THE ASTRONOMICAL ALMANAC

FOR THE YEAR

1992

Data for Astronomy, Space Sciences, Geodesy, Surveying, Navigation and other applications

WASHINGTON

Issued by the
Nautical Almanac Office
United States
Naval Observatory
by direction of the
Secretary of the Navy
and under the
authority of Congress

LONDON

Issued by
Her Majesty's
Nautical Almanac Office
Royal Greenwich Observatory
on behalf of the
Science and Engineering
Research
Council

WASHINGTON: U.S. GOVERNMENT PRINTING OFFICE
LONDON: HER MAJESTY'S STATIONERY OFFICE

ISBN 0 11 886942 6

UNITED STATES

For sale by the
Superintendent of Documents
U.S. GOVERNMENT PRINTING OFFICE
Washington, D.C., 20402

UNITED KINGDOM

For sale by

HER MAJESTY'S STATIONERY OFFICE

HMSO publications are available from:

HMSO Publications Centre
(Mail and telephone orders only)
PO Box 276, London, SW8 5DT
Telephone orders 071-873 9090
General enquiries 071-873 0011
(queuing system in operation for both numbers)

HMSO Bookshops
49 High Holborn, London, WC1V 6HB 071-873 0011 (Counter service only)
258 Broad Street, Birmingham, B1 2HE 021-643 3740
Southey House, 33 Wine Street, Bristol BS1 2BQ (0272) 264306
9–21 Princess Street, Manchester, M60 8AS 061-834 7201
80 Chichester Street, Belfast, BT1 4JY (0232) 238451
71–73 Lothian Road, Edinburgh, EH3 9AZ 031-228 4181

HMSO's Accredited Agents
(see Yellow Pages)

And through good booksellers

Overseas Orders to:
The Government Bookshop
PO Box 276, London SW8 5DT

NOTE

Every care is taken to prevent errors in the production of this publication. As a final precaution it is recommended that the sequence of pages in this copy be examined on receipt. If faulty it should be returned for replacement.

Printed in the United States of America
by the U.S. Government Printing Office

Beginning with the edition for 1981, the title *The Astronomical Almanac* replaced both the title *The American Ephemeris and Nautical Almanac* and the title *The Astronomical Ephemeris*. The changes in title symbolise the unification of the two series, which until 1980 were published separately in the United States of America since 1855 and in the United Kingdom since 1767. *The Astronomical Almanac* is prepared jointly by the Nautical Almanac Office, United States Naval Observatory, and H.M. Nautical Almanac Office, Royal Greenwich Observatory, and is published jointly by the United States Government Printing Office and Her Majesty's Stationery Office; it is printed only in the United States of America but some of the reproducible material that is used is prepared in the United Kingdom.

The principal ephemerides in this Almanac have been computed from fundamental ephemerides of the planets and the Moon prepared at the Jet Propulsion Laboratory, California, in cooperation with the U.S. Naval Observatory. They are in general accord with the recommendations of the International Astronomical Union and are consistent with the IAU (1976) system of astronomical constants apart from minor modifications introduced to permit a better fit to observations; in particular, dynamical time-scales and the standard reference system of J2000·0 are used where appropriate. A brief description of the use of each ephemeris is given with it, and the bases and additional notes are given in the Explanation at the end of the volume. Additional information about the IAU recommendations and the ephemerides is given in the *Supplement to the Astronomical Almanac for 1984*. A new Explanatory Supplement to the Astronomical Almanac is in preparation and will be published as soon as possible.

By international agreement the tasks of computation and publication of astronomical ephemerides are shared between the ephemeris offices of a number of countries. The sources of the basic data for this Almanac are indicated in the list of contributors on page vii. This volume was designed in consultation with other astronomers of many countries, and is intended to provide current, accurate astronomical data for use in the making and reduction of observations and for general purposes. (The other publications listed on pages viii–ix give astronomical data for particular applications, such as navigation and surveying.) Any changes introduced since the previous volume are listed on page iv. Suggestions for further improvement of this Almanac would be welcomed; they should be sent to the Director, Nautical Almanac Office, United States Naval Observatory or to the Head of H.M. Nautical Almanac Office, Royal Greenwich Observatory.

JAMES B. HAGEN
Captain, U.S. Navy,
Superintendent, U.S. Naval Observatory,
Washington, D.C. 20390,
U.S.A.

ALEXANDER BOKSENBERG,
Director,
Royal Greenwich Observatory,
Madingley Road,
Cambridge, CB3 0EZ, England

May 1990

Astronomical Almanac 1990

Page F69 The units for a/Δ should be arc seconds (″) rather than arc minutes (′).

Astronomical Almanac 1991

Pages H2–H31 In the column of star names, the Bayer designation τ should be replaced by σ for the following B.S. numbers:

68	2646	3616	5447	6498
740	2878	4386	5603	7121
847	2973	4743	6084	7462
1052	3418	5015	6093	7228
1931	3575	5425	6168	8143

For the following stars, the Bayer designation σ should be replaced by ρ:

82	1698	3576	5429	7724
708	2852	4133	5453	8252
921	3185	4638	5928	9045
1444	3492	4828	7340	

CONTENTS, 1992

The pagination within each section is given in full on the first page of each section.

U. S. NAVAL OBSERVATORY

Captain James B. Hagen, *U.S.N., Superintendent*

ASTRONOMICAL COUNCIL

Captain James B. Hagen, *U.S.N., Superintendent*
Lt. Commander Thomas D. Lage, *U.S.N., Deputy Superintendent*
Gart Westerhout, *Scientific Director*
Gernot M. R. Winkler, *Director, Time Service Department*
P. Kenneth Seidelmann, *Director, Nautical Almanac Office*
James A. Hughes, *Director, Astrometry Department*

NAUTICAL ALMANAC OFFICE

P. Kenneth Seidelmann, *Director*

Paul M. Janiczek, *Chief, Ephemerides Division*
Alan D. Fiala, *Chief, Astronomical Data Division*
LeRoy E. Doggett, *Head, Publications Branch*
Marta R. Goldblatt, *Head, ADP Branch*

Dan Pascu
John A. Bangert
Peter C. Kammeyer
David A. Nutile
Marie R. Lukac
Tim S. Carroll
James L. Hilton
James R. Rohde
William T. Harris
Stephen P. Panossian
William J. Tangren
Jennifer J. Weeks
QMC (SW) Donald A. Dias, *U.S.N.*
Carla G. Lee
Charlotte S. James
Candice P. Baines
Yvette Holley

ROYAL GREENWICH OBSERVATORY

Alexander Boksenberg, B.Sc. Ph.D., F.R.S., *Director*

ASTRONOMY SUPPORT AND RESEARCH DIVISION

P. G. Murdin, B.A., Ph.D., *Division Head*

HER MAJESTY'S NAUTICAL ALMANAC OFFICE

B. D. Yallop, B.Sc., Ph.D., *Department Head*

D. B. Taylor, B.Sc., Ph.D.
Miss C. Y. Hohenkerk, B.Sc.
Mrs. J. P. Hamblyn

In addition, the following persons have assisted in the preparation and proofreading of the publications of the Office:

Mrs. P. V. Long and Mrs. R. A. Yallop.

May 1990

The data in this volume have been prepared as follows:—

By H. M. Nautical Almanac Office, Royal Greenwich Observatory:

Section A—phenomena, rising and setting of Sun and Moon; B—ephemerides and tables relating to time-scales and coordinate reference frames; D—physical ephemerides, geocentric coordinates and daily polynomial coefficients of the Moon; G—geocentric positions of minor planets; K—tables and data.

By the Nautical Almanac Office, United States Naval Observatory:

Section A—eclipses of Sun and Moon; C—physical ephemerides, geocentric and rectangular coordinates of the Sun; E—physical ephemerides, geocentric coordinates and transit times of the major planets; F—ephemerides of satellites, except Jupiter I–IV; H–data for lists of bright stars, lists of photometric standard stars, bright galaxies, radio source positions, radio flux calibrators, X-ray sources, radial velocity standard stars, variable stars, quasars and pulsars; J—information on observatories; L—explanation; M—glossary; N–index.

By the Service des Calculs, Bureau des Longitudes, Paris:

Section F—ephemerides of satellites I–IV of Jupiter.

By the Institute of Theoretical Astronomy, Leningrad:

Section G—orbital elements of minor planets.

In general the Office responsible for the preparation of the data has drafted the related explanatory notes and auxiliary material, but both have contributed to the final form of the material. The preliminaries, part of Section A and Sections B, D, G and K have been composed in the United Kingdom, while the rest of the material has been composed in the United States. The work of proofreading has been shared, but no attempt has been made to eliminate the differences in spelling and style between the contributions of the two Offices.

Joint publications of the Royal Greenwich Observatory and the United States Naval Observatory

Except for the *Explanatory Supplement*, these publications are available from Her Majesty's Stationery Office at the addresses listed on page ii of this volume and from the Superintendent of Documents, U.S. Government Printing Office, Washington, D.C. 20402.

The Nautical Almanac contains ephemerides at an interval of one hour and auxiliary astronomical data for marine navigation.

The Air Almanac contains ephemerides at an interval of ten minutes and auxiliary astronomical data for air navigation.

Astronomical Phenomena contains extracts from *The Astronomical Almanac* and is published annually in advance of the main volume. It contains the dates and times of planetary and lunar phenomena and other astronomical data of general interest.

Planetary and Lunar Coordinates, 1984–2000 provides low-precision astronomical data for use in advance of the annual ephemerides and for other purposes. It contains heliocentric, geocentric, spherical and rectangular coordinates of the Sun, Moon and planets, eclipse data, and auxiliary data, such as orbital elements and precessional constants.

Explanatory Supplement to The Astronomical Ephemeris and The American Ephemeris and Nautical Almanac contains detailed explanations of the basis and derivation of each ephemeris in the *AE* in the edition for 1960; it also contains other useful material that is relevant to positional and dynamical astronomy and to chronology. Footnotes indicate the changes that have been introduced since 1960 and it contains a reprint of *The Supplement to A.E. 1968*, which gives an account of the introduction of the IAU (1964) system of astronomical constants. It was published by Her Majesty's Stationery Office but is now out of print. A new Explanatory Supplement to the Astronomical Almanac is in preparation.

Other publications of the United States Naval Observatory

Almanac for Computers contains short mathematical series which are used to represent the positions of the Sun, Moon and planets for efficient evaluation with small computers or programmable calculators. Data for both astronomical and navigational applications are included.

Astronomical Papers of the American Ephemeris are issued irregularly and contain reports of research in celestial mechanics with particular relevance to ephemerides.

U.S. Naval Observatory Circulars are issued irregularly to disseminate astronomical data concerning ephemerides or astronomical phenomena.

The Floppy Almanac is an integrated package of software and astronomical data on a floppy diskette. *The Floppy Almanac* will produce to full precision most of the data in *The Astronomical Almanac* including both positional and physical data interpolated to any date and time within the appropriate year. Versions are available for microcomputers compatible with the IBM PC, XT or AT, running under MS-DOS with at least 256k bytes of memory.

Other publications of the Royal Greenwich Observatory

The Star Almanac for Land Surveyors contains tabulations of *R*, declination and *E* for the Sun for every 6 hours, and right ascension to $0^{s}{\cdot}1$ and declination to 1″ of all stars brighter than magnitude 4·0 for each month. In addition the ephemerides of *R*, declination and *E* for the Sun are represented by polynomial series for each month. This publication is available from Her Majesty's Stationery Office and from Bernan-UNIPUB, 4611/F Assembly Drive, Lanham, MD 20706-4391, U.S.A.

Compact Data for Navigation and Astronomy for 1991 to 1995 contains data, which are mainly in the form of polynomial coefficients, for use by navigators and astronomers to calculate the positions of the Sun, Moon, navigational planets and bright stars using a small programmable calculator or personal computer. This publication is available from Cambridge University Press, The Edinburgh Building, Shaftesbury Road, Cambridge, CB2 2RU.

Publications of other countries

Apparent Places of Fundamental Stars is prepared annually by the Astronomisches Rechen-Institut in Heidelberg and contains mean and apparent coordinates of 1535 stars of the *Fifth Fundamental Catalogue* (FK5). This volume is available from Verlag G. Braun, Karl-Friedrich-Strasse, 14–18, Karlsruhe, Germany.

Ephemerides of Minor Planets is prepared annually by the Institute of Theoretical Astronomy, and published by the Academy of Sciences of the U.S.S.R. Included in this volume are elements, opposition dates and opposition ephemerides of all numbered minor planets. This volume is available from the Institute of Theoretical Astronomy, Naberezhuaga Kutozova 10, 191187 Leningrad, U.S.S.R.

CONTENTS OF SECTION A

NOTE: All the times in this section are expressed in universal time (UT).

THE SUN

		d h			d h m		d h m
Perigee ...	Jan.	3 15	Equinoxes ...	Mar.	20 08 48 ...	... Sept.	22 18 43
Apogee ...	July	3 12	Solstices ...	June	21 03 14 ...	... Dec.	21 14 43

PHASES OF THE MOON

Lunation	New Moon		First Quarter		Full Moon		Last Quarter	
		d h m		d h m		d h m		d h m
854	Jan.	4 23 10	Jan.	13 02 32	Jan.	19 21 28	Jan.	26 15 27
855	Feb.	3 19 00	Feb.	11 16 15	Feb.	18 08 04	Feb.	25 07 56
856	Mar.	4 13 22	Mar.	12 02 36	Mar.	18 18 18	Mar.	26 02 30
857	Apr.	3 05 01	Apr.	10 10 06	Apr.	17 04 42	Apr.	24 21 40
858	May	2 17 44	May	9 15 43	May	16 16 03	May	24 15 53
859	June	1 03 57	June	7 20 47	June	15 04 50	June	23 08 11
860	June	30 12 18	July	7 02 43	July	14 19 06	July	22 22 12
861	July	29 19 35	Aug.	5 10 58	Aug.	13 10 27	Aug.	21 10 01
862	Aug.	28 02 42	Sept.	3 22 39	Sept.	12 02 17	Sept.	19 19 53
863	Sept.	26 10 40	Oct.	3 14 12	Oct.	11 18 03	Oct.	19 04 12
864	Oct.	25 20 34	Nov.	2 09 11	Nov.	10 09 20	Nov.	17 11 39
865	Nov.	24 09 11	Dec.	2 06 17	Dec.	9 23 41	Dec.	16 19 13
866	Dec.	24 00 43						

ECLIPSES

Annular eclipse of the Sun	Jan. 4–5	Oceania, Philippines, Japan, extreme coast of N.E. Asia, N. Australia, W. coast of N. America
Partial eclipse of the Moon	June 15	Antarctica, E. Africa, S. tip of Greenland, S. America, N. America except N.W., Central America, E. New Zealand
Total eclipse of the Sun	June 30	Central S. America, S.W. Africa
Total eclipse of the Moon	Dec. 9–10	Asia except extreme E., Europe including the British Isles, Africa, Iceland, Greenland, S. America except S., Central America, N. America except W. coast
Partial eclipse of the Sun	Dec. 23–24	E. China, Korea, Japan, extreme E. of U.S.S.R., S.W. Alaska

MOON AT PERIGEE

	d h		d h		d h
Jan.	19 22	June	4 02	Oct.	23 05
Feb.	17 11	July	2 01	Nov.	19 00
Mar.	16 18	July	30 08	Dec.	13 21
Apr.	13 07	Aug.	27 18		
May	8 12	Sept.	25 03		

MOON AT APOGEE

	d h		d h		d h
Jan.	6 12	May	23 05	Oct.	7 06
Feb.	2 12	June	19 22	Nov.	3 23
Feb.	29 21	July	17 11	Dec.	1 20
Mar.	28 14	Aug.	13 16	Dec.	29 17
Apr.	25 10	Sept.	9 19		

OCCULTATIONS OF PLANETS AND BRIGHT STARS BY THE MOON

Date d h	Body	Areas of Visibility
Jan. 3 10	*Mars*	Extreme S. of S. America, Antarctica
Jan. 31 17	*Venus*	Most of Antarctica
Feb. 1 09	*Uranus*	N. Africa, Europe except extreme N. but including the British Isles, W. of U.S.S.R., N. of Middle East
Feb. 1 12	*Neptune*	Central S. America, S. Atlantic Ocean, Central Africa
Feb. 28 21	*Neptune*	E. Java, E. Borneo, New Guinea, N. Australia, Pacific Ocean, S. Hawaiian Islands
Mar. 27 05	*Neptune*	Central Africa, S.W. Asia
Apr. 23 14	*Neptune*	N.E. Pacific Ocean, N. America except extreme N. and E.
May 20 21	*Neptune*	Central Asia
June 17 04	*Neptune*	E. of N. America, S. Greenland, Iceland, W. Europe including the British Isles
July 14 09	*Neptune*	Pacific Ocean, E. of N. America
Aug. 10 14	*Neptune*	E. of Asia except extreme S., Japan
Sept. 6 19	*Neptune*	N. Africa, Europe except extreme N. but including British Isles, N. of Middle East, W. of U.S.S.R.
Sept. 20 09	*Mars*	S. of S. America, extreme tip of Antarctic Peninsula, S.W. Atlantic Ocean
Oct. 4 02	*Neptune*	N.W. Canada
Oct. 27 15	*Mercury*	E. coast of N. America, West Indies, central Atlantic Ocean, N.E. Africa, tip of Italy, Greece
Oct. 28 15	*Venus*	N. coast of S. America, central Atlantic, central Africa, Saudi Arabia

No occultations of bright stars in 1992

OCCULTATIONS OF X-RAY SOURCES BY THE MOON

Occultations occur at intervals of a lunar month between the dates given below:

Source	Dates	Source	Dates	Source	Dates
4U1621–23	Jan. 1–Dec. 22	V348SGR	Jan. 4–Oct. 3	IC443	Mar. 12–Dec. 11
A1704 – 25	Jan. 2–Aug. 8	NGC918	Jan. 14–Dec. 7	4U1438–18	Mar. 21–Dec. 20
GX1 + 4	Jan. 3–Sept. 4	4U1203 – 06	Jan. 24–Dec. 17	GXO + 9	Mar. 24–Dec. 22
GX5 – 1	Jan. 3–Feb. 27	4U1221 – 08	Jan. 24–Dec. 17	CRAB	Oct. 16–Dec. 10
MX1803–24	Jan. 3–Apr. 22	MKN372	Mar. 9–Dec. 7	4U1730 – 22	Oct. 29–Dec. 23
GX10–6	Jan. 4–Oct. 3				

AVAILABILITY OF PREDICTIONS OF LUNAR OCCULTATIONS

The International Lunar Occultation Centre, Geodesy and Geophysics Division, Hydrographic Department, Tsukiji-5, Chuo-ku, Tokyo, 104 JAPAN is responsible for the predictions and for the reductions of timings of occultations of stars by the Moon.

GEOCENTRIC PHENOMENA

MERCURY

	d h	d h	d h
Superior conjunction ...	Feb. 12 09	May 31 16	Sept. 15 04
Greatest elongation East	Mar. 9 21 (18°)	July 6 01 (26°)	Oct. 31 16 (24°)
Stationary	Mar. 16 15	July 19 03	Nov. 11 14
Inferior conjunction ...	Mar. 26 15	Aug. 2 21	Nov. 21 22
Stationary	Apr. 8 01	Aug. 12 13	Dec. 1 06
Greatest elongation West	Apr. 23 15 (27°)	Aug. 21 02 (18°)	Dec. 9 14 (21°)

VENUS

		d h
Superior conjunction ...	June	13 16

SUPERIOR PLANETS

	Conjunction	Stationary	Opposition	Stationary
	d h	d h	d h	d h
Mars	—	—	—	Nov. 29 16
Jupiter	Sept. 17 19	—	Feb. 29 01	May 1 05
Saturn	Jan. 29 22	May 29 03	Aug. 7 10	Oct. 16 04
Uranus	Jan. 5 01	Apr. 22 01	July 7 23	Sept. 23 00
Neptune	Jan. 7 13	Apr. 20 11	July 9 13	Sept. 27 16
Pluto	Nov. 15 00	Feb. 28 02	May 12 01	Aug. 4 09

OCCULTATIONS BY PLANETS AND SATELLITES

Details of predictions of occultations of stars by planets, minor planets and satellites are given in *The Handbook of the British Astronomical Association*.

HELIOCENTRIC PHENOMENA

	Aphelion	Perihelion	Descending Node	Greatest Lat. South	Ascending Node	Greatest Lat. North
Mercury	Jan. 21	Mar. 5	Jan. 11	Feb. 11	Mar. 1	Mar. 16
	Apr. 18	June 1	Apr. 8	May 9	May 28	June 11
	July 15	Aug. 28	July 5	Aug. 4	Aug. 24	Sept. 7
	Oct. 11	Nov. 24	Oct. 1	Oct. 31	Nov. 20	Dec. 4
	—	—	Dec. 28	—	—	—
Venus	Mar. 23	July 13	Feb. 17	Apr. 14	June 9	Aug. 4
	Nov. 3	—	Sept. 29	Nov. 25	—	—
Mars	—	May 17	—	Apr. 21	Sept. 15	—

Jupiter, Saturn, Uranus, Neptune, Pluto: None in 1992

ELONGATIONS AND MAGNITUDES OF PLANETS AT 0^h UT

Date	Mercury		Venus	
	Elong.	Mag.	Elong.	Mag.
	°		°	
Jan. –1	W. 22	−0·3	W. 40	−4·1
4	21	0·3	39	4·1
9	19	0·3	38	4·1
14	17	0·3	37	4·0
19	15	0·4	36	4·0
24	W. 13	−0·5	W. 35	−4·0
29	10	0·7	34	4·0
Feb. 3	7	0·9	33	4·0
8	W. 4	1·2	32	4·0
13	E. 2	1·5	31	4·0
18	E. 5	−1·5	W. 29	−4·0
23	9	1·3	28	3·9
28	13	1·2	27	3·9
Mar. 4	16	0·9	26	3·9
9	18	−0·4	25	3·9
14	E. 17	+0·5	W. 24	−3·9
19	13	2·1	22	3·9
24	E. 6	4·2	21	3·9
29	W. 5	4·6	20	3·9
Apr. 3	13	2·8	19	3·9
8	W. 20	+1·7	W. 18	−3·9
13	24	1·1	16	3·9
18	26	0·7	15	3·9
23	27	0·5	14	3·9
28	27	0·3	13	3·9
May 3	W. 25	+0·1	W. 11	−3·9
8	23	−0·2	10	3·9
13	20	0·4	9	3·9
18	15	0·8	7	3·9
23	10	1·2	6	3·9
28	W. 5	−1·8	W. 5	−3·9
June 2	E. 2	2·2	3	3·9
7	8	1·5	W. 2	3·9
12	13	1·0	0	3·9
17	18	0·6	E. 1	3·9
22	E. 22	−0·2	E. 2	−3·9
27	24	+0·1	4	3·9
July 2	E. 26	+0·3	E. 5	−3·9

Date	Mercury		Venus	
	Elong.	Mag.	Elong.	Mag.
	°		°	
July 2	E. 26	+0·3	E. 5	−3·9
7	26	0·6	6	3·9
12	25	0·9	8	3·9
17	23	1·4	9	3·9
22	18	2·1	11	3·9
27	E. 12	+3·2	E. 12	−3·9
Aug. 1	E. 6	4·5	13	3·9
6	W. 7	4·1	15	3·9
11	13	2·4	16	3·9
16	17	+1·0	17	3·9
21	W. 18	0·0	E. 19	−3·9
26	17	−0·7	20	3·9
31	14	1·1	21	3·9
Sept. 5	10	1·4	23	3·9
10	5	1·6	24	3·9
15	W. 2	−1·7	E. 25	−3·9
20	E. 4	1·3	26	3·9
25	8	0·9	28	3·9
30	11	0·6	29	3·9
Oct. 5	14	0·4	30	3·9
10	E. 17	−0·3	E. 31	−3·9
15	19	0·2	32	4·0
20	21	0·2	33	4·0
25	23	0·2	35	4·0
30	24	0·2	36	4·0
Nov. 4	E. 23	−0·1	E. 37	−4·0
9	21	+0·2	38	4·0
14	16	1·1	39	4·0
19	E. 7	3·4	40	4·1
24	W. 5	3·9	41	4·1
29	W. 15	+1·0	E. 42	−4·1
Dec. 4	19	−0·1	42	4·1
9	21	0·4	43	4·2
14	20	0·5	44	4·2
19	19	0·5	45	4·2
24	W. 17	−0·5	E. 45	−4·2
29	14	0·5	46	4·3
34	W. 12	−0·5	E. 46	−4·3

MINOR PLANETS

	Stationary	Opposition	Stationary	Conjunction
Ceres ...	June 7	July 25	Sept. 16	—
Pallas ...	Apr. 24	June 18	Aug. 14	—
Juno ...	Nov. 17	Dec. 28	—	Mar. 8
Vesta ...	Jan. 29	Mar. 9	Apr. 27	Dec. 4

ELONGATIONS AND MAGNITUDES OF PLANETS AT 0^h UT

Date	Mars Elong.	Mars Mag.	Jupiter Elong.	Jupiter Mag.	Saturn Elong.	Saturn Mag.	Uranus Elong.	Neptune Elong.	Pluto Elong.
	°		°		°		°	°	°
Jan. 0	W. 15	+1·4	W. 113	−2·2	E. 28	+0·7	E. 6	E. 8	W. 48
9	18	1·4	123	2·3	19	0·6	W. 4	W. 2	57
19	21	1·4	134	2·4	10	0·6	13	11	66
29	24	1·4	145	2·4	E. 1	0·6	23	21	76
Feb. 8	26	1·3	156	2·5	W. 8	0·6	33	31	86
18	W. 29	+1·3	W. 167	−2·5	W. 17	+0·7	W. 42	W. 41	W. 95
28	31	1·3	W. 178	2·5	26	0·7	52	50	105
Mar. 9	33	1·2	E. 170	2·5	35	0·8	61	60	115
19	36	1·2	159	2·5	44	0·8	71	70	124
29	38	1·2	148	2·4	53	0·8	81	80	134
Apr. 8	W. 40	+1·1	E. 137	−2·4	W. 62	+0·8	W. 90	W. 89	W. 143
18	42	1·1	127	2·3	71	0·8	100	99	152
28	44	1·1	117	2·2	80	0·8	110	109	159
May 8	46	1·1	107	2·2	90	0·7	120	119	W. 164
18	48	1·0	98	2·1	99	0·7	130	129	E. 164
28	W. 50	+1·0	E. 89	−2·0	W. 108	+0·7	W. 139	W. 138	E. 158
June 7	52	1·0	80	2·0	118	0·6	149	148	150
17	54	0·9	72	1·9	128	0·5	159	158	142
27	57	0·9	64	1·9	138	0·5	169	168	133
July 7	59	0·9	56	1·8	148	0·4	W. 179	W. 177	124
17	W. 61	+0·8	E. 48	−1·8	W. 158	+0·3	E. 171	E. 173	E. 115
27	64	0·8	40	1·7	168	0·3	161	163	105
Aug. 6	67	0·7	32	1·7	W. 178	0·2	151	153	96
16	70	0·7	25	1·7	E. 171	0·2	141	143	87
26	73	0·6	17	1·7	161	0·3	131	133	78
Sept. 5	W. 77	+0·6	E. 10	−1·7	E. 150	+0·3	E. 121	E. 124	E. 69
15	81	0·5	E. 2	1·7	140	0·4	112	114	59
25	85	0·4	W. 6	1·7	130	0·4	102	104	50
Oct. 5	90	0·3	13	1·7	120	0·5	92	94	41
15	95	+0·1	21	1·7	110	0·5	82	84	33
25	W. 101	0·0	W. 29	−1·7	E. 100	+0·6	E. 73	E. 75	E. 24
Nov. 4	108	−0·2	37	1·7	90	0·6	63	65	18
14	116	0·4	45	1·8	81	0·7	53	55	E. 14
24	124	0·6	53	1·8	71	0·7	44	45	W. 17
Dec. 4	135	0·8	62	1·8	62	0·7	34	35	23
14	W. 146	−1·1	W. 71	−1·9	E. 52	+0·7	E. 24	E. 25	W. 31
24	159	1·3	80	2·0	43	0·7	15	16	40
34	W. 172	−1·4	W. 89	−2·0	E. 34	+0·8	E. 5	E. 6	W. 50

Magnitudes at opposition: Uranus 5·6 Neptune 7·9 Pluto 13·7

VISUAL MAGNITUDES OF MINOR PLANETS

	Jan. 9	Feb. 18	Mar. 29	May 8	June 17	July 27	Sept. 5	Oct. 15	Nov. 24	Dec. 34
Ceres	9·0	9.2	9·1	8·8	8·1	7·5	8·2	8·9	9·2	9·3
Pallas	10·0	10·0	9·8	9·6	9·4	9·7	10·1	10·4	10·5	10·5
Juno	10·4	10·1	9·9	9·9	9·8	9·6	9·3	8·8	8·1	7·8
Vesta	7·2	6·3	6·2	6·9	7·4	7·7	7·8	7·8	7·5	7·7

VISIBILITY OF PLANETS

The planet diagram on page A7 shows, in graphical form for any date during the year, the local mean times of meridian passage of the Sun, of the five planets, Mercury, Venus, Mars, Jupiter and Saturn, and of every 2^h of right ascension. Intermediate lines, corresponding to particular stars, may be drawn in by the user if he so desires. The diagram is intended to provide a general picture of the availability of planets and stars for observation during the year.

On each side of the line marking the time of meridian passage of the Sun, a band 45^m wide is shaded to indicate that planets and most stars crossing the meridian within 45^m of the Sun are generally too close to the Sun for observation.

For any date the diagram provides immediately the local mean time of meridian passage of the Sun, planets and stars, and thus the following information:
(a) whether a planet or star is too close to the Sun for observation;
(b) visibility of a planet or star in the morning or evening;
(c) location of a planet or star during twilight;
(d) proximity of planets to stars or other planets.

When the meridian passage of a body occurs at midnight, it is close to opposition to the Sun and is visible all night, and may be observed in both morning and evening twilights. As the time of meridian passage decreases, the body ceases to be observable in the morning, but its altitude above the eastern horizon during evening twilight gradually increases until it is on the meridian at evening twilight. From then onwards the body is observable above the western horizon, its altitude at evening twilight gradually decreasing, until it becomes too close to the Sun for observation. When it again becomes visible, it is seen in the morning twilight, low in the east. Its altitude at morning twilight gradually increases until meridian passage occurs at the time of morning twilight, then as the time of meridian passage decreases to 0^h, the body is observable in the west in the morning twilight with a gradually decreasing altitude, until it once again reaches opposition.

Notes on the visibility of the principal planets, except Pluto, are given on page A8. Further information on the visibility of planets may be obtained from the diagram below which shows, in graphical form for any date during the year, the declinations of the bodies plotted on the planet diagram on page A7.

DECLINATIONS OF SUN AND PLANETS, 1992

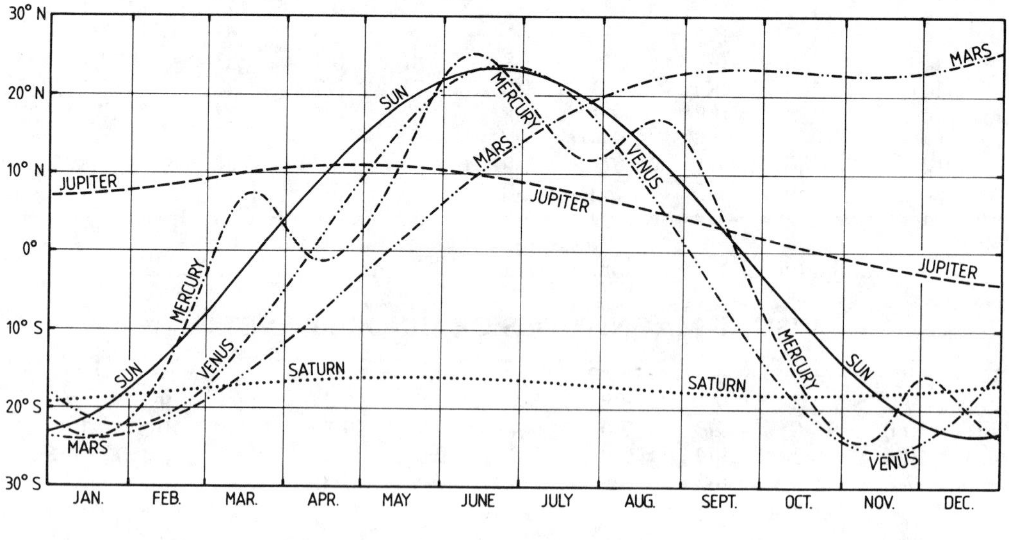

LOCAL MEAN TIME OF MERIDIAN PASSAGE

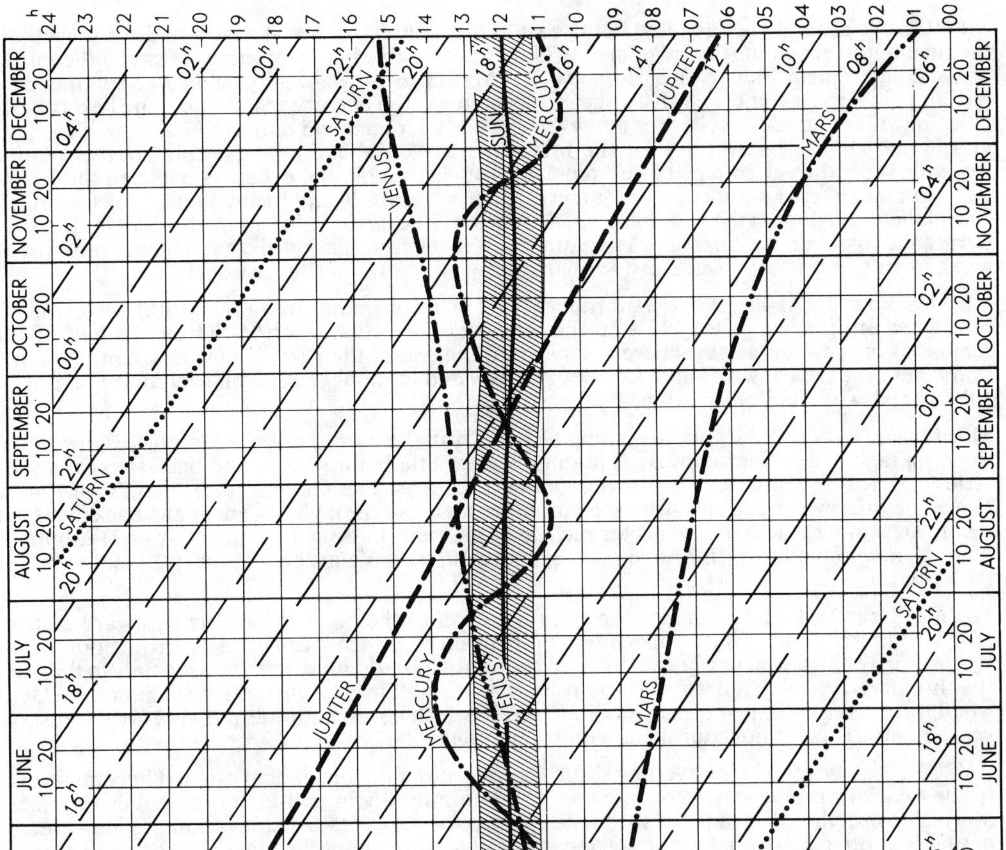

LOCAL MEAN TIME OF MERIDIAN PASSAGE

VISIBILITY OF PLANETS

MERCURY can only be seen low in the east before sunrise, or low in the west after sunset (about the time of the beginning or end of civil twilight). It is visible in the mornings between the following approximate dates: January 1 to January 30, April 3 to May 24, August 11 to September 6 and November 28 to December 31. The planet is brighter at the end of each period, (the best conditions in low northern latitudes occur for a few days in early January and during the third week in August and in most northern latitudes for the first three weeks in December, and in southern latitudes from the second week in April until mid-May). It is visible in the evenings between the following approximate dates: February 23 to March 19, June 8 to July 26 and September 27 to November 16. The planet is brighter at the beginning of each period, (the best conditions in northern latitudes occur during the first half of March and in southern latitudes from the third week in June until the third week in July and second week in October until just before mid-November).

VENUS is a brilliant object in the morning sky from the beginning of the year until the second week in May when it becomes too close to the Sun for observation. During the second half of July it reappears in the evening sky where it stays until the end of the year. Venus is in conjunction with Mars on February 19, with Saturn on February 29 and December 21, with Mercury on April 5 and July 25 and with Jupiter on August 23.

MARS can be seen at the very beginning of January in the morning sky in Ophiuchus then it passes through Sagittarius, Capricornus, Aquarius, Pisces, briefly into Cetus and back into Pisces, on to Aries and Taurus (passing 5° N. of *Aldebaran* on August 11), Gemini (passing 5° S. of *Pollux* on November 4 by which time it can be seen for more than half the night), Cancer and back into Gemini in mid December where it remains for the rest of the year (passing 3° S. of *Pollux* on December 22). Mars is in conjunction with Mercury on January 10, with Venus on February 19 and with Saturn on March 6.

JUPITER can be seen in January for more than half of the night in Leo, its westward elongation gradually increases until it is at opposition on February 29 when it can be seen throughout the night. Its eastward elongation then gradually decreases until by early June it can be seen only in the evening sky. In early September it becomes too close to the Sun for observation until the beginning of October when it reappears in the morning sky in Virgo in which constellation it remains for the rest of the year. Jupiter is in conjunction with Venus on August 23.

SATURN can be seen in the evening sky in Capricornus until mid-January then it becomes too close to the Sun for observation. It reappears in the morning sky in mid-February still in Capricornus, in which constellation it remains throughout the year. Its westward elongation gradually increases until it is at opposition on August 7 when it is visible throughout the night. Its eastward elongation then gradually decreases until from early November it can only be seen in the evening sky. Saturn is in conjunction with Venus on February 29 and December 21 and with Mars on March 6.

URANUS is too close to the Sun for observation until towards the end of January when it appears in the morning sky in Sagittarius, in which constellation it remains throughout the year. It is at opposition on July 7 when it can be seen throughout the night, after which its eastward elongation gradually decreases. From early October it can only be seen in the evening sky until the second half of December when it again becomes too close to the Sun for observation.

NEPTUNE is too close to the Sun for observation until late January when it can be seen in the morning sky shortly before sunrise in Sagittarius, in which constellation it remains throughout the year. It is at opposition on July 9 when it can be seen throughout the night. From mid-October it can only be seen in the evening sky until mid-December when it again becomes too close to the Sun for observation.

DO NOT CONFUSE (1) Mercury with Mars during the first half of January, the reddish tint of Mars should assist in its identification. (2) Venus with Mars from mid-February to early March, with Saturn from late February to early March and mid-December to late December, with Mercury in early April and in the third week of May and with Jupiter late August, in all occasions Venus is the brighter object. (3) Mars with Saturn for the first half of March when Mars is the brighter object.

VISIBILITY OF PLANETS IN MORNING AND EVENING TWILIGHT

	MORNING	EVENING
VENUS	January 1 –May 7	July 20 –December 31
MARS	January 1 –December 31	
JUPITER	January 1 –February 29 October 1 –December 31	February 29–September 4
SATURN	February 16–August 7	January 1 –January 13 August 7 –December 31

CONFIGURATIONS OF SUN, MOON AND PLANETS

	d h		
Jan.	1 12	Venus 5° N. of Moon	
	3 01	Mercury 3° N. of Moon	
	3 10	Mars 0°·8 N. of Moon	Occn.
	3 15	Earth at perihelion	
	4 23	NEW MOON	Eclipse
	5 01	Uranus in conjunction with Sun	
	6 12	Moon at apogee	
	6 23	Saturn 3° S. of Moon	
	7 13	Neptune in conjunction with Sun	
	7 19	Venus 7° N. of Antares	
	7 23	Juno 0°·2 S. of Moon	Occn.
	10 20	Mercury 0°·6 N. of Mars	
	13 03	FIRST QUARTER	
	19 21	FULL MOON	
	19 22	Moon at perigee	
	20 03	Mercury 0°·6 S. of Uranus	
	21 11	Mercury 1°·9 S. of Neptune	
	23 01	Jupiter 7° N. of Moon	
	26 15	LAST QUARTER	
	29 20	Vesta stationary	
	29 21	Mars 0°·4 S. of Uranus	
	29 22	Saturn in conjunction with Sun	
	31 17	Venus 1°·0 N. of Moon	Occn.
Feb.	1 08	Mars 1°·5 S. of Neptune	
	1 09	Uranus 0°·9 S. of Moon	Occn.
	1 12	Neptune 0°·03 N. of Moon	Occn.
	1 13	Mars 1°·5 S. of Moon	
	2 12	Moon at apogee	
	3 19	NEW MOON	
	7 07	Venus 0°·9 N. of Uranus	
	8 15	Venus 0°·3 S. of Neptune	
	11 16	FIRST QUARTER	
	12 09	Mercury in superior conjunction	
	17 11	Moon at perigee	
	18 08	FULL MOON	
	19 07	Jupiter 6° N. of Moon	
	19 22	Venus 0°·9 N. of Mars	
	25 08	LAST QUARTER	
	28 02	Pluto stationary	
	28 18	Uranus 1°·2 S. of Moon	
	28 21	Neptune 0°·2 S. of Moon	Occn.
	29 01	Jupiter at opposition	
	29 02	Venus 0°·1 N. of Saturn	
	29 21	Moon at apogee	
Mar.	1 18	Mars 4° S. of Moon	
	2 01	Saturn 4° S. of Moon	
	2 06	Venus 4° S. of Moon	
	4 13	NEW MOON	
	6 06	Mercury 4° S. of Moon	
	6 13	Mars 0°·4 S. of Saturn	
Mar.	8 12	Juno in conjunction with Sun	
	9 04	Vesta at opposition	
	9 21	Mercury greatest elong. E. (18°)	
	12 03	FIRST QUARTER	
	16 15	Mercury stationary	
	16 18	Moon at perigee	
	17 12	Jupiter 6° N. of Moon	
	18 18	FULL MOON	
	20 09	Equinox	
	26 02	LAST QUARTER	
	26 15	Mercury in inferior conjunction	
	27 03	Uranus 1°·5 S. of Moon	
	27 05	Neptune 0°·5 S. of Moon	Occn.
	28 14	Moon at apogee	
	29 14	Saturn 4° S. of Moon	
	31 01	Mars 6° S. of Moon	
Apr.	1 19	Venus 7° S. of Moon	
	2 08	Mercury 4° S. of Moon	
	3 05	NEW MOON	
	5 23	Mercury 2° N. of Venus	
	8 01	Mercury stationary	
	10 10	FIRST QUARTER	
	13 07	Moon at perigee	
	13 16	Jupiter 6° N. of Moon	
	17 05	FULL MOON	
	20 11	Neptune stationary	
	22 01	Uranus stationary	
	23 12	Uranus 1°·8 S. of Moon	
	23 14	Neptune 0°·7 S. of Moon	Occn.
	23 15	Mercury greatest elong. W. (27°)	
	24 04	Pallas stationary	
	24 22	LAST QUARTER	
	25 10	Moon at apogee	
	26 02	Saturn 5° S. of Moon	
	27 07	Vesta stationary	
	29 07	Mars 7° S. of Moon	
	30 21	Mercury 8° S. of Moon	
May	1 05	Jupiter stationary	
	2 18	NEW MOON	
	8 12	Moon at perigee	
	9 16	FIRST QUARTER	
	10 22	Jupiter 6° N. of Moon	
	12 01	Pluto at opposition	
	16 16	FULL MOON	
	20 20	Uranus 2° S. of Moon	
	20 21	Neptune 0°·9 S. of Moon	Occn.
	23 05	Moon at apogee	
	23 12	Saturn 5° S. of Moon	
	24 16	LAST QUARTER	

CONFIGURATIONS OF SUN, MOON AND PLANETS

	d h	
May	28 09	Mars 7° S. of Moon
	29 03	Saturn stationary
	31 16	Mercury in superior conjunction
June	1 04	NEW MOON
	4 02	Moon at perigee
	7 07	Jupiter 7° N. of Moon
	7 12	Ceres stationary
	7 21	FIRST QUARTER
	13 16	Venus in superior conjunction
	15 05	FULL MOON Eclipse
	17 01	Uranus 1°·9 S. of Moon
	17 04	Neptune 0°·9 S. of Moon Occn.
	18 23	Pallas at opposition
	19 19	Saturn 5° S. of Moon
	19 22	Moon at apogee
	21 03	Solstice
	23 02	Mercury 5° S. of Pollux
	23 08	LAST QUARTER
	26 08	Mars 6° S. of Moon
	30 12	NEW MOON Eclipse
July	2 01	Moon at perigee
	2 10	Mercury 4° N. of Moon
	3 12	Earth at aphelion
	4 20	Jupiter 7° N. of Moon
	6 01	Mercury greatest elong. E. (26°)
	7 03	FIRST QUARTER
	7 23	Uranus at opposition
	9 13	Neptune at opposition
	14 06	Uranus 1°·8 S. of Moon
	14 09	Neptune 0°·8 S. of Moon Occn.
	14 19	FULL MOON
	16 23	Saturn 5° S. of Moon
	17 11	Moon at apogee
	19 03	Mercury stationary
	22 22	LAST QUARTER
	25 04	Mars 4° S. of Moon
	25 15	Mercury 6° S. of Venus
	25 22	Ceres at opposition
	29 20	NEW MOON
	30 08	Moon at perigee
Aug.	1 12	Jupiter 7° N. of Moon
	2 21	Mercury in inferior conjunction
	4 09	Pluto stationary
	5 11	FIRST QUARTER
	6 18	Venus 1°·1 N. of Regulus
	7 10	Saturn at opposition
	10 10	Uranus 1°·8 S. of Moon
	10 14	Neptune 0°·8 S. of Moon Occn.
	11 09	Mars 5° N. of Aldebaran
Aug.	12 13	Mercury stationary
	13 01	Saturn 5° S. of Moon
	13 10	FULL MOON
	13 16	Moon at apogee
	14 01	Pallas stationary
	21 02	Mercury greatest elong. W. (18°)
	21 10	LAST QUARTER
	22 21	Mars 1°·4 S. of Moon
	23 03	Venus 0°·3 N. of Jupiter
	27 01	Mercury 5° N. of Moon
	27 18	Moon at perigee
	28 03	NEW MOON
	29 19	Venus 7° N. of Moon
Sept.	2 23	Mercury 1°·2 N. of Regulus
	3 23	FIRST QUARTER
	6 15	Uranus 1°·9 S. of Moon
	6 19	Neptune 0°·9 S. of Moon Occn.
	9 03	Saturn 5° S. of Moon
	9 19	Moon at apogee
	12 02	FULL MOON
	15 04	Mercury in superior conjunction
	16 12	Ceres stationary
	17 19	Jupiter in conjunction with Sun
	19 05	Venus 3° N. of Spica
	19 20	LAST QUARTER
	20 09	Mars 0°·9 N. of Moon Occn.
	22 19	Equinox
	23 00	Uranus stationary
	25 03	Moon at perigee
	26 11	NEW MOON
	27 16	Neptune stationary
	28 15	Venus 4° N. of Moon
Oct.	2 22	Mercury 2° N. of Spica
	3 14	FIRST QUARTER
	3 22	Uranus 2° S. of Moon
	4 02	Neptune 1°·2 S. of Moon Occn.
	6 08	Saturn 5° S. of Moon
	7 06	Moon at apogee
	11 18	FULL MOON
	16 04	Saturn stationary
	18 15	Mars 3° N. of Moon
	19 04	LAST QUARTER
	23 05	Moon at perigee
	24 00	Jupiter 7° N. of Moon
	25 21	NEW MOON
	27 07	Venus 3° N. of Antares
	27 15	Mercury 0°·5 S. of Moon Occn.
	28 15	Venus 0°·4 S. of Moon Occn.

CONFIGURATIONS OF SUN, MOON AND PLANETS

	d	h	
Oct.	31	07	Uranus 2° S. of Moon
	31	10	Neptune 1°·4 S. of Moon
	31	16	Mercury greatest elong. E. (24°)
Nov.	2	09	FIRST QUARTER
	2	16	Saturn 5° S. of Moon
	3	23	Moon at apogee
	4	21	Mars 5° S. of Pollux
	10	09	FULL MOON
	11	14	Mercury stationary
	15	00	Pluto in conjunction with Sun
	15	12	Mars 5° N. of Moon
	17	04	Juno stationary
	17	12	LAST QUARTER
	19	00	Moon at perigee
	20	16	Jupiter 7° N. of Moon
	21	22	Mercury in inferior conjunction
	24	09	NEW MOON
	26	12	Venus 1°·9 S. of Uranus
	27	13	Venus 3° S. of Neptune
	27	18	Uranus 3° S. of Moon
	27	21	Neptune 1°·6 S. of Moon
	27	21	Venus 5° S. of Moon
Nov.	29	16	Mars stationary
	30	03	Saturn 5° S. of Moon
Dec.	1	06	Mercury stationary
	1	20	Moon at apogee
	2	06	FIRST QUARTER
	4	10	Vesta in conjunction with Sun
	9	14	Mercury greatest elong. W.(21°)
	10	00	FULL MOON Eclipse
	12	19	Mars 6° N. of Moon
	13	21	Moon at perigee
	16	19	LAST QUARTER
	18	05	Jupiter 7° N. of Moon
	19	03	Mercury 6° N. of Antares
	21	15	Solstice
	21	16	Venus 1°·1 S. of Saturn
	22	14	Mercury 1°·5 N. of Moon
	22	21	Mars 3° S. of Pollux
	24	01	NEW MOON Eclipse
	27	17	Saturn 6° S. of Moon
	28	02	Juno at opposition
	28	07	Venus 7° S. of Moon
	29	17	Moon at apogee

Arrangement and basis of the tabulations

The tabulations of risings, settings and twilights on pages A14–A77 refer to the instants when the true geocentric zenith distance of the central point of the disk of the Sun or Moon takes the value indicated in the following table. The tabular times are in universal time (UT) for selected latitudes on the meridian of Greenwich; the times for other latitudes and longitudes may be obtained by interpolation as described below and as exemplified on page A13.

	Phenomena	*Zenith distance*	*Pages*
SUN (interval 4 days):	sunrise and sunset	90° 50′	A14–A21
	civil twilight	96°	A22–A29
	nautical twilight	102°	A30–A37
	astronomical twilight	108°	A38–A45
MOON (interval 1 day):	moonrise and moonset	$90° 34' + s - \pi$	A46–A77
(s = semidiameter, π = horizontal parallax)			

The zenith distance at the times for rising and setting is such that under normal conditions the upper limb of the Sun and Moon appears to be on the horizon of an observer at sea-level. The parallax of the Sun is ignored. The observed time may differ from the tabular time because of a variation of the atmospheric refraction from the adopted value (34′) and because of a difference in height of the observer and the actual horizon.

Use of tabulations

The following procedure may be used to obtain times of the phenomena for a non-tabular place and date.

Step 1: Interpolate linearly for latitude. The differences between adjacent values are usually small and so the required interpolates can often be obtained by inspection.

Step 2: Interpolate linearly for date and longitude in order to obtain the local mean times of the phenomena at the longitude concerned. For the Sun the variations with longitude of the local mean times of the phenomena are small, but to obtain better precision the interpolation factor for date should be increased by

west longitude in degrees / 1440

since the interval of tabulation is 4 days. For the Moon, the interpolating factor to be used is simply

west longitude in degrees / 360

since the interval of tabulation is 1 day; backward interpolation should be carried out for east longitudes.

Step 3: Convert the times so obtained (which are on the scale of local mean time for the local meridian to universal time (UT) or to the appropriate clock time, which may differ from the time of the nearest standard meridian according to the customs of the country concerned. The UT of the phenomenon is obtained from the local mean time by applying the longitude expressed in time measure (1 hour for each 15° of longitude), adding for west longitudes and subtracting for east longitudes. The times so obtained may require adjustment by 24^h; if so, the corresponding date must be changed accordingly.

Approximate formulae for direct calculation

The approximate UT of rising or setting of a body with right ascension α and declination δ at latitude ϕ and *east* longitude λ may be calculated from

$$\mathrm{UT} = 0{\cdot}997\,27\,\{\alpha - \lambda \pm \cos^{-1}(-\tan\phi\tan\delta) - (\mathrm{GMST\ at\ 0^hUT})\}$$

where each term is expressed in time measure and the GMST at 0^hUT is given in the tabulations on pages B8–B15. The negative sign corresponds to rising and the positive sign to setting. The formula ignores refraction, semi-diameter and any changes in α and δ during the day. If $\tan\phi\tan\delta$ is numerically greater than 1, there is no phenomenon.

Examples

The following examples of the calculations of the times of rising and setting phenomena use the procedure described on page A12.

1. To find the times of sunrise and sunset for Paris on 1992 July 20. Paris is at latitude N 48° 52′ (= +48°·87), longitude E 2° 20′ (=E 2°·33=E 0^h 09^m), and in the summer the clocks are kept two hours in advance of UT. The relevant portions of the tabulation on page A19 and the results of the interpolation for latitude are as follows, where the interpolation factor is (48·87 − 48)/2 = 0·44:

	Sunrise			Sunset		
	+48°	+50°	+48°·87	+48°	+50°	+48°·87
	h m	h m	h m	h m	h m	h m
July 17	04 18	04 10	04 14	19 53	20 02	19 57
July 21	04 23	04 15	04 19	19 49	19 57	19 53

The interpolation factor for date and longitude is (20 − 17)/4 − 2·33/1440 = 0·75

	Sunrise	Sunset
	d h m	d h m
Interpolate to obtain local mean time:	20 04 18	20 19 54
Subtract 0^h 09^m to obtain universal time:	20 04 09	20 19 45
Add 2^h to obtain clock time:	20 06 09	20 21 45

2. To find the times of beginning and end of astronomical twilight for Canberra, Australia on 1992 November 15. Canberra is at latitude S 35° 18′ (= −35°·30), longitude E 149° 08′ (=E 149°·13=E 9^h 57^m), and in the summer the clocks are kept eleven hours in advance of UT. The relevant portions of the tabulation on page A44 and the results of the interpolation for latitude are as follows, where the interpolation factor is (−35·30 − (−40))/5 = 0·94:

	Astronomical Twilight					
	beginning			end		
	−40°	−35°	−35°·3	−40°	−35°	−35°·3
	h m	h m	h m	h m	h m	h m
Nov. 14	02 46	03 09	03 08	20 44	20 21	20 22
Nov. 18	02 41	03 05	03 04	20 51	20 27	20 28

The interpolation factor for date and longitude is (15 − 14)/4 − 149·13/1440 = 0·15

	Astronomical Twilight	
	beginning	end
	d h m	d h m
Interpolation to obtain local mean time:	15 03 07	15 20 23
Subtract 9^h 57^m to obtain universal time:	14 17 10	15 10 26
Add 11^h to obtain clock time:	15 04 10	15 21 26

3. To find the times of moonrise and moonset for Washington, D.C. on 1992 February 3. Washington is at latitude N 38° 55′ (= +38°·92), longitude W 77° 00′ (=W 77°·0=W 5^h 08^m), and in the winter the clocks are kept five hours behind UT. The relevant portions of the tabulation on page A48 and the results of the interpolation for latitude are as follows, where the interpolation factor is (38·92 − 35)/5 = 0·78:

	Moonrise			Moonset		
	+35°	+40°	+38°·92	+35°	+40°	+38°·92
	h m	h m	h m	h m	h m	h m
Feb. 3	06 35	06 44	06 42	17 27	17 18	17 20
Feb. 4	07 04	07 11	07 09	18 23	18 18	18 19

The interpolation factor for longitude is 77·0/360 = 0·21

	Moonrise	Moonset
	d h m	d h m
Interpolate to obtain local mean time:	3 06 48	3 17 32
Add 5^h 08^m to obtain universal time:	3 11 56	3 22 40
Subtract 5^h to obtain clock time:	3 06 56	3 17 40

SUNRISE AND SUNSET, 1992

UNIVERSAL TIME FOR MERIDIAN OF GREENWICH

SUNRISE

Lat.	−55°	−50°	−45°	−40°	−35°	−30°	−20°	−10°	0°	+10°	+20°	+30°	+35°	+40°
	h m	h m	h m	h m	h m	h m	h m	h m	h m	h m	h m	h m	h m	h m
Jan. −1	3 23	3 53	4 15	4 33	4 48	5 01	5 22	5 41	5 59	6 16	6 34	6 55	7 07	7 21
3	3 28	3 57	4 18	4 36	4 51	5 03	5 25	5 43	6 00	6 18	6 36	6 56	7 08	7 22
7	3 33	4 01	4 23	4 40	4 54	5 06	5 27	5 46	6 02	6 19	6 37	6 57	7 09	7 22
11	3 39	4 07	4 27	4 44	4 58	5 10	5 30	5 48	6 04	6 20	6 37	6 57	7 08	7 22
15	3 46	4 12	4 32	4 48	5 01	5 13	5 33	5 50	6 06	6 21	6 38	6 57	7 08	7 20
19	3 54	4 18	4 37	4 52	5 05	5 16	5 35	5 52	6 07	6 22	6 38	6 56	7 07	7 19
23	4 02	4 25	4 43	4 57	5 09	5 20	5 38	5 54	6 08	6 22	6 38	6 55	7 05	7 16
27	4 10	4 32	4 48	5 02	5 13	5 24	5 41	5 55	6 09	6 23	6 37	6 53	7 03	7 14
31	4 19	4 39	4 54	5 07	5 18	5 27	5 43	5 57	6 10	6 23	6 36	6 51	7 00	7 10
Feb. 4	4 27	4 46	5 00	5 12	5 22	5 31	5 45	5 58	6 10	6 22	6 35	6 49	6 57	7 06
8	4 36	4 53	5 06	5 17	5 26	5 34	5 48	6 00	6 11	6 22	6 33	6 46	6 54	7 02
12	4 45	5 00	5 12	5 22	5 30	5 37	5 50	6 01	6 11	6 21	6 31	6 43	6 50	6 58
16	4 53	5 07	5 18	5 26	5 34	5 41	5 52	6 02	6 11	6 20	6 29	6 40	6 46	6 53
20	5 02	5 14	5 23	5 31	5 38	5 44	5 54	6 02	6 10	6 18	6 27	6 36	6 41	6 47
24	5 11	5 21	5 29	5 36	5 42	5 47	5 55	6 03	6 10	6 17	6 24	6 32	6 37	6 42
28	5 19	5 28	5 35	5 40	5 45	5 50	5 57	6 03	6 09	6 15	6 21	6 28	6 32	6 36
Mar. 3	5 27	5 34	5 40	5 45	5 49	5 52	5 59	6 04	6 09	6 13	6 18	6 24	6 27	6 30
7	5 35	5 41	5 46	5 49	5 52	5 55	6 00	6 04	6 08	6 11	6 15	6 19	6 21	6 24
11	5 44	5 48	5 51	5 54	5 56	5 58	6 01	6 04	6 07	6 09	6 12	6 14	6 16	6 18
15	5 52	5 54	5 56	5 58	5 59	6 00	6 02	6 04	6 06	6 07	6 08	6 10	6 11	6 11
19	5 59	6 00	6 01	6 02	6 02	6 03	6 04	6 04	6 04	6 05	6 05	6 05	6 05	6 05
23	6 07	6 07	6 06	6 06	6 06	6 05	6 05	6 04	6 03	6 02	6 01	6 00	5 59	5 58
27	6 15	6 13	6 12	6 10	6 09	6 08	6 06	6 04	6 02	6 00	5 58	5 55	5 54	5 52
31	6 23	6 19	6 17	6 14	6 12	6 10	6 07	6 04	6 01	5 58	5 54	5 50	5 48	5 45
Apr. 4	6 31	6 26	6 22	6 18	6 15	6 13	6 08	6 04	6 00	5 55	5 51	5 46	5 43	5 39

SUNSET

Lat.	−55°	−50°	−45°	−40°	−35°	−30°	−20°	−10°	0°	+10°	+20°	+30°	+35°	+40°
	h m	h m	h m	h m	h m	h m	h m	h m	h m	h m	h m	h m	h m	h m
Jan. −1	20 41	20 12	19 50	19 32	19 17	19 04	18 42	18 23	18 06	17 49	17 31	17 10	16 57	16 43
3	20 40	20 11	19 50	19 32	19 18	19 05	18 44	18 25	18 08	17 51	17 33	17 12	17 00	16 47
7	20 38	20 10	19 49	19 32	19 18	19 06	18 45	18 26	18 10	17 53	17 36	17 15	17 04	16 50
11	20 35	20 08	19 48	19 31	19 18	19 06	18 45	18 28	18 11	17 55	17 38	17 19	17 07	16 54
15	20 31	20 05	19 46	19 30	19 17	19 05	18 46	18 29	18 13	17 57	17 41	17 22	17 11	16 58
19	20 26	20 02	19 43	19 28	19 15	19 04	18 46	18 29	18 14	17 59	17 43	17 25	17 15	17 03
23	20 20	19 58	19 40	19 26	19 14	19 03	18 45	18 30	18 15	18 01	17 46	17 29	17 19	17 07
27	20 14	19 53	19 36	19 23	19 11	19 01	18 44	18 30	18 16	18 03	17 48	17 32	17 23	17 12
31	20 07	19 47	19 32	19 19	19 09	18 59	18 43	18 30	18 17	18 04	17 51	17 36	17 27	17 17
Feb. 4	19 59	19 41	19 27	19 15	19 05	18 57	18 42	18 29	18 17	18 06	17 53	17 39	17 31	17 22
8	19 51	19 35	19 22	19 11	19 02	18 54	18 40	18 29	18 18	18 07	17 55	17 42	17 35	17 27
12	19 42	19 28	19 16	19 06	18 58	18 51	18 38	18 28	18 18	18 08	17 57	17 46	17 39	17 31
16	19 34	19 20	19 10	19 01	18 54	18 47	18 36	18 26	18 18	18 09	17 59	17 49	17 43	17 36
20	19 24	19 13	19 03	18 56	18 49	18 43	18 34	18 25	18 17	18 09	18 01	17 52	17 47	17 41
24	19 15	19 05	18 57	18 50	18 44	18 39	18 31	18 24	18 17	18 10	18 03	17 55	17 50	17 45
28	19 05	18 57	18 50	18 44	18 39	18 35	18 28	18 22	18 16	18 10	18 04	17 58	17 54	17 50
Mar. 3	18 55	18 48	18 43	18 38	18 34	18 31	18 25	18 20	18 15	18 11	18 06	18 01	17 58	17 54
7	18 45	18 40	18 36	18 32	18 29	18 26	18 22	18 18	18 14	18 11	18 07	18 03	18 01	17 59
11	18 35	18 31	18 28	18 26	18 24	18 22	18 18	18 16	18 13	18 11	18 09	18 06	18 04	18 03
15	18 25	18 23	18 21	18 19	18 18	18 17	18 15	18 13	18 12	18 11	18 10	18 08	18 08	18 07
19	18 15	18 14	18 13	18 13	18 12	18 12	18 12	18 11	18 11	18 11	18 11	18 11	18 11	18 11
23	18 05	18 05	18 06	18 06	18 07	18 07	18 08	18 09	18 10	18 11	18 12	18 13	18 14	18 15
27	17 54	17 57	17 58	18 00	18 01	18 02	18 05	18 07	18 09	18 11	18 13	18 16	18 17	18 19
31	17 44	17 48	17 51	17 53	17 56	17 58	18 01	18 04	18 07	18 11	18 14	18 18	18 21	18 23
Apr. 4	17 34	17 39	17 44	17 47	17 50	17 53	17 58	18 02	18 06	18 10	18 15	18 21	18 24	18 27

UNIVERSAL TIME FOR MERIDIAN OF GREENWICH

SUNRISE

Lat.	+40°	+42°	+44°	+46°	+48°	+50°	+52°	+54°	+56°	+58°	+60°	+62°	+64°	+66°
	h m	h m	h m	h m	h m	h m	h m	h m	h m	h m	h m	h m	h m	h m
Jan. −1	7 21	7 28	7 35	7 42	7 50	7 59	8 08	8 19	8 32	8 46	9 03	9 24	9 52	10 32
3	7 22	7 28	7 35	7 42	7 50	7 58	8 08	8 19	8 31	8 45	9 02	9 22	9 48	10 25
7	7 22	7 28	7 35	7 42	7 49	7 58	8 07	8 17	8 29	8 43	8 59	9 18	9 43	10 17
11	7 22	7 27	7 34	7 40	7 48	7 56	8 05	8 15	8 26	8 39	8 54	9 13	9 36	10 07
15	7 20	7 26	7 32	7 38	7 46	7 53	8 02	8 11	8 22	8 34	8 49	9 06	9 27	9 55
19	7 19	7 24	7 30	7 36	7 43	7 50	7 58	8 07	8 17	8 29	8 42	8 58	9 18	9 43
23	7 16	7 21	7 27	7 33	7 39	7 46	7 54	8 02	8 12	8 22	8 35	8 50	9 07	9 30
27	7 14	7 18	7 23	7 29	7 35	7 41	7 49	7 56	8 05	8 15	8 27	8 40	8 56	9 16
31	7 10	7 15	7 19	7 25	7 30	7 36	7 43	7 50	7 58	8 07	8 18	8 30	8 45	9 02
Feb. 4	7 06	7 11	7 15	7 20	7 25	7 30	7 36	7 43	7 51	7 59	8 09	8 20	8 33	8 48
8	7 02	7 06	7 10	7 14	7 19	7 24	7 30	7 36	7 42	7 50	7 59	8 08	8 20	8 34
12	6 58	7 01	7 05	7 09	7 13	7 17	7 22	7 28	7 34	7 41	7 48	7 57	8 07	8 19
16	6 53	6 56	6 59	7 02	7 06	7 10	7 15	7 19	7 25	7 31	7 38	7 45	7 54	8 05
20	6 47	6 50	6 53	6 56	6 59	7 03	7 07	7 11	7 16	7 21	7 27	7 33	7 41	7 50
24	6 42	6 44	6 47	6 49	6 52	6 55	6 58	7 02	7 06	7 10	7 15	7 21	7 27	7 35
28	6 36	6 38	6 40	6 42	6 45	6 47	6 50	6 53	6 56	7 00	7 04	7 08	7 14	7 20
Mar. 3	6 30	6 32	6 33	6 35	6 37	6 39	6 41	6 43	6 46	6 49	6 52	6 56	7 00	7 05
7	6 24	6 25	6 26	6 28	6 29	6 30	6 32	6 34	6 36	6 38	6 40	6 43	6 46	6 49
11	6 18	6 18	6 19	6 20	6 21	6 22	6 23	6 24	6 25	6 27	6 28	6 30	6 32	6 34
15	6 11	6 12	6 12	6 13	6 13	6 13	6 14	6 14	6 15	6 16	6 16	6 17	6 18	6 19
19	6 05	6 05	6 05	6 05	6 05	6 05	6 05	6 05	6 04	6 04	6 04	6 04	6 04	6 04
23	5 58	5 58	5 58	5 57	5 57	5 56	5 55	5 55	5 54	5 53	5 52	5 51	5 50	5 48
27	5 52	5 51	5 50	5 49	5 48	5 47	5 46	5 45	5 43	5 42	5 40	5 38	5 36	5 33
31	5 45	5 44	5 43	5 42	5 40	5 39	5 37	5 35	5 33	5 30	5 28	5 25	5 21	5 17
Apr. 4	5 39	5 37	5 36	5 34	5 32	5 30	5 28	5 25	5 22	5 19	5 16	5 12	5 07	5 02

SUNSET

Lat.	+40°	+42°	+44°	+46°	+48°	+50°	+52°	+54°	+56°	+58°	+60°	+62°	+64°	+66°
	h m	h m	h m	h m	h m	h m	h m	h m	h m	h m	h m	h m	h m	h m
Jan. −1	16 43	16 37	16 30	16 23	16 15	16 06	15 57	15 46	15 33	15 19	15 02	14 40	14 13	13 33
3	16 47	16 40	16 34	16 27	16 19	16 10	16 01	15 50	15 38	15 24	15 07	14 47	14 21	13 43
7	16 50	16 44	16 38	16 31	16 23	16 15	16 06	15 55	15 43	15 30	15 14	14 54	14 30	13 56
11	16 54	16 48	16 42	16 35	16 28	16 20	16 11	16 01	15 50	15 37	15 22	15 03	14 40	14 09
15	16 58	16 53	16 47	16 40	16 33	16 26	16 17	16 08	15 57	15 44	15 30	15 13	14 52	14 24
19	17 03	16 57	16 52	16 46	16 39	16 32	16 23	16 14	16 04	15 53	15 39	15 23	15 04	14 39
23	17 07	17 02	16 57	16 51	16 45	16 38	16 30	16 22	16 12	16 02	15 49	15 34	15 17	14 54
27	17 12	17 07	17 02	16 57	16 51	16 44	16 37	16 30	16 21	16 11	15 59	15 46	15 30	15 10
31	17 17	17 13	17 08	17 03	16 57	16 51	16 45	16 37	16 29	16 20	16 10	15 57	15 43	15 25
Feb. 4	17 22	17 18	17 13	17 09	17 04	16 58	16 52	16 45	16 38	16 30	16 20	16 09	15 56	15 41
8	17 27	17 23	17 19	17 15	17 10	17 05	17 00	16 53	16 47	16 39	16 31	16 21	16 09	15 56
12	17 31	17 28	17 24	17 21	17 16	17 12	17 07	17 02	16 56	16 49	16 41	16 33	16 22	16 10
16	17 36	17 33	17 30	17 26	17 23	17 19	17 14	17 10	17 04	16 58	16 52	16 44	16 35	16 25
20	17 41	17 38	17 35	17 32	17 29	17 26	17 22	17 18	17 13	17 08	17 02	16 56	16 48	16 39
24	17 45	17 43	17 41	17 38	17 35	17 32	17 29	17 26	17 22	17 17	17 13	17 07	17 01	16 53
28	17 50	17 48	17 46	17 44	17 42	17 39	17 36	17 34	17 30	17 27	17 23	17 18	17 13	17 07
Mar. 3	17 54	17 53	17 51	17 50	17 48	17 46	17 44	17 41	17 39	17 36	17 33	17 29	17 25	17 21
7	17 59	17 57	17 56	17 55	17 54	17 52	17 51	17 49	17 47	17 45	17 43	17 40	17 38	17 34
11	18 03	18 02	18 01	18 01	18 00	17 59	17 58	17 57	17 56	17 54	17 53	17 51	17 50	17 47
15	18 07	18 07	18 06	18 06	18 06	18 05	18 05	18 04	18 04	18 03	18 03	18 02	18 01	18 01
19	18 11	18 11	18 11	18 11	18 12	18 12	18 12	18 12	18 12	18 12	18 13	18 13	18 13	18 14
23	18 15	18 16	18 16	18 17	18 17	18 18	18 19	18 19	18 20	18 21	18 22	18 24	18 25	18 27
27	18 19	18 20	18 21	18 22	18 23	18 24	18 26	18 27	18 29	18 30	18 32	18 34	18 37	18 40
31	18 23	18 25	18 26	18 27	18 29	18 31	18 32	18 34	18 37	18 39	18 42	18 45	18 49	18 53
Apr. 4	18 27	18 29	18 31	18 33	18 35	18 37	18 39	18 42	18 45	18 48	18 52	18 56	19 01	19 06

UNIVERSAL TIME FOR MERIDIAN OF GREENWICH

SUNRISE

Lat.	−55°	−50°	−45°	−40°	−35°	−30°	−20°	−10°	0°	+10°	+20°	+30°	+35°	+40°
	h m	h m	h m	h m	h m	h m	h m	h m	h m	h m	h m	h m	h m	h m
Mar. 31	6 23	6 19	6 17	6 14	6 12	6 10	6 07	6 04	6 01	5 58	5 54	5 50	5 48	5 45
Apr. 4	6 31	6 26	6 22	6 18	6 15	6 13	6 08	6 04	6 00	5 55	5 51	5 46	5 43	5 39
8	6 38	6 32	6 27	6 22	6 18	6 15	6 09	6 04	5 58	5 53	5 48	5 41	5 37	5 33
12	6 46	6 38	6 32	6 26	6 22	6 17	6 10	6 04	5 57	5 51	5 44	5 36	5 32	5 27
16	6 53	6 44	6 37	6 30	6 25	6 20	6 11	6 04	5 56	5 49	5 41	5 32	5 27	5 21
20	7 01	6 50	6 41	6 34	6 28	6 22	6 12	6 04	5 55	5 47	5 38	5 28	5 22	5 15
24	7 09	6 56	6 46	6 38	6 31	6 25	6 14	6 04	5 55	5 45	5 35	5 24	5 17	5 09
28	7 16	7 02	6 51	6 42	6 34	6 27	6 15	6 04	5 54	5 44	5 33	5 20	5 12	5 04
May 2	7 24	7 08	6 56	6 46	6 37	6 30	6 16	6 05	5 53	5 42	5 30	5 16	5 08	4 59
6	7 31	7 14	7 01	6 50	6 41	6 32	6 18	6 05	5 53	5 41	5 28	5 13	5 04	4 54
10	7 38	7 20	7 06	6 54	6 44	6 35	6 19	6 06	5 53	5 40	5 26	5 10	5 00	4 50
14	7 45	7 26	7 10	6 58	6 47	6 37	6 21	6 06	5 53	5 39	5 24	5 07	4 57	4 46
18	7 52	7 31	7 15	7 01	6 50	6 40	6 22	6 07	5 53	5 38	5 23	5 05	4 54	4 42
22	7 58	7 36	7 19	7 05	6 53	6 42	6 24	6 08	5 53	5 38	5 22	5 03	4 52	4 39
26	8 04	7 41	7 23	7 08	6 55	6 44	6 26	6 09	5 53	5 38	5 21	5 01	4 50	4 36
30	8 09	7 45	7 26	7 11	6 58	6 47	6 27	6 10	5 54	5 38	5 20	5 00	4 48	4 34
June 3	8 14	7 49	7 30	7 14	7 00	6 49	6 29	6 11	5 55	5 38	5 20	4 59	4 47	4 32
7	8 19	7 53	7 33	7 16	7 03	6 51	6 30	6 12	5 55	5 38	5 20	4 58	4 46	4 31
11	8 22	7 55	7 35	7 18	7 05	6 52	6 31	6 13	5 56	5 39	5 20	4 58	4 45	4 31
15	8 25	7 58	7 37	7 20	7 06	6 54	6 33	6 14	5 57	5 39	5 20	4 58	4 45	4 30
19	8 26	7 59	7 38	7 22	7 07	6 55	6 34	6 15	5 58	5 40	5 21	4 59	4 46	4 31
23	8 27	8 00	7 39	7 22	7 08	6 56	6 35	6 16	5 59	5 41	5 22	5 00	4 47	4 32
27	8 27	8 00	7 39	7 23	7 09	6 56	6 35	6 17	5 59	5 42	5 23	5 01	4 48	4 33
July 1	8 26	7 59	7 39	7 23	7 09	6 57	6 36	6 17	6 00	5 43	5 24	5 02	4 50	4 35
5	8 24	7 58	7 38	7 22	7 08	6 56	6 36	6 18	6 01	5 44	5 26	5 04	4 52	4 37

SUNSET

Lat.	−55°	−50°	−45°	−40°	−35°	−30°	−20°	−10°	0°	+10°	+20°	+30°	+35°	+40°
	h m	h m	h m	h m	h m	h m	h m	h m	h m	h m	h m	h m	h m	h m
Mar. 31	17 44	17 48	17 51	17 53	17 56	17 58	18 01	18 04	18 07	18 11	18 14	18 18	18 21	18 23
Apr. 4	17 34	17 39	17 44	17 47	17 50	17 53	17 58	18 02	18 06	18 10	18 15	18 21	18 24	18 27
8	17 24	17 31	17 36	17 41	17 45	17 48	17 54	18 00	18 05	18 10	18 16	18 23	18 27	18 32
12	17 15	17 23	17 29	17 35	17 39	17 44	17 51	17 58	18 04	18 10	18 17	18 25	18 30	18 36
16	17 05	17 15	17 22	17 29	17 34	17 39	17 48	17 56	18 03	18 11	18 19	18 28	18 33	18 40
20	16 56	17 07	17 16	17 23	17 29	17 35	17 45	17 54	18 02	18 11	18 20	18 30	18 37	18 44
24	16 47	16 59	17 09	17 17	17 25	17 31	17 42	17 52	18 01	18 11	18 21	18 33	18 40	18 48
28	16 38	16 52	17 03	17 12	17 20	17 27	17 40	17 51	18 01	18 11	18 23	18 36	18 43	18 52
May 2	16 29	16 45	16 57	17 07	17 16	17 24	17 37	17 49	18 00	18 12	18 24	18 38	18 46	18 56
6	16 22	16 38	16 52	17 03	17 12	17 21	17 35	17 48	18 00	18 12	18 25	18 41	18 50	19 00
10	16 14	16 32	16 47	16 59	17 09	17 18	17 33	17 47	18 00	18 13	18 27	18 43	18 53	19 04
14	16 07	16 27	16 42	16 55	17 06	17 15	17 32	17 46	18 00	18 14	18 29	18 46	18 56	19 08
18	16 00	16 21	16 38	16 51	17 03	17 13	17 30	17 46	18 00	18 15	18 30	18 48	18 59	19 11
22	15 55	16 17	16 34	16 48	17 00	17 11	17 29	17 45	18 00	18 15	18 32	18 51	19 02	19 15
26	15 50	16 13	16 31	16 46	16 58	17 09	17 28	17 45	18 01	18 16	18 33	18 53	19 05	19 18
30	15 45	16 10	16 28	16 44	16 57	17 08	17 28	17 45	18 01	18 17	18 35	18 55	19 08	19 21
June 3	15 42	16 07	16 26	16 42	16 56	17 07	17 28	17 45	18 02	18 19	18 37	18 58	19 10	19 24
7	15 39	16 05	16 25	16 41	16 55	17 07	17 28	17 46	18 03	18 20	18 38	19 00	19 12	19 27
11	15 37	16 04	16 24	16 41	16 55	17 07	17 28	17 46	18 03	18 21	18 39	19 01	19 14	19 29
15	15 36	16 03	16 24	16 41	16 55	17 07	17 28	17 47	18 04	18 22	18 41	19 03	19 16	19 31
19	15 36	16 04	16 24	16 41	16 55	17 08	17 29	17 48	18 05	18 23	18 42	19 04	19 17	19 32
23	15 37	16 04	16 25	16 42	16 56	17 09	17 30	17 48	18 06	18 23	18 42	19 05	19 18	19 33
27	15 39	16 06	16 27	16 43	16 58	17 10	17 31	17 49	18 07	18 24	18 43	19 05	19 18	19 33
July 1	15 42	16 08	16 29	16 45	16 59	17 11	17 32	17 50	18 08	18 25	18 43	19 05	19 18	19 33
5	15 45	16 11	16 31	16 47	17 01	17 13	17 33	17 51	18 08	18 25	18 44	19 05	19 17	19 32

UNIVERSAL TIME FOR MERIDIAN OF GREENWICH

SUNRISE

Lat.	+40°	+42°	+44°	+46°	+48°	+50°	+52°	+54°	+56°	+58°	+60°	+62°	+64°	+66°
	h m	h m	h m	h m	h m	h m	h m	h m	h m	h m	h m	h m	h m	h m
Mar. 31	5 45	5 44	5 43	5 42	5 40	5 39	5 37	5 35	5 33	5 30	5 28	5 25	5 21	5 17
Apr. 4	5 39	5 37	5 36	5 34	5 32	5 30	5 28	5 25	5 22	5 19	5 16	5 12	5 07	5 02
8	5 33	5 31	5 29	5 26	5 24	5 21	5 19	5 15	5 12	5 08	5 04	4 59	4 53	4 46
12	5 27	5 24	5 22	5 19	5 16	5 13	5 10	5 06	5 02	4 57	4 52	4 46	4 39	4 31
16	5 21	5 18	5 15	5 12	5 08	5 05	5 01	4 56	4 52	4 46	4 40	4 33	4 25	4 15
20	5 15	5 12	5 08	5 05	5 01	4 57	4 52	4 47	4 42	4 35	4 28	4 20	4 11	4 00
24	5 09	5 06	5 02	4 58	4 54	4 49	4 44	4 38	4 32	4 25	4 17	4 08	3 57	3 44
28	5 04	5 00	4 56	4 51	4 47	4 42	4 36	4 30	4 23	4 15	4 06	3 55	3 43	3 28
May 2	4 59	4 55	4 50	4 45	4 40	4 34	4 28	4 21	4 13	4 05	3 55	3 43	3 29	3 13
6	4 54	4 49	4 45	4 39	4 34	4 28	4 21	4 13	4 05	3 55	3 44	3 31	3 16	2 57
10	4 50	4 45	4 40	4 34	4 28	4 21	4 14	4 06	3 56	3 46	3 34	3 20	3 03	2 41
14	4 46	4 40	4 35	4 29	4 22	4 15	4 07	3 59	3 49	3 37	3 24	3 09	2 50	2 25
18	4 42	4 37	4 31	4 24	4 17	4 10	4 01	3 52	3 41	3 29	3 15	2 58	2 37	2 09
22	4 39	4 33	4 27	4 20	4 13	4 05	3 56	3 46	3 35	3 22	3 07	2 48	2 25	1 53
26	4 36	4 30	4 24	4 17	4 09	4 01	3 51	3 41	3 29	3 15	2 59	2 39	2 13	1 37
30	4 34	4 28	4 21	4 14	4 06	3 57	3 47	3 36	3 24	3 09	2 52	2 30	2 02	1 20
June 3	4 32	4 26	4 19	4 12	4 03	3 54	3 44	3 33	3 20	3 04	2 46	2 23	1 53	1 02
7	4 31	4 25	4 18	4 10	4 02	3 52	3 42	3 30	3 16	3 01	2 41	2 17	1 44	0 44
11	4 31	4 24	4 17	4 09	4 00	3 51	3 40	3 28	3 14	2 58	2 38	2 13	1 37	0 19
15	4 30	4 24	4 16	4 09	4 00	3 50	3 39	3 27	3 13	2 56	2 36	2 10	1 33	** **
19	4 31	4 24	4 17	4 09	4 00	3 50	3 39	3 27	3 13	2 56	2 35	2 09	1 31	** **
23	4 32	4 25	4 18	4 10	4 01	3 51	3 40	3 28	3 14	2 57	2 36	2 10	1 32	** **
27	4 33	4 26	4 19	4 11	4 02	3 53	3 42	3 30	3 16	2 59	2 39	2 13	1 35	** **
July 1	4 35	4 28	4 21	4 13	4 05	3 55	3 44	3 32	3 18	3 02	2 42	2 17	1 41	0 21
5	4 37	4 30	4 23	4 16	4 07	3 58	3 48	3 36	3 22	3 06	2 47	2 23	1 50	0 48

SUNSET

Lat.	+40°	+42°	+44°	+46°	+48°	+50°	+52°	+54°	+56°	+58°	+60°	+62°	+64°	+66°
	h m	h m	h m	h m	h m	h m	h m	h m	h m	h m	h m	h m	h m	h m
Mar. 31	18 23	18 25	18 26	18 27	18 29	18 31	18 32	18 34	18 37	18 39	18 42	18 45	18 49	18 53
Apr. 4	18 27	18 29	18 31	18 33	18 35	18 37	18 39	18 42	18 45	18 48	18 52	18 56	19 01	19 06
8	18 32	18 34	18 36	18 38	18 40	18 43	18 46	18 49	18 53	18 57	19 01	19 07	19 12	19 19
12	18 36	18 38	18 41	18 43	18 46	18 49	18 53	18 57	19 01	19 06	19 11	19 17	19 25	19 33
16	18 40	18 42	18 45	18 49	18 52	18 56	19 00	19 04	19 09	19 15	19 21	19 28	19 37	19 47
20	18 44	18 47	18 50	18 54	18 58	19 02	19 07	19 12	19 17	19 24	19 31	19 39	19 49	20 00
24	18 48	18 51	18 55	18 59	19 03	19 08	19 13	19 19	19 26	19 33	19 41	19 50	20 01	20 15
28	18 52	18 56	19 00	19 04	19 09	19 14	19 20	19 27	19 34	19 42	19 51	20 02	20 14	20 29
May 2	18 56	19 00	19 05	19 09	19 15	19 21	19 27	19 34	19 42	19 51	20 01	20 13	20 27	20 44
6	19 00	19 04	19 09	19 15	19 20	19 27	19 33	19 41	19 50	20 00	20 11	20 24	20 40	20 59
10	19 04	19 09	19 14	19 20	19 26	19 33	19 40	19 48	19 58	20 08	20 21	20 35	20 53	21 15
14	19 08	19 13	19 18	19 24	19 31	19 38	19 46	19 55	20 05	20 17	20 30	20 46	21 06	21 31
18	19 11	19 17	19 23	19 29	19 36	19 44	19 52	20 02	20 13	20 25	20 39	20 57	21 18	21 47
22	19 15	19 21	19 27	19 34	19 41	19 49	19 58	20 08	20 20	20 33	20 48	21 07	21 31	22 04
26	19 18	19 24	19 31	19 38	19 45	19 54	20 03	20 14	20 26	20 40	20 57	21 17	21 43	22 21
30	19 21	19 28	19 34	19 42	19 50	19 58	20 08	20 19	20 32	20 47	21 04	21 26	21 55	22 39
June 3	19 24	19 31	19 38	19 45	19 53	20 02	20 13	20 24	20 37	20 53	21 11	21 34	22 06	22 58
7	19 27	19 33	19 40	19 48	19 57	20 06	20 16	20 28	20 42	20 58	21 17	21 42	22 15	23 19
11	19 29	19 36	19 43	19 51	19 59	20 09	20 20	20 32	20 46	21 02	21 22	21 47	22 23	23 52
15	19 31	19 37	19 45	19 53	20 01	20 11	20 22	20 34	20 48	21 05	21 25	21 51	22 29	** **
19	19 32	19 39	19 46	19 54	20 03	20 12	20 23	20 36	20 50	21 07	21 27	21 54	22 32	** **
23	19 33	19 39	19 47	19 55	20 04	20 13	20 24	20 36	20 51	21 07	21 28	21 54	22 32	** **
27	19 33	19 40	19 47	19 55	20 04	20 13	20 24	20 36	20 50	21 07	21 27	21 53	22 30	** **
July 1	19 33	19 39	19 47	19 54	20 03	20 12	20 23	20 35	20 49	21 05	21 25	21 50	22 25	23 39
5	19 32	19 38	19 45	19 53	20 02	20 11	20 21	20 33	20 46	21 02	21 21	21 45	22 18	23 16

(** **) indicates Sun continuously above horizon.

SUNRISE AND SUNSET, 1992

UNIVERSAL TIME FOR MERIDIAN OF GREENWICH

SUNRISE

Lat.	−55°	−50°	−45°	−40°	−35°	−30°	−20°	−10°	0°	+10°	+20°	+30°	+35°	+40°
	h m	h m	h m	h m	h m	h m	h m	h m	h m	h m	h m	h m	h m	h m
July 1	8 26	7 59	7 39	7 23	7 09	6 57	6 36	6 17	6 00	5 43	5 24	5 02	4 50	4 35
5	8 24	7 58	7 38	7 22	7 08	6 56	6 36	6 18	6 01	5 44	5 26	5 04	4 52	4 37
9	8 21	7 56	7 37	7 21	7 07	6 56	6 36	6 18	6 02	5 45	5 27	5 06	4 54	4 40
13	8 17	7 53	7 34	7 19	7 06	6 55	6 35	6 18	6 02	5 46	5 28	5 08	4 56	4 42
17	8 13	7 50	7 32	7 17	7 04	6 53	6 35	6 18	6 03	5 47	5 30	5 10	4 59	4 45
21	8 07	7 45	7 28	7 14	7 02	6 52	6 34	6 18	6 03	5 48	5 31	5 12	5 01	4 49
25	8 01	7 41	7 24	7 11	7 00	6 50	6 32	6 17	6 03	5 48	5 33	5 15	5 04	4 52
29	7 55	7 35	7 20	7 08	6 57	6 47	6 31	6 17	6 03	5 49	5 34	5 17	5 07	4 56
Aug. 2	7 48	7 30	7 15	7 04	6 53	6 45	6 29	6 16	6 03	5 50	5 36	5 20	5 10	4 59
6	7 40	7 23	7 10	6 59	6 50	6 41	6 27	6 14	6 02	5 50	5 37	5 22	5 13	5 03
10	7 32	7 17	7 04	6 54	6 46	6 38	6 25	6 13	6 02	5 51	5 38	5 24	5 16	5 07
14	7 23	7 10	6 59	6 49	6 41	6 34	6 22	6 11	6 01	5 51	5 40	5 27	5 19	5 11
18	7 14	7 02	6 52	6 44	6 37	6 30	6 19	6 10	6 00	5 51	5 41	5 29	5 22	5 14
22	7 05	6 54	6 46	6 38	6 32	6 26	6 16	6 08	5 59	5 51	5 42	5 31	5 25	5 18
26	6 56	6 46	6 39	6 32	6 27	6 22	6 13	6 06	5 58	5 51	5 43	5 34	5 28	5 22
30	6 46	6 38	6 32	6 26	6 22	6 17	6 10	6 03	5 57	5 51	5 44	5 36	5 31	5 26
Sept. 3	6 36	6 30	6 25	6 20	6 16	6 13	6 07	6 01	5 56	5 51	5 45	5 38	5 34	5 30
7	6 26	6 21	6 17	6 14	6 11	6 08	6 03	5 59	5 55	5 50	5 46	5 40	5 37	5 33
11	6 16	6 13	6 10	6 07	6 05	6 03	6 00	5 56	5 53	5 50	5 47	5 42	5 40	5 37
15	6 06	6 04	6 02	6 01	5 59	5 58	5 56	5 54	5 52	5 50	5 47	5 45	5 43	5 41
19	5 56	5 55	5 55	5 54	5 54	5 53	5 52	5 51	5 50	5 49	5 48	5 47	5 46	5 45
23	5 46	5 46	5 47	5 47	5 48	5 48	5 49	5 49	5 49	5 49	5 49	5 49	5 49	5 49
27	5 35	5 37	5 39	5 41	5 42	5 43	5 45	5 46	5 48	5 49	5 50	5 51	5 52	5 52
Oct. 1	5 25	5 29	5 32	5 34	5 36	5 38	5 41	5 44	5 46	5 49	5 51	5 53	5 55	5 56
5	5 15	5 20	5 24	5 28	5 31	5 33	5 38	5 42	5 45	5 48	5 52	5 56	5 58	6 00

SUNSET

Lat.	−55°	−50°	−45°	−40°	−35°	−30°	−20°	−10°	0°	+10°	+20°	+30°	+35°	+40°
	h m	h m	h m	h m	h m	h m	h m	h m	h m	h m	h m	h m	h m	h m
July 1	15 42	16 08	16 29	16 45	16 59	17 11	17 32	17 50	18 08	18 25	18 43	19 05	19 18	19 33
5	15 45	16 11	16 31	16 47	17 01	17 13	17 33	17 51	18 08	18 25	18 44	19 05	19 17	19 32
9	15 50	16 15	16 34	16 50	17 03	17 15	17 35	17 52	18 09	18 25	18 43	19 04	19 16	19 31
13	15 55	16 19	16 37	16 53	17 06	17 17	17 36	17 53	18 09	18 26	18 43	19 03	19 15	19 29
17	16 00	16 23	16 41	16 56	17 08	17 19	17 38	17 54	18 10	18 25	18 42	19 02	19 13	19 26
21	16 06	16 28	16 45	16 59	17 11	17 21	17 39	17 55	18 10	18 25	18 41	19 00	19 11	19 24
25	16 12	16 33	16 49	17 02	17 14	17 23	17 41	17 56	18 10	18 24	18 40	18 58	19 08	19 20
29	16 19	16 38	16 53	17 06	17 16	17 26	17 42	17 56	18 10	18 24	18 38	18 55	19 05	19 17
Aug. 2	16 25	16 43	16 58	17 09	17 19	17 28	17 43	17 57	18 10	18 23	18 36	18 52	19 02	19 12
6	16 32	16 49	17 02	17 13	17 22	17 31	17 45	17 57	18 09	18 21	18 34	18 49	18 58	19 08
10	16 39	16 55	17 07	17 17	17 25	17 33	17 46	17 58	18 09	18 20	18 32	18 46	18 54	19 03
14	16 47	17 00	17 11	17 20	17 28	17 35	17 47	17 58	18 08	18 18	18 29	18 42	18 49	18 58
18	16 54	17 06	17 16	17 24	17 31	17 37	17 48	17 58	18 07	18 16	18 26	18 38	18 45	18 52
22	17 01	17 12	17 21	17 28	17 34	17 40	17 49	17 58	18 06	18 14	18 23	18 34	18 40	18 47
26	17 08	17 18	17 25	17 32	17 37	17 42	17 50	17 58	18 05	18 12	18 20	18 29	18 35	18 41
30	17 16	17 24	17 30	17 35	17 40	17 44	17 51	17 58	18 04	18 10	18 17	18 25	18 29	18 34
Sept. 3	17 23	17 29	17 35	17 39	17 43	17 46	17 52	17 57	18 03	18 08	18 13	18 20	18 24	18 28
7	17 30	17 35	17 39	17 43	17 46	17 48	17 53	17 57	18 01	18 05	18 10	18 15	18 18	18 22
11	17 38	17 41	17 44	17 46	17 49	17 50	17 54	17 57	18 00	18 03	18 06	18 10	18 12	18 15
15	17 45	17 47	17 49	17 50	17 51	17 53	17 55	17 56	17 58	18 00	18 03	18 05	18 07	18 09
19	17 53	17 53	17 54	17 54	17 54	17 55	17 55	17 56	17 57	17 58	17 59	18 00	18 01	18 02
23	18 00	17 59	17 58	17 58	17 57	17 57	17 56	17 56	17 55	17 55	17 55	17 55	17 55	17 55
27	18 08	18 05	18 03	18 02	18 00	17 59	17 57	17 56	17 54	17 53	17 52	17 50	17 49	17 49
Oct. 1	18 15	18 11	18 08	18 06	18 03	18 01	17 58	17 55	17 53	17 50	17 48	17 45	17 44	17 42
5	18 23	18 18	18 13	18 10	18 07	18 04	17 59	17 55	17 52	17 48	17 44	17 40	17 38	17 36

UNIVERSAL TIME FOR MERIDIAN OF GREENWICH

SUNRISE

Lat.	+40°	+42°	+44°	+46°	+48°	+50°	+52°	+54°	+56°	+58°	+60°	+62°	+64°	+66°
	h m	h m	h m	h m	h m	h m	h m	h m	h m	h m	h m	h m	h m	h m
July 1	4 35	4 28	4 21	4 13	4 05	3 55	3 44	3 32	3 18	3 02	2 42	2 17	1 41	0 21
5	4 37	4 30	4 23	4 16	4 07	3 58	3 48	3 36	3 22	3 06	2 47	2 23	1 50	0 48
9	4 40	4 33	4 26	4 19	4 11	4 01	3 51	3 40	3 27	3 12	2 53	2 30	1 59	1 09
13	4 42	4 36	4 29	4 22	4 14	4 05	3 56	3 45	3 32	3 18	3 00	2 38	2 10	1 27
17	4 45	4 39	4 33	4 26	4 18	4 10	4 01	3 50	3 38	3 24	3 08	2 48	2 22	1 45
21	4 49	4 43	4 37	4 30	4 23	4 15	4 06	3 56	3 45	3 31	3 16	2 57	2 34	2 02
25	4 52	4 47	4 41	4 34	4 27	4 20	4 11	4 02	3 51	3 39	3 25	3 08	2 46	2 18
29	4 56	4 50	4 45	4 39	4 32	4 25	4 17	4 09	3 59	3 47	3 34	3 18	2 59	2 34
Aug. 2	4 59	4 54	4 49	4 44	4 38	4 31	4 23	4 15	4 06	3 56	3 43	3 29	3 12	2 50
6	5 03	4 59	4 54	4 48	4 43	4 37	4 30	4 22	4 14	4 04	3 53	3 40	3 24	3 05
10	5 07	5 03	4 58	4 53	4 48	4 42	4 36	4 29	4 21	4 13	4 02	3 51	3 37	3 20
14	5 11	5 07	5 03	4 58	4 54	4 48	4 43	4 36	4 29	4 21	4 12	4 02	3 49	3 34
18	5 14	5 11	5 07	5 03	4 59	4 54	4 49	4 43	4 37	4 30	4 22	4 12	4 02	3 48
22	5 18	5 15	5 12	5 08	5 04	5 00	4 56	4 51	4 45	4 39	4 31	4 23	4 14	4 02
26	5 22	5 19	5 16	5 13	5 10	5 06	5 02	4 58	4 53	4 47	4 41	4 34	4 26	4 16
30	5 26	5 24	5 21	5 18	5 15	5 12	5 09	5 05	5 01	4 56	4 50	4 44	4 37	4 29
Sept. 3	5 30	5 28	5 26	5 23	5 21	5 18	5 15	5 12	5 08	5 04	5 00	4 55	4 49	4 42
7	5 33	5 32	5 30	5 28	5 26	5 24	5 22	5 19	5 16	5 13	5 09	5 05	5 01	4 55
11	5 37	5 36	5 35	5 33	5 32	5 30	5 28	5 26	5 24	5 22	5 19	5 16	5 12	5 08
15	5 41	5 40	5 39	5 38	5 37	5 36	5 35	5 33	5 32	5 30	5 28	5 26	5 23	5 21
19	5 45	5 44	5 44	5 43	5 43	5 42	5 41	5 40	5 40	5 39	5 38	5 36	5 35	5 33
23	5 49	5 48	5 48	5 48	5 48	5 48	5 48	5 48	5 47	5 47	5 47	5 47	5 46	5 46
27	5 52	5 53	5 53	5 53	5 54	5 54	5 54	5 55	5 55	5 56	5 56	5 57	5 58	5 58
Oct. 1	5 56	5 57	5 58	5 59	5 59	6 00	6 01	6 02	6 03	6 04	6 06	6 07	6 09	6 11
5	6 00	6 01	6 03	6 04	6 05	6 06	6 08	6 09	6 11	6 13	6 15	6 18	6 21	6 24

SUNSET

Lat.	+40°	+42°	+44°	+46°	+48°	+50°	+52°	+54°	+56°	+58°	+60°	+62°	+64°	+66°
	h m	h m	h m	h m	h m	h m	h m	h m	h m	h m	h m	h m	h m	h m
July 1	19 33	19 39	19 47	19 54	20 03	20 12	20 23	20 35	20 49	21 05	21 25	21 50	22 25	23 39
5	19 32	19 38	19 45	19 53	20 02	20 11	20 21	20 33	20 46	21 02	21 21	21 45	22 18	23 16
9	19 31	19 37	19 44	19 51	19 59	20 08	20 19	20 30	20 43	20 58	21 16	21 39	22 09	22 58
13	19 29	19 35	19 42	19 49	19 57	20 05	20 15	20 26	20 38	20 53	21 10	21 31	21 59	22 40
17	19 26	19 32	19 39	19 46	19 53	20 02	20 11	20 21	20 33	20 47	21 03	21 23	21 48	22 24
21	19 24	19 29	19 35	19 42	19 49	19 57	20 06	20 16	20 27	20 40	20 55	21 13	21 36	22 07
25	19 20	19 26	19 32	19 38	19 45	19 52	20 01	20 10	20 20	20 32	20 46	21 03	21 24	21 51
29	19 17	19 22	19 27	19 33	19 40	19 47	19 54	20 03	20 13	20 24	20 37	20 53	21 11	21 35
Aug. 2	19 12	19 17	19 22	19 28	19 34	19 41	19 48	19 56	20 05	20 15	20 27	20 41	20 58	21 19
6	19 08	19 12	19 17	19 22	19 28	19 34	19 41	19 48	19 57	20 06	20 17	20 30	20 45	21 04
10	19 03	19 07	19 12	19 16	19 22	19 27	19 33	19 40	19 48	19 56	20 06	20 18	20 31	20 48
14	18 58	19 02	19 06	19 10	19 15	19 20	19 25	19 32	19 39	19 46	19 55	20 06	20 18	20 32
18	18 52	18 56	18 59	19 03	19 08	19 12	19 17	19 23	19 29	19 36	19 44	19 53	20 04	20 16
22	18 47	18 50	18 53	18 56	19 00	19 04	19 09	19 14	19 19	19 26	19 33	19 41	19 50	20 01
26	18 41	18 43	18 46	18 49	18 53	18 56	19 00	19 04	19 09	19 15	19 21	19 28	19 36	19 45
30	18 34	18 37	18 39	18 42	18 45	18 48	18 51	18 55	18 59	19 04	19 09	19 15	19 22	19 30
Sept. 3	18 28	18 30	18 32	18 34	18 37	18 39	18 42	18 45	18 49	18 53	18 57	19 02	19 08	19 14
7	18 22	18 23	18 25	18 27	18 29	18 31	18 33	18 36	18 38	18 41	18 45	18 49	18 53	18 59
11	18 15	18 16	18 18	18 19	18 20	18 22	18 24	18 26	18 28	18 30	18 33	18 36	18 39	18 43
15	18 09	18 09	18 10	18 11	18 12	18 13	18 14	18 16	18 17	18 19	18 21	18 23	18 25	18 28
19	18 02	18 02	18 03	18 03	18 04	18 04	18 05	18 06	18 07	18 07	18 08	18 10	18 11	18 12
23	17 55	17 55	17 55	17 55	17 55	17 56	17 56	17 56	17 56	17 56	17 56	17 56	17 57	17 57
27	17 49	17 48	17 48	17 48	17 47	17 47	17 46	17 46	17 45	17 45	17 44	17 43	17 43	17 42
Oct. 1	17 42	17 41	17 41	17 40	17 39	17 38	17 37	17 36	17 35	17 33	17 32	17 30	17 28	17 26
5	17 36	17 35	17 33	17 32	17 31	17 29	17 28	17 26	17 24	17 22	17 20	17 17	17 14	17 11

UNIVERSAL TIME FOR MERIDIAN OF GREENWICH

SUNRISE

Lat.	−55°	−50°	−45°	−40°	−35°	−30°	−20°	−10°	0°	+10°	+20°	+30°	+35°	+40°
	h m	h m	h m	h m	h m	h m	h m	h m	h m	h m	h m	h m	h m	h m
Oct. 1	5 25	5 29	5 32	5 34	5 36	5 38	5 41	5 44	5 46	5 49	5 51	5 53	5 55	5 56
5	5 15	5 20	5 24	5 28	5 31	5 33	5 38	5 42	5 45	5 48	5 52	5 56	5 58	6 00
9	5 05	5 11	5 17	5 21	5 25	5 29	5 34	5 39	5 44	5 48	5 53	5 58	6 01	6 04
13	4 55	5 03	5 10	5 15	5 20	5 24	5 31	5 37	5 43	5 48	5 54	6 01	6 04	6 08
17	4 45	4 55	5 03	5 09	5 15	5 20	5 28	5 35	5 42	5 49	5 55	6 03	6 08	6 13
21	4 35	4 47	4 56	5 03	5 10	5 15	5 25	5 33	5 41	5 49	5 57	6 06	6 11	6 17
25	4 26	4 39	4 49	4 58	5 05	5 11	5 22	5 32	5 41	5 49	5 58	6 09	6 15	6 21
29	4 17	4 32	4 43	4 53	5 01	5 08	5 20	5 31	5 40	5 50	6 00	6 12	6 18	6 26
Nov. 2	4 08	4 24	4 37	4 48	4 57	5 04	5 18	5 29	5 40	5 51	6 02	6 15	6 22	6 30
6	4 00	4 18	4 32	4 43	4 53	5 01	5 16	5 28	5 40	5 52	6 04	6 18	6 26	6 35
10	3 52	4 11	4 27	4 39	4 49	4 59	5 14	5 28	5 40	5 53	6 06	6 21	6 30	6 40
14	3 45	4 06	4 22	4 35	4 47	4 56	5 13	5 28	5 41	5 54	6 08	6 24	6 34	6 44
18	3 38	4 01	4 18	4 32	4 44	4 54	5 12	5 27	5 42	5 56	6 11	6 28	6 38	6 49
22	3 32	3 56	4 14	4 29	4 42	4 53	5 12	5 28	5 43	5 57	6 13	6 31	6 41	6 53
26	3 27	3 52	4 12	4 27	4 40	4 52	5 11	5 28	5 44	5 59	6 16	6 34	6 45	6 58
30	3 23	3 49	4 09	4 26	4 39	4 51	5 12	5 29	5 45	6 01	6 18	6 38	6 49	7 02
Dec. 4	3 19	3 47	4 08	4 25	4 39	4 51	5 12	5 30	5 47	6 03	6 21	6 41	6 52	7 06
8	3 17	3 46	4 07	4 25	4 39	4 52	5 13	5 31	5 48	6 05	6 23	6 44	6 56	7 09
12	3 16	3 45	4 07	4 25	4 40	4 52	5 14	5 33	5 50	6 07	6 26	6 46	6 59	7 13
16	3 15	3 45	4 08	4 26	4 41	4 54	5 16	5 35	5 52	6 09	6 28	6 49	7 01	7 15
20	3 16	3 47	4 09	4 27	4 42	4 55	5 17	5 36	5 54	6 11	6 30	6 51	7 04	7 18
24	3 19	3 49	4 11	4 29	4 44	4 57	5 19	5 38	5 56	6 13	6 32	6 53	7 06	7 20
28	3 22	3 52	4 14	4 32	4 47	5 00	5 22	5 41	5 58	6 15	6 34	6 55	7 07	7 21
32	3 26	3 55	4 17	4 35	4 50	5 02	5 24	5 43	6 00	6 17	6 35	6 56	7 08	7 22
36	3 31	4 00	4 21	4 38	4 53	5 05	5 27	5 45	6 02	6 19	6 36	6 57	7 09	7 22

SUNSET

	h m	h m	h m	h m	h m	h m	h m	h m	h m	h m	h m	h m	h m	h m
Oct. 1	18 15	18 11	18 08	18 06	18 03	18 01	17 58	17 55	17 53	17 50	17 48	17 45	17 44	17 42
5	18 23	18 18	18 13	18 10	18 07	18 04	17 59	17 55	17 52	17 48	17 44	17 40	17 38	17 36
9	18 31	18 24	18 18	18 14	18 10	18 06	18 00	17 55	17 50	17 46	17 41	17 36	17 33	17 29
13	18 39	18 30	18 24	18 18	18 13	18 09	18 02	17 55	17 49	17 44	17 38	17 31	17 27	17 23
17	18 47	18 37	18 29	18 22	18 16	18 11	18 03	17 55	17 49	17 42	17 35	17 27	17 22	17 17
21	18 55	18 43	18 34	18 27	18 20	18 14	18 04	17 56	17 48	17 40	17 32	17 23	17 18	17 12
25	19 03	18 50	18 40	18 31	18 24	18 17	18 06	17 56	17 47	17 39	17 29	17 19	17 13	17 06
29	19 12	18 57	18 45	18 36	18 27	18 20	18 08	17 57	17 47	17 37	17 27	17 15	17 09	17 01
Nov. 2	19 20	19 04	18 51	18 40	18 31	18 23	18 10	17 58	17 47	17 36	17 25	17 12	17 05	16 56
6	19 29	19 11	18 57	18 45	18 35	18 26	18 12	17 59	17 47	17 35	17 23	17 09	17 01	16 52
10	19 37	19 17	19 02	18 50	18 39	18 30	18 14	18 00	17 47	17 35	17 22	17 07	16 58	16 48
14	19 45	19 24	19 08	18 54	18 43	18 33	18 16	18 02	17 48	17 35	17 20	17 04	16 55	16 44
18	19 54	19 31	19 13	18 59	18 47	18 36	18 19	18 03	17 49	17 35	17 20	17 03	16 53	16 41
22	20 01	19 37	19 19	19 04	18 51	18 40	18 21	18 05	17 50	17 35	17 19	17 01	16 51	16 39
26	20 09	19 43	19 24	19 08	18 55	18 43	18 24	18 07	17 51	17 36	17 19	17 00	16 49	16 37
30	20 16	19 49	19 29	19 12	18 58	18 47	18 26	18 09	17 52	17 36	17 19	17 00	16 49	16 36
Dec. 4	20 22	19 54	19 33	19 16	19 02	18 50	18 29	18 11	17 54	17 38	17 20	17 00	16 48	16 35
8	20 28	19 59	19 37	19 20	19 05	18 53	18 31	18 13	17 56	17 39	17 21	17 00	16 48	16 35
12	20 33	20 03	19 41	19 23	19 08	18 55	18 34	18 15	17 58	17 40	17 22	17 01	16 49	16 35
16	20 36	20 06	19 44	19 26	19 11	18 58	18 36	18 17	18 00	17 42	17 24	17 02	16 50	16 36
20	20 39	20 09	19 46	19 28	19 13	19 00	18 38	18 19	18 02	17 44	17 25	17 04	16 52	16 38
24	20 41	20 11	19 48	19 30	19 15	19 02	18 40	18 21	18 04	17 46	17 28	17 06	16 54	16 40
28	20 41	20 12	19 49	19 31	19 17	19 04	18 42	18 23	18 05	17 48	17 30	17 09	16 56	16 42
32	20 41	20 12	19 50	19 32	19 17	19 05	18 43	18 25	18 07	17 50	17 32	17 11	16 59	16 46
36	20 39	20 11	19 49	19 32	19 18	19 05	18 44	18 26	18 09	17 52	17 35	17 14	17 03	16 49

SUNRISE AND SUNSET, 1992

UNIVERSAL TIME FOR MERIDIAN OF GREENWICH

SUNRISE

Lat.		+40°	+42°	+44°	+46°	+48°	+50°	+52°	+54°	+56°	+58°	+60°	+62°	+64°	+66°
		h m	h m	h m	h m	h m	h m	h m	h m	h m	h m	h m	h m	h m	h m
Oct.	1	5 56	5 57	5 58	5 59	5 59	6 00	6 01	6 02	6 03	6 04	6 06	6 07	6 09	6 11
	5	6 00	6 01	6 03	6 04	6 05	6 06	6 08	6 09	6 11	6 13	6 15	6 18	6 21	6 24
	9	6 04	6 06	6 07	6 09	6 11	6 13	6 15	6 17	6 19	6 22	6 25	6 29	6 32	6 37
	13	6 08	6 10	6 12	6 14	6 17	6 19	6 22	6 24	6 27	6 31	6 35	6 39	6 44	6 50
	17	6 13	6 15	6 17	6 20	6 22	6 25	6 28	6 32	6 36	6 40	6 45	6 50	6 56	7 03
	21	6 17	6 20	6 22	6 25	6 28	6 32	6 36	6 40	6 44	6 49	6 55	7 01	7 08	7 17
	25	6 21	6 24	6 27	6 31	6 34	6 38	6 43	6 47	6 53	6 58	7 05	7 12	7 21	7 31
	29	6 26	6 29	6 33	6 36	6 41	6 45	6 50	6 55	7 01	7 08	7 15	7 24	7 33	7 45
Nov.	2	6 30	6 34	6 38	6 42	6 47	6 52	6 57	7 03	7 10	7 17	7 25	7 35	7 46	8 00
	6	6 35	6 39	6 43	6 48	6 53	6 58	7 04	7 11	7 18	7 26	7 36	7 46	7 59	8 14
	10	6 40	6 44	6 49	6 54	6 59	7 05	7 11	7 19	7 27	7 36	7 46	7 58	8 12	8 29
	14	6 44	6 49	6 54	6 59	7 05	7 11	7 18	7 26	7 35	7 45	7 56	8 09	8 25	8 44
	18	6 49	6 54	6 59	7 05	7 11	7 18	7 25	7 34	7 43	7 54	8 06	8 20	8 38	8 59
	22	6 53	6 59	7 04	7 10	7 17	7 24	7 32	7 41	7 51	8 02	8 16	8 31	8 50	9 15
	26	6 58	7 03	7 09	7 15	7 22	7 30	7 39	7 48	7 59	8 11	8 25	8 42	9 02	9 30
	30	7 02	7 08	7 14	7 20	7 28	7 36	7 45	7 54	8 06	8 18	8 33	8 51	9 14	9 44
Dec.	4	7 06	7 12	7 18	7 25	7 33	7 41	7 50	8 00	8 12	8 25	8 41	9 00	9 25	9 58
	8	7 09	7 15	7 22	7 29	7 37	7 45	7 55	8 06	8 18	8 32	8 48	9 08	9 34	10 11
	12	7 13	7 19	7 26	7 33	7 41	7 50	7 59	8 10	8 23	8 37	8 54	9 15	9 42	10 21
	16	7 15	7 22	7 29	7 36	7 44	7 53	8 03	8 14	8 26	8 41	8 58	9 20	9 48	10 29
	20	7 18	7 24	7 31	7 39	7 47	7 56	8 05	8 17	8 29	8 44	9 02	9 23	9 52	10 34
	24	7 20	7 26	7 33	7 40	7 48	7 57	8 07	8 18	8 31	8 46	9 03	9 25	9 53	10 36
	28	7 21	7 27	7 34	7 42	7 50	7 58	8 08	8 19	8 32	8 46	9 04	9 25	9 52	10 33
	32	7 22	7 28	7 35	7 42	7 50	7 59	8 08	8 19	8 31	8 45	9 02	9 23	9 49	10 28
	36	7 22	7 28	7 35	7 42	7 49	7 58	8 07	8 18	8 30	8 43	9 00	9 19	9 45	10 20

SUNSET

Lat.		+40°	+42°	+44°	+46°	+48°	+50°	+52°	+54°	+56°	+58°	+60°	+62°	+64°	+66°
		h m	h m	h m	h m	h m	h m	h m	h m	h m	h m	h m	h m	h m	h m
Oct.	1	17 42	17 41	17 41	17 40	17 39	17 38	17 37	17 36	17 35	17 33	17 32	17 30	17 28	17 26
	5	17 36	17 35	17 33	17 32	17 31	17 29	17 28	17 26	17 24	17 22	17 20	17 17	17 14	17 11
	9	17 29	17 28	17 26	17 25	17 23	17 21	17 19	17 17	17 14	17 11	17 08	17 05	17 01	16 56
	13	17 23	17 21	17 19	17 17	17 15	17 13	17 10	17 07	17 04	17 00	16 56	16 52	16 47	16 41
	17	17 17	17 15	17 13	17 10	17 07	17 04	17 01	16 58	16 54	16 50	16 45	16 39	16 33	16 26
	21	17 12	17 09	17 06	17 03	17 00	16 57	16 53	16 49	16 44	16 39	16 33	16 27	16 19	16 11
	25	17 06	17 03	17 00	16 57	16 53	16 49	16 45	16 40	16 35	16 29	16 22	16 15	16 06	15 56
	29	17 01	16 58	16 54	16 50	16 46	16 42	16 37	16 32	16 26	16 19	16 11	16 03	15 53	15 41
Nov.	2	16 56	16 53	16 49	16 44	16 40	16 35	16 29	16 23	16 17	16 09	16 01	15 51	15 40	15 26
	6	16 52	16 48	16 44	16 39	16 34	16 28	16 22	16 16	16 09	16 00	15 51	15 40	15 27	15 12
	10	16 48	16 44	16 39	16 34	16 28	16 22	16 16	16 09	16 01	15 52	15 41	15 29	15 15	14 58
	14	16 44	16 40	16 35	16 29	16 23	16 17	16 10	16 02	15 53	15 44	15 32	15 19	15 03	14 44
	18	16 41	16 36	16 31	16 25	16 19	16 12	16 05	15 56	15 47	15 36	15 24	15 09	14 52	14 30
	22	16 39	16 34	16 28	16 22	16 15	16 08	16 00	15 51	15 41	15 30	15 16	15 01	14 41	14 17
	26	16 37	16 31	16 25	16 19	16 12	16 04	15 56	15 46	15 36	15 24	15 10	14 53	14 32	14 05
	30	16 36	16 30	16 24	16 17	16 10	16 02	15 53	15 43	15 32	15 19	15 04	14 46	14 23	13 53
Dec.	4	16 35	16 29	16 22	16 15	16 08	16 00	15 50	15 40	15 28	15 15	14 59	14 40	14 16	13 42
	8	16 35	16 28	16 22	16 15	16 07	15 58	15 49	15 38	15 26	15 12	14 56	14 36	14 10	13 33
	12	16 35	16 29	16 22	16 15	16 07	15 58	15 48	15 37	15 25	15 11	14 54	14 33	14 06	13 26
	16	16 36	16 30	16 23	16 16	16 07	15 59	15 49	15 38	15 25	15 10	14 53	14 32	14 04	13 22
	20	16 38	16 31	16 24	16 17	16 09	16 00	15 50	15 39	15 26	15 11	14 54	14 32	14 04	13 21
	24	16 40	16 33	16 27	16 19	16 11	16 02	15 52	15 41	15 29	15 14	14 56	14 35	14 06	13 24
	28	16 42	16 36	16 29	16 22	16 14	16 05	15 55	15 44	15 32	15 17	15 00	14 39	14 11	13 30
	32	16 46	16 39	16 33	16 25	16 18	16 09	15 59	15 49	15 36	15 22	15 05	14 45	14 18	13 40
	36	16 49	16 43	16 36	16 29	16 22	16 13	16 04	15 53	15 42	15 28	15 12	14 52	14 27	13 52

UNIVERSAL TIME FOR MERIDIAN OF GREENWICH

MORNING CIVIL TWILIGHT

Lat.	−55°	−50°	−45°	−40°	−35°	−30°	−20°	−10°	0°	+10°	+20°	+30°	+35°	+40°
	h m	h m	h m	h m	h m	h m	h m	h m	h m	h m	h m	h m	h m	h m
Jan. −1	2 26	3 08	3 38	4 00	4 18	4 33	4 58	5 18	5 36	5 53	6 10	6 29	6 39	6 51
3	2 31	3 13	3 41	4 03	4 21	4 36	5 00	5 20	5 38	5 55	6 12	6 30	6 40	6 52
7	2 38	3 18	3 46	4 07	4 25	4 39	5 03	5 23	5 40	5 56	6 13	6 31	6 41	6 52
11	2 46	3 24	3 51	4 12	4 28	4 43	5 06	5 25	5 42	5 58	6 14	6 31	6 41	6 52
15	2 54	3 30	3 56	4 16	4 32	4 46	5 09	5 27	5 43	5 59	6 14	6 31	6 40	6 51
19	3 03	3 37	4 02	4 21	4 37	4 50	5 11	5 29	5 45	6 00	6 14	6 30	6 39	6 49
23	3 13	3 45	4 08	4 26	4 41	4 54	5 14	5 31	5 46	6 00	6 14	6 29	6 38	6 47
27	3 23	3 52	4 14	4 31	4 45	4 57	5 17	5 33	5 47	6 01	6 14	6 28	6 36	6 45
31	3 33	4 00	4 21	4 37	4 50	5 01	5 20	5 35	5 48	6 01	6 13	6 26	6 34	6 42
Feb. 4	3 43	4 08	4 27	4 42	4 54	5 05	5 22	5 36	5 49	6 00	6 12	6 24	6 31	6 38
8	3 53	4 16	4 33	4 47	4 59	5 09	5 25	5 38	5 49	6 00	6 10	6 21	6 28	6 34
12	4 03	4 24	4 40	4 53	5 03	5 12	5 27	5 39	5 50	5 59	6 09	6 19	6 24	6 30
16	4 13	4 32	4 46	4 58	5 07	5 16	5 29	5 40	5 50	5 58	6 07	6 15	6 20	6 25
20	4 22	4 39	4 52	5 03	5 12	5 19	5 31	5 41	5 49	5 57	6 04	6 12	6 16	6 20
24	4 31	4 47	4 58	5 08	5 16	5 22	5 33	5 42	5 49	5 56	6 02	6 08	6 11	6 15
28	4 41	4 54	5 04	5 13	5 19	5 25	5 35	5 42	5 48	5 54	5 59	6 04	6 06	6 09
Mar. 3	4 50	5 01	5 10	5 17	5 23	5 28	5 36	5 43	5 48	5 52	5 56	6 00	6 01	6 03
7	4 58	5 08	5 16	5 22	5 27	5 31	5 38	5 43	5 47	5 50	5 53	5 55	5 56	5 57
11	5 07	5 15	5 21	5 26	5 30	5 34	5 39	5 43	5 46	5 48	5 50	5 51	5 51	5 51
15	5 15	5 22	5 27	5 31	5 34	5 36	5 40	5 43	5 45	5 46	5 46	5 46	5 45	5 44
19	5 23	5 28	5 32	5 35	5 37	5 39	5 42	5 43	5 44	5 44	5 43	5 41	5 40	5 38
23	5 31	5 35	5 37	5 39	5 40	5 41	5 43	5 43	5 43	5 41	5 39	5 36	5 34	5 31
27	5 39	5 41	5 42	5 43	5 44	5 44	5 44	5 43	5 41	5 39	5 36	5 31	5 28	5 25
31	5 47	5 47	5 47	5 47	5 47	5 46	5 45	5 43	5 40	5 37	5 32	5 26	5 23	5 18
Apr. 4	5 54	5 53	5 52	5 51	5 50	5 49	5 46	5 43	5 39	5 34	5 29	5 22	5 17	5 12

EVENING CIVIL TWILIGHT

Lat.	−55°	−50°	−45°	−40°	−35°	−30°	−20°	−10°	0°	+10°	+20°	+30°	+35°	+40°
	h m	h m	h m	h m	h m	h m	h m	h m	h m	h m	h m	h m	h m	h m
Jan. −1	21 39	20 56	20 27	20 04	19 46	19 31	19 07	18 46	18 29	18 12	17 54	17 36	17 25	17 14
3	21 37	20 55	20 27	20 05	19 47	19 32	19 08	18 48	18 30	18 14	17 57	17 39	17 28	17 17
7	21 33	20 53	20 26	20 04	19 47	19 33	19 09	18 49	18 32	18 16	17 59	17 41	17 31	17 20
11	21 29	20 51	20 24	20 03	19 47	19 33	19 09	18 50	18 34	18 18	18 02	17 45	17 35	17 24
15	21 23	20 47	20 22	20 02	19 46	19 32	19 10	18 51	18 35	18 20	18 04	17 48	17 38	17 28
19	21 16	20 43	20 18	19 59	19 44	19 31	19 09	18 52	18 36	18 22	18 07	17 51	17 42	17 32
23	21 09	20 37	20 15	19 57	19 42	19 29	19 09	18 52	18 37	18 23	18 09	17 54	17 46	17 37
27	21 01	20 32	20 10	19 53	19 39	19 27	19 08	18 52	18 38	18 25	18 12	17 58	17 50	17 41
31	20 52	20 25	20 05	19 49	19 36	19 25	19 07	18 52	18 39	18 26	18 14	18 01	17 54	17 46
Feb. 4	20 43	20 18	20 00	19 45	19 33	19 22	19 05	18 51	18 39	18 27	18 16	18 04	17 57	17 50
8	20 34	20 11	19 54	19 40	19 29	19 19	19 03	18 50	18 39	18 29	18 18	18 07	18 01	17 55
12	20 24	20 03	19 48	19 35	19 25	19 16	19 01	18 49	18 39	18 29	18 20	18 10	18 05	17 59
16	20 14	19 55	19 41	19 30	19 20	19 12	18 59	18 48	18 39	18 30	18 22	18 13	18 09	18 04
20	20 04	19 47	19 34	19 24	19 15	19 08	18 56	18 47	18 38	18 31	18 24	18 16	18 12	18 08
24	19 54	19 39	19 27	19 18	19 10	19 04	18 53	18 45	18 38	18 31	18 25	18 19	18 16	18 13
28	19 43	19 30	19 20	19 12	19 05	19 00	18 50	18 43	18 37	18 32	18 27	18 22	18 20	18 17
Mar. 3	19 33	19 22	19 13	19 06	19 00	18 55	18 47	18 41	18 36	18 32	18 28	18 25	18 23	18 21
7	19 22	19 13	19 05	18 59	18 54	18 50	18 44	18 39	18 35	18 32	18 29	18 27	18 26	18 26
11	19 12	19 04	18 58	18 53	18 49	18 46	18 40	18 37	18 34	18 32	18 31	18 30	18 30	18 30
15	19 01	18 55	18 50	18 46	18 43	18 41	18 37	18 34	18 33	18 32	18 32	18 32	18 33	18 34
19	18 51	18 46	18 43	18 40	18 38	18 36	18 34	18 32	18 32	18 32	18 33	18 35	18 36	18 38
23	18 41	18 37	18 35	18 33	18 32	18 31	18 30	18 30	18 30	18 32	18 34	18 37	18 40	18 42
27	18 30	18 29	18 28	18 27	18 26	18 26	18 27	18 28	18 29	18 32	18 35	18 40	18 43	18 47
31	18 20	18 20	18 20	18 20	18 21	18 21	18 23	18 25	18 28	18 32	18 36	18 42	18 46	18 51
Apr. 4	18 10	18 12	18 13	18 14	18 15	18 17	18 20	18 23	18 27	18 32	18 37	18 45	18 49	18 55

UNIVERSAL TIME FOR MERIDIAN OF GREENWICH

MORNING CIVIL TWILIGHT

Lat.	+40°	+42°	+44°	+46°	+48°	+50°	+52°	+54°	+56°	+58°	+60°	+62°	+64°	+66°
	h m	h m	h m	h m	h m	h m	h m	h m	h m	h m	h m	h m	h m	h m
Jan. −1	6 51	6 56	7 01	7 07	7 13	7 20	7 28	7 36	7 45	7 55	8 06	8 19	8 35	8 54
3	6 52	6 57	7 02	7 08	7 14	7 20	7 28	7 35	7 44	7 54	8 05	8 18	8 33	8 52
7	6 52	6 57	7 02	7 07	7 13	7 20	7 27	7 34	7 43	7 52	8 03	8 16	8 30	8 48
11	6 52	6 56	7 01	7 07	7 12	7 18	7 25	7 32	7 41	7 50	8 00	8 12	8 26	8 43
15	6 51	6 55	7 00	7 05	7 10	7 16	7 23	7 30	7 37	7 46	7 56	8 07	8 20	8 36
19	6 49	6 53	6 58	7 03	7 08	7 13	7 20	7 26	7 33	7 42	7 51	8 01	8 13	8 28
23	6 47	6 51	6 55	7 00	7 05	7 10	7 16	7 22	7 29	7 36	7 45	7 54	8 05	8 19
27	6 45	6 48	6 52	6 56	7 01	7 06	7 11	7 17	7 23	7 30	7 38	7 47	7 57	8 09
31	6 42	6 45	6 49	6 52	6 57	7 01	7 06	7 11	7 17	7 23	7 30	7 38	7 47	7 58
Feb. 4	6 38	6 41	6 44	6 48	6 52	6 56	7 00	7 05	7 10	7 16	7 22	7 29	7 37	7 47
8	6 34	6 37	6 40	6 43	6 46	6 50	6 54	6 58	7 02	7 07	7 13	7 19	7 26	7 35
12	6 30	6 32	6 35	6 38	6 41	6 44	6 47	6 51	6 55	6 59	7 04	7 09	7 15	7 22
16	6 25	6 27	6 29	6 32	6 34	6 37	6 40	6 43	6 46	6 50	6 54	6 58	7 03	7 09
20	6 20	6 22	6 24	6 26	6 28	6 30	6 32	6 35	6 37	6 40	6 43	6 47	6 51	6 56
24	6 15	6 16	6 17	6 19	6 21	6 22	6 24	6 26	6 28	6 30	6 33	6 35	6 38	6 42
28	6 09	6 10	6 11	6 12	6 13	6 15	6 16	6 17	6 19	6 20	6 22	6 23	6 25	6 28
Mar. 3	6 03	6 04	6 04	6 05	6 06	6 07	6 07	6 08	6 09	6 10	6 10	6 11	6 12	6 13
7	5 57	5 57	5 58	5 58	5 58	5 58	5 58	5 59	5 59	5 59	5 59	5 59	5 59	5 58
11	5 51	5 51	5 51	5 50	5 50	5 50	5 49	5 49	5 48	5 48	5 47	5 46	5 45	5 43
15	5 44	5 44	5 43	5 43	5 42	5 41	5 40	5 39	5 38	5 37	5 35	5 33	5 31	5 28
19	5 38	5 37	5 36	5 35	5 34	5 33	5 31	5 29	5 27	5 25	5 23	5 20	5 16	5 12
23	5 31	5 30	5 29	5 27	5 26	5 24	5 22	5 19	5 17	5 14	5 10	5 06	5 02	4 57
27	5 25	5 23	5 21	5 19	5 17	5 15	5 12	5 09	5 06	5 02	4 58	4 53	4 47	4 40
31	5 18	5 16	5 14	5 11	5 09	5 06	5 03	4 59	4 55	4 50	4 45	4 39	4 32	4 24
Apr. 4	5 12	5 09	5 07	5 04	5 00	4 57	4 53	4 49	4 44	4 39	4 32	4 25	4 17	4 07

EVENING CIVIL TWILIGHT

Lat.	+40°	+42°	+44°	+46°	+48°	+50°	+52°	+54°	+56°	+58°	+60°	+62°	+64°	+66°
	h m	h m	h m	h m	h m	h m	h m	h m	h m	h m	h m	h m	h m	h m
Jan. −1	17 14	17 09	17 03	16 58	16 51	16 45	16 37	16 29	16 20	16 10	15 59	15 45	15 30	15 10
3	17 17	17 12	17 07	17 01	16 55	16 48	16 41	16 33	16 25	16 15	16 04	15 51	15 35	15 17
7	17 20	17 16	17 10	17 05	16 59	16 53	16 46	16 38	16 30	16 20	16 09	15 57	15 42	15 24
11	17 24	17 19	17 14	17 09	17 04	16 57	16 51	16 43	16 35	16 26	16 16	16 04	15 50	15 33
15	17 28	17 24	17 19	17 14	17 08	17 03	16 56	16 49	16 42	16 33	16 23	16 12	15 59	15 43
19	17 32	17 28	17 24	17 19	17 14	17 08	17 02	16 56	16 48	16 40	16 31	16 21	16 09	15 54
23	17 37	17 33	17 28	17 24	17 19	17 14	17 08	17 02	16 55	16 48	16 40	16 30	16 19	16 06
27	17 41	17 37	17 34	17 29	17 25	17 20	17 15	17 09	17 03	16 56	16 48	16 40	16 29	16 17
31	17 46	17 42	17 39	17 35	17 31	17 26	17 22	17 16	17 11	17 04	16 57	16 50	16 40	16 30
Feb. 4	17 50	17 47	17 44	17 40	17 37	17 33	17 28	17 24	17 19	17 13	17 07	17 00	16 52	16 42
8	17 55	17 52	17 49	17 46	17 43	17 39	17 35	17 31	17 27	17 22	17 16	17 10	17 03	16 55
12	17 59	17 57	17 54	17 52	17 49	17 46	17 42	17 39	17 35	17 31	17 26	17 21	17 15	17 08
16	18 04	18 02	18 00	17 57	17 55	17 52	17 49	17 46	17 43	17 40	17 36	17 31	17 26	17 21
20	18 08	18 07	18 05	18 03	18 01	17 59	17 56	17 54	17 51	17 49	17 45	17 42	17 38	17 34
24	18 13	18 11	18 10	18 08	18 07	18 05	18 03	18 02	18 00	17 57	17 55	17 53	17 50	17 46
28	18 17	18 16	18 15	18 14	18 13	18 12	18 10	18 09	18 08	18 06	18 05	18 03	18 01	17 59
Mar. 3	18 21	18 21	18 20	18 19	18 19	18 18	18 18	18 17	18 16	18 15	18 15	18 14	18 13	18 12
7	18 26	18 25	18 25	18 25	18 25	18 25	18 25	18 24	18 24	18 24	18 25	18 25	18 25	18 25
11	18 30	18 30	18 30	18 30	18 31	18 31	18 32	18 32	18 33	18 33	18 34	18 36	18 37	18 38
15	18 34	18 35	18 35	18 36	18 37	18 38	18 39	18 40	18 41	18 43	18 44	18 46	18 49	18 52
19	18 38	18 39	18 40	18 41	18 43	18 44	18 46	18 47	18 49	18 52	18 54	18 57	19 01	19 05
23	18 42	18 44	18 45	18 47	18 48	18 50	18 53	18 55	18 58	19 01	19 04	19 08	19 13	19 19
27	18 47	18 48	18 50	18 52	18 54	18 57	19 00	19 03	19 06	19 10	19 15	19 20	19 26	19 33
31	18 51	18 53	18 55	18 58	19 00	19 03	19 07	19 11	19 15	19 19	19 25	19 31	19 38	19 47
Apr. 4	18 55	18 57	19 00	19 03	19 06	19 10	19 14	19 18	19 23	19 29	19 35	19 43	19 51	20 02

UNIVERSAL TIME FOR MERIDIAN OF GREENWICH

MORNING CIVIL TWILIGHT

Lat.	−55°	−50°	−45°	−40°	−35°	−30°	−20°	−10°	0°	+10°	+20°	+30°	+35°	+40°
	h m	h m	h m	h m	h m	h m	h m	h m	h m	h m	h m	h m	h m	h m
Mar. 31	5 47	5 47	5 47	5 47	5 47	5 46	5 45	5 43	5 40	5 37	5 32	5 26	5 23	5 18
Apr. 4	5 54	5 53	5 52	5 51	5 50	5 49	5 46	5 43	5 39	5 34	5 29	5 22	5 17	5 12
8	6 02	5 59	5 57	5 55	5 53	5 51	5 47	5 42	5 38	5 32	5 25	5 17	5 11	5 05
12	6 09	6 05	6 02	5 59	5 56	5 53	5 48	5 42	5 37	5 30	5 22	5 12	5 06	4 59
16	6 16	6 11	6 07	6 03	5 59	5 56	5 49	5 42	5 35	5 28	5 19	5 07	5 01	4 52
20	6 24	6 17	6 11	6 07	6 02	5 58	5 50	5 42	5 34	5 26	5 15	5 03	4 55	4 46
24	6 31	6 23	6 16	6 10	6 05	6 00	5 51	5 42	5 33	5 24	5 12	4 59	4 50	4 40
28	6 38	6 28	6 21	6 14	6 08	6 03	5 52	5 43	5 33	5 22	5 10	4 55	4 45	4 35
May 2	6 45	6 34	6 25	6 18	6 11	6 05	5 54	5 43	5 32	5 20	5 07	4 51	4 41	4 29
6	6 51	6 39	6 30	6 21	6 14	6 07	5 55	5 43	5 32	5 19	5 05	4 47	4 37	4 24
10	6 58	6 45	6 34	6 25	6 17	6 10	5 56	5 44	5 31	5 18	5 03	4 44	4 33	4 19
14	7 04	6 50	6 38	6 28	6 20	6 12	5 58	5 44	5 31	5 17	5 01	4 41	4 29	4 15
18	7 10	6 55	6 42	6 32	6 23	6 14	5 59	5 45	5 31	5 16	4 59	4 38	4 26	4 11
22	7 16	6 59	6 46	6 35	6 25	6 16	6 00	5 46	5 31	5 15	4 58	4 36	4 23	4 08
26	7 21	7 04	6 50	6 38	6 28	6 19	6 02	5 47	5 31	5 15	4 57	4 34	4 21	4 05
30	7 26	7 08	6 53	6 41	6 30	6 21	6 03	5 47	5 32	5 15	4 56	4 33	4 19	4 02
June 3	7 30	7 11	6 56	6 44	6 33	6 23	6 05	5 48	5 32	5 15	4 55	4 32	4 17	4 00
7	7 34	7 14	6 59	6 46	6 35	6 24	6 06	5 49	5 33	5 15	4 55	4 31	4 16	3 59
11	7 37	7 17	7 01	6 48	6 36	6 26	6 07	5 50	5 33	5 16	4 55	4 31	4 16	3 58
15	7 39	7 19	7 03	6 50	6 38	6 27	6 09	5 51	5 34	5 16	4 56	4 31	4 16	3 58
19	7 41	7 21	7 04	6 51	6 39	6 29	6 10	5 52	5 35	5 17	4 56	4 32	4 16	3 58
23	7 42	7 21	7 05	6 52	6 40	6 29	6 11	5 53	5 36	5 18	4 57	4 32	4 17	3 59
27	7 42	7 22	7 06	6 52	6 40	6 30	6 11	5 54	5 37	5 19	4 58	4 34	4 18	4 00
July 1	7 41	7 21	7 05	6 52	6 41	6 30	6 12	5 55	5 38	5 20	5 00	4 35	4 20	4 02
5	7 39	7 20	7 05	6 52	6 40	6 30	6 12	5 55	5 39	5 21	5 01	4 37	4 22	4 04

EVENING CIVIL TWILIGHT

Lat.	−55°	−50°	−45°	−40°	−35°	−30°	−20°	−10°	0°	+10°	+20°	+30°	+35°	+40°
	h m	h m	h m	h m	h m	h m	h m	h m	h m	h m	h m	h m	h m	h m
Mar. 31	18 20	18 20	18 20	18 20	18 21	18 21	18 23	18 25	18 28	18 32	18 36	18 42	18 46	18 51
Apr. 4	18 10	18 12	18 13	18 14	18 15	18 17	18 20	18 23	18 27	18 32	18 37	18 45	18 49	18 55
8	18 01	18 03	18 06	18 08	18 10	18 12	18 16	18 21	18 26	18 32	18 39	18 47	18 53	18 59
12	17 51	17 55	17 59	18 02	18 05	18 08	18 13	18 19	18 25	18 32	18 40	18 50	18 56	19 03
16	17 42	17 47	17 52	17 56	18 00	18 04	18 10	18 17	18 24	18 32	18 41	18 53	19 00	19 08
20	17 33	17 40	17 46	17 51	17 55	17 59	18 07	18 15	18 23	18 32	18 43	18 55	19 03	19 12
24	17 25	17 33	17 39	17 45	17 51	17 56	18 05	18 14	18 23	18 33	18 44	18 58	19 07	19 17
28	17 16	17 26	17 34	17 40	17 46	17 52	18 02	18 12	18 22	18 33	18 45	19 01	19 10	19 21
May 2	17 09	17 19	17 28	17 36	17 42	17 49	18 00	18 11	18 22	18 34	18 47	19 04	19 14	19 25
6	17 01	17 13	17 23	17 31	17 39	17 46	17 58	18 10	18 22	18 34	18 49	19 06	19 17	19 30
10	16 54	17 07	17 18	17 27	17 35	17 43	17 56	18 09	18 22	18 35	18 50	19 09	19 21	19 34
14	16 48	17 02	17 14	17 24	17 32	17 40	17 55	18 08	18 22	18 36	18 52	19 12	19 24	19 38
18	16 42	16 57	17 10	17 21	17 30	17 38	17 54	18 08	18 22	18 37	18 54	19 15	19 27	19 42
22	16 37	16 53	17 07	17 18	17 28	17 37	17 53	18 07	18 22	18 38	18 56	19 17	19 31	19 46
26	16 33	16 50	17 04	17 16	17 26	17 35	17 52	18 07	18 23	18 39	18 58	19 20	19 34	19 50
30	16 29	16 47	17 01	17 14	17 25	17 34	17 52	18 08	18 23	18 40	18 59	19 22	19 37	19 53
June 3	16 26	16 45	17 00	17 12	17 24	17 34	17 51	18 08	18 24	18 41	19 01	19 25	19 39	19 57
7	16 24	16 43	16 59	17 12	17 23	17 33	17 51	18 08	18 25	18 43	19 03	19 27	19 42	19 59
11	16 22	16 42	16 58	17 11	17 23	17 33	17 52	18 09	18 26	18 44	19 04	19 29	19 44	20 02
15	16 22	16 42	16 58	17 11	17 23	17 34	17 52	18 10	18 27	18 45	19 05	19 30	19 45	20 03
19	16 22	16 42	16 58	17 12	17 24	17 34	17 53	18 10	18 28	18 46	19 06	19 31	19 46	20 05
23	16 23	16 43	16 59	17 13	17 24	17 35	17 54	18 11	18 28	18 47	19 07	19 32	19 47	20 05
27	16 25	16 45	17 01	17 14	17 26	17 36	17 55	18 12	18 29	18 47	19 08	19 32	19 48	20 06
July 1	16 27	16 47	17 02	17 16	17 27	17 38	17 56	18 13	18 30	18 48	19 08	19 32	19 47	20 05
5	16 30	16 49	17 05	17 18	17 29	17 39	17 57	18 14	18 31	18 48	19 08	19 32	19 47	20 04

UNIVERSAL TIME FOR MERIDIAN OF GREENWICH

MORNING CIVIL TWILIGHT

Lat.	+40°	+42°	+44°	+46°	+48°	+50°	+52°	+54°	+56°	+58°	+60°	+62°	+64°	+66°
	h m	h m	h m	h m	h m	h m	h m	h m	h m	h m	h m	h m	h m	h m
Mar. 31	5 18	5 16	5 14	5 11	5 09	5 06	5 03	4 59	4 55	4 50	4 45	4 39	4 32	4 24
Apr. 4	5 12	5 09	5 07	5 04	5 00	4 57	4 53	4 49	4 44	4 39	4 32	4 25	4 17	4 07
8	5 05	5 02	4 59	4 56	4 52	4 48	4 44	4 39	4 33	4 27	4 19	4 11	4 01	3 50
12	4 59	4 55	4 52	4 48	4 44	4 39	4 34	4 28	4 22	4 15	4 06	3 57	3 45	3 32
16	4 52	4 49	4 45	4 40	4 36	4 31	4 25	4 18	4 11	4 03	3 53	3 42	3 29	3 13
20	4 46	4 42	4 38	4 33	4 28	4 22	4 16	4 08	4 00	3 51	3 40	3 28	3 13	2 54
24	4 40	4 36	4 31	4 26	4 20	4 14	4 06	3 59	3 50	3 39	3 27	3 13	2 56	2 34
28	4 35	4 30	4 25	4 19	4 12	4 05	3 58	3 49	3 39	3 28	3 14	2 58	2 38	2 12
May 2	4 29	4 24	4 18	4 12	4 05	3 58	3 49	3 39	3 29	3 16	3 01	2 43	2 20	1 48
6	4 24	4 18	4 12	4 06	3 58	3 50	3 41	3 30	3 18	3 04	2 48	2 27	2 00	1 19
10	4 19	4 13	4 07	4 00	3 52	3 43	3 33	3 22	3 09	2 53	2 35	2 11	1 38	0 38
14	4 15	4 09	4 02	3 54	3 45	3 36	3 25	3 13	2 59	2 42	2 22	1 54	1 14	// //
18	4 11	4 04	3 57	3 49	3 40	3 30	3 18	3 05	2 50	2 32	2 09	1 37	0 40	// //
22	4 08	4 00	3 53	3 44	3 35	3 24	3 12	2 58	2 42	2 22	1 56	1 18	// //	// //
26	4 05	3 57	3 49	3 40	3 30	3 19	3 06	2 52	2 34	2 12	1 43	0 57	// //	// //
30	4 02	3 54	3 46	3 37	3 26	3 15	3 01	2 46	2 27	2 04	1 31	0 28	// //	// //
June 3	4 00	3 52	3 44	3 34	3 23	3 11	2 57	2 41	2 21	1 56	1 20	// //	// //	// //
7	3 59	3 51	3 42	3 32	3 21	3 09	2 54	2 37	2 16	1 50	1 09	// //	// //	// //
11	3 58	3 50	3 41	3 31	3 19	3 07	2 52	2 34	2 13	1 45	1 00	// //	// //	// //
15	3 58	3 49	3 40	3 30	3 19	3 06	2 51	2 33	2 11	1 41	0 53	// //	// //	** **
19	3 58	3 50	3 40	3 30	3 19	3 06	2 50	2 32	2 10	1 40	0 49	// //	// //	** **
23	3 59	3 50	3 41	3 31	3 20	3 06	2 51	2 33	2 11	1 41	0 50	// //	// //	** **
27	4 00	3 52	3 43	3 33	3 21	3 08	2 53	2 35	2 13	1 44	0 55	// //	// //	** **
July 1	4 02	3 54	3 45	3 35	3 24	3 11	2 56	2 39	2 17	1 49	1 04	// //	// //	// //
5	4 04	3 56	3 48	3 38	3 27	3 14	3 00	2 43	2 22	1 55	1 14	// //	// //	// //

EVENING CIVIL TWILIGHT

Lat.	+40°	+42°	+44°	+46°	+48°	+50°	+52°	+54°	+56°	+58°	+60°	+62°	+64°	+66°
	h m	h m	h m	h m	h m	h m	h m	h m	h m	h m	h m	h m	h m	h m
Mar. 31	18 51	18 53	18 55	18 58	19 00	19 03	19 07	19 11	19 15	19 19	19 25	19 31	19 38	19 47
Apr. 4	18 55	18 57	19 00	19 03	19 06	19 10	19 14	19 18	19 23	19 29	19 35	19 43	19 51	20 02
8	18 59	19 02	19 05	19 09	19 13	19 17	19 21	19 26	19 32	19 39	19 46	19 55	20 05	20 17
12	19 03	19 07	19 10	19 14	19 19	19 23	19 29	19 34	19 41	19 48	19 57	20 07	20 19	20 33
16	19 08	19 12	19 16	19 20	19 25	19 30	19 36	19 43	19 50	19 58	20 08	20 19	20 33	20 50
20	19 12	19 16	19 21	19 26	19 31	19 37	19 43	19 51	19 59	20 09	20 19	20 32	20 48	21 07
24	19 17	19 21	19 26	19 31	19 37	19 44	19 51	19 59	20 08	20 19	20 31	20 46	21 04	21 27
28	19 21	19 26	19 31	19 37	19 44	19 51	19 59	20 07	20 18	20 29	20 43	21 00	21 20	21 48
May 2	19 25	19 31	19 37	19 43	19 50	19 58	20 06	20 16	20 27	20 40	20 55	21 14	21 38	22 12
6	19 30	19 35	19 42	19 49	19 56	20 04	20 14	20 24	20 36	20 51	21 08	21 29	21 58	22 42
10	19 34	19 40	19 47	19 54	20 02	20 11	20 21	20 33	20 46	21 02	21 21	21 45	22 19	23 33
14	19 38	19 45	19 52	20 00	20 08	20 18	20 28	20 41	20 55	21 12	21 34	22 02	22 45	// //
18	19 42	19 49	19 57	20 05	20 14	20 24	20 36	20 49	21 04	21 23	21 47	22 20	23 25	// //
22	19 46	19 53	20 01	20 10	20 19	20 30	20 42	20 56	21 13	21 33	22 00	22 39	// //	// //
26	19 50	19 57	20 06	20 15	20 25	20 36	20 49	21 04	21 21	21 44	22 13	23 03	// //	// //
30	19 53	20 01	20 10	20 19	20 29	20 41	20 55	21 10	21 29	21 53	22 26	23 39	// //	// //
June 3	19 57	20 05	20 13	20 23	20 34	20 46	21 00	21 16	21 36	22 02	22 39	// //	// //	// //
7	19 59	20 07	20 16	20 26	20 37	20 50	21 04	21 21	21 42	22 10	22 51	// //	// //	// //
11	20 02	20 10	20 19	20 29	20 40	20 53	21 08	21 25	21 47	22 16	23 02	// //	// //	// //
15	20 03	20 12	20 21	20 31	20 43	20 56	21 11	21 28	21 51	22 20	23 10	// //	// //	** **
19	20 05	20 13	20 22	20 33	20 44	20 57	21 12	21 30	21 53	22 23	23 14	// //	// //	** **
23	20 05	20 14	20 23	20 33	20 45	20 58	21 13	21 31	21 53	22 23	23 14	// //	// //	** **
27	20 06	20 14	20 23	20 33	20 45	20 58	21 13	21 30	21 52	22 22	23 10	// //	// //	** **
July 1	20 05	20 14	20 23	20 33	20 44	20 56	21 11	21 29	21 50	22 18	23 02	// //	// //	// //
5	20 04	20 12	20 21	20 31	20 42	20 54	21 09	21 26	21 46	22 13	22 52	// //	// //	// //

(** **) indicates Sun continuously above horizon.

(// //) indicates continuous twilight.

CIVIL TWILIGHT, 1992

UNIVERSAL TIME FOR MERIDIAN OF GREENWICH

MORNING CIVIL TWILIGHT

Lat.	−55°	−50°	−45°	−40°	−35°	−30°	−20°	−10°	0°	+10°	+20°	+30°	+35°	+40°
	h m	h m	h m	h m	h m	h m	h m	h m	h m	h m	h m	h m	h m	h m
July 1	7 41	7 21	7 05	6 52	6 41	6 30	6 12	5 55	5 38	5 20	5 00	4 35	4 20	4 02
5	7 39	7 20	7 05	6 52	6 40	6 30	6 12	5 55	5 39	5 21	5 01	4 37	4 22	4 04
9	7 37	7 18	7 03	6 51	6 40	6 30	6 12	5 55	5 39	5 22	5 03	4 39	4 24	4 07
13	7 34	7 16	7 01	6 49	6 38	6 29	6 12	5 56	5 40	5 23	5 04	4 41	4 27	4 10
17	7 30	7 13	6 59	6 47	6 37	6 28	6 11	5 56	5 40	5 24	5 06	4 43	4 30	4 14
21	7 25	7 09	6 56	6 45	6 35	6 26	6 10	5 55	5 41	5 25	5 07	4 46	4 33	4 17
25	7 20	7 05	6 52	6 42	6 33	6 24	6 09	5 55	5 41	5 26	5 09	4 48	4 36	4 21
29	7 14	7 00	6 48	6 39	6 30	6 22	6 08	5 54	5 41	5 27	5 11	4 51	4 39	4 25
Aug. 2	7 07	6 55	6 44	6 35	6 27	6 19	6 06	5 54	5 41	5 28	5 12	4 54	4 42	4 29
6	7 00	6 49	6 39	6 31	6 23	6 16	6 04	5 53	5 41	5 28	5 14	4 56	4 46	4 33
10	6 53	6 42	6 34	6 26	6 19	6 13	6 02	5 51	5 40	5 29	5 15	4 59	4 49	4 37
14	6 45	6 36	6 28	6 21	6 15	6 10	6 00	5 50	5 40	5 29	5 17	5 02	4 52	4 42
18	6 37	6 29	6 22	6 16	6 11	6 06	5 57	5 48	5 39	5 29	5 18	5 04	4 56	4 46
22	6 28	6 21	6 16	6 11	6 06	6 02	5 54	5 46	5 38	5 30	5 19	5 07	4 59	4 50
26	6 19	6 14	6 09	6 05	6 01	5 58	5 51	5 44	5 37	5 30	5 20	5 09	5 02	4 54
30	6 10	6 06	6 02	5 59	5 56	5 53	5 48	5 42	5 36	5 30	5 22	5 11	5 05	4 58
Sept. 3	6 00	5 57	5 55	5 53	5 51	5 49	5 45	5 40	5 35	5 29	5 23	5 14	5 08	5 02
7	5 50	5 49	5 48	5 47	5 45	5 44	5 41	5 38	5 34	5 29	5 24	5 16	5 11	5 06
11	5 40	5 40	5 40	5 40	5 40	5 39	5 38	5 35	5 33	5 29	5 24	5 18	5 15	5 10
15	5 30	5 32	5 33	5 34	5 34	5 34	5 34	5 33	5 31	5 29	5 25	5 21	5 18	5 14
19	5 20	5 23	5 25	5 27	5 28	5 29	5 30	5 30	5 30	5 28	5 26	5 23	5 21	5 18
23	5 09	5 14	5 18	5 20	5 23	5 24	5 27	5 28	5 28	5 28	5 27	5 25	5 24	5 22
27	4 59	5 05	5 10	5 14	5 17	5 19	5 23	5 25	5 27	5 28	5 28	5 27	5 27	5 25
Oct. 1	4 48	4 56	5 02	5 07	5 11	5 14	5 19	5 23	5 26	5 28	5 29	5 30	5 30	5 29
5	4 38	4 47	4 55	5 00	5 05	5 09	5 16	5 21	5 24	5 27	5 30	5 32	5 33	5 33

EVENING CIVIL TWILIGHT

Lat.	−55°	−50°	−45°	−40°	−35°	−30°	−20°	−10°	0°	+10°	+20°	+30°	+35°	+40°
	h m	h m	h m	h m	h m	h m	h m	h m	h m	h m	h m	h m	h m	h m
July 1	16 27	16 47	17 02	17 16	17 27	17 38	17 56	18 13	18 30	18 48	19 08	19 32	19 47	20 05
5	16 30	16 49	17 05	17 18	17 29	17 39	17 57	18 14	18 31	18 48	19 08	19 32	19 47	20 04
9	16 34	16 52	17 07	17 20	17 31	17 41	17 59	18 15	18 31	18 48	19 08	19 31	19 46	20 03
13	16 38	16 56	17 10	17 23	17 33	17 43	18 00	18 16	18 32	18 48	19 07	19 30	19 44	20 01
17	16 43	17 00	17 14	17 25	17 36	17 45	18 01	18 17	18 32	18 48	19 06	19 28	19 42	19 58
21	16 48	17 04	17 17	17 28	17 38	17 47	18 03	18 17	18 32	18 48	19 05	19 26	19 39	19 55
25	16 54	17 09	17 21	17 31	17 41	17 49	18 04	18 18	18 32	18 47	19 04	19 24	19 37	19 51
29	17 00	17 13	17 25	17 35	17 43	17 51	18 05	18 19	18 32	18 46	19 02	19 21	19 33	19 47
Aug. 2	17 06	17 18	17 29	17 38	17 46	17 53	18 07	18 19	18 31	18 45	19 00	19 18	19 29	19 43
6	17 12	17 24	17 33	17 41	17 49	17 55	18 08	18 19	18 31	18 43	18 57	19 15	19 25	19 38
10	17 18	17 29	17 37	17 45	17 52	17 58	18 09	18 19	18 30	18 42	18 55	19 11	19 21	19 32
14	17 25	17 34	17 42	17 48	17 54	18 00	18 10	18 19	18 29	18 40	18 52	19 07	19 16	19 27
18	17 32	17 40	17 46	17 52	17 57	18 02	18 11	18 19	18 28	18 38	18 49	19 03	19 11	19 21
22	17 39	17 45	17 51	17 55	18 00	18 04	18 12	18 19	18 27	18 36	18 46	18 58	19 06	19 15
26	17 45	17 51	17 55	17 59	18 03	18 06	18 13	18 19	18 26	18 34	18 43	18 54	19 01	19 09
30	17 52	17 56	18 00	18 03	18 05	18 08	18 13	18 19	18 25	18 31	18 39	18 49	18 55	19 02
Sept. 3	18 00	18 02	18 04	18 06	18 08	18 10	18 14	18 19	18 23	18 29	18 36	18 44	18 49	18 56
7	18 07	18 08	18 09	18 10	18 11	18 12	18 15	18 18	18 22	18 26	18 32	18 39	18 44	18 49
11	18 14	18 13	18 13	18 13	18 14	18 14	18 16	18 18	18 20	18 24	18 28	18 34	18 38	18 42
15	18 21	18 19	18 18	18 17	18 17	18 16	18 17	18 17	18 19	18 21	18 25	18 29	18 32	18 36
19	18 29	18 25	18 23	18 21	18 20	18 19	18 17	18 17	18 18	18 19	18 21	18 24	18 26	18 29
23	18 36	18 31	18 28	18 25	18 23	18 21	18 18	18 17	18 16	18 16	18 17	18 19	18 20	18 22
27	18 44	18 38	18 33	18 29	18 26	18 23	18 19	18 17	18 15	18 14	18 14	18 14	18 15	18 16
Oct. 1	18 52	18 44	18 38	18 33	18 29	18 25	18 20	18 16	18 14	18 11	18 10	18 09	18 09	18 09
5	19 00	18 51	18 43	18 37	18 32	18 28	18 21	18 16	18 12	18 09	18 07	18 04	18 03	18 03

UNIVERSAL TIME FOR MERIDIAN OF GREENWICH
MORNING CIVIL TWILIGHT

Lat.	+40°	+42°	+44°	+46°	+48°	+50°	+52°	+54°	+56°	+58°	+60°	+62°	+64°	+66°
	h m	h m	h m	h m	h m	h m	h m	h m	h m	h m	h m	h m	h m	h m
July 1	4 02	3 54	3 45	3 35	3 24	3 11	2 56	2 39	2 17	1 49	1 04	// //	// //	// //
5	4 04	3 56	3 48	3 38	3 27	3 14	3 00	2 43	2 22	1 55	1 14	// //	// //	// //
9	4 07	3 59	3 51	3 41	3 30	3 18	3 04	2 48	2 28	2 03	1 26	// //	// //	// //
13	4 10	4 03	3 54	3 45	3 35	3 23	3 10	2 54	2 35	2 12	1 39	0 33	// //	// //
17	4 14	4 06	3 58	3 49	3 39	3 28	3 15	3 01	2 43	2 21	1 52	1 04	// //	// //
21	4 17	4 10	4 02	3 54	3 44	3 34	3 22	3 08	2 51	2 31	2 05	1 27	// //	// //
25	4 21	4 14	4 07	3 59	3 50	3 40	3 28	3 15	3 00	2 41	2 18	1 46	0 46	// //
29	4 25	4 19	4 12	4 04	3 55	3 46	3 35	3 23	3 09	2 52	2 31	2 03	1 21	// //
Aug. 2	4 29	4 23	4 16	4 09	4 01	3 52	3 42	3 31	3 18	3 03	2 44	2 20	1 46	0 41
6	4 33	4 28	4 21	4 15	4 07	3 59	3 50	3 39	3 27	3 13	2 56	2 35	2 07	1 25
10	4 37	4 32	4 26	4 20	4 13	4 05	3 57	3 47	3 36	3 23	3 08	2 50	2 26	1 53
14	4 42	4 37	4 31	4 26	4 19	4 12	4 04	3 55	3 45	3 34	3 20	3 04	2 43	2 17
18	4 46	4 41	4 36	4 31	4 25	4 19	4 12	4 04	3 54	3 44	3 32	3 17	3 00	2 37
22	4 50	4 46	4 41	4 36	4 31	4 25	4 19	4 12	4 03	3 54	3 43	3 30	3 15	2 56
26	4 54	4 50	4 46	4 42	4 37	4 32	4 26	4 19	4 12	4 04	3 54	3 43	3 29	3 13
30	4 58	4 55	4 51	4 47	4 43	4 38	4 33	4 27	4 21	4 13	4 05	3 55	3 43	3 29
Sept. 3	5 02	4 59	4 56	4 53	4 49	4 45	4 40	4 35	4 29	4 23	4 15	4 07	3 57	3 45
7	5 06	5 03	5 01	4 58	4 55	4 51	4 47	4 43	4 38	4 32	4 26	4 18	4 10	4 00
11	5 10	5 08	5 05	5 03	5 00	4 57	4 54	4 50	4 46	4 41	4 36	4 30	4 22	4 14
15	5 14	5 12	5 10	5 08	5 06	5 03	5 01	4 58	4 54	4 50	4 46	4 41	4 35	4 28
19	5 18	5 16	5 15	5 13	5 12	5 10	5 07	5 05	5 02	4 59	4 56	4 51	4 47	4 41
23	5 22	5 21	5 20	5 18	5 17	5 16	5 14	5 12	5 10	5 08	5 05	5 02	4 59	4 54
27	5 25	5 25	5 24	5 24	5 23	5 22	5 21	5 20	5 18	5 17	5 15	5 13	5 10	5 07
Oct. 1	5 29	5 29	5 29	5 29	5 28	5 28	5 27	5 27	5 26	5 25	5 24	5 23	5 22	5 20
5	5 33	5 34	5 34	5 34	5 34	5 34	5 34	5 34	5 34	5 34	5 34	5 34	5 33	5 33

EVENING CIVIL TWILIGHT

Lat.	+40°	+42°	+44°	+46°	+48°	+50°	+52°	+54°	+56°	+58°	+60°	+62°	+64°	+66°
	h m	h m	h m	h m	h m	h m	h m	h m	h m	h m	h m	h m	h m	h m
July 1	20 05	20 14	20 23	20 33	20 44	20 56	21 11	21 29	21 50	22 18	23 02	// //	// //	// //
5	20 04	20 12	20 21	20 31	20 42	20 54	21 09	21 26	21 46	22 13	22 52	// //	// //	// //
9	20 03	20 11	20 19	20 29	20 39	20 51	21 05	21 21	21 41	22 06	22 42	// //	// //	// //
13	20 01	20 08	20 17	20 26	20 36	20 48	21 01	21 16	21 35	21 58	22 30	23 29	// //	// //
17	19 58	20 05	20 13	20 22	20 32	20 43	20 56	21 10	21 28	21 49	22 18	23 02	// //	// //
21	19 55	20 02	20 10	20 18	20 27	20 38	20 50	21 04	21 20	21 40	22 05	22 42	// //	// //
25	19 51	19 58	20 05	20 13	20 22	20 32	20 43	20 56	21 11	21 29	21 52	22 23	23 16	// //
29	19 47	19 53	20 00	20 08	20 16	20 26	20 36	20 48	21 02	21 19	21 39	22 06	22 45	// //
Aug. 2	19 43	19 49	19 55	20 02	20 10	20 19	20 29	20 40	20 53	21 08	21 26	21 49	22 21	23 18
6	19 38	19 43	19 49	19 56	20 03	20 12	20 21	20 31	20 43	20 57	21 13	21 33	22 00	22 40
10	19 32	19 38	19 43	19 50	19 56	20 04	20 12	20 22	20 33	20 45	21 00	21 18	21 41	22 12
14	19 27	19 32	19 37	19 43	19 49	19 56	20 04	20 12	20 22	20 33	20 47	21 03	21 22	21 48
18	19 21	19 25	19 30	19 35	19 41	19 48	19 55	20 02	20 11	20 22	20 34	20 48	21 05	21 26
22	19 15	19 19	19 23	19 28	19 33	19 39	19 45	19 53	20 01	20 10	20 20	20 33	20 48	21 06
26	19 09	19 12	19 16	19 21	19 25	19 30	19 36	19 43	19 50	19 58	20 07	20 18	20 31	20 47
30	19 02	19 05	19 09	19 13	19 17	19 22	19 27	19 32	19 39	19 46	19 54	20 04	20 15	20 29
Sept. 3	18 56	18 59	19 02	19 05	19 09	19 13	19 17	19 22	19 28	19 34	19 41	19 50	19 59	20 11
7	18 49	18 52	18 54	18 57	19 00	19 04	19 08	19 12	19 17	19 22	19 28	19 35	19 44	19 54
11	18 42	18 44	18 47	18 49	18 52	18 55	18 58	19 02	19 06	19 10	19 15	19 21	19 28	19 37
15	18 36	18 37	18 39	18 41	18 43	18 46	18 48	18 51	18 55	18 58	19 03	19 08	19 13	19 20
19	18 29	18 30	18 32	18 33	18 35	18 37	18 39	18 41	18 44	18 47	18 50	18 54	18 59	19 04
23	18 22	18 23	18 24	18 25	18 26	18 28	18 29	18 31	18 33	18 35	18 38	18 41	18 44	18 48
27	18 16	18 16	18 17	18 17	18 18	18 19	18 20	18 21	18 22	18 24	18 25	18 27	18 30	18 32
Oct. 1	18 09	18 09	18 09	18 10	18 10	18 10	18 11	18 11	18 12	18 12	18 13	18 14	18 16	18 17
5	18 03	18 02	18 02	18 02	18 02	18 02	18 01	18 01	18 01	18 01	18 01	18 02	18 02	18 02

(// //) indicates continuous twilight.

UNIVERSAL TIME FOR MERIDIAN OF GREENWICH

MORNING CIVIL TWILIGHT

Lat.	−55°	−50°	−45°	−40°	−35°	−30°	−20°	−10°	0°	+10°	+20°	+30°	+35°	+40°
	h m	h m	h m	h m	h m	h m	h m	h m	h m	h m	h m	h m	h m	h m
Oct. 1	4 48	4 56	5 02	5 07	5 11	5 14	5 19	5 23	5 26	5 28	5 29	5 30	5 30	5 29
5	4 38	4 47	4 55	5 00	5 05	5 09	5 16	5 21	5 24	5 27	5 30	5 32	5 33	5 33
9	4 27	4 38	4 47	4 54	5 00	5 04	5 12	5 18	5 23	5 27	5 31	5 34	5 36	5 37
13	4 17	4 30	4 39	4 47	4 54	5 00	5 09	5 16	5 22	5 27	5 32	5 37	5 39	5 41
17	4 06	4 21	4 32	4 41	4 49	4 55	5 06	5 14	5 21	5 27	5 33	5 39	5 42	5 45
21	3 56	4 12	4 25	4 35	4 44	4 51	5 03	5 12	5 20	5 28	5 35	5 42	5 45	5 49
25	3 46	4 04	4 18	4 29	4 39	4 47	5 00	5 10	5 20	5 28	5 36	5 44	5 49	5 54
29	3 36	3 56	4 11	4 24	4 34	4 43	4 57	5 09	5 19	5 28	5 38	5 47	5 52	5 58
Nov. 2	3 26	3 48	4 05	4 18	4 30	4 39	4 55	5 08	5 19	5 29	5 39	5 50	5 56	6 02
6	3 16	3 41	3 59	4 14	4 26	4 36	4 53	5 07	5 19	5 30	5 41	5 53	5 59	6 07
10	3 07	3 33	3 53	4 09	4 22	4 33	4 51	5 06	5 19	5 31	5 43	5 56	6 03	6 11
14	2 58	3 27	3 48	4 05	4 19	4 30	4 50	5 05	5 19	5 32	5 45	5 59	6 07	6 15
18	2 50	3 21	3 44	4 01	4 16	4 28	4 48	5 05	5 20	5 34	5 47	6 02	6 10	6 20
22	2 42	3 16	3 40	3 58	4 14	4 26	4 48	5 05	5 21	5 35	5 50	6 05	6 14	6 24
26	2 36	3 11	3 36	3 56	4 12	4 25	4 47	5 06	5 22	5 37	5 52	6 09	6 18	6 28
30	2 30	3 07	3 33	3 54	4 10	4 24	4 47	5 06	5 23	5 39	5 54	6 12	6 21	6 32
Dec. 4	2 25	3 04	3 31	3 53	4 10	4 24	4 48	5 07	5 24	5 41	5 57	6 15	6 25	6 36
8	2 21	3 02	3 30	3 52	4 10	4 24	4 49	5 08	5 26	5 43	5 59	6 17	6 28	6 39
12	2 19	3 01	3 30	3 52	4 10	4 25	4 50	5 10	5 28	5 45	6 02	6 20	6 31	6 42
16	2 18	3 01	3 30	3 53	4 11	4 26	4 51	5 11	5 30	5 47	6 04	6 23	6 33	6 45
20	2 18	3 02	3 32	3 54	4 13	4 28	4 53	5 13	5 31	5 49	6 06	6 25	6 36	6 47
24	2 20	3 04	3 34	3 56	4 15	4 30	4 55	5 15	5 33	5 51	6 08	6 27	6 37	6 49
28	2 24	3 07	3 37	3 59	4 17	4 32	4 57	5 17	5 35	5 53	6 10	6 28	6 39	6 51
32	2 29	3 11	3 40	4 02	4 20	4 35	5 00	5 20	5 37	5 54	6 11	6 30	6 40	6 52
36	2 36	3 16	3 44	4 06	4 24	4 38	5 02	5 22	5 39	5 56	6 13	6 31	6 41	6 52

EVENING CIVIL TWILIGHT

Lat.	−55°	−50°	−45°	−40°	−35°	−30°	−20°	−10°	0°	+10°	+20°	+30°	+35°	+40°
	h m	h m	h m	h m	h m	h m	h m	h m	h m	h m	h m	h m	h m	h m
Oct. 1	18 52	18 44	18 38	18 33	18 29	18 25	18 20	18 16	18 14	18 11	18 10	18 09	18 09	18 09
5	19 00	18 51	18 43	18 37	18 32	18 28	18 21	18 16	18 12	18 09	18 07	18 04	18 03	18 03
9	19 09	18 57	18 48	18 41	18 35	18 30	18 23	18 16	18 11	18 07	18 03	18 00	17 58	17 56
13	19 17	19 04	18 54	18 46	18 39	18 33	18 24	18 16	18 10	18 05	18 00	17 55	17 53	17 50
17	19 26	19 11	18 59	18 50	18 42	18 36	18 25	18 17	18 10	18 03	17 57	17 51	17 48	17 45
21	19 35	19 18	19 05	18 55	18 46	18 39	18 27	18 17	18 09	18 02	17 54	17 47	17 43	17 39
25	19 44	19 25	19 11	19 00	18 50	18 42	18 29	18 18	18 09	18 00	17 52	17 43	17 39	17 34
29	19 54	19 33	19 17	19 05	18 54	18 45	18 31	18 19	18 08	17 59	17 50	17 40	17 35	17 29
Nov. 2	20 03	19 40	19 23	19 10	18 58	18 49	18 33	18 20	18 08	17 58	17 48	17 37	17 31	17 24
6	20 13	19 48	19 29	19 15	19 02	18 52	18 35	18 21	18 09	17 57	17 46	17 34	17 27	17 20
10	20 23	19 56	19 36	19 20	19 07	18 56	18 37	18 22	18 09	17 57	17 45	17 32	17 24	17 17
14	20 32	20 03	19 42	19 25	19 11	18 59	18 40	18 24	18 10	17 57	17 44	17 30	17 22	17 13
18	20 42	20 11	19 48	19 30	19 15	19 03	18 42	18 26	18 11	17 57	17 43	17 28	17 20	17 11
22	20 52	20 18	19 54	19 35	19 19	19 06	18 45	18 27	18 12	17 57	17 43	17 27	17 18	17 08
26	21 01	20 25	19 59	19 40	19 24	19 10	18 48	18 29	18 13	17 58	17 43	17 26	17 17	17 07
30	21 09	20 32	20 05	19 44	19 28	19 13	18 50	18 31	18 15	17 59	17 43	17 26	17 16	17 06
Dec. 4	21 17	20 38	20 10	19 49	19 31	19 17	18 53	18 34	18 16	18 00	17 44	17 26	17 16	17 05
8	21 24	20 43	20 14	19 52	19 35	19 20	18 56	18 36	18 18	18 02	17 45	17 27	17 16	17 05
12	21 30	20 47	20 18	19 56	19 38	19 23	18 58	18 38	18 20	18 03	17 46	17 27	17 17	17 05
16	21 34	20 51	20 21	19 59	19 41	19 25	19 01	18 40	18 22	18 05	17 48	17 29	17 18	17 07
20	21 37	20 54	20 24	20 01	19 43	19 28	19 03	18 42	18 24	18 07	17 50	17 31	17 20	17 08
24	21 39	20 55	20 26	20 03	19 45	19 30	19 05	18 44	18 26	18 09	17 52	17 33	17 22	17 10
28	21 39	20 56	20 27	20 04	19 46	19 31	19 06	18 46	18 28	18 11	17 54	17 35	17 25	17 13
32	21 37	20 56	20 27	20 05	19 47	19 32	19 08	18 48	18 30	18 13	17 56	17 38	17 27	17 16
36	21 34	20 54	20 26	20 05	19 47	19 33	19 09	18 49	18 32	18 15	17 59	17 41	17 30	17 19

UNIVERSAL TIME FOR MERIDIAN OF GREENWICH

MORNING CIVIL TWILIGHT

Lat.	+40°	+42°	+44°	+46°	+48°	+50°	+52°	+54°	+56°	+58°	+60°	+62°	+64°	+66°
	h m	h m	h m	h m	h m	h m	h m	h m	h m	h m	h m	h m	h m	h m
Oct. 1	5 29	5 29	5 29	5 29	5 28	5 28	5 27	5 27	5 26	5 25	5 24	5 23	5 22	5 20
5	5 33	5 34	5 34	5 34	5 34	5 34	5 34	5 34	5 34	5 34	5 34	5 34	5 33	5 33
9	5 37	5 38	5 38	5 39	5 40	5 40	5 41	5 41	5 42	5 43	5 43	5 44	5 45	5 45
13	5 41	5 42	5 43	5 44	5 45	5 46	5 48	5 49	5 50	5 51	5 53	5 54	5 56	5 58
17	5 45	5 47	5 48	5 49	5 51	5 53	5 54	5 56	5 58	6 00	6 02	6 05	6 07	6 11
21	5 49	5 51	5 53	5 55	5 57	5 59	6 01	6 03	6 06	6 09	6 12	6 15	6 19	6 23
25	5 54	5 56	5 58	6 00	6 02	6 05	6 08	6 11	6 14	6 17	6 21	6 25	6 30	6 36
29	5 58	6 00	6 03	6 05	6 08	6 11	6 14	6 18	6 22	6 26	6 30	6 36	6 41	6 48
Nov. 2	6 02	6 05	6 08	6 11	6 14	6 18	6 21	6 25	6 30	6 34	6 40	6 46	6 53	7 01
6	6 07	6 10	6 13	6 16	6 20	6 24	6 28	6 32	6 37	6 43	6 49	6 56	7 04	7 13
10	6 11	6 14	6 18	6 22	6 26	6 30	6 35	6 40	6 45	6 51	6 58	7 06	7 15	7 25
14	6 15	6 19	6 23	6 27	6 31	6 36	6 41	6 47	6 53	7 00	7 07	7 16	7 26	7 37
18	6 20	6 23	6 28	6 32	6 37	6 42	6 47	6 53	7 00	7 08	7 16	7 25	7 36	7 49
22	6 24	6 28	6 32	6 37	6 42	6 48	6 54	7 00	7 07	7 15	7 24	7 34	7 46	8 00
26	6 28	6 32	6 37	6 42	6 47	6 53	6 59	7 06	7 14	7 22	7 32	7 43	7 56	8 11
30	6 32	6 36	6 41	6 47	6 52	6 58	7 05	7 12	7 20	7 29	7 39	7 51	8 05	8 21
Dec. 4	6 36	6 40	6 45	6 51	6 57	7 03	7 10	7 18	7 26	7 35	7 46	7 58	8 13	8 30
8	6 39	6 44	6 49	6 55	7 01	7 07	7 15	7 22	7 31	7 41	7 52	8 05	8 20	8 38
12	6 42	6 47	6 53	6 58	7 05	7 11	7 19	7 27	7 35	7 45	7 57	8 10	8 26	8 45
16	6 45	6 50	6 56	7 01	7 08	7 14	7 22	7 30	7 39	7 49	8 01	8 14	8 30	8 50
20	6 47	6 52	6 58	7 04	7 10	7 17	7 24	7 33	7 42	7 52	8 04	8 17	8 34	8 53
24	6 49	6 54	7 00	7 06	7 12	7 19	7 26	7 35	7 44	7 54	8 06	8 19	8 35	8 55
28	6 51	6 56	7 01	7 07	7 13	7 20	7 27	7 35	7 45	7 55	8 06	8 20	8 35	8 55
32	6 52	6 57	7 02	7 08	7 14	7 20	7 28	7 36	7 44	7 54	8 06	8 19	8 34	8 53
36	6 52	6 57	7 02	7 08	7 14	7 20	7 27	7 35	7 43	7 53	8 04	8 17	8 31	8 49

EVENING CIVIL TWILIGHT

Lat.	+40°	+42°	+44°	+46°	+48°	+50°	+52°	+54°	+56°	+58°	+60°	+62°	+64°	+66°
	h m	h m	h m	h m	h m	h m	h m	h m	h m	h m	h m	h m	h m	h m
Oct. 1	18 09	18 09	18 09	18 10	18 10	18 10	18 11	18 11	18 12	18 12	18 13	18 14	18 16	18 17
5	18 03	18 02	18 02	18 02	18 02	18 02	18 01	18 01	18 01	18 01	18 01	18 02	18 02	18 02
9	17 56	17 56	17 55	17 55	17 54	17 53	17 53	17 52	17 51	17 50	17 50	17 49	17 48	17 47
13	17 50	17 49	17 48	17 47	17 46	17 45	17 44	17 43	17 41	17 40	17 38	17 37	17 35	17 33
17	17 45	17 43	17 42	17 40	17 39	17 37	17 35	17 34	17 32	17 30	17 27	17 25	17 22	17 19
21	17 39	17 37	17 36	17 34	17 32	17 30	17 27	17 25	17 22	17 19	17 16	17 13	17 09	17 05
25	17 34	17 32	17 30	17 27	17 25	17 22	17 20	17 17	17 13	17 10	17 06	17 02	16 57	16 51
29	17 29	17 27	17 24	17 21	17 18	17 15	17 12	17 09	17 05	17 01	16 56	16 51	16 45	16 38
Nov. 2	17 24	17 22	17 19	17 16	17 12	17 09	17 05	17 01	16 57	16 52	16 46	16 40	16 33	16 25
6	17 20	17 17	17 14	17 11	17 07	17 03	16 59	16 54	16 49	16 44	16 37	16 30	16 22	16 13
10	17 17	17 13	17 10	17 06	17 02	16 57	16 53	16 48	16 42	16 36	16 29	16 21	16 12	16 02
14	17 13	17 10	17 06	17 02	16 57	16 52	16 47	16 42	16 36	16 29	16 21	16 12	16 02	15 51
18	17 11	17 07	17 02	16 58	16 53	16 48	16 42	16 36	16 30	16 22	16 14	16 04	15 53	15 41
22	17 08	17 04	17 00	16 55	16 50	16 44	16 38	16 32	16 25	16 17	16 08	15 57	15 45	15 31
26	17 07	17 02	16 57	16 52	16 47	16 41	16 35	16 28	16 20	16 12	16 02	15 51	15 38	15 23
30	17 06	17 01	16 56	16 51	16 45	16 39	16 32	16 25	16 17	16 08	15 58	15 46	15 32	15 16
Dec. 4	17 05	17 00	16 55	16 50	16 44	16 37	16 30	16 23	16 14	16 05	15 54	15 42	15 27	15 10
8	17 05	17 00	16 55	16 49	16 43	16 36	16 29	16 21	16 13	16 03	15 52	15 39	15 24	15 05
12	17 05	17 00	16 55	16 49	16 43	16 36	16 29	16 21	16 12	16 02	15 51	15 37	15 22	15 03
16	17 07	17 01	16 56	16 50	16 44	16 37	16 30	16 21	16 12	16 02	15 51	15 37	15 21	15 02
20	17 08	17 03	16 58	16 52	16 45	16 39	16 31	16 23	16 14	16 03	15 52	15 38	15 22	15 02
24	17 10	17 05	17 00	16 54	16 48	16 41	16 33	16 25	16 16	16 06	15 54	15 40	15 24	15 05
28	17 13	17 08	17 02	16 57	16 50	16 44	16 36	16 28	16 19	16 09	15 57	15 44	15 28	15 09
32	17 16	17 11	17 06	17 00	16 54	16 47	16 40	16 32	16 23	16 13	16 02	15 49	15 33	15 15
36	17 19	17 14	17 09	17 04	16 58	16 51	16 44	16 36	16 28	16 18	16 07	15 55	15 40	15 22

UNIVERSAL TIME FOR MERIDIAN OF GREENWICH

BEGINNING NAUTICAL TWILIGHT

Lat.	−55°	−50°	−45°	−40°	−35°	−30°	−20°	−10°	0°	+10°	+20°	+30°	+35°	+40°
	h m	h m	h m	h m	h m	h m	h m	h m	h m	h m	h m	h m	h m	h m
Jan. −1	// //	2 04	2 48	3 18	3 41	4 00	4 29	4 51	5 10	5 27	5 43	5 59	6 08	6 17
3	0 21	2 09	2 53	3 22	3 45	4 03	4 31	4 53	5 12	5 29	5 44	6 00	6 09	6 18
7	0 48	2 16	2 58	3 26	3 48	4 06	4 34	4 56	5 14	5 30	5 46	6 01	6 09	6 18
11	1 07	2 24	3 03	3 31	3 53	4 10	4 37	4 58	5 16	5 32	5 46	6 02	6 09	6 18
15	1 25	2 32	3 10	3 36	3 57	4 14	4 40	5 00	5 18	5 33	5 47	6 02	6 09	6 17
19	1 42	2 41	3 17	3 42	4 02	4 18	4 43	5 03	5 19	5 34	5 47	6 01	6 08	6 16
23	1 58	2 51	3 24	3 48	4 07	4 22	4 46	5 05	5 21	5 35	5 47	6 00	6 07	6 14
27	2 13	3 00	3 31	3 54	4 11	4 26	4 49	5 07	5 22	5 35	5 47	5 59	6 05	6 12
31	2 27	3 10	3 38	4 00	4 16	4 30	4 52	5 09	5 23	5 35	5 47	5 58	6 03	6 09
Feb. 4	2 41	3 19	3 46	4 06	4 21	4 34	4 55	5 11	5 24	5 35	5 46	5 56	6 01	6 06
8	2 54	3 29	3 53	4 12	4 26	4 38	4 57	5 12	5 24	5 35	5 44	5 53	5 58	6 02
12	3 07	3 38	4 00	4 17	4 31	4 42	5 00	5 14	5 25	5 34	5 43	5 50	5 54	5 58
16	3 19	3 47	4 08	4 23	4 36	4 46	5 02	5 15	5 25	5 33	5 41	5 47	5 50	5 53
20	3 31	3 56	4 14	4 29	4 40	4 50	5 05	5 16	5 25	5 32	5 39	5 44	5 46	5 48
24	3 42	4 05	4 21	4 34	4 45	4 53	5 07	5 17	5 25	5 31	5 36	5 40	5 42	5 43
28	3 53	4 13	4 28	4 39	4 49	4 57	5 09	5 17	5 24	5 29	5 33	5 36	5 37	5 38
Mar. 3	4 03	4 21	4 34	4 45	4 53	5 00	5 10	5 18	5 24	5 28	5 30	5 32	5 32	5 32
7	4 13	4 28	4 40	4 49	4 57	5 03	5 12	5 18	5 23	5 26	5 27	5 28	5 27	5 26
11	4 22	4 36	4 46	4 54	5 01	5 06	5 13	5 19	5 22	5 24	5 24	5 23	5 22	5 19
15	4 31	4 43	4 52	4 59	5 04	5 08	5 15	5 19	5 21	5 22	5 21	5 18	5 16	5 13
19	4 40	4 50	4 57	5 03	5 08	5 11	5 16	5 19	5 20	5 19	5 17	5 13	5 10	5 06
23	4 48	4 57	5 03	5 07	5 11	5 14	5 17	5 19	5 19	5 17	5 14	5 08	5 05	5 00
27	4 57	5 03	5 08	5 12	5 14	5 16	5 18	5 19	5 17	5 15	5 10	5 03	4 59	4 53
31	5 05	5 10	5 13	5 16	5 17	5 19	5 19	5 18	5 16	5 12	5 07	4 58	4 53	4 46
Apr. 4	5 12	5 16	5 18	5 20	5 21	5 21	5 20	5 18	5 15	5 10	5 03	4 53	4 47	4 39

ENDING NAUTICAL TWILIGHT

Lat.	−55°	−50°	−45°	−40°	−35°	−30°	−20°	−10°	0°	+10°	+20°	+30°	+35°	+40°
	h m	h m	h m	h m	h m	h m	h m	h m	h m	h m	h m	h m	h m	h m
Jan. −1	// //	22 00	21 16	20 46	20 23	20 05	19 36	19 14	18 55	18 38	18 22	18 06	17 57	17 48
3	23 40	21 58	21 15	20 46	20 23	20 05	19 37	19 15	18 57	18 40	18 24	18 08	18 00	17 51
7	23 20	21 55	21 14	20 45	20 23	20 06	19 38	19 16	18 58	18 42	18 27	18 11	18 03	17 54
11	23 04	21 50	21 11	20 44	20 22	20 05	19 38	19 17	19 00	18 44	18 29	18 14	18 06	17 58
15	22 50	21 45	21 08	20 41	20 21	20 04	19 38	19 18	19 01	18 46	18 31	18 17	18 10	18 02
19	22 36	21 38	21 03	20 38	20 19	20 03	19 38	19 18	19 02	18 47	18 34	18 20	18 13	18 06
23	22 23	21 31	20 59	20 35	20 16	20 01	19 37	19 18	19 03	18 49	18 36	18 23	18 17	18 10
27	22 10	21 23	20 53	20 31	20 13	19 59	19 36	19 18	19 03	18 50	18 38	18 26	18 20	18 14
31	21 57	21 15	20 47	20 26	20 10	19 56	19 34	19 18	19 04	18 52	18 40	18 30	18 24	18 18
Feb. 4	21 44	21 07	20 41	20 21	20 06	19 53	19 33	19 17	19 04	18 53	18 42	18 33	18 28	18 23
8	21 31	20 58	20 34	20 16	20 01	19 49	19 31	19 16	19 04	18 54	18 44	18 36	18 31	18 27
12	21 19	20 49	20 27	20 10	19 57	19 46	19 28	19 15	19 04	18 54	18 46	18 39	18 35	18 31
16	21 07	20 39	20 20	20 04	19 52	19 42	19 26	19 13	19 03	18 55	18 48	18 41	18 38	18 36
20	20 55	20 30	20 12	19 58	19 47	19 37	19 23	19 12	19 03	18 55	18 49	18 44	18 42	18 40
24	20 43	20 21	20 04	19 52	19 41	19 33	19 20	19 10	19 02	18 56	18 51	18 47	18 45	18 44
28	20 31	20 11	19 56	19 45	19 36	19 28	19 16	19 08	19 01	18 56	18 52	18 50	18 49	18 48
Mar. 3	20 19	20 02	19 49	19 38	19 30	19 23	19 13	19 06	19 00	18 56	18 54	18 52	18 52	18 53
7	20 08	19 52	19 41	19 32	19 24	19 19	19 10	19 03	18 59	18 56	18 55	18 55	18 56	18 57
11	19 56	19 43	19 33	19 25	19 19	19 14	19 06	19 01	18 58	18 56	18 56	18 58	18 59	19 01
15	19 45	19 33	19 25	19 18	19 13	19 09	19 03	18 59	18 57	18 56	18 57	19 00	19 02	19 06
19	19 34	19 24	19 17	19 11	19 07	19 04	18 59	18 57	18 56	18 56	18 58	19 03	19 06	19 10
23	19 23	19 15	19 09	19 05	19 01	18 59	18 56	18 54	18 54	18 56	19 00	19 05	19 09	19 14
27	19 13	19 06	19 02	18 58	18 56	18 54	18 52	18 52	18 53	18 56	19 01	19 08	19 13	19 19
31	19 02	18 57	18 54	18 52	18 50	18 49	18 49	18 50	18 52	18 56	19 02	19 10	19 16	19 23
Apr. 4	18 52	18 49	18 47	18 45	18 45	18 44	18 45	18 48	18 51	18 56	19 03	19 13	19 20	19 28

(// //) indicates continuous twilight.

UNIVERSAL TIME FOR MERIDIAN OF GREENWICH
BEGINNING NAUTICAL TWILIGHT

Lat.	+40°	+42°	+44°	+46°	+48°	+50°	+52°	+54°	+56°	+58°	+60°	+62°	+64°	+66°
	h m	h m	h m	h m	h m	h m	h m	h m	h m	h m	h m	h m	h m	h m
Jan. −1	6 17	6 21	6 25	6 29	6 34	6 39	6 44	6 50	6 56	7 02	7 10	7 18	7 27	7 38
3	6 18	6 22	6 26	6 30	6 34	6 39	6 44	6 50	6 55	7 02	7 09	7 17	7 26	7 37
7	6 18	6 22	6 26	6 30	6 34	6 39	6 44	6 49	6 55	7 01	7 08	7 15	7 24	7 34
11	6 18	6 22	6 25	6 29	6 33	6 38	6 42	6 47	6 53	6 59	7 05	7 12	7 21	7 30
15	6 17	6 21	6 24	6 28	6 32	6 36	6 40	6 45	6 50	6 56	7 02	7 08	7 16	7 25
19	6 16	6 19	6 22	6 26	6 30	6 33	6 37	6 42	6 47	6 52	6 57	7 04	7 11	7 19
23	6 14	6 17	6 20	6 23	6 27	6 30	6 34	6 38	6 42	6 47	6 52	6 58	7 04	7 11
27	6 12	6 15	6 17	6 20	6 23	6 26	6 30	6 33	6 37	6 41	6 46	6 51	6 57	7 03
31	6 09	6 11	6 14	6 17	6 19	6 22	6 25	6 28	6 32	6 35	6 39	6 43	6 48	6 54
Feb. 4	6 06	6 08	6 10	6 12	6 15	6 17	6 20	6 22	6 25	6 28	6 32	6 35	6 39	6 44
8	6 02	6 04	6 06	6 08	6 10	6 12	6 14	6 16	6 18	6 21	6 23	6 26	6 29	6 33
12	5 58	5 59	6 01	6 02	6 04	6 06	6 07	6 09	6 11	6 13	6 14	6 17	6 19	6 21
16	5 53	5 54	5 56	5 57	5 58	5 59	6 00	6 01	6 03	6 04	6 05	6 06	6 08	6 09
20	5 48	5 49	5 50	5 51	5 52	5 52	5 53	5 54	5 54	5 55	5 55	5 56	5 56	5 56
24	5 43	5 44	5 44	5 44	5 45	5 45	5 45	5 45	5 45	5 45	5 45	5 44	5 44	5 43
28	5 38	5 38	5 38	5 38	5 37	5 37	5 37	5 36	5 36	5 35	5 34	5 32	5 31	5 29
Mar. 3	5 32	5 31	5 31	5 31	5 30	5 29	5 28	5 27	5 26	5 24	5 22	5 20	5 17	5 14
7	5 26	5 25	5 24	5 23	5 22	5 21	5 19	5 18	5 16	5 13	5 11	5 07	5 04	4 59
11	5 19	5 18	5 17	5 16	5 14	5 12	5 10	5 08	5 05	5 02	4 58	4 54	4 49	4 43
15	5 13	5 12	5 10	5 08	5 06	5 04	5 01	4 58	4 54	4 51	4 46	4 41	4 34	4 27
19	5 06	5 05	5 02	5 00	4 58	4 55	4 51	4 48	4 43	4 39	4 33	4 27	4 19	4 10
23	5 00	4 57	4 55	4 52	4 49	4 45	4 42	4 37	4 32	4 26	4 20	4 12	4 03	3 52
27	4 53	4 50	4 47	4 44	4 40	4 36	4 32	4 26	4 21	4 14	4 06	3 57	3 46	3 33
31	4 46	4 43	4 39	4 36	4 31	4 27	4 21	4 15	4 09	4 01	3 52	3 42	3 29	3 13
Apr. 4	4 39	4 36	4 32	4 27	4 22	4 17	4 11	4 04	3 57	3 48	3 38	3 25	3 11	2 52

ENDING NAUTICAL TWILIGHT

Lat.	+40°	+42°	+44°	+46°	+48°	+50°	+52°	+54°	+56°	+58°	+60°	+62°	+64°	+66°
	h m	h m	h m	h m	h m	h m	h m	h m	h m	h m	h m	h m	h m	h m
Jan. −1	17 48	17 44	17 40	17 35	17 31	17 26	17 21	17 15	17 09	17 03	16 55	16 47	16 38	16 27
3	17 51	17 47	17 43	17 39	17 34	17 30	17 25	17 19	17 13	17 07	17 00	16 52	16 43	16 32
7	17 54	17 50	17 47	17 42	17 38	17 34	17 29	17 24	17 18	17 12	17 05	16 57	16 49	16 39
11	17 58	17 54	17 50	17 47	17 42	17 38	17 34	17 29	17 23	17 17	17 11	17 04	16 55	16 46
15	18 02	17 58	17 55	17 51	17 47	17 43	17 39	17 34	17 29	17 23	17 17	17 11	17 03	16 54
19	18 06	18 02	17 59	17 56	17 52	17 48	17 44	17 40	17 35	17 30	17 24	17 18	17 11	17 03
23	18 10	18 07	18 04	18 01	17 57	17 54	17 50	17 46	17 42	17 37	17 32	17 27	17 20	17 13
27	18 14	18 11	18 08	18 06	18 03	17 59	17 56	17 53	17 49	17 45	17 40	17 35	17 30	17 23
31	18 18	18 16	18 13	18 11	18 08	18 05	18 02	17 59	17 56	17 52	17 49	17 44	17 40	17 34
Feb. 4	18 23	18 20	18 18	18 16	18 14	18 11	18 09	18 06	18 03	18 00	17 57	17 54	17 50	17 45
8	18 27	18 25	18 23	18 21	18 19	18 18	18 15	18 13	18 11	18 09	18 06	18 03	18 00	17 57
12	18 31	18 30	18 28	18 27	18 25	18 24	18 22	18 21	18 19	18 17	18 15	18 13	18 11	18 09
16	18 36	18 34	18 33	18 32	18 31	18 30	18 29	18 28	18 27	18 26	18 24	18 23	18 22	18 21
20	18 40	18 39	18 38	18 38	18 37	18 36	18 36	18 35	18 35	18 34	18 34	18 34	18 33	18 33
24	18 44	18 44	18 43	18 43	18 43	18 43	18 43	18 43	18 43	18 43	18 43	18 44	18 45	18 46
28	18 48	18 48	18 48	18 49	18 49	18 49	18 50	18 50	18 51	18 52	18 53	18 55	18 56	18 59
Mar. 3	18 53	18 53	18 53	18 54	18 55	18 56	18 57	18 58	18 59	19 01	19 03	19 05	19 08	19 12
7	18 57	18 58	18 59	19 00	19 01	19 02	19 04	19 05	19 08	19 10	19 13	19 16	19 20	19 25
11	19 01	19 02	19 04	19 05	19 07	19 09	19 11	19 13	19 16	19 19	19 23	19 28	19 33	19 39
15	19 06	19 07	19 09	19 11	19 13	19 15	19 18	19 21	19 25	19 29	19 34	19 39	19 46	19 53
19	19 10	19 12	19 14	19 16	19 19	19 22	19 25	19 29	19 34	19 39	19 44	19 51	19 59	20 08
23	19 14	19 17	19 19	19 22	19 25	19 29	19 33	19 37	19 43	19 48	19 55	20 03	20 13	20 24
27	19 19	19 21	19 24	19 28	19 32	19 36	19 40	19 46	19 52	19 59	20 07	20 16	20 27	20 41
31	19 23	19 26	19 30	19 34	19 38	19 43	19 48	19 54	20 01	20 09	20 18	20 29	20 42	20 58
Apr. 4	19 28	19 31	19 35	19 40	19 45	19 50	19 56	20 03	20 11	20 20	20 31	20 43	20 59	21 18

UNIVERSAL TIME FOR MERIDIAN OF GREENWICH
BEGINNING NAUTICAL TWILIGHT

Lat.	−55°	−50°	−45°	−40°	−35°	−30°	−20°	−10°	0°	+10°	+20°	+30°	+35°	+40°
	h m	h m	h m	h m	h m	h m	h m	h m	h m	h m	h m	h m	h m	h m
Mar. 31	5 05	5 10	5 13	5 16	5 17	5 19	5 19	5 18	5 16	5 12	5 07	4 58	4 53	4 46
Apr. 4	5 12	5 16	5 18	5 20	5 21	5 21	5 20	5 18	5 15	5 10	5 03	4 53	4 47	4 39
8	5 20	5 22	5 23	5 24	5 24	5 23	5 21	5 18	5 13	5 07	4 59	4 48	4 41	4 32
12	5 27	5 28	5 28	5 27	5 27	5 25	5 22	5 18	5 12	5 05	4 56	4 43	4 35	4 25
16	5 34	5 34	5 33	5 31	5 30	5 28	5 23	5 18	5 11	5 03	4 52	4 38	4 29	4 19
20	5 41	5 39	5 37	5 35	5 32	5 30	5 24	5 18	5 10	5 01	4 49	4 34	4 24	4 12
24	5 48	5 45	5 42	5 38	5 35	5 32	5 25	5 18	5 09	4 59	4 46	4 29	4 18	4 06
28	5 55	5 50	5 46	5 42	5 38	5 34	5 26	5 18	5 08	4 57	4 43	4 25	4 13	3 59
May 2	6 01	5 56	5 50	5 46	5 41	5 36	5 27	5 18	5 07	4 55	4 40	4 21	4 08	3 53
6	6 08	6 01	5 55	5 49	5 44	5 39	5 28	5 18	5 06	4 53	4 37	4 17	4 04	3 48
10	6 14	6 06	5 59	5 52	5 47	5 41	5 30	5 18	5 06	4 52	4 35	4 13	3 59	3 42
14	6 19	6 10	6 03	5 56	5 49	5 43	5 31	5 19	5 06	4 51	4 33	4 10	3 55	3 37
18	6 25	6 15	6 07	5 59	5 52	5 45	5 32	5 19	5 05	4 50	4 31	4 07	3 51	3 33
22	6 30	6 19	6 10	6 02	5 54	5 47	5 34	5 20	5 05	4 49	4 29	4 04	3 48	3 29
26	6 35	6 23	6 14	6 05	5 57	5 49	5 35	5 21	5 05	4 48	4 28	4 02	3 45	3 25
30	6 39	6 27	6 17	6 07	5 59	5 51	5 36	5 21	5 06	4 48	4 27	4 00	3 43	3 22
June 3	6 43	6 30	6 20	6 10	6 01	5 53	5 38	5 22	5 06	4 48	4 27	3 59	3 41	3 19
7	6 47	6 33	6 22	6 12	6 03	5 55	5 39	5 23	5 07	4 48	4 26	3 58	3 40	3 18
11	6 49	6 36	6 24	6 14	6 05	5 56	5 40	5 24	5 07	4 49	4 26	3 58	3 39	3 17
15	6 52	6 38	6 26	6 16	6 06	5 58	5 41	5 25	5 08	4 49	4 27	3 58	3 39	3 16
19	6 53	6 39	6 27	6 17	6 07	5 59	5 42	5 26	5 09	4 50	4 27	3 58	3 39	3 16
23	6 54	6 40	6 28	6 18	6 08	6 00	5 43	5 27	5 10	4 51	4 28	3 59	3 40	3 17
27	6 54	6 40	6 28	6 18	6 09	6 00	5 44	5 28	5 11	4 52	4 29	4 00	3 42	3 19
July 1	6 53	6 40	6 28	6 18	6 09	6 00	5 44	5 28	5 12	4 53	4 31	4 02	3 44	3 21
5	6 52	6 39	6 28	6 18	6 09	6 00	5 45	5 29	5 12	4 54	4 32	4 04	3 46	3 23

ENDING NAUTICAL TWILIGHT

Lat.	−55°	−50°	−45°	−40°	−35°	−30°	−20°	−10°	0°	+10°	+20°	+30°	+35°	+40°
	h m	h m	h m	h m	h m	h m	h m	h m	h m	h m	h m	h m	h m	h m
Mar. 31	19 02	18 57	18 54	18 52	18 50	18 49	18 49	18 50	18 52	18 56	19 02	19 10	19 16	19 23
Apr. 4	18 52	18 49	18 47	18 45	18 45	18 44	18 45	18 48	18 51	18 56	19 03	19 13	19 20	19 28
8	18 43	18 41	18 40	18 39	18 39	18 40	18 42	18 45	18 50	18 56	19 05	19 16	19 23	19 32
12	18 33	18 33	18 33	18 33	18 34	18 36	18 39	18 43	18 49	18 57	19 06	19 19	19 27	19 37
16	18 24	18 25	18 26	18 28	18 29	18 31	18 36	18 42	18 48	18 57	19 08	19 22	19 31	19 42
20	18 15	18 18	18 20	18 22	18 25	18 27	18 33	18 40	18 48	18 57	19 09	19 25	19 35	19 46
24	18 07	18 10	18 14	18 17	18 20	18 24	18 31	18 38	18 47	18 58	19 11	19 28	19 38	19 51
28	17 59	18 04	18 08	18 12	18 16	18 20	18 28	18 37	18 47	18 58	19 13	19 31	19 42	19 56
May 2	17 52	17 58	18 03	18 08	18 12	18 17	18 26	18 36	18 47	18 59	19 14	19 34	19 46	20 01
6	17 45	17 52	17 58	18 04	18 09	18 14	18 24	18 35	18 47	19 00	19 16	19 37	19 50	20 06
10	17 38	17 46	17 53	18 00	18 06	18 12	18 23	18 34	18 47	19 01	19 18	19 40	19 54	20 11
14	17 33	17 42	17 49	17 56	18 03	18 09	18 22	18 34	18 47	19 02	19 20	19 43	19 58	20 16
18	17 27	17 37	17 46	17 54	18 01	18 07	18 20	18 33	18 47	19 03	19 22	19 47	20 02	20 21
22	17 23	17 33	17 43	17 51	17 59	18 06	18 20	18 33	18 48	19 04	19 24	19 50	20 06	20 25
26	17 19	17 30	17 40	17 49	17 57	18 05	18 19	18 33	18 49	19 06	19 26	19 52	20 09	20 30
30	17 15	17 28	17 38	17 47	17 56	18 04	18 19	18 34	18 49	19 07	19 28	19 55	20 12	20 34
June 3	17 13	17 26	17 37	17 46	17 55	18 03	18 19	18 34	18 50	19 08	19 30	19 58	20 15	20 37
7	17 11	17 24	17 36	17 45	17 54	18 03	18 19	18 35	18 51	19 10	19 32	20 00	20 18	20 40
11	17 10	17 23	17 35	17 45	17 54	18 03	18 19	18 35	18 52	19 11	19 33	20 02	20 20	20 43
15	17 09	17 23	17 35	17 45	17 55	18 03	18 20	18 36	18 53	19 12	19 34	20 03	20 22	20 45
19	17 10	17 24	17 35	17 46	17 55	18 04	18 21	18 37	18 54	19 13	19 35	20 05	20 23	20 46
23	17 11	17 25	17 36	17 47	17 56	18 05	18 21	18 38	18 55	19 14	19 36	20 05	20 24	20 47
27	17 12	17 26	17 38	17 48	17 57	18 06	18 22	18 39	18 55	19 14	19 37	20 06	20 24	20 47
July 1	17 14	17 28	17 40	17 50	17 59	18 07	18 24	18 39	18 56	19 15	19 37	20 06	20 24	20 47
5	17 17	17 31	17 42	17 52	18 01	18 09	18 25	18 40	18 57	19 15	19 37	20 05	20 23	20 45

UNIVERSAL TIME FOR MERIDIAN OF GREENWICH
BEGINNING NAUTICAL TWILIGHT

Lat.	+40°	+42°	+44°	+46°	+48°	+50°	+52°	+54°	+56°	+58°	+60°	+62°	+64°	+66°
	h m	h m	h m	h m	h m	h m	h m	h m	h m	h m	h m	h m	h m	h m
Mar. 31	4 46	4 43	4 39	4 36	4 31	4 27	4 21	4 15	4 09	4 01	3 52	3 42	3 29	3 13
Apr. 4	4 39	4 36	4 32	4 27	4 22	4 17	4 11	4 04	3 57	3 48	3 38	3 25	3 11	2 52
8	4 32	4 28	4 24	4 19	4 14	4 08	4 01	3 53	3 45	3 34	3 23	3 09	2 51	2 29
12	4 25	4 21	4 16	4 11	4 05	3 58	3 50	3 42	3 32	3 21	3 07	2 51	2 30	2 03
16	4 19	4 14	4 08	4 02	3 56	3 48	3 40	3 30	3 19	3 07	2 51	2 32	2 07	1 31
20	4 12	4 07	4 01	3 54	3 47	3 39	3 29	3 19	3 07	2 52	2 34	2 12	1 41	0 44
24	4 06	4 00	3 53	3 46	3 38	3 29	3 19	3 07	2 53	2 37	2 16	1 49	1 07	// //
28	3 59	3 53	3 46	3 38	3 30	3 20	3 08	2 56	2 40	2 21	1 57	1 22	// //	// //
May 2	3 53	3 47	3 39	3 31	3 21	3 10	2 58	2 44	2 27	2 05	1 36	0 46	// //	// //
6	3 48	3 40	3 32	3 23	3 13	3 01	2 48	2 32	2 13	1 47	1 10	// //	// //	// //
10	3 42	3 34	3 26	3 16	3 05	2 53	2 38	2 20	1 58	1 28	0 33	// //	// //	// //
14	3 37	3 29	3 20	3 09	2 58	2 44	2 28	2 09	1 43	1 06	// //	// //	// //	// //
18	3 33	3 24	3 14	3 03	2 51	2 36	2 19	1 57	1 28	0 36	// //	// //	// //	// //
22	3 29	3 19	3 09	2 57	2 44	2 28	2 09	1 45	1 11	// //	// //	// //	// //	// //
26	3 25	3 15	3 04	2 52	2 38	2 21	2 01	1 34	0 51	// //	// //	// //	// //	// //
30	3 22	3 12	3 01	2 48	2 33	2 15	1 53	1 23	0 25	// //	// //	// //	// //	// //
June 3	3 19	3 09	2 57	2 44	2 29	2 10	1 46	1 13	// //	// //	// //	// //	// //	// //
7	3 18	3 07	2 55	2 41	2 25	2 06	1 40	1 03	// //	// //	// //	// //	// //	// //
11	3 17	3 06	2 53	2 39	2 23	2 02	1 36	0 55	// //	// //	// //	// //	// //	// //
15	3 16	3 05	2 53	2 38	2 21	2 00	1 33	0 48	// //	// //	// //	// //	// //	** **
19	3 16	3 05	2 53	2 38	2 21	2 00	1 32	0 45	// //	// //	// //	// //	// //	** **
23	3 17	3 06	2 53	2 39	2 22	2 01	1 33	0 46	// //	// //	// //	// //	// //	** **
27	3 19	3 08	2 55	2 41	2 24	2 03	1 36	0 51	// //	// //	// //	// //	// //	** **
July 1	3 21	3 10	2 58	2 43	2 27	2 07	1 40	0 58	// //	// //	// //	// //	// //	// //
5	3 23	3 13	3 01	2 47	2 31	2 11	1 46	1 08	// //	// //	// //	// //	// //	// //

ENDING NAUTICAL TWILIGHT

Lat.	+40°	+42°	+44°	+46°	+48°	+50°	+52°	+54°	+56°	+58°	+60°	+62°	+64°	+66°
	h m	h m	h m	h m	h m	h m	h m	h m	h m	h m	h m	h m	h m	h m
Mar. 31	19 23	19 26	19 30	19 34	19 38	19 43	19 48	19 54	20 01	20 09	20 18	20 29	20 42	20 58
Apr. 4	19 28	19 31	19 35	19 40	19 45	19 50	19 56	20 03	20 11	20 20	20 31	20 43	20 59	21 18
8	19 32	19 36	19 41	19 46	19 51	19 57	20 04	20 12	20 21	20 31	20 43	20 58	21 16	21 39
12	19 37	19 41	19 46	19 52	19 58	20 05	20 13	20 21	20 31	20 43	20 57	21 14	21 35	22 05
16	19 42	19 47	19 52	19 58	20 05	20 13	20 21	20 31	20 42	20 55	21 11	21 31	21 57	22 36
20	19 46	19 52	19 58	20 05	20 12	20 21	20 30	20 41	20 53	21 08	21 27	21 50	22 23	23 34
24	19 51	19 57	20 04	20 11	20 19	20 29	20 39	20 51	21 05	21 22	21 43	22 12	22 59	// //
28	19 56	20 03	20 10	20 18	20 27	20 37	20 48	21 01	21 17	21 37	22 02	22 39	// //	// //
May 2	20 01	20 08	20 16	20 25	20 34	20 45	20 58	21 12	21 30	21 52	22 23	23 21	// //	// //
6	20 06	20 14	20 22	20 31	20 42	20 53	21 07	21 23	21 43	22 10	22 50	// //	// //	// //
10	20 11	20 19	20 28	20 38	20 49	21 02	21 17	21 35	21 58	22 30	23 36	// //	// //	// //
14	20 16	20 25	20 34	20 44	20 56	21 10	21 26	21 46	22 13	22 52	// //	// //	// //	// //
18	20 21	20 30	20 40	20 51	21 04	21 18	21 36	21 58	22 29	23 28	// //	// //	// //	// //
22	20 25	20 35	20 45	20 57	21 11	21 26	21 46	22 10	22 47	// //	// //	// //	// //	// //
26	20 30	20 39	20 50	21 03	21 17	21 34	21 55	22 22	23 08	// //	// //	// //	// //	// //
30	20 34	20 44	20 55	21 08	21 23	21 41	22 04	22 34	23 41	// //	// //	// //	// //	// //
June 3	20 37	20 48	21 00	21 13	21 29	21 48	22 12	22 46	// //	// //	// //	// //	// //	// //
7	20 40	20 51	21 03	21 17	21 33	21 53	22 19	22 57	// //	// //	// //	// //	// //	// //
11	20 43	20 54	21 06	21 20	21 37	21 58	22 24	23 07	// //	// //	// //	// //	// //	// //
15	20 45	20 56	21 09	21 23	21 40	22 01	22 28	23 14	// //	// //	// //	// //	// //	** **
19	20 46	20 58	21 10	21 25	21 42	22 03	22 31	23 18	// //	// //	// //	// //	// //	** **
23	20 47	20 58	21 11	21 25	21 42	22 03	22 31	23 18	// //	// //	// //	// //	// //	** **
27	20 47	20 58	21 11	21 25	21 42	22 03	22 30	23 14	// //	// //	// //	// //	// //	** **
July 1	20 47	20 58	21 10	21 24	21 40	22 00	22 27	23 07	// //	// //	// //	// //	// //	// //
5	20 45	20 56	21 08	21 22	21 38	21 57	22 22	22 59	// //	// //	// //	// //	// //	// //

(** **) indicates Sun continuously above horizon.
(// //) indicates continuous twilight.

UNIVERSAL TIME FOR MERIDIAN OF GREENWICH

BEGINNING NAUTICAL TWILIGHT

Lat.	−55°	−50°	−45°	−40°	−35°	−30°	−20°	−10°	0°	+10°	+20°	+30°	+35°	+40°
	h m	h m	h m	h m	h m	h m	h m	h m	h m	h m	h m	h m	h m	h m
July 1	6 53	6 40	6 28	6 18	6 09	6 00	5 44	5 28	5 12	4 53	4 31	4 02	3 44	3 21
5	6 52	6 39	6 28	6 18	6 09	6 00	5 45	5 29	5 12	4 54	4 32	4 04	3 46	3 23
9	6 50	6 37	6 26	6 17	6 08	6 00	5 45	5 29	5 13	4 55	4 34	4 06	3 48	3 27
13	6 47	6 35	6 25	6 16	6 07	5 59	5 44	5 30	5 14	4 56	4 35	4 08	3 51	3 30
17	6 44	6 32	6 23	6 14	6 06	5 58	5 44	5 30	5 15	4 58	4 37	4 11	3 54	3 34
21	6 40	6 29	6 20	6 12	6 04	5 57	5 43	5 30	5 15	4 59	4 39	4 14	3 58	3 38
25	6 35	6 25	6 17	6 09	6 02	5 55	5 42	5 29	5 15	5 00	4 41	4 17	4 01	3 43
29	6 29	6 21	6 13	6 06	5 59	5 53	5 41	5 29	5 16	5 01	4 43	4 20	4 05	3 47
Aug. 2	6 23	6 16	6 09	6 02	5 56	5 51	5 40	5 28	5 16	5 02	4 45	4 23	4 09	3 52
6	6 17	6 10	6 04	5 58	5 53	5 48	5 38	5 27	5 16	5 02	4 46	4 26	4 13	3 57
10	6 10	6 04	5 59	5 54	5 49	5 45	5 36	5 26	5 15	5 03	4 48	4 29	4 16	4 01
14	6 02	5 58	5 53	5 49	5 45	5 41	5 33	5 25	5 15	5 04	4 50	4 32	4 20	4 06
18	5 54	5 51	5 48	5 44	5 41	5 38	5 31	5 23	5 15	5 04	4 51	4 34	4 24	4 11
22	5 46	5 44	5 41	5 39	5 37	5 34	5 28	5 22	5 14	5 04	4 53	4 37	4 27	4 16
26	5 37	5 36	5 35	5 33	5 32	5 30	5 25	5 20	5 13	5 05	4 54	4 40	4 31	4 20
30	5 28	5 28	5 28	5 28	5 27	5 26	5 22	5 18	5 12	5 05	4 55	4 43	4 35	4 25
Sept. 3	5 18	5 20	5 21	5 22	5 21	5 21	5 19	5 16	5 11	5 05	4 56	4 45	4 38	4 29
7	5 08	5 12	5 14	5 15	5 16	5 16	5 15	5 13	5 10	5 05	4 58	4 48	4 41	4 33
11	4 58	5 03	5 07	5 09	5 10	5 11	5 12	5 11	5 08	5 04	4 59	4 50	4 45	4 38
15	4 48	4 54	4 59	5 02	5 05	5 07	5 08	5 08	5 07	5 04	5 00	4 53	4 48	4 42
19	4 37	4 45	4 51	4 56	4 59	5 01	5 05	5 06	5 06	5 04	5 01	4 55	4 51	4 46
23	4 26	4 36	4 43	4 49	4 53	4 56	5 01	5 03	5 04	5 04	5 01	4 57	4 54	4 50
27	4 15	4 27	4 35	4 42	4 47	4 51	4 57	5 01	5 03	5 03	5 02	4 59	4 57	4 54
Oct. 1	4 04	4 17	4 27	4 35	4 41	4 46	4 54	4 58	5 02	5 03	5 03	5 02	5 00	4 58
5	3 53	4 08	4 19	4 28	4 35	4 41	4 50	4 56	5 00	5 03	5 04	5 04	5 03	5 02

ENDING NAUTICAL TWILIGHT

Lat.	−55°	−50°	−45°	−40°	−35°	−30°	−20°	−10°	0°	+10°	+20°	+30°	+35°	+40°
	h m	h m	h m	h m	h m	h m	h m	h m	h m	h m	h m	h m	h m	h m
July 1	17 14	17 28	17 40	17 50	17 59	18 07	18 24	18 39	18 56	19 15	19 37	20 06	20 24	20 47
5	17 17	17 31	17 42	17 52	18 01	18 09	18 25	18 40	18 57	19 15	19 37	20 05	20 23	20 45
9	17 21	17 33	17 44	17 54	18 02	18 11	18 26	18 41	18 57	19 15	19 37	20 04	20 22	20 43
13	17 25	17 37	17 47	17 56	18 04	18 12	18 27	18 42	18 57	19 15	19 36	20 03	20 20	20 41
17	17 29	17 40	17 50	17 59	18 07	18 14	18 28	18 43	18 58	19 15	19 35	20 01	20 17	20 38
21	17 34	17 44	17 53	18 02	18 09	18 16	18 30	18 43	18 58	19 14	19 33	19 58	20 14	20 34
25	17 39	17 48	17 57	18 04	18 11	18 18	18 31	18 44	18 57	19 13	19 32	19 56	20 11	20 30
29	17 44	17 53	18 01	18 07	18 14	18 20	18 32	18 44	18 57	19 12	19 30	19 53	20 07	20 25
Aug. 2	17 50	17 57	18 04	18 11	18 16	18 22	18 33	18 44	18 57	19 11	19 27	19 49	20 03	20 20
6	17 56	18 02	18 08	18 14	18 19	18 24	18 34	18 44	18 56	19 09	19 25	19 45	19 58	20 14
10	18 02	18 07	18 12	18 17	18 22	18 26	18 35	18 45	18 55	19 07	19 22	19 41	19 53	20 08
14	18 08	18 12	18 16	18 20	18 24	18 28	18 36	18 44	18 54	19 05	19 19	19 37	19 48	20 02
18	18 14	18 17	18 21	18 24	18 27	18 30	18 37	18 44	18 53	19 03	19 16	19 32	19 43	19 56
22	18 21	18 23	18 25	18 27	18 29	18 32	18 38	18 44	18 52	19 01	19 12	19 28	19 37	19 49
26	18 28	18 28	18 29	18 31	18 32	18 34	18 38	18 44	18 50	18 59	19 09	19 23	19 32	19 42
30	18 34	18 34	18 34	18 34	18 35	18 36	18 39	18 43	18 49	18 56	19 05	19 18	19 26	19 35
Sept. 3	18 41	18 39	18 38	18 38	18 38	18 38	18 40	18 43	18 48	18 54	19 02	19 13	19 20	19 29
7	18 49	18 45	18 43	18 41	18 40	18 40	18 41	18 43	18 46	18 51	18 58	19 08	19 14	19 22
11	18 56	18 51	18 47	18 45	18 43	18 42	18 41	18 42	18 45	18 48	18 54	19 02	19 08	19 15
15	19 04	18 57	18 52	18 49	18 46	18 44	18 42	18 42	18 43	18 46	18 50	18 57	19 02	19 08
19	19 12	19 03	18 57	18 53	18 49	18 46	18 43	18 41	18 42	18 43	18 46	18 52	18 56	19 01
23	19 20	19 10	19 02	18 57	18 52	18 49	18 44	18 41	18 40	18 41	18 43	18 47	18 50	18 54
27	19 28	19 16	19 07	19 01	18 55	18 51	18 45	18 41	18 39	18 38	18 39	18 42	18 44	18 47
Oct. 1	19 37	19 23	19 13	19 05	18 59	18 54	18 46	18 41	18 38	18 36	18 35	18 37	18 38	18 40
5	19 46	19 30	19 18	19 09	19 02	18 56	18 47	18 41	18 36	18 34	18 32	18 32	18 33	18 34

NAUTICAL TWILIGHT, 1992

UNIVERSAL TIME FOR MERIDIAN OF GREENWICH
BEGINNING NAUTICAL TWILIGHT

Lat.	+40°	+42°	+44°	+46°	+48°	+50°	+52°	+54°	+56°	+58°	+60°	+62°	+64°	+66°
	h m	h m	h m	h m	h m	h m	h m	h m	h m	h m	h m	h m	h m	h m
July 1	3 21	3 10	2 58	2 43	2 27	2 07	1 40	0 58	// //	// //	// //	// //	// //	// //
5	3 23	3 13	3 01	2 47	2 31	2 11	1 46	1 08	// //	// //	// //	// //	// //	// //
9	3 27	3 16	3 04	2 51	2 36	2 17	1 53	1 19	// //	// //	// //	// //	// //	// //
13	3 30	3 20	3 09	2 56	2 41	2 23	2 01	1 31	0 31	// //	// //	// //	// //	// //
17	3 34	3 24	3 13	3 01	2 47	2 30	2 10	1 43	0 59	// //	// //	// //	// //	// //
21	3 38	3 29	3 19	3 07	2 54	2 38	2 19	1 54	1 19	// //	// //	// //	// //	// //
25	3 43	3 34	3 24	3 13	3 00	2 46	2 28	2 06	1 37	0 42	// //	// //	// //	// //
29	3 47	3 39	3 30	3 19	3 07	2 54	2 38	2 18	1 52	1 14	// //	// //	// //	// //
Aug. 2	3 52	3 44	3 35	3 26	3 15	3 02	2 47	2 29	2 07	1 36	0 37	// //	// //	// //
6	3 57	3 49	3 41	3 32	3 22	3 10	2 56	2 40	2 21	1 55	1 16	// //	// //	// //
10	4 01	3 55	3 47	3 38	3 29	3 18	3 06	2 51	2 34	2 12	1 42	0 48	// //	// //
14	4 06	4 00	3 53	3 45	3 36	3 26	3 15	3 02	2 46	2 27	2 02	1 26	// //	// //
18	4 11	4 05	3 58	3 51	3 43	3 34	3 24	3 12	2 58	2 41	2 20	1 52	1 06	// //
22	4 16	4 10	4 04	3 57	3 50	3 42	3 32	3 22	3 09	2 54	2 36	2 13	1 40	0 36
26	4 20	4 15	4 10	4 04	3 57	3 49	3 41	3 31	3 20	3 07	2 51	2 31	2 06	1 27
30	4 25	4 20	4 15	4 10	4 04	3 57	3 49	3 40	3 30	3 19	3 05	2 48	2 27	1 58
Sept. 3	4 29	4 25	4 20	4 16	4 10	4 04	3 57	3 49	3 40	3 30	3 18	3 04	2 46	2 22
7	4 33	4 30	4 26	4 21	4 16	4 11	4 05	3 58	3 50	3 41	3 30	3 18	3 03	2 43
11	4 38	4 34	4 31	4 27	4 23	4 18	4 12	4 06	3 59	3 52	3 42	3 31	3 18	3 02
15	4 42	4 39	4 36	4 33	4 29	4 25	4 20	4 15	4 09	4 02	3 54	3 44	3 33	3 20
19	4 46	4 44	4 41	4 38	4 35	4 31	4 27	4 23	4 17	4 11	4 05	3 57	3 47	3 36
23	4 50	4 48	4 46	4 43	4 41	4 38	4 34	4 30	4 26	4 21	4 15	4 08	4 01	3 51
27	4 54	4 52	4 51	4 49	4 47	4 44	4 41	4 38	4 35	4 30	4 26	4 20	4 13	4 05
Oct. 1	4 58	4 57	4 55	4 54	4 52	4 50	4 48	4 46	4 43	4 39	4 36	4 31	4 26	4 19
5	5 02	5 01	5 00	4 59	4 58	4 57	4 55	4 53	4 51	4 48	4 45	4 42	4 38	4 33

ENDING NAUTICAL TWILIGHT

Lat.	+40°	+42°	+44°	+46°	+48°	+50°	+52°	+54°	+56°	+58°	+60°	+62°	+64°	+66°
	h m	h m	h m	h m	h m	h m	h m	h m	h m	h m	h m	h m	h m	h m
July 1	20 47	20 58	21 10	21 24	21 40	22 00	22 27	23 07	// //	// //	// //	// //	// //	// //
5	20 45	20 56	21 08	21 22	21 38	21 57	22 22	22 59	// //	// //	// //	// //	// //	// //
9	20 43	20 54	21 05	21 19	21 34	21 53	22 16	22 49	// //	// //	// //	// //	// //	// //
13	20 41	20 51	21 02	21 15	21 29	21 47	22 09	22 38	23 32	// //	// //	// //	// //	// //
17	20 38	20 47	20 58	21 10	21 24	21 41	22 01	22 27	23 08	// //	// //	// //	// //	// //
21	20 34	20 43	20 53	21 05	21 18	21 33	21 52	22 16	22 49	// //	// //	// //	// //	// //
25	20 30	20 38	20 48	20 59	21 11	21 26	21 43	22 04	22 33	23 21	// //	// //	// //	// //
29	20 25	20 33	20 42	20 52	21 04	21 18	21 33	21 53	22 17	22 53	// //	// //	// //	// //
Aug. 2	20 20	20 27	20 36	20 46	20 56	21 09	21 23	21 41	22 02	22 32	23 23	// //	// //	// //
6	20 14	20 21	20 29	20 38	20 49	21 00	21 13	21 29	21 48	22 13	22 49	// //	// //	// //
10	20 08	20 15	20 23	20 31	20 40	20 51	21 03	21 17	21 34	21 56	22 24	23 11	// //	// //
14	20 02	20 08	20 15	20 23	20 32	20 41	20 53	21 05	21 21	21 39	22 03	22 37	// //	// //
18	19 56	20 01	20 08	20 15	20 23	20 32	20 42	20 54	21 07	21 24	21 44	22 11	22 52	// //
22	19 49	19 54	20 00	20 07	20 14	20 22	20 31	20 42	20 54	21 09	21 26	21 48	22 19	23 13
26	19 42	19 47	19 53	19 59	20 05	20 13	20 21	20 30	20 41	20 54	21 09	21 28	21 53	22 28
30	19 35	19 40	19 45	19 50	19 56	20 03	20 10	20 19	20 29	20 40	20 53	21 10	21 30	21 57
Sept. 3	19 29	19 33	19 37	19 42	19 47	19 53	20 00	20 07	20 16	20 26	20 38	20 52	21 09	21 31
7	19 22	19 25	19 29	19 33	19 38	19 43	19 49	19 56	20 04	20 13	20 23	20 35	20 50	21 08
11	19 15	19 18	19 21	19 25	19 29	19 34	19 39	19 45	19 52	20 00	20 08	20 19	20 32	20 47
15	19 08	19 10	19 13	19 17	19 20	19 24	19 29	19 34	19 40	19 47	19 54	20 03	20 14	20 27
19	19 01	19 03	19 05	19 08	19 11	19 15	19 19	19 23	19 28	19 34	19 41	19 48	19 58	20 09
23	18 54	18 56	18 58	19 00	19 03	19 06	19 09	19 13	19 17	19 22	19 27	19 34	19 42	19 51
27	18 47	18 49	18 50	18 52	18 54	18 57	18 59	19 02	19 06	19 10	19 14	19 20	19 26	19 34
Oct. 1	18 40	18 42	18 43	18 44	18 46	18 48	18 50	18 52	18 55	18 58	19 02	19 06	19 11	19 17
5	18 34	18 35	18 36	18 37	18 38	18 39	18 40	18 42	18 44	18 47	18 50	18 53	18 57	19 02

(// //) indicates continuous twilight.

NAUTICAL TWILIGHT, 1992

UNIVERSAL TIME FOR MERIDIAN OF GREENWICH
BEGINNING NAUTICAL TWILIGHT

Lat.	−55°	−50°	−45°	−40°	−35°	−30°	−20°	−10°	0°	+10°	+20°	+30°	+35°	+40°
	h m	h m	h m	h m	h m	h m	h m	h m	h m	h m	h m	h m	h m	h m
Oct. 1	4 04	4 17	4 27	4 35	4 41	4 46	4 54	4 58	5 02	5 03	5 03	5 02	5 00	4 58
5	3 53	4 08	4 19	4 28	4 35	4 41	4 50	4 56	5 00	5 03	5 04	5 04	5 03	5 02
9	3 41	3 58	4 11	4 21	4 29	4 36	4 46	4 54	4 59	5 03	5 05	5 06	5 06	5 06
13	3 29	3 49	4 03	4 15	4 24	4 31	4 43	4 51	4 58	5 03	5 06	5 09	5 10	5 10
17	3 17	3 39	3 55	4 08	4 18	4 26	4 39	4 49	4 57	5 03	5 08	5 11	5 13	5 14
21	3 05	3 30	3 47	4 01	4 12	4 22	4 36	4 47	4 56	5 03	5 09	5 14	5 16	5 18
25	2 53	3 20	3 40	3 55	4 07	4 17	4 33	4 45	4 55	5 03	5 10	5 16	5 19	5 22
29	2 41	3 11	3 32	3 49	4 02	4 13	4 30	4 44	4 54	5 03	5 12	5 19	5 23	5 26
Nov. 2	2 29	3 02	3 25	3 43	3 57	4 09	4 28	4 42	4 54	5 04	5 13	5 22	5 26	5 30
6	2 16	2 53	3 18	3 37	3 53	4 05	4 25	4 41	4 54	5 05	5 15	5 24	5 29	5 34
10	2 03	2 44	3 12	3 32	3 49	4 02	4 23	4 40	4 54	5 06	5 17	5 27	5 33	5 38
14	1 50	2 36	3 05	3 27	3 45	3 59	4 22	4 39	4 54	5 07	5 19	5 30	5 36	5 43
18	1 37	2 28	3 00	3 23	3 42	3 57	4 20	4 39	4 54	5 08	5 21	5 33	5 40	5 47
22	1 24	2 20	2 55	3 19	3 39	3 55	4 19	4 39	4 55	5 09	5 23	5 36	5 43	5 51
26	1 10	2 14	2 50	3 16	3 37	3 53	4 19	4 39	4 56	5 11	5 25	5 39	5 47	5 55
30	0 56	2 08	2 47	3 14	3 35	3 52	4 19	4 40	4 57	5 13	5 27	5 42	5 50	5 58
Dec. 4	0 41	2 03	2 44	3 12	3 34	3 51	4 19	4 40	4 58	5 14	5 30	5 45	5 53	6 02
8	0 23	1 59	2 42	3 11	3 33	3 51	4 19	4 41	5 00	5 16	5 32	5 48	5 56	6 05
12	// //	1 57	2 41	3 11	3 33	3 52	4 20	4 43	5 01	5 18	5 34	5 50	5 59	6 08
16	// //	1 56	2 41	3 11	3 34	3 53	4 22	4 44	5 03	5 20	5 36	5 53	6 02	6 11
20	// //	1 56	2 42	3 13	3 36	3 54	4 24	4 46	5 05	5 22	5 38	5 55	6 04	6 13
24	// //	1 58	2 44	3 15	3 38	3 57	4 26	4 48	5 07	5 24	5 40	5 57	6 06	6 15
28	// //	2 02	2 47	3 17	3 40	3 59	4 28	4 50	5 09	5 26	5 42	5 59	6 07	6 17
32	// //	2 07	2 51	3 21	3 44	4 02	4 30	4 53	5 11	5 28	5 44	6 00	6 08	6 18
36	0 41	2 14	2 56	3 25	3 47	4 05	4 33	4 55	5 13	5 30	5 45	6 01	6 09	6 18

ENDING NAUTICAL TWILIGHT

Lat.	−55°	−50°	−45°	−40°	−35°	−30°	−20°	−10°	0°	+10°	+20°	+30°	+35°	+40°
	h m	h m	h m	h m	h m	h m	h m	h m	h m	h m	h m	h m	h m	h m
Oct. 1	19 37	19 23	19 13	19 05	18 59	18 54	18 46	18 41	18 38	18 36	18 35	18 37	18 38	18 40
5	19 46	19 30	19 18	19 09	19 02	18 56	18 47	18 41	18 36	18 34	18 32	18 32	18 33	18 34
9	19 55	19 37	19 24	19 14	19 06	18 59	18 49	18 41	18 35	18 31	18 29	18 27	18 27	18 28
13	20 05	19 45	19 30	19 19	19 09	19 02	18 50	18 41	18 35	18 30	18 26	18 23	18 22	18 22
17	20 15	19 53	19 36	19 24	19 13	19 05	18 52	18 42	18 34	18 28	18 23	18 19	18 17	18 16
21	20 26	20 01	19 43	19 29	19 17	19 08	18 53	18 42	18 33	18 26	18 20	18 15	18 13	18 11
25	20 37	20 10	19 50	19 34	19 22	19 12	18 55	18 43	18 33	18 25	18 18	18 11	18 08	18 06
29	20 49	20 18	19 56	19 40	19 26	19 15	18 58	18 44	18 33	18 24	18 16	18 08	18 04	18 01
Nov. 2	21 01	20 27	20 03	19 45	19 31	19 19	19 00	18 45	18 33	18 23	18 14	18 05	18 01	17 56
6	21 14	20 36	20 10	19 51	19 35	19 23	19 02	18 47	18 34	18 23	18 12	18 03	17 58	17 53
10	21 27	20 46	20 18	19 57	19 40	19 26	19 05	18 48	18 34	18 22	18 11	18 00	17 55	17 49
14	21 41	20 55	20 25	20 02	19 45	19 30	19 08	18 50	18 35	18 22	18 10	17 58	17 52	17 46
18	21 56	21 04	20 32	20 08	19 50	19 34	19 10	18 52	18 36	18 23	18 10	17 57	17 50	17 44
22	22 12	21 14	20 39	20 14	19 54	19 38	19 13	18 54	18 38	18 23	18 10	17 56	17 49	17 42
26	22 28	21 23	20 46	20 19	19 59	19 42	19 16	18 56	18 39	18 24	18 10	17 55	17 48	17 40
30	22 45	21 31	20 52	20 24	20 03	19 46	19 19	18 58	18 41	18 25	18 10	17 55	17 47	17 39
Dec. 4	23 04	21 39	20 58	20 29	20 07	19 50	19 22	19 00	18 42	18 26	18 11	17 56	17 47	17 39
8	23 27	21 46	21 03	20 34	20 11	19 53	19 25	19 03	18 44	18 28	18 12	17 56	17 48	17 39
12	// //	21 52	21 07	20 37	20 15	19 56	19 27	19 05	18 46	18 29	18 14	17 57	17 49	17 39
16	// //	21 56	21 11	20 41	20 17	19 59	19 30	19 07	18 48	18 31	18 15	17 59	17 50	17 41
20	// //	21 59	21 14	20 43	20 20	20 01	19 32	19 09	18 50	18 33	18 17	18 00	17 52	17 42
24	// //	22 01	21 15	20 45	20 22	20 03	19 34	19 11	18 52	18 35	18 19	18 03	17 54	17 44
28	// //	22 01	21 16	20 46	20 23	20 04	19 35	19 13	18 54	18 37	18 21	18 05	17 56	17 47
32	23 49	21 59	21 16	20 46	20 23	20 05	19 37	19 15	18 56	18 39	18 24	18 07	17 59	17 50
36	23 25	21 56	21 14	20 45	20 23	20 06	19 38	19 16	18 58	18 41	18 26	18 10	18 02	17 53

(// //) indicates continuous twilight.

UNIVERSAL TIME FOR MERIDIAN OF GREENWICH

BEGINNING NAUTICAL TWILIGHT

Lat.	+40°	+42°	+44°	+46°	+48°	+50°	+52°	+54°	+56°	+58°	+60°	+62°	+64°	+66°
	h m	h m	h m	h m	h m	h m	h m	h m	h m	h m	h m	h m	h m	h m
Oct. 1	4 58	4 57	4 55	4 54	4 52	4 50	4 48	4 46	4 43	4 39	4 36	4 31	4 26	4 19
5	5 02	5 01	5 00	4 59	4 58	4 57	4 55	4 53	4 51	4 48	4 45	4 42	4 38	4 33
9	5 06	5 06	5 05	5 04	5 04	5 03	5 02	5 01	4 59	4 57	4 55	4 53	4 50	4 46
13	5 10	5 10	5 10	5 10	5 09	5 09	5 08	5 08	5 07	5 06	5 05	5 03	5 01	4 59
17	5 14	5 14	5 15	5 15	5 15	5 15	5 15	5 15	5 15	5 14	5 14	5 13	5 12	5 11
21	5 18	5 19	5 19	5 20	5 21	5 21	5 22	5 22	5 23	5 23	5 23	5 23	5 23	5 23
25	5 22	5 23	5 24	5 25	5 26	5 27	5 28	5 29	5 30	5 31	5 32	5 33	5 34	5 35
29	5 26	5 27	5 29	5 30	5 32	5 33	5 35	5 36	5 38	5 40	5 41	5 43	5 45	5 47
Nov. 2	5 30	5 32	5 34	5 35	5 37	5 39	5 41	5 43	5 45	5 48	5 50	5 53	5 56	5 59
6	5 34	5 36	5 38	5 41	5 43	5 45	5 48	5 50	5 53	5 56	5 59	6 02	6 06	6 10
10	5 38	5 41	5 43	5 46	5 48	5 51	5 54	5 57	6 00	6 04	6 07	6 11	6 16	6 21
14	5 43	5 45	5 48	5 51	5 54	5 57	6 00	6 03	6 07	6 11	6 15	6 20	6 26	6 32
18	5 47	5 49	5 52	5 56	5 59	6 02	6 06	6 10	6 14	6 18	6 23	6 29	6 35	6 42
22	5 51	5 54	5 57	6 00	6 04	6 08	6 12	6 16	6 20	6 25	6 31	6 37	6 44	6 52
26	5 55	5 58	6 01	6 05	6 09	6 13	6 17	6 22	6 27	6 32	6 38	6 45	6 52	7 01
30	5 58	6 02	6 05	6 09	6 13	6 18	6 22	6 27	6 32	6 38	6 45	6 52	7 00	7 09
Dec. 4	6 02	6 06	6 09	6 13	6 18	6 22	6 27	6 32	6 38	6 44	6 51	6 58	7 07	7 17
8	6 05	6 09	6 13	6 17	6 22	6 26	6 31	6 37	6 42	6 49	6 56	7 04	7 13	7 23
12	6 08	6 12	6 16	6 20	6 25	6 30	6 35	6 41	6 47	6 53	7 00	7 09	7 18	7 29
16	6 11	6 15	6 19	6 23	6 28	6 33	6 38	6 44	6 50	6 57	7 04	7 12	7 22	7 33
20	6 13	6 17	6 21	6 26	6 30	6 35	6 41	6 46	6 53	6 59	7 07	7 15	7 25	7 36
24	6 15	6 19	6 23	6 28	6 32	6 37	6 43	6 48	6 54	7 01	7 09	7 17	7 27	7 38
28	6 17	6 21	6 25	6 29	6 34	6 39	6 44	6 49	6 55	7 02	7 10	7 18	7 27	7 38
32	6 18	6 21	6 26	6 30	6 34	6 39	6 44	6 50	6 56	7 02	7 09	7 17	7 27	7 37
36	6 18	6 22	6 26	6 30	6 34	6 39	6 44	6 49	6 55	7 01	7 08	7 16	7 25	7 35

ENDING NAUTICAL TWILIGHT

Lat.	+40°	+42°	+44°	+46°	+48°	+50°	+52°	+54°	+56°	+58°	+60°	+62°	+64°	+66°
	h m	h m	h m	h m	h m	h m	h m	h m	h m	h m	h m	h m	h m	h m
Oct. 1	18 40	18 42	18 43	18 44	18 46	18 48	18 50	18 52	18 55	18 58	19 02	19 06	19 11	19 17
5	18 34	18 35	18 36	18 37	18 38	18 39	18 40	18 42	18 44	18 47	18 50	18 53	18 57	19 02
9	18 28	18 28	18 29	18 29	18 30	18 31	18 32	18 33	18 34	18 36	18 38	18 40	18 43	18 46
13	18 22	18 22	18 22	18 22	18 22	18 22	18 23	18 23	18 24	18 25	18 26	18 28	18 30	18 32
17	18 16	18 16	18 15	18 15	18 15	18 15	18 15	18 15	18 15	18 15	18 15	18 16	18 17	18 18
21	18 11	18 10	18 09	18 08	18 08	18 07	18 07	18 06	18 05	18 05	18 05	18 04	18 04	18 04
25	18 06	18 04	18 03	18 02	18 01	18 00	17 59	17 58	17 57	17 56	17 55	17 53	17 52	17 51
29	18 01	17 59	17 58	17 56	17 55	17 53	17 52	17 50	17 49	17 47	17 45	17 43	17 41	17 39
Nov. 2	17 56	17 55	17 53	17 51	17 49	17 47	17 45	17 43	17 41	17 38	17 36	17 33	17 30	17 27
6	17 53	17 50	17 48	17 46	17 44	17 41	17 39	17 36	17 34	17 31	17 28	17 24	17 20	17 16
10	17 49	17 47	17 44	17 42	17 39	17 36	17 33	17 30	17 27	17 24	17 20	17 16	17 11	17 06
14	17 46	17 43	17 41	17 38	17 35	17 32	17 28	17 25	17 21	17 17	17 13	17 08	17 02	16 56
18	17 44	17 41	17 38	17 34	17 31	17 28	17 24	17 20	17 16	17 11	17 06	17 01	16 55	16 48
22	17 42	17 38	17 35	17 32	17 28	17 24	17 20	17 16	17 11	17 06	17 01	16 55	16 48	16 40
26	17 40	17 37	17 33	17 30	17 26	17 22	17 17	17 13	17 08	17 02	16 56	16 50	16 42	16 33
30	17 39	17 36	17 32	17 28	17 24	17 20	17 15	17 10	17 05	16 59	16 52	16 45	16 37	16 28
Dec. 4	17 39	17 35	17 31	17 27	17 23	17 18	17 13	17 08	17 03	16 56	16 50	16 42	16 33	16 24
8	17 39	17 35	17 31	17 27	17 22	17 18	17 13	17 07	17 01	16 55	16 48	16 40	16 31	16 21
12	17 39	17 36	17 31	17 27	17 23	17 18	17 13	17 07	17 01	16 54	16 47	16 39	16 30	16 19
16	17 41	17 37	17 32	17 28	17 23	17 19	17 13	17 08	17 02	16 55	16 47	16 39	16 29	16 18
20	17 42	17 38	17 34	17 30	17 25	17 20	17 15	17 09	17 03	16 56	16 49	16 40	16 31	16 19
24	17 44	17 40	17 36	17 32	17 27	17 22	17 17	17 11	17 05	16 58	16 51	16 42	16 33	16 22
28	17 47	17 43	17 39	17 35	17 30	17 25	17 20	17 14	17 08	17 01	16 54	16 46	16 36	16 25
32	17 50	17 46	17 42	17 38	17 33	17 28	17 23	17 18	17 12	17 05	16 58	16 50	16 41	16 30
36	17 53	17 49	17 45	17 41	17 37	17 32	17 27	17 22	17 16	17 10	17 03	16 55	16 47	16 36

UNIVERSAL TIME FOR MERIDIAN OF GREENWICH
BEGINNING ASTRONOMICAL TWILIGHT

Lat.	−55°	−50°	−45°	−40°	−35°	−30°	−20°	−10°	0°	+10°	+20°	+30°	+35°	+40°
	h m	h m	h m	h m	h m	h m	h m	h m	h m	h m	h m	h m	h m	h m
Jan. −1	// //	// //	1 43	2 30	3 01	3 24	3 58	4 24	4 44	5 01	5 16	5 30	5 37	5 44
3	// //	// //	1 49	2 34	3 05	3 28	4 01	4 26	4 46	5 02	5 17	5 31	5 38	5 45
7	// //	// //	1 56	2 39	3 09	3 31	4 04	4 28	4 48	5 04	5 19	5 32	5 39	5 45
11	// //	0 20	2 04	2 45	3 14	3 35	4 07	4 31	4 50	5 06	5 20	5 33	5 39	5 45
15	// //	0 55	2 13	2 51	3 19	3 39	4 11	4 33	4 52	5 07	5 20	5 33	5 39	5 45
19	// //	1 16	2 22	2 58	3 24	3 44	4 14	4 36	4 53	5 08	5 21	5 32	5 38	5 44
23	// //	1 34	2 31	3 05	3 29	3 48	4 17	4 38	4 55	5 09	5 21	5 32	5 37	5 42
27	// //	1 50	2 40	3 12	3 35	3 53	4 20	4 41	4 56	5 10	5 21	5 31	5 35	5 40
31	// //	2 05	2 50	3 19	3 41	3 58	4 24	4 43	4 58	5 10	5 20	5 29	5 33	5 37
Feb. 4	0 54	2 19	2 59	3 26	3 46	4 02	4 27	4 45	4 59	5 10	5 19	5 27	5 31	5 34
8	1 28	2 32	3 08	3 33	3 52	4 07	4 30	4 47	4 59	5 10	5 18	5 25	5 28	5 30
12	1 52	2 45	3 17	3 40	3 57	4 11	4 33	4 48	5 00	5 09	5 17	5 22	5 25	5 26
16	2 11	2 56	3 25	3 46	4 03	4 16	4 35	4 50	5 00	5 09	5 15	5 19	5 21	5 22
20	2 28	3 08	3 34	3 53	4 08	4 20	4 38	4 51	5 00	5 08	5 13	5 16	5 17	5 17
24	2 44	3 18	3 41	3 59	4 13	4 24	4 40	4 52	5 00	5 06	5 10	5 12	5 12	5 12
28	2 58	3 28	3 49	4 05	4 17	4 27	4 42	4 53	5 00	5 05	5 08	5 08	5 08	5 06
Mar. 3	3 10	3 37	3 56	4 11	4 22	4 31	4 44	4 53	4 59	5 03	5 05	5 04	5 03	5 00
7	3 22	3 46	4 03	4 16	4 26	4 34	4 46	4 54	4 59	5 01	5 02	5 00	4 58	4 54
11	3 34	3 55	4 10	4 21	4 30	4 37	4 47	4 54	4 58	4 59	4 59	4 55	4 52	4 48
15	3 44	4 03	4 16	4 26	4 34	4 40	4 49	4 54	4 57	4 57	4 55	4 50	4 46	4 41
19	3 54	4 10	4 22	4 31	4 38	4 43	4 50	4 54	4 56	4 55	4 52	4 45	4 41	4 34
23	4 04	4 18	4 28	4 35	4 41	4 46	4 52	4 54	4 55	4 53	4 48	4 40	4 35	4 27
27	4 13	4 25	4 33	4 40	4 45	4 48	4 53	4 54	4 53	4 50	4 44	4 35	4 28	4 20
31	4 21	4 31	4 39	4 44	4 48	4 51	4 54	4 54	4 52	4 48	4 41	4 30	4 22	4 13
Apr. 4	4 29	4 38	4 44	4 48	4 51	4 53	4 55	4 54	4 51	4 45	4 37	4 24	4 16	4 06

ENDING ASTRONOMICAL TWILIGHT

Lat.	−55°	−50°	−45°	−40°	−35°	−30°	−20°	−10°	0°	+10°	+20°	+30°	+35°	+40°
	h m	h m	h m	h m	h m	h m	h m	h m	h m	h m	h m	h m	h m	h m
Jan. −1	// //	// //	22 21	21 34	21 03	20 40	20 06	19 41	19 21	19 04	18 49	18 35	18 28	18 21
3	// //	// //	22 18	21 33	21 03	20 41	20 07	19 42	19 23	19 06	18 51	18 37	18 31	18 24
7	// //	// //	22 15	21 32	21 03	20 41	20 08	19 44	19 24	19 08	18 54	18 40	18 34	18 27
11	// //	23 44	22 10	21 29	21 01	20 40	20 08	19 44	19 26	19 10	18 56	18 43	18 37	18 30
15	// //	23 19	22 04	21 26	20 59	20 39	20 08	19 45	19 27	19 12	18 58	18 46	18 40	18 34
19	// //	23 01	21 58	21 22	20 57	20 37	20 07	19 45	19 28	19 13	19 00	18 49	18 43	18 38
23	// //	22 46	21 51	21 17	20 53	20 34	20 06	19 45	19 28	19 15	19 03	18 52	18 47	18 42
27	// //	22 32	21 43	21 12	20 49	20 32	20 04	19 44	19 29	19 16	19 05	18 55	18 50	18 46
31	// //	22 19	21 35	21 07	20 45	20 28	20 03	19 44	19 29	19 17	19 07	18 58	18 54	18 50
Feb. 4	23 24	22 06	21 27	21 01	20 41	20 25	20 01	19 43	19 29	19 18	19 09	19 01	18 57	18 54
8	22 55	21 53	21 19	20 54	20 36	20 21	19 58	19 42	19 29	19 19	19 10	19 04	19 01	18 59
12	22 32	21 41	21 10	20 48	20 30	20 16	19 55	19 40	19 28	19 19	19 12	19 07	19 04	19 03
16	22 13	21 30	21 01	20 41	20 25	20 12	19 53	19 39	19 28	19 20	19 14	19 09	19 08	19 07
20	21 56	21 18	20 53	20 34	20 19	20 07	19 49	19 37	19 27	19 20	19 15	19 12	19 11	19 11
24	21 40	21 07	20 44	20 27	20 13	20 02	19 46	19 35	19 26	19 20	19 17	19 15	19 15	19 16
28	21 25	20 56	20 35	20 19	20 07	19 57	19 43	19 33	19 25	19 21	19 18	19 17	19 18	19 20
Mar. 3	21 11	20 45	20 26	20 12	20 01	19 52	19 39	19 30	19 24	19 21	19 19	19 20	19 22	19 24
7	20 57	20 34	20 18	20 05	19 55	19 47	19 36	19 28	19 23	19 21	19 20	19 23	19 25	19 29
11	20 44	20 24	20 09	19 58	19 49	19 42	19 32	19 26	19 22	19 21	19 22	19 25	19 29	19 33
15	20 32	20 14	20 01	19 51	19 43	19 37	19 28	19 23	19 21	19 21	19 23	19 28	19 32	19 37
19	20 19	20 04	19 52	19 44	19 37	19 32	19 25	19 21	19 20	19 21	19 24	19 31	19 36	19 42
23	20 08	19 54	19 44	19 37	19 31	19 27	19 21	19 19	19 18	19 21	19 25	19 33	19 39	19 47
27	19 56	19 45	19 36	19 30	19 25	19 22	19 18	19 16	19 17	19 21	19 27	19 36	19 43	19 51
31	19 45	19 35	19 28	19 23	19 20	19 17	19 14	19 14	19 16	19 21	19 28	19 39	19 47	19 56
Apr. 4	19 35	19 27	19 21	19 17	19 14	19 12	19 11	19 12	19 15	19 21	19 29	19 42	19 51	20 01

(// //) indicates continuous twilight.

UNIVERSAL TIME FOR MERIDIAN OF GREENWICH

BEGINNING ASTRONOMICAL TWILIGHT

Lat.	+40°	+42°	+44°	+46°	+48°	+50°	+52°	+54°	+56°	+58°	+60°	+62°	+64°	+66°
	h m	h m	h m	h m	h m	h m	h m	h m	h m	h m	h m	h m	h m	h m
Jan. −1	5 44	5 47	5 50	5 53	5 56	5 59	6 03	6 06	6 10	6 14	6 18	6 23	6 28	6 33
3	5 45	5 48	5 51	5 54	5 57	6 00	6 03	6 07	6 10	6 14	6 18	6 22	6 27	6 33
7	5 45	5 48	5 51	5 54	5 57	6 00	6 03	6 06	6 09	6 13	6 17	6 21	6 25	6 30
11	5 45	5 48	5 51	5 53	5 56	5 59	6 02	6 05	6 08	6 11	6 15	6 18	6 22	6 27
15	5 45	5 47	5 50	5 52	5 55	5 57	6 00	6 03	6 05	6 08	6 12	6 15	6 19	6 22
19	5 44	5 46	5 48	5 50	5 53	5 55	5 57	6 00	6 02	6 05	6 08	6 10	6 14	6 17
23	5 42	5 44	5 46	5 48	5 50	5 52	5 54	5 56	5 58	6 00	6 03	6 05	6 08	6 10
27	5 40	5 42	5 43	5 45	5 47	5 48	5 50	5 52	5 54	5 55	5 57	5 59	6 01	6 02
31	5 37	5 39	5 40	5 41	5 43	5 44	5 46	5 47	5 48	5 49	5 50	5 52	5 53	5 54
Feb. 4	5 34	5 35	5 36	5 37	5 38	5 39	5 40	5 41	5 42	5 43	5 43	5 44	5 44	5 44
8	5 30	5 31	5 32	5 33	5 34	5 34	5 35	5 35	5 35	5 35	5 35	5 35	5 34	5 34
12	5 26	5 27	5 27	5 28	5 28	5 28	5 28	5 28	5 28	5 27	5 26	5 25	5 24	5 22
16	5 22	5 22	5 22	5 22	5 22	5 22	5 21	5 21	5 20	5 19	5 17	5 15	5 13	5 10
20	5 17	5 17	5 17	5 16	5 16	5 15	5 14	5 13	5 11	5 09	5 07	5 04	5 01	4 57
24	5 12	5 11	5 11	5 10	5 09	5 08	5 06	5 04	5 02	4 59	4 56	4 53	4 48	4 43
28	5 06	5 05	5 04	5 03	5 02	5 00	4 58	4 55	4 52	4 49	4 45	4 40	4 35	4 28
Mar. 3	5 00	4 59	4 58	4 56	4 54	4 52	4 49	4 46	4 42	4 38	4 33	4 27	4 20	4 12
7	4 54	4 53	4 51	4 48	4 46	4 43	4 40	4 36	4 32	4 27	4 21	4 14	4 05	3 55
11	4 48	4 46	4 43	4 41	4 38	4 34	4 30	4 26	4 21	4 15	4 08	3 59	3 49	3 37
15	4 41	4 39	4 36	4 33	4 29	4 25	4 20	4 15	4 09	4 02	3 54	3 44	3 32	3 18
19	4 34	4 31	4 28	4 24	4 20	4 15	4 10	4 04	3 57	3 49	3 39	3 28	3 14	2 57
23	4 27	4 24	4 20	4 16	4 11	4 06	3 59	3 53	3 44	3 35	3 24	3 11	2 54	2 33
27	4 20	4 16	4 12	4 07	4 02	3 55	3 49	3 41	3 31	3 21	3 08	2 52	2 33	2 06
31	4 13	4 08	4 04	3 58	3 52	3 45	3 37	3 28	3 18	3 06	2 51	2 32	2 08	1 33
Apr. 4	4 06	4 01	3 55	3 49	3 42	3 35	3 26	3 16	3 04	2 50	2 32	2 10	1 39	0 38

ENDING ASTRONOMICAL TWILIGHT

Lat.	+40°	+42°	+44°	+46°	+48°	+50°	+52°	+54°	+56°	+58°	+60°	+62°	+64°	+66°
	h m	h m	h m	h m	h m	h m	h m	h m	h m	h m	h m	h m	h m	h m
Jan. −1	18 21	18 18	18 15	18 12	18 09	18 05	18 02	17 59	17 55	17 51	17 47	17 42	17 37	17 31
3	18 24	18 21	18 18	18 15	18 12	18 09	18 06	18 02	17 59	17 55	17 51	17 46	17 42	17 36
7	18 27	18 24	18 21	18 19	18 16	18 13	18 10	18 06	18 03	18 00	17 56	17 52	17 47	17 42
11	18 30	18 28	18 25	18 23	18 20	18 17	18 14	18 11	18 08	18 05	18 01	17 58	17 54	17 49
15	18 34	18 32	18 29	18 27	18 24	18 22	18 19	18 16	18 14	18 11	18 08	18 04	18 01	17 57
19	18 38	18 36	18 34	18 31	18 29	18 27	18 24	18 22	18 20	18 17	18 14	18 11	18 08	18 05
23	18 42	18 40	18 38	18 36	18 34	18 32	18 30	18 28	18 26	18 24	18 22	18 19	18 17	18 14
27	18 46	18 44	18 43	18 41	18 39	18 38	18 36	18 34	18 33	18 31	18 29	18 28	18 26	18 24
31	18 50	18 49	18 47	18 46	18 45	18 43	18 42	18 41	18 40	18 38	18 37	18 36	18 35	18 34
Feb. 4	18 54	18 53	18 52	18 51	18 50	18 49	18 48	18 47	18 47	18 46	18 46	18 45	18 45	18 45
8	18 59	18 58	18 57	18 56	18 56	18 55	18 55	18 54	18 54	18 54	18 54	18 55	18 55	18 56
12	19 03	19 02	19 02	19 02	19 01	19 01	19 01	19 01	19 02	19 02	19 03	19 05	19 06	19 08
16	19 07	19 07	19 07	19 07	19 07	19 07	19 08	19 09	19 10	19 11	19 13	19 15	19 17	19 20
20	19 11	19 11	19 12	19 12	19 13	19 14	19 15	19 16	19 18	19 20	19 22	19 25	19 29	19 33
24	19 16	19 16	19 17	19 18	19 19	19 20	19 22	19 24	19 26	19 29	19 32	19 36	19 40	19 46
28	19 20	19 21	19 22	19 23	19 25	19 27	19 29	19 31	19 34	19 38	19 42	19 47	19 53	20 00
Mar. 3	19 24	19 26	19 27	19 29	19 31	19 33	19 36	19 39	19 43	19 47	19 52	19 58	20 06	20 14
7	19 29	19 30	19 32	19 35	19 37	19 40	19 44	19 47	19 52	19 57	20 03	20 10	20 19	20 30
11	19 33	19 35	19 38	19 40	19 43	19 47	19 51	19 56	20 01	20 07	20 14	20 23	20 33	20 46
15	19 37	19 40	19 43	19 46	19 50	19 54	19 59	20 04	20 11	20 18	20 26	20 36	20 48	21 03
19	19 42	19 45	19 48	19 52	19 57	20 01	20 07	20 13	20 20	20 29	20 39	20 50	21 05	21 23
23	19 47	19 50	19 54	19 58	20 03	20 09	20 15	20 22	20 31	20 40	20 52	21 05	21 22	21 45
27	19 51	19 55	20 00	20 05	20 10	20 17	20 24	20 32	20 41	20 52	21 06	21 22	21 42	22 10
31	19 56	20 01	20 06	20 11	20 18	20 25	20 33	20 42	20 53	21 05	21 21	21 40	22 06	22 44
Apr. 4	20 01	20 06	20 12	20 18	20 25	20 33	20 42	20 52	21 05	21 19	21 37	22 01	22 35	// //

UNIVERSAL TIME FOR MERIDIAN OF GREENWICH

BEGINNING ASTRONOMICAL TWILIGHT

Lat.	−55°	−50°	−45°	−40°	−35°	−30°	−20°	−10°	0°	+10°	+20°	+30°	+35°	+40°
	h m	h m	h m	h m	h m	h m	h m	h m	h m	h m	h m	h m	h m	h m
Mar. 31	4 21	4 31	4 39	4 44	4 48	4 51	4 54	4 54	4 52	4 48	4 41	4 30	4 22	4 13
Apr. 4	4 29	4 38	4 44	4 48	4 51	4 53	4 55	4 54	4 51	4 45	4 37	4 24	4 16	4 06
8	4 37	4 44	4 49	4 52	4 54	4 55	4 56	4 54	4 49	4 43	4 33	4 19	4 10	3 58
12	4 45	4 50	4 54	4 56	4 57	4 58	4 57	4 53	4 48	4 40	4 29	4 14	4 04	3 51
16	4 52	4 56	4 59	5 00	5 00	5 00	4 57	4 53	4 47	4 38	4 26	4 09	3 57	3 43
20	4 59	5 02	5 03	5 03	5 03	5 02	4 58	4 53	4 45	4 35	4 22	4 03	3 51	3 36
24	5 06	5 07	5 08	5 07	5 06	5 04	4 59	4 53	4 44	4 33	4 19	3 58	3 45	3 29
28	5 13	5 13	5 12	5 11	5 09	5 06	5 00	4 53	4 43	4 31	4 15	3 54	3 39	3 22
May 2	5 19	5 18	5 16	5 14	5 11	5 08	5 01	4 53	4 42	4 29	4 12	3 49	3 34	3 15
6	5 25	5 23	5 20	5 17	5 14	5 10	5 02	4 53	4 41	4 27	4 09	3 45	3 28	3 08
10	5 31	5 28	5 24	5 21	5 17	5 12	5 03	4 53	4 41	4 26	4 07	3 41	3 23	3 02
14	5 37	5 32	5 28	5 24	5 19	5 15	5 05	4 53	4 40	4 24	4 04	3 37	3 19	2 56
18	5 42	5 37	5 32	5 27	5 22	5 17	5 06	4 54	4 40	4 23	4 02	3 33	3 14	2 50
22	5 47	5 41	5 35	5 30	5 24	5 18	5 07	4 54	4 40	4 22	4 00	3 30	3 10	2 45
26	5 51	5 45	5 38	5 32	5 26	5 20	5 08	4 55	4 40	4 21	3 59	3 28	3 07	2 40
30	5 55	5 48	5 41	5 35	5 29	5 22	5 09	4 55	4 40	4 21	3 58	3 26	3 04	2 36
June 3	5 59	5 51	5 44	5 37	5 31	5 24	5 11	4 56	4 40	4 21	3 57	3 24	3 02	2 33
7	6 02	5 54	5 46	5 39	5 32	5 26	5 12	4 57	4 40	4 21	3 56	3 23	3 00	2 30
11	6 05	5 56	5 48	5 41	5 34	5 27	5 13	4 58	4 41	4 21	3 56	3 22	2 59	2 28
15	6 07	5 58	5 50	5 43	5 35	5 28	5 14	4 59	4 42	4 22	3 56	3 22	2 59	2 28
19	6 09	6 00	5 51	5 44	5 37	5 29	5 15	5 00	4 43	4 22	3 57	3 22	2 59	2 28
23	6 09	6 00	5 52	5 45	5 37	5 30	5 16	5 01	4 43	4 23	3 58	3 23	3 00	2 28
27	6 10	6 01	5 53	5 45	5 38	5 31	5 17	5 01	4 44	4 24	3 59	3 25	3 01	2 30
July 1	6 09	6 00	5 53	5 45	5 38	5 31	5 17	5 02	4 45	4 25	4 00	3 26	3 03	2 33
5	6 08	6 00	5 52	5 45	5 38	5 31	5 17	5 03	4 46	4 27	4 02	3 29	3 06	2 36

ENDING ASTRONOMICAL TWILIGHT

Lat.	−55°	−50°	−45°	−40°	−35°	−30°	−20°	−10°	0°	+10°	+20°	+30°	+35°	+40°
	h m	h m	h m	h m	h m	h m	h m	h m	h m	h m	h m	h m	h m	h m
Mar. 31	19 45	19 35	19 28	19 23	19 20	19 17	19 14	19 14	19 16	19 21	19 28	19 39	19 47	19 56
Apr. 4	19 35	19 27	19 21	19 17	19 14	19 12	19 11	19 12	19 15	19 21	19 29	19 42	19 51	20 01
8	19 25	19 18	19 14	19 11	19 09	19 08	19 08	19 10	19 14	19 21	19 31	19 45	19 55	20 06
12	19 15	19 10	19 07	19 05	19 04	19 03	19 05	19 08	19 14	19 22	19 33	19 48	19 59	20 12
16	19 06	19 02	19 00	18 59	18 59	18 59	19 02	19 06	19 13	19 22	19 34	19 52	20 03	20 17
20	18 57	18 55	18 54	18 54	18 54	18 55	18 59	19 05	19 12	19 23	19 36	19 55	20 07	20 23
24	18 49	18 48	18 48	18 48	18 50	18 52	18 57	19 03	19 12	19 23	19 38	19 58	20 12	20 28
28	18 41	18 41	18 42	18 44	18 46	18 48	18 54	19 02	19 12	19 24	19 40	20 02	20 16	20 34
May 2	18 34	18 35	18 37	18 39	18 42	18 45	18 52	19 01	19 12	19 25	19 42	20 05	20 21	20 40
6	18 27	18 29	18 32	18 35	18 39	18 42	18 51	19 00	19 12	19 26	19 44	20 09	20 26	20 46
10	18 21	18 24	18 28	18 32	18 36	18 40	18 49	19 00	19 12	19 27	19 47	20 13	20 30	20 52
14	18 15	18 20	18 24	18 29	18 33	18 38	18 48	18 59	19 13	19 29	19 49	20 16	20 35	20 58
18	18 10	18 16	18 21	18 26	18 31	18 36	18 47	18 59	19 13	19 30	19 51	20 20	20 39	21 04
22	18 06	18 12	18 18	18 23	18 29	18 35	18 46	18 59	19 14	19 31	19 53	20 23	20 44	21 10
26	18 02	18 09	18 15	18 21	18 27	18 33	18 46	18 59	19 15	19 33	19 56	20 27	20 48	21 15
30	17 59	18 07	18 13	18 20	18 26	18 33	18 46	19 00	19 15	19 34	19 58	20 30	20 51	21 20
June 3	17 57	18 05	18 12	18 19	18 26	18 32	18 46	19 00	19 16	19 36	20 00	20 33	20 55	21 24
7	17 55	18 03	18 11	18 18	18 25	18 32	18 46	19 01	19 17	19 37	20 02	20 35	20 58	21 28
11	17 54	18 03	18 11	18 18	18 25	18 32	18 46	19 01	19 18	19 38	20 03	20 37	21 01	21 31
15	17 54	18 03	18 11	18 18	18 26	18 33	18 47	19 02	19 19	19 39	20 05	20 39	21 03	21 34
19	17 54	18 03	18 11	18 19	18 26	18 33	18 48	19 03	19 20	19 40	20 06	20 40	21 04	21 35
23	17 55	18 04	18 12	18 20	18 27	18 34	18 49	19 04	19 21	19 41	20 07	20 41	21 05	21 36
27	17 57	18 05	18 14	18 21	18 28	18 35	18 50	19 05	19 22	19 42	20 07	20 41	21 05	21 36
July 1	17 59	18 07	18 15	18 23	18 30	18 37	18 51	19 06	19 22	19 42	20 07	20 41	21 04	21 35
5	18 02	18 10	18 17	18 24	18 31	18 38	18 52	19 06	19 23	19 42	20 07	20 40	21 03	21 33

UNIVERSAL TIME FOR MERIDIAN OF GREENWICH
BEGINNING ASTRONOMICAL TWILIGHT

Lat.	+40°	+42°	+44°	+46°	+48°	+50°	+52°	+54°	+56°	+58°	+60°	+62°	+64°	+66°
	h m	h m	h m	h m	h m	h m	h m	h m	h m	h m	h m	h m	h m	h m
Mar. 31	4 13	4 08	4 04	3 58	3 52	3 45	3 37	3 28	3 18	3 06	2 51	2 32	2 08	1 33
Apr. 4	4 06	4 01	3 55	3 49	3 42	3 35	3 26	3 16	3 04	2 50	2 32	2 10	1 39	0 38
8	3 58	3 53	3 47	3 40	3 32	3 24	3 14	3 02	2 49	2 33	2 12	1 44	0 57	// //
12	3 51	3 45	3 38	3 31	3 22	3 13	3 02	2 49	2 33	2 14	1 49	1 10	// //	// //
16	3 43	3 37	3 29	3 21	3 12	3 01	2 49	2 34	2 17	1 54	1 21	// //	// //	// //
20	3 36	3 29	3 21	3 12	3 02	2 50	2 36	2 19	1 58	1 30	0 39	// //	// //	// //
24	3 29	3 21	3 12	3 02	2 51	2 38	2 22	2 03	1 38	0 59	// //	// //	// //	// //
28	3 22	3 13	3 04	2 53	2 41	2 26	2 08	1 46	1 14	// //	// //	// //	// //	// //
May 2	3 15	3 06	2 55	2 44	2 30	2 14	1 54	1 27	0 41	// //	// //	// //	// //	// //
6	3 08	2 58	2 47	2 34	2 19	2 01	1 38	1 03	// //	// //	// //	// //	// //	// //
10	3 02	2 51	2 39	2 25	2 09	1 48	1 20	0 30	// //	// //	// //	// //	// //	// //
14	2 56	2 44	2 31	2 16	1 58	1 34	1 00	// //	// //	// //	// //	// //	// //	// //
18	2 50	2 38	2 24	2 08	1 47	1 20	0 33	// //	// //	// //	// //	// //	// //	// //
22	2 45	2 32	2 17	1 59	1 37	1 05	// //	// //	// //	// //	// //	// //	// //	// //
26	2 40	2 27	2 11	1 52	1 27	0 47	// //	// //	// //	// //	// //	// //	// //	// //
30	2 36	2 22	2 05	1 44	1 16	0 23	// //	// //	// //	// //	// //	// //	// //	// //
June 3	2 33	2 18	2 00	1 38	1 07	// //	// //	// //	// //	// //	// //	// //	// //	// //
7	2 30	2 15	1 56	1 33	0 58	// //	// //	// //	// //	// //	// //	// //	// //	// //
11	2 28	2 13	1 54	1 29	0 50	// //	// //	// //	// //	// //	// //	// //	// //	// //
15	2 28	2 11	1 52	1 26	0 45	// //	// //	// //	// //	// //	// //	// //	// //	** **
19	2 28	2 11	1 51	1 25	0 42	// //	// //	// //	// //	// //	// //	// //	// //	** **
23	2 28	2 12	1 52	1 26	0 42	// //	// //	// //	// //	// //	// //	// //	// //	** **
27	2 30	2 14	1 54	1 29	0 47	// //	// //	// //	// //	// //	// //	// //	// //	** **
July 1	2 33	2 17	1 58	1 33	0 54	// //	// //	// //	// //	// //	// //	// //	// //	// //
5	2 36	2 21	2 02	1 38	1 03	// //	// //	// //	// //	// //	// //	// //	// //	// //

ENDING ASTRONOMICAL TWILIGHT

Lat.	+40°	+42°	+44°	+46°	+48°	+50°	+52°	+54°	+56°	+58°	+60°	+62°	+64°	+66°
	h m	h m	h m	h m	h m	h m	h m	h m	h m	h m	h m	h m	h m	h m
Mar. 31	19 56	20 01	20 06	20 11	20 18	20 25	20 33	20 42	20 53	21 05	21 21	21 40	22 06	22 44
Apr. 4	20 01	20 06	20 12	20 18	20 25	20 33	20 42	20 52	21 05	21 19	21 37	22 01	22 35	// //
8	20 06	20 12	20 18	20 25	20 33	20 42	20 52	21 03	21 17	21 34	21 56	22 26	23 21	// //
12	20 12	20 18	20 25	20 32	20 41	20 51	21 02	21 15	21 31	21 51	22 18	23 01	// //	// //
16	20 17	20 24	20 31	20 40	20 49	21 00	21 13	21 28	21 46	22 10	22 46	// //	// //	// //
20	20 23	20 30	20 38	20 47	20 58	21 10	21 24	21 41	22 03	22 34	23 37	// //	// //	// //
24	20 28	20 36	20 45	20 55	21 07	21 20	21 36	21 56	22 23	23 06	// //	// //	// //	// //
28	20 34	20 43	20 53	21 04	21 16	21 31	21 49	22 13	22 47	// //	// //	// //	// //	// //
May 2	20 40	20 50	21 00	21 12	21 26	21 43	22 04	22 32	23 25	// //	// //	// //	// //	// //
6	20 46	20 56	21 08	21 21	21 36	21 55	22 19	22 56	// //	// //	// //	// //	// //	// //
10	20 52	21 03	21 15	21 29	21 46	22 08	22 37	23 38	// //	// //	// //	// //	// //	// //
14	20 58	21 10	21 23	21 38	21 57	22 21	22 58	// //	// //	// //	// //	// //	// //	// //
18	21 04	21 16	21 30	21 47	22 08	22 36	23 30	// //	// //	// //	// //	// //	// //	// //
22	21 10	21 22	21 37	21 56	22 19	22 52	// //	// //	// //	// //	// //	// //	// //	// //
26	21 15	21 28	21 44	22 04	22 30	23 12	// //	// //	// //	// //	// //	// //	// //	// //
30	21 20	21 34	21 51	22 12	22 41	23 42	// //	// //	// //	// //	// //	// //	// //	// //
June 3	21 24	21 39	21 57	22 20	22 52	// //	// //	// //	// //	// //	// //	// //	// //	// //
7	21 28	21 44	22 02	22 26	23 02	// //	// //	// //	// //	// //	// //	// //	// //	// //
11	21 31	21 47	22 06	22 31	23 11	// //	// //	// //	// //	// //	// //	// //	// //	// //
15	21 34	21 50	22 09	22 35	23 18	// //	// //	// //	// //	// //	// //	// //	// //	** **
19	21 35	21 51	22 11	22 38	23 21	// //	// //	// //	// //	// //	// //	// //	// //	** **
23	21 36	21 52	22 12	22 38	23 21	// //	// //	// //	// //	// //	// //	// //	// //	** **
27	21 36	21 52	22 11	22 37	23 18	// //	// //	// //	// //	// //	// //	// //	// //	** **
July 1	21 35	21 50	22 09	22 34	23 12	// //	// //	// //	// //	// //	// //	// //	// //	// //
5	21 33	21 48	22 06	22 30	23 04	// //	// //	// //	// //	// //	// //	// //	// //	// //

(** **) indicates Sun continuously above horizon.
(// //) indicates continuous twilight.

UNIVERSAL TIME FOR MERIDIAN OF GREENWICH
BEGINNING ASTRONOMICAL TWILIGHT

Lat.		−55°	−50°	−45°	−40°	−35°	−30°	−20°	−10°	0°	+10°	+20°	+30°	+35°	+40°
		h m	h m	h m	h m	h m	h m	h m	h m	h m	h m	h m	h m	h m	h m
July	1	6 09	6 00	5 53	5 45	5 38	5 31	5 17	5 02	4 45	4 25	4 00	3 26	3 03	2 33
	5	6 08	6 00	5 52	5 45	5 38	5 31	5 17	5 03	4 46	4 27	4 02	3 29	3 06	2 36
	9	6 06	5 58	5 51	5 44	5 38	5 31	5 18	5 03	4 47	4 28	4 04	3 31	3 09	2 40
	13	6 04	5 56	5 49	5 43	5 37	5 30	5 18	5 04	4 48	4 29	4 06	3 34	3 12	2 44
	17	6 00	5 54	5 47	5 41	5 35	5 29	5 17	5 04	4 49	4 31	4 08	3 37	3 16	2 49
	21	5 56	5 50	5 45	5 39	5 34	5 28	5 17	5 04	4 49	4 32	4 10	3 40	3 20	2 54
	25	5 52	5 47	5 42	5 37	5 32	5 27	5 16	5 04	4 50	4 33	4 12	3 43	3 24	3 00
	29	5 47	5 42	5 38	5 34	5 29	5 25	5 15	5 03	4 50	4 34	4 14	3 47	3 29	3 05
Aug.	2	5 41	5 38	5 34	5 30	5 26	5 22	5 13	5 03	4 51	4 35	4 16	3 50	3 33	3 11
	6	5 35	5 32	5 30	5 27	5 23	5 20	5 12	5 02	4 51	4 36	4 18	3 54	3 37	3 17
	10	5 28	5 26	5 25	5 22	5 20	5 17	5 10	5 01	4 51	4 37	4 20	3 57	3 42	3 23
	14	5 20	5 20	5 19	5 18	5 16	5 14	5 08	5 00	4 50	4 38	4 22	4 01	3 46	3 28
	18	5 12	5 13	5 14	5 13	5 12	5 10	5 05	4 58	4 50	4 39	4 24	4 04	3 51	3 34
	22	5 04	5 06	5 07	5 08	5 07	5 06	5 02	4 57	4 49	4 39	4 26	4 07	3 55	3 39
	26	4 55	4 59	5 01	5 02	5 02	5 02	5 00	4 55	4 49	4 40	4 27	4 10	3 59	3 45
	30	4 46	4 51	4 54	4 56	4 57	4 58	4 57	4 53	4 48	4 40	4 29	4 13	4 03	3 50
Sept.	3	4 36	4 43	4 47	4 50	4 52	4 53	4 53	4 51	4 47	4 40	4 30	4 16	4 07	3 55
	7	4 26	4 34	4 40	4 44	4 47	4 49	4 50	4 49	4 46	4 40	4 31	4 19	4 10	4 00
	11	4 15	4 25	4 32	4 37	4 41	4 44	4 46	4 47	4 44	4 40	4 33	4 22	4 14	4 04
	15	4 04	4 16	4 24	4 31	4 35	4 39	4 43	4 44	4 43	4 40	4 34	4 24	4 17	4 09
	19	3 53	4 06	4 16	4 24	4 29	4 34	4 39	4 42	4 42	4 40	4 35	4 27	4 21	4 13
	23	3 41	3 57	4 08	4 17	4 23	4 28	4 35	4 39	4 40	4 39	4 36	4 29	4 24	4 18
	27	3 29	3 47	4 00	4 09	4 17	4 23	4 31	4 37	4 39	4 39	4 37	4 32	4 27	4 22
Oct.	1	3 16	3 36	3 51	4 02	4 11	4 18	4 28	4 34	4 38	4 39	4 38	4 34	4 31	4 26
	5	3 03	3 26	3 42	3 55	4 05	4 12	4 24	4 31	4 36	4 39	4 39	4 36	4 34	4 30

ENDING ASTRONOMICAL TWILIGHT

Lat.		−55°	−50°	−45°	−40°	−35°	−30°	−20°	−10°	0°	+10°	+20°	+30°	+35°	+40°
		h m	h m	h m	h m	h m	h m	h m	h m	h m	h m	h m	h m	h m	h m
July	1	17 59	18 07	18 15	18 23	18 30	18 37	18 51	19 06	19 22	19 42	20 07	20 41	21 04	21 35
	5	18 02	18 10	18 17	18 24	18 31	18 38	18 52	19 06	19 23	19 42	20 07	20 40	21 03	21 33
	9	18 05	18 13	18 20	18 26	18 33	18 40	18 53	19 07	19 23	19 42	20 06	20 39	21 01	21 30
	13	18 08	18 16	18 22	18 29	18 35	18 41	18 54	19 08	19 23	19 42	20 05	20 37	20 59	21 26
	17	18 12	18 19	18 25	18 31	18 37	18 43	18 55	19 08	19 23	19 41	20 04	20 35	20 56	21 22
	21	18 17	18 23	18 28	18 34	18 39	18 45	18 56	19 09	19 23	19 41	20 03	20 32	20 52	21 17
	25	18 22	18 27	18 32	18 37	18 42	18 47	18 57	19 09	19 23	19 40	20 01	20 29	20 48	21 12
	29	18 27	18 31	18 35	18 39	18 44	18 48	18 58	19 09	19 22	19 38	19 58	20 25	20 43	21 06
Aug.	2	18 32	18 35	18 39	18 42	18 46	18 50	18 59	19 10	19 22	19 37	19 56	20 21	20 39	21 00
	6	18 38	18 40	18 43	18 46	18 49	18 52	19 00	19 10	19 21	19 35	19 53	20 17	20 33	20 54
	10	18 44	18 45	18 47	18 49	18 51	18 54	19 01	19 10	19 20	19 33	19 50	20 13	20 28	20 47
	14	18 50	18 50	18 50	18 52	18 54	18 56	19 02	19 09	19 19	19 31	19 47	20 08	20 22	20 40
	18	18 56	18 55	18 55	18 55	18 56	18 58	19 03	19 09	19 18	19 29	19 43	20 03	20 16	20 32
	22	19 03	19 00	18 59	18 58	18 59	19 00	19 03	19 09	19 16	19 26	19 39	19 58	20 10	20 25
	26	19 10	19 06	19 03	19 02	19 01	19 02	19 04	19 08	19 15	19 24	19 36	19 53	20 04	20 18
	30	19 17	19 11	19 08	19 05	19 04	19 04	19 05	19 08	19 13	19 21	19 32	19 47	19 57	20 10
Sept.	3	19 24	19 17	19 12	19 09	19 07	19 06	19 05	19 08	19 12	19 18	19 28	19 42	19 51	20 03
	7	19 32	19 23	19 17	19 13	19 10	19 08	19 06	19 07	19 10	19 16	19 24	19 36	19 45	19 55
	11	19 39	19 29	19 22	19 16	19 13	19 10	19 07	19 07	19 09	19 13	19 20	19 31	19 38	19 48
	15	19 48	19 36	19 27	19 20	19 16	19 12	19 08	19 06	19 07	19 10	19 16	19 25	19 32	19 40
	19	19 56	19 42	19 32	19 25	19 19	19 14	19 09	19 06	19 06	19 08	19 12	19 20	19 26	19 33
	23	20 05	19 49	19 38	19 29	19 22	19 17	19 10	19 06	19 04	19 05	19 08	19 15	19 20	19 26
	27	20 15	19 57	19 43	19 33	19 25	19 19	19 11	19 05	19 03	19 03	19 05	19 10	19 14	19 19
Oct.	1	20 25	20 04	19 49	19 38	19 29	19 22	19 12	19 05	19 02	19 00	19 01	19 05	19 08	19 12
	5	20 36	20 12	19 56	19 43	19 33	19 25	19 13	19 05	19 01	18 58	18 58	19 00	19 02	19 06

UNIVERSAL TIME FOR MERIDIAN OF GREENWICH
BEGINNING ASTRONOMICAL TWILIGHT

Lat.	+40°	+42°	+44°	+46°	+48°	+50°	+52°	+54°	+56°	+58°	+60°	+62°	+64°	+66°
	h m	h m	h m	h m	h m	h m	h m	h m	h m	h m	h m	h m	h m	h m
July 1	2 33	2 17	1 58	1 33	0 54	// //	// //	// //	// //	// //	// //	// //	// //	// //
5	2 36	2 21	2 02	1 38	1 03	// //	// //	// //	// //	// //	// //	// //	// //	// //
9	2 40	2 25	2 07	1 45	1 13	// //	// //	// //	// //	// //	// //	// //	// //	// //
13	2 44	2 30	2 13	1 52	1 24	0 29	// //	// //	// //	// //	// //	// //	// //	// //
17	2 49	2 36	2 20	2 00	1 35	0 55	// //	// //	// //	// //	// //	// //	// //	// //
21	2 54	2 42	2 27	2 09	1 46	1 13	// //	// //	// //	// //	// //	// //	// //	// //
25	3 00	2 48	2 34	2 17	1 57	1 29	0 39	// //	// //	// //	// //	// //	// //	// //
29	3 05	2 54	2 41	2 26	2 07	1 44	1 08	// //	// //	// //	// //	// //	// //	// //
Aug. 2	3 11	3 01	2 49	2 34	2 18	1 57	1 28	0 34	// //	// //	// //	// //	// //	// //
6	3 17	3 07	2 56	2 43	2 28	2 09	1 45	1 10	// //	// //	// //	// //	// //	// //
10	3 23	3 13	3 03	2 51	2 37	2 21	2 00	1 33	0 44	// //	// //	// //	// //	// //
14	3 28	3 20	3 10	2 59	2 47	2 32	2 14	1 51	1 18	// //	// //	// //	// //	// //
18	3 34	3 26	3 17	3 07	2 56	2 42	2 27	2 07	1 41	1 00	// //	// //	// //	// //
22	3 39	3 32	3 24	3 15	3 04	2 52	2 38	2 21	2 00	1 30	0 32	// //	// //	// //
26	3 45	3 38	3 31	3 22	3 13	3 02	2 49	2 34	2 16	1 52	1 18	// //	// //	// //
30	3 50	3 44	3 37	3 29	3 21	3 11	3 00	2 47	2 31	2 11	1 45	1 03	// //	// //
Sept. 3	3 55	3 49	3 43	3 36	3 29	3 20	3 10	2 58	2 44	2 27	2 06	1 36	0 43	// //
7	4 00	3 55	3 49	3 43	3 36	3 28	3 19	3 09	2 57	2 42	2 24	2 01	1 27	// //
11	4 04	4 00	3 55	3 49	3 43	3 36	3 28	3 19	3 08	2 55	2 40	2 21	1 55	1 17
15	4 09	4 05	4 00	3 56	3 50	3 44	3 37	3 28	3 19	3 08	2 55	2 38	2 18	1 50
19	4 13	4 10	4 06	4 02	3 57	3 51	3 45	3 38	3 29	3 20	3 08	2 54	2 37	2 15
23	4 18	4 15	4 11	4 07	4 03	3 58	3 53	3 46	3 39	3 31	3 21	3 09	2 55	2 36
27	4 22	4 19	4 16	4 13	4 09	4 05	4 00	3 55	3 49	3 41	3 33	3 23	3 10	2 55
Oct. 1	4 26	4 24	4 22	4 19	4 16	4 12	4 08	4 03	3 58	3 52	3 44	3 36	3 25	3 12
5	4 30	4 29	4 27	4 24	4 22	4 19	4 15	4 11	4 07	4 01	3 55	3 48	3 39	3 28

ENDING ASTRONOMICAL TWILIGHT

Lat.	+40°	+42°	+44°	+46°	+48°	+50°	+52°	+54°	+56°	+58°	+60°	+62°	+64°	+66°
	h m	h m	h m	h m	h m	h m	h m	h m	h m	h m	h m	h m	h m	h m
July 1	21 35	21 50	22 09	22 34	23 12	// //	// //	// //	// //	// //	// //	// //	// //	// //
5	21 33	21 48	22 06	22 30	23 04	// //	// //	// //	// //	// //	// //	// //	// //	// //
9	21 30	21 45	22 02	22 24	22 55	// //	// //	// //	// //	// //	// //	// //	// //	// //
13	21 26	21 40	21 57	22 18	22 45	23 35	// //	// //	// //	// //	// //	// //	// //	// //
17	21 22	21 36	21 51	22 10	22 35	23 13	// //	// //	// //	// //	// //	// //	// //	// //
21	21 17	21 30	21 45	22 02	22 25	22 56	// //	// //	// //	// //	// //	// //	// //	// //
25	21 12	21 24	21 38	21 54	22 14	22 40	23 25	// //	// //	// //	// //	// //	// //	// //
29	21 06	21 17	21 30	21 45	22 03	22 26	23 00	// //	// //	// //	// //	// //	// //	// //
Aug. 2	21 00	21 11	21 22	21 36	21 53	22 13	22 40	23 27	// //	// //	// //	// //	// //	// //
6	20 54	21 03	21 14	21 27	21 42	22 00	22 23	22 56	// //	// //	// //	// //	// //	// //
10	20 47	20 56	21 06	21 18	21 31	21 47	22 07	22 32	23 17	// //	// //	// //	// //	// //
14	20 40	20 48	20 57	21 08	21 20	21 35	21 52	22 14	22 45	// //	// //	// //	// //	// //
18	20 32	20 40	20 49	20 59	21 10	21 23	21 38	21 57	22 22	22 59	// //	// //	// //	// //
22	20 25	20 32	20 40	20 49	20 59	21 11	21 25	21 41	22 02	22 30	23 18	// //	// //	// //
26	20 18	20 24	20 31	20 40	20 49	20 59	21 12	21 26	21 44	22 07	22 39	// //	// //	// //
30	20 10	20 16	20 23	20 30	20 39	20 48	20 59	21 12	21 27	21 46	22 11	22 49	// //	// //
Sept. 3	20 03	20 08	20 14	20 21	20 28	20 37	20 47	20 58	21 12	21 28	21 48	22 16	23 02	// //
7	19 55	20 00	20 05	20 12	20 18	20 26	20 35	20 45	20 57	21 11	21 28	21 50	22 22	23 21
11	19 48	19 52	19 57	20 02	20 09	20 15	20 23	20 32	20 43	20 55	21 10	21 28	21 52	22 28
15	19 40	19 44	19 49	19 53	19 59	20 05	20 12	20 20	20 29	20 40	20 53	21 08	21 28	21 54
19	19 33	19 36	19 40	19 45	19 49	19 55	20 01	20 08	20 16	20 25	20 36	20 50	21 06	21 28
23	19 26	19 29	19 32	19 36	19 40	19 45	19 50	19 56	20 03	20 12	20 21	20 33	20 47	21 04
27	19 19	19 21	19 24	19 28	19 31	19 35	19 40	19 45	19 51	19 58	20 07	20 16	20 28	20 43
Oct. 1	19 12	19 14	19 17	19 19	19 22	19 26	19 30	19 34	19 40	19 46	19 53	20 01	20 11	20 23
5	19 06	19 07	19 09	19 11	19 14	19 17	19 20	19 24	19 28	19 34	19 40	19 47	19 55	20 05

(// //) indicates continuous twilight.

UNIVERSAL TIME FOR MERIDIAN OF GREENWICH

BEGINNING ASTRONOMICAL TWILIGHT

Lat.	−55°	−50°	−45°	−40°	−35°	−30°	−20°	−10°	0°	+10°	+20°	+30°	+35°	+40°
	h m	h m	h m	h m	h m	h m	h m	h m	h m	h m	h m	h m	h m	h m
Oct. 1	3 16	3 36	3 51	4 02	4 11	4 18	4 28	4 34	4 38	4 39	4 38	4 34	4 31	4 26
5	3 03	3 26	3 42	3 55	4 05	4 12	4 24	4 31	4 36	4 39	4 39	4 36	4 34	4 30
9	2 50	3 15	3 34	3 48	3 58	4 07	4 20	4 29	4 35	4 38	4 40	4 39	4 37	4 34
13	2 35	3 04	3 25	3 40	3 52	4 02	4 16	4 26	4 34	4 38	4 41	4 41	4 40	4 38
17	2 20	2 53	3 16	3 33	3 46	3 57	4 13	4 24	4 32	4 38	4 42	4 43	4 43	4 42
21	2 05	2 42	3 07	3 26	3 40	3 52	4 09	4 22	4 31	4 38	4 43	4 46	4 47	4 47
25	1 48	2 30	2 58	3 18	3 34	3 47	4 06	4 20	4 30	4 38	4 44	4 48	4 50	4 51
29	1 29	2 19	2 49	3 12	3 29	3 42	4 03	4 18	4 30	4 39	4 46	4 51	4 53	4 55
Nov. 2	1 07	2 07	2 41	3 05	3 23	3 38	4 00	4 16	4 29	4 39	4 47	4 54	4 56	4 59
6	0 40	1 54	2 32	2 58	3 18	3 34	3 57	4 15	4 29	4 40	4 49	4 56	5 00	5 03
10	// //	1 41	2 24	2 52	3 13	3 30	3 55	4 14	4 28	4 40	4 50	4 59	5 03	5 07
14	// //	1 28	2 16	2 46	3 09	3 26	3 53	4 13	4 28	4 41	4 52	5 02	5 06	5 11
18	// //	1 14	2 08	2 41	3 05	3 23	3 52	4 12	4 29	4 42	4 54	5 05	5 10	5 14
22	// //	0 59	2 01	2 36	3 01	3 21	3 50	4 12	4 29	4 44	4 56	5 07	5 13	5 18
26	// //	0 43	1 54	2 32	2 58	3 19	3 49	4 12	4 30	4 45	4 58	5 10	5 16	5 22
30	// //	0 21	1 48	2 28	2 56	3 17	3 49	4 12	4 31	4 47	5 00	5 13	5 19	5 26
Dec. 4	// //	// //	1 43	2 25	2 54	3 16	3 49	4 13	4 32	4 48	5 03	5 16	5 22	5 29
8	// //	// //	1 39	2 24	2 53	3 16	3 49	4 14	4 34	4 50	5 05	5 19	5 25	5 32
12	// //	// //	1 36	2 23	2 53	3 16	3 50	4 15	4 35	4 52	5 07	5 21	5 28	5 35
16	// //	// //	1 35	2 23	2 54	3 17	3 52	4 17	4 37	4 54	5 09	5 24	5 31	5 38
20	// //	// //	1 36	2 24	2 55	3 19	3 53	4 19	4 39	4 56	5 11	5 26	5 33	5 40
24	// //	// //	1 38	2 26	2 57	3 21	3 55	4 21	4 41	4 58	5 13	5 28	5 35	5 42
28	// //	// //	1 42	2 29	3 00	3 23	3 58	4 23	4 43	5 00	5 15	5 29	5 36	5 44
32	// //	// //	1 47	2 33	3 03	3 26	4 00	4 25	4 45	5 02	5 17	5 31	5 38	5 45
36	// //	// //	1 54	2 38	3 07	3 30	4 03	4 28	4 47	5 04	5 18	5 32	5 38	5 45

ENDING ASTRONOMICAL TWILIGHT

Lat.	−55°	−50°	−45°	−40°	−35°	−30°	−20°	−10°	0°	+10°	+20°	+30°	+35°	+40°
	h m	h m	h m	h m	h m	h m	h m	h m	h m	h m	h m	h m	h m	h m
Oct. 1	20 25	20 04	19 49	19 38	19 29	19 22	19 12	19 05	19 02	19 00	19 01	19 05	19 08	19 12
5	20 36	20 12	19 56	19 43	19 33	19 25	19 13	19 05	19 01	18 58	18 58	19 00	19 02	19 06
9	20 47	20 21	20 02	19 48	19 37	19 28	19 15	19 06	19 00	18 56	18 54	18 55	18 57	18 59
13	21 00	20 30	20 09	19 53	19 41	19 31	19 16	19 06	18 59	18 54	18 51	18 51	18 52	18 53
17	21 13	20 39	20 16	19 59	19 45	19 35	19 18	19 07	18 58	18 52	18 48	18 47	18 47	18 47
21	21 28	20 49	20 24	20 05	19 50	19 38	19 20	19 07	18 58	18 51	18 46	18 43	18 42	18 42
25	21 44	21 00	20 31	20 11	19 55	19 42	19 23	19 08	18 58	18 50	18 44	18 39	18 38	18 37
29	22 03	21 11	20 40	20 17	20 00	19 46	19 25	19 10	18 58	18 49	18 42	18 36	18 34	18 32
Nov. 2	22 25	21 23	20 48	20 24	20 05	19 50	19 28	19 11	18 58	18 48	18 40	18 33	18 30	18 28
6	22 56	21 36	20 57	20 30	20 10	19 54	19 30	19 13	18 59	18 48	18 38	18 31	18 27	18 24
10	// //	21 49	21 06	20 37	20 16	19 59	19 33	19 14	19 00	18 48	18 37	18 29	18 25	18 21
14	// //	22 04	21 15	20 44	20 21	20 03	19 36	19 16	19 01	18 48	18 37	18 27	18 22	18 18
18	// //	22 20	21 24	20 51	20 27	20 08	19 39	19 18	19 02	18 48	18 36	18 26	18 21	18 16
22	// //	22 37	21 33	20 58	20 32	20 12	19 42	19 21	19 03	18 49	18 36	18 25	18 19	18 14
26	// //	22 57	21 42	21 04	20 37	20 17	19 46	19 23	19 05	18 50	18 36	18 24	18 18	18 12
30	// //	23 25	21 51	21 10	20 42	20 21	19 49	19 25	19 07	18 51	18 37	18 24	18 18	18 12
Dec. 4	// //	// //	21 59	21 16	20 47	20 25	19 52	19 28	19 09	18 52	18 38	18 25	18 18	18 11
8	// //	// //	22 06	21 21	20 51	20 28	19 55	19 30	19 11	18 54	18 39	18 25	18 19	18 12
12	// //	// //	22 12	21 26	20 55	20 32	19 58	19 33	19 13	18 56	18 41	18 27	18 20	18 12
16	// //	// //	22 17	21 29	20 58	20 35	20 00	19 35	19 15	18 58	18 42	18 28	18 21	18 14
20	// //	// //	22 20	21 32	21 00	20 37	20 02	19 37	19 17	19 00	18 44	18 30	18 23	18 15
24	// //	// //	22 21	21 33	21 02	20 39	20 04	19 39	19 19	19 02	18 46	18 32	18 25	18 17
28	// //	// //	22 21	21 34	21 03	20 40	20 06	19 41	19 20	19 04	18 48	18 34	18 27	18 20
32	// //	// //	22 19	21 34	21 03	20 41	20 07	19 42	19 22	19 06	18 51	18 37	18 30	18 23
36	// //	// //	22 16	21 32	21 03	20 41	20 08	19 43	19 24	19 07	18 53	18 39	18 33	18 26

(// //) indicates continuous twilight.

UNIVERSAL TIME FOR MERIDIAN OF GREENWICH
BEGINNING ASTRONOMICAL TWILIGHT

Lat.	+40°	+42°	+44°	+46°	+48°	+50°	+52°	+54°	+56°	+58°	+60°	+62°	+64°	+66°
	h m	h m	h m	h m	h m	h m	h m	h m	h m	h m	h m	h m	h m	h m
Oct. 1	4 26	4 24	4 22	4 19	4 16	4 12	4 08	4 03	3 58	3 52	3 44	3 36	3 25	3 12
5	4 30	4 29	4 27	4 24	4 22	4 19	4 15	4 11	4 07	4 01	3 55	3 48	3 39	3 28
9	4 34	4 33	4 31	4 30	4 27	4 25	4 22	4 19	4 15	4 11	4 06	3 59	3 52	3 43
13	4 38	4 37	4 36	4 35	4 33	4 31	4 29	4 27	4 23	4 20	4 16	4 11	4 05	3 57
17	4 42	4 42	4 41	4 40	4 39	4 38	4 36	4 34	4 32	4 29	4 25	4 21	4 17	4 11
21	4 47	4 46	4 46	4 45	4 45	4 44	4 43	4 41	4 40	4 37	4 35	4 32	4 28	4 24
25	4 51	4 51	4 51	4 51	4 50	4 50	4 49	4 48	4 47	4 46	4 44	4 42	4 39	4 36
29	4 55	4 55	4 55	4 56	4 56	4 56	4 56	4 55	4 55	4 54	4 53	4 52	4 50	4 48
Nov. 2	4 59	4 59	5 00	5 01	5 01	5 02	5 02	5 02	5 02	5 02	5 02	5 01	5 01	4 59
6	5 03	5 04	5 05	5 06	5 07	5 07	5 08	5 09	5 10	5 10	5 10	5 11	5 11	5 10
10	5 07	5 08	5 09	5 11	5 12	5 13	5 14	5 15	5 17	5 18	5 19	5 20	5 20	5 21
14	5 11	5 12	5 14	5 15	5 17	5 19	5 20	5 22	5 23	5 25	5 26	5 28	5 30	5 31
18	5 14	5 16	5 18	5 20	5 22	5 24	5 26	5 28	5 30	5 32	5 34	5 36	5 38	5 41
22	5 18	5 20	5 23	5 25	5 27	5 29	5 31	5 34	5 36	5 39	5 41	5 44	5 47	5 50
26	5 22	5 24	5 27	5 29	5 32	5 34	5 37	5 39	5 42	5 45	5 48	5 51	5 55	5 58
30	5 26	5 28	5 31	5 33	5 36	5 39	5 42	5 44	5 48	5 51	5 54	5 58	6 02	6 06
Dec. 4	5 29	5 32	5 34	5 37	5 40	5 43	5 46	5 49	5 53	5 56	6 00	6 04	6 08	6 13
8	5 32	5 35	5 38	5 41	5 44	5 47	5 50	5 54	5 57	6 01	6 05	6 09	6 14	6 19
12	5 35	5 38	5 41	5 44	5 47	5 50	5 54	5 57	6 01	6 05	6 09	6 14	6 19	6 24
16	5 38	5 41	5 44	5 47	5 50	5 53	5 57	6 00	6 04	6 08	6 13	6 17	6 22	6 28
20	5 40	5 43	5 46	5 49	5 53	5 56	5 59	6 03	6 07	6 11	6 15	6 20	6 25	6 31
24	5 42	5 45	5 48	5 51	5 55	5 58	6 01	6 05	6 09	6 13	6 17	6 22	6 27	6 33
28	5 44	5 47	5 50	5 53	5 56	5 59	6 03	6 06	6 10	6 14	6 18	6 23	6 28	6 34
32	5 45	5 48	5 51	5 54	5 57	6 00	6 03	6 07	6 10	6 14	6 18	6 23	6 28	6 33
36	5 45	5 48	5 51	5 54	5 57	6 00	6 03	6 06	6 10	6 13	6 17	6 21	6 26	6 31

ENDING ASTRONOMICAL TWILIGHT

Lat.	+40°	+42°	+44°	+46°	+48°	+50°	+52°	+54°	+56°	+58°	+60°	+62°	+64°	+66°
	h m	h m	h m	h m	h m	h m	h m	h m	h m	h m	h m	h m	h m	h m
Oct. 1	19 12	19 14	19 17	19 19	19 22	19 26	19 30	19 34	19 40	19 46	19 53	20 01	20 11	20 23
5	19 06	19 07	19 09	19 11	19 14	19 17	19 20	19 24	19 28	19 34	19 40	19 47	19 55	20 05
9	18 59	19 00	19 02	19 04	19 06	19 08	19 11	19 14	19 18	19 22	19 27	19 33	19 40	19 49
13	18 53	18 54	18 55	18 56	18 58	19 00	19 02	19 05	19 07	19 11	19 15	19 20	19 26	19 33
17	18 47	18 48	18 49	18 50	18 51	18 52	18 54	18 55	18 58	19 00	19 04	19 07	19 12	19 18
21	18 42	18 42	18 42	18 43	18 44	18 44	18 45	18 47	18 48	18 50	18 53	18 56	18 59	19 04
25	18 37	18 37	18 37	18 37	18 37	18 37	18 38	18 39	18 40	18 41	18 43	18 45	18 47	18 50
29	18 32	18 32	18 31	18 31	18 31	18 31	18 31	18 31	18 31	18 32	18 33	18 34	18 36	18 38
Nov. 2	18 28	18 27	18 26	18 26	18 25	18 25	18 24	18 24	18 24	18 24	18 24	18 24	18 25	18 26
6	18 24	18 23	18 22	18 21	18 20	18 19	18 18	18 18	18 17	18 16	18 16	18 16	18 15	18 15
10	18 21	18 19	18 18	18 17	18 15	18 14	18 13	18 12	18 11	18 09	18 08	18 07	18 06	18 06
14	18 18	18 16	18 15	18 13	18 11	18 10	18 08	18 06	18 05	18 03	18 02	18 00	17 58	17 57
18	18 16	18 14	18 12	18 10	18 08	18 06	18 04	18 02	18 00	17 58	17 56	17 53	17 51	17 49
22	18 14	18 12	18 09	18 07	18 05	18 03	18 00	17 58	17 56	17 53	17 51	17 48	17 45	17 42
26	18 12	18 10	18 08	18 05	18 03	18 00	17 58	17 55	17 52	17 49	17 46	17 43	17 39	17 36
30	18 12	18 09	18 07	18 04	18 01	17 58	17 56	17 53	17 50	17 46	17 43	17 39	17 35	17 31
Dec. 4	18 11	18 09	18 06	18 03	18 00	17 57	17 54	17 51	17 48	17 44	17 40	17 36	17 32	17 27
8	18 12	18 09	18 06	18 03	18 00	17 57	17 54	17 50	17 47	17 43	17 39	17 35	17 30	17 25
12	18 12	18 09	18 07	18 03	18 00	17 57	17 54	17 50	17 47	17 43	17 38	17 34	17 29	17 23
16	18 14	18 11	18 08	18 05	18 01	17 58	17 55	17 51	17 47	17 43	17 39	17 34	17 29	17 23
20	18 15	18 12	18 09	18 06	18 03	18 00	17 56	17 52	17 49	17 45	17 40	17 35	17 30	17 24
24	18 17	18 14	18 11	18 08	18 05	18 02	17 58	17 55	17 51	17 47	17 42	17 38	17 32	17 27
28	18 20	18 17	18 14	18 11	18 08	18 04	18 01	17 58	17 54	17 50	17 46	17 41	17 36	17 30
32	18 23	18 20	18 17	18 14	18 11	18 08	18 04	18 01	17 57	17 54	17 49	17 45	17 40	17 35
36	18 26	18 23	18 20	18 17	18 15	18 12	18 08	18 05	18 02	17 58	17 54	17 50	17 45	17 40

UNIVERSAL TIME FOR MERIDIAN OF GREENWICH

MOONRISE

Lat.	−55°	−50°	−45°	−40°	−35°	−30°	−20°	−10°	0°	+10°	+20°	+30°	+35°	+40°
	h m	h m	h m	h m	h m	h m	h m	h m	h m	h m	h m	h m	h m	h m
Jan. 0	.. ..	0 21	0 39	0 54	1 06	1 17	1 36	1 53	2 08	2 24	2 41	3 00	3 12	3 25
1	0 25	0 53	1 15	1 32	1 47	2 00	2 22	2 41	2 58	3 16	3 36	3 58	4 11	4 26
2	1 02	1 33	1 57	2 16	2 32	2 46	3 10	3 30	3 49	4 09	4 29	4 53	5 07	5 23
3	1 50	2 22	2 46	3 06	3 22	3 36	4 00	4 21	4 40	5 00	5 20	5 44	5 59	6 15
4	2 48	3 19	3 42	4 00	4 16	4 29	4 52	5 12	5 30	5 49	6 09	6 31	6 45	7 00
5	3 55	4 21	4 42	4 58	5 12	5 24	5 45	6 02	6 19	6 35	6 53	7 13	7 25	7 39
6	5 06	5 27	5 44	5 57	6 09	6 19	6 36	6 51	7 05	7 19	7 34	7 51	8 01	8 12
7	6 18	6 34	6 47	6 57	7 06	7 13	7 27	7 38	7 49	8 00	8 12	8 25	8 32	8 41
8	7 30	7 41	7 49	7 56	8 02	8 07	8 16	8 24	8 32	8 39	8 47	8 56	9 01	9 07
9	8 42	8 47	8 51	8 55	8 58	9 00	9 05	9 09	9 13	9 17	9 21	9 25	9 28	9 31
10	9 54	9 54	9 54	9 54	9 54	9 54	9 54	9 54	9 54	9 54	9 54	9 54	9 54	9 54
11	11 08	11 02	10 58	10 54	10 51	10 48	10 43	10 39	10 35	10 32	10 28	10 23	10 21	10 18
12	12 23	12 12	12 03	11 56	11 50	11 44	11 35	11 27	11 19	11 11	11 03	10 54	10 49	10 43
13	13 43	13 25	13 12	13 01	12 51	12 43	12 29	12 17	12 05	11 54	11 42	11 29	11 21	11 12
14	15 05	14 41	14 23	14 08	13 56	13 45	13 27	13 11	12 56	12 41	12 26	12 08	11 57	11 46
15	16 27	15 58	15 36	15 18	15 03	14 50	14 28	14 09	13 51	13 34	13 15	12 54	12 41	12 27
16	17 44	17 10	16 46	16 26	16 10	15 56	15 32	15 11	14 52	14 32	14 12	13 48	13 34	13 18
17	18 46	18 14	17 49	17 30	17 13	16 59	16 35	16 14	15 55	15 35	15 15	14 51	14 37	14 20
18	19 32	19 04	18 43	18 25	18 11	17 58	17 36	17 17	16 59	16 41	16 22	16 00	15 47	15 32
19	20 04	19 43	19 26	19 12	19 00	18 50	18 32	18 16	18 02	17 47	17 31	17 13	17 02	16 50
20	20 27	20 13	20 01	19 51	19 43	19 35	19 23	19 11	19 01	18 50	18 39	18 25	18 18	18 09
21	20 45	20 37	20 30	20 25	20 20	20 16	20 09	20 02	19 56	19 50	19 43	19 36	19 31	19 27
22	21 00	20 58	20 56	20 55	20 54	20 53	20 51	20 49	20 48	20 47	20 45	20 43	20 42	20 41
23	21 14	21 18	21 21	21 24	21 26	21 28	21 32	21 35	21 38	21 41	21 44	21 48	21 51	21 53
24	21 28	21 38	21 46	21 52	21 58	22 03	22 12	22 19	22 27	22 34	22 42	22 51	22 57	23 03

MOONSET

Lat.	−55°	−50°	−45°	−40°	−35°	−30°	−20°	−10°	0°	+10°	+20°	+30°	+35°	+40°
	h m	h m	h m	h m	h m	h m	h m	h m	h m	h m	h m	h m	h m	h m
Jan. 0	16 59	16 32	16 11	15 55	15 40	15 28	15 07	14 49	14 32	14 16	13 58	13 37	13 25	13 11
1	18 07	17 36	17 13	16 54	16 38	16 25	16 02	15 42	15 23	15 04	14 44	14 21	14 08	13 52
2	19 05	18 33	18 08	17 49	17 32	17 18	16 54	16 33	16 14	15 55	15 34	15 10	14 56	14 39
3	19 51	19 19	18 56	18 37	18 21	18 07	17 44	17 24	17 05	16 46	16 25	16 02	15 48	15 31
4	20 25	19 57	19 36	19 19	19 05	18 52	18 30	18 12	17 54	17 36	17 18	16 56	16 43	16 28
5	20 50	20 27	20 09	19 55	19 43	19 32	19 13	18 57	18 42	18 26	18 10	17 51	17 39	17 27
6	21 09	20 51	20 37	20 26	20 16	20 07	19 52	19 39	19 27	19 14	19 01	18 45	18 36	18 26
7	21 23	21 11	21 01	20 53	20 46	20 40	20 29	20 19	20 10	20 01	19 51	19 40	19 33	19 25
8	21 36	21 29	21 23	21 18	21 13	21 10	21 03	20 57	20 52	20 46	20 40	20 33	20 29	20 24
9	21 47	21 45	21 43	21 41	21 40	21 39	21 36	21 34	21 32	21 31	21 29	21 26	21 25	21 23
10	21 58	22 01	22 03	22 05	22 06	22 07	22 10	22 12	22 14	22 16	22 18	22 20	22 21	22 23
11	22 10	22 17	22 24	22 29	22 33	22 37	22 44	22 50	22 56	23 02	23 08	23 15	23 19	23 24
12	22 23	22 36	22 47	22 56	23 03	23 10	23 21	23 31	23 41	23 50	.. ..	.. ..	.. ..	.. ..
13	22 41	22 59	23 14	23 26	23 37	23 46	.. ..	.. ..	.. ..	.. ..	0 01	0 12	0 19	0 27
14	23 04	23 29	23 48	.. ..	.. ..	.. ..	0 02	0 16	0 29	0 42	0 56	1 13	1 22	1 33
15	23 38	.. ..	.. ..	0 04	0 17	0 28	0 48	1 06	1 22	1 38	1 56	2 16	2 28	2 41
16	.. ..	0 08	0 31	0 49	1 05	1 18	1 41	2 01	2 20	2 38	2 58	3 21	3 35	3 50
17	0 28	1 01	1 26	1 46	2 02	2 17	2 41	3 02	3 21	3 41	4 02	4 26	4 40	4 57
18	1 38	2 10	2 34	2 53	3 09	3 23	3 46	4 07	4 25	4 44	5 04	5 27	5 41	5 56
19	3 04	3 31	3 52	4 08	4 22	4 34	4 55	5 13	5 29	5 45	6 03	6 23	6 34	6 48
20	4 40	4 59	5 15	5 27	5 38	5 47	6 03	6 17	6 30	6 43	6 56	7 12	7 20	7 30
21	6 16	6 28	6 38	6 46	6 53	6 59	7 10	7 19	7 27	7 36	7 44	7 54	8 00	8 07
22	7 49	7 54	7 59	8 02	8 06	8 08	8 13	8 17	8 21	8 25	8 29	8 33	8 36	8 38
23	9 18	9 17	9 17	9 16	9 15	9 15	9 14	9 13	9 12	9 11	9 10	9 09	9 08	9 08
24	10 45	10 37	10 32	10 27	10 22	10 19	10 12	10 07	10 01	9 56	9 50	9 44	9 40	9 36

(.. ..) indicates phenomenon will occur the next day.

UNIVERSAL TIME FOR MERIDIAN OF GREENWICH

MOONRISE

Lat.	+40°	+42°	+44°	+46°	+48°	+50°	+52°	+54°	+56°	+58°	+60°	+62°	+64°	+66°
	h m	h m	h m	h m	h m	h m	h m	h m	h m	h m	h m	h m	h m	h m
Jan. 0	3 25	3 31	3 37	3 44	3 51	3 59	4 09	4 19	4 30	4 44	5 00	5 19	5 43	6 16
1	4 26	4 33	4 40	4 48	4 57	5 07	5 17	5 30	5 44	6 00	6 20	6 46	7 21	8 28
2	5 23	5 31	5 39	5 47	5 57	6 07	6 19	6 32	6 48	7 06	7 29	7 59	8 45	-- --
3	6 15	6 22	6 30	6 39	6 48	6 59	7 11	7 24	7 39	7 58	8 20	8 50	9 35	-- --
4	7 00	7 07	7 14	7 22	7 31	7 41	7 52	8 04	8 18	8 34	8 54	9 19	9 52	10 51
5	7 39	7 45	7 51	7 58	8 06	8 14	8 24	8 34	8 46	8 59	9 15	9 34	9 58	10 30
6	8 12	8 17	8 22	8 28	8 34	8 41	8 48	8 57	9 06	9 16	9 28	9 42	9 59	10 20
7	8 41	8 45	8 49	8 53	8 58	9 03	9 08	9 14	9 21	9 29	9 37	9 47	9 59	10 12
8	9 07	9 09	9 12	9 15	9 18	9 21	9 25	9 29	9 34	9 38	9 44	9 50	9 58	10 06
9	9 31	9 32	9 33	9 35	9 36	9 38	9 40	9 42	9 44	9 47	9 49	9 52	9 56	10 00
10	9 54	9 54	9 54	9 54	9 54	9 54	9 54	9 54	9 54	9 54	9 54	9 54	9 54	9 54
11	10 18	10 16	10 15	10 14	10 12	10 10	10 08	10 06	10 04	10 02	9 59	9 56	9 53	9 49
12	10 43	10 40	10 38	10 35	10 32	10 28	10 24	10 20	10 16	10 11	10 05	9 59	9 51	9 43
13	11 12	11 08	11 04	10 59	10 54	10 49	10 43	10 37	10 30	10 22	10 13	10 03	9 51	9 37
14	11 46	11 40	11 35	11 29	11 22	11 15	11 07	10 58	10 49	10 37	10 25	10 10	9 52	9 29
15	12 27	12 21	12 14	12 06	11 58	11 49	11 39	11 28	11 16	11 01	10 44	10 23	9 56	9 19
16	13 18	13 11	13 03	12 55	12 45	12 35	12 24	12 11	11 56	11 38	11 17	10 50	10 11	** **
17	14 20	14 13	14 05	13 56	13 47	13 36	13 25	13 11	12 56	12 37	12 15	11 45	11 01	** **
18	15 32	15 25	15 18	15 10	15 02	14 52	14 41	14 29	14 16	14 00	13 40	13 16	12 43	11 47
19	16 50	16 44	16 39	16 32	16 25	16 18	16 09	16 00	15 49	15 37	15 23	15 06	14 45	14 17
20	18 09	18 05	18 01	17 57	17 52	17 46	17 41	17 34	17 27	17 19	17 10	16 59	16 46	16 31
21	19 27	19 24	19 22	19 20	19 17	19 14	19 11	19 07	19 03	18 59	18 54	18 48	18 42	18 34
22	20 41	20 41	20 40	20 40	20 39	20 38	20 38	20 37	20 36	20 35	20 34	20 33	20 31	20 30
23	21 53	21 54	21 56	21 57	21 58	22 00	22 01	22 03	22 05	22 08	22 10	22 13	22 16	22 20
24	23 03	23 06	23 08	23 12	23 15	23 19	23 23	23 27	23 32	23 37	23 43	23 51	23 59	.. ..

MOONSET

Lat.	+40°	+42°	+44°	+46°	+48°	+50°	+52°	+54°	+56°	+58°	+60°	+62°	+64°	+66°
	h m	h m	h m	h m	h m	h m	h m	h m	h m	h m	h m	h m	h m	h m
Jan. 0	13 11	13 05	12 58	12 51	12 43	12 35	12 25	12 15	12 03	11 49	11 33	11 13	10 48	10 14
1	13 52	13 45	13 38	13 30	13 21	13 11	13 00	12 47	12 33	12 16	11 56	11 30	10 55	9 48
2	14 39	14 32	14 24	14 15	14 06	13 55	13 43	13 30	13 14	12 56	12 33	12 03	11 16	-- --
3	15 31	15 24	15 16	15 08	14 58	14 48	14 36	14 23	14 08	13 49	13 27	12 57	12 13	-- --
4	16 28	16 21	16 14	16 06	15 58	15 48	15 37	15 25	15 12	14 56	14 36	14 12	13 39	12 41
5	17 27	17 21	17 15	17 08	17 01	16 53	16 44	16 34	16 22	16 09	15 54	15 36	15 12	14 41
6	18 26	18 22	18 17	18 11	18 06	17 59	17 52	17 45	17 36	17 26	17 15	17 01	16 45	16 26
7	19 25	19 22	19 19	19 15	19 11	19 06	19 01	18 56	18 50	18 43	18 35	18 26	18 16	18 03
8	20 24	20 22	20 20	20 18	20 15	20 13	20 10	20 06	20 03	19 59	19 54	19 49	19 43	19 36
9	21 23	21 23	21 22	21 21	21 20	21 19	21 18	21 17	21 16	21 15	21 13	21 11	21 09	21 07
10	22 23	22 23	22 24	22 25	22 26	22 27	22 27	22 29	22 30	22 31	22 32	22 34	22 36	22 38
11	23 24	23 26	23 28	23 30	23 33	23 35	23 38	23 42	23 45	23 49	23 54	23 59	.. ..	.. ..
12	.. ..	.. ..	.. ..	.. ..	.. ..	.. ..	.. ..	.. ..	.. ..	.. ..	.. ..	.. ..	0 05	0 12
13	0 27	0 30	0 34	0 38	0 42	0 47	0 52	0 58	1 04	1 11	1 19	1 28	1 39	1 52
14	1 33	1 38	1 43	1 48	1 55	2 01	2 09	2 17	2 26	2 36	2 48	3 03	3 20	3 41
15	2 41	2 47	2 54	3 01	3 09	3 17	3 27	3 38	3 50	4 04	4 21	4 41	5 07	5 44
16	3 50	3 57	4 05	4 13	4 22	4 32	4 44	4 56	5 11	5 28	5 49	6 16	6 54	** **
17	4 57	5 04	5 12	5 21	5 30	5 41	5 53	6 06	6 22	6 40	7 02	7 32	8 16	** **
18	5 56	6 03	6 11	6 19	6 28	6 38	6 49	7 01	7 15	7 32	7 51	8 16	8 50	9 46
19	6 48	6 53	7 00	7 07	7 14	7 22	7 31	7 41	7 52	8 05	8 20	8 38	9 00	9 28
20	7 30	7 35	7 40	7 45	7 50	7 56	8 03	8 10	8 18	8 27	8 37	8 49	9 03	9 19
21	8 07	8 10	8 13	8 16	8 19	8 23	8 27	8 32	8 37	8 42	8 48	8 55	9 03	9 13
22	8 38	8 40	8 41	8 42	8 44	8 46	8 47	8 49	8 51	8 54	8 56	8 59	9 03	9 07
23	9 08	9 07	9 07	9 07	9 06	9 06	9 05	9 05	9 04	9 04	9 03	9 02	9 02	9 01
24	9 36	9 34	9 32	9 30	9 28	9 26	9 23	9 20	9 17	9 14	9 10	9 05	9 00	8 55

(.. ..) indicates phenomenon will occur the next day.
(-- --) indicates Moon continuously below horizon.
(** **) indicates Moon continuously above horizon.

UNIVERSAL TIME FOR MERIDIAN OF GREENWICH

MOONRISE

Lat.	−55°	−50°	−45°	−40°	−35°	−30°	−20°	−10°	0°	+10°	+20°	+30°	+35°	+40°
	h m	h m	h m	h m	h m	h m	h m	h m	h m	h m	h m	h m	h m	h m
Jan. 23	21 14	21 18	21 21	21 24	21 26	21 28	21 32	21 35	21 38	21 41	21 44	21 48	21 51	21 53
24	21 28	21 38	21 46	21 52	21 58	22 03	22 12	22 19	22 27	22 34	22 42	22 51	22 57	23 03
25	21 44	22 00	22 12	22 22	22 31	22 39	22 52	23 04	23 16	23 27	23 39	23 53	.. ..	.. ..
26	22 04	22 25	22 42	22 55	23 07	23 17	23 35	23 50	.. ..	.. ..	.. ..	.. ..	0 01	0 11
27	22 29	22 55	23 16	23 32	23 46	23 59	.. ..	.. ..	0 05	0 20	0 35	0 53	1 04	1 16
28	23 02	23 33	23 56	.. ..	.. ..	.. ..	0 20	0 38	0 55	1 12	1 31	1 52	2 05	2 19
29	23 46	.. ..	.. ..	0 15	0 30	0 44	1 07	1 27	1 46	2 05	2 25	2 48	3 02	3 18
30	.. ..	0 19	0 43	1 02	1 19	1 33	1 57	2 17	2 37	2 56	3 17	3 41	3 55	4 12
31	0 41	1 12	1 36	1 55	2 11	2 25	2 48	3 08	3 27	3 46	4 06	4 29	4 43	4 59
Feb. 1	1 45	2 13	2 34	2 52	3 06	3 19	3 40	3 58	4 16	4 33	4 51	5 13	5 25	5 39
2	2 54	3 18	3 36	3 50	4 03	4 13	4 32	4 48	5 03	5 17	5 33	5 51	6 02	6 14
3	4 06	4 24	4 38	4 50	4 59	5 08	5 23	5 36	5 47	5 59	6 12	6 26	6 35	6 44
4	5 19	5 31	5 41	5 49	5 56	6 02	6 13	6 22	6 31	6 39	6 48	6 59	7 04	7 11
5	6 31	6 38	6 44	6 48	6 52	6 56	7 02	7 07	7 12	7 17	7 23	7 29	7 32	7 36
6	7 43	7 45	7 46	7 47	7 48	7 49	7 51	7 52	7 53	7 55	7 56	7 58	7 58	7 59
7	8 56	8 52	8 50	8 47	8 45	8 43	8 40	8 37	8 35	8 32	8 30	8 27	8 25	8 23
8	10 10	10 01	9 54	9 48	9 43	9 38	9 30	9 24	9 17	9 11	9 04	8 57	8 53	8 48
9	11 27	11 12	11 01	10 51	10 43	10 35	10 23	10 12	10 02	9 52	9 41	9 29	9 23	9 15
10	12 46	12 25	12 09	11 56	11 45	11 35	11 18	11 03	10 50	10 36	10 22	10 06	9 57	9 46
11	14 06	13 39	13 19	13 02	12 48	12 36	12 16	11 58	11 42	11 25	11 08	10 48	10 36	10 23
12	15 23	14 51	14 28	14 09	13 53	13 39	13 16	12 56	12 38	12 19	11 59	11 36	11 23	11 08
13	16 30	15 57	15 32	15 12	14 56	14 42	14 17	13 57	13 37	13 18	12 57	12 33	12 19	12 03
14	17 22	16 51	16 28	16 10	15 54	15 41	15 17	14 57	14 39	14 20	14 00	13 37	13 23	13 07
15	18 00	17 35	17 15	17 00	16 46	16 35	16 14	15 57	15 40	15 24	15 06	14 46	14 34	14 20
16	18 27	18 09	17 54	17 42	17 32	17 23	17 07	16 53	16 40	16 27	16 13	15 57	15 48	15 37

MOONSET

Lat.	−55°	−50°	−45°	−40°	−35°	−30°	−20°	−10°	0°	+10°	+20°	+30°	+35°	+40°
	h m	h m	h m	h m	h m	h m	h m	h m	h m	h m	h m	h m	h m	h m
Jan. 23	9 18	9 17	9 17	9 16	9 15	9 15	9 14	9 13	9 12	9 11	9 10	9 09	9 08	9 08
24	10 45	10 37	10 32	10 27	10 22	10 19	10 12	10 07	10 01	9 56	9 50	9 44	9 40	9 36
25	12 09	11 55	11 44	11 35	11 28	11 21	11 10	10 59	10 50	10 41	10 31	10 19	10 13	10 05
26	13 30	13 10	12 55	12 42	12 32	12 22	12 06	11 52	11 39	11 26	11 12	10 56	10 47	10 37
27	14 48	14 22	14 03	13 47	13 34	13 22	13 02	12 45	12 29	12 13	11 56	11 36	11 25	11 12
28	15 59	15 29	15 06	14 48	14 33	14 20	13 57	13 38	13 19	13 01	12 42	12 20	12 07	11 51
29	17 00	16 28	16 04	15 44	15 28	15 14	14 50	14 30	14 10	13 51	13 31	13 07	12 53	12 37
30	17 50	17 18	16 54	16 35	16 19	16 05	15 41	15 20	15 01	14 42	14 21	13 57	13 43	13 27
31	18 27	17 58	17 36	17 18	17 03	16 50	16 28	16 09	15 51	15 33	15 13	14 50	14 37	14 22
Feb. 1	18 55	18 30	18 12	17 56	17 43	17 32	17 12	16 55	16 39	16 22	16 05	15 45	15 33	15 20
2	19 15	18 56	18 41	18 29	18 18	18 09	17 52	17 38	17 25	17 11	16 57	16 40	16 30	16 19
3	19 32	19 18	19 06	18 57	18 49	18 42	18 30	18 19	18 08	17 58	17 47	17 34	17 27	17 18
4	19 45	19 36	19 29	19 23	19 17	19 13	19 05	18 58	18 51	18 44	18 37	18 28	18 23	18 18
5	19 57	19 53	19 49	19 47	19 44	19 42	19 39	19 35	19 32	19 29	19 26	19 22	19 19	19 17
6	20 08	20 09	20 10	20 10	20 11	20 11	20 12	20 13	20 13	20 14	20 14	20 15	20 16	20 16
7	20 19	20 25	20 30	20 34	20 37	20 41	20 46	20 51	20 55	20 59	21 04	21 10	21 13	21 16
8	20 32	20 43	20 52	21 00	21 06	21 12	21 22	21 30	21 38	21 47	21 55	22 05	22 11	22 18
9	20 48	21 04	21 17	21 28	21 38	21 46	22 00	22 13	22 25	22 36	22 49	23 04	23 12	23 22
10	21 08	21 30	21 48	22 02	22 14	22 25	22 43	22 59	23 14	23 29	23 45	.. ..	.. ..	.. ..
11	21 36	22 04	22 25	22 43	22 57	23 10	23 31	23 50	.. ..	.. ..	.. ..	0 04	0 15	0 27
12	22 17	22 49	23 13	23 32	23 48	.. ..	.. ..	.. ..	0 08	0 26	0 44	1 06	1 19	1 34
13	23 15	23 48	.. ..	.. ..	.. ..	0 02	0 26	0 46	1 06	1 25	1 45	2 09	2 23	2 39
14	.. ..	.. ..	0 12	0 32	0 48	1 02	1 26	1 47	2 06	2 26	2 46	3 10	3 24	3 40
15	0 31	1 00	1 23	1 41	1 56	2 09	2 31	2 50	3 08	3 26	3 45	4 06	4 19	4 34
16	2 00	2 23	2 42	2 56	3 09	3 20	3 38	3 54	4 09	4 24	4 40	4 57	5 08	5 20

(.. ..) indicates phenomenon will occur the next day.

UNIVERSAL TIME FOR MERIDIAN OF GREENWICH

MOONRISE

Lat.	+40°	+42°	+44°	+46°	+48°	+50°	+52°	+54°	+56°	+58°	+60°	+62°	+64°	+66°
	h m	h m	h m	h m	h m	h m	h m	h m	h m	h m	h m	h m	h m	h m
Jan. 23	21 53	21 54	21 56	21 57	21 58	22 00	22 01	22 03	22 05	22 08	22 10	22 13	22 16	22 20
24	23 03	23 06	23 08	23 12	23 15	23 19	23 23	23 27	23 32	23 37	23 43	23 51	23 59	.. ..
25	.. ..	.. ..	.. ..	.. ..	.. ..	.. ..	.. ..	.. ..	.. ..	.. ..	.. ..	.. ..	.. ..	0 09
26	0 11	0 15	0 19	0 24	0 29	0 35	0 41	0 48	0 56	1 05	1 15	1 27	1 41	1 58
27	1 16	1 22	1 28	1 34	1 41	1 49	1 57	2 07	2 17	2 29	2 44	3 01	3 22	3 50
28	2 19	2 26	2 33	2 41	2 49	2 58	3 08	3 20	3 33	3 49	4 08	4 31	5 02	5 51
29	3 18	3 25	3 33	3 42	3 51	4 01	4 13	4 26	4 41	4 59	5 21	5 50	6 32	-- --
30	4 12	4 19	4 27	4 36	4 45	4 56	5 08	5 21	5 37	5 55	6 18	6 48	7 35	-- --
31	4 59	5 06	5 13	5 21	5 31	5 41	5 52	6 04	6 19	6 36	6 57	7 24	8 01	9 27
Feb. 1	5 39	5 45	5 52	6 00	6 08	6 17	6 26	6 37	6 50	7 04	7 22	7 42	8 09	8 47
2	6 14	6 19	6 25	6 31	6 38	6 45	6 53	7 02	7 12	7 24	7 37	7 53	8 12	8 35
3	6 44	6 48	6 53	6 58	7 03	7 09	7 15	7 22	7 29	7 38	7 48	7 59	8 12	8 28
4	7 11	7 14	7 17	7 21	7 24	7 28	7 33	7 37	7 43	7 48	7 55	8 02	8 11	8 21
5	7 36	7 38	7 39	7 41	7 43	7 46	7 48	7 51	7 54	7 57	8 01	8 05	8 10	8 15
6	7 59	8 00	8 00	8 01	8 01	8 02	8 03	8 03	8 04	8 05	8 06	8 07	8 08	8 10
7	8 23	8 22	8 21	8 20	8 19	8 18	8 17	8 16	8 14	8 13	8 11	8 09	8 07	8 04
8	8 48	8 46	8 43	8 41	8 38	8 35	8 32	8 29	8 25	8 21	8 17	8 11	8 05	7 59
9	9 15	9 11	9 08	9 04	8 59	8 55	8 50	8 44	8 38	8 31	8 24	8 15	8 05	7 53
10	9 46	9 41	9 36	9 31	9 25	9 18	9 11	9 03	8 55	8 45	8 34	8 21	8 05	7 46
11	10 23	10 17	10 11	10 04	9 56	9 48	9 39	9 29	9 17	9 04	8 49	8 31	8 08	7 39
12	11 08	11 01	10 54	10 46	10 37	10 27	10 16	10 04	9 50	9 34	9 15	8 50	8 18	7 24
13	12 03	11 55	11 47	11 39	11 29	11 19	11 07	10 54	10 39	10 21	9 58	9 30	8 47	** **
14	13 07	13 00	12 53	12 44	12 35	12 25	12 14	12 01	11 46	11 29	11 08	10 41	10 02	** **
15	14 20	14 14	14 07	14 00	13 52	13 44	13 34	13 23	13 11	12 57	12 40	12 19	11 52	11 14
16	15 37	15 33	15 27	15 22	15 16	15 09	15 02	14 54	14 45	14 35	14 23	14 09	13 52	13 31

MOONSET

Lat.	+40°	+42°	+44°	+46°	+48°	+50°	+52°	+54°	+56°	+58°	+60°	+62°	+64°	+66°
	h m	h m	h m	h m	h m	h m	h m	h m	h m	h m	h m	h m	h m	h m
Jan. 23	9 08	9 07	9 07	9 07	9 06	9 06	9 05	9 05	9 04	9 04	9 03	9 02	9 02	9 01
24	9 36	9 34	9 32	9 30	9 28	9 26	9 23	9 20	9 17	9 14	9 10	9 05	9 00	8 55
25	10 05	10 02	9 59	9 55	9 51	9 47	9 42	9 37	9 31	9 24	9 17	9 09	8 59	8 48
26	10 37	10 32	10 27	10 22	10 16	10 10	10 03	9 56	9 47	9 38	9 27	9 14	8 59	8 41
27	11 12	11 06	11 00	10 53	10 46	10 38	10 29	10 19	10 08	9 55	9 40	9 22	9 00	8 32
28	11 51	11 45	11 37	11 30	11 21	11 11	11 01	10 49	10 35	10 19	10 00	9 37	9 05	8 15
29	12 37	12 29	12 21	12 13	12 03	11 53	11 41	11 28	11 13	10 55	10 32	10 04	9 21	-- --
30	13 27	13 20	13 12	13 03	12 53	12 43	12 31	12 18	12 02	11 44	11 21	10 51	10 04	-- --
31	14 22	14 15	14 07	13 59	13 50	13 41	13 30	13 17	13 03	12 46	12 25	11 59	11 22	9 56
Feb. 1	15 20	15 14	15 07	15 00	14 52	14 44	14 34	14 23	14 11	13 57	13 40	13 20	12 54	12 17
2	16 19	16 14	16 09	16 03	15 56	15 50	15 42	15 33	15 24	15 13	15 00	14 45	14 27	14 04
3	17 18	17 15	17 11	17 06	17 02	16 56	16 51	16 45	16 38	16 30	16 21	16 11	15 58	15 44
4	18 18	18 15	18 13	18 10	18 07	18 03	18 00	17 56	17 51	17 46	17 41	17 34	17 27	17 18
5	19 17	19 16	19 14	19 13	19 12	19 10	19 09	19 07	19 05	19 02	19 00	18 57	18 54	18 50
6	20 16	20 16	20 17	20 17	20 17	20 17	20 18	20 18	20 18	20 19	20 19	20 20	20 20	20 21
7	21 16	21 18	21 19	21 21	21 23	21 25	21 28	21 30	21 33	21 36	21 39	21 43	21 48	21 53
8	22 18	22 21	22 24	22 27	22 31	22 35	22 39	22 44	22 50	22 55	23 02	23 10	23 19	23 30
9	23 22	23 26	23 31	23 36	23 41	23 47	23 53	.. ..	.. ..	.. ..	.. ..	.. ..	.. ..	.. ..
10	.. ..	.. ..	.. ..	.. ..	.. ..	.. ..	.. ..	0 01	0 09	0 18	0 28	0 40	0 55	1 13
11	0 27	0 33	0 39	0 46	0 53	1 00	1 09	1 19	1 30	1 42	1 57	2 14	2 36	3 05
12	1 34	1 41	1 48	1 56	2 04	2 14	2 24	2 36	2 50	3 05	3 25	3 48	4 21	5 14
13	2 39	2 46	2 54	3 03	3 12	3 23	3 34	3 47	4 03	4 21	4 43	5 11	5 54	** **
14	3 40	3 47	3 55	4 03	4 13	4 23	4 34	4 47	5 02	5 20	5 41	6 09	6 48	** **
15	4 34	4 40	4 47	4 55	5 03	5 12	5 22	5 33	5 46	6 01	6 18	6 39	7 07	7 45
16	5 20	5 25	5 31	5 37	5 43	5 50	5 58	6 07	6 17	6 28	6 40	6 55	7 13	7 35

(.. ..) indicates phenomenon will occur the next day.
(-- --) indicates Moon continuously below horizon.
(** **) indicates Moon continuously above horizon.

UNIVERSAL TIME FOR MERIDIAN OF GREENWICH

MOONRISE

Lat.	−55°	−50°	−45°	−40°	−35°	−30°	−20°	−10°	0°	+10°	+20°	+30°	+35°	+40°
	h m	h m	h m	h m	h m	h m	h m	h m	h m	h m	h m	h m	h m	h m
Feb. 15	18 00	17 35	17 15	17 00	16 46	16 35	16 14	15 57	15 40	15 24	15 06	14 46	14 34	14 20
16	18 27	18 09	17 54	17 42	17 32	17 23	17 07	16 53	16 40	16 27	16 13	15 57	15 48	15 37
17	18 48	18 36	18 26	18 19	18 12	18 06	17 55	17 46	17 37	17 29	17 19	17 09	17 03	16 56
18	19 04	18 59	18 55	18 51	18 48	18 45	18 40	18 36	18 32	18 28	18 23	18 18	18 16	18 12
19	19 19	19 20	19 20	19 21	19 21	19 22	19 23	19 23	19 24	19 25	19 25	19 26	19 27	19 27
20	19 34	19 41	19 46	19 51	19 55	19 58	20 04	20 10	20 15	20 20	20 26	20 32	20 36	20 40
21	19 50	20 02	20 13	20 21	20 28	20 35	20 46	20 56	21 06	21 15	21 25	21 37	21 44	21 52
22	20 09	20 27	20 42	20 54	21 04	21 14	21 29	21 43	21 56	22 10	22 24	22 40	22 50	23 01
23	20 32	20 56	21 15	21 30	21 44	21 55	22 15	22 32	22 48	23 04	23 21	23 42	23 53	.. ..
24	21 03	21 32	21 54	22 12	22 27	22 40	23 02	23 21	23 40	23 58	.. ..	.. ..	.. ..	0 07
25	21 44	22 15	22 39	22 58	23 14	23 28	23 52	.. ..	.. ..	.. ..	0 17	0 40	0 53	1 09
26	22 35	23 07	23 31	23 50	.. ..	.. ..	.. ..	0 12	0 31	0 51	1 11	1 35	1 49	2 05
27	23 36	.. ..	.. ..	.. ..	0 06	0 19	0 43	1 03	1 22	1 41	2 02	2 25	2 39	2 55
28	.. ..	0 05	0 27	0 45	1 00	1 13	1 35	1 54	2 12	2 30	2 49	3 10	3 23	3 38
29	0 44	1 08	1 28	1 43	1 56	2 07	2 27	2 44	2 59	3 15	3 32	3 51	4 02	4 15
Mar. 1	1 55	2 14	2 30	2 42	2 53	3 02	3 18	3 32	3 45	3 58	4 11	4 27	4 36	4 46
2	3 07	3 21	3 32	3 41	3 49	3 56	4 08	4 19	4 28	4 38	4 48	5 00	5 07	5 15
3	4 19	4 28	4 35	4 41	4 46	4 50	4 58	5 04	5 11	5 17	5 24	5 31	5 35	5 40
4	5 32	5 35	5 38	5 40	5 42	5 44	5 47	5 50	5 52	5 55	5 58	6 01	6 02	6 04
5	6 45	6 43	6 41	6 40	6 39	6 38	6 37	6 35	6 34	6 33	6 31	6 30	6 29	6 28
6	7 59	7 52	7 46	7 41	7 37	7 33	7 27	7 22	7 17	7 12	7 06	7 00	6 57	6 53
7	9 15	9 02	8 52	8 44	8 36	8 30	8 19	8 10	8 01	7 52	7 43	7 32	7 26	7 20
8	10 34	10 15	10 00	9 48	9 38	9 29	9 14	9 00	8 48	8 36	8 23	8 08	7 59	7 50
9	11 53	11 28	11 09	10 54	10 41	10 30	10 10	9 54	9 38	9 23	9 06	8 48	8 37	8 24
10	13 09	12 40	12 17	11 59	11 44	11 31	11 09	10 50	10 32	10 14	9 55	9 34	9 21	9 06

MOONSET

Lat.	−55°	−50°	−45°	−40°	−35°	−30°	−20°	−10°	0°	+10°	+20°	+30°	+35°	+40°
	h m	h m	h m	h m	h m	h m	h m	h m	h m	h m	h m	h m	h m	h m
Feb. 15	0 31	1 00	1 23	1 41	1 56	2 09	2 31	2 50	3 08	3 26	3 45	4 06	4 19	4 34
16	2 00	2 23	2 42	2 56	3 09	3 20	3 38	3 54	4 09	4 24	4 40	4 57	5 08	5 20
17	3 34	3 51	4 04	4 14	4 23	4 31	4 45	4 57	5 08	5 19	5 30	5 43	5 50	5 59
18	5 09	5 19	5 26	5 32	5 38	5 42	5 50	5 57	6 03	6 10	6 16	6 24	6 28	6 33
19	6 42	6 45	6 47	6 48	6 50	6 51	6 53	6 55	6 57	6 58	7 00	7 02	7 03	7 04
20	8 12	8 08	8 05	8 02	8 00	7 58	7 54	7 51	7 48	7 45	7 42	7 38	7 36	7 34
21	9 40	9 30	9 22	9 15	9 09	9 03	8 54	8 46	8 39	8 32	8 24	8 15	8 10	8 04
22	11 06	10 49	10 36	10 25	10 16	10 07	9 53	9 41	9 30	9 18	9 06	8 53	8 45	8 36
23	12 28	12 05	11 47	11 33	11 21	11 10	10 52	10 36	10 21	10 06	9 51	9 33	9 22	9 10
24	13 44	13 16	12 54	12 37	12 23	12 10	11 49	11 30	11 13	10 55	10 37	10 16	10 03	9 49
25	14 50	14 19	13 55	13 36	13 21	13 07	12 44	12 23	12 05	11 46	11 26	11 02	10 49	10 33
26	15 45	15 13	14 49	14 30	14 14	14 00	13 36	13 15	12 56	12 37	12 16	11 52	11 38	11 22
27	16 27	15 57	15 34	15 16	15 01	14 47	14 25	14 05	13 46	13 28	13 08	12 45	12 31	12 16
28	16 58	16 32	16 12	15 56	15 42	15 30	15 10	14 52	14 35	14 18	14 00	13 39	13 27	13 13
29	17 21	17 00	16 44	16 30	16 19	16 08	15 51	15 36	15 21	15 07	14 52	14 34	14 23	14 11
Mar. 1	17 39	17 23	17 10	17 00	16 51	16 43	16 29	16 17	16 06	15 55	15 42	15 28	15 20	15 11
2	17 53	17 42	17 34	17 27	17 20	17 15	17 05	16 57	16 49	16 41	16 32	16 22	16 17	16 10
3	18 06	18 00	17 55	17 51	17 48	17 45	17 40	17 35	17 31	17 26	17 22	17 16	17 13	17 09
4	18 17	18 16	18 16	18 15	18 15	18 14	18 14	18 13	18 12	18 12	18 11	18 10	18 10	18 09
5	18 29	18 33	18 37	18 39	18 42	18 44	18 48	18 51	18 54	18 57	19 01	19 05	19 07	19 09
6	18 42	18 51	18 58	19 05	19 10	19 15	19 23	19 31	19 38	19 45	19 52	20 00	20 05	20 11
7	18 57	19 11	19 23	19 33	19 41	19 48	20 01	20 13	20 23	20 34	20 45	20 58	21 06	21 14
8	19 16	19 36	19 52	20 05	20 16	20 26	20 43	20 58	21 12	21 26	21 41	21 58	22 08	22 19
9	19 41	20 07	20 27	20 43	20 57	21 08	21 29	21 47	22 04	22 20	22 38	22 59	23 11	23 25
10	20 17	20 47	21 10	21 29	21 44	21 57	22 20	22 40	22 59	23 18	23 37	.. ..	.. ..	.. ..

(.. ..) indicates phenomenon will occur the next day.

UNIVERSAL TIME FOR MERIDIAN OF GREENWICH

MOONRISE

Lat.	+40°	+42°	+44°	+46°	+48°	+50°	+52°	+54°	+56°	+58°	+60°	+62°	+64°	+66°
	h m	h m	h m	h m	h m	h m	h m	h m	h m	h m	h m	h m	h m	h m
Feb. 15	14 20	14 14	14 07	14 00	13 52	13 44	13 34	13 23	13 11	12 57	12 40	12 19	11 52	11 14
16	15 37	15 33	15 27	15 22	15 16	15 09	15 02	14 54	14 45	14 35	14 23	14 09	13 52	13 31
17	16 56	16 52	16 49	16 45	16 42	16 37	16 33	16 28	16 22	16 15	16 08	16 00	15 50	15 39
18	18 12	18 11	18 10	18 08	18 06	18 04	18 02	18 00	17 57	17 55	17 51	17 48	17 44	17 39
19	19 27	19 28	19 28	19 28	19 29	19 29	19 29	19 30	19 30	19 31	19 31	19 32	19 33	19 34
20	20 40	20 42	20 44	20 47	20 49	20 52	20 54	20 57	21 01	21 05	21 09	21 14	21 20	21 26
21	21 52	21 55	21 59	22 03	22 07	22 12	22 17	22 23	22 29	22 36	22 45	22 54	23 05	23 18
22	23 01	23 06	23 11	23 17	23 23	23 30	23 37	23 45	23 55	.. ..	.. ..	.. ..	.. ..	.. ..
23	.. ..	.. ..	.. ..	.. ..	.. ..	.. ..	.. ..	.. ..	.. ..	0 05	0 18	0 32	0 50	1 12
24	0 07	0 13	0 20	0 27	0 35	0 43	0 53	1 03	1 16	1 30	1 46	2 07	2 33	3 10
25	1 09	1 16	1 23	1 31	1 40	1 50	2 01	2 14	2 28	2 45	3 06	3 32	4 09	5 40
26	2 05	2 13	2 20	2 29	2 38	2 49	3 01	3 14	3 29	3 48	4 10	4 40	5 25	-- --
27	2 55	3 02	3 10	3 18	3 27	3 38	3 49	4 02	4 17	4 34	4 56	5 23	6 03	-- --
28	3 38	3 44	3 51	3 59	4 08	4 17	4 27	4 39	4 52	5 07	5 25	5 48	6 17	7 01
29	4 15	4 20	4 26	4 33	4 40	4 48	4 57	5 06	5 17	5 30	5 44	6 01	6 22	6 49
Mar. 1	4 46	4 51	4 56	5 01	5 07	5 13	5 20	5 28	5 36	5 45	5 56	6 09	6 24	6 41
2	5 15	5 18	5 22	5 25	5 30	5 34	5 39	5 45	5 51	5 57	6 05	6 14	6 23	6 35
3	5 40	5 42	5 45	5 47	5 50	5 53	5 56	5 59	6 03	6 07	6 12	6 17	6 23	6 30
4	6 04	6 05	6 06	6 07	6 08	6 09	6 11	6 12	6 14	6 15	6 17	6 19	6 22	6 24
5	6 28	6 28	6 27	6 27	6 27	6 26	6 25	6 25	6 24	6 23	6 22	6 22	6 20	6 19
6	6 53	6 51	6 49	6 47	6 45	6 43	6 41	6 38	6 35	6 32	6 28	6 24	6 20	6 14
7	7 20	7 17	7 13	7 10	7 06	7 02	6 58	6 53	6 48	6 42	6 35	6 28	6 19	6 09
8	7 50	7 45	7 41	7 36	7 30	7 25	7 18	7 11	7 03	6 55	6 45	6 33	6 20	6 04
9	8 24	8 19	8 13	8 07	8 00	7 52	7 44	7 35	7 24	7 12	6 59	6 42	6 23	5 58
10	9 06	9 00	8 53	8 45	8 37	8 28	8 18	8 06	7 53	7 38	7 20	6 59	6 30	5 50

MOONSET

Lat.	+40°	+42°	+44°	+46°	+48°	+50°	+52°	+54°	+56°	+58°	+60°	+62°	+64°	+66°
	h m	h m	h m	h m	h m	h m	h m	h m	h m	h m	h m	h m	h m	h m
Feb. 15	4 34	4 40	4 47	4 55	5 03	5 12	5 22	5 33	5 46	6 01	6 18	6 39	7 07	7 45
16	5 20	5 25	5 31	5 37	5 43	5 50	5 58	6 07	6 17	6 28	6 40	6 55	7 13	7 35
17	5 59	6 03	6 07	6 11	6 16	6 21	6 26	6 32	6 39	6 46	6 54	7 04	7 15	7 28
18	6 33	6 35	6 38	6 40	6 43	6 45	6 49	6 52	6 56	7 00	7 04	7 09	7 15	7 22
19	7 04	7 05	7 05	7 06	7 07	7 07	7 08	7 09	7 10	7 11	7 12	7 13	7 15	7 16
20	7 34	7 33	7 32	7 31	7 29	7 28	7 27	7 25	7 23	7 21	7 19	7 17	7 14	7 11
21	8 04	8 01	7 59	7 56	7 53	7 49	7 46	7 42	7 37	7 32	7 27	7 20	7 13	7 05
22	8 36	8 32	8 27	8 23	8 18	8 12	8 07	8 00	7 53	7 45	7 36	7 25	7 13	6 59
23	9 10	9 05	8 59	8 53	8 46	8 39	8 31	8 22	8 12	8 01	7 48	7 33	7 14	6 51
24	9 49	9 43	9 36	9 28	9 20	9 11	9 02	8 50	8 38	8 23	8 06	7 45	7 18	6 41
25	10 33	10 26	10 18	10 10	10 01	9 51	9 40	9 27	9 12	8 55	8 34	8 08	7 30	6 00
26	11 22	11 15	11 07	10 58	10 49	10 38	10 27	10 13	9 58	9 40	9 17	8 47	8 03	-- --
27	12 16	12 09	12 01	11 53	11 44	11 34	11 22	11 10	10 55	10 37	10 16	9 49	9 09	-- --
28	13 13	13 06	12 59	12 52	12 44	12 35	12 25	12 14	12 01	11 46	11 28	11 06	10 37	9 53
29	14 11	14 06	14 00	13 54	13 47	13 40	13 32	13 22	13 12	13 00	12 46	12 30	12 09	11 43
Mar. 1	15 11	15 07	15 02	14 57	14 52	14 46	14 40	14 33	14 25	14 16	14 06	13 55	13 41	13 24
2	16 10	16 07	16 04	16 01	15 57	15 53	15 49	15 44	15 39	15 33	15 26	15 19	15 10	14 59
3	17 09	17 08	17 06	17 04	17 02	17 00	16 58	16 55	16 52	16 49	16 46	16 42	16 37	16 32
4	18 09	18 09	18 08	18 08	18 08	18 08	18 07	18 07	18 06	18 06	18 05	18 05	18 04	18 03
5	19 09	19 10	19 12	19 13	19 14	19 16	19 17	19 19	19 21	19 23	19 26	19 28	19 32	19 35
6	20 11	20 13	20 16	20 19	20 22	20 25	20 29	20 33	20 37	20 42	20 48	20 54	21 02	21 11
7	21 14	21 18	21 22	21 27	21 32	21 37	21 43	21 49	21 56	22 04	22 13	22 24	22 36	22 51
8	22 19	22 25	22 30	22 36	22 43	22 50	22 58	23 06	23 16	23 27	23 41	23 56	.. ..	.. ..
9	23 25	23 31	23 38	23 45	23 53	.. ..	.. ..	.. ..	.. ..	.. ..	.. ..	.. ..	0 15	0 39
10	.. ..	.. ..	.. ..	.. ..	.. ..	0 02	0 12	0 23	0 36	0 50	1 08	1 29	1 57	2 37

(.. ..) indicates phenomenon will occur the next day.
(-- --) indicates Moon continuously below horizon.
(** **) indicates Moon continuously above horizon.

UNIVERSAL TIME FOR MERIDIAN OF GREENWICH

MOONRISE

Lat.	−55°	−50°	−45°	−40°	−35°	−30°	−20°	−10°	0°	+10°	+20°	+30°	+35°	+40°
	h m	h m	h m	h m	h m	h m	h m	h m	h m	h m	h m	h m	h m	h m
Mar. 9	11 53	11 28	11 09	10 54	10 41	10 30	10 10	9 54	9 38	9 23	9 06	8 48	8 37	8 24
10	13 09	12 40	12 17	11 59	11 44	11 31	11 09	10 50	10 32	10 14	9 55	9 34	9 21	9 06
11	14 18	13 46	13 22	13 02	12 46	12 32	12 09	11 48	11 29	11 10	10 50	10 26	10 12	9 56
12	15 14	14 43	14 19	14 00	13 44	13 31	13 07	12 47	12 28	12 09	11 49	11 25	11 11	10 56
13	15 56	15 29	15 08	14 51	14 37	14 25	14 03	13 45	13 27	13 10	12 51	12 30	12 17	12 03
14	16 27	16 05	15 49	15 35	15 23	15 13	14 56	14 40	14 26	14 11	13 56	13 38	13 27	13 15
15	16 49	16 35	16 23	16 13	16 04	15 57	15 44	15 33	15 22	15 11	15 00	14 47	14 39	14 30
16	17 07	16 59	16 52	16 46	16 41	16 37	16 29	16 23	16 16	16 10	16 03	15 55	15 51	15 46
17	17 23	17 21	17 19	17 17	17 16	17 15	17 12	17 10	17 09	17 07	17 05	17 03	17 02	17 00
18	17 38	17 42	17 45	17 47	17 49	17 51	17 54	17 57	18 00	18 03	18 06	18 09	18 12	18 14
19	17 54	18 03	18 11	18 17	18 23	18 28	18 36	18 44	18 51	18 59	19 06	19 15	19 21	19 27
20	18 12	18 27	18 40	18 50	18 59	19 06	19 20	19 32	19 43	19 54	20 07	20 21	20 29	20 38
21	18 34	18 55	19 12	19 26	19 37	19 48	20 05	20 21	20 36	20 50	21 06	21 24	21 35	21 47
22	19 02	19 29	19 49	20 06	20 20	20 32	20 53	21 12	21 29	21 46	22 05	22 26	22 39	22 53
23	19 40	20 10	20 33	20 51	21 07	21 20	21 43	22 03	22 22	22 41	23 01	23 24	23 38	23 53
24	20 28	20 59	21 23	21 42	21 58	22 12	22 35	22 55	23 14	23 33	23 54	.. ..	.. ..	.. ..
25	21 26	21 56	22 19	22 37	22 52	23 05	23 28	23 47	.. ..	.. ..	.. ..	0 17	0 31	0 47
26	22 32	22 58	23 18	23 34	23 48	.. ..	.. ..	.. ..	0 05	0 23	0 43	1 05	1 18	1 33
27	23 42	.. ..	.. ..	.. ..	.. ..	0 00	0 20	0 37	0 54	1 10	1 28	1 48	1 59	2 13
28	.. ..	0 03	0 19	0 33	0 44	0 54	1 11	1 26	1 40	1 54	2 09	2 26	2 35	2 46
29	0 53	1 09	1 22	1 32	1 41	1 49	2 02	2 14	2 24	2 35	2 47	3 00	3 07	3 16
30	2 05	2 16	2 24	2 31	2 37	2 43	2 52	3 00	3 07	3 15	3 23	3 32	3 37	3 43
31	3 17	3 23	3 27	3 30	3 34	3 36	3 41	3 45	3 49	3 53	3 57	4 02	4 04	4 07
Apr. 1	4 30	4 30	4 30	4 30	4 30	4 30	4 31	4 31	4 31	4 31	4 31	4 31	4 31	4 32
2	5 44	5 39	5 35	5 31	5 28	5 26	5 21	5 17	5 13	5 10	5 06	5 02	4 59	4 56

MOONSET

Lat.	−55°	−50°	−45°	−40°	−35°	−30°	−20°	−10°	0°	+10°	+20°	+30°	+35°	+40°
	h m	h m	h m	h m	h m	h m	h m	h m	h m	h m	h m	h m	h m	h m
Mar. 9	19 41	20 07	20 27	20 43	20 57	21 08	21 29	21 47	22 04	22 20	22 38	22 59	23 11	23 25
10	20 17	20 47	21 10	21 29	21 44	21 57	22 20	22 40	22 59	23 18	23 37	.. ..	.. ..	.. ..
11	21 07	21 39	22 04	22 23	22 39	22 53	23 17	23 38	23 57	.. ..	.. ..	0 00	0 14	0 30
12	22 14	22 45	23 08	23 26	23 42	23 55	.. ..	.. ..	.. ..	0 16	0 37	1 00	1 14	1 30
13	23 34	.. ..	.. ..	.. ..	.. ..	.. ..	0 18	0 38	0 56	1 15	1 34	1 57	2 10	2 25
14	.. ..	0 01	0 20	0 37	0 50	1 02	1 22	1 39	1 55	2 11	2 28	2 48	2 59	3 12
15	1 03	1 23	1 38	1 51	2 01	2 11	2 26	2 40	2 53	3 05	3 19	3 34	3 43	3 53
16	2 35	2 48	2 58	3 06	3 13	3 19	3 30	3 39	3 48	3 57	4 06	4 16	4 22	4 28
17	4 07	4 12	4 17	4 21	4 25	4 28	4 33	4 37	4 41	4 45	4 50	4 54	4 57	5 00
18	5 37	5 36	5 36	5 35	5 35	5 35	5 34	5 34	5 33	5 33	5 32	5 31	5 31	5 30
19	7 06	6 59	6 53	6 49	6 45	6 41	6 35	6 29	6 24	6 19	6 14	6 08	6 04	6 00
20	8 34	8 20	8 10	8 01	7 53	7 47	7 35	7 25	7 16	7 07	6 57	6 45	6 39	6 32
21	9 59	9 40	9 24	9 11	9 01	8 51	8 35	8 21	8 08	7 55	7 41	7 25	7 16	7 06
22	11 20	10 55	10 35	10 19	10 06	9 54	9 34	9 17	9 01	8 45	8 28	8 08	7 57	7 44
23	12 33	12 03	11 41	11 23	11 07	10 54	10 32	10 12	9 54	9 36	9 17	8 55	8 42	8 27
24	13 34	13 02	12 39	12 20	12 04	11 50	11 27	11 06	10 47	10 28	10 08	9 45	9 31	9 15
25	14 21	13 51	13 28	13 10	12 54	12 41	12 18	11 58	11 39	11 20	11 00	10 37	10 23	10 08
26	14 57	14 30	14 09	13 53	13 38	13 26	13 05	12 46	12 29	12 12	11 53	11 31	11 19	11 04
27	15 23	15 01	14 43	14 29	14 17	14 06	13 48	13 32	13 16	13 01	12 45	12 26	12 15	12 02
28	15 43	15 26	15 12	15 01	14 51	14 42	14 27	14 14	14 02	13 49	13 36	13 21	13 12	13 01
29	15 59	15 47	15 37	15 29	15 21	15 15	15 04	14 54	14 45	14 36	14 26	14 15	14 08	14 01
30	16 13	16 05	15 59	15 54	15 50	15 46	15 39	15 33	15 27	15 22	15 16	15 08	15 04	15 00
31	16 25	16 22	16 20	16 18	16 17	16 16	16 13	16 11	16 09	16 07	16 05	16 02	16 01	15 59
Apr. 1	16 37	16 39	16 41	16 43	16 44	16 45	16 47	16 49	16 51	16 53	16 55	16 57	16 58	16 59
2	16 50	16 57	17 03	17 08	17 12	17 16	17 23	17 29	17 34	17 40	17 46	17 53	17 57	18 01

(.. ..) indicates phenomenon will occur the next day.

UNIVERSAL TIME FOR MERIDIAN OF GREENWICH

MOONRISE

Lat.	+40°	+42°	+44°	+46°	+48°	+50°	+52°	+54°	+56°	+58°	+60°	+62°	+64°	+66°
	h m	h m	h m	h m	h m	h m	h m	h m	h m	h m	h m	h m	h m	h m
Mar. 9	8 24	8 19	8 13	8 07	8 00	7 52	7 44	7 35	7 24	7 12	6 59	6 42	6 23	5 58
10	9 06	9 00	8 53	8 45	8 37	8 28	8 18	8 06	7 53	7 38	7 20	6 59	6 30	5 50
11	9 56	9 49	9 42	9 33	9 24	9 14	9 03	8 50	8 35	8 18	7 57	7 30	6 52	** **
12	10 56	10 48	10 41	10 32	10 23	10 13	10 02	9 49	9 34	9 16	8 55	8 27	7 47	** **
13	12 03	11 56	11 49	11 42	11 33	11 24	11 14	11 02	10 49	10 33	10 15	9 52	9 22	8 34
14	13 15	13 10	13 04	12 58	12 51	12 44	12 35	12 26	12 16	12 04	11 50	11 34	11 13	10 47
15	14 30	14 26	14 22	14 18	14 13	14 08	14 02	13 55	13 48	13 40	13 31	13 20	13 07	12 52
16	15 46	15 43	15 41	15 38	15 36	15 32	15 29	15 25	15 21	15 17	15 12	15 06	14 59	14 51
17	17 00	17 00	16 59	16 58	16 58	16 57	16 56	16 55	16 54	16 53	16 51	16 50	16 48	16 46
18	18 14	18 15	18 16	18 17	18 18	18 20	18 21	18 23	18 25	18 27	18 29	18 32	18 35	18 39
19	19 27	19 29	19 32	19 35	19 38	19 42	19 46	19 50	19 55	20 00	20 06	20 13	20 21	20 31
20	20 38	20 42	20 47	20 51	20 57	21 02	21 09	21 16	21 23	21 32	21 42	21 54	22 08	22 25
21	21 47	21 53	21 59	22 05	22 12	22 20	22 28	22 38	22 49	23 01	23 15	23 33	23 54	.. ..
22	22 53	23 00	23 07	23 14	23 23	23 32	23 42	23 54	.. ..	.. ..	.. ..	.. ..	.. ..	0 22
23	23 53	.. ..	.. ..	.. ..	.. ..	.. ..	.. ..	.. ..	0 07	0 23	0 41	1 05	1 36	2 25
24	.. ..	0 01	0 08	0 17	0 26	0 36	0 47	1 00	1 15	1 33	1 54	2 22	3 02	-- --
25	0 47	0 54	1 02	1 10	1 20	1 30	1 41	1 54	2 09	2 27	2 49	3 16	3 57	-- --
26	1 33	1 40	1 47	1 55	2 04	2 13	2 24	2 36	2 50	3 06	3 25	3 48	4 20	5 12
27	2 13	2 19	2 25	2 32	2 39	2 48	2 57	3 07	3 19	3 32	3 47	4 06	4 29	5 00
28	2 46	2 51	2 57	3 02	3 09	3 15	3 23	3 31	3 40	3 50	4 02	4 16	4 33	4 53
29	3 16	3 20	3 24	3 28	3 33	3 38	3 44	3 50	3 56	4 04	4 13	4 23	4 34	4 48
30	3 43	3 45	3 48	3 51	3 54	3 57	4 01	4 05	4 10	4 15	4 20	4 27	4 34	4 43
31	4 07	4 09	4 10	4 12	4 13	4 15	4 17	4 19	4 21	4 24	4 27	4 30	4 34	4 38
Apr. 1	4 32	4 32	4 32	4 32	4 32	4 32	4 32	4 32	4 32	4 32	4 33	4 33	4 33	4 33
2	4 56	4 55	4 54	4 52	4 51	4 49	4 47	4 46	4 43	4 41	4 39	4 36	4 32	4 29

MOONSET

Lat.	+40°	+42°	+44°	+46°	+48°	+50°	+52°	+54°	+56°	+58°	+60°	+62°	+64°	+66°
	h m	h m	h m	h m	h m	h m	h m	h m	h m	h m	h m	h m	h m	h m
Mar. 9	23 25	23 31	23 38	23 45	23 53	.. ..	.. ..	.. ..	.. ..	.. ..	.. ..	.. ..	0 15	0 39
10	.. ..	.. ..	.. ..	.. ..	.. ..	0 02	0 12	0 23	0 36	0 50	1 08	1 29	1 57	2 37
11	0 30	0 37	0 44	0 52	1 01	1 11	1 23	1 35	1 50	2 07	2 28	2 55	3 32	** **
12	1 30	1 37	1 45	1 54	2 03	2 13	2 25	2 38	2 53	3 10	3 32	4 00	4 40	** **
13	2 25	2 32	2 39	2 47	2 55	3 05	3 15	3 27	3 41	3 57	4 15	4 39	5 10	5 57
14	3 12	3 18	3 24	3 31	3 38	3 46	3 55	4 04	4 15	4 28	4 42	5 00	5 21	5 48
15	3 53	3 57	4 02	4 07	4 12	4 18	4 25	4 32	4 40	4 49	4 59	5 11	5 25	5 41
16	4 28	4 31	4 34	4 38	4 41	4 45	4 49	4 54	4 59	5 05	5 11	5 18	5 26	5 36
17	5 00	5 02	5 03	5 05	5 06	5 08	5 10	5 12	5 14	5 17	5 20	5 23	5 27	5 31
18	5 30	5 30	5 30	5 30	5 29	5 29	5 29	5 29	5 28	5 28	5 27	5 27	5 26	5 26
19	6 00	5 59	5 57	5 55	5 53	5 50	5 48	5 45	5 42	5 39	5 35	5 31	5 26	5 21
20	6 32	6 28	6 25	6 21	6 17	6 13	6 08	6 03	5 57	5 51	5 44	5 36	5 26	5 15
21	7 06	7 01	6 56	6 51	6 45	6 39	6 32	6 24	6 16	6 06	5 55	5 43	5 28	5 10
22	7 44	7 38	7 31	7 25	7 17	7 09	7 00	6 50	6 39	6 26	6 11	5 53	5 31	5 02
23	8 27	8 20	8 13	8 05	7 56	7 47	7 36	7 24	7 10	6 55	6 36	6 12	5 41	4 51
24	9 15	9 08	9 00	8 51	8 42	8 32	8 20	8 07	7 52	7 35	7 13	6 45	6 05	-- --
25	10 08	10 01	9 53	9 44	9 35	9 25	9 14	9 01	8 46	8 28	8 07	7 39	6 59	-- --
26	11 04	10 57	10 50	10 43	10 34	10 25	10 15	10 03	9 49	9 34	9 15	8 51	8 20	7 29
27	12 02	11 57	11 51	11 44	11 37	11 29	11 20	11 10	10 59	10 46	10 31	10 13	9 50	9 20
28	13 01	12 57	12 52	12 47	12 41	12 35	12 28	12 20	12 11	12 02	11 50	11 37	11 21	11 02
29	14 01	13 57	13 54	13 50	13 46	13 41	13 36	13 31	13 25	13 18	13 10	13 01	12 50	12 38
30	15 00	14 58	14 55	14 53	14 51	14 48	14 45	14 41	14 38	14 34	14 29	14 24	14 18	14 10
31	15 59	15 58	15 58	15 57	15 56	15 55	15 54	15 53	15 51	15 50	15 48	15 46	15 44	15 42
Apr. 1	16 59	17 00	17 01	17 01	17 02	17 03	17 04	17 05	17 06	17 07	17 08	17 10	17 12	17 14
2	18 01	18 03	18 05	18 07	18 10	18 13	18 15	18 19	18 22	18 26	18 30	18 36	18 41	18 48

(.. ..) indicates phenomenon will occur the next day.
(-- --) indicates Moon continuously below horizon.
(** **) indicates Moon continuously above horizon.

MOONRISE AND MOONSET, 1992

UNIVERSAL TIME FOR MERIDIAN OF GREENWICH

MOONRISE

Lat.	−55°	−50°	−45°	−40°	−35°	−30°	−20°	−10°	0°	+10°	+20°	+30°	+35°	+40°
	h m	h m	h m	h m	h m	h m	h m	h m	h m	h m	h m	h m	h m	h m
Apr. 1	4 30	4 30	4 30	4 30	4 30	4 30	4 31	4 31	4 31	4 31	4 31	4 31	4 31	4 32
2	5 44	5 39	5 35	5 31	5 28	5 26	5 21	5 17	5 13	5 10	5 06	5 02	4 59	4 56
3	7 00	6 50	6 41	6 34	6 28	6 22	6 13	6 05	5 58	5 50	5 42	5 34	5 28	5 23
4	8 19	8 02	7 49	7 39	7 29	7 21	7 08	6 56	6 45	6 33	6 22	6 08	6 01	5 52
5	9 39	9 17	8 59	8 45	8 33	8 22	8 04	7 49	7 35	7 20	7 05	6 48	6 38	6 26
6	10 58	10 30	10 08	9 51	9 37	9 25	9 03	8 45	8 28	8 11	7 53	7 32	7 20	7 06
7	12 09	11 38	11 15	10 56	10 40	10 26	10 03	9 43	9 24	9 06	8 46	8 23	8 10	7 54
8	13 09	12 38	12 14	11 55	11 39	11 25	11 02	10 42	10 23	10 04	9 44	9 20	9 07	8 51
9	13 55	13 27	13 05	12 48	12 33	12 20	11 58	11 39	11 22	11 04	10 45	10 23	10 10	9 55
10	14 28	14 05	13 47	13 33	13 20	13 09	12 51	12 34	12 19	12 04	11 47	11 28	11 17	11 05
11	14 52	14 36	14 22	14 11	14 02	13 53	13 39	13 26	13 14	13 02	12 50	12 35	12 26	12 17
12	15 12	15 01	14 52	14 45	14 39	14 33	14 24	14 15	14 08	14 00	13 51	13 41	13 36	13 29
13	15 28	15 23	15 19	15 16	15 13	15 11	15 06	15 02	14 59	14 55	14 51	14 47	14 44	14 42
14	15 43	15 44	15 45	15 45	15 46	15 46	15 47	15 48	15 49	15 50	15 51	15 52	15 52	15 53
15	15 58	16 05	16 10	16 15	16 19	16 22	16 29	16 34	16 39	16 44	16 50	16 57	17 00	17 05
16	16 15	16 28	16 38	16 46	16 53	17 00	17 11	17 21	17 30	17 39	17 50	18 01	18 08	18 16
17	16 35	16 54	17 08	17 20	17 30	17 40	17 55	18 09	18 22	18 35	18 49	19 05	19 15	19 26
18	17 01	17 25	17 43	17 59	18 12	18 23	18 42	18 59	19 15	19 31	19 49	20 09	20 20	20 34
19	17 35	18 03	18 25	18 43	18 57	19 10	19 32	19 51	20 09	20 27	20 47	21 09	21 22	21 38
20	18 19	18 50	19 13	19 32	19 48	20 01	20 24	20 44	21 03	21 22	21 42	22 06	22 19	22 35
21	19 14	19 45	20 07	20 26	20 41	20 55	21 18	21 37	21 56	22 14	22 34	22 57	23 10	23 25
22	20 18	20 45	21 06	21 23	21 37	21 50	22 11	22 29	22 46	23 03	23 21	23 42	23 54	.. ..
23	21 27	21 50	22 08	22 22	22 34	22 45	23 03	23 19	23 34	23 48	.. ..	.. ..	.. ..	0 08
24	22 38	22 56	23 10	23 21	23 31	23 40	23 54	.. ..	.. ..	.. ..	0 04	0 22	0 32	0 44
25	23 50	.. ..	.. ..	.. ..	.. ..	.. ..	.. ..	0 07	0 19	0 31	0 43	0 58	1 06	1 16

MOONSET

Lat.	−55°	−50°	−45°	−40°	−35°	−30°	−20°	−10°	0°	+10°	+20°	+30°	+35°	+40°
	h m	h m	h m	h m	h m	h m	h m	h m	h m	h m	h m	h m	h m	h m
Apr. 1	16 37	16 39	16 41	16 43	16 44	16 45	16 47	16 49	16 51	16 53	16 55	16 57	16 58	16 59
2	16 50	16 57	17 03	17 08	17 12	17 16	17 23	17 29	17 34	17 40	17 46	17 53	17 57	18 01
3	17 04	17 17	17 27	17 36	17 43	17 49	18 01	18 10	18 20	18 29	18 39	18 50	18 57	19 05
4	17 22	17 41	17 55	18 07	18 17	18 26	18 42	18 55	19 08	19 21	19 35	19 50	20 00	20 10
5	17 46	18 10	18 29	18 44	18 57	19 08	19 27	19 44	20 00	20 16	20 32	20 52	21 03	21 17
6	18 19	18 48	19 10	19 28	19 42	19 55	20 18	20 37	20 55	21 13	21 32	21 54	22 07	22 22
7	19 05	19 37	20 01	20 20	20 36	20 50	21 13	21 33	21 52	22 11	22 31	22 55	23 09	23 24
8	20 07	20 38	21 02	21 20	21 36	21 49	22 13	22 32	22 51	23 09	23 29	23 52	.. ..	.. ..
9	21 23	21 50	22 11	22 27	22 41	22 54	23 15	23 32	23 49	.. ..	.. ..	.. ..	0 05	0 20
10	22 47	23 08	23 25	23 39	23 50	.. ..	.. ..	.. ..	.. ..	0 06	0 24	0 44	0 56	1 09
11	.. ..	.. ..	.. ..	.. ..	.. ..	0 00	0 17	0 32	0 46	0 59	1 14	1 30	1 40	1 51
12	0 15	0 30	0 41	0 51	0 59	1 07	1 19	1 30	1 40	1 50	2 00	2 12	2 19	2 27
13	1 43	1 51	1 58	2 04	2 08	2 13	2 20	2 26	2 32	2 38	2 44	2 51	2 55	2 59
14	3 11	3 13	3 14	3 16	3 17	3 18	3 20	3 21	3 23	3 24	3 26	3 27	3 28	3 29
15	4 37	4 33	4 30	4 27	4 25	4 23	4 19	4 16	4 13	4 10	4 07	4 03	4 01	3 58
16	6 04	5 54	5 45	5 38	5 33	5 27	5 18	5 11	5 03	4 56	4 48	4 39	4 34	4 29
17	7 30	7 13	7 00	6 49	6 40	6 32	6 18	6 06	5 55	5 43	5 31	5 18	5 10	5 01
18	8 53	8 31	8 13	7 59	7 46	7 36	7 18	7 02	6 47	6 33	6 17	5 59	5 49	5 37
19	10 11	9 43	9 22	9 05	8 50	8 38	8 17	7 58	7 41	7 24	7 06	6 45	6 32	6 18
20	11 18	10 47	10 24	10 06	9 50	9 37	9 14	8 54	8 35	8 17	7 57	7 34	7 21	7 05
21	12 12	11 42	11 18	11 00	10 44	10 31	10 07	9 47	9 29	9 10	8 50	8 27	8 13	7 57
22	12 53	12 25	12 04	11 46	11 32	11 19	10 57	10 38	10 20	10 02	9 43	9 21	9 08	8 53
23	13 23	13 00	12 41	12 26	12 13	12 02	11 42	11 25	11 09	10 53	10 36	10 16	10 05	9 51
24	13 46	13 27	13 12	12 59	12 49	12 39	12 23	12 09	11 56	11 42	11 28	11 11	11 02	10 51
25	14 04	13 50	13 38	13 29	13 21	13 14	13 01	12 50	12 40	12 30	12 18	12 06	11 58	11 50

(.. ..) indicates phenomenon will occur the next day.

UNIVERSAL TIME FOR MERIDIAN OF GREENWICH

MOONRISE

Lat.	+40°	+42°	+44°	+46°	+48°	+50°	+52°	+54°	+56°	+58°	+60°	+62°	+64°	+66°
	h m	h m	h m	h m	h m	h m	h m	h m	h m	h m	h m	h m	h m	h m
Apr. 1	4 32	4 32	4 32	4 32	4 32	4 32	4 32	4 32	4 32	4 32	4 33	4 33	4 33	4 33
2	4 56	4 55	4 54	4 52	4 51	4 49	4 47	4 46	4 43	4 41	4 39	4 36	4 32	4 29
3	5 23	5 20	5 17	5 15	5 11	5 08	5 04	5 00	4 56	4 51	4 46	4 39	4 32	4 24
4	5 52	5 48	5 44	5 40	5 35	5 30	5 24	5 18	5 11	5 03	4 55	4 45	4 33	4 19
5	6 26	6 21	6 15	6 10	6 03	5 56	5 49	5 40	5 31	5 20	5 08	4 53	4 36	4 15
6	7 06	7 00	6 53	6 46	6 38	6 30	6 20	6 10	5 58	5 44	5 28	5 08	4 43	4 10
7	7 54	7 47	7 40	7 32	7 23	7 13	7 02	6 50	6 36	6 20	6 00	5 35	5 01	4 03
8	8 51	8 44	8 36	8 28	8 19	8 08	7 57	7 44	7 30	7 12	6 51	6 24	5 45	** **
9	9 55	9 48	9 41	9 33	9 25	9 15	9 05	8 53	8 39	8 23	8 04	7 40	7 08	6 15
10	11 05	10 59	10 53	10 46	10 39	10 31	10 22	10 12	10 01	9 48	9 33	9 15	8 52	8 22
11	12 17	12 12	12 08	12 03	11 57	11 51	11 45	11 37	11 29	11 20	11 09	10 57	10 42	10 24
12	13 29	13 26	13 23	13 20	13 17	13 13	13 09	13 04	12 59	12 53	12 46	12 39	12 30	12 20
13	14 42	14 40	14 39	14 37	14 36	14 34	14 32	14 30	14 28	14 26	14 23	14 20	14 16	14 12
14	15 53	15 54	15 54	15 54	15 55	15 55	15 56	15 56	15 57	15 57	15 58	15 59	16 00	16 01
15	17 05	17 06	17 08	17 11	17 13	17 16	17 18	17 21	17 25	17 29	17 33	17 38	17 44	17 50
16	18 16	18 19	18 23	18 27	18 31	18 36	18 41	18 46	18 53	19 00	19 08	19 17	19 28	19 41
17	19 26	19 31	19 36	19 42	19 48	19 54	20 02	20 10	20 19	20 30	20 42	20 56	21 14	21 35
18	20 34	20 40	20 46	20 53	21 01	21 10	21 19	21 30	21 42	21 56	22 12	22 32	22 58	23 34
19	21 38	21 44	21 52	22 00	22 09	22 18	22 29	22 41	22 56	23 12	23 32	23 58	.. ..	.. ..
20	22 35	22 42	22 50	22 58	23 08	23 18	23 29	23 42	23 57	.. ..	.. ..	.. ..	0 33	1 40
21	23 25	23 32	23 40	23 48	23 57	.. ..	.. ..	.. ..	.. ..	0 14	0 36	1 03	1 43	-- --
22	.. ..	.. ..	.. ..	.. ..	.. ..	0 06	0 17	0 29	0 44	1 00	1 20	1 45	2 19	3 21
23	0 08	0 14	0 21	0 28	0 36	0 45	0 55	1 05	1 18	1 32	1 48	2 09	2 34	3 10
24	0 44	0 50	0 55	1 01	1 08	1 15	1 23	1 32	1 42	1 54	2 07	2 22	2 41	3 04
25	1 16	1 20	1 24	1 29	1 34	1 40	1 46	1 53	2 01	2 09	2 19	2 30	2 43	2 59

MOONSET

Lat.	+40°	+42°	+44°	+46°	+48°	+50°	+52°	+54°	+56°	+58°	+60°	+62°	+64°	+66°
	h m	h m	h m	h m	h m	h m	h m	h m	h m	h m	h m	h m	h m	h m
Apr. 1	16 59	17 00	17 01	17 01	17 02	17 03	17 04	17 05	17 06	17 07	17 08	17 10	17 12	17 14
2	18 01	18 03	18 05	18 07	18 10	18 13	18 15	18 19	18 22	18 26	18 30	18 36	18 41	18 48
3	19 05	19 08	19 12	19 15	19 20	19 24	19 29	19 35	19 41	19 48	19 56	20 05	20 15	20 28
4	20 10	20 15	20 20	20 25	20 31	20 38	20 45	20 53	21 02	21 12	21 23	21 37	21 53	22 14
5	21 17	21 23	21 29	21 36	21 43	21 51	22 01	22 11	22 22	22 36	22 52	23 11	23 35	.. ..
6	22 22	22 29	22 36	22 44	22 53	23 03	23 13	23 25	23 39	23 55	.. ..	.. ..	.. ..	0 08
7	23 24	23 32	23 39	23 48	23 57	.. ..	.. ..	.. ..	.. ..	.. ..	0 15	0 40	1 13	2 11
8	.. ..	.. ..	.. ..	.. ..	.. ..	0 07	0 18	0 31	0 46	1 03	1 24	1 51	2 30	** **
9	0 20	0 27	0 35	0 43	0 51	1 01	1 12	1 24	1 38	1 54	2 14	2 38	3 10	4 04
10	1 09	1 15	1 22	1 29	1 36	1 45	1 54	2 04	2 16	2 29	2 45	3 03	3 27	3 58
11	1 51	1 56	2 01	2 06	2 12	2 19	2 26	2 34	2 43	2 53	3 04	3 18	3 33	3 53
12	2 27	2 30	2 34	2 38	2 42	2 47	2 52	2 57	3 03	3 10	3 18	3 26	3 36	3 48
13	2 59	3 01	3 03	3 05	3 08	3 10	3 13	3 16	3 19	3 23	3 27	3 32	3 37	3 43
14	3 29	3 30	3 30	3 30	3 31	3 32	3 32	3 33	3 34	3 35	3 35	3 36	3 38	3 39
15	3 58	3 57	3 56	3 55	3 54	3 52	3 51	3 49	3 47	3 45	3 43	3 41	3 38	3 34
16	4 29	4 26	4 23	4 20	4 17	4 14	4 10	4 06	4 02	3 57	3 52	3 45	3 38	3 30
17	5 01	4 57	4 53	4 48	4 43	4 38	4 32	4 26	4 19	4 11	4 02	3 51	3 39	3 25
18	5 37	5 32	5 26	5 20	5 14	5 07	4 59	4 50	4 40	4 29	4 16	4 01	3 43	3 20
19	6 18	6 12	6 05	5 58	5 50	5 41	5 31	5 20	5 08	4 54	4 37	4 16	3 50	3 14
20	7 05	6 58	6 51	6 42	6 33	6 24	6 13	6 00	5 46	5 29	5 09	4 43	4 08	3 01
21	7 57	7 50	7 42	7 34	7 25	7 14	7 03	6 50	6 35	6 18	5 56	5 29	4 49	-- --
22	8 53	8 46	8 39	8 31	8 22	8 13	8 02	7 50	7 36	7 20	7 00	6 35	6 01	5 00
23	9 51	9 45	9 39	9 32	9 24	9 16	9 07	8 56	8 44	8 30	8 14	7 54	7 29	6 54
24	10 51	10 46	10 40	10 35	10 28	10 21	10 14	10 05	9 56	9 45	9 33	9 18	9 00	8 37
25	11 50	11 46	11 42	11 38	11 33	11 28	11 22	11 16	11 09	11 01	10 52	10 42	10 29	10 15

(.. ..) indicates phenomenon will occur the next day.
(-- --) indicates Moon continuously below horizon.
(** **) indicates Moon continuously above horizon.

UNIVERSAL TIME FOR MERIDIAN OF GREENWICH

MOONRISE

Lat.	−55°	−50°	−45°	−40°	−35°	−30°	−20°	−10°	0°	+10°	+20°	+30°	+35°	+40°
	h m	h m	h m	h m	h m	h m	h m	h m	h m	h m	h m	h m	h m	h m
Apr. 24	22 38	22 56	23 10	23 21	23 31	23 40	23 54	.. ..	.. ..	.. ..	0 04	0 22	0 32	0 44
25	23 50	.. ..	.. ..	.. ..	.. ..	.. ..	.. ..	0 07	0 19	0 31	0 43	0 58	1 06	1 16
26	.. ..	0 02	0 12	0 20	0 27	0 33	0 44	0 53	1 02	1 11	1 20	1 30	1 36	1 43
27	1 01	1 08	1 14	1 19	1 23	1 27	1 33	1 39	1 44	1 49	1 55	2 01	2 05	2 09
28	2 13	2 15	2 17	2 18	2 19	2 20	2 22	2 24	2 26	2 27	2 29	2 31	2 32	2 33
29	3 26	3 23	3 20	3 18	3 16	3 15	3 12	3 10	3 08	3 05	3 03	3 01	2 59	2 58
30	4 41	4 33	4 26	4 20	4 15	4 11	4 04	3 57	3 51	3 45	3 39	3 32	3 28	3 23
May 1	5 59	5 45	5 34	5 24	5 16	5 09	4 57	4 47	4 37	4 28	4 17	4 06	3 59	3 52
2	7 20	6 59	6 44	6 31	6 20	6 10	5 54	5 40	5 27	5 14	5 00	4 44	4 35	4 24
3	8 40	8 14	7 55	7 39	7 25	7 13	6 53	6 36	6 20	6 04	5 47	5 27	5 16	5 03
4	9 57	9 26	9 04	8 45	8 30	8 17	7 54	7 35	7 17	6 59	6 39	6 17	6 04	5 49
5	11 02	10 31	10 07	9 48	9 32	9 18	8 55	8 35	8 16	7 57	7 37	7 14	7 00	6 44
6	11 53	11 24	11 02	10 44	10 29	10 16	9 53	9 34	9 16	8 58	8 38	8 16	8 03	7 48
7	12 30	12 06	11 47	11 31	11 18	11 07	10 48	10 30	10 14	9 58	9 41	9 21	9 10	8 57
8	12 57	12 39	12 24	12 12	12 02	11 53	11 37	11 24	11 11	10 58	10 44	10 28	10 19	10 08
9	13 17	13 05	12 55	12 47	12 40	12 34	12 23	12 13	12 04	11 55	11 45	11 34	11 28	11 20
10	13 34	13 28	13 22	13 18	13 14	13 11	13 05	13 00	12 55	12 50	12 45	12 39	12 35	12 32
11	13 49	13 48	13 48	13 47	13 47	13 46	13 45	13 45	13 44	13 44	13 43	13 42	13 42	13 42
12	14 04	14 09	14 13	14 16	14 19	14 21	14 25	14 29	14 33	14 37	14 41	14 45	14 48	14 51
13	14 20	14 30	14 38	14 45	14 51	14 57	15 06	15 14	15 22	15 30	15 39	15 48	15 54	16 00
14	14 38	14 54	15 07	15 18	15 27	15 35	15 49	16 01	16 13	16 24	16 37	16 51	17 00	17 09
15	15 01	15 23	15 40	15 54	16 06	16 16	16 34	16 50	17 05	17 20	17 36	17 54	18 05	18 17
16	15 31	15 58	16 18	16 35	16 49	17 01	17 22	17 41	17 58	18 15	18 34	18 55	19 08	19 22
17	16 11	16 41	17 04	17 22	17 38	17 51	18 14	18 33	18 52	19 10	19 30	19 53	20 07	20 23
18	17 02	17 33	17 56	18 15	18 30	18 44	19 07	19 27	19 45	20 04	20 24	20 47	21 00	21 16

MOONSET

Lat.	−55°	−50°	−45°	−40°	−35°	−30°	−20°	−10°	0°	+10°	+20°	+30°	+35°	+40°
	h m	h m	h m	h m	h m	h m	h m	h m	h m	h m	h m	h m	h m	h m
Apr. 24	13 46	13 27	13 12	12 59	12 49	12 39	12 23	12 09	11 56	11 42	11 28	11 11	11 02	10 51
25	14 04	13 50	13 38	13 29	13 21	13 14	13 01	12 50	12 40	12 30	12 18	12 06	11 58	11 50
26	14 18	14 09	14 01	13 55	13 50	13 45	13 37	13 29	13 22	13 15	13 08	12 59	12 54	12 49
27	14 31	14 26	14 23	14 20	14 17	14 15	14 11	14 07	14 04	14 01	13 57	13 53	13 50	13 47
28	14 43	14 43	14 44	14 44	14 44	14 45	14 45	14 45	14 46	14 46	14 46	14 46	14 47	14 47
29	14 56	15 01	15 05	15 09	15 12	15 15	15 20	15 24	15 28	15 32	15 37	15 42	15 44	15 48
30	15 10	15 20	15 29	15 36	15 42	15 47	15 57	16 05	16 13	16 21	16 29	16 39	16 44	16 51
May 1	15 27	15 43	15 55	16 06	16 15	16 23	16 37	16 49	17 00	17 12	17 24	17 38	17 46	17 56
2	15 49	16 10	16 27	16 41	16 53	17 03	17 21	17 37	17 52	18 06	18 22	18 40	18 51	19 03
3	16 19	16 46	17 07	17 23	17 37	17 50	18 11	18 29	18 47	19 04	19 22	19 44	19 56	20 10
4	17 01	17 32	17 55	18 14	18 29	18 43	19 06	19 26	19 45	20 03	20 23	20 46	21 00	21 15
5	17 59	18 31	18 54	19 13	19 29	19 43	20 06	20 26	20 44	21 03	21 23	21 46	21 59	22 15
6	19 12	19 41	20 02	20 20	20 34	20 47	21 08	21 27	21 44	22 01	22 19	22 40	22 53	23 07
7	20 35	20 58	21 16	21 30	21 43	21 53	22 11	22 27	22 42	22 56	23 11	23 29	23 39	23 51
8	22 02	22 19	22 32	22 43	22 52	23 00	23 13	23 25	23 36	23 47	23 59	.. ..	.. ..	.. ..
9	23 29	23 40	23 48	23 54	.. ..	.. ..	.. ..	.. ..	.. ..	.. ..	.. ..	0 12	0 20	0 28
10	.. ..	.. ..	.. ..	.. ..	0 00	0 05	0 14	0 21	0 28	0 35	0 43	0 51	0 56	1 01
11	0 55	0 59	1 02	1 05	1 07	1 09	1 13	1 16	1 18	1 21	1 24	1 27	1 29	1 31
12	2 20	2 18	2 16	2 15	2 14	2 13	2 11	2 09	2 07	2 06	2 04	2 02	2 01	2 00
13	3 44	3 36	3 30	3 24	3 20	3 15	3 08	3 02	2 56	2 51	2 44	2 37	2 33	2 29
14	5 08	4 54	4 43	4 33	4 25	4 18	4 06	3 56	3 46	3 36	3 26	3 14	3 07	3 00
15	6 31	6 11	5 55	5 42	5 31	5 21	5 05	4 51	4 37	4 24	4 10	3 54	3 44	3 34
16	7 50	7 24	7 05	6 49	6 35	6 24	6 04	5 46	5 30	5 14	4 57	4 37	4 25	4 12
17	9 02	8 32	8 10	7 52	7 37	7 24	7 01	6 42	6 24	6 06	5 47	5 24	5 11	4 56
18	10 01	9 31	9 07	8 49	8 33	8 20	7 57	7 36	7 18	6 59	6 39	6 16	6 02	5 46

(.. ..) indicates phenomenon will occur the next day.

UNIVERSAL TIME FOR MERIDIAN OF GREENWICH

MOONRISE

Lat.	+40°	+42°	+44°	+46°	+48°	+50°	+52°	+54°	+56°	+58°	+60°	+62°	+64°	+66°
	h m	h m	h m	h m	h m	h m	h m	h m	h m	h m	h m	h m	h m	h m
Apr. 24	0 44	0 50	0 55	1 01	1 08	1 15	1 23	1 32	1 42	1 54	2 07	2 22	2 41	3 04
25	1 16	1 20	1 24	1 29	1 34	1 40	1 46	1 53	2 01	2 09	2 19	2 30	2 43	2 59
26	1 43	1 46	1 50	1 53	1 57	2 01	2 05	2 10	2 15	2 21	2 28	2 35	2 44	2 54
27	2 09	2 11	2 12	2 14	2 17	2 19	2 22	2 24	2 28	2 31	2 35	2 39	2 44	2 50
28	2 33	2 34	2 34	2 35	2 36	2 36	2 37	2 38	2 39	2 40	2 41	2 42	2 44	2 46
29	2 58	2 57	2 56	2 55	2 54	2 53	2 52	2 51	2 50	2 49	2 47	2 46	2 44	2 41
30	3 23	3 21	3 19	3 17	3 14	3 12	3 09	3 06	3 02	2 58	2 54	2 49	2 44	2 37
May 1	3 52	3 48	3 45	3 41	3 37	3 32	3 28	3 22	3 17	3 10	3 03	2 54	2 45	2 33
2	4 24	4 20	4 15	4 09	4 04	3 57	3 51	3 43	3 35	3 25	3 14	3 02	2 47	2 29
3	5 03	4 57	4 51	4 44	4 37	4 29	4 20	4 10	3 59	3 47	3 32	3 15	2 53	2 25
4	5 49	5 43	5 35	5 28	5 19	5 10	4 59	4 48	4 34	4 19	4 00	3 37	3 07	2 22
5	6 44	6 37	6 30	6 21	6 12	6 02	5 51	5 38	5 24	5 07	4 46	4 20	3 43	2 23
6	7 48	7 41	7 33	7 25	7 17	7 07	6 56	6 44	6 30	6 14	5 54	5 29	4 55	3 55
7	8 57	8 51	8 44	8 37	8 30	8 21	8 12	8 02	7 50	7 36	7 20	7 01	6 36	6 02
8	10 08	10 04	9 59	9 53	9 47	9 41	9 34	9 26	9 17	9 07	8 55	8 41	8 25	8 04
9	11 20	11 17	11 14	11 10	11 06	11 01	10 57	10 51	10 45	10 39	10 31	10 22	10 12	10 00
10	12 32	12 30	12 28	12 26	12 24	12 22	12 19	12 16	12 13	12 10	12 06	12 02	11 57	11 51
11	13 42	13 41	13 41	13 41	13 41	13 41	13 40	13 40	13 40	13 39	13 39	13 39	13 38	13 38
12	14 51	14 52	14 54	14 56	14 57	14 59	15 01	15 03	15 06	15 08	15 11	15 15	15 19	15 23
13	16 00	16 03	16 06	16 10	16 13	16 17	16 21	16 26	16 31	16 37	16 44	16 51	17 00	17 10
14	17 09	17 14	17 18	17 23	17 29	17 35	17 41	17 48	17 56	18 05	18 16	18 28	18 43	19 00
15	18 17	18 23	18 29	18 35	18 42	18 50	18 59	19 08	19 19	19 32	19 46	20 04	20 26	20 54
16	19 22	19 29	19 36	19 44	19 52	20 01	20 11	20 23	20 36	20 52	21 10	21 34	22 04	22 53
17	20 23	20 30	20 37	20 45	20 55	21 05	21 16	21 28	21 43	22 00	22 21	22 48	23 26	-- --
18	21 16	21 23	21 31	21 39	21 48	21 58	22 09	22 22	22 36	22 53	23 14	23 40	.. ..	-- --

MOONSET

Lat.	+40°	+42°	+44°	+46°	+48°	+50°	+52°	+54°	+56°	+58°	+60°	+62°	+64°	+66°
	h m	h m	h m	h m	h m	h m	h m	h m	h m	h m	h m	h m	h m	h m
Apr. 24	10 51	10 46	10 40	10 35	10 28	10 21	10 14	10 05	9 56	9 45	9 33	9 18	9 00	8 37
25	11 50	11 46	11 42	11 38	11 33	11 28	11 22	11 16	11 09	11 01	10 52	10 42	10 29	10 15
26	12 49	12 46	12 43	12 40	12 37	12 34	12 30	12 26	12 22	12 16	12 11	12 04	11 56	11 48
27	13 47	13 46	13 45	13 43	13 42	13 40	13 38	13 36	13 34	13 32	13 29	13 26	13 22	13 18
28	14 47	14 47	14 47	14 47	14 47	14 47	14 47	14 48	14 48	14 48	14 48	14 48	14 48	14 49
29	15 48	15 49	15 51	15 52	15 54	15 56	15 58	16 00	16 03	16 06	16 09	16 12	16 17	16 21
30	16 51	16 53	16 56	17 00	17 03	17 07	17 11	17 15	17 21	17 26	17 33	17 40	17 48	17 59
May 1	17 56	18 00	18 04	18 09	18 14	18 20	18 26	18 33	18 41	18 50	19 00	19 12	19 25	19 42
2	19 03	19 08	19 14	19 21	19 28	19 35	19 43	19 53	20 03	20 15	20 29	20 46	21 07	21 34
3	20 10	20 17	20 24	20 31	20 40	20 49	20 59	21 10	21 23	21 39	21 57	22 19	22 49	23 34
4	21 15	21 22	21 30	21 38	21 47	21 57	22 08	22 21	22 36	22 53	23 13	23 40	.. ..	.. ..
5	22 15	22 22	22 29	22 37	22 46	22 56	23 07	23 19	23 34	23 50	.. ..	.. ..	0 16	1 36
6	23 07	23 13	23 20	23 27	23 35	23 44	23 53	.. ..	.. ..	.. ..	0 10	0 35	1 09	2 09
7	23 51	23 56	.. ..	.. ..	.. ..	.. ..	.. ..	0 04	0 16	0 30	0 47	1 07	1 32	2 07
8	.. ..	.. ..	0 01	0 07	0 14	0 21	0 28	0 37	0 47	0 57	1 10	1 24	1 42	2 03
9	0 28	0 32	0 36	0 41	0 45	0 50	0 56	1 02	1 09	1 16	1 25	1 35	1 46	1 59
10	1 01	1 04	1 06	1 09	1 12	1 15	1 18	1 22	1 26	1 30	1 35	1 41	1 48	1 55
11	1 31	1 32	1 33	1 34	1 35	1 36	1 38	1 39	1 40	1 42	1 44	1 46	1 49	1 51
12	2 00	1 59	1 59	1 58	1 57	1 57	1 56	1 55	1 54	1 53	1 52	1 50	1 49	1 47
13	2 29	2 27	2 25	2 22	2 20	2 17	2 14	2 11	2 08	2 04	2 00	1 55	1 49	1 43
14	3 00	2 56	2 53	2 49	2 44	2 40	2 35	2 29	2 23	2 17	2 09	2 00	1 50	1 39
15	3 34	3 29	3 24	3 18	3 12	3 06	2 59	2 51	2 42	2 33	2 21	2 08	1 53	1 34
16	4 12	4 06	4 00	3 53	3 46	3 38	3 29	3 19	3 07	2 54	2 39	2 21	1 59	1 29
17	4 56	4 50	4 42	4 35	4 26	4 17	4 06	3 54	3 41	3 25	3 06	2 43	2 11	1 23
18	5 46	5 39	5 32	5 23	5 14	5 04	4 53	4 40	4 25	4 08	3 47	3 20	2 42	-- --

(.. ..) indicates phenomenon will occur the next day.
(-- --) indicates Moon continuously below horizon.
(** **) indicates Moon continuously above horizon.

UNIVERSAL TIME FOR MERIDIAN OF GREENWICH

MOONRISE

Lat.	−55°	−50°	−45°	−40°	−35°	−30°	−20°	−10°	0°	+10°	+20°	+30°	+35°	+40°
	h m	h m	h m	h m	h m	h m	h m	h m	h m	h m	h m	h m	h m	h m
May 17	16 11	16 41	17 04	17 22	17 38	17 51	18 14	18 33	18 52	19 10	19 30	19 53	20 07	20 23
18	17 02	17 33	17 56	18 15	18 30	18 44	19 07	19 27	19 45	20 04	20 24	20 47	21 00	21 16
19	18 03	18 32	18 54	19 11	19 26	19 39	20 00	20 19	20 37	20 55	21 13	21 35	21 48	22 02
20	19 11	19 36	19 55	20 10	20 23	20 34	20 54	21 10	21 26	21 42	21 58	22 17	22 29	22 41
21	20 22	20 42	20 57	21 10	21 20	21 30	21 46	22 00	22 13	22 26	22 39	22 55	23 04	23 15
22	21 34	21 48	21 59	22 09	22 17	22 24	22 36	22 47	22 57	23 07	23 17	23 29	23 36	23 44
23	22 45	22 54	23 01	23 07	23 13	23 17	23 25	23 32	23 39	23 45	23 52	.. ..	.. ..	.. ..
24	23 56	.. ..	.. ..	.. ..	.. ..	.. ..	.. ..	.. ..	.. ..	.. ..	.. ..	0 00	0 05	0 10
25	.. ..	0 00	0 03	0 06	0 08	0 10	0 14	0 17	0 20	0 23	0 26	0 30	0 32	0 34
26	1 07	1 06	1 05	1 05	1 04	1 04	1 03	1 02	1 01	1 01	1 00	0 59	0 59	0 58
27	2 20	2 14	2 09	2 05	2 01	1 58	1 53	1 48	1 44	1 39	1 35	1 30	1 27	1 23
28	3 36	3 25	3 15	3 07	3 01	2 55	2 45	2 36	2 28	2 20	2 12	2 02	1 56	1 50
29	4 55	4 38	4 24	4 12	4 03	3 54	3 40	3 27	3 16	3 04	2 52	2 38	2 30	2 21
30	6 16	5 53	5 35	5 20	5 07	4 57	4 38	4 22	4 07	3 52	3 37	3 19	3 08	2 56
31	7 36	7 07	6 46	6 28	6 13	6 01	5 39	5 20	5 03	4 46	4 27	4 06	3 54	3 40
June 1	8 48	8 16	7 53	7 34	7 18	7 04	6 41	6 21	6 02	5 44	5 24	5 01	4 47	4 32
2	9 46	9 16	8 53	8 34	8 19	8 05	7 42	7 22	7 04	6 45	6 25	6 02	5 49	5 34
3	10 29	10 03	9 43	9 26	9 13	9 01	8 40	8 22	8 05	7 48	7 30	7 09	6 57	6 43
4	11 00	10 40	10 24	10 11	9 59	9 50	9 33	9 18	9 04	8 50	8 35	8 18	8 07	7 56
5	11 23	11 09	10 58	10 48	10 40	10 33	10 21	10 10	9 59	9 49	9 38	9 26	9 18	9 10
6	11 41	11 33	11 26	11 21	11 16	11 12	11 05	10 58	10 52	10 46	10 39	10 32	10 28	10 23
7	11 57	11 54	11 52	11 51	11 49	11 48	11 46	11 44	11 42	11 40	11 38	11 36	11 35	11 34
8	12 11	12 14	12 17	12 19	12 21	12 23	12 26	12 28	12 31	12 33	12 36	12 39	12 41	12 43
9	12 27	12 35	12 42	12 48	12 53	12 58	13 06	13 13	13 19	13 26	13 33	13 41	13 46	13 52
10	12 44	12 58	13 09	13 19	13 27	13 34	13 47	13 58	14 08	14 19	14 30	14 43	14 51	15 00

MOONSET

Lat.	−55°	−50°	−45°	−40°	−35°	−30°	−20°	−10°	0°	+10°	+20°	+30°	+35°	+40°
	h m	h m	h m	h m	h m	h m	h m	h m	h m	h m	h m	h m	h m	h m
May 17	9 02	8 32	8 10	7 52	7 37	7 24	7 01	6 42	6 24	6 06	5 47	5 24	5 11	4 56
18	10 01	9 31	9 07	8 49	8 33	8 20	7 57	7 36	7 18	6 59	6 39	6 16	6 02	5 46
19	10 48	10 19	9 57	9 39	9 24	9 11	8 48	8 29	8 10	7 52	7 33	7 10	6 57	6 41
20	11 23	10 57	10 37	10 21	10 08	9 56	9 35	9 18	9 01	8 44	8 26	8 06	7 53	7 39
21	11 48	11 27	11 11	10 58	10 46	10 36	10 18	10 03	9 49	9 34	9 19	9 01	8 51	8 39
22	12 08	11 52	11 39	11 29	11 20	11 12	10 58	10 46	10 34	10 23	10 10	9 56	9 48	9 38
23	12 23	12 12	12 04	11 56	11 50	11 44	11 34	11 25	11 17	11 09	11 00	10 50	10 44	10 37
24	12 37	12 31	12 26	12 21	12 18	12 14	12 09	12 04	11 59	11 54	11 49	11 43	11 40	11 36
25	12 49	12 48	12 46	12 45	12 44	12 44	12 42	12 41	12 40	12 39	12 37	12 36	12 35	12 34
26	13 01	13 05	13 07	13 10	13 12	13 13	13 16	13 19	13 22	13 24	13 27	13 30	13 32	13 34
27	13 14	13 23	13 29	13 35	13 40	13 44	13 52	13 58	14 05	14 11	14 18	14 25	14 30	14 35
28	13 30	13 43	13 54	14 03	14 11	14 18	14 30	14 41	14 50	15 00	15 11	15 23	15 30	15 38
29	13 49	14 09	14 24	14 36	14 47	14 56	15 12	15 26	15 40	15 53	16 07	16 24	16 33	16 44
30	14 16	14 41	15 00	15 15	15 28	15 40	16 00	16 17	16 33	16 50	17 07	17 27	17 39	17 52
31	14 53	15 22	15 45	16 03	16 18	16 31	16 53	17 13	17 31	17 49	18 09	18 31	18 44	19 00
June 1	15 46	16 17	16 41	17 00	17 15	17 29	17 53	18 13	18 32	18 50	19 11	19 34	19 47	20 03
2	16 55	17 25	17 47	18 05	18 20	18 34	18 56	19 15	19 33	19 51	20 10	20 32	20 45	20 59
3	18 17	18 42	19 01	19 17	19 30	19 42	20 01	20 18	20 33	20 49	21 05	21 24	21 35	21 48
4	19 46	20 05	20 19	20 31	20 41	20 50	21 05	21 18	21 31	21 43	21 56	22 10	22 19	22 28
5	21 15	21 27	21 37	21 45	21 52	21 57	22 08	22 17	22 25	22 33	22 42	22 51	22 57	23 03
6	22 43	22 48	22 53	22 57	23 00	23 03	23 08	23 12	23 16	23 20	23 24	23 29	23 31	23 34
7	.. ..	.. ..	.. ..	.. ..	.. ..	.. ..	.. ..	.. ..	.. ..	.. ..	.. ..	.. ..	.. ..	.. ..
8	0 08	0 08	0 07	0 07	0 07	0 06	0 06	0 06	0 05	0 05	0 05	0 04	0 04	0 03
9	1 32	1 25	1 20	1 16	1 12	1 09	1 03	0 58	0 54	0 49	0 44	0 39	0 36	0 32
10	2 55	2 42	2 32	2 24	2 17	2 11	2 00	1 51	1 43	1 34	1 25	1 14	1 09	1 02

(.. ..) indicates phenomenon will occur the next day.

UNIVERSAL TIME FOR MERIDIAN OF GREENWICH

MOONRISE

Lat.	+40°	+42°	+44°	+46°	+48°	+50°	+52°	+54°	+56°	+58°	+60°	+62°	+64°	+66°
	h m	h m	h m	h m	h m	h m	h m	h m	h m	h m	h m	h m	h m	h m
May 17	20 23	20 30	20 37	20 45	20 55	21 05	21 16	21 28	21 43	22 00	22 21	22 48	23 26	-- --
18	21 16	21 23	21 31	21 39	21 48	21 58	22 09	22 22	22 36	22 53	23 14	23 40	.. ..	-- --
19	22 02	22 09	22 16	22 23	22 32	22 41	22 51	23 02	23 15	23 30	23 48	.. ..	0 17	1 34
20	22 41	22 47	22 53	23 00	23 07	23 15	23 23	23 33	23 44	23 56	.. ..	0 10	0 39	1 21
21	23 15	23 19	23 24	23 30	23 35	23 42	23 48	23 56	.. ..	.. ..	0 10	0 27	0 48	1 15
22	23 44	23 47	23 51	23 55	23 59	.. ..	.. ..	.. ..	0 04	0 14	0 25	0 38	0 52	1 11
23	.. ..	.. ..	.. ..	.. ..	.. ..	0 04	0 09	0 14	0 20	0 27	0 35	0 44	0 54	1 06
24	0 10	0 12	0 15	0 17	0 20	0 23	0 26	0 30	0 34	0 38	0 43	0 48	0 55	1 02
25	0 34	0 35	0 37	0 38	0 39	0 40	0 42	0 43	0 45	0 47	0 49	0 52	0 55	0 58
26	0 58	0 58	0 58	0 58	0 58	0 57	0 57	0 57	0 56	0 56	0 55	0 55	0 54	0 54
27	1 23	1 22	1 20	1 19	1 17	1 15	1 13	1 10	1 08	1 05	1 02	0 58	0 54	0 50
28	1 50	1 47	1 44	1 41	1 38	1 34	1 30	1 26	1 21	1 16	1 10	1 03	0 55	0 46
29	2 21	2 16	2 12	2 07	2 02	1 57	1 51	1 45	1 37	1 29	1 20	1 09	0 57	0 42
30	2 56	2 51	2 45	2 39	2 33	2 25	2 18	2 09	1 59	1 48	1 35	1 20	1 01	0 38
31	3 40	3 33	3 26	3 19	3 11	3 02	2 53	2 42	2 29	2 15	1 58	1 37	1 11	0 35
June 1	4 32	4 25	4 17	4 09	4 00	3 51	3 40	3 27	3 13	2 56	2 36	2 11	1 36	0 35
2	5 34	5 27	5 19	5 11	5 02	4 52	4 41	4 28	4 14	3 57	3 37	3 11	2 35	1 22
3	6 43	6 36	6 30	6 22	6 14	6 05	5 55	5 44	5 31	5 17	4 59	4 38	4 10	3 29
4	7 56	7 51	7 45	7 39	7 33	7 26	7 18	7 09	6 59	6 48	6 35	6 19	6 00	5 36
5	9 10	9 06	9 02	8 58	8 53	8 48	8 43	8 37	8 30	8 22	8 13	8 03	7 51	7 37
6	10 23	10 21	10 18	10 16	10 13	10 10	10 07	10 03	10 00	9 55	9 50	9 45	9 38	9 31
7	11 34	11 33	11 32	11 32	11 31	11 30	11 29	11 28	11 27	11 26	11 25	11 23	11 21	11 19
8	12 43	12 44	12 45	12 46	12 47	12 48	12 50	12 51	12 53	12 55	12 57	12 59	13 02	13 05
9	13 52	13 54	13 57	13 59	14 02	14 06	14 09	14 13	14 18	14 22	14 28	14 34	14 42	14 50
10	15 00	15 03	15 08	15 12	15 17	15 22	15 28	15 34	15 41	15 49	15 59	16 09	16 22	16 37

MOONSET

Lat.	+40°	+42°	+44°	+46°	+48°	+50°	+52°	+54°	+56°	+58°	+60°	+62°	+64°	+66°
	h m	h m	h m	h m	h m	h m	h m	h m	h m	h m	h m	h m	h m	h m
May 17	4 56	4 50	4 42	4 35	4 26	4 17	4 06	3 54	3 41	3 25	3 06	2 43	2 11	1 23
18	5 46	5 39	5 32	5 23	5 14	5 04	4 53	4 40	4 25	4 08	3 47	3 20	2 42	-- --
19	6 41	6 34	6 27	6 19	6 10	6 00	5 49	5 36	5 22	5 05	4 45	4 19	3 42	2 25
20	7 39	7 33	7 26	7 19	7 11	7 02	6 52	6 41	6 28	6 13	5 56	5 34	5 06	4 21
21	8 39	8 33	8 28	8 21	8 15	8 07	7 59	7 50	7 39	7 27	7 14	6 57	6 37	6 10
22	9 38	9 34	9 29	9 25	9 19	9 14	9 07	9 00	8 52	8 43	8 33	8 21	8 07	7 50
23	10 37	10 34	10 31	10 28	10 24	10 20	10 15	10 10	10 05	9 59	9 52	9 44	9 35	9 24
24	11 36	11 34	11 32	11 30	11 28	11 26	11 23	11 20	11 17	11 14	11 10	11 05	11 00	10 55
25	12 34	12 34	12 33	12 33	12 32	12 32	12 31	12 30	12 29	12 29	12 28	12 27	12 25	12 24
26	13 34	13 34	13 35	13 36	13 37	13 39	13 40	13 41	13 43	13 45	13 46	13 49	13 51	13 54
27	14 35	14 37	14 39	14 42	14 45	14 48	14 51	14 54	14 58	15 03	15 08	15 13	15 20	15 28
28	15 38	15 42	15 46	15 50	15 54	15 59	16 04	16 10	16 17	16 24	16 33	16 42	16 54	17 07
29	16 44	16 49	16 55	17 00	17 06	17 13	17 21	17 29	17 38	17 49	18 01	18 16	18 33	18 55
30	17 52	17 58	18 05	18 12	18 20	18 28	18 38	18 48	19 00	19 14	19 30	19 50	20 16	20 51
31	19 00	19 06	19 14	19 22	19 31	19 40	19 51	20 03	20 17	20 34	20 54	21 19	21 53	22 55
June 1	20 03	20 10	20 18	20 26	20 35	20 45	20 56	21 09	21 23	21 40	22 01	22 27	23 03	.. ..
2	20 59	21 06	21 13	21 21	21 29	21 38	21 48	22 00	22 13	22 28	22 46	23 08	23 36	0 15
3	21 48	21 53	21 59	22 06	22 12	22 20	22 28	22 38	22 48	23 00	23 14	23 30	23 50	0 18
4	22 28	22 33	22 37	22 42	22 47	22 53	22 59	23 06	23 14	23 22	23 32	23 43	23 56	0 15
5	23 03	23 06	23 09	23 12	23 16	23 19	23 23	23 28	23 33	23 38	23 44	23 51	23 59	0 12
6	23 34	23 36	23 37	23 39	23 40	23 42	23 44	23 46	23 48	23 50	23 53	23 56	.. ..	0 08
7	.. ..	.. ..	.. ..	.. ..	.. ..	.. ..	.. ..	.. ..	.. ..	.. ..	.. ..	.. ..	0 00	0 04
8	0 03	0 03	0 03	0 03	0 03	0 03	0 02	0 02	0 02	0 02	0 01	0 01	0 00	0 00
9	0 32	0 30	0 29	0 27	0 25	0 23	0 21	0 18	0 15	0 12	0 09	0 05	0 01	23 52
10	1 02	0 59	0 56	0 52	0 48	0 45	0 40	0 35	0 30	0 24	0 18	0 10	0 02	23 47

(.. ..) indicates phenomenon will occur the next day.
(-- --) indicates Moon continuously below horizon.
(** **) indicates Moon continuously above horizon.

UNIVERSAL TIME FOR MERIDIAN OF GREENWICH

MOONRISE

Lat.	−55°	−50°	−45°	−40°	−35°	−30°	−20°	−10°	0°	+10°	+20°	+30°	+35°	+40°
	h m	h m	h m	h m	h m	h m	h m	h m	h m	h m	h m	h m	h m	h m
June 8	12 11	12 14	12 17	12 19	12 21	12 23	12 26	12 28	12 31	12 33	12 36	12 39	12 41	12 43
9	12 27	12 35	12 42	12 48	12 53	12 58	13 06	13 13	13 19	13 26	13 33	13 41	13 46	13 52
10	12 44	12 58	13 09	13 19	13 27	13 34	13 47	13 58	14 08	14 19	14 30	14 43	14 51	15 00
11	13 05	13 25	13 40	13 53	14 04	14 14	14 30	14 45	14 59	15 13	15 28	15 45	15 55	16 07
12	13 32	13 57	14 16	14 32	14 45	14 57	15 17	15 34	15 51	16 07	16 25	16 46	16 58	17 12
13	14 07	14 36	14 58	15 16	15 31	15 44	16 06	16 26	16 44	17 02	17 22	17 44	17 58	18 13
14	14 53	15 24	15 47	16 06	16 22	16 35	16 58	17 18	17 37	17 56	18 16	18 39	18 53	19 08
15	15 51	16 20	16 43	17 01	17 16	17 29	17 52	18 11	18 29	18 47	19 07	19 29	19 42	19 57
16	16 56	17 23	17 43	17 59	18 13	18 25	18 45	19 03	19 19	19 36	19 53	20 14	20 25	20 39
17	18 07	18 28	18 45	18 59	19 10	19 20	19 38	19 53	20 07	20 21	20 36	20 53	21 03	21 14
18	19 18	19 35	19 48	19 58	20 07	20 15	20 29	20 41	20 52	21 03	21 15	21 28	21 36	21 45
19	20 30	20 41	20 50	20 57	21 03	21 09	21 19	21 27	21 35	21 43	21 51	22 00	22 06	22 12
20	21 40	21 46	21 51	21 55	21 59	22 02	22 07	22 12	22 16	22 21	22 25	22 30	22 33	22 37
21	22 51	22 52	22 53	22 53	22 54	22 55	22 55	22 56	22 57	22 58	22 59	23 00	23 00	23 01
22	.. ..	23 58	23 55	23 52	23 50	23 48	23 44	23 41	23 38	23 35	23 32	23 29	23 27	23 25
23	0 02	.. ..	.. ..	.. ..	.. ..	.. ..	.. ..	.. ..	.. ..	.. ..	.. ..	.. ..	23 55	23 50
24	1 16	1 06	0 59	0 52	0 47	0 42	0 34	0 27	0 21	0 14	0 07	0 00	.. ..	.. ..
25	2 32	2 17	2 05	1 55	1 47	1 39	1 27	1 16	1 06	0 56	0 45	0 33	0 26	0 18
26	3 50	3 30	3 13	3 00	2 49	2 39	2 22	2 08	1 55	1 41	1 27	1 11	1 01	0 51
27	5 10	4 44	4 23	4 07	3 54	3 42	3 21	3 04	2 47	2 31	2 14	1 54	1 43	1 30
28	6 26	5 55	5 33	5 14	4 59	4 45	4 23	4 03	3 45	3 27	3 07	2 45	2 32	2 17
29	7 31	7 00	6 37	6 18	6 02	5 48	5 25	5 05	4 46	4 27	4 07	3 44	3 30	3 14
30	8 22	7 54	7 32	7 15	7 00	6 47	6 25	6 06	5 48	5 30	5 11	4 49	4 36	4 21
July 1	8 59	8 36	8 18	8 04	7 51	7 40	7 21	7 05	6 50	6 34	6 18	5 58	5 47	5 34
2	9 26	9 10	8 56	8 45	8 36	8 27	8 13	8 00	7 48	7 37	7 24	7 09	7 01	6 51

MOONSET

Lat.	−55°	−50°	−45°	−40°	−35°	−30°	−20°	−10°	0°	+10°	+20°	+30°	+35°	+40°
	h m	h m	h m	h m	h m	h m	h m	h m	h m	h m	h m	h m	h m	h m
June 8	0 08	0 08	0 07	0 07	0 07	0 06	0 06	0 06	0 05	0 05	0 05	0 04	0 04	0 03
9	1 32	1 25	1 20	1 16	1 12	1 09	1 03	0 58	0 54	0 49	0 44	0 39	0 36	0 32
10	2 55	2 42	2 32	2 24	2 17	2 11	2 00	1 51	1 43	1 34	1 25	1 14	1 09	1 02
11	4 16	3 58	3 44	3 32	3 22	3 13	2 58	2 45	2 32	2 20	2 07	1 52	1 44	1 34
12	5 35	5 11	4 53	4 38	4 25	4 14	3 55	3 39	3 24	3 08	2 52	2 34	2 23	2 10
13	6 49	6 20	5 59	5 41	5 27	5 14	4 53	4 34	4 16	3 59	3 40	3 19	3 06	2 52
14	7 52	7 22	6 59	6 40	6 25	6 11	5 48	5 28	5 10	4 51	4 31	4 08	3 55	3 39
15	8 44	8 13	7 51	7 32	7 17	7 04	6 41	6 21	6 02	5 44	5 24	5 01	4 47	4 32
16	9 22	8 55	8 34	8 18	8 03	7 51	7 30	7 11	6 54	6 36	6 18	5 56	5 43	5 29
17	9 51	9 28	9 11	8 56	8 44	8 33	8 14	7 58	7 43	7 27	7 11	6 52	6 41	6 28
18	10 13	9 55	9 41	9 29	9 19	9 10	8 55	8 42	8 29	8 16	8 03	7 47	7 38	7 28
19	10 30	10 17	10 06	9 58	9 50	9 44	9 32	9 22	9 13	9 03	8 53	8 41	8 35	8 27
20	10 44	10 36	10 29	10 24	10 19	10 15	10 07	10 01	9 55	9 49	9 42	9 35	9 30	9 25
21	10 56	10 53	10 50	10 48	10 46	10 44	10 41	10 38	10 36	10 33	10 31	10 27	10 25	10 23
22	11 08	11 10	11 11	11 12	11 13	11 13	11 15	11 16	11 17	11 18	11 19	11 20	11 21	11 22
23	11 21	11 27	11 32	11 36	11 40	11 43	11 49	11 54	11 58	12 03	12 08	12 14	12 17	12 21
24	11 35	11 46	11 55	12 02	12 09	12 15	12 25	12 34	12 42	12 50	12 59	13 09	13 15	13 22
25	11 52	12 08	12 21	12 32	12 42	12 50	13 04	13 17	13 29	13 40	13 53	14 08	14 16	14 26
26	12 14	12 36	12 53	13 08	13 20	13 30	13 48	14 04	14 19	14 34	14 50	15 09	15 20	15 32
27	12 45	13 12	13 33	13 50	14 05	14 17	14 38	14 57	15 14	15 32	15 50	16 12	16 25	16 39
28	13 30	14 00	14 24	14 42	14 58	15 12	15 35	15 55	16 14	16 32	16 52	17 15	17 29	17 45
29	14 31	15 02	15 26	15 44	16 00	16 14	16 37	16 57	17 15	17 34	17 54	18 16	18 30	18 45
30	15 49	16 17	16 38	16 55	17 09	17 22	17 43	18 01	18 18	18 34	18 52	19 13	19 25	19 38
July 1	17 18	17 40	17 57	18 10	18 22	18 32	18 49	19 04	19 18	19 32	19 46	20 03	20 12	20 23
2	18 51	19 06	19 17	19 27	19 35	19 42	19 55	20 05	20 15	20 25	20 36	20 47	20 54	21 02

(.. ..) indicates phenomenon will occur the next day.

UNIVERSAL TIME FOR MERIDIAN OF GREENWICH

MOONRISE

Lat.	+40°	+42°	+44°	+46°	+48°	+50°	+52°	+54°	+56°	+58°	+60°	+62°	+64°	+66°
	h m	h m	h m	h m	h m	h m	h m	h m	h m	h m	h m	h m	h m	h m
June 8	12 43	12 44	12 45	12 46	12 47	12 48	12 50	12 51	12 53	12 55	12 57	12 59	13 02	13 05
9	13 52	13 54	13 57	13 59	14 02	14 06	14 09	14 13	14 18	14 22	14 28	14 34	14 42	14 50
10	15 00	15 03	15 08	15 12	15 17	15 22	15 28	15 34	15 41	15 49	15 59	16 09	16 22	16 37
11	16 07	16 12	16 17	16 23	16 30	16 37	16 45	16 54	17 04	17 15	17 28	17 44	18 03	18 27
12	17 12	17 18	17 25	17 32	17 40	17 49	17 58	18 09	18 22	18 36	18 54	19 15	19 42	20 22
13	18 13	18 20	18 27	18 35	18 44	18 54	19 05	19 17	19 32	19 49	20 09	20 35	21 11	22 24
14	19 08	19 16	19 23	19 31	19 41	19 51	20 02	20 15	20 29	20 47	21 08	21 35	22 13	-- --
15	19 57	20 04	20 11	20 19	20 28	20 37	20 48	21 00	21 13	21 29	21 48	22 12	22 44	23 35
16	20 39	20 45	20 51	20 58	21 06	21 14	21 23	21 34	21 45	21 59	22 15	22 34	22 57	23 29
17	21 14	21 19	21 25	21 30	21 37	21 44	21 51	22 00	22 09	22 19	22 32	22 46	23 03	23 24
18	21 45	21 49	21 53	21 57	22 02	22 08	22 13	22 20	22 27	22 34	22 43	22 53	23 05	23 20
19	22 12	22 15	22 18	22 21	22 24	22 28	22 32	22 36	22 41	22 46	22 52	22 59	23 06	23 15
20	22 37	22 38	22 40	22 42	22 44	22 46	22 48	22 50	22 53	22 55	22 59	23 02	23 06	23 11
21	23 01	23 01	23 01	23 02	23 02	23 02	23 03	23 03	23 04	23 04	23 05	23 05	23 06	23 07
22	23 25	23 24	23 23	23 22	23 21	23 19	23 18	23 16	23 15	23 13	23 11	23 09	23 06	23 03
23	23 50	23 48	23 46	23 43	23 40	23 37	23 34	23 31	23 27	23 23	23 18	23 12	23 06	22 59
24	.. ..	.. ..	.. ..	.. ..	.. ..	23 58	23 53	23 47	23 41	23 34	23 27	23 18	23 07	22 55
25	0 18	0 15	0 11	0 07	0 03	.. ..	.. ..	.. ..	.. ..	23 50	23 39	23 26	23 10	22 52
26	0 51	0 46	0 41	0 35	0 30	0 23	0 16	0 08	0 00	.. ..	23 57	23 39	23 17	22 48
27	1 30	1 24	1 17	1 11	1 03	0 55	0 46	0 36	0 25	0 12	.. ..	.. ..	23 33	22 46
28	2 17	2 10	2 03	1 55	1 47	1 37	1 27	1 15	1 01	0 46	0 27	0 04	.. ..	22 54
29	3 14	3 07	3 00	2 51	2 42	2 32	2 21	2 08	1 54	1 37	1 16	0 50	0 13	.. ..
30	4 21	4 14	4 07	3 59	3 51	3 41	3 30	3 18	3 05	2 49	2 29	2 05	1 33	0 39
July 1	5 34	5 29	5 23	5 16	5 09	5 01	4 52	4 42	4 30	4 18	4 02	3 44	3 21	2 51
2	6 51	6 46	6 42	6 37	6 31	6 25	6 19	6 11	6 03	5 54	5 44	5 31	5 16	4 58

MOONSET

Lat.	+40°	+42°	+44°	+46°	+48°	+50°	+52°	+54°	+56°	+58°	+60°	+62°	+64°	+66°
	h m	h m	h m	h m	h m	h m	h m	h m	h m	h m	h m	h m	h m	h m
June 8	0 03	0 03	0 03	0 03	0 03	0 03	0 02	0 02	0 02	0 02	0 01	0 01	0 00	0 00
9	0 32	0 30	0 29	0 27	0 25	0 23	0 21	0 18	0 15	0 12	0 09	0 05	0 01	23 52
10	1 02	0 59	0 56	0 52	0 48	0 45	0 40	0 35	0 30	0 24	0 18	0 10	0 02	23 47
11	1 34	1 30	1 25	1 20	1 15	1 09	1 03	0 55	0 48	0 39	0 29	0 17	0 04	23 43
12	2 10	2 05	1 59	1 52	1 46	1 38	1 30	1 20	1 10	0 58	0 44	0 28	0 08	23 37
13	2 52	2 45	2 38	2 31	2 23	2 14	2 03	1 52	1 39	1 24	1 07	0 45	0 17	23 26
14	3 39	3 32	3 24	3 16	3 07	2 57	2 46	2 34	2 19	2 02	1 42	1 16	0 39	-- --
15	4 32	4 25	4 17	4 09	4 00	3 50	3 38	3 26	3 11	2 54	2 33	2 06	1 28	-- --
16	5 29	5 22	5 15	5 07	4 59	4 49	4 39	4 27	4 14	3 58	3 39	3 16	2 44	1 53
17	6 28	6 22	6 16	6 09	6 02	5 54	5 45	5 35	5 24	5 11	4 55	4 37	4 14	3 43
18	7 28	7 23	7 18	7 12	7 07	7 00	6 53	6 45	6 36	6 26	6 15	6 01	5 45	5 25
19	8 27	8 23	8 20	8 16	8 11	8 07	8 01	7 56	7 49	7 42	7 34	7 25	7 14	7 01
20	9 25	9 23	9 21	9 18	9 15	9 12	9 09	9 06	9 02	8 57	8 52	8 47	8 40	8 32
21	10 23	10 22	10 21	10 20	10 19	10 18	10 17	10 15	10 13	10 12	10 10	10 07	10 04	10 01
22	11 22	11 22	11 22	11 23	11 23	11 24	11 24	11 25	11 25	11 26	11 27	11 28	11 29	11 30
23	12 21	12 23	12 24	12 26	12 28	12 31	12 33	12 36	12 39	12 42	12 46	12 50	12 55	13 00
24	13 22	13 25	13 28	13 32	13 35	13 40	13 44	13 49	13 54	14 00	14 07	14 15	14 24	14 35
25	14 26	14 30	14 35	14 40	14 45	14 51	14 58	15 05	15 13	15 22	15 32	15 45	15 59	16 17
26	15 32	15 37	15 43	15 50	15 57	16 05	16 13	16 23	16 33	16 46	17 00	17 18	17 39	18 07
27	16 39	16 46	16 53	17 00	17 09	17 18	17 28	17 40	17 53	18 08	18 27	18 50	19 20	20 07
28	17 45	17 52	17 59	18 07	18 16	18 26	18 38	18 50	19 05	19 22	19 42	20 09	20 46	22 04
29	18 45	18 52	18 59	19 07	19 16	19 26	19 37	19 49	20 03	20 19	20 39	21 03	21 36	22 30
30	19 38	19 44	19 51	19 58	20 05	20 14	20 23	20 34	20 45	20 59	21 15	21 33	21 57	22 28
July 1	20 23	20 28	20 33	20 39	20 45	20 51	20 59	21 07	21 15	21 25	21 37	21 50	22 06	22 25
2	21 02	21 05	21 09	21 13	21 17	21 21	21 26	21 32	21 37	21 44	21 52	22 00	22 10	22 21

(.. ..) indicates phenomenon will occur the next day.
(-- --) indicates Moon continuously below horizon.
(** **) indicates Moon continuously above horizon.

MOONRISE AND MOONSET, 1992

UNIVERSAL TIME FOR MERIDIAN OF GREENWICH

MOONRISE

Lat.	−55°	−50°	−45°	−40°	−35°	−30°	−20°	−10°	0°	+10°	+20°	+30°	+35°	+40°
	h m	h m	h m	h m	h m	h m	h m	h m	h m	h m	h m	h m	h m	h m
July 1	8 59	8 36	8 18	8 04	7 51	7 40	7 21	7 05	6 50	6 34	6 18	5 58	5 47	5 34
2	9 26	9 10	8 56	8 45	8 36	8 27	8 13	8 00	7 48	7 37	7 24	7 09	7 01	6 51
3	9 47	9 36	9 28	9 21	9 15	9 09	9 00	8 52	8 44	8 36	8 28	8 19	8 13	8 07
4	10 04	9 59	9 56	9 53	9 50	9 48	9 44	9 40	9 37	9 33	9 30	9 26	9 23	9 21
5	10 19	10 20	10 21	10 22	10 23	10 24	10 25	10 26	10 27	10 28	10 30	10 31	10 32	10 33
6	10 34	10 41	10 47	10 52	10 56	10 59	11 06	11 11	11 17	11 22	11 28	11 35	11 38	11 43
7	10 51	11 04	11 14	11 22	11 29	11 36	11 47	11 57	12 06	12 16	12 26	12 37	12 44	12 52
8	11 11	11 29	11 43	11 55	12 05	12 14	12 30	12 43	12 56	13 09	13 23	13 39	13 48	13 59
9	11 35	11 59	12 17	12 32	12 45	12 56	13 15	13 32	13 47	14 03	14 20	14 40	14 51	15 05
10	12 08	12 35	12 57	13 14	13 29	13 41	14 03	14 22	14 40	14 57	15 16	15 39	15 52	16 07
11	12 50	13 20	13 43	14 02	14 17	14 31	14 54	15 14	15 32	15 51	16 11	16 34	16 48	17 03
12	13 43	14 13	14 36	14 54	15 10	15 23	15 46	16 06	16 24	16 43	17 02	17 25	17 38	17 54
13	14 45	15 13	15 34	15 51	16 05	16 18	16 39	16 57	17 15	17 32	17 50	18 11	18 23	18 37
14	15 54	16 17	16 35	16 50	17 02	17 13	17 32	17 48	18 03	18 18	18 34	18 52	19 03	19 15
15	17 05	17 23	17 38	17 50	18 00	18 08	18 23	18 36	18 49	19 01	19 14	19 29	19 37	19 47
16	18 17	18 30	18 40	18 49	18 56	19 02	19 14	19 23	19 32	19 41	19 51	20 02	20 08	20 15
17	19 28	19 35	19 42	19 47	19 52	19 56	20 02	20 09	20 14	20 20	20 26	20 33	20 36	20 41
18	20 38	20 41	20 43	20 45	20 47	20 48	20 51	20 53	20 55	20 57	20 59	21 02	21 03	21 05
19	21 48	21 46	21 44	21 43	21 42	21 41	21 39	21 37	21 36	21 34	21 33	21 31	21 30	21 29
20	23 00	22 52	22 47	22 42	22 38	22 34	22 28	22 22	22 17	22 12	22 07	22 00	21 57	21 53
21	.. ..	.. ..	23 51	23 42	23 35	23 29	23 18	23 09	23 00	22 52	22 42	22 32	22 26	22 19
22	0 13	0 01	.. ..	.. ..	.. ..	.. ..	.. ..	23 58	23 46	23 34	23 21	23 07	22 59	22 49
23	1 29	1 11	0 56	0 45	0 35	0 26	0 11	.. ..	.. ..	.. ..	.. ..	23 47	23 36	23 24
24	2 46	2 23	2 04	1 49	1 37	1 26	1 07	0 51	0 36	0 21	0 05	.. ..	.. ..	.. ..
25	4 02	3 34	3 12	2 55	2 40	2 27	2 06	1 47	1 30	1 12	0 54	0 33	0 20	0 06

MOONSET

Lat.	−55°	−50°	−45°	−40°	−35°	−30°	−20°	−10°	0°	+10°	+20°	+30°	+35°	+40°
	h m	h m	h m	h m	h m	h m	h m	h m	h m	h m	h m	h m	h m	h m
July 1	17 18	17 40	17 57	18 10	18 22	18 32	18 49	19 04	19 18	19 32	19 46	20 03	20 12	20 23
2	18 51	19 06	19 17	19 27	19 35	19 42	19 55	20 05	20 15	20 25	20 36	20 47	20 54	21 02
3	20 22	20 30	20 37	20 42	20 47	20 51	20 58	21 04	21 09	21 15	21 21	21 27	21 31	21 35
4	21 51	21 53	21 54	21 55	21 56	21 57	21 59	22 00	22 01	22 02	22 03	22 04	22 05	22 06
5	23 17	23 13	23 09	23 06	23 04	23 01	22 58	22 54	22 51	22 48	22 44	22 40	22 38	22 35
6	.. ..	.. ..	.. ..	.. ..	.. ..	.. ..	23 56	23 48	23 40	23 33	23 25	23 16	23 11	23 05
7	0 42	0 31	0 23	0 16	0 10	0 05	.. ..	.. ..	.. ..	.. ..	.. ..	23 53	23 45	23 37
8	2 04	1 48	1 35	1 24	1 15	1 07	0 53	0 41	0 30	0 19	0 07	.. ..	.. ..	.. ..
9	3 24	3 02	2 45	2 31	2 19	2 08	1 51	1 35	1 21	1 06	0 51	0 33	0 23	0 11
10	4 39	4 12	3 51	3 35	3 21	3 08	2 47	2 29	2 12	1 56	1 37	1 17	1 05	0 51
11	5 46	5 15	4 53	4 34	4 19	4 06	3 43	3 23	3 05	2 47	2 27	2 04	1 51	1 36
12	6 40	6 10	5 47	5 28	5 13	4 59	4 36	4 16	3 57	3 39	3 19	2 55	2 42	2 26
13	7 23	6 54	6 33	6 15	6 01	5 48	5 26	5 07	4 49	4 31	4 12	3 49	3 36	3 21
14	7 54	7 30	7 11	6 56	6 43	6 31	6 12	5 54	5 38	5 22	5 05	4 45	4 33	4 19
15	8 18	7 59	7 43	7 30	7 20	7 10	6 53	6 39	6 25	6 12	5 57	5 40	5 30	5 19
16	8 37	8 22	8 10	8 01	7 52	7 45	7 32	7 21	7 10	6 59	6 48	6 35	6 27	6 18
17	8 52	8 42	8 34	8 27	8 22	8 17	8 08	8 00	7 53	7 45	7 37	7 28	7 23	7 17
18	9 05	9 00	8 56	8 52	8 49	8 46	8 42	8 38	8 34	8 30	8 26	8 21	8 18	8 15
19	9 17	9 16	9 16	9 16	9 16	9 15	9 15	9 15	9 14	9 14	9 14	9 13	9 13	9 13
20	9 29	9 33	9 37	9 40	9 42	9 45	9 49	9 52	9 55	9 59	10 02	10 06	10 09	10 11
21	9 42	9 51	9 59	10 05	10 10	10 15	10 23	10 31	10 37	10 44	10 52	11 00	11 05	11 11
22	9 57	10 11	10 23	10 32	10 41	10 48	11 00	11 11	11 22	11 32	11 43	11 56	12 03	12 12
23	10 17	10 36	10 52	11 04	11 15	11 25	11 41	11 56	12 09	12 23	12 38	12 54	13 04	13 15
24	10 43	11 07	11 27	11 42	11 56	12 07	12 27	12 45	13 01	13 17	13 35	13 55	14 07	14 20
25	11 19	11 48	12 11	12 28	12 44	12 57	13 19	13 39	13 57	14 15	14 34	14 57	15 10	15 25

(.. ..) indicates phenomenon will occur the next day.

UNIVERSAL TIME FOR MERIDIAN OF GREENWICH

MOONRISE

Lat.	+40°	+42°	+44°	+46°	+48°	+50°	+52°	+54°	+56°	+58°	+60°	+62°	+64°	+66°
	h m	h m	h m	h m	h m	h m	h m	h m	h m	h m	h m	h m	h m	h m
July 1	5 34	5 29	5 23	5 16	5 09	5 01	4 52	4 42	4 30	4 18	4 02	3 44	3 21	2 51
2	6 51	6 46	6 42	6 37	6 31	6 25	6 19	6 11	6 03	5 54	5 44	5 31	5 16	4 58
3	8 07	8 04	8 01	7 58	7 54	7 51	7 47	7 42	7 37	7 31	7 25	7 18	7 09	6 59
4	9 21	9 20	9 18	9 17	9 16	9 14	9 12	9 11	9 08	9 06	9 04	9 01	8 57	8 53
5	10 33	10 33	10 34	10 34	10 35	10 35	10 36	10 36	10 37	10 38	10 39	10 40	10 41	10 43
6	11 43	11 45	11 47	11 49	11 52	11 54	11 57	12 00	12 04	12 08	12 12	12 17	12 23	12 30
7	12 52	12 55	12 59	13 03	13 07	13 12	13 17	13 22	13 29	13 36	13 44	13 53	14 04	14 17
8	13 59	14 04	14 09	14 15	14 21	14 27	14 35	14 43	14 52	15 02	15 14	15 28	15 45	16 06
9	15 05	15 10	15 17	15 24	15 31	15 40	15 49	15 59	16 11	16 25	16 41	17 00	17 25	17 59
10	16 07	16 13	16 21	16 28	16 37	16 47	16 57	17 09	17 23	17 39	17 59	18 23	18 57	19 55
11	17 03	17 10	17 18	17 26	17 35	17 46	17 57	18 10	18 24	18 42	19 03	19 30	20 09	-- --
12	17 54	18 01	18 08	18 16	18 25	18 35	18 46	18 58	19 12	19 29	19 49	20 14	20 49	21 52
13	18 37	18 44	18 51	18 58	19 06	19 15	19 25	19 36	19 48	20 02	20 19	20 40	21 06	21 44
14	19 15	19 20	19 26	19 32	19 39	19 47	19 55	20 04	20 14	20 26	20 39	20 55	21 14	21 38
15	19 47	19 51	19 56	20 01	20 06	20 12	20 19	20 26	20 33	20 42	20 52	21 04	21 17	21 34
16	20 15	20 18	20 22	20 25	20 29	20 33	20 38	20 43	20 49	20 55	21 02	21 10	21 19	21 30
17	20 41	20 43	20 45	20 47	20 49	20 52	20 55	20 58	21 01	21 05	21 09	21 14	21 19	21 26
18	21 05	21 06	21 06	21 07	21 08	21 09	21 10	21 11	21 13	21 14	21 16	21 17	21 19	21 22
19	21 29	21 28	21 28	21 27	21 26	21 26	21 25	21 24	21 24	21 23	21 22	21 20	21 19	21 18
20	21 53	21 51	21 49	21 48	21 45	21 43	21 41	21 38	21 35	21 32	21 28	21 24	21 19	21 14
21	22 19	22 16	22 13	22 10	22 06	22 02	21 58	21 53	21 48	21 42	21 36	21 28	21 20	21 10
22	22 49	22 45	22 40	22 36	22 30	22 25	22 19	22 12	22 04	21 56	21 46	21 35	21 22	21 06
23	23 24	23 19	23 13	23 07	23 00	22 53	22 44	22 36	22 25	22 14	22 01	21 45	21 26	21 03
24	.. ..	.. ..	23 53	23 45	23 37	23 29	23 19	23 08	22 55	22 41	22 24	22 03	21 37	21 00
25	0 06	0 00	.. ..	.. ..	.. ..	.. ..	.. ..	23 52	23 38	23 22	23 02	22 36	22 02	21 01

MOONSET

Lat.	+40°	+42°	+44°	+46°	+48°	+50°	+52°	+54°	+56°	+58°	+60°	+62°	+64°	+66°
	h m	h m	h m	h m	h m	h m	h m	h m	h m	h m	h m	h m	h m	h m
July 1	20 23	20 28	20 33	20 39	20 45	20 51	20 59	21 07	21 15	21 25	21 37	21 50	22 06	22 25
2	21 02	21 05	21 09	21 13	21 17	21 21	21 26	21 32	21 37	21 44	21 52	22 00	22 10	22 21
3	21 35	21 37	21 39	21 41	21 44	21 46	21 49	21 52	21 55	21 58	22 02	22 07	22 12	22 18
4	22 06	22 06	22 07	22 07	22 07	22 08	22 08	22 09	22 10	22 10	22 11	22 12	22 13	22 14
5	22 35	22 34	22 33	22 32	22 30	22 29	22 27	22 25	22 24	22 21	22 19	22 16	22 13	22 10
6	23 05	23 02	23 00	22 57	22 54	22 50	22 47	22 43	22 38	22 33	22 28	22 21	22 14	22 06
7	23 37	23 33	23 28	23 24	23 19	23 14	23 08	23 02	22 55	22 47	22 38	22 28	22 16	22 01
8	.. ..	.. ..	.. ..	23 55	23 48	23 41	23 33	23 25	23 15	23 04	22 52	22 37	22 19	21 57
9	0 11	0 06	0 01	.. ..	.. ..	.. ..	.. ..	23 54	23 42	23 28	23 11	22 52	22 26	21 52
10	0 51	0 45	0 38	0 31	0 23	0 14	0 05	.. ..	.. ..	.. ..	23 42	23 17	22 43	21 44
11	1 36	1 29	1 21	1 13	1 05	0 55	0 44	0 32	0 18	0 01	.. ..	23 59	23 21	-- --
12	2 26	2 19	2 11	2 03	1 54	1 44	1 33	1 20	1 05	0 48	0 27	.. ..	.. ..	23 25
13	3 21	3 14	3 07	2 59	2 50	2 41	2 30	2 18	2 04	1 47	1 28	1 03	0 28	.. ..
14	4 19	4 13	4 07	4 00	3 52	3 43	3 34	3 23	3 11	2 57	2 40	2 20	1 54	1 18
15	5 19	5 14	5 08	5 02	4 56	4 49	4 41	4 32	4 23	4 12	3 59	3 44	3 25	3 01
16	6 18	6 14	6 10	6 06	6 01	5 55	5 49	5 43	5 36	5 28	5 18	5 07	4 55	4 39
17	7 17	7 14	7 11	7 08	7 05	7 01	6 58	6 53	6 48	6 43	6 37	6 30	6 22	6 12
18	8 15	8 14	8 12	8 11	8 09	8 07	8 05	8 03	8 00	7 57	7 54	7 51	7 47	7 42
19	9 13	9 13	9 13	9 13	9 12	9 12	9 12	9 12	9 12	9 11	9 11	9 11	9 10	9 10
20	10 11	10 12	10 14	10 15	10 16	10 18	10 20	10 21	10 23	10 26	10 28	10 31	10 35	10 38
21	11 11	11 13	11 16	11 19	11 22	11 25	11 29	11 32	11 37	11 42	11 47	11 54	12 01	12 10
22	12 12	12 16	12 20	12 24	12 29	12 34	12 39	12 46	12 53	13 00	13 09	13 19	13 32	13 46
23	13 15	13 20	13 26	13 32	13 38	13 45	13 52	14 01	14 10	14 21	14 34	14 49	15 07	15 30
24	14 20	14 26	14 33	14 40	14 48	14 56	15 06	15 16	15 29	15 43	15 59	16 19	16 45	17 21
25	15 25	15 32	15 39	15 47	15 56	16 06	16 17	16 29	16 43	16 59	17 19	17 44	18 18	19 19

(.. ..) indicates phenomenon will occur the next day.
(-- --) indicates Moon continuously below horizon.
(** **) indicates Moon continuously above horizon.

UNIVERSAL TIME FOR MERIDIAN OF GREENWICH

MOONRISE

Lat.	−55°	−50°	−45°	−40°	−35°	−30°	−20°	−10°	0°	+10°	+20°	+30°	+35°	+40°
	h m	h m	h m	h m	h m	h m	h m	h m	h m	h m	h m	h m	h m	h m
July 24	2 46	2 23	2 04	1 49	1 37	1 26	1 07	0 51	0 36	0 21	0 05	.. ..	.. ..	.. ..
25	4 02	3 34	3 12	2 55	2 40	2 27	2 06	1 47	1 30	1 12	0 54	0 33	0 20	0 06
26	5 12	4 41	4 17	3 59	3 43	3 29	3 06	2 46	2 27	2 09	1 49	1 26	1 12	0 57
27	6 10	5 40	5 17	4 58	4 43	4 29	4 07	3 47	3 28	3 10	2 50	2 27	2 14	1 58
28	6 54	6 28	6 08	5 51	5 37	5 25	5 05	4 47	4 30	4 13	3 55	3 34	3 22	3 08
29	7 25	7 06	6 50	6 37	6 26	6 16	5 59	5 44	5 31	5 17	5 02	4 45	4 35	4 24
30	7 49	7 36	7 25	7 16	7 08	7 01	6 49	6 39	6 29	6 19	6 09	5 56	5 49	5 41
31	8 09	8 01	7 55	7 51	7 46	7 42	7 36	7 30	7 25	7 19	7 13	7 07	7 03	6 58
Aug. 1	8 25	8 24	8 23	8 22	8 21	8 21	8 20	8 19	8 18	8 17	8 16	8 15	8 14	8 14
2	8 41	8 46	8 50	8 53	8 55	8 58	9 02	9 06	9 09	9 13	9 17	9 22	9 24	9 27
3	8 58	9 08	9 17	9 24	9 30	9 35	9 44	9 53	10 01	10 08	10 17	10 27	10 32	10 39
4	9 17	9 33	9 46	9 57	10 06	10 14	10 28	10 40	10 52	11 03	11 16	11 31	11 39	11 49
5	9 40	10 02	10 19	10 33	10 45	10 55	11 13	11 29	11 44	11 58	12 14	12 33	12 44	12 56
6	10 10	10 37	10 57	11 14	11 28	11 40	12 01	12 19	12 36	12 53	13 12	13 33	13 45	14 00
7	10 49	11 19	11 41	11 59	12 15	12 28	12 51	13 10	13 29	13 47	14 07	14 30	14 43	14 59
8	11 39	12 09	12 32	12 50	13 06	13 19	13 42	14 02	14 21	14 39	14 59	15 22	15 36	15 51
9	12 38	13 06	13 28	13 46	14 00	14 13	14 35	14 54	15 11	15 29	15 48	16 09	16 22	16 37
10	13 44	14 09	14 28	14 44	14 57	15 08	15 28	15 44	16 00	16 16	16 33	16 52	17 03	17 16
11	14 54	15 14	15 30	15 43	15 54	16 03	16 19	16 33	16 47	17 00	17 14	17 30	17 39	17 50
12	16 06	16 20	16 32	16 42	16 50	16 57	17 10	17 21	17 31	17 41	17 52	18 04	18 11	18 19
13	17 17	17 26	17 34	17 40	17 46	17 51	17 59	18 06	18 13	18 20	18 27	18 35	18 40	18 45
14	18 27	18 32	18 35	18 38	18 41	18 43	18 47	18 51	18 54	18 58	19 01	19 05	19 07	19 10
15	19 37	19 37	19 37	19 36	19 36	19 36	19 36	19 35	19 35	19 35	19 35	19 34	19 34	19 34
16	20 48	20 43	20 38	20 35	20 32	20 29	20 24	20 20	20 16	20 12	20 08	20 04	20 01	19 58
17	22 00	21 50	21 41	21 34	21 28	21 23	21 14	21 06	20 58	20 51	20 43	20 34	20 29	20 24

MOONSET

Lat.	−55°	−50°	−45°	−40°	−35°	−30°	−20°	−10°	0°	+10°	+20°	+30°	+35°	+40°
	h m	h m	h m	h m	h m	h m	h m	h m	h m	h m	h m	h m	h m	h m
July 24	10 43	11 07	11 27	11 42	11 56	12 07	12 27	12 45	13 01	13 17	13 35	13 55	14 07	14 20
25	11 19	11 48	12 11	12 28	12 44	12 57	13 19	13 39	13 57	14 15	14 34	14 57	15 10	15 25
26	12 11	12 42	13 05	13 24	13 40	13 54	14 17	14 37	14 56	15 15	15 35	15 58	16 12	16 27
27	13 20	13 49	14 12	14 30	14 45	14 58	15 20	15 40	15 58	16 15	16 34	16 56	17 09	17 24
28	14 44	15 09	15 28	15 43	15 56	16 07	16 27	16 43	16 59	17 14	17 31	17 50	18 00	18 13
29	16 16	16 34	16 48	17 00	17 10	17 19	17 34	17 47	17 59	18 11	18 23	18 38	18 46	18 55
30	17 50	18 01	18 11	18 18	18 24	18 30	18 40	18 48	18 56	19 03	19 11	19 21	19 26	19 32
31	19 23	19 28	19 31	19 34	19 37	19 39	19 43	19 47	19 50	19 53	19 57	20 00	20 02	20 05
Aug. 1	20 53	20 52	20 50	20 49	20 48	20 47	20 45	20 44	20 42	20 41	20 40	20 38	20 37	20 36
2	22 22	22 13	22 07	22 02	21 57	21 53	21 46	21 40	21 34	21 28	21 22	21 15	21 11	21 06
3	23 47	23 33	23 22	23 12	23 04	22 57	22 45	22 35	22 25	22 15	22 05	21 53	21 46	21 38
4	.. ..	.. ..	.. ..	.. ..	.. ..	.. ..	23 44	23 30	23 16	23 03	22 49	22 33	22 23	22 13
5	1 10	0 50	0 34	0 21	0 10	0 01	.. ..	.. ..	.. ..	23 53	23 35	23 16	23 04	22 51
6	2 28	2 03	1 43	1 27	1 14	1 02	0 42	0 25	0 09	.. ..	.. ..	.. ..	23 49	23 34
7	3 38	3 09	2 46	2 29	2 14	2 01	1 38	1 19	1 01	0 43	0 24	0 02	.. ..	.. ..
8	4 37	4 06	3 43	3 24	3 09	2 55	2 32	2 12	1 54	1 35	1 15	0 52	0 39	0 23
9	5 23	4 54	4 31	4 13	3 58	3 45	3 23	3 04	2 45	2 27	2 08	1 45	1 32	1 16
10	5 57	5 32	5 12	4 56	4 42	4 30	4 10	3 52	3 35	3 18	3 00	2 40	2 27	2 13
11	6 24	6 02	5 46	5 32	5 21	5 10	4 53	4 37	4 23	4 08	3 53	3 35	3 24	3 12
12	6 44	6 27	6 14	6 04	5 54	5 46	5 32	5 20	5 08	4 56	4 44	4 29	4 21	4 11
13	7 00	6 48	6 39	6 32	6 25	6 19	6 09	6 00	5 51	5 43	5 34	5 23	5 17	5 10
14	7 14	7 07	7 02	6 57	6 53	6 50	6 44	6 38	6 33	6 28	6 23	6 16	6 13	6 08
15	7 26	7 24	7 22	7 21	7 20	7 19	7 17	7 15	7 14	7 12	7 11	7 09	7 08	7 06
16	7 38	7 41	7 43	7 45	7 47	7 48	7 50	7 53	7 55	7 57	7 59	8 01	8 03	8 04
17	7 51	7 58	8 04	8 10	8 14	8 18	8 25	8 31	8 36	8 42	8 48	8 55	8 59	9 03

(.. ..) indicates phenomenon will occur the next day.

UNIVERSAL TIME FOR MERIDIAN OF GREENWICH

MOONRISE

Lat.	+40°	+42°	+44°	+46°	+48°	+50°	+52°	+54°	+56°	+58°	+60°	+62°	+64°	+66°
	h m	h m	h m	h m	h m	h m	h m	h m	h m	h m	h m	h m	h m	h m
July 24	.. ..	.. ..	23 53	23 45	23 37	23 29	23 19	23 08	22 55	22 41	22 24	22 03	21 37	21 00
25	0 06	0 00	.. ..	.. ..	.. ..	.. ..	.. ..	23 52	23 38	23 22	23 02	22 36	22 02	21 01
26	0 57	0 50	0 43	0 34	0 26	0 16	0 05	.. ..	.. ..	.. ..	.. ..	23 36	23 00	21 50
27	1 58	1 51	1 44	1 35	1 26	1 17	1 06	0 53	0 39	0 22	0 02	.. ..	.. ..	23 55
28	3 08	3 02	2 55	2 47	2 39	2 31	2 21	2 10	1 57	1 42	1 25	1 03	0 36	.. ..
29	4 24	4 18	4 13	4 07	4 01	3 54	3 46	3 37	3 27	3 16	3 03	2 48	2 29	2 06
30	5 41	5 38	5 34	5 30	5 25	5 20	5 15	5 09	5 03	4 55	4 47	4 37	4 26	4 12
31	6 58	6 56	6 54	6 52	6 50	6 47	6 44	6 41	6 38	6 34	6 30	6 25	6 19	6 12
Aug. 1	8 14	8 14	8 13	8 13	8 13	8 12	8 12	8 11	8 11	8 10	8 10	8 09	8 08	8 07
2	9 27	9 29	9 30	9 32	9 33	9 35	9 37	9 39	9 41	9 44	9 47	9 50	9 54	9 59
3	10 39	10 42	10 45	10 48	10 52	10 56	11 00	11 05	11 10	11 16	11 22	11 30	11 39	11 49
4	11 49	11 53	11 58	12 03	12 08	12 14	12 20	12 28	12 36	12 45	12 55	13 08	13 22	13 40
5	12 56	13 02	13 08	13 14	13 21	13 29	13 37	13 47	13 58	14 10	14 25	14 42	15 04	15 33
6	14 00	14 06	14 13	14 21	14 29	14 39	14 49	15 00	15 13	15 29	15 47	16 10	16 40	17 27
7	14 59	15 06	15 13	15 21	15 30	15 40	15 51	16 04	16 19	16 36	16 56	17 23	18 00	19 29
8	15 51	15 58	16 06	16 14	16 23	16 33	16 44	16 56	17 11	17 28	17 48	18 14	18 50	20 03
9	16 37	16 43	16 50	16 58	17 06	17 15	17 25	17 37	17 50	18 05	18 23	18 45	19 14	19 56
10	17 16	17 22	17 28	17 34	17 41	17 49	17 58	18 08	18 19	18 31	18 46	19 03	19 24	19 52
11	17 50	17 54	17 59	18 05	18 10	18 17	18 24	18 31	18 40	18 50	19 01	19 14	19 29	19 48
12	18 19	18 23	18 26	18 30	18 35	18 39	18 45	18 50	18 57	19 04	19 12	19 21	19 31	19 44
13	18 45	18 48	18 50	18 53	18 56	18 59	19 02	19 06	19 10	19 15	19 20	19 26	19 32	19 40
14	19 10	19 11	19 12	19 14	19 15	19 17	19 18	19 20	19 22	19 24	19 27	19 29	19 33	19 36
15	19 34	19 34	19 34	19 34	19 34	19 34	19 33	19 33	19 33	19 33	19 33	19 33	19 33	19 33
16	19 58	19 57	19 55	19 54	19 52	19 51	19 49	19 47	19 45	19 42	19 39	19 36	19 33	19 29
17	20 24	20 21	20 18	20 16	20 12	20 09	20 05	20 02	19 57	19 52	19 47	19 41	19 34	19 25

MOONSET

Lat.	+40°	+42°	+44°	+46°	+48°	+50°	+52°	+54°	+56°	+58°	+60°	+62°	+64°	+66°
	h m	h m	h m	h m	h m	h m	h m	h m	h m	h m	h m	h m	h m	h m
July 24	14 20	14 26	14 33	14 40	14 48	14 56	15 06	15 16	15 29	15 43	15 59	16 19	16 45	17 21
25	15 25	15 32	15 39	15 47	15 56	16 06	16 17	16 29	16 43	16 59	17 19	17 44	18 18	19 19
26	16 27	16 34	16 42	16 50	16 59	17 09	17 20	17 33	17 47	18 04	18 24	18 50	19 26	20 36
27	17 24	17 30	17 37	17 45	17 53	18 02	18 12	18 24	18 37	18 52	19 10	19 32	20 00	20 41
28	18 13	18 18	18 24	18 31	18 38	18 45	18 53	19 03	19 13	19 25	19 38	19 55	20 14	20 39
29	18 55	18 59	19 04	19 08	19 14	19 19	19 25	19 32	19 39	19 48	19 57	20 08	20 20	20 35
30	19 32	19 34	19 37	19 40	19 43	19 47	19 51	19 55	19 59	20 04	20 10	20 16	20 23	20 32
31	20 05	20 06	20 07	20 08	20 09	20 11	20 12	20 14	20 16	20 18	20 20	20 22	20 25	20 28
Aug. 1	20 36	20 35	20 35	20 34	20 33	20 33	20 32	20 31	20 30	20 30	20 28	20 27	20 26	20 24
2	21 06	21 04	21 02	21 00	20 57	20 55	20 52	20 49	20 45	20 42	20 37	20 32	20 27	20 21
3	21 38	21 35	21 31	21 27	21 23	21 18	21 13	21 08	21 02	20 55	20 47	20 39	20 29	20 17
4	22 13	22 08	22 03	21 57	21 51	21 45	21 38	21 30	21 21	21 11	21 00	20 47	20 32	20 13
5	22 51	22 45	22 39	22 32	22 25	22 17	22 08	21 58	21 46	21 33	21 18	21 00	20 38	20 08
6	23 34	23 28	23 21	23 13	23 04	22 55	22 44	22 33	22 19	22 04	21 45	21 22	20 51	20 04
7	.. ..	.. ..	.. ..	.. ..	23 51	23 41	23 30	23 17	23 03	22 46	22 25	21 58	21 21	19 52
8	0 23	0 16	0 08	0 00	.. ..	.. ..	.. ..	.. ..	23 58	23 41	23 21	22 55	22 19	21 06
9	1 16	1 09	1 02	0 54	0 45	0 35	0 24	0 12	.. ..	.. ..	.. ..	.. ..	23 39	22 57
10	2 13	2 07	2 00	1 53	1 45	1 36	1 26	1 15	1 02	0 47	0 29	0 08	.. ..	.. ..
11	3 12	3 07	3 01	2 54	2 48	2 40	2 32	2 22	2 12	2 00	1 46	1 29	1 08	0 42
12	4 11	4 07	4 02	3 57	3 52	3 46	3 40	3 32	3 24	3 15	3 05	2 53	2 38	2 20
13	5 10	5 07	5 04	5 00	4 56	4 52	4 48	4 42	4 37	4 31	4 23	4 15	4 06	3 54
14	6 08	6 07	6 05	6 03	6 00	5 58	5 55	5 52	5 49	5 45	5 41	5 36	5 31	5 25
15	7 06	7 06	7 05	7 05	7 04	7 03	7 02	7 01	7 00	6 59	6 58	6 57	6 55	6 53
16	8 04	8 05	8 06	8 07	8 08	8 08	8 10	8 11	8 12	8 13	8 15	8 17	8 19	8 21
17	9 03	9 05	9 07	9 10	9 12	9 15	9 18	9 21	9 24	9 28	9 33	9 38	9 44	9 51

(.. ..) indicates phenomenon will occur the next day.
(-- --) indicates Moon continuously below horizon.
(** **) indicates Moon continuously above horizon.

UNIVERSAL TIME FOR MERIDIAN OF GREENWICH

MOONRISE

Lat.	−55°	−50°	−45°	−40°	−35°	−30°	−20°	−10°	0°	+10°	+20°	+30°	+35°	+40°
	h m	h m	h m	h m	h m	h m	h m	h m	h m	h m	h m	h m	h m	h m
Aug. 16	20 48	20 43	20 38	20 35	20 32	20 29	20 24	20 20	20 16	20 12	20 08	20 04	20 01	19 58
17	22 00	21 50	21 41	21 34	21 28	21 23	21 14	21 06	20 58	20 51	20 43	20 34	20 29	20 24
18	23 14	22 58	22 45	22 35	22 26	22 18	22 05	21 53	21 43	21 32	21 21	21 08	21 00	20 52
19	.. ..	.. ..	23 51	23 38	23 26	23 16	22 59	22 44	22 30	22 16	22 01	21 45	21 35	21 24
20	0 29	0 08	.. ..	.. ..	.. ..	.. ..	23 55	23 37	23 21	23 04	22 47	22 27	22 15	22 02
21	1 44	1 18	0 57	0 41	0 27	0 15	.. ..	.. ..	.. ..	23 57	23 38	23 16	23 03	22 48
22	2 55	2 25	2 02	1 44	1 28	1 15	0 53	0 33	0 15	.. ..	.. ..	.. ..	23 58	23 42
23	3 56	3 25	3 02	2 43	2 28	2 14	1 51	1 31	1 13	0 54	0 34	0 11	.. ..	.. ..
24	4 45	4 17	3 55	3 38	3 23	3 11	2 49	2 30	2 12	1 54	1 36	1 14	1 01	0 46
25	5 22	4 59	4 41	4 26	4 14	4 03	3 44	3 27	3 12	2 57	2 40	2 21	2 10	1 57
26	5 49	5 32	5 19	5 08	4 58	4 50	4 35	4 23	4 11	3 59	3 46	3 31	3 22	3 13
27	6 11	6 00	5 52	5 45	5 39	5 33	5 24	5 15	5 07	5 00	4 51	4 42	4 36	4 30
28	6 29	6 25	6 21	6 18	6 16	6 13	6 09	6 06	6 02	5 59	5 56	5 52	5 49	5 47
29	6 46	6 48	6 49	6 50	6 51	6 52	6 53	6 55	6 56	6 57	6 59	7 00	7 01	7 03
30	7 03	7 10	7 17	7 22	7 26	7 30	7 37	7 43	7 49	7 55	8 01	8 08	8 12	8 17
31	7 22	7 35	7 46	7 55	8 03	8 09	8 21	8 32	8 42	8 52	9 02	9 15	9 22	9 30
Sept. 1	7 44	8 03	8 18	8 31	8 42	8 51	9 07	9 22	9 35	9 49	10 03	10 20	10 30	10 41
2	8 12	8 37	8 56	9 11	9 24	9 36	9 55	10 13	10 29	10 45	11 03	11 23	11 35	11 48
3	8 49	9 17	9 39	9 56	10 11	10 24	10 46	11 05	11 23	11 41	12 00	12 22	12 35	12 50
4	9 35	10 05	10 28	10 46	11 02	11 15	11 38	11 57	12 16	12 34	12 54	13 17	13 30	13 46
5	10 32	11 01	11 23	11 41	11 56	12 09	12 31	12 50	13 08	13 25	13 45	14 07	14 19	14 34
6	11 36	12 02	12 22	12 38	12 51	13 03	13 23	13 41	13 57	14 13	14 31	14 51	15 02	15 16
7	12 45	13 06	13 23	13 36	13 48	13 58	14 15	14 30	14 44	14 58	15 13	15 30	15 40	15 51
8	13 55	14 12	14 25	14 35	14 44	14 52	15 06	15 18	15 29	15 40	15 52	16 05	16 13	16 22
9	15 06	15 17	15 26	15 34	15 40	15 46	15 55	16 04	16 12	16 20	16 28	16 38	16 43	16 49

MOONSET

Lat.	−55°	−50°	−45°	−40°	−35°	−30°	−20°	−10°	0°	+10°	+20°	+30°	+35°	+40°
	h m	h m	h m	h m	h m	h m	h m	h m	h m	h m	h m	h m	h m	h m
Aug. 16	7 38	7 41	7 43	7 45	7 47	7 48	7 50	7 53	7 55	7 57	7 59	8 01	8 03	8 04
17	7 51	7 58	8 04	8 10	8 14	8 18	8 25	8 31	8 36	8 42	8 48	8 55	8 59	9 03
18	8 05	8 18	8 28	8 36	8 43	8 50	9 01	9 10	9 19	9 28	9 38	9 49	9 56	10 03
19	8 23	8 41	8 54	9 06	9 16	9 25	9 39	9 53	10 05	10 17	10 31	10 46	10 55	11 05
20	8 46	9 09	9 26	9 41	9 53	10 04	10 22	10 39	10 54	11 09	11 25	11 44	11 55	12 08
21	9 17	9 44	10 05	10 22	10 37	10 49	11 10	11 29	11 46	12 04	12 22	12 44	12 56	13 11
22	10 00	10 31	10 54	11 12	11 28	11 41	12 04	12 24	12 42	13 01	13 21	13 43	13 57	14 12
23	10 59	11 30	11 53	12 11	12 27	12 40	13 03	13 23	13 41	13 59	14 19	14 41	14 54	15 10
24	12 14	12 41	13 02	13 19	13 33	13 45	14 06	14 24	14 41	14 57	15 15	15 35	15 47	16 01
25	13 40	14 02	14 19	14 32	14 44	14 54	15 11	15 26	15 40	15 54	16 08	16 25	16 35	16 46
26	15 13	15 28	15 39	15 49	15 57	16 05	16 17	16 28	16 38	16 48	16 58	17 10	17 17	17 25
27	16 46	16 54	17 01	17 06	17 11	17 15	17 22	17 28	17 34	17 39	17 45	17 52	17 56	18 00
28	18 19	18 21	18 22	18 23	18 24	18 24	18 26	18 27	18 28	18 29	18 30	18 31	18 32	18 32
29	19 50	19 45	19 41	19 38	19 35	19 33	19 28	19 25	19 21	19 17	19 14	19 09	19 07	19 04
30	21 20	21 09	21 00	20 52	20 46	20 40	20 30	20 22	20 14	20 06	19 58	19 48	19 42	19 36
31	22 47	22 29	22 16	22 04	21 55	21 46	21 32	21 19	21 07	20 55	20 43	20 28	20 20	20 11
Sept. 1	.. ..	23 46	23 28	23 14	23 01	22 50	22 32	22 16	22 01	21 46	21 30	21 12	21 01	20 49
2	0 10	.. ..	.. ..	.. ..	.. ..	23 52	23 30	23 12	22 55	22 38	22 19	21 58	21 46	21 31
3	1 25	0 57	0 36	0 19	0 04	.. ..	.. ..	.. ..	23 48	23 30	23 10	22 48	22 35	22 19
4	2 29	1 59	1 36	1 18	1 02	0 49	0 26	0 07	.. ..	.. ..	.. ..	23 40	23 27	23 12
5	3 19	2 50	2 28	2 10	1 54	1 41	1 19	0 59	0 41	0 23	0 03	.. ..	.. ..	.. ..
6	3 58	3 32	3 11	2 54	2 40	2 28	2 07	1 49	1 32	1 15	0 56	0 35	0 22	0 08
7	4 27	4 05	3 47	3 33	3 21	3 10	2 51	2 35	2 20	2 05	1 49	1 30	1 19	1 06
8	4 49	4 32	4 18	4 06	3 56	3 47	3 32	3 19	3 06	2 53	2 40	2 24	2 15	2 05
9	5 07	4 54	4 44	4 35	4 27	4 21	4 09	3 59	3 50	3 40	3 30	3 18	3 11	3 04

(.. ..) indicates phenomenon will occur the next day.

UNIVERSAL TIME FOR MERIDIAN OF GREENWICH

MOONRISE

Lat.	+40°	+42°	+44°	+46°	+48°	+50°	+52°	+54°	+56°	+58°	+60°	+62°	+64°	+66°
	h m	h m	h m	h m	h m	h m	h m	h m	h m	h m	h m	h m	h m	h m
Aug. 16	19 58	19 57	19 55	19 54	19 52	19 51	19 49	19 47	19 45	19 42	19 39	19 36	19 33	19 29
17	20 24	20 21	20 18	20 16	20 12	20 09	20 05	20 02	19 57	19 52	19 47	19 41	19 34	19 25
18	20 52	20 48	20 44	20 40	20 35	20 30	20 25	20 19	20 12	20 05	19 56	19 47	19 35	19 22
19	21 24	21 19	21 14	21 08	21 02	20 55	20 48	20 40	20 31	20 21	20 09	19 55	19 39	19 19
20	22 02	21 56	21 50	21 43	21 35	21 27	21 18	21 08	20 57	20 43	20 28	20 10	19 47	19 17
21	22 48	22 41	22 34	22 26	22 18	22 08	21 58	21 46	21 32	21 17	20 58	20 35	20 05	19 18
22	23 42	23 35	23 28	23 20	23 11	23 01	22 50	22 37	22 23	22 06	21 46	21 20	20 45	19 38
23	.. ..	.. ..	.. ..	.. ..	.. ..	.. ..	23 56	23 44	23 31	23 15	22 56	22 33	22 01	21 11
24	0 46	0 39	0 32	0 24	0 16	0 07	.. ..	.. ..	.. ..	.. ..	.. ..	.. ..	23 44	23 14
25	1 57	1 51	1 45	1 39	1 31	1 23	1 14	1 05	0 53	0 40	0 25	0 07	.. ..	.. ..
26	3 13	3 08	3 04	2 58	2 53	2 47	2 40	2 33	2 25	2 16	2 05	1 52	1 38	1 19
27	4 30	4 27	4 24	4 21	4 17	4 13	4 09	4 05	4 00	3 54	3 48	3 40	3 32	3 21
28	5 47	5 46	5 44	5 43	5 42	5 40	5 38	5 36	5 34	5 32	5 30	5 27	5 23	5 20
29	7 03	7 03	7 04	7 04	7 05	7 06	7 06	7 07	7 08	7 09	7 10	7 11	7 13	7 15
30	8 17	8 19	8 22	8 24	8 27	8 30	8 33	8 36	8 40	8 44	8 49	8 55	9 01	9 08
31	9 30	9 34	9 38	9 42	9 47	9 52	9 57	10 03	10 10	10 17	10 26	10 36	10 48	11 02
Sept. 1	10 41	10 46	10 51	10 57	11 04	11 11	11 18	11 27	11 37	11 48	12 00	12 15	12 34	12 57
2	11 48	11 54	12 01	12 08	12 16	12 25	12 34	12 45	12 57	13 11	13 28	13 48	14 15	14 52
3	12 50	12 57	13 05	13 13	13 21	13 31	13 42	13 54	14 08	14 24	14 44	15 09	15 43	16 43
4	13 46	13 53	14 00	14 09	14 18	14 27	14 38	14 51	15 05	15 22	15 43	16 08	16 45	17 57
5	14 34	14 41	14 48	14 56	15 04	15 14	15 24	15 36	15 49	16 05	16 23	16 46	17 17	18 04
6	15 16	15 22	15 28	15 35	15 42	15 51	16 00	16 10	16 21	16 35	16 50	17 09	17 32	18 02
7	15 51	15 56	16 01	16 07	16 13	16 20	16 28	16 36	16 45	16 56	17 08	17 22	17 39	18 00
8	16 22	16 26	16 30	16 34	16 39	16 44	16 50	16 57	17 04	17 11	17 20	17 30	17 42	17 57
9	16 49	16 52	16 55	16 58	17 01	17 05	17 09	17 13	17 18	17 24	17 30	17 36	17 44	17 53

MOONSET

Lat.	+40°	+42°	+44°	+46°	+48°	+50°	+52°	+54°	+56°	+58°	+60°	+62°	+64°	+66°
	h m	h m	h m	h m	h m	h m	h m	h m	h m	h m	h m	h m	h m	h m
Aug. 16	8 04	8 05	8 06	8 07	8 08	8 08	8 10	8 11	8 12	8 13	8 15	8 17	8 19	8 21
17	9 03	9 05	9 07	9 10	9 12	9 15	9 18	9 21	9 24	9 28	9 33	9 38	9 44	9 51
18	10 03	10 06	10 10	10 14	10 18	10 22	10 27	10 32	10 38	10 45	10 53	11 01	11 12	11 24
19	11 05	11 09	11 14	11 19	11 25	11 31	11 38	11 46	11 54	12 04	12 15	12 28	12 43	13 02
20	12 08	12 13	12 19	12 26	12 33	12 41	12 50	12 59	13 10	13 23	13 38	13 56	14 18	14 47
21	13 11	13 17	13 24	13 32	13 40	13 50	14 00	14 11	14 25	14 40	14 58	15 21	15 51	16 38
22	14 12	14 19	14 27	14 35	14 44	14 54	15 05	15 17	15 31	15 48	16 08	16 34	17 09	18 17
23	15 10	15 16	15 24	15 32	15 40	15 50	16 00	16 12	16 26	16 42	17 01	17 25	17 57	18 47
24	16 01	16 07	16 13	16 20	16 28	16 36	16 46	16 56	17 08	17 21	17 37	17 55	18 19	18 50
25	16 46	16 51	16 56	17 01	17 07	17 14	17 21	17 29	17 38	17 48	18 00	18 13	18 29	18 48
26	17 25	17 28	17 32	17 36	17 40	17 45	17 49	17 55	18 01	18 08	18 15	18 24	18 34	18 45
27	18 00	18 02	18 04	18 06	18 08	18 11	18 13	18 16	18 19	18 23	18 27	18 31	18 36	18 42
28	18 32	18 33	18 33	18 33	18 34	18 34	18 34	18 35	18 35	18 36	18 37	18 37	18 38	18 39
29	19 04	19 03	19 01	19 00	18 58	18 57	18 55	18 53	18 51	18 48	18 46	18 43	18 39	18 36
30	19 36	19 33	19 31	19 27	19 24	19 20	19 16	19 12	19 07	19 02	18 56	18 49	18 41	18 32
31	20 11	20 07	20 02	19 57	19 52	19 47	19 40	19 34	19 26	19 18	19 08	18 57	18 44	18 29
Sept. 1	20 49	20 43	20 38	20 31	20 25	20 17	20 09	20 00	19 50	19 38	19 25	19 09	18 50	18 26
2	21 31	21 25	21 18	21 11	21 03	20 54	20 44	20 33	20 21	20 06	19 49	19 28	19 01	18 24
3	22 19	22 12	22 05	21 57	21 48	21 38	21 27	21 15	21 01	20 45	20 25	20 00	19 25	18 24
4	23 12	23 05	22 57	22 49	22 40	22 30	22 19	22 07	21 53	21 36	21 16	20 50	20 14	19 01
5	.. ..	.. ..	23 54	23 47	23 38	23 29	23 19	23 07	22 54	22 39	22 21	21 58	21 28	20 41
6	0 08	0 01	.. ..	.. ..	.. ..	.. ..	.. ..	.. ..	.. ..	23 50	23 35	23 17	22 54	22 24
7	1 06	1 00	0 54	0 47	0 40	0 32	0 24	0 14	0 03	.. ..	.. ..	.. ..	.. ..	.. ..
8	2 05	2 00	1 55	1 50	1 44	1 38	1 31	1 23	1 14	1 04	0 52	0 39	0 23	0 03
9	3 04	3 00	2 56	2 52	2 48	2 43	2 38	2 32	2 26	2 19	2 11	2 01	1 50	1 37

(.. ..) indicates phenomenon will occur the next day.
(-- --) indicates Moon continuously below horizon.
(** **) indicates Moon continuously above horizon.

MOONRISE AND MOONSET, 1992

UNIVERSAL TIME FOR MERIDIAN OF GREENWICH

MOONRISE

Lat.	−55°	−50°	−45°	−40°	−35°	−30°	−20°	−10°	0°	+10°	+20°	+30°	+35°	+40°
	h m	h m	h m	h m	h m	h m	h m	h m	h m	h m	h m	h m	h m	h m
Sept. 8	13 55	14 12	14 25	14 35	14 44	14 52	15 06	15 18	15 29	15 40	15 52	16 05	16 13	16 22
9	15 06	15 17	15 26	15 34	15 40	15 46	15 55	16 04	16 12	16 20	16 28	16 38	16 43	16 49
10	16 17	16 23	16 28	16 32	16 36	16 39	16 44	16 49	16 53	16 58	17 03	17 08	17 11	17 15
11	17 27	17 28	17 29	17 30	17 31	17 31	17 32	17 33	17 34	17 35	17 36	17 38	17 38	17 39
12	18 38	18 34	18 31	18 28	18 26	18 24	18 21	18 18	18 16	18 13	18 10	18 07	18 05	18 03
13	19 50	19 41	19 34	19 28	19 23	19 18	19 11	19 04	18 58	18 52	18 45	18 38	18 33	18 29
14	21 03	20 49	20 38	20 28	20 20	20 13	20 02	19 51	19 42	19 32	19 22	19 10	19 04	18 56
15	22 17	21 58	21 43	21 30	21 20	21 10	20 54	20 41	20 28	20 15	20 02	19 46	19 37	19 27
16	23 32	23 07	22 48	22 33	22 20	22 09	21 49	21 33	21 17	21 02	20 45	20 26	20 16	20 03
17	.. ..	.. ..	23 52	23 35	23 20	23 07	22 46	22 27	22 10	21 52	21 34	21 12	21 00	20 46
18	0 43	0 14	.. ..	.. ..	.. ..	.. ..	23 43	23 23	23 05	22 46	22 27	22 04	21 51	21 36
19	1 46	1 15	0 53	0 34	0 19	0 05	.. ..	.. ..	.. ..	23 44	23 25	23 02	22 49	22 34
20	2 38	2 09	1 47	1 29	1 14	1 01	0 39	0 20	0 02	.. ..	.. ..	.. ..	23 54	23 40
21	3 18	2 53	2 34	2 18	2 05	1 53	1 33	1 16	0 59	0 43	0 26	0 05	.. ..	.. ..
22	3 48	3 29	3 13	3 01	2 50	2 41	2 24	2 10	1 56	1 43	1 28	1 12	1 02	0 51
23	4 12	3 58	3 48	3 39	3 31	3 24	3 12	3 02	2 52	2 42	2 32	2 20	2 13	2 05
24	4 31	4 24	4 18	4 13	4 09	4 05	3 58	3 52	3 47	3 41	3 35	3 28	3 24	3 20
25	4 49	4 47	4 46	4 45	4 44	4 44	4 42	4 41	4 40	4 39	4 38	4 37	4 36	4 35
26	5 06	5 10	5 14	5 17	5 20	5 22	5 26	5 30	5 33	5 37	5 40	5 45	5 47	5 50
27	5 24	5 35	5 43	5 50	5 56	6 01	6 11	6 19	6 27	6 34	6 43	6 53	6 58	7 05
28	5 46	6 02	6 15	6 25	6 35	6 43	6 57	7 09	7 21	7 33	7 45	8 00	8 08	8 18
29	6 12	6 34	6 51	7 05	7 17	7 27	7 45	8 01	8 16	8 31	8 47	9 06	9 16	9 29
30	6 46	7 12	7 33	7 49	8 03	8 15	8 36	8 54	9 12	9 29	9 47	10 08	10 21	10 35
Oct. 1	7 29	7 59	8 21	8 39	8 54	9 07	9 29	9 48	10 07	10 25	10 44	11 07	11 20	11 35
2	8 23	8 53	9 15	9 33	9 48	10 01	10 23	10 42	11 00	11 18	11 37	12 00	12 13	12 28

MOONSET

Lat.	−55°	−50°	−45°	−40°	−35°	−30°	−20°	−10°	0°	+10°	+20°	+30°	+35°	+40°
	h m	h m	h m	h m	h m	h m	h m	h m	h m	h m	h m	h m	h m	h m
Sept. 8	4 49	4 32	4 18	4 06	3 56	3 47	3 32	3 19	3 06	2 53	2 40	2 24	2 15	2 05
9	5 07	4 54	4 44	4 35	4 27	4 21	4 09	3 59	3 50	3 40	3 30	3 18	3 11	3 04
10	5 22	5 13	5 07	5 01	4 56	4 52	4 45	4 38	4 32	4 26	4 19	4 11	4 07	4 02
11	5 35	5 31	5 28	5 26	5 24	5 22	5 19	5 16	5 13	5 10	5 08	5 04	5 02	5 00
12	5 47	5 48	5 49	5 50	5 51	5 51	5 52	5 53	5 54	5 55	5 56	5 57	5 57	5 58
13	6 00	6 06	6 11	6 15	6 18	6 21	6 27	6 31	6 36	6 40	6 45	6 50	6 53	6 57
14	6 14	6 25	6 34	6 41	6 47	6 53	7 02	7 11	7 19	7 26	7 35	7 45	7 50	7 57
15	6 31	6 47	7 00	7 10	7 19	7 27	7 40	7 52	8 03	8 15	8 27	8 41	8 49	8 58
16	6 53	7 13	7 30	7 43	7 55	8 05	8 22	8 37	8 51	9 05	9 21	9 38	9 48	10 00
17	7 21	7 46	8 06	8 22	8 36	8 47	9 08	9 25	9 42	9 58	10 16	10 37	10 49	11 02
18	7 59	8 28	8 50	9 08	9 23	9 36	9 58	10 18	10 36	10 54	11 13	11 35	11 48	12 03
19	8 51	9 21	9 44	10 02	10 18	10 31	10 54	11 13	11 32	11 50	12 10	12 32	12 45	13 01
20	9 57	10 26	10 47	11 05	11 19	11 32	11 53	12 12	12 29	12 46	13 05	13 26	13 38	13 52
21	11 16	11 40	11 58	12 13	12 26	12 37	12 55	13 12	13 27	13 42	13 58	14 16	14 26	14 38
22	12 43	13 00	13 14	13 26	13 36	13 44	13 59	14 11	14 23	14 35	14 47	15 01	15 09	15 18
23	14 13	14 24	14 33	14 40	14 47	14 52	15 02	15 10	15 18	15 26	15 34	15 43	15 48	15 54
24	15 44	15 48	15 52	15 56	15 58	16 01	16 05	16 09	16 12	16 15	16 19	16 23	16 25	16 27
25	17 15	17 13	17 12	17 11	17 10	17 09	17 08	17 06	17 05	17 04	17 03	17 01	17 00	16 59
26	18 45	18 37	18 31	18 26	18 21	18 17	18 10	18 04	17 58	17 53	17 47	17 40	17 36	17 31
27	20 15	20 00	19 49	19 40	19 32	19 25	19 13	19 02	18 52	18 42	18 32	18 20	18 13	18 05
28	21 42	21 21	21 05	20 52	20 41	20 31	20 15	20 00	19 47	19 34	19 19	19 03	18 54	18 43
29	23 02	22 37	22 17	22 01	21 48	21 36	21 16	20 59	20 43	20 26	20 09	19 49	19 38	19 25
30	.. ..	23 44	23 22	23 05	22 50	22 37	22 15	21 56	21 38	21 20	21 01	20 39	20 26	20 12
Oct. 1	0 13	.. ..	.. ..	.. ..	23 46	23 33	23 10	22 51	22 33	22 14	21 55	21 32	21 19	21 04
2	1 11	0 41	0 19	0 01	.. ..	.. ..	.. ..	23 43	23 25	23 08	22 49	22 27	22 14	22 00

(.. ..) indicates phenomenon will occur the next day.

UNIVERSAL TIME FOR MERIDIAN OF GREENWICH

MOONRISE

Lat.	+40°	+42°	+44°	+46°	+48°	+50°	+52°	+54°	+56°	+58°	+60°	+62°	+64°	+66°
	h m	h m	h m	h m	h m	h m	h m	h m	h m	h m	h m	h m	h m	h m
Sept. 8	16 22	16 26	16 30	16 34	16 39	16 44	16 50	16 57	17 04	17 11	17 20	17 30	17 42	17 57
9	16 49	16 52	16 55	16 58	17 01	17 05	17 09	17 13	17 18	17 24	17 30	17 36	17 44	17 53
10	17 15	17 16	17 18	17 20	17 22	17 24	17 26	17 28	17 31	17 34	17 37	17 41	17 45	17 50
11	17 39	17 39	17 40	17 40	17 41	17 41	17 41	17 42	17 43	17 43	17 44	17 45	17 46	17 47
12	18 03	18 02	18 01	18 00	17 59	17 58	17 57	17 56	17 54	17 52	17 51	17 49	17 46	17 43
13	18 29	18 27	18 24	18 22	18 19	18 17	18 14	18 10	18 07	18 03	17 58	17 53	17 47	17 40
14	18 56	18 53	18 49	18 45	18 41	18 37	18 32	18 27	18 21	18 14	18 07	17 59	17 49	17 38
15	19 27	19 23	19 18	19 13	19 07	19 01	18 54	18 47	18 39	18 30	18 19	18 07	17 53	17 35
16	20 03	19 58	19 52	19 45	19 38	19 31	19 22	19 13	19 02	18 50	18 36	18 20	18 00	17 34
17	20 46	20 39	20 32	20 25	20 17	20 08	19 58	19 47	19 34	19 20	19 02	18 41	18 14	17 36
18	21 36	21 29	21 22	21 14	21 05	20 55	20 45	20 32	20 19	20 02	19 43	19 18	18 45	17 50
19	22 34	22 28	22 20	22 12	22 04	21 54	21 44	21 32	21 18	21 02	20 43	20 19	19 46	18 52
20	23 40	23 34	23 27	23 20	23 13	23 04	22 55	22 44	22 32	22 18	22 02	21 42	21 16	20 40
21	.. ..	.. ..	.. ..	.. ..	.. ..	.. ..	.. ..	.. ..	23 57	23 46	23 33	23 19	23 01	22 39
22	0 51	0 46	0 41	0 35	0 29	0 22	0 14	0 06	.. ..	.. ..	.. ..	.. ..	.. ..	.. ..
23	2 05	2 01	1 58	1 53	1 49	1 44	1 39	1 33	1 27	1 19	1 11	1 02	0 50	0 37
24	3 20	3 18	3 16	3 14	3 11	3 08	3 06	3 02	2 59	2 55	2 51	2 45	2 40	2 33
25	4 35	4 35	4 34	4 34	4 33	4 33	4 33	4 32	4 31	4 31	4 30	4 29	4 28	4 27
26	5 50	5 51	5 53	5 54	5 56	5 57	5 59	6 01	6 04	6 06	6 09	6 12	6 16	6 20
27	7 05	7 07	7 11	7 14	7 17	7 21	7 25	7 30	7 35	7 41	7 48	7 55	8 04	8 15
28	8 18	8 22	8 27	8 32	8 38	8 43	8 50	8 57	9 05	9 15	9 25	9 38	9 52	10 10
29	9 29	9 35	9 41	9 47	9 54	10 02	10 11	10 20	10 31	10 44	10 58	11 16	11 38	12 07
30	10 35	10 42	10 49	10 56	11 05	11 14	11 24	11 35	11 49	12 04	12 22	12 45	13 15	14 00
Oct. 1	11 35	11 42	11 50	11 58	12 06	12 16	12 27	12 39	12 53	13 10	13 30	13 55	14 30	15 33
2	12 28	12 34	12 42	12 49	12 58	13 08	13 18	13 30	13 44	14 00	14 19	14 42	15 14	16 05

MOONSET

Lat.	+40°	+42°	+44°	+46°	+48°	+50°	+52°	+54°	+56°	+58°	+60°	+62°	+64°	+66°
	h m	h m	h m	h m	h m	h m	h m	h m	h m	h m	h m	h m	h m	h m
Sept. 8	2 05	2 00	1 55	1 50	1 44	1 38	1 31	1 23	1 14	1 04	0 52	0 39	0 23	0 03
9	3 04	3 00	2 56	2 52	2 48	2 43	2 38	2 32	2 26	2 19	2 11	2 01	1 50	1 37
10	4 02	4 00	3 57	3 55	3 52	3 49	3 46	3 42	3 38	3 33	3 28	3 23	3 16	3 08
11	5 00	4 59	4 58	4 57	4 56	4 54	4 53	4 51	4 49	4 48	4 45	4 43	4 40	4 37
12	5 58	5 58	5 59	5 59	5 59	6 00	6 00	6 01	6 01	6 02	6 02	6 03	6 04	6 05
13	6 57	6 58	7 00	7 02	7 04	7 06	7 08	7 11	7 13	7 17	7 20	7 24	7 28	7 34
14	7 57	8 00	8 03	8 06	8 09	8 13	8 17	8 22	8 27	8 33	8 39	8 47	8 55	9 06
15	8 58	9 02	9 06	9 11	9 16	9 22	9 28	9 35	9 42	9 51	10 01	10 12	10 25	10 42
16	10 00	10 05	10 11	10 17	10 24	10 31	10 39	10 48	10 58	11 09	11 23	11 39	11 58	12 23
17	11 02	11 09	11 15	11 23	11 30	11 39	11 49	12 00	12 12	12 26	12 43	13 04	13 30	14 08
18	12 03	12 10	12 17	12 25	12 34	12 44	12 54	13 06	13 20	13 36	13 56	14 20	14 53	15 48
19	13 01	13 07	13 15	13 23	13 31	13 41	13 52	14 04	14 18	14 34	14 53	15 17	15 50	16 44
20	13 52	13 59	14 06	14 13	14 21	14 30	14 39	14 50	15 03	15 17	15 34	15 54	16 20	16 56
21	14 38	14 44	14 49	14 55	15 02	15 09	15 17	15 26	15 36	15 48	16 01	16 16	16 35	16 58
22	15 18	15 22	15 27	15 31	15 37	15 42	15 48	15 54	16 02	16 10	16 19	16 30	16 42	16 57
23	15 54	15 57	16 00	16 03	16 06	16 09	16 13	16 17	16 22	16 27	16 32	16 39	16 46	16 55
24	16 27	16 28	16 30	16 31	16 32	16 34	16 35	16 37	16 39	16 41	16 43	16 46	16 49	16 52
25	16 59	16 59	16 58	16 58	16 57	16 57	16 56	16 55	16 55	16 54	16 53	16 52	16 51	16 49
26	17 31	17 29	17 27	17 25	17 23	17 20	17 17	17 14	17 11	17 07	17 03	16 58	16 53	16 47
27	18 05	18 02	17 58	17 54	17 50	17 46	17 41	17 35	17 29	17 22	17 15	17 06	16 56	16 44
28	18 43	18 38	18 33	18 27	18 21	18 15	18 08	18 00	17 51	17 41	17 30	17 17	17 01	16 42
29	19 25	19 19	19 12	19 06	18 58	18 50	18 41	18 31	18 19	18 06	17 51	17 33	17 11	16 41
30	20 12	20 05	19 58	19 50	19 42	19 32	19 22	19 10	18 57	18 41	18 23	18 00	17 30	16 44
Oct. 1	21 04	20 57	20 49	20 41	20 33	20 23	20 12	20 00	19 45	19 29	19 09	18 44	18 09	17 06
2	22 00	21 53	21 46	21 38	21 30	21 20	21 10	20 58	20 45	20 29	20 10	19 47	19 15	18 24

(.. ..) indicates phenomenon will occur the next day.
(-- --) indicates Moon continuously below horizon.
(** **) indicates Moon continuously above horizon.

UNIVERSAL TIME FOR MERIDIAN OF GREENWICH

MOONRISE

Lat.	−55°	−50°	−45°	−40°	−35°	−30°	−20°	−10°	0°	+10°	+20°	+30°	+35°	+40°
	h m	h m	h m	h m	h m	h m	h m	h m	h m	h m	h m	h m	h m	h m
Oct. 1	7 29	7 59	8 21	8 39	8 54	9 07	9 29	9 48	10 07	10 25	10 44	11 07	11 20	11 35
2	8 23	8 53	9 15	9 33	9 48	10 01	10 23	10 42	11 00	11 18	11 37	12 00	12 13	12 28
3	9 26	9 53	10 13	10 30	10 44	10 56	11 17	11 35	11 51	12 08	12 26	12 47	12 59	13 12
4	10 34	10 57	11 14	11 29	11 41	11 51	12 09	12 25	12 40	12 54	13 10	13 28	13 38	13 50
5	11 44	12 02	12 16	12 27	12 37	12 46	13 01	13 14	13 26	13 38	13 50	14 05	14 13	14 23
6	12 55	13 07	13 18	13 26	13 33	13 40	13 51	14 00	14 09	14 18	14 28	14 38	14 45	14 52
7	14 05	14 13	14 19	14 24	14 29	14 33	14 39	14 45	14 51	14 57	15 03	15 09	15 13	15 18
8	15 15	15 18	15 20	15 22	15 24	15 25	15 28	15 30	15 32	15 34	15 37	15 39	15 41	15 42
9	16 25	16 23	16 22	16 20	16 19	16 18	16 16	16 15	16 13	16 12	16 11	16 09	16 08	16 07
10	17 37	17 30	17 24	17 20	17 16	17 12	17 06	17 01	16 56	16 51	16 45	16 39	16 36	16 32
11	18 50	18 38	18 28	18 20	18 13	18 07	17 57	17 48	17 39	17 31	17 22	17 12	17 06	16 59
12	20 05	19 48	19 34	19 22	19 13	19 04	18 50	18 37	18 25	18 14	18 01	17 47	17 39	17 30
13	21 20	20 58	20 40	20 25	20 13	20 03	19 45	19 29	19 14	19 00	18 44	18 27	18 16	18 05
14	22 33	22 06	21 45	21 28	21 14	21 02	20 41	20 23	20 06	19 49	19 32	19 11	18 59	18 45
15	23 39	23 09	22 47	22 29	22 13	22 00	21 38	21 19	21 01	20 43	20 24	20 01	19 49	19 34
16	.. ..	.. ..	23 42	23 24	23 09	22 56	22 34	22 15	21 57	21 39	21 20	20 57	20 44	20 29
17	0 34	0 04	.. ..	.. ..	.. ..	23 49	23 28	23 10	22 53	22 37	22 19	21 58	21 46	21 32
18	1 16	0 51	0 31	0 14	0 01	.. ..	.. ..	.. ..	23 49	23 35	23 20	23 02	22 52	22 40
19	1 49	1 28	1 12	0 58	0 46	0 36	0 19	0 04	.. ..	.. ..	.. ..	.. ..	23 59	23 50
20	2 14	1 59	1 47	1 36	1 28	1 20	1 06	0 55	0 44	0 33	0 21	0 07	.. ..	.. ..
21	2 35	2 25	2 17	2 11	2 05	2 00	1 51	1 44	1 37	1 29	1 22	1 13	1 08	1 02
22	2 52	2 48	2 45	2 42	2 40	2 38	2 35	2 31	2 28	2 25	2 22	2 19	2 17	2 14
23	3 09	3 11	3 12	3 14	3 15	3 15	3 17	3 19	3 20	3 21	3 23	3 25	3 26	3 27
24	3 27	3 34	3 40	3 45	3 50	3 53	4 00	4 06	4 12	4 18	4 24	4 31	4 35	4 40
25	3 47	4 00	4 10	4 19	4 27	4 33	4 45	4 55	5 05	5 15	5 26	5 38	5 45	5 53

MOONSET

Lat.	−55°	−50°	−45°	−40°	−35°	−30°	−20°	−10°	0°	+10°	+20°	+30°	+35°	+40°
	h m	h m	h m	h m	h m	h m	h m	h m	h m	h m	h m	h m	h m	h m
Oct. 1	0 13	.. ..	.. ..	.. ..	23 46	23 33	23 10	22 51	22 33	22 14	21 55	21 32	21 19	21 04
2	1 11	0 41	0 19	0 01	.. ..	.. ..	.. ..	23 43	23 25	23 08	22 49	22 27	22 14	22 00
3	1 55	1 28	1 07	0 50	0 35	0 23	0 01	.. ..	.. ..	23 59	23 42	23 23	23 11	22 58
4	2 28	2 04	1 46	1 31	1 18	1 07	0 48	0 31	0 15	.. ..	.. ..	.. ..	.. ..	23 57
5	2 53	2 34	2 18	2 06	1 55	1 46	1 30	1 16	1 02	0 49	0 34	0 18	0 08	.. ..
6	3 12	2 58	2 46	2 37	2 28	2 21	2 08	1 57	1 47	1 36	1 25	1 12	1 04	0 56
7	3 28	3 18	3 10	3 04	2 58	2 53	2 44	2 37	2 30	2 22	2 14	2 05	2 00	1 54
8	3 42	3 37	3 33	3 29	3 26	3 24	3 19	3 15	3 11	3 07	3 03	2 58	2 55	2 52
9	3 55	3 54	3 54	3 54	3 53	3 53	3 53	3 52	3 52	3 52	3 51	3 51	3 50	3 50
10	4 08	4 12	4 16	4 18	4 21	4 23	4 27	4 30	4 34	4 37	4 40	4 44	4 46	4 49
11	4 22	4 31	4 38	4 44	4 50	4 54	5 02	5 10	5 16	5 23	5 30	5 38	5 43	5 49
12	4 38	4 52	5 04	5 13	5 21	5 28	5 40	5 51	6 01	6 11	6 22	6 34	6 42	6 50
13	4 59	5 18	5 33	5 45	5 56	6 05	6 21	6 35	6 48	7 02	7 16	7 32	7 42	7 53
14	5 25	5 49	6 08	6 23	6 36	6 47	7 06	7 23	7 39	7 55	8 12	8 31	8 42	8 56
15	6 01	6 29	6 50	7 07	7 22	7 34	7 56	8 15	8 32	8 50	9 08	9 30	9 43	9 57
16	6 49	7 18	7 41	7 59	8 14	8 27	8 50	9 09	9 27	9 46	10 05	10 27	10 40	10 56
17	7 50	8 19	8 41	8 58	9 13	9 26	9 48	10 06	10 24	10 41	11 00	11 22	11 34	11 48
18	9 04	9 29	9 48	10 04	10 17	10 28	10 48	11 05	11 20	11 36	11 53	12 12	12 23	12 35
19	10 26	10 45	11 00	11 13	11 24	11 33	11 49	12 03	12 15	12 28	12 42	12 57	13 06	13 16
20	11 51	12 05	12 15	12 24	12 32	12 39	12 50	13 00	13 09	13 18	13 28	13 39	13 45	13 52
21	13 18	13 26	13 32	13 36	13 41	13 44	13 51	13 56	14 01	14 06	14 12	14 18	14 21	14 25
22	14 46	14 47	14 48	14 49	14 50	14 50	14 51	14 52	14 53	14 54	14 54	14 55	14 56	14 56
23	16 14	16 09	16 05	16 02	15 59	15 56	15 52	15 48	15 44	15 41	15 37	15 33	15 30	15 27
24	17 42	17 31	17 22	17 15	17 08	17 03	16 53	16 45	16 37	16 29	16 21	16 12	16 06	16 00
25	19 10	18 52	18 39	18 28	18 18	18 10	17 55	17 43	17 31	17 19	17 07	16 53	16 45	16 36

(.. ..) indicates phenomenon will occur the next day.

UNIVERSAL TIME FOR MERIDIAN OF GREENWICH

MOONRISE

Lat.	+40°	+42°	+44°	+46°	+48°	+50°	+52°	+54°	+56°	+58°	+60°	+62°	+64°	+66°
	h m	h m	h m	h m	h m	h m	h m	h m	h m	h m	h m	h m	h m	h m
Oct. 1	11 35	11 42	11 50	11 58	12 06	12 16	12 27	12 39	12 53	13 10	13 30	13 55	14 30	15 33
2	12 28	12 34	12 42	12 49	12 58	13 08	13 18	13 30	13 44	14 00	14 19	14 42	15 14	16 05
3	13 12	13 18	13 25	13 32	13 40	13 49	13 58	14 09	14 21	14 35	14 51	15 11	15 36	16 10
4	13 50	13 55	14 01	14 07	14 14	14 21	14 29	14 38	14 48	14 59	15 12	15 28	15 46	16 09
5	14 23	14 27	14 32	14 36	14 42	14 47	14 54	15 01	15 08	15 17	15 27	15 38	15 51	16 07
6	14 52	14 55	14 58	15 02	15 05	15 09	15 14	15 19	15 24	15 31	15 37	15 45	15 54	16 05
7	15 18	15 20	15 22	15 24	15 26	15 29	15 32	15 35	15 38	15 42	15 46	15 51	15 56	16 02
8	15 42	15 43	15 44	15 45	15 46	15 47	15 48	15 49	15 50	15 52	15 53	15 55	15 57	15 59
9	16 07	16 07	16 06	16 05	16 05	16 04	16 04	16 03	16 02	16 01	16 00	15 59	15 58	15 57
10	16 32	16 31	16 29	16 27	16 25	16 23	16 20	16 18	16 15	16 11	16 08	16 04	15 59	15 54
11	16 59	16 56	16 53	16 50	16 47	16 43	16 38	16 34	16 29	16 23	16 17	16 10	16 01	15 52
12	17 30	17 26	17 21	17 17	17 11	17 06	17 00	16 53	16 46	16 38	16 28	16 18	16 05	15 50
13	18 05	17 59	17 54	17 48	17 41	17 34	17 26	17 18	17 08	16 57	16 44	16 29	16 12	15 49
14	18 45	18 39	18 33	18 26	18 18	18 10	18 00	17 50	17 38	17 24	17 08	16 49	16 24	15 52
15	19 34	19 27	19 20	19 12	19 04	18 54	18 44	18 32	18 19	18 03	17 44	17 21	16 51	16 03
16	20 29	20 23	20 15	20 08	19 59	19 49	19 39	19 27	19 13	18 57	18 38	18 14	17 42	16 49
17	21 32	21 26	21 19	21 12	21 04	20 55	20 45	20 34	20 22	20 07	19 50	19 29	19 02	18 24
18	22 40	22 34	22 29	22 23	22 16	22 09	22 01	21 52	21 41	21 30	21 16	21 00	20 40	20 15
19	23 50	23 46	23 42	23 37	23 32	23 27	23 21	23 14	23 07	22 58	22 49	22 38	22 24	22 09
20	.. ..	.. ..	.. ..	.. ..	.. ..	.. ..	.. ..	.. ..	.. ..	.. ..	.. ..	.. ..	.. ..	.. ..
21	1 02	1 00	0 57	0 54	0 51	0 47	0 43	0 39	0 35	0 30	0 24	0 17	0 09	0 00
22	2 14	2 13	2 12	2 11	2 10	2 08	2 07	2 05	2 03	2 01	1 59	1 57	1 54	1 50
23	3 27	3 27	3 28	3 29	3 29	3 30	3 31	3 32	3 33	3 34	3 35	3 36	3 38	3 40
24	4 40	4 42	4 44	4 47	4 49	4 52	4 55	4 59	5 02	5 07	5 11	5 17	5 23	5 30
25	5 53	5 57	6 00	6 05	6 09	6 14	6 19	6 25	6 32	6 40	6 48	6 58	7 09	7 23

MOONSET

Lat.	+40°	+42°	+44°	+46°	+48°	+50°	+52°	+54°	+56°	+58°	+60°	+62°	+64°	+66°
	h m	h m	h m	h m	h m	h m	h m	h m	h m	h m	h m	h m	h m	h m
Oct. 1	21 04	20 57	20 49	20 41	20 33	20 23	20 12	20 00	19 45	19 29	19 09	18 44	18 09	17 06
2	22 00	21 53	21 46	21 38	21 30	21 20	21 10	20 58	20 45	20 29	20 10	19 47	19 15	18 24
3	22 58	22 52	22 46	22 39	22 31	22 23	22 14	22 03	21 52	21 38	21 22	21 03	20 38	20 05
4	23 57	23 52	23 47	23 41	23 35	23 28	23 20	23 12	23 02	22 52	22 39	22 24	22 06	21 44
5	.. ..	.. ..	.. ..	.. ..	.. ..	.. ..	.. ..	.. ..	.. ..	.. ..	23 57	23 46	23 34	23 19
6	0 56	0 52	0 48	0 43	0 39	0 33	0 28	0 21	0 14	0 06	.. ..	.. ..	.. ..	.. ..
7	1 54	1 52	1 49	1 46	1 42	1 39	1 35	1 31	1 26	1 21	1 15	1 08	1 00	0 50
8	2 52	2 51	2 49	2 48	2 46	2 44	2 42	2 40	2 37	2 34	2 31	2 28	2 24	2 19
9	3 50	3 50	3 50	3 50	3 49	3 49	3 49	3 49	3 49	3 48	3 48	3 48	3 47	3 47
10	4 49	4 50	4 51	4 52	4 54	4 55	4 57	4 59	5 01	5 03	5 05	5 08	5 11	5 15
11	5 49	5 51	5 54	5 56	5 59	6 03	6 06	6 10	6 14	6 19	6 24	6 31	6 38	6 46
12	6 50	6 54	6 57	7 02	7 06	7 11	7 17	7 23	7 29	7 37	7 46	7 56	8 07	8 21
13	7 53	7 57	8 03	8 08	8 14	8 21	8 28	8 37	8 46	8 56	9 08	9 22	9 40	10 01
14	8 56	9 01	9 08	9 15	9 22	9 30	9 39	9 49	10 01	10 14	10 30	10 49	11 13	11 45
15	9 57	10 04	10 11	10 19	10 27	10 36	10 46	10 58	11 11	11 27	11 45	12 08	12 38	13 25
16	10 56	11 02	11 10	11 17	11 26	11 36	11 46	11 58	12 12	12 28	12 47	13 11	13 44	14 37
17	11 48	11 55	12 02	12 09	12 17	12 26	12 36	12 48	13 00	13 15	13 33	13 54	14 21	15 00
18	12 35	12 41	12 47	12 53	13 00	13 08	13 17	13 26	13 37	13 49	14 03	14 20	14 40	15 06
19	13 16	13 21	13 25	13 30	13 36	13 42	13 49	13 56	14 04	14 13	14 24	14 36	14 50	15 07
20	13 52	13 55	13 59	14 02	14 06	14 10	14 15	14 20	14 25	14 31	14 38	14 46	14 55	15 06
21	14 25	14 27	14 29	14 31	14 33	14 35	14 37	14 40	14 43	14 46	14 50	14 54	14 58	15 04
22	14 56	14 57	14 57	14 57	14 57	14 58	14 58	14 58	14 59	14 59	15 00	15 00	15 01	15 01
23	15 27	15 26	15 25	15 23	15 22	15 20	15 19	15 17	15 14	15 12	15 09	15 06	15 03	14 59
24	16 00	15 57	15 54	15 51	15 48	15 44	15 40	15 36	15 31	15 26	15 20	15 14	15 06	14 57
25	16 36	16 31	16 27	16 22	16 17	16 12	16 06	15 59	15 52	15 43	15 34	15 23	15 10	14 55

(.. ..) indicates phenomenon will occur the next day.
(-- --) indicates Moon continuously below horizon.
(** **) indicates Moon continuously above horizon.

UNIVERSAL TIME FOR MERIDIAN OF GREENWICH

MOONRISE

Lat.	−55°	−50°	−45°	−40°	−35°	−30°	−20°	−10°	0°	+10°	+20°	+30°	+35°	+40°
	h m	h m	h m	h m	h m	h m	h m	h m	h m	h m	h m	h m	h m	h m
Oct. 24	3 27	3 34	3 40	3 45	3 50	3 53	4 00	4 06	4 12	4 18	4 24	4 31	4 35	4 40
25	3 47	4 00	4 10	4 19	4 27	4 33	4 45	4 55	5 05	5 15	5 26	5 38	5 45	5 53
26	4 10	4 29	4 44	4 57	5 07	5 16	5 33	5 47	6 00	6 13	6 28	6 44	6 54	7 05
27	4 41	5 05	5 24	5 39	5 52	6 03	6 23	6 40	6 56	7 12	7 29	7 49	8 01	8 15
28	5 21	5 49	6 10	6 27	6 42	6 54	7 16	7 35	7 52	8 10	8 29	8 51	9 04	9 19
29	6 11	6 40	7 03	7 20	7 35	7 48	8 11	8 30	8 48	9 06	9 25	9 48	10 01	10 16
30	7 12	7 39	8 01	8 18	8 32	8 45	9 06	9 24	9 41	9 59	10 17	10 38	10 51	11 05
31	8 19	8 43	9 02	9 17	9 30	9 41	10 00	10 17	10 32	10 47	11 04	11 23	11 34	11 46
Nov. 1	9 29	9 49	10 04	10 17	10 27	10 37	10 53	11 07	11 20	11 32	11 46	12 02	12 11	12 21
2	10 40	10 55	11 06	11 16	11 24	11 31	11 43	11 54	12 04	12 14	12 25	12 37	12 44	12 52
3	11 51	12 00	12 08	12 14	12 20	12 25	12 33	12 40	12 47	12 54	13 01	13 09	13 14	13 19
4	13 00	13 05	13 09	13 12	13 15	13 17	13 21	13 25	13 28	13 32	13 35	13 39	13 42	13 44
5	14 10	14 10	14 10	14 10	14 10	14 10	14 09	14 09	14 09	14 09	14 09	14 09	14 09	14 09
6	15 21	15 16	15 12	15 08	15 05	15 03	14 58	14 54	14 51	14 47	14 43	14 39	14 37	14 34
7	16 34	16 24	16 15	16 08	16 03	15 58	15 49	15 41	15 34	15 27	15 19	15 11	15 06	15 00
8	17 48	17 33	17 21	17 10	17 02	16 54	16 41	16 30	16 19	16 09	15 58	15 45	15 38	15 30
9	19 04	18 43	18 27	18 14	18 03	17 53	17 36	17 21	17 08	16 54	16 40	16 23	16 14	16 03
10	20 19	19 54	19 34	19 18	19 04	18 53	18 33	18 16	17 59	17 43	17 26	17 07	16 56	16 43
11	21 29	21 00	20 38	20 20	20 06	19 53	19 31	19 12	18 54	18 37	18 18	17 56	17 44	17 29
12	22 29	21 59	21 37	21 19	21 04	20 51	20 29	20 09	19 51	19 33	19 14	18 52	18 39	18 24
13	23 16	22 49	22 28	22 12	21 57	21 45	21 24	21 06	20 49	20 32	20 13	19 52	19 40	19 25
14	23 51	23 29	23 12	22 57	22 45	22 35	22 16	22 00	21 45	21 30	21 14	20 56	20 45	20 32
15	.. ..	.. ..	23 48	23 37	23 28	23 19	23 05	22 52	22 40	22 28	22 15	22 01	21 52	21 42
16	0 18	0 02	.. ..	.. ..	.. ..	.. ..	23 50	23 41	23 33	23 24	23 16	23 05	23 00	22 53
17	0 40	0 29	0 20	0 12	0 06	0 00	.. ..	.. ..	.. ..	.. ..	.. ..	.. ..	.. ..	.. ..

MOONSET

Lat.	−55°	−50°	−45°	−40°	−35°	−30°	−20°	−10°	0°	+10°	+20°	+30°	+35°	+40°
	h m	h m	h m	h m	h m	h m	h m	h m	h m	h m	h m	h m	h m	h m
Oct. 24	17 42	17 31	17 22	17 15	17 08	17 03	16 53	16 45	16 37	16 29	16 21	16 12	16 06	16 00
25	19 10	18 52	18 39	18 28	18 18	18 10	17 55	17 43	17 31	17 19	17 07	16 53	16 45	16 36
26	20 34	20 11	19 53	19 39	19 26	19 16	18 57	18 41	18 27	18 12	17 56	17 38	17 27	17 15
27	21 51	21 24	21 03	20 46	20 32	20 19	19 58	19 40	19 23	19 06	18 48	18 27	18 15	18 01
28	22 56	22 27	22 05	21 47	21 32	21 19	20 57	20 37	20 19	20 01	19 42	19 20	19 07	18 52
29	23 47	23 19	22 57	22 40	22 25	22 13	21 51	21 32	21 14	20 56	20 37	20 15	20 02	19 47
30	.. ..	.. ..	23 41	23 25	23 12	23 00	22 40	22 23	22 06	21 50	21 32	21 12	21 00	20 46
31	0 25	0 00	.. ..	.. ..	23 52	23 42	23 25	23 10	22 55	22 41	22 26	22 08	21 58	21 46
Nov. 1	0 54	0 33	0 17	0 03	.. ..	.. ..	.. ..	23 53	23 41	23 30	23 17	23 03	22 55	22 45
2	1 15	0 59	0 47	0 36	0 27	0 19	0 05	.. ..	.. ..	.. ..	.. ..	23 57	23 51	23 44
3	1 33	1 21	1 12	1 05	0 58	0 52	0 42	0 33	0 25	0 17	0 07	.. ..	.. ..	.. ..
4	1 47	1 41	1 35	1 31	1 27	1 23	1 17	1 12	1 07	1 02	0 56	0 50	0 46	0 42
5	2 01	1 59	1 57	1 56	1 54	1 53	1 51	1 50	1 48	1 46	1 44	1 42	1 41	1 40
6	2 14	2 16	2 18	2 20	2 22	2 23	2 25	2 27	2 29	2 31	2 33	2 35	2 36	2 38
7	2 28	2 35	2 41	2 46	2 50	2 54	3 00	3 06	3 11	3 17	3 22	3 29	3 33	3 37
8	2 44	2 56	3 05	3 13	3 20	3 26	3 37	3 46	3 55	4 04	4 14	4 25	4 31	4 38
9	3 03	3 20	3 33	3 44	3 54	4 03	4 17	4 30	4 42	4 54	5 07	5 22	5 31	5 41
10	3 27	3 49	4 06	4 21	4 33	4 43	5 01	5 17	5 32	5 47	6 03	6 21	6 32	6 44
11	4 00	4 26	4 47	5 03	5 17	5 29	5 50	6 08	6 25	6 42	7 01	7 22	7 34	7 48
12	4 45	5 14	5 36	5 54	6 09	6 22	6 44	7 03	7 21	7 39	7 59	8 21	8 34	8 49
13	5 43	6 12	6 34	6 52	7 07	7 20	7 42	8 01	8 19	8 36	8 55	9 17	9 30	9 44
14	6 54	7 20	7 40	7 57	8 10	8 22	8 42	9 00	9 16	9 32	9 49	10 09	10 20	10 34
15	8 14	8 36	8 52	9 05	9 17	9 27	9 43	9 58	10 12	10 25	10 40	10 56	11 05	11 16
16	9 39	9 54	10 06	10 16	10 24	10 32	10 44	10 55	11 05	11 16	11 26	11 38	11 45	11 53
17	11 04	11 13	11 20	11 26	11 32	11 36	11 44	11 51	11 57	12 03	12 10	12 18	12 22	12 27

(.. ..) indicates phenomenon will occur the next day.

MOONRISE AND MOONSET, 1992

UNIVERSAL TIME FOR MERIDIAN OF GREENWICH

MOONRISE

Lat.	+40°	+42°	+44°	+46°	+48°	+50°	+52°	+54°	+56°	+58°	+60°	+62°	+64°	+66°
	h m	h m	h m	h m	h m	h m	h m	h m	h m	h m	h m	h m	h m	h m
Oct. 24	4 40	4 42	4 44	4 47	4 49	4 52	4 55	4 59	5 02	5 07	5 11	5 17	5 23	5 30
25	5 53	5 57	6 00	6 05	6 09	6 14	6 19	6 25	6 32	6 40	6 48	6 58	7 09	7 23
26	7 05	7 10	7 16	7 21	7 28	7 34	7 42	7 50	8 00	8 11	8 23	8 38	8 56	9 19
27	8 15	8 21	8 27	8 34	8 42	8 51	9 00	9 11	9 23	9 36	9 53	10 13	10 39	11 14
28	9 19	9 26	9 33	9 41	9 49	9 59	10 09	10 21	10 35	10 51	11 10	11 34	12 06	12 59
29	10 16	10 23	10 30	10 38	10 47	10 56	11 07	11 19	11 33	11 49	12 08	12 33	13 05	14 00
30	11 05	11 11	11 18	11 26	11 34	11 43	11 52	12 04	12 16	12 31	12 48	13 09	13 36	14 15
31	11 46	11 52	11 58	12 04	12 11	12 19	12 28	12 37	12 48	13 00	13 14	13 31	13 51	14 18
Nov. 1	12 21	12 26	12 31	12 36	12 42	12 48	12 55	13 03	13 11	13 21	13 32	13 44	13 59	14 17
2	12 52	12 55	12 59	13 03	13 07	13 12	13 17	13 23	13 29	13 36	13 44	13 53	14 03	14 16
3	13 19	13 21	13 24	13 27	13 29	13 33	13 36	13 40	13 44	13 48	13 53	13 59	14 06	14 13
4	13 44	13 46	13 47	13 48	13 50	13 51	13 53	13 55	13 57	13 59	14 01	14 04	14 07	14 11
5	14 09	14 09	14 09	14 09	14 09	14 09	14 09	14 09	14 09	14 09	14 09	14 09	14 09	14 09
6	14 34	14 33	14 31	14 30	14 28	14 27	14 25	14 23	14 21	14 19	14 16	14 13	14 10	14 06
7	15 00	14 58	14 55	14 52	14 49	14 46	14 43	14 39	14 35	14 30	14 25	14 19	14 12	14 04
8	15 30	15 26	15 22	15 18	15 13	15 09	15 03	14 57	14 51	14 44	14 35	14 26	14 15	14 02
9	16 03	15 58	15 53	15 48	15 42	15 35	15 28	15 20	15 11	15 01	14 50	14 37	14 21	14 02
10	16 43	16 37	16 31	16 24	16 17	16 09	16 00	15 50	15 39	15 26	15 11	14 54	14 32	14 04
11	17 29	17 23	17 16	17 08	17 00	16 51	16 41	16 29	16 16	16 01	15 44	15 22	14 54	14 12
12	18 24	18 17	18 10	18 02	17 53	17 44	17 33	17 21	17 08	16 52	16 33	16 09	15 37	14 46
13	19 25	19 19	19 12	19 05	18 56	18 47	18 37	18 26	18 13	17 58	17 41	17 19	16 50	16 08
14	20 32	20 27	20 21	20 14	20 07	20 00	19 51	19 42	19 31	19 18	19 04	18 47	18 25	17 57
15	21 42	21 38	21 33	21 28	21 23	21 17	21 10	21 03	20 55	20 45	20 35	20 22	20 08	19 50
16	22 53	22 50	22 47	22 43	22 40	22 36	22 31	22 26	22 21	22 15	22 08	22 00	21 51	21 40
17	.. ..	.. ..	.. ..	23 59	23 57	23 55	23 52	23 50	23 47	23 44	23 41	23 37	23 33	23 28

MOONSET

Lat.	+40°	+42°	+44°	+46°	+48°	+50°	+52°	+54°	+56°	+58°	+60°	+62°	+64°	+66°
	h m	h m	h m	h m	h m	h m	h m	h m	h m	h m	h m	h m	h m	h m
Oct. 24	16 00	15 57	15 54	15 51	15 48	15 44	15 40	15 36	15 31	15 26	15 20	15 14	15 06	14 57
25	16 36	16 31	16 27	16 22	16 17	16 12	16 06	15 59	15 52	15 43	15 34	15 23	15 10	14 55
26	17 15	17 10	17 04	16 58	16 51	16 44	16 36	16 27	16 17	16 06	15 52	15 37	15 18	14 55
27	18 01	17 54	17 48	17 40	17 32	17 24	17 14	17 03	16 51	16 36	16 20	15 59	15 33	14 57
28	18 52	18 45	18 37	18 30	18 21	18 11	18 01	17 49	17 35	17 19	17 00	16 36	16 03	15 10
29	19 47	19 40	19 33	19 25	19 17	19 07	18 57	18 45	18 31	18 15	17 56	17 32	16 59	16 04
30	20 46	20 40	20 33	20 26	20 18	20 09	20 00	19 49	19 36	19 22	19 05	18 44	18 17	17 39
31	21 46	21 40	21 35	21 29	21 22	21 14	21 06	20 57	20 47	20 35	20 22	20 05	19 45	19 20
Nov. 1	22 45	22 41	22 37	22 32	22 26	22 21	22 14	22 07	21 59	21 50	21 40	21 28	21 14	20 57
2	23 44	23 41	23 38	23 34	23 30	23 26	23 22	23 17	23 11	23 05	22 58	22 50	22 40	22 29
3	.. ..	.. ..	.. ..	.. ..	.. ..	.. ..	.. ..	.. ..	.. ..	.. ..	.. ..	.. ..	.. ..	23 58
4	0 42	0 40	0 38	0 36	0 34	0 31	0 29	0 26	0 23	0 19	0 15	0 10	0 05	.. ..
5	1 40	1 39	1 39	1 38	1 37	1 36	1 35	1 35	1 33	1 32	1 31	1 29	1 28	1 26
6	2 38	2 39	2 39	2 40	2 41	2 42	2 43	2 44	2 45	2 46	2 47	2 49	2 51	2 53
7	3 37	3 39	3 41	3 43	3 46	3 48	3 51	3 54	3 57	4 01	4 06	4 10	4 16	4 23
8	4 38	4 41	4 45	4 48	4 52	4 56	5 01	5 06	5 12	5 19	5 26	5 34	5 44	5 56
9	5 41	5 45	5 50	5 55	6 00	6 06	6 13	6 20	6 29	6 38	6 49	7 01	7 16	7 34
10	6 44	6 50	6 56	7 02	7 09	7 17	7 25	7 35	7 46	7 58	8 12	8 29	8 50	9 18
11	7 48	7 54	8 01	8 08	8 17	8 25	8 35	8 46	8 59	9 14	9 31	9 53	10 21	11 02
12	8 49	8 55	9 03	9 11	9 19	9 29	9 39	9 51	10 05	10 21	10 40	11 03	11 35	12 27
13	9 44	9 51	9 58	10 06	10 14	10 23	10 33	10 45	10 58	11 13	11 31	11 53	12 22	13 04
14	10 34	10 39	10 46	10 53	11 00	11 08	11 17	11 27	11 38	11 51	12 06	12 24	12 46	13 15
15	11 16	11 21	11 26	11 32	11 38	11 44	11 51	11 59	12 08	12 18	12 29	12 42	12 58	13 17
16	11 53	11 57	12 01	12 05	12 09	12 14	12 19	12 24	12 31	12 37	12 45	12 54	13 04	13 17
17	12 27	12 29	12 31	12 33	12 36	12 39	12 42	12 45	12 49	12 53	12 57	13 03	13 08	13 15

(.. ..) indicates phenomenon will occur the next day.
(-- --) indicates Moon continuously below horizon.
(** **) indicates Moon continuously above horizon.

UNIVERSAL TIME FOR MERIDIAN OF GREENWICH

MOONRISE

Lat.	−55°	−50°	−45°	−40°	−35°	−30°	−20°	−10°	0°	+10°	+20°	+30°	+35°	+40°
	h m	h m	h m	h m	h m	h m	h m	h m	h m	h m	h m	h m	h m	h m
Nov. 16	0 18	0 02	.. ..	.. ..	.. ..	.. ..	23 50	23 41	23 33	23 24	23 16	23 05	23 00	22 53
17	0 40	0 29	0 20	0 12	0 06	0 00	.. ..	.. ..	.. ..	.. ..	.. ..	.. ..	.. ..	.. ..
18	0 58	0 52	0 48	0 44	0 41	0 38	0 32	0 28	0 24	0 19	0 15	0 10	0 07	0 03
19	1 15	1 14	1 14	1 14	1 14	1 14	1 14	1 14	1 14	1 14	1 14	1 14	1 14	1 14
20	1 31	1 37	1 41	1 44	1 48	1 50	1 55	2 00	2 04	2 08	2 12	2 18	2 21	2 24
21	1 49	2 00	2 09	2 16	2 23	2 28	2 38	2 47	2 55	3 03	3 12	3 22	3 28	3 35
22	2 11	2 27	2 40	2 51	3 01	3 09	3 23	3 36	3 47	3 59	4 12	4 27	4 36	4 45
23	2 38	3 00	3 17	3 31	3 43	3 53	4 11	4 27	4 42	4 57	5 13	5 32	5 42	5 55
24	3 13	3 39	3 59	4 16	4 30	4 42	5 03	5 21	5 38	5 55	6 13	6 34	6 47	7 01
25	3 59	4 27	4 49	5 07	5 22	5 35	5 57	6 16	6 34	6 52	7 11	7 34	7 47	8 02
26	4 55	5 24	5 45	6 03	6 18	6 31	6 52	7 11	7 29	7 47	8 06	8 27	8 40	8 55
27	6 01	6 27	6 46	7 02	7 16	7 28	7 48	8 05	8 22	8 38	8 55	9 15	9 27	9 40
28	7 11	7 33	7 49	8 03	8 15	8 25	8 42	8 57	9 11	9 25	9 40	9 57	10 07	10 18
29	8 23	8 39	8 53	9 03	9 12	9 20	9 34	9 46	9 57	10 09	10 21	10 34	10 42	10 51
30	9 34	9 46	9 55	10 02	10 09	10 15	10 25	10 33	10 41	10 49	10 58	11 08	11 13	11 20
Dec. 1	10 44	10 51	10 56	11 00	11 04	11 08	11 13	11 18	11 23	11 28	11 33	11 39	11 42	11 46
2	11 54	11 55	11 57	11 58	11 59	12 00	12 01	12 03	12 04	12 05	12 07	12 08	12 09	12 10
3	13 03	13 00	12 58	12 56	12 54	12 52	12 50	12 47	12 45	12 43	12 40	12 38	12 36	12 35
4	14 14	14 06	14 00	13 54	13 50	13 46	13 39	13 33	13 27	13 21	13 15	13 08	13 05	13 00
5	15 27	15 14	15 04	14 55	14 47	14 41	14 30	14 20	14 11	14 02	13 52	13 41	13 35	13 28
6	16 43	16 24	16 09	15 57	15 47	15 38	15 23	15 10	14 58	14 45	14 32	14 17	14 09	13 59
7	17 58	17 35	17 16	17 01	16 49	16 38	16 19	16 03	15 48	15 33	15 17	14 59	14 48	14 36
8	19 12	18 44	18 23	18 05	17 51	17 39	17 17	16 59	16 42	16 25	16 07	15 46	15 34	15 20
9	20 17	19 48	19 25	19 07	18 52	18 39	18 16	17 57	17 39	17 21	17 02	16 40	16 27	16 12
10	21 11	20 43	20 21	20 04	19 49	19 36	19 15	18 56	18 38	18 20	18 01	17 40	17 27	17 12

MOONSET

Lat.	−55°	−50°	−45°	−40°	−35°	−30°	−20°	−10°	0°	+10°	+20°	+30°	+35°	+40°
	h m	h m	h m	h m	h m	h m	h m	h m	h m	h m	h m	h m	h m	h m
Nov. 16	9 39	9 54	10 06	10 16	10 24	10 32	10 44	10 55	11 05	11 16	11 26	11 38	11 45	11 53
17	11 04	11 13	11 20	11 26	11 32	11 36	11 44	11 51	11 57	12 03	12 10	12 18	12 22	12 27
18	12 29	12 32	12 35	12 37	12 39	12 40	12 43	12 45	12 47	12 50	12 52	12 54	12 56	12 57
19	13 54	13 51	13 49	13 47	13 46	13 44	13 42	13 39	13 37	13 35	13 33	13 31	13 29	13 28
20	15 19	15 10	15 03	14 58	14 53	14 48	14 41	14 34	14 28	14 22	14 15	14 08	14 03	13 58
21	16 44	16 30	16 18	16 08	16 00	15 53	15 41	15 30	15 20	15 10	14 59	14 47	14 40	14 32
22	18 08	17 48	17 32	17 19	17 08	16 58	16 41	16 27	16 13	16 00	15 45	15 29	15 19	15 09
23	19 28	19 02	18 43	18 27	18 14	18 02	17 42	17 25	17 09	16 52	16 35	16 16	16 04	15 51
24	20 38	20 10	19 48	19 31	19 16	19 03	18 41	18 22	18 05	17 47	17 28	17 06	16 54	16 39
25	21 36	21 07	20 45	20 28	20 13	20 00	19 38	19 19	19 01	18 43	18 23	18 01	17 48	17 33
26	22 20	21 54	21 33	21 17	21 03	20 51	20 30	20 12	19 55	19 37	19 19	18 58	18 45	18 31
27	22 53	22 31	22 13	21 59	21 46	21 36	21 17	21 01	20 46	20 30	20 14	19 55	19 44	19 31
28	23 18	23 00	22 46	22 34	22 24	22 15	22 00	21 46	21 34	21 21	21 07	20 52	20 43	20 32
29	23 37	23 24	23 13	23 05	22 57	22 50	22 39	22 29	22 19	22 09	21 59	21 47	21 40	21 32
30	23 53	23 44	23 38	23 32	23 27	23 22	23 15	23 08	23 02	22 55	22 48	22 40	22 36	22 31
Dec. 1	.. ..	.. ..	.. ..	23 57	23 55	23 53	23 49	23 46	23 43	23 40	23 37	23 33	23 31	23 28
2	0 07	0 03	0 00	.. ..	.. ..	.. ..	.. ..	.. ..	.. ..	.. ..	.. ..	.. ..	.. ..	.. ..
3	0 20	0 20	0 21	0 21	0 22	0 22	0 23	0 23	0 24	0 24	0 25	0 25	0 25	0 26
4	0 33	0 38	0 43	0 46	0 49	0 52	0 57	1 01	1 05	1 09	1 13	1 18	1 21	1 24
5	0 48	0 58	1 06	1 13	1 18	1 23	1 32	1 40	1 48	1 55	2 03	2 12	2 17	2 23
6	1 05	1 20	1 32	1 42	1 50	1 58	2 11	2 22	2 33	2 44	2 55	3 08	3 16	3 25
7	1 27	1 47	2 02	2 15	2 26	2 36	2 53	3 07	3 21	3 35	3 50	4 07	4 17	4 28
8	1 56	2 21	2 40	2 55	3 09	3 20	3 40	3 57	4 13	4 30	4 47	5 07	5 19	5 32
9	2 36	3 04	3 26	3 43	3 58	4 10	4 32	4 51	5 09	5 27	5 46	6 07	6 20	6 35
10	3 29	3 59	4 21	4 39	4 54	5 07	5 30	5 49	6 07	6 25	6 44	7 06	7 19	7 34

(.. ..) indicates phenomenon will occur the next day.

UNIVERSAL TIME FOR MERIDIAN OF GREENWICH

MOONRISE

Lat.	+40°	+42°	+44°	+46°	+48°	+50°	+52°	+54°	+56°	+58°	+60°	+62°	+64°	+66°
	h m	h m	h m	h m	h m	h m	h m	h m	h m	h m	h m	h m	h m	h m
Nov. 16	22 53	22 50	22 47	22 43	22 40	22 36	22 31	22 26	22 21	22 15	22 08	22 00	21 51	21 40
17	.. ..	.. ..	.. ..	23 59	23 57	23 55	23 52	23 50	23 47	23 44	23 41	23 37	23 33	23 28
18	0 03	0 02	0 00	.. ..	.. ..	.. ..	.. ..	.. ..	.. ..	.. ..	.. ..	.. ..	.. ..	.. ..
19	1 14	1 14	1 14	1 14	1 14	1 14	1 14	1 14	1 14	1 14	1 14	1 14	1 14	1 14
20	2 24	2 26	2 27	2 29	2 31	2 33	2 35	2 37	2 40	2 43	2 47	2 50	2 55	3 00
21	3 35	3 38	3 41	3 44	3 48	3 52	3 57	4 02	4 07	4 13	4 20	4 28	4 37	4 48
22	4 45	4 50	4 55	5 00	5 05	5 11	5 18	5 25	5 33	5 43	5 54	6 06	6 21	6 39
23	5 55	6 01	6 07	6 13	6 20	6 28	6 37	6 46	6 57	7 10	7 24	7 42	8 04	8 33
24	7 01	7 08	7 15	7 22	7 30	7 39	7 50	8 01	8 14	8 29	8 47	9 09	9 39	10 23
25	8 02	8 08	8 16	8 24	8 32	8 42	8 53	9 05	9 19	9 35	9 54	10 19	10 52	11 48
26	8 55	9 01	9 08	9 16	9 25	9 34	9 44	9 56	10 09	10 24	10 43	11 05	11 35	12 20
27	9 40	9 46	9 52	9 59	10 07	10 15	10 24	10 34	10 46	10 59	11 15	11 33	11 56	12 27
28	10 18	10 23	10 29	10 34	10 41	10 48	10 55	11 04	11 13	11 23	11 36	11 50	12 07	12 28
29	10 51	10 55	10 59	11 04	11 09	11 14	11 20	11 26	11 33	11 41	11 50	12 00	12 13	12 27
30	11 20	11 22	11 25	11 29	11 32	11 36	11 40	11 44	11 49	11 55	12 01	12 08	12 16	12 25
Dec. 1	11 46	11 47	11 49	11 51	11 53	11 55	11 57	12 00	12 03	12 06	12 09	12 13	12 18	12 23
2	12 10	12 11	12 11	12 12	12 12	12 13	12 14	12 14	12 15	12 16	12 17	12 18	12 19	12 21
3	12 35	12 34	12 33	12 32	12 31	12 31	12 29	12 28	12 27	12 26	12 24	12 22	12 20	12 18
4	13 00	12 58	12 56	12 54	12 52	12 49	12 46	12 43	12 40	12 36	12 32	12 28	12 22	12 16
5	13 28	13 25	13 21	13 18	13 14	13 10	13 05	13 00	12 55	12 49	12 42	12 34	12 25	12 14
6	13 59	13 55	13 50	13 45	13 40	13 34	13 28	13 21	13 13	13 04	12 54	12 43	12 29	12 13
7	14 36	14 31	14 25	14 19	14 12	14 04	13 56	13 47	13 37	13 26	13 12	12 57	12 38	12 14
8	15 20	15 13	15 07	15 00	14 52	14 43	14 33	14 23	14 10	13 56	13 40	13 20	12 54	12 19
9	16 12	16 05	15 58	15 50	15 42	15 32	15 22	15 10	14 56	14 41	14 22	13 58	13 27	12 39
10	17 12	17 06	16 59	16 51	16 42	16 33	16 23	16 11	15 58	15 42	15 24	15 01	14 30	13 43

MOONSET

Lat.	+40°	+42°	+44°	+46°	+48°	+50°	+52°	+54°	+56°	+58°	+60°	+62°	+64°	+66°
	h m	h m	h m	h m	h m	h m	h m	h m	h m	h m	h m	h m	h m	h m
Nov. 16	11 53	11 57	12 01	12 05	12 09	12 14	12 19	12 24	12 31	12 37	12 45	12 54	13 04	13 17
17	12 27	12 29	12 31	12 33	12 36	12 39	12 42	12 45	12 49	12 53	12 57	13 03	13 08	13 15
18	12 57	12 58	12 59	13 00	13 01	13 02	13 03	13 04	13 05	13 06	13 08	13 09	13 11	13 13
19	13 28	13 27	13 26	13 25	13 24	13 23	13 22	13 21	13 20	13 19	13 17	13 15	13 13	13 11
20	13 58	13 56	13 54	13 52	13 49	13 46	13 43	13 40	13 36	13 32	13 27	13 22	13 16	13 09
21	14 32	14 28	14 24	14 20	14 16	14 11	14 06	14 00	13 54	13 47	13 39	13 30	13 20	13 07
22	15 09	15 04	14 59	14 53	14 47	14 41	14 33	14 25	14 17	14 07	13 55	13 42	13 26	13 06
23	15 51	15 45	15 39	15 32	15 24	15 16	15 07	14 57	14 46	14 33	14 18	13 59	13 37	13 07
24	16 39	16 32	16 25	16 18	16 09	16 00	15 50	15 38	15 25	15 10	14 51	14 29	13 59	13 15
25	17 33	17 26	17 19	17 11	17 02	16 53	16 42	16 30	16 16	16 00	15 40	15 16	14 43	13 47
26	18 31	18 25	18 18	18 10	18 02	17 53	17 43	17 31	17 18	17 03	16 45	16 23	15 53	15 08
27	19 31	19 26	19 20	19 13	19 06	18 58	18 49	18 39	18 28	18 15	18 00	17 42	17 19	16 49
28	20 32	20 27	20 22	20 17	20 11	20 05	19 58	19 50	19 41	19 31	19 19	19 06	18 49	18 29
29	21 32	21 28	21 25	21 21	21 16	21 11	21 06	21 00	20 54	20 47	20 38	20 29	20 17	20 04
30	22 31	22 28	22 26	22 23	22 20	22 17	22 14	22 10	22 06	22 01	21 56	21 50	21 43	21 35
Dec. 1	23 28	23 27	23 26	23 25	23 23	23 22	23 20	23 19	23 17	23 14	23 12	23 09	23 06	23 02
2	.. ..	.. ..	.. ..	.. ..	.. ..	.. ..	.. ..	.. ..	.. ..	.. ..	.. ..	.. ..	.. ..	.. ..
3	0 26	0 26	0 26	0 26	0 26	0 27	0 27	0 27	0 27	0 28	0 28	0 28	0 29	0 29
4	1 24	1 25	1 27	1 28	1 30	1 32	1 34	1 36	1 39	1 41	1 44	1 48	1 52	1 57
5	2 23	2 26	2 29	2 32	2 35	2 39	2 43	2 47	2 52	2 57	3 03	3 10	3 18	3 27
6	3 25	3 28	3 33	3 37	3 42	3 47	3 53	4 00	4 07	4 15	4 24	4 35	4 47	5 02
7	4 28	4 33	4 38	4 44	4 51	4 58	5 05	5 14	5 23	5 34	5 47	6 02	6 20	6 44
8	5 32	5 38	5 44	5 52	5 59	6 08	6 17	6 27	6 39	6 53	7 09	7 29	7 54	8 29
9	6 35	6 42	6 49	6 57	7 05	7 14	7 25	7 36	7 50	8 05	8 24	8 47	9 18	10 07
10	7 34	7 41	7 48	7 56	8 05	8 14	8 24	8 36	8 50	9 05	9 24	9 47	10 18	11 05

(.. ..) indicates phenomenon will occur the next day.
(-- --) indicates Moon continuously below horizon.
(** **) indicates Moon continuously above horizon.

UNIVERSAL TIME FOR MERIDIAN OF GREENWICH

MOONRISE

Lat.	−55°	−50°	−45°	−40°	−35°	−30°	−20°	−10°	0°	+10°	+20°	+30°	+35°	+40°
	h m	h m	h m	h m	h m	h m	h m	h m	h m	h m	h m	h m	h m	h m
Dec. 9	20 17	19 48	19 25	19 07	18 52	18 39	18 16	17 57	17 39	17 21	17 02	16 40	16 27	16 12
10	21 11	20 43	20 21	20 04	19 49	19 36	19 15	18 56	18 38	18 20	18 01	17 40	17 27	17 12
11	21 52	21 27	21 09	20 53	20 40	20 29	20 10	19 53	19 37	19 21	19 04	18 44	18 33	18 19
12	22 22	22 03	21 49	21 36	21 26	21 17	21 01	20 47	20 34	20 21	20 07	19 51	19 41	19 31
13	22 46	22 33	22 22	22 14	22 06	22 00	21 48	21 38	21 29	21 19	21 09	20 57	20 51	20 43
14	23 05	22 58	22 52	22 47	22 43	22 39	22 32	22 26	22 21	22 15	22 10	22 03	21 59	21 55
15	23 22	23 20	23 19	23 18	23 17	23 16	23 14	23 13	23 11	23 10	23 09	23 07	23 06	23 06
16	23 38	23 42	23 45	23 48	23 50	23 52	23 55	23 58	.. ..	.. ..	.. ..	.. ..	.. ..	.. ..
17	23 56	.. ..	.. ..	.. ..	.. ..	.. ..	.. ..	.. ..	0 01	0 04	0 07	0 11	0 13	0 15
18	.. ..	0 05	0 12	0 18	0 24	0 28	0 37	0 44	0 51	0 58	1 05	1 14	1 19	1 25
19	0 15	0 30	0 42	0 51	1 00	1 07	1 20	1 31	1 42	1 53	2 04	2 17	2 25	2 34
20	0 39	0 59	1 15	1 28	1 39	1 49	2 06	2 20	2 34	2 48	3 03	3 21	3 31	3 42
21	1 10	1 35	1 54	2 10	2 23	2 35	2 55	3 12	3 28	3 45	4 02	4 23	4 35	4 48
22	1 51	2 19	2 40	2 58	3 12	3 25	3 47	4 06	4 23	4 41	5 00	5 22	5 35	5 50
23	2 42	3 11	3 33	3 51	4 06	4 19	4 41	5 00	5 18	5 36	5 55	6 18	6 31	6 46
24	3 44	4 11	4 32	4 49	5 03	5 15	5 36	5 54	6 11	6 28	6 47	7 08	7 20	7 34
25	4 53	5 16	5 34	5 49	6 02	6 12	6 31	6 47	7 02	7 17	7 33	7 52	8 03	8 15
26	6 05	6 23	6 38	6 50	7 00	7 09	7 24	7 38	7 50	8 03	8 16	8 31	8 40	8 50
27	7 17	7 30	7 41	7 50	7 58	8 04	8 16	8 26	8 35	8 45	8 55	9 06	9 13	9 20
28	8 28	8 36	8 43	8 49	8 54	8 58	9 06	9 12	9 18	9 24	9 31	9 38	9 43	9 47
29	9 37	9 41	9 44	9 47	9 49	9 51	9 54	9 57	10 00	10 02	10 05	10 08	10 10	10 13
30	10 47	10 46	10 45	10 44	10 43	10 43	10 42	10 41	10 40	10 39	10 39	10 38	10 37	10 37
31	11 56	11 50	11 45	11 41	11 38	11 35	11 30	11 25	11 21	11 17	11 13	11 08	11 05	11 01
32	13 07	12 56	12 47	12 40	12 34	12 29	12 19	12 11	12 03	11 56	11 48	11 39	11 34	11 28
33	14 20	14 04	13 51	13 41	13 32	13 24	13 10	12 59	12 48	12 37	12 26	12 13	12 05	11 57

MOONSET

Lat.	−55°	−50°	−45°	−40°	−35°	−30°	−20°	−10°	0°	+10°	+20°	+30°	+35°	+40°
	h m	h m	h m	h m	h m	h m	h m	h m	h m	h m	h m	h m	h m	h m
Dec. 9	2 36	3 04	3 26	3 43	3 58	4 10	4 32	4 51	5 09	5 27	5 46	6 07	6 20	6 35
10	3 29	3 59	4 21	4 39	4 54	5 07	5 30	5 49	6 07	6 25	6 44	7 06	7 19	7 34
11	4 38	5 05	5 26	5 43	5 58	6 10	6 31	6 49	7 06	7 23	7 41	8 02	8 14	8 28
12	5 58	6 21	6 38	6 53	7 05	7 16	7 34	7 50	8 04	8 19	8 34	8 52	9 02	9 14
13	7 23	7 40	7 54	8 05	8 14	8 23	8 37	8 49	9 00	9 12	9 24	9 37	9 45	9 54
14	8 50	9 01	9 10	9 17	9 23	9 29	9 38	9 46	9 54	10 01	10 09	10 18	10 23	10 29
15	10 16	10 21	10 25	10 29	10 31	10 34	10 38	10 42	10 45	10 48	10 52	10 56	10 58	11 01
16	11 41	11 40	11 40	11 39	11 38	11 38	11 37	11 36	11 35	11 34	11 33	11 32	11 32	11 31
17	13 06	12 59	12 53	12 49	12 45	12 41	12 35	12 30	12 25	12 20	12 15	12 09	12 05	12 01
18	14 29	14 16	14 06	13 58	13 51	13 44	13 33	13 24	13 15	13 06	12 57	12 46	12 40	12 33
19	15 52	15 33	15 19	15 07	14 57	14 48	14 32	14 19	14 07	13 54	13 41	13 26	13 17	13 08
20	17 11	16 47	16 29	16 14	16 01	15 50	15 32	15 15	15 00	14 45	14 29	14 10	13 59	13 47
21	18 24	17 56	17 35	17 18	17 04	16 52	16 30	16 12	15 55	15 37	15 19	14 58	14 46	14 31
22	19 26	18 57	18 35	18 17	18 02	17 49	17 27	17 08	16 50	16 32	16 12	15 50	15 37	15 22
23	20 15	19 48	19 26	19 09	18 55	18 42	18 20	18 02	17 44	17 26	17 08	16 46	16 33	16 18
24	20 52	20 28	20 09	19 54	19 41	19 29	19 10	18 52	18 36	18 20	18 03	17 43	17 31	17 17
25	21 20	21 00	20 45	20 32	20 21	20 11	19 54	19 40	19 26	19 12	18 57	18 40	18 30	18 18
26	21 42	21 27	21 15	21 04	20 56	20 48	20 35	20 23	20 12	20 01	19 50	19 36	19 28	19 19
27	21 59	21 49	21 40	21 33	21 27	21 22	21 12	21 04	20 56	20 49	20 40	20 30	20 25	20 19
28	22 13	22 08	22 03	21 59	21 56	21 53	21 47	21 43	21 38	21 34	21 29	21 24	21 20	21 17
29	22 27	22 26	22 25	22 24	22 23	22 22	22 21	22 20	22 19	22 18	22 17	22 16	22 15	22 14
30	22 40	22 43	22 46	22 48	22 50	22 52	22 55	22 57	23 00	23 02	23 05	23 08	23 10	23 12
31	22 54	23 02	23 08	23 13	23 18	23 22	23 29	23 35	23 41	23 47	23 53	.. ..	.. ..	.. ..
32	23 09	23 22	23 32	23 41	23 48	23 54	.. ..	.. ..	.. ..	.. ..	.. ..	0 01	0 05	0 10
33	23 28	23 46	.. ..	.. ..	.. ..	.. ..	0 05	0 15	0 24	0 34	0 44	0 55	1 01	1 09

(.. ..) indicates phenomenon will occur the next day.

UNIVERSAL TIME FOR MERIDIAN OF GREENWICH

MOONRISE

Lat.	+40°	+42°	+44°	+46°	+48°	+50°	+52°	+54°	+56°	+58°	+60°	+62°	+64°	+66°
	h m	h m	h m	h m	h m	h m	h m	h m	h m	h m	h m	h m	h m	h m
Dec. 9	16 12	16 05	15 58	15 50	15 42	15 32	15 22	15 10	14 56	14 41	14 22	13 58	13 27	12 39
10	17 12	17 06	16 59	16 51	16 42	16 33	16 23	16 11	15 58	15 42	15 24	15 01	14 30	13 43
11	18 19	18 14	18 07	18 00	17 53	17 45	17 35	17 25	17 13	17 00	16 44	16 25	16 01	15 28
12	19 31	19 26	19 21	19 15	19 09	19 02	18 55	18 47	18 38	18 28	18 16	18 02	17 45	17 24
13	20 43	20 40	20 36	20 32	20 28	20 23	20 18	20 12	20 06	19 59	19 51	19 42	19 31	19 18
14	21 55	21 53	21 51	21 49	21 46	21 44	21 41	21 38	21 34	21 30	21 26	21 21	21 15	21 09
15	23 06	23 05	23 05	23 04	23 04	23 03	23 02	23 02	23 01	23 00	22 59	22 58	22 57	22 56
16	.. ..	.. ..	.. ..	.. ..	.. ..	.. ..	.. ..	.. ..	.. ..	.. ..	.. ..	.. ..	.. ..	.. ..
17	0 15	0 16	0 18	0 19	0 20	0 22	0 23	0 25	0 27	0 29	0 31	0 34	0 37	0 41
18	1 25	1 27	1 30	1 33	1 36	1 40	1 43	1 48	1 52	1 57	2 03	2 10	2 17	2 26
19	2 34	2 38	2 42	2 47	2 52	2 57	3 03	3 10	3 17	3 25	3 34	3 45	3 58	4 14
20	3 42	3 48	3 53	3 59	4 06	4 13	4 21	4 30	4 40	4 51	5 04	5 20	5 39	6 04
21	4 48	4 55	5 01	5 08	5 16	5 25	5 34	5 45	5 58	6 12	6 29	6 49	7 16	7 53
22	5 50	5 57	6 04	6 12	6 20	6 30	6 40	6 52	7 06	7 22	7 41	8 05	8 38	9 31
23	6 46	6 52	6 59	7 07	7 16	7 25	7 36	7 48	8 02	8 18	8 37	9 00	9 32	10 24
24	7 34	7 40	7 47	7 54	8 02	8 11	8 20	8 31	8 44	8 58	9 15	9 35	10 01	10 37
25	8 15	8 20	8 26	8 32	8 39	8 47	8 55	9 04	9 14	9 26	9 40	9 56	10 15	10 40
26	8 50	8 54	8 59	9 04	9 10	9 16	9 22	9 29	9 37	9 46	9 57	10 09	10 23	10 40
27	9 20	9 24	9 27	9 31	9 35	9 39	9 44	9 49	9 55	10 02	10 09	10 17	10 27	10 38
28	9 47	9 49	9 52	9 54	9 57	10 00	10 03	10 06	10 10	10 14	10 18	10 23	10 29	10 36
29	10 13	10 13	10 14	10 16	10 17	10 18	10 19	10 21	10 22	10 24	10 26	10 28	10 31	10 34
30	10 37	10 37	10 36	10 36	10 36	10 35	10 35	10 35	10 34	10 34	10 33	10 33	10 32	10 32
31	11 01	11 00	10 59	10 57	10 55	10 53	10 51	10 49	10 47	10 44	10 41	10 38	10 34	10 29
32	11 28	11 25	11 22	11 19	11 16	11 13	11 09	11 05	11 00	10 55	10 50	10 43	10 36	10 27
33	11 57	11 53	11 49	11 44	11 40	11 35	11 29	11 23	11 16	11 09	11 00	10 51	10 39	10 26

MOONSET

Lat.	+40°	+42°	+44°	+46°	+48°	+50°	+52°	+54°	+56°	+58°	+60°	+62°	+64°	+66°
	h m	h m	h m	h m	h m	h m	h m	h m	h m	h m	h m	h m	h m	h m
Dec. 9	6 35	6 42	6 49	6 57	7 05	7 14	7 25	7 36	7 50	8 05	8 24	8 47	9 18	10 07
10	7 34	7 41	7 48	7 56	8 05	8 14	8 24	8 36	8 50	9 05	9 24	9 47	10 18	11 05
11	8 28	8 34	8 40	8 48	8 55	9 04	9 13	9 24	9 36	9 50	10 06	10 26	10 50	11 23
12	9 14	9 19	9 25	9 31	9 37	9 44	9 52	10 01	10 10	10 21	10 34	10 48	11 06	11 28
13	9 54	9 58	10 02	10 07	10 11	10 17	10 22	10 29	10 36	10 44	10 52	11 03	11 15	11 29
14	10 29	10 32	10 34	10 37	10 40	10 44	10 47	10 51	10 56	11 01	11 06	11 12	11 19	11 28
15	11 01	11 02	11 03	11 04	11 06	11 07	11 09	11 11	11 12	11 15	11 17	11 20	11 23	11 26
16	11 31	11 31	11 30	11 30	11 30	11 29	11 29	11 28	11 28	11 27	11 27	11 26	11 25	11 24
17	12 01	11 59	11 58	11 56	11 54	11 51	11 49	11 46	11 43	11 40	11 36	11 32	11 27	11 22
18	12 33	12 30	12 26	12 23	12 19	12 15	12 11	12 06	12 00	11 54	11 47	11 40	11 31	11 20
19	13 08	13 03	12 58	12 53	12 48	12 42	12 36	12 28	12 20	12 11	12 01	11 49	11 36	11 19
20	13 47	13 41	13 35	13 29	13 22	13 14	13 06	12 57	12 46	12 34	12 20	12 04	11 44	11 19
21	14 31	14 25	14 18	14 11	14 03	13 54	13 44	13 33	13 20	13 06	12 49	12 28	12 01	11 23
22	15 22	15 15	15 08	15 00	14 51	14 42	14 31	14 19	14 06	13 49	13 30	13 06	12 34	11 40
23	16 18	16 11	16 04	15 56	15 48	15 39	15 28	15 16	15 03	14 47	14 28	14 04	13 33	12 41
24	17 17	17 11	17 05	16 58	16 50	16 42	16 32	16 22	16 10	15 56	15 39	15 19	14 53	14 18
25	18 18	18 13	18 08	18 02	17 55	17 48	17 40	17 32	17 22	17 10	16 57	16 42	16 23	15 59
26	19 19	19 15	19 11	19 06	19 01	18 55	18 49	18 43	18 35	18 27	18 17	18 06	17 53	17 37
27	20 19	20 16	20 13	20 09	20 06	20 02	19 58	19 53	19 48	19 42	19 36	19 29	19 20	19 10
28	21 17	21 15	21 13	21 12	21 10	21 08	21 05	21 03	21 00	20 57	20 53	20 49	20 44	20 39
29	22 14	22 14	22 14	22 13	22 13	22 12	22 12	22 11	22 10	22 10	22 09	22 08	22 07	22 06
30	23 12	23 12	23 13	23 14	23 15	23 17	23 18	23 19	23 21	23 22	23 24	23 26	23 29	23 32
31	.. ..	.. ..	.. ..	.. ..	.. ..	.. ..	.. ..	.. ..	.. ..	.. ..	.. ..	.. ..	.. ..	.. ..
32	0 10	0 12	0 14	0 16	0 19	0 22	0 25	0 28	0 32	0 36	0 41	0 46	0 52	1 00
33	1 09	1 12	1 16	1 20	1 24	1 28	1 33	1 39	1 45	1 52	1 59	2 08	2 19	2 31

(.. ..) indicates phenomenon will occur the next day.
(-- --) indicates Moon continuously below horizon.
(** **) indicates Moon continuously above horizon.

There are five eclipses, three of the Sun and two of the Moon.

I	January 4-5	Annular eclipse of the Sun
II	June 15	Partial eclipse of the Moon
III	June 30	Total eclipse of the Sun
IV	December 9-10	Total eclipse of the Moon
V	December 23-24	Partial eclipse of the Sun

A standard correction of +0″.5 has been applied to the tabular longitude of the Moon, and of –0″.25 has been applied to the tabular latitude of the Moon, to help correct for the difference between the center of figure and the center of mass when using the LE200 lunar ephemeris.

The arguments are given provisionally in Universal Time, using $\Delta T(\mathrm{A}) = +58^{\mathrm{s}}$. Once the value of ΔT is known, the data on these pages may be expressed in Universal Time as follows:

Define $\delta T = \Delta T - \Delta T(\mathrm{A})$.

Convert all arguments in preliminary Universal Time by subtracting δT.

Apply the correction $1.0027\,\delta T$ to μ and the longitudes in such a way that if δT is positive, μ decreases and the longitudes shift to the east.

Leave all other quantities unchanged.

Longitude is positive to the east, and negative to the west.

I.—*Annular Eclipse of the Sun*, January 4-5.

ELEMENTS OF THE ECLIPSE

U.T. of geocentric conjunction in right ascension January $4^{\mathrm{d}}\ 23^{\mathrm{h}}\ 14^{\mathrm{m}}\ 43^{\mathrm{s}}.499$

Julian Day No. = 2448626.4685590107

	h m s		s s
R.A. of Sun and Moon	19 00 11.850	Hourly motions	10.995 and 126.011
	° ′ ″		′ ″
Declination of Sun	−22 43 10.39	Hourly motion	+ 0 15.80
Declination of Moon	−22 20 39.72	Hourly motion	+ 5 46.57
Equatorial hor. par. of Sun	8.94	True semidiameter of Sun	16 15.9
Equatorial hor. par. of Moon	54 02.75	True semidiameter of Moon	14 43.6

CIRCUMSTANCES OF THE ECLIPSE

		U.T.	Longitude	Latitude
		d h m	° ′	° ′
Eclipse begins	January	4 20 03.6	+151 38.1	+ 3 08.6
Central eclipse begins		4 21 16.0	+137 07.3	+11 32.7
Central eclipse at local apparent noon		4 23 14.7	−167 27.2	+ 1 58.9
Central eclipse ends		5 0 53.1	−117 45.2	+32 53.2
Eclipse ends		5 2 05.6	−131 16.4	+24 42.5

II.—*Partial Eclipse of the Moon*, June 15; the beginning of the umbral phase visible in South America, Central America, the United States except the northwest, southeastern Canada, Antarctica, the eastern South Pacific Ocean, extreme western Europe, Africa except the northeast, and the Atlantic Ocean; the end visible in New Zealand, Hawaii, South America, Central America, North America except Alaska and northern Canada, Antarctica, extreme western Africa, the South Pacific Ocean, the southeastern North Pacific Ocean, and most of the Atlantic Ocean except the northeast.

ELEMENTS OF THE ECLIPSE

U.T. of geocentric opposition in right ascension June $15^d\ 04^h\ 52^m\ 51^s.466$

Julian Day No. = 2448788.7033734503

	h m s		s
R.A. of Sun	5 35 20.235	Hourly motion	10.387
R.A. of Moon	17 35 20.235	Hourly motion	136.006
	° ′ ″		′ ″
Declination of Sun	+23 19 08.68	Hourly motion	+ 0 06.12
Declination of Moon	−23 54 00.16	Hourly motion	+ 1 31.29
Equatorial hor. par. of Sun	8.66	True semidiameter of Sun	15 44.7
Equatorial hor. par. of Moon	55 20.39	True semidiameter of Moon	15 04.8

CIRCUMSTANCES OF THE ECLIPSE

		d	h m	
Moon enters penumbra	June	15	2 09.1	
Moon enters umbra		15	3 26.6	
Middle of eclipse		15	4 57.0	U.T.
Moon leaves umbra		15	6 27.2	
Moon leaves penumbra		15	7 44.9	

Contacts of Umbra with Limb of Moon	Position Angles from the North Point	The Moon being in the Zenith in Longitude	Latitude
	°	° ′	° ′
First	48.0 to East	− 52 18.4	− 23 56.1
Last	54.4 to West	− 95 52.3	− 23 51.5

Magnitude of the eclipse: 0.687

III.—*Total Eclipse of the Sun*, June 30.

ELEMENTS OF THE ECLIPSE

U.T. of geocentric conjunction in right ascension June $30^d\ 12^h\ 23^m\ 22^s.989$

Julian Day No. = 2448804.0162382999

	h m s		s		s
R.A. of Sun and Moon	6 38 57.402	Hourly motions	10.350	and	158.436

	° ′ ″		′ ″
Declination of Sun	+23 08 17.09	Hourly motion	− 0 09.63
Declination of Moon	+22 22 19.96	Hourly motion	− 5 51.42
Equatorial hor. par. of Sun	8.65	True semidiameter of Sun	15 43.9
Equatorial hor. par. of Moon	60 29.24	True semidiameter of Moon	16 28.9

Total Eclipse of the Sun, June 30 (continued).

CIRCUMSTANCES OF THE ECLIPSE

		U.T.	Longitude	Latitude
		d h m	° ′	° ′
Eclipse begins	June	30 09 50.9	− 48 39.1	− 18 22.2
Central eclipse begins		30 11 01.7	− 56 32.7	− 35 49.2
Central eclipse at local apparent noon		30 12 23.4	− 4 55.7	− 26 39.5
Central eclipse ends		30 13 18.9	+ 38 36.9	− 51 33.9
Eclipse ends		30 14 29.7	+ 35 52.1	− 35 19.6

IV.—*Total Eclipse of the Moon*, December 9-10; the beginning of the umbral phase visible in northeastern South America, eastern and northern North America, Greenland, the Arctic regions, Africa, Europe, most of Asia, most of the Atlantic Ocean, and the Indian Ocean; the end visible in South America, Central America, most of North America, Greenland, the Arctic regions, Africa, Europe, northern and western Asia, the eastern Pacific Ocean, the Atlantic Ocean, and the western Indian Ocean.

ELEMENTS OF THE ECLIPSE

U.T. of geocentric opposition in right ascension December $9^d\ 23^h\ 43^m\ 43^s.663$

Julian Day No. = 2448966.4886998031

	h m s		s
R.A. of Sun	17 08 34.398	Hourly motion	10.986
R.A. of Moon	5 08 34.398	Hourly motion	150.781
	° ′ ″		′ ″
Declination of Sun	−22 54 48.21	Hourly motion	− 0 13.49
Declination of Moon	+23 13 09.62	Hourly motion	− 0 07.20
Equatorial hor. par. of Sun	8.93	True semidiameter of Sun	16 14.5
Equatorial hor. par. of Moon	58 24.18	True semidiameter of Moon	15 54.9

CIRCUMSTANCES OF THE ECLIPSE

		d h m	
Moon enters penumbra	December	9 20 55.4	
Moon enters umbra		9 21 59.4	
Moon enters totality		9 23 06.8	
Middle of eclipse		9 23 44.1	U.T.
Moon leaves totality		10 0 21.5	
Moon leaves umbra		10 1 28.8	
Moon leaves penumbra		10 2 32.7	

Contacts of Umbra with Limb of Moon	Position Angles from the North Point	The Moon being in the Zenith in Longitude	Latitude
	°	° ′	° ′
First	108.7 to East	+ 27 18.5	+ 23 13.1
Last	107.5 to West	− 22 59.9	+ 23 12.7

Magnitude of the eclipse: 1.276

V.—*Partial Eclipse of the Sun*, December 23-24.

ELEMENTS OF THE ECLIPSE

U.T. of geocentric conjunction in right ascension December $24^d\ 00^h\ 45^m\ 09^s.177$

Julian Day No. = 2448980.5313562183

	h m s		s s
R.A. of Sun and Moon	18 10 44.453	Hourly motions	11.104 and 137.162
	° ′ ″		′ ″
Declination of Sun	−23 25 00.08	Hourly motion	+ 0 02.85
Declination of Moon	−22 24 45.87	Hourly motion	+ 3 28.17
Equatorial hor. par. of Sun	8.94	True semidiameter of Sun	16 15.6
Equatorial hor. par. of Moon	56 00.29	True semidiameter of Moon	15 15.7

CIRCUMSTANCES OF THE ECLIPSE

		U.T.	Longitude	Latitude
		d h m	° ′	° ′
Eclipse begins	December	23 22 20.7	+131 11.6	+33 15.3
Greatest eclipse		24 00 30.7	+155 44.6	+65 41.6
Eclipse ends		24 02 40.7	−155 45.9	+44 48.8

Magnitude of greatest eclipse: 0.843

ANNULAR SOLAR ECLIPSE OF 4-5 JANUARY 1992

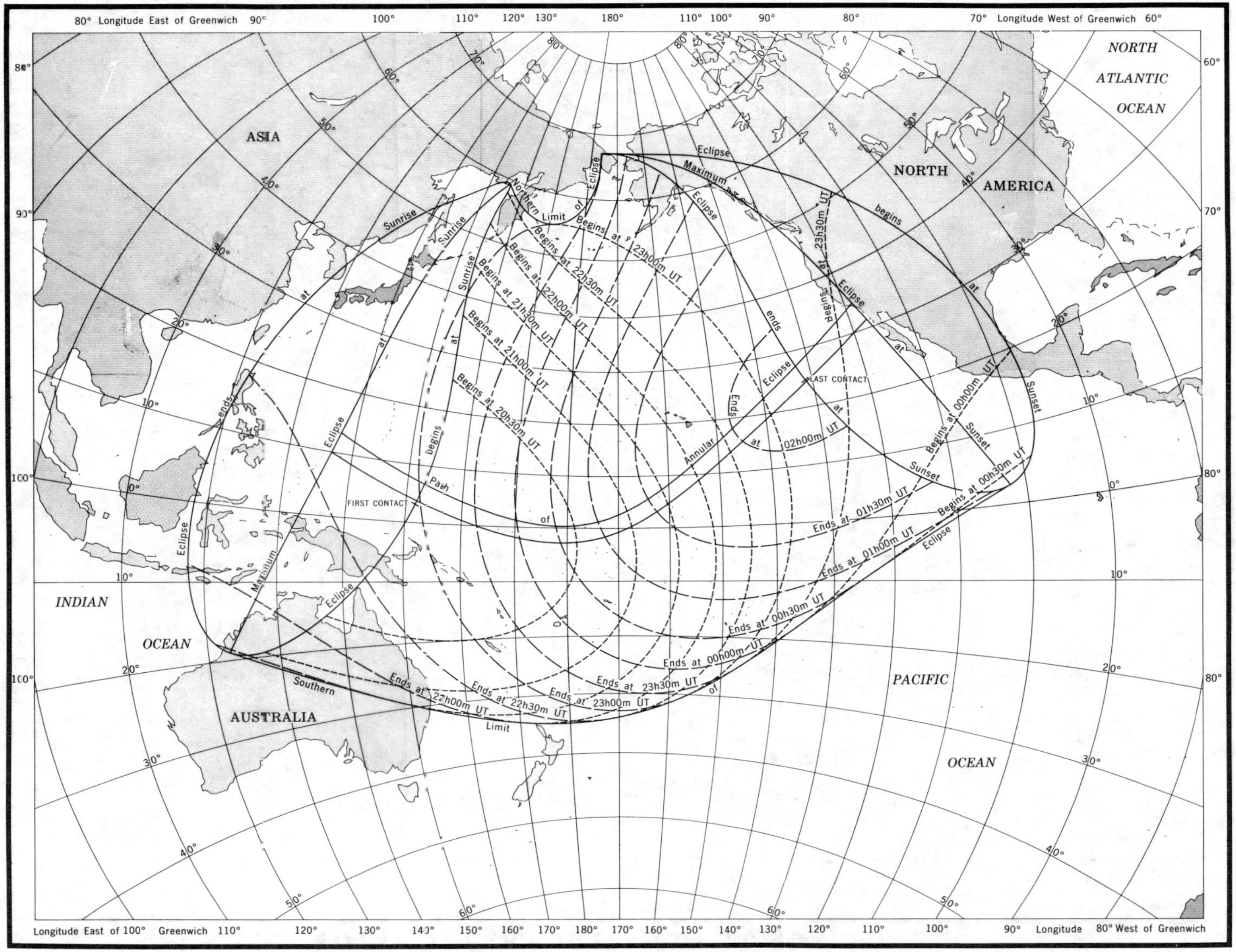

TOTAL SOLAR ECLIPSE OF 30 JUNE 1992

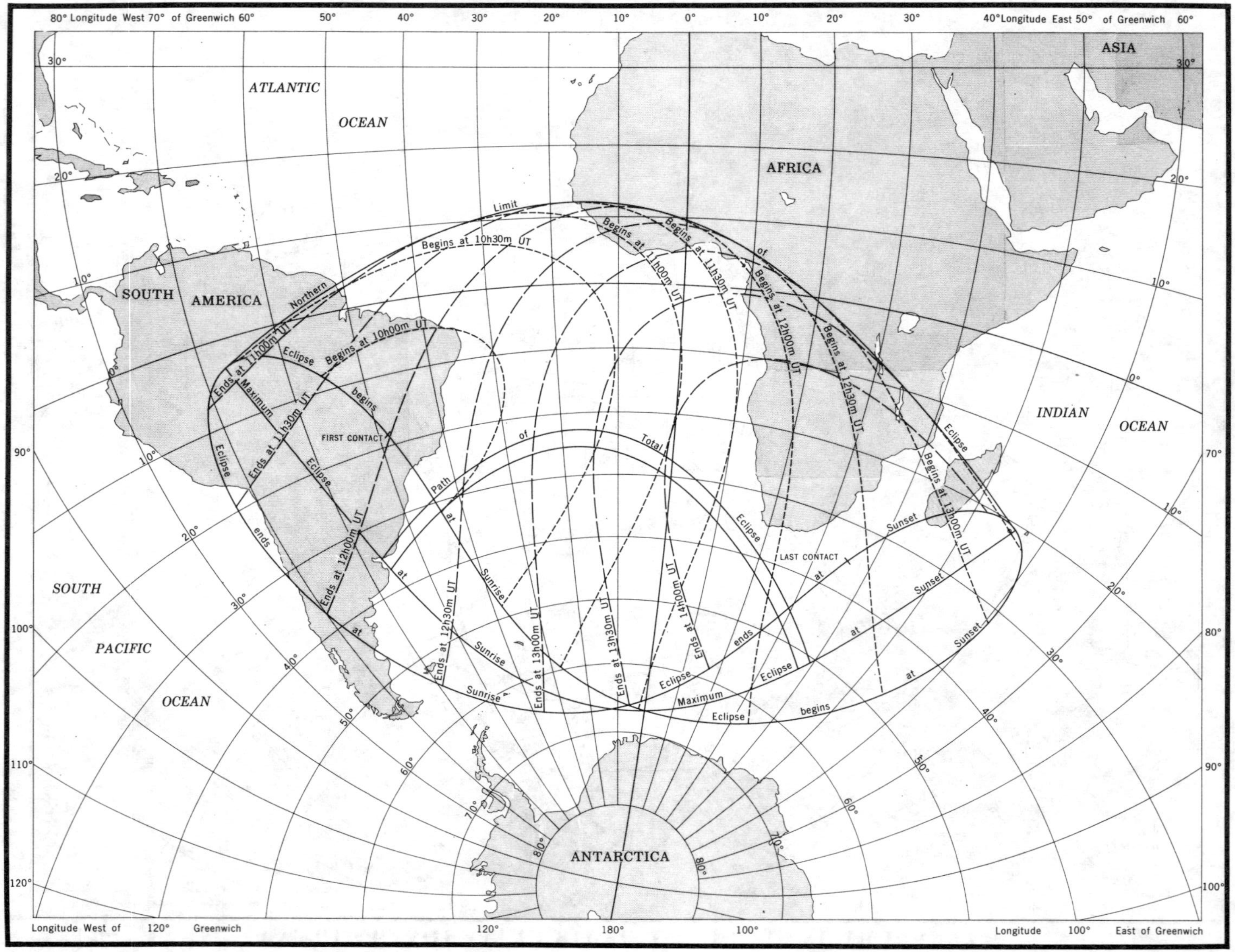

PARTIAL SOLAR ECLIPSE OF 23-24 DECEMBER 1992

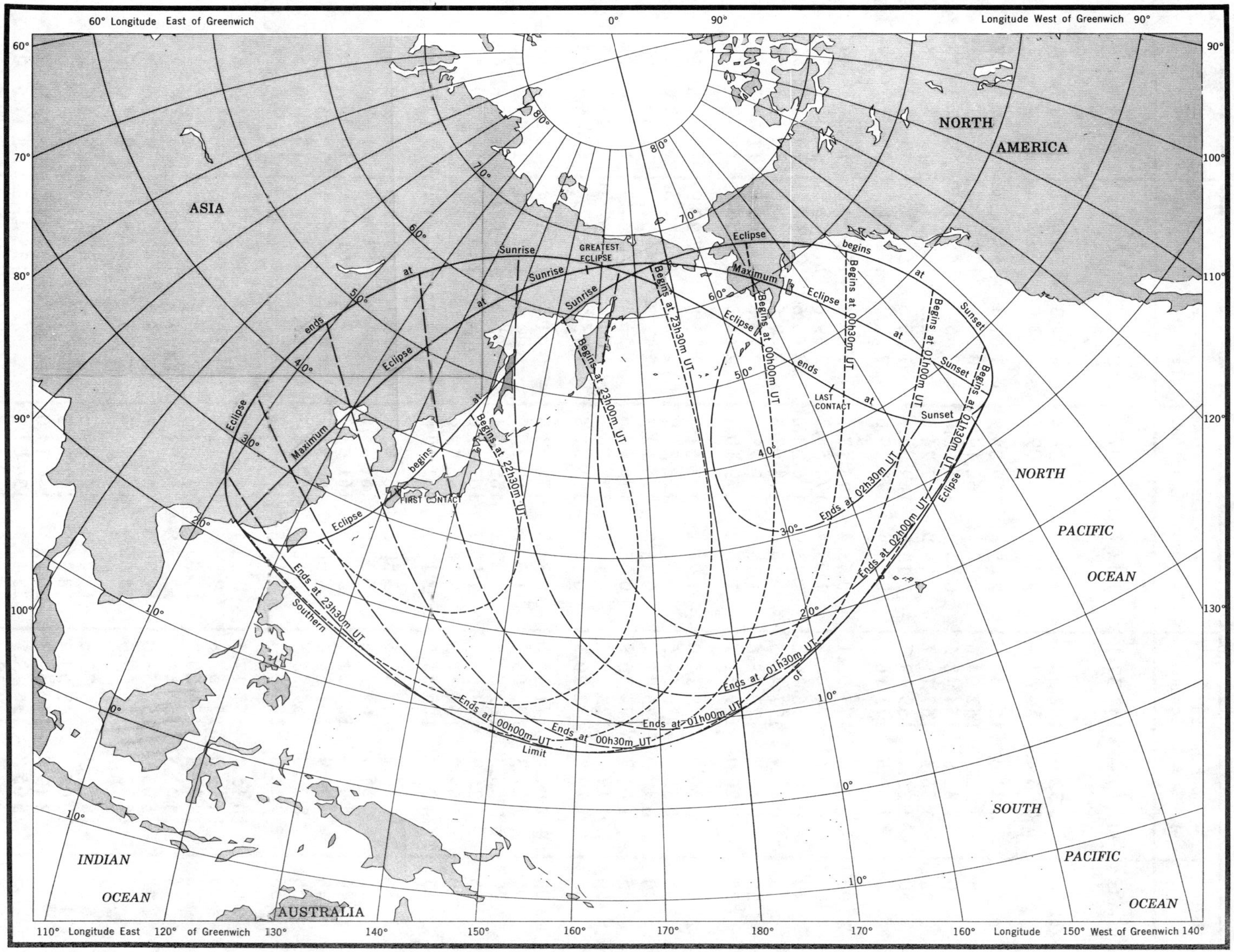

PATH OF CENTRAL PHASE: ANNULAR SOLAR ECLIPSE OF JANUARY 4-5

U.T.	Northern Limit Latitude	Northern Limit Longitude	Central Line Latitude	Central Line Longitude	Southern Limit Latitude	Southern Limit Longitude	Central Line Duration	Alt.
	° ′	° ′	° ′	° ′	° ′	° ′	m s	°
Limits	+13 10.4	+137 37.1	+11 32.7	+137 07.3	+ 9 56.2	+136 36.9	7 14.1	–
h m								
21 20	+ 8 32.0	+148 34.4	+ 6 23.7	+149 30.4	+ 4 24.3	+150 09.1	7 56.7	14
25	6 04.8	154 42.4	4 11.6	155 09.6	2 23.0	155 29.2	8 22.9	21
21 30	+ 4 33.0	+158 49.7	+ 2 45.6	+159 08.2	+ 1 01.4	+159 21.8	8 44.4	27
35	3 26.5	162 04.6	1 42.4	162 19.5	+ 0 00.9	162 30.8	9 03.6	31
40	2 35.8	164 48.6	0 53.9	165 02.0	– 0 45.7	165 12.4	9 21.3	35
45	1 56.3	167 11.7	+ 0 16.0	167 24.6	1 22.2	167 35.2	9 37.8	38
50	1 25.4	169 19.7	– 0 13.6	169 32.7	1 50.6	169 43.9	9 53.2	42
55	1 01.5	171 16.0	0 36.4	171 29.6	2 12.4	171 41.7	10 07.7	44
22 00	+ 0 43.6	+173 03.1	– 0 53.4	+173 17.6	– 2 28.6	+173 30.8	10 21.1	47
05	0 30.6	174 42.7	1 05.4	174 58.3	2 39.8	175 12.8	10 33.6	50
10	0 22.2	176 16.2	1 13.0	176 33.0	2 46.6	176 48.9	10 45.1	52
15	0 17.9	177 44.6	1 16.6	178 02.6	2 49.5	178 20.0	10 55.5	54
20	0 17.2	+179 08.6	1 16.5	+179 28.0	2 48.6	+179 46.8	11 04.7	56
25	0 20.0	–179 31.1	1 13.0	–179 10.3	2 44.3	–178 50.0	11 12.9	58
22 30	+ 0 26.0	–178 14.0	– 1 06.2	–177 51.8	– 2 36.8	–177 29.9	11 19.8	60
35	0 35.0	176 59.5	0 56.4	176 35.9	2 26.2	176 12.5	11 25.6	61
40	0 47.0	175 47.2	0 43.7	175 22.3	2 12.7	174 57.4	11 30.1	63
45	1 01.7	174 36.9	0 28.2	174 10.6	1 56.4	173 44.3	11 33.4	64
50	1 19.1	173 28.1	0 10.0	173 00.5	1 37.4	172 32.8	11 35.4	65
55	1 39.2	172 20.5	0 11.0	171 51.6	1 15.7	171 22.6	11 36.2	65
23 00	+ 2 02.0	–171 14.0	+ 0 34.5	–170 43.8	– 0 51.4	–170 13.4	11 35.8	66
05	2 27.3	170 08.0	1 00.6	169 36.7	– 0 24.4	169 05.1	11 34.3	66
10	2 55.3	169 02.5	1 29.3	168 30.1	+ 0 05.0	167 57.3	11 31.6	66
15	3 25.9	167 57.1	2 00.7	167 23.6	0 37.1	166 49.7	11 27.9	65
20	3 59.2	166 51.5	2 34.6	166 17.0	1 11.8	165 42.1	11 23.1	65
25	4 35.3	165 45.5	3 11.3	165 10.0	1 49.1	164 34.1	11 17.4	64
23 30	+ 5 14.1	–164 38.6	+ 3 50.8	–164 02.3	+ 2 29.1	–163 25.5	11 10.9	63
35	5 55.9	163 30.6	4 33.0	162 53.6	3 11.9	162 15.9	11 03.5	61
40	6 40.8	162 21.1	5 18.3	161 43.3	3 57.6	161 04.9	10 55.4	60
45	7 28.9	161 09.6	6 06.7	160 31.2	4 46.4	159 52.1	10 46.5	58
50	8 20.5	159 55.5	6 58.5	159 16.7	5 38.3	158 37.1	10 37.0	56
55	9 15.8	158 38.4	7 53.8	157 59.2	6 33.8	157 19.1	10 27.0	54
0 00	+10 15.2	–157 17.5	+ 8 53.1	–156 38.0	+ 7 32.9	–155 57.6	10 16.3	52
05	11 19.0	155 51.8	9 56.6	155 12.2	8 36.2	154 31.6	10 05.2	49
10	12 27.9	154 20.2	11 05.0	153 40.7	9 44.1	153 00.1	9 53.5	47
15	13 42.5	152 41.3	12 18.8	152 02.2	10 57.3	151 21.8	9 41.3	44
20	15 03.9	150 53.0	13 39.0	150 14.7	12 16.5	149 34.7	9 28.5	41
25	16 33.5	148 52.5	15 06.9	148 15.4	13 43.0	147 36.4	9 15.2	38
0 30	+18 13.4	–146 35.6	+16 44.4	–146 00.6	+15 18.4	–145 23.1	9 01.1	34
35	20 06.8	143 55.4	18 34.5	143 23.8	17 05.7	142 49.1	8 46.2	30
40	22 19.8	140 39.3	20 42.4	140 14.0	19 09.2	139 44.1	8 29.9	26
45	25 05.8	136 19.3	23 19.1	136 07.2	21 38.6	135 47.3	8 11.5	20
50	29 12.2	129 13.0	26 59.1	129 47.7	25 00.7	129 58.3	7 48.2	13
Limits	+34 27.6	–118 31.1	+32 53.2	–117 45.2	+31 19.6	–117 00.3	7 15.9	–

BESSELIAN ELEMENTS: ANNULAR SOLAR ECLIPSE OF JANUARY 4-5

U.T.	Intersection of Axis of Shadow with Fundamental Plane		Direction of Axis of Shadow			Radius of Shadow on Fundamental Plane	
	x	y	sin d	cos d	μ	Penumbra	Umbra
h m					°		
19 40	−1.766179	+0.053218	−0.386475	0.922300	113.78516	0.574830	+0.028303
50	1.683928	0.070113	.386464	.922305	116.28459	.574846	.028318
20 00	−1.601675	+0.087015	−0.386453	0.922309	118.78401	0.574862	+0.028334
10	1.519422	0.103925	.386442	.922314	121.28344	.574877	.028349
20	1.437168	0.120843	.386431	.922319	123.78287	.574891	.028363
30	1.354913	0.137769	.386420	.922323	126.28230	.574905	.028377
40	1.272658	0.154702	.386409	.922328	128.78173	.574918	.028390
50	1.190402	0.171644	.386398	.922332	131.28116	.574931	.028403
21 00	−1.108146	+0.188593	−0.386387	0.922337	133.78059	0.574943	+0.028415
10	1.025890	0.205551	.386375	.922342	136.28002	.574955	.028426
20	0.943633	0.222515	.386364	.922346	138.77945	.574966	.028437
30	0.861377	0.239488	.386353	.922351	141.27888	.574976	.028448
40	0.779122	0.256468	.386342	.922355	143.77831	.574986	.028458
50	0.696866	0.273456	.386331	.922360	146.27774	.574996	.028467
22 00	−0.614612	+0.290451	−0.386320	0.922365	148.77717	0.575005	+0.028476
10	0.532358	0.307454	.386309	.922369	151.27660	.575013	.028484
20	0.450105	0.324465	.386298	.922374	153.77603	.575021	.028492
30	0.367853	0.341482	.386287	.922379	156.27546	.575028	.028499
40	0.285603	0.358508	.386276	.922383	158.77489	.575035	.028506
50	0.203354	0.375541	.386265	.922388	161.27432	.575041	.028512
23 00	−0.121106	+0.392581	−0.386254	0.922393	163.77375	0.575046	+0.028518
10	−0.038860	0.409628	.386243	.922397	166.27319	.575051	.028523
20	+0.043384	0.426683	.386232	.922402	168.77262	.575056	.028527
30	0.125626	0.443745	.386221	.922406	171.27205	.575060	.028531
40	0.207866	0.460814	.386209	.922411	173.77148	.575063	.028534
50	0.290103	0.477891	.386198	.922416	176.27091	.575066	.028537
0 00	+0.372338	+0.494974	−0.386187	0.922420	178.77034	0.575068	+0.028539
10	0.454571	0.512065	.386176	.922425	181.26977	.575070	.028541
20	0.536800	0.529163	.386165	.922430	183.76920	.575071	.028542
30	0.619027	0.546268	.386154	.922434	186.26863	.575072	.028543
40	0.701251	0.563379	.386143	.922439	188.76807	.575072	.028543
50	0.783471	0.580498	.386132	.922444	191.26750	.575071	.028542
1 00	+0.865688	+0.597624	−0.386120	0.922448	193.76693	0.575070	+0.028541
10	0.947902	0.614757	.386109	.922453	196.26636	.575069	.028540
20	1.030111	0.631896	.386098	.922458	198.76579	.575067	.028538
30	1.112317	0.649043	.386087	.922462	201.26522	.575064	.028535
40	1.194519	0.666196	.386076	.922467	203.76466	.575061	.028532
50	1.276717	0.683356	.386065	.922472	206.26409	.575057	.028528
2 00	+1.358911	+0.700523	−0.386053	0.922476	208.76352	0.575052	+0.028524
10	1.441100	0.717696	.386042	.922481	211.26295	.575048	.028519
20	1.523285	0.734876	.386031	.922486	213.76238	.575042	.028514
2 30	+1.605465	+0.752063	−0.386020	0.922490	216.26182	0.575036	+0.028508

tan f_1	0.004756	
tan f_2	0.004733	
μ'	0.261740	radians per hour
d'	+0.000072	radians per hour

PATH OF CENTRAL PHASE: TOTAL SOLAR ECLIPSE OF JUNE 30

U.T.	Northern Limit Latitude	Northern Limit Longitude	Central Line Latitude	Central Line Longitude	Southern Limit Latitude	Southern Limit Longitude	Central Line Duration	Central Line Alt.
	° ′	° ′	° ′	° ′	° ′	° ′	m s	°
Limits	−34 53.4	− 56 51.4	−35 49.2	− 56 32.7	−36 46.1	− 56 13.3	2 57.8	–
h m								
11 02	−31 08.3	− 48 27.8	−34 13.3	− 52 58.6	——	——	3 07.0	3
04	29 27.2	44 29.8	31 29.1	46 39.7	−34 00.4	− 49 55.8	3 25.5	10
06	28 18.6	41 40.4	30 05.1	43 15.1	32 03.2	45 15.8	3 36.8	13
08	27 25.6	39 23.3	29 04.4	40 40.0	30 50.3	42 12.4	3 46.0	16
11 10	−26 42.3	− 37 25.8	−28 16.4	− 38 31.3	−29 55.6	− 39 47.8	3 54.0	18
12	26 05.9	35 41.8	27 36.9	36 39.5	29 11.8	37 45.6	4 01.2	20
14	25 34.8	34 07.9	27 03.4	34 59.7	28 35.4	35 58.4	4 07.9	22
16	25 07.9	32 41.7	26 34.8	33 29.0	28 04.5	34 22.0	4 14.0	24
18	24 44.4	31 21.7	26 10.0	32 05.4	27 38.0	32 54.0	4 19.7	25
11 20	−24 23.9	− 30 06.9	−25 48.4	− 30 47.5	−27 15.1	− 31 32.6	4 25.1	27
25	23 43.1	27 17.1	25 05.6	27 52.2	26 30.1	28 30.6	4 37.3	30
11 30	−23 14.3	− 24 45.7	−24 35.7	− 25 17.0	−25 58.6	− 25 51.0	4 47.9	32
35	22 55.2	22 27.7	24 15.7	22 56.2	25 37.7	23 26.9	4 57.0	34
40	22 44.1	20 19.6	24 04.0	20 46.0	25 25.4	21 14.1	5 04.8	36
45	22 40.1	18 19.0	23 59.6	18 43.7	25 20.5	19 09.9	5 11.3	38
50	22 42.5	16 24.2	24 01.7	16 47.5	25 22.3	17 12.0	5 16.6	39
55	22 50.8	14 33.7	24 09.8	14 55.9	25 30.1	15 19.0	5 20.6	40
12 00	−23 04.7	− 12 46.5	−24 23.5	− 13 07.6	−25 43.6	− 13 29.4	5 23.5	41
05	23 24.0	11 01.4	24 42.7	11 21.5	26 02.7	11 42.1	5 25.1	41
10	23 48.6	9 17.7	25 07.2	9 36.8	26 27.2	9 56.2	5 25.6	41
15	24 18.5	7 34.3	25 37.1	7 52.4	26 57.1	8 10.7	5 24.8	41
20	24 53.7	5 50.5	26 12.4	6 07.5	27 32.5	6 24.5	5 22.9	41
25	25 34.6	4 05.2	26 53.4	4 21.1	28 13.8	4 36.7	5 19.9	40
30	26 21.3	2 17.6	27 40.4	2 32.0	29 01.1	2 46.0	5 15.7	39
12 35	−27 14.4	− 0 26.4	−28 33.9	− 0 39.2	−29 55.2	− 0 51.3	5 10.4	38
40	28 14.4	+ 1 29.9	29 34.5	+ 1 19.1	30 56.6	+ 1 09.2	5 04.0	36
45	29 22.3	3 32.9	30 43.4	3 24.7	32 06.6	3 17.7	4 56.4	35
50	30 39.3	5 45.3	32 01.8	5 40.3	33 26.6	5 37.0	4 47.6	32
55	32 07.3	8 10.5	33 31.8	8 09.8	34 59.1	8 11.5	4 37.5	30
13 00	−33 49.4	+ 10 53.7	−35 16.8	+ 10 59.1	−36 47.7	+ 11 08.0	4 26.0	27
02	34 35.1	12 05.8	36 04.2	12 14.5	37 37.1	12 27.2	4 21.0	26
04	35 24.3	13 23.0	36 55.4	13 35.6	38 30.8	13 53.0	4 15.6	24
06	36 17.5	14 46.4	37 51.2	15 03.8	39 29.7	15 27.2	4 09.8	23
08	37 15.6	16 17.5	38 52.6	16 41.1	40 35.3	17 12.2	4 03.7	21
13 10	−38 19.9	+ 17 58.5	−40 01.2	+ 18 30.2	−41 49.6	+ 19 11.9	3 57.0	19
12	39 32.1	19 52.8	41 19.5	20 36.0	43 16.3	21 33.3	3 49.6	17
14	40 55.4	22 06.1	42 52.1	23 06.8	45 03.0	24 30.6	3 41.1	14
16	42 35.5	24 49.4	44 49.0	26 21.7	47 31.2	28 45.2	3 31.0	11
18	44 47.1	+ 28 30.5	47 46.3	31 30.3	——	——	3 16.8	6
Limits	−50 43.7	+ 39 22.1	−51 33.9	+ 38 36.9	−52 24.7	+ 37 48.9	3 00.4	–

BESSELIAN ELEMENTS: TOTAL SOLAR ECLIPSE OF JUNE 30

U.T.	Intersection of Axis of Shadow with Fundamental Plane		Direction of Axis of Shadow			Radius of Shadow on Fundamental Plane	
	x	y	sin d	cos d	μ	Penumbra	Umbra
h m					°		
9 20	−1.733994	−0.474861	+0.393097	0.919497	319.08427	0.533977	−0.012347
30	1.639457	0.490430	.393091	.919500	321.58418	.533976	.012348
40	1.544917	0.506007	.393084	.919503	324.08408	.533974	.012350
50	1.450373	0.521591	.393078	.919505	326.58399	.533971	.012353
10 00	−1.355826	−0.537183	+0.393071	0.919508	329.08389	0.533968	−0.012356
10	1.261277	0.552782	.393065	.919511	331.58380	.533964	.012360
20	1.166725	0.568388	.393058	.919514	334.08370	.533959	.012365
30	1.072170	0.584002	.393052	.919516	336.58360	.533954	.012370
40	0.977614	0.599623	.393045	.919519	339.08351	.533948	.012376
50	0.883056	0.615251	.393039	.919522	341.58341	.533941	.012383
11 00	−0.788496	−0.630886	+0.393032	0.919525	344.08332	0.533933	−0.012390
10	0.693935	0.646528	.393026	.919528	346.58322	.533925	.012398
20	0.599373	0.662177	.393019	.919530	349.08313	.533916	.012407
30	0.504810	0.677834	.393013	.919533	351.58303	.533907	.012417
40	0.410246	0.693497	.393006	.919536	354.08294	.533897	.012427
50	0.315682	0.709167	.393000	.919539	356.58284	.533886	.012438
12 00	−0.221119	−0.724844	+0.392993	0.919541	359.08274	0.533874	−0.012449
10	0.126555	0.740528	.392986	.919544	1.58265	.533862	.012461
20	−0.031992	0.756219	.392980	.919547	4.08255	.533849	.012474
30	+0.062570	0.771916	.392973	.919550	6.58246	.533835	.012488
40	0.157132	0.787620	.392967	.919553	9.08236	.533821	.012502
50	0.251692	0.803331	.392960	.919555	11.58227	.533806	.012517
13 00	+0.346250	−0.819048	+0.392954	0.919558	14.08217	0.533790	−0.012533
10	0.440807	0.834772	.392947	.919561	16.58208	.533774	.012549
20	0.535362	0.850503	.392941	.919564	19.08198	.533756	.012566
30	0.629914	0.866239	.392934	.919567	21.58189	.533739	.012584
40	0.724464	0.881983	.392927	.919570	24.08180	.533720	.012602
50	0.819011	0.897732	.392921	.919572	26.58170	.533701	.012621
14 00	+0.913555	−0.913488	+0.392914	0.919575	29.08161	0.533681	−0.012641
10	1.008096	0.929250	.392908	.919578	31.58151	.533660	.012662
20	1.102633	0.945018	.392901	.919581	34.08142	.533639	.012683
30	1.197166	0.960793	.392894	.919584	36.58132	.533617	.012705
40	1.291695	0.976573	.392888	.919586	39.08123	.533595	.012727
50	1.386220	0.992360	.392881	.919589	41.58113	.533571	.012750
15 00	+1.480741	−1.008152	+0.392874	0.919592	44.08104	0.533547	−0.012774

tan f_1	0.004598	
tan f_2	0.004575	
μ'	0.261789	radians per hour
d'	−0.000043	radians per hour

BESSELIAN ELEMENTS: PARTIAL SOLAR ECLIPSE OF DECEMBER 23-24

U.T.	Intersection of Axis of Shadow with Fundamental Plane		Direction of Axis of Shadow			Radius of Penumbra on Fundamental Plane
	x	*y*	sin *d*	cos *d*	μ	Plane
h m					°	
22 00	−1.435928	+0.909577	−0.397485	0.917608	150.12845	0.563953
10	1.348982	0.919750	.397484	.917609	152.62781	.563980
20	1.262036	0.929929	.397482	.917610	155.12718	.564007
30	1.175089	0.940115	.397481	.917611	157.62655	.564033
40	1.088141	0.950307	.397479	.917611	160.12592	.564059
50	1.001194	0.960506	.397477	.917612	162.62529	.564083
23 00	−0.914246	+0.970712	−0.397476	0.917613	165.12466	0.564108
10	0.827298	0.980924	.397474	.917613	167.62403	.564131
20	0.740350	0.991144	.397472	.917614	170.12340	.564154
30	0.653403	1.001369	.397471	.917615	172.62277	.564177
40	0.566456	1.011601	.397469	.917616	175.12213	.564199
50	0.479510	1.021840	.397467	.917616	177.62150	.564220
0 00	−0.392565	+1.032085	−0.397466	0.917617	180.12087	0.564241
10	0.305622	1.042337	.397464	.917618	182.62024	.564261
20	0.218679	1.052595	.397462	.917618	185.11961	.564280
30	0.131738	1.062860	.397461	.917619	187.61898	.564299
40	−0.044799	1.073131	.397459	.917620	190.11835	.564317
50	+0.042139	1.083408	.397457	.917621	192.61772	.564335
1 00	+0.129074	+1.093692	−0.397456	0.917621	195.11709	0.564352
10	0.216007	1.103982	.397454	.917622	197.61646	.564369
20	0.302938	1.114278	.397452	.917623	200.11582	.564385
30	0.389866	1.124581	.397450	.917624	202.61519	.564400
40	0.476792	1.134890	.397449	.917624	205.11456	.564415
50	0.563714	1.145205	.397447	.917625	207.61393	.564429
2 00	+0.650634	+1.155527	−0.397445	0.917626	210.11330	0.564442
10	0.737550	1.165854	.397443	.917627	212.61267	.564455
20	0.824462	1.176188	.397442	.917627	215.11204	.564467
30	0.911371	1.186528	.397440	.917628	217.61141	.564479
40	0.998276	1.196874	.397438	.917629	220.11078	.564490
50	1.085177	1.207226	.397436	.917630	222.61015	.564501
3 00	+1.172074	+1.217584	−0.397435	0.917631	225.10951	0.564511
3 10	+1.258966	+1.227948	−0.397433	0.917631	227.60888	0.564520

tan f_1	0.004755	
μ'	0.261733	radians per hour
d'	+0.000011	radians per hour

CONTENTS OF SECTION B

NOTE

The tables and formulae in this section were revised in 1984 to bring them into accordance with the recommendations of the International Astronomical Union at its General Assemblies in 1976, 1979 and 1982. They are intended for use with the new dynamical time-scales, the new FK5 celestial reference system and the new standard epoch of J2000·0, and formulae are given for the computation of relativistic effects in the reduction from mean to apparent place. Except when the highest precision is required it is possible, however, to continue to use the classical methods (e.g. to use day numbers), but the catalogue position should be reduced to the FK5 system and to the standard equinox of J2000·0 as explained in the Supplement to the Almanac for 1984, and precession should be applied to give the mean place for the *middle* of the year as the starting point for the reduction from mean to apparent place when day numbers are to be used.

Background information about the new time and coordinate reference systems, and about the changes in the procedures, are given in the Explanation and in the Supplement to the Almanac for 1984.

CALENDAR, 1992

Day of Month	JANUARY Day of Week	JANUARY Day of Year	FEBRUARY Day of Week	FEBRUARY Day of Year	MARCH Day of Week	MARCH Day of Year	APRIL Day of Week	APRIL Day of Year	MAY Day of Week	MAY Day of Year	JUNE Day of Week	JUNE Day of Year
1	Wed.	1	Sat.	32	Sun.	61	Wed.	92	Fri.	122	Mon.	153
2	Thu.	2	Sun.	33	Mon.	62	Thu.	93	Sat.	123	Tue.	154
3	Fri.	3	Mon.	34	Tue.	63	Fri.	94	Sun.	124	Wed.	155
4	Sat.	4	Tue.	35	Wed.	64	Sat.	95	Mon.	125	Thu.	156
5	Sun.	5	Wed.	36	Thu.	65	Sun.	96	Tue.	126	Fri.	157
6	Mon.	6	Thu.	37	Fri.	66	Mon.	97	Wed.	127	Sat.	158
7	Tue.	7	Fri.	38	Sat.	67	Tue.	98	Thu.	128	Sun.	159
8	Wed.	8	Sat.	39	Sun.	68	Wed.	99	Fri.	129	Mon.	160
9	Thu.	9	Sun.	40	Mon.	69	Thu.	100	Sat.	130	Tue.	161
10	Fri.	10	Mon.	41	Tue.	70	Fri.	101	Sun.	131	Wed.	162
11	Sat.	11	Tue.	42	Wed.	71	Sat.	102	Mon.	132	Thu.	163
12	Sun.	12	Wed.	43	Thu.	72	Sun.	103	Tue.	133	Fri.	164
13	Mon.	13	Thu.	44	Fri.	73	Mon.	104	Wed.	134	Sat.	165
14	Tue.	14	Fri.	45	Sat.	74	Tue.	105	Thu.	135	Sun.	166
15	Wed.	15	Sat.	46	Sun.	75	Wed.	106	Fri.	136	Mon.	167
16	Thu.	16	Sun.	47	Mon.	76	Thu.	107	Sat.	137	Tue.	168
17	Fri.	17	Mon.	48	Tue.	77	Fri.	108	Sun.	138	Wed.	169
18	Sat.	18	Tue.	49	Wed.	78	Sat.	109	Mon.	139	Thu.	170
19	Sun.	19	Wed.	50	Thu.	79	Sun.	110	Tue.	140	Fri.	171
20	Mon.	20	Thu.	51	Fri.	80	Mon.	111	Wed.	141	Sat.	172
21	Tue.	21	Fri.	52	Sat.	81	Tue.	112	Thu.	142	Sun.	173
22	Wed.	22	Sat.	53	Sun.	82	Wed.	113	Fri.	143	Mon.	174
23	Thu.	23	Sun.	54	Mon.	83	Thu.	114	Sat.	144	Tue.	175
24	Fri.	24	Mon.	55	Tue.	84	Fri.	115	Sun.	145	Wed.	176
25	Sat.	25	Tue.	56	Wed.	85	Sat.	116	Mon.	146	Thu.	177
26	Sun.	26	Wed.	57	Thu.	86	Sun.	117	Tue.	147	Fri.	178
27	Mon.	27	Thu.	58	Fri.	87	Mon.	118	Wed.	148	Sat.	179
28	Tue.	28	Fri.	59	Sat.	88	Tue.	119	Thu.	149	Sun.	180
29	Wed.	29	Sat.	60	Sun.	89	Wed.	120	Fri.	150	Mon.	181
30	Thu.	30			Mon.	90	Thu.	121	Sat.	151	Tue.	182
31	Fri.	31			Tue.	91			Sun.	152		

CHRONOLOGICAL CYCLES AND ERAS

Dominical Letter	ED	Julian Period (year of)	6705
Epact	25*	Roman Indiction	15
Golden Number (Lunar Cycle) ...	XVII	Solar Cycle	13

* Second Epact 25 to be used as if it were 26.

All dates are given in terms of the Gregorian calendar in which 1992 January 14 corresponds to 1992 January 1 of the Julian calendar.

ERA	YEAR	BEGINS	ERA	YEAR	BEGINS
Byzantine	7501	Sept. 14	Japanese	2652	Jan. 1
Jewish (A.M.)*	5753	Sept. 27	Grecian (Seleucidæ) ...	2304	Sept. 14 (or Oct. 14)
Chinese	(4629)	Feb. 4			
Roman (A.U.C.)	2745	Jan. 14	Indian (Saka)	1914	Mar. 21
Nabonassar	2741	Apr. 25	Diocletian	1709	Sept. 11
			Islamic (Hegira)*	1413	July 1

* Year begins at sunset.

Day of Month	JULY		AUGUST		SEPTEMBER		OCTOBER		NOVEMBER		DECEMBER	
	Day of Week	Day of Year	Day of Week	Day of Year	Day of Week	Day of Year	Day of Week	Day of Year	Day of Week	Day of Year	Day of Week	Day of Year
1	Wed.	183	Sat.	214	Tue.	245	Thu.	275	Sun.	306	Tue.	336
2	Thu.	184	Sun.	215	Wed.	246	Fri.	276	Mon.	307	Wed.	337
3	Fri.	185	Mon.	216	Thu.	247	Sat.	277	Tue.	308	Thu.	338
4	Sat.	186	Tue.	217	Fri.	248	Sun.	278	Wed.	309	Fri.	339
5	Sun.	187	Wed.	218	Sat.	249	Mon.	279	Thu.	310	Sat.	340
6	Mon.	188	Thu.	219	Sun.	250	Tue.	280	Fri.	311	Sun.	341
7	Tue.	189	Fri.	220	Mon.	251	Wed.	281	Sat.	312	Mon.	342
8	Wed.	190	Sat.	221	Tue.	252	Thu.	282	Sun.	313	Tue.	343
9	Thu.	191	Sun.	222	Wed.	253	Fri.	283	Mon.	314	Wed.	344
10	Fri.	192	Mon.	223	Thu.	254	Sat.	284	Tue.	315	Thu.	345
11	Sat.	193	Tue.	224	Fri.	255	Sun.	285	Wed.	316	Fri.	346
12	Sun.	194	Wed.	225	Sat.	256	Mon.	286	Thu.	317	Sat.	347
13	Mon.	195	Thu.	226	Sun.	257	Tue.	287	Fri.	318	Sun.	348
14	Tue.	196	Fri.	227	Mon.	258	Wed.	288	Sat.	319	Mon.	349
15	Wed.	197	Sat.	228	Tue.	259	Thu.	289	Sun.	320	Tue.	350
16	Thu.	198	Sun.	229	Wed.	260	Fri.	290	Mon.	321	Wed.	351
17	Fri.	199	Mon.	230	Thu.	261	Sat.	291	Tue.	322	Thu.	352
18	Sat.	200	Tue.	231	Fri.	262	Sun.	292	Wed.	323	Fri.	353
19	Sun.	201	Wed.	232	Sat.	263	Mon.	293	Thu.	324	Sat.	354
20	Mon.	202	Thu.	233	Sun.	264	Tue.	294	Fri.	325	Sun.	355
21	Tue.	203	Fri.	234	Mon.	265	Wed.	295	Sat.	326	Mon.	356
22	Wed.	204	Sat.	235	Tue.	266	Thu.	296	Sun.	327	Tue.	357
23	Thu.	205	Sun.	236	Wed.	267	Fri.	297	Mon.	328	Wed.	358
24	Fri.	206	Mon.	237	Thu.	268	Sat.	298	Tue.	329	Thu.	359
25	Sat.	207	Tue.	238	Fri.	269	Sun.	299	Wed.	330	Fri.	360
26	Sun.	208	Wed.	239	Sat.	270	Mon.	300	Thu.	331	Sat.	361
27	Mon.	209	Thu.	240	Sun.	271	Tue.	301	Fri.	332	Sun.	362
28	Tue.	210	Fri.	241	Mon.	272	Wed.	302	Sat.	333	Mon.	363
29	Wed.	211	Sat.	242	Tue.	273	Thu.	303	Sun.	334	Tue.	364
30	Thu.	212	Sun.	243	Wed.	274	Fri.	304	Mon.	335	Wed.	365
31	Fri.	213	Mon.	244			Sat.	305			Thu.	366

RELIGIOUS CALENDARS

Epiphany ... ... Jan. 6
Ash Wednesday ... ... Mar. 4
Palm Sunday ... ... Apr. 12
Good Friday ... ... Apr. 17
Easter Day ... ... Apr. 19

First Day of Passover (Pesach) ... Apr. 18
Feast of Weeks (Shavuot) ... June 7
Jewish New Year (tabular) ...
(Rosh Hashanah) ... ... Sept. 28

Islamic New Year ... ... July 2
(tabular) ... ...

Ascension Day ... ... May 28
Whit Sunday—Pentecost ... ... June 7
Trinity Sunday ... ... June 14
First Sunday in Advent ... ... Nov. 29
Christmas Day (Friday) ... ... Dec. 25

Day of Atonement
(Yom Kippur) ... ... Oct. 7
First day of Tabernacles
(Succoth) ... ... Oct. 12

First day of Ramadân ... ... Mar. 5
(tabular)

The Jewish and Islamic dates above are tabular dates, which begin at sunset on the previous evening and end at sunset on the date tabulated. In practice, the dates of Islamic fasts and festivals are determined by an actual sighting of the appropriate new Moon.

Julian date

A tabulation of Julian date (JD) at 0^h UT against calendar date is given with the ephemeris of universal and sidereal times on pages B8–B15. The following relationship holds during 1992:

Julian date = 244 8621·5 + day of year + fraction of day from 0^h UT

where the day of the year for the current year of the Gregorian calendar is given on pages B2–B3. The following table gives the Julian dates at day 0 of each month of 1992:

0^h UT	JD	0^h UT	JD	0^h UT	JD	0^h UT	JD
Jan. 0	244 8621·5	Apr. 0	244 8712·5	July 0	244 8803·5	Oct. 0	244 8895·5
Feb. 0	244 8652·5	May 0	244 8742·5	Aug. 0	244 8834·5	Nov. 0	244 8926·5
Mar. 0	244 8681·5	June 0	244 8773·5	Sept. 0	244 8865·5	Dec. 0	244 8956·5

Tabulations of Julian date against calendar date for other years are given on pages K2–K4. Other relevant dates are:

400-day date, JD 244 8800·5 = 1992 June 27

Standard epoch, 1900 January 0, 12^h UT = JD 241 5020·000
B1950·0 = 1950 Jan. 0·923 = JD 243 3282·423
B1992·0 = 1992 Jan. 1·096 = JD 244 8622·596
J1992·5 = 1992 July 2·125 = JD 244 8805·625
J2000·0 = 2000 Jan. 1·5 = JD 245 1545·0

The fraction of the year from 1992·5 is tabulated with the Besselian day numbers on pages B24–B31.

The "*modified Julian date*" (MJD) is the Julian date minus 240 0000·5 and in 1992 is given by: MJD = 48621·0 + day of year + fraction of day from 0^h UT.

A date may also be expressed in years as a Julian epoch, or for some purposes as a Besselian epoch, using:

Julian epoch = J[2000·0 + (JD − 245 1545·0)/365·25]
Besselian epoch = B[1900·0 + (JD − 241 5020·313 52)/365·242 198 781]

where JD is the Julian date; the prefixes J and B may be omitted only where the context, or precision, make them superfluous.

Notation for time-scales

A summary of the notation for time-scales and related quantities used in this Almanac is given below. Additional information is given in the Glossary, in the Explanation and in the Supplement to the Almanac for 1984.

UT	= UT1; universal time; counted from 0^h at midnight; unit is mean solar day
UT0	local approximation to universal time; not corrected for polar motion
GMST	Greenwich mean sidereal time; GHA of mean equinox of date
GAST	Greenwich apparent sidereal time; GHA of true equinox of date
TAI	international atomic time; unit is the SI second
UTC	coordinated universal time; differs from TAI by an integral number of seconds, and is the basis of most radio time signals and legal time systems
ΔUT	= UT − UTC; increment to be applied to UTC to give UT
DUT	= predicted value of ΔUT, rounded to $0^s{\cdot}1$, given in some radio time signals

Notation for time-scales (continued)

ET ephemeris time; was used in dynamical theories and in the Almanac from 1960–1983; but is now replaced by TDT and TDB

TDT terrestrial dynamical time; used as time-scale of ephemerides for observations from the Earth's surface. TDT = TAI + $32^s \cdot 184$

TDB barycentric dynamical time; used as time-scale of ephemerides referred to the barycentre of the solar system

ΔT = ET − UT (prior to 1984); increment to be applied to UT to give ET

ΔT = TDT − UT (1984 onwards); increment to be applied to UT to give TDT

ΔT = TAI + $32^s \cdot 184$ − UT

ΔAT = TAI − UTC; increment to be applied to UTC to give TAI

ΔET = ET − UTC; increment to be applied to UTC to give ET

ΔTT = TDT − UTC; increment to be applied to UTC to give TDT

For most purposes, ET up to 1983 December 31 and TDT from 1984 January 1 can be regarded as a continuous time-scale. Values of ΔT for the years 1620 onwards are given on pages K8–K9.

The name Greenwich mean time (GMT) is not used in this Almanac since it is ambiguous and is now used, although not in astronomy, in the sense of UTC in addition to the earlier sense of UT; prior to 1925 it was reckoned for astronomical purposes from Greenwich mean noon (12^h UT).

Relationships between time-scales

The relationships between universal and sidereal times are described on page B6 and a daily ephemeris is given on pages B8–B15; examples of the use of the ephemeris are given on page B7.

The scale of coordinated universal time (UTC) contains step adjustments of exactly one second (leap seconds) so that universal time (UT) may be obtained directly from it with an accuracy of 1 second or better and so that international atomic time (TAI) may be obtained by the addition of an integral number of seconds. The step adjustments are usually inserted after the 60th second of the last minute of December 31 or June 30. Values of the differences ΔAT for 1972 onwards are given on page K9. Accurate values of the increment ΔUT to be applied to UTC to give UT are derived from observations, but predicted values are transmitted in code in some time signals.

The differences between the terrestrial and barycentric dynamical time-scales (due to the variations in gravitational potential around the Earth's orbit) are given by:

$$\mathrm{TDB} = \mathrm{TDT} + 0^s \cdot 001\,658 \sin g + 0^s \cdot 000\,014 \sin 2g$$
$$g = 357^\circ \cdot 53 + 0^\circ \cdot 985\,600\,28\,(\mathrm{JD} - 245\,1545 \cdot 0)$$

where higher-order terms are neglected and g is the mean anomaly of the Earth in its orbit around the Sun. For the current year

$$g = 356^\circ \cdot 13 + 0^\circ \cdot 985\,600\,28d$$

where d is the day of the year tabulated on pages B2–B3.

Relationships between universal and sidereal time

The ephemeris of universal and sidereal times on pages B8–B15 is primarily intended to facilitate the conversion of universal time to local apparent sidereal time, and vice versa, for use in the computation and reduction of quantities dependent on local hour angle. Numerical examples of such conversions using the ephemeris and other tables are given opposite on page B7. Alternatively, such conversions may be carried out using the basic formulae and numerical coefficients given below.

Universal time is defined in terms of Greenwich mean sidereal time, (i.e. the Greenwich hour angle (GHA) of the mean equinox of date), by:

$$\text{GMST at } 0^{h}\text{ UT} = 24\,110^{s}{\cdot}548\,41 + 8640\,184^{s}{\cdot}812\,866\, T_{U} + 0^{s}{\cdot}093\,104\, T_{U}^{2} - 6^{s}{\cdot}2 \times 10^{-6}\, T_{U}^{3}$$

where $T_{U} = (\text{JD} - 245\,1545{\cdot}0)/36\,525$

T_U is the interval of time, measured in Julian centuries of 36 525 days of universal time (mean solar days), elapsed since the epoch 2000 January $1^{d}\ 12^{h}$ UT.

The following relationship holds during 1992:

on day of year d at t^{h} UT, $\text{GMST} = 6^{h}{\cdot}594\,7030 + 0^{h}{\cdot}065\,709\,8243\, d + 1^{h}{\cdot}002\,737\,91\, t$

where the day of year d is tabulated on pages B2–B3. Add or subtract multiples of 24^{h} as necessary.

In 1992:

1 mean solar day $= 1{\cdot}002\,737\,909\,35$ mean sidereal days
$= 24^{h}\ 03^{m}\ 56^{s}{\cdot}555\,37$ of mean sidereal time

1 mean sidereal day $= 0{\cdot}997\,269\,566\,33$ mean solar days
$= 23^{h}\ 56^{m}\ 04^{s}{\cdot}090\,53$ of mean solar time

Greenwich apparent sidereal time (i.e. the Greenwich hour angle of the true equinox of date) is given by:

$$\text{GAST} = \text{GMST} + \text{equation of equinoxes}$$

The equation of the equinoxes is tabulated on pages B8–B15 at 0^{h} UT for each day and should be interpolated to the required time if full precision is required; it is equal to the total nutation in longitude multiplied by the cosine of the true obliquity of the ecliptic.

Relationships with local time and hour angle

The following general relationships are used:

local mean solar time = universal time + east longitude
local mean sidereal time = Greenwich mean sidereal time + east longitude
local apparent sidereal time = local mean sidereal time + equation of equinoxes
= Greenwich apparent sidereal time + east longitude
local hour angle = local apparent sidereal time − apparent right ascension
= local mean sidereal time
− (apparent right ascension − equation of equinoxes)

A further small correction for the effect of polar motion is required in the reduction of very precise observations; for details see page B60.

Examples of the use of the ephemeris of universal and sidereal times

1. *Conversion of universal time to local sidereal time*

To find the local apparent sidereal time at $09^h\ 44^m\ 30^s$ UT on 1992 July 8 in longitude 80° 22′ 55″·79 west.

	h m s
Greenwich mean sidereal time on July 8 at 0^h UT is (page B12)	19 04 46·4508
Add the equivalent mean sidereal time interval from 0^h to $09^h\ 44^m\ 30^s$ UT (multiply UT interval by 1·002 737 9093)	9 46 06·0185
Greenwich mean sidereal time at required UT:	4 50 52·4693
Add equation of equinoxes, interpolated using second-order differences to approximate UT$=0^d$·41	+1·0841
Greenwich apparent sidereal time:	4 50 53·5534
Subtract west longitude (add east longitude)	5 21 31·7193
Local apparent sidereal time:	23 29 21·8341

The calculation for local mean sidereal time is similar, but omit the step which allows for the equation of the equinoxes.

2. *Conversion of local sidereal time to universal time*

To find the universal time at $23^h\ 29^m\ 21^s$·8341 local apparent sidereal time on 1992 July 8 in longitude 80° 22′ 55″·79 west.

	h m s
Local apparent sidereal time:	23 29 21·8341
Add west longitude (subtract east longitude)	5 21 31·7193
Greenwich apparent sidereal time:	4 50 53·5534
Subtract equation of equinoxes, interpolated using second-order differences to approximate UT$=0^d$·41	+1·0841
Greenwich mean sidereal time:	4 50 52·4693
Subtract Greenwich mean sidereal time at 0^h UT	19 04 46·4508
Mean sidereal time interval from 0^h UT:	9 46 06·0185
Equivalent UT interval (multiply mean sidereal time interval by 0·997 269 5663)	9 44 30·0000

The conversion of mean sidereal time to universal time is carried out by a similar procedure; omit the step which allows for the equation of the equinoxes.

Date 0^h UT		Julian Date	G. SIDEREAL TIME (GHA of the Equinox) Apparent	Mean	Equation of Equinoxes at 0^h UT	GSD at 0^h GMST	UT at 0^h GMST (Greenwich Transit of the Mean Equinox)	
		244	h m s	s	s	245		h m s
Jan.	0	**8621·5**	6 35 41·9770	40·9309	+1·0460	**5327·0**	Jan.	0 17 21 27·9826
	1	**8622·5**	6 39 38·5369	37·4863	1·0506	**5328·0**		1 17 17 32·0732
	2	**8623·5**	6 43 35·0989	34·0417	1·0572	**5329·0**		2 17 13 36·1637
	3	**8624·5**	6 47 31·6619	30·5970	1·0649	**5330·0**		3 17 09 40·2542
	4	**8625·5**	6 51 28·2252	27·1524	1·0728	**5331·0**		4 17 05 44·3448
	5	**8626·5**	6 55 24·7877	23·7078	+1·0800	**5332·0**		5 17 01 48·4353
	6	**8627·5**	6 59 21·3487	20·2631	1·0856	**5333·0**		6 16 57 52·5258
	7	**8628·5**	7 03 17·9076	16·8185	1·0891	**5334·0**		7 16 53 56·6164
	8	**8629·5**	7 07 14·4641	13·3739	1·0902	**5335·0**		8 16 50 00·7069
	9	**8630·5**	7 11 11·0183	09·9292	1·0891	**5336·0**		9 16 46 04·7974
	10	**8631·5**	7 15 07·5708	06·4846	+1·0862	**5337·0**		10 16 42 08·8880
	11	**8632·5**	7 19 04·1222	03·0400	1·0822	**5338·0**		11 16 38 12·9785
	12	**8633·5**	7 22 60·6735	59·5953	1·0782	**5339·0**		12 16 34 17·0690
	13	**8634·5**	7 26 57·2260	56·1507	1·0753	**5340·0**		13 16 30 21·1595
	14	**8635·5**	7 30 53·7806	52·7061	1·0746	**5341·0**		14 16 26 25·2501
	15	**8636·5**	7 34 50·3385	49·2614	+1·0771	**5342·0**		15 16 22 29·3406
	16	**8637·5**	7 38 46·9001	45·8168	1·0833	**5343·0**		16 16 18 33·4311
	17	**8638·5**	7 42 43·4649	42·3722	1·0927	**5344·0**		17 16 14 37·5217
	18	**8639·5**	7 46 40·0316	38·9275	1·1040	**5345·0**		18 16 10 41·6122
	19	**8640·5**	7 50 36·5979	35·4829	1·1150	**5346·0**		19 16 06 45·7027
	20	**8641·5**	7 54 33·1616	32·0383	+1·1233	**5347·0**		20 16 02 49·7933
	21	**8642·5**	7 58 29·7209	28·5936	1·1272	**5348·0**		21 15 58 53·8838
	22	**8643·5**	8 02 26·2756	25·1490	1·1266	**5349·0**		22 15 54 57·9743
	23	**8644·5**	8 06 22·8271	21·7044	1·1227	**5350·0**		23 15 51 02·0649
	24	**8645·5**	8 10 19·3771	18·2597	1·1174	**5351·0**		24 15 47 06·1554
	25	**8646·5**	8 14 15·9277	14·8151	+1·1126	**5352·0**		25 15 43 10·2459
	26	**8647·5**	8 18 12·4803	11·3705	1·1098	**5353·0**		26 15 39 14·3365
	27	**8648·5**	8 22 09·0353	07·9258	1·1095	**5354·0**		27 15 35 18·4270
	28	**8649·5**	8 26 05·5929	04·4812	1·1117	**5355·0**		28 15 31 22·5175
	29	**8650·5**	8 30 02·1525	01·0366	1·1159	**5356·0**		29 15 27 26·6080
	30	**8651·5**	8 33 58·7133	57·5920	+1·1214	**5357·0**		30 15 23 30·6986
	31	**8652·5**	8 37 55·2745	54·1473	1·1272	**5358·0**		31 15 19 34·7891
Feb.	1	**8653·5**	8 41 51·8351	50·7027	1·1324	**5359·0**	Feb.	1 15 15 38·8796
	2	**8654·5**	8 45 48·3942	47·2581	1·1362	**5360·0**		2 15 11 42·9702
	3	**8655·5**	8 49 44·9513	43·8134	1·1379	**5361·0**		3 15 07 47·0607
	4	**8656·5**	8 53 41·5060	40·3688	+1·1373	**5362·0**		4 15 03 51·1512
	5	**8657·5**	8 57 38·0584	36·9242	1·1343	**5363·0**		5 14 59 55·2418
	6	**8658·5**	9 01 34·6088	33·4795	1·1293	**5364·0**		6 14 55 59·3323
	7	**8659·5**	9 05 31·1579	30·0349	1·1230	**5365·0**		7 14 52 03·4228
	8	**8660·5**	9 09 27·7066	26·5903	1·1164	**5366·0**		8 14 48 07·5134
	9	**8661·5**	9 13 24·2561	23·1456	+1·1105	**5367·0**		9 14 44 11·6039
	10	**8662·5**	9 17 20·8074	19·7010	1·1064	**5368·0**		10 14 40 15·6944
	11	**8663·5**	9 21 17·3614	16·2564	1·1050	**5369·0**		11 14 36 19·7850
	12	**8664·5**	9 25 13·9187	12·8117	1·1069	**5370·0**		12 14 32 23·8755
	13	**8665·5**	9 29 10·4791	09·3671	1·1120	**5371·0**		13 14 28 27·9660
	14	**8666·5**	9 33 07·0418	05·9225	+1·1193	**5372·0**		14 14 24 32·0565
	15	**8667·5**	9 37 03·6051	02·4778	+1·1273	**5373·0**		15 14 20 36·1471

Date 0ʰ UT	Julian Date	G. SIDEREAL TIME (GHA of the Equinox) Apparent	Mean	Equation of Equinoxes at 0ʰ UT	GSD at 0ʰ GMST	UT at 0ʰ GMST (Greenwich Transit of the Mean Equinox)
	244	h m s	s	s	**245**	h m s
Feb. 15	**8667·5**	9 37 03·6051	02·4778	+1·1273	**5373·0**	Feb. 15 14 20 36·1471
16	**8668·5**	9 40 60·1670	59·0332	1·1338	**5374·0**	16 14 16 40·2376
17	**8669·5**	9 44 56·7255	55·5886	1·1369	**5375·0**	17 14 12 44·3281
18	**8670·5**	9 48 53·2796	52·1439	1·1357	**5376·0**	18 14 08 48·4187
19	**8671·5**	9 52 49·8297	48·6993	1·1304	**5377·0**	19 14 04 52·5092
20	**8672·5**	9 56 46·3774	45·2547	+1·1227	**5378·0**	20 14 00 56·5997
21	**8673·5**	10 00 42·9248	41·8100	1·1148	**5379·0**	21 13 57 00·6903
22	**8674·5**	10 04 39·4738	38·3654	1·1084	**5380·0**	22 13 53 04·7808
23	**8675·5**	10 08 36·0254	34·9208	1·1046	**5381·0**	23 13 49 08·8713
24	**8676·5**	10 12 32·5797	31·4761	1·1036	**5382·0**	24 13 45 12·9619
25	**8677·5**	10 16 29·1364	28·0315	+1·1049	**5383·0**	25 13 41 17·0524
26	**8678·5**	10 20 25·6946	24·5869	1·1077	**5384·0**	26 13 37 21·1429
27	**8679·5**	10 24 22·2533	21·1422	1·1111	**5385·0**	27 13 33 25·2335
28	**8680·5**	10 28 18·8117	17·6976	1·1141	**5386·0**	28 13 29 29·3240
29	**8681·5**	10 32 15·3688	14·2530	1·1158	**5387·0**	29 13 25 33·4145
Mar. 1	**8682·5**	10 36 11·9240	10·8083	+1·1157	**5388·0**	Mar. 1 13 21 37·5050
2	**8683·5**	10 40 08·4770	07·3637	1·1132	**5389·0**	2 13 17 41·5956
3	**8684·5**	10 44 05·0275	03·9191	1·1085	**5390·0**	3 13 13 45·6861
4	**8685·5**	10 48 01·5761	00·4744	1·1016	**5391·0**	4 13 09 49·7766
5	**8686·5**	10 51 58·1231	57·0298	1·0933	**5392·0**	5 13 05 53·8672
6	**8687·5**	10 55 54·6696	53·5852	+1·0844	**5393·0**	6 13 01 57·9577
7	**8688·5**	10 59 51·2166	50·1405	1·0760	**5394·0**	7 12 58 02·0482
8	**8689·5**	11 03 47·7652	46·6959	1·0693	**5395·0**	8 12 54 06·1388
9	**8690·5**	11 07 44·3164	43·2513	1·0651	**5396·0**	9 12 50 10·2293
10	**8691·5**	11 11 40·8707	39·8067	1·0640	**5397·0**	10 12 46 14·3198
11	**8692·5**	11 15 37·4280	36·3620	+1·0660	**5398·0**	11 12 42 18·4104
12	**8693·5**	11 19 33·9877	32·9174	1·0703	**5399·0**	12 12 38 22·5009
13	**8694·5**	11 23 30·5485	29·4728	1·0757	**5400·0**	13 12 34 26·5914
14	**8695·5**	11 27 27·1085	26·0281	1·0804	**5401·0**	14 12 30 30·6820
15	**8696·5**	11 31 23·6661	22·5835	1·0826	**5402·0**	15 12 26 34·7725
16	**8697·5**	11 35 20·2201	19·1389	+1·0812	**5403·0**	16 12 22 38·8630
17	**8698·5**	11 39 16·7701	15·6942	1·0759	**5404·0**	17 12 18 42·9535
18	**8699·5**	11 43 13·3172	12·2496	1·0676	**5405·0**	18 12 14 47·0441
19	**8700·5**	11 47 09·8631	08·8050	1·0581	**5406·0**	19 12 10 51·1346
20	**8701·5**	11 51 06·4099	05·3603	1·0495	**5407·0**	20 12 06 55·2251
21	**8702·5**	11 55 02·9591	01·9157	+1·0434	**5408·0**	21 12 02 59·3157
22	**8703·5**	11 58 59·5113	58·4711	1·0402	**5409·0**	22 11 59 03·4062
23	**8704·5**	12 02 56·0663	55·0264	1·0399	**5410·0**	23 11 55 07·4967
24	**8705·5**	12 06 52·6234	51·5818	1·0416	**5411·0**	24 11 51 11·5873
25	**8706·5**	12 10 49·1815	48·1372	1·0443	**5412·0**	25 11 47 15·6778
26	**8707·5**	12 14 45·7394	44·6925	+1·0469	**5413·0**	26 11 43 19·7683
27	**8708·5**	12 18 42·2964	41·2479	1·0485	**5414·0**	27 11 39 23·8589
28	**8709·5**	12 22 38·8517	37·8033	1·0484	**5415·0**	28 11 35 27·9494
29	**8710·5**	12 26 35·4048	34·3586	1·0462	**5416·0**	29 11 31 32·0399
30	**8711·5**	12 30 31·9557	30·9140	1·0416	**5417·0**	30 11 27 36·1305
31	**8712·5**	12 34 28·5044	27·4694	+1·0350	**5418·0**	31 11 23 40·2210
Apr. 1	**8713·5**	12 38 25·0516	24·0247	+1·0268	**5419·0**	Apr. 1 11 19 44·3115

Date 0h UT		Julian Date	G. SIDEREAL TIME (GHA of the Equinox) Apparent			Mean	Equation of Equinoxes at 0h UT	GSD at 0h GMST	UT at 0h GMST (Greenwich Transit of the Mean Equinox)				
		244	h	m	s	s	s	245			h	m	s
Apr.	1	8713·5	12	38	25·0516	24·0247	+1·0268	5419·0	Apr.	1	11	19	44·3115
	2	8714·5	12	42	21·5979	20·5801	1·0178	5420·0		2	11	15	48·4020
	3	8715·5	12	46	18·1446	17·1355	1·0091	5421·0		3	11	11	52·4926
	4	8716·5	12	50	14·6927	13·6908	1·0019	5422·0		4	11	07	56·5831
	5	8717·5	12	54	11·2434	10·2462	0·9972	5423·0		5	11	04	00·6736
	6	8718·5	12	58	07·7971	06·8016	+0·9956	5424·0		6	11	00	04·7642
	7	8719·5	13	02	04·3541	03·3569	0·9972	5425·0		7	10	56	08·8547
	8	8720·5	13	05	60·9137	59·9123	1·0014	5426·0		8	10	52	12·9452
	9	8721·5	13	09	57·4745	56·4677	1·0068	5427·0		9	10	48	17·0358
	10	8722·5	13	13	54·0350	53·0230	1·0120	5428·0		10	10	44	21·1263
	11	8723·5	13	17	50·5935	49·5784	+1·0151	5429·0		11	10	40	25·2168
	12	8724·5	13	21	47·1488	46·1338	1·0151	5430·0		12	10	36	29·3074
	13	8725·5	13	25	43·7006	42·6891	1·0114	5431·0		13	10	32	33·3979
	14	8726·5	13	29	40·2493	39·2445	1·0047	5432·0		14	10	28	37·4884
	15	8727·5	13	33	36·7963	35·7999	0·9965	5433·0		15	10	24	41·5790
	16	8728·5	13	37	33·3436	32·3552	+0·9884	5434·0		16	10	20	45·6695
	17	8729·5	13	41	29·8928	28·9106	·9822	5435·0		17	10	16	49·7600
	18	8730·5	13	45	26·4449	25·4660	·9789	5436·0		18	10	12	53·8505
	19	8731·5	13	49	23·0001	22·0214	·9788	5437·0		19	10	08	57·9411
	20	8732·5	13	53	19·5579	18·5767	·9812	5438·0		20	10	05	02·0316
	21	8733·5	13	57	16·1172	15·1321	+0·9851	5439·0		21	10	01	06·1221
	22	8734·5	14	01	12·6770	11·6875	·9895	5440·0		22	9	57	10·2127
	23	8735·5	14	05	09·2360	08·2428	·9932	5441·0		23	9	53	14·3032
	24	8736·5	14	09	05·7935	04·7982	·9953	5442·0		24	9	49	18·3937
	25	8737·5	14	13	02·3490	01·3536	·9955	5443·0		25	9	45	22·4843
	26	8738·5	14	16	58·9023	57·9089	+0·9933	5444·0		26	9	41	26·5748
	27	8739·5	14	20	55·4534	54·4643	·9891	5445·0		27	9	37	30·6653
	28	8740·5	14	24	52·0028	51·0197	·9831	5446·0		28	9	33	34·7559
	29	8741·5	14	28	48·5512	47·5750	·9761	5447·0		29	9	29	38·8464
	30	8742·5	14	32	45·0995	44·1304	·9691	5448·0		30	9	25	42·9369
May	1	8743·5	14	36	41·6490	40·6858	+0·9632	5449·0	May	1	9	21	47·0275
	2	8744·5	14	40	38·2008	37·2411	·9596	5450·0		2	9	17	51·1180
	3	8745·5	14	44	34·7557	33·7965	·9592	5451·0		3	9	13	55·2085
	4	8746·5	14	48	31·3140	30·3519	·9622	5452·0		4	9	09	59·2990
	5	8747·5	14	52	27·8754	26·9072	·9682	5453·0		5	9	06	03·3896
	6	8748·5	14	56	24·4385	23·4626	+0·9759	5454·0		6	9	02	07·4801
	7	8749·5	15	00	21·0016	20·0180	·9836	5455·0		7	8	58	11·5706
	8	8750·5	15	04	17·5629	16·5733	·9895	5456·0		8	8	54	15·6612
	9	8751·5	15	08	14·1211	13·1287	·9924	5457·0		9	8	50	19·7517
	10	8752·5	15	12	10·6757	09·6841	·9917	5458·0		10	8	46	23·8422
	11	8753·5	15	16	07·2273	06·2394	+0·9878	5459·0		11	8	42	27·9328
	12	8754·5	15	20	03·7769	02·7948	·9821	5460·0		12	8	38	32·0233
	13	8755·5	15	23	60·3262	59·3502	·9761	5461·0		13	8	34	36·1138
	14	8756·5	15	27	56·8770	55·9055	·9714	5462·0		14	8	30	40·2044
	15	8757·5	15	31	53·4303	52·4609	·9694	5463·0		15	8	26	44·2949
	16	8758·5	15	35	49·9867	49·0163	+0·9704	5464·0		16	8	22	48·3854
	17	8759·5	15	39	46·5459	45·5716	+0·9742	5465·0		17	8	18	52·4760

Date 0ʰ UT		Julian Date	G. SIDEREAL TIME (GHA of the Equinox) Apparent			Mean	Equation of Equinoxes at 0ʰ UT	GSD at 0ʰ GMST	UT at 0ʰ GMST (Greenwich Transit of the Mean Equinox)				
		244	h	m	s	s	s	**245**			h	m	s
May	17	**8759·5**	15	39	46·5459	45·5716	+0·9742	**5465·0**	May	17	8	18	52·4760
	18	**8760·5**	15	43	43·1070	42·1270	·9800	**5466·0**		18	8	14	56·5665
	19	**8761·5**	15	47	39·6691	38·6824	·9867	**5467·0**		19	8	11	00·6570
	20	**8762·5**	15	51	36·2308	35·2377	·9931	**5468·0**		20	8	07	04·7475
	21	**8763·5**	15	55	32·7913	31·7931	0·9982	**5469·0**		21	8	03	08·8381
	22	**8764·5**	15	59	29·3498	28·3485	+1·0013	**5470·0**		22	7	59	12·9286
	23	**8765·5**	16	03	25·9061	24·9038	1·0023	**5471·0**		23	7	55	17·0191
	24	**8766·5**	16	07	22·4601	21·4592	1·0009	**5472·0**		24	7	51	21·1097
	25	**8767·5**	16	11	19·0123	18·0146	0·9977	**5473·0**		25	7	47	25·2002
	26	**8768·5**	16	15	15·5632	14·5699	·9932	**5474·0**		26	7	43	29·2907
	27	**8769·5**	16	19	12·1137	11·1253	+0·9884	**5475·0**		27	7	39	33·3813
	28	**8770·5**	16	23	08·6649	07·6807	·9842	**5476·0**		28	7	35	37·4718
	29	**8771·5**	16	27	05·2179	04·2361	·9818	**5477·0**		29	7	31	41·5623
	30	**8772·5**	16	31	01·7737	00·7914	·9823	**5478·0**		30	7	27	45·6529
	31	**8773·5**	16	34	58·3330	57·3468	·9862	**5479·0**		31	7	23	49·7434
June	1	**8774·5**	16	38	54·8957	53·9022	+0·9936	**5480·0**	June	1	7	19	53·8339
	2	**8775·5**	16	42	51·4609	50·4575	1·0034	**5481·0**		2	7	15	57·9245
	3	**8776·5**	16	46	48·0267	47·0129	1·0138	**5482·0**		3	7	12	02·0150
	4	**8777·5**	16	50	44·5911	43·5683	1·0228	**5483·0**		4	7	08	06·1055
	5	**8778·5**	16	54	41·1524	40·1236	1·0288	**5484·0**		5	7	04	10·1960
	6	**8779·5**	16	58	37·7100	36·6790	+1·0310	**5485·0**		6	7	00	14·2866
	7	**8780·5**	17	02	34·2640	33·2344	1·0297	**5486·0**		7	6	56	18·3771
	8	**8781·5**	17	06	30·8157	29·7897	1·0260	**5487·0**		8	6	52	22·4676
	9	**8782·5**	17	10	27·3667	26·3451	1·0216	**5488·0**		9	6	48	26·5582
	10	**8783·5**	17	14	23·9187	22·9005	1·0182	**5489·0**		10	6	44	30·6487
	11	**8784·5**	17	18	20·4729	19·4558	+1·0171	**5490·0**		11	6	40	34·7392
	12	**8785·5**	17	22	17·0299	16·0112	1·0187	**5491·0**		12	6	36	38·8298
	13	**8786·5**	17	26	13·5898	12·5666	1·0232	**5492·0**		13	6	32	42·9203
	14	**8787·5**	17	30	10·1517	09·1219	1·0298	**5493·0**		14	6	28	47·0108
	15	**8788·5**	17	34	06·7149	05·6773	1·0376	**5494·0**		15	6	24	51·1014
	16	**8789·5**	17	38	03·2781	02·2327	+1·0454	**5495·0**		16	6	20	55·1919
	17	**8790·5**	17	41	59·8402	58·7880	1·0522	**5496·0**		17	6	16	59·2824
	18	**8791·5**	17	45	56·4006	55·3434	1·0572	**5497·0**		18	6	13	03·3730
	19	**8792·5**	17	49	52·9587	51·8988	1·0600	**5498·0**		19	6	09	07·4635
	20	**8793·5**	17	53	49·5145	48·4541	1·0604	**5499·0**		20	6	05	11·5540
	21	**8794·5**	17	57	46·0682	45·0095	+1·0587	**5500·0**		21	6	01	15·6445
	22	**8795·5**	18	01	42·6204	41·5649	1·0555	**5501·0**		22	5	57	19·7351
	23	**8796·5**	18	05	39·1718	38·1202	1·0515	**5502·0**		23	5	53	23·8256
	24	**8797·5**	18	09	35·7234	34·6756	1·0478	**5503·0**		24	5	49	27·9161
	25	**8798·5**	18	13	32·2763	31·2310	1·0454	**5504·0**		25	5	45	32·0067
	26	**8799·5**	18	17	28·8316	27·7863	+1·0453	**5505·0**		26	5	41	36·0972
	27	**8800·5**	18	21	25·3900	24·3417	1·0483	**5506·0**		27	5	37	40·1877
	28	**8801·5**	18	25	21·9518	20·8971	1·0548	**5507·0**		28	5	33	44·2783
	29	**8802·5**	18	29	18·5166	17·4524	1·0642	**5508·0**		29	5	29	48·3688
	30	**8803·5**	18	33	15·0830	14·0078	1·0752	**5509·0**		30	5	25	52·4593
July	1	**8804·5**	18	37	11·6489	10·5632	+1·0857	**5510·0**	July	1	5	21	56·5499
	2	**8805·5**	18	41	08·2122	07·1185	+1·0936	**5511·0**		2	5	18	00·6404

Date 0h UT		Julian Date	G. SIDEREAL TIME (GHA of the Equinox) Apparent	Mean	Equation of Equinoxes at 0h UT	GSD at 0h GMST	UT at 0h GMST (Greenwich Transit of the Mean Equinox)	
		244	h m s	s	s	245		h m s
July	2	**8805·5**	18 41 08·2122	07·1185	+1·0936	**5511·0**	July 2	5 18 00·6404
	3	**8806·5**	18 45 04·7715	03·6739	1·0976	**5512·0**	3	5 14 04·7309
	4	**8807·5**	18 49 01·3268	00·2293	1·0975	**5513·0**	4	5 10 08·8215
	5	**8808·5**	18 52 57·8790	56·7847	1·0943	**5514·0**	5	5 06 12·9120
	6	**8809·5**	18 56 54·4299	53·3400	1·0899	**5515·0**	6	5 02 17·0025
	7	**8810·5**	19 00 50·9814	49·8954	+1·0860	**5516·0**	7	4 58 21·0930
	8	**8811·5**	19 04 47·5348	46·4508	1·0841	**5517·0**	8	4 54 25·1836
	9	**8812·5**	19 08 44·0909	43·0061	1·0848	**5518·0**	9	4 50 29·2741
	10	**8813·5**	19 12 40·6497	39·5615	1·0882	**5519·0**	10	4 46 33·3646
	11	**8814·5**	19 16 37·2107	36·1169	1·0938	**5520·0**	11	4 42 37·4552
	12	**8815·5**	19 20 33·7730	32·6722	+1·1007	**5521·0**	12	4 38 41·5457
	13	**8816·5**	19 24 30·3355	29·2276	1·1079	**5522·0**	13	4 34 45·6362
	14	**8817·5**	19 28 26·8973	25·7830	1·1143	**5523·0**	14	4 30 49·7268
	15	**8818·5**	19 32 23·4574	22·3383	1·1191	**5524·0**	15	4 26 53·8173
	16	**8819·5**	19 36 20·0153	18·8937	1·1216	**5525·0**	16	4 22 57·9078
	17	**8820·5**	19 40 16·5708	15·4491	+1·1218	**5526·0**	17	4 19 01·9984
	18	**8821·5**	19 44 13·1241	12·0044	1·1197	**5527·0**	18	4 15 06·0889
	19	**8822·5**	19 48 09·6756	08·5598	1·1158	**5528·0**	19	4 11 10·1794
	20	**8823·5**	19 52 06·2261	05·1152	1·1109	**5529·0**	20	4 07 14·2699
	21	**8824·5**	19 56 02·7764	01·6705	1·1058	**5530·0**	21	4 03 18·3605
	22	**8825·5**	19 59 59·3275	58·2259	+1·1016	**5531·0**	22	3 59 22·4510
	23	**8826·5**	20 03 55·8805	54·7813	1·0993	**5532·0**	23	3 55 26·5415
	24	**8827·5**	20 07 52·4362	51·3366	1·0996	**5533·0**	24	3 51 30·6321
	25	**8828·5**	20 11 48·9951	47·8920	1·1031	**5534·0**	25	3 47 34·7226
	26	**8829·5**	20 15 45·5570	44·4474	1·1096	**5535·0**	26	3 43 38·8131
	27	**8830·5**	20 19 42·1211	41·0027	+1·1183	**5536·0**	27	3 39 42·9037
	28	**8831·5**	20 23 38·6857	37·5581	1·1276	**5537·0**	28	3 35 46·9942
	29	**8832·5**	20 27 35·2488	34·1135	1·1353	**5538·0**	29	3 31 51·0847
	30	**8833·5**	20 31 31·8084	30·6688	1·1396	**5539·0**	30	3 27 55·1753
	31	**8834·5**	20 35 28·3638	27·2242	1·1396	**5540·0**	31	3 23 59·2658
Aug.	1	**8835·5**	20 39 24·9153	23·7796	+1·1357	**5541·0**	Aug. 1	3 20 03·3563
	2	**8836·5**	20 43 21·4646	20·3349	1·1296	**5542·0**	2	3 16 07·4469
	3	**8837·5**	20 47 18·0138	16·8903	1·1235	**5543·0**	3	3 12 11·5374
	4	**8838·5**	20 51 14·5647	13·4457	1·1190	**5544·0**	4	3 08 15·6279
	5	**8839·5**	20 55 11·1182	10·0010	1·1172	**5545·0**	5	3 04 19·7184
	6	**8840·5**	20 59 07·6745	06·5564	+1·1181	**5546·0**	6	3 00 23·8090
	7	**8841·5**	21 03 04·2331	03·1118	1·1213	**5547·0**	7	2 56 27·8995
	8	**8842·5**	21 06 60·7932	59·6671	1·1260	**5548·0**	8	2 52 31·9900
	9	**8843·5**	21 10 57·3536	56·2225	1·1311	**5549·0**	9	2 48 36·0806
	10	**8844·5**	21 14 53·9135	52·7779	1·1356	**5550·0**	10	2 44 40·1711
	11	**8845·5**	21 18 50·4719	49·3332	+1·1386	**5551·0**	11	2 40 44·2616
	12	**8846·5**	21 22 47·0282	45·8886	1·1395	**5552·0**	12	2 36 48·3522
	13	**8847·5**	21 26 43·5821	42·4440	1·1381	**5553·0**	13	2 32 52·4427
	14	**8848·5**	21 30 40·1337	38·9994	1·1343	**5554·0**	14	2 28 56·5332
	15	**8849·5**	21 34 36·6833	35·5547	1·1286	**5555·0**	15	2 25 00·6238
	16	**8850·5**	21 38 33·2317	32·1101	+1·1216	**5556·0**	16	2 21 04·7143
	17	**8851·5**	21 42 29·7797	28·6655	+1·1143	**5557·0**	17	2 17 08·8048

Date 0h UT		Julian Date	G. SIDEREAL TIME (GHA of the Equinox) Apparent			Mean	Equation of Equinoxes at 0h UT	GSD at 0h GMST	UT at 0h GMST (Greenwich Transit of the Mean Equinox)				
		244	h	m	s	s	s	245			h	m	s
Aug.	17	**8851·5**	21	42	29·7797	28·6655	+1·1143	**5557·0**	Aug.	17	2	17	08·8048
	18	**8852·5**	21	46	26·3283	25·2208	1·1075	**5558·0**		18	2	13	12·8954
	19	**8853·5**	21	50	22·8783	21·7762	1·1022	**5559·0**		19	2	09	16·9859
	20	**8854·5**	21	54	19·4307	18·3316	1·0992	**5560·0**		20	2	05	21·0764
	21	**8855·5**	21	58	15·9860	14·8869	1·0991	**5561·0**		21	2	01	25·1669
	22	**8856·5**	22	02	12·5442	11·4423	+1·1019	**5562·0**		22	1	57	29·2575
	23	**8857·5**	22	06	09·1047	07·9977	1·1070	**5563·0**		23	1	53	33·3480
	24	**8858·5**	22	10	05·6664	04·5530	1·1133	**5564·0**		24	1	49	37·4385
	25	**8859·5**	22	14	02·2275	01·1084	1·1191	**5565·0**		25	1	45	41·5291
	26	**8860·5**	22	17	58·7863	57·6638	1·1225	**5566·0**		26	1	41	45·6196
	27	**8861·5**	22	21	55·3412	54·2191	+1·1221	**5567·0**		27	1	37	49·7101
	28	**8862·5**	22	25	51·8920	50·7745	1·1175	**5568·0**		28	1	33	53·8007
	29	**8863·5**	22	29	48·4398	47·3299	1·1100	**5569·0**		29	1	29	57·8912
	30	**8864·5**	22	33	44·9866	43·8852	1·1013	**5570·0**		30	1	26	01·9817
	31	**8865·5**	22	37	41·5344	40·4406	1·0938	**5571·0**		31	1	22	06·0723
Sept.	1	**8866·5**	22	41	38·0849	36·9960	+1·0889	**5572·0**	Sept.	1	1	18	10·1628
	2	**8867·5**	22	45	34·6384	33·5513	1·0870	**5573·0**		2	1	14	14·2533
	3	**8868·5**	22	49	31·1945	30·1067	1·0878	**5574·0**		3	1	10	18·3439
	4	**8869·5**	22	53	27·7524	26·6621	1·0903	**5575·0**		4	1	06	22·4344
	5	**8870·5**	22	57	24·3110	23·2174	1·0935	**5576·0**		5	1	02	26·5249
	6	**8871·5**	23	01	20·8691	19·7728	+1·0963	**5577·0**		6	0	58	30·6154
	7	**8872·5**	23	05	17·4260	16·3282	1·0978	**5578·0**		7	0	54	34·7060
	8	**8873·5**	23	09	13·9809	12·8835	1·0974	**5579·0**		8	0	50	38·7965
	9	**8874·5**	23	13	10·5335	09·4389	1·0946	**5580·0**		9	0	46	42·8870
	10	**8875·5**	23	17	07·0838	05·9943	1·0895	**5581·0**		10	0	42	46·9776
	11	**8876·5**	23	21	03·6321	02·5496	+1·0824	**5582·0**		11	0	38	51·0681
	12	**8877·5**	23	24	60·1790	59·1050	1·0740	**5583·0**		12	0	34	55·1586
	13	**8878·5**	23	28	56·7253	55·6604	1·0649	**5584·0**		13	0	30	59·2492
	14	**8879·5**	23	32	53·2720	52·2157	1·0562	**5585·0**		14	0	27	03·3397
	15	**8880·5**	23	36	49·8200	48·7711	1·0489	**5586·0**		15	0	23	07·4302
	16	**8881·5**	23	40	46·3702	45·3265	+1·0437	**5587·0**		16	0	19	11·5208
	17	**8882·5**	23	44	42·9231	41·8818	1·0413	**5588·0**		17	0	15	15·6113
	18	**8883·5**	23	48	39·4788	38·4372	1·0416	**5589·0**		18	0	11	19·7018
	19	**8884·5**	23	52	36·0370	34·9926	1·0444	**5590·0**		19	0	07	23·7924
	20	**8885·5**	23	56	32·5965	31·5480	+1·0486	**5591·0**		20	0	03	27·8829
								5592·0		20	23	59	31·9734
	21	**8886·5**	0	00	29·1561	28·1033	+1·0528	**5593·0**		21	23	55	36·0639
	22	**8887·5**	0	04	25·7141	24·6587	1·0554	**5594·0**		22	23	51	40·1545
	23	**8888·5**	0	08	22·2690	21·2141	1·0550	**5595·0**		23	23	47	44·2450
	24	**8889·5**	0	12	18·8203	17·7694	1·0509	**5596·0**		24	23	43	48·3355
	25	**8890·5**	0	16	15·3682	14·3248	+1·0435	**5597·0**		25	23	39	52·4261
	26	**8891·5**	0	20	11·9143	10·8802	1·0342	**5598·0**		26	23	35	56·5166
	27	**8892·5**	0	24	08·4606	07·4355	1·0251	**5599·0**		27	23	32	00·6071
	28	**8893·5**	0	28	05·0091	03·9909	1·0182	**5600·0**		28	23	28	04·6977
	29	**8894·5**	0	32	01·5608	00·5463	1·0145	**5601·0**		29	23	24	08·7882
	30	**8895·5**	0	35	58·1157	57·1016	+1·0140	**5602·0**		30	23	20	12·8787
Oct.	1	**8896·5**	0	39	54·6729	53·6570	+1·0159	**5603·0**	Oct.	1	23	16	16·9693

Date 0h UT		Julian Date	G. SIDEREAL TIME (GHA of the Equinox) Apparent	Mean	Equation of Equinoxes at 0h UT	GSD at 0h GMST	UT at 0h GMST (Greenwich Transit of the Mean Equinox)	
		244	h m s	s	s	245		h m s
Oct.	1	**8896·5**	0 39 54·6729	53·6570	+1·0159	**5603·0**	Oct.	1 23 16 16·9693
	2	**8897·5**	0 43 51·2313	50·2124	1·0189	**5604·0**		2 23 12 21·0598
	3	**8898·5**	0 47 47·7896	46·7677	1·0219	**5605·0**		3 23 08 25·1503
	4	**8899·5**	0 51 44·3468	43·3231	1·0238	**5606·0**		4 23 04 29·2409
	5	**8900·5**	0 55 40·9023	39·8785	1·0238	**5607·0**		5 23 00 33·3314
	6	**8901·5**	0 59 37·4555	36·4338	+1·0216	**5608·0**		6 22 56 37·4219
	7	**8902·5**	1 03 34·0064	32·9892	1·0172	**5609·0**		7 22 52 41·5124
	8	**8903·5**	1 07 30·5553	29·5446	1·0107	**5610·0**		8 22 48 45·6030
	9	**8904·5**	1 11 27·1027	26·0999	1·0027	**5611·0**		9 22 44 49·6935
	10	**8905·5**	1 15 23·6493	22·6553	0·9940	**5612·0**		10 22 40 53·7840
	11	**8906·5**	1 19 20·1962	19·2107	+0·9855	**5613·0**		11 22 36 57·8746
	12	**8907·5**	1 23 16·7442	15·7660	·9782	**5614·0**		12 22 33 01·9651
	13	**8908·5**	1 27 13·2944	12·3214	·9730	**5615·0**		13 22 29 06·0556
	14	**8909·5**	1 31 09·8473	08·8768	·9705	**5616·0**		14 22 25 10·1462
	15	**8910·5**	1 35 06·4031	05·4321	·9710	**5617·0**		15 22 21 14·2367
	16	**8911·5**	1 39 02·9614	01·9875	+0·9739	**5618·0**		16 22 17 18·3272
	17	**8912·5**	1 42 59·5213	58·5429	·9785	**5619·0**		17 22 13 22·4178
	18	**8913·5**	1 46 56·0816	55·0982	·9833	**5620·0**		18 22 09 26·5083
	19	**8914·5**	1 50 52·6406	51·6536	·9870	**5621·0**		19 22 05 30·5988
	20	**8915·5**	1 54 49·1972	48·2090	·9882	**5622·0**		20 22 01 34·6893
	21	**8916·5**	1 58 45·7504	44·7643	+0·9861	**5623·0**		21 21 57 38·7799
	22	**8917·5**	2 02 42·3005	41·3197	·9808	**5624·0**		22 21 53 42·8704
	23	**8918·5**	2 06 38·8484	37·8751	·9733	**5625·0**		23 21 49 46·9609
	24	**8919·5**	2 10 35·3958	34·4304	·9654	**5626·0**		24 21 45 51·0515
	25	**8920·5**	2 14 31·9447	30·9858	·9589	**5627·0**		25 21 41 55·1420
	26	**8921·5**	2 18 28·4964	27·5412	+0·9553	**5628·0**		26 21 37 59·2325
	27	**8922·5**	2 22 25·0516	24·0965	·9551	**5629·0**		27 21 34 03·3231
	28	**8923·5**	2 26 21·6099	20·6519	·9579	**5630·0**		28 21 30 07·4136
	29	**8924·5**	2 30 18·1700	17·2073	·9627	**5631·0**		29 21 26 11·5041
	30	**8925·5**	2 34 14·7305	13·7627	·9679	**5632·0**		30 21 22 15·5947
	31	**8926·5**	2 38 11·2904	10·3180	+0·9723	**5633·0**		31 21 18 19·6852
Nov.	1	**8927·5**	2 42 07·8486	06·8734	·9752	**5634·0**	Nov.	1 21 14 23·7757
	2	**8928·5**	2 46 04·4046	03·4288	·9758	**5635·0**		2 21 10 27·8663
	3	**8929·5**	2 49 60·9583	59·9841	·9742	**5636·0**		3 21 06 31·9568
	4	**8930·5**	2 53 57·5099	56·5395	·9704	**5637·0**		4 21 02 36·0473
	5	**8931·5**	2 57 54·0599	53·0949	+0·9650	**5638·0**		5 20 58 40·1378
	6	**8932·5**	3 01 50·6089	49·6502	·9587	**5639·0**		6 20 54 44·2284
	7	**8933·5**	3 05 47·1579	46·2056	·9523	**5640·0**		7 20 50 48·3189
	8	**8934·5**	3 09 43·7078	42·7610	·9469	**5641·0**		8 20 46 52·4094
	9	**8935·5**	3 13 40·2597	39·3163	·9433	**5642·0**		9 20 42 56·5000
	10	**8936·5**	3 17 36·8142	35·8717	+0·9425	**5643·0**		10 20 39 00·5905
	11	**8937·5**	3 21 33·3718	32·4271	·9447	**5644·0**		11 20 35 04·6810
	12	**8938·5**	3 25 29·9322	28·9824	·9498	**5645·0**		12 20 31 08·7716
	13	**8939·5**	3 29 26·4946	25·5378	·9568	**5646·0**		13 20 27 12·8621
	14	**8940·5**	3 33 23·0576	22·0932	·9644	**5647·0**		14 20 23 16·9526
	15	**8941·5**	3 37 19·6197	18·6485	+0·9711	**5648·0**		15 20 19 21·0432
	16	**8942·5**	3 41 16·1793	15·2039	+0·9754	**5649·0**		16 20 15 25·1337

Date 0h UT	Julian Date	G. SIDEREAL TIME (GHA of the Equinox) Apparent	Mean	Equation of Equinoxes at 0h UT	GSD at 0h GMST	UT at 0h GMST (Greenwich Transit of the Mean Equinox)
	244	h m s	s	s	245	h m s
Nov. 16	8942·5	3 41 16·1793	15·2039	+0·9754	5649·0	Nov. 16 20 15 25·1337
17	8943·5	3 45 12·7358	11·7593	·9766	5650·0	17 20 11 29·2242
18	8944·5	3 49 09·2891	08·3146	·9745	5651·0	18 20 07 33·3148
19	8945·5	3 53 05·8401	04·8700	·9701	5652·0	19 20 03 37·4053
20	8946·5	3 57 02·3901	01·4254	·9648	5653·0	20 19 59 41·4958
21	8947·5	4 00 58·9410	57·9807	+0·9603	5654·0	21 19 55 45·5863
22	8948·5	4 04 55·4943	54·5361	·9582	5655·0	22 19 51 49·6769
23	8949·5	4 08 52·0507	51·0915	·9593	5656·0	23 19 47 53·7674
24	8950·5	4 12 48·6104	47·6468	·9635	5657·0	24 19 43 57·8579
25	8951·5	4 16 45·1724	44·2022	·9702	5658·0	25 19 40 01·9485
26	8952·5	4 20 41·7357	40·7576	+0·9781	5659·0	26 19 36 06·0390
27	8953·5	4 24 38·2986	37·3129	·9857	5660·0	27 19 32 10·1295
28	8954·5	4 28 34·8603	33·8683	·9919	5661·0	28 19 28 14·2201
29	8955·5	4 32 31·4198	30·4237	·9961	5662·0	29 19 24 18·3106
30	8956·5	4 36 27·9769	26·9790	·9979	5663·0	30 19 20 22·4011
Dec. 1	8957·5	4 40 24·5318	23·5344	+0·9974	5664·0	Dec. 1 19 16 26·4917
2	8958·5	4 44 21·0848	20·0898	·9950	5665·0	2 19 12 30·5822
3	8959·5	4 48 17·6365	16·6451	·9914	5666·0	3 19 08 34·6727
4	8960·5	4 52 14·1879	13·2005	·9874	5667·0	4 19 04 38·7633
5	8961·5	4 56 10·7398	09·7559	·9839	5668·0	5 19 00 42·8538
6	8962·5	5 00 07·2932	06·3113	+0·9820	5669·0	6 18 56 46·9443
7	8963·5	5 04 03·8490	02·8666	·9824	5670·0	7 18 52 51·0348
8	8964·5	5 07 60·4078	59·4220	·9858	5671·0	8 18 48 55·1254
9	8965·5	5 11 56·9697	55·9774	0·9923	5672·0	9 18 44 59·2159
10	8966·5	5 15 53·5340	52·5327	1·0013	5673·0	10 18 41 03·3064
11	8967·5	5 19 50·0996	49·0881	+1·0115	5674·0	11 18 37 07·3970
12	8968·5	5 23 46·6646	45·6435	1·0211	5675·0	12 18 33 11·4875
13	8969·5	5 27 43·2274	42·1988	1·0286	5676·0	13 18 29 15·5780
14	8970·5	5 31 39·7868	38·7542	1·0326	5677·0	14 18 25 19·6686
15	8971·5	5 35 36·3428	35·3096	1·0332	5678·0	15 18 21 23·7591
16	8972·5	5 39 32·8959	31·8649	+1·0310	5679·0	16 18 17 27·8496
17	8973·5	5 43 29·4477	28·4203	1·0274	5680·0	17 18 13 31·9402
18	8974·5	5 47 25·9999	24·9757	1·0242	5681·0	18 18 09 36·0307
19	8975·5	5 51 22·5540	21·5310	1·0230	5682·0	19 18 05 40·1212
20	8976·5	5 55 19·1109	18·0864	1·0245	5683·0	20 18 01 44·2118
21	8977·5	5 59 15·6708	14·6418	+1·0291	5684·0	21 17 57 48·3023
22	8978·5	6 03 12·2333	11·1971	1·0362	5685·0	22 17 53 52·3928
23	8979·5	6 07 08·7972	07·7525	1·0447	5686·0	23 17 49 56·4833
24	8980·5	6 11 05·3614	04·3079	1·0535	5687·0	24 17 46 00·5739
25	8981·5	6 15 01·9246	00·8632	1·0613	5688·0	25 17 42 04·6644
26	8982·5	6 18 58·4859	57·4186	+1·0673	5689·0	26 17 38 08·7549
27	8983·5	6 22 55·0447	53·9740	1·0708	5690·0	27 17 34 12·8455
28	8984·5	6 26 51·6011	50·5293	1·0718	5691·0	28 17 30 16·9360
29	8985·5	6 30 48·1554	47·0847	1·0707	5692·0	29 17 26 21·0265
30	8986·5	6 34 44·7081	43·6401	1·0680	5693·0	30 17 22 25·1171
31	8987·5	6 38 41·2600	40·1954	+1·0645	5694·0	31 17 18 29·2076
32	8988·5	6 42 37·8120	36·7508	+1·0612	5695·0	32 17 14 33·2981

Purpose and arrangement

The formulae, tables and ephemerides in the remainder of this section are mainly intended to provide for the reduction of celestial coordinates (especially of right ascension and declination) from one reference system to another (especially for stars from catalogue (barycentric) place to apparent (geocentric) place) but some of the data may be used for other purposes. Formulae and numerical values are given on pages B16–B21 for the separate steps in such reductions (i.e. for proper motion, aberration, light-deflection, parallax, precession and nutation). Formulae, examples and ephemerides are given for approximate reductions using the day-number technique on pages B22–B35 and for full-precision reductions using vectors and the rotation-matrix technique on pages B36–B59. Finally, formulae and numerical values are given for the reduction from geocentric to topocentric place on pages B60 and B61. Background information is given in the Glossary and the Explanation.

Notation and units

- t — an epoch expressed in terms of the Julian year (see page B4); the difference between two epochs represents a time-interval expressed in Julian years; subscripts zero and one are used to indicate the epoch of a catalogue place, usually the standard epoch of J2000·0, and the epoch of the middle of a Julian year (here shortened to "epoch of year"), respectively.
- τ — fraction of year measured from the epoch of year; $\tau = t - t_1$
- T — an interval of time expressed in Julian centuries of 36 525 days; usually measured from J2000·0, i.e. from JD 245 1545·0.
- α, δ, π — right ascension, declination and annual parallax; in the formulae for computation, right ascension and related quantities are expressed in time-measure ($1^h = 15°$, etc.), while declination and related quantities, including annual parallax, are expressed in sexagesimal angular measure, unless the contrary is indicated.
- μ_α, μ_δ — components of *centennial* proper motion in right ascension and declination.
- λ, β — ecliptic longitude and latitude.
- Ω, i, ω — orbital elements referred to the ecliptic; longitude of ascending node, inclination, argument of perihelion.
- X, Y, Z — rectangular coordinates of the Earth with respect to the barycentre of the solar system, referred to the mean equinox and equator of J2000·0, and expressed in astronomical units (au).
- $\dot{X}, \dot{Y}, \dot{Z}$ — first derivatives of X, Y, Z with respect to time expressed in days.

Approximate reduction for proper motion

In its simplest form the reduction for the proper motion is given by:

$$\alpha = \alpha_0 + (t - t_0)\,\mu_\alpha/100 \qquad \delta = \delta_0 + (t - t_0)\,\mu_\delta/100$$

In some cases it is necessary to allow also for second-order terms, radial velocity and orbital motion, but appropriate formulae are usually given in the catalogue.

Approximate reduction for annual parallax

The reduction for annual parallax from the catalogue place (α_0, δ_0) to the geocentric place (α, δ) is given by:

$$\alpha = \alpha_0 + (\pi/15 \cos\delta_0)\,(X \sin\alpha_0 - Y \cos\alpha_0)$$
$$\delta = \delta_0 + \pi(X \cos\alpha_0 \sin\delta_0 + Y \sin\alpha_0 \sin\delta_0 - Z \cos\delta_0)$$

where X, Y, Z are the coordinates of the Earth tabulated on pages B44 onwards. Expressions for X, Y, Z may be obtained from page C24, since $X = -x$, $Y = -y$, $Z = -z$. The correction may be applied with the correction for annual aberration using the C and D day numbers, (see page B22).

The times of reception of periodic phenomena, such as pulsar signals, may be reduced to a common origin at the barycentre by adding the light-time corresponding to the component of the Earth's position vector along the direction to the object; that is by adding to the observed times $(X \cos\alpha \cos\delta + Y \sin\alpha \cos\delta + Z \sin\delta)/c$, where the velocity of light, $c = 173{\cdot}14$ au/d, and the light time for 1 au, $1/c = 0^d{\cdot}005\,7755$.

Approximate reduction for annual aberration

The reduction for annual aberration from a geometric geocentric place (α_0, δ_0) to an apparent geocentric place (α, δ) is given by:

$$\alpha = \alpha_0 + (-\dot{X} \sin \alpha_0 + \dot{Y} \cos \alpha_0)/(c \cos \delta_0)$$

$$\delta = \delta_0 + (-\dot{X} \cos \alpha_0 \sin \delta_0 - \dot{Y} \sin \alpha_0 \sin \delta_0 + \dot{Z} \cos \delta_0)/c$$

where $c = 173{\cdot}14$ au/d, and $\dot{X}$, $\dot{Y}$, $\dot{Z}$ are the velocity components of the Earth given on pages **B44** onwards. Alternatively, but to lower precision, it is possible to use the expressions

$$\dot{X} = +0{\cdot}0172 \sin \lambda \qquad \dot{Y} = -0{\cdot}0158 \cos \lambda \qquad \dot{Z} = -0{\cdot}0068 \cos \lambda$$

where the apparent longitude of the Sun λ is given by the expression on page C24. The reduction may also be carried out by using the day-number technique (see page B22) or the rotation-matrix technique (see page B39) when full precision is required.

Measurements of radial velocity may be reduced to a common origin at the barycentre by adding the component of the Earth's velocity in the direction of the object; that is by adding

$$\dot{X} \cos \alpha_0 \cos \delta_0 + \dot{Y} \sin \alpha_0 \cos \delta_0 + \dot{Z} \sin \delta_0$$

Classical reduction for planetary aberration

In the case of a body in the solar system the apparent direction at the instant of observation (t) differs from the geometric direction at that instant because of (a) the motion of the body during the light-time and (b) the relative motion of the Earth and the light. The reduction may be carried out in two stages: (*i*) by combining the barycentric position of the body at time $t - \Delta t$, where Δt is the light-time, with the barycentric position of the Earth at time t, and then (*ii*) by applying the correction for annual aberration as described above. Alternatively it is possible to interpolate the geometric (geocentric) ephemeris of the body to the time $t - \Delta t$; it is usually sufficient to subtract the product of the light-time and the first derivative of the coordinate. The light-time Δt in days is given by the distance in au between the body and the Earth, multiplied by 0·005 7755; strictly, the light-time corresponds to the distance from the position of the Earth at time t to the position of the body at time $t - \Delta t$, but it is usually sufficient to use the geocentric distance at time t.

Approximate reduction for light-deflection

The apparent direction of a star or of a body in the solar system may be significantly affected by the deflection of light in the gravitational field of the Sun. The elongation (E) from the centre of the Sun is increased by an amount (ΔE) that, for a star, depends on the elongation in the following manner:

$$\Delta E = 0''{\cdot}00407/\tan (E/2)$$

E	0°·25	0°·5	1°	2°	5°	10°	20°	50°	90°
ΔE	1″·866	0″·933	0″·466	0″·233	0″·093	0″·047	0″·023	0″·009	0″·004

The body disappears behind the Sun when E is less than the limiting grazing value of about 0°·25. The effects in right ascension and declination may be calculated approximately from:

$$\cos E = \sin \delta \sin \delta_0 + \cos \delta \cos \delta_0 \cos (\alpha - \alpha_0)$$

$$\Delta\alpha = 0^{s}{\cdot}000\,271 \cos \delta_0 \sin (\alpha - \alpha_0)/(1 - \cos E) \cos \delta$$

$$\Delta\delta = 0''{\cdot}004\,07 [\sin \delta \cos \delta_0 \cos (\alpha - \alpha_0) - \cos \delta \sin \delta_0]/(1 - \cos E)$$

where α, δ refer to the star, and α_0, δ_0 to the Sun. See also page B39.

Reduction for precession—rigorous formulae

Rigorous formulae for the reduction of mean equatorial positions from an initial epoch t_0 to epoch of date t, and vice versa, are as follows:

For right ascension and declination:

$$\begin{aligned}
\sin(\alpha - z_A)\cos\delta &= \sin(\alpha_0+\zeta_A)\cos\delta_0 \\
\cos(\alpha - z_A)\cos\delta &= \cos(\alpha_0+\zeta_A)\cos\theta_A\cos\delta_0 - \sin\theta_A\sin\delta_0 \\
\sin\delta &= \cos(\alpha_0+\zeta_A)\sin\theta_A\cos\delta_0 + \cos\theta_A\sin\delta_0
\end{aligned}$$

$$\begin{aligned}
\sin(\alpha_0+\zeta_A)\cos\delta_0 &= \sin(\alpha - z_A)\cos\delta \\
\cos(\alpha_0+\zeta_A)\cos\delta_0 &= \cos(\alpha - z_A)\cos\theta_A\cos\delta + \sin\theta_A\sin\delta \\
\sin\delta_0 &= -\cos(\alpha - z_A)\sin\theta_A\cos\delta + \cos\theta_A\sin\delta
\end{aligned}$$

where ζ_A, z_A, θ_A are angles that serve to specify the position of the mean equinox and equator of date with respect to the mean equinox and equator of the initial epoch.

For reduction with respect to the standard epoch $t_0 = \text{J2000·0}$

$$\begin{aligned}
\zeta_A &= 0^\circ{\cdot}640\,6161\ T + 0^\circ{\cdot}000\,0839\ T^2 + 0^\circ{\cdot}000\,0050\ T^3 \\
z_A &= 0^\circ{\cdot}640\,6161\ T + 0^\circ{\cdot}000\,3041\ T^2 + 0^\circ{\cdot}000\,0051\ T^3 \\
\theta_A &= 0^\circ{\cdot}556\,7530\ T - 0^\circ{\cdot}000\,1185\ T^2 - 0^\circ{\cdot}000\,0116\ T^3
\end{aligned}$$

where $T = (t - 2000{\cdot}0)/100 = (\text{JD} - 245\,1545{\cdot}0)/36\,525$

For equatorial rectangular coordinates (or direction cosines):

$$\mathbf{r} = \mathbf{P}\,\mathbf{r}_0 \qquad \mathbf{r}_0 = \mathbf{P}^{-1}\,\mathbf{r} = \mathbf{P}'\,\mathbf{r} \qquad \text{where } \mathbf{r} \text{ is the position vector } (x, y, z).$$

The inverse of the rotation matrix $\mathbf{P}$ is equal to its transpose, i.e. $\mathbf{P}^{-1} = \mathbf{P}'$. The elements of $\mathbf{P}$ may be expressed in terms of ζ_A, z_A, θ_A as follows:

$$\begin{matrix}
\cos\zeta_A\cos\theta_A\cos z_A - \sin\zeta_A\sin z_A & -\sin\zeta_A\cos\theta_A\cos z_A - \cos\zeta_A\sin z_A & -\sin\theta_A\cos z_A \\
\cos\zeta_A\cos\theta_A\sin z_A + \sin\zeta_A\cos z_A & -\sin\zeta_A\cos\theta_A\sin z_A + \cos\zeta_A\cos z_A & -\sin\theta_A\sin z_A \\
\cos\zeta_A\sin\theta_A & -\sin\zeta_A\sin\theta_A & \cos\theta_A
\end{matrix}$$

Values of the angles ζ_A, z_A, θ_A and of the elements of $\mathbf{P}$ for reduction from the standard epoch J2000·0 to epoch of year are as follows:

Epoch J1992·5	Rotation matrix **P** for reduction to epoch J1992·5		
$\zeta_A = -172''{\cdot}96 = -0^\circ{\cdot}048\,046$	+0·999 998 33	+0·001 677 09	+0·000 728 80
$z_A = -172''{\cdot}96 = -0^\circ{\cdot}048\,045$	−0·001 677 09	+0·999 998 59	−0·000 000 61
$\theta_A = -150''{\cdot}33 = -0^\circ{\cdot}041\,757$	−0·000 728 80	−0·000 000 61	+0·999 999 73

The obliquity of the ecliptic of date (with respect to the mean equator of date) is given by:

$$\begin{aligned}
\varepsilon &= 23^\circ\,26'\,21''{\cdot}45 - 46''{\cdot}815\ T - 0''{\cdot}0006\ T^2 + 0''{\cdot}001\,81\ T^3 \\
\varepsilon &= 23^\circ{\cdot}439\,291 - 0^\circ{\cdot}013\,0042\ T - 0^\circ{\cdot}000\,000\,16\ T^2 + 0^\circ{\cdot}000\,000\,504\ T^3
\end{aligned}$$

The precessional motion of the ecliptic is specified by the inclination (π_A) and longitude of the node (Π_A) of the ecliptic of date with respect to the ecliptic and equinox of J2000·0; they are given by:

$$\begin{aligned}
\pi_A\sin\Pi_A &= +\ 4''{\cdot}198\ T + 0''{\cdot}1945\ T^2 - 0''{\cdot}000\,18\ T^3 \\
\pi_A\cos\Pi_A &= -46''{\cdot}815\ T + 0''{\cdot}0506\ T^2 + 0''{\cdot}000\,34\ T^3
\end{aligned}$$

For epoch J1992·5

$$\begin{aligned}
\varepsilon &= 23^\circ\,26'\,24''{\cdot}96 = 23^\circ{\cdot}440\,266 \\
\pi_A &= -3''{\cdot}525 = -0^\circ{\cdot}000\,9793 \\
\Pi_A &= 174^\circ\,53'{\cdot}7 = 174^\circ{\cdot}895
\end{aligned}$$

Reduction for precession—approximate formulae

Approximate formulae for the reduction of coordinates and orbital elements referred to the mean equinox and equator or ecliptic of date (t) are as follows:

For reduction to J2000·0

$$\alpha_0 = \alpha - M - N \sin\alpha_m \tan\delta_m$$
$$\delta_0 = \delta - N \cos\alpha_m$$
$$\lambda_0 = \lambda - a + b\cos(\lambda + c')\tan\beta_0$$
$$\beta_0 = \beta - b\sin(\lambda + c')$$
$$\Omega_0 = \Omega - a + b\sin(\Omega + c')\cot i_0$$
$$i_0 = i - b\cos(\Omega + c')$$
$$\omega_0 = \omega - b\sin(\Omega + c')\operatorname{cosec} i_0$$

For reduction from J2000·0

$$\alpha = \alpha_0 + M + N \sin\alpha_m \tan\delta_m$$
$$\delta = \delta_0 + N \cos\alpha_m$$
$$\lambda = \lambda_0 + a - b\cos(\lambda_0 + c)\tan\beta$$
$$\beta = \beta_0 + b\sin(\lambda_0 + c)$$
$$\Omega = \Omega_0 + a - b\sin(\Omega_0 + c)\cot i$$
$$i = i_0 + b\cos(\Omega_0 + c)$$
$$\omega = \omega_0 + b\sin(\Omega_0 + c)\operatorname{cosec} i$$

where the subscript zero refers to epoch J2000·0 and α_m, δ_m refer to the mean epoch; with sufficient accuracy:

$$\alpha_m = \alpha - \tfrac{1}{2}(M + N\sin\alpha\tan\delta)$$
$$\delta_m = \delta - \tfrac{1}{2}N\cos\alpha_m$$

or

$$\alpha_m = \alpha_0 + \tfrac{1}{2}(M + N\sin\alpha_0\tan\delta_0)$$
$$\delta_m = \delta_0 + \tfrac{1}{2}N\cos\alpha_m$$

The precessional constants M, N, etc., are given by:

$$M = 1^\circ{\cdot}281\,2323\,T + 0^\circ{\cdot}000\,3879\,T^2 + 0^\circ{\cdot}000\,0101\,T^3$$
$$N = 0^\circ{\cdot}556\,7530\,T - 0^\circ{\cdot}000\,1185\,T^2 - 0^\circ{\cdot}000\,0116\,T^3$$
$$a = 1^\circ{\cdot}396\,971\,T + 0^\circ{\cdot}000\,3086\,T^2$$
$$b = 0^\circ{\cdot}013\,056\,T - 0^\circ{\cdot}000\,0092\,T^2$$
$$c = 5^\circ{\cdot}123\,62 + 0^\circ{\cdot}241\,614\,T + 0^\circ{\cdot}000\,1122\,T^2$$
$$c' = 5^\circ{\cdot}123\,62 - 1^\circ{\cdot}155\,358\,T - 0^\circ{\cdot}000\,1964\,T^2$$

where $T = (t - 2000{\cdot}0)/100 = (\text{JD} - 245\,1545{\cdot}0)/36\,525$

Formulae for the reduction from the mean equinox and equator or ecliptic of the middle of year (t_1) to date (t) are as follows:

$$\alpha = \alpha_1 + \tau\,(m + n\sin\alpha_1\tan\delta_1)$$
$$\delta = \delta_1 + \tau\,n\cos\alpha_1$$
$$\lambda = \lambda_1 + \tau\,(p - \pi\cos(\lambda_1 + 6^\circ)\tan\beta)$$
$$\beta = \beta_1 + \tau\,\pi\sin(\lambda_1 + 6^\circ)$$
$$\Omega = \Omega_1 + \tau\,(p - \pi\sin(\Omega_1 + 6^\circ)\cot i)$$
$$i = i_1 + \tau\,\pi\cos(\Omega_1 + 6^\circ)$$
$$\omega = \omega_1 + \tau\,\pi\sin(\Omega_1 + 6^\circ)\operatorname{cosec} i$$

where $\tau = t - t_1$ and π is the annual rate of rotation of the ecliptic. The precessional constants p, m, etc. are as follows:

	Epoch J1992·5
Annual general precession	$p = +0^\circ{\cdot}013\,9692$
Annual precession in R.A.	$m = +0^\circ{\cdot}012\,8117$
Annual precession in Dec.	$n = +0^\circ{\cdot}005\,5677$
Annual rate of rotation	$\pi = +0^\circ{\cdot}000\,1306$
Longitude of axis	$\Pi = +174^\circ{\cdot}8079$
	$\gamma = 180^\circ - \Pi = +5^\circ{\cdot}1921$

where Π is the longitude of the instantaneous rotation axis of the ecliptic, measured from the mean equinox of date.

Approximate reduction for nutation

To first order, the contributions of the nutations in longitude ($\Delta\psi$) and in obliquity ($\Delta\varepsilon$) to the reduction from mean place to true place are given by:

$$\Delta\alpha = (\cos\varepsilon + \sin\varepsilon \sin\alpha \tan\delta)\,\Delta\psi - \cos\alpha \tan\delta\,\Delta\varepsilon \qquad \Delta\lambda = \Delta\psi$$
$$\Delta\delta = \sin\varepsilon \cos\alpha\,\Delta\psi + \sin\alpha\,\Delta\varepsilon \qquad \Delta\beta = 0$$

Daily values of $\Delta\psi$ and $\Delta\varepsilon$ during 1992 are tabulated on pages B24–B31. The following formulae may be used to compute $\Delta\psi$ and $\Delta\varepsilon$ to a precision of about 0°·0002 (1″) during 1992.

$$\Delta\psi = -0°{\cdot}0048 \sin(279°{\cdot}9 - 0{\cdot}053\,d) - 0°{\cdot}0004 \sin(197°{\cdot}9 + 1{\cdot}971\,d)$$
$$\Delta\varepsilon = +0°{\cdot}0026 \cos(279°{\cdot}9 - 0{\cdot}053\,d) + 0°{\cdot}0002 \cos(197°{\cdot}9 + 1{\cdot}971\,d)$$

where d = JD − 2448 256·5; for this precision

$$\varepsilon = 23°{\cdot}44 \qquad \cos\varepsilon = 0{\cdot}917 \qquad \sin\varepsilon = 0{\cdot}398$$

The corrections to be added to the mean rectangular coordinates (x, y, z) to produce the true rectangular coordinates are given by:

$$\Delta x = -(y\cos\varepsilon + z\sin\varepsilon)\,\Delta\psi \qquad \Delta y = +x\cos\varepsilon\,\Delta\psi - z\,\Delta\varepsilon \qquad \Delta z = +x\sin\varepsilon\,\Delta\psi + y\,\Delta\varepsilon$$

where $\Delta\psi$ and $\Delta\varepsilon$ are expressed in radians.

The elements of the corresponding rotation matrix are:

$$\begin{matrix} 1 & -\Delta\psi\cos\varepsilon & -\Delta\psi\sin\varepsilon \\ +\Delta\psi\cos\varepsilon & 1 & -\Delta\varepsilon \\ +\Delta\psi\sin\varepsilon & +\Delta\varepsilon & 1 \end{matrix}$$

The full series for nutation in $\Delta\psi$ and $\Delta\varepsilon$ are given in *The Astronomical Almanac 1984* on pages S23–S26.

Approximate reduction for precession and nutation

The following formulae and table may be used for the approximate reduction from the standard equinox and equator of J2000·0 to the true equinox and equator of date during 1992:

$$\alpha = \alpha_0 + f + g\sin(G + \alpha_0)\tan\delta_0$$
$$\delta = \delta_0 + g\cos(G + \alpha_0)$$

where the units of the correction to α_0 and δ_0 are seconds of time and minutes of arc, respectively.

Date 1992	f (s)	g (s)	g (′)	G (h m)	Date 1992	f (s)	g (s)	g (′)	G (h m)
Jan. −1	−23·6	10·2	2·56	12 01	**July 7**	−21·9	9·5	2·38	11 59
9	23·4	10·2	2·55	12 02	**17**	21·8	9·5	2·37	11 59
19*	23·3	10·1	2·54	12 01	**27**	21·7	9·4	2·36	11 59
29	23·3	10·1	2·53	12 02	**Aug. 6***	21·6	9·4	2·35	11 59
Feb. 8	23·2	10·1	2·52	12 02	**16**	21·6	9·4	2·34	12 00
18	−23·1	10·0	2·51	12 02	**26**	−21·5	9·3	2·33	12 00
28*	23·0	10·0	2·50	12 02	**Sept. 5**	21·4	9·3	2·33	12 00
Mar. 9	23·0	10·0	2·50	12 02	**15***	21.4	9·3	2·32	12 00
19	22·9	9.9	2·49	12 02	**25**	21·3	9·3	2·31	12 00
29	22·8	9·9	2·48	12 02	**Oct. 5**	21·2	9·2	2·31	11 59
Apr. 8*	−22·8	9·9	2·47	12 02	**15**	−21·2	9·2	2·30	11 59
18	22·7	9·9	2·47	12 01	**25***	21·1	9·2	2·30	11 59
28	22·6	9·8	2·46	12 01	**Nov. 4**	21·0	9·1	2·29	11 58
May 8	22·5	9·8	2·45	12 01	**14**	21·0	9·1	2·28	11 58
18*	22·5	9·8	2·44	12 00	**24**	20·9	9·1	2·27	11 57
28	−22·4	9·7	2·43	12 00	**Dec. 4***	−20·8	9·0	2·26	11 57
June 7	22·2	9·7	2·42	12 00	**14**	20·6	9·0	2·24	11 57
17	22·1	9·6	2·41	11 59	**24**	20·5	8·9	2·23	11 56
27*†	22·1	9·6	2·40	11 59	**34**	20·4	8·9	2·22	11 56
July 7	21·9	9·5	2·38	11 59					

*40-day date †400-day date for osculation epoch

Differential precession and nutation

The corrections for differential precession and nutation are given below. These are to be added to the observed differences of the right ascension and declination, $\Delta\alpha$ and $\Delta\delta$, of an object relative to a comparison star to obtain the differences in the mean place for a standard epoch (e.g. J2000·0 or the beginning of the year). The differences $\Delta\alpha$ and $\Delta\delta$ are measured in the sense "object − comparison star", and the corrections are in the same units as $\Delta\alpha$ and $\Delta\delta$. In the correction to right ascension the same units must be used for $\Delta\alpha$ and $\Delta\delta$.

correction to right ascension	$e \tan\delta\ \Delta\alpha - f \sec^2\delta\ \Delta\delta$
correction to declination	$f\,\Delta\alpha$

where $e = -\cos\alpha\,(nt + \sin\varepsilon\,\Delta\psi) - \sin\alpha\,\Delta\varepsilon$

$f = +\sin\alpha\,(nt + \sin\varepsilon\,\Delta\psi) - \cos\alpha\,\Delta\varepsilon$

and $\varepsilon = 23°\!\cdot\!44, \qquad \sin\varepsilon = 0\cdot3978$

$n = 0\cdot000\,0972$ radians for epoch J1992·5

t is the time in years *from* the standard epoch *to* the time of observation.

$\Delta\psi$, $\Delta\varepsilon$ are nutations in longitude and obliquity at the time of observation, *expressed in radians.* ($1'' = 0\cdot000\,004\,8481$ rad).

The errors in arc units caused by using these formulae are of order $10^{-8}\,t^2 \sec^2\delta$ multiplied by the displacement in arc from the comparison star.

Differential aberration

The corrections for differential annual aberration to be added to the observed differences (in the sense moving object minus star) of right ascension and declination to give the true differences are:

in right ascension	$a\,\Delta\alpha + b\,\Delta\delta$	in units of $0^s\!\cdot\!001$
in declination	$c\,\Delta\alpha + d\,\Delta\delta$	in units of $0''\!\cdot\!01$

where $\Delta\alpha$, $\Delta\delta$ are the observed differences in units of 1^m and $1'$ respectively, and where a, b, c, d are coefficients defined by:

$a = -5\cdot701 \cos(H+\alpha) \sec\delta \qquad b = -0\cdot380 \sin(H+\alpha) \sec\delta \tan\delta$

$c = +8\cdot552 \sin(H+\alpha) \sin\delta \qquad d = -0\cdot570 \cos(H+\alpha) \cos\delta$

$H^h = 23\cdot4 - (\text{day of year} / 15\cdot2)$

The day of year is tabulated on pages B2–B3.

Astrometric positions

An astrometric position of a body in the solar system is formed by applying the correction for the barycentric motion of the body during the light-time to the geometric geocentric position referred to the equator and equinox of the standard epoch of J2000·0. Such a position is then directly comparable with the astrometric positions of stars formed by applying the corrections for proper motion and annual parallax to the catalogue positions for the standard epoch of J2000·0. The deflection of light has been ignored.

Formulae using day numbers

For stars and other objects outside the solar system the usual procedure for the computation of apparent positions from catalogue data is as follows, but the techniques described on pages B39–B41 should be used if full precision is required.

From	To	Step	Correction
Catalogue epoch	current epoch	*i*	proper motion
catalogue equinox	mean equinox of year	*ii*	precession
mean equinox of year	mean equinox of date	*iii*	precession
mean equinox of date	true equinox of date	*iv*	nutation
true (heliocentric) position	apparent (geocentric) position	*v*	aberration (annual)
		vi	parallax (annual)

Star catalogues usually provide coefficients for steps *i* and *ii* for the reduction from catalogue position (α_0, δ_0) to the position for the mean equinox of another epoch. Besselian day numbers (A to E), which provide for steps *iii* to *v* for the reductions from the position (α_1, δ_1) for the mean equinox of the middle of the year to the apparent geocentric position (α, δ), are given on pages B24–B31; for high declinations, the second-order day numbers (J, J') given on pages B32–B35 may be required. The formulae to be used are:

$$\alpha = \alpha_1 + Aa + Bb + Cc + Dd + E + J\tan^2\delta_1$$
$$\delta = \delta_1 + Aa' + Bb' + Cc' + Dd' + J'\tan\delta_1$$

where the Besselian star constants are given by:

$$a = (m/n) + \sin\alpha_1 \tan\delta_1 \qquad a' = \cos\alpha_1$$
$$b = \cos\alpha_1 \tan\delta_1 \qquad b' = -\sin\alpha_1$$
$$c = \cos\alpha_1 \sec\delta_1 \qquad c' = \tan\varepsilon\cos\delta_1 - \sin\alpha_1 \sin\delta_1$$
$$d = \sin\alpha_1 \sec\delta_1 \qquad d' = \cos\alpha_1 \sin\delta_1$$

where α and δ are in arc units. For 1992·5, $m/n = 2{\cdot}301\,08$ and $\tan\varepsilon = 0{\cdot}433\,57$.

The additional corrections for the proper motion (centennial components μ_α, μ_δ) during the fraction of year (τ) and for annual parallax (π) are given by:

$$\Delta\alpha = \tau\mu_\alpha/100 + \pi(dX - cY) \qquad \Delta\delta = \tau\mu_\delta/100 + \pi(d'X - c'Y)$$

where X, Y are the coordinates of the Earth with respect to the solar-system barycentre given on pages B44–B59. Strictly, this parallax correction should be computed using the coordinates of the Earth referred to the mean equinox of the middle of the year, or using star constants computed for the standard epoch of J2000·0.

The corrections for annual parallax may be included with the corrections for annual aberration by substituting $C - \pi Y$ for C and $D + \pi X$ for D in the formulae given above. Alternatively if the annual parallax is small enough it is possible to make the substitutions

$$c + 0{\cdot}0532\, d\pi \text{ for } c \qquad d - 0{\cdot}0448\, c\pi \text{ for } d$$
$$c' + 0{\cdot}0532\, d'\pi \text{ for } c' \qquad d' - 0{\cdot}0448\, c'\pi \text{ for } d'$$

The error in this approximate method is negligible if the parallax of the star is less than about 0″·2.

A further correction to allow for the deflection of the light in the gravitational field of the Sun may also be required—appropriate formulae are given on page B17.

The day-number technique may also be used for objects within the solar system but steps *i* and *vi* are omitted and step *v* is replaced by forming the geocentric position by combining the barycentric position of the body at time $t - \Delta t$, where Δt is the light-time, with the barycentric position of the Earth at time t.

Example of day-number technique

To calculate the apparent place of a star at 0^h TDT at Greenwich on 1992 January 1 from the mean place for J1992·5 using day numbers.

Step 1. From a fundamental star catalogue, such as the FK5, calculate for epoch and equinox J1992·5 the mean right ascension and declination (α_1, δ_1), the centennial proper motion (μ_α, μ_δ) and the parallax (π).

Assume the following fictitious values for the calculation:

$\alpha_1 = 14^h\ 39^m\ 05^s{\cdot}238$ $\qquad \delta_1 = -60°\ 48'\ 16''{\cdot}94$ $\qquad \pi = 0''{\cdot}752$

$\mu_\alpha = -49^s{\cdot}419$ per century $\qquad \mu_\delta = +69''{\cdot}84$ per century

Step 2. Form the star constants as follows:

$$a = \tfrac{1}{15}((m/n) + \sin\alpha_1 \tan\delta_1) = +0{\cdot}229\ 73 \qquad a' = \cos\alpha_1 = -0{\cdot}768\ 60$$

$$b = \tfrac{1}{15}\cos\alpha_1 \tan\delta_1 = +0{\cdot}091\ 70 \qquad b' = -\sin\alpha_1 = +0{\cdot}639\ 73$$

$$c = \tfrac{1}{15}\cos\alpha_1 \sec\delta_1 = -0{\cdot}105\ 05 \qquad c' = \tan\varepsilon\cos\delta_1 - \sin\alpha_1 \sin\delta_1 = -0{\cdot}346\ 97$$

$$d = \tfrac{1}{15}\sin\alpha_1 \sec\delta_1 = -0{\cdot}087\ 43 \qquad d' = \cos\alpha_1 \sin\delta_1 = +0{\cdot}670\ 96$$

Step 3. Extract the day numbers for pages B24, B34 and B35. In general, linear interpolation is required and second differences may be significant for *A* and *B*. The values for 1992 January 1 at 0^h TDT are:

$A = -3''{\cdot}216$ $\qquad C = -3''{\cdot}266$ $\qquad E = +0^s{\cdot}0024$ $\qquad J = +0^s{\cdot}000\ 09$

$B = -1''{\cdot}043$ $\qquad D = +20''{\cdot}534$ $\qquad \tau = -0{\cdot}5014$ $\qquad J' = -0''{\cdot}0004$

Step 4. Extract the values of the Earth's rectangular coordinates from page B44 (the values for J2000·0 are of sufficient accuracy for computing the parallax correction). The values are:

$$X = -0{\cdot}167 \qquad Y = +0{\cdot}891$$

Step 5. Calculate the corrections for light-deflection, $\Delta\alpha$ and $\Delta\delta$.

For the Sun for 1992 January 1 at 0^h TDT, $\alpha_0 = 18^h\ 42^m{\cdot}7$, $\delta_0 = -23°\ 05'$. Using the formulae on page B17, cos (elongation) = +0·5604 and the corrections for light-deflection are $\Delta\alpha = -0^s{\cdot}001$ and $\Delta\delta = 0''{\cdot}00$.

Step 6. Compute the apparent position as follows:

Mean position 1992·5,	$\alpha_1 = 14^h\ 39^m\ 05^s{\cdot}238$		$\delta_1 = -60°\ 48'\ 16''{\cdot}94$
$Aa + Bb + Cc + Dd + E$	$= -2^s{\cdot}284$	$Aa' + Bb' + Cc' + Dd'$	$= +16''{\cdot}72$
$J \tan^2 \delta_1$	$= 0^s{\cdot}000$	$J' \tan \delta_1$	$= 0''{\cdot}00$
$\tau\mu_\alpha/100$	$= +0^s{\cdot}248$	$\tau\mu_\delta/100$	$= -0''{\cdot}35$
$\pi\,(dX - cY)$	$= +0^s{\cdot}081$	$\pi(d'X - c'Y)$	$= +0''{\cdot}15$
$\Delta\alpha$	$= -0^s{\cdot}001$	$\Delta\delta$	$= 0''{\cdot}00$
Apparent position	$\alpha = 14^h\ 39^m\ 03^s{\cdot}282$		$\delta = -60°\ 48'\ 00''{\cdot}43$

FOR 0h DYNAMICAL TIME

Date 0h TDT		Nutation in Long.	Nutation in Obl.	Obl. of Ecliptic 23° 26′	Besselian Day Numbers for Mean Equinox J1992·5 A	B	C	D	E	Fraction of Year τ
		″	″	″	″	″	″	″	(0″·0001)	
Jan.	0	+17·102	+1·084	26·280	− 3·301	−1·084	− 2·935	+20·592	+24	−0·5041
	1	17·177	1·043	26·237	3·216	1·043	3·266	20·534	24	·5014
	2	17·284	1·013	26·206	3·119	1·013	3·597	20·469	25	·4986
	3	17·410	1·000	26·191	3·014	1·000	3·926	20·398	25	·4959
	4	17·540	1·004	26·194	2·908	1·004	4·255	20·320	25	·4932
	5	+17·657	+1·025	26·214	− 2·806	−1·025	− 4·582	+20·235	+25	−0·4904
	6	17·748	1·059	26·247	2·715	1·059	4·907	20·144	25	·4877
	7	17·805	1·101	26·287	2·637	1·101	5·231	20·046	25	·4849
	8	17·824	1·145	26·330	2·575	1·145	5·553	19·942	25	·4822
	9	17·806	1·183	26·367	2·527	1·183	5·873	19·831	25	·4795
	10	+17·758	+1·211	26·394	− 2·491	−1·211	− 6·191	+19·714	+25	−0·4767
	11	17·694	1·224	26·405	2·462	1·224	6·506	19·590	25	·4740
	12	17·628	1·218	26·398	2·433	1·218	6·820	19·461	25	·4713
	13	17·580	1·196	26·374	2·398	1·196	7·130	19·325	25	·4685
	14	17·569	1·159	26·336	2·347	1·159	7·438	19·184	25	·4658
	15	+17·610	+1·115	26·291	− 2·276	−1·115	− 7·744	+19·036	+25	−0·4630
	16	17·711	1·076	26·250	2·181	1·076	8·046	18·883	25	·4603
	17	17·865	1·052	26·226	2·065	1·052	8·346	18·725	25	·4576
	18	18·050	1·056	26·228	1·936	1·056	8·642	18·562	26	·4548
	19	18·230	1·089	26·260	1·810	1·089	8·936	18·393	26	·4521
	20	+18·365	+1·146	26·315	− 1·701	−1·146	− 9·227	+18·219	+26	−0·4493
	21	18·429	1·212	26·381	1·621	1·212	9·516	18·040	26	·4466
	22	18·419	1·272	26·439	1·570	1·272	9·802	17·856	26	·4439
	23	18·355	1·310	26·476	1·541	1·310	10·085	17·667	26	·4411
	24	18·268	1·322	26·486	1·520	1·322	10·366	17·472	26	·4384
	25	+18·191	+1·309	26·472	− 1·496	−1·309	−10·643	+17·272	+26	−0·4357
	26	18·144	1·281	26·442	1·460	1·281	10·918	17·067	26	·4329
	27	18·139	1·246	26·407	1·407	1·246	11·190	16·856	26	·4302
	28	18·175	1·215	26·374	1·338	1·215	11·458	16·639	26	·4274
	29	18·244	1·194	26·352	1·255	1·194	11·723	16·417	26	·4247
	30	+18·334	+1·189	26·345	− 1·165	−1·189	−11·984	+16·190	+26	−0·4220
	31	18·428	1·200	26·356	1·072	1·200	12·241	15·957	26	·4192
Feb.	1	18·514	1·229	26·383	0·984	1·229	12·495	15·720	26	·4165
	2	18·575	1·271	26·424	0·904	1·271	12·745	15·477	26	·4138
	3	18·604	1·321	26·473	0·838	1·321	12·990	15·229	26	·4110
	4	+18·593	+1·375	26·525	− 0·787	−1·375	−13·231	+14·976	+26	−0·4083
	5	18·544	1·424	26·573	0·752	1·424	13·468	14·718	26	·4055
	6	18·463	1·463	26·611	0·729	1·463	13·700	14·456	26	·4028
	7	18·360	1·488	26·634	0·715	1·488	13·927	14·189	26	·4001
	8	18·252	1·494	26·639	0·704	1·494	14·150	13·918	26	·3973
	9	+18·155	+1·483	26·627	− 0·687	−1·483	−14·368	+13·643	+26	−0·3946
	10	18·089	1·457	26·599	0·659	1·457	14·581	13·364	26	·3919
	11	18·066	1·422	26·563	0·613	1·422	14·789	13·081	26	·3891
	12	18·097	1·388	26·528	0·545	1·388	14·992	12·795	26	·3864
	13	18·180	1·363	26·502	0·458	1·363	15·190	12·505	26	·3836
	14	+18·300	+1·360	26·497	− 0·355	−1·360	−15·382	+12·212	+26	−0·3809
	15	+18·430	+1·383	26·519	− 0·248	−1·383	−15·570	+11·915	+26	−0·3782

FOR 0ʰ DYNAMICAL TIME

Date 0ʰ TDT		Nutation in Long.	Nutation in Obl.	Obl. of Ecliptic 23° 26′	Besselian Day Numbers for Mean Equinox J1992·5 *A*	*B*	*C*	*D*	*E*	Fraction of Year τ
		″	″	″	″	″	″	″	(0ˢ0001)	
Feb.	**15**	+18·430	+1·383	26·519	− 0·248	−1·383	−15·570	+11·915	+26	−0·3782
	16	18·537	1·431	26·566	0·151	1·431	15·754	11·616	26	·3754
	17	18·588	1·495	26·628	0·076	1·495	15·932	11·314	26	·3727
	18	18·568	1·560	26·692	0·029	1·560	16·106	11·009	26	·3700
	19	18·481	1·609	26·740	0·009	1·609	16·275	10·702	26	·3672
	20	+18·356	+1·632	26·762	− 0·004	−1·632	−16·439	+10·391	+26	−0·3645
	21	18·226	1·627	26·756	0·000	1·627	16·599	10·077	26	·3617
	22	18·122	1·600	26·727	+ 0·013	1·600	16·754	9·761	26	·3590
	23	18·060	1·562	26·687	0·043	1·562	16·904	9·441	26	·3563
	24	18·043	1·523	26·647	0·091	1·523	17·050	9·119	26	·3535
	25	+18·064	+1·493	26·616	+ 0·155	−1·493	−17·190	+ 8·793	+26	−0·3508
	26	18·110	1·477	26·599	0·228	1·477	17·325	8·465	26	·3480
	27	18·166	1·478	26·599	0·305	1·478	17·455	8·134	26	·3453
	28	18·214	1·496	26·615	0·379	1·496	17·579	7·800	26	·3426
	29	18·243	1·527	26·646	0·445	1·527	17·698	7·464	26	·3398
Mar.	**1**	+18·240	+1·569	26·686	+ 0·499	−1·569	−17·811	+ 7·125	+26	−0·3371
	2	18·201	1·614	26·729	0·538	1·614	17·919	6·785	26	·3344
	3	18·123	1·656	26·770	0·562	1·656	18·021	6·442	26	·3316
	4	18·011	1·689	26·802	0·572	1·689	18·117	6·097	26	·3289
	5	17·875	1·708	26·820	0·573	1·708	18·207	5·751	25	·3261
	6	+17·729	+1·709	26·819	+ 0·570	−1·709	−18·291	+ 5·403	+25	−0·3234
	7	17·592	1·691	26·800	0·571	1·691	18·369	5·053	25	·3207
	8	17·482	1·657	26·765	0·582	1·657	18·442	4·703	25	·3179
	9	17·414	1·612	26·719	0·609	1·612	18·508	4·351	25	·3152
	10	17·396	1·566	26·671	0·657	1·566	18·568	3·999	25	·3125
	11	+17·428	+1·527	26·631	+ 0·725	−1·527	−18·623	+ 3·646	+25	−0·3097
	12	17·499	1·505	26·608	0·808	1·505	18·671	3·292	25	·3070
	13	17·587	1·506	26·608	0·898	1·506	18·714	2·938	25	·3042
	14	17·664	1·531	26·631	0·983	1·531	18·751	2·585	25	·3015
	15	17·700	1·574	26·673	1·053	1·574	18·782	2·231	25	·2988
	16	+17·677	+1·623	26·721	+ 1·098	−1·623	−18·808	+ 1·877	+25	−0·2960
	17	17·590	1·663	26·760	1·118	1·663	18·829	1·524	25	·2933
	18	17·454	1·683	26·778	1·119	1·683	18·844	1·171	25	·2906
	19	17·300	1·674	26·768	1·113	1·674	18·854	0·817	25	·2878
	20	17·159	1·639	26·732	1·112	1·639	18·859	0·464	24	·2851
	21	+17·058	+1·587	26·678	+ 1·127	−1·587	−18·859	+ 0·110	+24	−0·2823
	22	17·007	1·529	26·619	1·161	1·529	18·854	− 0·243	24	·2796
	23	17·002	1·476	26·565	1·214	1·476	18·843	0·596	24	·2769
	24	17·029	1·436	26·523	1·280	1·436	18·827	0·950	24	·2741
	25	17·073	1·413	26·500	1·352	1·413	18·806	1·303	24	·2714
	26	+17·116	+1·408	26·493	+ 1·424	−1·408	−18·778	− 1·656	+24	−0·2687
	27	17·143	1·419	26·502	1·489	1·419	18·746	2·008	24	·2659
	28	17·141	1·440	26·522	1·544	1·440	18·707	2·360	24	·2632
	29	17·104	1·466	26·547	1·584	1·466	18·663	2·711	24	·2604
	30	17·030	1·492	26·571	1·609	1·492	18·613	3·061	24	·2577
	31	+16·922	+1·510	26·588	+ 1·621	−1·510	−18·557	− 3·411	+24	−0·2550
Apr.	**1**	+16·788	+1·515	26·592	+ 1·623	−1·515	−18·496	− 3·759	+24	−0·2522

FOR 0^h DYNAMICAL TIME

Date		Nutation		Obl. of	Besselian Day Numbers					Fraction
		in	in	Ecliptic	for Mean Equinox J1992·5					of Year
0^h TDT		Long.	Obl.	23° 26′	*A*	*B*	*C*	*D*	*E*	τ
		″	″	″	″	″	″	″	($0^{s}0001$)	
Apr.	**1**	+16·788	+1·515	26·592	+ 1·623	−1·515	−18·496	− 3·759	+24	−0·2522
	2	16·641	1·503	26·579	1·619	1·503	18·429	4·106	24	·2495
	3	16·499	1·472	26·546	1·617	1·472	18·356	4·451	23	·2467
	4	16·380	1·423	26·496	1·625	1·423	18·278	4·795	23	·2440
	5	16·303	1·362	26·434	1·649	1·362	18·193	5·137	23	·2413
	6	+16·277	+1·296	26·367	+ 1·694	−1·296	−18·103	− 5·477	+23	−0·2385
	7	16·303	1·237	26·307	1·759	1·237	18·008	5·814	23	·2358
	8	16·371	1·194	26·262	1·841	1·194	17·907	6·150	23	·2331
	9	16·461	1·172	26·239	1·932	1·172	17·800	6·482	23	·2303
	10	16·545	1·173	26·239	2·020	1·173	17·689	6·812	24	·2276
	11	+16·596	+1·194	26·258	+ 2·095	−1·194	−17·572	− 7·140	+24	−0·2248
	12	16·595	1·223	26·286	2·150	1·223	17·451	7·464	24	·2221
	13	16·536	1·249	26·311	2·181	1·249	17·325	7·785	23	·2194
	14	16·427	1·258	26·319	2·192	1·258	17·194	8·104	23	·2166
	15	16·291	1·243	26·302	2·193	1·243	17·059	8·420	23	·2139
	16	+16·159	+1·202	26·260	+ 2·196	−1·202	−16·919	− 8·733	+23	−0·2112
	17	16·058	1·140	26·197	2·210	1·140	16·775	9·044	23	·2084
	18	16·004	1·068	26·124	2·244	1·068	16·627	9·351	23	·2057
	19	16·002	0·997	26·051	2·298	0·997	16·474	9·657	23	·2029
	20	16·041	0·936	25·989	2·368	0·936	16·317	9·959	23	·2002
	21	+16·106	+0·892	25·943	+ 2·449	−0·892	−16·155	−10·259	+23	−0·1975
	22	16·178	0·866	25·916	2·532	0·866	15·989	10·556	23	·1947
	23	16·238	0·858	25·907	2·611	0·858	15·818	10·850	23	·1920
	24	16·273	0·863	25·910	2·680	0·863	15·643	11·140	23	·1893
	25	16·275	0·875	25·922	2·736	0·875	15·463	11·428	23	·1865
	26	+16·240	+0·889	25·935	+ 2·777	−0·889	−15·279	−11·713	+23	−0·1838
	27	16·171	0·898	25·942	2·804	0·898	15·090	11·994	23	·1810
	28	16·073	0·896	25·939	2·820	0·896	14·896	12·272	23	·1783
	29	15·959	0·879	25·920	2·829	0·879	14·698	12·546	23	·1756
	30	15·845	0·843	25·883	2·839	0·843	14·496	12·816	23	·1728
May	**1**	+15·748	+0·790	25·828	+ 2·855	−0·790	−14·289	−13·082	+22	−0·1701
	2	15·689	0·721	25·759	2·887	0·721	14·078	13·344	22	·1674
	3	15·682	0·646	25·682	2·939	0·646	13·863	13·602	22	·1646
	4	15·731	0·574	25·609	3·013	0·574	13·643	13·856	22	·1619
	5	15·829	0·517	25·551	3·107	0·517	13·419	14·105	22	·1591
	6	+15·955	+0·482	25·514	+ 3·212	−0·482	−13·192	−14·349	+23	−0·1564
	7	16·081	0·471	25·502	3·317	0·471	12·961	14·588	23	·1537
	8	16·178	0·482	25·511	3·410	0·482	12·726	14·823	23	·1509
	9	16·225	0·504	25·532	3·484	0·504	12·488	15·053	23	·1482
	10	16·213	0·525	25·552	3·534	0·525	12·247	15·277	23	·1454
	11	+16·150	+0·534	25·560	+ 3·564	−0·534	−12·002	−15·497	+23	−0·1427
	12	16·056	0·521	25·546	3·581	0·521	11·755	15·713	23	·1400
	13	15·958	0·484	25·507	3·597	0·484	11·506	15·923	23	·1372
	14	15·882	0·426	25·448	3·622	0·426	11·253	16·129	23	·1345
	15	15·849	0·355	25·376	3·664	0·355	10·998	16·330	23	·1318
	16	+15·865	+0·282	25·301	+ 3·725	−0·282	−10·740	−16·527	+23	−0·1290
	17	+15·928	+0·216	25·234	+ 3·805	−0·216	−10·480	−16·719	+23	−0·1263

FOR 0^h DYNAMICAL TIME

Date 0^h TDT		Nutation in Long.	Nutation in Obl.	Obl. of Ecliptic 23° 26′	Besselian Day Numbers for Mean Equinox J1992·5 *A*	*B*	*C*	*D*	*E*	Fraction of Year τ
		″	″	″	″	″	″	″	(0″0001)	
May	**17**	+15·928	+0·216	25·234	+ 3·805	−0·216	−10·480	−16·719	+23	−0·1263
	18	16·023	0·165	25·182	3·897	0·165	10·217	16·907	23	·1235
	19	16·132	0·133	25·148	3·996	0·133	9·951	17·090	23	·1208
	20	16·236	0·119	25·134	4·092	0·119	9·683	17·269	23	·1181
	21	16·319	0·121	25·134	4·180	0·121	9·412	17·443	23	·1153
	22	+16·371	+0·134	25·146	+ 4·256	−0·134	− 9·138	−17·612	+23	−0·1126
	23	16·386	0·150	25·161	4·316	0·150	8·862	17·777	23	·1099
	24	16·365	0·164	25·173	4·363	0·164	8·583	17·936	23	·1071
	25	16·312	0·169	25·177	4·397	0·169	8·302	18·091	23	·1044
	26	16·239	0·161	25·168	4·422	0·161	8·018	18·240	23	·1016
	27	+16·159	+0·136	25·142	+ 4·446	−0·136	− 7·731	−18·385	+23	−0·0989
	28	16·091	0·094	25·098	4·473	0·094	7·442	18·524	23	·0962
	29	16·052	+0·036	25·039	4·513	−0·036	7·151	18·657	23	·0934
	30	16·060	−0·032	24·970	4·571	+0·032	6·858	18·785	23	·0907
	31	16·124	0·100	24·900	4·651	0·100	6·562	18·908	23	·0880
June	**1**	+16·244	−0·158	24·841	+ 4·754	+0·158	− 6·264	−19·025	+23	−0·0852
	2	16·404	0·195	24·803	4·872	0·195	5·965	19·135	23	·0825
	3	16·574	0·205	24·792	4·995	0·205	5·663	19·240	24	·0797
	4	16·722	0·190	24·805	5·109	0·190	5·361	19·339	24	·0770
	5	16·820	0·160	24·834	5·202	0·160	5·057	19·431	24	·0743
	6	+16·856	−0·127	24·866	+ 5·272	+0·127	− 4·752	−19·518	+24	−0·0715
	7	16·834	0·104	24·887	5·318	0·104	4·446	19·599	24	·0688
	8	16·774	0·102	24·888	5·349	0·102	4·139	19·674	24	·0661
	9	16·703	0·122	24·866	5·375	0·122	3·832	19·744	24	·0633
	10	16·647	0·165	24·823	5·408	0·165	3·525	19·807	24	·0606
	11	+16·628	−0·221	24·765	+ 5·455	+0·221	− 3·216	−19·866	+24	−0·0578
	12	16·656	0·282	24·703	5·521	0·282	2·908	19·919	24	·0551
	13	16·729	0·337	24·647	5·605	0·337	2·599	19·967	24	·0524
	14	16·837	0·379	24·603	5·703	0·379	2·289	20·009	24	·0496
	15	16·964	0·404	24·577	5·808	0·404	1·979	20·046	24	·0469
	16	+17·091	−0·409	24·570	+ 5·914	+0·409	− 1·669	−20·078	+24	−0·0441
	17	17·203	0·398	24·580	6·013	0·398	1·358	20·105	24	·0414
	18	17·285	0·375	24·603	6·101	0·375	1·047	20·126	25	·0387
	19	17·330	0·345	24·631	6·173	0·345	0·735	20·142	25	·0359
	20	17·337	0·315	24·660	6·231	0·315	0·424	20·153	25	·0332
	21	+17·309	−0·292	24·682	+ 6·275	+0·292	− 0·111	−20·158	+25	−0·0305
	22	17·257	0·280	24·692	6·309	0·280	+ 0·201	20·157	25	·0277
	23	17·192	0·284	24·687	6·338	0·284	0·514	20·151	24	·0250
	24	17·131	0·304	24·665	6·369	0·304	0·827	20·140	24	·0222
	25	17·091	0·340	24·628	6·408	0·340	1·140	20·122	24	·0195
	26	+17·089	−0·388	24·579	+ 6·462	+0·388	+ 1·453	−20·099	+24	−0·0168
	27	17·139	0·441	24·525	6·536	0·441	1·766	20·070	24	·0140
	28	17·244	0·488	24·476	6·633	0·488	2·078	20·035	24	·0113
	29	17·399	0·519	24·444	6·750	0·519	2·391	19·994	25	·0086
	30	17·578	0·524	24·438	6·876	0·524	2·702	19·947	25	·0058
July	**1**	+17·750	−0·502	24·459	+ 6·999	+0·502	+ 3·014	−19·894	+25	−0·0031
	2	+17·880	−0·458	24·501	+ 7·106	+0·458	+ 3·324	−19·835	+25	−0·0003

FOR 0ʰ DYNAMICAL TIME

Date 0ʰ TDT		Nutation in Long.	Nutation in Obl.	Obl. of Ecliptic 23° 26′	Besselian Day Numbers for Mean Equinox J1992·5 A	B	C	D	E	Fraction of Year τ
		″	″	″	″	″	″	″	(0″0001)	
July	**1**	+17·750	−0·502	24·459	+ 6·999	+0·502	+ 3·014	−19·894	+25	−0·0031
	2	17·880	0·458	24·501	7·106	0·458	3·324	19·835	25	− ·0003
	3	17·945	0·406	24·552	7·186	0·406	3·633	19·769	25	+ ·0024
	4	17·943	0·361	24·596	7·241	0·361	3·940	19·698	25	·0051
	5	17·892	0·334	24·621	7·275	0·334	4·246	19·620	25	·0079
	6	+17·819	−0·333	24·621	+ 7·301	+0·333	+ 4·550	−19·537	+25	+0·0106
	7	17·756	0·355	24·598	7·331	0·355	4·853	19·449	25	·0133
	8	17·724	0·392	24·559	7·373	0·392	5·154	19·355	25	·0161
	9	17·735	0·437	24·514	7·432	0·437	5·453	19·256	25	·0188
	10	17·791	0·477	24·472	7·509	0·477	5·750	19·153	25	·0216
	11	+17·883	−0·507	24·441	+ 7·601	+0·507	+ 6·045	−19·044	+25	+0·0243
	12	17·996	0·520	24·427	7·701	0·520	6·339	18·930	26	·0270
	13	18·114	0·514	24·431	7·802	0·514	6·630	18·811	26	·0298
	14	18·218	0·491	24·453	7·899	0·491	6·920	18·687	26	·0325
	15	18·296	0·455	24·487	7·984	0·455	7·208	18·558	26	·0352
	16	+18·337	−0·411	24·530	+ 8·056	+0·411	+ 7·495	−18·424	+26	+0·0380
	17	18·340	0·366	24·574	8·112	0·366	7·779	18·286	26	·0407
	18	18·306	0·325	24·613	8·153	0·325	8·062	18·142	26	·0435
	19	18·243	0·295	24·642	8·183	0·295	8·342	17·993	26	·0462
	20	18·163	0·279	24·657	8·206	0·279	8·621	17·840	26	·0489
	21	+18·080	−0·280	24·655	+ 8·228	+0·280	+ 8·897	−17·681	+26	+0·0517
	22	18·011	0·296	24·637	8·255	0·296	9·172	17·518	26	·0544
	23	17·972	0·325	24·607	8·295	0·325	9·444	17·349	26	·0572
	24	17·978	0·362	24·569	8·352	0·362	9·714	17·175	26	·0599
	25	18·035	0·397	24·533	8·429	0·397	9·982	16·997	26	·0626
	26	+18·141	−0·421	24·507	+ 8·527	+0·421	+10·247	−16·812	+26	+0·0654
	27	18·284	0·425	24·502	8·638	0·425	10·510	16·623	26	·0681
	28	18·435	0·403	24·523	8·753	0·403	10·769	16·428	26	·0708
	29	18·561	0·355	24·569	8·858	0·355	11·026	16·228	26	·0736
	30	18·632	0·293	24·631	8·941	0·293	11·280	16·023	26	·0763
	31	+18·631	−0·230	24·692	+ 8·996	+0·230	+11·530	−15·813	+26	+0·0791
Aug.	**1**	18·568	0·184	24·737	9·026	0·184	11·777	15·598	26	·0818
	2	18·469	0·163	24·757	9·041	0·163	12·020	15·378	26	·0845
	3	18·369	0·169	24·750	9·056	0·169	12·259	15·154	26	·0873
	4	18·295	0·195	24·722	9·082	0·195	12·494	14·926	26	·0900
	5	+18·265	−0·231	24·684	+ 9·125	+0·231	+12·726	−14·693	+26	+0·0927
	6	18·280	0·267	24·648	9·186	0·267	12·953	14·457	26	·0955
	7	18·333	0·292	24·621	9·261	0·292	13·177	14·217	26	·0982
	8	18·409	0·302	24·610	9·347	0·302	13·397	13·972	26	·1010
	9	18·493	0·294	24·617	9·435	0·294	13·613	13·725	26	·1037
	10	+18·566	−0·269	24·641	+ 9·519	+0·269	+13·825	−13·473	+26	+0·1064
	11	18·616	0·229	24·679	9·593	0·229	14·034	13·219	26	·1092
	12	18·631	0·182	24·725	9·654	0·182	14·239	12·960	26	·1119
	13	18·607	0·132	24·774	9·700	0·132	14·440	12·698	26	·1146
	14	18·545	0·085	24·819	9·730	0·085	14·638	12·433	26	·1174
	15	+18·452	−0·048	24·855	+ 9·748	+0·048	+14·831	−12·164	+26	+0·1201
	16	+18·338	−0·025	24·877	+ 9·757	+0·025	+15·021	−11·891	+26	+0·1229

FOR 0^h DYNAMICAL TIME

Date 0^h TDT		Nutation in Long.	Nutation in Obl.	Obl. of Ecliptic 23° 26′	Besselian Day Numbers for Mean Equinox J1992·5 *A*	*B*	*C*	*D*	*E*	Fraction of Year *τ*
		″	″	″	″	″	″	″	(0″0001)	
Aug.	**16**	+18·338	−0·025	24·877	+ 9·757	+0·025	+15·021	−11·891	+26	+0·1229
	17	18·218	0·018	24·882	9·764	0·018	15·207	11·615	26	·1256
	18	18·106	0·027	24·872	9·775	0·027	15·389	11·336	26	·1283
	19	18·019	0·051	24·847	9·795	0·051	15·567	11·053	26	·1311
	20	17·971	0·083	24·814	9·831	0·083	15·741	10·766	26	·1338
	21	+17·969	−0·116	24·779	+ 9·885	+0·116	+15·912	−10·477	+26	+0·1366
	22	18·015	0·143	24·751	9·958	0·143	16·077	10·183	26	·1393
	23	18·099	0·154	24·739	10·046	0·154	16·239	9·887	26	·1420
	24	18·202	0·143	24·748	10·142	0·143	16·396	9·586	26	·1448
	25	18·297	0·108	24·782	10·235	0·108	16·549	9·282	26	·1475
	26	+18·352	−0·054	24·835	+10·312	+0·054	+16·697	− 8·975	+26	+0·1502
	27	18·345	+0·007	24·895	10·364	−0·007	16·840	8·665	26	·1530
	28	18·271	0·060	24·946	10·389	0·060	16·978	8·351	26	·1557
	29	18·147	0·089	24·974	10·395	0·089	17·111	8·035	26	·1585
	30	18·006	0·090	24·973	10·394	0·090	17·238	7·716	26	·1612
	31	+17·883	+0·064	24·947	+10·400	−0·064	+17·360	− 7·395	+25	+0·1639
Sept.	**1**	17·803	+0·023	24·904	10·422	−0·023	17·476	7·071	25	·1667
	2	17·772	−0·022	24·857	10·465	+0·022	17·587	6·746	25	·1694
	3	17·785	0·060	24·819	10·525	0·060	17·693	6·418	25	·1721
	4	17·826	0·082	24·795	10·596	0·082	17·794	6·089	25	·1749
	5	+17·878	−0·086	24·790	+10·672	+0·086	+17·890	− 5·759	+25	+0·1776
	6	17·924	0·073	24·801	10·745	0·073	17·980	5·427	25	·1804
	7	17·948	0·046	24·827	10·810	0·046	18·066	5·093	25	·1831
	8	17·941	−0·009	24·863	10·861	+0·009	18·146	4·758	25	·1858
	9	17·896	+0·031	24·902	10·898	−0·031	18·221	4·422	25	·1886
	10	+17·813	+0·069	24·939	+10·920	−0·069	+18·292	− 4·085	+25	+0·1913
	11	17·697	0·098	24·967	10·929	0·098	18·357	3·746	25	·1940
	12	17·558	0·114	24·981	10·929	0·114	18·418	3·406	25	·1968
	13	17·410	0·114	24·980	10·925	0·114	18·473	3·065	25	·1995
	14	17·268	0·097	24·961	10·923	0·097	18·523	2·723	25	·2023
	15	+17·149	+0·065	24·928	+10·930	−0·065	+18·569	− 2·379	+24	+0·2050
	16	17·064	+0·022	24·884	10·952	−0·022	18·609	2·035	24	·2077
	17	17·024	−0·024	24·837	10·991	+0·024	18·645	1·689	24	·2105
	18	17·030	0·065	24·795	11·048	0·065	18·675	1·343	24	·2132
	19	17·075	0·093	24·765	11·121	0·093	18·700	0·995	24	·2159
	20	+17·143	−0·102	24·755	+11·203	+0·102	+18·720	− 0·646	+24	+0·2187
	21	17·212	0·090	24·766	11·285	0·090	18·734	− 0·297	24	·2214
	22	17·255	0·058	24·796	11·357	0·058	18·743	+ 0·054	24	·2242
	23	17·248	−0·015	24·838	11·409	+0·015	18·746	0·405	24	·2269
	24	17·181	+0·025	24·877	11·437	−0·025	18·744	0·757	24	·2296
	25	+17·060	+0·049	24·900	+11·444	−0·049	+18·735	+ 1·109	+24	+0·2324
	26	16·908	0·047	24·896	11·438	0·047	18·721	1·462	24	·2351
	27	16·760	+0·016	24·864	11·434	−0·016	18·700	1·814	24	·2379
	28	16·647	−0·037	24·809	11·444	+0·037	18·674	2·166	24	·2406
	29	16·586	0·100	24·745	11·475	0·100	18·641	2·517	24	·2433
	30	+16·579	−0·160	24·684	+11·527	+0·160	+18·603	+ 2·868	+24	+0·2461
Oct.	**1**	+16·610	−0·207	24·636	+11·594	+0·207	+18·559	+ 3·217	+24	+0·2488

FOR 0^h DYNAMICAL TIME

Date 0^h TDT		Nutation in Long.	Nutation in Obl.	Obl. of Ecliptic 23° 26′	Besselian Day Numbers for Mean Equinox J1992·5 *A*	*B*	*C*	*D*	*E*	Fraction of Year *τ*
		″	″	″	″	″	″	″	(0^s0001)	
Oct.	**1**	+16·610	−0·207	24·636	+11·594	+0·207	+18·559	+ 3·217	+24	+0·2488
	2	16·659	0·235	24·606	11·669	0·235	18·509	3·566	24	·2515
	3	16·707	0·244	24·596	11·743	0·244	18·454	3·913	24	·2543
	4	16·738	0·238	24·601	11·810	0·238	18·393	4·259	24	·2570
	5	16·738	0·220	24·617	11·865	0·220	18·327	4·604	24	·2598
	6	+16·703	−0·198	24·639	+11·906	+0·198	+18·256	+ 4·947	+24	+0·2625
	7	16·631	0·176	24·659	11·932	0·176	18·179	5·289	24	·2652
	8	16·525	0·162	24·672	11·944	0·162	18·097	5·629	23	·2680
	9	16·394	0·159	24·673	11·947	0·159	18·010	5·968	23	·2707
	10	16·252	0·173	24·658	11·946	0·173	17·918	6·306	23	·2734
	11	+16·112	−0·203	24·626	+11·945	+0·203	+17·821	+ 6·641	+23	+0·2762
	12	15·993	0·250	24·579	11·952	0·250	17·719	6·975	23	·2789
	13	15·908	0·308	24·519	11·973	0·308	17·611	7·308	23	·2817
	14	15·867	0·371	24·455	12·012	0·371	17·499	7·638	23	·2844
	15	15·874	0·431	24·394	12·070	0·431	17·382	7·967	23	·2871
	16	+15·922	−0·480	24·344	+12·144	+0·480	+17·260	+ 8·294	+23	+0·2899
	17	15·997	0·510	24·312	12·228	0·510	17·132	8·620	23	·2926
	18	16·077	0·519	24·302	12·315	0·519	17·000	8·943	23	·2953
	19	16·137	0·509	24·310	12·394	0·509	16·862	9·265	23	·2981
	20	16·156	0·487	24·331	12·456	0·487	16·719	9·585	23	·3008
	21	+16·122	−0·463	24·354	+12·498	+0·463	+16·570	+ 9·902	+23	+0·3036
	22	16·035	0·450	24·366	12·518	0·450	16·417	10·217	23	·3063
	23	15·913	0·458	24·357	12·524	0·458	16·257	10·530	23	·3090
	24	15·783	0·493	24·320	12·528	0·493	16·092	10·839	22	·3118
	25	15·677	0·552	24·260	12·540	0·552	15·922	11·146	22	·3145
	26	+15·618	−0·626	24·184	+12·571	+0·626	+15·746	+11·449	+22	+0·3172
	27	15·615	0·703	24·106	12·625	0·703	15·565	11·749	22	·3200
	28	15·662	0·770	24·038	12·699	0·770	15·379	12·045	22	·3227
	29	15·739	0·820	23·986	12·784	0·820	15·188	12·337	22	·3255
	30	15·824	0·850	23·955	12·873	0·850	14·992	12·626	22	·3282
	31	+15·897	−0·862	23·943	+12·957	+0·862	+14·791	+12·910	+23	+0·3309
Nov.	**1**	15·943	0·859	23·944	13·030	0·859	14·586	13·190	23	·3337
	2	15·954	0·849	23·953	13·089	0·849	14·376	13·466	23	·3364
	3	15·927	0·837	23·963	13·134	0·837	14·162	13·738	23	·3392
	4	15·866	0·831	23·968	13·164	0·831	13·944	14·006	23	·3419
	5	+15·778	−0·835	23·963	+13·184	+0·835	+13·722	+14·269	+22	+0·3446
	6	15·674	0·853	23·944	13·198	0·853	13·496	14·529	22	·3474
	7	15·570	0·888	23·908	13·211	0·888	13·265	14·783	22	·3501
	8	15·481	0·938	23·856	13·230	0·938	13·031	15·034	22	·3528
	9	15·423	1·002	23·790	13·262	1·002	12·793	15·280	22	·3556
	10	+15·409	−1·073	23·718	+13·312	+1·073	+12·552	+15·523	+22	+0·3583
	11	15·446	1·143	23·647	13·381	1·143	12·307	15·760	22	·3611
	12	15·528	1·202	23·586	13·469	1·202	12·058	15·994	22	·3638
	13	15·643	1·244	23·544	13·569	1·244	11·805	16·223	22	·3665
	14	15·768	1·263	23·523	13·674	1·263	11·549	16·448	22	·3693
	15	+15·877	−1·262	23·523	+13·772	+1·262	+11·289	+16·668	+23	+0·3720
	16	+15·948	−1·245	23·538	+13·855	+1·245	+11·026	+16·884	+23	+0·3747

FOR 0ʰ DYNAMICAL TIME

Date 0ʰ TDT		Nutation in Long.	Nutation in Obl.	Obl. of Ecliptic 23° 26′	Besselian Day Numbers for Mean Equinox J1992·5 A	B	C	D	E	Fraction of Year τ
		″	″	″	″	″	″	″	(0ˢ0001)	
Nov.	16	+15·948	−1·245	23·538	+13·855	+1·245	+11·026	+16·884	+23	+0·3747
	17	15·966	1·225	23·558	13·917	1·225	10·758	17·096	23	·3775
	18	15·933	1·211	23·570	13·959	1·211	10·487	17·302	23	·3802
	19	15·860	1·216	23·564	13·985	1·216	10·212	17·504	22	·3830
	20	15·773	1·244	23·534	14·005	1·244	9·933	17·701	22	·3857
	21	+15·700	−1·296	23·481	+14·031	+1·296	+ 9·651	+17·892	+22	+0·3884
	22	15·665	1·365	23·411	14·072	1·365	9·364	18·078	22	·3912
	23	15·683	1·440	23·335	14·134	1·440	9·075	18·258	22	·3939
	24	15·753	1·509	23·264	14·217	1·509	8·782	18·432	22	·3966
	25	15·863	1·565	23·208	14·315	1·565	8·486	18·600	23	·3994
	26	+15·991	−1·600	23·171	+14·421	+1·600	+ 8·187	+18·762	+23	+0·4021
	27	16·115	1·615	23·155	14·525	1·615	7·886	18·918	23	·4049
	28	16·217	1·613	23·156	14·621	1·613	7·582	19·068	23	·4076
	29	16·286	1·599	23·168	14·703	1·599	7·275	19·211	23	·4103
	30	16·315	1·582	23·184	14·769	1·582	6·967	19·349	23	·4131
Dec.	1	+16·306	−1·567	23·197	+14·821	+1·567	+ 6·656	+19·480	+23	+0·4158
	2	16·267	1·561	23·202	14·860	1·561	6·344	19·605	23	·4185
	3	16·208	1·567	23·195	14·892	1·567	6·029	19·724	23	·4213
	4	16·143	1·588	23·172	14·921	1·588	5·713	19·837	23	·4240
	5	16·086	1·625	23·134	14·953	1·625	5·395	19·944	23	·4268
	6	+16·054	−1·676	23·082	+14·995	+1·676	+ 5·076	+20·044	+23	+0·4295
	7	16·062	1·735	23·022	15·053	1·735	4·756	20·139	23	·4322
	8	16·118	1·797	22·959	15·130	1·797	4·434	20·228	23	·4350
	9	16·224	1·851	22·903	15·227	1·851	4·111	20·311	23	·4377
	10	16·370	1·889	22·864	15·340	1·889	3·787	20·388	23	·4405
	11	+16·537	−1·904	22·847	+15·461	+1·904	+ 3·462	+20·459	+23	+0·4432
	12	16·694	1·896	22·855	15·579	1·896	3·136	20·524	24	·4459
	13	16·816	1·868	22·881	15·682	1·868	2·808	20·584	24	·4487
	14	16·883	1·833	22·914	15·764	1·833	2·480	20·637	24	·4514
	15	16·892	1·803	22·943	15·822	1·803	2·150	20·685	24	·4541
	16	+16·856	−1·789	22·957	+15·863	+1·789	+ 1·819	+20·726	+24	+0·4569
	17	16·797	1·797	22·947	15·894	1·797	1·486	20·762	24	·4596
	18	16·746	1·827	22·915	15·929	1·827	1·153	20·790	24	·4624
	19	16·724	1·876	22·865	15·975	1·876	0·819	20·812	24	·4651
	20	16·750	1·933	22·807	16·040	1·933	0·484	20·828	24	·4678
	21	+16·824	−1·987	22·752	+16·124	+1·987	+ 0·149	+20·837	+24	+0·4706
	22	16·940	2·030	22·708	16·225	2·030	− 0·187	20·838	24	·4733
	23	17·080	2·054	22·682	16·336	2·054	0·523	20·833	24	·4760
	24	17·224	2·058	22·677	16·448	2·058	0·859	20·821	24	·4788
	25	17·352	2·044	22·690	16·554	2·044	1·194	20·802	25	·4815
	26	+17·449	−2·016	22·716	+16·647	+2·016	− 1·530	+20·776	+25	+0·4843
	27	17·506	1·982	22·750	16·725	1·982	1·864	20·744	25	·4870
	28	17·523	1·947	22·783	16·787	1·947	2·198	20·705	25	·4897
	29	17·505	1·918	22·810	16·834	1·918	2·531	20·659	25	·4925
	30	17·461	1·901	22·826	16·872	1·901	2·863	20·606	25	·4952
	31	+17·404	−1·898	22·828	+16·904	+1·898	− 3·193	+20·547	+25	+0·4979
	32	+17·349	−1·910	22·815	+16·937	+1·910	− 3·522	+20·481	+25	+0·5007

J for NORTHERN DECLINATIONS
FOR 0^h TDT AND EQUINOX J1992·5

Right Ascension

Date	0^h 12^h	1^h 13^h	2^h 14^h	3^h 15^h	4^h 16^h	5^h 17^h	6^h 18^h	7^h 19^h	8^h 20^h	9^h 21^h	10^h 22^h	11^h 23^h	12^h 24^h
Jan. −1	− 2	0	+ 3	+ 5	+ 5	+ 4	+ 2	0	− 3	− 5	− 5	− 4	− 2
9	− 4	− 1	+ 2	+ 4	+ 5	+ 5	+ 4	+ 1	− 2	− 4	− 5	− 5	− 4
19	− 5	− 3	0	+ 3	+ 5	+ 6	+ 5	+ 3	0	− 3	− 5	− 6	− 5
29	− 6	− 5	− 2	+ 1	+ 4	+ 6	+ 6	+ 5	+ 2	− 1	− 4	− 6	− 6
Feb. 8	− 7	− 6	− 4	− 1	+ 2	+ 5	+ 7	+ 6	+ 4	+ 1	− 2	− 5	− 7
18	− 6	− 7	− 6	− 3	0	+ 4	+ 6	+ 7	+ 6	+ 3	0	− 4	− 6
28	− 5	− 7	− 7	− 5	− 2	+ 2	+ 5	+ 7	+ 7	+ 5	+ 2	− 2	− 5
Mar. 9	− 3	− 6	− 7	− 6	− 4	0	+ 3	+ 6	+ 7	+ 6	+ 4	0	− 3
19	− 1	− 4	− 6	− 7	− 5	− 2	+ 1	+ 4	+ 6	+ 7	+ 5	+ 2	− 1
29	+ 1	− 3	− 5	− 7	− 6	− 4	− 1	+ 3	+ 5	+ 7	+ 6	+ 4	+ 1
Apr. 8	+ 3	0	− 4	− 6	− 6	− 5	− 3	0	+ 4	+ 6	+ 6	+ 5	+ 3
18	+ 4	+ 1	− 2	− 4	− 6	− 6	− 4	− 1	+ 2	+ 4	+ 6	+ 6	+ 4
28	+ 5	+ 3	0	− 3	− 5	− 5	− 5	− 3	0	+ 3	+ 5	+ 5	+ 5
May 8	+ 5	+ 4	+ 2	− 1	− 3	− 5	− 5	− 4	− 2	+ 1	+ 3	+ 5	+ 5
18	+ 4	+ 4	+ 3	+ 1	− 1	− 3	− 4	− 4	− 3	− 1	+ 1	+ 3	+ 4
28	+ 3	+ 4	+ 4	+ 2	0	− 2	− 3	− 4	− 4	− 2	0	+ 2	+ 3
June 7	+ 2	+ 3	+ 4	+ 3	+ 2	0	− 2	− 3	− 4	− 3	− 2	0	+ 2
17	0	+ 2	+ 3	+ 3	+ 3	+ 1	0	− 2	− 3	− 3	− 3	− 1	0
27	− 1	+ 1	+ 2	+ 3	+ 3	+ 2	+ 1	− 1	− 2	− 3	− 3	− 2	− 1
July 7	− 2	− 1	+ 1	+ 2	+ 3	+ 3	+ 2	+ 1	− 1	− 2	− 3	− 3	− 2
17	− 3	− 2	− 1	+ 1	+ 2	+ 3	+ 3	+ 2	+ 1	− 1	− 2	− 3	− 3
27	− 3	− 3	− 2	− 1	+ 1	+ 2	+ 3	+ 3	+ 2	+ 1	− 1	− 2	− 3
Aug. 6	− 2	− 3	− 3	− 2	− 1	+ 1	+ 2	+ 3	+ 3	+ 2	+ 1	− 1	− 2
16	− 1	− 3	− 4	− 4	− 3	− 1	+ 1	+ 3	+ 4	+ 4	+ 3	+ 1	− 1
26	+ 1	− 2	− 4	− 5	− 4	− 3	− 1	+ 2	+ 4	+ 5	+ 4	+ 3	+ 1
Sept. 5	+ 3	0	− 3	− 5	− 6	− 5	− 3	0	+ 3	+ 5	+ 6	+ 5	+ 3
15	+ 5	+ 2	− 1	− 4	− 6	− 7	− 5	− 2	+ 1	+ 4	+ 6	+ 7	+ 5
25	+ 8	+ 5	+ 1	− 3	− 6	− 8	− 8	− 5	− 1	+ 3	+ 6	+ 8	+ 8
Oct. 5	+10	+ 8	+ 4	− 1	− 6	− 9	−10	− 8	− 4	+ 1	+ 6	+ 9	+10
15	+12	+11	+ 7	+ 1	− 5	− 9	−12	−11	− 7	− 1	+ 5	+ 9	+12
25	+13	+13	+10	+ 5	− 2	− 9	−13	−13	−10	− 5	+ 2	+ 9	+13
Nov. 4	+13	+15	+14	+ 8	+ 1	− 7	−13	−15	−14	− 8	− 1	+ 7	+13
14	+12	+17	+17	+12	+ 4	− 5	−12	−17	−17	−12	− 4	+ 5	+12
24	+11	+17	+19	+16	+ 8	− 2	−11	−17	−19	−16	− 8	+ 2	+11
Dec. 4	+ 8	+16	+20	+19	+12	+ 2	− 8	−16	−20	−19	−12	− 2	+ 8
14	+ 5	+15	+21	+21	+16	+ 6	− 5	−15	−21	−21	−16	− 6	+ 5
24	+ 1	+12	+20	+22	+19	+10	− 1	−12	−20	−22	−19	−10	+ 1
34	− 3	+ 9	+18	+22	+21	+14	+ 3	− 9	−18	−22	−21	−14	− 3

The second-order day number J is given in this table in units of $0\overset{s}{.}000\,01$.
The apparent right ascension of a star is given by:

$$\alpha = \alpha_1 + \tau\mu_\alpha / 100 + Aa + Bb + Cc + Dd + E + J \tan^2 \delta_1$$

where the position (α_1, δ_1) and centennial proper motion in right ascension (μ_α) are referred to the mean equator and equinox of J1992·5

J' for NORTHERN DECLINATIONS FOR 0^h TDT AND EQUINOX J1992·5

Right Ascension

Date	0^h 12^h	1^h 13^h	2^h 14^h	3^h 15^h	4^h 16^h	5^h 17^h	6^h 18^h	7^h 19^h	8^h 20^h	9^h 21^h	10^h 22^h	11^h 23^h	12^h 24^h
Jan. −1	0	0	− 1	− 2	− 4	− 6	− 7	− 8	− 7	− 5	− 3	− 2	0
9	− 1	0	0	− 1	− 3	− 5	− 7	− 8	− 8	− 7	− 5	− 3	− 1
19	− 2	− 1	0	− 1	− 2	− 4	− 7	− 8	− 9	− 9	− 7	− 5	− 2
29	− 4	− 2	0	0	− 1	− 3	− 6	− 8	− 9	−10	− 9	− 7	− 4
Feb. 8	− 6	− 3	− 1	0	0	− 2	− 4	− 7	− 9	−10	−10	− 8	− 6
18	− 8	− 5	− 2	− 1	0	− 1	− 3	− 6	− 8	−10	−10	−10	− 8
28	− 9	− 6	− 4	− 1	0	0	− 2	− 4	− 7	− 9	−10	−10	− 9
Mar. 9	−10	− 8	− 5	− 3	− 1	0	− 1	− 2	− 5	− 8	−10	−10	−10
19	−10	− 9	− 7	− 4	− 2	0	0	− 1	− 3	− 6	− 9	−10	−10
29	−10	− 9	− 8	− 5	− 3	− 1	0	0	− 2	− 4	− 7	− 9	−10
Apr. 8	− 9	− 9	− 8	− 7	− 4	− 2	0	0	− 1	− 3	− 5	− 7	− 9
18	− 8	− 9	− 9	− 7	− 5	− 3	− 1	0	0	− 1	− 3	− 6	− 8
28	− 6	− 8	− 8	− 8	− 6	− 4	− 2	− 1	0	0	− 2	− 4	− 6
May 8	− 4	− 6	− 7	− 7	− 7	− 5	− 3	− 1	0	0	− 1	− 2	− 4
18	− 3	− 4	− 6	− 7	− 7	− 6	− 4	− 2	− 1	0	0	− 1	− 3
28	− 1	− 3	− 4	− 6	− 6	− 6	− 5	− 3	− 2	− 1	0	0	− 1
June 7	0	− 2	− 3	− 4	− 5	− 5	− 5	− 4	− 3	− 1	0	0	0
17	0	− 1	− 2	− 3	− 4	− 5	− 5	− 4	− 3	− 2	− 1	0	0
27	0	0	− 1	− 2	− 3	− 4	− 4	− 5	− 4	− 3	− 2	− 1	0
July 7	− 1	0	0	− 1	− 2	− 3	− 4	− 4	− 4	− 4	− 3	− 2	− 1
17	− 2	− 1	0	0	− 1	− 1	− 3	− 3	− 4	− 4	− 4	− 3	− 2
27	− 3	− 2	− 1	0	0	− 1	− 2	− 3	− 4	− 4	− 4	− 4	− 3
Aug. 6	− 4	− 3	− 2	− 1	0	0	− 1	− 2	− 3	− 4	− 5	− 5	− 4
16	− 5	− 5	− 3	− 2	− 1	0	0	− 1	− 2	− 4	− 5	− 6	− 5
26	− 7	− 7	− 6	− 4	− 2	− 1	0	0	− 1	− 3	− 5	− 6	− 7
Sept. 5	− 8	− 8	− 8	− 6	− 4	− 2	− 1	0	− 1	− 2	− 4	− 6	− 8
15	− 8	−10	−10	− 9	− 7	− 4	− 2	0	0	− 1	− 3	− 6	− 8
25	− 8	−11	−12	−12	−10	− 7	− 4	− 1	0	0	− 2	− 5	− 8
Oct. 5	− 8	−12	−14	−15	−13	−10	− 7	− 3	− 1	0	− 1	− 5	− 8
15	− 8	−12	−16	−17	−17	−14	−10	− 5	− 2	0	− 1	− 4	− 8
25	− 7	−12	−17	−20	−20	−18	−14	− 8	− 4	− 1	0	− 2	− 7
Nov. 4	− 5	−11	−17	−21	−23	−22	−18	−12	− 6	− 2	0	− 1	− 5
14	− 4	−10	−17	−22	−26	−25	−22	−16	− 9	− 4	0	− 1	− 4
24	− 3	− 8	−15	−22	−27	−28	−26	−20	−13	− 6	− 1	0	− 3
Dec. 4	− 1	− 6	−14	−21	−28	−30	−29	−24	−17	− 9	− 3	0	− 1
14	0	− 4	−12	−20	−27	−32	−32	−28	−21	−12	− 5	− 1	0
24	0	− 3	− 9	−18	−26	−32	−34	−31	−24	−16	− 8	− 2	0
34	0	− 1	− 7	−15	−24	−31	−34	−33	−27	−19	−10	− 3	0

The second-order day number J' is given in this table in units of 0″0001.
The apparent declination of a star is given by:

$$\delta = \delta_1 + \tau\mu_\delta / 100 + Aa' + Bb' + Cc' + J' \tan \delta_1$$

where the declination (δ_1) and centennial proper motion in declination (μ_δ) are referred to the mean equator and equinox of J1992·5

J for SOUTHERN DECLINATIONS FOR 0ʰ TDT AND EQUINOX J1992·5

Right Ascension

Date	0ʰ 12ʰ	1ʰ 13ʰ	2ʰ 14ʰ	3ʰ 15ʰ	4ʰ 16ʰ	5ʰ 17ʰ	6ʰ 18ʰ	7ʰ 19ʰ	8ʰ 20ʰ	9ʰ 21ʰ	10ʰ 22ʰ	11ʰ 23ʰ	12ʰ 24ʰ
Jan. −1	− 1	+ 4	+ 7	+ 9	+ 9	+ 6	+ 1	− 4	− 7	− 9	− 9	− 6	− 1
9	− 3	+ 1	+ 5	+ 8	+ 8	+ 7	+ 3	− 1	− 5	− 8	− 8	− 7	− 3
19	− 5	− 2	+ 2	+ 6	+ 7	+ 7	+ 5	+ 2	− 2	− 6	− 7	− 7	− 5
29	− 6	− 4	0	+ 3	+ 6	+ 7	+ 6	+ 4	0	− 3	− 6	− 7	− 6
Feb. 8	− 6	− 5	− 2	+ 1	+ 4	+ 6	+ 6	+ 5	+ 2	− 1	− 4	− 6	− 6
18	− 5	− 5	− 4	− 1	+ 1	+ 4	+ 5	+ 5	+ 4	+ 1	− 1	− 4	− 5
28	− 4	− 5	− 5	− 3	− 1	+ 2	+ 4	+ 5	+ 5	+ 3	+ 1	− 2	− 4
Mar. 9	− 2	− 4	− 5	− 4	− 3	0	+ 2	+ 4	+ 5	+ 4	+ 3	0	− 2
19	0	− 2	− 4	− 5	− 4	− 3	0	+ 2	+ 4	+ 5	+ 4	+ 3	0
29	+ 2	0	− 3	− 4	− 5	− 4	− 2	0	+ 3	+ 4	+ 5	+ 4	+ 2
Apr. 8	+ 4	+ 2	− 1	− 3	− 5	− 5	− 4	− 2	+ 1	+ 3	+ 5	+ 5	+ 4
18	+ 6	+ 4	+ 1	− 2	− 4	− 6	− 6	− 4	− 1	+ 2	+ 4	+ 6	+ 6
28	+ 7	+ 6	+ 4	+ 1	− 3	− 6	− 7	− 6	− 4	− 1	+ 3	+ 6	+ 7
May 8	+ 7	+ 8	+ 6	+ 3	− 1	− 5	− 7	− 8	− 6	− 3	+ 1	+ 5	+ 7
18	+ 7	+ 9	+ 8	+ 5	+ 1	− 3	− 7	− 9	− 8	− 5	− 1	+ 3	+ 7
28	+ 5	+ 9	+ 9	+ 8	+ 4	− 1	− 5	− 9	− 9	− 8	− 4	+ 1	+ 5
June 7	+ 4	+ 8	+10	+10	+ 7	+ 2	− 4	− 8	−10	−10	− 7	− 2	+ 4
17	+ 1	+ 7	+10	+11	+ 9	+ 4	− 1	− 7	−10	−11	− 9	− 4	+ 1
27	− 1	+ 5	+ 9	+11	+10	+ 7	+ 1	− 5	− 9	−11	−10	− 7	− 1
July 7	− 4	+ 2	+ 8	+11	+12	+ 9	+ 4	− 2	− 8	−11	−12	− 9	− 4
17	− 6	0	+ 6	+10	+12	+11	+ 6	0	− 6	−10	−12	−11	− 6
27	− 8	− 3	+ 3	+ 9	+12	+11	+ 8	+ 3	− 3	− 9	−12	−11	− 8
Aug. 6	−10	− 5	+ 1	+ 6	+10	+12	+10	+ 5	− 1	− 6	−10	−12	−10
16	−10	− 7	− 2	+ 4	+ 9	+11	+10	+ 7	+ 2	− 4	− 9	−11	−10
26	−10	− 8	− 4	+ 2	+ 7	+10	+10	+ 8	+ 4	− 2	− 7	−10	−10
Sept. 5	− 9	− 9	− 5	− 1	+ 4	+ 8	+ 9	+ 9	+ 5	+ 1	− 4	− 8	− 9
15	− 8	− 8	− 6	− 3	+ 2	+ 6	+ 8	+ 8	+ 6	+ 3	− 2	− 6	− 8
25	− 6	− 7	− 7	− 4	0	+ 3	+ 6	+ 7	+ 7	+ 4	0	− 3	− 6
Oct. 5	− 4	− 6	− 6	− 4	− 2	+ 1	+ 4	+ 6	+ 6	+ 4	+ 2	− 1	− 4
15	− 2	− 4	− 5	− 4	− 3	0	+ 2	+ 4	+ 5	+ 4	+ 3	0	− 2
25	− 1	− 2	− 4	− 4	− 3	− 1	+ 1	+ 2	+ 4	+ 4	+ 3	+ 1	− 1
Nov. 4	0	− 1	− 2	− 3	− 3	− 2	0	+ 1	+ 2	+ 3	+ 3	+ 2	0
14	+ 1	0	− 1	− 2	− 2	− 2	− 1	0	+ 1	+ 2	+ 2	+ 2	+ 1
24	+ 1	+ 1	0	− 1	− 1	− 1	− 1	− 1	0	+ 1	+ 1	+ 1	+ 1
Dec. 4	+ 1	+ 1	0	0	0	− 1	− 1	− 1	0	0	0	+ 1	+ 1
14	0	0	0	0	0	0	0	0	0	0	0	0	0
24	0	0	0	0	0	0	0	0	0	0	0	0	0
34	− 1	− 1	− 1	0	0	0	+ 1	+ 1	+ 1	0	0	0	− 1

The second-order day number J is given in this table in units of 0ˢ.000 01. The apparent right ascension of a star is given by:

$$\alpha = \alpha_1 + \tau\mu_\alpha / 100 + Aa + Bb + Cc + Dd + E + J \tan^2 \delta_1$$

where the position (α_1, δ_1) and centennial proper motion in right ascension (μ_α) are referred to the mean equator and equinox of J1992·5

J' for SOUTHERN DECLINATIONS
FOR 0^h TDT AND EQUINOX J1992·5

Right Ascension

Date	0^h / 12^h	1^h / 13^h	2^h / 14^h	3^h / 15^h	4^h / 16^h	5^h / 17^h	6^h / 18^h	7^h / 19^h	8^h / 20^h	9^h / 21^h	10^h / 22^h	11^h / 23^h	12^h / 24^h
Jan. −1	0	− 1	− 3	− 6	−10	−13	−14	−13	−11	− 8	− 4	− 1	0
9	− 1	0	− 1	− 4	− 7	−10	−12	−13	−11	− 9	− 6	− 3	− 1
19	− 1	0	0	− 2	− 4	− 7	−10	−11	−11	−10	− 7	− 4	− 1
29	− 3	− 1	0	− 1	− 2	− 5	− 8	− 9	−10	−10	− 8	− 5	− 3
Feb. 8	− 4	− 2	0	0	− 1	− 3	− 5	− 7	− 9	− 9	− 8	− 6	− 4
18	− 5	− 3	− 1	0	0	− 1	− 3	− 5	− 7	− 8	− 8	− 7	− 5
28	− 6	− 4	− 3	− 1	0	0	− 1	− 3	− 5	− 7	− 8	− 7	− 6
Mar. 9	− 7	− 6	− 4	− 2	− 1	0	0	− 2	− 3	− 5	− 7	− 7	− 7
19	− 7	− 7	− 5	− 4	− 2	− 1	0	0	− 2	− 3	− 5	− 7	− 7
29	− 7	− 8	− 7	− 6	− 4	− 2	0	0	− 1	− 2	− 4	− 6	− 7
Apr. 8	− 7	− 8	− 8	− 7	− 6	− 4	− 2	0	0	− 1	− 3	− 5	− 7
18	− 6	− 8	− 9	− 9	− 8	− 6	− 3	− 1	0	0	− 1	− 4	− 6
28	− 5	− 7	− 9	−10	−10	− 8	− 6	− 3	− 1	0	− 1	− 2	− 5
May 8	− 4	− 7	− 9	−11	−12	−10	− 8	− 5	− 2	0	0	− 1	− 4
18	− 2	− 6	− 9	−12	−13	−12	−10	− 7	− 4	− 1	0	0	− 2
28	− 1	− 4	− 8	−11	−13	−14	−13	−10	− 6	− 3	− 1	0	− 1
June 7	− 1	− 3	− 7	−11	−14	−15	−15	−13	− 9	− 5	− 2	0	− 1
17	0	− 2	− 5	− 9	−13	−16	−17	−15	−11	− 7	− 3	− 1	0
27	0	− 1	− 4	− 8	−12	−16	−17	−16	−14	− 9	− 5	− 2	0
July 7	0	0	− 2	− 6	−11	−15	−17	−18	−16	−12	− 7	− 3	0
17	− 1	0	− 1	− 4	− 9	−13	−17	−18	−17	−14	− 9	− 5	− 1
27	− 2	0	0	− 3	− 7	−12	−15	−18	−18	−15	−11	− 6	− 2
Aug. 6	− 4	− 1	0	− 1	− 5	− 9	−14	−17	−17	−16	−13	− 8	− 4
16	− 5	− 2	0	− 1	− 3	− 7	−11	−15	−17	−16	−14	−10	− 5
26	− 7	− 3	− 1	0	− 2	− 5	− 9	−13	−15	−16	−14	−11	− 7
Sept. 5	− 8	− 4	− 1	0	− 1	− 3	− 7	−10	−13	−14	−14	−11	− 8
15	− 8	− 5	− 2	0	0	− 2	− 4	− 8	−11	−12	−13	−11	− 8
25	− 9	− 6	− 3	− 1	0	− 1	− 3	− 5	8	10	11	11	9
Oct. 5	− 8	− 6	− 4	− 1	0	0	− 1	− 3	− 6	− 8	− 9	− 9	− 8
15	− 7	− 6	− 4	− 2	− 1	0	0	− 2	− 4	− 5	− 7	− 7	− 7
25	− 6	− 5	− 4	− 2	− 1	0	0	− 1	− 2	− 3	− 5	− 6	− 6
Nov. 4	− 4	− 4	− 3	− 2	− 1	0	0	0	− 1	− 2	− 3	− 4	− 4
14	− 3	− 3	− 3	− 2	− 1	− 1	0	0	0	− 1	− 1	− 2	− 3
24	− 1	− 2	− 2	− 2	− 1	− 1	0	0	0	0	0	− 1	− 1
Dec. 4	0	− 1	− 1	− 1	− 1	− 1	− 1	0	0	0	0	0	0
14	0	0	0	0	− 1	− 1	− 1	0	0	0	0	0	0
24	0	0	0	0	0	0	0	− 1	− 1	− 1	− 1	0	0
34	− 1	− 1	0	0	0	0	0	− 1	− 1	− 1	− 1	− 1	− 1

The second-order day number J' is given in this table in units of 0.″0001.
The apparent declination of a star is given by:

$$\delta = \delta_1 + \tau\mu_\delta / 100 + Aa' + Bb' + Cc' + J' \tan \delta_1$$

where the declination (δ_1) and centennial proper motion in declination (μ_δ) are referred to the mean equator and equinox of J1992·5

Planetary reduction

Data and formulae are provided for the precise computation for an object within the solar system of apparent geocentric right ascension and declination at an epoch in terrestrial dynamical time, from a barycentric ephemeris in rectangular coordinates and barycentric dynamical time referred to the standard equator and equinox of J2000·0. The stages in the reduction may be summarised as follows:

1. Convert from terrestrial dynamical time TDT (proper time) to barycentric dynamical time TDB (coordinate time).
2. Calculate the geocentric rectangular coordinates of the planet from barycentric ephemerides of the planet and the Earth for the standard equator and equinox of J2000·0 and coordinate time argument TDB, allowing for light time calculated from heliocentric coordinates.
3. Calculate the direction of the planet relative to the natural frame (i.e. the geocentric inertial frame that is instantaneously stationary in the space time reference frame of the solar system), allowing for light deflection due to solar gravitation.
4. Calculate the direction of the planet relative to the geocentric proper frame by applying the correction for the Earth's orbital velocity about the barycentre (i.e. annual aberration). The resulting direction is for the standard equator and equinox of J2000·0.
5. Apply precession and nutation to convert to the true equator and equinox of date.
6. Convert to spherical coordinates.

Formulae and method for planetary reduction

Step 1. The apparent place is required for a time in TDT whilst the barycentric ephemeris is referred to TDB. For calculating an apparent place the following approximate formulae are sufficient for converting from TDT to TDB:

$$\text{TDB} = \text{TDT} + 0^{s}{\cdot}001\,658 \sin g + 0^{s}{\cdot}000\,014 \sin 2g$$

where $g = 357^{\circ}{\cdot}53 + 0^{\circ}{\cdot}985\,6003\,(\text{JD} - 245\,1545{\cdot}0)$
and JD = Julian date to two decimals of a day.

Step 2. Obtain the Earth's barycentric position $\mathbf{E_B}(t)$ in au and velocity $\dot{\mathbf{E}}_\mathbf{B}(t)$ in au / d, at coordinate time t = TDB referred to the equator and equinox of J2000·0.

Using an ephemeris, obtain the barycentric position of the planet $\mathbf{Q_B}$ in au at time $(t - \tau)$ for the equator and equinox of J2000·0 where τ is the light time, so that light emitted by the planet at the event $\mathbf{Q_B}(t - \tau)$ arrives at the Earth at the event $\mathbf{E_B}(t)$.

The light time equation is solved iteratively using the heliocentric position of the Earth (**E**) and the planet (**Q**), starting with the approximation $\tau = 0$, as follows:

Form **P**, the vector from the Earth to the planet from the equation:

$$\mathbf{P} = \mathbf{Q_B}(t - \tau) - \mathbf{E_B}(t)$$

Form **E** and **Q** from the equations:

$$\mathbf{E} = \mathbf{E_B}(t) - \mathbf{S_B}(t)$$
$$\mathbf{Q} = \mathbf{Q_B}(t - \tau) - \mathbf{S_B}(t - \tau)$$

where $\mathbf{S_B}$ is the barycentric position of the Sun.

Calculate τ from: $c\tau = P + (2\mu / c^2) \ln [(E + P + Q) / (E - P + Q)]$

where the light time (τ) includes the effect of gravitational retardation due to the Sun, and

$\mu = GM_0$ $\quad c$ = velocity of light = 173·1446 au / d
G = the gravitational constant $\quad \mu / c^2 = 9.87 \times 10^{-9}$ au
M_0 = mass of Sun $\quad P = |\mathbf{P}|, \quad Q = |\mathbf{Q}|, \quad E = |\mathbf{E}|$

where | | means calculate the square root of the sum of the squares of the components.

Formulae and method for planetary reduction (continued)

After convergence, form unit vectors **p**, **q**, **e** by dividing **P**, **Q**, **E** by P, Q, E respectively.

Step 3. Calculate the geocentric direction ($\mathbf{p}_1$) of the planet, corrected for light deflection in the natural frame, from:

$$\mathbf{p}_1 = \mathbf{p} + (2\mu/c^2E)\,((\mathbf{p}\,.\,\mathbf{q})\,\mathbf{e} - (\mathbf{e}\,.\,\mathbf{p})\,\mathbf{q})/(1 + \mathbf{q}\,.\,\mathbf{e})$$

where the dot indicates a scalar product. (The scalar product of two vectors is the sum of the products of their corresponding components in the same reference frame.)

The vector $\mathbf{p}_1$ is a unit vector to order μ/c^2.

Step 4. Calculate the proper direction of the planet ($\mathbf{p}_2$) in the geocentric inertial frame that is moving with the instantaneous velocity (**V**) of the Earth relative to the natural frame from:

$$\mathbf{p}_2 = (\beta^{-1}\mathbf{p}_1 + (1 + (\mathbf{p}_1\,.\,\mathbf{V})/(1 + \beta^{-1}))\,\mathbf{V})/(1 + \mathbf{p}_1\,.\,\mathbf{V})$$

where $\mathbf{V} = \dot{\mathbf{E}}_\mathrm{B}/c = 0{\cdot}005\ 7755\ \dot{\mathbf{E}}_\mathrm{B}$ and $\beta = (1 - V^2)^{-1/2}$; the velocity (**V**) is expressed in units of the velocity of light and is equal to the Earth's velocity in the barycentric frame to order V^2.

Step 5. Apply precession and nutation to the proper direction ($\mathbf{p}_2$) by multiplying by the rotation matrix **R** given on the odd pages B45 to B59 to obtain the apparent direction $\mathbf{p}_3$ from:

$$\mathbf{p}_3 = \mathbf{R}\,\mathbf{p}_2$$

using row by column multiplication.

Step 6. Convert to spherical coordinates α, δ using: $\alpha = \tan^{-1}(\eta/\xi)$, $\delta = \sin^{-1}\zeta$ where $\mathbf{p}_3 = (\xi, \eta, \zeta)$ and the quadrant of α is determined by the signs of ξ and η.

Example of planetary reduction

Calculate the apparent place of Venus on 1992 June 13 at 0^h TDT:

Step 1. From page B11, JD = 244 8786·5,
hence $g = 158^\circ{\cdot}75$ and TDB − TDT = $6{\cdot}8 \times 10^{-9}$ days.
The difference between TDB and TDT may be neglected in this example.

Step 2. Tabular values, taken from the JPL DE200/LE200 barycentric ephemeris, referred to J2000·0, which are required for the calculation, are as follows:

Vector	Julian date (TDB)	Rectangular components x	y	z
$\mathbf{E}_\mathrm{B}$	244 8786·5	−0·132 942 635	−0·920 444 691	−0·399 167 011
$\dot{\mathbf{E}}_\mathrm{B}$	244 8786·5	+0·016 764 861	−0·002 153 368	−0·000 934 211
$\mathbf{Q}_\mathrm{B}$	244 8784·5	+0·143 824 125	+0·650 506 337	+0·283 548 296
	244 8785·5	+0·123 870 626	+0·653 327 817	+0·286 080 387
	244 8786·5	+0·103 820 934	+0·655 634 351	+0·288 386 913
	244 8787·5	+0·083 690 892	+0·657 423 888	+0·290 465 948
	244 8788·5	+0·063 496 419	+0·658 694 791	+0·292 315 748
$\mathbf{S}_\mathrm{B}$	244 8785·5	+0·002 404 014	+0·003 026 335	+0·001 223 984
	244 8786·5	+0·002 404 245	+0·003 031 403	+0·001 226 182
	244 8787·5	+0·002 404 467	+0·003 036 473	+0·001 228 382

Example of planetary reduction (continued)

Hence on JD 244 8786·5

$\mathbf{E} = (-0{\cdot}135\,346\,880,\quad -0{\cdot}923\,476\,094,\quad -0{\cdot}400\,393\,193)\qquad E = 1{\cdot}015\,599\,125$

The first iteration, with $\tau = 0$, gives:

$\mathbf{P} = (+0{\cdot}236\,763\,569,\quad +1{\cdot}576\,079\,042,\quad +0{\cdot}687\,553\,924)\qquad P = 1{\cdot}735\,745\,526$
$\mathbf{Q} = (+0{\cdot}101\,416\,689,\quad +0{\cdot}652\,602\,948,\quad +0{\cdot}287\,160\,731)\qquad Q = 0{\cdot}720\,164\,730$
$\tau = +0^{\mathrm{d}}{\cdot}010\,024\,83$

The second iteration, with $\tau = 0^{\mathrm{d}}{\cdot}010\,024\,83$ using Stirling's central-difference formula up to δ^4 to interpolate $\mathbf{Q}_{\mathrm{B}}$, and up to δ^2 to interpolate $\mathbf{S}_{\mathrm{B}}$, gives:

$\mathbf{P} = (+0{\cdot}236\,964\,989,\quad +1{\cdot}576\,058\,482,\quad +0{\cdot}687\,531\,927)\qquad P = 1{\cdot}735\,745\,631$
$\mathbf{Q} = (+0{\cdot}101\,618\,111,\quad +0{\cdot}652\,582\,439,\quad +0{\cdot}287\,138\,756)\qquad Q = 0{\cdot}720\,165\,776$
$\tau = 0^{\mathrm{d}}{\cdot}010\,024\,83$

Iterate until P changes by less than 10^{-9}.

The unit vectors are:

$$\mathbf{p} = (+0{\cdot}136\,520\,573,\quad +0{\cdot}908\,000\,835,\quad +0{\cdot}396\,101\,776)$$
$$\mathbf{q} = (+0{\cdot}141\,103\,777,\quad +0{\cdot}906\,155\,861,\quad +0{\cdot}398\,712\,027)$$
$$\mathbf{e} = (-0{\cdot}133\,268\,015,\quad -0{\cdot}909\,291\,935,\quad -0{\cdot}394\,243\,342)$$

Step 3. Calculate the scalar products:

$\mathbf{p}.\mathbf{q} = +0{\cdot}999\,984\,388\quad \mathbf{e}.\mathbf{p} = -0{\cdot}999\,992\,150\quad \mathbf{q}.\mathbf{e} = -0{\cdot}999\,954\,398$ then

$(2\mu/c^2E)((\mathbf{p}.\mathbf{q})\mathbf{e}-(\mathbf{e}.\mathbf{p})\mathbf{q})/(1+\mathbf{q}.\mathbf{e}) = (+0{\cdot}000\,003\,340,\ -0{\cdot}000\,001\,334,\ +0{\cdot}000\,001\,906)$ and $\mathbf{p}_1 = (+0{\cdot}136\,523\,914,\ +0{\cdot}907\,999\,501,\ +0{\cdot}396\,103\,682)$

Step 4. Take $\dot{\mathbf{E}}_{\mathrm{B}}$ from the table in *Step* 2 and calculate:

$$\mathbf{V} = 0{\cdot}005\,7755\,\dot{\mathbf{E}}_{\mathrm{B}} = (+0{\cdot}000\,096\,826,\quad -0{\cdot}000\,012\,437,\quad -0{\cdot}000\,005\,396)$$

Then $V = 0{\cdot}000\,097\,770$, $\beta = 1{\cdot}000\,000\,005$ and $\beta^{-1} = 0{\cdot}999\,999\,995$

Calculate the scalar product $\mathbf{p}_1 . \mathbf{V} = -0{\cdot}000\,000\,211$

Then $1 + (\mathbf{p}_1 . \mathbf{V})/(1+\beta^{-1}) = 0{\cdot}999\,999\,895$

Hence $\mathbf{p}_2 = (+0{\cdot}136\,620\,768,\ +0{\cdot}907\,987\,251,\ +0{\cdot}396\,098\,368)$

Step 5. From page B51, the precession and nutation matrix $\mathbf{R}$ is given by:

$$\mathbf{R} = \begin{bmatrix} +0{\cdot}999\,998\,45 & +0{\cdot}001\,614\,39 & +0{\cdot}000\,701\,63 \\ -0{\cdot}001\,614\,39 & +0{\cdot}999\,998\,70 & +0{\cdot}000\,001\,07 \\ -0{\cdot}000\,701\,62 & -0{\cdot}000\,002\,20 & +0{\cdot}999\,999\,75 \end{bmatrix}$$

Hence $\mathbf{p}_3 = \mathbf{R}\,\mathbf{p}_2 = (+0{\cdot}138\,364\,31,\quad +0{\cdot}907\,765\,93,\quad +0{\cdot}396\,000\,42)$

Step 6. Converting to spherical coordinates

$$\alpha = 5^{\mathrm{h}}\,25^{\mathrm{m}}\,20^{\mathrm{s}}{\cdot}04 \qquad \delta = +23^\circ\,19'\,42''{\cdot}2$$

The geometric distance between the Earth and Venus at time t = JD 244 8786·5 is the value of $P = 1{\cdot}735\,745\,526$ au in the first iteration in *Step* 2, where $\tau = 0$. The light path distance between the Earth at time t and Venus at time $(t-\tau)$ is the value of $P = 1{\cdot}735\,745\,631$ au in the final iteration in *Step* 2, where $\tau = 0^{\mathrm{d}}{\cdot}010\,024\,83$.

Solar reduction

The method for solar reduction is identical to the method for planetary reduction, except for the following differences:

In *Step* 2 set $\mathbf{Q_B} = \mathbf{S_B}$ and hence $\mathbf{P} = \mathbf{S_B}(t - \tau) - \mathbf{E_B}(t)$. Calculate the light time (τ) by iteration from $\tau = P / c$ and form the unit vector $\mathbf{p}$ only.

In *Step* 3 set $\mathbf{p}_1 = \mathbf{p}$ since there is no light deflection from the centre of the Sun's disk.

Stellar reduction

The method for planetary reduction may be applied with some modification to the calculation of the apparent places of stars.

The barycentric direction of a star at epoch TDB is calculated from its right ascension, declination and space motion for the standard equator and equinox of J2000·0 on the FK5 system. A concise method of conversion from B1950·0 on the FK4 system to J2000·0 on the FK5 system is given on page B42.

The main modifications to the planetary reduction in the stellar case are: in *Step* 1, the distinction between TDB and TDT is not significant; in *Step* 2, the space motion of the star is included but light time is ignored; in *Step* 3, the relativity term for light deflection is modified to the asymptotic case where the star is assumed to be at infinity.

Formulae and method for stellar reduction

The steps in the stellar reduction are as follows:

Step 1. Set TDB = TDT

Step 2. Obtain the Earth's barycentric position $\mathbf{E_B}$ in au and velocity $\dot{\mathbf{E}}_\mathbf{B}$ in au / d, at coordinate time t = TDB referred to the equator and equinox of J2000·0.

The barycentric direction ($\mathbf{q}$) of a star at epoch J2000·0, referred to the standard equator and equinox of J2000·0, is given by:

$$\mathbf{q} = (\cos\alpha_0 \cos\delta_0,\ \sin\alpha_0 \cos\delta_0,\ \sin\delta_0)$$

where α_0 and δ_0 are the right ascension and declination for the equator, equinox and epoch of J2000·0.

The space motion vector $\mathbf{m} = (m_x, m_y, m_z)$ of the star expressed in radians per century, is given by:

$$\begin{aligned}
m_x &= -\mu_\alpha \cos\delta_0 \sin\alpha_0 - \mu_\delta \sin\delta_0 \cos\alpha_0 + v\pi \cos\delta_0 \cos\alpha_0\\
m_y &= \mu_\alpha \cos\delta_0 \cos\alpha_0 - \mu_\delta \sin\delta_0 \sin\alpha_0 + v\pi \cos\delta_0 \sin\alpha_0\\
m_z &= \mu_\delta \cos\delta_0 + v\pi \sin\delta_0
\end{aligned}$$

where these expressions take into account radial velocity (v) in au / century (1 km / s = 21·095 au / century), measured positively away from the Earth, as well as proper motion (μ_α, μ_δ) in right ascension and declination in radians / century, and π is the parallax in radians.

Calculate $\mathbf{P}$, the geocentric vector of the star at the required epoch, from:

$$\mathbf{P} = \mathbf{q} + T\mathbf{m} - \pi\, \mathbf{E_B}$$

where T = (JD − 245 1545·0) / 36 525, which is the interval in Julian centuries from J2000·0, and JD is the Julian date to one decimal of a day.

Form the heliocentric position of the Earth ($\mathbf{E}$) from:

$$\mathbf{E} = \mathbf{E_B} - \mathbf{S_B}$$

where $\mathbf{S_B}$ is the barycentric position of the Sun at time t.

Form the geocentric direction ($\mathbf{p}$) of the star and the unit vector ($\mathbf{e}$) from $\mathbf{p} = \mathbf{P} / |\mathbf{P}|$ and $\mathbf{e} = \mathbf{E} / |\mathbf{E}|$.

Formulae and method for stellar reduction (continued)

Step 3. Calculate the geocentric direction ($\mathbf{p}_1$) of the star, corrected for light deflection in the natural frame, from:

$$\mathbf{p}_1 = \mathbf{p} + (2\mu/c^2E)(\mathbf{e} - (\mathbf{p}\,.\,\mathbf{e})\mathbf{p})/(1 + \mathbf{p}\,.\,\mathbf{e})$$

where the dot indicates a scalar product, $\mu/c^2 = 9{\cdot}87 \times 10^{-9}$ au and $E = |\mathbf{E}|$. Note that the expression is derived from the planetary case by substituting $\mathbf{q} = \mathbf{p}$ in the small term which allows for light deflection.

The vector $\mathbf{p}_1$ is a unit vector to order μ/c^2.

Step 4. Calculate the proper direction ($\mathbf{p}_2$) in the geocentric inertial frame, that is moving with the instantaneous velocity ($\mathbf{V}$) of the Earth relative to the natural frame, from:

$$\mathbf{p}_2 = (\beta^{-1}\mathbf{p}_1 + (1 + (\mathbf{p}_1\,.\,\mathbf{V})/(1 + \beta^{-1}))\mathbf{V})/(1 + \mathbf{p}_1\,.\,\mathbf{V})$$

where $\mathbf{V} = \dot{\mathbf{E}}_B/c = 0{\cdot}005\,7755\,\dot{\mathbf{E}}_B$ and $\beta = (1 - V^2)^{-1/2}$; the velocity ($\mathbf{V}$) is expressed in units of velocity of light and is equal to the Earth's velocity in the barycentric frame to order V^2.

Step 5. Apply precession and nutation to the proper direction ($\mathbf{p}_2$) by multiplying by the rotation matrix ($\mathbf{R}$), given on the odd pages B45 to B59, to obtain the apparent direction ($\mathbf{p}_3$) from:

$$\mathbf{p}_3 = \mathbf{R}\,\mathbf{p}_2$$

using row by column multiplication.

Step 6. Convert to spherical coordinates (α, δ) using: $\alpha = \tan^{-1}(\eta/\xi)$, $\delta = \sin^{-1}\zeta$ where $\mathbf{p}_3 = (\xi, \eta, \zeta)$ and the quadrant of α is determined by the signs of ξ and η.

Example of stellar reduction

Calculate the apparent position of a fictitious star on 1992 January 1 at 0^h TDT. The mean right ascension (α_0), declination (δ_0), centennial proper motions (μ_α, μ_δ), parallax (π) and radial velocity (v) of the star at the standard equator and equinox of J2000·0 are given by:

$\alpha = 14^h\,39^m\,36^s{\cdot}087$ $\quad\delta_0 = -60°\,50'\,07''{\cdot}14$ $\quad\pi = 0''{\cdot}752 = 3{\cdot}6458 \times 10^{-6}$ rad
$\mu_\alpha = -49{\cdot}486$ s/cy $\quad\mu_\delta = +69{\cdot}60''$/cy $\quad v = -22{\cdot}2$ km/s
$= -0{\cdot}003\,598\,723$ rad/cy, $= +0{\cdot}000\,337\,430$ rad/cy, $v\pi = -0{\cdot}001\,707\,357$ rad/cy

Step 1. TDB = TDT = JD244 8622·5

Step 2. Tabular values of $\mathbf{E}_B$, $\dot{\mathbf{E}}_B$ and $\mathbf{S}_B$, taken from the JPL DE200/LE200 barycentric ephemeris referred to J2000·0, are:

Vector	Julian date (TDB)	Rectangular components x	y	z
$\mathbf{E}_B$	244 8622·5	−0·167 197 870	+0·890 911 185	+0·386 192 545
$\dot{\mathbf{E}}_B$	244 8622·5	−0·017 231 180	−0·002 770 779	−0·001 201 858
$\mathbf{S}_B$	244 8622·5	+0·002 253 657	+0·002 220 725	+0·000 876 542

From the positional data, calculate:

$$\mathbf{q} = (-0{\cdot}373\,854\,098,\ -0{\cdot}312\,594\,565,\ -0{\cdot}873\,222\,624)$$
$$\mathbf{m} = (-0{\cdot}000\,712\,685,\ +0{\cdot}001\,690\,102,\ +0{\cdot}001\,655\,339)$$

Example of stellar reduction (continued)

Form $\mathbf{P} = \mathbf{q} + T\mathbf{m} - \pi\mathbf{E}_B = (-0{\cdot}373\ 796\ 464,\ -0{\cdot}312\ 733\ 044,\ -0{\cdot}873\ 356\ 482)$

where $T = (244\ 8622{\cdot}5 - 245\ 1545{\cdot}0)/36\ 525 = -0{\cdot}080\ 013\ 689,$

and form $\mathbf{E} = \mathbf{E}_B - \mathbf{S}_B = (-0{\cdot}169\ 451\ 527,\ +0{\cdot}888\ 690\ 460,\ +0{\cdot}385\ 316\ 003),$

$E = 0{\cdot}983\ 337\ 671$

Hence the unit vectors are:

$$\mathbf{p} = (-0{\cdot}373\ 744\ 648,\ -0{\cdot}312\ 689\ 693,\ -0{\cdot}873\ 235\ 417)$$
$$\mathbf{e} = (-0{\cdot}172\ 322\ 827,\ +0{\cdot}903\ 749\ 023,\ +0{\cdot}391\ 845\ 054)$$

Step 3. Calculate the scalar product $\mathbf{p.e} = -0{\cdot}560\ 361\ 250$

then

$(2\mu/c^2E)\,(\mathbf{e} - (\mathbf{p.e})\,\mathbf{p})/(1 + \mathbf{p.e}) = (-0{\cdot}000\ 000\ 017,\ +0{\cdot}000\ 000\ 033,\ -0{\cdot}000\ 000\ 004)$
and $\mathbf{p}_1 = (-0{\cdot}373\ 744\ 666,\ -0{\cdot}312\ 689\ 660,\ -0{\cdot}873\ 235\ 422)$

Step 4.

Calculate $\mathbf{V} = 0{\cdot}005\ 7755\ \dot{\mathbf{E}}_B = (-0{\cdot}000\ 099\ 519,\ -0{\cdot}000\ 016\ 003,\ -0{\cdot}000\ 006\ 941)$

where $\dot{\mathbf{E}}_B$ is taken from the table in *Step* 2.

Then $V = 0{\cdot}000\ 101\ 036$, $\beta = 1{\cdot}000\ 000\ 005$ and $\beta^{-1} = 0{\cdot}999\ 999\ 995$

Calculate the scalar product $\mathbf{p}_1.\mathbf{V} = +0{\cdot}000\ 048\ 260$

Then $1 + (\mathbf{p}_1.\mathbf{V})/(1 + \beta^{-1}) = 1{\cdot}000\ 024\ 130$

Hence $\mathbf{p}_2 = (-0{\cdot}373\ 826\ 144,\ -0{\cdot}312\ 690\ 571,\ -0{\cdot}873\ 200\ 218)$

Step 5. From page B45, the precession and nutation matrix **R** is given by:

$$\mathbf{R} = \begin{bmatrix} +0{\cdot}999\ 998\ 26 & +0{\cdot}001\ 712\ 80 & +0{\cdot}000\ 744\ 39 \\ -0{\cdot}001\ 712\ 79 & +0{\cdot}999\ 998\ 53 & -0{\cdot}000\ 005\ 69 \\ -0{\cdot}000\ 744\ 40 & +0{\cdot}000\ 004\ 42 & +0{\cdot}999\ 999\ 72 \end{bmatrix}$$

Hence $\mathbf{p}_3 = \mathbf{R}\,\mathbf{p}_2 = (-0{\cdot}375\ 011\ 07,\ -0{\cdot}312\ 044\ 85,\ -0{\cdot}872\ 923\ 08)$

Step 6. Converting to spherical coordinates:

$$\alpha = 14^{h}\ 39^{m}\ 03^{s}{\cdot}282 \qquad \delta = -60^{\circ}\ 48'\ 00''{\cdot}42$$

Conversion of stellar positions and proper motions from the standard epoch B1950·0 to the standard epoch J2000·0

A matrix method for calculating the mean place of a star at J2000·0 on the FK5 system from the mean place at B1950·0 on the FK4 system, ignoring the systematic corrections FK5–FK4 and individual star corrections to the FK5, is as follows:

1. From a star catalogue obtain the FK4 position (α_0, δ_0), proper motion $(\mu_{\alpha 0}, \mu_{\delta 0})$ in seconds of arc per tropical century, parallax (π_0) in seconds of arc and radial velocity (v_0) in km / s for B1950·0. If π_0 or v_0 are unspecified, set them both equal to zero.

2. Calculate the rectangular components of the position vector $\mathbf{r}_0$ and velocity vector $\dot{\mathbf{r}}_0$ from:

$$\mathbf{r}_0 = \begin{bmatrix} \cos\alpha_0 \cos\delta_0 \\ \sin\alpha_0 \cos\delta_0 \\ \sin\delta_0 \end{bmatrix} \qquad \dot{\mathbf{r}}_0 = \begin{bmatrix} -\mu_{\alpha 0} \sin\alpha_0 \cos\delta_0 - \mu_{\delta 0} \cos\alpha_0 \sin\delta_0 \\ \mu_{\alpha 0} \cos\alpha_0 \cos\delta_0 - \mu_{\delta 0} \sin\alpha_0 \sin\delta_0 \\ \mu_{\delta 0} \cos\delta_0 \end{bmatrix} + 21{\cdot}095\, v_0\, \pi_0\, \mathbf{r}_0$$

3. Remove the effects of the E-terms of aberration to form $\mathbf{r}_1$ and $\dot{\mathbf{r}}_1$ from:

$$\mathbf{r}_1 = \mathbf{r}_0 - \mathbf{A} + (\mathbf{r}_0' \mathbf{A})\, \mathbf{r}_0$$

$$\dot{\mathbf{r}}_1 = \dot{\mathbf{r}}_0 - \dot{\mathbf{A}} + (\mathbf{r}_0' \dot{\mathbf{A}})\, \mathbf{r}_0$$

where $$\mathbf{A} = 10^{-6} \begin{bmatrix} -1{\cdot}625\,57 \\ -0{\cdot}319\,19 \\ -0{\cdot}138\,43 \end{bmatrix} \qquad \dot{\mathbf{A}} = 10^{-3} \begin{bmatrix} +1{\cdot}244 \\ -1{\cdot}579 \\ -0{\cdot}660 \end{bmatrix}$$

and $\mathbf{r}_0'$ is the transpose of $\mathbf{r}_0$. (The terms $\mathbf{r}_0' \mathbf{A}$ and $\mathbf{r}_0' \dot{\mathbf{A}}$ are scalar products).

4. Form the vector $\mathbf{R}_1 = \begin{bmatrix} \mathbf{r}_1 \\ \dot{\mathbf{r}}_1 \end{bmatrix}$ and calculate the vector $\mathbf{R} = \begin{bmatrix} \mathbf{r} \\ \dot{\mathbf{r}} \end{bmatrix}$ from:

$$\mathbf{R} = \mathbf{M}\, \mathbf{R}_1$$

where $\mathbf{M}$ is a constant 6 × 6 matrix given by:

$$\begin{bmatrix} +0{\cdot}999\,925\,6782 & -0{\cdot}011\,182\,0611 & -0{\cdot}004\,857\,9477 & +0{\cdot}000\,002\,423\,950\,18 & -0{\cdot}000\,000\,027\,106\,63 & -0{\cdot}000\,000\,011\,776\,56 \\ +0{\cdot}011\,182\,0610 & +0{\cdot}999\,937\,4784 & -0{\cdot}000\,027\,1765 & +0{\cdot}000\,000\,027\,106\,63 & +0{\cdot}000\,002\,423\,978\,78 & -0{\cdot}000\,000\,000\,065\,87 \\ +0{\cdot}004\,857\,9479 & -0{\cdot}000\,027\,1474 & +0{\cdot}999\,988\,1997 & +0{\cdot}000\,000\,011\,776\,56 & -0{\cdot}000\,000\,000\,065\,82 & +0{\cdot}000\,002\,424\,101\,73 \\ -0{\cdot}000\,551 & -0{\cdot}238\,565 & +0{\cdot}435\,739 & +0{\cdot}999\,947\,04 & -0{\cdot}011\,182\,51 & -0{\cdot}004\,857\,67 \\ +0{\cdot}238\,514 & -0{\cdot}002\,667 & -0{\cdot}008\,541 & +0{\cdot}011\,182\,51 & +0{\cdot}999\,958\,83 & -0{\cdot}000\,027\,18 \\ -0{\cdot}435\,623 & +0{\cdot}012\,254 & +0{\cdot}002\,117 & +0{\cdot}004\,857\,67 & -0{\cdot}000\,027\,14 & +1{\cdot}000\,009\,56 \end{bmatrix}$$

and set $(x, y, z, \dot{x}, \dot{y}, \dot{z}) = \mathbf{R}'$

5. Calculate the FK5 mean position (α_1, δ_1), proper motion $(\mu_{\alpha 1}, \mu_{\delta 1})$ in seconds of arc per Julian century, parallax (π_1) in seconds of arc and radial velocity (v_1) in km / s for J2000·0 from:

$$\cos\alpha_1 \cos\delta_1 = x / r \qquad \sin\alpha_1 \cos\delta_1 = y / r \qquad \sin\delta_1 = z / r$$

$$\mu_{\alpha 1} = (x\dot{y} - y\dot{x}) / (x^2 + y^2) \qquad \mu_{\delta 1} = [\dot{z}(x^2 + y^2) - z(x\dot{x} + y\dot{y})] / [r^2 (x^2 + y^2)^{1/2}]$$

$$v_1 = (x\dot{x} + y\dot{y} + z\dot{z}) / (21{\cdot}095\, \pi_0\, r) \qquad \pi_1 = \pi_0 / r$$

where $r = (x^2 + y^2 + z^2)^{1/2}$

If π_0 is zero, set $v_1 = v_0$

References.

Standish, E. M., (1982) *Astron. Astrophys.*, **115**, 20–22.
Aoki, S., Sôma, M., Kinoshita, H., Inoue, K., (1983) *Astron. Astrophys.*, **128**, 263–267.

Conversion of stellar positions and proper motions from the standard epoch J2000·0 to the standard epoch B1950·0

A matrix method for calculating the mean place of a star at B1950·0 on the FK4 system from the mean place at J2000·0 on the FK5 system, ignoring the systematic corrections FK4–FK5 and individual star corrections to the FK4, is as follows:

1. From a star catalogue obtain the FK5 position (α_0, δ_0), proper motion $(\mu_{\alpha 0}, \mu_{\delta 0})$ in seconds of arc per Julian century, parallax (π_0) in seconds of arc and radial velocity (v_0) in km/s for J2000·0. If π_0 or v_0 are unspecified, set them both equal to zero.

2. Calculate the rectangular components of the position vector $\mathbf{r}_0$ and velocity vector $\dot{\mathbf{r}}_0$ from:

$$\mathbf{r}_0 = \begin{bmatrix} \cos\alpha_0 \cos\delta_0 \\ \sin\alpha_0 \cos\delta_0 \\ \sin\delta_0 \end{bmatrix} \qquad \dot{\mathbf{r}}_0 = \begin{bmatrix} -\mu_{\alpha 0} \sin\alpha_0 \cos\delta_0 - \mu_{\delta 0} \cos\alpha_0 \sin\delta_0 \\ \mu_{\alpha 0} \cos\alpha_0 \cos\delta_0 - \mu_{\delta 0} \sin\alpha_0 \sin\delta_0 \\ \mu_{\delta 0} \cos\delta_0 \end{bmatrix} + 21{\cdot}095\, v_0\, \pi_0\, \mathbf{r}_0$$

3. Form the vector $\mathbf{R}_0 = \begin{bmatrix} \mathbf{r}_0 \\ \dot{\mathbf{r}}_0 \end{bmatrix}$ and calculate the vector $\mathbf{R}_1 = \begin{bmatrix} \mathbf{r}_1 \\ \dot{\mathbf{r}}_1 \end{bmatrix}$ from:

$$\mathbf{R}_1 = \mathbf{M}^{-1}\, \mathbf{R}_0$$

where $\mathbf{M}^{-1}$ is a constant 6×6 matrix given by:

$$\begin{bmatrix}
+0{\cdot}999\,925\,6795 & +0{\cdot}011\,181\,4828 & +0{\cdot}004\,859\,0039 & -0{\cdot}000\,002\,423\,898\,40 & -0{\cdot}000\,000\,027\,105\,44 & -0{\cdot}000\,000\,011\,777\,42 \\
-0{\cdot}011\,181\,4828 & +0{\cdot}999\,937\,4849 & -0{\cdot}000\,027\,1771 & +0{\cdot}000\,000\,027\,105\,44 & -0{\cdot}000\,002\,423\,927\,02 & +0{\cdot}000\,000\,000\,065\,85 \\
-0{\cdot}004\,859\,0040 & -0{\cdot}000\,027\,1557 & +0{\cdot}999\,988\,1946 & +0{\cdot}000\,000\,011\,777\,42 & +0{\cdot}000\,000\,000\,065\,85 & -0{\cdot}000\,002\,424\,049\,95 \\
-0{\cdot}000\,551 & +0{\cdot}238\,509 & -0{\cdot}435\,614 & +0{\cdot}999\,904\,32 & +0{\cdot}011\,181\,45 & +0{\cdot}004\,858\,52 \\
-0{\cdot}238\,560 & -0{\cdot}002\,667 & +0{\cdot}012\,254 & -0{\cdot}011\,181\,45 & +0{\cdot}999\,916\,13 & -0{\cdot}000\,027\,17 \\
+0{\cdot}435\,730 & -0{\cdot}008\,541 & +0{\cdot}002\,117 & -0{\cdot}004\,858\,52 & -0{\cdot}000\,027\,16 & +0{\cdot}999\,966\,84
\end{bmatrix}$$

4. Include the effects of the E-terms of aberration as follows:

Form $\mathbf{s}_1 = \mathbf{r}_1/r_1$ and $\dot{\mathbf{s}}_1 = \dot{\mathbf{r}}_1/r_1$ where $r_1 = (x_1^2 + y_1^2 + z_1^2)^{1/2}$

Set $\mathbf{s} = \mathbf{s}_1$ and calculate $\mathbf{r}$ from $\mathbf{r} = \mathbf{s}_1 + \mathbf{A} - (\mathbf{s}'\,\mathbf{A})\,\mathbf{s}$

where
$$\mathbf{A} = 10^{-6} \begin{bmatrix} -1{\cdot}625\,57 \\ -0{\cdot}319\,19 \\ -0{\cdot}138\,43 \end{bmatrix}$$

($\mathbf{s}'$ is the transpose of $\mathbf{s}$, so that $\mathbf{s}'\mathbf{A}$ is a scalar product.)

Set $\mathbf{s} = \mathbf{r}/r$ and iterate the expression for $\mathbf{r}$ once or twice until a consistent value for $\mathbf{r}$ is obtained, then calculate:

$$\dot{\mathbf{r}} = \dot{\mathbf{s}}_1 + \dot{\mathbf{A}} - (\mathbf{s}'\dot{\mathbf{A}})\,\mathbf{s} \qquad \text{where} \quad \dot{\mathbf{A}} = 10^{-3} \begin{bmatrix} +1{\cdot}244 \\ -1{\cdot}579 \\ -0.660 \end{bmatrix}$$

5. Calculate the FK4 mean position (α_1, δ_1), proper motion $(\mu_{\alpha 1}, \mu_{\delta 1})$ in seconds of arc per tropical century, parallax (π_1) in seconds of arc and radial velocity (v_1) in km/s for B1950·0, as follows:

Set $\quad (x, y, z) = \mathbf{r}' \qquad (\dot{x}, \dot{y}, \dot{z}) = \dot{\mathbf{r}}' \qquad$ and $\qquad r = (x^2 + y^2 + z^2)^{1/2}$

Then $\quad \cos\alpha_1 \cos\delta_1 = x/r \qquad \sin\alpha_1 \cos\delta_1 = y/r \qquad \sin\delta_1 = z/r$

$$\mu_{\alpha 1} = (x\dot{y} - y\dot{x})/(x^2 + y^2) \qquad \mu_{\delta 1} = [\dot{z}(x^2 + y^2) - z(x\dot{x} + y\dot{y})]/[r^2(x^2 + y^2)^{1/2}]$$

In step 4 set

$$(x_1, y_1, z_1) = \mathbf{r}_1' \qquad (\dot{x}_1, \dot{y}_1, \dot{z}_1) = \dot{\mathbf{r}}_1' \qquad \text{and} \qquad r_1 = (x_1^2 + y_1^2 + z_1^2)^{1/2}$$

then $\quad v_1 = (x_1\dot{x}_1 + y_1\dot{y}_1 + z_1\dot{z}_1)/(21{\cdot}095\,\pi_0\, r_1) \qquad \pi_1 = \pi_0/r_1$

If π_0 is zero, set $v_1 = v_0$

ORIGIN AT SOLAR SYSTEM BARYCENTRE

MEAN EQUATOR AND EQUINOX J2000·0

Date 0^h TDB		X	Y	Z	$\dot{X}$	$\dot{Y}$	$\dot{Z}$
Jan.	**0**	−0·149 941 700	+0·893 543 000	+0·387 334 093	−1728 0219	− 249 2747	− 108 1189
	1	·167 197 870	·890 911 185	·386 192 545	1723 1180	277 0779	120 1858
	2	·184 402 175	·888 001 673	·384 930 486	1717 6487	304 8117	132 2199
	3	·201 548 960	·884 815 232	·383 548 278	1711 6141	332 4612	144 2147
	4	·218 632 576	·881 352 778	·382 046 346	1705 0153	360 0119	156 1635
	5	−0·235 647 394	+0·877 615 372	+0·380 425 182	−1697 8550	− 387 4492	− 168 0600
	6	·252 587 816	·873 604 220	·378 685 341	1690 1369	414 7588	179 8980
	7	·269 448 289	·869 320 667	·376 827 437	1681 8659	441 9270	191 6718
	8	·286 223 310	·864 766 196	·374 852 139	1673 0477	468 9404	203 3756
	9	·302 907 441	·859 942 419	·372 760 174	1663 6890	495 7861	215 0044
	10	−0·319 495 314	+0·854 851 073	+0·370 552 317	−1653 7973	− 522 4520	− 226 5532
	11	·335 981 638	·849 494 018	·368 229 393	1643 3809	548 9262	238 0172
	12	·352 361 212	·843 873 226	·365 792 270	1632 4487	575 1976	249 3922
	13	·368 628 927	·837 990 777	·363 241 859	1621 0110	601 2557	260 6742
	14	·384 779 784	·831 848 855	·360 579 108	1609 0790	627 0909	271 8598
	15	−0·400 808 902	+0·825 449 729	+0·357 804 993	−1596 6653	− 652 6950	− 282 9464
	16	·416 711 529	·818 795 745	·354 920 516	1583 7832	678 0621	293 9322
	17	·432 483 052	·811 889 289	·351 926 686	1570 4466	703 1889	304 8170
	18	·448 118 989	·804 732 768	·348 824 510	1556 6681	728 0757	315 6018
	19	·463 614 974	·797 328 564	·345 614 976	1542 4574	752 7261	326 2889
	20	−0·478 966 714	+0·789 679 015	+0·342 299 049	−1527 8196	− 777 1457	− 336 8810
	21	·494 169 941	·781 786 401	·338 877 666	1512 7543	801 3400	347 3802
	22	·509 220 358	·773 652 956	·335 351 753	1497 2564	825 3120	357 7870
	23	·524 113 601	·765 280 904	·331 722 240	1481 3181	849 0606	368 0996
	24	·538 845 224	·756 672 507	·327 990 087	1464 9312	872 5800	378 3143
	25	−0·553 410 707	+0·747 830 099	+0·324 156 298	−1448 0893	− 895 8608	− 388 4257
	26	·567 805 481	·738 756 127	·320 221 938	1430 7888	918 8909	398 4276
	27	·582 024 952	·729 453 160	·316 188 133	1413 0288	941 6574	408 3135
	28	·596 064 530	·719 923 903	·312 056 076	1394 8108	964 1468	418 0770
	29	·609 919 652	·710 171 192	·307 827 021	1376 1382	986 3458	427 7120
	30	−0·623 585 795	+0·700 197 997	+0·303 502 282	−1357 0159	−1008 2417	− 437 2128
	31	·637 058 492	·690 007 410	·299 083 232	1337 4500	1029 8220	446 5736
Feb.	**1**	·650 333 340	·679 602 649	·294 571 294	1317 4475	1051 0746	455 7893
	2	·663 406 012	·668 987 049	·289 967 947	1297 0162	1071 9878	464 8547
	3	·676 272 264	·658 164 062	·285 274 716	1276 1650	1092 5502	473 7652
	4	−0·688 927 946	+0·647 137 252	+0·280 493 176	−1254 9037	−1112 7508	− 482 5159
	5	·701 369 008	·635 910 288	·275 624 944	1233 2431	1132 5791	491 1028
	6	·713 591 515	·624 486 944	·270 671 679	1211 1947	1152 0253	499 5220
	7	·725 591 652	·612 871 087	·265 635 075	1188 7712	1171 0802	507 7699
	8	·737 365 732	·601 066 670	·260 516 862	1165 9857	1189 7358	515 8435
	9	−0·748 910 206	+0·589 077 726	+0·255 318 796	−1142 8523	−1207 9848	− 523 7403
	10	·760 221 667	·576 908 350	·250 042 653	1119 3856	1225 8210	531 4583
	11	·771 296 859	·564 562 698	·244 690 230	1095 6010	1243 2395	538 9962
	12	·782 132 680	·552 044 962	·239 263 331	1071 5143	1260 2373	546 3534
	13	·792 726 191	·539 359 359	·233 763 763	1047 1416	1276 8131	553 5303
	14	−0·803 074 610	+0·526 510 103	+0·228 193 322	−1022 4984	−1292 9682	− 560 5282
	15	−0·813 175 305	+0·513 501 383	+0·222 553 787	− 997 5989	−1308 7068	− 567 3497

$\dot{X}$, $\dot{Y}$, $\dot{Z}$ are in units of 10^{-9} au / d.

MATRIX ELEMENTS FOR CONVERSION FROM MEAN EQUINOX OF J2000·0 TO TRUE EQUINOX OF DATE

Julian Date	$R_{11}-1$	R_{12}	R_{13}	R_{21}	$R_{22}-1$	R_{23}	R_{31}	R_{32}	$R_{33}-1$
244									
8621·5	− 175	+171 374	+ 74 480	−171 374	− 147	− 590	− 74 481	+ 462	− 28
8622·5	174	171 280	74 439	171 279	147	569	74 440	442	28
8623·5	174	171 171	74 392	171 170	147	555	74 393	427	28
8624·5	174	171 053	74 341	171 053	146	548	74 342	421	28
8625·5	174	170 935	74 290	170 934	146	550	74 290	423	28
8626·5	− 173	+170 821	+ 74 240	−170 821	− 146	− 560	− 74 241	+ 434	− 28
8627·5	173	170 719	74 196	170 719	146	577	74 197	450	28
8628·5	173	170 633	74 158	170 632	146	597	74 159	471	27
8629·5	173	170 563	74 128	170 563	145	618	74 129	492	27
8630·5	173	170 510	74 105	170 510	145	637	74 106	511	27
8631·5	− 173	+170 470	+ 74 088	−170 470	− 145	− 650	− 74 089	+ 524	− 27
8632·5	173	170 438	74 074	170 437	145	656	74 075	530	27
8633·5	173	170 406	74 060	170 405	145	654	74 061	528	27
8634·5	173	170 366	74 042	170 365	145	643	74 043	517	27
8635·5	172	170 310	74 018	170 309	145	625	74 019	499	27
8636·5	− 172	+170 230	+ 73 983	−170 230	− 145	− 604	− 73 984	+ 478	− 27
8637·5	172	170 124	73 937	170 124	145	584	73 938	459	27
8638·5	172	169 994	73 881	169 994	144	573	73 882	447	27
8639·5	171	169 850	73 819	169 850	144	574	73 819	449	27
8640·5	171	169 709	73 757	169 709	144	590	73 758	465	27
8641·5	− 171	+169 588	+ 73 705	−169 588	− 144	− 618	− 73 706	+ 493	− 27
8642·5	171	169 498	73 666	169 498	144	650	73 667	525	27
8643·5	171	169 441	73 641	169 441	144	679	73 642	554	27
8644·5	171	169 409	73 627	169 408	143	697	73 628	573	27
8645·5	171	169 386	73 617	169 386	143	703	73 618	578	27
8646·5	− 171	+169 359	+ 73 605	−169 359	− 143	− 697	− 73 606	+ 572	− 27
8647·5	170	169 319	73 588	169 318	143	683	73 589	559	27
8648·5	170	169 260	73 562	169 259	143	666	73 563	542	27
8649·5	170	169 183	73 529	169 182	143	651	73 530	527	27
8650·5	170	169 091	73 489	169 090	143	641	73 490	517	27
8651·5	− 170	+168 990	+ 73 445	−168 989	− 143	− 638	− 73 446	+ 514	− 27
8652·5	170	168 886	73 400	168 886	143	644	73 401	520	27
8653·5	169	168 787	73 357	168 787	142	658	73 358	534	27
8654·5	169	168 699	73 318	168 698	142	678	73 319	554	27
8655·5	169	168 625	73 286	168 624	142	702	73 287	579	27
8656·5	− 169	+168 568	+ 73 262	−168 568	− 142	− 728	− 73 263	+ 605	− 27
8657·5	169	168 529	73 244	168 528	142	752	73 246	629	27
8658·5	169	168 504	73 233	168 503	142	771	73 235	648	27
8659·5	169	168 488	73 227	168 488	142	783	73 228	660	27
8660·5	169	168 475	73 221	168 475	142	786	73 222	663	27
8661·5	− 169	+168 457	+ 73 213	−168 456	− 142	− 781	− 73 214	+ 657	− 27
8662·5	169	168 425	73 199	168 425	142	768	73 200	645	27
8663·5	169	168 374	73 177	168 373	142	751	73 178	628	27
8664·5	168	168 299	73 144	168 298	142	734	73 146	611	27
8665·5	168	168 201	73 102	168 200	141	723	73 103	600	27
8666·5	− 168	+168 086	+ 73 052	−168 086	− 141	− 721	− 73 053	+ 598	− 27
8667·5	− 168	+167 967	+ 73 000	−167 967	− 141	− 732	− 73 001	+ 609	− 27

Values are in units of 10^{-8}.

ORIGIN AT SOLAR SYSTEM BARYCENTRE

MEAN EQUATOR AND EQUINOX J2000·0

Date 0^h TDB	X	Y	Z	$\dot{X}$	$\dot{Y}$	$\dot{Z}$
Feb. 15	−0·813 175 305	+0·513 501 383	+0·222 553 787	− 997 5989	−1308 7068	− 567 3497
16	·823 025 772	·500 337 333	·216 846 905	972 4545	1324 0357	573 9982
17	·832 623 602	·487 022 008	·211 074 389	947 0722	1338 9630	580 4773
18	·841 966 431	·473 559 385	·205 237 914	921 4542	1353 4968	586 7900
19	·851 051 893	·459 953 365	·199 339 137	895 5981	1367 6428	592 9381
20	−0·859 877 582	+0·446 207 815	+0·193 379 704	− 869 4988	−1381 4027	− 598 9208
21	·868 441 040	·432 326 608	·187 361 280	843 1510	1394 7735	604 7356
22	·876 739 762	·418 313 666	·181 285 566	816 5513	1407 7482	610 3781
23	·884 771 224	·404 172 995	·175 154 308	789 6990	1420 3175	615 8434
24	·892 532 910	·389 908 702	·168 969 307	762 5970	1432 4708	621 1260
25	−0·900 022 349	+0·375 525 000	+0·162 732 413	− 735 2506	−1444 1977	− 626 2211
26	·907 237 131	·361 026 203	·156 445 526	707 6671	1455 4882	631 1240
27	·914 174 930	·346 416 721	·150 110 587	679 8555	1466 3332	635 8307
28	·920 833 512	·331 701 052	·143 729 578	651 8253	1476 7244	640 3375
29	·927 210 742	·316 883 772	·137 304 514	623 5870	1486 6541	644 6412
Mar. 1	−0·933 304 594	+0·301 969 534	+0·130 837 441	− 595 1515	−1496 1149	− 648 7388
2	·939 113 153	·286 963 059	·124 330 435	566 5304	1505 1001	652 6273
3	·944 634 625	·271 869 139	·117 785 599	537 7361	1513 6031	656 3044
4	·949 867 342	·256 692 625	·111 205 060	508 7817	1521 6179	659 7677
5	·954 809 773	·241 438 426	·104 590 965	479 6814	1529 1392	663 0153
6	−0·959 460 534	+0·226 111 502	+0·097 945 478	− 450 4502	−1536 1624	− 666 0458
7	·963 818 393	·210 716 849	·091 270 775	421 1038	1542 6843	668 8585
8	·967 882 282	·195 259 493	·084 569 035	391 6588	1548 7029	671 4531
9	·971 651 296	·179 744 470	·077 842 437	362 1317	1554 2178	673 8303
10	·975 124 700	·164 176 813	·071 093 150	332 5395	1559 2301	675 9913
11	−0·978 301 924	+0·148 561 532	+0·064 323 325	− 302 8986	−1563 7432	− 677 9383
12	·981 182 562	·132 903 597	·057 535 088	273 2247	1567 7620	679 6742
13	·983 766 356	·117 207 916	·050 730 532	243 5322	1571 2937	681 2027
14	·986 053 184	·101 479 316	·043 911 712	213 8334	1574 3474	682 5280
15	·988 043 034	·085 722 527	·037 080 633	184 1377	1576 9335	683 6550
16	−0·989 735 968	+0·069 942 168	+0·030 239 257	− 154 4509	−1579 0632	− 684 5882
17	·991 132 087	·054 142 750	·023 389 501	124 7745	1580 7468	685 3316
18	·992 231 488	·038 328 690	·016 533 250	95 1069	1581 9925	685 8875
19	·993 034 240	·022 504 344	·009 672 373	65 4440	1582 8046	686 2567
20	·993 540 369	+ ·006 674 042	+ ·002 808 741	35 7817	1583 1835	686 4382
21	−0·993 749 867	−0·009 157 870	−0·004 055 756	− 6 1175	−1583 1259	− 686 4292
22	·993 662 714	·024 987 001	·010 919 196	+ 23 5481	1582 6260	686 2264
23	·993 278 911	·040 808 894	·017 779 625	53 2118	1581 6772	685 8262
24	·992 598 506	·056 619 026	·024 635 051	82 8671	1580 2727	685 2253
25	·991 621 623	·072 412 809	·031 483 453	112 5059	1578 4064	684 4211
26	−0·990 348 475	−0·088 185 598	−0·038 322 787	+ 142 1185	−1576 0733	− 683 4113
27	·988 779 377	·103 932 705	·045 150 988	171 6943	1573 2694	682 1942
28	·986 914 749	·119 649 406	·051 965 976	201 2224	1569 9914	680 7687
29	·984 755 125	·135 330 944	·058 765 663	230 6916	1566 2365	679 1336
30	·982 301 153	·150 972 539	·065 547 948	260 0901	1562 0025	677 2884
31	−0·979 553 599	−0·166 569 389	−0·072 310 729	+ 289 4057	−1557 2873	− 675 2325
Apr. 1	−0·976 513 358	−0·182 116 675	−0·079 051 895	+ 318 6254	−1552 0893	− 672 9655

$\dot{X}$, $\dot{Y}$, $\dot{Z}$ are in units of 10^{-9} au / d.

MATRIX ELEMENTS FOR CONVERSION FROM MEAN EQUINOX OF J2000·0 TO TRUE EQUINOX OF DATE

Julian Date	$R_{11}-1$	R_{12}	R_{13}	R_{21}	$R_{22}-1$	R_{23}	R_{31}	R_{32}	$R_{33}-1$
244									
8667·5	− 168	+167 967	+ 73 000	−167 967	− 141	− 732	− 73 001	+ 609	− 27
8668·5	167	167 858	72 953	167 858	141	755	72 954	632	27
8669·5	167	167 774	72 917	167 774	141	786	72 918	664	27
8670·5	167	167 722	72 894	167 722	141	817	72 895	695	27
8671·5	167	167 700	72 884	167 699	141	841	72 885	719	27
8672·5	− 167	+167 694	+ 72 882	−167 694	− 141	− 852	− 72 883	+ 730	− 27
8673·5	167	167 691	72 880	167 690	141	850	72 881	728	27
8674·5	167	167 676	72 874	167 675	141	837	72 875	715	27
8675·5	167	167 642	72 859	167 642	141	818	72 860	696	27
8676·5	167	167 589	72 836	167 588	140	799	72 837	677	27
8677·5	− 167	+167 518	+ 72 805	−167 517	− 140	− 785	− 72 806	+ 663	− 27
8678·5	167	167 436	72 769	167 436	140	777	72 771	655	26
8679·5	166	167 350	72 732	167 350	140	777	72 733	656	26
8680·5	166	167 267	72 696	167 267	140	786	72 697	664	26
8681·5	166	167 193	72 664	167 193	140	801	72 665	680	26
8682·5	− 166	+167 133	+ 72 638	−167 133	− 140	− 821	− 72 639	+ 700	− 26
8683·5	166	167 090	72 619	167 089	140	843	72 620	722	26
8684·5	166	167 063	72 607	167 063	140	863	72 609	742	26
8685·5	166	167 052	72 602	167 051	140	880	72 604	758	26
8686·5	166	167 051	72 602	167 051	140	889	72 603	767	26
8687·5	− 166	+167 055	+ 72 603	−167 054	− 140	− 889	− 72 605	+ 768	− 26
8688·5	166	167 054	72 603	167 054	140	880	72 605	759	26
8689·5	166	167 042	72 598	167 041	140	864	72 599	743	26
8690·5	166	167 011	72 584	167 011	139	842	72 586	721	26
8691·5	166	166 958	72 561	166 957	139	820	72 563	699	26
8692·5	− 166	+166 882	+ 72 529	−166 882	− 139	− 801	− 72 530	+ 680	− 26
8693·5	165	166 790	72 488	166 789	139	790	72 489	669	26
8694·5	165	166 689	72 445	166 689	139	791	72 446	670	26
8695·5	165	166 594	72 403	166 593	139	803	72 405	682	26
8696·5	165	166 516	72 370	166 516	139	823	72 371	703	26
8697·5	− 165	+166 466	+ 72 347	−166 465	− 139	− 847	− 72 349	+ 727	− 26
8698·5	165	166 443	72 338	166 443	139	867	72 339	746	26
8699·5	165	166 442	72 337	166 442	139	876	72 339	756	26
8700·5	165	166 450	72 340	166 449	139	872	72 342	751	26
8701·5	165	166 451	72 341	166 450	139	855	72 342	734	26
8702·5	− 165	+166 435	+ 72 334	−166 434	− 139	− 830	− 72 335	+ 709	− 26
8703·5	165	166 396	72 317	166 396	138	801	72 318	681	26
8704·5	164	166 338	72 292	166 337	138	776	72 293	655	26
8705·5	164	166 264	72 260	166 263	138	756	72 261	636	26
8706·5	164	166 183	72 224	166 183	138	745	72 226	625	26
8707·5	− 164	+166 103	+ 72 190	−166 102	− 138	− 743	− 72 191	+ 623	− 26
8708·5	164	166 030	72 158	166 029	138	748	72 159	628	26
8709·5	164	165 969	72 132	165 969	138	758	72 133	638	26
8710·5	164	165 925	72 112	165 924	138	771	72 113	651	26
8711·5	164	165 896	72 100	165 896	138	783	72 101	663	26
8712·5	− 164	+165 883	+ 72 094	−165 883	− 138	− 792	− 72 095	+ 672	− 26
8713·5	− 164	+165 882	+ 72 093	−165 881	− 138	− 794	− 72 095	+ 675	− 26

Values are in units of 10^{-8}.

ORIGIN AT SOLAR SYSTEM BARYCENTRE

MEAN EQUATOR AND EQUINOX J2000·0

Date 0^h TDB	X	Y	Z	$\dot{X}$	$\dot{Y}$	$\dot{Z}$
Apr. 1	−0·976 513 358	−0·182 116 675	−0·079 051 895	+ 318 6254	−1552 0893	− 672 9655
2	·973 181 456	·197 609 563	·085 769 336	347 7354	1546 4077	670 4876
3	·969 559 065	·213 043 215	·092 460 946	376 7208	1540 2422	667 7994
4	·965 647 508	·228 412 798	·099 124 626	405 5659	1533 5942	664 9020
5	·961 448 269	·243 713 500	·105 758 296	434 2544	1526 4666	661 7977
6	−0·956 962 996	−0·258 940 547	−0·112 359 900	+ 462 7701	−1518 8643	− 658 4892
7	·952 193 497	·274 089 227	·118 927 414	491 0971	1510 7944	654 9806
8	·947 141 731	·289 154 907	·125 458 860	519 2210	1502 2661	651 2764
9	·941 809 797	·304 133 058	·131 952 308	547 1289	1493 2904	647 3820
10	·936 199 909	·319 019 265	·138 405 884	574 8103	1483 8797	643 3030
11	−0·930 314 371	−0·333 809 245	−0·144 817 772	+ 602 2577	−1474 0472	− 639 0453
12	·924 155 552	·348 498 850	·151 186 213	629 4662	1463 8067	634 6145
13	·917 725 849	·363 084 062	·157 509 503	656 4343	1453 1709	630 0158
14	·911 027 664	·377 560 989	·163 785 983	683 1632	1442 1514	625 2532
15	·904 063 371	·391 925 841	·170 014 031	709 6565	1430 7570	620 3297
16	−0·896 835 304	−0·406 174 900	−0·176 192 045	+ 735 9187	−1418 9936	− 615 2465
17	·889 345 753	·420 304 492	·182 318 429	761 9541	1406 8638	610 0036
18	·881 596 969	·434 310 953	·188 391 582	787 7655	1394 3671	604 5999
19	·873 591 188	·448 190 603	·194 409 886	813 3533	1381 5011	599 0336
20	·865 330 656	·461 939 734	·200 371 704	838 7150	1368 2625	593 3025
21	−0·856 817 658	−0·475 554 600	−0·206 275 382	+ 863 8456	−1354 6479	− 587 4052
22	·848 054 537	·489 031 429	·212 119 249	888 7383	1340 6546	581 3402
23	·839 043 713	·502 366 423	·217 901 626	913 3848	1326 2807	575 1071
24	·829 787 692	·515 555 773	·223 620 830	937 7761	1311 5257	568 7056
25	·820 289 074	·528 595 666	·229 275 178	961 9027	1296 3895	562 1361
26	−0·810 550 553	−0·541 482 295	−0·234 862 993	+ 985 7551	−1280 8728	− 555 3991
27	·800 574 920	·554 211 858	·240 382 604	1009 3232	1264 9769	548 4955
28	·790 365 070	·566 780 572	·245 832 350	1032 5968	1248 7030	541 4263
29	·779 924 002	·579 184 665	·251 210 582	1055 5650	1232 0531	534 1927
30	·769 254 828	·591 420 388	·256 515 662	1078 2159	1215 0294	526 7964
May 1	−0·758 360 784	−0·603 484 018	−0·261 745 974	+1100 5367	−1197 6351	− 519 2393
2	·747 245 240	·615 371 870	·266 899 922	1122 5137	1179 8747	511 5243
3	·735 911 705	·627 080 314	·271 975 945	1144 1325	1161 7548	503 6550
4	·724 363 832	·638 605 798	·276 972 524	1165 3789	1143 2844	495 6362
5	·712 605 413	·649 944 871	·281 888 190	1186 2397	1124 4749	487 4737
6	−0·700 640 361	−0·661 094 210	−0·286 721 540	+1206 7038	−1105 3401	− 479 1740
7	·688 472 687	·672 050 639	·291 471 237	1226 7629	1085 8955	470 7443
8	·676 106 469	·682 811 142	·296 136 017	1246 4121	1066 1575	462 1916
9	·663 545 817	·693 372 865	·300 714 683	1265 6497	1046 1421	453 5227
10	·650 794 842	·703 733 111	·305 206 102	1284 4772	1025 8643	444 7433
11	−0·637 857 628	−0·713 889 321	−0·309 609 197	+1302 8984	−1005 3372	− 435 8584
12	·624 738 212	·723 839 060	·313 922 932	1320 9186	984 5716	426 8717
13	·611 440 572	·733 579 986	·318 146 300	1338 5441	963 5757	417 7856
14	·597 968 625	·743 109 825	·322 278 318	1355 7811	942 3551	408 6016
15	·584 326 226	·752 426 350	·326 318 007	1372 6350	920 9129	399 3199
16	−0·570 517 188	−0·761 527 349	−0·330 264 390	+1389 1096	− 899 2501	− 389 9403
17	−0·556 545 294	−0·770 410 615	−0·334 116 484	+1405 2063	− 877 3663	− 380 4619

$\dot{X}$, $\dot{Y}$, $\dot{Z}$ are in units of 10^{-9} au / d.

MATRIX ELEMENTS FOR CONVERSION FROM MEAN EQUINOX OF J2000·0 TO TRUE EQUINOX OF DATE

Julian Date	$R_{11}-1$	R_{12}	R_{13}	R_{21}	$R_{22}-1$	R_{23}	R_{31}	R_{32}	$R_{33}-1$
244									
8713·5	− 164	+165 882	+ 72 093	−165 881	− 138	− 794	− 72 095	+ 675	− 26
8714·5	164	165 886	72 095	165 885	138	788	72 096	669	26
8715·5	164	165 888	72 096	165 887	138	773	72 097	654	26
8716·5	164	165 879	72 092	165 879	138	750	72 093	630	26
8717·5	164	165 852	72 080	165 852	138	720	72 081	600	26
8718·5	− 163	+165 803	+ 72 059	−165 802	− 137	− 688	− 72 060	+ 569	− 26
8719·5	163	165 730	72 027	165 729	137	660	72 028	540	26
8720·5	163	165 638	71 987	165 638	137	638	71 988	519	26
8721·5	163	165 537	71 943	165 537	137	628	71 944	509	26
8722·5	163	165 439	71 901	165 438	137	628	71 902	509	26
8723·5	− 163	+165 355	+ 71 864	−165 354	− 137	− 638	− 71 865	+ 519	− 26
8724·5	162	165 294	71 838	165 293	137	652	71 839	534	26
8725·5	162	165 259	71 823	165 259	137	665	71 824	546	26
8726·5	162	165 246	71 817	165 246	137	669	71 818	551	26
8727·5	162	165 245	71 817	165 245	137	662	71 818	543	26
8728·5	− 162	+165 243	+ 71 815	−165 242	− 137	− 642	− 71 816	+ 523	− 26
8729·5	162	165 227	71 808	165 226	137	612	71 809	494	26
8730·5	162	165 189	71 792	165 189	136	577	71 793	459	26
8731·5	162	165 129	71 766	165 129	136	543	71 767	424	26
8732·5	162	165 050	71 732	165 050	136	513	71 732	395	26
8733·5	− 162	+164 960	+ 71 693	−164 960	− 136	− 491	− 71 693	+ 373	− 26
8734·5	162	164 867	71 652	164 867	136	479	71 653	361	26
8735·5	161	164 780	71 614	164 779	136	475	71 615	357	26
8736·5	161	164 703	71 581	164 702	136	477	71 581	359	26
8737·5	161	164 640	71 554	164 640	136	483	71 554	366	26
8738·5	− 161	+164 595	+ 71 534	−164 594	− 135	− 490	− 71 534	+ 372	− 26
8739·5	161	164 564	71 521	164 564	135	494	71 521	377	26
8740·5	161	164 547	71 513	164 546	135	493	71 514	376	26
8741·5	161	164 536	71 508	164 536	135	485	71 509	367	26
8742·5	161	164 526	71 504	164 526	135	468	71 504	350	26
8743·5	− 161	+164 507	+ 71 496	−164 507	− 135	− 442	− 71 496	+ 324	− 26
8744·5	161	164 472	71 480	164 472	135	408	71 481	291	26
8745·5	161	164 415	71 455	164 414	135	372	71 456	254	26
8746·5	161	164 331	71 419	164 331	135	337	71 420	220	26
8747·5	160	164 227	71 374	164 227	135	309	71 374	192	25
8748·5	− 160	+164 109	+ 71 323	−164 109	− 135	− 292	− 71 323	+ 175	− 25
8749·5	160	163 992	71 272	163 992	134	287	71 272	170	25
8750·5	160	163 888	71 227	163 888	134	292	71 227	175	25
8751·5	159	163 806	71 191	163 806	134	303	71 191	186	25
8752·5	159	163 750	71 167	163 750	134	313	71 167	196	25
8753·5	− 159	+163 716	+ 71 152	−163 716	− 134	− 317	− 71 152	+ 201	− 25
8754·5	159	163 697	71 144	163 697	134	311	71 144	194	25
8755·5	159	163 680	71 136	163 679	134	293	71 136	176	25
8756·5	159	163 652	71 124	163 652	134	265	71 124	148	25
8757·5	159	163 606	71 104	163 605	134	230	71 104	114	25
8758·5	− 159	+163 537	+ 71 074	−163 537	− 134	− 195	− 71 074	+ 78	− 25
8759·5	− 159	+163 448	+ 71 035	−163 448	− 134	− 163	− 71 036	+ 47	− 25

Values are in units of 10^{-8}.

ORIGIN AT SOLAR SYSTEM BARYCENTRE

MEAN EQUATOR AND EQUINOX J2000·0

Date 0ʰ TDB		X	Y	Z	$\dot{X}$	$\dot{Y}$	$\dot{Z}$
May	17	−0·556 545 294	−0·770 410 615	−0·334 116 484	+1405 2063	− 877 3663	− 380 4619
	18	·542 414 323	·779 073 935	·337 873 296	1420 9247	855 2605	370 8839
	19	·528 128 071	·787 515 082	·341 533 829	1436 2620	832 9317	361 2058
	20	·513 690 368	·795 731 823	·345 097 079	1451 2138	810 3793	351 4274
	21	·499 105 099	·803 721 924	·348 562 044	1465 7744	787 6037	341 5490
	22	−0·484 376 206	−0·811 483 158	−0·351 927 728	+1479 9374	− 764 6063	− 331 5715
	23	·469 507 698	·819 013 316	·355 193 146	1493 6960	741 3890	321 4960
	24	·454 503 657	·826 310 214	·358 357 326	1507 0432	717 9546	311 3241
	25	·439 368 231	·833 371 695	·361 419 313	1519 9715	694 3063	301 0577
	26	·424 105 648	·840 195 638	·364 378 171	1532 4735	670 4476	290 6986
	27	−0·408 720 210	−0·846 779 958	−0·367 232 984	+1544 5410	− 646 3823	− 280 2491
	28	·393 216 306	·853 122 611	·369 982 860	1556 1651	622 1149	269 7117
	29	·377 598 419	·859 221 599	·372 626 935	1567 3360	597 6505	259 0894
	30	·361 871 134	·865 074 985	·375 164 376	1578 0427	572 9956	248 3857
	31	·346 039 150	·870 680 904	·377 594 394	1588 2739	548 1588	237 6054
June	1	−0·330 107 279	−0·876 037 592	−0·379 916 249	+1598 0182	− 523 1515	− 226 7543
	2	·314 080 443	·881 143 413	·382 129 267	1607 2657	497 9880	215 8394
	3	·297 963 646	·885 996 886	·384 232 849	1616 0093	472 6850	204 8684
	4	·281 761 948	·890 596 709	·386 226 476	1624 2457	447 2609	193 8496
	5	·265 480 421	·894 941 763	·388 109 708	1631 9758	421 7343	182 7907
	6	−0·249 124 104	−0·899 031 111	−0·389 882 179	+1639 2045	− 396 1224	− 171 6985
	7	·232 697 976	·902 863 976	·391 543 584	1645 9395	370 4400	160 5784
	8	·216 206 928	·906 439 715	·393 093 668	1652 1901	344 6987	149 4346
	9	·199 655 756	·909 757 785	·394 532 206	1657 9659	318 9075	138 2697
	10	·183 049 161	·912 817 719	·395 858 998	1663 2761	293 0725	127 0854
	11	−0·166 391 758	−0·915 619 103	−0·397 073 852	+1668 1288	− 267 1978	− 115 8824
	12	·149 688 088	·918 161 550	·398 176 585	1672 5305	241 2855	104 6610
	13	·132 942 635	·920 444 691	·399 167 011	1676 4861	215 3368	93 4211
	14	·116 159 842	·922 468 166	·400 044 943	1679 9987	189 3521	82 1623
	15	·099 344 132	·924 231 614	·400 810 193	1683 0696	163 3314	70 8845
	16	−0·082 499 924	−0·925 734 675	−0·401 462 570	+1685 6983	− 137 2749	− 59 5878
	17	·065 631 645	·926 976 995	·402 001 888	1687 8832	111 1833	48 2726
	18	·048 743 749	·927 958 229	·402 427 963	1689 6214	85 0580	36 9397
	19	·031 840 717	·928 678 050	·402 740 626	1690 9095	58 9010	25 5903
	20	− ·014 927 072	·929 136 154	·402 939 720	1691 7435	32 7154	14 2261
	21	+0·001 992 624	−0·929 332 273	−0·403 025 105	+1692 1190	− 6 5045	− 2 8490
	22	·018 913 766	·929 266 173	·402 996 664	1692 0318	+ 19 7276	+ 8 5387
	23	·035 831 701	·928 937 665	·402 854 304	1691 4770	45 9764	19 9346
	24	·052 741 730	·928 346 606	·402 597 953	1690 4494	72 2370	31 3362
	25	·069 639 095	·927 492 904	·402 227 572	1688 9433	98 5041	42 7403
	26	+0·086 518 978	−0·926 376 523	−0·401 743 150	+1686 9519	+ 124 7715	+ 54 1436
	27	·103 376 491	·924 997 497	·401 144 717	1684 4679	151 0317	65 5419
	28	·120 206 666	·923 355 945	·400 432 345	1681 4829	177 2750	76 9304
	29	·137 004 454	·921 452 094	·399 606 163	1677 9894	203 4892	88 3026
	30	·153 764 737	·919 286 309	·398 666 372	1673 9809	229 6589	99 6512
July	1	+0·170 482 346	−0·916 859 125	−0·397 613 246	+1669 4545	+ 255 7658	+ 110 9678
	2	+0·187 152 106	−0·914 171 267	−0·396 447 149	+1664 4117	+ 281 7905	+ 122 2441

$\dot{X}$, $\dot{Y}$, $\dot{Z}$ are in units of 10^{-9} au / d.

MATRIX ELEMENTS FOR CONVERSION FROM MEAN EQUINOX OF J2000·0 TO TRUE EQUINOX OF DATE

Julian Date	$R_{11}-1$	R_{12}	R_{13}	R_{21}	$R_{22}-1$	R_{23}	R_{31}	R_{32}	$R_{33}-1$
244									
8759·5	− 159	+163 448	+ 71 035	−163 448	− 134	− 163	− 71 036	+ 47	− 25
8760·5	159	163 345	70 990	163 345	133	138	70 991	22	25
8761·5	158	163 235	70 943	163 235	133	122	70 943	+ 6	25
8762·5	158	163 127	70 896	163 127	133	116	70 896	0	25
8763·5	158	163 029	70 853	163 029	133	117	70 853	+ 1	25
8764·5	− 158	+162 945	+ 70 817	−162 945	− 133	− 123	− 70 817	+ 7	− 25
8765·5	158	162 877	70 787	162 877	133	130	70 787	15	25
8766·5	158	162 825	70 765	162 825	133	137	70 765	22	25
8767·5	158	162 787	70 748	162 787	132	140	70 749	24	25
8768·5	157	162 759	70 736	162 759	132	136	70 736	21	25
8769·5	− 157	+162 733	+ 70 725	−162 733	− 132	− 124	− 70 725	+ 9	− 25
8770·5	157	162 702	70 711	162 702	132	103	70 711	− 12	25
8771·5	157	162 658	70 692	162 658	132	75	70 692	40	25
8772·5	157	162 594	70 664	162 594	132	42	70 664	73	25
8773·5	157	162 504	70 625	162 504	132	− 9	70 625	106	25
8774·5	− 157	+162 389	+ 70 575	−162 389	− 132	+ 19	− 70 575	− 134	− 25
8775·5	156	162 257	70 518	162 257	132	37	70 518	152	25
8776·5	156	162 120	70 458	162 120	131	42	70 458	156	25
8777·5	156	161 993	70 403	161 993	131	35	70 403	149	25
8778·5	156	161 888	70 358	161 888	131	21	70 358	135	25
8779·5	− 156	+161 811	+ 70 324	−161 811	− 131	+ 5	− 70 324	− 119	− 25
8780·5	156	161 759	70 302	161 759	131	− 6	70 302	107	25
8781·5	155	161 725	70 287	161 725	131	− 8	70 287	106	25
8782·5	155	161 695	70 274	161 695	131	+ 3	70 274	116	25
8783·5	155	161 659	70 258	161 659	131	23	70 258	137	25
8784·5	− 155	+161 606	+ 70 235	−161 606	− 131	+ 50	− 70 235	− 164	− 25
8785·5	155	161 533	70 203	161 533	130	80	70 203	193	25
8786·5	155	161 439	70 163	161 439	130	107	70 162	220	25
8787·5	155	161 330	70 115	161 330	130	127	70 115	240	25
8788·5	154	161 212	70 064	161 212	130	139	70 064	252	25
8789·5	− 154	+161 094	+ 70 013	−161 094	− 130	+ 142	− 70 012	− 255	− 25
8790·5	154	160 983	69 965	160 983	130	137	69 964	249	24
8791·5	154	160 886	69 922	160 886	129	125	69 922	238	24
8792·5	154	160 804	69 887	160 804	129	111	69 887	223	24
8793·5	154	160 740	69 859	160 740	129	97	69 859	209	24
8794·5	− 153	+160 691	+ 69 838	−160 691	− 129	+ 85	− 69 838	− 198	− 24
8795·5	153	160 653	69 821	160 653	129	80	69 821	192	24
8796·5	153	160 621	69 807	160 621	129	82	69 807	194	24
8797·5	153	160 587	69 792	160 587	129	92	69 792	204	24
8798·5	153	160 543	69 773	160 543	129	109	69 773	221	24
8799·5	− 153	+160 483	+ 69 747	−160 483	− 129	+ 132	− 69 747	− 244	− 24
8800·5	153	160 399	69 711	160 400	129	158	69 711	270	24
8801·5	153	160 291	69 664	160 291	128	181	69 664	293	24
8802·5	152	160 161	69 608	160 162	128	196	69 607	307	24
8803·5	152	160 020	69 546	160 020	128	198	69 546	310	24
8804·5	− 152	+159 882	+ 69 487	−159 883	− 128	+ 188	− 69 486	− 299	− 24
8805·5	− 152	+159 764	+ 69 435	−159 764	− 128	+ 167	− 69 435	− 278	− 24

Values are in units of 10^{-8}.

ORIGIN AT SOLAR SYSTEM BARYCENTRE

MEAN EQUATOR AND EQUINOX J2000·0

Date 0ʰ TDB		X	Y	Z	$\dot{X}$	$\dot{Y}$	$\dot{Z}$
July	**1**	+0·170 482 346	−0·916 859 125	−0·397 613 246	+1669 4545	+ 255 7658	+ 110 9678
	2	·187 152 106	·914 171 267	·396 447 149	1664 4117	281 7905	122 2441
	3	·203 768 879	·911 223 654	·395 168 525	1658 8588	307 7136	133 4721
	4	·220 327 618	·908 017 390	·393 777 890	1652 8069	333 5182	144 6454
	5	·236 823 398	·904 553 729	·392 275 815	1646 2692	359 1911	155 7595
	6	+0·253 251 432	−0·900 834 040	−0·390 662 908	+1639 2602	+ 384 7227	+ 166 8115
	7	·269 607 077	·896 859 766	·388 939 797	1631 7934	410 1073	177 8000
	8	·285 885 813	·892 632 395	·387 107 120	1623 8804	435 3416	188 7248
	9	·302 083 230	·888 153 441	·385 165 514	1615 5311	460 4239	199 5858
	10	·318 195 007	·883 424 425	·383 115 615	1606 7533	485 3539	210 3834
	11	+0·334 216 889	−0·878 446 872	−0·380 958 055	+1597 5530	+ 510 1312	+ 221 1180
	12	·350 144 673	·873 222 309	·378 693 465	1587 9344	534 7560	231 7897
	13	·365 974 194	·867 752 261	·376 322 471	1577 9008	559 2281	242 3985
	14	·381 701 310	·862 038 258	·373 845 704	1567 4538	583 5470	252 9443
	15	·397 321 894	·856 081 833	·371 263 798	1556 5942	607 7121	263 4264
	16	+0·412 831 819	−0·849 884 533	−0·368 577 392	+1545 3221	+ 631 7219	+ 273 8440
	17	·428 226 958	·843 447 921	·365 787 138	1533 6366	655 5743	284 1958
	18	·443 503 170	·836 773 581	·362 893 701	1521 5367	679 2666	294 4802
	19	·458 656 304	·829 863 132	·359 897 765	1509 0207	702 7957	304 6953
	20	·473 682 190	·822 718 225	·356 800 033	1496 0868	726 1575	314 8390
	21	+0·488 576 641	−0·815 340 555	−0·353 601 232	+1482 7331	+ 749 3475	+ 324 9087
	22	·503 335 444	·807 731 863	·350 302 114	1468 9570	772 3610	334 9018
	23	·517 954 364	·799 893 943	·346 903 462	1454 7559	795 1922	344 8151
	24	·532 429 133	·791 828 648	·343 406 090	1440 1261	817 8347	354 6451
	25	·546 755 443	·783 537 902	·339 810 850	1425 0636	840 2808	364 3878
	26	+0·560 928 948	−0·775 023 719	−0·336 118 642	+1409 5642	+ 862 5206	+ 374 0380
	27	·574 945 257	·766 288 218	·332 330 419	1393 6239	884 5420	383 5897
	28	·588 799 949	·757 333 657	·328 447 200	1377 2406	906 3298	393 0357
	29	·602 488 595	·748 162 458	·324 470 084	1360 4153	927 8664	402 3680
	30	·616 006 800	·738 777 230	·320 400 246	1343 1540	949 1327	411 5786
	31	+0·629 350 257	−0·729 180 767	−0·316 238 942	+1325 4679	+ 970 1103	+ 420 6601
Aug.	**1**	·642 514 794	·719 376 037	·311 987 492	1307 3727	990 7839	429 6071
	2	·655 496 410	·709 366 139	·307 647 260	1288 8869	1011 1425	438 4161
	3	·668 291 292	·699 154 258	·303 219 634	1270 0289	1031 1799	447 0859
	4	·680 895 804	·688 743 620	·298 706 006	1250 8154	1050 8937	455 6166
	5	+0·693 306 463	−0·678 137 461	−0·294 107 762	+1231 2605	+1070 2843	+ 464 0092
	6	·705 519 914	·667 339 007	·289 426 277	1211 3755	1089 3531	472 2651
	7	·717 532 901	·656 351 465	·284 662 912	1191 1691	1108 1020	480 3854
	8	·729 342 248	·645 178 028	·279 819 018	1170 6484	1126 5327	488 3710
	9	·740 944 839	·633 821 868	·274 895 939	1149 8189	1144 6465	496 2226
	10	+0·752 337 611	−0·622 286 150	−0·269 895 013	+1128 6852	+1162 4444	+ 503 9403
	11	·763 517 541	·610 574 031	·264 817 578	1107 2509	1179 9267	511 5244
	12	·774 481 639	·598 688 667	·259 664 972	1085 5192	1197 0935	518 9743
	13	·785 226 941	·586 633 214	·254 438 539	1063 4923	1213 9443	526 2898
	14	·795 750 507	·574 410 837	·249 139 628	1041 1720	1230 4782	533 4698
	15	+0·806 049 408	−0·562 024 711	−0·243 769 597	+1018 5596	+1246 6936	+ 540 5134
	16	+0·816 120 727	−0·549 478 032	−0·238 329 819	+ 995 6559	+1262 5885	+ 547 4191

$\dot{X}$, $\dot{Y}$, $\dot{Z}$ are in units of 10^{-9} au / d.

MATRIX ELEMENTS FOR CONVERSION FROM MEAN EQUINOX OF J2000·0 TO TRUE EQUINOX OF DATE

Julian Date	$R_{11}-1$	R_{12}	R_{13}	R_{21}	$R_{22}-1$	R_{23}	R_{31}	R_{32}	$R_{33}-1$
244									
8804·5	− 152	+159 882	+ 69 487	−159 883	− 128	+ 188	− 69 486	− 299	− 24
8805·5	152	159 764	69 435	159 764	128	167	69 435	278	24
8806·5	152	159 673	69 396	159 674	127	141	69 396	252	24
8807·5	151	159 613	69 370	159 613	127	119	69 369	230	24
8808·5	151	159 575	69 353	159 575	127	107	69 353	217	24
8809·5	− 151	+159 546	+ 69 340	−159 546	− 127	+ 106	− 69 340	− 217	− 24
8810·5	151	159 513	69 326	159 513	127	117	69 326	227	24
8811·5	151	159 466	69 306	159 466	127	135	69 305	246	24
8812·5	151	159 399	69 277	159 400	127	156	69 276	267	24
8813·5	151	159 313	69 239	159 314	127	176	69 239	287	24
8814·5	− 151	+159 211	+ 69 195	−159 211	− 127	+ 191	− 69 195	− 301	− 24
8815·5	150	159 100	69 147	159 100	127	197	69 146	307	24
8816·5	150	158 986	69 097	158 986	126	194	69 097	304	24
8817·5	150	158 879	69 051	158 879	126	183	69 050	293	24
8818·5	150	158 783	69 009	158 783	126	166	69 009	275	24
8819·5	− 150	+158 703	+ 68 974	−158 703	− 126	+ 145	− 68 974	− 254	− 24
8820·5	150	158 641	68 947	158 641	126	123	68 947	232	24
8821·5	150	158 595	68 927	158 595	126	103	68 927	212	24
8822·5	149	158 561	68 913	158 562	126	88	68 913	198	24
8823·5	149	158 536	68 902	158 536	126	81	68 901	190	24
8824·5	− 149	+158 512	+ 68 891	−158 512	− 126	+ 81	− 68 891	− 190	− 24
8825·5	149	158 481	68 878	158 481	126	89	68 877	198	24
8826·5	149	158 437	68 859	158 437	126	103	68 858	212	24
8827·5	149	158 373	68 831	158 373	125	121	68 831	230	24
8828·5	149	158 287	68 793	158 287	125	138	68 793	247	24
8829·5	− 149	+158 178	+ 68 746	−158 178	− 125	+ 150	− 68 746	− 259	− 24
8830·5	148	158 054	68 692	158 054	125	152	68 692	260	24
8831·5	148	157 925	68 636	157 925	125	141	68 636	249	24
8832·5	148	157 808	68 585	157 808	125	118	68 585	226	24
8833·5	148	157 715	68 545	157 715	124	88	68 545	196	23
8834·5	− 148	+157 654	+ 68 519	−157 654	− 124	+ 58	− 68 518	− 166	− 23
8835·5	148	157 621	68 504	157 621	124	35	68 504	143	23
8836·5	148	157 604	68 497	157 604	124	25	68 497	133	23
8837·5	148	157 587	68 489	157 587	124	28	68 489	136	23
8838·5	148	157 559	68 477	157 559	124	41	68 477	149	23
8839·5	− 147	+157 511	+ 68 456	−157 511	− 124	+ 58	− 68 456	− 166	− 23
8840·5	147	157 443	68 427	157 443	124	75	68 426	183	23
8841·5	147	157 358	68 390	157 358	124	88	68 390	195	23
8842·5	147	157 263	68 349	157 263	124	93	68 348	200	23
8843·5	147	157 165	68 306	157 165	124	89	68 306	196	23
8844·5	− 147	+157 071	+ 68 265	−157 071	− 123	+ 77	− 68 265	− 184	− 23
8845·5	147	156 988	68 229	156 988	123	58	68 229	165	23
8846·5	146	156 920	68 199	156 920	123	35	68 199	142	23
8847·5	146	156 869	68 177	156 869	123	+ 10	68 177	117	23
8848·5	146	156 835	68 163	156 835	123	− 12	68 163	95	23
8849·5	− 146	+156 815	+ 68 154	−156 815	− 123	− 30	− 68 154	− 77	− 23
8850·5	− 146	+156 805	+ 68 149	−156 805	− 123	− 41	− 68 149	− 65	− 23

Values are in units of 10^{-8}.

ORIGIN AT SOLAR SYSTEM BARYCENTRE

MEAN EQUATOR AND EQUINOX J2000·0

Date 0h TDB	X	Y	Z	$\dot{X}$	$\dot{Y}$	$\dot{Z}$
Aug. 16	+0·816 120 727	−0·549 478 032	−0·238 329 819	+ 995 6559	+1262 5885	+ 547 4191
17	·825 961 556	·536 774 019	·232 821 682	972 4615	1278 1600	554 1850
18	·835 568 988	·523 915 921	·227 246 591	948 9765	1293 4047	560 8093
19	·844 940 118	·510 907 027	·221 605 977	925 2011	1308 3185	567 2893
20	·854 072 040	·497 750 671	·215 901 294	901 1348	1322 8963	573 6224
21	+0·862 961 844	−0·484 450 240	−0·210 134 029	+ 876 7773	+1337 1322	+ 579 8053
22	·871 606 613	·471 009 191	·204 305 702	852 1279	1351 0188	585 8340
23	·880 003 426	·457 431 058	·198 417 878	827 1860	1364 5472	591 7041
24	·888 149 360	·443 719 479	·192 472 168	801 9521	1377 7061	597 4100
25	·896 041 502	·429 878 213	·186 470 245	776 4284	1390 4820	602 9456
26	+0·903 676 979	−0·415 911 168	−0·180 413 848	+ 750 6202	+1402 8593	+ 608 3039
27	·911 052 990	·401 822 410	·174 304 781	724 5375	1414 8217	613 4782
28	·918 166 863	·387 616 169	·168 144 916	698 1953	1426 3536	618 4627
29	·925 016 095	·373 296 814	·161 936 174	671 6129	1437 4428	623 2532
30	·931 598 393	·358 868 815	·155 680 505	644 8120	1448 0816	627 8478
31	+0·937 911 679	−0·344 336 692	−0·149 379 869	+ 617 8140	+1458 2676	+ 632 2470
Sept. 1	·943 954 080	·329 704 967	·143 036 211	590 6380	1468 0024	636 4525
2	·949 723 899	·314 978 134	·136 651 455	563 2998	1477 2903	640 4670
3	·955 219 577	·300 160 634	·130 227 498	535 8117	1486 1364	644 2932
4	·960 439 665	·285 256 861	·123 766 210	508 1833	1494 5459	647 9335
5	+0·965 382 802	−0·270 271 156	−0·117 269 440	+ 480 4225	+1502 5233	+ 651 3899
6	·970 047 694	·255 207 823	·110 739 020	452 5354	1510 0721	654 6638
7	·974 433 109	·240 071 132	·104 176 770	424 5278	1517 1955	657 7562
8	·978 537 866	·224 865 324	·097 584 499	396 4047	1523 8958	660 6679
9	·982 360 834	·209 594 620	·090 964 013	368 1707	1530 1748	663 3994
10	+0·985 900 925	−0·194 263 226	−0·084 317 111	+ 339 8299	+1536 0343	+ 665 9510
11	·989 157 089	·178 875 329	·077 645 591	311 3860	1541 4754	668 3229
12	·992 128 310	·163 435 110	·070 951 253	282 8419	1546 4988	670 5147
13	·994 813 601	·147 946 744	·064 235 899	254 2002	1551 1048	672 5261
14	·997 211 997	·132 414 407	·057 501 337	225 4631	1555 2928	674 3561
15	+0·999 322 552	−0·116 842 285	−0·050 749 385	+ 196 6324	+1559 0614	+ 676 0037
16	1·001 144 337	·101 234 584	·043 981 877	167 7095	1562 4083	677 4670
17	1·002 676 439	·085 595 537	·037 200 666	138 6960	1565 3299	678 7440
18	1·003 917 961	·069 929 419	·030 407 626	109 5937	1567 8217	679 8321
19	1·004 868 024	·054 240 556	·023 604 663	80 4047	1569 8776	680 7281
20	+1·005 525 776	−0·038 533 344	−0·016 793 717	+ 51 1320	+1571 4903	+ 681 4282
21	1·005 890 399	·022 812 258	·009 976 766	+ 21 7799	1572 6506	681 9283
22	1·005 961 130	− ·007 081 874	− ·003 155 834	− 7 6453	1573 3479	682 2235
23	1·005 737 280	+ ·008 653 118	+ ·003 667 005	37 1341	1573 5705	682 3090
24	1·005 218 279	·024 387 912	·010 489 631	66 6732	1573 3063	682 1801
25	+1·004 403 706	+0·040 117 587	+0·017 309 881	− 96 2450	+1572 5450	+ 681 8333
26	1·003 293 338	·055 837 130	·024 125 564	125 8286	1571 2791	681 2666
27	1·001 887 171	·071 541 477	·030 934 480	155 4011	1569 5058	680 4801
28	1·000 185 428	·087 225 560	·037 734 441	184 9402	1567 2269	679 4761
29	0·998 188 545	·102 884 350	·044 523 288	214 4260	1564 4485	678 2581
30	+0·995 897 140	+0·118 512 894	+0·051 298 903	− 243 8422	+1561 1791	+ 676 8305
Oct. 1	+0·993 311 975	+0·134 106 326	+0·058 059 213	− 273 1761	+1557 4279	+ 675 1976

$\dot{X}$, $\dot{Y}$, $\dot{Z}$ are in units of 10^{-9} au / d.

MATRIX ELEMENTS FOR CONVERSION FROM MEAN EQUINOX OF J2000·0 TO TRUE EQUINOX OF DATE

Julian Date	$R_{11}-1$	R_{12}	R_{13}	R_{21}	$R_{22}-1$	R_{23}	R_{31}	R_{32}	$R_{33}-1$
244									
8850·5	− 146	+156 805	+ 68 149	−156 805	− 123	− 41	− 68 149	− 65	− 23
8851·5	146	156 797	68 146	156 797	123	45	68 146	62	23
8852·5	146	156 786	68 141	156 786	123	40	68 141	67	23
8853·5	146	156 763	68 131	156 763	123	29	68 131	78	23
8854·5	146	156 723	68 114	156 723	123	− 13	68 114	93	23
8855·5	− 146	+156 663	+ 68 088	−156 663	− 123	+ 3	− 68 087	− 110	− 23
8856·5	146	156 581	68 052	156 581	123	16	68 052	123	23
8857·5	146	156 483	68 009	156 483	122	22	68 009	128	23
8858·5	145	156 376	67 963	156 376	122	+ 16	67 963	123	23
8859·5	145	156 272	67 918	156 272	122	− 1	67 918	105	23
8860·5	− 145	+156 186	+ 67 881	−156 186	− 122	− 27	− 67 881	− 79	− 23
8861·5	145	156 128	67 855	156 128	122	57	67 855	49	23
8862·5	145	156 100	67 843	156 100	122	82	67 843	24	23
8863·5	145	156 094	67 840	156 094	122	96	67 840	10	23
8864·5	145	156 095	67 841	156 095	122	96	67 841	9	23
8865·5	− 145	+156 089	+ 67 838	−156 089	− 122	− 84	− 67 838	− 22	− 23
8866·5	145	156 064	67 827	156 064	122	64	67 827	42	23
8867·5	145	156 016	67 806	156 016	122	42	67 806	64	23
8868·5	145	155 949	67 777	155 949	122	24	67 777	82	23
8869·5	144	155 869	67 743	155 869	121	13	67 743	92	23
8870·5	− 144	+155 785	+ 67 706	−155 785	− 121	− 11	− 67 706	− 95	− 23
8871·5	144	155 704	67 671	155 704	121	17	67 671	88	23
8872·5	144	155 631	67 639	155 631	121	30	67 639	75	23
8873·5	144	155 574	67 614	155 574	121	48	67 614	57	23
8874·5	144	155 532	67 596	155 532	121	68	67 596	37	23
8875·5	− 144	+155 508	+ 67 586	−155 508	− 121	− 86	− 67 586	− 19	− 23
8876·5	144	155 498	67 581	155 498	121	100	67 581	− 5	23
8877·5	144	155 499	67 581	155 499	121	108	67 582	+ 3	23
8878·5	144	155 503	67 583	155 503	121	108	67 584	+ 3	23
8879·5	144	155 505	67 584	155 505	121	99	67 584	− 6	23
8880·5	− 144	+155 497	+ 67 581	−155 497	− 121	− 84	− 67 581	− 21	− 23
8881·5	144	155 474	67 570	155 474	121	63	67 570	42	23
8882·5	144	155 430	67 551	155 430	121	41	67 551	64	23
8883·5	143	155 367	67 524	155 367	121	21	67 524	84	23
8884·5	143	155 285	67 488	155 285	121	7	67 488	97	23
8885·5	− 143	+155 194	+ 67 449	−155 194	− 120	− 3	− 67 449	− 102	− 23
8886·5	143	155 102	67 409	155 102	120	9	67 409	96	23
8887·5	143	155 022	67 374	155 022	120	24	67 374	80	23
8888·5	143	154 963	67 349	154 963	120	45	67 349	60	23
8889·5	143	154 932	67 335	154 932	120	64	67 335	40	23
8890·5	− 143	+154 925	+ 67 332	−154 925	− 120	− 76	− 67 332	− 28	− 23
8891·5	143	154 931	67 334	154 931	120	75	67 335	29	23
8892·5	143	154 936	67 336	154 936	120	60	67 336	44	23
8893·5	143	154 925	67 332	154 925	120	34	67 332	70	23
8894·5	143	154 890	67 317	154 890	120	− 4	67 317	101	23
8895·5	− 143	+154 833	+ 67 292	−154 833	− 120	+ 26	− 67 291	− 130	− 23
8896·5	− 142	+154 758	+ 67 259	−154 758	− 120	+ 48	− 67 259	− 152	− 23

Values are in units of 10^{-8}.

ORIGIN AT SOLAR SYSTEM BARYCENTRE

MEAN EQUATOR AND EQUINOX J2000·0

Date 0ʰ TDB		X	Y	Z	$\dot{X}$	$\dot{Y}$	$\dot{Z}$
Oct.	1	+0·993 311 975	+0·134 106 326	+0·058 059 213	− 273 1761	+1557 4279	+ 675 1976
	2	·990 433 923	·149 659 877	·064 802 185	302 4182	1553 2042	673 3635
	3	·987 263 943	·165 168 861	·071 525 823	331 5608	1548 5159	671 3315
	4	·983 803 060	·180 628 670	·078 228 164	360 5976	1543 3702	669 1043
	5	·980 052 362	·196 034 761	·084 907 266	389 5230	1537 7732	666 6843
	6	+0·976 012 988	+0·211 382 648	+0·091 561 212	− 418 3318	+1531 7303	+ 664 0733
	7	·971 686 130	·226 667 897	·098 188 102	447 0191	1525 2464	661 2733
	8	·967 073 027	·241 886 123	·104 786 053	475 5801	1518 3262	658 2856
	9	·962 174 962	·257 032 982	·111 353 195	504 0109	1510 9741	655 1118
	10	·956 993 257	·272 104 178	·117 887 673	532 3077	1503 1940	651 7531
	11	+0·951 529 264	+0·287 095 447	+0·124 387 643	− 560 4678	+1494 9894	+ 648 2102
	12	·945 784 365	·302 002 560	·130 851 266	588 4888	1486 3630	644 4839
	13	·939 759 958	·316 821 307	·137 276 709	616 3689	1477 3165	640 5740
	14	·933 457 462	·331 547 490	·143 662 133	644 1066	1467 8502	636 4802
	15	·926 878 308	·346 176 908	·150 005 695	671 6999	1457 9631	632 2012
	16	+0·920 023 953	+0·360 705 342	+0·156 305 535	− 699 1465	+1447 6527	+ 627 7356
	17	·912 895 879	·375 128 539	·162 559 777	726 4427	1436 9152	623 0810
	18	·905 495 615	·389 442 204	·168 766 518	753 5837	1425 7452	618 2352
	19	·897 824 747	·403 641 982	·174 923 834	780 5622	1414 1368	613 1953
	20	·889 884 943	·417 723 455	·181 029 769	807 3690	1402 0831	607 9587
	21	+0·881 677 981	+0·431 682 137	+0·187 082 344	− 833 9917	+1389 5775	+ 602 5229
	22	·873 205 774	·445 513 478	·193 079 557	860 4150	1376 6140	596 8861
	23	·864 470 405	·459 212 877	·199 019 393	886 6209	1363 1886	591 0476
	24	·855 474 149	·472 775 707	·204 899 840	912 5889	1349 3005	585 0084
	25	·846 219 492	·486 197 353	·210 718 900	938 2976	1334 9525	578 7708
	26	+0·836 709 128	+0·499 473 247	+0·216 474 609	− 963 7268	+1320 1517	+ 572 3392
	27	·826 945 948	·512 598 911	·222 165 055	988 8584	1304 9084	565 7189
	28	·816 933 002	·525 569 982	·227 788 378	1013 6776	1289 2354	558 9159
	29	·806 673 473	·538 382 231	·233 342 784	1038 1735	1273 1461	551 9363
	30	·796 170 637	·551 031 561	·238 826 534	1062 3378	1256 6536	544 7856
	31	+0·785 427 841	+0·563 514 001	+0·244 237 943	−1086 1647	+1239 7702	+ 537 4689
Nov.	1	·774 448 483	·575 825 697	·249 575 372	1109 6497	1222 5065	529 9904
	2	·763 236 000	·587 962 897	·254 837 223	1132 7888	1204 8726	522 3539
	3	·751 793 871	·599 921 945	·260 021 935	1155 5787	1186 8774	514 5629
	4	·740 125 603	·611 699 269	·265 127 978	1178 0159	1168 5292	506 6206
	5	+0·728 234 737	+0·623 291 380	+0·270 153 853	−1200 0976	+1149 8363	+ 498 5300
	6	·716 124 844	·634 694 871	·275 098 093	1221 8211	1130 8063	490 2940
	7	·703 799 516	·645 906 409	·279 959 257	1243 1845	1111 4470	481 9152
	8	·691 262 359	·656 922 736	·284 735 930	1264 1866	1091 7653	473 3961
	9	·678 516 989	·667 740 661	·289 426 719	1284 8271	1071 7674	464 7388
	10	+0·665 567 021	+0·678 357 048	+0·294 030 250	−1305 1062	+1051 4585	+ 455 9447
	11	·652 416 067	·688 768 805	·298 545 159	1325 0246	1030 8419	447 0145
	12	·639 067 732	·698 972 866	·302 970 087	1344 5824	1009 9192	437 9484
	13	·625 525 623	·708 966 169	·307 303 672	1363 7789	988 6902	428 7457
	14	·611 793 367	·718 745 643	·311 544 542	1382 6114	967 1530	419 4053
	15	+0·597 874 625	+0·728 308 192	+0·315 691 315	−1401 0750	+ 945 3047	+ 409 9260
	16	+0·583 773 122	+0·737 650 690	+0·319 742 595	−1419 1620	+ 923 1422	+ 400 3067

$\dot{X}$, $\dot{Y}$, $\dot{Z}$ are in units of 10^{-9} au / d.

MATRIX ELEMENTS FOR CONVERSION FROM MEAN EQUINOX OF J2000·0 TO TRUE EQUINOX OF DATE

Julian Date	$R_{11}-1$	R_{12}	R_{13}	R_{21}	$R_{22}-1$	R_{23}	R_{31}	R_{32}	$R_{33}-1$
244									
8896·5	− 142	+154 758	+ 67 259	−154 758	− 120	+ 48	− 67 259	− 152	− 23
8897·5	142	154 674	67 223	154 675	120	62	67 223	166	23
8898·5	142	154 592	67 187	154 592	119	67	67 187	170	23
8899·5	142	154 517	67 154	154 517	119	63	67 154	167	23
8900·5	142	154 455	67 128	154 456	119	55	67 128	159	23
8901·5	− 142	+154 410	+ 67 108	−154 410	− 119	+ 44	− 67 108	− 148	− 23
8902·5	142	154 381	67 095	154 381	119	34	67 095	137	23
8903·5	142	154 367	67 089	154 367	119	27	67 089	130	23
8904·5	142	154 364	67 088	154 364	119	26	67 088	129	23
8905·5	142	154 366	67 089	154 366	119	32	67 088	136	23
8906·5	− 142	+154 367	+ 67 089	−154 367	− 119	+ 47	− 67 089	− 150	− 23
8907·5	142	154 359	67 085	154 359	119	69	67 085	173	23
8908·5	142	154 335	67 075	154 335	119	98	67 075	201	23
8909·5	142	154 292	67 056	154 292	119	128	67 056	232	22
8910·5	141	154 228	67 028	154 228	119	157	67 028	261	22
8911·5	− 141	+154 145	+ 66 992	−154 145	− 119	+ 181	− 66 992	− 284	− 22
8912·5	141	154 051	66 951	154 051	119	196	66 951	299	22
8913·5	141	153 954	66 909	153 954	119	200	66 909	303	22
8914·5	141	153 866	66 871	153 866	118	196	66 871	298	22
8915·5	141	153 796	66 841	153 796	118	185	66 841	287	22
8916·5	− 141	+153 750	+ 66 821	−153 750	− 118	+ 173	− 66 821	− 276	− 22
8917·5	140	153 727	66 811	153 728	118	167	66 811	269	22
8918·5	140	153 721	66 808	153 721	118	171	66 808	273	22
8919·5	140	153 717	66 806	153 717	118	188	66 806	290	22
8920·5	140	153 703	66 800	153 703	118	216	66 800	319	22
8921·5	− 140	+153 668	+ 66 785	−153 669	− 118	+ 252	− 66 785	− 355	− 22
8922·5	140	153 608	66 759	153 608	118	290	66 759	392	22
8923·5	140	153 526	66 723	153 527	118	322	66 723	425	22
8924·5	140	153 431	66 682	153 431	118	347	66 681	449	22
8925·5	140	153 332	66 639	153 332	118	361	66 638	463	22
8926·5	− 140	+153 238	+ 66 598	−153 238	− 117	+ 367	− 66 598	− 469	− 22
8927·5	139	153 156	66 563	153 156	117	365	66 562	467	22
8928·5	139	153 090	66 534	153 090	117	360	66 533	462	22
8929·5	139	153 041	66 513	153 041	117	355	66 512	457	22
8930·5	139	153 007	66 498	153 007	117	352	66 497	454	22
8931·5	− 139	+152 985	+ 66 488	−152 985	− 117	+ 354	− 66 488	− 456	− 22
8932·5	139	152 970	66 482	152 970	117	363	66 481	464	22
8933·5	139	152 955	66 475	152 955	117	379	66 474	481	22
8934·5	139	152 933	66 466	152 934	117	404	66 465	506	22
8935·5	139	152 898	66 450	152 898	117	435	66 450	537	22
8936·5	− 139	+152 843	+ 66 426	−152 843	− 117	+ 470	− 66 425	− 571	− 22
8937·5	139	152 765	66 393	152 766	117	503	66 392	605	22
8938·5	139	152 667	66 350	152 668	117	532	66 349	634	22
8939·5	138	152 555	66 301	152 555	116	552	66 300	654	22
8940·5	138	152 438	66 251	152 439	116	562	66 250	663	22
8941·5	− 138	+152 329	+ 66 203	−152 329	− 116	+ 561	− 66 202	− 662	− 22
8942·5	− 138	+152 236	+ 66 163	−152 236	− 116	+ 553	− 66 162	− 654	− 22

Values are in units of 10^{-8}.

ORIGIN AT SOLAR SYSTEM BARYCENTRE
MEAN EQUATOR AND EQUINOX J2000·0

Date 0^h TDB		X	Y	Z	$\dot{X}$	$\dot{Y}$	$\dot{Z}$
Nov.	16	+0·583 773 122	+0·737 650 690	+0·319 742 595	−1419 1620	+ 923 1422	+ 400 3067
	17	·569 492 675	·746 769 980	·323 696 979	1436 8620	900 6627	390 5467
	18	·555 037 214	·755 662 882	·327 553 060	1454 1623	877 8647	380 6461
	19	·540 410 812	·764 326 213	·331 309 437	1471 0478	854 7486	370 6062
	20	·525 617 696	·772 756 804	·334 964 727	1487 5022	831 3172	360 4292
	21	+0·510 662 262	+0·780 951 526	+0·338 517 576	−1503 5084	+ 807 5763	+ 350 1187
	22	·495 549 079	·788 907 326	·341 966 672	1519 0496	783 5346	339 6797
	23	·480 282 873	·796 621 255	·345 310 762	1534 1103	759 2041	329 1183
	24	·464 868 517	·804 090 493	·348 548 653	1548 6776	734 5990	318 4413
	25	·449 310 999	·811 312 373	·351 679 226	1562 7413	709 7353	307 6559
	26	+0·433 615 393	+0·818 284 391	+0·354 701 433	−1576 2944	+ 684 6293	+ 296 7694
	27	·417 786 827	·825 004 205	·357 614 299	1589 3326	659 2973	285 7886
	28	·401 830 465	·831 469 633	·360 416 911	1601 8537	633 7543	274 7196
	29	·385 751 480	·837 678 636	·363 108 417	1613 8570	608 0145	263 5682
	30	·369 555 051	·843 629 309	·365 688 017	1625 3426	582 0904	252 3393
Dec.	1	+0·353 246 352	+0·849 319 870	+0·368 154 960	−1636 3111	+ 555 9941	+ 241 0375
	2	·336 830 551	·854 748 654	·370 508 538	1646 7631	529 7366	229 6671
	3	·320 312 807	·859 914 100	·372 748 087	1656 6998	503 3285	218 2322
	4	·303 698 268	·864 814 756	·374 872 981	1666 1224	476 7801	206 7368
	5	·286 992 066	·869 449 268	·376 882 633	1675 0330	450 1015	195 1845
	6	+0·270 199 306	+0·873 816 384	+0·378 776 494	−1683 4343	+ 423 3022	+ 183 5790
	7	·253 325 066	·877 914 939	·380 554 046	1691 3300	396 3908	171 9232
	8	·236 374 377	·881 743 851	·382 214 799	1698 7247	369 3747	160 2196
	9	·219 352 225	·885 302 102	·383 758 284	1705 6233	342 2593	148 4699
	10	·202 263 548	·888 588 717	·385 184 045	1712 0305	315 0478	136 6747
	11	+0·185 113 242	+0·891 602 742	+0·386 491 627	−1717 9495	+ 287 7412	+ 124 8342
	12	·167 906 182	·894 343 220	·387 680 574	1723 3810	260 3384	112 9475
	13	·150 647 253	·896 809 182	·388 750 422	1728 3226	232 8374	101 0142
	14	·133 341 382	·898 999 635	·389 700 703	1732 7683	205 2365	89 0341
	15	·115 993 570	·900 913 579	·390 530 950	1736 7090	177 5357	77 0079
	16	+0·098 608 923	+0·902 550 023	+0·391 240 714	−1740 1335	+ 149 7372	+ 64 9378
	17	·081 192 662	·903 908 015	·391 829 570	1743 0295	121 8462	52 8270
	18	·063 750 136	·904 986 667	·392 297 135	1745 3846	93 8709	40 6804
	19	·046 286 811	·905 785 186	·392 643 077	1747 1875	65 8219	28 5035
	20	·028 808 262	·906 302 902	·392 867 126	1748 4279	37 7124	16 3030
	21	+0·011 320 154	+0·906 539 283	+0·392 969 081	−1749 0979	+ 9 5575	+ 4 0859
	22	− ·006 171 776	·906 493 955	·392 948 813	1749 1918	− 18 6266	− 8 1403
	23	·023 661 749	·906 166 712	·392 806 270	1748 7061	46 8228	20 3679
	24	·041 143 965	·905 557 516	·392 541 475	1747 6403	75 0140	32 5897
	25	·058 612 626	·904 666 504	·392 154 520	1745 9957	103 1835	44 7986
	26	−0·076 061 960	+0·903 493 972	+0·391 645 568	−1743 7756	− 131 3154	− 56 9881
	27	·093 486 235	·902 040 370	·391 014 842	1740 9847	159 3952	69 1525
	28	·110 879 769	·900 306 288	·390 262 621	1737 6284	187 4092	81 2863
	29	·128 236 939	·898 292 446	·389 389 234	1733 7128	215 3452	93 3848
	30	·145 552 182	·895 999 683	·388 395 057	1729 2442	243 1914	105 4435
	31	−0·162 820 001	+0·893 428 953	+0·387 280 510	−1724 2290	− 270 9370	− 117 4583
	32	−0·180 034 964	+0·890 581 313	+0·386 046 052	−1718 6742	− 298 5716	− 129 4251

$\dot{X}$, $\dot{Y}$, $\dot{Z}$ are in units of 10^{-9} au / d.

MATRIX ELEMENTS FOR CONVERSION FROM MEAN EQUINOX OF J2000·0 TO TRUE EQUINOX OF DATE

Julian Date	$R_{11}-1$	R_{12}	R_{13}	R_{21}	$R_{22}-1$	R_{23}	R_{31}	R_{32}	$R_{33}-1$
244									
8942·5	− 138	+152 236	+ 66 163	−152 236	− 116	+ 553	− 66 162	− 654	− 22
8943·5	138	152 166	66 133	152 167	116	543	66 132	644	22
8944·5	138	152 120	66 113	152 121	116	537	66 112	638	22
8945·5	137	152 091	66 100	152 092	116	539	66 099	640	22
8946·5	137	152 069	66 090	152 069	116	553	66 089	654	22
8947·5	− 137	+152 040	+ 66 078	−152 040	− 116	+ 578	− 66 077	− 679	− 22
8948·5	137	151 994	66 058	151 995	116	611	66 057	712	22
8949·5	137	151 925	66 028	151 926	115	648	66 027	748	22
8950·5	137	151 833	65 987	151 833	115	682	65 986	782	22
8951·5	137	151 723	65 940	151 723	115	708	65 939	809	22
8952·5	− 137	+151 605	+ 65 888	−151 605	− 115	+ 726	− 65 887	− 826	− 22
8953·5	136	151 488	65 838	151 488	115	733	65 837	833	22
8954·5	136	151 381	65 792	151 382	115	732	65 790	832	22
8955·5	136	151 290	65 752	151 290	114	726	65 751	825	22
8956·5	136	151 216	65 720	151 216	114	717	65 718	817	22
8957·5	− 136	+151 158	+ 65 695	−151 159	− 114	+ 710	− 65 693	− 810	− 22
8958·5	136	151 114	65 675	151 115	114	707	65 674	806	22
8959·5	136	151 079	65 660	151 080	114	710	65 659	809	22
8960·5	136	151 047	65 646	151 048	114	720	65 645	820	22
8961·5	136	151 011	65 631	151 012	114	738	65 629	837	22
8962·5	− 135	+150 964	+ 65 610	−150 965	− 114	+ 763	− 65 609	− 862	− 22
8963·5	135	150 900	65 582	150 900	114	792	65 581	891	22
8964·5	135	150 813	65 545	150 814	114	822	65 543	921	21
8965·5	135	150 705	65 498	150 706	114	848	65 496	947	21
8966·5	135	150 579	65 443	150 579	113	867	65 441	965	21
8967·5	− 135	+150 443	+ 65 384	−150 444	− 113	+ 874	− 65 383	− 972	− 21
8968·5	134	150 312	65 327	150 313	113	870	65 326	968	21
8969·5	134	150 197	65 277	150 197	113	857	65 276	955	21
8970·5	134	150 106	65 238	150 106	113	840	65 236	938	21
8971·5	134	150 040	65 209	150 041	113	825	65 208	923	21
8972·5	− 134	+149 995	+ 65 190	−149 996	− 113	+ 818	− 65 188	− 916	− 21
8973·5	134	149 960	65 174	149 961	112	822	65 173	920	21
8974·5	134	149 922	65 158	149 923	112	837	65 156	935	21
8975·5	134	149 870	65 135	149 871	112	861	65 134	958	21
8976·5	133	149 798	65 104	149 798	112	888	65 102	986	21
8977·5	− 133	+149 703	+ 65 063	−149 704	− 112	+ 915	− 65 061	−1012	− 21
8978·5	133	149 590	65 014	149 591	112	935	65 012	1033	21
8979·5	133	149 467	64 960	149 468	112	947	64 958	1044	21
8980·5	133	149 342	64 906	149 342	112	949	64 904	1046	21
8981·5	132	149 224	64 854	149 224	111	943	64 853	1039	21
8982·5	− 132	+149 119	+ 64 809	−149 120	− 111	+ 929	− 64 808	−1026	− 21
8983·5	132	149 033	64 771	149 033	111	912	64 770	1009	21
8984·5	132	148 964	64 742	148 965	111	896	64 740	992	21
8985·5	132	148 911	64 719	148 912	111	882	64 717	978	21
8986·5	132	148 869	64 700	148 870	111	874	64 699	970	21
8987·5	− 132	+148 833	+ 64 685	−148 834	− 111	+ 872	− 64 683	− 968	− 21
8988·5	− 132	+148 796	+ 64 669	−148 797	− 111	+ 878	− 64 667	− 974	− 21

Values are in units of 10^{-8}.

Reduction for polar motion

The rotation of the Earth is represented by a diurnal rotation around a reference axis whose motion with respect to the inertial reference frame is represented by the theories of precession and nutation. This reference axis does not coincide with the axis of figure (maximum moment of inertia) of the Earth, but moves slowly (in a terrestrial reference frame) in a quasi-circular path around it. The reference axis is the celestial ephemeris pole (normal to the true equator) and its motion with respect to the terrestrial reference frame is known as polar motion. The maximum amplitude of the polar motion is typically about $0''.3$ (corresponding to a displacement of about 9 m on the surface of the Earth) and the principal periods are about 365 and 428 days. The motion is affected by unpredictable geophysical forces and is determined from observations of stars, of radio sources and of appropriate satellites of the Earth.

The pole and zero (Greenwich) meridian of the terrestrial reference frame are defined implicitly by the adoption of a set of coordinates for the instruments that are used to determine UT and polar motion from astronomical observations. (The pole of this system is known as the conventional international origin.) The position of the terrestrial reference frame with respect to the true equator and equinox of date is defined by successive rotations through two small angles x, y and the Greenwich apparent sidereal time θ. The angles x, y correspond to the coordinates of the celestial ephemeris pole with respect to the terrestrial pole measured along the meridians at longitudes 0° and 270° (90° west). Current values of the coordinates of the pole for use in the reduction of observations are published by the International Earth Rotation Service. Previous values from 1970 January 1 onwards are given on page K10. An 80-year long series of values on a consistent basis has been published by the International Polar Motion Service. The coordinates x and y are usually measured in seconds of arc.

Polar motion causes variations in the zenith distance and azimuth of the celestial ephemeris pole and hence in the values of terrestrial latitude (ϕ) and longitude (λ) that are determined from direct astronomical observations of latitude and time. To first order, the departures from the mean values ϕ_m, λ_m are given by:

$$\Delta\phi = x \cos \lambda_m - y \sin \lambda_m \quad \text{and} \quad \Delta\lambda = (x \sin \lambda_m + y \cos \lambda_m) \tan \phi_m$$

The variation in longitude must be taken into account in the determination of GMST, and hence of UT, from observations.

The rigorous transformation of a vector $\mathbf{p}_3$ with respect to the celestial frame of the true equator and equinox of date to the corresponding vector $\mathbf{p}_4$ with respect to the terrestrial frame is given by the formula:

$$\mathbf{p}_4 = \mathbf{R}_2(-x)\,\mathbf{R}_1(-y)\,\mathbf{R}_3(\theta)\,\mathbf{p}_3$$

and conversely,

$$\mathbf{p}_3 = \mathbf{R}_3(-\theta)\,\mathbf{R}_1(y)\,\mathbf{R}_2(x)\,\mathbf{p}_4$$

where $\mathbf{R}_1(\alpha)$, $\mathbf{R}_2(\alpha)$, $\mathbf{R}_3(\alpha)$ are, respectively, the matrices:

$$\begin{bmatrix} 1 & 0 & 0 \\ 0 & \cos\alpha & \sin\alpha \\ 0 & -\sin\alpha & \cos\alpha \end{bmatrix} \qquad \begin{bmatrix} \cos\alpha & 0 & -\sin\alpha \\ 0 & 1 & 0 \\ \sin\alpha & 0 & \cos\alpha \end{bmatrix} \qquad \begin{bmatrix} \cos\alpha & \sin\alpha & 0 \\ -\sin\alpha & \cos\alpha & 0 \\ 0 & 0 & 1 \end{bmatrix}$$

corresponding to rotations α about the x, y and z axes. The vector $\mathbf{p}$ could represent, for example, the coordinates of a point on the Earth's surface or of a satellite in orbit around the Earth.

Reduction for diurnal parallax and diurnal aberration

The computation of diurnal parallax and aberration due to the displacement of the observer from the centre of the Earth requires a knowledge of the geocentric coordinates (ρ, geocentric distance in units of the Earth's equatorial radius, and ϕ', geocentric latitude, see page K5) of the place of observation and the local sidereal time (θ_0) of the observation (see page B6).

For bodies whose equatorial horizontal parallax (π) normally amounts to only a few seconds of arc the corrections for diurnal parallax in right ascension and declination (in the sense geocentric place *minus* topocentric place) are given by:

$$\Delta\alpha = \pi(\rho \cos\phi' \sin h \sec\delta)$$
$$\Delta\delta = \pi(\rho \sin\phi' \cos\delta - \rho\cos\phi' \cos h \sin\delta)$$

where h is the local hour angle ($\theta_0 - \alpha$) and π may be calculated from 8″·794 divided by the geocentric distance of the body (in au). For the Moon (and other very close bodies) more precise formulae are required (see page D3).

The corrections for diurnal aberration in right ascension and declination (in the sense apparent place *minus* mean place) are given by:

$$\Delta\alpha = 0^{s}\!\cdot\!0213\ \rho\cos\phi'\cos h\sec\delta \qquad \Delta\delta = 0''\!\cdot\!319\ \rho\cos\phi'\sin h\sin\delta$$

For a body at transit the local hour angle (h) is zero and so $\Delta\delta$ is zero, but

$$\Delta\alpha = \pm 0^{s}\!\cdot\!0213\ \rho\cos\phi'\sec\delta$$

where the plus and minus signs are used for the upper and lower transits, respectively; this may be regarded as a correction to the time of transit.

Alternatively, the effects may be computed in rectangular coordinates using the following expressions for the geocentric coordinates and velocity components of the observer with respect to the celestial equatorial reference frame:

$$\text{position:}\quad (a\rho\cos\phi'\cos\theta_0,\ a\rho\cos\phi'\sin\theta_0,\ a\rho\sin\phi')$$
$$\text{velocity:}\quad (-a\omega\rho\cos\phi'\sin\theta_0,\ a\omega\rho\cos\phi'\cos\theta_0,\ 0)$$

where θ_0 is the local sidereal time (mean or apparent as appropriate), a is the equatorial radius of the Earth and ω the angular velocity of the Earth.

$$\theta_0 = \text{Greenwich sidereal time} + \text{east longitude}$$
$$a\omega = 0\!\cdot\!464 \text{ km/s} = 0\!\cdot\!268 \times 10^{-3} \text{ au/d} \qquad c = 2\!\cdot\!998 \times 10^{5} \text{ km/s} = 173\!\cdot\!14 \text{ au/d}$$
$$a\omega/c = 1\!\cdot\!55 \times 10^{-6} \text{ rad} = 0''\!\cdot\!319 = 0^{s}\!\cdot\!0213$$

These geocentric position and velocity vectors of the observer are added to the barycentric position and velocity of the Earth's centre, respectively, to obtain the corresponding barycentric vectors of the observer.

Conversion to altitude and azimuth

It is convenient to use the local hour angle (h) as an intermediary in the conversion from the apparent right ascension (α) and declination (δ) to the azimuth (A) and altitude (a). The local apparent sidereal time (θ_0) corresponding to the UT of the observation must be determined first (see page B6). The formulae are:

$$\theta_0 = \text{GMST} + \lambda + \text{equation of equinoxes}$$
$$h = \theta_0 - \alpha$$
$$\cos a \sin A = -\cos\delta\sin h$$
$$\cos a\cos A = \sin\delta\cos\phi - \cos\delta\cos h\sin\phi$$
$$\sin a = \sin\delta\sin\phi + \cos\delta\cos h\cos\phi$$

where azimuth (A) is measured from the north through east in the plane of the horizon, altitude (a) is measured perpendicular to the horizon, and λ, ϕ are the astronomical values of the east longitude and latitude of the place of observation. The plane of

Conversion to altitude and azimuth (continued)

the horizon is defined to be perpendicular to the apparent direction of gravity. Zenith distance is given by $z = 90° - a$.

For most purposes the values of the geodetic longitude and latitude may be used but in some cases the effects of local gravity anomalies and polar motion must be included. For full precision, the values of α, δ must be corrected for diurnal parallax and diurnal aberration. The inverse formulae are:

$$\begin{aligned} \cos\delta \sin h &= -\cos a \sin A \\ \cos\delta \cos h &= \sin a \cos\phi - \cos a \cos A \sin\phi \\ \sin\delta &= \sin a \sin\phi + \cos a \cos A \cos\phi \end{aligned}$$

Correction for refraction

For most astronomical purposes the effect of refraction in the Earth's atmosphere is to decrease the zenith distance (computed by the formulae of the previous section) by an amount R that depends on the zenith distance and on the meteorological conditions at the site. A simple expression for R for zenith distances less than 75° (altitudes greater than 15°) is:

$$\begin{aligned} R &= 0°{\cdot}004\,52\, P \tan z \,/\, (273 + T) \\ &= 0°{\cdot}004\,52\, P \,/\, ((273 + T) \tan a) \end{aligned}$$

where T is the temperature (°C) and P is the barometric pressure (millibars). This formula is usually accurate to about 0′·1 for altitudes above 15°, but the error increases rapidly at lower altitudes, especially in abnormal meteorological conditions. For altitudes below 15° use the approximate formula:

$$R = P(0{\cdot}1594 + 0{\cdot}0196\, a + 0{\cdot}000\,02\, a^2) \,/\, [(273 + T)(1 + 0{\cdot}505\, a + 0{\cdot}0845\, a^2)]$$

where a is the apparent altitude in degrees.

DETERMINATION OF LATITUDE AND AZIMUTH

Use of the Polaris Table

The table on pages B64–B67 gives data for obtaining latitude from an observed altitude of Polaris (suitably corrected for instrumental errors and refraction) and the azimuth of this star (measured from north, positive to the east and negative to the west), for all hour angles and northern latitudes. The six tabulated quantities, each given to a precision of 0′·1, are a_0, a_1, a_2, referring to the correction to altitude, and b_0, b_1, b_2, to the azimuth.

$$\begin{aligned} \text{latitude} &= \text{corrected observed altitude} + a_0 + a_1 + a_2 \\ \text{azimuth} &= (b_0 + b_1 + b_2)/\cos\,(\text{latitude}) \end{aligned}$$

The table is to be entered with the local sidereal time of observation (LST), and gives the values of a_0, b_0 directly; interpolation, with maximum differences of 0′·7, can be done mentally. In the same vertical column, the values of a_1, b_1 are found with the latitude, and those of a_2, b_2 with the date, as argument. Thus all six quantities can, if desired, be extracted together. The errors due to the adoption of a mean value of the local sidereal time for each of the subsidiary tables have been reduced to a minimum, and the total error is not likely to exceed 0′·2. Interpolation between columns should not be attempted.

The observed altitude must be corrected for refraction before being used to determine the astronomical latitude of the place of observation. Both the latitude and the azimuth so obtained are affected by local gravity anomalies.

Pole Star formulae

The formulae below provide a method for obtaining latitude from the observed altitude of one of the pole stars, *Polaris* or σ Octantis, and an assumed *east* longitude of the observer λ. In addition, the azimuth of a pole star may be calculated from an assumed *east* longitude λ and the observed altitude a, or from λ and an assumed latitude ϕ. An error of $0°{\cdot}002$ in a or $0°{\cdot}1$ in λ will produce an error of about $0°{\cdot}002$ in the calculated latitude. Likewise an error of $0°{\cdot}03$ in λ, a or ϕ will produce an error of about $0°{\cdot}002$ in the calculated azimuth for latitudes below 70°.

Step 1. Calculate the Greenwich hour angle GHA and polar distance p, in degrees, from expressions of the form:

$$\mathrm{GHA} = a_0 + a_1 L + a_2 \sin L + a_3 \cos L + 15\,t$$
$$p = a_0 + a_1 L + a_2 \sin L + a_3 \cos L$$

where
$$L = 0°{\cdot}985\,65\,d$$
$$d = \text{day of year (from pages B2–B3)} + t/24$$

and where the coefficients a_0, a_1, a_2, a_3 are given in the table below, t is the universal time in hours, d is the interval in days from 1992 January 0 at 0^h UT to the time of observation, and the quantity L is in degrees. In the above formulae d is required to two decimals of a day, L to two decimals of a degree and t to three decimals of an hour.

Step 2. Calculate the local hour angle LHA from:

$$\mathrm{LHA} = \mathrm{GHA} + \lambda \quad \text{(add or subtract multiples of 360°)}$$

where λ is the assumed longitude measured east from the Greenwich meridian.

Form the quantities: $S = p \sin(\mathrm{LHA})$ $\qquad C = p \cos(\mathrm{LHA})$

Step 3. The latitude of the place of observation, in degrees, is given by:

$$\text{latitude} = a - C + 0{\cdot}0087\,S^2 \tan a$$

where a is the observed altitude of the pole star after correction for instrument error and atmospheric refraction.

Step 4. The azimuth of the pole star, in degrees, is given by:

$$\text{azimuth of } \textit{Polaris} = -S/\cos a$$
$$\text{azimuth of } \sigma \text{ Octantis} = 180° + S/\cos a$$

where azimuth is measured eastwards around the horizon from north.

In step 4, if a has not been observed, use the quantity:

$$a = \phi + C - 0{\cdot}0087\,S^2 \tan \phi$$

where ϕ is an assumed latitude, taken to be positive in either hemisphere.

POLE STAR COEFFICIENTS FOR 1992

	Polaris		σ Octantis	
	GHA	p	GHA	p
a_0	63°06	0°7695	143°45	1°0124
a_1	0·999 11	−0·00000 98	0·999 47	0·0000 118
a_2	0·35	−0·0021	0·18	0·0041
a_3	−0·20	−0·0050	0·26	−0·0035

POLARIS TABLE, 1992

LST	0^h a_0	0^h b_0	1^h a_0	1^h b_0	2^h a_0	2^h b_0	3^h a_0	3^h b_0	4^h a_0	4^h b_0	5^h a_0	5^h b_0
m	′	′	′	′	′	′	′	′	′	′	′	′
0	−37·1	+27·4	−42·9	+16·8	−45·7	+ 4·9	−45·4	− 7·3	−42·0	−18·9	−35·6	−29·3
3	37·4	26·9	43·1	16·2	45·8	4·3	45·3	7·9	41·7	19·5	35·2	29·7
6	37·8	26·4	43·3	15·6	45·9	3·7	45·2	8·5	41·5	20·1	34·8	30·2
9	38·1	25·9	43·5	15·0	45·9	3·1	45·1	9·1	41·2	20·6	34·5	30·7
12	38·4	25·4	43·7	14·5	45·9	2·5	45·0	9·7	40·9	21·1	34·0	31·1
15	−38·8	+24·9	−43·9	+13·9	−46·0	+ 1·9	−44·9	−10·3	−40·6	−21·7	−33·6	−31·6
18	39·1	24·4	44·1	13·3	46·0	1·3	44·7	10·9	40·4	22·2	33·2	32·0
21	39·4	23·9	44·2	12·7	46·0	+ 0·7	44·6	11·5	40·1	22·8	32·8	32·4
24	39·7	23·4	44·4	12·1	46·0	0·0	44·4	12·1	39·8	23·3	32·4	32·9
27	40·0	22·8	44·6	11·5	46·0	− 0·6	44·3	12·6	39·5	23·8	31·9	33·3
30	−40·3	+22·3	−44·7	+10·9	−46·0	− 1·2	−44·1	−13·2	−39·1	−24·3	−31·5	−33·7
33	40·6	21·8	44·8	10·3	46·0	1·8	43·9	13·8	38·8	24·8	31·1	34·1
36	40·9	21·2	45·0	9·8	45·9	2·4	43·7	14·4	38·5	25·4	30·6	34·5
39	41·2	20·7	45·1	9·2	45·9	3·0	43·5	15·0	38·2	25·9	30·2	34·9
42	41·4	20·1	45·2	8·6	45·9	3·6	43·3	15·5	37·8	26·4	29·7	35·3
45	−41·7	+19·6	−45·3	+ 8·0	−45·8	− 4·2	−43·1	−16·1	−37·5	−26·9	−29·2	−35·7
48	41·9	19·0	45·4	7·3	45·7	4·8	42·9	16·7	37·1	27·4	28·8	36·1
51	42·2	18·5	45·5	6·7	45·7	5·5	42·7	17·3	36·8	27·8	28·3	36·5
54	42·4	17·9	45·6	6·1	45·6	6·1	42·5	17·8	36·4	28·3	27·8	36·8
57	42·7	17·3	45·7	5·5	45·5	6·7	42·2	18·4	36·0	28·8	27·3	37·2
60	−42·9	+16·8	−45·7	+ 4·9	−45·4	− 7·3	−42·0	−18·9	−35·6	−29·3	−26·8	−37·5

Lat.	a_1	b_1	a_1	b_1	a_1	b_1	a_1	b_1	a_1	b_1	a_1	b_1
°												
0	−·1	−·3	·0	−·2	·0	·0	·0	+·2	−·1	+·3	−·2	+·4
10	−·1	−·3	·0	−·1	·0	·0	·0	+·2	−·1	+·3	−·2	+·3
20	−·1	−·2	·0	−·1	·0	·0	·0	+·1	−·1	+·2	−·1	+·3
30	·0	−·2	·0	−·1	·0	·0	·0	+·1	−·1	+·2	−·1	+·2
40	·0	−·1	·0	·0	·0	·0	·0	+·1	·0	+·1	−·1	+·1
45	·0	·0	·0	·0	·0	·0	·0	·0	·0	+·1	·0	+·1
50	·0	·0	·0	·0	·0	·0	·0	·0	·0	·0	·0	·0
55	·0	+·1	·0	·0	·0	·0	·0	·0	·0	−·1	·0	−·1
60	·0	+·1	·0	+·1	·0	·0	·0	−·1	·0	−·1	+·1	−·2
62	·0	+·2	·0	+·1	·0	·0	·0	−·1	+·1	−·2	+·1	−·2
64	+·1	+·2	·0	+·1	·0	·0	·0	−·1	+·1	−·2	+·1	−·3
66	+·1	+·3	·0	+·1	·0	·0	·0	−·2	+·1	−·3	+·2	−·3

Month	a_2	b_2	a_2	b_2	a_2	b_2	a_2	b_2	a_2	b_2	a_2	b_2
Jan.	+·1	−·1	+·1	−·1	+·1	·0	+·2	·0	+·2	·0	+·1	+·1
Feb.	·0	−·2	+·1	−·2	+·2	−·2	+·2	−·1	+·2	−·1	+·3	·0
Mar.	−·1	−·3	·0	−·3	+·1	−·3	+·2	−·3	+·2	−·2	+·3	−·2
Apr.	−·2	−·3	−·2	−·4	−·1	−·4	·0	−·4	+·1	−·4	+·2	−·3
May	−·4	−·2	−·3	−·3	−·2	−·3	−·1	−·4	·0	−·4	+·1	−·4
June	−·4	·0	−·4	−·1	−·3	−·2	−·2	−·3	−·2	−·4	−·1	−·4
July	−·4	+·1	−·4	·0	−·4	−·1	−·3	−·2	−·3	−·2	−·2	−·3
Aug.	−·2	+·2	−·3	+·2	−·3	+·1	−·3	·0	−·3	−·1	−·3	−·2
Sept.	−·1	+·3	−·1	+·3	−·2	+·2	−·3	+·2	−·3	+·1	−·3	·0
Oct.	+·1	+·3	·0	+·4	−·1	+·4	−·1	+·3	−·2	+·3	−·3	+·2
Nov.	+·3	+·3	+·2	+·3	+·1	+·4	·0	+·4	−·1	+·4	−·2	+·4
Dec.	+·4	+·1	+·4	+·2	+·3	+·3	+·2	+·4	+·1	+·4	·0	+·4

Latitude = Corrected observed altitude of *Polaris* + $a_0 + a_1 + a_2$
Azimuth of *Polaris* = $(b_0 + b_1 + b_2)$ / cos (latitude)

LST	6h		7h		8h		9h		10h		11h	
	a_0	b_0	a_0	b_0	a_0	b_0	a_0	b_0	a_0	b_0	a_0	b_0
m	′	′	′	′	′	′	′	′	′	′	′	′
0	−26·8	−37·5	−16·2	−43·2	− 4·5	−45·8	+ 7·5	−45·3	+19·0	−41·8	+29·1	−35·4
3	26·3	37·9	15·6	43·4	3·9	45·9	8·1	45·2	19·5	41·5	29·6	35·0
6	25·8	38·2	15·1	43·6	3·3	45·9	8·7	45·1	20·1	41·3	30·1	34·6
9	25·3	38·6	14·5	43·8	2·7	45·9	9·3	45·0	20·6	41·0	30·5	34·2
12	24·8	38·9	13·9	44·0	2·1	46·0	9·9	44·9	21·1	40·7	31·0	33·8
15	−24·3	−39·2	−13·3	−44·1	− 1·5	−46·0	+10·5	−44·7	+21·7	−40·4	+31·4	−33·4
18	23·8	39·5	12·8	44·3	0·9	46·0	11·0	44·6	22·2	40·1	31·8	33·0
21	23·3	39·8	12·2	44·5	− 0·3	46·0	11·6	44·4	22·7	39·8	32·3	32·6
24	22·8	40·1	11·6	44·6	+ 0·3	46·0	12·2	44·3	23·2	39·5	32·7	32·2
27	22·2	40·4	11·0	44·8	0·9	46·0	12·8	44·1	23·8	39·2	33·1	31·8
30	−21·7	−40·7	−10·4	−44·9	+ 1·5	−46·0	+13·4	−43·9	+24·3	−38·9	+33·5	−31·3
33	21·2	41·0	9·8	45·0	2·1	45·9	13·9	43·7	24·8	38·6	33·9	30·9
36	20·6	41·3	9·3	45·1	2·7	45·9	14·5	43·5	25·3	38·3	34·3	30·4
39	20·1	41·5	8·7	45·2	3·3	45·9	15·1	43·3	25·8	37·9	34·7	30·0
42	19·5	41·8	8·1	45·4	3·9	45·8	15·6	43·1	26·3	37·6	35·1	29·5
45	−19·0	−42·0	− 7·5	−45·5	+ 4·5	−45·7	+16·2	−42·9	+26·8	−37·2	+35·5	−29·1
48	18·4	42·3	6·9	45·5	5·1	45·7	16·8	42·7	27·2	36·9	35·9	28·6
51	17·9	42·5	6·3	45·6	5·7	45·6	17·3	42·5	27·7	36·5	36·2	28·2
54	17·3	42·7	5·7	45·7	6·3	45·5	17·9	42·3	28·2	36·2	36·6	27·7
57	16·8	43·0	5·1	45·8	6·9	45·4	18·4	42·0	28·7	35·8	37·0	27·2
60	−16·2	−43·2	− 4·5	−45·8	+ 7·5	−45·3	+19·0	−41·8	+29·1	−35·4	+37·3	−26·7

Lat.	a_1	b_1	a_1	b_1	a_1	b_1	a_1	b_1	a_1	b_1	a_1	b_1
°												
0	−·3	+·3	−·3	+·2	−·4	·0	−·3	−·2	−·3	−·3	−·2	−·4
10	−·2	+·3	−·3	+·1	−·3	·0	−·3	−·2	−·2	−·3	−·1	−·3
20	−·2	+·2	−·2	+·1	−·3	·0	−·2	−·1	−·2	−·2	−·1	−·3
30	−·1	+·2	−·2	+·1	−·2	·0	−·2	−·1	−·1	−·2	−·1	−·2
40	−·1	+·1	−·1	·0	−·1	·0	−·1	−·1	−·1	−·1	−·1	−·1
45	·0	·0	−·1	·0	−·1	·0	−·1	·0	·0	−·1	·0	−·1
50	·0	·0	·0	·0	·0	·0	·0	·0	·0	·0	·0	·0
55	+·1	−·1	+·1	·0	+·1	·0	+·1	·0	+·1	+·1	·0	+·1
60	+·1	−·1	+·2	−·1	+·2	·0	+·2	+·1	+·1	+·1	+·1	+·2
62	+·2	−·2	+·2	−·1	+·2	·0	+·2	+·1	+·2	+·2	+·1	+·2
64	+·2	−·2	+·2	−·1	+·3	·0	+·2	+·1	+·2	+·2	+·1	+·3
66	+·3	−·3	+·3	−·1	+·3	·0	+·3	+·2	+·2	+·3	+·2	+·3

Month	a_2	b_2	a_2	b_2	a_2	b_2	a_2	b_2	a_2	b_2	a_2	b_2
Jan.	+·1	+·1	+·1	+·1	·0	+·1	·0	+·2	·0	+·2	−·1	+·1
Feb.	+·2	·0	+·2	+·1	+·2	+·2	+·1	+·2	+·1	+·2	·0	+·3
Mar.	+·3	−·1	+·3	·0	+·3	+·1	+·3	+·2	+·2	+·2	+·2	+·3
Apr.	+·3	−·2	+·4	−·2	+·4	−·1	+·4	·0	+·4	+·1	+·3	+·2
May	+·2	−·4	+·3	−·3	+·3	−·2	+·4	−·1	+·4	·0	+·4	+·1
June	·0	−·4	+·1	−·4	+·2	−·3	+·3	−·2	+·4	−·2	+·4	−·1
July	−·1	−·4	·0	−·4	+·1	−·4	+·2	−·3	+·2	−·3	+·3	−·2
Aug.	−·2	−·2	−·2	−·3	−·1	−·3	·0	−·3	+·1	−·3	+·2	−·3
Sept.	−·3	−·1	−·3	−·1	−·2	−·2	−·2	−·3	−·1	−·3	·0	−·3
Oct.	−·3	+·1	−·4	·0	−·4	−·1	−·3	−·1	−·3	−·2	−·2	−·3
Nov.	−·3	+·3	−·3	+·2	−·4	+·1	−·4	·0	−·4	−·1	−·4	−·2
Dec.	−·1	+·4	−·2	+·4	−·3	+·3	−·4	+·2	−·4	+·1	−·4	·0

Latitude = Corrected observed altitude of *Polaris* + a_0 + a_1 + a_2

Azimuth of *Polaris* = (b_0 + b_1 + b_2) / cos (latitude)

LST	12h		13h		14h		15h		16h		17h	
	a_0	b_0	a_0	b_0	a_0	b_0	a_0	b_0	a_0	b_0	a_0	b_0
m	′	′	′	′	′	′	′	′	′	′	′	′
0	+37·3	−26·7	+43·0	−16·3	+45·7	− 4·8	+45·4	+ 7·0	+42·1	+18·4	+35·9	+28·6
3	37·7	26·2	43·2	15·7	45·8	4·2	45·4	7·6	41·9	18·9	35·5	29·0
6	38·0	25·7	43·4	15·2	45·9	3·6	45·2	8·2	41·6	19·5	35·2	29·5
9	38·3	25·2	43·6	14·6	45·9	3·0	45·1	8·8	41·3	20·0	34·8	29·9
12	38·7	24·8	43·8	14·0	45·9	2·4	45·0	9·4	41·1	20·6	34·4	30·4
15	+39·0	−24·2	+44·0	−13·5	+46·0	− 1·8	+44·9	+10·0	+40·8	+21·1	+34·0	+30·8
18	39·3	23·7	44·1	12·9	46·0	1·2	44·8	10·5	40·5	21·6	33·6	31·3
21	39·6	23·2	44·3	12·3	46·0	− 0·6	44·6	11·1	40·2	22·1	33·2	31·7
24	39·9	22·7	44·4	11·8	46·0	0·0	44·5	11·7	39·9	22·6	32·7	32·1
27	40·2	22·2	44·6	11·2	46·0	+ 0·6	44·3	12·3	39·6	23·2	32·3	32·6
30	+40·5	−21·7	+44·7	−10·6	+46·0	+ 1·1	+44·1	+12·8	+39·3	+23·7	+31·9	+33·0
33	40·8	21·2	44·9	10·0	46·0	1·7	44·0	13·4	39·0	24·2	31·5	33·4
36	41·0	20·6	45·0	9·5	45·9	2·3	43·8	14·0	38·7	24·7	31·0	33·8
39	41·3	20·1	45·1	8·9	45·9	2·9	43·6	14·5	38·4	25·2	30·6	34·2
42	41·6	19·6	45·2	8·3	45·9	3·5	43·4	15·1	38·0	25·7	30·1	34·6
45	+41·8	−19·0	+45·3	− 7·7	+45·8	+ 4·1	+43·2	+15·6	+37·7	+26·2	+29·7	+35·0
48	42·1	18·5	45·4	7·1	45·8	4·7	43·0	16·2	37·4	26·7	29·2	35·4
51	42·3	17·9	45·5	6·5	45·7	5·3	42·8	16·8	37·0	27·1	28·7	35·7
54	42·5	17·4	45·6	5·9	45·6	5·9	42·6	17·3	36·7	27·6	28·3	36·1
57	42·8	16·8	45·7	5·4	45·5	6·5	42·3	17·9	36·3	28·1	27·8	36·5
60	+43·0	−16·3	+45·7	− 4·8	+45·4	+ 7·0	+42·1	+18·4	+35·9	+28·6	+27·3	+36·8
Lat.	a_1	b_1	a_1	b_1	a_1	b_1	a_1	b_1	a_1	b_1	a_1	b_1
°												
0	−·1	−·3	·0	−·2	·0	·0	·0	+·2	−·1	+·3	−·2	+·4
10	−·1	−·3	·0	−·1	·0	·0	·0	+·2	−·1	+·3	−·2	+·3
20	−·1	−·2	·0	−·1	·0	·0	·0	+·1	−·1	+·2	−·1	+·3
30	·0	−·2	·0	−·1	·0	·0	·0	+·1	−·1	+·2	−·1	+·2
40	·0	−·1	·0	·0	·0	·0	·0	+·1	·0	+·1	−·1	+·1
45	·0	·0	·0	·0	·0	·0	·0	·0	·0	+·1	·0	+·1
50	·0	·0	·0	·0	·0	·0	·0	·0	·0	·0	·0	·0
55	·0	+·1	·0	·0	·0	·0	·0	·0	·0	−·1	·0	−·1
60	·0	+·1	·0	+·1	·0	·0	·0	−·1	·0	−·1	+·1	−·2
62	·0	+·2	·0	+·1	·0	·0	·0	−·1	+·1	−·2	+·1	−·2
64	+·1	+·2	·0	+·1	·0	·0	·0	−·1	+·1	−·2	+·1	−·3
66	+·1	+·3	·0	+·1	·0	·0	·0	−·2	+·1	−·3	+·2	−·3
Month	a_2	b_2	a_2	b_2	a_2	b_2	a_2	b_2	a_2	b_2	a_2	b_2
Jan.	−·1	+·1	−·1	+·1	−·1	·0	−·2	·0	−·2	·0	−·1	−·1
Feb.	·0	+·2	−·1	+·2	−·2	+·2	−·2	+·1	−·2	+·1	−·3	·0
Mar.	+·1	+·3	·0	+·3	−·1	+·3	−·2	+·3	−·2	+·2	−·3	+·2
Apr.	+·2	+·3	+·2	+·4	+·1	+·4	·0	+·4	−·1	+·4	−·2	+·3
May	+·4	+·2	+·3	+·3	+·2	+·3	+·1	+·4	·0	+·4	−·1	+·4
June	+·4	·0	+·4	+·1	+·3	+·2	+·2	+·3	+·2	+·4	+·1	+·4
July	+·4	−·1	+·4	·0	+·4	+·1	+·3	+·2	+·3	+·2	+·2	+·3
Aug.	+·2	−·2	+·3	−·2	+·3	−·1	+·3	·0	+·3	+·1	+·3	+·2
Sept.	+·1	−·3	+·1	−·3	+·2	−·2	+·3	−·2	+·3	−·1	+·3	·0
Oct.	−·1	−·3	·0	−·4	+·1	−·4	+·1	−·3	+·2	−·3	+·3	−·2
Nov.	−·3	−·3	−·2	−·3	−·1	−·4	·0	−·4	+·1	−·4	+·2	−·4
Dec.	−·4	−·1	−·4	−·2	−·3	−·3	−·2	−·4	−·1	−·4	·0	−·4

Latitude = Corrected observed altitude of *Polaris* + a_0 + a_1 + a_2

Azimuth of *Polaris* = (b_0 + b_1 + b_2) / cos (latitude)

LST	18h a_0	18h b_0	19h a_0	19h b_0	20h a_0	20h b_0	21h a_0	21h b_0	22h a_0	22h b_0	23h a_0	23h b_0
m	′	′	′	′	′	′	′	′	′	′	′	′
0	+27·3	+36·8	+16·8	+42·7	+ 5·2	+45·7	− 6·8	+45·6	−18·4	+42·3	−28·7	+36·1
3	26·8	37·2	16·3	42·9	4·6	45·7	7·4	45·5	18·9	42·1	29·2	35·8
6	26·3	37·5	15·7	43·1	4·0	45·8	8·0	45·4	19·5	41·8	29·6	35·4
9	25·8	37·9	15·2	43·3	3·4	45·9	8·6	45·3	20·0	41·6	30·1	35·0
12	25·3	38·2	14·6	43·5	2·8	45·9	9·2	45·2	20·6	41·3	30·5	34·6
15	+24·8	+38·6	+14·0	+43·7	+ 2·2	+45·9	− 9·8	+45·0	−21·1	+41·0	−31·0	+34·2
18	24·3	38·9	13·4	43·9	1·6	46·0	10·4	44·9	21·6	40·8	31·4	33·8
21	23·8	39·2	12·9	44·1	1·0	46·0	10·9	44·8	22·2	40·5	31·9	33·3
24	23·3	39·5	12·3	44·2	+ 0·4	46·0	11·5	44·6	22·7	40·2	32·3	32·9
27	22·8	39·8	11·7	44·4	− 0·2	46·0	12·1	44·5	23·2	39·9	32·7	32·5
30	+22·3	+40·1	+11·1	+44·6	− 0·8	+46·0	−12·7	+44·3	−23·7	+39·6	−33·2	+32·1
33	21·7	40·4	10·5	44·7	1·4	46·0	13·3	44·1	24·3	39·3	33·6	31·6
36	21·2	40·7	10·0	44·8	2·0	46·0	13·8	44·0	24·8	38·9	34·0	31·2
39	20·7	40·9	9·4	45·0	2·6	46·0	14·4	43·8	25·3	38·6	34·4	30·7
42	20·1	41·2	8·8	45·1	3·2	45·9	15·0	43·6	25·8	38·3	34·8	30·3
45	+19·6	+41·5	+ 8·2	+45·2	− 3·8	+45·9	−15·6	+43·4	−26·3	+37·9	−35·2	+29·8
48	19·1	41·7	7·6	45·3	4·4	45·8	16·1	43·2	26·8	37·6	35·6	29·3
51	18·5	42·0	7·0	45·4	5·0	45·8	16·7	43·0	27·3	37·2	36·0	28·9
54	18·0	42·2	6·4	45·5	5·6	45·7	17·3	42·8	27·7	36·9	36·3	28·4
57	17·4	42·5	5·8	45·6	6·2	45·6	17·8	42·5	28·2	36·5	36·7	27·9
60	+16·8	+42·7	+ 5·2	+45·7	− 6·8	+45·6	−18·4	+42·3	−28·7	+36·1	−37·1	+27·4
Lat.	a_1	b_1	a_1	b_1	a_1	b_1	a_1	b_1	a_1	b_1	a_1	b_1
°												
0	−·3	+·3	−·3	+·2	−·4	·0	−·3	−·2	−·3	−·3	−·2	−·4
10	−·2	+·3	−·3	+·1	−·3	·0	−·3	−·2	−·2	−·3	−·1	−·3
20	−·2	+·2	−·2	+·1	−·3	·0	−·2	−·1	−·2	−·2	−·1	−·3
30	−·1	+·2	−·2	+·1	−·2	·0	−·2	−·1	−·1	−·2	−·1	−·2
40	−·1	+·1	−·1	·0	−·1	·0	−·1	−·1	−·1	−·1	−·1	−·1
45	·0	·0	−·1	·0	−·1	·0	−·1	·0	·0	−·1	·0	−·1
50	·0	·0	·0	·0	·0	·0	·0	·0	·0	·0	·0	·0
55	+·1	−·1	+·1	·0	+·1	·0	+·1	·0	+·1	+·1	·0	+·1
60	+·1	−·1	+·2	−·1	+·2	·0	+·2	+·1	+·1	+·1	+·1	+·2
62	+·2	−·2	+·2	−·1	+·2	·0	+·2	+·1	+·2	+·2	+·1	+·2
64	+·2	−·2	+·2	−·1	+·3	·0	+·2	+·1	+·2	+·2	+·1	+·3
66	+·3	−·3	+·3	−·1	+·3	·0	+·3	+·2	+·2	+·3	+·2	+·3
Month	a_2	b_2	a_2	b_2	a_2	b_2	a_2	b_2	a_2	b_2	a_2	b_2
Jan.	−·1	−·1	−·1	−·1	·0	−·1	·0	−·2	·0	−·2	+·1	−·1
Feb.	−·2	·0	−·2	−·1	−·2	−·2	−·1	−·2	−·1	−·2	·0	−·3
Mar.	−·3	+·1	−·3	·0	−·3	−·1	−·3	−·2	−·2	−·2	−·2	−·3
Apr.	−·3	+·2	−·4	+·2	−·4	+·1	−·4	·0	−·4	−·1	−·3	−·2
May	−·2	+·4	−·3	+·3	−·3	+·2	−·4	+·1	−·4	·0	−·4	−·1
June	·0	+·4	−·1	+·4	−·2	+·3	−·3	+·2	−·4	+·2	−·4	+·1
July	+·1	+·4	·0	+·4	−·1	+·4	−·2	+·3	−·2	+·3	−·3	+·2
Aug.	+·2	+·2	+·2	+·3	+·1	+·3	·0	+·3	−·1	+·3	−·2	+·3
Sept.	+·3	+·1	+·3	+·1	+·2	+·2	+·2	+·3	+·1	+·3	·0	+·3
Oct.	+·3	−·1	+·4	·0	+·4	+·1	+·3	+·1	+·3	+·2	+·2	+·3
Nov.	+·3	−·3	+·3	−·2	+·4	−·1	+·4	·0	+·4	+·1	+·4	+·2
Dec.	+·1	−·4	+·2	−·4	+·3	−·3	+·4	−·2	+·4	−·1	+·4	·0

Latitude = Corrected observed altitude of *Polaris* + a_0 + a_1 + a_2

Azimuth of *Polaris* = $(b_0 + b_1 + b_2)$ / cos (latitude)

CONTENTS OF SECTION C

NOTES AND FORMULAS

Mean orbital elements of the Sun

Mean elements of the orbit of the Sun, referred to the mean equinox and ecliptic of date, are given by the following expressions. The time argument d is the interval in days from 1992 January 0, 0^h TDT. These expressions are intended for use only during the year of this volume.

d = JD − 244 8621.5 = day of year (from B2–B3) + fraction of day from 0^h TDT.

Geometric mean longitude: $278°.926\,017 + 0.985\,647\,36\, d$

Mean longitude of perigee: $282°.800\,722 + 0.000\,047\,07\, d$

Mean anomaly: $356°.125\,295 + 0.985\,600\,28\, d$

Eccentricity: $0.016\,711\,70 - 0.000\,000\,0012\, d$

Mean obliquity of the ecliptic with respect to the mean equator of date:

$$23°.440\,332 - 0.000\,000\,36\, d$$

The position of the ecliptic of date with respect to the ecliptic of the standard epoch is given by formulas on page B18.

Accurate osculating elements of the Earth/Moon barycenter are given on pages E3 and E4.

Lengths of principal years

The lengths of the principal years at 1992.0 as derived from the Sun's mean motion are:

		d	d h m s
tropical year	(equinox to equinox)	365.242 190	365 05 48 45.3
sidereal year	(fixed star to fixed star)	365.256 363	365 06 09 09.8
anomalistic year	(perigee to perigee)	365.259 635	365 06 13 52.5
eclipse year	(node to node)	346.620 073	346 14 52 54.3

NOTES AND FORMULAS

Apparent ecliptic coordinates of the Sun

The apparent longitude may be computed from the geometric longitude tabulated on pages C4–C18 using:

$$\text{apparent longitude} = \text{tabulated longitude} + \text{nutation in longitude } (\Delta\psi) - 20''.496/R$$

where $\Delta\psi$ is tabulated on pages B24–B31 and R is the true distance; the tabulated longitude is the geometric longitude with respect to the mean equinox of date. The apparent latitude is equal to the geometric latitude to the precision of tabulation.

Time of transit of the Sun

The quantity tabulated as "Ephemeris Transit" on pages C5–C19 is the TDT of transit of the Sun over the ephemeris meridian, which is at the longitude 1.002 738 ΔT east of the prime (Greenwich) meridian; in this expression ΔT is the difference TDT – UT. The TDT of transit of the Sun over a local meridian is obtained by interpolation where the first differences are about 24 hours. The interpolation factor p is given by:

$$p = -\lambda + 1.002\,738\,\Delta T$$

where λ is the east longitude and the right-hand side is expressed in days. (Divide longitude in degrees by 360 and ΔT in seconds by 86 400). During 1992 it is expected that ΔT will be about 58 seconds, so that the second term is about +0.000 66 days.

The UT of transit is obtained by subtracting ΔT from the TDT of transit obtained by interpolation.

Equation of time

The equation of time is defined so that:

local mean solar time = local apparent solar – equation of time.

To obtain the equation of time to a precision of about 1 second it is sufficient to use:

equation of time at 12^h UT = 12^h – tabulated value of TDT of ephemeris transit.

Alternatively it may be calculated for any instant during 1992 in seconds of time to a precision of about 3 seconds directly from the expression:

$$\begin{aligned}\text{equation of time} = &-106.2 \sin L + 596.2 \sin 2L + 4.4 \sin 3L - 12.7 \sin 4L \\ &- 428.9 \cos L - 2.1 \cos 2L + 19.3 \cos 3L\end{aligned}$$

where L is the mean longitude of the Sun, given by:

$$L = 278^\circ.926 + 0.985\,647\,d$$

and where d is the interval in days from 1992 January 0 at 0^h UT, given by:

$$d = \text{day of year (from B2–B3)} + \text{fraction of day from } 0^h \text{ UT}$$

Geocentric rectangular coordinates of the Sun

The geocentric equatorial rectangular coordinates of the Sun are given, in au, on pages C20–C23 and are referred to the mean equator and equinox of J2000.0. The x-axis is directed towards the equinox, the y-axis towards the point on the equator at right ascension 6^h, and the z-axis towards the north pole of the equator.

These geocentric rectangular coordinates (x, y, z) may be used to convert an object's heliocentric rectangular coordinates (x_0, y_0, z_0) to the corresponding geometric geocentric rectangular coordinates (ξ_0, η_0, ζ_0) by means of the formulas:

$$\xi_0 = x_0 + x \qquad \eta_0 = y_0 + y \qquad \zeta_0 = z_0 + z$$

See pages B36–B39 for a rigorous method of forming an apparent place of an object in the solar system.

NOTES AND FORMULAS

Elements of the rotation of Sun

The mean elements of the rotation of the Sun during 1992 are given by:

Longitude of the ascending node of the solar equator:

on the ecliptic of date, 75°.65 on the mean equator of date, 16°.11

Inclination of the solar equator:

on the ecliptic of date, 7°.25 on the mean equator of date, 26°.14

The mean position of the pole of the solar equator is at:

right ascension, 286°.11 declination, 63°.86

Sidereal period of rotation of the prime meridian is 25.38 days.

Mean synodic period of rotation of the prime meridian is 27.2753 days.

These data are derived from elements given by R. C. Carrington (*Observations of the Spots on the Sun*, p. 244, 1863).

Heliographic coordinates

The values of P (position angle of the northern extremity of the axis of rotation, measured eastwards from the north point of the disk), B_0 and L_0 (the heliographic latitude and longitude of the central point of the disk) are for 0^h UT; they may be interpolated linearly. The horizontal parallax and semidiameter are given for 0^h TDT, but may be regarded as being for 0^h UT.

If ρ_1, θ are the observed angular distance and position angle of a sunspot from the center of the disk of the Sun as seen from the Earth, and ρ is the heliocentric angular distance of the spot on the solar surface from the center of the Sun's disk, then

$$\sin(\rho + \rho_1) = \rho_1 / S$$

where S is the semidiameter of the Sun. The position angle is measured from the north point of the disk towards the east.

The formulas for the computation of the heliographic coordinates (L, B) of a sunspot (or other feature on the surface of the Sun) from (ρ, θ) are as follows:

$$\begin{aligned}
\sin B &= \sin B_0 \cos \rho + \cos B_0 \sin \rho \cos (P - \theta) \\
\cos B \sin (L - L_0) &= \sin \rho \sin (P - \theta) \\
\cos B \cos (L - L_0) &= \cos \rho \cos B_0 - \sin B_0 \sin \rho \cos (P - \theta)
\end{aligned}$$

where B is measured positive to the north of the solar equator and L is measured from 0° to 360° in the direction of rotation of the Sun, i.e., westwards on the apparent disk as seen from the Earth.

SYNODIC ROTATION NUMBERS, 1992

Number	Date of Commencement	Number	Date of Commencement	Number	Date of Commencement
1850	1991 Dec. 8.84	1855	1992 Apr. 23.43	1860	1992 Sept. 6.53
1851	1992 Jan. 5.17	1856	May 20.67	1861	Oct. 3.80
1852	Feb. 1.51	1857	June 16.87	1862	Oct. 31.09
1853	Feb. 28.85	1858	July 14.07	1863	Nov. 27.40
1854	Mar. 27.16	1859	Aug. 10.28	1864	Dec. 24.72

At the date of commencement of each synodic rotation period the value of L_0 is zero; that is, the prime meridian passes through the central point of the disk.

FOR 0^h DYNAMICAL TIME

Date		Julian Date	Ecliptic Long. (for Mean Equinox of Date)	Ecliptic Lat. (for Mean Equinox of Date)	Apparent Right Ascension	Apparent Declination	True Geocentric Distance
		244	° ′ ″	″	h m s	° ′ ″	
Jan.	0	8621.5	278 47 30.03	−0.59	18 38 16.16	−23 08 57.5	0.983 3503
	1	8622.5	279 48 40.05	−0.51	18 42 41.54	23 04 40.8	.983 3377
	2	8623.5	280 49 50.37	−0.41	18 47 06.64	22 59 56.5	.983 3292
	3	8624.5	281 51 00.91	−0.30	18 51 31.44	22 54 44.6	.983 3247
	4	8625.5	282 52 11.59	−0.18	18 55 55.89	22 49 05.2	.983 3242
	5	8626.5	283 53 22.31	−0.05	19 00 19.97	−22 42 58.7	0.983 3276
	6	8627.5	284 54 32.99	+0.07	19 04 43.63	22 36 25.1	.983 3349
	7	8628.5	285 55 43.54	+0.19	19 09 06.85	22 29 24.7	.983 3461
	8	8629.5	286 56 53.88	+0.29	19 13 29.60	22 21 57.7	.983 3613
	9	8630.5	287 58 03.92	+0.37	19 17 51.83	22 14 04.3	.983 3805
	10	8631.5	288 59 13.59	+0.44	19 22 13.54	−22 05 44.7	0.983 4039
	11	8632.5	290 00 22.81	+0.47	19 26 34.68	21 56 59.2	.983 4316
	12	8633.5	291 01 31.51	+0.48	19 30 55.24	21 47 48.1	.983 4639
	13	8634.5	292 02 39.64	+0.45	19 35 15.19	21 38 11.6	.983 5010
	14	8635.5	293 03 47.13	+0.40	19 39 34.51	21 28 10.0	.983 5431
	15	8636.5	294 04 53.95	+0.31	19 43 53.18	−21 17 43.7	0.983 5904
	16	8637.5	295 06 00.06	+0.20	19 48 11.18	21 06 52.8	.983 6433
	17	8638.5	296 07 05.46	+0.07	19 52 28.50	20 55 37.7	.983 7021
	18	8639.5	297 08 10.15	−0.08	19 56 45.11	20 43 58.8	.983 7670
	19	8640.5	298 09 14.16	−0.23	20 01 01.01	20 31 56.3	.983 8383
	20	8641.5	299 10 17.52	−0.38	20 05 16.18	−20 19 30.6	0.983 9162
	21	8642.5	300 11 20.29	−0.52	20 09 30.61	20 06 42.0	.984 0005
	22	8643.5	301 12 22.52	−0.63	20 13 44.29	19 53 30.7	.984 0913
	23	8644.5	302 13 24.24	−0.72	20 17 57.22	19 39 57.1	.984 1883
	24	8645.5	303 14 25.48	−0.77	20 22 09.40	19 26 01.5	.984 2912
	25	8646.5	304 15 26.24	−0.79	20 26 20.81	−19 11 44.3	0.984 3998
	26	8647.5	305 16 26.51	−0.78	20 30 31.45	18 57 05.8	.984 5135
	27	8648.5	306 17 26.26	−0.73	20 34 41.32	18 42 06.4	.984 6322
	28	8649.5	307 18 25.45	−0.66	20 38 50.41	18 26 46.5	.984 7554
	29	8650.5	308 19 24.02	−0.57	20 42 58.71	18 11 06.5	.984 8829
	30	8651.5	309 20 21.92	−0.46	20 47 06.22	−17 55 06.7	0.985 0144
	31	8652.5	310 21 19.08	−0.34	20 51 12.92	17 38 47.7	.985 1497
Feb.	1	8653.5	311 22 15.42	−0.22	20 55 18.82	17 22 09.8	.985 2886
	2	8654.5	312 23 10.87	−0.09	20 59 23.91	17 05 13.4	.985 4309
	3	8655.5	313 24 05.36	+0.03	21 03 28.19	16 47 58.9	.985 5764
	4	8656.5	314 24 58.81	+0.13	21 07 31.64	−16 30 26.8	0.985 7252
	5	8657.5	315 25 51.13	+0.22	21 11 34.28	16 12 37.5	.985 8770
	6	8658.5	316 26 42.24	+0.29	21 15 36.10	15 54 31.5	.986 0320
	7	8659.5	317 27 32.07	+0.33	21 19 37.11	15 36 09.1	.986 1900
	8	8660.5	318 28 20.53	+0.35	21 23 37.30	15 17 30.7	.986 3512
	9	8661.5	319 29 07.55	+0.33	21 27 36.69	−14 58 36.9	0.986 5157
	10	8662.5	320 29 53.07	+0.29	21 31 35.26	14 39 28.0	.986 6835
	11	8663.5	321 30 37.02	+0.22	21 35 33.05	14 20 04.4	.986 8548
	12	8664.5	322 31 19.35	+0.12	21 39 30.04	14 00 26.6	.987 0300
	13	8665.5	323 32 00.02	+0.00	21 43 26.25	13 40 35.1	.987 2091
	14	8666.5	324 32 39.00	−0.13	21 47 21.68	−13 20 30.2	0.987 3925
	15	8667.5	325 33 16.29	−0.28	21 51 16.36	−13 00 12.3	0.987 5804

FOR 0^h DYNAMICAL TIME

Date		Position Angle of Axis P	Heliographic Latitude B_0	Heliographic Longitude L_0	H. P.	Semi-Diameter	Ephemeris Transit
		°	°	°	″	′ ″	h m s
Jan.	0	+ 2.87	−2.83	68.08	8.94	16 15.89	12 02 48.68
	1	2.38	2.95	54.91	8.94	16 15.91	12 03 17.38
	2	1.90	3.07	41.74	8.94	16 15.91	12 03 45.78
	3	1.41	3.18	28.57	8.94	16 15.92	12 04 13.85
	4	0.92	3.30	15.40	8.94	16 15.92	12 04 41.57
	5	+ 0.44	−3.41	2.23	8.94	16 15.92	12 05 08.89
	6	− 0.05	3.53	349.06	8.94	16 15.91	12 05 35.78
	7	0.53	3.64	335.89	8.94	16 15.90	12 06 02.21
	8	1.01	3.75	322.72	8.94	16 15.88	12 06 28.15
	9	1.49	3.86	309.55	8.94	16 15.86	12 06 53.58
	10	− 1.98	−3.97	296.39	8.94	16 15.84	12 07 18.46
	11	2.45	4.07	283.22	8.94	16 15.81	12 07 42.77
	12	2.93	4.18	270.05	8.94	16 15.78	12 08 06.48
	13	3.41	4.28	256.88	8.94	16 15.74	12 08 29.56
	14	3.88	4.39	243.71	8.94	16 15.70	12 08 52.01
	15	− 4.35	−4.49	230.55	8.94	16 15.66	12 09 13.79
	16	4.82	4.59	217.38	8.94	16 15.60	12 09 34.89
	17	5.29	4.69	204.21	8.94	16 15.54	12 09 55.29
	18	5.75	4.78	191.04	8.94	16 15.48	12 10 14.98
	19	6.21	4.88	177.87	8.94	16 15.41	12 10 33.95
	20	− 6.67	−4.97	164.71	8.94	16 15.33	12 10 52.18
	21	7.13	5.06	151.54	8.94	16 15.25	12 11 09.68
	22	7.58	5.16	138.37	8.94	16 15.16	12 11 26.43
	23	8.03	5.24	125.21	8.94	16 15.06	12 11 42.43
	24	8.47	5.33	112.04	8.93	16 14.96	12 11 57.67
	25	− 8.92	−5.42	98.87	8.93	16 14.85	12 12 12.14
	26	9.35	5.50	85.71	8.93	16 14.74	12 12 25.84
	27	9.79	5.59	72.54	8.93	16 14.62	12 12 38.76
	28	10.22	5.67	59.37	8.93	16 14.50	12 12 50.89
	29	10.65	5.75	46.21	8.93	16 14.37	12 13 02.23
	30	−11.07	−5.82	33.04	8.93	16 14.24	12 13 12.77
	31	11.49	5.90	19.87	8.93	16 14.11	12 13 22.51
Feb.	1	11.90	5.97	6.71	8.93	16 13.97	12 13 31.44
	2	12.31	6.04	353.54	8.92	16 13.83	12 13 39.55
	3	12.71	6.11	340.38	8.92	16 13.69	12 13 46.85
	4	−13.11	−6.18	327.21	8.92	16 13.54	12 13 53.34
	5	13.51	6.25	314.04	8.92	16 13.39	12 13 59.01
	6	13.90	6.31	300.88	8.92	16 13.24	12 14 03.87
	7	14.28	6.37	287.71	8.92	16 13.08	12 14 07.91
	8	14.66	6.43	274.55	8.92	16 12.92	12 14 11.14
	9	−15.04	−6.49	261.38	8.91	16 12.76	12 14 13.56
	10	15.41	6.54	248.21	8.91	16 12.60	12 14 15.18
	11	15.77	6.60	235.05	8.91	16 12.43	12 14 16.01
	12	16.13	6.65	221.88	8.91	16 12.26	12 14 16.04
	13	16.48	6.70	208.71	8.91	16 12.08	12 14 15.30
	14	−16.83	−6.75	195.54	8.91	16 11.90	12 14 13.78
	15	−17.17	−6.79	182.37	8.90	16 11.71	12 14 11.51

FOR 0h DYNAMICAL TIME

Date		Julian Date	Ecliptic Long. (for Mean Equinox of Date)	Ecliptic Lat. (for Mean Equinox of Date)	Apparent Right Ascension	Apparent Declination	True Geocentric Distance
		244	° ′ ″	″	h m s	° ′ ″	
Feb.	15	8667.5	325 33 16.29	−0.28	21 51 16.36	−13 00 12.3	0.987 5804
	16	8668.5	326 33 51.90	−0.42	21 55 10.28	12 39 41.9	.987 7732
	17	8669.5	327 34 25.86	−0.56	21 59 03.46	12 18 59.3	.987 9709
	18	8670.5	328 34 58.21	−0.67	22 02 55.93	11 58 05.0	.988 1738
	19	8671.5	329 35 29.03	−0.76	22 06 47.68	11 36 59.2	.988 3818
	20	8672.5	330 35 58.36	−0.82	22 10 38.75	−11 15 42.5	0.988 5949
	21	8673.5	331 36 26.25	−0.85	22 14 29.16	10 54 15.1	.988 8128
	22	8674.5	332 36 52.76	−0.84	22 18 18.92	10 32 37.4	.989 0352
	23	8675.5	333 37 17.89	−0.81	22 22 08.06	10 10 49.8	.989 2618
	24	8676.5	334 37 41.66	−0.74	22 25 56.60	9 48 52.7	.989 4923
	25	8677.5	335 38 04.05	−0.65	22 29 44.54	− 9 26 46.6	0.989 7263
	26	8678.5	336 38 25.04	−0.55	22 33 31.91	9 04 31.8	.989 9634
	27	8679.5	337 38 44.61	−0.43	22 37 18.71	8 42 08.7	.990 2033
	28	8680.5	338 39 02.71	−0.31	22 41 04.98	8 19 37.9	.990 4457
	29	8681.5	339 39 19.31	−0.18	22 44 50.71	7 56 59.6	.990 6903
Mar.	1	8682.5	340 39 34.35	−0.06	22 48 35.93	− 7 34 14.3	0.990 9369
	2	8683.5	341 39 47.77	+0.04	22 52 20.64	7 11 22.5	.991 1852
	3	8684.5	342 39 59.53	+0.13	22 56 04.88	6 48 24.5	.991 4350
	4	8685.5	343 40 09.55	+0.20	22 59 48.65	6 25 20.7	.991 6862
	5	8686.5	344 40 17.78	+0.25	23 03 31.97	6 02 11.6	.991 9386
	6	8687.5	345 40 24.13	+0.27	23 07 14.85	− 5 38 57.6	0.992 1921
	7	8688.5	346 40 28.55	+0.26	23 10 57.32	5 15 39.1	.992 4467
	8	8689.5	347 40 30.97	+0.23	23 14 39.39	4 52 16.4	.992 7023
	9	8690.5	348 40 31.31	+0.16	23 18 21.07	4 28 50.0	.992 9590
	10	8691.5	349 40 29.53	+0.07	23 22 02.39	4 05 20.2	.993 2170
	11	8692.5	350 40 25.56	−0.04	23 25 43.37	− 3 41 47.5	0.993 4762
	12	8693.5	351 40 19.36	−0.16	23 29 24.01	3 18 12.3	.993 7370
	13	8694.5	352 40 10.92	−0.29	23 33 04.35	2 54 34.9	.993 9996
	14	8695.5	353 40 00.21	−0.43	23 36 44.39	2 30 55.8	.994 2642
	15	8696.5	354 39 47.24	−0.56	23 40 24.16	2 07 15.2	.994 5311
	16	8697.5	355 39 32.04	−0.67	23 44 03.67	− 1 43 33.6	0.994 8005
	17	8698.5	356 39 14.65	−0.76	23 47 42.95	1 19 51.2	.995 0726
	18	8699.5	357 38 55.14	−0.82	23 51 22.03	0 56 08.5	.995 3475
	19	8700.5	358 38 33.56	−0.85	23 55 00.94	0 32 25.8	.995 6253
	20	8701.5	359 38 10.01	−0.85	23 58 39.69	− 0 08 43.2	.995 9059
	21	8702.5	0 37 44.53	−0.81	0 02 18.32	+ 0 14 58.7	0.996 1890
	22	8703.5	1 37 17.18	−0.75	0 05 56.85	0 38 39.7	.996 4744
	23	8704.5	2 36 48.02	−0.66	0 09 35.30	1 02 19.4	.996 7619
	24	8705.5	3 36 17.06	−0.56	0 13 13.70	1 25 57.5	.997 0510
	25	8706.5	4 35 44.33	−0.44	0 16 52.07	1 49 33.6	.997 3414
	26	8707.5	5 35 09.83	−0.32	0 20 30.43	+ 2 13 07.4	0.997 6329
	27	8708.5	6 34 33.56	−0.19	0 24 08.79	2 36 38.4	.997 9249
	28	8709.5	7 33 55.51	−0.07	0 27 47.18	3 00 06.4	.998 2173
	29	8710.5	8 33 15.66	+0.04	0 31 25.61	3 23 30.9	.998 5097
	30	8711.5	9 32 33.98	+0.13	0 35 04.10	3 46 51.5	.998 8018
	31	8712.5	10 31 50.46	+0.21	0 38 42.68	+ 4 10 08.0	0.999 0934
Apr.	1	8713.5	11 31 05.04	+0.26	0 42 21.35	+ 4 33 20.0	0.999 3842

FOR 0^h DYNAMICAL TIME

Date		Position Angle of Axis P	Heliographic Latitude B_0	Heliographic Longitude L_0	H. P.	Semi-Diameter	Ephemeris Transit
		°	°	°	″	′ ″	h m s
Feb.	15	−17.17	−6.79	182.37	8.90	16 11.71	12 14 11.51
	16	17.51	6.83	169.21	8.90	16 11.52	12 14 08.49
	17	17.84	6.87	156.04	8.90	16 11.33	12 14 04.74
	18	18.17	6.91	142.87	8.90	16 11.13	12 14 00.29
	19	18.49	6.95	129.70	8.90	16 10.93	12 13 55.14
	20	−18.80	−6.98	116.53	8.90	16 10.72	12 13 49.32
	21	19.11	7.02	103.36	8.89	16 10.50	12 13 42.85
	22	19.41	7.05	90.19	8.89	16 10.28	12 13 35.75
	23	19.71	7.07	77.02	8.89	16 10.06	12 13 28.03
	24	20.00	7.10	63.85	8.89	16 09.84	12 13 19.70
	25	−20.28	−7.12	50.68	8.89	16 09.61	12 13 10.80
	26	20.56	7.14	37.51	8.88	16 09.37	12 13 01.32
	27	20.83	7.16	24.34	8.88	16 09.14	12 12 51.29
	28	21.10	7.18	11.17	8.88	16 08.90	12 12 40.72
	29	21.36	7.19	357.99	8.88	16 08.66	12 12 29.64
Mar.	1	−21.61	−7.21	344.82	8.87	16 08.42	12 12 18.04
	2	21.86	7.22	331.65	8.87	16 08.18	12 12 05.96
	3	22.10	7.23	318.48	8.87	16 07.94	12 11 53.41
	4	22.34	7.23	305.30	8.87	16 07.69	12 11 40.40
	5	22.56	7.23	292.13	8.87	16 07.44	12 11 26.95
	6	−22.79	−7.24	278.95	8.86	16 07.20	12 11 13.08
	7	23.00	7.23	265.78	8.86	16 06.95	12 10 58.80
	8	23.21	7.23	252.60	8.86	16 06.70	12 10 44.12
	9	23.41	7.23	239.43	8.86	16 06.45	12 10 29.07
	10	23.61	7.22	226.25	8.85	16 06.20	12 10 13.66
	11	−23.80	−7.21	213.07	8.85	16 05.95	12 09 57.92
	12	23.98	7.20	199.89	8.85	16 05.69	12 09 41.84
	13	24.15	7.18	186.71	8.85	16 05.44	12 09 25.47
	14	24.32	7.16	173.53	8.84	16 05.18	12 09 08.81
	15	24.49	7.15	160.35	8.84	16 04.92	12 08 51.89
	16	−24.64	−7.13	147.17	8.84	16 04.66	12 08 34.74
	17	24.79	7.10	133.99	8.84	16 04.40	12 08 17.37
	18	24.93	7.08	120.81	8.84	16 04.13	12 07 59.81
	19	25.07	7.05	107.63	8.83	16 03.86	12 07 42.09
	20	25.19	7.02	94.44	8.83	16 03.59	12 07 24.24
	21	−25.31	−6.99	81.26	8.83	16 03.32	12 07 06.27
	22	25.43	6.96	68.07	8.83	16 03.04	12 06 48.21
	23	25.54	6.92	54.89	8.82	16 02.76	12 06 30.08
	24	25.64	6.88	41.70	8.82	16 02.48	12 06 11.91
	25	25.73	6.84	28.51	8.82	16 02.20	12 05 53.72
	26	−25.82	−6.80	15.32	8.82	16 01.92	12 05 35.52
	27	25.89	6.76	2.14	8.81	16 01.64	12 05 17.34
	28	25.97	6.71	348.95	8.81	16 01.36	12 04 59.20
	29	26.03	6.66	335.76	8.81	16 01.08	12 04 41.11
	30	26.09	6.61	322.57	8.80	16 00.80	12 04 23.09
	31	−26.14	−6.56	309.37	8.80	16 00.52	12 04 05.17
Apr.	1	−26.18	−6.51	296.18	8.80	16 00.24	12 03 47.35

FOR 0^h DYNAMICAL TIME

Date		Julian Date	Ecliptic Long. (for Mean Equinox of Date)	Ecliptic Lat. (for Mean Equinox of Date)	Apparent Right Ascension	Apparent Declination	True Geocentric Distance
		244	° ′ ″	″	h m s	° ′ ″	
Apr.	1	8713.5	11 31 05.04	+0.26	0 42 21.35	+ 4 33 20.0	0.999 3842
	2	8714.5	12 30 17.70	+0.28	0 46 00.14	4 56 27.0	.999 6739
	3	8715.5	13 29 28.38	+0.28	0 49 39.06	5 19 28.7	0.999 9624
	4	8716.5	14 28 37.03	+0.24	0 53 18.13	5 42 24.8	1.000 2495
	5	8717.5	15 27 43.59	+0.18	0 56 57.37	6 05 14.9	.000 5351
	6	8718.5	16 26 48.01	+0.10	1 00 36.79	+ 6 27 58.6	1.000 8192
	7	8719.5	17 25 50.23	−0.01	1 04 16.40	6 50 35.6	.001 1018
	8	8720.5	18 24 50.20	−0.13	1 07 56.23	7 13 05.6	.001 3831
	9	8721.5	19 23 47.89	−0.25	1 11 36.27	7 35 28.0	.001 6630
	10	8722.5	20 22 43.26	−0.38	1 15 16.56	7 57 42.7	.001 9420
	11	8723.5	21 21 36.31	−0.51	1 18 57.09	+ 8 19 49.2	1.002 2201
	12	8724.5	22 20 27.03	−0.62	1 22 37.89	8 41 47.3	.002 4977
	13	8725.5	23 19 15.45	−0.71	1 26 18.97	9 03 36.5	.002 7750
	14	8726.5	24 18 01.62	−0.77	1 30 00.35	9 25 16.6	.003 0523
	15	8727.5	25 16 45.59	−0.80	1 33 42.06	9 46 47.2	.003 3297
	16	8728.5	26 15 27.43	−0.80	1 37 24.10	+10 08 08.1	1.003 6074
	17	8729.5	27 14 07.22	−0.77	1 41 06.51	10 29 19.0	.003 8853
	18	8730.5	28 12 45.05	−0.70	1 44 49.31	10 50 19.5	.004 1636
	19	8731.5	29 11 20.98	−0.62	1 48 32.50	11 11 09.4	.004 4419
	20	8732.5	30 09 55.10	−0.51	1 52 16.12	11 31 48.3	.004 7201
	21	8733.5	31 08 27.47	−0.39	1 56 00.17	+11 52 15.9	1.004 9980
	22	8734.5	32 06 58.13	−0.26	1 59 44.67	12 12 31.8	.005 2752
	23	8735.5	33 05 27.12	−0.13	2 03 29.64	12 32 35.9	.005 5515
	24	8736.5	34 03 54.48	−0.01	2 07 15.07	12 52 27.6	.005 8266
	25	8737.5	35 02 20.24	+0.11	2 11 01.00	13 12 06.7	.006 1000
	26	8738.5	36 00 44.39	+0.20	2 14 47.43	+13 31 32.8	1.006 3716
	27	8739.5	36 59 06.97	+0.28	2 18 34.36	13 50 45.7	.006 6410
	28	8740.5	37 57 27.96	+0.34	2 22 21.81	14 09 44.9	.006 9080
	29	8741.5	38 55 47.36	+0.37	2 26 09.79	14 28 30.1	.007 1721
	30	8742.5	39 54 05.16	+0.37	2 29 58.30	14 47 01.0	.007 4331
May	1	8743.5	40 52 21.33	+0.34	2 33 47.36	+15 05 17.3	1.007 6909
	2	8744.5	41 50 35.85	+0.29	2 37 36.96	15 23 18.6	.007 9450
	3	8745.5	42 48 48.67	+0.21	2 41 27.12	15 41 04.7	.008 1954
	4	8746.5	43 46 59.75	+0.10	2 45 17.82	15 58 35.1	.008 4420
	5	8747.5	44 45 09.04	−0.02	2 49 09.08	16 15 49.6	.008 6846
	6	8748.5	45 43 16.49	−0.14	2 53 00.90	+16 32 47.9	1.008 9233
	7	8749.5	46 41 22.06	−0.28	2 56 53.26	16 49 29.6	.009 1583
	8	8750.5	47 39 25.73	−0.40	3 00 46.17	17 05 54.4	.009 3897
	9	8751.5	48 37 27.47	−0.51	3 04 39.63	17 22 02.0	.009 6177
	10	8752.5	49 35 27.29	−0.60	3 08 33.63	17 37 52.1	.009 8426
	11	8753.5	50 33 25.21	−0.67	3 12 28.18	+17 53 24.4	1.010 0648
	12	8754.5	51 31 21.26	−0.70	3 16 23.28	18 08 38.6	.010 2844
	13	8755.5	52 29 15.51	−0.71	3 20 18.94	18 23 34.5	.010 5017
	14	8756.5	53 27 08.02	−0.68	3 24 15.15	18 38 11.8	.010 7170
	15	8757.5	54 24 58.86	−0.62	3 28 11.92	18 52 30.2	.010 9303
	16	8758.5	55 22 48.12	−0.53	3 32 09.26	+19 06 29.5	1.011 1418
	17	8759.5	56 20 35.88	−0.43	3 36 07.16	+19 20 09.4	1.011 3513

FOR 0^h DYNAMICAL TIME

Date		Position Angle of Axis P	Heliographic Latitude B_0	Heliographic Longitude L_0	H. P.	Semi-Diameter	Ephemeris Transit
		°	°	°	″	′ ″	h m s
Apr.	1	−26.18	−6.51	296.18	8.80	16 00.24	12 03 47.35
	2	26.22	6.45	282.99	8.80	15 59.96	12 03 29.66
	3	26.25	6.40	269.80	8.79	15 59.68	12 03 12.11
	4	26.27	6.34	256.60	8.79	15 59.41	12 02 54.72
	5	26.29	6.28	243.41	8.79	15 59.13	12 02 37.50
	6	−26.29	−6.21	230.21	8.79	15 58.86	12 02 20.47
	7	26.29	6.15	217.01	8.78	15 58.59	12 02 03.63
	8	26.29	6.08	203.82	8.78	15 58.32	12 01 47.01
	9	26.27	6.01	190.62	8.78	15 58.05	12 01 30.62
	10	26.25	5.94	177.42	8.78	15 57.78	12 01 14.47
	11	−26.22	−5.87	164.22	8.77	15 57.52	12 00 58.57
	12	26.19	5.80	151.02	8.77	15 57.25	12 00 42.96
	13	26.14	5.72	137.81	8.77	15 56.99	12 00 27.64
	14	26.09	5.65	124.61	8.77	15 56.72	12 00 12.64
	15	26.03	5.57	111.41	8.76	15 56.46	11 59 57.96
	16	−25.97	−5.49	98.20	8.76	15 56.20	11 59 43.64
	17	25.89	5.41	85.00	8.76	15 55.93	11 59 29.70
	18	25.81	5.33	71.79	8.76	15 55.67	11 59 16.14
	19	25.73	5.24	58.58	8.76	15 55.40	11 59 02.99
	20	25.63	5.16	45.37	8.75	15 55.14	11 58 50.26
	21	−25.53	−5.07	32.17	8.75	15 54.87	11 58 37.98
	22	25.42	4.98	18.96	8.75	15 54.61	11 58 26.15
	23	25.30	4.89	5.75	8.75	15 54.35	11 58 14.79
	24	25.18	4.80	352.54	8.74	15 54.09	11 58 03.92
	25	25.05	4.71	339.32	8.74	15 53.83	11 57 53.54
	26	−24.91	−4.61	326.11	8.74	15 53.57	11 57 43.66
	27	24.76	4.52	312.90	8.74	15 53.31	11 57 34.30
	28	24.61	4.42	299.68	8.73	15 53.06	11 57 25.47
	29	24.45	4.32	286.47	8.73	15 52.81	11 57 17.17
	30	24.28	4.22	273.25	8.73	15 52.56	11 57 09.40
May	1	−24.10	−4.12	260.04	8.73	15 52.32	11 57 02.18
	2	23.92	4.02	246.82	8.72	15 52.08	11 56 55.51
	3	23.73	3.92	233.60	8.72	15 51.84	11 56 49.38
	4	23.53	3.82	220.39	8.72	15 51.61	11 56 43.81
	5	23.33	3.71	207.17	8.72	15 51.38	11 56 38.78
	6	−23.12	−3.61	193.95	8.72	15 51.16	11 56 34.30
	7	22.90	3.50	180.73	8.71	15 50.94	11 56 30.38
	8	22.67	3.39	167.51	8.71	15 50.72	11 56 27.00
	9	22.44	3.28	154.29	8.71	15 50.50	11 56 24.17
	10	22.20	3.18	141.06	8.71	15 50.29	11 56 21.89
	11	−21.96	−3.07	127.84	8.71	15 50.08	11 56 20.16
	12	21.70	2.95	114.62	8.70	15 49.88	11 56 18.99
	13	21.44	2.84	101.39	8.70	15 49.67	11 56 18.37
	14	21.18	2.73	88.17	8.70	15 49.47	11 56 18.31
	15	20.91	2.62	74.94	8.70	15 49.27	11 56 18.81
	16	−20.63	−2.50	61.71	8.70	15 49.07	11 56 19.87
	17	−20.34	−2.39	48.49	8.70	15 48.87	11 56 21.49

SUN, 1992

FOR 0h DYNAMICAL TIME

Date		Julian Date	Ecliptic Long. (for Mean Equinox of Date)	Ecliptic Lat. (for Mean Equinox of Date)	Apparent Right Ascension	Apparent Declination	True Geocentric Distance
		244	° ′ ″	″	h m s	° ′ ″	
May	17	8759.5	56 20 35.88	−0.43	3 36 07.16	+19 20 09.4	1.011 3513
	18	8760.5	57 18 22.24	−0.31	3 40 05.61	19 33 29.7	.011 5589
	19	8761.5	58 16 07.28	−0.17	3 44 04.63	19 46 30.1	.011 7643
	20	8762.5	59 13 51.06	−0.04	3 48 04.21	19 59 10.4	.011 9675
	21	8763.5	60 11 33.65	+0.09	3 52 04.33	20 11 30.3	.012 1682
	22	8764.5	61 09 15.12	+0.21	3 56 05.00	+20 23 29.5	1.012 3661
	23	8765.5	62 06 55.51	+0.32	4 00 06.21	20 35 07.7	.012 5610
	24	8766.5	63 04 34.86	+0.40	4 04 07.94	20 46 24.8	.012 7527
	25	8767.5	64 02 13.22	+0.47	4 08 10.20	20 57 20.5	.012 9408
	26	8768.5	64 59 50.62	+0.51	4 12 12.97	21 07 54.6	.013 1250
	27	8769.5	65 57 27.07	+0.52	4 16 16.24	+21 18 06.7	1.013 3051
	28	8770.5	66 55 02.59	+0.50	4 20 20.01	21 27 56.8	.013 4808
	29	8771.5	67 52 37.18	+0.45	4 24 24.24	21 37 24.5	.013 6517
	30	8772.5	68 50 10.84	+0.37	4 28 28.94	21 46 29.8	.013 8176
	31	8773.5	69 47 43.55	+0.28	4 32 34.08	21 55 12.3	.013 9783
June	1	8774.5	70 45 15.27	+0.16	4 36 39.65	+22 03 32.0	1.014 1335
	2	8775.5	71 42 45.97	+0.03	4 40 45.61	22 11 28.5	.014 2831
	3	8776.5	72 40 15.60	−0.11	4 44 51.95	22 19 01.9	.014 4271
	4	8777.5	73 37 44.12	−0.24	4 48 58.65	22 26 11.8	.014 5655
	5	8778.5	74 35 11.48	−0.36	4 53 05.67	22 32 58.2	.014 6984
	6	8779.5	75 32 37.67	−0.46	4 57 12.99	+22 39 20.9	1.014 8262
	7	8780.5	76 30 02.68	−0.53	5 01 20.60	22 45 19.6	.014 9490
	8	8781.5	77 27 26.52	−0.57	5 05 28.47	22 50 54.4	.015 0672
	9	8782.5	78 24 49.21	−0.58	5 09 36.58	22 56 05.1	.015 1811
	10	8783.5	79 22 10.81	−0.56	5 13 44.92	23 00 51.6	.015 2910
	11	8784.5	80 19 31.37	−0.50	5 17 53.47	+23 05 13.8	1.015 3972
	12	8785.5	81 16 50.96	−0.42	5 22 02.21	23 09 11.6	.015 4998
	13	8786.5	82 14 09.66	−0.32	5 26 11.12	23 12 45.0	.015 5991
	14	8787.5	83 11 27.57	−0.20	5 30 20.18	23 15 53.9	.015 6951
	15	8788.5	84 08 44.75	−0.07	5 34 29.37	23 18 38.2	.015 7879
	16	8789.5	85 06 01.31	+0.06	5 38 38.68	+23 20 57.8	1.015 8775
	17	8790.5	86 03 17.32	+0.20	5 42 48.08	23 22 52.8	.015 9637
	18	8791.5	87 00 32.87	+0.32	5 46 57.55	23 24 23.1	.016 0466
	19	8792.5	87 57 48.03	+0.44	5 51 07.07	23 25 28.5	.016 1259
	20	8793.5	88 55 02.86	+0.53	5 55 16.62	23 26 09.2	.016 2015
	21	8794.5	89 52 17.44	+0.60	5 59 26.18	+23 26 25.1	1.016 2732
	22	8795.5	90 49 31.82	+0.65	6 03 35.73	23 26 16.1	.016 3408
	23	8796.5	91 46 46.05	+0.67	6 07 45.24	23 25 42.3	.016 4040
	24	8797.5	92 44 00.18	+0.66	6 11 54.70	23 24 43.7	.016 4626
	25	8798.5	93 41 14.24	+0.63	6 16 04.08	23 23 20.2	.016 5163
	26	8799.5	94 38 28.27	+0.56	6 20 13.36	+23 21 32.1	1.016 5649
	27	8800.5	95 35 42.27	+0.47	6 24 22.51	23 19 19.2	.016 6080
	28	8801.5	96 32 56.24	+0.36	6 28 31.52	23 16 41.8	.016 6453
	29	8802.5	97 30 10.18	+0.23	6 32 40.35	23 13 39.8	.016 6767
	30	8803.5	98 27 24.06	+0.10	6 36 48.97	23 10 13.3	.016 7019
July	1	8804.5	99 24 37.84	−0.04	6 40 57.37	+23 06 22.6	1.016 7207
	2	8805.5	100 21 51.46	−0.16	6 45 05.50	+23 02 07.6	1.016 7331

FOR 0^h DYNAMICAL TIME

Date		Position Angle of Axis *P*	Heliographic Latitude B_0	Heliographic Longitude L_0	H. P.	Semi-Diameter	Ephemeris Transit
		°	°	°	″	′ ″	h m s
May	17	−20.34	−2.39	48.49	8.70	15 48.87	11 56 21.49
	18	20.05	2.28	35.26	8.69	15 48.68	11 56 23.66
	19	19.75	2.16	22.03	8.69	15 48.49	11 56 26.40
	20	19.45	2.04	8.80	8.69	15 48.30	11 56 29.68
	21	19.14	1.93	355.57	8.69	15 48.11	11 56 33.52
	22	−18.82	−1.81	342.34	8.69	15 47.92	11 56 37.90
	23	18.50	1.69	329.11	8.69	15 47.74	11 56 42.81
	24	18.17	1.58	315.88	8.68	15 47.56	11 56 48.26
	25	17.83	1.46	302.65	8.68	15 47.38	11 56 54.22
	26	17.49	1.34	289.42	8.68	15 47.21	11 57 00.69
	27	−17.15	−1.22	276.19	8.68	15 47.04	11 57 07.66
	28	16.80	1.10	262.96	8.68	15 46.88	11 57 15.11
	29	16.44	0.98	249.73	8.68	15 46.72	11 57 23.03
	30	16.08	0.86	236.49	8.67	15 46.57	11 57 31.39
	31	15.71	0.74	223.26	8.67	15 46.42	11 57 40.19
June	1	−15.34	−0.62	210.03	8.67	15 46.27	11 57 49.39
	2	14.96	0.50	196.79	8.67	15 46.13	11 57 58.98
	3	14.58	0.38	183.56	8.67	15 46.00	11 58 08.94
	4	14.20	0.26	170.33	8.67	15 45.87	11 58 19.24
	5	13.81	0.14	157.09	8.67	15 45.74	11 58 29.85
	6	−13.41	−0.02	143.86	8.67	15 45.62	11 58 40.77
	7	13.01	+0.10	130.62	8.66	15 45.51	11 58 51.95
	8	12.61	0.22	117.39	8.66	15 45.40	11 59 03.40
	9	12.20	0.34	104.15	8.66	15 45.29	11 59 15.08
	10	11.79	0.46	90.91	8.66	15 45.19	11 59 26.97
	11	−11.38	+0.58	77.68	8.66	15 45.09	11 59 39.06
	12	10.96	0.70	64.44	8.66	15 45.00	11 59 51.33
	13	10.54	0.82	51.21	8.66	15 44.91	12 00 03.76
	14	10.11	0.94	37.97	8.66	15 44.82	12 00 16.33
	15	9.69	1.06	24.73	8.66	15 44.73	12 00 29.03
	16	− 9.26	+1.18	11.49	8.66	15 44.65	12 00 41.82
	17	8.82	1.30	358.26	8.66	15 44.57	12 00 54.70
	18	8.39	1.42	345.02	8.66	15 44.49	12 01 07.64
	19	7.95	1.53	331.78	8.65	15 44.42	12 01 20.62
	20	7.51	1.65	318.55	8.65	15 44.34	12 01 33.62
	21	− 7.07	+1.77	305.31	8.65	15 44.28	12 01 46.63
	22	6.63	1.88	292.07	8.65	15 44.22	12 01 59.61
	23	6.18	2.00	278.83	8.65	15 44.16	12 02 12.55
	24	5.73	2.12	265.60	8.65	15 44.10	12 02 25.42
	25	5.29	2.23	252.36	8.65	15 44.05	12 02 38.20
	26	− 4.84	+2.34	239.12	8.65	15 44.01	12 02 50.87
	27	4.38	2.46	225.89	8.65	15 43.97	12 03 03.39
	28	3.93	2.57	212.65	8.65	15 43.93	12 03 15.75
	29	3.48	2.68	199.41	8.65	15 43.90	12 03 27.92
	30	3.03	2.79	186.18	8.65	15 43.88	12 03 39.87
July	1	− 2.57	+2.90	172.94	8.65	15 43.86	12 03 51.57
	2	− 2.12	+3.01	159.71	8.65	15 43.85	12 04 03.00

SUN, 1992

FOR 0^h DYNAMICAL TIME

Date		Julian Date	Ecliptic Long. (for Mean Equinox of Date)	Ecliptic Lat. (for Mean Equinox of Date)	Apparent Right Ascension	Apparent Declination	True Geocentric Distance
		244	° ′ ″	″	h m s	° ′ ″	
July	1	8804.5	99 24 37.84	−0.04	6 40 57.37	+23 06 22.6	1.016 7207
	2	8805.5	100 21 51.46	−0.16	6 45 05.50	23 02 07.6	.016 7331
	3	8806.5	101 19 04.89	−0.27	6 49 13.34	22 57 28.5	.016 7392
	4	8807.5	102 16 18.09	−0.35	6 53 20.86	22 52 25.4	.016 7393
	5	8808.5	103 13 31.02	−0.40	6 57 28.04	22 46 58.4	.016 7335
	6	8809.5	104 10 43.70	−0.42	7 01 34.86	+22 41 07.7	1.016 7222
	7	8810.5	105 07 56.12	−0.41	7 05 41.30	22 34 53.5	.016 7058
	8	8811.5	106 05 08.32	−0.37	7 09 47.34	22 28 15.8	.016 6845
	9	8812.5	107 02 20.34	−0.29	7 13 52.97	22 21 14.9	.016 6588
	10	8813.5	107 59 32.25	−0.20	7 17 58.17	22 13 50.9	.016 6288
	11	8814.5	108 56 44.11	−0.08	7 22 02.93	+22 06 04.0	1.016 5949
	12	8815.5	109 53 55.99	+0.04	7 26 07.23	21 57 54.4	.016 5572
	13	8816.5	110 51 07.98	+0.18	7 30 11.07	21 49 22.2	.016 5159
	14	8817.5	111 48 20.15	+0.31	7 34 14.42	21 40 27.8	.016 4710
	15	8818.5	112 45 32.59	+0.44	7 38 17.27	21 31 11.1	.016 4227
	16	8819.5	113 42 45.39	+0.55	7 42 19.62	+21 21 32.5	1.016 3709
	17	8820.5	114 39 58.61	+0.65	7 46 21.46	21 11 32.1	.016 3156
	18	8821.5	115 37 12.34	+0.73	7 50 22.78	21 01 10.2	.016 2568
	19	8822.5	116 34 26.65	+0.78	7 54 23.56	20 50 26.9	.016 1943
	20	8823.5	117 31 41.61	+0.81	7 58 23.81	20 39 22.4	.016 1281
	21	8824.5	118 28 57.30	+0.81	8 02 23.52	+20 27 57.1	1.016 0579
	22	8825.5	119 26 13.75	+0.78	8 06 22.67	20 16 11.0	.015 9835
	23	8826.5	120 23 31.03	+0.73	8 10 21.28	20 04 04.5	.015 9049
	24	8827.5	121 20 49.19	+0.65	8 14 19.32	19 51 37.9	.015 8216
	25	8828.5	122 18 08.25	+0.54	8 18 16.80	19 38 51.2	.015 7335
	26	8829.5	123 15 28.23	+0.42	8 22 13.71	+19 25 45.0	1.015 6403
	27	8830.5	124 12 49.14	+0.29	8 26 10.05	19 12 19.3	.015 5418
	28	8831.5	125 10 10.98	+0.16	8 30 05.79	18 58 34.6	.015 4375
	29	8832.5	126 07 33.70	+0.03	8 34 00.95	18 44 31.1	.015 3274
	30	8833.5	127 04 57.26	−0.09	8 37 55.49	18 30 09.1	.015 2113
	31	8834.5	128 02 21.62	−0.18	8 41 49.43	+18 15 28.9	1.015 0892
Aug.	1	8835.5	128 59 46.72	−0.24	8 45 42.74	18 00 30.9	.014 9612
	2	8836.5	129 57 12.50	−0.27	8 49 35.44	17 45 15.2	.014 8275
	3	8837.5	130 54 38.96	−0.27	8 53 27.51	17 29 42.4	.014 6884
	4	8838.5	131 52 06.06	−0.24	8 57 18.97	17 13 52.5	.014 5443
	5	8839.5	132 49 33.83	−0.17	9 01 09.80	+16 57 46.0	1.014 3956
	6	8840.5	133 47 02.28	−0.09	9 05 00.02	16 41 23.1	.014 2426
	7	8841.5	134 44 31.46	+0.02	9 08 49.63	16 24 44.2	.014 0856
	8	8842.5	135 42 01.41	+0.14	9 12 38.63	16 07 49.5	.013 9250
	9	8843.5	136 39 32.19	+0.27	9 16 27.03	15 50 39.3	.013 7610
	10	8844.5	137 37 03.87	+0.40	9 20 14.84	+15 33 14.0	1.013 5940
	11	8845.5	138 34 36.50	+0.52	9 24 02.07	15 15 33.9	.013 4239
	12	8846.5	139 32 10.17	+0.64	9 27 48.72	14 57 39.1	.013 2511
	13	8847.5	140 29 44.95	+0.73	9 31 34.80	14 39 30.1	.013 0756
	14	8848.5	141 27 20.90	+0.81	9 35 20.33	14 21 07.0	.012 8976
	15	8849.5	142 24 58.11	+0.87	9 39 05.32	+14 02 30.3	1.012 7170
	16	8850.5	143 22 36.65	+0.90	9 42 49.78	+13 43 40.1	1.012 5339

FOR 0^h DYNAMICAL TIME

Date		Position Angle of Axis P	Heliographic Latitude B_0	Heliographic Longitude L_0	H. P.	Semi-Diameter	Ephemeris Transit
		°	°	°	″	′ ″	h m s
July	1	− 2.57	+2.90	172.94	8.65	15 43.86	12 03 51.57
	2	2.12	3.01	159.71	8.65	15 43.85	12 04 03.00
	3	1.66	3.12	146.47	8.65	15 43.85	12 04 14.12
	4	1.21	3.23	133.24	8.65	15 43.85	12 04 24.92
	5	0.76	3.34	120.00	8.65	15 43.85	12 04 35.37
	6	− 0.30	+3.44	106.77	8.65	15 43.86	12 04 45.45
	7	+ 0.15	3.55	93.53	8.65	15 43.88	12 04 55.14
	8	0.60	3.65	80.30	8.65	15 43.90	12 05 04.43
	9	1.05	3.76	67.06	8.65	15 43.92	12 05 13.29
	10	1.51	3.86	53.83	8.65	15 43.95	12 05 21.71
	11	+ 1.95	+3.96	40.59	8.65	15 43.98	12 05 29.68
	12	2.40	4.06	27.36	8.65	15 44.01	12 05 37.19
	13	2.85	4.16	14.12	8.65	15 44.05	12 05 44.22
	14	3.30	4.26	0.89	8.65	15 44.09	12 05 50.76
	15	3.74	4.35	347.66	8.65	15 44.14	12 05 56.80
	16	+ 4.18	+4.45	334.42	8.65	15 44.19	12 06 02.34
	17	4.62	4.54	321.19	8.65	15 44.24	12 06 07.36
	18	5.06	4.64	307.96	8.65	15 44.29	12 06 11.85
	19	5.50	4.73	294.73	8.65	15 44.35	12 06 15.82
	20	5.93	4.82	281.50	8.65	15 44.41	12 06 19.24
	21	+ 6.36	+4.91	268.27	8.66	15 44.48	12 06 22.12
	22	6.79	5.00	255.04	8.66	15 44.55	12 06 24.45
	23	7.22	5.08	241.81	8.66	15 44.62	12 06 26.22
	24	7.64	5.17	228.58	8.66	15 44.70	12 06 27.43
	25	8.06	5.25	215.35	8.66	15 44.78	12 06 28.06
	26	+ 8.48	+5.33	202.12	8.66	15 44.87	12 06 28.12
	27	8.90	5.41	188.89	8.66	15 44.96	12 06 27.59
	28	9.31	5.49	175.66	8.66	15 45.06	12 06 26.48
	29	9.72	5.57	162.44	8.66	15 45.16	12 06 24.76
	30	10.12	5.65	149.21	8.66	15 45.27	12 06 22.44
	31	+10.53	+5.72	135.99	8.66	15 45.38	12 06 19.51
Aug.	1	10.92	5.80	122.76	8.66	15 45.50	12 06 15.96
	2	11.32	5.87	109.53	8.67	15 45.62	12 06 11.79
	3	11.71	5.94	96.31	8.67	15 45.75	12 06 07.00
	4	12.10	6.01	83.09	8.67	15 45.89	12 06 01.59
	5	+12.48	+6.07	69.86	8.67	15 46.03	12 05 55.56
	6	12.86	6.14	56.64	8.67	15 46.17	12 05 48.92
	7	13.24	6.20	43.42	8.67	15 46.32	12 05 41.66
	8	13.61	6.26	30.19	8.67	15 46.47	12 05 33.80
	9	13.98	6.32	16.97	8.67	15 46.62	12 05 25.35
	10	+14.34	+6.38	3.75	8.68	15 46.77	12 05 16.30
	11	14.70	6.44	350.53	8.68	15 46.93	12 05 06.68
	12	15.05	6.49	337.31	8.68	15 47.09	12 04 56.49
	13	15.40	6.54	324.09	8.68	15 47.26	12 04 45.74
	14	15.75	6.59	310.87	8.68	15 47.43	12 04 34.45
	15	+16.09	+6.64	297.65	8.68	15 47.59	12 04 22.62
	16	+16.43	+6.69	284.43	8.69	15 47.77	12 04 10.28

SUN, 1992

FOR 0ʰ DYNAMICAL TIME

Date		Julian Date	Ecliptic Long. (for Mean Equinox of Date)	Ecliptic Lat. (for Mean Equinox of Date)	Apparent Right Ascension	Apparent Declination	True Geocentric Distance
		244	° ′ ″	″	h m s	° ′ ″	
Aug.	16	8850.5	143 22 36.65	+0.90	9 42 49.78	+13 43 40.1	1.012 5339
	17	8851.5	144 20 16.60	+0.90	9 46 33.73	13 24 36.8	.012 3483
	18	8852.5	145 17 58.01	+0.88	9 50 17.18	13 05 20.6	.012 1600
	19	8853.5	146 15 40.97	+0.83	9 54 00.14	12 45 52.0	.011 9690
	20	8854.5	147 13 25.51	+0.75	9 57 42.64	12 26 11.0	.011 7752
	21	8855.5	148 11 11.71	+0.65	10 01 24.67	+12 06 18.1	1.011 5783
	22	8856.5	149 08 59.60	+0.54	10 05 06.27	11 46 13.7	.011 3781
	23	8857.5	150 06 49.22	+0.41	10 08 47.44	11 25 57.9	.011 1745
	24	8858.5	151 04 40.57	+0.28	10 12 28.19	11 05 31.1	.010 9671
	25	8859.5	152 02 33.68	+0.15	10 16 08.53	10 44 53.8	.010 7556
	26	8860.5	153 00 28.51	+0.03	10 19 48.48	+10 24 06.1	1.010 5399
	27	8861.5	153 58 25.03	−0.07	10 23 28.04	10 03 08.6	.010 3197
	28	8862.5	154 56 23.19	−0.14	10 27 07.22	9 42 01.4	.010 0949
	29	8863.5	155 54 22.93	−0.18	10 30 46.04	9 20 45.1	.009 8655
	30	8864.5	156 52 24.17	−0.19	10 34 24.51	8 59 19.8	.009 6316
	31	8865.5	157 50 26.88	−0.16	10 38 02.64	+ 8 37 46.0	1.009 3935
Sept.	1	8866.5	158 48 31.01	−0.11	10 41 40.45	8 16 04.0	.009 1515
	2	8867.5	159 46 36.53	−0.03	10 45 17.94	7 54 14.2	.008 9058
	3	8868.5	160 44 43.45	+0.07	10 48 55.14	7 32 16.7	.008 6570
	4	8869.5	161 42 51.76	+0.19	10 52 32.07	7 10 12.1	.008 4053
	5	8870.5	162 41 01.50	+0.31	10 56 08.73	+ 6 48 00.6	1.008 1512
	6	8871.5	163 39 12.69	+0.44	10 59 45.15	6 25 42.6	.007 8949
	7	8872.5	164 37 25.38	+0.56	11 03 21.34	6 03 18.3	.007 6368
	8	8873.5	165 35 39.62	+0.67	11 06 57.32	5 40 48.0	.007 3771
	9	8874.5	166 33 55.44	+0.76	11 10 33.12	5 18 12.1	.007 1161
	10	8875.5	167 32 12.91	+0.84	11 14 08.75	+ 4 55 31.0	1.006 8540
	11	8876.5	168 30 32.09	+0.89	11 17 44.25	4 32 44.8	.006 5909
	12	8877.5	169 28 53.04	+0.92	11 21 19.62	4 09 53.9	.006 3271
	13	8878.5	170 27 15.82	+0.93	11 24 54.89	3 46 58.6	.006 0625
	14	8879.5	171 25 40.50	+0.90	11 28 30.09	3 23 59.2	.005 7974
	15	8880.5	172 24 07.15	+0.85	11 32 05.24	+ 3 00 56.0	1.005 5317
	16	8881.5	173 22 35.82	+0.77	11 35 40.36	2 37 49.3	.005 2654
	17	8882.5	174 21 06.57	+0.68	11 39 15.48	2 14 39.4	.004 9984
	18	8883.5	175 19 39.47	+0.56	11 42 50.62	1 51 26.7	.004 7306
	19	8884.5	176 18 14.55	+0.44	11 46 25.80	1 28 11.4	.004 4620
	20	8885.5	177 16 51.86	+0.30	11 50 01.05	+ 1 04 54.0	1.004 1922
	21	8886.5	178 15 31.41	+0.18	11 53 36.37	0 41 34.7	.003 9210
	22	8887.5	179 14 13.22	+0.05	11 57 11.80	+ 0 18 13.9	.003 6483
	23	8888.5	180 12 57.28	−0.05	12 00 47.35	− 0 05 08.0	.003 3737
	24	8889.5	181 11 43.56	−0.13	12 04 23.03	0 28 30.6	.003 0970
	25	8890.5	182 10 32.00	−0.18	12 07 58.87	− 0 51 53.7	1.002 8181
	26	8891.5	183 09 22.56	−0.20	12 11 34.88	1 15 16.8	.002 5367
	27	8892.5	184 08 15.14	−0.18	12 15 11.08	1 38 39.5	.002 2529
	28	8893.5	185 07 09.69	−0.13	12 18 47.49	2 02 01.6	.001 9669
	29	8894.5	186 06 06.13	−0.06	12 22 24.12	2 25 22.6	.001 6787
	30	8895.5	187 05 04.40	+0.04	12 26 01.00	− 2 48 42.1	1.001 3888
Oct.	1	8896.5	188 04 04.47	+0.16	12 29 38.13	− 3 11 59.8	1.001 0974

FOR 0^h DYNAMICAL TIME

Date		Position Angle of Axis P	Heliographic Latitude B_0	Heliographic Longitude L_0	H. P.	Semi-Diameter	Ephemeris Transit
		°	°	°	″	′ ″	h m s
Aug.	16	+16.43	+6.69	284.43	8.69	15 47.77	12 04 10.28
	17	16.76	6.74	271.21	8.69	15 47.94	12 03 57.42
	18	17.09	6.78	258.00	8.69	15 48.12	12 03 44.08
	19	17.41	6.82	244.78	8.69	15 48.29	12 03 30.26
	20	17.73	6.86	231.56	8.69	15 48.48	12 03 15.97
	21	+18.04	+6.90	218.35	8.69	15 48.66	12 03 01.23
	22	18.35	6.93	205.13	8.70	15 48.85	12 02 46.05
	23	18.65	6.97	191.92	8.70	15 49.04	12 02 30.45
	24	18.95	7.00	178.71	8.70	15 49.23	12 02 14.44
	25	19.24	7.03	165.49	8.70	15 49.43	12 01 58.03
	26	+19.53	+7.06	152.28	8.70	15 49.64	12 01 41.22
	27	19.81	7.08	139.07	8.70	15 49.84	12 01 24.04
	28	20.09	7.11	125.86	8.71	15 50.05	12 01 06.50
	29	20.36	7.13	112.65	8.71	15 50.27	12 00 48.59
	30	20.62	7.15	99.44	8.71	15 50.49	12 00 30.35
	31	+20.89	+7.17	86.23	8.71	15 50.71	12 00 11.77
Sept.	1	21.14	7.18	73.02	8.71	15 50.94	11 59 52.87
	2	21.39	7.20	59.81	8.72	15 51.17	11 59 33.67
	3	21.63	7.21	46.60	8.72	15 51.41	11 59 14.18
	4	21.87	7.22	33.39	8.72	15 51.65	11 58 54.41
	5	+22.10	+7.23	20.19	8.72	15 51.89	11 58 34.40
	6	22.33	7.23	6.98	8.73	15 52.13	11 58 14.15
	7	22.55	7.23	353.77	8.73	15 52.37	11 57 53.68
	8	22.77	7.23	340.57	8.73	15 52.62	11 57 33.02
	9	22.98	7.23	327.36	8.73	15 52.86	11 57 12.19
	10	+23.18	+7.23	314.16	8.73	15 53.11	11 56 51.20
	11	23.38	7.23	300.95	8.74	15 53.36	11 56 30.09
	12	23.57	7.22	287.75	8.74	15 53.61	11 56 08.86
	13	23.75	7.21	274.54	8.74	15 53.86	11 55 47.56
	14	23.93	7.20	261.34	8.74	15 54.11	11 55 26.19
	15	+24.11	+7.19	248.14	8.75	15 54.37	11 55 04.78
	16	24.27	7.17	234.94	8.75	15 54.62	11 54 43.35
	17	24.43	7.15	221.73	8.75	15 54.87	11 54 21.93
	18	24.59	7.13	208.53	8.75	15 55.13	11 54 00.54
	19	24.74	7.11	195.33	8.76	15 55.38	11 53 39.19
	20	+24.88	+7.09	182.13	8.76	15 55.64	11 53 17.92
	21	25.01	7.06	168.93	8.76	15 55.90	11 52 56.74
	22	25.14	7.03	155.73	8.76	15 56.16	11 52 35.67
	23	25.26	7.00	142.54	8.76	15 56.42	11 52 14.74
	24	25.38	6.97	129.34	8.77	15 56.68	11 51 53.95
	25	+25.49	+6.94	116.14	8.77	15 56.95	11 51 33.33
	26	25.59	6.90	102.94	8.77	15 57.22	11 51 12.89
	27	25.69	6.86	89.75	8.77	15 57.49	11 50 52.65
	28	25.77	6.82	76.55	8.78	15 57.76	11 50 32.62
	29	25.86	6.78	63.35	8.78	15 58.04	11 50 12.82
	30	+25.93	+6.73	50.16	8.78	15 58.31	11 49 53.27
Oct.	1	+26.00	+6.69	36.96	8.78	15 58.59	11 49 33.98

SUN, 1992

FOR 0^h DYNAMICAL TIME

Date		Julian Date	Ecliptic Long.	Ecliptic Lat.	Apparent Right Ascension	Apparent Declination	True Geocentric Distance
			for Mean Equinox of Date				
		244	° ′ ″	″	h m s	° ′ ″	
Oct.	1	8896.5	188 04 04.47	+0.16	12 29 38.13	− 3 11 59.8	1.001 0974
	2	8897.5	189 03 06.31	+0.28	12 33 15.53	3 35 15.4	.000 8050
	3	8898.5	190 02 09.91	+0.40	12 36 53.23	3 58 28.4	.000 5118
	4	8899.5	191 01 15.26	+0.52	12 40 31.24	4 21 38.5	1.000 2183
	5	8900.5	192 00 22.39	+0.63	12 44 09.58	4 44 45.4	0.999 9247
	6	8901.5	192 59 31.30	+0.73	12 47 48.27	− 5 07 48.7	0.999 6314
	7	8902.5	193 58 42.01	+0.80	12 51 27.33	5 30 48.1	.999 3387
	8	8903.5	194 57 54.56	+0.86	12 55 06.79	5 53 43.1	.999 0468
	9	8904.5	195 57 08.98	+0.88	12 58 46.66	6 16 33.5	.998 7561
	10	8905.5	196 56 25.30	+0.88	13 02 26.97	6 39 18.9	.998 4667
	11	8906.5	197 55 43.58	+0.86	13 06 07.74	− 7 01 59.0	0.998 1789
	12	8907.5	198 55 03.85	+0.80	13 09 48.99	7 24 33.4	.997 8928
	13	8908.5	199 54 26.18	+0.72	13 13 30.75	7 47 01.7	.997 6085
	14	8909.5	200 53 50.61	+0.62	13 17 13.04	8 09 23.7	.997 3261
	15	8910.5	201 53 17.20	+0.50	13 20 55.88	8 31 38.9	.997 0457
	16	8911.5	202 52 46.00	+0.37	13 24 39.29	− 8 53 47.0	0.996 7671
	17	8912.5	203 52 17.05	+0.23	13 28 23.29	9 15 47.7	.996 4903
	18	8913.5	204 51 50.39	+0.10	13 32 07.91	9 37 40.4	.996 2152
	19	8914.5	205 51 26.04	−0.03	13 35 53.15	9 59 24.9	.995 9415
	20	8915.5	206 51 04.02	−0.14	13 39 39.04	10 21 00.7	.995 6690
	21	8916.5	207 50 44.31	−0.22	13 43 25.59	−10 42 27.5	0.995 3974
	22	8917.5	208 50 26.89	−0.28	13 47 12.82	11 03 44.8	.995 1266
	23	8918.5	209 50 11.72	−0.31	13 51 00.73	11 24 52.2	.994 8561
	24	8919.5	210 49 58.74	−0.30	13 54 49.35	11 45 49.2	.994 5860
	25	8920.5	211 49 47.86	−0.26	13 58 38.69	12 06 35.6	.994 3160
	26	8921.5	212 49 39.01	−0.19	14 02 28.75	−12 27 10.9	0.994 0461
	27	8922.5	213 49 32.10	−0.10	14 06 19.55	12 47 34.6	.993 7764
	28	8923.5	214 49 27.04	+0.02	14 10 11.09	13 07 46.3	.993 5072
	29	8924.5	215 49 23.76	+0.14	14 14 03.38	13 27 45.7	.993 2385
	30	8925.5	216 49 22.21	+0.27	14 17 56.43	13 47 32.3	.992 9708
	31	8926.5	217 49 22.34	+0.39	14 21 50.24	−14 07 05.7	0.992 7043
Nov.	1	8927.5	218 49 24.10	+0.50	14 25 44.81	14 26 25.4	.992 4393
	2	8928.5	219 49 27.47	+0.60	14 29 40.17	14 45 31.2	.992 1762
	3	8929.5	220 49 32.43	+0.68	14 33 36.32	15 04 22.5	.991 9154
	4	8930.5	221 49 38.97	+0.74	14 37 33.26	15 22 59.0	.991 6570
	5	8931.5	222 49 47.08	+0.77	14 41 31.00	−15 41 20.3	0.991 4015
	6	8932.5	223 49 56.77	+0.77	14 45 29.56	15 59 25.9	.991 1491
	7	8933.5	224 50 08.05	+0.74	14 49 28.93	16 17 15.6	.990 9000
	8	8934.5	225 50 20.93	+0.69	14 53 29.13	16 34 48.8	.990 6546
	9	8935.5	226 50 35.43	+0.61	14 57 30.17	16 52 05.3	.990 4132
	10	8936.5	227 50 51.58	+0.50	15 01 32.06	−17 09 04.6	0.990 1758
	11	8937.5	228 51 09.43	+0.38	15 05 34.79	17 25 46.4	.989 9426
	12	8938.5	229 51 29.01	+0.24	15 09 38.37	17 42 10.2	.989 7138
	13	8939.5	230 51 50.37	+0.10	15 13 42.81	17 58 15.8	.989 4894
	14	8940.5	231 52 13.54	−0.05	15 17 48.11	18 14 02.6	.989 2694
	15	8941.5	232 52 38.56	−0.18	15 21 54.28	−18 29 30.4	0.989 0535
	16	8942.5	233 53 05.46	−0.30	15 26 01.30	−18 44 38.6	0.988 8417

FOR 0^h DYNAMICAL TIME

Date		Position Angle of Axis *P*	Heliographic Latitude B_0	Heliographic Longitude L_0	H. P.	Semi-Diameter	Ephemeris Transit
		°	°	°	″	′ ″	h m s
Oct.	1	+26.00	+6.69	36.96	8.78	15 58.59	11 49 33.98
	2	26.06	6.64	23.77	8.79	15 58.87	11 49 14.98
	3	26.11	6.59	10.57	8.79	15 59.15	11 48 56.27
	4	26.16	6.54	357.38	8.79	15 59.44	11 48 37.89
	5	26.20	6.48	344.18	8.79	15 59.72	11 48 19.85
	6	+26.23	+6.43	330.99	8.80	16 00.00	11 48 02.17
	7	26.26	6.37	317.80	8.80	16 00.28	11 47 44.87
	8	26.28	6.31	304.60	8.80	16 00.56	11 47 27.98
	9	26.29	6.24	291.41	8.81	16 00.84	11 47 11.52
	10	26.29	6.18	278.22	8.81	16 01.12	11 46 55.51
	11	+26.29	+6.11	265.02	8.81	16 01.40	11 46 39.97
	12	26.28	6.05	251.83	8.81	16 01.67	11 46 24.92
	13	26.26	5.98	238.64	8.82	16 01.95	11 46 10.39
	14	26.24	5.91	225.45	8.82	16 02.22	11 45 56.40
	15	26.20	5.83	212.26	8.82	16 02.49	11 45 42.96
	16	+26.16	+5.76	199.07	8.82	16 02.76	11 45 30.11
	17	26.11	5.68	185.88	8.83	16 03.02	11 45 17.85
	18	26.06	5.60	172.69	8.83	16 03.29	11 45 06.21
	19	25.99	5.52	159.50	8.83	16 03.56	11 44 55.21
	20	25.92	5.44	146.31	8.83	16 03.82	11 44 44.87
	21	+25.84	+5.36	133.12	8.83	16 04.08	11 44 35.19
	22	25.76	5.27	119.93	8.84	16 04.34	11 44 26.21
	23	25.66	5.19	106.74	8.84	16 04.61	11 44 17.92
	24	25.56	5.10	93.55	8.84	16 04.87	11 44 10.34
	25	25.45	5.01	80.36	8.84	16 05.13	11 44 03.48
	26	+25.33	+4.91	67.18	8.85	16 05.39	11 43 57.35
	27	25.21	4.82	53.99	8.85	16 05.65	11 43 51.95
	28	25.07	4.73	40.80	8.85	16 05.92	11 43 47.30
	29	24.93	4.63	27.62	8.85	16 06.18	11 43 43.40
	30	24.78	4.53	14.43	8.86	16 06.44	11 43 40.26
	31	+24.63	+4.43	1.24	8.86	16 06.70	11 43 37.88
Nov.	1	24.46	4.33	348.06	8.86	16 06.96	11 43 36.28
	2	24.29	4.23	334.87	8.86	16 07.21	11 43 35.47
	3	24.11	4.13	321.69	8.87	16 07.47	11 43 35.45
	4	23.92	4.02	308.50	8.87	16 07.72	11 43 36.23
	5	+23.72	+3.92	295.32	8.87	16 07.97	11 43 37.82
	6	23.52	3.81	282.13	8.87	16 08.21	11 43 40.22
	7	23.31	3.70	268.95	8.87	16 08.46	11 43 43.45
	8	23.09	3.59	255.76	8.88	16 08.70	11 43 47.51
	9	22.86	3.48	242.58	8.88	16 08.93	11 43 52.41
	10	+22.63	+3.37	229.39	8.88	16 09.17	11 43 58.15
	11	22.38	3.26	216.21	8.88	16 09.39	11 44 04.74
	12	22.13	3.15	203.02	8.89	16 09.62	11 44 12.19
	13	21.88	3.03	189.84	8.89	16 09.84	11 44 20.49
	14	21.61	2.92	176.66	8.89	16 10.05	11 44 29.65
	15	+21.34	+2.80	163.47	8.89	16 10.27	11 44 39.67
	16	+21.06	+2.68	150.29	8.89	16 10.47	11 44 50.55

SUN, 1992

FOR 0h DYNAMICAL TIME

Date	Julian Date	Ecliptic Long. for Mean Equinox of Date	Ecliptic Lat. for Mean Equinox of Date	Apparent Right Ascension	Apparent Declination	True Geocentric Distance
	244	° ′ ″	″	h m s	° ′ ″	
Nov. 16	8942.5	233 53 05.46	−0.30	15 26 01.30	−18 44 38.6	0.988 8417
17	8943.5	234 53 34.24	−0.40	15 30 09.17	18 59 27.0	.988 6338
18	8944.5	235 54 04.90	−0.46	15 34 17.91	19 13 55.1	.988 4294
19	8945.5	236 54 37.42	−0.50	15 38 27.48	19 28 02.5	.988 2283
20	8946.5	237 55 11.76	−0.50	15 42 37.91	19 41 48.8	.988 0302
21	8947.5	238 55 47.85	−0.47	15 46 49.16	−19 55 13.7	0.987 8348
22	8948.5	239 56 25.62	−0.41	15 51 01.24	20 08 16.9	.987 6421
23	8949.5	240 57 04.99	−0.32	15 55 14.12	20 20 57.8	.987 4518
24	8950.5	241 57 45.86	−0.21	15 59 27.80	20 33 16.2	.987 2638
25	8951.5	242 58 28.14	−0.09	16 03 42.24	20 45 11.8	.987 0783
26	8952.5	243 59 11.73	+0.04	16 07 57.44	−20 56 44.2	0.986 8953
27	8953.5	244 59 56.56	+0.17	16 12 13.37	21 07 53.0	.986 7150
28	8954.5	246 00 42.52	+0.29	16 16 30.02	21 18 38.0	.986 5375
29	8955.5	247 01 29.57	+0.39	16 20 47.34	21 28 58.8	.986 3632
30	8956.5	248 02 17.63	+0.48	16 25 05.34	21 38 55.1	.986 1922
Dec. 1	8957.5	249 03 06.66	+0.55	16 29 23.98	−21 48 26.6	0.986 0250
2	8958.5	250 03 56.61	+0.58	16 33 43.25	21 57 33.0	.985 8616
3	8959.5	251 04 47.45	+0.59	16 38 03.12	22 06 14.1	.985 7025
4	8960.5	252 05 39.13	+0.58	16 42 23.57	22 14 29.6	.985 5479
5	8961.5	253 06 31.65	+0.53	16 46 44.58	22 22 19.2	.985 3980
6	8962.5	254 07 24.98	+0.45	16 51 06.13	−22 29 42.7	0.985 2532
7	8963.5	255 08 19.12	+0.35	16 55 28.18	22 36 39.9	.985 1138
8	8964.5	256 09 14.07	+0.23	16 59 50.73	22 43 10.6	.984 9799
9	8965.5	257 10 09.85	+0.09	17 04 13.75	22 49 14.6	.984 8518
10	8966.5	258 11 06.49	−0.06	17 08 37.20	22 54 51.7	.984 7297
11	8967.5	259 12 04.00	−0.21	17 13 01.07	−23 00 01.6	0.984 6137
12	8968.5	260 13 02.44	−0.35	17 17 25.33	23 04 44.3	.984 5039
13	8969.5	261 14 01.84	−0.48	17 21 49.94	23 08 59.6	.984 4001
14	8970.5	262 15 02.23	−0.58	17 26 14.89	23 12 47.2	.984 3022
15	8971.5	263 16 03.62	−0.66	17 30 40.15	23 16 07.2	.984 2100
16	8972.5	264 17 06.03	−0.71	17 35 05.68	−23 18 59.2	0.984 1232
17	8973.5	265 18 09.42	−0.72	17 39 31.45	23 21 23.2	.984 0416
18	8974.5	266 19 13.77	−0.70	17 43 57.44	23 23 19.2	.983 9647
19	8975.5	267 20 19.03	−0.64	17 48 23.61	23 24 47.0	.983 8923
20	8976.5	268 21 25.13	−0.56	17 52 49.93	23 25 46.6	.983 8242
21	8977.5	269 22 31.99	−0.46	17 57 16.36	−23 26 17.9	0.983 7600
22	8978.5	270 23 39.53	−0.34	18 01 42.86	23 26 20.9	.983 6997
23	8979.5	271 24 47.64	−0.21	18 06 09.40	23 25 55.7	.983 6431
24	8980.5	272 25 56.23	−0.08	18 10 35.92	23 25 02.3	.983 5901
25	8981.5	273 27 05.20	+0.05	18 15 02.39	23 23 40.6	.983 5409
26	8982.5	274 28 14.47	+0.16	18 19 28.78	−23 21 50.6	0.983 4954
27	8983.5	275 29 23.94	+0.26	18 23 55.04	23 19 32.6	.983 4538
28	8984.5	276 30 33.52	+0.33	18 28 21.13	23 16 46.5	.983 4163
29	8985.5	277 31 43.15	+0.38	18 32 47.03	23 13 32.3	.983 3829
30	8986.5	278 32 52.76	+0.40	18 37 12.69	23 09 50.3	.983 3539
31	8987.5	279 34 02.28	+0.39	18 41 38.09	−23 05 40.4	0.983 3295
32	8988.5	280 35 11.66	+0.36	18 46 03.19	−23 01 02.9	0.983 3100

FOR 0^h DYNAMICAL TIME

Date		Position Angle of Axis P	Heliographic Latitude B_0	Heliographic Longitude L_0	H. P.	Semi-Diameter	Ephemeris Transit
		°	°	°	″	′ ″	h m s
Nov.	16	+21.06	+2.68	150.29	8.89	16 10.47	11 44 50.55
	17	20.77	2.56	137.11	8.90	16 10.68	11 45 02.29
	18	20.47	2.44	123.93	8.90	16 10.88	11 45 14.89
	19	20.17	2.32	110.74	8.90	16 11.08	11 45 28.33
	20	19.86	2.20	97.56	8.90	16 11.27	11 45 42.62
	21	+19.54	+2.08	84.38	8.90	16 11.46	11 45 57.73
	22	19.22	1.96	71.20	8.90	16 11.65	11 46 13.65
	23	18.89	1.84	58.02	8.91	16 11.84	11 46 30.37
	24	18.55	1.71	44.84	8.91	16 12.02	11 46 47.87
	25	18.20	1.59	31.66	8.91	16 12.21	11 47 06.13
	26	+17.85	+1.46	18.48	8.91	16 12.39	11 47 25.14
	27	17.49	1.34	5.30	8.91	16 12.57	11 47 44.86
	28	17.13	1.21	352.12	8.91	16 12.74	11 48 05.29
	29	16.76	1.09	338.94	8.92	16 12.91	11 48 26.39
	30	16.38	0.96	325.76	8.92	16 13.08	11 48 48.16
Dec.	1	+16.00	+0.83	312.59	8.92	16 13.25	11 49 10.56
	2	15.61	0.71	299.41	8.92	16 13.41	11 49 33.58
	3	15.21	0.58	286.23	8.92	16 13.56	11 49 57.19
	4	14.81	0.45	273.05	8.92	16 13.72	11 50 21.38
	5	14.41	0.33	259.87	8.92	16 13.87	11 50 46.11
	6	+13.99	+0.20	246.70	8.93	16 14.01	11 51 11.37
	7	13.58	+0.07	233.52	8.93	16 14.15	11 51 37.12
	8	13.16	−0.06	220.34	8.93	16 14.28	11 52 03.35
	9	12.73	0.19	207.16	8.93	16 14.41	11 52 30.03
	10	12.30	0.31	193.99	8.93	16 14.53	11 52 57.14
	11	+11.86	−0.44	180.81	8.93	16 14.64	11 53 24.65
	12	11.42	0.57	167.63	8.93	16 14.75	11 53 52.53
	13	10.98	0.70	154.46	8.93	16 14.85	11 54 20.76
	14	10.53	0.82	141.28	8.93	16 14.95	11 54 49.32
	15	10.08	0.95	128.11	8.94	16 15.04	11 55 18.16
	16	+ 9.62	−1.08	114.93	8.94	16 15.13	11 55 47.28
	17	9.17	1.20	101.76	8.94	16 15.21	11 56 16.62
	18	8.70	1.33	88.58	8.94	16 15.28	11 56 46.16
	19	8.24	1.46	75.41	8.94	16 15.36	11 57 15.87
	20	7.77	1.58	62.24	8.94	16 15.42	11 57 45.70
	21	+ 7.30	−1.71	49.06	8.94	16 15.49	11 58 15.62
	22	6.83	1.83	35.89	8.94	16 15.55	11 58 45.59
	23	6.35	1.96	22.72	8.94	16 15.60	11 59 15.57
	24	5.88	2.08	9.55	8.94	16 15.66	11 59 45.52
	25	5.40	2.20	356.37	8.94	16 15.70	12 00 15.40
	26	+ 4.92	−2.32	343.20	8.94	16 15.75	12 00 45.18
	27	4.43	2.44	330.03	8.94	16 15.79	12 01 14.81
	28	3.95	2.57	316.86	8.94	16 15.83	12 01 44.27
	29	3.47	2.68	303.69	8.94	16 15.86	12 02 13.51
	30	2.98	2.80	290.52	8.94	16 15.89	12 02 42.50
	31	+ 2.50	−2.92	277.35	8.94	16 15.91	12 03 11.20
	32	+ 2.01	−3.04	264.18	8.94	16 15.93	12 03 39.60

GEOCENTRIC RECTANGULAR COORDINATES
MEAN EQUATOR AND EQUINOX OF J2000.0

Date 0ʰTDT	x	y	z
Jan. 0	+0.152 1937	−0.891 3271	−0.386 4596
1	0.169 4515	0.888 6905	0.385 3160
2	0.186 6575	0.885 7762	0.384 0519
3	0.203 8060	0.882 5849	0.382 6676
4	0.220 8912	0.879 1177	0.381 1637
5	+0.237 9077	−0.875 3755	−0.379 5404
6	0.254 8497	0.871 3595	0.377 7985
7	0.271 7118	0.867 0712	0.375 9386
8	0.288 4885	0.862 5119	0.373 9612
9	0.305 1742	0.857 6833	0.371 8672
10	+0.321 7637	−0.852 5872	−0.369 6573
11	0.338 2516	0.847 2253	0.367 3323
12	0.354 6327	0.841 5997	0.364 8931
13	0.370 9020	0.835 7124	0.362 3406
14	0.387 0544	0.829 5656	0.359 6758
15	+0.403 0851	−0.823 1617	−0.356 8996
16	0.418 9892	0.816 5029	0.354 0131
17	0.434 7623	0.809 5916	0.351 0172
18	0.450 3997	0.802 4302	0.347 9129
19	0.465 8972	0.795 0212	0.344 7013
20	+0.481 2504	−0.787 3668	−0.341 3833
21	0.496 4551	0.779 4693	0.337 9598
22	0.511 5070	0.771 3310	0.334 4318
23	0.526 4017	0.762 9541	0.330 8002
24	0.541 1348	0.754 3409	0.327 0660
25	+0.555 7017	−0.745 4936	−0.323 2301
26	0.570 0979	0.736 4148	0.319 2937
27	0.584 3188	0.727 1070	0.315 2578
28	0.598 3598	0.717 5729	0.311 1236
29	0.612 2163	0.707 8153	0.306 8925
30	+0.625 8839	−0.697 8372	−0.302 5657
31	0.639 3579	0.687 6418	0.298 1445
Feb. 1	0.652 6342	0.677 2321	0.293 6305
2	0.665 7082	0.666 6117	0.289 0251
3	0.678 5758	0.655 7838	0.284 3297
4	+0.691 2328	−0.644 7521	−0.279 5461
5	0.703 6752	0.633 5203	0.274 6758
6	0.715 8990	0.622 0920	0.269 7204
7	0.727 9005	0.610 4713	0.264 6817
8	0.739 6759	0.598 6620	0.259 5614
9	+0.751 2216	−0.586 6682	−0.254 3612
10	0.762 5344	0.574 4939	0.249 0830
11	0.773 6109	0.562 1434	0.243 7285
12	0.784 4479	0.549 6207	0.238 2994
13	0.795 0427	0.536 9302	0.232 7978
14	+0.805 3924	−0.524 0761	−0.227 2252
15	+0.815 4943	−0.511 0625	−0.221 5836
Feb. 15	+0.815 4943	−0.511 0625	−0.221 5836
16	0.825 3460	0.497 8935	0.215 8746
17	0.834 9450	0.484 5733	0.210 1000
18	0.844 2891	0.471 1058	0.204 2614
19	0.853 3757	0.457 4949	0.198 3605
20	+0.862 2026	−0.443 7444	−0.192 3990
21	0.870 7673	0.429 8583	0.186 3784
22	0.879 0671	0.415 8404	0.180 3006
23	0.887 0998	0.401 6949	0.174 1672
24	0.894 8626	0.387 4257	0.167 9801
25	+0.902 3532	−0.373 0370	−0.161 7411
26	0.909 5691	0.358 5333	0.155 4521
27	0.916 5080	0.343 9189	0.149 1150
28	0.923 1677	0.329 1983	0.142 7319
29	0.929 5461	0.314 3762	0.136 3047
Mar. 1	+0.935 6410	−0.299 4570	−0.129 8355
2	0.941 4507	0.284 4456	0.123 3264
3	0.946 9732	0.269 3468	0.116 7795
4	0.952 2070	0.254 1653	0.110 1968
5	0.957 1505	0.238 9062	0.103 5806
6	+0.961 8023	−0.223 5743	−0.096 9330
7	0.966 1612	0.208 1747	0.090 2561
8	0.970 2262	0.192 7125	0.083 5523
9	0.973 9962	0.177 1925	0.076 8236
10	0.977 4706	0.161 6199	0.070 0721
11	+0.980 6489	−0.145 9997	−0.063 3002
12	0.983 5305	0.130 3368	0.056 5098
13	0.986 1153	0.114 6362	0.049 7031
14	0.988 4031	0.098 9027	0.042 8822
15	0.990 3939	0.083 1409	0.036 0490
16	+0.992 0878	−0.067 3556	−0.029 2055
17	0.993 4849	0.051 5513	0.022 3536
18	0.994 5852	0.035 7323	0.015 4952
19	0.995 3889	0.019 9030	0.008 6322
20	0.995 8960	−0.004 0677	−0.001 7664
21	+0.996 1064	+0.011 7692	+0.005 1002
22	0.996 0202	0.027 6032	0.011 9658
23	0.995 6373	0.043 4301	0.018 8284
24	0.994 9578	0.059 2452	0.025 6859
25	0.993 9818	0.075 0439	0.032 5365
26	+0.992 7095	+0.090 8217	+0.039 3779
27	0.991 1413	0.106 5737	0.046 2083
28	0.989 2775	0.122 2954	0.053 0254
29	0.987 1187	0.137 9819	0.059 8272
30	0.984 6656	0.153 6284	0.066 6117
31	+0.981 9189	+0.169 2303	+0.073 3766
Apr. 1	+0.978 8794	+0.184 7825	+0.080 1199

GEOCENTRIC RECTANGULAR COORDINATES
MEAN EQUATOR AND EQUINOX OF J2000.0

Date 0ʰTDT	x	y	z	Date 0ʰTDT	x	y	z
Apr. 1	+0.978 8794	+0.184 7825	+0.080 1199	May 17	+0.558 9403	+0.773 3059	+0.335 2837
2	0.975 5483	0.200 2804	0.086 8395	18	0.544 8098	0.781 9742	0.339 0427
3	0.971 9268	0.215 7190	0.093 5332	19	0.530 5240	0.790 4204	0.342 7054
4	0.968 0160	0.231 0935	0.100 1991	20	0.516 0867	0.798 6422	0.346 2708
5	0.963 8175	0.246 3992	0.106 8349	21	0.501 5019	0.806 6373	0.349 7379
6	+0.959 3331	+0.261 6312	+0.113 4386	22	+0.486 7734	+0.814 4035	+0.353 1058
7	0.954 5643	0.276 7848	0.120 0083	23	0.471 9053	0.821 9387	0.356 3734
8	0.949 5133	0.291 8555	0.126 5419	24	0.456 9017	0.829 2406	0.359 5398
9	0.944 1821	0.306 8386	0.133 0375	25	0.441 7666	0.836 3072	0.362 6039
10	0.938 5730	0.321 7298	0.139 4932	26	0.426 5044	0.843 1361	0.365 5650
11	+0.932 6882	+0.336 5248	+0.145 9072	27	+0.411 1194	+0.849 7255	+0.368 4219
12	0.926 5301	0.351 2193	0.152 2778	28	0.395 6158	0.856 0732	0.371 1740
13	0.920 1011	0.365 8095	0.158 6033	29	0.379 9983	0.862 1772	0.373 8203
14	0.913 4037	0.380 2914	0.164 8819	30	0.364 2714	0.868 0356	0.376 3599
15	0.906 4401	0.394 6613	0.171 1121	31	0.348 4397	0.873 6466	0.378 7921
16	+0.899 2127	+0.408 9153	+0.177 2923	June 1	+0.332 5082	+0.879 0083	+0.381 1161
17	0.891 7238	0.423 0499	0.183 4208	2	0.316 4817	0.884 1192	0.383 3313
18	0.883 9757	0.437 0613	0.189 4961	3	0.300 3652	0.888 9777	0.385 4371
19	0.875 9706	0.450 9459	0.195 5166	4	0.284 1638	0.893 5826	0.387 4329
20	0.867 7108	0.464 7000	0.201 4805	5	0.267 8826	0.897 9327	0.389 3183
21	+0.859 1984	+0.478 3199	+0.207 3864	6	+0.251 5265	+0.902 0271	+0.391 0930
22	0.850 4360	0.491 8017	0.213 2324	7	0.235 1007	0.905 8650	0.392 7566
23	0.841 4258	0.505 1417	0.219 0169	8	0.218 6099	0.909 4458	0.394 3089
24	0.832 1704	0.518 3360	0.224 7383	9	0.202 0590	0.912 7689	0.395 7496
25	0.822 6724	0.531 3809	0.230 3948	10	0.185 4527	0.915 8339	0.397 0786
26	+0.812 9345	+0.544 2725	+0.235 9848	11	+0.168 7955	+0.918 6404	+0.398 2956
27	0.802 9595	0.557 0071	0.241 5065	12	0.152 0921	0.921 1879	0.399 4006
28	0.792 7502	0.569 5808	0.246 9584	13	0.135 3469	0.923 4761	0.400 3932
29	0.782 3097	0.581 9899	0.252 3388	14	0.118 5643	0.925 5046	0.401 2733
30	0.771 6411	0.594 2306	0.257 6461	15	0.101 7488	0.927 2732	0.402 0408
May 1	+0.760 7477	+0.606 2992	+0.262 8785	16	+0.084 9048	+0.928 7813	+0.402 6954
2	0.749 6327	0.618 1921	0.268 0346	17	0.068 0367	0.930 0287	0.403 2369
3	0.738 2997	0.629 9055	0.273 1128	18	0.051 1490	0.931 0150	0.403 6652
4	0.726 7524	0.641 4360	0.278 1116	19	0.034 2462	0.931 7399	0.403 9800
5	0.714 9945	0.652 7800	0.283 0294	20	0.017 3327	0.932 2031	0.404 1813
6	+0.703 0300	+0.663 9344	+0.287 8649	21	+0.000 4131	+0.932 4043	+0.404 2689
7	0.690 8629	0.674 8958	0.292 6168	22	−0.016 5079	0.932 3433	0.404 2427
8	0.678 4972	0.685 6613	0.297 2837	23	0.033 4256	0.932 0198	0.404 1025
9	0.665 9370	0.696 2281	0.301 8645	24	0.050 3355	0.931 4339	0.403 8484
10	0.653 1866	0.706 5933	0.306 3581	25	0.067 2328	0.930 5853	0.403 4802
11	+0.640 2499	+0.716 7545	+0.310 7634	26	−0.084 1126	+0.929 4740	+0.402 9980
12	0.627 1309	0.726 7093	0.315 0793	27	0.100 9700	0.928 1000	0.402 4018
13	0.613 8338	0.736 4552	0.319 3048	28	0.117 8001	0.926 4636	0.401 6916
14	0.600 3623	0.745 9901	0.323 4390	29	0.134 5978	0.924 5648	0.400 8677
15	0.586 7204	0.755 3116	0.327 4809	30	0.151 3580	0.922 4041	0.399 9301
16	+0.572 9118	+0.764 4176	+0.331 4294	July 1	−0.168 0755	+0.919 9820	+0.398 8792
17	+0.558 9403	+0.773 3059	+0.335 2837	2	−0.184 7452	+0.917 2993	+0.397 7153

GEOCENTRIC RECTANGULAR COORDINATES
MEAN EQUATOR AND EQUINOX OF J2000.0

Date 0ʰTDT	x	y	z
July 1	−0.168 0755	+0.919 9820	+0.398 8792
2	0.184 7452	0.917 2993	0.397 7153
3	0.201 3619	0.914 3568	0.396 4389
4	0.217 9206	0.911 1556	0.395 0505
5	0.234 4164	0.907 6970	0.393 5506
6	−0.250 8444	+0.903 9824	+0.391 9399
7	0.267 2000	0.900 0133	0.390 2190
8	0.283 4788	0.895 7910	0.388 3886
9	0.299 6762	0.891 3171	0.386 4492
10	0.315 7880	0.886 5932	0.384 4015
11	−0.331 8099	+0.881 6208	+0.382 2462
12	0.347 7377	0.876 4013	0.379 9838
13	0.363 5673	0.870 9364	0.377 6150
14	0.379 2945	0.865 2275	0.375 1405
15	0.394 9151	0.859 2762	0.372 5608
16	−0.410 4251	+0.853 0840	+0.369 8766
17	0.425 8203	0.846 6525	0.367 0886
18	0.441 0967	0.839 9833	0.364 1974
19	0.456 2499	0.833 0779	0.361 2037
20	0.471 2759	0.825 9381	0.358 1082
21	−0.486 1705	+0.818 5656	+0.354 9116
22	0.500 9294	0.810 9620	0.351 6147
23	0.515 5485	0.803 1292	0.348 2183
24	0.530 0234	0.795 0690	0.344 7231
25	0.544 3499	0.786 7834	0.341 1301
26	−0.558 5235	+0.778 2743	+0.337 4402
27	0.572 5400	0.769 5439	0.333 6542
28	0.586 3949	0.760 5945	0.329 7732
29	0.600 0837	0.751 4284	0.325 7983
30	0.613 6022	0.742 0483	0.321 7307
31	−0.626 9458	+0.732 4570	+0.317 5716
Aug. 1	0.640 1106	0.722 6573	0.313 3224
2	0.653 0924	0.712 6526	0.308 9844
3	0.665 8876	0.702 4458	0.304 5590
4	0.678 4923	0.692 0403	0.300 0476
5	−0.690 9033	+0.681 4392	+0.295 4516
6	0.703 1170	0.670 6459	0.290 7724
7	0.715 1303	0.659 6635	0.286 0112
8	0.726 9399	0.648 4952	0.281 1696
9	0.738 5428	0.637 1441	0.276 2487
10	−0.749 9359	+0.625 6135	+0.271 2500
11	0.761 1161	0.613 9065	0.266 1748
12	0.772 0806	0.602 0263	0.261 0245
13	0.782 8262	0.589 9759	0.255 8003
14	0.793 3501	0.577 7587	0.250 5036
15	−0.803 6494	+0.565 3777	+0.245 1358
16	−0.813 7211	+0.552 8361	+0.239 6983
Aug. 16	−0.813 7211	+0.552 8361	+0.239 6983
17	0.823 5623	0.540 1372	0.234 1924
18	0.833 1701	0.527 2842	0.228 6195
19	0.842 5416	0.514 2805	0.222 9811
20	0.851 6739	0.501 1292	0.217 2787
21	−0.860 5641	+0.487 8339	+0.211 5137
22	0.869 2093	0.474 3980	0.205 6876
23	0.877 6066	0.460 8250	0.199 8020
24	0.885 7530	0.447 1185	0.193 8585
25	0.893 6455	0.433 2824	0.187 8588
26	−0.901 2815	+0.419 3204	+0.181 8047
27	0.908 6580	0.405 2368	0.175 6979
28	0.915 7723	0.391 0357	0.169 5402
29	0.922 6220	0.376 7214	0.163 3337
30	0.929 2048	0.362 2985	0.157 0803
31	−0.935 5186	+0.347 7715	+0.150 7819
Sept. 1	0.941 5615	0.333 1449	0.144 4405
2	0.947 3319	0.318 4232	0.138 0580
3	0.952 8281	0.303 6108	0.131 6363
4	0.958 0487	0.288 7122	0.125 1772
5	−0.962 9924	+0.273 7316	+0.118 6827
6	0.967 6578	0.258 6734	0.112 1545
7	0.972 0438	0.243 5418	0.105 5945
8	0.976 1491	0.228 3411	0.099 0045
9	0.979 9727	0.213 0755	0.092 3862
10	−0.983 5134	+0.197 7492	+0.085 7416
11	0.986 7702	0.182 3664	0.079 0723
12	0.989 7420	0.166 9313	0.072 3802
13	0.992 4279	0.151 4481	0.065 6671
14	0.994 8269	0.135 9208	0.058 9348
15	−0.996 9381	+0.120 3538	+0.052 1850
16	0.998 7606	0.104 7512	0.045 4198
17	1.000 2933	0.089 1173	0.038 6408
18	1.001 5355	0.073 4563	0.031 8500
19	1.002 4863	0.057 7725	0.025 0493
20	−1.003 1447	+0.042 0704	+0.018 2406
21	1.003 5100	0.026 3544	0.011 4259
22	1.003 5815	+0.010 6291	+0.004 6072
23	1.003 3583	−0.005 1008	−0.002 2134
24	1.002 8400	0.020 8305	0.009 0338
25	−1.002 0262	−0.036 5550	−0.015 8518
26	1.000 9166	0.052 2695	0.022 6653
27	0.999 5112	0.067 9687	0.029 4719
28	0.997 8102	0.083 6477	0.036 2697
29	0.995 8141	0.099 3014	0.043 0563
30	−0.993 5234	−0.114 9249	−0.049 8296
Oct. 1	−0.990 9390	−0.130 5132	−0.056 5877

GEOCENTRIC RECTANGULAR COORDINATES
MEAN EQUATOR AND EQUINOX OF J2000.0

Date 0ʰTDT	x	y	z	Date 0ʰTDT	x	y	z
Oct. 1	−0.990 9390	−0.130 5132	−0.056 5877	Nov. 16	−0.581 4446	−0.733 8249	−0.318 1686
2	0.988 0618	0.146 0617	0.063 3285	17	0.567 1653	0.742 9391	0.322 1208
3	0.984 8926	0.161 5656	0.070 0499	18	0.552 7109	0.751 8270	0.325 9746
4	0.981 4325	0.177 0203	0.076 7500	19	0.538 0857	0.760 4853	0.329 7288
5	0.977 6826	0.192 4213	0.083 4268	20	0.523 2937	0.768 9108	0.333 3818
6	−0.973 6441	−0.207 7641	−0.090 0785	21	−0.508 3394	−0.777 1005	−0.336 9325
7	0.969 3181	0.223 0443	0.096 7032	22	0.493 2274	0.785 0513	0.340 3793
8	0.964 7058	0.238 2574	0.103 2989	23	0.477 9624	0.792 7602	0.343 7212
9	0.959 8086	0.253 3992	0.109 8638	24	0.462 5492	0.800 2244	0.346 9569
10	0.954 6277	0.268 4653	0.116 3961	25	0.446 9929	0.807 4412	0.350 0852
11	−0.949 1646	−0.283 4515	−0.122 8938	26	−0.431 2985	−0.814 4082	−0.353 1052
12	0.943 4206	0.298 3536	0.129 3552	27	0.415 4711	0.821 1229	0.356 0158
13	0.937 3971	0.313 1673	0.135 7784	28	0.399 5159	0.827 5833	0.358 8162
14	0.931 0955	0.327 8884	0.142 1616	29	0.383 4382	0.833 7873	0.361 5055
15	0.924 5172	0.342 5127	0.148 5029	30	0.367 2430	0.839 7329	0.364 0829
16	−0.917 6637	−0.357 0361	−0.154 8005	Dec. 1	−0.350 9355	−0.845 4185	−0.366 5476
17	0.910 5366	0.371 4542	0.161 0525	2	0.334 5209	0.850 8422	0.368 8990
18	0.903 1372	0.385 7628	0.167 2571	3	0.318 0044	0.856 0026	0.371 1363
19	0.895 4673	0.399 9575	0.173 4121	4	0.301 3911	0.860 8982	0.373 2590
20	0.887 5284	0.414 0339	0.179 5158	5	0.284 6862	0.865 5277	0.375 2664
21	−0.879 3224	−0.427 9875	−0.185 5662	6	−0.267 8947	−0.869 8898	−0.377 1580
22	0.870 8511	0.441 8138	0.191 5612	7	0.251 0217	0.873 9833	0.378 9333
23	0.862 1167	0.455 5082	0.197 4988	8	0.234 0723	0.877 8071	0.380 5919
24	0.853 1214	0.469 0659	0.203 3770	9	0.217 0515	0.881 3603	0.382 1331
25	0.843 8678	0.482 4825	0.209 1938	10	0.199 9641	0.884 6419	0.383 5567
26	−0.834 3584	−0.495 7534	−0.214 9473	11	−0.182 8151	−0.887 6509	−0.384 8620
27	0.824 5962	0.508 8740	0.220 6355	12	0.165 6094	0.890 3863	0.386 0487
28	0.814 5842	0.521 8400	0.226 2566	13	0.148 3517	0.892 8472	0.387 1163
29	0.804 3257	0.534 6472	0.231 8088	14	0.131 0472	0.895 0327	0.388 0644
30	0.793 8239	0.547 2915	0.237 2903	15	0.113 7007	0.896 9416	0.388 8924
31	−0.783 0821	−0.559 7689	−0.242 6995	16	−0.096 3174	−0.898 5729	−0.389 5999
Nov. 1	0.772 1038	0.572 0755	0.248 0347	17	0.078 9025	0.899 9259	0.390 1866
2	0.760 8923	0.584 2077	0.253 2943	18	0.061 4613	0.900 9995	0.390 6519
3	0.749 4512	0.596 1617	0.258 4768	19	0.043 9994	0.901 7930	0.390 9956
4	0.737 7840	0.607 9339	0.263 5807	20	0.026 5222	0.902 3056	0.391 2174
5	−0.725 8942	−0.619 5210	−0.268 6043	21	−0.009 0355	−0.902 5370	−0.391 3172
6	0.713 7853	0.630 9194	0.273 5463	22	+0.008 4551	0.902 4866	0.391 2947
7	0.701 4611	0.642 1259	0.278 4053	23	0.025 9436	0.902 1543	0.391 1499
8	0.688 9250	0.653 1372	0.283 1797	24	0.043 4244	0.901 5401	0.390 8829
9	0.676 1807	0.663 9501	0.287 8683	25	0.060 8917	0.900 6440	0.390 4937
10	−0.663 2318	−0.674 5615	−0.292 4696	26	+0.078 3396	−0.899 4664	−0.389 9825
11	0.650 0819	0.684 9682	0.296 9823	27	0.095 7624	0.898 0078	0.389 3495
12	0.636 7347	0.695 1672	0.301 4050	28	0.113 1545	0.896 2686	0.388 5951
13	0.623 1937	0.705 1555	0.305 7363	29	0.130 5103	0.894 2497	0.387 7194
14	0.609 4626	0.714 9299	0.309 9750	30	0.147 8240	0.891 9519	0.386 7230
15	−0.595 5449	−0.724 4874	−0.314 1195	31	+0.165 0904	−0.889 3761	−0.385 6062
16	−0.581 4446	−0.733 8249	−0.318 1686	32	+0.182 3039	−0.886 5235	−0.384 3696

NOTES AND FORMULAS

Low precision formulas for the Sun's coordinates and the equation of time

The following formulas give the apparent coordinates of the Sun to a precision of $0^\circ.01$ and the equation of time to a precision of $0^{\mathrm{m}}.1$ between 1950 and 2050; on this page the time argument n is the number of days from J2000.0.

n = JD – 2451545.0 = –2923.5 + day of year (B2–B3) + fraction of day from 0^{h}UT

Mean longitude of Sun, corrected for aberration: $L = 280^\circ.460 + 0^\circ.985\,6474\,n$

Mean anomaly: $g = 357^\circ.528 + 0^\circ.985\,6003\,n$

Put L and g in the range 0° to 360° by adding multiples of 360°.

Ecliptic longitude: $\lambda = L + 1^\circ.915 \sin g + 0^\circ.020 \sin 2g$

Ecliptic latitude: $\beta = 0^\circ$

Obliquity of ecliptic: $\varepsilon = 23^\circ.439 - 0^\circ.000\,0004\,n$

Right ascension (in same quadrant as λ): $\alpha = \tan^{-1}(\cos\varepsilon \tan\lambda)$

Alternatively, α may be calculated directly from

$$\alpha = \lambda - f\,t \sin 2\lambda + (f/2)\,t^2 \sin 4\lambda$$

where $f = 180/\pi$ and $t = \tan^2 \varepsilon/2$

Declination: $\delta = \sin^{-1}(\sin\varepsilon \sin\lambda)$

Distance of Sun from Earth, in au: $R = 1.000\,14 - 0.016\,71 \cos g - 0.000\,14 \cos 2g$

Equatorial rectangular coordinates of the Sun, in au:

$$x = R\cos\lambda, \qquad y = R\cos\varepsilon\sin\lambda, \qquad z = R\sin\varepsilon\sin\lambda$$

Equation of time (apparent time minus mean time):

E, in minutes of time = $(L - \alpha)$, in degrees, multiplied by 4.

Horizontal parallax: $0^\circ.0024$

Semidiameter: $0^\circ.2666\,/\,R$

Light time: $0^{\mathrm{d}}.0058$

CONTENTS OF SECTION D

PHASES OF THE MOON

Lunation	New Moon d h m	First Quarter d h m	Full Moon d h m	Last Quarter d h m
854	Jan. 4 23 10	Jan. 13 02 32	Jan. 19 21 28	Jan. 26 15 27
855	Feb. 3 19 00	Feb. 11 16 15	Feb. 18 08 04	Feb. 25 07 56
856	Mar. 4 13 22	Mar. 12 02 36	Mar. 18 18 18	Mar. 26 02 30
857	Apr. 3 05 01	Apr. 10 10 06	Apr. 17 04 42	Apr. 24 21 40
858	May 2 17 44	May 9 15 43	May 16 16 03	May 24 15 53
859	June 1 03 57	June 7 20 47	June 15 04 50	June 23 08 11
860	June 30 12 18	July 7 02 43	July 14 19 06	July 22 22 12
861	July 29 19 35	Aug. 5 10 58	Aug. 13 10 27	Aug. 21 10 01
862	Aug. 28 02 42	Sept. 3 22 39	Sept. 12 02 17	Sept. 19 19 53
863	Sept. 26 10 40	Oct. 3 14 12	Oct. 11 18 03	Oct. 19 04 12
864	Oct. 25 20 34	Nov. 2 09 11	Nov. 10 09 20	Nov. 17 11 39
865	Nov. 24 09 11	Dec. 2 06 17	Dec. 9 23 41	Dec. 16 19 13
866	Dec. 24 00 43			

MOON AT PERIGEE

d h	d h	d h
Jan. 19 22	June 4 02	Oct. 23 05
Feb. 17 11	July 2 01	Nov. 19 00
Mar. 16 18	July 30 08	Dec. 13 21
Apr. 13 07	Aug. 27 18	
May 8 12	Sept. 25 03	

MOON AT APOGEE

d h	d h	d h
Jan. 6 12	May 23 05	Oct. 7 06
Feb. 2 12	June 19 22	Nov. 3 23
Feb. 29 21	July 17 11	Dec. 1 20
Mar. 28 14	Aug. 13 16	Dec. 29 17
Apr. 25 10	Sept. 9 19	

NOTES AND FORMULAE

Mean elements of the orbit of the Moon

The following expressions for the mean elements of the Moon are based on the fundamental arguments used in the IAU (1980) Theory of Nutation which are given in *The Astronomical Almanac 1984* on page S26. The angular elements are referred to the mean equinox and ecliptic of date. The time argument (d) is the interval in days from 1992 January 0 at 0^h TDT. These expressions are intended for use during 1992 only.

$$d = \text{JD} - 244\,8621{\cdot}5 = \text{day of year (from B2–B3)} + \text{fraction of day from } 0^h \text{ TDT}$$

Mean longitude of the Moon, measured in the ecliptic to the mean ascending node and then along the mean orbit:

$$L' = 217^\circ{\cdot}121\,325 + 13{\cdot}176\,396\,48\,d$$

Mean longitude of the lunar perigee, measured as for L':

$$\Gamma' = 117^\circ{\cdot}665\,169 + 0{\cdot}111\,403\,57\,d$$

Mean longitude of the mean ascending node of the lunar orbit on the ecliptic:

$$\Omega = 279^\circ{\cdot}854\,867 - 0{\cdot}052\,953\,77\,d$$

Mean elongation of the Moon from the Sun:

$$D = L' - L = 298^\circ{\cdot}195\,309 + 12{\cdot}190\,749\,12\,d$$

Mean inclination of the lunar orbit to the ecliptic: $5^\circ{\cdot}145\,3964$.

Mean elements of the rotation of the Moon

The following expressions give the mean elements of the mean equator of the Moon, referred to the true equator of the Earth, during 1992 to a precision of about $0^\circ{\cdot}001$; the time-argument d is as defined above for the orbital elements.

Inclination of the mean equator of the Moon to the true equator of the Earth:

$$i = 23^\circ{\cdot}2231 + 0{\cdot}001\,431\,d + 0{\cdot}000\,000\,041\,d^2$$

Arc of the mean equator of the Moon from its ascending node on the true equator of the Earth to its ascending node on the ecliptic of date:

$$\Delta = 96^\circ{\cdot}3184 - 0{\cdot}053\,293\,d + 0{\cdot}000\,001\,498\,d^2$$

Arc of the true equator of the Earth from the true equinox of date to the ascending node of the mean equator of the Moon:

$$\Omega' = +\,3^\circ{\cdot}8565 + 0{\cdot}000\,390\,d - 0{\cdot}000\,001\,634\,d^2$$

The inclination (I) of the mean lunar equator to the ecliptic: $1^\circ\,32'\,32''{\cdot}7$

The ascending node of the mean lunar equator on the ecliptic is at the descending node of the mean lunar orbit on the ecliptic, that is at longitude $\Omega + 180^\circ$.

Lengths of mean months

The lengths of the mean months at 1992·0, as derived from the mean orbital elements are:

		d	d h m s
synodic month	(new moon to new moon)	29·530 589	29 12 44 02·9
tropical month	(equinox to equinox)	27·321 582	27 07 43 04·7
sidereal month	(fixed star to fixed star)	27·321 662	27 07 43 11·6
anomalistic month	(perigee to perigee)	27·554 550	27 13 18 33·1
draconic month	(node to node)	27·212 221	27 05 05 35·9

NOTES AND FORMULAE

Geocentric coordinates

The apparent longitude (λ) and latitude (β) of the Moon given on pages D6–D20 are referred to the ecliptic of date: the apparent right ascension (α) and declination (δ) are referred to the true equator of date. These coordinates are primarily intended for planning purposes. The true distance (r) is expressed in Earth-radii and is derived, as is the semi-diameter (s), from the horizontal parallax (π).

The maximum errors which may result if Bessel's second-order interpolation formula is used are as follows:

λ	β	α	δ	r	π	s
$\pm 0°.02$	$\pm 0°.02$	$\pm 2^s.4$	$\pm 24''$	$\pm 0{\cdot}002$	$\pm 0''.07$	$\pm 0''.02$

More precise values of right ascension, declination and horizontal parallax may be obtained by using the polynomial coefficients given on pages D23–D45. Precise values of true distance and semi-diameter may be obtained from the parallax using:

$$r = 6\,378{\cdot}137 / \sin \pi \text{ km} \qquad \sin s = 0{\cdot}272\,493 \sin \pi$$

The tabulated values are all referred to the centre of the Earth, and may differ from the topocentric values by up to about 1 degree in angle and 2 per cent in distance.

Time of transit of the Moon

The TDT of upper (or lower) transit of the Moon over a local meridian may be obtained by interpolation in the tabulation of the time of upper (or lower) transit over the ephemeris meridian given on pages D6–D20, where the first differences are about 25 hours. The interpolation factor p is given by:

$$p = -\lambda + 1{\cdot}002\,738\,\Delta T$$

where λ is the *east* longitude and the right-hand side is expressed in days. (Divide longitude in degrees by 360 and ΔT in seconds by 86 400). During 1992 it is expected that ΔT will be about 58 seconds, so that the second term is about $+0{\cdot}000\,67$ days. In general, second-order differences are sufficient to give times to a few seconds, but higher-order differences must be taken into account if a precision of better than 1 second is required. The UT of transit is obtained by subtracting ΔT from the TDT of transit, which is obtained by interpolation.

Topocentric coordinates

The topocentric equatorial rectangular coordinates of the Moon (x', y', z'), referred to the true equinox of date, are equal to the geocentric equatorial rectangular coordinates of the Moon *minus* the geocentric equatorial rectangular coordinates of the observer. Hence, the topocentric right ascension (α'), declination (δ') and distance (r') of the Moon may be calculated from the formulae:

$$\begin{aligned} x' &= r' \cos \delta' \cos \alpha' = r \cos \delta \cos \alpha - \rho \cos \phi' \cos \theta_0 \\ y' &= r' \cos \delta' \sin \alpha' = r \cos \delta \sin \alpha - \rho \cos \phi' \sin \theta_0 \\ z' &= r' \sin \delta' \qquad\quad\;\; = r \sin \delta \qquad\qquad - \rho \sin \phi' \end{aligned}$$

where θ_0 is the local apparent sidereal time (see pages B6, B7) and ρ and ϕ' are the geocentric distance and latitude of the observer.

Then $\quad r'^2 = x'^2 + y'^2 + z'^2, \qquad \alpha' = \tan^{-1} (y' / x'), \qquad \delta' = \sin^{-1}(z' / r')$

The topocentric hour angle (h') may be calculated from $h' = \theta_0 - \alpha'$.

Physical ephemeris

See page D4 for notes on the physical ephemeris of the Moon on pages D7–D21.

NOTES AND FORMULAE

Appearance of the Moon

The quantities tabulated in the ephemeris for physical observations of the Moon on odd pages D7–D21 represent the geocentric aspect and illumination of the Moon's disk. For most purposes it is sufficient to regard the instant of tabulation as 0^h universal time. The age is the number of days elapsed since the previous new Moon; the fraction illuminated (or phase) is the ratio of the illuminated area to the total area of the lunar disk; it is also the fraction of the diameter illuminated perpendicular to the line of cusps. These quantities indicate the general aspect of the Moon, while the precise times of the four principal phases are given on pages A1 and D1; they are the times when the apparent longitudes of the Moon and Sun differ by 0°, 90°, 180° and 270°.

The position angle of the bright limb is measured anticlockwise around the disk from the north point (of the hour circle through the centre of the apparent disk) to the mid-point of the bright limb. Before full moon the morning terminator is visible and the position angle of the northern cusp is 90° greater than the position angle of the bright limb; after full moon the evening terminator is visible and the position angle of the northern cusp is 90° less than the position angle of the bright limb.

The brightness of the Moon is determined largely by the fraction illuminated, but it also depends on the distance of the Moon, on the nature of the part of the lunar surface that is illuminated, and on other factors. The integrated visual magnitude of the full Moon at mean distance is about −12·7. The crescent Moon is not normally visible to the naked eye when the phase is less than 0·01, but much depends on the conditions of observation.

Selenographic coordinates

The positions of points on the Moon's surface are specified by a system of selenographic coordinates, in which latitude is measured positively to the north from the equator of the pole of rotation, and longitude is measured positively to the east on the selenocentric celestial sphere from the lunar meridian through the mean centre of the apparent disk. Selenographic longitudes are measured positive to the west (towards Mare Crisium) on the apparent disk; this sign convention implies that the longitudes of the Sun and of the terminators are decreasing functions of time, and so for some purposes it is convenient to use colongitude which is 90° (or 450°) minus longitude.

The tabulated values of the Earth's selenographic longitude and latitude specify the sub-terrestrial point on the Moon's surface (that is, the centre of the apparent disk). The position angle of the axis of rotation is measured anticlockwise from the north point, and specifies the orientation of the lunar meridian through the sub-terrestrial point, which is the pole of the great circle that corresponds to the limb of the Moon.

The tabulated values of the Sun's selenographic colongitude and latitude specify the sub-solar point of the Moon's surface (that is at the pole of the great circle that bounds the illuminated hemisphere). The following relations hold approximately:

$$\text{longitude of morning terminator} = 360° - \text{colongitude of Sun}$$
$$\text{longitude of evening terminator} = 180° \text{ (or } 540°) - \text{colongitude of Sun}$$

The altitude (a) of the Sun above the lunar horizon at a point at selenographic longitude and latitude (l, b) may be calculated from:

$$\sin a = \sin b_0 \sin b + \cos b_0 \cos b \sin (c_0 + l)$$

where (c_0, b_0) are the Sun's colongitude and latitude at the time.

NOTES AND FORMULAE

Librations of the Moon

On average the same hemisphere of the Moon is always turned to the Earth but there is a periodic oscillation or libration of the apparent position of the lunar surface that allows about 59 per cent of the surface to be seen from the Earth. The libration is due partly to a physical libration, which is an oscillation of the actual rotational motion about its mean rotation, but mainly to the much larger geocentric optical libration, which results from the non-uniformity of the revolution of the Moon around the centre of the Earth. Both of these effects are taken into account in the computation of the Earth's selenographic longitude (l) and latitude (b) and of the position angle (C) of the axis of rotation. The contributions due to the physical libration are tabulated separately. There is a further contribution to the optical libration due to the difference between the viewpoints of the observer on the surface of the Earth and of the hypothetical observer at the centre of the Earth. These topocentric optical librations may be as much as 1° and have important effects on the apparent contour of the limb.

When the libration in longitude, that is the selenographic longitude of the Earth, is positive the mean centre of the disk is displaced eastwards on the celestial sphere, exposing to view a region on the west limb. When the libration in latitude, or selenographic latitude of the Earth, is positive the mean centre of the disk is displaced towards the south, and a region on the north limb is exposed to view. In a similar way the selenographic co-ordinates of the Sun show which regions of the lunar surface are illuminated.

Differential corrections to be applied to the tabular geocentric librations to form the topocentric librations may be computed from the following formulae:

$$\begin{aligned} \Delta l &= -\pi' \sin (Q - C) \sec b \\ \Delta b &= +\pi' \cos (Q - C) \\ \Delta C &= +\sin (b + \Delta b)\, \Delta l - \pi' \sin Q \tan \delta \end{aligned}$$

where Q is the geocentric parallactic angle of the Moon and π' is the topocentric horizontal parallax. The latter is obtained from the geocentric horizontal parallax (π), which is tabulated on even pages D6–D20 by using:

$$\pi' = \pi (\sin z + 0{\cdot}0084 \sin 2z)$$

where z is the geocentric zenith distance of the Moon. The values of z and Q may be calculated from the geocentric right ascension (α) and declination (δ) of the Moon by using:

$$\begin{aligned} \sin z \sin Q &= \cos \phi \sin h \\ \sin z \cos Q &= \cos \delta \sin \phi - \sin \delta \cos \phi \cos h \\ \cos z &= \sin \delta \sin \phi + \cos \delta \cos \phi \cos h \end{aligned}$$

where ϕ is the geocentric latitude of the observer and h is the local hour angle of the Moon, given by:

$$h = \text{local apparent sidereal time} - \alpha$$

Second differences must be taken into account in the interpolation of the tabular geocentric librations to the time of observation.

FOR 0ʰ DYNAMICAL TIME

Date 0ʰ TDT		Apparent Long.	Apparent Lat.	Apparent R.A.	Apparent Dec.	True Dist.	Horiz. Parallax	Semi-diameter	Ephemeris Transit for date Upper	Ephemeris Transit for date Lower
		°	°	h m s	° ′ ″		′ ″	′ ″	h	h
Jan.	0	223·72	−4·41	14 39 25·7	−20 09 21	61·708	55 42·72	15 10·83	08·3393	20·7562
	1	236·15	−3·67	15 31 35·5	−22 51 16	62·281	55 11·97	15 02·45	09·1775	21·6019
	2	248·38	−2·77	16 24 37·5	−24 26 32	62·755	54 46·95	14 55·64	10·0277	22·4530
	3	260·45	−1·77	17 17 52·6	−24 51 30	63·134	54 27·25	14 50·27	10·8757	23·2937
	4	272·40	−0·69	18 10 30·6	−24 06 34	63·420	54 12·50	14 46·25	11·7053	. . .
	5	284·26	+0·41	19 01 44·8	−22 16 22	63·615	54 02·51	14 43·53	12·5040	00·1091
	6	296·07	+1·48	19 51 05·7	−19 28 49	63·716	53 57·38	14 42·13	13·2661	00·8897
	7	307·86	+2·49	20 38 26·1	−15 53 45	63·714	53 57·48	14 42·16	13·9932	01·6337
	8	319·67	+3·40	21 23 59·9	−11 41 32	63·597	54 03·46	14 43·79	14·6929	02·3458
	9	331·55	+4·16	22 08 17·9	− 7 02 13	63·349	54 16·14	14 47·24	15·3771	03·0361
	10	343·54	+4·75	22 52 02·2	− 2 05 20	62·958	54 36·35	14 52·75	16·0607	03·7180
	11	355·70	+5·13	23 36 03·6	+ 2 59 54	62·417	55 04·79	15 00·50	16·7609	04·4076
	12	8·09	+5·29	0 21 18·7	+ 8 03 47	61·726	55 41·79	15 10·58	17·4961	05·1229
	13	20·78	+5·19	1 08 48·1	+12 55 09	60·901	56 27·06	15 22·91	18·2858	05·8830
	14	33·84	+4·83	1 59 31·6	+17 20 09	59·974	57 19·40	15 37·17	19·1471	06·7067
	15	47·31	+4·20	2 54 17·7	+21 01 22	58·995	58 16·46	15 52·72	20·0893	07·6081
	16	61·22	+3·30	3 53 25·3	+23 38 07	58·033	59 14·47	16 08·53	21·1059	08·5896
	17	75·58	+2·17	4 56 21·0	+24 49 15	57·165	60 08·42	16 23·22	22·1690	09·6342
	18	90·34	+0·87	6 01 30·3	+24 18 45	56·476	60 52·46	16 35·22	23·2353	10·7046
	19	105·42	−0·51	7 06 39·3	+22 02 20	56·038	61 20·97	16 42·99	. . .	11·7564
	20	120·66	−1·88	8 09 46·4	+18 10 31	55·903	61 29·90	16 45·42	00·2642	12·7568
	21	135·92	−3·11	9 09 42·8	+13 05 50	56·084	61 17·95	16 42·17	01·2334	13·6944
	22	151·01	−4·11	10 06 20·9	+ 7 16 27	56·561	60 46·94	16 33·72	02·1411	14·5754
	23	165·79	−4·81	11 00 15·4	+ 1 10 11	57·278	60 01·27	16 21·28	02·9995	15·4158
	24	180·17	−5·18	11 52 20·6	− 4 49 06	58·160	59 06·66	16 06·40	03·8265	16·2339
	25	194·07	−5·24	12 43 33·3	−10 22 19	59·124	58 08·84	15 50·64	04·6400	17·0464
	26	207·50	−5·00	13 34 42·3	−15 14 44	60·092	57 12·62	15 35·33	05·4547	17·8658
	27	220·50	−4·51	14 26 21·7	−19 15 00	61·001	56 21·47	15 21·39	06·2803	18·6984
	28	233·12	−3·81	15 18 45·7	−22 14 27	61·805	55 37·52	15 09·41	07·1196	19·5433
	29	245·43	−2·95	16 11 46·3	−24 07 01	62·474	55 01·77	14 59·67	07·9681	20·3927
	30	257·52	−1·98	17 04 55·1	−24 49 33	62·996	54 34·38	14 52·21	08·8153	21·2341
	31	269·44	−0·93	17 57 31·7	−24 22 19	63·372	54 14·96	14 46·92	09·6477	22·0547
Feb.	1	281·26	+0·14	18 48 55·9	−22 49 07	63·610	54 02·80	14 43·61	10·4539	22·8447
	2	293·05	+1·21	19 38 38·9	−20 16 43	63·722	53 57·10	14 42·05	11·2269	23·6006
	3	304·84	+2·22	20 26 30·1	−16 53 57	63·721	53 57·11	14 42·06	11·9663	. . .
	4	316·67	+3·14	21 12 37·4	−12 50 43	63·619	54 02·31	14 43·47	12·6773	00·3249
	5	328·58	+3·92	21 57 24·4	− 8 17 15	63·422	54 12·41	14 46·23	13·3692	01·0249
	6	340·58	+4·54	22 41 25·9	− 3 23 42	63·130	54 27·45	14 50·32	14·0543	01·7117
	7	352·72	+4·96	23 25 24·7	+ 1 39 52	62·741	54 47·70	14 55·84	14·7470	02·3987
	8	5·01	+5·15	0 10 09·5	+ 6 43 11	62·251	55 13·59	15 02·90	15·4632	03·1011
	9	17·49	+5·11	0 56 32·0	+11 35 08	61·657	55 45·52	15 11·60	16·2195	03·8353
	10	30·21	+4·82	1 45 24·5	+16 03 09	60·962	56 23·64	15 21·98	17·0314	04·6177
	11	43·21	+4·27	2 37 32·2	+19 52 31	60·181	57 07·57	15 33·95	17·9103	05·4621
	12	56·53	+3·49	3 33 22·4	+22 46 17	59·340	57 56·14	15 47·18	18·8578	06·3759
	13	70·22	+2·49	4 32 47·8	+24 26 32	58·483	58 47·08	16 01·06	19·8607	07·3538
	14	84·30	+1·31	5 34 55·3	+24 37 25	57·668	59 36·91	16 14·64	20·8906	08·3744
	15	98·77	+0·01	6 38 10·9	+23 09 55	56·966	60 21·04	16 26·66	21·9129	09·4047

EPHEMERIS FOR PHYSICAL OBSERVATIONS
FOR 0^h DYNAMICAL TIME

Date 0^h TDT		Age	The Earth's Selenographic Long.	The Earth's Selenographic Lat.	Physical Libration Lg.	Physical Libration Lt.	Physical Libration P.A.	The Sun's Selenographic Colong.	The Sun's Selenographic Lat.	Position Angle Axis	Position Angle Bright Limb	Fraction Illum.
		d	°	°	(0°001)			°	°	°	°	
Jan.	**0**	24·8	+6·676	+5·697	+10	+12	+17	208·20	−0·01	18·716	104·76	0·22
	1	25·8	5·938	4·744	9	12	16	220·38	+0·02	15·050	100·08	0·14
	2	26·8	4·981	3·588	7	12	16	232·56	0·04	10·600	94·91	0·08
	3	27·8	3·851	2·287	5	13	15	244·75	0·06	5·644	89·67	0·03
	4	28·8	2·591	+0·898	3	13	15	256·93	0·08	0·500	85·22	0·01
	5	0·0	+1·241	−0·519	+ 2	+13	+15	269·12	+0·10	355·474	216·30	0·00
	6	1·0	−0·160	1·906	0	14	16	281·31	0·12	350·809	251·85	0·01
	7	2·0	1·571	3·210	− 2	14	16	293·50	0·15	346·670	249·13	0·04
	8	3·0	2·950	4·377	4	14	17	305·68	0·17	343·158	246·69	0·08
	9	4·0	4·251	5·361	6	14	19	317·86	0·19	340·334	245·01	0·14
	10	5·0	−5·425	−6·120	− 8	+13	+20	330·04	+0·22	338·253	244·19	0·21
	11	6·0	6·416	6·616	9	13	21	342·21	0·25	336·981	244·28	0·30
	12	7·0	7·166	6·815	11	13	22	354·38	0·27	336·614	245·35	0·39
	13	8·0	7·615	6·689	11	12	23	6·54	0·30	337·279	247·45	0·49
	14	9·0	7·707	6·219	12	12	24	18·69	0·33	339·124	250·67	0·59
	15	10·0	−7·397	−5·397	−12	+11	+25	30·84	+0·37	342·269	255·00	0·70
	16	11·0	6·659	4·235	11	10	25	42·98	0·40	346·736	260·34	0·80
	17	12·0	5·493	2·775	10	10	25	55·11	0·44	352·342	266·27	0·88
	18	13·0	3·941	−1·092	9	9	25	67·24	0·48	358·649	271·88	0·95
	19	14·0	2·086	+0·699	7	9	25	79·37	0·51	5·043	274·28	0·99
	20	15·0	−0·052	+2·461	− 6	+ 8	+24	91·49	+0·55	10·920	153·88	1·00
	21	16·0	+2·005	4·047	4	8	23	103·62	0·59	15·852	117·96	0·98
	22	17·0	3·922	5·333	3	8	23	115·74	0·62	19·611	117·66	0·93
	23	18·0	5·551	6·235	2	8	22	127·88	0·65	22·108	117·72	0·86
	24	19·0	6·780	6·715	2	8	21	140·01	0·68	23·322	116·89	0·77
	25	20·0	+7·548	+6·779	− 2	+ 8	+20	152·16	+0·71	23·266	115·02	0·67
	26	21·0	7·840	6·465	2	8	19	164·31	0·73	21·976	112·13	0·57
	27	22·0	7·684	5·824	3	8	18	176·46	0·75	19·527	108·36	0·46
	28	23·0	7·135	4·915	4	8	17	188·63	0·77	16·053	103·87	0·37
	29	24·0	6·264	3·799	6	8	16	200·80	0·79	11·753	98·93	0·27
	30	25·0	+5·147	+2·534	− 7	+ 9	+15	212·97	+0·81	6·896	93·88	0·19
	31	26·0	3·862	+1·177	9	9	15	225·16	0·83	1·785	89·17	0·12
Feb.	**1**	27·0	2·476	−0·219	11	9	15	237·34	0·85	356·720	85·41	0·07
	2	28·0	+1·051	1·598	13	9	15	249·53	0·86	351·949	83·89	0·03
	3	29·0	−0·363	2·907	14	9	16	261·72	0·88	347·656	90·70	0·01
	4	0·2	−1·723	−4·092	−16	+10	+17	273·92	+0·90	343·959	198·44	0·00
	5	1·2	2·997	5·105	18	10	18	286·11	0·91	340·938	233·61	0·01
	6	2·2	4·156	5·900	20	9	19	298·30	0·93	338·655	237·91	0·05
	7	3·2	5·176	6·438	22	9	20	310·49	0·95	337·174	239·78	0·09
	8	4·2	6·030	6·687	23	9	21	322·68	0·97	336·579	241·62	0·16
	9	5·2	−6·688	−6·623	−24	+ 8	+22	334·86	+0·99	336·973	244·03	0·24
	10	6·2	7·117	6·235	24	8	23	347·03	1·01	338·468	247·26	0·33
	11	7·2	7·277	5·521	25	8	23	359·20	1·03	341·164	251·42	0·43
	12	8·2	7·125	4·494	24	8	24	11·37	1·05	345·092	256·48	0·54
	13	9·2	6·624	3·186	24	7	24	23·53	1·08	350·147	262·24	0·64
	14	10·2	−5·747	−1·651	−23	+ 7	+23	35·68	+1·11	356·031	268·26	0·75
	15	11·2	−4·493	+0·029	−21	+ 7	+23	47·82	+1·13	2·263	273·80	0·84

FOR 0ʰ DYNAMICAL TIME

Date 0ʰ TDT		Apparent Long.	Apparent Lat.	Apparent R.A.	Apparent Dec.	True Dist.	Horiz. Parallax	Semi-diameter	Ephemeris Transit for date Upper	Ephemeris Transit for date Lower
		°	°	h m s	° ′ ″		′ ″	′ ″	h	h
Feb.	**15**	98·77	+0·01	6 38 10·9	+23 09 55	56·966	60 21·04	16 26·66	21·9129	09·4047
	16	113·58	−1·31	7 40 49·0	+20 05 42	56·447	60 54·33	16 35·73	22·9001	10·4120
	17	128·64	−2·56	8 41 29·8	+15 38 04	56·175	61 12·05	16 40·56	23·8403	11·3762
	18	143·81	−3·63	9 39 39·5	+10 09 13	56·189	61 11·08	16 40·30	. . .	12·2933
	19	158·94	−4·45	10 35 28·1	+ 4 05 43	56·500	60 50·90	16 34·80	00·7366	13·1719
	20	173·84	−4·95	11 29 32·2	− 2 05 40	57·081	60 13·76	16 24·68	01·6011	14·0261
	21	188·38	−5·11	12 22 38·8	− 8 00 56	57·875	59 24·16	16 11·17	02·4487	14·8706
	22	202·46	−4·96	13 15 31·3	−13 20 15	58·807	58 27·64	15 55·77	03·2932	15·7174
	23	216·06	−4·52	14 08 40·9	−17 48 13	59·796	57 29·66	15 39·97	04·1438	16·5726
	24	229·17	−3·86	15 02 20·2	−21 13 50	60·762	56 34·80	15 25·02	05·0036	17·4359
	25	241·85	−3·03	15 56 21·0	−23 30 06	61·641	55 46·37	15 11·83	05·8686	18·3002
	26	254·18	−2·08	16 50 16·1	−24 34 03	62·385	55 06·48	15 00·96	06·7292	19·1541
	27	266·24	−1·05	17 43 28·3	−24 26 26	62·963	54 36·12	14 52·69	07·5733	19·9856
	28	278·12	0·00	18 35 21·3	−23 11 26	63·362	54 15·48	14 47·06	08·3901	20·7862
	29	289·91	+1·05	19 25 30·3	−20 55 45	63·584	54 04·13	14 43·97	09·1736	21·5525
Mar.	**1**	301·68	+2·04	20 13 47·3	−17 47 44	63·641	54 01·22	14 43·18	09·9234	22·2871
	2	313·49	+2·95	21 00 21·5	−13 56 36	63·553	54 05·68	14 44·39	10·6445	22·9968
	3	325·40	+3·74	21 45 35·9	− 9 31 54	63·345	54 16·32	14 47·29	11·3454	23·6919
	4	337·45	+4·36	22 30 03·5	− 4 43 25	63·041	54 32·05	14 51·58	12·0378	. . .
	5	349·64	+4·80	23 14 24·2	+ 0 18 42	62·661	54 51·89	14 56·98	12·7346	00·3848
	6	2·01	+5·01	23 59 22·3	+ 5 23 42	62·221	55 15·16	15 03·32	13·4502	01·0892
	7	14·55	+4·99	0 45 43·9	+10 19 52	61·732	55 41·43	15 10·48	14·1990	01·8195
	8	27·27	+4·72	1 34 14·1	+14 54 06	61·199	56 10·52	15 18·41	14·9943	02·5901
	9	40·19	+4·21	2 25 30·9	+18 51 51	60·626	56 42·38	15 27·09	15·8458	03·4127
	10	53·31	+3·47	3 19 56·4	+21 57 20	60·018	57 16·87	15 36·49	16·7557	04·2937
	11	66·67	+2·52	4 17 24·8	+23 54 37	59·384	57 53·56	15 46·48	17·7146	05·2301
	12	80·28	+1·42	5 17 14·7	+24 30 00	58·744	58 31·41	15 56·79	18·7016	06·2062
	13	94·17	+0·21	6 18 13·1	+23 35 04	58·129	59 08·59	16 06·92	19·6894	07·1970
	14	108·36	−1·04	7 18 56·1	+21 09 19	57·580	59 42·43	16 16·14	20·6544	08·1758
	15	122·82	−2·24	8 18 16·1	+17 21 01	57·147	60 09·54	16 23·53	21·5841	09·1239
	16	137·50	−3·31	9 15 40·6	+12 26 00	56·883	60 26·34	16 28·11	22·4786	10·0354
	17	152·33	−4·17	10 11 13·2	+ 6 45 10	56·829	60 29·76	16 29·04	23·3470	10·9153
	18	167·17	−4·74	11 05 23·7	+ 0 41 59	57·012	60 18·13	16 25·87	. . .	11·7755
	19	181·87	−5·00	11 58 53·8	− 5 19 43	57·431	59 51·68	16 18·66	00·2025	12·6297
	20	196·31	−4·93	12 52 24·9	−10 57 32	58·061	59 12·71	16 08·04	01·0584	13·4897
	21	210·38	−4·55	13 46 27·5	−15 51 59	58·851	58 25·02	15 55·05	01·9243	14·3623
	22	224·01	−3·93	14 41 13·4	−19 47 33	59·735	57 33·19	15 40·93	02·8035	15·2470
	23	237·19	−3·11	15 36 31·0	−22 33 34	60·639	56 41·70	15 26·90	03·6916	16·1357
	24	249·94	−2·16	16 31 47·5	−24 04 41	61·494	55 54·40	15 14·01	04·5774	17·0148
	25	262·33	−1·13	17 26 18·1	−24 20 47	62·239	55 14·19	15 03·06	05·4462	17·8700
	26	274·43	−0·07	18 19 19·8	−23 26 11	62·831	54 43·00	14 54·56	06·2852	18·6908
	27	286·34	+0·98	19 10 23·8	−21 28 17	63·238	54 21·84	14 48·80	07·0867	19·4729
	28	298·16	+1·97	19 59 21·5	−18 36 02	63·450	54 10·97	14 45·83	07·8500	20·2188
	29	309·95	+2·88	20 46 23·8	−14 58 48	63·468	54 10·04	14 45·58	08·5804	20·9361
	30	321·82	+3·67	21 31 56·8	−10 45 47	63·310	54 18·16	14 47·79	09·2873	21·6357
	31	333·82	+4·30	22 16 36·6	− 6 06 04	63·002	54 34·09	14 52·13	09·9830	22·3308
Apr.	**1**	346·00	+4·74	23 01 05·8	− 1 08 55	62·578	54 56·25	14 58·17	10·6811	23·0357

EPHEMERIS FOR PHYSICAL OBSERVATIONS
FOR 0ʰ DYNAMICAL TIME

Date 0ʰ TDT	Age	The Earth's Selenographic Long.	The Earth's Selenographic Lat.	Physical Libration Lg.	Physical Libration Lt.	Physical Libration P.A.	The Sun's Selenographic Colong.	The Sun's Selenographic Lat.	Position Angle Axis	Position Angle Bright Limb	Fraction Illum.
	d	°	°	(0°001)			°	°	°	°	
Feb. 15	11·2	−4·493	+0·029	−21	+ 7	+23	47·82	+1·13	2·263	273·80	0·84
16	12·2	2·897	1·743	20	7	22	59·96	1·16	8·296	277·74	0·92
17	13·2	−1·040	3·363	19	7	22	72·09	1·19	13·650	277·54	0·97
18	14·2	+0·949	4·757	17	7	21	84·23	1·22	17·993	251·83	1·00
19	15·2	2·907	5·811	16	6	20	96·36	1·24	21·135	137·08	0·99
20	16·2	+4·663	+6·453	−15	+ 6	+19	108·50	+1·27	22·976	124·67	0·96
21	17·2	6·066	6·659	15	6	19	120·63	1·29	23·471	120·22	0·90
22	18·2	7·013	6·449	15	6	18	132·78	1·30	22·628	116·37	0·82
23	19·2	7·457	5·877	15	6	17	144·93	1·32	20·511	112·10	0·73
24	20·2	7·405	5·011	15	6	16	157·08	1·33	17·259	107·27	0·63
25	21·2	+6·906	+3·922	−16	+ 6	+15	169·25	+1·34	13·090	102·01	0·53
26	22·2	6·039	2·679	17	6	15	181·42	1·35	8·289	96·57	0·44
27	23·2	4·894	+1·342	19	6	15	193·59	1·36	3·174	91·29	0·34
28	24·2	3·565	−0·032	20	6	15	205·78	1·37	358·052	86·50	0·26
29	25·2	2·144	1·393	22	6	15	217·97	1·38	353·179	82·55	0·18
Mar. 1	26·2	+0·710	−2·689	−23	+ 6	+15	230·16	+1·39	348·741	79·85	0·11
2	27·2	−0·669	3·871	25	6	16	242·36	1·39	344·867	79·11	0·06
3	28·2	1·939	4·892	26	6	17	254·56	1·40	341·643	82·45	0·02
4	29·2	3·064	5·705	28	6	18	266·76	1·41	339·139	103·27	0·00
5	0·4	4·022	6·267	29	6	19	278·97	1·41	337·430	203·10	0·00
6	1·4	−4·803	−6·544	−30	+ 5	+20	291·17	+1·42	336·602	229·87	0·02
7	2·4	5·405	6·510	31	5	21	303·38	1·42	336·758	237·62	0·06
8	3·4	5·828	6·155	32	4	22	315·58	1·43	338·004	242·86	0·12
9	4·4	6·068	5·482	32	4	22	327·78	1·44	340·424	247·93	0·19
10	5·4	6·114	4·510	32	4	23	339·97	1·45	344·040	253·44	0·28
11	6·4	−5·945	−3·276	−31	+ 4	+23	352·15	+1·46	348·753	259·45	0·38
12	7·4	5·534	1·834	30	4	22	4·33	1·47	354·305	265·73	0·49
13	8·4	4·853	−0·254	29	4	22	16·51	1·48	0·290	271·85	0·60
14	9·4	3·882	+1·372	28	4	21	28·67	1·49	6·235	277·24	0·71
15	10·4	2·627	2·941	27	5	20	40·84	1·50	11·697	281·30	0·81
16	11·4	−1·130	+4·340	−25	+ 5	+19	52·99	+1·52	16·343	283·28	0·89
17	12·4	+0·524	5·458	24	5	18	65·14	1·53	19·943	281·65	0·95
18	13·4	2·210	6·210	23	5	17	77·29	1·54	22·347	268·75	0·99
19	14·4	3·778	6·542	22	5	17	89·44	1·55	23·449	170·60	1·00
20	15·4	5·081	6·447	22	5	16	101·59	1·56	23·188	128·88	0·98
21	16·4	+6·003	+5·960	−21	+ 5	+15	113·74	+1·56	21·568	118·81	0·93
22	17·4	6·472	5·143	21	5	14	125·90	1·56	18·681	112·00	0·87
23	18·4	6·473	4·072	22	5	14	138·06	1·56	14·727	105·71	0·79
24	19·4	6·034	2·828	22	5	14	150·23	1·56	10·003	99·54	0·70
25	20·4	5·219	1·482	23	5	14	162·40	1·56	4·859	93·59	0·61
26	21·4	+4·115	+0·098	−24	+ 5	+14	174·58	+1·56	359·636	88·08	0·51
27	22·4	2·817	−1·268	25	5	14	186·77	1·55	354·619	83·26	0·42
28	23·4	1·424	2·567	26	5	15	198·96	1·55	350·013	79·32	0·33
29	24·4	+0·027	3·754	27	5	16	211·16	1·54	345·951	76·42	0·24
30	25·4	−1·290	4·783	28	4	17	223·37	1·54	342·518	74·77	0·17
31	26·4	−2·461	−5·612	−30	+ 4	+18	235·58	+1·53	339·781	74·70	0·10
Apr. 1	27·4	−3·436	−6·199	−31	+ 4	+19	247·80	+1·52	337·814	77·12	0·05

FOR 0^h DYNAMICAL TIME

Date 0^h TDT		Apparent Long.	Apparent Lat.	Apparent R.A.	Apparent Dec.	True Dist.	Horiz. Parallax	Semi-diameter	Ephemeris Transit for date Upper	Ephemeris Transit for date Lower
		°	°	h m s	° ′ ″		′ ″	′ ″	h	h
Apr.	1	346·00	+4·74	23 01 05·8	− 1 08 55	62·578	54 56·25	14 58·17	10·6811	23·0357
	2	358·39	+4·98	23 46 10·6	+ 3 55 31	62·076	55 22·93	15 05·44	11·3964	23·7652
	3	11·02	+4·97	0 32 38·2	+ 8 55 45	61·530	55 52·39	15 13·47	12·1436	. . .
	4	23·87	+4·71	1 21 14·0	+13 38 21	60·973	56 23·02	15 21·81	12·9361	00·5335
	5	36·93	+4·21	2 12 34·8	+17 47 59	60·429	56 53·52	15 30·12	13·7834	01·3526
	6	50·19	+3·47	3 07 00·1	+21 07 52	59·912	57 22·95	15 38·14	14·6873	02·2286
	7	63·62	+2·53	4 04 20·3	+23 21 23	59·432	57 50·75	15 45·71	15·6382	03·1579
	8	77·22	+1·43	5 03 50·0	+24 14 45	58·993	58 16·59	15 52·76	16·6148	04·1250
	9	90·97	+0·23	6 04 13·8	+23 40 01	58·597	58 40·21	15 59·19	17·5901	05·1043
	10	104·88	−0·99	7 04 08·0	+21 37 18	58·251	59 01·15	16 04·90	18·5404	06·0694
	11	118·95	−2·17	8 02 28·0	+18 14 36	57·966	59 18·57	16 09·64	19·4535	07·0018
	12	133·17	−3·23	8 58 45·1	+13 46 06	57·761	59 31·20	16 13·08	20·3296	07·8957
	13	147·50	−4·09	9 53 06·7	+ 8 29 42	57·661	59 37·40	16 14·77	21·1784	08·7566
	14	161·91	−4·70	10 46 05·8	+ 2 45 03	57·691	59 35·49	16 14·25	22·0142	09·5970
	15	176·30	−5·01	11 38 28·3	− 3 07 26	57·874	59 24·19	16 11·17	22·8521	10·4321
	16	190·59	−5·00	12 31 01·4	− 8 47 21	58·219	59 03·06	16 05·42	23·7045	11·2759
	17	204·68	−4·69	13 24 23·1	−13 55 07	58·721	58 32·81	15 57·18	. . .	12·1386
	18	218·49	−4·10	14 18 53·6	−18 12 59	59·354	57 55·33	15 46·96	00·5781	13·0226
	19	231·95	−3·30	15 14 27·6	−21 26 39	60·079	57 13·41	15 35·54	01·4711	13·9218
	20	245·04	−2·35	16 10 32·5	−23 26 43	60·842	56 30·31	15 23·80	02·3728	14·8216
	21	257·76	−1·29	17 06 15·9	−24 09 50	61·587	55 49·32	15 12·63	03·2660	15·7038
	22	270·16	−0·20	18 00 41·3	−23 38 30	62·255	55 13·36	15 02·83	04·1330	16·5524
	23	282·29	+0·88	18 53 05·5	−21 59 51	62·797	54 44·78	14 55·04	04·9611	17·3587
	24	294·24	+1·90	19 43 08·6	−19 23 34	63·172	54 25·25	14 49·72	05·7457	18·1225
	25	306·09	+2·84	20 30 56·0	−15 59 59	63·356	54 15·78	14 47·14	06·4902	18·8501
	26	317·92	+3·65	21 16 52·5	−11 58 58	63·338	54 16·71	14 47·40	07·2038	19·5530
	27	329·83	+4·31	22 01 36·1	− 7 29 35	63·123	54 27·80	14 50·42	07·8994	20·2450
	28	341·89	+4·79	22 45 53·0	− 2 40 31	62·732	54 48·17	14 55·97	08·5918	20·9418
	29	354·16	+5·05	23 30 33·6	+ 2 19 21	62·198	55 16·40	15 03·66	09·2971	21·6597
	30	6·70	+5·09	0 16 30·8	+ 7 19 50	61·565	55 50·53	15 12·96	10·0318	22·4152
May	1	19·54	+4·87	1 04 36·3	+12 08 41	60·881	56 28·13	15 23·20	10·8118	23·2229
	2	32·67	+4·39	1 55 35·2	+16 30 55	60·199	57 06·52	15 33·67	11·6497	. . .
	3	46·07	+3·66	2 49 56·1	+20 08 59	59·565	57 43·04	15 43·61	12·5512	00·0926
	4	59·72	+2·71	3 47 36·5	+22 44 15	59·015	58 15·32	15 52·41	13·5091	01·0241
	5	73·57	+1·58	4 47 52·2	+24 00 05	58·574	58 41·61	15 59·57	14·5020	02·0030
	6	87·56	+0·35	5 49 19·1	+23 46 02	58·253	59 01·00	16 04·86	15·4981	03·0017
	7	101·64	−0·92	6 50 17·2	+22 01 01	58·051	59 13·36	16 08·22	16·4677	03·9877
	8	115·79	−2·13	7 49 25·2	+18 53 28	57·956	59 19·14	16 09·80	17·3928	04·9363
	9	129·96	−3·22	8 46 03·0	+14 38 46	57·957	59 19·10	16 09·79	18·2710	05·8374
	10	144·13	−4·11	9 40 14·2	+ 9 35 36	58·041	59 13·96	16 08·39	19·1116	06·6951
	11	158·27	−4·75	10 32 33·5	+ 4 03 17	58·201	59 04·19	16 05·72	19·9300	07·5225
	12	172·35	−5·10	11 23 51·7	− 1 39 24	58·436	58 49·95	16 01·84	20·7436	08·3363
	13	186·33	−5·15	12 15 03·4	− 7 14 27	58·748	58 31·19	15 56·73	21·5681	09·1536
	14	200·17	−4·89	13 06 56·6	−12 24 29	59·140	58 07·91	15 50·39	22·4150	09·9883
	15	213·81	−4·36	14 00 04·9	−16 52 57	59·611	57 40·33	15 42·87	23·2880	10·8483
	16	227·21	−3·59	14 54 37·8	−20 24 57	60·153	57 09·16	15 34·38	. . .	11·7329
	17	240·34	−2·65	15 50 16·1	−22 48 43	60·746	56 35·67	15 25·26	00·1814	12·6314

EPHEMERIS FOR PHYSICAL OBSERVATIONS

FOR 0^h DYNAMICAL TIME

Date 0^h TDT		Age	The Earth's Selenographic Long.	The Earth's Selenographic Lat.	Physical Libration Lg.	Physical Libration Lt.	Physical Libration P.A.	The Sun's Selenographic Colong.	The Sun's Selenographic Lat.	Position Angle Axis	Position Angle Bright Limb	Fraction Illum.
		d	°	°	(0°.001)			°	°	°	°	
Apr.	**1**	27·4	−3·436	−6·199	−31	+4	+19	247·80	+1·52	337·814	77·12	0·05
	2	28·4	4·187	6·505	31	3	20	260·02	1·52	336·707	85·54	0·02
	3	29·4	4·706	6·501	32	2	21	272·24	1·51	336·573	130·22	0·00
	4	0·8	5·000	6·170	32	2	21	284·46	1·50	337·532	221·60	0·01
	5	1·8	5·092	5·512	32	2	22	296·68	1·50	339·687	239·92	0·04
	6	2·8	−5·004	−4·548	−32	+1	+22	308·90	+1·49	343·073	248·98	0·09
	7	3·8	4·757	3·319	31	1	22	321·11	1·48	347·598	256·48	0·15
	8	4·8	4·362	1·886	30	1	22	333·32	1·48	353·009	263·60	0·24
	9	5·8	3·819	−0·325	29	1	21	345·53	1·47	358·902	270·34	0·34
	10	6·8	3·120	+1·273	28	1	20	357·73	1·47	4·811	276·40	0·45
	11	7·8	−2·256	+2·814	−26	+2	+19	9·92	+1·46	10·311	281·41	0·57
	12	8·8	1·231	4·196	25	2	18	22·11	1·46	15·084	285·04	0·68
	13	9·8	−0·068	5·327	24	3	17	34·29	1·46	18·918	287·06	0·78
	14	10·8	+1·182	6·126	23	3	16	46·46	1·45	21·670	287·12	0·87
	15	11·8	2·435	6·537	22	4	15	58·63	1·45	23·226	284·47	0·94
	16	12·8	+3·590	+6·536	−21	+4	+14	70·80	+1·44	23·494	276·04	0·98
	17	13·8	4·538	6·135	21	4	13	82·96	1·43	22·420	230·45	1·00
	18	14·8	5·188	5·381	20	4	12	95·13	1·42	20·023	130·68	0·99
	19	15·8	5·475	4·341	20	4	12	107·30	1·41	16·434	113·09	0·96
	20	16·8	5·373	3·095	21	4	12	119·47	1·40	11·908	103·91	0·91
	21	17·8	+4·895	+1·726	−21	+4	+12	131·65	+1·39	6·798	96·53	0·84
	22	18·8	4·085	+0·305	21	4	12	143·83	1·37	1·483	90·06	0·77
	23	19·8	3·009	−1·101	22	4	13	156·02	1·35	356·298	84·43	0·68
	24	20·8	1·751	2·439	23	4	13	168·21	1·34	351·489	79·70	0·59
	25	21·8	+0·401	3·660	24	4	14	180·41	1·32	347·212	75·94	0·49
	26	22·8	−0·949	−4·724	−25	+4	+16	192·62	+1·31	343·557	73·20	0·40
	27	23·8	2·212	5·589	25	3	17	204·83	1·29	340·583	71·52	0·31
	28	24·8	3·311	6·218	26	3	18	217·05	1·27	338·350	70·97	0·22
	29	25·8	4·183	6·574	27	2	19	229·27	1·25	336·939	71·74	0·15
	30	26·8	4·784	6·625	27	2	20	241·50	1·24	336·458	74·24	0·08
May	**1**	27·8	−5·095	−6·347	−27	+1	+21	253·74	+1·22	337·041	79·71	0·04
	2	28·8	5·117	5·732	27	0	22	265·97	1·20	338·819	94·90	0·01
	3	0·3	4·875	4·791	26	0	22	278·21	1·18	341·876	204·65	0·00
	4	1·3	4·408	3·561	25	−1	22	290·45	1·16	346·173	248·04	0·02
	5	2·3	3·762	2·102	24	1	22	302·68	1·14	351·492	260·05	0·06
	6	3·3	−2·983	−0·501	−22	−1	+21	314·91	+1·12	357·424	268·55	0·13
	7	4·3	2·110	+1·145	21	1	20	327·14	1·11	3·463	275·61	0·21
	8	5·3	1·172	2·729	19	−1	19	339·36	1·09	9·135	281·40	0·31
	9	6·3	−0·194	4·149	18	0	18	351·58	1·07	14·094	285·80	0·43
	10	7·3	+0·802	5·316	16	0	17	3·79	1·05	18·129	288·76	0·54
	11	8·3	+1·787	+6·156	−15	+1	+15	15·99	+1·04	21·114	290·23	0·65
	12	9·3	2·725	6·622	14	1	14	28·19	1·02	22·962	290·19	0·76
	13	10·3	3·568	6·689	14	1	13	40·38	1·00	23·597	288·56	0·85
	14	11·3	4·260	6·364	13	2	12	52·57	0·98	22·956	285·08	0·92
	15	12·3	4·747	5·681	13	2	11	64·75	0·96	21·021	278·78	0·97
	16	13·3	+4·980	+4·693	−13	+2	+10	76·94	+0·94	17·857	263·18	0·99
	17	14·3	+4·926	+3·474	−13	+2	+10	89·12	+0·92	13·649	135·83	1·00

FOR 0^h DYNAMICAL TIME

Date 0^h TDT		Apparent Long.	Apparent Lat.	Apparent R.A.	Apparent Dec.	True Dist.	Horiz. Parallax	Semi-diameter	Ephemeris Transit for date Upper	Ephemeris Transit for date Lower
		°	°	h m s	° ′ ″		′ ″	′ ″	h	h
May	**17**	240·34	−2·65	15 50 16·1	−22 48 43	60·746	56 35·67	15 25·26	00·1814	12·6314
	18	253·18	−1·59	16 46 12·6	−23 57 28	61·362	56 01·58	15 15·97	01·0804	13·5257
	19	265·75	−0·47	17 41 25·5	−23 50 25	61·963	55 28·97	15 07·09	01·9650	14·3960
	20	278·07	+0·65	18 34 57·1	−22 32 36	62·507	55 00·01	14 59·19	02·8171	15·2272
	21	290·17	+1·73	19 26 10·6	−20 13 04	62·951	54 36·75	14 52·86	03·6260	16·0134
	22	302·11	+2·71	20 14 56·7	−17 02 45	63·255	54 21·00	14 48·57	04·3901	16·7571
	23	313·97	+3·58	21 01 31·2	−13 12 35	63·388	54 14·16	14 46·70	05·1157	17·4675
	24	325·82	+4·28	21 46 27·5	− 8 52 34	63·329	54 17·17	14 47·52	05·8142	18·1577
	25	337·75	+4·81	22 30 30·5	− 4 11 38	63·072	54 30·45	14 51·14	06·5002	18·8436
	26	349·82	+5·14	23 14 31·6	+ 0 41 46	62·625	54 53·81	14 57·51	07·1901	19·5420
	27	2·13	+5·23	23 59 26·7	+ 5 38 50	62·012	55 26·37	15 06·38	07·9015	20·2710
	28	14·72	+5·09	0 46 13·1	+10 29 14	61·272	56 06·51	15 17·31	08·6525	21·0482
	29	27·65	+4·68	1 35 45·8	+15 00 04	60·459	56 51·80	15 29·65	09·4598	21·8886
	30	40·95	+4·01	2 28 49·2	+18 55 10	59·634	57 39·03	15 42·52	10·3354	22·8000
	31	54·60	+3·10	3 25 42·4	+21 55 30	58·861	58 24·47	15 54·90	11·2811	23·7764
June	**1**	68·58	+1·98	4 26 02·0	+23 41 35	58·200	59 04·24	16 05·74	12·2822	. . .
	2	82·82	+0·72	5 28 33·3	+23 57 59	57·700	59 34·94	16 14·10	13·3075	00·7942
	3	97·25	−0·60	6 31 24·9	+22 38 25	57·391	59 54·23	16 19·36	14·3193	01·8172
	4	111·77	−1·90	7 32 47·6	+19 48 23	57·279	60 01·24	16 21·27	15·2883	02·8103
	5	126·29	−3·07	8 31 31·7	+15 43 25	57·352	59 56·63	16 20·01	16·2027	03·7523
	6	140·75	−4·04	9 27 19·8	+10 44 28	57·582	59 42·26	16 16·10	17·0668	04·6403
	7	155·07	−4·75	10 20 37·1	+ 5 13 31	57·933	59 20·58	16 10·19	17·8948	05·4842
	8	169·20	−5·17	11 12 13·1	− 0 28 57	58·368	58 54·06	16 02·96	18·7049	06·3009
	9	183·13	−5·27	12 03 05·4	− 6 04 44	58·855	58 24·78	15 54·99	19·5149	07·1089
	10	196·83	−5·07	12 54 08·7	−11 17 29	59·373	57 54·24	15 46·67	20·3394	07·9246
	11	210·30	−4·59	13 46 06·4	−15 52 16	59·905	57 23·37	15 38·25	21·1872	08·7601
	12	223·53	−3·87	14 39 22·0	−19 35 33	60·443	56 52·70	15 29·90	22·0588	09·6204
	13	236·52	−2·97	15 33 52·9	−22 15 49	60·981	56 22·58	15 21·69	22·9450	10·5010
	14	249·29	−1·93	16 29 07·6	−23 45 01	61·512	55 53·41	15 13·75	23·8289	11·3884
	15	261·83	−0·81	17 24 12·5	−23 59 57	62·023	55 25·77	15 06·21	. . .	12·2640
	16	274·17	+0·33	18 18 08·0	−23 02 58	62·497	55 00·51	14 59·33	00·6916	13·1099
	17	286·33	+1·43	19 10 06·3	−21 01 12	62·913	54 38·73	14 53·40	01·5178	13·9147
	18	298·34	+2·46	19 59 43·4	−18 04 54	63·241	54 21·73	14 48·77	02·3006	14·6759
	19	310·25	+3·38	20 47 01·7	−14 25 21	63·452	54 10·87	14 45·81	03·0415	15·3986
	20	322·10	+4·14	21 32 24·4	−10 13 29	63·518	54 07·48	14 44·88	03·7486	16·0933
	21	333·95	+4·73	22 16 29·4	− 5 39 10	63·416	54 12·69	14 46·30	04·4343	16·7737
	22	345·87	+5·11	23 00 03·9	− 0 51 19	63·132	54 27·36	14 50·30	05·1136	17·4560
	23	357·94	+5·28	23 44 00·8	+ 4 01 34	62·661	54 51·89	14 56·98	05·8033	18·1575
	24	10·22	+5·22	0 29 16·8	+ 8 50 24	62·017	55 26·10	15 06·30	06·5211	18·8963
	25	22·78	+4·90	1 16 49·8	+13 24 30	61·226	56 09·05	15 18·01	07·2854	19·6902
	26	35·70	+4·34	2 07 33·8	+17 30 27	60·334	56 58·88	15 31·58	08·1124	20·5532
	27	49·02	+3·52	3 02 08·2	+20 51 36	59·400	57 52·63	15 46·23	09·0128	21·4905
	28	62·75	+2·48	4 00 40·7	+23 08 44	58·496	58 46·32	16 00·85	09·9844	22·4915
	29	76·89	+1·26	5 02 30·3	+24 03 13	57·697	59 35·17	16 14·16	11·0074	23·5274
	30	91·39	−0·06	6 06 04·3	+23 22 10	57·072	60 14·28	16 24·82	12·0463	. . .
July	**1**	106·16	−1·41	7 09 23·6	+21 03 39	56·677	60 39·48	16 31·69	13·0630	00·5594
	2	121·08	−2·68	8 10 46·2	+17 18 24	56·540	60 48·32	16 34·09	14·0320	01·5543

EPHEMERIS FOR PHYSICAL OBSERVATIONS

FOR 0^h DYNAMICAL TIME

Date 0^h TDT		Age	The Earth's Selenographic Long.	The Earth's Selenographic Lat.	Physical Libration Lg.	Physical Libration Lt.	Physical Libration P.A.	The Sun's Selenographic Colong.	The Sun's Selenographic Lat.	Position Angle Axis	Position Angle Bright Limb	Fraction Illum.
		d	°	°	(0°001)			°	°	°	°	
May	17	14·3	+4·926	+3·474	−13	+ 2	+10	89·12	+0·92	13·649	135·83	1·00
	18	15·3	4·574	2·100	13	3	10	101·30	0·90	8·701	102·80	0·98
	19	16·3	3·934	+0·651	13	3	10	113·49	0·87	3·388	92·75	0·94
	20	17·3	3·035	−0·802	14	3	11	125·68	0·85	358·081	85·72	0·89
	21	18·3	1·928	2·196	14	3	12	137·87	0·82	353·078	80·15	0·82
	22	19·3	+0·674	−3·476	−15	+ 3	+13	150·07	+0·80	348·575	75·72	0·74
	23	20·3	−0·654	4·599	16	3	14	162·27	0·78	344·688	72·35	0·66
	24	21·3	1·976	5·524	16	2	15	174·48	0·75	341·478	70·00	0·56
	25	22·3	3·213	6·216	17	2	17	186·70	0·73	338·994	68·65	0·47
	26	23·3	4·286	6·643	18	2	18	198·92	0·70	337·296	68·31	0·37
	27	24·3	−5·125	−6·776	−18	+ 1	+19	211·15	+0·68	336·472	69·02	0·28
	28	25·3	5·671	6·590	18	0	21	223·38	0·65	336·644	70·90	0·20
	29	26·3	5·886	6·067	17	0	21	235·62	0·63	337·954	74·11	0·12
	30	27·3	5·751	5·208	17	− 1	22	247·87	0·60	340·527	79·01	0·06
	31	28·3	5·276	4·033	16	2	22	260·11	0·58	344·408	86·88	0·02
June	1	29·3	−4·494	−2·592	−14	− 2	+22	272·36	+0·55	349·473	123·21	0·00
	2	0·8	3·461	−0·962	12	3	21	284·61	0·52	355·378	263·21	0·01
	3	1·8	2·251	+0·750	11	3	21	296·86	0·50	1·601	274·43	0·05
	4	2·8	−0·941	2·426	9	3	20	309·10	0·47	7·591	281·44	0·11
	5	3·8	+0·388	3·947	7	3	18	321·35	0·45	12·907	286·58	0·19
	6	4·8	+1·663	+5·209	− 5	− 3	+17	333·58	+0·42	17·278	290·14	0·29
	7	5·8	2·824	6·134	4	2	16	345·81	0·40	20·563	292·20	0·40
	8	6·8	3·818	6·673	3	2	14	358·03	0·37	22·689	292·84	0·52
	9	7·8	4·610	6·809	2	2	13	10·25	0·35	23·605	292·16	0·63
	10	8·8	5·173	6·552	1	1	12	22·46	0·32	23·274	290·22	0·73
	11	9·8	+5·490	+5·936	− 1	− 1	+10	34·66	+0·30	21·684	287·08	0·82
	12	10·8	5·555	5·012	1	− 1	10	46·86	0·27	18·879	282·84	0·89
	13	11·8	5·366	3·843	1	0	9	59·06	0·24	14·997	277·58	0·95
	14	12·8	4·933	2·501	1	0	9	71·25	0·21	10·285	271·10	0·99
	15	13·8	4·270	+1·058	1	0	9	83·44	0·19	5·079	254·21	1·00
	16	14·8	+3·398	−0·411	− 2	0	+ 9	95·64	+0·16	359·747	86·16	0·99
	17	15·8	2·349	1·841	3	0	10	107·83	0·13	354·615	79·25	0·97
	18	16·8	+1·159	3·173	3	0	11	120·03	0·10	349·922	74·51	0·93
	19	17·8	−0·126	4·354	4	0	12	132·23	0·07	345·817	70·91	0·87
	20	18·8	1·454	5·343	5	0	14	144·43	0·05	342·382	68·31	0·80
	21	19·8	−2·764	−6·103	− 6	0	+15	156·64	+0·02	339·667	66·66	0·72
	22	20·8	3·994	6·606	6	0	17	168·85	−0·01	337·719	65·94	0·63
	23	21·8	5·074	6·824	7	− 1	18	181·07	0·03	336·604	66·16	0·53
	24	22·8	5·935	6·737	7	1	19	193·30	0·06	336·418	67·34	0·44
	25	23·8	6·512	6·330	7	1	20	205·53	0·08	337·282	69·54	0·34
	26	24·8	−6·750	−5·596	− 6	− 2	+21	217·77	−0·11	339·327	72·81	0·24
	27	25·8	6·606	4·542	5	2	21	230·01	0·13	342·649	77·12	0·16
	28	26·8	6·060	3·199	4	3	21	242·26	0·16	347·229	82·28	0·09
	29	27·8	5·124	−1·625	− 2	3	21	254·51	0·19	352·851	87·69	0·03
	30	28·8	3·839	+0·091	0	4	20	266·76	0·21	359·082	90·09	0·00
July	1	0·5	−2·286	+1·831	+ 2	− 4	+20	279·02	−0·24	5·350	288·57	0·00
	2	1·5	−0·570	+3·464	+ 4	− 4	+18	291·27	−0·27	11·114	289·46	0·03

FOR 0ʰ DYNAMICAL TIME

Date 0ʰ TDT		Apparent Long.	Apparent Lat.	Apparent R.A.	Apparent Dec.	True Dist.	Horiz. Parallax	Semi-diameter	Ephemeris Transit for date Upper	Ephemeris Transit for date Lower
		°	°	h m s	° ′ ″		′ ″	′ ″	h	h
July	**1**	106·16	−1·41	7 09 23·6	+21 03 39	56·677	60 39·48	16 31·69	13·0630	00·5594
	2	121·08	−2·68	8 10 46·2	+17 18 24	56·540	60 48·32	16 34·09	14·0320	01·5543
	3	136·02	−3·76	9 09 19·1	+12 26 46	56·660	60 40·59	16 31·99	14·9463	02·4958
	4	150·84	−4·58	10 05 01·4	+ 6 53 26	57·009	60 18·29	16 25·91	15·8138	03·3850
	5	165·43	−5·09	10 58 28·9	+ 1 02 47	57·539	59 44·98	16 16·84	16·6502	04·2348
	6	179·72	−5·27	11 50 34·5	− 4 43 36	58·191	59 04·79	16 05·89	17·4733	05·0623
	7	193·67	−5·14	12 42 13·4	−10 07 30	58·908	58 21·62	15 54·13	18·2990	05·8849
	8	207·26	−4·72	13 34 12·5	−14 53 39	59·641	57 38·60	15 42·40	19·1388	06·7167
	9	220·52	−4·05	14 27 03·0	−18 49 13	60·352	56 57·86	15 31·31	19·9971	07·5657
	10	233·47	−3·19	15 20 53·5	−21 43 43	61·016	56 20·68	15 21·17	20·8700	08·4324
	11	246·15	−2·19	16 15 26·5	−23 29 34	61·618	55 47·62	15 12·17	21·7456	09·3085
	12	258·61	−1·10	17 10 02·2	−24 03 07	62·153	55 18·82	15 04·32	22·6077	10·1794
	13	270·88	+0·02	18 03 49·2	−23 25 15	62·616	54 54·26	14 57·63	23·4410	11·0287
	14	283·00	+1·12	18 56 00·5	−21 41 20	63·005	54 33·93	14 52·09	. . .	11·8436
	15	295·00	+2·16	19 46 06·1	−19 00 11	63·313	54 17·99	14 47·75	00·2359	12·6180
	16	306·91	+3·10	20 33 58·9	−15 32 24	63·531	54 06·81	14 44·70	00·9903	13·3534
	17	318·77	+3·90	21 19 52·2	−11 29 03	63·644	54 01·03	14 43·12	01·7086	14·0571
	18	330·62	+4·53	22 04 15·5	− 7 00 39	63·636	54 01·44	14 43·24	02·4004	14·7403
	19	342·48	+4·97	22 47 48·3	− 2 16 55	63·489	54 08·96	14 45·29	03·0784	15·4166
	20	354·42	+5·19	23 31 17·1	+ 2 33 05	63·188	54 24·46	14 49·51	03·7570	16·1015
	21	6·49	+5·19	0 15 32·9	+ 7 20 22	62·723	54 48·63	14 56·09	04·4523	16·8115
	22	18·74	+4·95	1 01 29·2	+11 55 16	62·097	55 21·80	15 05·13	05·1812	17·5634
	23	31·26	+4·48	1 49 59·0	+16 06 26	61·323	56 03·73	15 16·56	05·9601	18·3729
	24	44·10	+3·77	2 41 48·0	+19 40 07	60·432	56 53·34	15 30·07	06·8029	19·2508
	25	57·33	+2·85	3 37 23·4	+22 19 58	59·471	57 48·49	15 45·10	07·7163	20·1982
	26	70·99	+1·73	4 36 37·3	+23 48 19	58·504	58 45·79	16 00·71	08·6940	21·2005
	27	85·11	+0·48	5 38 35·6	+23 49 33	57·608	59 40·65	16 15·66	09·7135	22·2286
	28	99·65	−0·84	6 41 43·6	+22 14 53	56·862	60 27·62	16 28·45	10·7413	23·2478
	29	114·55	−2·14	7 44 16·1	+19 06 37	56·340	61 01·26	16 37·62	11·7453	. . .
	30	129·69	−3·30	8 44 54·0	+14 38 41	56·094	61 17·33	16 42·00	12·7065	00·2318
	31	144·92	−4·23	9 43 04·7	+ 9 13 39	56·147	61 13·88	16 41·06	13·6224	01·1698
Aug.	**1**	160·05	−4·86	10 38 58·9	+ 3 17 52	56·486	60 51·81	16 35·04	14·5019	02·0658
	2	174·93	−5·15	11 33 14·4	− 2 42 51	57·068	60 14·52	16 24·89	15·3598	02·9326
	3	189·45	−5·10	12 26 38·7	− 8 25 47	57·829	59 26·98	16 11·93	16·2109	03·7853
	4	203·54	−4·74	13 19 56·4	−13 32 13	58·694	58 34·41	15 57·61	17·0670	04·6378
	5	217·17	−4·12	14 13 38·6	−17 47 29	59·592	57 41·46	15 43·18	17·9340	05·4991
	6	230·37	−3·29	15 07 57·2	−21 00 40	60·462	56 51·61	15 29·60	18·8103	06·3714
	7	243·19	−2·32	16 02 41·0	−23 04 34	61·261	56 07·13	15 17·48	19·6871	07·2494
	8	255·70	−1·27	16 57 18·5	−23 55 59	61·958	55 29·23	15 07·16	20·5512	08·1216
	9	267·96	−0·17	17 51 07·0	−23 35 52	62·539	54 58·30	14 58·73	21·3893	08·9742
	10	280·05	+0·91	18 43 25·7	−22 09 06	62·999	54 34·21	14 52·17	22·1920	09·7954
	11	292·01	+1·94	19 33 47·0	−19 43 33	63·341	54 16·54	14 47·35	22·9562	10·5788
	12	303·90	+2·88	20 22 02·3	−16 28 56	63·570	54 04·82	14 44·16	23·6847	11·3245
	13	315·75	+3·68	21 08 21·3	−12 35 44	63·691	53 58·65	14 42·48	. . .	12·0378
	14	327·60	+4·33	21 53 08·0	− 8 14 20	63·707	53 57·85	14 42·26	00·3852	12·7281
	15	339·47	+4·79	22 36 56·7	− 3 34 47	63·616	54 02·45	14 43·51	01·0683	13·4072
	16	351·40	+5·04	23 20 27·8	+ 1 13 19	63·414	54 12·78	14 46·33	01·7466	14·0883

EPHEMERIS FOR PHYSICAL OBSERVATIONS
FOR 0ʰ DYNAMICAL TIME

Date 0ʰ TDT		Age	The Earth's Selenographic Long.	The Earth's Selenographic Lat.	Physical Libration Lg.	Physical Libration Lt.	Physical Libration P.A.	The Sun's Selenographic Colong.	The Sun's Selenographic Lat.	Position Angle Axis	Position Angle Bright Limb	Fraction Illum.
		d	°	°	(0°001)			°	°	°	°	
July	1	0·5	−2·286	+1·831	+ 2	− 4	+20	279·02	−0·24	5·350	288·57	0·00
	2	1·5	−0·570	3·464	4	4	18	291·27	0·27	11·114	289·46	0·03
	3	2·5	+1·181	4·862	6	4	17	303·52	0·30	15·985	292·45	0·09
	4	3·5	2·838	5·921	7	4	16	315·77	0·32	19·741	294·51	0·17
	5	4·5	4·288	6·579	9	4	14	328·01	0·35	22·273	295·33	0·27
	6	5·5	+5·444	+6·811	+10	− 4	+13	340·24	−0·38	23·533	294·88	0·38
	7	6·5	6·254	6·632	11	4	12	352·47	0·40	23·501	293·25	0·49
	8	7·5	6·701	6·080	11	4	10	4·69	0·43	22·188	290·53	0·60
	9	8·5	6·794	5·214	11	4	9	16·91	0·45	19·652	286·87	0·70
	10	9·5	6·565	4·098	11	4	9	29·12	0·48	16·024	282·50	0·79
	11	10·5	+6·053	+2·802	+11	− 3	+ 8	41·32	−0·51	11·525	277·74	0·87
	12	11·5	5·304	+1·395	11	3	8	53·52	0·54	6·459	273·12	0·93
	13	12·5	4·362	−0·056	10	3	8	65·72	0·57	1·170	269·67	0·97
	14	13·5	3·271	1·485	10	3	9	77·92	0·59	355·984	271·68	0·99
	15	14·5	2·068	2·833	9	3	9	90·11	0·62	351·162	35·73	1·00
	16	15·5	+0·790	−4·046	+ 8	− 3	+10	102·31	−0·65	346·879	62·64	0·99
	17	16·5	−0·529	5·078	7	3	12	114·50	0·67	343·243	63·57	0·96
	18	17·5	1·852	5·889	6	3	13	126·70	0·69	340·317	63·12	0·91
	19	18·5	3·139	6·449	5	3	15	138·90	0·72	338·150	62·89	0·85
	20	19·5	4·347	6·733	4	3	16	151·11	0·74	336·797	63·28	0·77
	21	20·5	−5·424	−6·723	+ 4	− 4	+17	163·32	−0·76	336·333	64·42	0·69
	22	21·5	6·312	6·409	4	4	18	175·53	0·78	336·852	66·42	0·59
	23	22·5	6·951	5·787	4	4	19	187·75	0·80	338·463	69·34	0·49
	24	23·5	7·277	4·863	4	5	20	199·98	0·82	341·263	73·19	0·39
	25	24·5	7·231	3·657	5	5	20	212·21	0·84	345·280	77·90	0·29
	26	25·5	−6·768	−2·207	+ 7	− 5	+20	224·45	−0·86	350·409	83·20	0·19
	27	26·5	5·866	−0·581	8	5	19	236·69	0·88	356·354	88·46	0·11
	28	27·5	4·537	+1·130	10	5	18	248·94	0·90	2·634	92·37	0·05
	29	28·5	2·843	2·802	12	5	17	261·19	0·92	8·702	89·76	0·01
	30	0·2	−0·894	4·304	14	5	16	273·44	0·95	14·073	336·82	0·00
	31	1·2	+1·160	+5·507	+16	− 6	+15	285·69	−0·97	18·412	302·92	0·02
Aug.	1	2·2	3·144	6·316	17	6	13	297·94	0·99	21·518	300·00	0·07
	2	3·2	4·892	6·682	19	6	12	310·19	1·01	23·287	298·50	0·15
	3	4·2	6·275	6·606	20	6	11	322·43	1·03	23·678	296·48	0·24
	4	5·2	7·215	6·127	21	6	10	334·66	1·05	22·701	293·60	0·34
	5	6·2	+7·687	+5·311	+21	− 6	+ 9	346·89	−1·08	20·426	289·85	0·45
	6	7·2	7·713	4·233	22	6	8	359·11	1·10	17·000	285·39	0·56
	7	8·2	7·344	2·970	22	6	7	11·32	1·12	12·652	280·47	0·66
	8	9·2	6·650	1·594	21	6	7	23·53	1·14	7·684	275·44	0·75
	9	10·2	5·704	+0·171	21	6	7	35·73	1·16	2·433	270·73	0·83
	10	11·2	+4·577	−1·237	+20	− 6	+ 7	47·93	−1·18	357·219	266·89	0·90
	11	12·2	3·333	2·574	19	6	8	60·12	1·20	352·306	264·76	0·95
	12	13·2	2·025	3·788	18	6	9	72·31	1·22	347·884	266·79	0·98
	13	14·2	+0·695	4·833	17	6	10	84·50	1·24	344·074	290·75	1·00
	14	15·2	−0·624	5·668	16	6	12	96·69	1·26	340·955	35·10	1·00
	15	16·2	−1·906	−6·258	+15	− 6	+13	108·88	−1·27	338·584	52·86	0·98
	16	17·2	−3·125	−6·579	+14	− 7	+14	121·07	−1·29	337·019	57·46	0·94

FOR 0h DYNAMICAL TIME

Date 0h TDT		Apparent Long.	Apparent Lat.	Apparent R.A.	Apparent Dec.	True Dist.	Horiz. Parallax	Semi-diameter	Ephemeris Transit for date Upper	Ephemeris Transit for date Lower
		°	°	h m s	° ′ ″		′ ″	′ ″	h	h
Aug.	16	351·40	+5·04	23 20 27·8	+ 1 13 19	63·414	54 12·78	14 46·33	01·7466	14·0883
	17	3·41	+5·07	0 04 25·8	+ 6 00 27	63·094	54 29·31	14 50·83	02·4341	14·7859
	18	15·54	+4·87	0 49 37·4	+10 36 46	62·648	54 52·59	14 57·17	03·1454	15·5145
	19	27·84	+4·45	1 36 49·0	+14 51 31	62·073	55 23·08	15 05·48	03·8949	16·2882
	20	40·35	+3·81	2 26 42·5	+18 32 30	61·374	56 00·95	15 15·80	04·6956	17·1181
	21	53·14	+2·97	3 19 47·1	+21 25 50	60·565	56 45·82	15 28·02	05·5561	18·0093
	22	66·27	+1·95	4 16 08·8	+23 16 30	59·678	57 36·45	15 41·82	06·4767	18·9563
	23	79·79	+0·79	5 15 19·8	+23 50 10	58·759	58 30·55	15 56·56	07·4457	19·9415
	24	93·75	−0·45	6 16 17·7	+22 56 19	57·869	59 24·52	16 11·26	08·4404	20·9389
	25	108·15	−1·70	7 17 40·2	+20 31 46	57·082	60 13·65	16 24·65	09·4339	21·9230
	26	122·95	−2·86	8 18 12·2	+16 42 51	56·475	60 52·54	16 35·25	10·4047	22·8781
	27	138·05	−3·85	9 17 08·5	+11 45 10	56·113	61 16·10	16 41·66	11·3433	23·8008
	28	153·32	−4·57	10 14 21·0	+ 6 01 10	56·041	61 20·77	16 42·94	12·2519	. . .
	29	168·57	−4·97	11 10 11·2	− 0 03 21	56·273	61 05·60	16 38·80	13·1404	00·6979
	30	183·61	−5·02	12 05 15·5	− 6 02 29	56·786	60 32·52	16 29·79	14·0213	01·5811
	31	198·31	−4·73	13 00 11·5	−11 32 51	57·525	59 45·82	16 17·07	14·9052	02·4624
Sept.	1	212·54	−4·15	13 55 26·8	−16 15 08	58·418	58 51·01	16 02·13	15·7975	03·3502
	2	226·28	−3·35	14 51 10·2	−19 54 58	59·384	57 53·59	15 46·49	16·6965	04·2465
	3	239·54	−2·39	15 47 08·2	−22 23 17	60·346	56 58·20	15 31·40	17·5933	05·1459
	4	252·36	−1·34	16 42 47·8	−23 36 22	61·241	56 08·24	15 17·78	18·4748	06·0369
	5	264·83	−0·25	17 37 26·4	−23 35 27	62·022	55 25·83	15 06·23	19·3279	06·9056
	6	277·03	+0·82	18 30 24·7	−22 25 54	62·659	54 52·02	14 57·02	20·1435	07·7407
	7	289·03	+1·84	19 21 17·8	−20 15 52	63·138	54 27·03	14 50·21	20·9189	08·5362
	8	300·91	+2·77	20 09 59·7	−17 14 59	63·458	54 10·52	14 45·71	21·6572	09·2923
	9	312·75	+3·57	20 56 42·1	−13 33 18	63·629	54 01·80	14 43·34	22·3658	10·0146
	10	324·59	+4·22	21 41 50·1	− 9 20 46	63·664	54 00·01	14 42·85	23·0554	10·7123
	11	336·47	+4·68	22 25 57·3	− 4 47 02	63·580	54 04·29	14 44·01	23·7379	11·3968
	12	348·43	+4·94	23 09 42·8	− 0 01 38	63·393	54 13·91	14 46·63	. . .	12·0806
	13	0·48	+4·98	23 53 48·3	+ 4 45 42	63·113	54 28·33	14 50·56	00·4263	12·7769
	14	12·64	+4·80	0 38 56·4	+ 9 24 44	62·749	54 47·28	14 55·73	01·1339	13·4990
	15	24·94	+4·39	1 25 48·2	+13 44 20	62·305	55 10·73	15 02·11	01·8736	14·2590
	16	37·39	+3·77	2 14 59·5	+17 32 19	61·781	55 38·79	15 09·76	02·6565	15·0667
	17	50·02	+2·96	3 06 54·3	+20 35 29	61·180	56 11·59	15 18·70	03·4900	15·9264
	18	62·87	+1·97	4 01 37·3	+22 40 15	60·507	56 49·08	15 28·91	04·3750	16·8345
	19	75·99	+0·87	4 58 46·2	+23 34 06	59·777	57 30·74	15 40·26	05·3030	17·7781
	20	89·43	−0·31	5 57 32·3	+23 07 42	59·014	58 15·33	15 52·41	06·2570	18·7372
	21	103·23	−1·50	6 56 51·1	+21 17 12	58·259	59 00·65	16 04·76	07·2160	19·6915
	22	117·40	−2·64	7 55 42·4	+18 05 43	57·563	59 43·49	16 16·43	08·1620	20·6267
	23	131·94	−3·63	8 53 27·6	+13 43 34	56·987	60 19·72	16 26·30	09·0855	21·5386
	24	146·78	−4·39	9 49 58·2	+ 8 27 10	56·592	60 44·93	16 33·17	09·9870	22·4320
	25	161·82	−4·87	10 45 31·7	+ 2 37 29	56·433	60 55·22	16 35·98	10·8748	23·3172
	26	176·91	−5·01	11 40 41·2	− 3 21 48	56·540	60 48·30	16 34·09	11·7604	. . .
	27	191·87	−4·80	12 36 03·3	− 9 06 23	56·916	60 24·18	16 27·52	12·6547	00·2060
	28	206·54	−4·28	13 32 06·1	−14 13 22	57·533	59 45·31	16 16·93	13·5640	01·1074
	29	220·82	−3·50	14 28 58·8	−18 23 29	58·336	58 55·96	16 03·48	14·4869	02·0242
	30	234·63	−2·53	15 26 24·9	−21 22 55	59·253	58 01·26	15 48·58	15·4132	02·9505
Oct.	1	247·96	−1·46	16 23 44·2	−23 04 29	60·205	57 06·18	15 33·57	16·3270	03·8728

EPHEMERIS FOR PHYSICAL OBSERVATIONS
FOR 0ʰ DYNAMICAL TIME

Date 0ʰ TDT	Age	The Earth's Selenographic Long.	The Earth's Selenographic Lat.	Physical Libration Lg.	Physical Libration Lt.	Physical Libration P.A.	The Sun's Selenographic Colong.	The Sun's Selenographic Lat.	Position Angle Axis	Position Angle Bright Limb	Fraction Illum.
	d	°	°	(0°001)			°	°	°	°	
Aug. 16	17·2	−3·125	−6·579	+14	−7	+14	121·07	−1·29	337·019	57·46	0·94
17	18·2	4·257	6·612	13	7	15	133·26	1·30	336·327	60·25	0·89
18	19·2	5·271	6·349	13	7	17	145·45	1·31	336·590	63·01	0·82
19	20·2	6·129	5·791	13	7	17	157·65	1·32	337·895	66·25	0·74
20	21·2	6·781	4·950	13	7	18	169·86	1·33	340·318	70·20	0·65
21	22·2	−7·171	−3·845	+13	−7	+18	182·07	−1·34	343·889	74·89	0·55
22	23·2	7·238	2·511	14	7	18	194·28	1·35	348·543	80·21	0·44
23	24·2	6·922	−1·001	15	7	17	206·50	1·36	354·069	85·86	0·33
24	25·2	6·179	+0·614	16	7	17	218·73	1·37	0·103	91·33	0·23
25	26·2	4·993	2·238	18	7	16	230·96	1·38	6·176	95·84	0·14
26	27·2	−3·392	+3·755	+19	−7	+14	243·20	−1·39	11·815	98·12	0·07
27	28·2	−1·464	5·040	21	7	13	255·44	1·40	16·626	94·34	0·02
28	29·2	+0·647	5·980	23	6	12	267·68	1·41	20·328	40·46.	0·00
29	0·9	2·755	6·492	24	6	10	279·92	1·42	22·730	314·04	0·01
30	1·9	4·667	6·545	25	6	9	292·16	1·43	23·717	303·38	0·06
31	2·9	+6·218	+6·161	+26	−6	+8	304·40	−1·45	23·246	298·19	0·12
Sept. 1	3·9	7·300	5·400	27	6	7	316·63	1·46	21·354	293·49	0·21
2	4·9	7·869	4·347	28	6	6	328·85	1·47	18·184	288·46	0·30
3	5·9	7·937	3·090	28	7	6	341·07	1·48	13·976	283·07	0·40
4	6·9	7·560	1·715	28	7	5	353·28	1·49	9·058	277·54	0·51
5	7·9	+6·819	+0·294	+28	−7	+6	5·49	−1·50	3·788	272·19	0·61
6	8·9	5·801	−1·109	27	7	6	17·69	1·51	358·506	267·34	0·70
7	9·9	4·598	2·439	26	7	7	29·88	1·52	353·487	263·32	0·78
8	10·9	3·288	3·649	25	8	8	42·07	1·53	348·927	260·44	0·86
9	11·9	1·940	4·695	24	8	9	54·25	1·54	344·952	259·19	0·92
10	12·9	+0·609	−5·537	+23	−8	+10	66·43	−1·55	341·646	260·67	0·96
11	13·9	−0·665	6·140	22	8	11	78·61	1·55	339·070	269·48	0·99
12	14·9	1·858	6·479	21	9	13	90·79	1·55	337·289	325·00	1·00
13	15·9	2·950	6·531	20	9	14	102·96	1·55	336·374	40·30	0·99
14	16·9	3·932	6·289	19	9	15	115·14	1·55	336·407	54·67	0·96
15	17·9	−4·790	−5·753	+19	−10	+16	127·31	−1·55	337·472	61·36	0·92
16	18·9	5·507	4·939	18	10	16	139·49	1·55	339·636	66·75	0·86
17	19·9	6·055	3·871	18	10	16	151·67	1·54	342·921	72·14	0·78
18	20·9	6·394	2·589	19	10	16	163·86	1·54	347·260	77·85	0·69
19	21·9	6·477	−1·145	19	10	16	176·05	1·53	352·467	83·81	0·59
20	22·9	−6·250	+0·397	+20	−9	+15	188·25	−1·53	358·231	89·74	0·48
21	23·9	5·665	1·955	21	9	14	200·46	1·52	4·148	95·20	0·37
22	24·9	4·692	3·435	22	9	13	212·67	1·52	9·802	99·69	0·27
23	25·9	3·337	4·730	23	8	11	224·88	1·52	14·824	102·66	0·17
24	26·9	−1·657	5·735	25	8	10	237·10	1·51	18·922	103·28	0·09
25	27·9	+0·236	+6·358	+26	−7	+8	249·33	−1·51	21·869	99·35	0·03
26	28·9	2·184	6·541	27	7	7	261·55	1·51	23·490	74·90	0·00
27	0·6	4·006	6·273	28	7	6	273·78	1·51	23·663	324·87	0·01
28	1·6	5·533	5·592	29	7	5	286·01	1·51	22·343	302·30	0·04
29	2·6	6·641	4·573	29	7	4	298·23	1·50	19·603	293·53	0·09
30	3·6	+7·265	+3·314	+30	−7	+4	310·45	−1·50	15·650	286·63	0·16
Oct. 1	4·6	+7·396	+1·910	+30	−7	+3	322·66	−1·50	10·815	280·18	0·25

FOR 0^h^ DYNAMICAL TIME

Date 0ʰ TDT		Apparent Long.	Apparent Lat.	Apparent R.A.	Apparent Dec.	True Dist.	Horiz. Parallax	Semi-diameter	Ephemeris Transit for date Upper	Ephemeris Transit for date Lower
		°	°	h m s	° ′ ″		′ ″	′ ″	h	h
Oct.	1	247·96	−1·46	16 23 44·2	−23 04 29	60·205	57 06·18	15 33·57	16·3270	03·8728
	2	260·85	−0·34	17 20 03·3	−23 27 45	61·120	56 14·88	15 19·60	17·2112	04·7737
	3	273·36	+0·76	18 14 32·6	−22 37 59	61·936	55 30·44	15 07·49	18·0538	05·6382
	4	285·57	+1·81	19 06 40·1	−20 44 11	62·606	54 54·79	14 57·77	18·8502	06·4577
	5	297·57	+2·75	19 56 17·7	−17 56 56	63·102	54 28·87	14 50·71	19·6036	07·2319
	6	309·45	+3·57	20 43 38·3	−14 26 56	63·413	54 12·87	14 46·35	20·3221	07·9665
	7	321·29	+4·22	21 29 10·5	−10 24 11	63·540	54 06·34	14 44·57	21·0173	08·6719
	8	333·15	+4·70	22 13 31·9	− 5 57 58	63·499	54 08·43	14 45·14	21·7023	09·3603
	9	345·09	+4·97	22 57 25·0	− 1 17 15	63·314	54 17·96	14 47·74	22·3909	10·0453
	10	357·16	+5·02	23 41 34·2	+ 3 28 49	63·011	54 33·63	14 52·01	23·0970	10·7409
	11	9·37	+4·85	0 26 44·0	+ 8 10 16	62·620	54 54·08	14 57·58	23·8337	11·4607
	12	21·75	+4·45	1 13 36·2	+12 35 52	62·167	55 18·06	15 04·11	. . .	12·2171
	13	34·30	+3·83	2 02 46·0	+16 32 58	61·675	55 44·55	15 11·33	00·6122	13·0196
	14	47·02	+3·00	2 54 35·3	+19 47 42	61·158	56 12·79	15 19·03	01·4397	13·8723
	15	59·92	+2·01	3 49 05·1	+22 05 58	60·628	56 42·31	15 27·07	02·3165	14·7708
	16	73·01	+0·90	4 45 48·9	+23 15 11	60·089	57 12·84	15 35·39	03·2334	15·7017
	17	86·30	−0·28	5 43 54·3	+23 06 31	59·546	57 44·13	15 43·91	04·1731	16·6448
	18	99·82	−1·47	6 42 15·9	+21 36 53	59·007	58 15·74	15 52·52	05·1144	17·5797
	19	113·58	−2·59	7 39 55·1	+18 49 42	58·488	58 46·81	16 00·99	06·0394	18·4927
	20	127·61	−3·58	8 36 17·3	+14 54 19	58·011	59 15·81	16 08·89	06·9394	19·3800
	21	141·89	−4·37	9 31 18·0	+10 04 35	57·610	59 40·56	16 15·63	07·8155	20·2471
	22	156·38	−4·89	10 25 19·2	+ 4 37 31	57·324	59 58·38	16 20·49	08·6765	21·1056
	23	171·02	−5·10	11 18 59·0	− 1 07 34	57·196	60 06·48	16 22·70	09·5359	21·9694
	24	185·71	−4·97	12 13 01·3	− 6 49 47	57·257	60 02·62	16 21·64	10·4073	22·8510
	25	200·31	−4·52	13 08 04·2	−12 07 30	57·526	59 45·77	16 17·05	11·3011	23·7578
	26	214·72	−3·79	14 04 28·7	−16 39 54	57·999	59 16·53	16 09·09	12·2204	. . .
	27	228·82	−2·84	15 02 08·4	−20 09 09	58·648	58 37·15	15 58·36	13·1584	00·6879
	28	242·55	−1·75	16 00 25·8	−22 22 54	59·426	57 51·11	15 45·81	14·0980	01·6293
	29	255·86	−0·58	16 58 19·1	−23 16 07	60·270	57 02·50	15 32·57	15·0177	02·5617
	30	268·77	+0·58	17 54 41·0	−22 51 17	61·113	56 15·30	15 19·71	15·8981	03·4637
	31	281·33	+1·68	18 48 39·8	−21 16 46	61·889	55 32·95	15 08·17	16·7286	04·3198
Nov.	1	293·59	+2·68	19 39 52·1	−18 44 04	62·543	54 58·09	14 58·67	17·5083	05·1245
	2	305·63	+3·55	20 28 22·9	−15 25 19	63·032	54 32·50	14 51·70	18·2443	05·8811
	3	317·54	+4·25	21 14 39·0	−11 31 43	63·330	54 17·12	14 47·51	18·9483	06·5994
	4	329·39	+4·76	21 59 20·1	− 7 13 06	63·426	54 12·21	14 46·17	19·6346	07·2927
	5	341·28	+5·07	22 43 13·0	− 2 38 12	63·325	54 17·37	14 47·58	20·3185	07·9759
	6	353·27	+5·17	23 27 07·1	+ 2 04 33	63·048	54 31·68	14 51·48	21·0157	08·6645
	7	5·42	+5·03	0 11 52·7	+ 6 46 15	62·626	54 53·73	14 57·48	21·7414	09·3741
	8	17·77	+4·66	0 58 18·1	+11 16 41	62·099	55 21·71	15 05·11	22·5093	10·1193
	9	30·35	+4·06	1 47 05·8	+15 23 42	61·509	55 53·54	15 13·78	23·3296	10·9125
	10	43·18	+3·24	2 38 45·9	+18 53 04	60·901	56 27·04	15 22·91	. . .	11·7607
	11	56·24	+2·24	3 33 25·5	+21 29 24	60·311	57 00·15	15 31·93	00·2053	12·6620
	12	69·52	+1·10	4 30 40·3	+22 58 02	59·770	57 31·13	15 40·37	01·1288	13·6031
	13	83·00	−0·12	5 29 32·4	+23 08 06	59·295	57 58·76	15 47·90	02·0816	14·5610
	14	96·66	−1·35	6 28 43·6	+21 55 18	58·896	58 22·37	15 54·33	03·0382	15·5103
	15	110·48	−2·52	7 26 59·7	+19 23 08	58·571	58 41·79	15 59·62	03·9752	16·4317
	16	124·43	−3·55	8 23 33·7	+15 41 59	58·317	58 57·12	16 03·80	04·8792	17·3180

EPHEMERIS FOR PHYSICAL OBSERVATIONS
FOR 0^h DYNAMICAL TIME

Date 0^h TDT	Age	The Earth's Selenographic Long.	The Earth's Selenographic Lat.	Physical Libration Lg.	Physical Libration Lt.	Physical Libration P.A.	The Sun's Selenographic Colong.	The Sun's Selenographic Lat.	Position Angle Axis	Position Angle Bright Limb	Fraction Illum.
	d	°	°	(0°001)			°	°	°	°	
Oct. 1	4·6	+7·396	+1·910	+30	− 7	+ 3	322·66	−1·50	10·815	280·18	0·25
2	5·6	7·076	+0·450	30	7	3	334·87	1·50	5·497	274·07	0·34
3	6·6	6·374	−0·991	30	7	4	347·07	1·50	0·085	268·47	0·44
4	7·6	5·379	2·353	29	7	5	359·26	1·49	354·897	263·59	0·54
5	8·6	4·185	3·587	28	8	6	11·45	1·49	350·151	259·61	0·64
6	9·6	+2·880	−4·653	+27	− 8	+ 7	23·63	−1·49	345·982	256·64	0·72
7	10·6	1·545	5·513	26	8	8	35·81	1·48	342·473	254·82	0·80
8	11·6	+0·249	6·138	25	9	10	47·98	1·48	339·679	254·32	0·87
9	12·6	−0·957	6·499	24	9	11	60·14	1·47	337·662	255·64	0·93
10	13·6	2·037	6·576	23	10	12	72·30	1·46	336·494	260·23	0·97
11	14·6	−2·968	−6·354	+22	−10	+14	84·46	−1·44	336·262	275·98	0·99
12	15·6	3·743	5·833	21	11	14	96·62	1·43	337·063	10·27	1·00
13	16·6	4·359	5·023	21	11	15	108·78	1·41	338·978	55·37	0·98
14	17·6	4·816	3·953	20	11	15	120·94	1·39	342·038	67·15	0·95
15	18·6	5·109	2·666	21	11	15	133·10	1·37	346·184	75·00	0·89
16	19·6	−5·226	−1·219	+21	−11	+15	145·26	−1·35	351·233	81·98	0·82
17	20·6	5·144	+0·318	21	11	14	157·43	1·33	356·873	88·55	0·73
18	21·6	4·836	1·865	22	11	13	169·60	1·31	2·710	94·58	0·63
19	22·6	4·272	3·332	22	10	11	181·78	1·30	8·344	99·78	0·52
20	23·6	3·434	4·627	23	10	10	193·97	1·28	13·434	103·87	0·41
21	24·6	−2·325	+5·660	+24	− 9	+ 8	206·16	−1·26	17·714	106·59	0·30
22	25·6	−0·981	6·347	25	9	7	218·35	1·24	20·980	107·70	0·20
23	26·6	+0·521	6·628	25	8	5	230·56	1·23	23·062	106·84	0·11
24	27·6	2·071	6·475	26	8	4	242·77	1·21	23·810	103·02	0·05
25	28·6	3·534	5·900	27	7	3	254·98	1·20	23·120	91·26	0·01
26	0·1	+4·781	+4·954	+27	− 7	+ 2	267·19	−1·18	20·968	353·39	0·00
27	1·1	5·701	3·721	27	7	1	279·40	1·17	17·465	296·68	0·02
28	2·1	6·227	2·299	28	7	1	291·60	1·16	12·879	284·76	0·06
29	3·1	6·333	+0·789	28	7	1	303·81	1·14	7·601	276·77	0·12
30	4·1	6·035	−0·721	27	7	2	316·01	1·13	2·067	270·08	0·19
31	5·1	+5·381	−2·157	+27	− 7	+ 2	328·20	−1·11	356·663	264·35	0·28
Nov. 1	6·1	4·440	3·460	26	7	3	340·39	1·10	351·666	259·59	0·37
2	7·1	3·294	4·587	25	8	5	352·57	1·08	347·245	255·83	0·46
3	8·1	2·028	5·503	24	8	6	4·75	1·07	343·490	253·07	0·56
4	9·1	+0·725	6·179	23	8	8	16·92	1·05	340·453	251·32	0·65
5	10·1	−0·538	−6·591	+22	− 9	+ 9	29·08	−1·03	338·179	250·59	0·74
6	11·1	1·693	6·719	21	9	11	41·24	1·01	336·729	250·95	0·82
7	12·1	2·689	6·550	20	10	12	53·39	0·99	336·185	252·59	0·89
8	13·1	3·487	6·075	19	10	13	65·54	0·97	336·648	255·94	0·94
9	14·1	4·066	5·300	18	11	14	77·69	0·94	338·223	262·60	0·98
10	15·1	−4·419	−4·246	+18	−11	+15	89·83	−0·91	340·976	286·86	1·00
11	16·1	4·551	2·951	18	12	15	101·97	0·88	344·892	59·46	1·00
12	17·1	4·474	−1·474	18	12	14	114·11	0·85	349·819	78·23	0·97
13	18·1	4·204	+0·107	18	12	13	126·25	0·82	355·449	87·17	0·92
14	19·1	3·756	1·705	19	12	12	138·40	0·79	1·364	94·21	0·86
15	20·1	−3·143	+3·223	+19	−12	+11	150·55	−0·76	7·126	100·08	0·77
16	21·1	−2·378	+4·565	+20	−11	+ 9	162·70	−0·73	12·364	104·77	0·67

FOR 0ʰ DYNAMICAL TIME

Date 0ʰ TDT		Apparent Long.	Apparent Lat.	Apparent R.A.	Apparent Dec.	True Dist.	Horiz. Parallax	Semi-diameter	Ephemeris Transit for date Upper	Ephemeris Transit for date Lower
		°	°	h m s	° ′ ″		′ ″	′ ″	h	h
Nov.	16	124·43	−3·55	8 23 33·7	+15 41 59	58·317	58 57·12	16 03·80	04·8792	17·3180
	17	138·51	−4·38	9 18 14·6	+11 06 38	58·131	59 08·48	16 06·89	05·7489	18·1733
	18	152·67	−4·95	10 11 23·4	+ 5 54 03	58·012	59 15·71	16 08·86	06·5928	19·0095
	19	166·90	−5·22	11 03 41·5	+ 0 21 55	57·970	59 18·30	16 09·57	07·4252	19·8423
	20	181·15	−5·16	11 55 59·3	− 5 11 41	58·017	59 15·39	16 08·78	08·2624	20·6876
	21	195·37	−4·80	12 49 05·8	−10 28 20	58·171	59 06·00	16 06·22	09·1191	21·5580
	22	209·49	−4·14	13 43 37·6	−15 09 33	58·446	58 49·35	16 01·68	10·0047	22·4591
	23	223·44	−3·24	14 39 48·8	−18 57 44	58·847	58 25·26	15 55·12	10·9201	23·3860
	24	237·16	−2·17	15 37 21·1	−21 38 08	59·370	57 54·42	15 46·71	11·8545	. . .
	25	250·60	−1·00	16 35 23·8	−23 01 19	59·992	57 18·39	15 36·90	12·7879	00·3228
	26	263·74	+0·21	17 32 45·5	−23 05 03	60·677	56 39·53	15 26·31	13·6964	01·2466
	27	276·56	+1·37	18 28 16·9	−21 54 22	61·380	56 00·63	15 15·71	14·5615	02·1352
	28	289·09	+2·45	19 21 11·4	−19 39 33	62·046	55 24·53	15 05·88	15·3745	02·9746
	29	301·35	+3·38	20 11 14·6	−16 33 23	62·624	54 53·86	14 57·52	16·1370	03·7616
	30	313·42	+4·15	20 58 40·2	−12 48 33	63·065	54 30·80	14 51·24	16·8579	04·5018
Dec.	1	325·35	+4·74	21 44 02·2	− 8 36 22	63·333	54 16·96	14 47·47	17·5507	05·2069
	2	337·22	+5·11	22 28 05·5	− 4 06 32	63·403	54 13·36	14 46·48	18·2309	05·8914
	3	349·10	+5·27	23 11 41·0	+ 0 32 30	63·266	54 20·39	14 48·40	18·9149	06·5714
	4	1·09	+5·21	23 55 42·0	+ 5 12 35	62·930	54 37·83	14 53·15	19·6194	07·2635
	5	13·25	+4·91	0 41 02·4	+ 9 44 57	62·417	55 04·76	15 00·49	20·3610	07·9846
	6	25·65	+4·37	1 28 33·5	+13 59 11	61·766	55 39·61	15 09·98	21·1539	08·7503
	7	38·34	+3·62	2 18 58·7	+17 42 33	61·026	56 20·08	15 21·01	22·0076	09·5729
	8	51·35	+2·66	3 12 43·7	+20 39 53	60·256	57 03·29	15 32·78	22·9222	10·4578
	9	64·69	+1·53	4 09 43·4	+22 34 55	59·515	57 45·90	15 44·39	23·8842	11·3986
	10	78·35	+0·29	5 09 12·8	+23 13 08	58·859	58 24·56	15 54·93	. . .	12·3752
	11	92·28	−0·99	6 09 51·3	+22 25 44	58·331	58 56·30	16 03·58	00·8679	13·3582
	12	106·42	−2·23	7 10 05·5	+20 12 46	57·958	59 19·07	16 09·78	01·8430	14·3195
	13	120·72	−3·35	8 08 41·1	+16 43 36	57·748	59 31·97	16 13·29	02·7862	15·2421
	14	135·08	−4·26	9 05 03·3	+12 14 21	57·694	59 35·35	16 14·22	03·6875	16·1231
	15	149·43	−4·90	9 59 17·6	+ 7 04 26	57·772	59 30·48	16 12·89	04·5504	16·9712
	16	163·73	−5·24	10 51 58·8	+ 1 33 39	57·957	59 19·08	16 09·78	05·3874	17·8012
	17	177·91	−5·25	11 43 56·7	− 3 59 14	58·223	59 02·87	16 05·37	06·2147	18·6300
	18	191·94	−4·96	12 36 04·2	− 9 16 40	58·547	58 43·21	16 00·01	07·0489	19·4729
	19	205·80	−4·37	13 29 06·7	−14 02 10	58·920	58 20·95	15 53·94	07·9033	20·3406
	20	219·47	−3·55	14 23 33·0	−18 00 20	59·334	57 56·50	15 47·28	08·7848	21·2354
	21	232·94	−2·53	15 19 25·8	−20 57 30	59·789	57 30·02	15 40·07	09·6910	22·1497
	22	246·21	−1·40	16 16 16·6	−22 43 24	60·285	57 01·65	15 32·34	10·6090	23·0662
	23	259·25	−0·21	17 13 10·0	−23 12 56	60·815	56 31·81	15 24·21	11·5186	23·9636
	24	272·08	+0·97	18 08 58·9	−22 27 24	61·368	56 01·27	15 15·89	12·3990	. . .
	25	284·68	+2·08	19 02 46·1	−20 34 00	61·920	55 31·27	15 07·71	13·2356	00·8234
	26	297·07	+3·07	19 53 58·9	−17 43 59	62·443	55 03·38	15 00·11	14·0232	01·6355
	27	309·27	+3·90	20 42 33·6	−14 10 09	62·900	54 39·40	14 53·58	14·7654	02·3995
	28	321·31	+4·55	21 28 50·1	−10 04 56	63·252	54 21·15	14 48·61	15·4720	03·1224
	29	333·23	+5·00	22 13 24·4	− 5 39 23	63·462	54 10·32	14 45·66	16·1561	03·8159
	30	345·09	+5·23	22 57 01·4	− 1 03 03	63·501	54 08·34	14 45·12	16·8330	04·4945
	31	356·96	+5·23	23 40 31·2	+ 3 35 33	63·347	54 16·23	14 47·27	17·5188	05·1738
	32	8·91	+5·01	0 24 46·2	+ 8 08 13	62·993	54 34·56	14 52·26	18·2300	05·8702

EPHEMERIS FOR PHYSICAL OBSERVATIONS
FOR 0ʰ DYNAMICAL TIME

Date 0ʰ TDT		Age	The Earth's Selenographic Long.	The Earth's Selenographic Lat.	Physical Libration Lg.	Physical Libration Lt.	Physical Libration P.A.	The Sun's Selenographic Colong.	The Sun's Selenographic Lat.	Position Angle Axis	Position Angle Bright Limb	Fraction Illum.
		d	°	°	(0°001)			°	°	°	°	
Nov.	**16**	21·1	−2·378	+4·565	+20	−11	+ 9	162·70	−0·73	12·364	104·77	0·67
	17	22·1	1·476	5·644	20	11	8	174·86	0·70	16·811	108·17	0·56
	18	23·1	−0·462	6·386	20	10	6	187·03	0·67	20·283	110·22	0·44
	19	24·1	+0·628	6·741	21	9	4	199·21	0·65	22·643	110·90	0·33
	20	25·1	1·742	6·680	21	9	3	211·39	0·62	23·770	110·16	0·23
	21	26·1	+2·817	+6·209	+21	− 9	+ 2	223·58	−0·60	23·560	107·93	0·14
	22	27·1	3·778	5·363	22	8	0	235·77	0·57	21·955	104·02	0·07
	23	28·1	4·553	4·205	22	8	0	247·96	0·55	18·984	97·62	0·02
	24	29·1	5·081	2·820	22	8	− 1	260·16	0·53	14·815	79·17	0·00
	25	0·6	5·317	+1·302	21	8	1	272·36	0·50	9·763	285·64	0·00
	26	1·6	+5·239	−0·253	+21	− 8	− 1	284·55	−0·48	4·245	272·13	0·03
	27	2·6	4·850	1·760	21	8	0	296·74	0·46	358·688	264·96	0·07
	28	3·6	4·172	3·150	20	8	+ 1	308·93	0·44	353·440	259·45	0·14
	29	4·6	3·248	4·365	19	8	2	321·12	0·41	348·732	255·10	0·21
	30	5·6	2·136	5·366	18	8	4	333·30	0·39	344·690	251·82	0·29
Dec.	**1**	6·6	+0·901	−6·123	+16	− 8	+ 6	345·47	−0·37	341·377	249·52	0·38
	2	7·6	−0·382	6·614	15	9	7	357·64	0·34	338·831	248·17	0·48
	3	8·6	1·638	6·823	14	9	9	9·80	0·32	337·093	247·75	0·57
	4	9·6	2·794	6·738	13	10	10	21·96	0·29	336·224	248·25	0·66
	5	10·6	3·780	6·352	12	10	12	34·11	0·27	336·313	249·71	0·75
	6	11·6	−4·538	−5·666	+11	−10	+13	46·25	−0·24	337·463	252·18	0·83
	7	12·6	5·020	4·689	10	11	13	58·39	0·21	339·771	255·71	0·90
	8	13·6	5·197	3·449	10	11	14	70·53	0·17	343·283	260·34	0·95
	9	14·6	5·059	1·993	10	12	14	82·66	0·14	347·925	266·28	0·99
	10	15·6	4·617	−0·390	10	12	13	94·79	0·10	353·457	24·19	1·00
	11	16·6	−3·901	+1·267	+11	−12	+12	106·92	−0·07	359·476	95·24	0·99
	12	17·6	2·962	2·871	11	12	11	119·04	−0·03	5·494	101·41	0·95
	13	18·6	1·863	4·312	12	12	9	131·18	0·00	11·061	106·30	0·89
	14	19·6	−0·677	5·489	12	11	8	143·31	+0·04	15·841	109·97	0·80
	15	20·6	+0·526	6·322	13	11	6	155·46	0·07	19·618	112·35	0·70
	16	21·6	+1·677	+6·757	+13	−11	+ 4	167·61	+0·10	22·257	113·43	0·59
	17	22·6	2·720	6·776	13	10	3	179·76	0·13	23·667	113·24	0·48
	18	23·6	3·610	6·388	13	10	+ 1	191·92	0·16	23·776	111·84	0·37
	19	24·6	4·316	5·630	13	10	0	204·09	0·19	22·546	109·30	0·26
	20	25·6	4·817	4·559	13	9	− 1	216·27	0·21	19·994	105·75	0·17
	21	26·6	+5·100	+3·249	+13	− 9	− 2	228·45	+0·24	16·236	101·45	0·10
	22	27·6	5·160	1·783	12	9	2	240·64	0·27	11·510	96·91	0·04
	23	28·6	4·993	+0·247	12	9	2	252·82	0·29	6·170	93·65	0·01
	24	29·6	4·605	−1·276	11	9	2	265·01	0·32	0·618	158·90	0·00
	25	1·0	4·001	2·710	10	9	− 1	277·20	0·34	355·231	253·47	0·01
	26	2·0	+3·197	−3·991	+ 9	− 9	0	289·39	+0·36	350·290	251·72	0·04
	27	3·0	2·215	5·069	8	9	+ 2	301·58	0·39	345·975	249·14	0·09
	28	4·0	+1·086	5·907	6	10	3	313·76	0·41	342·381	247·05	0·15
	29	5·0	−0·146	6·479	5	10	5	325·94	0·43	339·559	245·69	0·22
	30	6·0	1·431	6·772	3	10	7	338·11	0·46	337·544	245·13	0·30
	31	7·0	−2·707	−6·775	+ 2	−10	+ 8	350·28	+0·48	336·379	245·38	0·39
	32	8·0	−3·905	−6·486	+ 1	−11	+10	2·44	+0·50	336·128	246·46	0·49

NOTES AND FORMULAE

Use of the polynomial coefficients for the lunar coordinates

On pages D23–D45 for each day of the year, the apparent right ascension (α) and declination (δ) of the Moon are represented by economised polynomials of the fifth degree, and the horizontal parallax (π) is represented by an economised polynomial of the fourth degree. The formulae to be evaluated are of the form:

$$a_0 + a_1 p + a_2 p^2 + a_3 p^3 + a_4 p^4 + a_5 p^5$$

where a_5 is zero for the parallax.

The time-interval from 0^h TDT is expressed as a fraction of a day to form the interpolation factor p, where $0 \leqslant p < 1$, and the polynomial is evaluated directly, or by re-expressing it in the nested form:

$$((((a_5 p + a_4)p + a_3)p + a_2)p + a_1)p + a_0$$

to avoid the separate formation of the powers of p. Alternatively this nested form for α and δ may be written as:

$$b_{n+1} = b_n p + a_{5-n}, \text{ for } n = 1 \text{ to } 5,$$

where $b_1 = a_5$ and b_6 is the required value. For the parallax a_5 is zero, so that:

$$b_{n+1} = b_n p + a_{4-n}, \text{ for } n = 1 \text{ to } 4,$$

where $b_1 = a_4$ and b_5 is the required value.

The polynomial coefficients are expressed in decimals of a degree, even for α, and the signs are given on the right-hand sides of the coefficients to facilitate their use with small calculators. Subtract 360° from α if it exceeds 360°. In order to obtain the full precision of the ephemeris the interpolating factor p must be evaluated to 8 decimal places (10^{-3} s); estimates of the precision of unrounded interpolated values are:

RA	Dec	HP
$\pm 0^s{\cdot}0003$	$\pm 0''{\cdot}003$	$\pm 0''{\cdot}0003$

Particular care must be taken to ensure that the coefficients are entered with the correct signs.

Example. To calculate the apparent right ascension (α) the declination (δ) and the horizontal parallax (π) for the Moon on 1992 January 21^d 13^h 23^m $48^s{\cdot}32$ UT, using an assumed value of $\Delta T = 58^s$.

TDT $= 13^h\ 24^m\ 46^s{\cdot}32$, hence $p = 0{\cdot}558\ 869\ 44$

	right ascension	declination	horizontal parallax
	°	°	°
b_1	$-0{\cdot}000\ 8655$	$-0{\cdot}000\ 4719$	$+0{\cdot}000\ 023\ 76$
b_2	$+0{\cdot}006\ 8511$	$-0{\cdot}001\ 1363$	$+0{\cdot}000\ 176\ 17$
b_3	$+0{\cdot}017\ 5938$	$+0{\cdot}080\ 6306$	$-0{\cdot}002\ 581\ 87$
b_4	$-0{\cdot}410\ 7718$	$-0{\cdot}326\ 7916$	$-0{\cdot}007\ 563\ 03$
b_5	$+14{\cdot}329\ 5157$	$-5{\cdot}713\ 5284$	$\pi = 1{\cdot}017\ 427\ 13$
b_6	$\alpha = 145{\cdot}436\ 5045$	$\delta = +9{\cdot}904\ 0146$	
	$= 9^h\ 41^m\ 44^s{\cdot}761$	$= +9°\ 54'\ 14''{\cdot}45$	$= 1°\ 01'\ 02''{\cdot}738$

DAILY POLYNOMIAL COEFFICIENTS

	Apparent Right Ascension	Apparent Declination	Horizontal Parallax
	January 0		
	°	°	°
a_0	219·8569 359+	20·1559 702−	0·9285 3341+
a_1	12·8995 890+	3·2187 247−	0·0093 6574−
a_2	1488 081+	4989 853+	8 3167+
a_3	28 768−	229 278+	441−
a_4	47 297−	14 265−	354−
a_5	981+	1 682−	
	January 1		
a_0	232·8978 247+	22·8543 766−	0·9199 9139+
a_1	13·1701 500+	2·1585 175−	0·0077 2978−
a_2	1127 787+	5575 211+	7 9808+
a_3	206 437−	155 556+	1838−
a_4	42 456−	23 125−	71−
a_5	4 002+	1 409−	
	January 2		
a_0	246·1562 642+	24·4422 708−	0·9130 4058+
a_1	13·3187 966+	1·0067 617−	0·0061 9163−
a_2	293 874+	5888 995+	7 3932+
a_3	335 238−	49 682+	2102−
a_4	21 552−	30 469−	146+
a_5	5 820+	144−	
	January 3		
a_0	259·4693 513+	24·8582 262−	0·9075 6872+
a_1	13·2712 886+	0·1736 840+	0·0047 7020−
a_2	782 800−	5853 803+	6 8546+
a_3	363 792−	72 784−	1494−
a_4	8 639+	31 071−	289+
a_5	4 860+	1 356+	
	January 4		
a_0	272·6273 306+	24·1094 117−	0·9034 7193+
a_1	13·0114 734+	1·3108 600+	0·0034 3252−
a_2	1773 711−	5462 643+	6 5819+
a_3	282 271−	183 044−	313−
a_4	33 203+	23 891−	354+
a_5	1 949+	2 159+	
	January 5		
a_0	285·4367 210+	22·2727 650−	0·9006 9801+
a_1	12·5863 026+	2·3399 984+	0·0021 1137−
a_2	2401 901−	4791 804+	6 6999+
a_3	131 479−	257 106−	1126+
a_4	42 406+	12 757−	342+
a_5	707−	1 985+	
	January 6		
a_0	297·7738 555+	19·4803 740−	0·8992 7130+
a_1	12·0830 862+	3·2171 164+	0·0007 2394−
a_2	2549 085−	3963 811+	7 2403+
a_3	30 378+	288 720−	2514+
a_4	38 176+	2 715−	262+
a_5	1 867−	1 216+	
	January 7		
a_0	309·6087 020+	15·8958 984−	0·8992 9916+
a_1	11·5967 197+	3·9227 835+	0·0008 1006+
a_2	2247 635−	3093 509+	8 1477+
a_3	164 445+	287 933−	3580+
a_4	28 413+	3 318+	126+
a_5	1 766−	332+	

	Apparent Right Ascension	Apparent Declination	Horizontal Parallax
	January 8		
	°	°	°
a_0	320·9997 675+	11·6921 924−	0·9009 6104+
a_1	11·2070 093+	4·4565 974+	0·0025 5202+
a_2	1601 505−	2252 918+	9 2913+
a_3	260 780+	271 783−	4097+
a_4	19 462+	4 885+	57−
a_5	1 162−	424−	
	January 9		
a_0	332·0745 342+	7·0370 354−	0·9044 8258+
a_1	10·9721 465+	4·8273 871+	0·0045 3089+
a_2	714 013−	1462 618+	10 4788+
a_3	327 302+	256 835−	3881+
a_4	13 720+	2 681+	283−
a_5	649−	1 021−	
	January 10		
a_0	343·0093 167+	2·0889 040−	0·9100 9732+
a_1	10·9326 979+	5·0434 216+	0·0067 3174+
a_2	343 745+	697 973+	11 4645+
a_3	375 716+	256 590−	2762+
a_4	10 676+	2 506−	555−
a_5	607−	1 485−	
	January 11		
a_0	354·0149 677+	2·9982 569+	0·9179 9759+
a_1	11·1181 280+	5·1042 943+	0·0090 8534+
a_2	1528 927+	101 706−	11 9506+
a_3	411 892+	281 607−	558+
a_4	8 001+	10 079−	872−
a_5	1 384−	1 745−	
	January 12		
a_0	5·3278 393+	8·0630 374+	0·9282 7486+
a_1	11·5499 871+	4·9945 673+	0·0114 5733+
a_2	2798 833+	1024 506−	11 5842+
a_3	428 893+	339 197−	2927−
a_4	1 576+	19 156−	1218−
a_5	3 373−	1 474−	
	January 13		
a_0	17·2004 193+	12·9191 714+	0·9408 4916+
a_1	12·2373 621+	4·6795 096+	0·0136 3762+
a_2	4061 267+	2171 878−	9 9659+
a_3	399 659+	429 657−	7828−
a_4	15 041−	27 233−	1526−
a_5	6 548−	34+	
	January 14		
a_0	29·8817 149+	17·3358 075+	0·9553 8982+
a_1	13·1602 199+	4·1053 644+	0·0153 3484+
a_2	5104 319+	3624 048−	6 6979+
a_3	272 753+	536 204−	1 4024−
a_4	49 002−	27 972−	1655−
a_5	8 936−	3 561+	
	January 15		
a_0	43·5738 482+	21·0227 056+	0·9712 3767+
a_1	14·2388 453+	3·2102 906+	0·0161 8744+
a_2	5538 636+	5364 885−	1 5069+
a_3	10 309−	609 741−	2 0810−
a_4	97 533−	10 191−	1382−
a_5	5 224−	8 444+	

Formula: Quantity in degrees $= a_0 + a_1 p + a_2 p^2 + a_3 p^3 + a_4 p^4 + a_5 p^5$
where p is the fraction of a day from 0^h TDT.

DAILY POLYNOMIAL COEFFICIENTS

	Apparent Right Ascension	Apparent Declination	Horizontal Parallax	Apparent Right Ascension	Apparent Declination	Horizontal Parallax
	January 16			**January 24**		
	°	°	°	°	°	°
a_0	58·3552 504+	23·6353 590+	0·9873 5389+	178·0858 230+	4·8183 979−	0·9851 8434+
a_1	15·3018 693+	1·9545 392+	0·0158 0914+	12·8744 475+	5·8172 933−	0·0158 6525−
a_2	4869 795+	7170 472−	5 5374−	1080 036−	2202 396+	4 3509−
a_3	445 161−	564 994−	2 6552−	381 053+	462 994+	2 5387+
a_4	126 944−	34 372+	499−	15 162−	30 524−	1466−
a_5	8 117+	10 547+		2 252−	1 505+	
	January 17			**January 25**		
a_0	74·0877 004+	24·8208 435+	1·0023 3879+	190·8886 308+	10·3720 542−	0·9691 2321+
a_1	16·0955 746+	0·3699 628+	0·0138 8507+	12·7655 660+	5·2493 738−	0·0160 3257−
a_2	2854 408+	8553 286−	13 7563−	50 377−	3423 329+	2 3792+
a_3	864 932−	325 103−	2 8699−	298 189+	355 681+	1 9367+
a_4	82 164−	90 361+	988+	26 404−	22 692−	1649−
a_5	20 770+	5 161+		1 777−	1 030+	
	January 18			**January 26**		
a_0	90·3760 833+	24·3125 196+	1·0145 7113+	203·6761 599+	15·2456 931−	0·9535 0573+
a_1	16·3844 916+	1·3995 114−	0·0103 1240+	12·8334 983+	4·4665 662−	0·0150 4174−
a_2	24 581−	8934 822−	21 7222−	667 939+	4364 540+	7 2050+
a_3	988 039−	82 482+	2 4690−	175 254+	274 709+	1 2693+
a_4	29 419+	116 145+	2652+	35 635−	17 459−	1473−
a_5	16 744+	4 674−		1 038−	135+	
	January 19			**January 27**		
a_0	106·6639 292+	22·0389 213+	1·0224 9094+	216·5903 101+	19·2500 669−	0·9392 9670+
a_1	16·1032 862+	3·1176 162−	0·0053 3352+	13·0048 907+	3·5181 629−	0·0132 7889−
a_2	2644 670−	8037 667−	27 5052−	969 424+	5085 238+	10 1391+
a_3	711 295−	497 126+	1 3767−	23 292+	205 784+	6779+
a_4	113 679+	89 785+	3728+	41 313−	16 930−	1147−
a_5	1 474+	10 247−		619+	631−	
	January 20			**January 28**		
a_0	122·4431 341+	18·1752 048+	1·0249 7355+	229·6904 031+	22·2408 836−	0·9270 8804+
a_1	15·4071 614+	4·5452 202−	0·0004 3120−	13·1895 503+	2·4464 677−	0·0110 9358−
a_2	4082 372−	6110 364−	29 4021−	797 587+	5594 657+	11 4953+
a_3	247 396−	754 411+	1562+	134 436−	131 664+	2196+
a_4	116 703+	36 388+	3613+	38 354−	20 387−	800−
a_5	8 129−	8 970−		2 956+	789−	
	January 21			**January 29**		
a_0	137·4281 760+	13·0971 311+	1·0216 5388+	242·9427 288+	24·1168 369−	0·9171 5795+
a_1	14·5590 835+	5·5308 946−	0·0061 2010−	13·2948 752+	1·2965 861−	0·0087 6065−
a_2	4206 044−	3718 536−	26 8033−	193 782+	5859 396+	11 6836+
a_3	137 649+	812 657+	1 6289+	257 350−	42 555+	991−
a_4	73 348+	8 726−	2376+	23 034−	24 591−	488−
a_5	8 655−	4 719−		4 634+	200−	
	January 22			**January 30**		
a_0	151·5868 893+	7·2743 042+	1·0130 4009+	256·2294 072+	24·8257 070−	0·9095 5089+
a_1	13·7841 843+	6·0366 501−	0·0108 9704−	13·2495 297+	0·1218 758−	0·0064 7314−
a_2	3439 620−	1380 011−	20 5424−	670 022−	5837 513+	11 1021+
a_3	346 090+	732 702+	2 5804+	303 311−	57 273−	2926−
a_4	29 722+	31 690−	694+	972+	25 602−	223−
a_5	5 764−	1 019−		4 345+	746+	
	January 23			**January 31**		
a_0	165·0641 164+	1·1696 523+	1·0003 5379+	269·3821 354+	24·3720 443−	0·9041 5648+
a_1	13·2090 969+	6·1060 249−	0·0142 0371−	13·0270 914+	1·0185 781+	0·0043 4938−
a_2	2280 601−	617 873+	12 4277−	1530 624−	5519 575+	10 0976+
a_3	408 720+	596 934+	2 8406+	257 115−	151 835−	3802−
a_4	1 354+	36 057−	702−	23 070+	21 659−	5−
a_5	3 376−	998+		2 310+	1 416+	

Formula: Quantity in degrees $= a_0 + a_1 p + a_2 p^2 + a_3 p^3 + a_4 p^4 + a_5 p^5$
where p is the fraction of a day from 0^h TDT.

DAILY POLYNOMIAL COEFFICIENTS

February 1

	Apparent Right Ascension	Apparent Declination	Horizontal Parallax
	°	°	°
a_0	282·2329 910+	22·8187 165−	0·9007 7878+
a_1	12·6542 127+	2·0689 871+	0·0024 4412−
a_2	2140 498−	4948 307+	8 9594+
a_3	143 031−	224 268−	3806−
a_4	34 335+	14 346−	167+
a_5	32+	1 484+	

February 2

	Apparent Right Ascension	Apparent Declination	Horizontal Parallax
a_0	294·6622 875+	20·2786 117−	0·8991 9421+
a_1	12·1969 522+	2·9863 713+	0·0007 5974−
a_2	2363 351−	4204 282+	7 9216+
a_3	6 144−	267 034−	3120−
a_4	33 945+	6 814−	288+
a_5	1 288−	1 087+	

February 3

	Apparent Right Ascension	Apparent Declination	Horizontal Parallax
a_0	306·6255 560+	16·8990 884−	0·8991 9831+
a_1	11·7353 726+	3·7449 347+	0·0007 4250+
a_2	2191 063−	3373 164+	7 1601+
a_3	116 596+	283 739−	1947−
a_4	27 087+	1 383−	348+
a_5	1 532−	527+	

February 4

	Apparent Right Ascension	Apparent Declination	Horizontal Parallax
a_0	318·1560 373+	12·8452 967−	0·9006 4083+
a_1	11·3422 079+	4·3341 558+	0·0021 3002+
a_2	1694 109−	2518 913+	6 7841+
a_3	209 809+	284 302−	533−
a_4	19 246+	1 199+	337+
a_5	1 188−	2+	

February 5

	Apparent Right Ascension	Apparent Declination	Horizontal Parallax
a_0	329·3516 210+	8·2875 598−	0·9034 4730+
a_1	11·0734 337+	4·7531 277+	0·0034 8435+
a_2	961 096−	1673 207+	6 8237+
a_3	275 145+	279 745−	839+
a_4	13 300+	1 150+	251+
a_5	771−	438−	

February 6

	Apparent Right Ascension	Apparent Declination	Horizontal Parallax
a_0	340·3577 124+	3·3950 147−	0·9076 2491+
a_1	10·9686 925+	5·0040 861+	0·0048 8431+
a_2	63 567−	836 478+	7 2210+
a_3	320 709+	279 736−	1866+
a_4	9 553+	1 098−	90+
a_5	623−	797−	

February 7

	Apparent Right Ascension	Apparent Declination	Horizontal Parallax
a_0	351·3530 121+	1·6645 562+	0·9132 5089+
a_1	11·0557 010+	5·0866 232+	0·0063 8811+
a_2	949 668+	17 302−	7 8276+
a_3	352 460+	292 236−	2247+
a_4	6 641+	5 175−	141−
a_5	1 002−	1 041−	

February 8

	Apparent Right Ascension	Apparent Declination	Horizontal Parallax
a_0	2·5394 898+	6·7196 040+	0·9204 4281+
a_1	11·3535 269+	4·9929 015+	0·0080 1541+
a_2	2036 913+	935 505−	8 4081+
a_3	368 337+	323 323−	1701+
a_4	1 926+	10 576−	433−
a_5	2 143−	1 009−	

February 9

	Apparent Right Ascension	Apparent Declination	Horizontal Parallax
	°	°	°
a_0	14·1335 199+	11·5854 642+	0·9293 1172+
a_1	11·8711 069+	4·7040 696+	0·0097 3078+
a_2	3132 064+	1979 077−	8 6485+
a_3	353 480+	375 310−	18−
a_4	8 588−	16 002−	770−
a_5	4 094−	335−	

February 10

	Apparent Right Ascension	Apparent Declination	Horizontal Parallax
a_0	26·3519 131+	16·0524 615+	0·9398 9948+
a_1	12·5980 794+	4·1890 951+	0·0114 2914+
a_2	4099 956+	3204 454−	8 1705+
a_3	277 123+	441 615−	3104−
a_4	29 512−	18 237−	1116−
a_5	6 021−	1 467+	

February 11

	Apparent Right Ascension	Apparent Declination	Horizontal Parallax
a_0	39·3841 472+	19·8752 726+	0·9521 0347+
a_1	13·4863 938+	3·4091 618+	0·0129 2547+
a_2	4693 761+	4624 102−	6 5613+
a_3	99 395+	498 193−	7609−
a_4	61 519−	11 249−	1392−
a_5	5 288−	4 442+	

February 12

	Apparent Right Ascension	Apparent Declination	Horizontal Parallax
a_0	53·3431 760+	22·7715 243+	0·9655 9506+
a_1	14·4277 207+	2·3326 077+	0·0139 5372+
a_2	4569 562+	6141 656−	3 4415+
a_3	195 761−	497 324−	1 3276−
a_4	90 643−	11 658+	1456−
a_5	1 271+	7 072+	

February 13

	Apparent Right Ascension	Apparent Declination	Horizontal Parallax
a_0	68·1993 396+	24·4421 070+	0·9797 4561+
a_1	15·2472 944+	0·9632 778+	0·0141 8542+
a_2	3451 155+	7492 685−	1 4039−
a_3	540 032−	380 424−	1 9261−
a_4	84 098−	48 912+	1114−
a_5	11 358+	6 363+	

February 14

	Apparent Right Ascension	Apparent Declination	Horizontal Parallax
a_0	83·7304 725+	24·6236 014+	0·9935 8690+
a_1	15·7475 598+	0·6266 458−	0·0132 8215+
a_2	1440 694+	8276 667−	7 8225−
a_3	760 843−	123 965−	2 3901−
a_4	22 757−	82 038+	216−
a_5	15 039+	1 348+	

February 15

	Apparent Right Ascension	Apparent Declination	Horizontal Parallax
a_0	99·5452 458+	23·1652 311+	1·0058 4564+
a_1	15·8058 547+	2·2856 855−	0·0109 9193+
a_2	827 513−	8142 951−	15 0797−
a_3	705 436−	214 444+	2 4871−
a_4	55 772+	88 047+	1165+
a_5	7 767+	4 372−	

February 16

	Apparent Right Ascension	Apparent Declination	Horizontal Parallax
a_0	115·2041 596+	20·0950 624+	1·0150 9256+
a_1	15·4549 030+	3·8169 126−	0·0072 7653+
a_2	2531 707−	7015 282−	21 7980−
a_3	410 194−	521 418+	2 0121−
a_4	93 302+	64 609+	2595+
a_5	2 077−	6 909−	

Formula: Quantity in degrees $= a_0 + a_1 p + a_2 p^2 + a_3 p^3 + a_4 p^4 + a_5 p^5$
where p is the fraction of a day from 0^h TDT.

DAILY POLYNOMIAL COEFFICIENTS

	Apparent Right Ascension	Apparent Declination	Horizontal Parallax
	February 17		
	°	°	°
a_0	130·3739 948+	15·6345 333+	1·0200 1403+
a_1	14·8617 800+	5·0411 538−	0·0024 1724+
a_2	3223 649−	5132 597−	26 2527−
a_3	60 426−	711 135+	9444−
a_4	80 256+	29 129+	3409+
a_5	6 581−	6 179−	
	February 18		
a_0	144·9147 348+	10·1535 282+	1·0197 4565+
a_1	14·2277 344+	5·8457 684−	0·0029 8007−
a_2	2989 402−	2886 236−	27 0481−
a_3	194 892+	767 077+	4549+
a_4	46 063+	1 969−	3159+
a_5	6 241−	4 036−	
	February 19		
a_0	158·8670 005+	4·0952 434+	1·0141 3784+
a_1	13·7036 284+	6·1956 955−	0·0081 2675−
a_2	2190 787−	637 144−	23 8239−
a_3	317 723+	720 103+	1 7399+
a_4	14 738+	21 958−	1983+
a_5	4 470−	1 851−	
	February 20		
a_0	172·3843 494+	2·0945 371−	1·0038 2251+
a_1	13·3644 499+	6·1168 003−	0·0122 9024−
a_2	1193 871−	1372 962+	17 4602−
a_3	332 790+	614 656+	2 5321+
a_4	7 423−	30 849−	489+
a_5	3 083−	299−	
	February 21		
a_0	185·6616 408+	8·0156 903−	0·9900 4434+
a_1	13·2210 036+	5·6702 993−	0·0150 0316−
a_2	270 871−	3028 907+	9 6076−
a_3	272 836+	488 674+	2 7120+
a_4	22 825−	31 968−	717−
a_5	2 147−	383+	
	February 22		
a_0	198·8803 436+	13·3373 899−	0·9743 4443+
a_1	13·2384 780+	4·9305 112−	0·0161 3988−
a_2	389 164+	4306 990+	1 9225−
a_3	160 730+	364 637+	2 4068+
a_4	33 849−	29 824−	1369−
a_5	1 020−	362+	
	February 23		
a_0	212·1703 242+	17·8036 846−	0·9582 3930+
a_1	13·3504 825+	3·9714 708−	0·0158 5715−
a_2	657 997+	5225 583+	4 4715+
a_3	16 189+	248 836+	1 8462+
a_4	39 322−	28 013−	1524−
a_5	813+	115+	
	February 24		
a_0	225·5843 743+	21·2305 033−	0·9429 9869+
a_1	13·4716 186+	2·8628 512−	0·0144 6998−
a_2	478 746+	5805 138+	9 1001+
a_3	131 721−	137 950+	1 2302+
a_4	35 276−	27 604−	1377−
a_5	3 021+	138+	
	February 25		
	°	°	°
a_0	239·0874 700+	23·5017 924−	0·9295 4798+
a_1	13·5152 534+	1·6714 107−	0·0123 3596−
a_2	97 787−	6054 727+	11 9731+
a_3	241 851−	29 170+	6779+
a_4	19 609−	27 044−	1108−
a_5	4 376+	588+	
	February 26		
a_0	252·5672 363+	24·5674 591−	0·9184 6605+
a_1	13·4174 845+	0·4622 377−	0·0097 8227−
a_2	897 139−	5985 856+	13 3509+
a_3	276 808−	72 812−	2354+
a_4	2 998+	24 051−	827−
a_5	3 875+	1 153+	
	February 27		
a_0	265·8680 134+	24·4406 822−	0·9100 3414+
a_1	13·1581 489+	0·7040 462+	0·0070 7454−
a_2	1670 795−	5634 669+	13 5687+
a_3	227 165−	157 351−	941−
a_4	22 634+	18 097−	579−
a_5	1 922+	1 391+	
	February 28		
a_0	278·8388 218+	23·1905 747−	0·9043 0127+
a_1	12·7658 523+	1·7772 315+	0·0044 1217−
a_2	2197 324−	5067 970+	12 9459+
a_3	118 572−	215 953−	3244−
a_4	31 940+	10 983−	368−
a_5	110−	1 163+	
	February 29		
a_0	291·3762 676+	20·9291 235−	0·9011 4757+
a_1	12·3035 359+	2·7222 272+	0·0019 3503−
a_2	2362 576−	4365 842+	11 7576+
a_3	7 425+	248 543−	4709−
a_4	30 901+	5 134−	185−
a_5	1 242−	665+	
	March 1		
a_0	303·4472 542+	17·7956 132−	0·9003 3938+
a_1	11·8449 872+	3·5191 115+	0·0002 6784+
a_2	2167 381−	3596 055+	10 2393+
a_3	118 470+	262 712−	5443−
a_4	24 325+	1 865−	18−
a_5	1 435−	168+	
	March 2		
a_0	315·0896 392+	13·9433 370−	0·9015 7655+
a_1	11·4560 646+	4·1588 467+	0·0021 5171+
a_2	1680 407−	2798 394+	8 6003+
a_3	201 573+	268 703−	5509−
a_4	16 989+	1 113−	132+
a_5	1 140−	192−	
	March 3		
a_0	326·3994 054+	9·5316 516−	0·9045 3452+
a_1	11·1866 812+	4·6373 737+	0·0037 1180+
a_2	985 163−	1983 680+	7 0309+
a_3	258 315+	275 197−	4967−
a_4	11 287+	2 143−	255+
a_5	796−	403−	

Formula: Quantity in degrees $= a_0 + a_1 p + a_2 p^2 + a_3 p^3 + a_4 p^4 + a_5 p^5$

where p is the fraction of a day from 0^h TDT.

DAILY POLYNOMIAL COEFFICIENTS

	Apparent Right Ascension	Apparent Declination	Horizontal Parallax	Apparent Right Ascension	Apparent Declination	Horizontal Parallax
	March 4			**March 12**		
	°	°	°	°	°	°
a_0	337·5144 509+	4·7236 842−	0·9089 0230+	79·3112 477+	24·5000 357+	0·9753 9083+
a_1	11·0712 604+	4·9504 917+	0·0049 7920+	15·1589 647+	0·1493 940−	0·0105 2523+
a_2	150 448−	1141 186+	5 6963+	1448 937+	7583 521−	8153−
a_3	295 553+	287 866−	3925−	593 162−	138 788−	1 0518−
a_4	7 390+	4 220−	329+	22 028−	59 372+	993−
a_5	690−	507−		10 975+	1 259+	
	March 5			**March 13**		
a_0	348·6008 919+	0·3116 667+	0·9144 1516+	94·5546 846+	23·5844 739+	0·9857 1942+
a_1	11·1324 476+	5·0904 274+	0·0060 1388+	15·2674 750+	1·6833 604−	0·0100 0683+
a_2	773 668+	247 181+	4 7161+	352 631−	7631 119−	4 5553−
a_3	318 016+	309 823−	2582−	573 728−	109 410+	1 4614−
a_4	4 089+	6 830−	329+	35 297+	65 303+	631−
a_5	1 010−	512−		6 965+	2 055−	
	March 6			**March 14**		
a_0	359·8428 158+	5·3950 958+	0·9208 7812+	109·7337 501+	21·1552 673+	0·9951 1828+
a_1	11·3837 157+	5·0439 292+	0·0068 9282+	15·0424 244+	3·1516 694−	0·0086 3206+
a_2	1742 167+	728 406−	4 1361+	1792 426−	6931 738−	9 2948−
a_3	323 753+	342 151−	1235−	366 766−	349 103+	1 7253−
a_4	786−	9 524−	237+	69 833+	54 198+	116+
a_5	1 881−	328−		29+	3 705−	
	March 7			**March 15**		
a_0	11·4328 568+	10·3309 842+	0·9281 7457+	124·5672 415+	17·3503 837+	1·0026 4948+
a_1	11·8280 187+	4·7916 301+	0·0076 9249+	14·6018 523+	4·4134 595−	0·0062 6009+
a_2	2689 898+	1815 319−	3 9021+	2473 687−	5596 352−	14 3706−
a_3	301 017+	383 183−	257−	89 639−	528 831+	1 6831−
a_4	10 138−	11 406−	51+	68 288+	35 234+	1121+
a_5	3 233−	253+		4 284−	3 667−	
	March 8			**March 16**		
a_0	23·5586 299+	14·9016 488+	0·9362 5520+	138·9191 616+	12·4333 288+	1·0073 1542+
a_1	12·4506 303+	4·3091 772+	0·0084 6722+	14·1053 953+	5·3618 197−	0·0029 2593+
a_2	3499 713+	3030 819−	3 8469+	2375 893−	3835 124−	18 7190−
a_3	227 515+	425 561−	34−	140 150+	633 436+	1 2255−
a_4	26 734−	10 451−	216−	45 719+	16 881+	2050+
a_5	4 365−	1 476+		5 106−	3 050−	
	March 9			**March 17**		
a_0	36·3788 730+	18·8642 906+	0·9451 0461+	152·8050 438+	6·7527 234+	1·0082 6740+
a_1	13·2059 522+	3·5719 052+	0·0092 2696+	13·6879 969+	5·9335 854−	0·0011 0342−
a_2	3978 000+	4355 463−	3 6977+	1732 244−	1864 009−	21 1522−
a_3	77 472+	451 580−	891−	272 367+	670 796+	3845−
a_4	49 877−	3 204−	527−	19 898+	1 707+	2492+
a_5	3 535−	3 261+		4 368−	2 480−	
	March 10			**March 18**		
a_0	49·9850 311+	21·9554 971+	0·9546 8716+	166·3486 061+	0·6997 393+	1·0050 3523+
a_1	14·0030 804+	2·5656 888+	0·0099 1870+	13·4290 346+	6·1057 052−	0·0053 4941−
a_2	3875 547+	5696 756−	3 1054+	839 437−	133 832+	20 8188−
a_3	154 739−	431 010−	3014−	308 743+	653 113+	6351+
a_4	69 242−	13 555+	823−	1 899−	10 650−	2223+
a_5	1 045+	4 672+		3 552−	1 995−	
	March 11			**March 19**		
a_0	64·3533 727+	23·9102 321+	0·9648 7803+	179·7240 262+	5·3285 359−	0·9976 8967+
a_1	14·7046 015+	1·3047 920+	0·0104 1640+	13·3512 353+	5·8882 621−	0·0092 3365−
a_2	3006 325+	6861 585−	1 7014+	39 850+	2009 329+	17 6055−
a_3	417 451−	330 398−	6357−	266 023+	590 821+	1 5375+
a_4	63 993−	37 959+	1018−	19 766−	20 572−	1377+
a_5	7 855+	4 140+		2 874−	1 556−	

Formula: Quantity in degrees $= a_0 + a_1 p + a_2 p^2 + a_3 p^3 + a_4 p^4 + a_5 p^5$
where p is the fraction of a day from 0^h TDT.

DAILY POLYNOMIAL COEFFICIENTS

	Apparent Right Ascension	Apparent Declination	Horizontal Parallax	Apparent Right Ascension	Apparent Declination	Horizontal Parallax
	March 20			**March 28**		
	°	°	°	°	°	°
a_0	193·1035 848+	10·9589 958−	0·9868 6298+	299·8394 519+	18·6006 444−	0·9030 4816+
a_1	13·4296 705+	5·3181 562−	0·0122 3843−	11·9865 738+	3·2716 449+	0·0016 0979−
a_2	690 512+	3642 814+	12 1988−	2430 152−	3747 719+	13 8529+
a_3	158 904+	493 204+	2 0875+	137 354+	259 880−	3028−
a_4	34 568−	28 305−	339+	26 167+	2 082+	521−
a_5	1 717−	1 178−		1 806−	49−	
	March 21			**March 29**		
a_0	206·6145 684+	15·8664 986−	0·9736 1680+	311·5991 819+	14·9800 122−	0·9027 8818+
a_1	13·6007 610+	4·4535 430−	0·0140 3843−	11·5513 136+	3·9440 326+	0·0010 4913+
a_2	942 556+	4940 805+	5 7592−	1879 184−	2980 062+	12 6359+
a_3	4 767+	368 468+	2 2132+	224 219+	252 348−	5111−
a_4	43 706−	34 254−	515−	16 991+	1 701+	403−
a_5	535+	734−		1 332−	580−	
	March 22			**March 30**		
a_0	220·3057 446+	19·7926 131−	0·9592 1861+	322·9865 647+	10·7630 961−	0·9050 4576+
a_1	13·7734 906+	3·3689 091−	0·0145 4698−	11·2488 729+	4·4647 307+	0·0034 0687+
a_2	699 954+	5833 328+	5557+	1117 914−	2227 401+	10 8651+
a_3	163 040−	224 568+	1 9951+	279 119+	251 498−	6729−
a_4	41 091−	38 078−	1012−	10 350+	1 323−	272−
a_5	3 463+	42+		869−	839−	
	March 23			**March 31**		
a_0	234·1291 637+	22·5595 361−	0·9449 1659+	334·1525 062+	6·1009 913−	0·9094 6914+
a_1	13·8498 662+	2·1500 818−	0·0138 7784−	11·1127 316+	4·8338 129+	0·0053 6714+
a_2	988−	6278 981+	5 9285+	227 137−	1456 562+	8 6881+
a_3	291 734−	73 307+	1 5811+	311 913+	265 197−	7828−
a_4	23 002−	37 917−	1173−	6 116+	5 625−	113−
a_5	5 331+	1 142+		722−	874−	
	March 24			**April 1**		
a_0	247·9479 907+	24·0780 667−	0·9317 7797+	345·2742 547+	1·1486 918−	0·9156 2568+
a_1	13·7556 127+	0·8868 885−	0·0122 6478−	11·1629 629+	5·0428 795+	0·0068 6539+
a_2	960 750−	6282 850+	9 9693+	738 092+	618 461+	6 2775+
a_3	330 799−	66 422−	1 1069+	328 953+	296 330−	8288−
a_4	4 672+	31 991−	1117−	2 675+	10 113−	72+
a_5	4 671+	2 058+		1 059−	694−	
	March 25			**April 2**		
a_0	261·5753 827+	24·3463 057−	0·9206 0963+	356·5440 837+	3·9253 202+	0·9230 3667+
a_1	13·4684 242+	0·3379 879+	0·0099 8357−	11·4098 066+	5·0732 814+	0·0078 7515+
a_2	1878 371−	5912 273+	12 6241+	1730 426+	338 171−	3 8403+
a_3	266 864−	173 695−	6584+	328 521+	343 445−	7993−
a_4	28 356+	21 347−	965−	2 441−	13 749−	261+
a_5	2 080+	2 239+		1 974−	224−	
	March 26			**April 3**		
a_0	274·8323 270+	23·4363 707−	0·9119 4466+	8·1593 435+	8·9290 427+	0·9312 1853+
a_1	13·0250 706+	1·4609 141+	0·0072 9983−	11·8524 834+	4·8970 033+	0·0084 1387+
a_2	2488 096−	5285 529+	14 0256+	2681 602+	1453 277−	1 6036+
a_3	134 096−	237 016−	2728+	298 240+	400 149−	6927−
a_4	38 308+	9 923−	796−	12 286−	15 109−	409+
a_5	474−	1 661+		3 339−	684+	
	March 27			**April 4**		
a_0	287·5989 618+	21·4714 315−	0·9060 6671+	20·3082 487+	13·6392 609+	0·9397 2757+
a_1	12·5023 074+	2·4437 757+	0·0044 4472−	12·4716 912+	4·4806 034+	0·0085 4314+
a_2	2665 376−	4531 560+	14 3710+	3469 137+	2737 572−	2277−
a_3	13 654+	260 607−	447−	215 166+	452 899−	5256−
a_4	35 289+	1 597−	647−	29 479−	11 920−	463+
a_5	1 740−	758+		4 346−	2 142+	

Formula: Quantity in degrees $= a_0 + a_1 p + a_2 p^2 + a_3 p^3 + a_4 p^4 + a_5 p^5$
where p is the fraction of a day from 0^h TDT.

DAILY POLYNOMIAL COEFFICIENTS

	Apparent Right Ascension	Apparent Declination	Horizontal Parallax	Apparent Right Ascension	Apparent Declination	Horizontal Parallax
	April 5			**April 13**		
	°	°	°	°	°	°
a_0	33·1449 877+	17·7998 394+	0·9482 0001+	148·2777 389+	8·4948 851+	0·9937 2271+
a_1	13·2161 056+	3·7935 240+	0·0083 5847+	13·3888 969+	5·5649 595−	0·0006 6863+
a_2	3894 091+	4146 370−	1 5287−	1739 346−	2357 292−	11 2986−
a_3	54 519+	478 197−	3360−	296 444+	562 360+	7667−
a_4	52 604−	1 174−	392+	21 059+	4 304+	742+
a_5	3 171−	3 852+		3 976−	856−	
	April 6			**April 14**		
a_0	46·7503 767+	21·1311 745+	0·9563 7593+	161·5240 540+	2·7507 772+	0·9931 9224+
a_1	13·9886 576+	2·8222 482+	0·0079 6762+	13·1363 974+	5·8664 168−	0·0017 9138−
a_2	3710 068+	5549 392−	2 3077−	763 409−	652 911−	13 1385−
a_3	184 776−	443 921−	1758−	341 242+	570 939+	4639−
a_4	70 070−	18 739+	196+	1 318+	285+	1230+
a_5	1 768+	4 701+		3 396−	1 003−	
	April 7			**April 15**		
a_0	61·0847 333+	23·3564 354+	0·9640 9715+	174·6180 269+	3·1239 087−	0·9900 5292+
a_1	14·6481 018+	1·5890 382+	0·0074 6116+	13·0849 175+	5·8261 057−	0·0045 0900−
a_2	2753 039+	6721 558−	2 7263−	234 251+	1051 603+	13 7859−
a_3	443 666−	322 766−	962−	312 655+	561 736+	391+
a_4	60 906−	43 299+	79−	15 656−	4 608−	1442+
a_5	8 426+	3 307+		3 239−	1 552−	
	April 8			**April 16**		
a_0	75·9585 245+	24·2457 018+	0·9712 7527+	187·7557 454+	8·7892 965−	0·9841 8366+
a_1	15·0454 629+	0·1668 660+	0·0068 8387+	13·2176 827+	5·4498 841−	0·0071 9671−
a_2	1141 255+	7396 935−	3 0708−	1045 817+	2693 643+	12 8082−
a_3	601 727−	118 407−	1294−	217 934+	527 448+	6274+
a_4	15 967−	60 249+	353−	32 266−	12 398−	1288+
a_5	10 805+	77−		2 798−	2 144−	
	April 9			**April 17**		
a_0	91·0574 241+	23·6670 508+	0·9778 3559+	201·0962 969+	13·9185 257−	0·9757 8175+
a_1	15·0922 065+	1·3239 855−	0·0062 1675+	13·4779 233+	4·7589 524−	0·0095 1857−
a_2	651 377−	7391 536−	3 6762−	1477 919+	4180 131+	10 1670−
a_3	560 037−	120 144+	2752−	61 985+	456 311+	1 1497+
a_4	40 233+	59 145+	532−	47 071−	23 355−	836+
a_5	6 278+	3 036−		966−	2 348−	
	April 10			**April 18**		
a_0	106·0331 403+	21·6215 370+	0·9836 5188+	214·7234 069+	18·2164 042−	0·9653 6980+
a_1	14·8131 448+	2·7441 102−	0·0053 7764+	13·7727 953+	3·7965 479−	0·0111 7361−
a_2	2027 395−	6706 729−	4 8207−	1371 697+	5385 399+	6 2337−
a_3	340 104−	325 953+	4950−	133 928−	339 819+	1 4849+
a_4	71 081+	42 999+	528−	52 484−	35 519−	265+
a_5	441−	3 708−		2 584+	1 665−	
	April 11			**April 19**		
a_0	120·6165 991+	18·2432 784+	0·9884 9268+	228·6149 891+	21·4441 488−	0·9537 2395+
a_1	14·3338 421+	3·9723 231−	0·0042 4385+	13·9872 582+	2·6325 609−	0·0119 6431−
a_2	2625 883−	5508 001−	6 6148−	680 914+	6175 043+	1 6359−
a_3	62 416−	461 444+	7139−	316 072−	182 069+	1 5866+
a_4	67 200+	24 107+	288−	39 004−	44 172−	245−
a_5	4 242−	2 654−		6 076+	47+	
	April 12			**April 20**		
a_0	134·6879 071+	13·7684 450+	0·9920 0078+	242·6354 387+	23·4454 110−	0·9417 5226+
a_1	13·8146 991+	4·9271 728−	0·0026 9518+	14·0160 563+	1·3605 747−	0·0118 2535−
a_2	2452 507−	4005 532−	8 9149−	440 363−	6456 707+	2 9662+
a_3	163 610+	532 038+	8349−	410 994−	6 975+	1 4827+
a_4	44 969+	11 041+	172+	7 119−	43 791−	589−
a_5	4 745−	1 417−		6 674+	2 039+	

Formula: Quantity in degrees $= a_0 + a_1 p + a_2 p^2 + a_3 p^3 + a_4 p^4 + a_5 p^5$

where p is the fraction of a day from 0^h TDT.

DAILY POLYNOMIAL COEFFICIENTS

	Apparent Right Ascension	Apparent Declination	Horizontal Parallax	Apparent Right Ascension	Apparent Declination	Horizontal Parallax
		April 21			April 29	
	°	°	°	°	°	°
a_0	256·5663 147+	24·1637 927−	0·9303 6591+	352·6401 864+	2·3224 820+	0·9212 2331+
a_1	13·8051 716+	0·0836 364−	0·0108 1088−	11·2919 700+	5·0335 200+	0·0087 6284+
a_2	1649 195−	6235 359+	7 0555+	1595 063+	67 256+	8 2364+
a_3	374 309−	147 186−	1 2420+	369 288+	305 242−	1 0370−
a_4	27 204+	33 031−	758−	187−	15 319−	443−
a_5	3 841+	3 107+		1 824−	958−	
		April 22			April 30	
a_0	270·1722 404+	23·6416 043−	0·9203 7720+	4·1283 904+	7·3305 757+	0·9307 0167+
a_1	13·3758 377+	1·1076 201+	0·0090 5751−	11·7207 801+	4·9487 933+	0·0100 8132+
a_2	2570 544−	5626 756+	10 3255+	2683 579+	950 015−	4 8681+
a_3	229 198−	248 415−	9358+	349 339+	375 511−	1 2177−
a_4	46 075+	16 991−	797−	9 117−	20 449−	183−
a_5	101+	2 777+		3 485−	33+	
		April 23			May 1	
a_0	283·2727 214+	21·9975 715−	0·9124 3785+	16·1512 021+	12·1447 747+	0·9411 4619+
a_1	12·8114 471+	2·1530 375+	0·0067 4356−	12·3569 063+	4·6379 758+	0·0106 8227+
a_2	2980 824−	4807 355+	12 6553+	3641 982+	2198 979−	1 1161+
a_3	45 177−	289 274−	6157+	277 053+	455 882−	1 2936−
a_4	45 665+	2 950−	771−	26 876−	20 698−	177+
a_5	2 110−	1 601+		5 225−	1 922+	
		April 24			May 2	
a_0	295·7859 239+	19·3928 608−	0·9070 1368+	28·8968 019+	16·5153 869+	0·9518 1249+
a_1	12·2189 397+	3·0273 454+	0·0040 5861−	13·1550 564+	4·0541 005+	0·0105 2449+
a_2	2863 573−	3937 824+	14 0413+	4259 405+	3671 609−	2 6464−
a_3	116 157+	285 777−	3075+	117 559+	517 977−	1 2222−
a_4	34 445+	4 947+	728−	54 520−	11 217−	570+
a_5	2 454−	363+		4 937−	4 505+	
		April 25			May 3	
a_0	307·7333 213+	15·9997 796−	0·9043 8267+	42·4836 090+	20·1498 576+	0·9619 5582+
a_1	11·6936 245+	3·7313 366+	0·0011 8721−	14·0179 346+	3·1621 532+	0·0096 5137+
a_2	2333 008−	3113 781+	14 5280+	4235 246+	5247 681−	5 9615−
a_3	229 746+	262 880−	168+	146 893−	516 789−	9895−
a_4	21 933+	6 585+	696−	81 558−	12 077+	881+
a_5	1 829−	538−		190+	6 340+	
		April 26			May 4	
a_0	319·2186 299+	11·9827 482−	0·9046 4298+	56·9022 421+	22·7374 055+	0·9709 2092+
a_1	11·3038 061+	4·2775 932+	0·0016 9558+	14·7883 976+	1·9655 796+	0·0081 9753+
a_2	1530 471−	2359 247+	14 1612+	3307 060+	6661 953−	8 3976−
a_3	299 597+	242 233−	2615−	466 262−	405 841−	6292−
a_4	12 807+	3 744+	678−	80 998−	45 403+	996+
a_5	1 099−	1 071−		9 069+	5 072+	
		April 27			May 5	
a_0	330·4005 194+	7·4931 863−	0·9077 2174+	71·9675 266+	24·0012 532+	0·9782 2573+
a_1	11·0921 649+	4·6777 347+	0·0044 2222+	15·2820 713+	0·5321 285+	0·0063 6913+
a_2	565 813−	1644 284+	12 9703+	1513 465+	7556 207−	9 6918−
a_3	340 050+	238 115−	5336−	696 984−	176 171−	2231−
a_4	7 452+	1 733−	655−	32 152−	71 692+	863+
a_5	716−	1 323−		13 786+	338+	
		April 28			May 6	
a_0	341·4707 816+	2·6751 402−	0·9133 8109+	87·3294 094+	23·7673 469+	0·9836 1200+
a_1	11·0836 396+	4·9338 025+	0·0068 3001+	15·3696 959+	1·0031 236−	0·0043 9840+
a_2	491 906+	906 287+	10 9787+	632 052−	7651 322−	9 8533−
a_3	362 578+	258 277−	7976−	690 408−	111 334+	1269+
a_4	4 074+	8 490−	590−	40 230+	72 432+	532+
a_5	905−	1 323−		8 862+	4 363−	

Formula: Quantity in degrees $= a_0 + a_1 p + a_2 p^2 + a_3 p^3 + a_4 p^4 + a_5 p^5$

where p is the fraction of a day from 0^h TDT.

DAILY POLYNOMIAL COEFFICIENTS

	Apparent Right Ascension	Apparent Declination	Horizontal Parallax	Apparent Right Ascension	Apparent Declination	Horizontal Parallax
	May 7			**May 15**		
	°	°	°	°	°	°
a_0	102·5717 685+	22·0170 314+	0·9870 4307+	210·0202 509+	16·8825 505−	0·9612 0139+
a_1	15·0566 762+	2·4731 977−	0·0024 8708+	13·4637 385+	4·0418 454−	0·0082 0425−
a_2	2373 347−	6926 582−	9 1659−	1818 075+	4728 435+	5 0439−
a_3	445 840−	356 689+	3397+	26 889−	382 667+	4554+
a_4	84 088+	49 021+	127+	55 152−	21 610−	641+
a_5	52−	5 599−		342+	2 713−	
	May 8			**May 16**		
a_0	117·3549 296+	18·8911 866+	0·9886 4879+	223·6576 270+	20·4157 181−	0·9525 4469+
a_1	14·4818 581+	3·7346 964−	0·0007 6086+	13·7974 018+	2·9913 579−	0·0090 5076−
a_2	3207 205−	5618 475−	8 0812−	1409 870+	5719 554+	3 2977−
a_3	112 996−	497 740+	3863+	241 653−	269 620+	7159+
a_4	81 499+	20 326+	211−	53 702−	35 805−	495+
a_5	5 175−	3 844−		4 624+	1 812−	
	May 9			**May 17**		
a_0	131·5124 000+	14·6460 649+	0·9886 3805+	237·5669 427+	22·8119 202−	0·9432 4071+
a_1	13·8365 296+	4·7028 585−	0·0007 4795−	13·9877 142+	1·7817 863−	0·0094 7570−
a_2	3109 171−	4041 708−	7 0539−	409 106+	6295 421+	8602−
a_3	160 909+	541 931+	2956+	408 617−	109 535+	9159+
a_4	54 192+	1 308+	379−	29 332−	45 199−	254+
a_5	5 610−	1 508−		7 510+	398+	
	May 10			**May 18**		
a_0	145·0589 617+	9·5932 086+	0·9872 1048+	251·5525 237+	23·9576 910−	0·9337 7312+
a_1	13·2818 416+	5·3488 499−	0·0020 8523−	13·9389 715+	0·5077 196−	0·0093 6282−
a_2	2357 435−	2423 071−	6 3932−	917 400−	6356 866+	2 0316+
a_3	322 324+	532 935+	1377+	451 495−	66 028−	1 0169+
a_4	25 907+	5 740−	343−	9 902+	42 855−	9−
a_5	4 247−	44−		6 339+	2 607+	
	May 11			**May 19**		
a_0	158·1394 582+	4·0547 667+	0·9844 9627+	265·3562 297+	23·8403 516−	0·9247 1506+
a_1	12·9152 923+	5·6759 014−	0·0033 3629−	13·6271 687+	0·7280 074+	0·0086 5180−
a_2	1277 462−	859 084−	6 1793−	2149 031−	5927 821+	5 0699+
a_3	384 085+	509 751+	39−	350 786−	210 910−	1 0115+
a_4	4 940+	5 531−	133−	42 011+	29 111−	233−
a_5	3 201−	325+		2 245+	3 410+	
	May 12			**May 20**		
a_0	170·9655 867+	1·6565 885−	0·9805 4034+	278·7378 422+	22·5432 231−	0·9166 6906+
a_1	12·7754 015+	5·6968 430−	0·0045 7864−	13·1100 492+	1·8403 588+	0·0073 4371−
a_2	127 548−	640 287+	6 2615−	2927 006−	5154 597+	7 9593+
a_3	371 968+	490 615+	585−	162 355−	293 647−	9161+
a_4	10 825−	3 600−	168+	52 400+	11 594−	396−
a_5	2 995−	149−		1 362−	2 682+	
	May 13			**May 21**		
a_0	183·7640 483+	7·2407 163−	0·9753 3138+	291·5440 590+	20·2176 605−	0·9102 0893+
a_1	12·8556 551+	5·4231 171−	0·0058 4175−	12·4962 187+	2·7798 861+	0·0054 9287−
a_2	893 432+	2089 065+	6 3272−	3113 446−	4230 923+	10 4666+
a_3	298 670+	474 153+	110+	32 797+	313 983−	7563+
a_4	25 927−	4 180−	455+	44 614+	1 871+	503−
a_5	3 115−	1 151−		2 734−	1 302+	
	May 14			**May 22**		
a_0	196·7360 093+	12·4080 447−	0·9688 6256+	303·7364 009+	17·0457 631−	0·9058 3333+
a_1	13·1120 149+	4·8653 068−	0·0070 8567−	11·8998 476+	3·5332 736+	0·0031 9278−
a_2	1602 611+	3474 932+	6 0158−	2774 793−	3313 191+	12 4314+
a_3	164 253+	445 311+	1978+	184 072+	294 203−	5546+
a_4	42 206−	9 991−	630+	30 435+	8 215+	578−
a_5	2 391−	2 243−		2 406−	56+	

Formula: Quantity in degrees $= a_0 + a_1 p + a_2 p^2 + a_3 p^3 + a_4 p^4 + a_5 p^5$
where p is the fraction of a day from 0^h TDT.

DAILY POLYNOMIAL COEFFICIENTS

	Apparent Right Ascension	Apparent Declination	Horizontal Parallax	Apparent Right Ascension	Apparent Declination	Horizontal Parallax
	May 23			May 31		
	°	°	°	°	°	°
a_0	315·3799 793+	13·2097 636−	0·9039 3336+	51·4267 137+	21·9250 617+	0·9734 6293+
a_1	11·4110 826+	4·1109 642+	0·0005 6326−	14·6859 997+	2·4384 353+	0·0120 0922+
a_2	2064 049−	2480 407+	13 7459+	4394 542+	6215 237−	7 9076−
a_3	282 253+	261 273−	3235+	337 980−	523 246−	1 7931−
a_4	18 312+	8 315+	650−	103 474−	27 195+	911+
a_5	1 520−	789−		5 044+	8 133+	
	May 24			June 1		
a_0	326·6145 615+	8·8761 334−	0·9047 7055+	66·5085 266+	23·6931 816+	0·9845 1119+
a_1	11·0895 145+	4·5315 944+	0·0022 5698+	15·4246 579+	1·0533 544+	0·0099 2624+
a_2	1122 607−	1738 561+	14 3237+	2810 496+	7540 145−	12 7201−
a_3	340 707+	236 207−	641+	695 759−	335 351−	1 4212−
a_4	10 821+	4 235+	736−	76 121−	70 155+	1575+
a_5	818−	1 300−		15 441+	4 262+	
	May 25			June 2		
a_0	337·6268 862+	4·1940 101−	0·9084 5895+	82·1385 902+	23·9664 280+	0·9930 3904+
a_1	10·9711 244+	4·8094 884+	0·0051 1153+	15·7553 007+	0·5250 952−	0·0070 1892+
a_2	43 726−	1042 335+	14 0714+	421 699+	8082 632−	16 0297−
a_3	375 902+	232 424−	2300−	846 063−	16 140−	7762−
a_4	6 929+	2 372−	837−	6 770+	91 603+	1876+
a_5	644−	1 580−		15 035+	2 956−	
	May 26			June 3		
a_0	348·6318 567+	0·6960 742+	0·9149 4625+	97·8536 351+	22·6403 203+	0·9983 9612+
a_1	11·0775 986+	4·9464 893+	0·0078 2333+	15·5960 355+	2·1113 053−	0·0036 5521+
a_2	1119 144+	315 005+	12 8766+	1925 267−	7611 296−	17 2391−
a_3	396 838+	257 719−	5658−	674 506−	318 186+	102−
a_4	3 996+	10 431−	927−	83 741+	74 688+	1678+
a_5	1 216−	1 609−		4 317+	7 389−	
	May 27			June 4		
a_0	359·8613 315+	5·6470 882+	0·9239 9139+	113·1984 990+	19·8064 340+	1·0003 4317+
a_1	11·4214 672+	4·9271 988+	0·0101 9180+	15·0442 747+	3·5119 270−	0·0002 7150+
a_2	2321 515+	536 870−	10 6224+	3403 569−	6282 745−	16 2814−
a_3	399 772+	315 238−	9400−	301 606−	543 396+	6702+
a_4	1 729−	18 783−	949−	102 609+	36 010+	1077+
a_5	2 747−	1 140−		4 848−	6 644−	
	May 28			June 5		
a_0	11·5544 797+	10·4870 839+	0·9351 4194+	127·8820 323+	15·7235 086+	0·9990 6432+
a_1	12·0036 339+	4·7171 719+	0·0119 9628+	14·3116 959+	4·5943 719−	0·0027 4065−
a_2	3482 995+	1606 763−	7 2371+	3741 596−	4502 972−	13 6477−
a_3	364 034+	400 909−	1 3259−	58 913+	622 784+	1 1007+
a_4	15 322−	25 016−	820−	75 814+	2 552+	333+
a_5	5 118−	312+		7 128−	3 446−	
	May 29			June 6		
a_0	23·9407 724+	15·0010 183+	0·9477 2114+	141·8323 285+	10·7410 285+	0·9950 7229+
a_1	12·8007 533+	4·2656 999+	0·0130 1309+	13·6078 141+	5·3088 304−	0·0051 2670−
a_2	4431 839+	2956 550−	2 7787+	3181 368−	2653 692−	10 1651−
a_3	250 626+	496 200−	1 6622−	291 728+	600 115+	1 2268+
a_4	41 804−	24 044−	459−	39 435+	14 150−	290−
a_5	6 879−	3 163+		5 544−	695−	
	May 30			June 7		
a_0	37·2049 039+	18·9193 551+	0·9608 4130+	155·1545 677+	5·2253 560+	0·9890 4885+
a_1	13·7421 510+	3·5174 978+	0·0130 5179+	13·0720 630+	5·6655 398−	0·0068 0335−
a_2	4863 704+	4557 770−	2 4641−	2124 992−	945 102−	6 6701−
a_3	16 167+	558 742−	1 8529−	395 090+	537 374+	1 1012+
a_4	78 885−	8 132−	153+	11 952+	17 034−	647−
a_5	4 397−	6 731+		3 674−	675+	

Formula: Quantity in degrees $= a_0 + a_1 p + a_2 p^2 + a_3 p^3 + a_4 p^4 + a_5 p^5$
where p is the fraction of a day from 0^h TDT.

DAILY POLYNOMIAL COEFFICIENTS

	Apparent Right Ascension	Apparent Declination	Horizontal Parallax	Apparent Right Ascension	Apparent Declination	Horizontal Parallax
	June 8			**June 16**		
	°	°	°	°	°	°
a_0	168·0544 682+	0·4825 926−	0·9816 8214+	274·5333 858+	23·0494 169−	0·9168 0846+
a_1	12·7685 360+	5·6998 243−	0·0078 3293−	13·2627 565+	1·5149 253+	0·0065 8999−
a_2	904 712−	571 628+	3 7570−	2485 867−	5418 812+	4 8148+
a_3	406 655+	476 087+	8344+	261 689−	256 179−	5723+
a_4	6 041−	13 221−	723−	46 783+	21 342−	159+
a_5	2 823−	835+		459+	2 990+	
	June 9			**June 17**		
a_0	180·7723 122+	6·0788 839−	0·9735 4973+	287·5261 109+	21·0200 635−	0·9107 5876+
a_1	12·7057 626+	5·4475 439−	0·0083 6296−	12·7060 163+	2·5147 910+	0·0054 4897−
a_2	250 790+	1928 965+	1 6835−	2985 791−	4552 160+	6 6229+
a_3	354 304+	431 173+	5400+	71 404−	312 146−	6365+
a_4	20 035−	8 748−	588−	48 146+	6 103−	17+
a_5	2 793−	160+		2 042−	2 097+	
	June 10			**June 18**		
a_0	193·5363 014+	11·2912 728−	0·9650 6653+	299·9310 182+	18·0816 716−	0·9060 3590+
a_1	12·8528 015+	4·9358 194−	0·0085 6121−	12·1056 740+	3·3301 852+	0·0039 3277−
a_2	1165 497+	3171 620+	4089−	2931 659−	3600 068+	8 5379+
a_3	246 350+	397 125+	3025+	100 452+	316 292−	6432+
a_4	34 368−	7 816−	340−	37 204+	4 355+	130−
a_5	2 650−	988−		2 528−	869+	
	June 11			**June 19**		
a_0	206·5265 858+	15·8710 980−	0·9564 9128+	311·7570 390+	14·4225 864−	0·9030 1993+
a_1	13·1447 353+	4·1859 792−	0·0085 6585−	11·5630 960+	3·9574 868+	0·0020 3744−
a_2	1671 703+	4306 204+	3028+	2432 410−	2685 992+	10 3848+
a_3	83 279+	355 370+	1669+	224 329+	290 773−	5909+
a_4	48 465−	12 908−	69−	24 271+	8 549+	275−
a_5	1 147−	2 071−		1 896−	153−	
	June 12			**June 20**		
a_0	219·8418 581+	19·5924 177−	0·9479 7170+	323·1015 644+	10·2247 380−	0·9020 7731+
a_1	13·4841 039+	3·2243 261−	0·0084 5798−	11·1526 740+	4·4107 956+	0·0002 0581+
a_2	1619 166+	5274 089+	7690+	1632 767−	1863 411+	11 9882+
a_3	120 091−	282 916+	1416+	302 910+	258 520−	4814+
a_4	54 919−	23 765−	156+	14 791+	7 639+	421−
a_5	2 272+	2 294−		1 082−	852−	
	June 13			**June 21**		
a_0	233·4706 047+	22·2636 492−	0·9396 0635+	334·1226 237+	5·6527 746−	0·9035 2586+
a_1	13·7510 822+	2·0952 853−	0·0082 5541−	10·9223 697+	4·7085 510+	0·0027 3102+
a_2	952 147+	5957 221+	1 2920+	646 098−	1125 148+	13 1745+
a_3	314 947−	165 605+	2074+	351 532+	236 751−	3137+
a_4	43 149−	35 811−	293+	9 518+	3 277+	583−
a_5	6 026+	1 088−		591−	1 307−	
	June 14			**June 22**		
a_0	247·2816 946+	23·7503 419−	0·9315 0381+	345·0164 295+	0·8551 869−	0·9075 9987+
a_1	13·8327 819+	0·8690 257−	0·0079 2305−	10·9021 211+	4·8632 125+	0·0054 3670+
a_2	191 122−	6228 272+	2 0912+	459 716+	421 473+	13 7596+
a_3	426 699−	12 692+	3278+	383 624+	236 871−	813+
a_4	11 502−	41 369−	328+	6 786+	3 355−	773−
a_5	7 071+	1 058+		687−	1 585−	
	June 15			**June 23**		
a_0	261·0522 513+	23·9993 023−	0·9238 2594+	356·0034 945+	4·0259 917+	0·9144 1293+
a_1	13·6654 792+	0·3644 196+	0·0073 9334−	11·1115 215+	4·8743 110+	0·0081 8210+
a_2	1469 371−	6028 782+	3 2698+	1644 477+	325 155−	13 5332+
a_3	403 457−	141 251−	4615+	403 355+	266 148−	2274−
a_4	24 973+	35 606−	274+	3 683+	11 459−	986−
a_5	4 408+	2 734+		1 608−	1 605−	

Formula: Quantity in degrees = $a_0 + a_1 p + a_2 p^2 + a_3 p^3 + a_4 p^4 + a_5 p^5$

where p is the fraction of a day from 0^h TDT.

DAILY POLYNOMIAL COEFFICIENTS

	Apparent Right Ascension	Apparent Declination	Horizontal Parallax	Apparent Right Ascension	Apparent Declination	Horizontal Parallax
		June 24			July 2	
	°	°	°	°	°	°
a_0	7·3200 066+	8·8398 661+	0·9239 1573+	122·6926 735+	17·3065 987+	1·0134 2106+
a_1	11·5620 901+	4·7240 504+	0·0107 8103+	15·0078 610+	4·3730 682−	0·0001 0869+
a_2	2860 601+	1208 458−	12 2528+	3632 875−	5560 440−	23 2853−
a_3	400 866+	327 675−	6235−	169 747−	665 450+	4979+
a_4	3 994−	19 858−	1188−	99 148+	26 991+	2324+
a_5	3 568−	1 022−		6 854−	6 795−	
		June 25			July 3	
a_0	19·2074 873+	13·4082 152+	0·9358 4782+	137·3295 016+	12·4460 512+	1·0112 7423+
a_1	12·2510 858+	4·3756 043+	0·0129 9700+	14·2665 931+	5·2781 183−	0·0043 0599−
a_2	4003 538+	2320 947−	9 6666+	3616 073−	3470 095−	20 4297−
a_3	347 741+	416 268−	1 1037−	157 993+	707 445+	1 4383+
a_4	21 893−	25 611−	1291−	62 892+	7 017−	1265+
a_5	6 177−	769+		7 182−	3 289−	
		June 26			July 4	
a_0	31·8908 939+	17·5076 138+	0·9496 8819+	151·2558 577+	6·8906 373+	1·0050 8174+
a_1	13·1442 690+	3·7766 786+	0·0145 4749+	13·6123 445+	5·7643 518−	0·0079 0988−
a_2	4853 397+	3715 817−	5 5849+	2836 609−	1422 662−	15 3911−
a_3	197 937+	509 069−	1 6303−	338 919+	648 090+	1 9393+
a_4	54 298−	22 380−	1161−	26 729+	22 881−	122+
a_5	7 176−	4 159+		5 053−	522−	
		June 27			July 5	
a_0	45·5341 489+	20·8599 817+	0·9646 1954+	164·6206 009+	1·0464 880+	0·9958 2790+
a_1	14·1490 279+	2·8739 261+	0·0151 2886+	13·1548 656+	5·8638 689−	0·0104 0149−
a_2	5049 171+	5335 653−	135+	1709 965−	379 197+	9 5257−
a_3	88 074−	554 902−	2 1086−	396 276+	552 127+	1 9748+
a_4	93 472−	1 158−	650−	1 820+	24 886−	715−
a_5	2 237−	7 930+		3 371−	800+	
		June 28			July 6	
a_0	60·1697 155+	23·1455 295+	0·9795 3239+	177·6439 425+	4·7266 571−	0·9846 6416+
a_1	15·0939 454+	1·6438 270+	0·0144 7292+	12·9307 991+	5·6319 459−	0·0117 4288−
a_2	4201 508+	6927 686−	6 6732−	543 892−	1894 328+	4 0430−
a_3	477 809−	480 083−	2 3810−	370 263+	460 640+	1 6750+
a_4	106 283−	40 731+	288+	14 769−	20 440−	1113−
a_5	9 436+	8 357+		2 692−	884+	
		June 29			July 7	
a_0	75·6263 461+	24·0534 884+	0·9931 0278+	190·5556 326+	10·1250 617−	0·9726 7334+
a_1	15·7531 181+	0·1347 294+	0·0124 3547+	12·9258 463+	5·1226 229−	0·0120 9357−
a_2	2225 345+	8039 663−	13 6073−	451 335+	3162 491+	3129+
a_3	804 089−	236 737−	2 2681−	284 469+	387 313+	1 2200+
a_4	54 806−	84 735+	1471+	28 345−	15 761−	1141−
a_5	17 749+	2 781+		2 389−	160+	
		June 30			July 8	
a_0	91·5178 843+	23·3693 293+	1·0039 6543+	203·5519 859+	14·8942 644−	0·9607 2165+
a_1	15·9439 067+	1·5089 485−	0·0090 9252+	13·0889 229+	4·3801 565−	0·0117 1068−
a_2	337 500−	8213 741−	19 4984−	1110 694+	4231 469+	3 2941+
a_3	848 688−	125 689+	1 6654−	147 793+	325 291+	7585+
a_4	39 393+	98 017+	2469+	40 844−	14 941−	942−
a_5	12 451+	4 962−		1 411−	861−	
		July 1			July 9	
a_0	107·3483 565+	21·0608 810+	1·0109 6626+	216·7625 320+	18·8203 249−	0·9494 0682+
a_1	15·6437 693+	3·0772 686−	0·0047 9212+	13·3383 594+	3·4426 826−	0·0108 6199−
a_2	2522 741−	7298 545−	23 0029−	1294 797+	5109 055+	5 0136+
a_3	573 380−	466 100+	6515−	28 299−	256 555+	3802+
a_4	101 371+	70 768+	2811+	48 561−	19 518−	645−
a_5	227+	8 459−		988+	1 519−	

Formula: Quantity in degrees = $a_0 + a_1 p + a_2 p^2 + a_3 p^3 + a_4 p^4 + a_5 p^5$
where p is the fraction of a day from 0^h TDT.

DAILY POLYNOMIAL COEFFICIENTS

	Apparent Right Ascension	Apparent Declination	Horizontal Parallax
July 10	°	°	°
a_0	230·2227 840+	21·7285 503−	0·9390 7776+
a_1	13·5699 025+	2·3524 716−	0·0097 7099−
a_2	928 406+	5746 354+	5 7769+
a_3	210 853−	163 479+	1229+
a_4	43 642−	27 575−	338−
a_5	4 183+	1 199−	
July 11			
a_0	243·8604 960+	23·4929 159−	0·9298 9338+
a_1	13·6769 648+	1·1657 848−	0·0085 9222−
a_2	75 956+	6059 315+	5 9515+
a_3	342 575−	41 945+	103−
a_4	21 756−	33 857−	71−
a_5	6 016+	126+	
July 12			
a_0	257·5092 250+	24·0519 479−	0·9218 9457+
a_1	13·5836 880+	0·0451 835+	0·0074 0784−
a_2	1021 992−	5983 288+	5 8843+
a_3	370 075−	91 369−	362−
a_4	9 460+	33 077−	129+
a_5	4 899+	1 635+	
July 13			
a_0	270·9551 423+	23·4207 167−	0·9150 7282+
a_1	13·2744 975+	1·2020 180+	0·0062 3670−
a_2	2026 437−	5527 131+	5 8565+
a_3	284 979−	206 899−	178+
a_4	34 194+	24 477−	249+
a_5	1 816+	2 372+	
July 14			
a_0	284·0020 993+	21·6888 861−	0·9094 2605+
a_1	12·7982 990+	2·2367 692+	0·0050 5007−
a_2	2658 145−	4783 337+	6 0608+
a_3	131 599−	281 228−	1200+
a_4	42 673+	12 283−	291+
a_5	903−	2 108+	
July 15			
a_0	296·5256 009+	19·0029 236−	0·9049 9697+
a_1	12·2438 068+	3·1052 076+	0·0037 9027−
a_2	2806 046−	3887 041+	6 5942+
a_3	29 392+	309 750−	2383+
a_4	37 433+	1 652−	261+
a_5	2 038−	1 285+	
July 16			
a_0	308·4952 817+	15·5400 236−	0·9018 9256+
a_1	11·7053 694+	3·7896 714+	0·0023 8950−
a_2	2513 723−	2960 714+	7 4630+
a_3	158 809+	304 015−	3443+
a_4	26 826+	4 706+	175+
a_5	1 896−	407+	
July 17			
a_0	319·9676 527+	11·4841 710−	0·9002 8554+
a_1	11·2600 506+	4·2926 950+	0·0007 8659−
a_2	1895 323−	2080 960+	8 5970+
a_3	247 511+	281 524−	4155+
a_4	17 239+	6 636+	47+
a_5	1 261−	289−	

	Apparent Right Ascension	Apparent Declination	Horizontal Parallax
July 18	°	°	°
a_0	331·0645 199+	7·0108 976−	0·9004 0068+
a_1	10·9615 052+	4·6269 394+	0·0010 5937+
a_2	1061 963−	1273 300+	9 8670+
a_3	304 170+	258 164−	4354+
a_4	10 988+	5 099+	113−
a_5	703−	788−	
July 19			
a_0	341·9512 742+	2·2820 136−	0·9024 8916+
a_1	10·8444 071+	4·8057 954+	0·0031 5886+
a_2	90 551−	521 508+	11 0992+
a_3	341 184+	245 853−	3911+
a_4	7 601+	1 080+	305−
a_5	525−	1 140−	
July 20			
a_0	352·8214 522+	2·5513 414+	0·9067 9400+
a_1	10·9314 293+	4·8362 030+	0·0054 8385+
a_2	973 377+	220 989−	12 0825+
a_3	366 104+	253 044−	2704+
a_4	5 191+	4 723−	533−
a_5	927−	1 340−	
July 21			
a_0	3·8872 561+	7·3395 347+	0·9135 0780+
a_1	11·2375 477+	4·7135 328+	0·0079 6014+
a_2	2093 603+	1021 894−	12 5652+
a_3	376 916+	285 265−	583+
a_4	854+	11 627−	803−
a_5	2 111−	1 234−	
July 22			
a_0	15·3717 301+	11·9210 655+	0·9227 2226+
a_1	11·7686 269+	4·4183 074+	0·0104 5856+
a_2	3208 385+	1959 859−	12 2491+
a_3	358 084+	343 647−	2625−
a_4	9 502	18 201	1102
a_5	4 099−	448−	
July 23			
a_0	27·4956 438+	16·1071 574+	0·9343 6846+
a_1	12·5118 767+	3·9157 392+	0·0127 8554+
a_2	4184 552+	3104 579−	10 7919+
a_3	278 030+	419 778−	7054−
a_4	30 474−	21 029−	1379−
a_5	6 023−	1 530+	
July 24			
a_0	40·4501 290+	19·6685 108+	0·9481 4885+
a_1	13·4169 965+	3·1612 465+	0·0146 7708+
a_2	4775 280+	4474 846−	7 8440+
a_3	96 485+	486 772−	1 2642−
a_4	62 525−	13 729−	1517−
a_5	5 205−	4 728+	
July 25			
a_0	54·3475 290+	22·3326 955+	0·9634 6874+
a_1	14·3733 931+	2·1171 212+	0·0158 0585+
a_2	4637 143+	5970 141−	3 1471+
a_3	201 802−	492 890−	1 8850−
a_4	91 217−	10 695+	1331−
a_5	1 483+	7 490+	

Formula: Quantity in degrees $= a_0 + a_1 p + a_2 p^2 + a_3 p^3 + a_4 p^4 + a_5 p^5$
where p is the fraction of a day from 0^h TDT.

DAILY POLYNOMIAL COEFFICIENTS

	Apparent Right Ascension	Apparent Declination	Horizontal Parallax	Apparent Right Ascension	Apparent Declination	Horizontal Parallax
	July 26			**August 3**		
	°	°	°	°	°	°
a_0	69·1554 828+	23·8053 323+	0·9793 8749+	186·6614 348+	8·4296 254−	0·9908 2681+
a_1	15·2045 468+	0·7832 484+	0·0158 1644+	13·3100 920+	5·4576 509−	0·0141 3957−
a_2	3499 299+	7309 439−	3 2844−	118 277−	3068 887+	6 8930−
a_3	546 270−	375 763−	2 4360−	277 292+	461 615+	2 3983+
a_4	83 515−	50 199+	629−	23 579−	28 051−	1169−
a_5	11 556+	6 604+		2 406−	484+	
	July 27			**August 4**		
a_0	84·6481 366+	23·8257 408+	0·9946 2561+	199·9848 298+	13·5369 829−	0·9762 2608+
a_1	15·7129 012+	0·7679 928−	0·0144 0351+	13·3589 910+	4·7163 680−	0·0148 4552−
a_2	1475 601+	8069 294−	10 9312−	548 002+	4290 295+	4091−
a_3	762 901−	112 016−	2 7029−	159 592+	354 010+	1 9151+
a_4	21 143−	84 599+	623+	35 975−	25 447−	1468−
a_5	15 005+	1 105+		1 273−	44+	
	July 28			**August 5**		
a_0	100·4316 941+	22·2481 873+	1·0076 7195+	213·4108 553+	17·7914 606−	0·9615 1649+
a_1	15·7781 885+	2·3810 713−	0·0114 3133+	13·5014 448+	3·7622 633−	0·0144 1155−
a_2	789 450−	7886 822−	18 6201−	798 119+	5200 069+	4 4581+
a_3	701 472−	233 976+	2 4528−	4 155+	252 388+	1 3190+
a_4	57 127+	89 254+	2123+	42 813−	25 328−	1381−
a_5	7 570+	5 020−		791+	448−	
	July 29			**August 6**		
a_0	116·0672 601+	19·1102 549+	1·0170 1723+	226·9883 253+	21·0110 558−	0·9476 6885+
a_1	15·4364 817+	3·8550 539−	0·0070 5652+	13·6455 883+	2·6568 884−	0·0131 7947−
a_2	2475 610−	6699 826−	24 6705−	561 603+	5800 744+	7 5952+
a_3	402 754−	539 343+	1 5801−	157 715−	146 631+	7632+
a_4	93 592+	62 349+	3243+	38 888−	27 854−	1102−
a_5	2 200−	7 482−		3 390+	396−	
	July 30			**August 7**		
a_0	131·2250 445+	14·6446 394+	1·0214 8112+	240·6707 526+	23·0760 318−	0·9353 1420+
a_1	14·8568 651+	5·0120 165−	0·0017 7830+	13·6967 359+	1·4640 891−	0·0114 7556−
a_2	3144 714−	4782 677−	27 4605−	110 881−	6069 523+	9 2337+
a_3	52 974−	714 580+	2466−	278 480−	31 673+	3220+
a_4	79 946+	23 888+	3394+	21 242−	30 046−	769−
a_5	6 569−	6 226−		4 982+	339+	
	July 31			**August 8**		
a_0	145·7694 785+	9·2275 794+	1·0205 2265+	254·3269 263+	23·9329 719−	0·9247 8652+
a_1	14·2407 239+	5·7477 326−	0·0036 5186−	13·5850 089+	0·2525 304−	0·0095 6298−
a_2	2889 845−	2557 882−	26 1912−	1023 839−	5987 665+	9 7479+
a_3	201 265+	749 406+	1 1395+	314 009−	84 589−	155+
a_4	45 849+	7 358−	2494+	4 556+	28 287−	454−
a_5	6 192−	3 514−		4 316+	1 280+	
	August 1			**August 9**		
a_0	159·7453 102+	3·2979 121+	1·0143 9054+	267·7790 376+	23·5978 954−	0·9161 9534+
a_1	13·7383 800+	6·0391 844−	0·0084 4846−	13·2900 162+	0·9089 519+	0·0076 2690−
a_2	2072 897−	388 892−	21 3209−	1895 338−	5577 016+	9 5304+
a_3	323 717+	686 211+	2 1438+	253 959−	184 646−	1644−
a_4	14 809+	24 555−	1034+	26 420+	21 623−	185−
a_5	4 491−	1 114−		1 954+	1 783+	
	August 2			**August 10**		
a_0	173·3098 040+	2·7141 073−	1·0040 3471+	280·8569 615+	22·1516 905−	0·9095 0320+
a_1	13·4245 953+	5·9214 769−	0·0120 2818−	12·8463 032+	1·9612 034+	0·0057 7750−
a_2	1057 781−	1511 348+	14 3105−	2479 225−	4911 202+	8 9332+
a_3	338 798+	577 630+	2 5455+	130 083−	253 388−	2364−
a_4	7 435−	29 622−	321−	35 790+	12 480−	29+
a_5	3 227−	232+		382−	1 632+	

Formula: Quantity in degrees $= a_0 + a_1 p + a_2 p^2 + a_3 p^3 + a_4 p^4 + a_5 p^5$

where p is the fraction of a day from 0^h TDT.

DAILY POLYNOMIAL COEFFICIENTS

	Apparent Right Ascension	Apparent Declination	Horizontal Parallax	Apparent Right Ascension	Apparent Declination	Horizontal Parallax
	August 11			**August 19**		
	°	°	°	°	°	°
a_0	293·4458 748+	19·7257 904−	0·9045 9567+	24·2043 329+	14·8585 005+	0·9230 7773+
a_1	12·3255 574+	2·8632 511+	0·0040 6059−	12·1095 269+	4·0007 186+	0·0094 9842+
a_2	2658 641−	4092 491+	8 2465+	3392 321+	2801 530−	10 3070+
a_3	8 554+	287 297−	2227−	263 553+	362 063−	64−
a_4	33 296+	4 246−	184+	19 848−	13 320−	795−
a_5	1 569−	1 081+		4 045−	598+	
	August 12			**August 20**		
a_0	305·5095 962+	16·4823 363−	0·9013 3930+	36·6770 580+	18·5415 875+	0·9335 9827+
a_1	11·8089 293+	3·5944 019+	0·0024 7074−	12·8570 955+	3·3267 667+	0·0115 2612+
a_2	2448 956−	3215 929+	7 6915+	4023 288+	3961 709−	9 8006+
a_3	125 971+	293 822−	1471−	143 684+	408 281−	3245−
a_4	25 051+	1 115+	276+	41 062−	10 651−	1130−
a_5	1 661−	469+		4 181−	2 477+	
	August 13			**August 21**		
a_0	317·0885 659+	12·5955 654−	0·8996 2576+	49·9463 262+	21·4305 378+	0·9460 6069+
a_1	11·3661 196+	4·1501 210+	0·0009 6552−	13·6863 463+	2·4089 209+	0·0133 4365+
a_2	1937 382−	2345 831+	7 4167+	4165 892+	5225 672−	8 1403+
a_3	209 792+	284 959−	346−	60 520−	424 909−	7799−
a_4	16 592+	3 384+	303+	63 702−	1 871+	1414−
a_5	1 230−	26−		1 025−	4 624+	
	August 14			**August 22**		
a_0	328·2834 627+	8·2390 215−	0·8994 0148+	64·0367 370+	23·2750 501+	0·9601 2624+
a_1	11·0476 032+	4·5351 393+	0·0005 1957+	14·4753 825+	1·2393 749+	0·0146 8112+
a_2	1220 760−	1510 978+	7 4936+	3591 736+	6442 799−	4 9493+
a_3	264 118+	271 903−	887+	321 899−	370 766−	1 3547−
a_4	10 443+	3 182+	265+	69 680−	25 929+	1508−
a_5	764−	387−		5 502+	5 423+	
	August 15			**August 23**		
a_0	339·2363 697+	3·5796 952−	0·9006 8194+	78·8326 854+	23·8362 036+	0·9751 5174+
a_1	10·8864 823+	4·7568 428+	0·0020 5551+	15·0720 449+	0·1473 340−	0·0152 0417+
a_2	373 382−	710 475+	7 9155+	2263 245+	7345 123−	100−
a_3	298 372+	263 204−	1966+	542 770−	213 958−	1 9738−
a_4	6 698+	1 186+	164+	40 266	54 227+	1211−
a_5	542−	650−		10 643+	3 435+	
	August 16			**August 24**		
a_0	350·1159 667+	1·2219 282+	0·9035 5029+	94·0738 157+	22·9387 277+	0·9901 4544+
a_1	10·9037 254+	4·8201 255+	0·0037 0413+	15·3510 767+	1·6571 421−	0·0145 6150+
a_2	556 509+	78 539−	8 5984+	500 197+	7627 240−	6 6311−
a_3	319 634+	265 061−	2638+	598 308−	35 088+	2 4776−
a_4	4 115+	2 136−	4+	15 959+	71 694+	350−
a_5	722−	822−		9 000+	487−	
	August 17			**August 25**		
a_0	1·1076 457+	6·0073 978+	0·9081 4068+	109·4175 773+	20·5294 912+	1·0037 9257+
a_1	11·1122 016+	4·7236 338+	0·0055 0313+	15·2825 002+	3·1436 331−	0·0124 7791+
a_2	1532 898+	894 783−	9 3856+	1108 877−	7096 772−	14 2312−
a_3	328 464+	281 824−	2672+	448 213−	315 103+	2 6304−
a_4	684+	6 370−	211−	61 658+	68 576+	1038+
a_5	1 424−	823−		2 265+	3 792−	
	August 18			**August 26**		
a_0	12·4059 095+	10·6126 516+	0·9146 0697+	124·5507 607+	16·7141 696+	1·0145 9472+
a_1	11·5168 806+	4·4571 709+	0·0074 5195+	14·9520 500+	4·4429 238−	0·0088 8414+
a_2	2508 164+	1786 743−	10 0519+	2061 140−	5778 042−	21 4532−
a_3	316 238+	315 311−	1843+	182 170−	550 703+	2 2083−
a_4	6 304−	10 717−	480−	71 502+	48 819+	2542+
a_5	2 669−	450−		3 410−	5 097−	

Formula: Quantity in degrees = $a_0 + a_1 p + a_2 p^2 + a_3 p^3 + a_4 p^4 + a_5 p^5$

where p is the fraction of a day from 0^h TDT.

DAILY POLYNOMIAL COEFFICIENTS

	Apparent Right Ascension	Apparent Declination	Horizontal Parallax
	August 27		
	°	°	°
a_0	139·2852 888+	11·7528 840+	1·0211 3814+
a_1	14·5120 640+	5·4163 422−	0·0040 3286+
a_2	2212 968−	3884 059−	26 5244−
a_3	68 557+	695 167+	1 1622−
a_4	52 985+	22 869+	3474+
a_5	5 361−	4 777−	
	August 28		
a_0	153·5876 742+	6·0194 619+	1·0224 3708+
a_1	14·1085 514+	5·9778 434−	0·0014 8150−
a_2	1743 093−	1709 137−	27 9310−
a_3	227 082+	739 538+	2650+
a_4	25 597+	1 202−	3321+
a_5	4 960−	3 606−	
	August 29		
a_0	167·5466 881+	0·0558 222−	1·0182 2219+
a_1	13·8358 173+	6·1000 915−	0·0068 5522−
a_2	957 888−	466 217+	25 1783−
a_3	280 431+	699 477+	1 6176+
a_4	704+	19 187−	2169+
a_5	3 985−	2 198−	
	August 30		
a_0	181·3144 317+	6·0414 828−	1·0090 3258+
a_1	13·7266 595+	5·8057 775−	0·0113 1883−
a_2	152 250−	2427 582+	19 0714−
a_3	244 011+	601 402+	2 4859+
a_4	19 363−	29 953−	628+
a_5	2 965−	1 066−	
	August 31		
a_0	195·0480 345+	11·5474 638−	0·9960 6146+
a_1	13·7601 871+	5·1523 539−	0·0143 6232−
a_2	433 893+	4041 445+	11 2768−
a_3	137 810+	471 298+	2 7211+
a_4	34 580−	35 066−	653−
a_5	1 463−	444−	
	September 1		
a_0	208·8617 876+	16·2520 944−	0·9808 3703+
a_1	13·8737 480+	4·2169 235−	0·0158 2759−
a_2	625 142+	5240 512+	3 5278−
a_3	13 746−	326 813+	2 4405+
a_4	42 317−	37 251−	1365−
a_5	945+	93−	
	September 2		
a_0	222·7925 381+	19·9160 198−	0·9648 8706+
a_1	13·9782 016+	3·0857 236−	0·0158 5565−
a_2	339 462+	5996 497+	2 9691+
a_3	172 019−	177 187+	1 8806+
a_4	37 493−	37 840−	1547−
a_5	3 662+	433+	
	September 3		
a_0	236·7841 008+	22·3881 156−	0·9495 0091+
a_1	13·9813 234+	1·8481 864−	0·0147 5955−
a_2	364 832−	6305 348+	7 6874+
a_3	284 593−	30 622+	1 2548+
a_4	18 380−	35 722−	1398−
a_5	5 056+	1 241+	

	Apparent Right Ascension	Apparent Declination	Horizontal Parallax
	September 4		
	°	°	°
a_0	250·6991 493+	23·6061 531−	0·9356 2161+
a_1	13·8181 540+	0·5915 979−	0·0129 0156−
a_2	1278 215−	6195 314+	10 6219+
a_3	308 098−	99 480−	6935+
a_4	7 767+	29 344−	1109−
a_5	4 077+	1 911+	
	September 5		
a_0	264·3598 565+	23·5909 109−	0·9238 4051+
a_1	13·4752 244+	0·6068 387+	0·0106 1347−
a_2	2115 118−	5739 958+	12 0467+
a_3	237 667−	197 699−	2504+
a_4	28 330+	19 514−	799−
a_5	1 576+	1 990+	
	September 6		
a_0	277·6027 929+	22·4315 987−	0·9144 4876+
a_1	12·9930 181+	1·6887 096+	0·0081 6096−
a_2	2642 456−	5049 703+	12 3273+
a_3	109 871−	256 135−	679−
a_4	35 740+	9 396−	520−
a_5	666−	1 486+	
	September 7		
a_0	290·3240 857+	20·2643 233−	0·9075 0854+
a_1	12·4455 273+	2·6187 934+	0·0057 3665−
a_2	2764 385−	4239 776+	11 8193+
a_3	25 821+	279 276−	2744−
a_4	31 825+	1 972−	283−
a_5	1 691−	761+	
	September 8		
a_0	302·4987 701+	17·2496 010−	0·9029 2356+
a_1	11·9122 813+	3·3825 565+	0·0034 6641−
a_2	2512 940−	3397 703+	10 8326+
a_3	136 207+	279 915−	3864−
a_4	23 006+	1 729+	85−
a_5	1 678−	145+	
	September 9		
a_0	314·1755 109+	13·5550 783−	0·9005 0092+
a_1	11·4589 193+	3·9788 861+	0·0014 1921−
a_2	1983 089−	2569 759+	9 6273+
a_3	211 711+	271 778−	4194−
a_4	14 492+	2 340+	78+
a_5	1 223−	243−	
	September 10		
a_0	325·4586 195+	9·3461 844−	0·9000 0329+
a_1	11·1310 011+	4·4121 189+	0·0003 8357+
a_2	1273 234−	1766 022+	8 4200+
a_3	257 705+	264 960−	3868−
a_4	8 394+	1 041+	204+
a_5	779−	433−	
	September 11		
a_0	336·4888 290+	4·7838 984−	0·9011 9221+
a_1	10·9566 339+	4·6860 353+	0·0019 5969+
a_2	457 546−	973 046+	7 3846+
a_3	283 598+	265 168−	3034−
a_4	4 565+	1 187−	284+
a_5	578−	499−	

Formula: Quantity in degrees $= a_0 + a_1 p + a_2 p^2 + a_3 p^3 + a_4 p^4 + a_5 p^5$

where p is the fraction of a day from 0^h TDT.

DAILY POLYNOMIAL COEFFICIENTS

September 12

	Apparent Right Ascension	Apparent Declination	Horizontal Parallax
	°	°	°
a_0	347·4284 668+	0·0272 439−	0·9038 6286+
a_1	10·9517 409+	4·8003 695+	0·0033 5697+
a_2	414 864+	165 413+	6 6455+
a_3	295 996+	274 896−	1876−
a_4	1 763+	3 747−	304+
a_5	726−	479−	

September 13

	Apparent Right Ascension	Apparent Declination	Horizontal Parallax
a_0	358·4513 973+	4·7617 547+	0·9078 6867+
a_1	11·1238 539+	4·7492 451+	0·0046 4198+
a_2	1306 178+	686 565−	6 2635+
a_3	295 489+	294 588−	636−
a_4	1 763−	6 235−	251+
a_5	1 273−	342−	

September 14

	Apparent Right Ascension	Apparent Declination	Horizontal Parallax
a_0	9·7351 143+	9·4122 268+	0·9131 3315+
a_1	11·4723 933+	4·5208 912+	0·0058 8567+
a_2	2169 333+	1611 182−	6 2194+
a_3	275 208+	322 721−	394+
a_4	8 081−	8 091−	120+
a_5	2 159−	31+	

September 15

	Apparent Right Ascension	Apparent Declination	Horizontal Parallax
a_0	21·4509 376+	13·7389 217+	0·9196 4590+
a_1	11·9845 093+	4·0986 184+	0·0071 4618+
a_2	2924 837+	2627 616−	6 4031+
a_3	220 824+	354 332−	897+
a_4	19 097−	8 138−	89−
a_5	2 995−	795+	

September 16

	Apparent Right Ascension	Apparent Declination	Horizontal Parallax
a_0	33·7478 039+	17·5386 110+	0·9274 4047+
a_1	12·6265 882+	3·4639 391+	0·0084 5018+
a_2	3442 663+	3731 517−	6 6105+
a_3	114 619+	378 262−	561+
a_4	34 813−	4 313−	364−
a_5	2 823−	1 972+	

September 17

	Apparent Right Ascension	Apparent Declination	Horizontal Parallax
a_0	46·7263 567+	20·5913 381+	0·9365 5368+
a_1	13·3341 727+	2·6034 193+	0·0097 7458+
a_2	3549 240+	4872 442−	6 5508+
a_3	51 464−	375 124−	885−
a_4	50 057−	5 684+	681−
a_5	431−	3 162+	

September 18

	Apparent Right Ascension	Apparent Declination	Horizontal Parallax
a_0	60·4052 581+	22·6708 854+	0·9469 6768+
a_1	14·0083 479+	1·5202 486+	0·0110 3094+
a_2	3090 105+	5932 013−	5 8670+
a_3	253 478−	320 605−	3622−
a_4	52 743−	22 037+	994−
a_5	4 037+	3 480+	

September 19

	Apparent Right Ascension	Apparent Declination	Horizontal Parallax
a_0	74·6923 982+	23·5684 239+	0·9585 3916+
a_1	14·5312 511+	0·2482 176+	0·0120 5591+
a_2	2053 740+	6726 724−	4 1773+
a_3	422 069−	198 349−	7646−
a_4	31 435−	40 043+	1217−
a_5	7 648+	2 279+	

September 20

	Apparent Right Ascension	Apparent Declination	Horizontal Parallax
	°	°	°
a_0	89·3844 377+	23·1283 665+	0·9709 2419+
a_1	14·8066 277+	1·1394 773−	0·0126 1329+
a_2	675 668+	7058 702−	1 1539+
a_3	471 677−	16 596−	1 2611−
a_4	8 733+	51 551+	1214−
a_5	7 023+	136+	

September 21

	Apparent Right Ascension	Apparent Declination	Horizontal Parallax
a_0	104·2130 401+	21·2865 281+	0·9835 1461+
a_1	14·8072 578+	2·5355 100−	0·0124 1709+
a_2	616 643−	6797 873−	3 3457−
a_3	368 939−	189 975+	1 7614−
a_4	44 590+	51 856+	828−
a_5	2 693+	1 597−	

September 22

	Apparent Right Ascension	Apparent Declination	Horizontal Parallax
a_0	118·9264 680+	18·0952 542+	0·9954 1272+
a_1	14·5924 253+	3·8181 492−	0·0111 8632+
a_2	1429 102−	5932 830−	9 1000−
a_3	166 156−	380 981+	2 1076−
a_4	57 319+	43 534+	33+
a_5	1 694−	2 337−	

September 23

	Apparent Right Ascension	Apparent Declination	Horizontal Parallax
a_0	133·3649 300+	13·7260 397+	1·0054 7863+
a_1	14·2788 362+	4·8741 764−	0·0087 3535+
a_2	1600 755−	4552 060−	15 3653−
a_3	44 855+	531 600+	2 1007−
a_4	47 834+	31 806+	1251+
a_5	3 915−	2 562−	

September 24

	Apparent Right Ascension	Apparent Declination	Horizontal Parallax
a_0	147·4925 681+	8·4527 417+	1·0124 7989+
a_1	13·9893 171+	5·6136 674−	0·0050 8217+
a_2	1218 425−	2792 032−	20 8812−
a_3	196 742+	633 065+	1 5899−
a_4	27 722+	19 067+	2417+
a_5	4 373−	2 799−	

September 25

	Apparent Right Ascension	Apparent Declination	Horizontal Parallax
a_0	161·3820 518+	2·6248 045+	1·0153 3912+
a_1	13·8135 570+	5·9759 271−	0·0005 2579+
a_2	505 637−	806 422−	24 1831−
a_3	263 991+	681 206+	5966−
a_4	5 687+	5 053+	2992+
a_5	4 206−	3 042−	

September 26

	Apparent Right Ascension	Apparent Declination	Horizontal Parallax
a_0	175·1715 922+	3·3634 431−	1·0134 1687+
a_1	13·7917 992+	5·9323 496−	0·0043 6996−
a_2	278 353+	1237 080+	24 1874−
a_3	244 944+	670 997+	6294+
a_4	15 573−	10 271−	2667+
a_5	3 778−	3 045−	

September 27

	Apparent Right Ascension	Apparent Declination	Horizontal Parallax
a_0	189·0137 861+	9·1063 166−	1·0067 1777+
a_1	13·9128 365+	5·4892 646−	0·0089 1185−
a_2	881 870+	3157 983+	20 7313−
a_3	145 628+	599 696+	1 7123+
a_4	35 063−	25 640−	1611+
a_5	2 517−	2 659−	

Formula: Quantity in degrees $= a_0 + a_1 p + a_2 p^2 + a_3 p^3 + a_4 p^4 + a_5 p^5$
where p is the fraction of a day from 0^h TDT.

DAILY POLYNOMIAL COEFFICIENTS

September 28

	Apparent Right Ascension	Apparent Declination	Horizontal Parallax
	°	°	°
a_0	203·0256 143+	14·2226 431−	0·9959 2013+
a_1	14·1176 183+	4·6893 435−	0·0124 7999−
a_2	1083 084+	4776 616+	14 6670−
a_3	18 156−	471 068+	2 3547+
a_4	48 398−	39 146−	336+
a_5	311+	1 777−	

September 29

	Apparent Right Ascension	Apparent Declination	Horizontal Parallax
a_0	217·2449 168+	18·3913 104−	0·9822 1226+
a_1	14·3095 890+	3·6092 446−	0·0146 9360−
a_2	741 319+	5937 149+	7 4325−
a_3	206 477−	297 593+	2 4757+
a_4	46 923−	48 262−	677−
a_5	4 115+	269−	

September 30

	Apparent Right Ascension	Apparent Declination	Horizontal Parallax
a_0	231·6037 092+	21·3819 339−	0·9670 1620+
a_1	14·3792 007+	2·3519 740−	0·0154 6457−
a_2	118 350−	6537 671+	4291−
a_3	351 699−	102 924+	2 1896+
a_4	25 247−	49 601−	1227−
a_5	6 467+	1 603+	

October 1

	Apparent Right Ascension	Apparent Declination	Horizontal Parallax
a_0	245·9340 270+	23·0746 483−	0·9517 1542+
a_1	14·2431 544+	1·0326 004−	0·0149 4263−
a_2	1260 076−	6564 919+	5 3992+
a_3	388 616−	78 674−	1 6880+
a_4	8 446+	41 181−	1363−
a_5	5 398+	2 975+	

October 2

	Apparent Right Ascension	Apparent Declination	Horizontal Parallax
a_0	260·0136 966+	23·4624 447−	0·9374 6788+
a_1	13·8806 281+	0·2417 968+	0·0134 1093−
a_2	2321 229−	6111 645+	9 6490+
a_3	302 783−	213 548−	1 1375+
a_4	35 763+	25 768−	1246−
a_5	1 960+	3 124+	

October 3

	Apparent Right Ascension	Apparent Declination	Horizontal Parallax
a_0	273·6356 958+	22·6331 026−	0·9251 2313+
a_1	13·3408 291+	1·3913 154+	0·0111 8977−
a_2	2995 509−	5347 678+	12 3203+
a_3	141 892−	285 885−	6374+
a_4	44 882+	9 861−	1026−
a_5	1 111−	2 214+	

October 4

	Apparent Right Ascension	Apparent Declination	Horizontal Parallax
a_0	286·6671 619+	20·7363 726−	0·9152 1886+
a_1	12·7165 557+	2·3722 468+	0·0085 7554−
a_2	3163 128−	4452 994+	13 6238+
a_3	25 777+	303 890−	2274+
a_4	38 511+	1 182+	795−
a_5	2 377−	981+	

October 5

	Apparent Right Ascension	Apparent Declination	Horizontal Parallax
a_0	299·0735 959+	17·9489 990−	0·9080 2049+
a_1	12·1058 790+	3·1726 408+	0·0058 1438−
a_2	2878 580−	3558 199+	13 8351+
a_3	156 125+	289 921−	897−
a_4	26 169+	5 911+	591−
a_5	2 207−	5−	

October 6

	Apparent Right Ascension	Apparent Declination	Horizontal Parallax
	°	°	°
a_0	310·9096 257+	14·4489 398−	0·9035 7474+
a_1	11·5863 656+	3·7996 655+	0·0030 9791−
a_2	2275 283−	2723 822+	13 2167+
a_3	239 131+	266 664−	3253−
a_4	15 018+	5 706+	417−
a_5	1 508−	581−	

October 7

	Apparent Right Ascension	Apparent Declination	Horizontal Parallax
a_0	322·2937 271+	10·4030 460−	0·9017 6180+
a_1	11·2083 022+	4·2664 221+	0·0005 6884−
a_2	1482 864−	1952 229+	11 9955+
a_3	284 479+	249 794−	4918−
a_4	7 525+	2 665+	261−
a_5	897−	821−	

October 8

	Apparent Right Ascension	Apparent Declination	Horizontal Parallax
a_0	333·3828 536+	5·9661 960−	0·9023 4073+
a_1	10·9996 349+	4·5825 851+	0·0016 7230+
a_2	593 237−	1210 607+	10 3686+
a_3	305 770+	247 354−	5960−
a_4	3 152+	1 536−	108−
a_5	625−	834−	

October 9

	Apparent Right Ascension	Apparent Declination	Horizontal Parallax
a_0	344·3539 945+	1·2875 227−	0·9049 8920+
a_1	10·9736 664+	4·7494 689+	0·0035 6289+
a_2	336 740+	450 971+	8 5204+
a_3	312 034+	261 747−	6393−
a_4	146+	5 798−	45+
a_5	771−	674−	

October 10

	Apparent Right Ascension	Apparent Declination	Horizontal Parallax
a_0	355·3924 757+	3·4802 214+	0·9093 4065+
a_1	11·1342 966+	4·7584 831+	0·0050 7699+
a_2	1266 011+	375 821−	6 6341+
a_3	304 562+	291 477−	6206−
a_4	3 599−	9 285−	190+
a_5	1 353−	323−	

October 11

	Apparent Right Ascension	Apparent Declination	Horizontal Parallax
a_0	6·6833 344+	8·1710 139+	0·9150 2090+
a_1	11·4767 502+	4·5920 010+	0·0062 2524+
a_2	2144 565+	1309 220−	4 8898+
a_3	276 116+	331 483−	5430−
a_4	10 352−	11 058−	303+
a_5	2 255−	305+	

October 12

	Apparent Right Ascension	Apparent Declination	Horizontal Parallax
a_0	18·4008 918+	12·5978 693+	0·9216 8385+
a_1	11·9832 286+	4·2264 424+	0·0070 5244+
a_2	2888 196+	2366 996−	3 4441+
a_3	211 743+	372 095−	4192−
a_4	21 921−	9 699−	351+
a_5	3 013−	1 285+	

October 13

	Apparent Right Ascension	Apparent Declination	Horizontal Parallax
a_0	30·6916 209+	16·5495 611+	0·9290 4229+
a_1	12·6141 163+	3·6381 788+	0·0076 2956+
a_2	3361 638+	3528 635−	2 3958+
a_3	94 213+	397 353−	2754−
a_4	37 798−	3 310−	307+
a_5	2 581−	2 494+	

Formula: Quantity in degrees $= a_0 + a_1 p + a_2 p^2 + a_3 p^3 + a_4 p^4 + a_5 p^5$

where p is the fraction of a day from 0^h TDT.

DAILY POLYNOMIAL COEFFICIENTS

	Apparent Right Ascension	Apparent Declination	Horizontal Parallax	Apparent Right Ascension	Apparent Declination	Horizontal Parallax
	October 14			**October 22**		
	°	°	°	°	°	°
a_0	43·6472 844+	19·7950 596+	0·9368 8696+	156·3299 144+	4·6253 657+	0·9995 5072+
a_1	13·2983 016+	2·8131 698+	0·0080 3841+	13·4303 617+	5·6574 628−	0·0037 2169+
a_2	3391 514+	4715 574−	1 7493+	453 724−	1513 703−	13 5499−
a_3	81 188−	385 172−	1493−	304 250+	562 727+	1 2786−
a_4	51 814−	9 459+	158+	9 539+	11 797+	1041+
a_5	169+	3 367+		3 832−	1 123−	
	October 15			**October 23**		
a_0	57·2714 540+	22·0994 375+	0·9450 8694+	169·7458 995+	1·1261 274−	1·0017 9998+
a_1	13·9316 116+	1·7599 703+	0·0083 4980+	13·4327 916+	5·7872 293−	0·0006 6980+
a_2	2838 696+	5780 574−	1 3887+	477 931+	234 062+	16 7395−
a_3	284 195−	313 877−	837−	303 985+	598 145+	8536−
a_4	51 292−	26 929+	82−	9 635−	6 382+	1755+
a_5	4 712+	3 024+		4 026−	2 073−	
	October 16			**October 24**		
a_0	71·4538 578+	23·2529 579+	0·9535 6642+	183·2555 167+	6·8297 052−	1·0007 2803+
a_1	14·3959 350+	0·5219 737+	0·0085 9916+	13·6137 071+	5·5594 588−	0·0028 6386−
a_2	1725 660+	6530 323−	1 0793+	1291 743+	2046 051+	18 2378−
a_3	440 511−	176 965−	1160−	225 324+	602 317+	1343−
a_4	26 403−	42 535+	373−	30 236−	4 020−	2055+
a_5	7 835+	1 186+		3 849−	3 174−	
	October 17			**October 25**		
a_0	85·9764 509+	23·1085 749+	0·9622 5817+	197·0175 220+	12·1250 467−	0·9960 4750+
a_1	14·6022 686+	0·8195 758−	0·0087 6528+	13·9256 364+	4·9727 492−	0·0064 6943−
a_2	324 310+	6794 169−	4990+	1747 644+	3797 092+	17 4158−
a_3	468 474−	3 768+	2671−	67 044+	554 204+	7052+
a_4	14 597+	48 290+	647−	50 553−	20 258−	1793+
a_5	6 569+	1 062−		1 920−	3 719−	
	October 18			**October 26**		
a_0	100·5664 198+	21·6146 817+	0·9710 4017+	211·1193 799+	16·6650 641−	0·9879 2495+
a_1	14·5357 070+	2·1584 956−	0·0087 5902+	14·2741 022+	4·0570 314−	0·0096 6924−
a_2	927 777−	6503 835−	6959−	1626 116+	5300 869+	14 2456−
a_3	346 878−	185 634+	5311−	151 851−	436 478+	1 4323+
a_4	47 910+	42 358+	814−	61 068−	39 531−	1093+
a_5	2 118+	2 221−		2 479+	2 886−	
	October 19			**October 27**		
a_0	114·9796 642+	18·8283 797+	0·9796 6834+	225·5350 498+	20·1526 025−	0·9769 8530+
a_1	14·2663 068+	3·3877 395−	0·0084 2788+	14·5305 881+	2·8831 668−	0·0120 4494−
a_2	1659 905−	5715 057−	2 7765−	829 032+	6344 180+	9 3185−
a_3	136 342−	332 971+	8653−	368 681−	250 912+	1 8689+
a_4	57 639+	30 824+	771−	48 046−	54 506−	254+
a_5	1 868−	1 974−		7 225+	405−	
	October 20			**October 28**		
a_0	129·0719 234+	14·9053 167+	0·9877 2434+	240·1075 910+	22·3817 513−	0·9641 9794+
a_1	13·9155 430+	4·4195 157−	0·0075 8212+	14·5701 856+	1·5610 589−	0·0133 3784−
a_2	1741 920−	4550 927−	5 8240−	492 750−	6765 858+	3 5810−
a_3	74 487+	436 969+	1 1839−	488 042−	30 528+	1 9627+
a_4	47 352+	20 979+	429−	9 865−	56 330−	442−
a_5	3 622−	1 184−		8 260+	2 585+	
	October 21			**October 29**		
a_0	142·8250 961+	10·0763 848+	0·9946 0137+	254·5795 369+	23·2685 459−	0·9506 9384+
a_1	13·6066 344+	5·1908 104−	0·0060 4492+	14·3254 032+	0·2199 662−	0·0134 8299−
a_2	1270 638−	3125 940−	9 6133−	1933 278−	6545 439+	2 0285+
a_3	227 498+	509 272+	1 3644−	446 925−	168 014−	1 7763+
a_4	28 839+	15 360+	219+	32 767+	42 551−	860−
a_5	3 860−	779−		4 591+	4 199+	

Formula: Quantity in degrees $= a_0 + a_1 p + a_2 p^2 + a_3 p^3 + a_4 p^4 + a_5 p^5$

where p is the fraction of a day from 0^h TDT.

DAILY POLYNOMIAL COEFFICIENTS

October 30

	Apparent Right Ascension	Apparent Declination	Horizontal Parallax
	°	°	°
a_0	268·6706 556+	22·8546 048−	0·9375 8273+
a_1	13·8200 670+	1·0237 961+	0·0125 7885−
a_2	3031 619−	5828 187+	6 8362+
a_3	272 675−	296 468−	1 4249+
a_4	55 231+	20 824−	1014−
a_5	250−	3 738+	

October 31

	Apparent Right Ascension	Apparent Declination	Horizontal Parallax
a_0	282·1657 914+	21·2793 454−	0·9258 1985+
a_1	13·1539 055+	2·0940 307+	0·0108 2471−
a_2	3520 938−	4851 245+	10 5033+
a_3	55 799−	343 295−	1 0155+
a_4	52 760+	1 951−	991−
a_5	2 912−	2 121+	

November 1

	Apparent Right Ascension	Apparent Declination	Horizontal Parallax
a_0	294·9670 079+	18·7345 028−	0·9161 3710+
a_1	12·4526 253+	2·9615 691+	0·0084 5908−
a_2	3401 032−	3830 830+	12 9578+
a_3	125 952+	330 809−	6176+
a_4	37 383+	8 464+	889−
a_5	3 126−	529+	

November 2

	Apparent Right Ascension	Apparent Declination	Horizontal Parallax
a_0	307·0955 510+	15·4220 323−	0·9090 2666+
a_1	11·8235 961+	3·6321 414+	0·0057 1783−
a_2	2830 179−	2894 437+	14 2804+
a_3	244 733+	292 275−	2620+
a_4	21 508+	10 850+	774−
a_5	2 210−	517−	

November 3

	Apparent Right Ascension	Apparent Declination	Horizontal Parallax
a_0	318·6625 323+	11·5286 414−	0·9047 5535+
a_1	11·3384 795+	4·1274 271+	0·0028 1407−
a_2	1989 033−	2077 510+	14 6054+
a_3	309 214+	254 358−	468−
a_4	10 511+	8 071+	675−
a_5	1 250−	1 045−	

November 4

	Apparent Right Ascension	Apparent Declination	Horizontal Parallax
a_0	329·8339 561+	7·2181 966−	0·9033 9039+
a_1	11·0370 172+	4·4693 274+	0·0000 6598+
a_2	1010 811−	1352 392+	14 0627+
a_3	339 076+	232 634−	3164−
a_4	4 410+	2 721+	594−
a_5	703−	1 235−	

November 5

	Apparent Right Ascension	Apparent Declination	Horizontal Parallax
a_0	340·8041 705+	2·6367 448−	0·9048 2507+
a_1	10·9379 903+	4·6704 863+	0·0027 5986+
a_2	25 861+	658 442+	12 7598+
a_3	349 685+	234 083−	5544−
a_4	1 067+	3 563−	513−
a_5	695−	1 200−	

November 6

	Apparent Right Ascension	Apparent Declination	Horizontal Parallax
a_0	351·7797 526+	2·0757 011+	0·9088 0034+
a_1	11·0481 464+	4·7299 249+	0·0051 2498+
a_2	1074 381+	77 212−	10 7925+
a_3	346 661+	260 172−	7609−
a_4	2 226−	9 708−	402−
a_5	1 263−	917−	

November 7

	Apparent Right Ascension	Apparent Declination	Horizontal Parallax
	°	°	°
a_0	2·9696 542+	6·7708 251+	0·9149 2446+
a_1	11·3654 976+	4·6320 903+	0·0070 3913+
a_2	2088 383+	925 177−	8 2741+
a_3	324 479+	307 774−	9236−
a_4	8 422−	14 516−	235−
a_5	2 367−	239−	

November 8

	Apparent Right Ascension	Apparent Declination	Horizontal Parallax
a_0	14·5753 591+	11·2781 449+	0·9226 9628+
a_1	11·8759 641+	4·3487 984+	0·0084 0744+
a_2	2987 582+	1938 026−	5 3691+
a_3	266 416+	367 494−	1 0197−
a_4	20 432−	15 991−	6−
a_5	3 611−	1 020+	

November 9

	Apparent Right Ascension	Apparent Declination	Horizontal Parallax
a_0	26·7743 187+	15·3948 942+	0·9315 3861+
a_1	12·5434 272+	3·8450 604+	0·0091 7513+
a_2	3627 996+	3126 284−	2 3151+
a_3	148 520+	420 231−	1 0224−
a_4	39 345−	11 056−	264+
a_5	3 794−	2 819+	

November 10

	Apparent Right Ascension	Apparent Declination	Horizontal Parallax
a_0	39·6910 836+	18·8844 794+	0·9408 4565+
a_1	13·2959 504+	3·0907 235+	0·0093 4200+
a_2	3799 310+	4425 073−	5860−
a_3	45 213−	435 351−	9149−
a_4	59 857−	3 354+	515+
a_5	1 111−	4 445+	

November 11

	Apparent Right Ascension	Apparent Declination	Horizontal Parallax
a_0	53·3563 469+	21·4899 405+	0·9500 4271+
a_1	14·0177 568+	2·0786 677+	0·0089 7095+
a_2	3293 287+	5666 414−	3 0171−
a_3	292 469−	377 514−	7046−
a_4	66 284−	26 536+	667+
a_5	4 676+	4 444+	

November 12

	Apparent Right Ascension	Apparent Declination	Horizontal Parallax
a_0	67·6680 247+	22·9673 133+	0·9586 4817+
a_1	14·5645 035+	0·8449 637+	0·0081 8287+
a_2	2065 138+	6595 176−	4 7309−
a_3	508 090−	228 365−	4320−
a_4	41 427−	49 684+	658+
a_5	9 673+	1 936+	

November 13

	Apparent Right Ascension	Apparent Declination	Horizontal Parallax
a_0	82·3850 575+	23·1350 850+	0·9663 2132+
a_1	14·8133 685+	0·5217 433−	0·0071 3342+
a_2	389 404+	6962 823−	5 6379−
a_3	577 497−	12 308−	1638−
a_4	9 623+	59 255+	474+
a_5	8 901+	1 684−	

November 14

	Apparent Right Ascension	Apparent Declination	Horizontal Parallax
a_0	97·1814 692+	21·9215 856+	0·9728 7931+
a_1	14·7262 937+	1·8951 430−	0·0059 7567+
a_2	1196 216−	6661 205−	5 8543−
a_3	453 168−	206 659+	284+
a_4	55 033+	49 831+	167+
a_5	3 167+	3 790−	

Formula: Quantity in degrees $= a_0 + a_1 p + a_2 p^2 + a_3 p^3 + a_4 p^4 + a_5 p^5$

where p is the fraction of a day from 0^h TDT.

DAILY POLYNOMIAL COEFFICIENTS

	Apparent Right Ascension	Apparent Declination	Horizontal Parallax	Apparent Right Ascension	Apparent Declination	Horizontal Parallax
	November 15			**November 23**		
	°	°	°	°	°	°
a_0	111·7486 446+	19·3855 920+	0·9782 7406+	219·9531 939+	18·9622 594−	0·9736 8459+
a_1	14·3746 910+	3·1473 487−	0·0048 2000+	14·2402 710+	3·2730 735−	0·0076 8268−
a_2	2194 026−	5780 263−	5 6791−	1752 240+	5686 153+	9 5059−
a_3	204 408−	368 235+	942+	246 448−	353 547+	5371+
a_4	69 688+	30 050+	164−	64 589−	38 535−	1034+
a_5	2 185−	3 460−		4 485+	2 991−	
	November 16			**November 24**		
a_0	125·8902 424+	15·6996 995+	0·9825 3393+	234·3380 337+	21·6355 155−	0·9651 1537+
a_1	13·9013 433+	4·1826 391−	0·0037 0585+	14·4931 963+	2·0466 847−	0·0093 8134−
a_2	2411 188−	4529 879−	5 5025−	670 395+	6485 592+	7 2837−
a_3	51 231+	454 727+	241+	457 547−	171 110+	9569+
a_4	57 384+	12 594+	409−	40 817−	54 111−	725+
a_5	4 289−	1 864−		8 840+	160−	
	November 17			**November 25**		
a_0	139·5608 994+	11·1106 182+	0·9856 8785+	248·8493 172+	23·0219 572−	0·9551 0859+
a_1	13·4552 839+	4·9480 896−	0·0025 9620+	14·4781 037+	0·7199 545−	0·0105 2202−
a_2	1956 176−	3108 731−	5 6775−	858 422−	6672 705+	3 9916−
a_3	237 930+	487 279+	1461−	532 824−	45 126−	1 2483+
a_4	35 371+	3 559+	478−	5 718+	54 546−	294+
a_5	4 120−	433−		8 099+	3 049+	
	November 18			**November 26**		
a_0	152·8474 837+	5·9006 962+	0·9876 9692+	263·1896 779+	23·0843 034−	0·9443 1518+
a_1	13·1475 165+	5·4224 439−	0·0013 9771+	14·1529 031+	0·5807 563+	0·0109 3412−
a_2	1071 367−	1629 799−	6 3978−	2341 491−	6240 681+	834−
a_3	338 524+	497 536+	3443−	431 772−	232 017−	1 3634+
a_4	14 822+	1 809+	326−	47 002+	38 305−	121−
a_5	3 546−	162+		3 053+	4 441+	
	November 19			**November 27**		
a_0	165·9228 435+	0·3652 230+	0·9884 1715+	277·0702 602+	21·9060 671−	0·9335 0784+
a_1	13·0389 560+	5·5983 389−	0·0000 0177+	13·5753 950+	1·7461 850+	0·0105 4667−
a_2	2 305−	124 660−	7 6157−	3324 429−	5359 308+	3 9242+
a_3	362 361+	506 192+	4801−	215 942−	341 269−	1 3107+
a_4	2 698−	3 008+	24+	61 224+	15 407−	430−
a_5	3 531−	183−		1 680−	3 620+	
	November 20			**November 28**		
a_0	178·9971 822+	5·1946 801−	0·9876 0958+	290·2975 723+	19·6592 569−	0·9234 8036+
a_1	13·1443 585+	5·4703 027−	0·0016 6443−	12·8693 738+	2·7113 112+	0·0093 8584−
a_2	1033 264+	1410 183+	9 0273−	3621 910−	4279 273+	7 5926+
a_3	316 055+	515 746+	4715−	11 071+	367 707−	1 1348+
a_4	20 416−	2 384+	480+	51 534+	2 775+	610−
a_5	3 927−	1 318−		3 490−	1 831+	
	November 21			**November 29**		
a_0	192·2740 383+	10·4722 832−	0·9850 0007+	302·8106 667+	16·5563 285−	0·9149 6116+
a_1	13·4356 986+	5·0332 492−	0·0035 9214−	12·1671 821+	3·4588 773+	0·0075 5127−
a_2	1819 550+	2958 557+	10 1415−	3314 495−	3211 074+	10 6288+
a_3	195 377+	511 252+	2755−	182 530+	339 210−	8884+
a_4	40 773−	4 169−	889+	33 447+	11 691+	687−
a_5	3 568−	2 822−		3 050−	253+	
	November 22			**November 30**		
a_0	205·9067 955+	15·1592 506−	0·9803 7511+	314·6676 921+	12·8090 704−	0·9085 5474+
a_1	13·8401 314+	4·2912 418−	0·0056 6750−	11·5708 975+	4·0041 309+	0·0051 8649−
a_2	2125 159+	4439 015+	10 4284−	2596 741−	2266 082+	12 8809+
a_3	1 774−	465 829+	882+	286 473+	290 487−	6125+
a_4	59 914−	18 732−	1100+	18 103+	12 700+	710−
a_5	801−	3 783−		1 904−	726−	

Formula: Quantity in degrees $= a_0 + a_1 p + a_2 p^2 + a_3 p^3 + a_4 p^4 + a_5 p^5$

where p is the fraction of a day from 0^h TDT.

DAILY POLYNOMIAL COEFFICIENTS

	Apparent Right Ascension	Apparent Declination	Horizontal Parallax
	December 1		
	°	°	°
a_0	326·0091 829+	8·6061 827−	0·9047 1049+
a_1	11·1437 820+	4·3749 173+	0·0024 5496−
a_2	1647 726−	1463 528+	14 2919+
a_3	340 386+	247 244−	3286+
a_4	8 714+	8 894+	721−
a_5	979−	1 221−	
	December 2		
a_0	337·0230 044+	4·1088 698−	0·9037 1037+
a_1	10·9193 490+	4·5963 965+	0·0004 7316+
a_2	584 058−	762 931+	14 8443+
a_3	365 671+	223 998−	408+
a_4	4 005+	2 679+	745−
a_5	596−	1 426−	
	December 3		
a_0	347·9208 557+	0·5415 453+	0·9056 6460+
a_1	10·9135 427+	4·6821 419+	0·0034 2446+
a_2	531 053+	92 732+	14 5184+
a_3	375 586+	227 537−	2570−
a_4	1 243+	4 563−	785−
a_5	848−	1 423−	
	December 4		
a_0	358·9251 018+	5·2096 081+	0·9105 0735+
a_1	11·1325 013+	4·6298 908+	0·0062 1965+
a_2	1656 816+	631 519−	13 2756+
a_3	371 522+	259 830−	5718−
a_4	2 761−	11 871−	817−
a_5	1 807−	1 114−	
	December 5		
a_0	10·2599 801+	9·7490 655+	0·9179 8921+
a_1	11·5733 113+	4·4203 338+	0·0086 7052+
a_2	2736 753+	1493 423−	11 0706+
a_3	341 498+	317 918−	9016−
a_4	11 673−	17 772−	793−
a_5	3 394−	216−	
	December 6		
a_0	22·1396 099+	13·9864 663+	0·9276 6870+
a_1	12·2167 434+	4·0190 588+	0·0105 8242+
a_2	3657 193+	2556 039−	7 8948+
a_3	260 054+	390 120−	1 2237−
a_4	29 060−	19 277−	647−
a_5	4 858−	1 591+	
	December 7		
a_0	34·7446 862+	17·7091 405+	0·9389 1175+
a_1	13·0121 465+	3·3839 024+	0·0117 6834+
a_2	4214 190+	3826 178−	3 8457+
a_3	95 703+	449 861−	1 4886−
a_4	54 844−	11 480−	322−
a_5	4 168−	4 154+	
	December 8		
a_0	48·1819 207+	20·6647 066+	0·9509 1258+
a_1	13·8596 796+	2·4811 959+	0·0120 7798+
a_2	4130 251+	5202 993−	7978−
a_3	162 381−	453 188−	1 6224−
a_4	77 735−	10 008+	184+
a_5	969+	6 071+	

	Apparent Right Ascension	Apparent Declination	Horizontal Parallax
	December 9		
	°	°	°
a_0	62·4307 107+	22·5818 923+	0·9627 5038+
a_1	14·6064 149+	1·3116 786+	0·0114 3904+
a_2	3186 376+	6441 569−	5 5364−
a_3	459 185−	353 068−	1 5493−
a_4	72 926−	41 928+	780+
a_5	8 949+	5 036+	
	December 10		
a_0	77·3034 470+	23·2188 036+	0·9734 8866+
a_1	15·0812 424+	0·0632 714−	0·0098 9821+
a_2	1461 206+	7198 707−	9 7008−
a_3	659 414−	137 483−	1 2309−
a_4	24 934−	68 063+	1283+
a_5	12 580+	613+	
	December 11		
a_0	92·4636 333+	22·4287 809+	0·9823 0654+
a_1	15·1719 707+	1·5167 308−	0·0076 4019+
a_2	540 434−	7196 753−	12 6168−
a_3	635 998−	138 296+	7058−
a_4	40 840+	70 347+	1497+
a_5	7 730+	4 017−	
	December 12		
a_0	107·5228 178+	20·2128 374+	0·9886 2944+
a_1	14·8932 766+	2·8884 641−	0·0049 6505+
a_2	2126 170−	6400 167−	13 8410−
a_3	399 799−	378 602+	947−
a_4	78 963+	48 772+	1326+
a_5	240−	5 554−	
	December 13		
a_0	122·1713 698+	16·7265 385+	0·9922 1417+
a_1	14·3795 630+	4·0381 836−	0·0022 2150+
a_2	2854 493−	5027 380−	13 3446−
a_3	88 979−	518 919+	4429+
a_4	75 714+	20 225+	840+
a_5	4 752−	4 130−	
	December 14		
a_0	136·2636 819+	12·2391 183+	0·9931 5389+
a_1	13·8098 800+	4·8819 564−	0·0002 8096−
a_2	2714 852−	3390 563−	11 5305−
a_3	166 032+	559 819+	7791+
a_4	50 713+	359−	236+
a_5	5 199−	1 834−	
	December 15		
a_0	149·8232 312+	7·0738 684+	0·9918 0015+
a_1	13·3344 058+	5·3931 816−	0·0023 4389−
a_2	1964 510−	1731 530−	9 0671−
a_3	317 483+	541 018+	8681+
a_4	24 522+	9 104−	277−
a_5	4 116−	153−	
	December 16		
a_0	162·9949 749+	1·5607 099+	0·9886 3358+
a_1	13·0445 003+	5·5808 993−	0·0039 0801−
a_2	906 065−	164 555−	6 6380−
a_3	374 848+	503 447+	7492+
a_4	4 183+	9 412−	568−
a_5	3 353−	481+	

Formula: Quantity in degrees $= a_0 + a_1 p + a_2 p^2 + a_3 p^3 + a_4 p^4 + a_5 p^5$
where p is the fraction of a day from 0^h TDT.

DAILY POLYNOMIAL COEFFICIENTS

	Apparent Right Ascension	Apparent Declination	Horizontal Parallax	Apparent Right Ascension	Apparent Declination	Horizontal Parallax
	December 17			**December 25**		
	°	°	°	°	°	°
a_0	175·9864 365+	3·9871 932−	0·9841 3101+	285·6920 554+	20·5666 363−	0·9253 5234+
a_1	12·9757 383+	5·4663 007−	0·0050 3362−	13·1373 682+	2·3964 280+	0·0081 1974−
a_2	210 060+	1294 185+	4 7326−	3267 629−	4728 765+	2 9088+
a_3	358 079+	470 409+	5142+	123 653−	350 654−	8096+
a_4	12 412−	6 618−	608−	54 006+	9 462−	143+
a_5	3 330−	131+		2 199−	2 810+	
	December 18			**December 26**		
a_0	189·0174 144+	9·2776 832−	0·9786 6947+	298·4954 761+	17·7330 622−	0·9176 0587+
a_1	13·1185 440+	5·0689 239−	0·0058 5020−	12·4672 482+	3·2346 037+	0·0072 8938−
a_2	1176 473+	2667 058+	3 5494−	3336 693−	3648 133+	5 4170+
a_3	275 118+	444 621+	2658+	69 852+	361 235−	8663+
a_4	29 328−	5 703−	440−	42 060+	4 595+	64−
a_5	3 411−	965−		3 036−	1 347+	
	December 19			**December 27**		
a_0	202·2778 436+	14·0361 062−	0·9724 8652+	310·6399 424+	14·1691 745−	0·9109 4419+
a_1	13·4229 385+	4·4048 917−	0·0064 9794−	11·8361 715+	3·8583 699+	0·0059 4865−
a_2	1791 608+	3957 043+	3 0069−	2905 208−	2605 438+	7 9719+
a_3	124 410+	411 394+	876+	208 081+	330 102−	8395+
a_4	47 266−	10 568−	151−	26 485+	11 139+	243−
a_5	2 251−	2 305−		2 380−	108+	
	December 20			**December 28**		
a_0	215·8874 322+	18·0054 415−	0·9656 9515+	322·2088 117+	10·0821 462−	0·9058 7424+
a_1	13·7985 555+	3·4954 452−	0·0070 7905−	11·3269 592+	4·2849 357+	0·0041 1216−
a_2	1858 581+	5104 698+	2 8248−	2145 870−	1683 016+	10 3400+
a_3	85 138−	345 732+	282+	290 769+	284 933−	7413+
a_4	59 546−	22 595−	160+	14 568+	11 486+	386−
a_5	1 240+	2 938−		1 407−	688−	
	December 21			**December 29**		
a_0	229·8575 015+	20·9583 969−	0·9583 3804+	333·3515 770+	5·6563 224−	0·9028 6635+
a_1	14·1215 371+	2·3812 915−	0·0076 2912−	10·9901 407+	4·5403 091+	0·0018 3722−
a_2	1258 293+	5976 839+	2 6364−	1200 212−	890 233+	12 3283+
a_3	308 285−	226 580+	957+	335 359+	246 129−	5865+
a_4	53 271−	38 046−	403+	7 658+	7 911+	507−
a_5	5 927+	1 898−		730−	1 131	
	December 22			**December 30**		
a_0	244·0693 050+	22·7233 410−	0·9504 5888+	344·2559 253+	1·0509 249−	0·9023 1553+
a_1	14·2623 676+	1·1341 148−	0·0081 1155−	10·8534 044+	4·6471 155+	0·0007 8409+
a_2	73 311+	6409 267+	2 1040−	155 470−	187 984+	13 7794+
a_3	460 846−	56 839+	2615+	358 780+	225 925−	3839+
a_4	21 959−	47 879−	517+	4 180+	2 158+	630−
a_5	8 168+	635+		569−	1 352−	
	December 23			**December 31**		
a_0	258·2915 400+	23·2155 696−	0·9421 6825+	355·1300 219+	3·5924 771+	0·9045 0966+
a_1	14·1340 735+	0·1459 589+	0·0084 3320−	10·9313 318+	4·6171 219+	0·0036 2998+
a_2	1359 076−	6298 927+	1 0103−	940 289+	490 385−	14 5490+
a_3	468 304−	126 989−	4726+	369 544+	230 824−	1330+
a_4	20 529+	44 237−	488+	1 553+	4 718−	773−
a_5	5 790+	3 004+		1 028−	1 377−	
	December 24			**December 32**		
a_0	272·2455 074+	22·4565 402−	0·9336 8616+	6·1923 896+	8·1368 684+	0·9096 0010+
a_1	13·7328 687+	1·3514 557+	0·0084 7394−	11·2303 591+	4·4472 221+	0·0065 4875+
a_2	2582 917−	5682 691+	6960+	2047 991+	1224 973−	14 4789+
a_3	330 906−	273 480−	6707+	364 803+	263 278−	1756−
a_4	49 500+	28 434−	345+	3 352−	11 817−	942−
a_5	1 117+	3 705+		2 193−	1 063−	

Formula: Quantity in degrees = $a_0 + a_1 p + a_2 p^2 + a_3 p^3 + a_4 p^4 + a_5 p^5$
where p is the fraction of a day from 0^h TDT.

NOTES AND FORMULAE

Low-precision formulae for geocentric coordinates of the Moon

The following formulae give approximate geocentric coordinates of the Moon. The errors will rarely exceed 0°·3 in ecliptic longitude (λ), 0°·2 in ecliptic latitude (β), 0°·003 in horizontal parallax (π), 0°·001 in semidiameter (SD), 0·2 Earth radii in distance (r), 0°·3 in right ascension (α) and 0°·2 in declination (δ).

On this page the time argument T is the number of Julian centuries from J2000·0.

$$T = (\text{JD} - 245\,1545{\cdot}0)\,/\,36\,525 = (\,-2923{\cdot}5 + \text{day of year} + \text{UT}\,/\,24)\,/\,36\,525$$

where day of year is given on pages B2–B3 and UT is the universal time in hours.

$$\begin{aligned}
\lambda = {} & 218^\circ{\cdot}32 + 481\,267^\circ{\cdot}883\,T \\
& +6^\circ{\cdot}29 \sin(134^\circ{\cdot}9 + 477\,198^\circ{\cdot}85\,T) - 1^\circ{\cdot}27 \sin(259^\circ{\cdot}2 - 413\,335^\circ{\cdot}38\,T) \\
& +0^\circ{\cdot}66 \sin(235^\circ{\cdot}7 + 890\,534^\circ{\cdot}23\,T) + 0^\circ{\cdot}21 \sin(269^\circ{\cdot}9 + 954\,397^\circ{\cdot}70\,T) \\
& -0^\circ{\cdot}19 \sin(357^\circ{\cdot}5 + 35\,999^\circ{\cdot}05\,T) - 0^\circ{\cdot}11 \sin(186^\circ{\cdot}6 + 966\,404^\circ{\cdot}05\,T) \\
\beta = {} & +5^\circ{\cdot}13 \sin(93^\circ{\cdot}3 + 483\,202^\circ{\cdot}03\,T) + 0^\circ{\cdot}28 \sin(228^\circ{\cdot}2 + 960\,400^\circ{\cdot}87\,T) \\
& -0^\circ{\cdot}28 \sin(318^\circ{\cdot}3 + 6\,003^\circ{\cdot}18\,T) - 0^\circ{\cdot}17 \sin(217^\circ{\cdot}6 - 407\,332^\circ{\cdot}20\,T) \\
\pi = {} & +0^\circ{\cdot}9508 \\
& +0^\circ{\cdot}0518 \cos(134^\circ{\cdot}9 + 477\,198^\circ{\cdot}85\,T) + 0^\circ{\cdot}0095 \cos(259^\circ{\cdot}2 - 413\,335^\circ{\cdot}38\,T) \\
& +0^\circ{\cdot}0078 \cos(235^\circ{\cdot}7 + 890\,534^\circ{\cdot}23\,T) + 0^\circ{\cdot}0028 \cos(269^\circ{\cdot}9 + 954\,397^\circ{\cdot}70\,T) \\
\text{SD} = {} & 0{\cdot}2725\,\pi \\
r = {} & 1\,/\sin\pi
\end{aligned}$$

Form the geocentric direction cosines (l, m, n) from:

$$\begin{aligned}
l &= \cos\beta\cos\lambda \\
m &= +0{\cdot}9175\cos\beta\sin\lambda - 0{\cdot}3978\sin\beta \\
n &= +0{\cdot}3978\cos\beta\sin\lambda + 0{\cdot}9175\sin\beta
\end{aligned}$$

where $l = \cos\delta\cos\alpha \qquad m = \cos\delta\sin\alpha \qquad n = \sin\delta$

Then $\alpha = \tan^{-1}(m\,/\,l)$ and $\delta = \sin^{-1}(n)$

where the quadrant of α is determined by the signs of l and m, and where α, δ are referred to the mean equator and equinox of date.

Low-precision formulae for topocentric coordinates of the Moon

The following formulae give approximate topocentric values of right ascension (α'), declination (δ'), distance (r'), parallax (π') and semi-diameter (SD$'$).

Form the geocentric rectangular coordinates (x, y, z) from:

$$\begin{aligned}
x &= rl = r\cos\delta\cos\alpha \\
y &= rm = r\cos\delta\sin\alpha \\
z &= rn = r\sin\delta
\end{aligned}$$

Form the topocentric rectangular coordinates (x', y', z') from:

$$\begin{aligned}
x' &= x - \cos\phi'\cos\theta_0 \\
y' &= y - \cos\phi'\sin\theta_0 \\
z' &= z - \sin\phi'
\end{aligned}$$

where ϕ' is the observer's geocentric latitude and θ_0 is the local sidereal time.

$$\theta_0 = 100^\circ{\cdot}46 + 36\,000^\circ{\cdot}77\,T + \lambda' + 15\,\text{UT}$$

where λ' is the observer's east longitude.

Then $r' = (x'^2 + y'^2 + z'^2)^{1/2} \quad \alpha' = \tan^{-1}(y'\,/\,x') \quad \delta' = \sin^{-1}(z'\,/\,r')$

$\pi' = \sin^{-1}(1\,/\,r') \qquad \text{SD}' = 0{\cdot}2725\,\pi'$

CONTENTS OF SECTION E

NOTES

1. Other data, explanatory notes and formulas are given on the following pages:

2. Other data on the planets are given on the following pages:

NOTES AND FORMULAS

Orbital elements

The heliocentric osculating orbital elements for the Earth given on pages E3–E4 refer to the Earth/Moon barycenter. In ecliptic rectangular coordinates, the correction from the Earth/Moon barycenter to the Earth's center is given by:

(Earth's center) = (Earth/Moon barycenter) – (0.000 0312 cos L, 0.000 0312 sin L, 0.0)

where $L = 218° + 481\,268°\,T$, with T in Julian centuries from JD 245 1545.0 to 5 decimal places; the coordinates are in au and are referred to the mean equinox and ecliptic of date.

Linear interpolation of the heliocentric osculating orbital elements usually leads to errors of about 1″ or 2″ in the geocentric positions of the Sun and planets: the errors may, however, reach about 7″ for Venus at inferior conjunction and about 3″ for Mars at opposition.

Heliocentric coordinates

The heliocentric ecliptic coordinates of the Earth may be obtained from the geocentric ecliptic coordinates of the Sun given on pages C4–C18 by adding ±180° to the longitude, and reversing the sign of the latitude.

Geocentric coordinates

Precise values of apparent semidiameter and horizontal parallax may be computed from the formulas and values given on page E43. Values of apparent diameter are tabulated in the ephemerides for physical observations on pages E52 onwards.

Times of transit, rising and setting

Formulas for obtaining the universal times of transit, rising and setting of the planets are given on page E43.

Ephemerides for physical observations

Full descriptions of the quantities tabulated in the ephemerides for physical observations of the planets are given in the Explanation (Section L).

Invariable plane of the solar system

Approximate coordinates of the north pole of the invariable plane for J2000.0 are:

$$\alpha_0 = 273^\circ.85 \qquad \delta_0 = 66^\circ.99$$

HELIOCENTRIC OSCULATING ORBITAL ELEMENTS REFERRED TO THE MEAN ECLIPTIC AND EQUINOX OF J2000.0

Julian Date 244	Inclin-ation i	Longitude Asc. Node Ω	Longitude Perihelion ϖ	Mean Distance a	Daily Motion n	Eccen-tricity e	Mean Longitude L
MERCURY	°	°	°		°		°
8680.5	7.005 42	48.3405	77.4438	0.387 0975	4.092 357	0.205 6268	49.746 69
8880.5	7.005 43	48.3399	77.4422	0.387 0979	4.092 350	0.205 6257	148.217 38
VENUS							
8680.5	3.394 59	76.7011	131.699	0.723 3260	1.602 151	0.006 7555	272.676 75
8880.5	3.394 59	76.7001	131.688	0.723 3269	1.602 148	0.006 7549	233.102 78
EARTH*							
8680.5	0.001 03	359.7	102.9287	1.000 0122	0.985 591 1	0.016 7250	157.187 35
8880.5	0.000 97	0.5	102.8141	1.000 0054	0.985 601 2	0.016 7119	354.308 64
MARS							
8680.5	1.850 34	49.5761	336.0691	1.523 6547	0.524 051 9	0.093 4974	294.352 42
8880.5	1.850 33	49.5753	336.0333	1.523 6512	0.524 053 7	0.093 5104	39.156 94
JUPITER							
8680.5	1.304 63	100.4687	15.6772	5.203 186	0.083 082 06	0.048 2538	156.366 56
8880.5	1.304 64	100.4687	15.6658	5.203 019	0.083 086 06	0.048 2876	172.982 39
SATURN							
8680.5	2.487 26	113.6774	93.6489	9.520 163	0.033 558 27	0.054 4734	314.107 15
8880.5	2.486 99	113.6753	93.5771	9.523 469	0.033 540 80	0.054 0387	320.789 14
URANUS							
8680.5	0.772 02	74.0108	171.3225	19.191 84	0.011 723 00	0.048 1286	279.512 73
8880.5	0.772 19	74.0185	172.3152	19.208 01	0.011 708 20	0.047 9195	281.842 05
NEPTUNE							
8680.5	1.772 03	131.7530	53.722	30.056 33	0.005 981 518	0.007 6826	287.578 41
8880.5	1.772 08	131.7523	47.854	30.087 03	0.005 972 364	0.006 9509	288.750 36
PLUTO							
8680.5	17.145 67	110.2586	223.7788	39.691 86	0.003 941 412	0.252 8372	227.382 05
8880.5	17.142 17	110.2788	223.8836	39.746 46	0.003 933 294	0.253 8339	228.204 78

HELIOCENTRIC COORDINATES AND VELOCITY COMPONENTS REFERRED TO THE MEAN EQUATOR AND EQUINOX OF J2000.0

	x	y	z	$\dot{x}$	$\dot{y}$	$\dot{z}$
MERCURY						
8680.5	+ 0.260 3940	+ 0.176 0307	+ 0.067 0142	− 0.021 999 15	+ 0.020 459 90	+ 0.013 210 70
8880.5	− 0.371 0643	+ 0.032 4824	+ 0.055 8427	− 0.009 747 59	− 0.023 940 70	− 0.011 776 73
VENUS						
8680.5	+ 0.039 7435	− 0.661 1896	− 0.299 9574	+ 0.020 060 91	+ 0.001 402 09	− 0.000 638 95
8880.5	− 0.426 6959	− 0.543 4086	− 0.217 4516	+ 0.016 200 95	− 0.010 583 77	− 0.005 786 60
EARTH*						
8680.5	− 0.923 1630	+ 0.329 1685	+ 0.142 7190	− 0.006 512 64	− 0.014 771 66	− 0.006 404 62
8880.5	+ 0.996 9673	− 0.120 3423	− 0.052 1774	+ 0.001 963 82	+ 0.015 591 32	+ 0.006 759 98
MARS						
8680.5	+ 0.405 5621	− 1.236 4320	− 0.578 0814	+ 0.013 944 12	+ 0.004 855 05	+ 0.001 849 65
8880.5	+ 0.960 9452	+ 1.021 3551	+ 0.442 4681	− 0.010 057 75	+ 0.009 300 63	+ 0.004 537 96
JUPITER						
8680.5	− 5.065 706	+ 1.676 754	+ 0.842 162	− 0.002 716 756	− 0.006 213 551	− 0.002 597 219
8880.5	− 5.414 519	+ 0.387 231	+ 0.297 913	− 0.000 755 930	− 0.006 600 401	− 0.002 810 817
SATURN						
8680.5	+ 6.412 183	− 6.901 778	− 3.126 276	+ 0.003 953 559	+ 0.003 378 972	+ 0.001 225 329
8880.5	+ 7.162 530	− 6.185 484	− 2.862 744	+ 0.003 541 520	+ 0.003 777 541	+ 0.001 407 692
URANUS						
8680.5	+ 4.935 42	− 17.273 00	− 7.634 84	+ 0.003 774 668	+ 0.000 758 884	+ 0.000 278 988
8880.5	+ 5.686 41	− 17.107 55	− 7.573 00	+ 0.003 734 247	+ 0.000 895 328	+ 0.000 339 303
NEPTUNE						
8680.5	+ 8.767 53	− 26.662 35	− 11.131 38	+ 0.002 982 380	+ 0.000 882 759	+ 0.000 286 949
8880.5	+ 9.362 25	− 26.480 09	− 11.071 61	+ 0.002 964 611	+ 0.000 939 808	+ 0.000 310 738
PLUTO						
8680.5	− 18.003 45	− 23.532 73	− 1.920 50	+ 0.002 585 865	− 0.001 962 502	− 0.001 392 448
8880.5	− 17.482 07	− 23.919 91	− 2.198 55	+ 0.002 627 764	− 0.001 909 015	− 0.001 387 915

*Values labelled for the Earth are actually for the Earth/Moon barycenter (see note on page E2).
The velocity components are expressed in astronomical units per day.

HELIOCENTRIC OSCULATING ORBITAL ELEMENTS REFERRED TO THE MEAN ECLIPTIC AND EQUINOX OF DATE

Date	Julian Date 244	Inclination i	Longitude Asc. Node Ω	Longitude Perihelion ϖ	Mean Distance a	Daily Motion n	Eccentricity e	Mean Longitude L
MERCURY		°	°	°		°		°
Jan. −21	8600.5	7.0048	48.235	77.332	0.387 098	4.092 34	0.205 620	82.2466
Jan. 19	8640.5	7.0048	48.236	77.333	0.387 098	4.092 36	0.205 628	245.9414
Feb. 28	8680.5	7.0048	48.238	77.334	0.387 098	4.092 36	0.205 627	49.6371
Apr. 8	8720.5	7.0048	48.239	77.335	0.387 097	4.092 36	0.205 631	213.3325
May 18	8760.5	7.0048	48.240	77.336	0.387 097	4.092 36	0.205 632	17.0287
June 27	8800.5	7.0049	48.242	77.337	0.387 097	4.092 36	0.205 631	180.7244
Aug. 6	8840.5	7.0049	48.243	77.338	0.387 098	4.092 34	0.205 627	344.4200
Sept. 15	8880.5	7.0049	48.244	77.340	0.387 098	4.092 35	0.205 626	148.1154
Oct. 25	8920.5	7.0048	48.245	77.343	0.387 100	4.092 32	0.205 620	311.8100
Dec. 4	8960.5	7.0048	48.246	77.345	0.387 099	4.092 34	0.205 617	115.5045
Dec. 44	9000.5	7.0049	48.248	77.347	0.387 099	4.092 34	0.205 620	279.1991
VENUS								
Jan. −21	8600.5	3.3944	76.606	131.63	0.723 329	1.602 14	0.006 761	144.3920
Jan. 19	8640.5	3.3944	76.607	131.59	0.723 325	1.602 15	0.006 759	208.4792
Feb. 28	8680.5	3.3944	76.609	131.59	0.723 326	1.602 15	0.006 755	272.5672
Apr. 8	8720.5	3.3944	76.610	131.63	0.723 331	1.602 13	0.006 748	336.6539
May 18	8760.5	3.3944	76.611	131.67	0.723 330	1.602 14	0.006 746	40.7404
June 27	8800.5	3.3944	76.612	131.69	0.723 329	1.602 14	0.006 747	104.8278
Aug. 6	8840.5	3.3944	76.613	131.66	0.723 332	1.602 13	0.006 753	168.9143
Sept. 15	8880.5	3.3945	76.614	131.59	0.723 327	1.602 15	0.006 755	233.0008
Oct. 25	8920.5	3.3945	76.615	131.55	0.723 326	1.602 15	0.006 753	297.0888
Dec. 4	8960.5	3.3945	76.616	131.56	0.723 327	1.602 15	0.006 750	1.1761
Dec. 44	9000.5	3.3945	76.618	131.60	0.723 325	1.602 15	0.006 749	65.2635
EARTH*								
Jan. −21	8600.5	0.0	—	102.823	1.000 001	0.985 607	0.016 709	78.2275
Jan. 19	8640.5	0.0	—	102.834	1.000 007	0.985 599	0.016 715	117.6536
Feb. 28	8680.5	0.0	—	102.819	1.000 012	0.985 591	0.016 725	157.0778
Apr. 8	8720.5	0.0	—	102.766	0.999 999	0.985 611	0.016 731	196.5022
May 18	8760.5	0.0	—	102.717	0.999 983	0.985 635	0.016 739	235.9286
June 27	8800.5	0.0	—	102.717	0.999 985	0.985 632	0.016 736	275.3556
Aug. 6	8840.5	0.0	—	102.717	0.999 997	0.985 613	0.016 723	314.7818
Sept. 15	8880.5	0.0	—	102.712	1.000 005	0.985 601	0.016 712	354.2067
Oct. 25	8920.5	0.0	—	102.711	1.000 006	0.985 600	0.016 706	33.6316
Dec. 4	8960.5	0.0	—	102.726	1.000 002	0.985 606	0.016 704	73.0573
Dec. 44	9000.5	0.0	—	102.750	1.000 003	0.985 605	0.016 705	112.4838
MARS								
Jan. −21	8600.5	1.8497	49.490	335.979	1.523 583	0.524 089	0.093 473	252.3130
Jan. 19	8640.5	1.8497	49.492	335.968	1.523 618	0.524 071	0.093 482	273.2784
Feb. 28	8680.5	1.8497	49.492	335.959	1.523 655	0.524 052	0.093 497	294.2429
Apr. 8	8720.5	1.8498	49.493	335.954	1.523 689	0.524 034	0.093 515	315.2062
May 18	8760.5	1.8498	49.494	335.952	1.523 706	0.524 026	0.093 524	336.1684
June 27	8800.5	1.8498	49.495	335.947	1.523 696	0.524 031	0.093 520	357.1301
Aug. 6	8840.5	1.8498	49.496	335.940	1.523 676	0.524 041	0.093 515	18.0921
Sept. 15	8880.5	1.8498	49.497	335.931	1.523 651	0.524 054	0.093 510	39.0550
Oct. 25	8920.5	1.8498	49.499	335.926	1.523 633	0.524 063	0.093 508	60.0190
Dec. 4	8960.5	1.8498	49.500	335.923	1.523 624	0.524 068	0.093 498	80.9843
Dec. 44	9000.5	1.8498	49.501	335.920	1.523 626	0.524 067	0.093 481	101.9507

*Values labelled for the Earth are actually for the Earth/Moon barycenter (see note on page E2).

FORMULAS

Mean anomaly, $M = L - \varpi$

Argument of perihelion, measured from node, $\omega = \varpi - \Omega$

True anomaly, $v = M + (2e - e^3/4)\sin M + (5e^2/4)\sin 2M + (13e^3/12)\sin 3M + \ldots$ in radians.

True distance, $r = a(1 - e^2)/(1 + e\cos v)$

Heliocentric rectangular coordinates, referred to the ecliptic of date, may be computed from these elements by:

$$x = r\{\cos(v+\omega)\cos\Omega - \sin(v+\omega)\cos i\sin\Omega\}$$
$$y = r\{\cos(v+\omega)\sin\Omega + \sin(v+\omega)\cos i\cos\Omega\}$$
$$z = r\sin(v+\omega)\sin i$$

HELIOCENTRIC OSCULATING ORBITAL ELEMENTS REFERRED TO THE MEAN ECLIPTIC AND EQUINOX OF DATE

Date	Julian Date 244	Inclin-ation i	Longitude Asc. Node Ω	Longitude Perihelion ϖ	Mean Distance a	Daily Motion n	Eccen-tricity e	Mean Longitude L
JUPITER		°	°	°		°		°
Jan. − 21	8600.5	1.3049	100.401	15.586	5.203 27	0.083 080 0	0.048 248	149.6062
Jan. 19	8640.5	1.3049	100.401	15.579	5.203 26	0.083 080 3	0.048 244	152.9319
Feb. 28	8680.5	1.3049	100.402	15.568	5.203 19	0.083 082 1	0.048 254	156.2570
Apr. 8	8720.5	1.3049	100.404	15.562	5.203 12	0.083 083 7	0.048 266	159.5816
May 18	8760.5	1.3049	100.406	15.565	5.203 08	0.083 084 5	0.048 275	162.9058
June 27	8800.5	1.3049	100.407	15.570	5.203 10	0.083 084 1	0.048 273	166.2304
Aug. 6	8840.5	1.3049	100.407	15.567	5.203 07	0.083 084 7	0.048 277	169.5555
Sept. 15	8880.5	1.3049	100.407	15.564	5.203 02	0.083 086 1	0.048 288	172.8805
Oct. 25	8920.5	1.3049	100.408	15.565	5.202 94	0.083 088 1	0.048 305	176.2049
Dec. 4	8960.5	1.3049	100.410	15.578	5.202 93	0.083 088 2	0.048 310	179.5286
Dec. 44	9000.5	1.3049	100.411	15.590	5.202 99	0.083 086 8	0.048 300	182.8529
SATURN								
Jan. − 21	8600.5	2.4879	113.587	93.527	9.518 87	0.033 565 1	0.054 664	311.3239
Jan. 19	8640.5	2.4878	113.588	93.544	9.519 45	0.033 562 0	0.054 570	312.6600
Feb. 28	8680.5	2.4878	113.589	93.539	9.520 16	0.033 558 3	0.054 473	313.9976
Apr. 8	8720.5	2.4877	113.589	93.530	9.520 84	0.033 554 7	0.054 384	315.3357
May 18	8760.5	2.4876	113.590	93.522	9.521 42	0.033 551 6	0.054 306	316.6742
June 27	8800.5	2.4875	113.591	93.521	9.521 97	0.033 548 7	0.054 225	318.0117
Aug. 6	8840.5	2.4875	113.592	93.504	9.522 69	0.033 544 9	0.054 132	319.3492
Sept. 15	8880.5	2.4875	113.593	93.475	9.523 47	0.033 540 8	0.054 039	320.6872
Oct. 25	8920.5	2.4874	113.593	93.437	9.524 25	0.033 536 7	0.053 953	322.0262
Dec. 4	8960.5	2.4873	113.594	93.415	9.524 82	0.033 533 7	0.053 885	323.3652
Dec. 44	9000.5	2.4873	113.594	93.400	9.525 36	0.033 530 8	0.053 811	324.7028
URANUS								
Jan. − 21	8600.5	0.7717	73.972	170.769	19.184 9	0.011 729 4	0.048 176	278.4724
Jan. 19	8640.5	0.7718	73.975	170.995	19.188 4	0.011 726 2	0.048 163	278.9366
Feb. 28	8680.5	0.7718	73.976	171.213	19.191 8	0.011 723 0	0.048 129	279.4032
Apr. 8	8720.5	0.7718	73.977	171.415	19.195 1	0.011 720 0	0.048 090	279.8705
May 18	8760.5	0.7718	73.977	171.596	19.198 0	0.011 717 4	0.048 055	280.3381
June 27	8800.5	0.7719	73.979	171.789	19.201 1	0.011 714 6	0.048 025	280.8045
Aug. 6	8840.5	0.7719	73.982	172.003	19.204 5	0.011 711 4	0.047 978	281.2718
Sept. 15	8880.5	0.7720	73.986	172.213	19.208 0	0.011 708 2	0.047 919	281.7401
Oct. 25	8920.5	0.7720	73.988	172.403	19.211 2	0.011 705 3	0.047 854	282.2099
Dec. 4	8960.5	0.7721	73.989	172.563	19.213 9	0.011 702 8	0.047 806	282.6787
Dec. 44	9000.5	0.7721	73.991	172.742	19.216 8	0.011 700 2	0.047 762	283.1461
NEPTUNE								
Jan. − 21	8600.5	1.7727	131.665	55.46	30.043 5	0.005 985 34	0.008 054	287.0030
Jan. 19	8640.5	1.7728	131.665	54.64	30.049 8	0.005 983 45	0.007 858	287.2344
Feb. 28	8680.5	1.7728	131.666	53.61	30.056 3	0.005 981 52	0.007 683	287.4688
Apr. 8	8720.5	1.7728	131.667	52.57	30.062 4	0.005 979 69	0.007 526	287.7044
May 18	8760.5	1.7727	131.669	51.58	30.067 9	0.005 978 06	0.007 389	287.9405
June 27	8800.5	1.7727	131.670	50.55	30.073 7	0.005 976 35	0.007 240	288.1752
Aug. 6	8840.5	1.7728	131.671	49.22	30.080 3	0.005 974 37	0.007 088	288.4111
Sept. 15	8880.5	1.7728	131.671	47.75	30.087 0	0.005 972 36	0.006 951	288.6484
Oct. 25	8920.5	1.7727	131.673	46.27	30.093 3	0.005 970 50	0.006 841	288.8876
Dec. 4	8960.5	1.7727	131.675	45.05	30.098 5	0.005 968 96	0.006 747	289.1259
Dec. 44	9000.5	1.7727	131.676	43.71	30.104 1	0.005 967 28	0.006 637	289.3625
PLUTO								
Jan. − 21	8600.5	17.1475	110.141	223.629	39.663 8	0.003 945 60	0.252 317	226.9422
Jan. 19	8640.5	17.1468	110.146	223.647	39.679 1	0.003 943 31	0.252 603	227.1064
Feb. 28	8680.5	17.1461	110.152	223.669	39.691 9	0.003 941 41	0.252 837	227.2724
Apr. 8	8720.5	17.1454	110.158	223.691	39.703 1	0.003 939 73	0.253 044	227.4383
May 18	8760.5	17.1448	110.163	223.711	39.713 3	0.003 938 22	0.253 231	227.6035
June 27	8800.5	17.1441	110.168	223.731	39.724 9	0.003 936 50	0.253 443	227.7684
Aug. 6	8840.5	17.1434	110.174	223.755	39.736 3	0.003 934 80	0.253 651	227.9351
Sept. 15	8880.5	17.1426	110.180	223.782	39.746 5	0.003 933 29	0.253 834	228.1027
Oct. 25	8920.5	17.1418	110.185	223.808	39.754 4	0.003 932 11	0.253 976	228.2705
Dec. 4	8960.5	17.1412	110.190	223.829	39.762 0	0.003 930 99	0.254 111	228.4363
Dec. 44	9000.5	17.1405	110.195	223.851	39.771 6	0.003 929 57	0.254 284	228.6018

HELIOCENTRIC POSITIONS FOR 0^h DYNAMICAL TIME

MEAN EQUINOX AND ECLIPTIC OF DATE

Date	Longitude	Latitude	Radius Vector	Date	Longitude	Latitude	Radius Vector
	° ′ ″	° ′ ″			° ′ ″	° ′ ″	
Jan. 0	191 14 35.3	+ 4 13 41.9	0.404 1392	Feb. 15	332 58 32.2	− 6 46 35.4	0.390 8008
1	194 51 40.1	3 52 00.6	.409 3807	16	336 59 15.9	6 38 11.2	.385 1298
2	198 23 09.8	3 29 58.5	.414 4556	17	341 07 04.7	6 27 30.6	.379 3844
3	201 49 28.0	3 07 42.0	.419 3489	18	345 22 20.9	6 14 25.0	.373 5923
4	205 10 57.6	2 45 16.6	.424 0472	19	349 45 26.0	5 58 46.0	.367 7843
5	208 28 00.4	+ 2 22 47.1	0.428 5388	20	354 16 40.4	− 5 40 25.9	0.361 9941
6	211 40 57.4	2 00 17.6	.432 8131	21	358 56 22.4	5 19 18.0	.356 2586
7	214 50 08.7	1 37 51.8	.436 8609	22	3 44 48.1	4 55 17.0	.350 6182
8	217 55 53.3	1 15 32.7	.440 6741	23	8 42 10.2	4 28 19.6	.345 1160
9	220 58 29.6	0 53 23.0	.444 2455	24	13 48 37.2	3 58 25.3	.339 7984
10	223 58 14.9	+ 0 31 24.8	0.447 5689	25	19 04 12.1	− 3 25 36.9	0.334 7144
11	226 55 25.9	+ 0 09 40.4	.450 6387	26	24 28 51.7	2 50 01.4	.329 9146
12	229 50 18.6	− 0 11 48.7	.453 4502	27	30 02 25.1	2 11 50.7	.325 4512
13	232 43 08.5	0 33 00.8	.455 9992	28	35 44 32.8	1 31 22.0	.321 3763
14	235 34 10.2	0 53 54.6	.458 2821	29	41 34 45.9	0 48 58.3	.317 7413
15	238 23 38.0	− 1 14 28.9	0.460 2959	Mar. 1	47 32 24.9	− 0 05 08.5	0.314 5947
16	241 11 45.8	1 34 42.3	.462 0380	2	53 36 39.8	+ 0 39 33.2	.311 9812
17	243 58 46.9	1 54 34.0	.463 5061	3	59 46 30.1	1 24 27.9	.309 9396
18	246 44 54.4	2 14 02.7	.464 6987	4	66 00 45.5	2 08 53.2	.308 5014
19	249 30 21.1	2 33 07.5	.465 6141	5	72 18 06.9	2 52 05.5	.307 6895
20	252 15 19.6	− 2 51 47.3	0.466 2514	6	78 37 08.8	+ 3 33 21.2	0.307 5172
21	255 00 02.3	3 10 01.2	.466 6099	7	84 56 21.5	4 11 59.4	.307 9872
22	257 44 41.4	3 27 47.9	.466 6891	8	91 14 13.8	4 47 23.8	.309 0919
23	260 29 29.3	3 45 06.3	.466 4890	9	97 29 16.4	5 19 04.3	.310 8135
24	263 14 37.9	4 01 55.2	.466 0097	10	103 40 04.2	5 46 38.1	.313 1249
25	266 00 19.7	− 4 18 13.3	0.465 2519	11	109 45 19.7	+ 6 09 50.2	0.315 9910
26	268 46 46.8	4 33 59.0	.464 2164	12	115 43 54.3	6 28 33.7	.319 3700
27	271 34 11.9	4 49 10.9	.462 9044	13	121 34 50.2	6 42 49.0	.323 2153
28	274 22 47.4	5 03 47.1	.461 3174	14	127 17 21.3	6 52 42.7	.327 4769
29	277 12 46.4	5 17 45.7	.459 4575	15	132 50 52.7	6 58 26.9	.332 1029
30	280 04 21.9	− 5 31 04.7	0.457 3268	16	138 15 00.9	+ 7 00 17.3	0.337 0410
31	282 57 47.5	5 43 41.7	.454 9283	17	143 29 32.7	6 58 32.4	.342 2397
Feb. 1	285 53 17.1	5 55 34.1	.452 2652	18	148 34 23.7	6 53 32.2	.347 6488
2	288 51 05.0	6 06 39.2	.449 3414	19	153 29 37.6	6 45 37.1	.353 2208
3	291 51 26.1	6 16 53.6	.446 1612	20	158 15 24.4	6 35 07.6	.358 9105
4	294 54 35.6	− 6 26 14.1	0.442 7297	21	162 51 59.3	+ 6 22 23.2	0.364 6762
5	298 00 49.6	6 34 36.8	.439 0528	22	167 19 41.5	6 07 42.3	.370 4791
6	301 10 24.6	6 41 57.6	.435 1371	23	171 38 52.8	5 51 22.1	.376 2838
7	304 23 37.8	6 48 11.8	.430 9902	24	175 49 57.3	5 33 38.1	.382 0578
8	307 40 47.1	6 53 14.6	.426 6206	25	179 53 20.2	5 14 44.2	.387 7720
9	311 02 11.1	− 6 57 00.4	0.422 0381	26	183 49 27.3	+ 4 54 53.1	0.393 4000
10	314 28 09.0	6 59 23.3	.417 2538	27	187 38 44.5	4 34 15.7	.398 9179
11	317 59 01.1	7 00 17.1	.412 2801	28	191 21 37.5	4 13 01.7	.404 3047
12	321 35 07.8	6 59 34.7	.407 1311	29	194 58 31.5	3 51 19.7	.409 5413
13	325 16 50.6	6 57 08.9	.401 8226	30	198 29 51.1	3 29 17.1	.414 6109
14	329 04 31.3	− 6 52 51.8	0.396 3725	31	201 55 59.9	+ 3 07 00.1	0.419 4984
15	332 58 32.2	− 6 46 35.4	0.390 8008	Apr. 1	205 17 20.8	+ 2 44 34.5	0.424 1906

HELIOCENTRIC POSITIONS FOR 0^h DYNAMICAL TIME
MEAN EQUINOX AND ECLIPTIC OF DATE

Date	Longitude	Latitude	Radius Vector	Date	Longitude	Latitude	Radius Vector
	° ′ ″	° ′ ″			° ′ ″	° ′ ″	
Apr. 1	205 17 20.8	+ 2 44 34.5	0.424 1906	May 17	349 53 55.4	− 5 58 14.3	0.367 6018
2	208 34 15.6	2 22 05.0	.428 6756	18	354 25 25.4	5 39 49.1	.361 8126
3	211 47 05.3	1 59 35.6	.432 9431	19	359 05 23.7	5 18 35.9	.356 0793
4	214 56 09.8	1 37 10.0	.436 9838	20	3 54 06.1	4 54 29.4	.350 4423
5	218 01 48.2	1 14 51.1	.440 7896	21	8 51 45.3	4 27 26.5	.344 9451
6	221 04 18.8	+ 0 52 41.7	0.444 3534	22	13 58 29.6	− 3 57 26.6	0.339 6340
7	224 03 59.0	0 30 44.0	.447 6690	23	19 14 21.8	3 24 32.9	.334 5579
8	227 01 05.4	+ 0 08 59.9	.450 7309	24	24 39 18.5	2 48 52.3	.329 7677
9	229 55 54.1	− 0 12 28.6	.453 5343	25	30 13 08.5	2 10 37.0	.325 3156
10	232 48 40.3	0 33 40.2	.456 0751	26	35 55 32.2	1 30 04.3	.321 2537
11	235 39 38.8	− 0 54 33.4	0.458 3497	27	41 46 00.1	− 0 47 37.4	0.317 6332
12	238 29 03.9	1 15 07.0	.460 3551	28	47 43 52.5	− 0 03 45.3	.314 5026
13	241 17 09.3	1 35 19.8	.462 0887	29	53 48 19.2	+ 0 40 57.4	.311 9064
14	244 04 08.5	1 55 10.7	.463 5483	30	59 58 19.2	1 25 51.9	.309 8832
15	246 50 14.5	2 14 38.7	.464 7322	31	66 12 41.9	2 10 15.7	.308 4644
16	249 35 40.1	− 2 33 42.8	0.465 6390	June 1	72 30 08.0	+ 2 53 25.0	0.307 6725
17	252 20 37.9	2 52 21.8	.466 2676	2	78 49 11.9	3 34 36.4	.307 5204
18	255 05 20.2	3 10 34.8	.466 6174	3	85 08 23.7	4 13 09.1	.308 0106
19	257 49 59.4	3 28 20.7	.466 6879	4	91 26 12.3	4 48 27.0	.309 1351
20	260 34 47.6	3 45 38.2	.466 4791	5	97 41 08.3	5 20 00.1	.310 8757
21	263 19 57.0	− 4 02 26.2	0.465 9911	6	103 51 47.0	+ 5 47 25.8	0.313 2052
22	266 05 40.0	4 18 43.3	.465 2246	7	109 56 51.0	6 10 29.6	.316 0882
23	268 52 08.7	4 34 28.0	.464 1805	8	115 55 12.2	6 29 04.7	.319 4828
24	271 39 35.7	4 49 38.7	.462 8598	9	121 45 53.0	6 43 11.6	.323 3421
25	274 28 13.6	5 04 13.8	.461 2643	10	127 28 07.7	6 52 57.4	.327 6161
26	277 18 15.3	− 5 18 11.3	0.459 3958	11	133 01 21.8	+ 6 58 34.0	0.332 2528
27	280 09 54.0	5 31 29.0	.457 2567	12	138 25 12.1	7 00 17.3	.337 2000
28	283 03 23.2	5 44 04.7	.454 8498	13	143 39 25.6	6 58 26.0	.342 4062
29	285 58 56.9	5 55 55.6	.452 1785	14	148 43 58.4	6 53 19.9	.347 8213
30	288 56 49.3	6 06 59.1	.449 2465	15	153 38 54.4	6 45 19.7	.353 3977
May 1	291 57 15.3	− 6 17 12.0	0.446 0583	16	158 24 23.6	+ 6 34 45.7	0.359 0906
2	295 00 30.3	6 26 30.7	.442 6190	17	163 00 41.4	6 21 57.3	.364 8582
3	298 06 50.3	6 34 51.6	.438 9345	18	167 28 07.1	6 07 13.1	.370 6618
4	301 16 31.8	6 42 10.3	.435 0114	19	171 47 02.8	5 50 50.1	.376 4660
5	304 29 52.1	6 48 22.5	.430 8573	20	175 57 52.4	5 33 03.7	.382 2387
6	307 47 09.0	− 6 53 22.9	0.426 4808	21	180 01 01.1	+ 5 14 07.8	0.387 9506
7	311 08 41.2	6 57 06.2	.421 8917	22	183 56 54.9	4 54 15.1	.393 5755
8	314 34 48.0	6 59 26.5	.417 1012	23	187 45 59.5	4 33 36.4	.399 0897
9	318 05 49.5	7 00 17.4	.412 1217	24	191 28 40.8	4 12 21.4	.404 4720
10	321 42 06.4	6 59 31.9	.406 9673	25	195 05 23.9	3 50 38.7	.409 7037
11	325 24 00.1	− 6 57 02.7	0.401 6540	26	198 36 33.3	+ 3 28 35.5	0.414 7678
12	329 11 52.4	6 52 42.1	.396 1997	27	202 02 32.7	3 06 18.2	.419 6494
13	333 06 05.5	6 46 21.9	.390 6244	28	205 23 44.9	2 43 52.4	.424 3352
14	337 07 02.1	6 37 53.5	.384 9505	29	208 40 31.7	2 21 22.8	.428 8136
15	341 15 04.6	6 27 08.5	.379 2031	30	211 53 13.9	1 58 53.5	.433 0740
16	345 30 35.3	− 6 13 58.2	0.373 4099	July 1	215 02 11.6	+ 1 36 28.0	0.437 1074
17	349 53 55.4	− 5 58 14.3	0.367 6018	2	218 07 43.9	+ 1 14 09.4	0.440 9058

HELIOCENTRIC POSITIONS FOR 0ʰ DYNAMICAL TIME
MEAN EQUINOX AND ECLIPTIC OF DATE

Date	Longitude	Latitude	Radius Vector	Date	Longitude	Latitude	Radius Vector
	° ′ ″	° ′ ″			° ′ ″	° ′ ″	
July 1	215 02 11.6	+ 1 36 28.0	0.437 1074	Aug. 16	4 03 26.9	− 4 53 41.3	0.350 2697
2	218 07 43.9	1 14 09.4	.440 9058	17	9 01 22.9	4 26 32.9	.344 7777
3	221 10 08.8	0 52 00.4	.444 4619	18	14 08 24.1	3 56 27.5	.339 4734
4	224 09 43.9	0 30 03.0	.447 7695	19	19 24 33.2	3 23 28.4	.334 4056
5	227 06 45.7	+ 0 08 19.5	.450 8233	20	24 49 46.6	2 47 42.9	.329 6254
6	230 01 30.3	− 0 13 08.6	0.453 6185	21	30 23 52.8	− 2 09 23.0	0.325 1848
7	232 54 12.8	0 34 19.6	.456 1510	22	36 06 31.8	1 28 46.3	.321 1361
8	235 45 08.2	0 55 12.2	.458 4171	23	41 57 13.9	0 46 16.3	.317 5302
9	238 34 30.5	1 15 45.2	.460 4140	24	47 55 19.2	− 0 02 22.2	.314 4158
10	241 22 33.7	1 35 57.3	.462 1389	25	53 59 56.9	+ 0 42 21.5	.311 8370
11	244 09 31.0	− 1 55 47.5	0.463 5899	26	60 10 05.9	+ 1 27 15.6	0.309 8323
12	246 55 35.5	2 15 14.8	.464 7651	27	66 24 35.2	2 11 37.7	.308 4327
13	249 41 00.0	2 34 18.1	.465 6631	28	72 42 05.4	2 54 43.9	.307 6606
14	252 25 57.1	2 52 56.3	.466 2830	29	79 01 10.6	3 35 51.0	.307 5285
15	255 10 39.2	3 11 08.5	.466 6239	30	85 20 20.8	4 14 18.1	.308 0386
16	257 55 18.5	− 3 28 53.5	0.466 6856	31	91 38 05.0	+ 4 49 29.4	0.309 1825
17	260 40 07.2	3 46 10.1	.466 4680	Sept. 1	97 52 53.9	5 20 55.1	.310 9418
18	263 25 17.5	4 02 57.2	.465 9713	2	104 03 23.1	5 48 12.8	.313 2890
19	266 11 01.7	4 19 13.3	.465 1960	3	110 08 15.3	6 11 08.3	.316 1884
20	268 57 32.1	4 34 56.9	.464 1431	4	116 06 22.7	6 29 35.0	.319 5980
21	271 45 01.1	− 4 50 06.6	0.462 8138	5	121 56 48.2	+ 6 43 33.7	0.323 4708
22	274 33 41.5	5 04 40.6	.461 2097	6	127 38 46.3	6 53 11.5	.327 7566
23	277 23 46.1	5 18 36.8	.459 3326	7	133 11 43.1	6 58 40.6	.332 4036
24	280 15 28.1	5 31 53.3	.457 1851	8	138 35 15.5	7 00 17.1	.337 3595
25	283 09 01.0	5 44 27.6	.454 7699	9	143 49 10.9	6 58 19.4	.342 5727
26	286 04 38.8	− 5 56 17.1	0.452 0903	10	148 53 25.6	+ 6 53 07.6	0.347 9933
27	289 02 35.8	6 07 19.1	.449 1503	11	153 48 03.7	6 45 02.3	.353 5738
28	292 03 07.0	6 17 30.3	.445 9542	12	158 33 15.5	6 34 23.8	.359 2695
29	295 06 27.6	6 26 47.3	.442 5071	13	163 09 16.5	6 21 31.7	.365 0385
30	298 12 53.6	6 35 06.3	.438 8150	14	167 36 26.0	6 06 44.1	.370 8425
31	301 22 41.8	− 6 42 23.0	0.434 8846	15	171 55 06.2	+ 5 50 18.3	0.376 6460
Aug. 1	304 36 09.2	6 48 33.0	.430 7234	16	176 05 41.1	5 32 29.5	.382 4171
2	307 53 33.8	6 53 31.1	.426 3402	17	180 08 36.0	5 13 31.7	.388 1265
3	311 15 14.4	6 57 11.9	.421 7448	18	184 04 16.7	4 53 37.4	.393 7482
4	314 41 30.2	6 59 29.5	.416 9482	19	187 53 09.2	4 32 57.4	.399 2585
5	318 12 41.3	− 7 00 17.5	0.411 9631	20	191 35 39.0	+ 4 11 41.5	0.404 6363
6	321 49 08.4	6 59 28.8	.406 8036	21	195 12 11.5	3 49 58.0	.409 8629
7	325 31 12.9	6 56 56.3	.401 4857	22	198 43 11.0	3 27 54.3	.414 9214
8	329 19 16.8	6 52 32.1	.396 0274	23	202 09 01.2	3 05 36.6	.419 7970
9	333 13 42.3	6 46 08.0	.390 4488	24	205 30 04.9	2 43 10.6	.424 4765
10	337 14 51.8	− 6 37 35.5	0.384 7724	25	208 46 43.8	+ 2 20 41.0	0.428 9482
11	341 23 08.0	6 26 46.0	.379 0232	26	211 59 18.9	1 58 11.7	.433 2017
12	345 38 53.0	6 13 31.1	.373 2293	27	215 08 10.0	1 35 46.4	.437 2279
13	350 02 28.1	5 57 42.2	.367 4214	28	218 13 36.2	1 13 28.1	.441 0187
14	354 34 13.7	5 39 11.9	.361 6335	29	221 15 55.7	0 51 19.3	.444 5672
15	359 14 28.0	− 5 17 53.3	0.355 9028	30	224 15 25.9	+ 0 29 22.4	0.447 8670
16	4 03 26.9	− 4 53 41.3	0.350 2697	Oct. 1	227 12 23.3	+ 0 07 39.3	0.450 9128

HELIOCENTRIC POSITIONS FOR 0^h DYNAMICAL TIME
MEAN EQUINOX AND ECLIPTIC OF DATE

Date	Longitude	Latitude	Radius Vector	Date	Longitude	Latitude	Radius Vector
	° ′ ″	° ′ ″			° ′ ″	° ′ ″	
Oct. 1	227 12 23.3	+ 0 07 39.3	0.450 9128	Nov. 16	25 00 04.2	− 2 46 34.3	0.329 4892
2	230 07 03.9	− 0 13 48.3	.453 6999	17	30 34 25.9	2 08 10.0	.325 0598
3	232 59 43.0	0 34 58.8	.456 2241	18	36 17 19.6	1 27 29.6	.321 0239
4	235 50 35.4	0 55 50.8	.458 4820	19	42 08 15.3	0 44 56.7	.317 4324
5	238 39 55.1	1 16 23.2	.460 4705	20	48 06 32.8	− 0 01 00.5	.314 3336
6	241 27 56.1	− 1 36 34.7	0.462 1870	21	54 11 21.0	+ 0 43 44.0	0.311 7717
7	244 14 51.6	1 56 24.2	.463 6295	22	60 21 38.5	1 28 37.7	.309 7849
8	247 00 54.8	2 15 50.7	.464 7961	23	66 36 13.9	2 12 58.1	.308 4040
9	249 46 18.4	2 34 53.3	.465 6856	24	72 53 47.8	2 56 01.2	.307 6511
10	252 31 14.9	2 53 30.7	.466 2969	25	79 12 53.9	3 37 03.9	.307 5384
11	255 15 56.8	− 3 11 42.1	0.466 6293	26	85 32 02.4	+ 4 15 25.5	0.308 0676
12	258 00 36.2	3 29 26.2	.466 6824	27	91 49 42.1	4 50 30.3	.309 2303
13	260 45 25.5	3 46 41.9	.466 4562	28	98 04 24.0	5 21 48.6	.311 0076
14	263 30 36.7	4 03 28.0	.465 9510	29	104 14 43.7	5 48 58.6	.313 3718
15	266 16 22.2	4 19 43.1	.465 1672	30	110 19 24.2	6 11 45.9	.316 2871
16	269 02 54.3	− 4 35 25.8	0.464 1059	Dec. 1	116 17 18.1	+ 6 30 04.4	0.319 7111
17	271 50 25.4	4 50 34.4	.462 7682	2	122 07 28.7	6 43 55.0	.323 5968
18	274 39 08.2	5 05 07.2	.461 1557	3	127 49 10.7	6 53 25.0	.327 8941
19	277 29 15.7	5 19 02.3	.459 2704	4	133 21 50.5	6 58 46.8	.332 5509
20	280 21 00.9	5 32 17.4	.457 1147	5	138 45 05.5	7 00 16.5	.337 5149
21	283 14 37.5	− 5 44 50.4	0.454 6915	6	143 58 43.3	+ 6 58 12.6	0.342 7348
22	286 10 19.4	5 56 38.5	.452 0040	7	149 02 40.4	6 52 55.2	.348 1607
23	289 08 21.0	6 07 38.9	.449 0561	8	153 57 01.1	6 44 45.0	.353 7451
24	292 08 57.1	6 17 48.5	.445 8524	9	158 41 56.1	6 34 02.2	.359 4432
25	295 12 23.2	6 27 03.7	.442 3979	10	163 17 40.7	6 21 06.2	.365 2136
26	298 18 55.3	− 6 35 20.8	0.438 6986	11	167 44 34.6	+ 6 06 15.4	0.371 0178
27	301 28 49.9	6 42 35.5	.434 7612	12	172 02 59.8	5 49 46.8	.376 8206
28	304 42 24.3	6 48 43.3	.430 5933	13	176 13 20.6	5 31 55.7	.382 5900
29	307 59 56.5	6 53 39.1	.426 2036	14	180 16 02.1	5 12 56.1	.388 2970
30	311 21 45.2	6 57 17.4	.421 6021	15	184 11 30.2	4 53 00.1	.393 9154
31	314 48 09.6	− 6 59 32.3	0.416 7998	16	188 00 10.8	+ 4 32 18.9	0.399 4219
Nov. 1	318 19 30.0	7 00 17.4	.411 8093	17	191 42 29.6	4 11 02.0	.404 7952
2	321 56 07.1	6 59 25.6	.406 6450	18	195 18 51.7	3 49 17.8	.410 0168
3	325 38 22.3	6 56 49.7	.401 3228	19	198 49 41.7	3 27 13.5	.415 0699
4	329 26 37.3	6 52 21.9	.395 8608	20	202 15 23.0	3 04 55.6	.419 9397
5	333 21 14.7	− 6 45 53.9	0.390 2791	21	205 36 18.5	+ 2 42 29.4	0.424 6130
6	337 22 36.8	6 37 17.3	.384 6003	22	208 52 49.9	2 19 59.6	.429 0781
7	341 31 06.2	6 26 23.5	.378 8496	23	212 05 18.1	1 57 30.4	.433 3248
8	345 47 05.0	6 13 03.9	.373 0551	24	215 14 02.9	1 35 05.3	.437 3440
9	350 10 54.6	5 57 10.2	.367 2475	25	218 19 23.3	1 12 47.2	.441 1275
10	354 42 55.1	− 5 38 34.8	0.361 4610	26	221 21 37.6	+ 0 50 38.7	0.444 6685
11	359 23 25.0	5 17 11.0	.355 7329	27	224 21 03.0	0 28 42.1	.447 9608
12	4 12 39.8	4 52 53.7	.350 1037	28	227 17 56.2	+ 0 06 59.5	.450 9988
13	9 10 52.0	4 25 39.8	.344 6170	29	230 12 33.0	− 0 14 27.6	.453 7780
14	14 18 09.5	3 55 29.1	.339 3193	30	233 05 08.8	0 35 37.6	.456 2943
15	19 34 34.9	− 3 22 24.8	0.334 2597	31	235 55 58.3	− 0 56 29.0	0.458 5441
16	25 00 04.2	− 2 46 34.3	0.329 4892	32	238 45 15.5	− 1 17 00.8	0.460 5244

HELIOCENTRIC POSITIONS FOR 0^h DYNAMICAL TIME
MEAN EQUINOX AND ECLIPTIC OF DATE

Date	Longitude	Latitude	Radius Vector	Date	Longitude	Latitude	Radius Vector
	° ′ ″	° ′ ″			° ′ ″	° ′ ″	
Jan. −1	177 00 09.5	+ 3 20 19.8	0.719 8757	Apr. 2	326 48 18.1	− 3 11 38.7	0.728 0396
1	180 14 31.4	3 17 56.4	.720 0756	4	329 58 22.0	3 15 09.5	.727 9607
3	183 28 45.5	3 14 55.1	.720 2858	6	333 08 29.9	3 18 04.7	.727 8676
5	186 42 51.4	3 11 16.8	.720 5055	8	336 18 41.9	3 20 23.6	.727 7604
7	189 56 48.5	3 07 02.1	.720 7340	10	339 28 58.3	3 22 05.8	.727 6395
9	193 10 36.5	+ 3 02 12.0	0.720 9706	12	342 39 19.2	− 3 23 10.9	0.727 5054
11	196 24 14.9	2 56 47.5	.721 2145	14	345 49 44.8	3 23 38.8	.727 3583
13	199 37 43.4	2 50 49.6	.721 4650	16	349 00 15.3	3 23 29.3	.727 1987
15	202 51 01.6	2 44 19.6	.721 7213	18	352 10 50.8	3 22 42.4	.727 0271
17	206 04 09.3	2 37 18.9	.721 9825	20	355 21 31.5	3 21 18.1	.726 8441
19	209 17 06.3	+ 2 29 48.7	0.722 2478	22	358 32 17.5	− 3 19 16.7	0.726 6501
21	212 29 52.5	2 21 50.6	.722 5164	24	1 43 08.8	3 16 38.5	.726 4459
23	215 42 27.8	2 13 26.2	.722 7874	26	4 54 05.6	3 13 23.9	.726 2319
25	218 54 52.1	2 04 37.2	.723 0600	28	8 05 08.0	3 09 33.4	.726 0089
27	222 07 05.6	1 55 25.1	.723 3334	30	11 16 16.0	3 05 07.7	.725 7775
29	225 19 08.2	+ 1 45 51.9	0.723 6066	May 2	14 27 29.8	− 3 00 07.5	0.725 5384
31	228 31 00.1	1 35 59.3	.723 8788	4	17 38 49.4	2 54 33.8	.725 2924
Feb. 2	231 42 41.5	1 25 49.3	.724 1491	6	20 50 14.9	2 48 27.4	.725 0402
4	234 54 12.7	1 15 23.7	.724 4168	8	24 01 46.3	2 41 49.5	.724 7826
6	238 05 33.9	1 04 44.6	.724 6810	10	27 13 23.9	2 34 41.1	.724 5204
8	241 16 45.4	+ 0 53 54.0	0.724 9408	12	30 25 07.6	− 2 27 03.6	0.724 2544
10	244 27 47.7	0 42 53.8	.725 1955	14	33 36 57.6	2 18 58.4	.723 9854
12	247 38 41.0	0 31 46.1	.725 4443	16	36 48 53.9	2 10 26.8	.723 7143
14	250 49 26.0	0 20 33.0	.725 6865	18	40 00 56.7	2 01 30.5	.723 4419
16	254 00 02.9	+ 0 09 16.6	.725 9212	20	43 13 06.0	1 52 11.0	.723 1690
18	257 10 32.4	− 0 02 01.1	0.726 1477	22	46 25 21.9	− 1 42 30.1	0.722 8966
20	260 20 55.0	0 13 18.1	.726 3655	24	49 37 44.5	1 32 29.5	.722 6253
22	263 31 11.1	0 24 32.2	.726 5737	26	52 50 13.9	1 22 11.0	.722 3562
24	266 41 21.4	0 35 41.4	.726 7718	28	56 02 50.1	1 11 36.7	.722 0901
26	269 51 26.4	0 46 43.7	.726 9592	30	59 15 33.4	1 00 48.4	.721 8277
28	273 01 26.7	− 0 57 37.1	0.727 1353	June 1	62 28 23.7	− 0 49 48.1	0.721 5699
Mar. 1	276 11 22.9	1 08 19.7	.727 2996	3	65 41 21.1	0 38 38.0	.721 3176
3	279 21 15.5	1 18 49.4	.727 4515	5	68 54 25.6	0 27 20.1	.721 0715
5	282 31 05.1	1 29 04.5	.727 5907	7	72 07 37.3	0 15 56.6	.720 8324
7	285 40 52.3	1 39 03.2	.727 7166	9	75 20 56.3	− 0 04 29.6	.720 6011
9	288 50 37.7	− 1 48 43.5	0.727 8290	11	78 34 22.4	+ 0 06 58.7	0.720 3783
11	292 00 21.7	1 58 03.8	.727 9275	13	81 47 55.8	0 18 26.1	.720 1647
13	295 10 05.0	2 07 02.4	.728 0118	15	85 01 36.3	0 29 50.4	.719 9611
15	298 19 48.0	2 15 37.7	.728 0816	17	88 15 23.9	0 41 09.3	.719 7680
17	301 29 31.3	2 23 48.2	.728 1367	19	91 29 18.4	0 52 20.8	.719 5861
19	304 39 15.3	− 2 31 32.4	0.728 1770	21	94 43 19.8	+ 1 03 22.6	0.719 4160
21	307 49 00.4	2 38 48.9	.728 2024	23	97 57 27.8	1 14 12.6	.719 2582
23	310 58 47.1	2 45 36.3	.728 2127	25	101 11 42.4	1 24 48.6	.719 1132
25	314 08 35.9	2 51 53.6	.728 2080	27	104 26 03.1	1 35 08.7	.718 9816
27	317 18 27.0	2 57 39.4	.728 1883	29	107 40 29.8	1 45 10.7	.718 8637
29	320 28 20.9	− 3 02 52.9	0.728 1535	July 1	110 55 02.2	+ 1 54 52.7	0.718 7599
31	323 38 17.8	− 3 07 32.9	0.728 1039	3	114 09 39.9	+ 2 04 12.8	0.718 6706

HELIOCENTRIC POSITIONS FOR 0^h DYNAMICAL TIME
MEAN EQUINOX AND ECLIPTIC OF DATE

Date	Longitude	Latitude	Radius Vector	Date	Longitude	Latitude	Radius Vector
	° ′ ″	° ′ ″			° ′ ″	° ′ ″	
July 1	110 55 02.2	+ 1 54 52.7	0.718 7599	Oct. 1	259 14 45.1	− 0 09 21.7	0.726 2914
3	114 09 39.9	2 04 12.8	.718 6706	3	262 25 03.4	0 20 37.0	.726 5029
5	117 24 22.5	2 13 09.2	.718 5960	5	265 35 15.6	0 31 48.1	.726 7045
7	120 39 09.5	2 21 40.0	.718 5364	7	268 45 22.4	0 42 53.1	.726 8955
9	123 54 00.6	2 29 43.7	.718 4920	9	271 55 24.2	0 53 49.8	.727 0755
11	127 08 55.2	+ 2 37 18.6	0.718 4629	11	275 05 21.8	− 1 04 36.4	0.727 2438
13	130 23 52.8	2 44 23.2	.718 4493	13	278 15 15.5	1 15 10.9	.727 3999
15	133 38 52.8	2 50 56.2	.718 4511	15	281 25 06.1	1 25 31.3	.727 5434
17	136 53 54.6	2 56 56.2	.718 4684	17	284 34 54.1	1 35 35.9	.727 6738
19	140 08 57.6	3 02 22.1	.718 5011	19	287 44 40.0	1 45 22.8	.727 7908
21	143 24 01.2	+ 3 07 12.8	0.718 5492	21	290 54 24.5	− 1 54 50.3	0.727 8939
23	146 39 04.6	3 11 27.4	.718 6124	23	294 04 08.0	2 03 56.7	.727 9830
25	149 54 07.2	3 15 05.0	.718 6905	25	297 13 51.1	2 12 40.4	.728 0576
27	153 09 08.3	3 18 05.1	.718 7834	27	300 23 34.2	2 20 59.7	.728 1176
29	156 24 07.2	3 20 26.9	.718 8907	29	303 33 17.9	2 28 53.3	.728 1628
31	159 39 03.2	+ 3 22 10.1	0.719 0120	31	306 43 02.6	− 2 36 19.6	0.728 1931
Aug. 2	162 53 55.5	3 23 14.4	.719 1470	Nov. 2	309 52 48.8	2 43 17.3	.728 2084
4	166 08 43.6	3 23 39.6	.719 2952	4	313 02 36.8	2 49 45.3	.728 2086
6	169 23 26.7	3 23 25.7	.719 4561	6	316 12 27.2	2 55 42.2	.728 1938
8	172 38 04.2	3 22 32.7	.719 6293	8	319 22 20.1	3 01 07.1	.728 1639
10	175 52 35.3	+ 3 21 01.0	0.719 8141	10	322 32 16.0	− 3 05 59.0	0.728 1191
12	179 06 59.5	3 18 50.8	.720 0101	12	325 42 15.2	3 10 16.8	.728 0595
14	182 21 16.3	3 16 02.7	.720 2164	14	328 52 17.9	3 14 00.0	.727 9853
16	185 35 25.0	3 12 37.3	.720 4326	16	332 02 24.5	3 17 07.6	.727 8968
18	188 49 25.1	3 08 35.2	.720 6578	18	335 12 35.2	3 19 39.2	.727 7941
20	192 03 16.3	+ 3 03 57.4	0.720 8914	20	338 22 50.1	− 3 21 34.3	0.727 6776
22	195 16 57.9	2 58 44.7	.721 1326	22	341 33 09.6	3 22 52.5	.727 5476
24	198 30 29.8	2 52 58.4	.721 3806	24	344 43 33.7	3 23 33.4	.727 4046
26	201 43 51.5	2 46 39.5	.721 6347	26	347 54 02.7	3 23 37.0	.727 2489
28	204 57 02.9	2 39 49.4	.721 8941	28	351 04 36.6	3 23 03.1	.727 0811
30	208 10 03.6	+ 2 32 29.3	0.722 1578	30	354 15 15.7	− 3 21 51.9	0.726 9017
Sept. 1	211 22 53.5	2 24 40.8	.722 4252	Dec. 2	357 26 00.0	3 20 03.4	.726 7111
3	214 35 32.5	2 16 25.5	.722 6953	4	0 36 49.7	3 17 38.0	.726 5100
5	217 48 00.6	2 07 44.8	.722 9673	6	3 47 44.8	3 14 36.1	.726 2990
7	221 00 17.8	1 58 40.6	.723 2403	8	6 58 45.5	3 10 58.0	.726 0787
9	224 12 24.2	+ 1 49 14.6	0.723 5134	10	10 09 51.8	− 3 06 44.6	0.725 8498
11	227 24 19.8	1 39 28.6	.723 7859	12	13 21 03.8	3 01 56.3	.725 6130
13	230 36 04.8	1 29 24.4	.724 0569	14	16 32 21.6	2 56 34.2	.725 3690
15	233 47 39.5	1 19 04.1	.724 3255	16	19 43 45.3	2 50 39.1	.725 1186
17	236 59 04.1	1 08 29.5	.724 5909	18	22 55 15.0	2 44 12.0	.724 8626
19	240 10 18.9	+ 0 57 42.6	0.724 8522	20	26 06 50.7	− 2 37 14.2	0.724 6017
21	243 21 24.3	0 46 45.5	.725 1087	22	29 18 32.6	2 29 46.7	.724 3367
23	246 32 20.7	0 35 40.2	.725 3595	24	32 30 20.7	2 21 51.0	.724 0685
25	249 43 08.5	0 24 28.8	.725 6040	26	35 42 15.2	2 13 28.4	.723 7978
27	252 53 48.2	0 13 13.3	.725 8412	28	38 54 16.0	2 04 40.5	.723 5256
29	256 04 20.2	+ 0 01 55.8	0.726 0706	30	42 06 23.3	− 1 55 28.9	0.723 2527
Oct. 1	259 14 45.1	− 0 09 21.7	0.726 2914	32	45 18 37.2	− 1 45 55.3	0.722 9798

HELIOCENTRIC POSITIONS FOR 0^h DYNAMICAL TIME
MEAN EQUINOX AND ECLIPTIC OF DATE

Date	Longitude	Latitude	Radius Vector	Date	Longitude	Latitude	Radius Vector
	° ′ ″	° ′ ″			° ′ ″	° ′ ″	
Jan. −1	252 14 01.7	− 0 42 55.1	1.495 0222	July 1	3 58 40.4	− 1 19 11.9	1.395 1821
3	254 24 30.5	0 46 46.3	.489 8318	5	6 27 45.4	1 15 45.2	.397 7412
7	256 35 54.0	0 50 35.1	.484 6707	9	8 56 16.1	1 12 10.8	.400 4955
11	258 48 12.7	0 54 20.9	.479 5467	13	11 24 10.2	1 08 29.2	.403 4392
15	261 01 26.5	0 58 03.5	.474 4676	17	13 51 25.6	1 04 41.0	.406 5661
19	263 15 35.6	− 1 01 42.3	1.469 4414	21	16 18 00.3	− 1 00 46.8	1.409 8696
23	265 30 40.0	1 05 16.8	.464 4761	25	18 43 52.5	0 56 47.1	.413 3429
27	267 46 39.5	1 08 46.8	.459 5799	29	21 09 00.4	0 52 42.6	.416 9789
31	270 03 33.7	1 12 11.6	.454 7609	Aug. 2	23 33 22.5	0 48 33.7	.420 7703
Feb. 4	272 21 22.5	1 15 30.8	.450 0273	6	25 56 57.3	0 44 21.1	.424 7097
8	274 40 05.2	− 1 18 44.0	1.445 3872	10	28 19 43.4	− 0 40 05.4	1.428 7893
12	276 59 41.4	1 21 50.6	.440 8490	14	30 41 39.7	0 35 46.9	.433 0013
16	279 20 10.2	1 24 50.2	.436 4207	18	33 02 45.2	0 31 26.4	.437 3378
20	281 41 30.7	1 27 42.4	.432 1105	22	35 22 58.8	0 27 04.4	.441 7907
24	284 03 42.1	1 30 26.6	.427 9262	26	37 42 19.8	0 22 41.3	.446 3519
28	286 26 43.2	− 1 33 02.4	1.423 8760	30	40 00 47.5	− 0 18 17.6	1.451 0133
Mar. 3	288 50 32.8	1 35 29.3	.419 9675	Sept. 3	42 18 21.4	0 13 53.9	.455 7666
7	291 15 09.4	1 37 46.9	.416 2085	7	44 35 00.9	0 09 30.6	.460 6037
11	293 40 31.6	1 39 54.8	.412 6063	11	46 50 45.9	0 05 08.1	.465 5164
15	296 06 37.6	1 41 52.6	.409 1683	15	49 05 36.0	− 0 00 47.0	.470 4965
19	298 33 25.8	− 1 43 39.8	1.405 9014	19	51 19 31.2	+ 0 03 32.5	1.475 5360
23	301 00 54.2	1 45 16.0	.402 8124	23	53 32 31.3	0 07 49.8	.480 6267
27	303 29 00.7	1 46 41.0	.399 9079	27	55 44 36.7	0 12 04.7	.485 7608
31	305 57 43.1	1 47 54.4	.397 1939	Oct. 1	57 55 47.3	0 16 16.8	.490 9303
Apr. 4	308 26 59.2	1 48 55.9	.394 6764	5	60 06 03.4	0 20 25.7	.496 1275
8	310 56 46.6	− 1 49 45.2	1.392 3607	9	62 15 25.4	+ 0 24 31.2	1.501 3449
12	313 27 02.7	1 50 22.1	.390 2520	13	64 23 53.8	0 28 32.9	.506 5747
16	315 57 44.9	1 50 46.5	.388 3548	17	66 31 28.9	0 32 30.5	.511 8098
20	318 28 50.5	1 50 58.1	.386 6736	21	68 38 11.3	0 36 23.9	.517 0428
24	321 00 16.7	1 50 56.8	.385 2119	25	70 44 01.7	0 40 12.7	.522 2666
28	323 32 00.6	− 1 50 42.6	1.383 9732	29	72 49 00.8	+ 0 43 56.7	1.527 4744
May 2	326 03 59.3	1 50 15.4	.382 9603	Nov. 2	74 53 09.1	0 47 35.8	.532 6594
6	328 36 09.8	1 49 35.2	.382 1753	6	76 56 27.6	0 51 09.8	.537 8149
10	331 08 28.9	1 48 42.1	.381 6203	10	78 58 57.1	0 54 38.4	.542 9347
14	333 40 53.8	1 47 36.2	.381 2964	14	81 00 38.4	0 58 01.5	.548 0123
18	336 13 21.2	− 1 46 17.5	1.381 2043	18	83 01 32.3	+ 1 01 18.9	1.553 0417
22	338 45 48.0	1 44 46.3	.381 3444	22	85 01 40.0	1 04 30.6	.558 0171
26	341 18 11.2	1 43 02.8	.381 7162	26	87 01 02.3	1 07 36.4	.562 9327
30	343 50 27.6	1 41 07.3	.382 3190	30	88 59 40.2	1 10 36.2	.567 7830
June 3	346 22 34.3	1 39 00.0	.383 1513	Dec. 4	90 57 34.8	1 13 29.9	.572 5626
7	348 54 28.1	− 1 36 41.3	1.384 2112	8	92 54 47.0	+ 1 16 17.5	1.577 2663
11	351 26 06.2	1 34 11.5	.385 4964	12	94 51 18.1	1 18 58.7	.581 8892
15	353 57 25.5	1 31 31.1	.387 0039	16	96 47 09.0	1 21 33.7	.586 4264
19	356 28 23.3	1 28 40.4	.388 7303	20	98 42 21.0	1 24 02.2	.590 8732
23	358 58 56.8	1 25 40.0	.390 6717	24	100 36 55.0	1 26 24.3	.595 2251
27	1 29 03.4	− 1 22 30.4	1.392 8239	28	102 30 52.3	+ 1 28 40.0	1.599 4780
July 1	3 58 40.4	− 1 19 11.9	1.395 1821	32	104 24 14.0	+ 1 30 49.2	1.603 6276

HELIOCENTRIC POSITIONS FOR 0^h DYNAMICAL TIME

MEAN EQUINOX AND ECLIPTIC OF DATE

Date	Longitude	Latitude	Radius Vector	Date	Longitude	Latitude	Radius Vector
	JUPITER				**SATURN**		
	° ′ ″	° ′ ″			° ′ ″	° ′ ″	
Jan. −1	154 57 42.0	+ 1 03 47.6	5.388 503	Jan. −1	308 17 12.5	− 0 37 54.0	9.936 598
9	155 44 08.0	1 04 24.0	.390 856	9	308 35 40.5	0 38 40.5	.934 852
19	156 30 31.5	1 04 59.7	.393 172	19	308 54 08.8	0 39 26.9	.933 094
29	157 16 52.7	1 05 34.6	.395 449	29	309 12 37.5	0 40 13.3	.931 324
Feb. 8	158 03 11.6	1 06 08.8	.397 688	Feb. 8	309 31 06.6	0 40 59.6	.929 541
18	158 49 28.2	+ 1 06 42.3	5.399 887	18	309 49 36.2	− 0 41 45.8	9.927 746
28	159 35 42.6	1 07 15.0	.402 048	28	310 08 06.2	0 42 32.1	.925 938
Mar. 9	160 21 54.8	1 07 46.9	.404 169	Mar. 9	310 26 36.6	0 43 18.2	.924 117
19	161 08 04.9	1 08 18.1	.406 250	19	310 45 07.4	0 44 04.3	.922 285
29	161 54 12.8	1 08 48.6	.408 290	29	311 03 38.7	0 44 50.3	.920 440
Apr. 8	162 40 18.7	+ 1 09 18.2	5.410 290	Apr. 8	311 22 10.4	− 0 45 36.3	9.918 582
18	163 26 22.6	1 09 47.1	.412 249	18	311 40 42.5	0 46 22.2	.916 712
28	164 12 24.6	1 10 15.2	.414 167	28	311 59 15.1	0 47 08.1	.914 830
May 8	164 58 24.6	1 10 42.6	.416 043	May 8	312 17 48.1	0 47 53.9	.912 936
18	165 44 22.7	1 11 09.2	.417 877	18	312 36 21.5	0 48 39.6	.911 030
28	166 30 19.0	+ 1 11 35.0	5.419 670	28	312 54 55.4	− 0 49 25.2	9.909 111
June 7	167 16 13.5	1 12 00.0	.421 419	June 7	313 13 29.7	0 50 10.8	.907 181
17	168 02 06.3	1 12 24.2	.423 126	17	313 32 04.5	0 50 56.3	.905 238
27	168 47 57.4	1 12 47.7	.424 790	27	313 50 39.8	0 51 41.8	.903 284
July 7	169 33 46.8	1 13 10.3	.426 410	July 7	314 09 15.5	0 52 27.1	.901 319
17	170 19 34.6	+ 1 13 32.2	5.427 987	17	314 27 51.6	− 0 53 12.4	9.899 341
27	171 05 20.9	1 13 53.2	.429 520	27	314 46 28.3	0 53 57.6	.897 352
Aug. 6	171 51 05.7	1 14 13.5	.431 009	Aug. 6	315 05 05.4	0 54 42.8	.895 351
16	172 36 48.9	1 14 33.0	.432 453	16	315 23 43.0	0 55 27.9	.893 338
26	173 22 30.8	1 14 51.7	.433 853	26	315 42 21.0	0 56 12.8	.891 314
Sept. 5	174 08 11.2	+ 1 15 09.5	5.435 207	Sept. 5	316 00 59.5	− 0 56 57.8	9.889 279
15	174 53 50.3	1 15 26.6	.436 517	15	316 19 38.6	0 57 42.6	.887 232
25	175 39 28.2	1 15 42.9	.437 781	25	316 38 18.1	0 58 27.3	.885 174
Oct. 5	176 25 04.8	1 15 58.3	.439 000	Oct. 5	316 56 58.0	0 59 12.0	.883 104
15	177 10 40.2	1 16 13.0	.440 174	15	317 15 38.5	0 59 56.6	.881 023
25	177 56 14.4	+ 1 16 26.8	5.441 302	25	317 34 19.5	− 1 00 41.0	9.878 930
Nov. 4	178 41 47.5	1 16 39.8	.442 383	Nov. 4	317 53 01.0	1 01 25.4	.876 826
14	179 27 19.6	1 16 52.0	.443 419	14	318 11 42.9	1 02 09.7	.874 711
24	180 12 50.6	1 17 03.4	.444 409	24	318 30 25.4	1 02 54.0	.872 585
Dec. 4	180 58 20.7	1 17 14.0	.445 352	Dec. 4	318 49 08.4	1 03 38.1	.870 447
14	181 43 49.9	+ 1 17 23.8	5.446 249	14	319 07 51.8	− 1 04 22.1	9.868 299
24	182 29 18.2	1 17 32.8	.447 099	24	319 26 35.8	1 05 06.1	.866 140
Jan. 3	183 14 45.7	1 17 40.9	.447 902	Jan. 3	319 45 20.3	1 05 49.9	.863 970
13	184 00 12.5	+ 1 17 48.2	5.448 659	13	320 04 05.3	− 1 06 33.6	9.861 789

Date	Longitude	Latitude	Radius Vector	Date	Longitude	Latitude	Radius Vector
	URANUS				**NEPTUNE**		
Jan. −21	283 37 39.8	− 0 22 54.7	19.505 43	Jan. −21	286 17 43.5	+ 0 45 35.0	30.195 23
Jan. 19	284 04 56.8	0 23 13.8	19.512 41	Jan. 19	286 32 02.0	0 45 11.1	30.194 45
Feb. 28	284 32 12.8	0 23 32.9	19.519 37	Feb. 28	286 46 20.7	0 44 47.1	30.193 67
Apr. 8	284 59 27.7	0 23 51.8	19.526 31	Apr. 8	287 00 39.5	0 44 23.1	30.192 90
May 18	285 26 41.7	0 24 10.6	19.533 23	May 18	287 14 58.4	0 43 59.0	30.192 14
June 27	285 53 54.6	− 0 24 29.3	19.540 13	June 27	287 29 17.5	+ 0 43 34.9	30.191 37
Aug. 6	286 21 06.5	0 24 47.9	19.547 01	Aug. 6	287 43 36.7	0 43 10.7	30.190 61
Sept. 15	286 48 17.5	0 25 06.4	19.553 87	Sept. 15	287 57 56.0	0 42 46.5	30.189 86
Oct. 25	287 15 27.4	0 25 24.8	19.560 71	Oct. 25	288 12 15.4	0 42 22.2	30.189 10
Dec. 4	287 42 36.4	0 25 43.1	19.567 52	Dec. 4	288 26 35.0	0 41 57.9	30.188 34
Dec. 44	288 09 44.3	− 0 26 01.2	19.574 31	Dec. 44	288 40 54.7	+ 0 41 33.6	30.187 59

GEOCENTRIC COORDINATES FOR 0^h DYNAMICAL TIME

Date	Apparent Right Ascension	Apparent Declination	True Geocentric Distance	Date	Apparent Right Ascension	Apparent Declination	True Geocentric Distance
	h m s	° ′ ″			h m s	° ′ ″	
Jan. 0	17 03 35.825	−21 13 00.66	1.078 9858	Feb. 15	22 02 01.802	−14 05 59.51	1.374 0895
1	17 08 35.222	21 28 01.70	.097 8410	16	22 08 58.749	13 24 37.42	.366 4894
2	17 13 44.825	21 42 39.69	.116 1012	17	22 15 55.219	12 41 53.86	.358 0088
3	17 19 03.663	21 56 48.06	.133 7601	18	22 22 51.014	11 57 50.98	.348 6083
4	17 24 30.876	22 10 20.99	.150 8152	19	22 29 45.893	11 12 31.42	.338 2475
5	17 30 05.703	−22 23 13.35	1.167 2673	20	22 36 39.563	−10 25 58.42	1.326 8856
6	17 35 47.468	22 35 20.55	.183 1191	21	22 43 31.671	9 38 15.91	.314 4821
7	17 41 35.570	22 46 38.51	.198 3755	22	22 50 21.790	8 49 28.63	.300 9984
8	17 47 29.472	22 57 03.61	.213 0426	23	22 57 09.408	7 59 42.23	.286 3984
9	17 53 28.694	23 06 32.59	.227 1274	24	23 03 53.914	7 09 03.41	.270 6510
10	17 59 32.806	−23 15 02.53	1.240 6376	25	23 10 34.589	− 6 17 39.99	1.253 7312
11	18 05 41.420	23 22 30.81	.253 5814	26	23 17 10.594	5 25 41.06	.235 6227
12	18 11 54.184	23 28 55.05	.265 9670	27	23 23 40.961	4 33 17.03	.216 3200
13	18 18 10.782	23 34 13.11	.277 8026	28	23 30 04.587	3 40 39.73	.195 8310
14	18 24 30.924	23 38 23.04	.289 0965	29	23 36 20.226	2 48 02.39	.174 1793
15	18 30 54.345	−23 41 23.05	1.299 8565	Mar. 1	23 42 26.496	− 1 55 39.66	1.151 4060
16	18 37 20.801	23 43 11.54	.310 0899	2	23 48 21.882	1 03 47.52	.127 5722
17	18 43 50.069	23 43 47.00	.319 8039	3	23 54 04.749	− 0 12 43.15	.102 7595
18	18 50 21.939	23 43 08.07	.329 0048	4	23 59 33.364	+ 0 37 15.28	.077 0708
19	18 56 56.219	23 41 13.46	.337 6982	5	0 04 45.921	1 25 48.82	.050 6296
20	19 03 32.728	−23 38 01.99	1.345 8892	6	0 09 40.571	+ 2 12 38.06	1.023 5782
21	19 10 11.299	23 33 32.55	.353 5818	7	0 14 15.463	2 57 23.43	0.996 0759
22	19 16 51.777	23 27 44.09	.360 7793	8	0 18 28.781	3 39 45.52	.968 2947
23	19 23 34.019	23 20 35.65	.367 4839	9	0 22 18.785	4 19 25.35	.940 4165
24	19 30 17.891	23 12 06.33	.373 6970	10	0 25 43.857	4 56 04.67	.912 6284
25	19 37 03.266	−23 02 15.30	1.379 4190	11	0 28 42.536	+ 5 29 26.23	0.885 1185
26	19 43 50.024	22 51 01.82	.384 6495	12	0 31 13.568	5 59 13.94	.858 0727
27	19 50 38.053	22 38 25.17	.389 3868	13	0 33 15.939	6 25 13.12	.831 6710
28	19 57 27.246	22 24 24.70	.393 6284	14	0 34 48.921	6 47 10.68	.806 0846
29	20 04 17.502	22 08 59.81	.397 3706	15	0 35 52.112	7 04 55.38	.781 4741
30	20 11 08.729	−21 52 09.93	1.400 6084	16	0 36 25.474	+ 7 18 18.10	0.757 9877
31	20 18 00.837	21 33 54.55	.403 3359	17	0 36 29.370	7 27 12.14	.735 7599
Feb. 1	20 24 53.745	21 14 13.20	.405 5454	18	0 36 04.602	7 31 33.64	.714 9107
2	20 31 47.376	20 53 05.44	.407 2283	19	0 35 12.430	7 31 21.99	.695 5451
3	20 38 41.659	20 30 30.90	.408 3742	20	0 33 54.588	7 26 40.25	.677 7521
4	20 45 36.530	−20 06 29.25	1.408 9712	21	0 32 13.279	+ 7 17 35.54	0.661 6046
5	20 52 31.924	19 41 00.20	.409 0058	22	0 30 11.155	7 04 19.40	.647 1587
6	20 59 27.786	19 14 03.53	.408 4628	23	0 27 51.272	6 47 07.90	.634 4538
7	21 06 24.057	18 45 39.09	.407 3250	24	0 25 17.024	6 26 21.62	.623 5120
8	21 13 20.685	18 15 46.82	.405 5735	25	0 22 32.061	6 02 25.39	.614 3383
9	21 20 17.614	−17 44 26.72	1.403 1871	26	0 19 40.184	+ 5 35 47.76	0.606 9209
10	21 27 14.786	17 11 38.94	.400 1427	27	0 16 45.233	5 07 00.13	.601 2321
11	21 34 12.141	16 37 23.71	.396 4152	28	0 13 50.970	4 36 35.86	.597 2292
12	21 41 09.608	16 01 41.46	.391 9769	29	0 11 00.977	4 05 09.08	.594 8560
13	21 48 07.107	15 24 32.75	.386 7982	30	0 08 18.555	3 33 13.62	.594 0450
14	21 55 04.542	−14 45 58.41	1.380 8471	31	0 05 46.656	+ 3 01 21.91	0.594 7191
15	22 02 01.802	−14 05 59.51	1.374 0895	Apr. 1	0 03 27.824	+ 2 30 04.10	0.596 7940

GEOCENTRIC COORDINATES FOR 0^h DYNAMICAL TIME

Date	Apparent Right Ascension	Apparent Declination	True Geocentric Distance	Date	Apparent Right Ascension	Apparent Declination	True Geocentric Distance
	h m s	° ′ ″			h m s	° ′ ″	
Apr. 1	0 03 27.824	+2 30 04.10	0.596 7940	May 17	2 33 26.688	+13 08 47.60	1.205 6118
2	0 01 24.175	1 59 47.36	.600 1808	18	2 40 33.479	13 51 10.41	.219 0917
3	23 59 37.387	1 30 55.42	.604 7876	19	2 47 50.467	14 33 39.67	.232 0918
4	23 58 08.720	1 03 48.33	.610 5222	20	2 55 17.869	15 16 08.28	.244 5426
5	23 56 59.037	0 38 42.45	.617 2934	21	3 02 55.862	15 58 28.46	.256 3690
6	23 56 08.854	+0 15 50.60	0.625 0127	22	3 10 44.570	+16 40 31.70	1.267 4912
7	23 55 38.378	−0 04 37.66	.633 5951	23	3 18 44.049	17 22 08.80	.277 8247
8	23 55 27.558	0 22 35.70	.642 9605	24	3 26 54.265	18 03 09.78	.287 2819
9	23 55 36.134	0 37 59.43	.653 0338	25	3 35 15.082	18 43 23.99	.295 7739
10	23 56 03.680	0 50 46.87	.663 7452	26	3 43 46.239	19 22 40.15	.303 2115
11	23 56 49.646	−1 00 57.82	0.675 0306	27	3 52 27.334	+20 00 46.45	1.309 5085
12	23 57 53.392	1 08 33.47	.686 8315	28	4 01 17.812	20 37 30.72	.314 5840
13	23 59 14.220	1 13 36.10	.699 0945	29	4 10 16.956	21 12 40.66	.318 3649
14	0 00 51.400	1 16 08.83	.711 7714	30	4 19 23.881	21 46 04.10	.320 7898
15	0 02 44.186	1 16 15.38	.724 8188	31	4 28 37.544	22 17 29.29	.321 8106
16	0 04 51.838	−1 13 59.93	0.738 1976	June 1	4 37 56.767	+22 46 45.15	1.321 3960
17	0 07 13.628	1 09 26.89	.751 8727	2	4 47 20.205	23 13 41.41	.319 5323
18	0 09 48.854	1 02 40.91	.765 8129	3	4 56 46.449	23 38 09.54	.316 2250
19	0 12 36.842	0 53 46.67	.779 9900	4	5 06 14.019	24 00 02.23	.311 4983
20	0 15 36.956	0 42 48.89	.794 3792	5	5 15 41.402	24 19 13.87	.305 3944
21	0 18 48.598	−0 29 52.24	0.808 9582	6	5 25 07.099	+24 35 40.67	1.297 9716
22	0 22 11.214	−0 15 01.31	.823 7070	7	5 34 29.654	24 49 20.61	.289 3019
23	0 25 44.294	+0 01 39.45	.838 6078	8	5 43 47.685	25 00 13.40	.279 4680
24	0 29 27.374	0 20 05.67	.853 6446	9	5 52 59.909	25 08 20.39	.268 5604
25	0 33 20.035	0 40 13.17	.868 8029	10	6 02 05.155	25 13 44.34	.256 6743
26	0 37 21.901	+1 01 57.89	0.884 0692	11	6 11 02.375	+25 16 29.29	1.243 9067
27	0 41 32.644	1 25 15.91	.899 4313	12	6 19 50.648	25 16 40.31	.230 3541
28	0 45 51.976	1 50 03.46	.914 8776	13	6 28 29.177	25 14 23.30	.216 1105
29	0 50 19.654	2 16 16.88	.930 3968	14	6 36 57.287	25 09 44.82	.201 2659
30	0 54 55.474	2 43 52.64	.945 9781	15	6 45 14.411	25 02 51.88	.185 9052
May 1	0 59 39.273	+3 12 47.29	0.961 6105	16	6 53 20.085	+24 53 51.82	1.170 1073
2	1 04 30.925	3 42 57.50	.977 2828	17	7 01 13.934	24 42 52.14	.153 9450
3	1 09 30.343	4 14 19.97	0.992 9835	18	7 08 55.660	24 30 00.43	.137 4846
4	1 14 37.473	4 46 51.47	1.008 7000	19	7 16 25.032	24 15 24.29	.120 7863
5	1 19 52.298	5 20 28.81	.024 4191	20	7 23 41.873	23 59 11.23	.103 9043
6	1 25 14.833	+5 55 08.77	1.040 1260	21	7 30 46.051	+23 41 28.68	1.086 8871
7	1 30 45.127	6 30 48.15	.055 8046	22	7 37 37.467	23 22 23.91	.069 7778
8	1 36 23.262	7 07 23.67	.071 4366	23	7 44 16.049	23 02 04.05	.052 6151
9	1 42 09.350	7 44 52.01	.087 0016	24	7 50 41.744	22 40 36.03	.035 4332
10	1 48 03.533	8 23 09.73	.102 4766	25	7 56 54.506	22 18 06.65	.018 2623
11	1 54 05.983	+9 02 13.26	1.117 8356	26	8 02 54.297	+21 54 42.53	1.001 1294
12	2 00 16.898	9 41 58.84	.133 0490	27	8 08 41.073	21 30 30.15	0.984 0586
13	2 06 36.497	10 22 22.50	.148 0836	28	8 14 14.787	21 05 35.86	.967 0715
14	2 13 05.020	11 03 19.99	.162 9018	29	8 19 35.376	20 40 05.88	.950 1874
15	2 19 42.718	11 44 46.72	.177 4616	30	8 24 42.762	20 14 06.37	.933 4242
16	2 26 29.854	+12 26 37.73	1.191 7157	July 1	8 29 36.849	+19 47 43.37	0.916 7982
17	2 33 26.688	+13 08 47.60	1.205 6118	2	8 34 17.517	+19 21 02.90	0.900 3250

GEOCENTRIC COORDINATES FOR 0h DYNAMICAL TIME

Date	Apparent Right Ascension	Apparent Declination	True Geocentric Distance
	h m s	° ′ ″	
July 1	8 29 36.849	+19 47 43.37	0.916 7982
2	8 34 17.517	19 21 02.90	.900 3250
3	8 38 44.623	18 54 10.92	.884 0195
4	8 42 58.002	18 27 13.38	.867 8963
5	8 46 57.461	18 00 16.25	.851 9702
6	8 50 42.783	+17 33 25.53	0.836 2560
7	8 54 13.723	17 06 47.27	.820 7693
8	8 57 30.007	16 40 27.62	.805 5264
9	9 00 31.335	16 14 32.82	.790 5447
10	9 03 17.382	15 49 09.22	.775 8427
11	9 05 47.796	+15 24 23.33	0.761 4407
12	9 08 02.211	15 00 21.78	.747 3608
13	9 10 00.241	14 37 11.36	.733 6270
14	9 11 41.496	14 14 59.00	.720 2658
15	9 13 05.587	13 53 51.79	.707 3062
16	9 14 12.135	+13 33 56.90	0.694 7802
17	9 15 00.789	13 15 21.60	.682 7225
18	9 15 31.238	12 58 13.21	.671 1715
19	9 15 43.234	12 42 38.98	.660 1687
20	9 15 36.609	12 28 46.06	.649 7592
21	9 15 11.305	+12 16 41.33	0.639 9916
22	9 14 27.399	12 06 31.31	.630 9182
23	9 13 25.133	11 58 21.94	.622 5945
24	9 12 04.948	11 52 18.46	.615 0790
25	9 10 27.510	11 48 25.12	.608 4333
26	9 08 33.744	+11 46 45.03	0.602 7205
27	9 06 24.855	11 47 19.89	.598 0058
28	9 04 02.349	11 50 09.83	.594 3543
29	9 01 28.042	11 55 13.15	.591 8313
30	8 58 44.063	12 02 26.23	.590 5001
31	8 55 52.836	+12 11 43.44	0.590 4215
Aug. 1	8 52 57.059	12 22 57.10	.591 6522
2	8 49 59.661	12 35 57.58	.594 2438
3	8 47 03.744	12 50 33.43	.598 2411
4	8 44 12.519	13 06 31.66	.603 6816
5	8 41 29.234	+13 23 37.95	0.610 5941
6	8 38 57.095	13 41 37.07	.618 9982
7	8 36 39.194	14 00 13.13	.628 9036
8	8 34 38.443	14 19 09.95	.640 3102
9	8 32 57.517	14 38 11.36	.653 2076
10	8 31 38.811	+14 57 01.37	0.667 5755
11	8 30 44.406	15 15 24.45	.683 3835
12	8 30 16.056	15 33 05.56	.700 5914
13	8 30 15.178	15 49 50.26	.719 1495
14	8 30 42.859	16 05 24.76	.738 9980
15	8 31 39.861	+16 19 35.85	0.760 0674
16	8 33 06.636	+16 32 10.95	0.782 2782

Date	Apparent Right Ascension	Apparent Declination	True Geocentric Distance
	h m s	° ′ ″	
Aug. 16	8 33 06.636	+16 32 10.95	0.782 2782
17	8 35 03.345	16 42 58.06	.805 5408
18	8 37 29.867	16 51 45.78	.829 7545
19	8 40 25.820	16 58 23.32	.854 8083
20	8 43 50.571	17 02 40.59	.880 5798
21	8 47 43.243	+17 04 28.28	0.906 9361
22	8 52 02.733	17 03 38.00	.933 7340
23	8 56 47.716	17 00 02.48	.960 8212
24	9 01 56.663	16 53 35.71	0.988 0378
25	9 07 27.855	16 44 13.21	1.015 2189
26	9 13 19.409	+16 31 52.14	1.042 1972
27	9 19 29.306	16 16 31.54	.068 8070
28	9 25 55.430	15 58 12.34	.094 8875
29	9 32 35.605	15 36 57.49	.120 2870
30	9 39 27.647	15 12 51.82	.144 8664
31	9 46 29.400	+14 46 01.99	1.168 5026
Sept. 1	9 53 38.790	14 16 36.23	.191 0908
2	10 00 53.853	13 44 44.12	.212 5459
3	10 08 12.777	13 10 36.24	.232 8032
4	10 15 33.920	12 34 23.87	.251 8179
5	10 22 55.827	+11 56 18.68	1.269 5640
6	10 30 17.238	11 16 32.44	.286 0325
7	10 37 37.085	10 35 16.76	.301 2292
8	10 44 54.482	9 52 42.90	.315 1723
9	10 52 08.718	9 09 01.61	.327 8899
10	10 59 19.237	+ 8 24 23.01	1.339 4176
11	11 06 25.621	7 38 56.54	.349 7968
12	11 13 27.575	6 52 50.93	.359 0721
13	11 20 24.906	6 06 14.18	.367 2904
14	11 27 17.511	5 19 13.56	.374 4993
15	11 34 05.358	+ 4 31 55.69	1.380 7460
16	11 40 48.473	3 44 26.49	.386 0767
17	11 47 26.927	2 56 51.41	.390 5357
18	11 54 00.832	2 09 15.30	.394 1654
19	12 00 30.329	1 21 42.54	.397 0055
20	12 06 55.578	+ 0 34 17.05	1.399 0935
21	12 13 16.754	− 0 12 57.63	.400 4637
22	12 19 34.041	0 59 58.28	.401 1482
23	12 25 47.632	1 46 42.02	.401 1762
24	12 31 57.719	2 33 06.22	.400 5744
25	12 38 04.500	− 3 19 08.47	1.399 3670
26	12 44 08.166	4 04 46.61	.397 5760
27	12 50 08.909	4 49 58.64	.395 2212
28	12 56 06.912	5 34 42.70	.392 3205
29	13 02 02.354	6 18 57.09	.388 8897
30	13 07 55.402	− 7 02 40.20	1.384 9430
Oct. 1	13 13 46.216	− 7 45 50.51	1.380 4926

GEOCENTRIC COORDINATES FOR 0^h DYNAMICAL TIME

Date	Apparent Right Ascension	Apparent Declination	True Geocentric Distance	Date	Apparent Right Ascension	Apparent Declination	True Geocentric Distance
	h m s	° ′ ″			h m s	° ′ ″	
Oct. 1	13 13 46.216	− 7 45 50.51	1.380 4926	Nov. 16	16 18 22.689	−22 40 47.28	0.718 6629
2	13 19 34.946	8 28 26.58	.375 5495	17	16 15 00.179	22 14 59.36	.705 5274
3	13 25 21.733	9 10 27.04	.370 1229	18	16 10 59.904	21 45 56.88	.694 5804
4	13 31 06.708	9 51 50.57	.364 2208	19	16 06 27.032	21 13 58.25	.686 1178
5	13 36 49.989	10 32 35.86	.357 8495	20	16 01 28.712	20 39 33.76	.680 4049
6	13 42 31.684	−11 12 41.67	1.351 0145	21	15 56 13.854	−20 03 26.13	0.677 6565
7	13 48 11.889	11 52 06.75	.343 7200	22	15 50 52.637	19 26 29.39	.678 0192
8	13 53 50.684	12 30 49.86	.335 9689	23	15 45 35.827	18 49 45.68	.681 5575
9	13 59 28.135	13 08 49.76	.327 7634	24	15 40 33.978	18 14 20.57	.688 2460
10	14 05 04.295	13 46 05.21	.319 1047	25	15 35 56.663	17 41 17.40	.697 9705
11	14 10 39.196	−14 22 34.94	1.309 9930	26	15 31 51.867	−17 11 32.01	0.710 5366
12	14 16 12.855	14 58 17.65	.300 4280	27	15 28 25.611	16 45 48.63	.725 6848
13	14 21 45.265	15 33 12.04	.290 4085	28	15 25 41.842	16 24 37.65	.743 1090
14	14 27 16.399	16 07 16.73	.279 9327	29	15 23 42.544	16 08 15.28	.762 4766
15	14 32 46.205	16 40 30.31	.268 9981	30	15 22 28.006	15 56 44.72	.783 4468
16	14 38 14.603	−17 12 51.32	1.257 6019	Dec. 1	15 21 57.174	−15 49 58.41	0.805 6860
17	14 43 41.481	17 44 18.23	.245 7408	2	15 22 08.018	15 47 40.61	.828 8796
18	14 49 06.698	18 14 49.43	.233 4112	3	15 22 57.866	15 49 29.97	.852 7397
19	14 54 30.072	18 44 23.24	.220 6091	4	15 24 23.687	15 55 01.86	.877 0094
20	14 59 51.381	19 12 57.89	.207 3308	5	15 26 22.315	16 03 50.18	.901 4650
21	15 05 10.361	−19 40 31.52	1.193 5723	6	15 28 50.608	−16 15 28.77	0.925 9150
22	15 10 26.694	20 07 02.17	.179 3300	7	15 31 45.556	16 29 32.32	.950 1987
23	15 15 40.008	20 32 27.78	.164 6008	8	15 35 04.346	16 45 37.04	.974 1836
24	15 20 49.867	20 56 46.16	.149 3822	9	15 38 44.403	17 03 20.95	0.997 7620
25	15 25 55.762	21 19 55.01	.133 6729	10	15 42 43.401	17 22 24.08	1.020 8482
26	15 30 57.106	−21 41 51.88	1.117 4727	11	15 46 59.257	−17 42 28.42	1.043 3751
27	15 35 53.221	22 02 34.17	.100 7834	12	15 51 30.128	18 03 17.92	.065 2918
28	15 40 43.326	22 21 59.08	.083 6089	13	15 56 14.387	18 24 38.28	.086 5600
29	15 45 26.530	22 40 03.60	.065 9558	14	16 01 10.607	18 46 16.88	.107 1530
30	15 50 01.821	22 56 44.52	.047 8341	15	16 06 17.540	19 08 02.57	.127 0525
31	15 54 28.046	−23 11 58.32	1.029 2580	16	16 11 34.096	−19 29 45.51	1.146 2478
Nov. 1	15 58 43.902	23 25 41.22	1.010 2464	17	16 16 59.326	19 51 17.07	.164 7338
2	16 02 47.922	23 37 49.07	0.990 8247	18	16 22 32.402	20 12 29.62	.182 5102
3	16 06 38.456	23 48 17.31	.971 0250	19	16 28 12.599	20 33 16.44	.199 5801
4	16 10 13.658	23 57 00.93	.950 8886	20	16 33 59.286	20 53 31.58	.215 9495
5	16 13 31.479	−24 03 54.38	0.930 4669	21	16 39 51.907	−21 13 09.79	1.231 6265
6	16 16 29.655	24 08 51.48	.909 8238	22	16 45 49.975	21 32 06.37	.246 6206
7	16 19 05.711	24 11 45.37	.889 0374	23	16 51 53.060	21 50 17.17	.260 9426
8	16 21 16.976	24 12 28.39	.868 2030	24	16 58 00.781	22 07 38.42	.274 6037
9	16 23 00.612	24 10 52.08	.847 4349	25	17 04 12.802	22 24 06.77	.287 6156
10	16 24 13.669	−24 06 47.11	0.826 8698	26	17 10 28.823	−22 39 39.18	1.299 9901
11	16 24 53.177	24 00 03.41	.806 6687	27	17 16 48.575	22 54 12.89	.311 7390
12	16 24 56.271	23 50 30.36	.787 0194	28	17 23 11.818	23 07 45.38	.322 8736
13	16 24 20.377	23 37 57.26	.768 1375	29	17 29 38.333	23 20 14.35	.333 4049
14	16 23 03.446	23 22 14.02	.750 2670	30	17 36 07.922	23 31 37.69	.343 3435
15	16 21 04.245	−23 03 12.33	0.733 6781	31	17 42 40.403	−23 41 53.45	1.352 6993
16	16 18 22.689	−22 40 47.28	0.718 6629	32	17 49 15.607	−23 50 59.82	1.361 4815

GEOCENTRIC COORDINATES FOR 0^h DYNAMICAL TIME

Date	Apparent Right Ascension	Apparent Declination	True Geocentric Distance	Date	Apparent Right Ascension	Apparent Declination	True Geocentric Distance
	h m s	° ′ ″			h m s	° ′ ″	
Jan. 0	15 50 01.657	−17 53 14.00	1.111 6497	Feb. 15	19 49 32.689	−20 55 23.75	1.384 8630
1	15 54 57.399	18 09 56.74	.118 2285	16	19 54 45.915	20 44 29.70	.390 0995
2	15 59 54.364	18 26 13.35	.124 7797	17	19 59 58.398	20 32 58.95	.395 3063
3	16 04 52.536	18 42 03.09	.131 3033	18	20 05 10.106	20 20 51.86	.400 4831
4	16 09 51.895	18 57 25.25	.137 7990	19	20 10 21.007	20 08 08.86	.405 6297
5	16 14 52.421	−19 12 19.08	1.144 2670	20	20 15 31.075	−19 54 50.37	1.410 7460
6	16 19 54.092	19 26 43.90	.150 7070	21	20 20 40.284	19 40 56.84	.415 8313
7	16 24 56.885	19 40 39.01	.157 1192	22	20 25 48.610	19 26 28.77	.420 8854
8	16 30 00.774	19 54 03.73	.163 5035	23	20 30 56.029	19 11 26.67	.425 9078
9	16 35 05.733	20 06 57.42	.169 8599	24	20 36 02.519	18 55 51.09	.430 8981
10	16 40 11.737	−20 19 19.43	1.176 1886	25	20 41 08.058	−18 39 42.58	1.435 8559
11	16 45 18.755	20 31 09.16	.182 4896	26	20 46 12.624	18 23 01.72	.440 7808
12	16 50 26.758	20 42 25.99	.188 7630	27	20 51 16.200	18 05 49.11	.445 6726
13	16 55 35.716	20 53 09.36	.195 0090	28	20 56 18.769	17 48 05.35	.450 5310
14	17 00 45.595	21 03 18.72	.201 2275	29	21 01 20.317	17 29 51.06	.455 3557
15	17 05 56.361	−21 12 53.54	1.207 4188	Mar. 1	21 06 20.831	−17 11 06.88	1.460 1466
16	17 11 07.980	21 21 53.31	.213 5829	2	21 11 20.304	16 51 53.43	.464 9033
17	17 16 20.412	21 30 17.55	.219 7199	3	21 16 18.727	16 32 11.37	.469 6259
18	17 21 33.618	21 38 05.80	.225 8297	4	21 21 16.097	16 12 01.35	.474 3141
19	17 26 47.556	21 45 17.60	.231 9123	5	21 26 12.411	15 51 24.05	.478 9678
20	17 32 02.182	−21 51 52.52	1.237 9676	6	21 31 07.672	−15 30 20.13	1.483 5871
21	17 37 17.450	21 57 50.14	.243 9951	7	21 36 01.881	15 08 50.28	.488 1718
22	17 42 33.315	22 03 10.06	.249 9945	8	21 40 55.045	14 46 55.17	.492 7220
23	17 47 49.731	22 07 51.93	.255 9654	9	21 45 47.172	14 24 35.50	.497 2377
24	17 53 06.647	22 11 55.42	.261 9073	10	21 50 38.269	14 01 51.96	.501 7190
25	17 58 24.016	−22 15 20.24	1.267 8197	11	21 55 28.349	−13 38 45.27	1.506 1661
26	18 03 41.782	22 18 06.13	.273 7023	12	22 00 17.424	13 15 16.11	.510 5790
27	18 08 59.891	22 20 12.89	.279 5547	13	22 05 05.508	12 51 25.19	.514 9579
28	18 14 18.285	22 21 40.33	.285 3766	14	22 09 52.617	12 27 13.22	.519 3028
29	18 19 36.907	22 22 28.29	.291 1678	15	22 14 38.769	12 02 40.89	.523 6139
30	18 24 55.697	−22 22 36.65	1.296 9281	16	22 19 23.985	−11 37 48.90	1.527 8912
31	18 30 14.595	22 22 05.32	.302 6572	17	22 24 08.288	11 12 37.93	.532 1346
Feb. 1	18 35 33.542	22 20 54.24	.308 3552	18	22 28 51.704	10 47 08.66	.536 3441
2	18 40 52.479	22 19 03.38	.314 0219	19	22 33 34.262	10 21 21.76	.540 5192
3	18 46 11.347	22 16 32.73	.319 6572	20	22 38 15.993	9 55 17.93	.544 6598
4	18 51 30.087	−22 13 22.34	1.325 2612	21	22 42 56.926	− 9 28 57.85	1.548 7653
5	18 56 48.644	22 09 32.25	.330 8339	22	22 47 37.092	9 02 22.24	.552 8353
6	19 02 06.961	22 05 02.57	.336 3754	23	22 52 16.519	8 35 31.79	.556 8694
7	19 07 24.986	21 59 53.41	.341 8856	24	22 56 55.237	8 08 27.24	.560 8669
8	19 12 42.664	21 54 04.94	.347 3648	25	23 01 33.277	7 41 09.29	.564 8274
9	19 17 59.947	−21 47 37.32	1.352 8131	26	23 06 10.667	− 7 13 38.68	1.568 7504
10	19 23 16.784	21 40 30.79	.358 2306	27	23 10 47.438	6 45 56.13	.572 6355
11	19 28 33.129	21 32 45.58	.363 6175	28	23 15 23.623	6 18 02.35	.576 4822
12	19 33 48.935	21 24 21.96	.368 9741	29	23 19 59.253	5 49 58.06	.580 2901
13	19 39 04.159	21 15 20.23	.374 3004	30	23 24 34.362	5 21 43.99	.584 0589
14	19 44 18.757	−21 05 40.71	1.379 5966	31	23 29 08.984	− 4 53 20.85	1.587 7881
15	19 49 32.689	−20 55 23.75	1.384 8630	Apr. 1	23 33 43.154	− 4 24 49.35	1.591 4774

GEOCENTRIC COORDINATES FOR 0^h DYNAMICAL TIME

Date	Apparent Right Ascension	Apparent Declination	True Geocentric Distance	Date	Apparent Right Ascension	Apparent Declination	True Geocentric Distance
	h m s	° ′ ″			h m s	° ′ ″	
Apr. 1	23 33 43.154	− 4 24 49.35	1.591 4774	May 17	3 06 40.995	+16 34 58.58	1.714 0531
2	23 38 16.908	3 56 10.20	.595 1264	18	3 11 35.370	16 56 45.85	.715 5801
3	23 42 50.282	3 27 24.13	.598 7350	19	3 16 30.854	17 18 07.32	.717 0548
4	23 47 23.312	2 58 31.83	.602 3027	20	3 21 27.454	17 39 02.26	.718 4768
5	23 51 56.036	2 29 34.01	.605 8295	21	3 26 25.175	17 59 29.98	.719 8455
6	23 56 28.490	− 2 00 31.39	1.609 3152	22	3 31 24.023	+18 19 29.74	1.721 1606
7	0 01 00.713	1 31 24.67	.612 7596	23	3 36 23.998	18 39 00.86	.722 4216
8	0 05 32.740	1 02 14.55	.616 1628	24	3 41 25.100	18 58 02.62	.723 6280
9	0 10 04.608	0 33 01.76	.619 5246	25	3 46 27.327	19 16 34.34	.724 7793
10	0 14 36.356	− 0 03 46.98	.622 8452	26	3 51 30.674	19 34 35.33	.725 8751
11	0 19 08.020	+ 0 25 29.07	1.626 1244	27	3 56 35.133	+19 52 04.92	1.726 9149
12	0 23 39.640	0 54 45.70	.629 3625	28	4 01 40.695	20 09 02.46	.727 8982
13	0 28 11.256	1 24 02.22	.632 5592	29	4 06 47.347	20 25 27.28	.728 8246
14	0 32 42.910	1 53 17.95	.635 7147	30	4 11 55.074	20 41 18.78	.729 6937
15	0 37 14.646	2 22 32.22	.638 8287	31	4 17 03.855	20 56 36.32	.730 5050
16	0 41 46.509	+ 2 51 44.36	1.641 9010	June 1	4 22 13.669	+21 11 19.30	1.731 2582
17	0 46 18.542	3 20 53.69	.644 9313	2	4 27 24.489	21 25 27.14	.731 9531
18	0 50 50.788	3 49 59.53	.647 9193	3	4 32 36.284	21 38 59.26	.732 5894
19	0 55 23.290	4 19 01.19	.650 8646	4	4 37 49.021	21 51 55.09	.733 1671
20	0 59 56.087	4 47 57.98	.653 7665	5	4 43 02.666	22 04 14.09	.733 6862
21	1 04 29.220	+ 5 16 49.20	1.656 6246	6	4 48 17.183	+22 15 55.73	1.734 1468
22	1 09 02.727	5 45 34.13	.659 4383	7	4 53 32.537	22 26 59.49	.734 5490
23	1 13 36.646	6 14 12.06	.662 2071	8	4 58 48.690	22 37 24.91	.734 8930
24	1 18 11.014	6 42 42.27	.664 9304	9	5 04 05.605	22 47 11.51	.735 1790
25	1 22 45.869	7 11 04.04	.667 6076	10	5 09 23.242	22 56 18.89	.735 4071
26	1 27 21.247	+ 7 39 16.65	1.670 2383	11	5 14 41.560	+23 04 46.64	1.735 5775
27	1 31 57.185	8 07 19.38	.672 8218	12	5 20 00.511	23 12 34.43	.735 6903
28	1 36 33.718	8 35 11.50	.675 3578	13	5 25 20.044	23 19 42.18	.735 7455
29	1 41 10.881	9 02 52.29	.677 8455	14	5 30 40.228	23 26 09.36	.735 7432
30	1 45 48.709	9 30 21.03	.680 2847	15	5 36 00.824	23 31 54.35	.735 6833
May 1	1 50 27.235	+ 9 57 36.98	1.682 6747	16	5 41 21.856	+23 36 58.62	1.735 5658
2	1 55 06.492	10 24 39.43	.685 0152	17	5 46 43.285	23 41 21.43	.735 3905
3	1 59 46.510	10 51 27.66	.687 3059	18	5 52 05.054	23 45 02.49	.735 1573
4	2 04 27.319	11 18 00.92	.689 5463	19	5 57 27.107	23 48 01.58	.734 8660
5	2 09 08.946	11 44 18.48	.691 7362	20	6 02 49.389	23 50 18.52	.734 5165
6	2 13 51.417	+12 10 19.62	1.693 8755	21	6 08 11.842	+23 51 53.17	1.734 1086
7	2 18 34.756	12 36 03.60	.695 9641	22	6 13 34.410	23 52 45.40	.733 6419
8	2 23 18.986	13 01 29.66	.698 0018	23	6 18 57.036	23 52 55.14	.733 1164
9	2 28 04.130	13 26 37.09	.699 9888	24	6 24 19.663	23 52 22.34	.732 5318
10	2 32 50.213	13 51 25.13	.701 9250	25	6 29 42.234	23 51 06.97	.731 8879
11	2 37 37.258	+14 15 53.08	1.703 8104	26	6 35 04.689	+23 49 09.07	1.731 1843
12	2 42 25.289	14 40 00.20	.705 6451	27	6 40 26.972	23 46 28.69	.730 4210
13	2 47 14.330	15 03 45.78	.707 4289	28	6 45 49.023	23 43 05.92	.729 5976
14	2 52 04.401	15 27 09.12	.709 1618	29	6 51 10.783	23 39 00.88	.728 7139
15	2 56 55.524	15 50 09.50	.710 8436	30	6 56 32.190	23 34 13.74	.727 7697
16	3 01 47.717	+16 12 46.23	1.712 4742	July 1	7 01 53.185	+23 28 44.67	1.726 7651
17	3 06 40.995	+16 34 58.58	1.714 0531	2	7 07 13.705	+23 22 33.90	1.725 6999

GEOCENTRIC COORDINATES FOR 0^h DYNAMICAL TIME

Date	Apparent Right Ascension	Apparent Declination	True Geocentric Distance
	h m s	° ′ ″	
July 1	7 01 53.185	+23 28 44.67	1.726 7651
2	7 07 13.705	23 22 33.90	.725 6999
3	7 12 33.694	23 15 41.65	.724 5743
4	7 17 53.096	23 08 08.19	.723 3887
5	7 23 11.859	22 59 53.81	.722 1432
6	7 28 29.934	+22 50 58.82	1.720 8383
7	7 33 47.275	22 41 23.57	.719 4744
8	7 39 03.840	22 31 08.44	.718 0520
9	7 44 19.585	22 20 13.83	.716 5714
10	7 49 34.470	22 08 40.16	.715 0332
11	7 54 48.458	+21 56 27.89	1.713 4378
12	8 00 01.512	21 43 37.48	.711 7853
13	8 05 13.598	21 30 09.40	.710 0763
14	8 10 24.684	21 16 04.17	.708 3110
15	8 15 34.743	21 01 22.30	.706 4896
16	8 20 43.748	+20 46 04.32	1.704 6125
17	8 25 51.676	20 30 10.78	.702 6797
18	8 30 58.506	20 13 42.24	.700 6914
19	8 36 04.221	19 56 39.28	.698 6479
20	8 41 08.806	19 39 02.50	.696 5493
21	8 46 12.249	+19 20 52.50	1.694 3956
22	8 51 14.539	19 02 09.90	.692 1869
23	8 56 15.670	18 42 55.35	.689 9233
24	9 01 15.635	18 23 09.49	.687 6049
25	9 06 14.431	18 02 52.98	.685 2316
26	9 11 12.054	+17 42 06.52	1.682 8035
27	9 16 08.504	17 20 50.79	.680 3205
28	9 21 03.780	16 59 06.50	.677 7826
29	9 25 57.881	16 36 54.38	.675 1898
30	9 30 50.809	16 14 15.13	.672 5423
31	9 35 42.568	+15 51 09.48	1.669 8401
Aug. 1	9 40 33.167	15 27 38.16	.667 0837
2	9 45 22.615	15 03 41.88	.664 2734
3	9 50 10.927	14 39 21.37	.661 4098
4	9 54 58.119	14 14 37.36	.658 4934
5	9 59 44.208	+13 49 30.57	1.655 5248
6	10 04 29.214	13 24 01.74	.652 5047
7	10 09 13.156	12 58 11.61	.649 4337
8	10 13 56.056	12 32 00.91	.646 3125
9	10 18 37.938	12 05 30.35	.643 1416
10	10 23 18.828	+11 38 40.68	1.639 9215
11	10 27 58.752	11 11 32.62	.636 6530
12	10 32 37.738	10 44 06.88	.633 3364
13	10 37 15.817	10 16 24.19	.629 9722
14	10 41 53.022	9 48 25.25	.626 5610
15	10 46 29.385	+ 9 20 10.79	1.623 1032
16	10 51 04.941	+ 8 51 41.50	1.619 5991
Aug. 16	10 51 04.941	+ 8 51 41.50	1.619 5991
17	10 55 39.727	8 22 58.10	.616 0492
18	11 00 13.779	7 54 01.30	.612 4538
19	11 04 47.136	7 24 51.80	.608 8131
20	11 09 19.837	6 55 30.31	.605 1275
21	11 13 51.920	+ 6 25 57.54	1.601 3972
22	11 18 23.425	5 56 14.21	.597 6223
23	11 22 54.389	5 26 21.03	.593 8030
24	11 27 24.853	4 56 18.74	.589 9395
25	11 31 54.853	4 26 08.06	.586 0317
26	11 36 24.428	+ 3 55 49.72	1.582 0797
27	11 40 53.614	3 25 24.46	.578 0836
28	11 45 22.451	2 54 53.02	.574 0435
29	11 49 50.977	2 24 16.11	.569 9596
30	11 54 19.234	1 53 34.47	.565 8322
31	11 58 47.261	+ 1 22 48.81	1.561 6616
Sept. 1	12 03 15.101	0 51 59.86	.557 4485
2	12 07 42.792	+ 0 21 08.33	.553 1934
3	12 12 10.376	− 0 09 45.05	.548 8969
4	12 16 37.891	0 40 39.56	.544 5596
5	12 21 05.380	− 1 11 34.49	1.540 1822
6	12 25 32.882	1 42 29.10	.535 7652
7	12 30 00.439	2 13 22.70	.531 3094
8	12 34 28.094	2 44 14.57	.526 8153
9	12 38 55.888	3 15 03.98	.522 2835
10	12 43 23.865	− 3 45 50.25	1.517 7145
11	12 47 52.069	4 16 32.65	.513 1088
12	12 52 20.543	4 47 10.48	.508 4671
13	12 56 49.331	5 17 43.03	.503 7898
14	13 01 18.477	5 48 09.59	.499 0773
15	13 05 48.027	− 6 18 29.47	1.494 3300
16	13 10 18.022	6 48 41.93	.489 5483
17	13 14 48.506	7 18 46.28	.484 7326
18	13 19 19.522	7 48 41.78	.479 8831
19	13 23 51.110	8 18 27.72	.475 0000
20	13 28 23.311	− 8 48 03.35	1.470 0835
21	13 32 56.162	9 17 27.93	.465 1336
22	13 37 29.701	9 46 40.71	.460 1505
23	13 42 03.964	10 15 40.93	.455 1341
24	13 46 38.988	10 44 27.83	.450 0844
25	13 51 14.805	−11 13 00.63	1.445 0014
26	13 55 51.451	11 41 18.58	.439 8851
27	14 00 28.958	12 09 20.90	.434 7354
28	14 05 07.357	12 37 06.83	.429 5527
29	14 09 46.677	13 04 35.61	.424 3370
30	14 14 26.943	−13 31 46.46	1.419 0889
Oct. 1	14 19 08.180	−13 58 38.62	1.413 8086

GEOCENTRIC COORDINATES FOR 0^h DYNAMICAL TIME

Date	Apparent Right Ascension	Apparent Declination	True Geocentric Distance	Date	Apparent Right Ascension	Apparent Declination	True Geocentric Distance
	h m s	° ′ ″			h m s	° ′ ″	
Oct. 1	14 19 08.180	−13 58 38.62	1.413 8086	Nov.16	18 13 07.314	−25 29 59.21	1.141 8266
2	14 23 50.410	14 25 11.30	.408 4966	17	18 18 26.252	25 29 34.98	.135 3598
3	14 28 33.656	14 51 23.73	.403 1534	18	18 23 44.895	25 28 26.42	.128 8730
4	14 33 17.939	15 17 15.13	.397 7795	19	18 29 03.174	25 26 33.56	.122 3658
5	14 38 03.278	15 42 44.73	.392 3753	20	18 34 21.020	25 23 56.51	.115 8383
6	14 42 49.694	−16 07 51.75	1.386 9415	21	18 39 38.364	−25 20 35.40	1.109 2902
7	14 47 37.205	16 32 35.42	.381 4784	22	18 44 55.138	25 16 30.40	.102 7213
8	14 52 25.827	16 56 54.99	.375 9865	23	18 50 11.269	25 11 41.73	.096 1315
9	14 57 15.576	17 20 49.68	.370 4664	24	18 55 26.686	25 06 09.64	.089 5206
10	15 02 06.466	17 44 18.74	.364 9184	25	19 00 41.317	24 59 54.42	.082 8885
11	15 06 58.510	−18 07 21.41	1.359 3431	26	19 05 55.091	−24 52 56.40	1.076 2352
12	15 11 51.719	18 29 56.95	.353 7407	27	19 11 07.935	24 45 15.93	.069 5608
13	15 16 46.103	18 52 04.61	.348 1119	28	19 16 19.781	24 36 53.38	.062 8653
14	15 21 41.667	19 13 43.65	.342 4568	29	19 21 30.562	24 27 49.17	.056 1490
15	15 26 38.415	19 34 53.35	.336 7759	30	19 26 40.214	24 18 03.74	.049 4120
16	15 31 36.350	−19 55 32.97	1.331 0693	Dec. 1	19 31 48.674	−24 07 37.54	1.042 6546
17	15 36 35.469	20 15 41.77	.325 3373	2	19 36 55.885	23 56 31.07	.035 8771
18	15 41 35.768	20 35 19.05	.319 5800	3	19 42 01.791	23 44 44.85	.029 0797
19	15 46 37.240	20 54 24.07	.313 7974	4	19 47 06.337	23 32 19.43	.022 2628
20	15 51 39.873	21 12 56.12	.307 9897	5	19 52 09.474	23 19 15.35	.015 4267
21	15 56 43.655	−21 30 54.49	1.302 1566	6	19 57 11.154	−23 05 33.23	1.008 5719
22	16 01 48.570	21 48 18.47	.296 2982	7	20 02 11.331	22 51 13.66	1.001 6986
23	16 06 54.600	22 05 07.40	.290 4142	8	20 07 09.964	22 36 17.28	0.994 8074
24	16 12 01.723	22 21 20.59	.284 5045	9	20 12 07.012	22 20 44.75	.987 8987
25	16 17 09.914	22 36 57.42	.278 5689	10	20 17 02.439	22 04 36.71	.980 9730
26	16 22 19.144	−22 51 57.26	1.272 6074	11	20 21 56.209	−21 47 53.86	0.974 0306
27	16 27 29.376	23 06 19.50	.266 6200	12	20 26 48.293	21 30 36.87	.967 0721
28	16 32 40.574	23 20 03.56	.260 6068	13	20 31 38.660	21 12 46.44	.960 0977
29	16 37 52.692	23 33 08.87	.254 5677	14	20 36 27.285	20 54 23.26	.953 1077
30	16 43 05.686	23 45 34.88	.248 5032	15	20 41 14.149	20 35 28.04	.946 1023
31	16 48 19.508	−23 57 21.06	1.242 4135	16	20 45 59.231	−20 16 01.50	0.939 0816
Nov. 1	16 53 34.107	24 08 26.92	.236 2987	17	20 50 42.516	19 56 04.35	.932 0456
2	16 58 49.433	24 18 51.96	.230 1594	18	20 55 23.991	19 35 37.35	.924 9944
3	17 04 05.430	24 28 35.73	.223 9958	19	21 00 03.641	19 14 41.25	.917 9279
4	17 09 22.044	24 37 37.82	.217 8083	20	21 04 41.454	18 53 16.84	.910 8459
5	17 14 39.217	−24 45 57.83	1.211 5972	21	21 09 17.418	−18 31 24.91	0.903 7485
6	17 19 56.892	24 53 35.39	.205 3629	22	21 13 51.518	18 09 06.27	.896 6356
7	17 25 15.008	25 00 30.16	.199 1058	23	21 18 23.742	17 46 21.75	.889 5070
8	17 30 33.503	25 06 41.86	.192 8263	24	21 22 54.077	17 23 12.17	.882 3630
9	17 35 52.314	25 12 10.20	.186 5248	25	21 27 22.511	16 59 38.36	.875 2034
10	17 41 11.378	−25 16 54.97	1.180 2017	26	21 31 49.032	−16 35 41.16	0.868 0285
11	17 46 30.629	25 20 55.95	.173 8573	27	21 36 13.633	16 11 21.41	.860 8384
12	17 51 50.000	25 24 12.97	.167 4920	28	21 40 36.303	15 46 39.94	.853 6334
13	17 57 09.423	25 26 45.90	.161 1060	29	21 44 57.036	15 21 37.60	.846 4137
14	18 02 28.830	25 28 34.64	.154 6997	30	21 49 15.825	14 56 15.21	.839 1797
15	18 07 48.150	−25 29 39.09	1.148 2732	31	21 53 32.666	−14 30 33.63	0.831 9318
16	18 13 07.314	−25 29 59.21	1.141 8266	32	21 57 47.554	−14 04 33.69	0.824 6703

GEOCENTRIC COORDINATES FOR 0^h DYNAMICAL TIME

Date	Apparent Right Ascension	Apparent Declination	True Geocentric Distance	Date	Apparent Right Ascension	Apparent Declination	True Geocentric Distance
	h m s	° ′ ″			h m s	° ′ ″	
Jan. 0	17 29 38.418	−23 42 35.61	2.416 1973	Feb. 15	19 59 17.739	−21 32 05.01	2.232 6351
1	17 32 50.811	23 45 24.03	.412 6702	16	20 02 31.099	21 23 26.28	.228 3284
2	17 36 03.533	23 47 57.92	.409 1147	17	20 05 44.191	21 14 33.79	.224 0168
3	17 39 16.569	23 50 17.21	.405 5312	18	20 08 57.009	21 05 27.62	.219 7004
4	17 42 29.904	23 52 21.81	.401 9201	19	20 12 09.545	20 56 07.87	.215 3797
5	17 45 43.523	−23 54 11.66	2.398 2819	20	20 15 21.797	−20 46 34.64	2.211 0546
6	17 48 57.409	23 55 46.69	.394 6171	21	20 18 33.761	20 36 48.03	.206 7253
7	17 52 11.549	23 57 06.82	.390 9263	22	20 21 45.432	20 26 48.17	.202 3918
8	17 55 25.928	23 58 12.00	.387 2102	23	20 24 56.806	20 16 35.20	.198 0540
9	17 58 40.530	23 59 02.17	.383 4693	24	20 28 07.875	20 06 09.25	.193 7120
10	18 01 55.342	−23 59 37.27	2.379 7044	25	20 31 18.632	−19 55 30.47	2.189 3657
11	18 05 10.350	23 59 57.24	.375 9161	26	20 34 29.070	19 44 39.02	.185 0153
12	18 08 25.541	24 00 02.05	.372 1052	27	20 37 39.181	19 33 35.04	.180 6609
13	18 11 40.903	23 59 51.67	.368 2724	28	20 40 48.957	19 22 18.70	.176 3027
14	18 14 56.424	23 59 26.05	.364 4185	29	20 43 58.391	19 10 50.14	.171 9408
15	18 18 12.089	−23 58 45.17	2.360 5442	Mar. 1	20 47 07.476	−18 59 09.54	2.167 5754
16	18 21 27.888	23 57 49.02	.356 6504	2	20 50 16.206	18 47 17.04	.163 2069
17	18 24 43.807	23 56 37.58	.352 7377	3	20 53 24.575	18 35 12.80	.158 8354
18	18 27 59.832	23 55 10.85	.348 8069	4	20 56 32.577	18 22 57.00	.154 4614
19	18 31 15.950	23 53 28.80	.344 8585	5	20 59 40.210	18 10 29.78	.150 0853
20	18 34 32.145	−23 51 31.43	2.340 8930	6	21 02 47.468	−17 57 51.33	2.145 7073
21	18 37 48.404	23 49 18.70	.336 9108	7	21 05 54.350	17 45 01.79	.141 3280
22	18 41 04.716	23 46 50.60	.332 9121	8	21 09 00.853	17 32 01.34	.136 9478
23	18 44 21.069	23 44 07.12	.328 8971	9	21 12 06.976	17 18 50.16	.132 5672
24	18 47 37.454	23 41 08.24	.324 8658	10	21 15 12.716	17 05 28.42	.128 1869
25	18 50 53.857	−23 37 53.98	2.320 8184	11	21 18 18.073	−16 51 56.30	2.123 8073
26	18 54 10.268	23 34 24.37	.316 7548	12	21 21 23.044	16 38 13.97	.119 4291
27	18 57 26.673	23 30 39.43	.312 6754	13	21 24 27.629	16 24 21.62	.115 0527
28	19 00 43.056	23 26 39.20	.308 5802	14	21 27 31.826	16 10 19.41	.110 6787
29	19 03 59.403	23 22 23.73	.304 4694	15	21 30 35.635	15 56 07.54	.106 3075
30	19 07 15.698	−23 17 53.06	2.300 3435	16	21 33 39.056	−15 41 46.15	2.101 9396
31	19 10 31.926	23 13 07.24	.296 2026	17	21 36 42.092	15 27 15.40	.097 5753
Feb. 1	19 13 48.070	23 08 06.31	.292 0471	18	21 39 44.747	15 12 35.46	.093 2146
2	19 17 04.115	23 02 50.34	.287 8775	19	21 42 47.027	14 57 46.47	.088 8577
3	19 20 20.047	22 57 19.39	.283 6942	20	21 45 48.936	14 42 48.60	.084 5044
4	19 23 35.850	−22 51 33.50	2.279 4977	21	21 48 50.480	−14 27 42.02	2.080 1547
5	19 26 51.511	22 45 32.75	.275 2885	22	21 51 51.662	14 12 26.92	.075 8085
6	19 30 07.016	22 39 17.20	.271 0672	23	21 54 52.485	13 57 03.49	.071 4654
7	19 33 22.354	22 32 46.93	.266 8343	24	21 57 52.949	13 41 31.92	.067 1255
8	19 36 37.512	22 26 02.01	.262 5905	25	22 00 53.057	13 25 52.42	.062 7884
9	19 39 52.481	−22 19 02.53	2.258 3364	26	22 03 52.808	−13 10 05.18	2.058 4542
10	19 43 07.249	22 11 48.57	.254 0727	27	22 06 52.204	12 54 10.41	.054 1227
11	19 46 21.806	22 04 20.22	.249 8001	28	22 09 51.246	12 38 08.31	.049 7939
12	19 49 36.144	21 56 37.58	.245 5194	29	22 12 49.937	12 21 59.07	.045 4677
13	19 52 50.252	21 48 40.77	.241 2312	30	22 15 48.276	12 05 42.90	.041 1442
14	19 56 04.120	−21 40 29.87	2.236 9362	31	22 18 46.268	−11 49 20.00	2.036 8234
15	19 59 17.739	−21 32 05.01	2.232 6351	Apr. 1	22 21 43.915	−11 32 50.56	2.032 5053

GEOCENTRIC COORDINATES FOR 0^h DYNAMICAL TIME

Date	Apparent Right Ascension	Apparent Declination	True Geocentric Distance	Date	Apparent Right Ascension	Apparent Declination	True Geocentric Distance
	h m s	° ′ ″			h m s	° ′ ″	
Apr. 1	22 21 43.915	−11 32 50.56	2.032 5053	May 17	0 33 16.613	+ 2 08 12.97	1.838 2418
2	22 24 41.220	11 16 14.77	.028 1901	18	0 36 04.482	2 26 14.72	.834 0861
3	22 27 38.187	10 59 32.84	.023 8779	19	0 38 52.303	2 44 14.07	.829 9308
4	22 30 34.819	10 42 44.97	.019 5688	20	0 41 40.082	3 02 10.86	.825 7754
5	22 33 31.122	10 25 51.35	.015 2632	21	0 44 27.824	3 20 04.90	.821 6194
6	22 36 27.099	−10 08 52.18	2.010 9614	22	0 47 15.533	+ 3 37 56.00	1.817 4623
7	22 39 22.754	9 51 47.67	.006 6635	23	0 50 03.216	3 55 44.00	.813 3034
8	22 42 18.091	9 34 38.00	2.002 3701	24	0 52 50.875	4 13 28.70	.809 1425
9	22 45 13.114	9 17 23.40	1.998 0814	25	0 55 38.517	4 31 09.93	.804 9789
10	22 48 07.826	9 00 04.04	.993 7979	26	0 58 26.145	4 48 47.51	.800 8122
11	22 51 02.233	− 8 42 40.13	1.989 5198	27	1 01 13.764	+ 5 06 21.28	1.796 6421
12	22 53 56.339	8 25 11.85	.985 2474	28	1 04 01.379	5 23 51.05	.792 4680
13	22 56 50.153	8 07 39.38	.980 9810	29	1 06 48.993	5 41 16.67	.788 2897
14	22 59 43.681	7 50 02.89	.976 7208	30	1 09 36.610	5 58 37.95	.784 1067
15	23 02 36.934	7 32 22.54	.972 4666	31	1 12 24.233	6 15 54.74	.779 9190
16	23 05 29.922	− 7 14 38.49	1.968 2185	June 1	1 15 11.863	+ 6 33 06.85	1.775 7263
17	23 08 22.654	6 56 50.93	.963 9763	2	1 17 59.501	6 50 14.12	.771 5285
18	23 11 15.141	6 39 00.02	.959 7396	3	1 20 47.147	7 07 16.38	.767 3256
19	23 14 07.391	6 21 05.96	.955 5083	4	1 23 34.802	7 24 13.45	.763 1176
20	23 16 59.409	6 03 08.94	.951 2819	5	1 26 22.466	7 41 05.17	.758 9047
21	23 19 51.202	− 5 45 09.14	1.947 0601	6	1 29 10.144	+ 7 57 51.38	1.754 6868
22	23 22 42.775	5 27 06.79	.942 8426	7	1 31 57.839	8 14 31.94	.750 4640
23	23 25 34.133	5 09 02.07	.938 6288	8	1 34 45.557	8 31 06.71	.746 2364
24	23 28 25.282	4 50 55.19	.934 4186	9	1 37 33.306	8 47 35.55	.742 0038
25	23 31 16.227	4 32 46.36	.930 2117	10	1 40 21.092	9 03 58.35	.737 7660
26	23 34 06.972	− 4 14 35.76	1.926 0077	11	1 43 08.921	+ 9 20 14.98	1.733 5227
27	23 36 57.524	3 56 23.60	.921 8063	12	1 45 56.798	9 36 25.32	.729 2736
28	23 39 47.887	3 38 10.07	.917 6074	13	1 48 44.729	9 52 29.22	.725 0184
29	23 42 38.069	3 19 55.38	.913 4108	14	1 51 32.718	10 08 26.56	.720 7565
30	23 45 28.074	3 01 39.70	.909 2162	15	1 54 20.765	10 24 17.19	.716 4873
May 1	23 48 17.909	− 2 43 23.25	1.905 0235	16	1 57 08.875	+10 40 00.97	1.712 2104
2	23 51 07.580	2 25 06.20	.900 8327	17	1 59 57.048	10 55 37.76	.707 9251
3	23 53 57.091	2 06 48.76	.896 6436	18	2 02 45.285	11 11 07.41	.703 6309
4	23 56 46.449	1 48 31.12	.892 4565	19	2 05 33.588	11 26 29.78	.699 3271
5	23 59 35.657	1 30 13.46	.888 2713	20	2 08 21.957	11 41 44.72	.695 0131
6	0 02 24.719	1 11 56.00	1.884 0882	21	2 11 10.394	+11 56 52.09	1.690 6883
7	0 05 13.640	0 53 38.91	.879 9074	22	2 13 58.898	12 11 51.76	.686 3522
8	0 08 02.425	0 35 22.39	.875 7290	23	2 16 47.470	12 26 43.59	.682 0042
9	0 10 51.078	− 0 17 06.62	.871 5533	24	2 19 36.109	12 41 27.45	.677 6437
10	0 13 39.606	+ 0 01 08.22	.867 3804	25	2 22 24.815	12 56 03.21	.673 2702
11	0 16 28.019	+ 0 19 21.98	1.863 2103	26	2 25 13.585	+13 10 30.73	1.668 8832
12	0 19 16.326	0 37 34.49	.859 0430	27	2 28 02.418	13 24 49.91	.664 4823
13	0 22 04.537	0 55 45.61	.854 8784	28	2 30 51.310	13 39 00.61	.660 0671
14	0 24 52.662	1 13 55.17	.850 7163	29	2 33 40.256	13 53 02.70	.655 6373
15	0 27 40.710	1 32 03.03	.846 5564	30	2 36 29.249	14 06 56.08	.651 1927
16	0 30 28.692	+ 1 50 09.02	1.842 3984	July 1	2 39 18.283	+14 20 40.61	1.646 7331
17	0 33 16.613	+ 2 08 12.97	1.838 2418	2	2 42 07.352	+14 34 16.15	1.642 2586

GEOCENTRIC COORDINATES FOR 0^h DYNAMICAL TIME

Date	Apparent Right Ascension	Apparent Declination	True Geocentric Distance	Date	Apparent Right Ascension	Apparent Declination	True Geocentric Distance
	h m s	° ′ ″			h m s	° ′ ″	
July 1	2 39 18.283	+14 20 40.61	1.646 7331	Aug. 16	4 48 16.672	+21 50 48.55	1.421 1102
2	2 42 07.352	14 34 16.15	.642 2586	17	4 51 00.956	21 56 22.57	.415 6504
3	2 44 56.449	14 47 42.60	.637 7691	18	4 53 44.900	22 01 45.81	.410 1625
4	2 47 45.572	15 00 59.85	.633 2648	19	4 56 28.488	22 06 58.33	.404 6461
5	2 50 34.720	15 14 07.78	.628 7457	20	4 59 11.707	22 12 00.18	.399 1009
6	2 53 23.892	+15 27 06.33	1.624 2119	21	5 01 54.539	+22 16 51.42	1.393 5265
7	2 56 13.089	15 39 55.40	.619 6631	22	5 04 36.968	22 21 32.11	.387 9225
8	2 59 02.312	15 52 34.93	.615 0993	23	5 07 18.975	22 26 02.32	.382 2887
9	3 01 51.559	16 05 04.85	.610 5202	24	5 10 00.538	22 30 22.14	.376 6250
10	3 04 40.829	16 17 25.07	.605 9256	25	5 12 41.637	22 34 31.62	.370 9312
11	3 07 30.119	+16 29 35.53	1.601 3151	26	5 15 22.249	+22 38 30.85	1.365 2073
12	3 10 19.426	16 41 36.14	.596 6881	27	5 18 02.353	22 42 19.90	.359 4536
13	3 13 08.746	16 53 26.81	.592 0444	28	5 20 41.929	22 45 58.84	.353 6702
14	3 15 58.075	17 05 07.48	.587 3832	29	5 23 20.959	22 49 27.75	.347 8575
15	3 18 47.407	17 16 38.05	.582 7042	30	5 25 59.429	22 52 46.74	.342 0159
16	3 21 36.737	+17 27 58.44	1.578 0067	31	5 28 37.325	+22 55 55.90	1.336 1459
17	3 24 26.061	17 39 08.58	.573 2902	Sept. 1	5 31 14.634	22 58 55.35	.330 2477
18	3 27 15.372	17 50 08.38	.568 5541	2	5 33 51.341	23 01 45.22	.324 3216
19	3 30 04.664	18 00 57.79	.563 7977	3	5 36 27.433	23 04 25.62	.318 3679
20	3 32 53.931	18 11 36.72	.559 0206	4	5 39 02.894	23 06 56.68	.312 3867
21	3 35 43.166	+18 22 05.11	1.554 2221	5	5 41 37.709	+23 09 18.50	1.306 3781
22	3 38 32.361	18 32 22.91	.549 4018	6	5 44 11.863	23 11 31.22	.300 3422
23	3 41 21.508	18 42 30.07	.544 5591	7	5 46 45.340	23 13 34.95	.294 2789
24	3 44 10.598	18 52 26.53	.539 6935	8	5 49 18.125	23 15 29.81	.288 1882
25	3 46 59.620	19 02 12.24	.534 8045	9	5 51 50.202	23 17 15.93	.282 0701
26	3 49 48.563	+19 11 47.17	1.529 8918	10	5 54 21.555	+23 18 53.43	1.275 9244
27	3 52 37.412	19 21 11.27	.524 9550	11	5 56 52.169	23 20 22.46	.269 7512
28	3 55 26.153	19 30 24.51	.519 9939	12	5 59 22.027	23 21 43.13	.263 5502
29	3 58 14.771	19 39 26.83	.515 0085	13	6 01 51.113	23 22 55.62	.257 3213
30	4 01 03.250	19 48 18.20	.509 9986	14	6 04 19.410	23 24 00.05	.251 0646
31	4 03 51.577	+19 56 58.58	1.504 9645	15	6 06 46.901	+23 24 56.59	1.244 7797
Aug. 1	4 06 39.742	20 05 27.93	.499 9061	16	6 09 13.566	23 25 45.41	.238 4667
2	4 09 27.734	20 13 46.25	.494 8238	17	6 11 39.385	23 26 26.68	.232 1255
3	4 12 15.548	20 21 53.53	.489 7177	18	6 14 04.338	23 27 00.59	.225 7561
4	4 15 03.174	20 29 49.78	.484 5877	19	6 16 28.401	23 27 27.31	.219 3585
5	4 17 50.607	+20 37 35.00	1.479 4340	20	6 18 51.550	+23 27 47.04	1.212 9327
6	4 20 37.837	20 45 09.22	.474 2564	21	6 21 13.759	23 27 59.97	.206 4790
7	4 23 24.854	20 52 32.44	.469 0548	22	6 23 35.002	23 28 06.30	.199 9975
8	4 26 11.649	20 59 44.68	.463 8290	23	6 25 55.251	23 28 06.22	.193 4888
9	4 28 58.211	21 06 45.94	.458 5786	24	6 28 14.479	23 27 59.92	.186 9532
10	4 31 44.529	+21 13 36.24	1.453 3035	25	6 30 32.662	+23 27 47.61	1.180 3915
11	4 34 30.592	21 20 15.58	.448 0032	26	6 32 49.777	23 27 29.48	.173 8043
12	4 37 16.389	21 26 43.98	.442 6773	27	6 35 05.803	23 27 05.74	.167 1926
13	4 40 01.907	21 33 01.45	.437 3255	28	6 37 20.719	23 26 36.61	.160 5571
14	4 42 47.135	21 39 08.02	.431 9473	29	6 39 34.507	23 26 02.31	.153 8987
15	4 45 32.061	+21 45 03.71	1.426 5424	30	6 41 47.144	+23 25 23.08	1.147 2182
16	4 48 16.672	+21 50 48.55	1.421 1102	Oct. 1	6 43 58.613	+23 24 39.13	1.140 5164

GEOCENTRIC COORDINATES FOR 0h DYNAMICAL TIME

Date	Apparent Right Ascension	Apparent Declination	True Geocentric Distance	Date	Apparent Right Ascension	Apparent Declination	True Geocentric Distance
	h m s	° ′ ″			h m s	° ′ ″	
Oct. 1	6 43 58.613	+23 24 39.13	1.140 5164	Nov. 16	7 55 45.026	+22 41 12.47	0.823 5531
2	6 46 08.890	23 23 50.69	.133 7940	17	7 56 27.366	22 41 53.11	.816 9934
3	6 48 17.956	23 22 57.98	.127 0516	18	7 57 06.799	22 42 42.70	.810 4797
4	6 50 25.789	23 22 01.22	.120 2897	19	7 57 43.269	22 43 41.49	.804 0149
5	6 52 32.368	23 21 00.65	.113 5091	20	7 58 16.722	22 44 49.72	.797 6025
6	6 54 37.671	+23 19 56.48	1.106 7103	21	7 58 47.103	+22 46 07.62	0.791 2458
7	6 56 41.677	23 18 48.95	.099 8937	22	7 59 14.362	22 47 35.40	.784 9485
8	6 58 44.365	23 17 38.28	.093 0599	23	7 59 38.449	22 49 13.23	.778 7142
9	7 00 45.710	23 16 24.72	.086 2095	24	7 59 59.316	22 51 01.29	.772 5466
10	7 02 45.691	23 15 08.52	.079 3429	25	8 00 16.917	22 52 59.73	.766 4497
11	7 04 44.284	+23 13 49.91	1.072 4606	26	8 00 31.207	+22 55 08.64	0.760 4270
12	7 06 41.465	23 12 29.16	.065 5633	27	8 00 42.145	22 57 28.12	.754 4825
13	7 08 37.206	23 11 06.54	.058 6513	28	8 00 49.690	22 59 58.22	.748 6200
14	7 10 31.482	23 09 42.31	.051 7254	29	8 00 53.806	23 02 38.97	.742 8433
15	7 12 24.261	23 08 16.78	.044 7860	30	8 00 54.458	23 05 30.36	.737 1562
16	7 14 15.514	+23 06 50.22	1.037 8339	Dec. 1	8 00 51.615	+23 08 32.38	0.731 5625
17	7 16 05.205	23 05 22.93	.030 8698	2	8 00 45.248	23 11 44.97	.726 0661
18	7 17 53.299	23 03 55.23	.023 8944	3	8 00 35.330	23 15 08.03	.720 6709
19	7 19 39.759	23 02 27.40	.016 9088	4	8 00 21.838	23 18 41.46	.715 3807
20	7 21 24.546	23 00 59.76	.009 9140	5	8 00 04.754	23 22 25.13	.710 1996
21	7 23 07.621	+22 59 32.61	1.002 9112	6	7 59 44.061	+23 26 18.85	0.705 1314
22	7 24 48.946	22 58 06.24	0.995 9017	7	7 59 19.745	23 30 22.42	.700 1801
23	7 26 28.482	22 56 40.97	.988 8870	8	7 58 51.798	23 34 35.63	.695 3497
24	7 28 06.194	22 55 17.10	.981 8686	9	7 58 20.213	23 38 58.21	.690 6442
25	7 29 42.046	22 53 54.93	.974 8484	10	7 57 44.986	23 43 29.87	.686 0676
26	7 31 16.003	+22 52 34.77	0.967 8281	11	7 57 06.116	+23 48 10.31	0.681 6240
27	7 32 48.030	22 51 16.95	.960 8095	12	7 56 23.606	23 52 59.15	.677 3177
28	7 34 18.092	22 50 01.76	.953 7944	13	7 55 37.467	23 57 55.99	.673 1530
29	7 35 46.151	22 48 49.52	.946 7847	14	7 54 47.712	24 03 00.39	.669 1341
30	7 37 12.173	22 47 40.52	.939 7821	15	7 53 54.366	24 08 11.83	.665 2658
31	7 38 36.121	+22 46 35.08	0.932 7883	16	7 52 57.463	+24 13 29.78	0.661 5526
Nov. 1	7 39 57.958	22 45 33.47	.925 8051	17	7 51 57.049	24 18 53.62	.657 9992
2	7 41 17.647	22 44 35.99	.918 8343	18	7 50 53.180	24 24 22.71	.654 6103
3	7 42 35.151	22 43 42.95	.911 8775	19	7 49 45.923	24 29 56.37	.651 3906
4	7 43 50.431	22 42 54.62	.904 9365	20	7 48 35.358	24 35 33.88	.648 3447
5	7 45 03.449	+22 42 11.31	0.898 0131	21	7 47 21.576	+24 41 14.48	0.645 4773
6	7 46 14.166	22 41 33.31	.891 1090	22	7 46 04.678	24 46 57.37	.642 7926
7	7 47 22.540	22 41 00.92	.884 2261	23	7 44 44.779	24 52 41.74	.640 2949
8	7 48 28.530	22 40 34.44	.877 3662	24	7 43 22.003	24 58 26.73	.637 9882
9	7 49 32.093	22 40 14.17	.870 5311	25	7 41 56.486	25 04 11.48	.635 8762
10	7 50 33.186	+22 40 00.42	0.863 7229	26	7 40 28.376	+25 09 55.11	0.633 9625
11	7 51 31.759	22 39 53.50	.856 9434	27	7 38 57.830	25 15 36.73	.632 2503
12	7 52 27.765	22 39 53.74	.850 1948	28	7 37 25.014	25 21 15.49	.630 7427
13	7 53 21.150	22 40 01.44	.843 4791	29	7 35 50.105	25 26 50.50	.629 4424
14	7 54 11.860	22 40 16.93	.836 7986	30	7 34 13.286	25 32 20.91	.628 3518
15	7 54 59.838	+22 40 40.50	0.830 1558	31	7 32 34.748	+25 37 45.91	0.627 4732
16	7 55 45.026	+22 41 12.47	0.823 5531	32	7 30 54.690	+25 43 04.71	0.626 8086

GEOCENTRIC COORDINATES FOR 0^h DYNAMICAL TIME

Date	Apparent Right Ascension	Apparent Declination	True Geocentric Distance	Date	Apparent Right Ascension	Apparent Declination	True Geocentric Distance
	h m s	° ′ ″			h m s	° ′ ″	
Jan. 0	11 05 11.636	+ 7 07 46.87	4.911 0241	Feb. 15	10 53 43.407	+ 8 30 22.94	4.442 9225
1	11 05 11.633	7 08 04.93	.896 5750	16	10 53 16.177	8 33 19.11	.438 7011
2	11 05 10.904	7 08 27.50	.882 2456	17	10 52 48.659	8 36 16.55	.434 7820
3	11 05 09.449	7 08 54.60	.868 0413	18	10 52 20.870	8 39 15.16	.431 1665
4	11 05 07.265	7 09 26.21	.853 9673	19	10 51 52.831	8 42 14.79	.427 8560
5	11 05 04.354	+ 7 10 02.34	4.840 0289	20	10 51 24.559	+ 8 45 15.33	4.424 8519
6	11 05 00.715	7 10 42.97	.826 2314	21	10 50 56.078	8 48 16.62	.422 1555
7	11 04 56.348	7 11 28.08	.812 5800	22	10 50 27.407	8 51 18.53	.419 7682
8	11 04 51.256	7 12 17.65	.799 0798	23	10 49 58.569	8 54 20.94	.417 6913
9	11 04 45.441	7 13 11.66	.785 7361	24	10 49 29.583	8 57 23.71	.415 9260
10	11 04 38.907	+ 7 14 10.07	4.772 5538	25	10 49 00.470	+ 9 00 26.71	4.414 4732
11	11 04 31.657	7 15 12.85	.759 5380	26	10 48 31.250	9 03 29.83	.413 3341
12	11 04 23.697	7 16 19.94	.746 6934	27	10 48 01.944	9 06 32.92	.412 5091
13	11 04 15.031	7 17 31.31	.734 0250	28	10 47 32.572	9 09 35.86	.411 9990
14	11 04 05.667	7 18 46.91	.721 5375	29	10 47 03.155	9 12 38.51	.411 8040
15	11 03 55.610	+ 7 20 06.69	4.709 2354	Mar. 1	10 46 33.714	+ 9 15 40.75	4.411 9244
16	11 03 44.867	7 21 30.61	.697 1233	2	10 46 04.272	9 18 42.42	.412 3602
17	11 03 33.443	7 22 58.62	.685 2055	3	10 45 34.850	9 21 43.41	.413 1111
18	11 03 21.343	7 24 30.69	.673 4863	4	10 45 05.470	9 24 43.56	.414 1768
19	11 03 08.570	7 26 06.79	.661 9699	5	10 44 36.157	9 27 42.74	.415 5566
20	11 02 55.130	+ 7 27 46.87	4.650 6603	6	10 44 06.932	+ 9 30 40.80	4.417 2497
21	11 02 41.026	7 29 30.90	.639 5619	7	10 43 37.819	9 33 37.61	.419 2550
22	11 02 26.264	7 31 18.83	.628 6788	8	10 43 08.842	9 36 33.03	.421 5714
23	11 02 10.853	7 33 10.58	.618 0153	9	10 42 40.022	9 39 26.94	.424 1973
24	11 01 54.802	7 35 06.08	.607 5758	10	10 42 11.383	9 42 19.20	.427 1310
25	11 01 38.121	+ 7 37 05.27	4.597 3648	11	10 41 42.944	+ 9 45 09.71	4.430 3706
26	11 01 20.820	7 39 08.06	.587 3868	12	10 41 14.727	9 47 58.35	.433 9141
27	11 01 02.908	7 41 14.39	.577 6462	13	10 40 46.750	9 50 45.02	.437 7593
28	11 00 44.396	7 43 24.18	.568 1474	14	10 40 19.031	9 53 29.63	.441 9037
29	11 00 25.295	7 45 37.35	.558 8948	15	10 39 51.586	9 56 12.08	.446 3449
30	11 00 05.614	+ 7 47 53.83	4.549 8927	16	10 39 24.433	+ 9 58 52.28	4.451 0804
31	10 59 45.366	7 50 13.52	.541 1452	17	10 38 57.589	10 01 30.13	.456 1076
Feb. 1	10 59 24.561	7 52 36.34	.532 6565	18	10 38 31.071	10 04 05.53	.461 4239
2	10 59 03.213	7 55 02.20	.524 4306	19	10 38 04.899	10 06 38.38	.467 0268
3	10 58 41.334	7 57 30.99	.516 4711	20	10 37 39.091	10 09 08.58	.472 9137
4	10 58 18.940	+ 8 00 02.61	4.508 7819	21	10 37 13.665	+10 11 36.03	4.479 0822
5	10 57 56.044	8 02 36.95	.501 3664	22	10 36 48.638	10 14 00.66	.485 5296
6	10 57 32.664	8 05 13.89	.494 2281	23	10 36 24.025	10 16 22.37	.492 2532
7	10 57 08.817	8 07 53.31	.487 3701	24	10 35 59.843	10 18 41.10	.499 2504
8	10 56 44.519	8 10 35.09	.480 7955	25	10 35 36.105	10 20 56.77	.506 5183
9	10 56 19.790	+ 8 13 19.10	4.474 5072	26	10 35 12.826	+10 23 09.31	4.514 0538
10	10 55 54.648	8 16 05.21	.468 5077	27	10 34 50.021	10 25 18.65	.521 8538
11	10 55 29.113	8 18 53.29	.462 7995	28	10 34 27.703	10 27 24.70	.529 9151
12	10 55 03.203	8 21 43.23	.457 3849	29	10 34 05.886	10 29 27.40	.538 2342
13	10 54 36.937	8 24 34.89	.452 2660	30	10 33 44.585	10 31 26.68	.546 8077
14	10 54 10.332	+ 8 27 28.16	4.447 4446	31	10 33 23.813	+10 33 22.47	4.555 6319
15	10 53 43.407	+ 8 30 22.94	4.442 9225	Apr. 1	10 33 03.583	+10 35 14.69	4.564 7030

GEOCENTRIC COORDINATES FOR 0ʰ DYNAMICAL TIME

Date	Apparent Right Ascension	Apparent Declination	True Geocentric Distance
	h m s	° ′ ″	
Apr. 1	10 33 03.583	+10 35 14.69	4.564 7030
2	10 32 43.910	10 37 03.28	.574 0170
3	10 32 24.808	10 38 48.17	.583 5700
4	10 32 06.288	10 40 29.31	.593 3577
5	10 31 48.363	10 42 06.63	.603 3758
6	10 31 31.046	+10 43 40.09	4.613 6199
7	10 31 14.345	10 45 09.65	.624 0853
8	10 30 58.270	10 46 35.29	.634 7675
9	10 30 42.827	10 47 56.97	.645 6618
10	10 30 28.023	10 49 14.68	.656 7632
11	10 30 13.864	+10 50 28.40	4.668 0672
12	10 30 00.352	10 51 38.12	.679 5689
13	10 29 47.495	10 52 43.82	.691 2637
14	10 29 35.296	10 53 45.48	.703 1469
15	10 29 23.762	10 54 43.07	.715 2141
16	10 29 12.899	+10 55 36.57	4.727 4608
17	10 29 02.711	10 56 25.97	.739 8827
18	10 28 53.204	10 57 11.25	.752 4755
19	10 28 44.380	10 57 52.41	.765 2350
20	10 28 36.244	10 58 29.44	.778 1569
21	10 28 28.796	+10 59 02.36	4.791 2370
22	10 28 22.038	10 59 31.15	.804 4711
23	10 28 15.971	10 59 55.82	.817 8548
24	10 28 10.598	11 00 16.36	.831 3837
25	10 28 05.918	11 00 32.78	.845 0534
26	10 28 01.934	+11 00 45.07	4.858 8595
27	10 27 58.647	11 00 53.23	.872 7975
28	10 27 56.059	11 00 57.25	.886 8627
29	10 27 54.170	11 00 57.12	.901 0506
30	10 27 52.983	11 00 52.86	.915 3565
May 1	10 27 52.499	+11 00 44.44	4.929 7758
2	10 27 52.718	11 00 31.89	.944 3036
3	10 27 53.640	11 00 15.20	.958 9352
4	10 27 55.267	10 59 54.39	.973 6655
5	10 27 57.594	10 59 29.48	4.988 4899
6	10 28 00.620	+10 59 00.50	5.003 4033
7	10 28 04.339	10 58 27.49	.018 4008
8	10 28 08.746	10 57 50.47	.033 4776
9	10 28 13.837	10 57 09.48	.048 6291
10	10 28 19.607	10 56 24.55	.063 8504
11	10 28 26.051	+10 55 35.68	5.079 1373
12	10 28 33.166	10 54 42.91	.094 4852
13	10 28 40.950	10 53 46.26	.109 8901
14	10 28 49.397	10 52 45.73	.125 3477
15	10 28 58.506	10 51 41.35	.140 8542
16	10 29 08.272	+10 50 33.16	5.156 4056
17	10 29 18.690	+10 49 21.19	5.171 9981
May 17	10 29 18.690	+10 49 21.19	5.171 9981
18	10 29 29.754	10 48 05.46	.187 6280
19	10 29 41.458	10 46 46.02	.203 2914
20	10 29 53.798	10 45 22.89	.218 9846
21	10 30 06.766	10 43 56.09	.234 7039
22	10 30 20.357	+10 42 25.66	5.250 4454
23	10 30 34.568	10 40 51.62	.266 2054
24	10 30 49.391	10 39 13.99	.281 9801
25	10 31 04.824	10 37 32.79	.297 7656
26	10 31 20.862	10 35 48.03	.313 5581
27	10 31 37.500	+10 33 59.74	5.329 3538
28	10 31 54.736	10 32 07.94	.345 1487
29	10 32 12.564	10 30 12.64	.360 9391
30	10 32 30.980	10 28 13.88	.376 7209
31	10 32 49.981	10 26 11.66	.392 4901
June 1	10 33 09.560	+10 24 06.04	5.408 2429
2	10 33 29.710	10 21 57.05	.423 9751
3	10 33 50.425	10 19 44.74	.439 6828
4	10 34 11.694	10 17 29.15	.455 3620
5	10 34 33.509	10 15 10.33	.471 0090
6	10 34 55.862	+10 12 48.31	5.486 6200
7	10 35 18.747	10 10 23.13	.502 1914
8	10 35 42.156	10 07 54.82	.517 7199
9	10 36 06.084	10 05 23.39	.533 2022
10	10 36 30.526	10 02 48.89	.548 6352
11	10 36 55.477	+10 00 11.33	5.564 0158
12	10 37 20.929	9 57 30.76	.579 3412
13	10 37 46.876	9 54 47.20	.594 6085
14	10 38 13.311	9 52 00.70	.609 8150
15	10 38 40.226	9 49 11.30	.624 9579
16	10 39 07.615	+ 9 46 19.03	5.640 0345
17	10 39 35.469	9 43 23.92	.655 0421
18	10 40 03.783	9 40 26.00	.669 9779
19	10 40 32.550	9 37 25.31	.684 8392
20	10 41 01.764	9 34 21.87	.699 6233
21	10 41 31.419	+ 9 31 15.70	5.714 3274
22	10 42 01.510	9 28 06.83	.728 9487
23	10 42 32.033	9 24 55.27	.743 4845
24	10 43 02.982	9 21 41.05	.757 9320
25	10 43 34.354	9 18 24.18	.772 2883
26	10 44 06.144	+ 9 15 04.70	5.786 5505
27	10 44 38.347	9 11 42.62	.800 7159
28	10 45 10.957	9 08 17.98	.814 7814
29	10 45 43.970	9 04 50.81	.828 7441
30	10 46 17.376	9 01 21.16	.842 6011
July 1	10 46 51.169	+ 8 57 49.08	5.856 3494
2	10 47 25.339	+ 8 54 14.60	5.869 9861

GEOCENTRIC COORDINATES FOR 0h DYNAMICAL TIME

Date	Apparent Right Ascension	Apparent Declination	True Geocentric Distance	Date	Apparent Right Ascension	Apparent Declination	True Geocentric Distance
	h m s	° ′ ″			h m s	° ′ ″	
July 1	10 46 51.169	+ 8 57 49.08	5.856 3494	Aug. 16	11 18 01.477	+ 5 40 22.16	6.335 1528
2	10 47 25.339	8 54 14.60	.869 9861	17	11 18 47.057	5 35 30.48	.341 5789
3	10 47 59.878	8 50 37.77	.883 5084	18	11 19 32.773	5 30 37.81	.347 8147
4	10 48 34.779	8 46 58.63	.896 9138	19	11 20 18.622	5 25 44.16	.353 8591
5	10 49 10.037	8 43 17.20	.910 1996	20	11 21 04.600	5 20 49.56	.359 7108
6	10 49 45.645	+ 8 39 33.51	5.923 3637	21	11 21 50.705	+ 5 15 54.04	6.365 3688
7	10 50 21.600	8 35 47.58	.936 4037	22	11 22 36.933	5 10 57.62	.370 8318
8	10 50 57.896	8 31 59.44	.949 3177	23	11 23 23.279	5 06 00.34	.376 0985
9	10 51 34.527	8 28 09.12	.962 1038	24	11 24 09.738	5 01 02.24	.381 1676
10	10 52 11.488	8 24 16.66	.974 7599	25	11 24 56.304	4 56 03.36	.386 0380
11	10 52 48.773	+ 8 20 22.09	5.987 2843	26	11 25 42.970	+ 4 51 03.74	6.390 7082
12	10 53 26.373	8 16 25.45	5.999 6751	27	11 26 29.731	4 46 03.43	.395 1772
13	10 54 04.284	8 12 26.78	6.011 9307	28	11 27 16.581	4 41 02.47	.399 4437
14	10 54 42.498	8 08 26.10	.024 0491	29	11 28 03.516	4 36 00.87	.403 5068
15	10 55 21.010	8 04 23.46	.036 0287	30	11 28 50.533	4 30 58.67	.407 3658
16	10 55 59.813	+ 8 00 18.87	6.047 8676	31	11 29 37.628	+ 4 25 55.89	6.411 0199
17	10 56 38.903	7 56 12.37	.059 5641	Sept. 1	11 30 24.799	4 20 52.56	.414 4687
18	10 57 18.274	7 52 03.98	.071 1163	2	11 31 12.041	4 15 48.71	.417 7118
19	10 57 57.922	7 47 53.72	.082 5224	3	11 31 59.349	4 10 44.38	.420 7488
20	10 58 37.844	7 43 41.60	.093 7805	4	11 32 46.718	4 05 39.60	.423 5796
21	10 59 18.035	+ 7 39 27.66	6.104 8889	5	11 33 34.141	+ 4 00 34.41	6.426 2039
22	10 59 58.492	7 35 11.91	.115 8455	6	11 34 21.615	3 55 28.86	.428 6215
23	11 00 39.211	7 30 54.36	.126 6484	7	11 35 09.134	3 50 22.97	.430 8322
24	11 01 20.188	7 26 35.06	.137 2958	8	11 35 56.693	3 45 16.78	.432 8357
25	11 02 01.419	7 22 14.01	.147 7855	9	11 36 44.289	3 40 10.31	.434 6318
26	11 02 42.901	+ 7 17 51.26	6.158 1156	10	11 37 31.916	+ 3 35 03.61	6.436 2203
27	11 03 24.626	7 13 26.84	.168 2841	11	11 38 19.573	3 29 56.68	.437 6010
28	11 04 06.588	7 09 00.79	.178 2887	12	11 39 07.256	3 24 49.57	.438 7736
29	11 04 48.781	7 04 33.16	.188 1276	13	11 39 54.962	3 19 42.28	.439 7378
30	11 05 31.196	7 00 03.99	.197 7986	14	11 40 42.688	3 14 34.85	.440 4934
31	11 06 13.827	+ 6 55 33.32	6.207 2998	15	11 41 30.433	+ 3 09 27.30	6.441 0399
Aug. 1	11 06 56.669	6 51 01.18	.216 6296	16	11 42 18.195	3 04 19.69	.441 3771
2	11 07 39.718	6 46 27.59	.225 7863	17	11 43 05.966	2 59 12.08	.441 5046
3	11 08 22.969	6 41 52.58	.234 7685	18	11 43 53.733	2 54 04.45	.441 4219
4	11 09 06.419	6 37 16.17	.243 5748	19	11 44 41.502	2 48 56.57	.441 1286
5	11 09 50.063	+ 6 32 38.39	6.252 2042	20	11 45 29.281	+ 2 43 48.69	6.440 6242
6	11 10 33.897	6 27 59.27	.260 6555	21	11 46 17.056	2 38 40.94	.439 9083
7	11 11 17.916	6 23 18.86	.268 9277	22	11 47 04.819	2 33 33.32	.438 9804
8	11 12 02.112	6 18 37.18	.277 0197	23	11 47 52.564	2 28 25.88	.437 8401
9	11 12 46.481	6 13 54.28	.284 9306	24	11 48 40.285	2 23 18.64	.436 4870
10	11 13 31.018	+ 6 09 10.18	6.292 6594	25	11 49 27.977	+ 2 18 11.64	6.434 9207
11	11 14 15.716	6 04 24.92	.300 2052	26	11 50 15.638	2 13 04.92	.433 1414
12	11 15 00.571	5 59 38.53	.307 5668	27	11 51 03.263	2 07 58.48	.431 1488
13	11 15 45.578	5 54 51.03	.314 7435	28	11 51 50.850	2 02 52.37	.428 9433
14	11 16 30.734	5 50 02.46	.321 7340	29	11 52 38.394	1 57 46.61	.426 5252
15	11 17 16.035	+ 5 45 12.83	6.328 5375	30	11 53 25.891	+ 1 52 41.24	6.423 8950
16	11 18 01.477	+ 5 40 22.16	6.335 1528	Oct. 1	11 54 13.335	+ 1 47 36.30	6.421 0532

GEOCENTRIC COORDINATES FOR 0^h DYNAMICAL TIME

Date	Apparent Right Ascension	Apparent Declination	True Geocentric Distance	Date	Apparent Right Ascension	Apparent Declination	True Geocentric Distance
	h m s	° ′ ″			h m s	° ′ ″	
Oct. 1	11 54 13.335	+ 1 47 36.30	6.421 0532	Nov. 16	12 28 16.073	− 1 48 11.46	6.074 1184
2	11 55 00.720	1 42 31.83	.418 0004	17	12 28 55.529	1 52 15.65	.062 2960
3	11 55 48.040	1 37 27.88	.414 7372	18	12 29 34.684	1 56 17.63	.050 3158
4	11 56 35.289	1 32 24.48	.411 2644	19	12 30 13.530	2 00 17.36	.038 1799
5	11 57 22.464	1 27 21.68	.407 5825	20	12 30 52.064	2 04 14.79	.025 8901
6	11 58 09.558	+ 1 22 19.50	6.403 6922	21	12 31 30.279	− 2 08 09.89	6.013 4487
7	11 58 56.568	1 17 17.98	.399 5941	22	12 32 08.169	2 12 02.63	6.000 8578
8	11 59 43.490	1 12 17.15	.395 2889	23	12 32 45.729	2 15 52.97	5.988 1201
9	12 00 30.319	1 07 17.04	.390 7773	24	12 33 22.951	2 19 40.87	.975 2380
10	12 01 17.054	1 02 17.67	.386 0598	25	12 33 59.828	2 23 26.28	.962 2144
11	12 02 03.690	+ 0 57 19.06	6.381 1371	26	12 34 36.350	− 2 27 09.15	5.949 0518
12	12 02 50.226	0 52 21.26	.376 0097	27	12 35 12.511	2 30 49.44	.935 7533
13	12 03 36.656	0 47 24.27	.370 6784	28	12 35 48.301	2 34 27.10	.922 3218
14	12 04 22.979	0 42 28.13	.365 1435	29	12 36 23.715	2 38 02.08	.908 7602
15	12 05 09.190	0 37 32.87	.359 4055	30	12 36 58.745	2 41 34.34	.895 0714
16	12 05 55.284	+ 0 32 38.53	6.353 4651	Dec. 1	12 37 33.385	− 2 45 03.85	5.881 2584
17	12 06 41.256	0 27 45.14	.347 3226	2	12 38 07.630	2 48 30.56	.867 3242
18	12 07 27.100	0 22 52.75	.340 9786	3	12 38 41.473	2 51 54.46	.853 2717
19	12 08 12.810	0 18 01.42	.334 4334	4	12 39 14.910	2 55 15.50	.839 1040
20	12 08 58.379	0 13 11.17	.327 6876	5	12 39 47.934	2 58 33.66	.824 8241
21	12 09 43.802	+ 0 08 22.07	6.320 7417	6	12 40 20.542	− 3 01 48.91	5.810 4348
22	12 10 29.073	+ 0 03 34.15	.313 5965	7	12 40 52.728	3 05 01.21	.795 9393
23	12 11 14.188	− 0 01 12.55	.306 2526	8	12 41 24.487	3 08 10.54	.781 3404
24	12 11 59.141	0 05 58.00	.298 7109	9	12 41 55.812	3 11 16.87	.766 6410
25	12 12 43.930	0 10 42.18	.290 9725	10	12 42 26.696	3 14 20.15	.751 8441
26	12 13 28.549	− 0 15 25.03	6.283 0386	11	12 42 57.133	− 3 17 20.34	5.736 9526
27	12 14 12.993	0 20 06.54	.274 9106	12	12 43 27.113	3 20 17.40	.721 9691
28	12 14 57.256	0 24 46.66	.266 5900	13	12 43 56.630	3 23 11.26	.706 8967
29	12 15 41.331	0 29 25.33	.258 0783	14	12 44 25.673	3 26 01.89	.691 7381
30	12 16 25.211	0 34 02.52	.249 3772	15	12 44 54.237	3 28 49.23	.676 4965
31	12 17 08.889	− 0 38 38.18	6.240 4883	16	12 45 22.314	− 3 31 33.24	5.661 1748
Nov. 1	12 17 52.359	0 43 12.26	.231 4134	17	12 45 49.899	3 34 13.89	.645 7763
2	12 18 35.615	0 47 44.74	.222 1541	18	12 46 16.985	3 36 51.15	.630 3044
3	12 19 18.652	0 52 15.56	.212 7121	19	12 46 43.567	3 39 24.98	.614 7627
4	12 20 01.464	0 56 44.69	.203 0893	20	12 47 09.637	3 41 55.35	.599 1547
5	12 20 44.048	− 1 01 12.11	6.193 2872	21	12 47 35.188	− 3 44 22.22	5.583 4843
6	12 21 26.399	1 05 37.78	.183 3076	22	12 48 00.211	3 46 45.55	.567 7554
7	12 22 08.512	1 10 01.68	.173 1522	23	12 48 24.699	3 49 05.29	.551 9721
8	12 22 50.383	1 14 23.76	.162 8227	24	12 48 48.643	3 51 21.40	.536 1385
9	12 23 32.009	1 18 44.01	.152 3209	25	12 49 12.035	3 53 33.85	.520 2587
10	12 24 13.385	− 1 23 02.40	6.141 6483	26	12 49 34.867	− 3 55 42.57	5.504 3371
11	12 24 54.506	1 27 18.89	.130 8067	27	12 49 57.133	3 57 47.55	.488 3777
12	12 25 35.367	1 31 33.45	.119 7976	28	12 50 18.826	3 59 48.74	.472 3850
13	12 26 15.961	1 35 46.04	.108 6227	29	12 50 39.940	4 01 46.11	.456 3630
14	12 26 56.281	1 39 56.60	.097 2834	30	12 51 00.469	4 03 39.65	.440 3162
15	12 27 36.321	− 1 44 05.09	6.085 7815	31	12 51 20.410	− 4 05 29.31	5.424 2488
16	12 28 16.073	− 1 48 11.46	6.074 1184	32	12 51 39.757	− 4 07 15.09	5.408 1650

GEOCENTRIC COORDINATES FOR 0^h DYNAMICAL TIME

Date	Apparent Right Ascension	Apparent Declination	True Geocentric Distance	Date	Apparent Right Ascension	Apparent Declination	True Geocentric Distance
	h m s	° ′ ″			h m s	° ′ ″	
Jan. 0	20 33 03.635	−19 24 02.34	10.802 8962	Feb. 15	20 55 09.234	−18 02 13.65	10.881 7259
1	20 33 31.113	19 22 26.22	.810 2816	16	20 55 37.736	18 00 22.43	.877 4073
2	20 33 58.726	19 20 49.38	.817 4295	17	20 56 06.147	17 58 31.36	.872 8402
3	20 34 26.468	19 19 11.84	.824 3380	18	20 56 34.460	17 56 40.46	.868 0258
4	20 34 54.333	19 17 33.63	.831 0053	19	20 57 02.673	17 54 49.72	.862 9653
5	20 35 22.314	−19 15 54.77	10.837 4295	20	20 57 30.783	−17 52 59.17	10.857 6597
6	20 35 50.406	19 14 15.26	.843 6091	21	20 57 58.789	17 51 08.80	.852 1100
7	20 36 18.601	19 12 35.13	.849 5426	22	20 58 26.689	17 49 18.63	.846 3173
8	20 36 46.896	19 10 54.38	.855 2286	23	20 58 54.479	17 47 28.68	.840 2824
9	20 37 15.284	19 09 13.04	.860 6659	24	20 59 22.156	17 45 38.99	.834 0064
10	20 37 43.760	−19 07 31.11	10.865 8532	25	20 59 49.714	−17 43 49.58	10.827 4906
11	20 38 12.321	19 05 48.61	.870 7895	26	21 00 17.150	17 42 00.49	.820 7362
12	20 38 40.962	19 04 05.54	.875 4740	27	21 00 44.457	17 40 11.75	.813 7445
13	20 39 09.679	19 02 21.91	.879 9058	28	21 01 11.629	17 38 23.38	.806 5170
14	20 39 38.469	19 00 37.75	.884 0842	29	21 01 38.661	17 36 35.41	.799 0552
15	20 40 07.327	−18 58 53.07	10.888 0085	Mar. 1	21 02 05.548	−17 34 47.87	10.791 3608
16	20 40 36.250	18 57 07.89	.891 6785	2	21 02 32.284	17 33 00.78	.783 4355
17	20 41 05.233	18 55 22.25	.895 0935	3	21 02 58.865	17 31 14.15	.775 2813
18	20 41 34.269	18 53 36.16	.898 2534	4	21 03 25.286	17 29 28.01	.766 9001
19	20 42 03.351	18 51 49.66	.901 1577	5	21 03 51.544	17 27 42.38	.758 2939
20	20 42 32.473	−18 50 02.77	10.903 8061	6	21 04 17.634	−17 25 57.27	10.749 4650
21	20 43 01.628	18 48 15.49	.906 1983	7	21 04 43.553	17 24 12.70	.740 4157
22	20 43 30.813	18 46 27.84	.908 3336	8	21 05 09.299	17 22 28.70	.731 1484
23	20 44 00.022	18 44 39.80	.910 2114	9	21 05 34.867	17 20 45.28	.721 6655
24	20 44 29.256	18 42 51.39	.911 8312	10	21 06 00.255	17 19 02.48	.711 9698
25	20 44 58.511	−18 41 02.61	10.913 1923	11	21 06 25.458	−17 17 20.32	10.702 0638
26	20 45 27.785	18 39 13.48	.914 2941	12	21 06 50.471	17 15 38.83	.691 9503
27	20 45 57.074	18 37 24.04	.915 1359	13	21 07 15.290	17 13 58.05	.681 6320
28	20 46 26.375	18 35 34.31	.915 7174	14	21 07 39.909	17 12 18.00	.671 1118
29	20 46 55.687	18 33 44.43	.916 0381	15	21 08 04.322	17 10 38.71	.660 3923
30	20 47 24.965	−18 31 54.69	10.916 0977	16	21 08 28.524	−17 09 00.20	10.649 4764
31	20 47 54.233	18 30 03.78	.915 8960	17	21 08 52.512	17 07 22.48	.638 3666
Feb. 1	20 48 23.518	18 28 12.95	.915 4330	18	21 09 16.282	17 05 45.56	.627 0655
2	20 48 52.781	18 26 22.05	.914 7087	19	21 09 39.833	17 04 09.43	.615 5756
3	20 49 22.019	18 24 31.02	.913 7231	20	21 10 03.165	17 02 34.12	.603 8993
4	20 49 51.225	−18 22 39.88	10.912 4766	21	21 10 26.275	−17 00 59.65	10.592 0390
5	20 50 20.396	18 20 48.62	.910 9695	22	21 10 49.160	16 59 26.03	.579 9971
6	20 50 49.526	18 18 57.26	.909 2023	23	21 11 11.816	16 57 53.31	.567 7762
7	20 51 18.612	18 17 05.82	.907 1756	24	21 11 34.239	16 56 21.51	.555 3787
8	20 51 47.649	18 15 14.32	.904 8900	25	21 11 56.425	16 54 50.67	.542 8074
9	20 52 16.634	−18 13 22.76	10.902 3465	26	21 12 18.366	−16 53 20.82	10.530 0648
10	20 52 45.564	18 11 31.18	.899 5459	27	21 12 40.060	16 51 51.98	.517 1540
11	20 53 14.434	18 09 39.58	.896 4892	28	21 13 01.500	16 50 24.18	.504 0779
12	20 53 43.241	18 07 48.00	.893 1776	29	21 13 22.682	16 48 57.43	.490 8394
13	20 54 11.980	18 05 56.47	.889 6124	30	21 13 43.602	16 47 31.77	.477 4416
14	20 54 40.647	−18 04 05.00	10.885 7947	31	21 14 04.256	−16 46 07.21	10.463 8879
15	20 55 09.234	−18 02 13.65	10.881 7259	Apr. 1	21 14 24.640	−16 44 43.76	10.450 1815

GEOCENTRIC COORDINATES FOR 0^h DYNAMICAL TIME

Date	Apparent Right Ascension	Apparent Declination	True Geocentric Distance	Date	Apparent Right Ascension	Apparent Declination	True Geocentric Distance
	h m s	° ′ ″			h m s	° ′ ″	
Apr. 1	21 14 24.640	−16 44 43.76	10.450 1815	May 17	21 24 18.929	−16 06 19.54	9.720 3089
2	21 14 44.751	16 43 21.45	.436 3259	18	21 24 23.442	16 06 08.48	.703 8165
3	21 15 04.587	16 42 00.29	.422 3245	19	21 24 27.573	16 05 59.24	.687 3662
4	21 15 24.145	16 40 40.30	.408 1809	20	21 24 31.318	16 05 51.83	.670 9622
5	21 15 43.422	16 39 21.50	.393 8990	21	21 24 34.676	16 05 46.26	.654 6088
6	21 16 02.416	−16 38 03.92	10.379 4824	22	21 24 37.645	−16 05 42.55	9.638 3103
7	21 16 21.123	16 36 47.57	.364 9351	23	21 24 40.224	16 05 40.68	.622 0713
8	21 16 39.539	16 35 32.50	.350 2610	24	21 24 42.412	16 05 40.66	.605 8962
9	21 16 57.660	16 34 18.73	.335 4640	25	21 24 44.209	16 05 42.48	.589 7896
10	21 17 15.481	16 33 06.28	.320 5481	26	21 24 45.614	16 05 46.15	.573 7561
11	21 17 32.997	−16 31 55.18	10.305 5172	27	21 24 46.629	−16 05 51.64	9.557 8003
12	21 17 50.205	16 30 45.45	.290 3751	28	21 24 47.253	16 05 58.97	.541 9269
13	21 18 07.100	16 29 37.09	.275 1257	29	21 24 47.489	16 06 08.12	.526 1408
14	21 18 23.680	16 28 30.11	.259 7727	30	21 24 47.338	16 06 19.09	.510 4468
15	21 18 39.946	16 27 24.52	.244 3198	31	21 24 46.800	16 06 31.88	.494 8497
16	21 18 55.894	−16 26 20.32	10.228 7704	June 1	21 24 45.876	−16 06 46.49	9.479 3545
17	21 19 11.526	16 25 17.52	.213 1282	2	21 24 44.566	16 07 02.92	.463 9661
18	21 19 26.840	16 24 16.15	.197 3967	3	21 24 42.870	16 07 21.18	.448 6894
19	21 19 41.833	16 23 16.22	.181 5793	4	21 24 40.786	16 07 41.27	.433 5292
20	21 19 56.501	16 22 17.77	.165 6796	5	21 24 38.315	16 08 03.18	.418 4903
21	21 20 10.842	−16 21 20.82	10.149 7012	6	21 24 35.456	−16 08 26.89	9.403 5773
22	21 20 24.849	16 20 25.39	.133 6479	7	21 24 32.213	16 08 52.38	.388 7946
23	21 20 38.521	16 19 31.52	.117 5235	8	21 24 28.588	16 09 19.61	.374 1466
24	21 20 51.852	16 18 39.21	.101 3317	9	21 24 24.586	16 09 48.58	.359 6373
25	21 21 04.838	16 17 48.50	.085 0766	10	21 24 20.210	16 10 19.25	.345 2711
26	21 21 17.477	−16 16 59.38	10.068 7622	11	21 24 15.465	−16 10 51.61	9.331 0518
27	21 21 29.766	16 16 11.88	.052 3927	12	21 24 10.354	16 11 25.65	.316 9833
28	21 21 41.701	16 15 26.01	.035 9721	13	21 24 04.879	16 12 01.36	.303 0697
29	21 21 53.282	16 14 41.78	.019 5049	14	21 23 59.041	16 12 38.73	.289 3148
30	21 22 04.505	16 13 59.19	10.002 9953	15	21 23 52.843	16 13 17.75	.275 7226
May 1	21 22 15.371	−16 13 18.26	9.986 4478	16	21 23 46.286	−16 13 58.43	9.262 2968
2	21 22 25.877	16 12 39.00	.969 8670	17	21 23 39.371	16 14 40.73	.249 0415
3	21 22 36.022	16 12 01.43	.953 2575	18	21 23 32.100	16 15 24.66	.235 9606
4	21 22 45.805	16 11 25.55	.936 6240	19	21 23 24.474	16 16 10.19	.223 0581
5	21 22 55.223	16 10 51.39	.919 9711	20	21 23 16.498	16 16 57.29	.210 3379
6	21 23 04.272	−16 10 18.98	9.903 3038	21	21 23 08.173	−16 17 45.96	9.197 8041
7	21 23 12.950	16 09 48.32	.886 6266	22	21 22 59.504	16 18 36.15	.185 4608
8	21 23 21.253	16 09 19.44	.869 9444	23	21 22 50.495	16 19 27.84	.173 3118
9	21 23 29.179	16 08 52.35	.853 2616	24	21 22 41.151	16 20 21.00	.161 3613
10	21 23 36.725	16 08 27.04	.836 5830	25	21 22 31.476	16 21 15.62	.149 6134
11	21 23 43.890	−16 08 03.51	9.819 9128	26	21 22 21.475	−16 22 11.65	9.138 0720
12	21 23 50.676	16 07 41.75	.803 2555	27	21 22 11.154	16 23 09.07	.126 7414
13	21 23 57.083	16 07 21.76	.786 6153	28	21 22 00.518	16 24 07.87	.115 6254
14	21 24 03.111	16 07 03.54	.769 9964	29	21 21 49.571	16 25 08.02	.104 7283
15	21 24 08.762	16 06 47.09	.753 4030	30	21 21 38.316	16 26 09.51	.094 0539
16	21 24 14.035	−16 06 32.42	9.736 8391	July 1	21 21 26.758	−16 27 12.31	9.083 6062
17	21 24 18.929	−16 06 19.54	9.720 3089	2	21 21 14.899	−16 28 16.40	9.073 3890

GEOCENTRIC COORDINATES FOR 0^h DYNAMICAL TIME

Date	Apparent Right Ascension	Apparent Declination	True Geocentric Distance	Date	Apparent Right Ascension	Apparent Declination	True Geocentric Distance
	h m s	° ′ ″			h m s	° ′ ″	
July 1	21 21 26.758	−16 27 12.31	9.083 6062	Aug. 16	21 09 06.561	−17 28 01.36	8.891 8699
2	21 21 14.899	16 28 16.40	.073 3890	17	21 08 49.003	17 29 22.09	.894 5366
3	21 21 02.744	16 29 21.73	.063 4058	18	21 08 31.525	17 30 42.23	.897 4982
4	21 20 50.298	16 30 28.28	.053 6601	19	21 08 14.137	17 32 01.73	.900 7538
5	21 20 37.570	16 31 35.98	.044 1551	20	21 07 56.848	17 33 20.55	.904 3028
6	21 20 24.566	−16 32 44.79	9.034 8937	21	21 07 39.667	−17 34 38.67	8.908 1443
7	21 20 11.297	16 33 54.67	.025 8788	22	21 07 22.605	17 35 56.04	.912 2773
8	21 19 57.770	16 35 05.57	.017 1132	23	21 07 05.670	17 37 12.65	.916 7008
9	21 19 43.992	16 36 17.46	.008 5993	24	21 06 48.868	17 38 28.45	.921 4136
10	21 19 29.971	16 37 30.32	9.000 3398	25	21 06 32.209	17 39 43.44	.926 4146
11	21 19 15.712	−16 38 44.11	8.992 3371	26	21 06 15.698	−17 40 57.57	8.931 7024
12	21 19 01.222	16 39 58.81	.984 5936	27	21 05 59.343	17 42 10.80	.937 2752
13	21 18 46.505	16 41 14.39	.977 1117	28	21 05 43.154	17 43 23.09	.943 1312
14	21 18 31.568	16 42 30.82	.969 8937	29	21 05 27.139	17 44 34.39	.949 2684
15	21 18 16.417	16 43 48.07	.962 9421	30	21 05 11.311	17 45 44.64	.955 6843
16	21 18 01.056	−16 45 06.09	8.956 2590	31	21 04 55.680	−17 46 53.81	8.962 3765
17	21 17 45.494	16 46 24.86	.949 8467	Sept. 1	21 04 40.257	17 48 01.87	.969 3423
18	21 17 29.737	16 47 44.32	.943 7075	2	21 04 25.051	17 49 08.78	.976 5787
19	21 17 13.793	16 49 04.43	.937 8435	3	21 04 10.068	17 50 14.54	.984 0830
20	21 16 57.670	16 50 25.15	.932 2571	4	21 03 55.317	17 51 19.13	.991 8522
21	21 16 41.377	−16 51 46.44	8.926 9502	5	21 03 40.802	−17 52 22.52	8.999 8834
22	21 16 24.921	16 53 08.25	.921 9251	6	21 03 26.531	17 53 24.70	9.008 1735
23	21 16 08.313	16 54 30.54	.917 1837	7	21 03 12.509	17 54 25.65	.016 7195
24	21 15 51.562	16 55 53.26	.912 7282	8	21 02 58.741	17 55 25.34	.025 5183
25	21 15 34.675	16 57 16.39	.908 5603	9	21 02 45.235	17 56 23.75	.034 5670
26	21 15 17.661	−16 58 39.87	8.904 6822	10	21 02 31.997	−17 57 20.84	9.043 8624
27	21 15 00.529	17 00 03.69	.901 0955	11	21 02 19.033	17 58 16.61	.053 4014
28	21 14 43.285	17 01 27.81	.897 8020	12	21 02 06.349	17 59 11.01	.063 1809
29	21 14 25.936	17 02 52.18	.894 8034	13	21 01 53.954	18 00 04.03	.073 1977
30	21 14 08.490	17 04 16.78	.892 1009	14	21 01 41.853	18 00 55.63	.083 4487
31	21 13 50.955	−17 05 41.54	8.889 6957	15	21 01 30.054	−18 01 45.80	9.093 9305
Aug. 1	21 13 33.340	17 07 06.40	.887 5887	16	21 01 18.564	18 02 34.52	.104 6401
2	21 13 15.658	17 08 31.31	.885 7806	17	21 01 07.389	18 03 21.76	.115 5741
3	21 12 57.920	17 09 56.21	.884 2717	18	21 00 56.535	18 04 07.53	.126 7292
4	21 12 40.137	17 11 21.05	.883 0624	19	21 00 46.007	18 04 51.79	.138 1021
5	21 12 22.321	−17 12 45.79	8.882 1527	20	21 00 35.810	−18 05 34.56	9.149 6894
6	21 12 04.480	17 14 10.39	.881 5426	21	21 00 25.948	18 06 15.80	.161 4874
7	21 11 46.624	17 15 34.83	.881 2321	22	21 00 16.424	18 06 55.53	.173 4926
8	21 11 28.761	17 16 59.06	.881 2208	23	21 00 07.242	18 07 33.70	.185 7011
9	21 11 10.898	17 18 23.06	.881 5088	24	20 59 58.407	18 08 10.31	.198 1091
10	21 10 53.044	−17 19 46.79	8.882 0955	25	20 59 49.924	−18 08 45.32	9.210 7125
11	21 10 35.207	17 21 10.22	.882 9808	26	20 59 41.799	18 09 18.71	.223 5068
12	21 10 17.394	17 22 33.30	.884 1642	27	20 59 34.040	18 09 50.44	.236 4876
13	21 09 59.614	17 23 56.00	.885 6453	28	20 59 26.652	18 10 20.51	.249 6504
14	21 09 41.876	17 25 18.27	.887 4236	29	20 59 19.643	18 10 48.89	.262 9902
15	21 09 24.188	−17 26 40.07	8.889 4987	30	20 59 13.014	−18 11 15.60	9.276 5025
16	21 09 06.561	−17 28 01.36	8.891 8699	Oct. 1	20 59 06.770	−18 11 40.64	9.290 1823

GEOCENTRIC COORDINATES FOR 0^h DYNAMICAL TIME

Date	Apparent Right Ascension	Apparent Declination	True Geocentric Distance	Date	Apparent Right Ascension	Apparent Declination	True Geocentric Distance
	h m s	° ′ ″			h m s	° ′ ″	
Oct. 1	20 59 06.770	−18 11 40.64	9.290 1823	Nov. 16	21 01 33.497	−18 00 06.73	10.019 6751
2	20 59 00.911	18 12 04.00	.304 0249	17	21 01 45.982	17 59 12.37	.035 9225
3	20 58 55.439	18 12 25.70	.318 0256	18	21 01 58.829	17 58 16.43	.052 1130
4	20 58 50.356	18 12 45.72	.332 1797	19	21 02 12.037	17 57 18.93	.068 2421
5	20 58 45.663	18 13 04.06	.346 4825	20	21 02 25.603	17 56 19.85	.084 3051
6	20 58 41.361	−18 13 20.71	9.360 9293	21	21 02 39.529	−17 55 19.19	10.100 2974
7	20 58 37.452	18 13 35.67	.375 5155	22	21 02 53.811	17 54 16.97	.116 2143
8	20 58 33.939	18 13 48.93	.390 2366	23	21 03 08.448	17 53 13.19	.132 0511
9	20 58 30.823	18 14 00.48	.405 0879	24	21 03 23.438	17 52 07.88	.147 8030
10	20 58 28.107	18 14 10.31	.420 0650	25	21 03 38.774	17 51 01.05	.163 4655
11	20 58 25.793	−18 14 18.42	9.435 1633	26	21 03 54.453	−17 49 52.74	10.179 0340
12	20 58 23.884	18 14 24.79	.450 3782	27	21 04 10.469	17 48 42.96	.194 5041
13	20 58 22.381	18 14 29.42	.465 7055	28	21 04 26.816	17 47 31.72	.209 8713
14	20 58 21.287	18 14 32.32	.481 1405	29	21 04 43.490	17 46 19.05	.225 1316
15	20 58 20.602	18 14 33.49	.496 6790	30	21 05 00.485	17 45 04.94	.240 2807
16	20 58 20.329	−18 14 32.93	9.512 3164	Dec. 1	21 05 17.798	−17 43 49.42	10.255 3147
17	20 58 20.465	18 14 30.65	.528 0483	2	21 05 35.424	17 42 32.49	.270 2296
18	20 58 21.012	18 14 26.65	.543 8701	3	21 05 53.360	17 41 14.15	.285 0217
19	20 58 21.969	18 14 20.94	.559 7774	4	21 06 11.602	17 39 54.42	.299 6872
20	20 58 23.334	18 14 13.52	.575 7654	5	21 06 30.147	17 38 33.31	.314 2225
21	20 58 25.107	−18 14 04.37	9.591 8294	6	21 06 48.991	−17 37 10.82	10.328 6241
22	20 58 27.289	18 13 53.49	.607 9646	7	21 07 08.132	17 35 46.98	.342 8884
23	20 58 29.881	18 13 40.87	.624 1659	8	21 07 27.565	17 34 21.79	.357 0123
24	20 58 32.886	18 13 26.49	.640 4283	9	21 07 47.286	17 32 55.27	.370 9923
25	20 58 36.305	18 13 10.35	.656 7465	10	21 08 07.292	17 31 27.45	.384 8253
26	20 58 40.141	−18 12 52.45	9.673 1153	11	21 08 27.576	−17 29 58.35	10.398 5081
27	20 58 44.392	18 12 32.79	.689 5294	12	21 08 48.132	17 28 27.99	.412 0376
28	20 58 49.059	18 12 11.40	.705 9834	13	21 09 08.955	17 26 56.38	.425 4106
29	20 58 54.139	18 11 48.29	.722 4722	14	21 09 30.040	17 25 23.54	.438 6239
30	20 58 59.629	18 11 23.48	.738 9905	15	21 09 51.381	17 23 49.47	.451 6741
31	20 59 05.525	−18 10 56.97	9.755 5332	16	21 10 12.976	−17 22 14.17	10.464 5581
Nov. 1	20 59 11.826	18 10 28.77	.772 0955	17	21 10 34.823	17 20 37.64	.477 2723
2	20 59 18.529	18 09 58.90	.788 6722	18	21 10 56.919	17 18 59.89	.489 8133
3	20 59 25.631	18 09 27.34	.805 2586	19	21 11 19.261	17 17 20.93	.502 1777
4	20 59 33.131	18 08 54.12	.821 8499	20	21 11 41.847	17 15 40.76	.514 3620
5	20 59 41.026	−18 08 19.22	9.838 4413	21	21 12 04.672	−17 13 59.43	10.526 3629
6	20 59 49.315	18 07 42.66	.855 0283	22	21 12 27.732	17 12 16.95	.538 1769
7	20 59 57.997	18 07 04.42	.871 6062	23	21 12 51.020	17 10 33.34	.549 8009
8	21 00 07.070	18 06 24.53	.888 1706	24	21 13 14.531	17 08 48.64	.561 2316
9	21 00 16.534	18 05 42.98	.904 7170	25	21 13 38.258	17 07 02.86	.572 4661
10	21 00 26.385	−18 04 59.78	9.921 2409	26	21 14 02.194	−17 05 16.03	10.583 5014
11	21 00 36.623	18 04 14.94	.937 7382	27	21 14 26.335	17 03 28.17	.594 3347
12	21 00 47.245	18 03 28.48	.954 2046	28	21 14 50.676	17 01 39.28	.604 9634
13	21 00 58.247	18 02 40.42	.970 6357	29	21 15 15.210	16 59 49.39	.615 3848
14	21 01 09.626	18 01 50.76	.987 0273	30	21 15 39.934	16 57 58.50	.625 5966
15	21 01 21.378	−18 00 59.53	10.003 3752	31	21 16 04.843	−16 56 06.62	10.635 5965
16	21 01 33.497	−18 00 06.73	10.019 6751	32	21 16 29.934	−16 54 13.79	10.645 3822

GEOCENTRIC COORDINATES FOR 0^h DYNAMICAL TIME

Date	Apparent Right Ascension	Apparent Declination	True Geocentric Distance	Date	Apparent Right Ascension	Apparent Declination	True Geocentric Distance
	h m s	° ′ ″			h m s	° ′ ″	
Jan. 0	18 59 20.390	−23 06 29.27	20.488 755	Feb. 15	19 10 35.582	−22 49 41.47	20.270 980
1	18 59 35.871	23 06 07.68	.490 230	16	19 10 48.302	22 49 21.32	.260 282
2	18 59 51.368	23 05 46.01	.491 419	17	19 11 00.887	22 49 01.36	.249 369
3	19 00 06.879	23 05 24.27	.492 323	18	19 11 13.335	22 48 41.58	.238 243
4	19 00 22.405	23 05 02.54	.492 941	19	19 11 25.643	22 48 21.99	.226 907
5	19 00 37.905	−23 04 41.75	20.493 274	20	19 11 37.811	−22 48 02.56	20.215 365
6	19 00 53.371	23 04 18.81	.493 320	21	19 11 49.838	22 47 43.30	.203 619
7	19 01 08.888	23 03 56.67	.493 081	22	19 12 01.725	22 47 24.21	.191 671
8	19 01 24.387	23 03 34.58	.492 555	23	19 12 13.469	22 47 05.32	.179 526
9	19 01 39.869	23 03 12.44	.491 745	24	19 12 25.070	22 46 46.63	.167 186
10	19 01 55.333	−23 02 50.25	20.490 649	25	19 12 36.523	−22 46 28.17	20.154 653
11	19 02 10.776	23 02 28.01	.489 269	26	19 12 47.826	22 46 09.94	.141 933
12	19 02 26.198	23 02 05.70	.487 605	27	19 12 58.974	22 45 51.96	.129 027
13	19 02 41.595	23 01 43.34	.485 659	28	19 13 09.964	22 45 34.24	.115 940
14	19 02 56.967	23 01 20.93	.483 431	29	19 13 20.793	22 45 16.78	.102 675
15	19 03 12.310	−23 00 58.49	20.480 922	Mar. 1	19 13 31.456	−22 44 59.58	20.089 236
16	19 03 27.623	23 00 36.02	.478 133	2	19 13 41.952	22 44 42.65	.075 627
17	19 03 42.902	23 00 13.54	.475 066	3	19 13 52.276	22 44 25.99	.061 851
18	19 03 58.142	22 59 51.08	.471 721	4	19 14 02.429	22 44 09.60	.047 914
19	19 04 13.339	22 59 28.64	.468 101	5	19 14 12.407	22 43 53.47	.033 819
20	19 04 28.485	−22 59 06.22	20.464 205	6	19 14 22.209	−22 43 37.61	20.019 569
21	19 04 43.577	22 58 43.83	.460 035	7	19 14 31.835	22 43 22.03	20.005 171
22	19 04 58.612	22 58 21.45	.455 593	8	19 14 41.283	22 43 06.72	19.990 628
23	19 05 13.587	22 57 59.06	.450 879	9	19 14 50.554	22 42 51.69	.975 944
24	19 05 28.501	22 57 36.67	.445 893	10	19 14 59.645	22 42 36.96	.961 124
25	19 05 43.355	−22 57 14.28	20.440 638	11	19 15 08.555	−22 42 22.55	19.946 174
26	19 05 58.146	22 56 51.90	.435 114	12	19 15 17.281	22 42 08.46	.931 096
27	19 06 12.873	22 56 29.55	.429 322	13	19 15 25.821	22 41 54.71	.915 896
28	19 06 27.532	22 56 07.24	.423 264	14	19 15 34.170	22 41 41.30	.900 579
29	19 06 42.119	22 55 44.99	.416 942	15	19 15 42.326	22 41 28.24	.885 148
30	19 06 56.630	−22 55 22.80	20.410 356	16	19 15 50.285	−22 41 15.53	19.869 609
31	19 07 11.060	22 55 00.69	.403 508	17	19 15 58.046	22 41 03.15	.853 964
Feb. 1	19 07 25.407	22 54 38.67	.396 401	18	19 16 05.608	22 40 51.10	.838 220
2	19 07 39.664	22 54 16.75	.389 037	19	19 16 12.972	22 40 39.36	.822 379
3	19 07 53.829	22 53 54.92	.381 416	20	19 16 20.139	22 40 27.94	.806 446
4	19 08 07.897	−22 53 33.18	20.373 543	21	19 16 27.109	−22 40 16.85	19.790 425
5	19 08 21.866	22 53 11.55	.365 419	22	19 16 33.883	22 40 06.10	.774 321
6	19 08 35.732	22 52 50.01	.357 046	23	19 16 40.457	22 39 55.70	.758 136
7	19 08 49.494	22 52 28.57	.348 428	24	19 16 46.831	22 39 45.68	.741 876
8	19 09 03.148	22 52 07.24	.339 567	25	19 16 53.000	22 39 36.04	.725 546
9	19 09 16.694	−22 51 46.01	20.330 465	26	19 16 58.964	−22 39 26.78	19.709 149
10	19 09 30.131	22 51 24.89	.321 126	27	19 17 04.718	22 39 17.92	.692 691
11	19 09 43.455	22 51 03.90	.311 553	28	19 17 10.261	22 39 09.45	.676 176
12	19 09 56.665	22 50 43.05	.301 749	29	19 17 15.591	22 39 01.36	.659 608
13	19 10 09.759	22 50 22.34	.291 716	30	19 17 20.707	22 38 53.67	.642 993
14	19 10 22.733	−22 50 01.81	20.281 459	31	19 17 25.607	−22 38 46.37	19.626 336
15	19 10 35.582	−22 49 41.47	20.270 980	Apr. 1	19 17 30.291	−22 38 39.44	19.609 642

GEOCENTRIC COORDINATES FOR 0^h DYNAMICAL TIME

Date	Apparent Right Ascension	Apparent Declination	True Geocentric Distance	Date	Apparent Right Ascension	Apparent Declination	True Geocentric Distance
	h m s	° ′ ″			h m s	° ′ ″	
Apr. 1	19 17 30.291	−22 38 39.44	19.609 642	May 17	19 17 12.749	−22 40 28.61	18.886 240
2	19 17 34.759	22 38 32.90	.592 915	18	19 17 07.548	22 40 39.65	.873 243
3	19 17 39.012	22 38 26.74	.576 162	19	19 17 02.166	22 40 51.00	.860 431
4	19 17 43.049	22 38 20.95	.559 386	20	19 16 56.605	22 41 02.68	.847 807
5	19 17 46.872	22 38 15.55	.542 593	21	19 16 50.865	22 41 14.68	.835 376
6	19 17 50.480	−22 38 10.55	19.525 789	22	19 16 44.948	−22 41 26.98	18.823 140
7	19 17 53.874	22 38 05.94	.508 979	23	19 16 38.855	22 41 39.58	.811 104
8	19 17 57.052	22 38 01.75	.492 167	24	19 16 32.588	22 41 52.47	.799 271
9	19 18 00.012	22 37 57.98	.475 360	25	19 16 26.150	22 42 05.63	.787 646
10	19 18 02.754	22 37 54.65	.458 561	26	19 16 19.545	22 42 19.04	.776 232
11	19 18 05.274	−22 37 51.73	19.441 777	27	19 16 12.776	−22 42 32.71	18.765 032
12	19 18 07.571	22 37 49.24	.425 011	28	19 16 05.847	22 42 46.62	.754 051
13	19 18 09.645	22 37 47.16	.408 269	29	19 15 58.762	22 43 00.76	.743 292
14	19 18 11.497	22 37 45.48	.391 556	30	19 15 51.525	22 43 15.13	.732 760
15	19 18 13.128	22 37 44.18	.374 875	31	19 15 44.141	22 43 29.73	.722 456
16	19 18 14.541	−22 37 43.26	19.358 232	June 1	19 15 36.611	−22 43 44.56	18.712 386
17	19 18 15.737	22 37 42.72	.341 631	2	19 15 28.939	22 43 59.63	.702 552
18	19 18 16.719	22 37 42.56	.325 076	3	19 15 21.126	22 44 14.94	.692 958
19	19 18 17.486	22 37 42.81	.308 571	4	19 15 13.172	22 44 30.47	.683 607
20	19 18 18.038	22 37 43.46	.292 122	5	19 15 05.081	22 44 46.23	.674 502
21	19 18 18.374	−22 37 44.52	19.275 733	6	19 14 56.853	−22 45 02.18	18.665 646
22	19 18 18.493	22 37 46.01	.259 408	7	19 14 48.494	22 45 18.31	.657 041
23	19 18 18.395	22 37 47.92	.243 152	8	19 14 40.007	22 45 34.59	.648 689
24	19 18 18.077	22 37 50.24	.226 970	9	19 14 31.399	22 45 51.02	.640 593
25	19 18 17.540	22 37 52.98	.210 866	10	19 14 22.676	22 46 07.58	.632 755
26	19 18 16.784	−22 37 56.12	19.194 846	11	19 14 13.842	−22 46 24.27	18.625 176
27	19 18 15.810	22 37 59.66	.178 915	12	19 14 04.902	22 46 41.09	.617 860
28	19 18 14.618	22 38 03.60	.163 076	13	19 13 55.859	22 46 58.04	.610 808
29	19 18 13.210	22 38 07.91	.147 336	14	19 13 46.716	22 47 15.13	.604 021
30	19 18 11.588	22 38 12.60	.131 699	15	19 13 37.475	22 47 32.34	.597 503
May 1	19 18 09.754	−22 38 17.66	19.116 170	16	19 13 28.140	−22 47 49.69	18.591 253
2	19 18 07.711	22 38 23.10	.100 754	17	19 13 18.712	22 48 07.16	.585 276
3	19 18 05.460	22 38 28.90	.085 456	18	19 13 09.193	22 48 24.74	.579 572
4	19 18 03.004	22 38 35.08	.070 280	19	19 12 59.587	22 48 42.42	.574 143
5	19 18 00.343	22 38 41.64	.055 233	20	19 12 49.898	22 49 00.18	.568 991
6	19 17 57.478	−22 38 48.60	19.040 318	21	19 12 40.129	−22 49 18.02	18.564 118
7	19 17 54.408	22 38 55.95	.025 540	22	19 12 30.285	22 49 35.91	.559 525
8	19 17 51.133	22 39 03.68	19.010 904	23	19 12 20.371	22 49 53.86	.555 215
9	19 17 47.653	22 39 11.79	18.996 415	24	19 12 10.391	22 50 11.83	.551 189
10	19 17 43.970	22 39 20.26	.982 075	25	19 12 00.351	22 50 29.84	.547 449
11	19 17 40.085	−22 39 29.08	18.967 890	26	19 11 50.256	−22 50 47.86	18.543 996
12	19 17 36.002	22 39 38.22	.953 864	27	19 11 40.112	22 51 05.90	.540 831
13	19 17 31.726	22 39 47.67	.939 999	28	19 11 29.923	22 51 23.96	.537 957
14	19 17 27.259	22 39 57.44	.926 301	29	19 11 19.693	22 51 42.04	.535 374
15	19 17 22.606	22 40 07.51	.912 773	30	19 11 09.424	22 52 00.15	.533 084
16	19 17 17.768	−22 40 17.90	18.899 418	July 1	19 10 59.120	−22 52 18.27	18.531 087
17	19 17 12.749	−22 40 28.61	18.886 240	2	19 10 48.782	−22 52 36.41	18.529 384

GEOCENTRIC COORDINATES FOR 0^h DYNAMICAL TIME

Date	Apparent Right Ascension	Apparent Declination	True Geocentric Distance	Date	Apparent Right Ascension	Apparent Declination	True Geocentric Distance
	h m s	° ′ ″			h m s	° ′ ″	
July 1	19 10 59.120	−22 52 18.27	18.531 087	Aug. 16	19 03 39.965	−23 04 06.91	18.749 029
2	19 10 48.782	22 52 36.41	.529 384	17	19 03 32.679	23 04 17.53	.759 966
3	19 10 38.413	22 52 54.53	.527 977	18	19 03 25.545	23 04 27.87	.771 128
4	19 10 28.018	22 53 12.63	.526 864	19	19 03 18.569	23 04 37.91	.782 513
5	19 10 17.602	22 53 30.67	.526 047	20	19 03 11.755	23 04 47.67	.794 116
6	19 10 07.171	−22 53 48.65	18.525 525	21	19 03 05.105	−23 04 57.15	18.805 936
7	19 09 56.733	22 54 06.54	.525 298	22	19 02 58.624	23 05 06.35	.817 969
8	19 09 46.293	22 54 24.35	.525 365	23	19 02 52.312	23 05 15.27	.830 211
9	19 09 35.856	22 54 42.08	.525 727	24	19 02 46.173	23 05 23.93	.842 659
10	19 09 25.428	22 54 59.73	.526 383	25	19 02 40.206	23 05 32.33	.855 311
11	19 09 15.010	−22 55 17.30	18.527 333	26	19 02 34.413	−23 05 40.45	18.868 161
12	19 09 04.608	22 55 34.80	.528 576	27	19 02 28.794	23 05 48.29	.881 207
13	19 08 54.222	22 55 52.22	.530 111	28	19 02 23.352	23 05 55.83	.894 443
14	19 08 43.857	22 56 09.56	.531 939	29	19 02 18.091	23 06 03.05	.907 867
15	19 08 33.515	22 56 26.80	.534 058	30	19 02 13.015	23 06 09.95	.921 473
16	19 08 23.199	−22 56 43.94	18.536 468	31	19 02 08.130	−23 06 16.51	18.935 257
17	19 08 12.914	22 57 00.97	.539 168	Sept. 1	19 02 03.439	23 06 22.74	.949 214
18	19 08 02.663	22 57 17.87	.542 158	2	19 01 58.945	23 06 28.66	.963 340
19	19 07 52.451	22 57 34.64	.545 438	3	19 01 54.649	23 06 34.27	.977 631
20	19 07 42.283	22 57 51.25	.549 006	4	19 01 50.552	23 06 39.58	18.992 082
21	19 07 32.163	−22 58 07.71	18.552 862	5	19 01 46.655	−23 06 44.60	19.006 688
22	19 07 22.098	22 58 24.00	.557 004	6	19 01 42.957	23 06 49.32	.021 445
23	19 07 12.093	22 58 40.12	.561 433	7	19 01 39.460	23 06 53.75	.036 349
24	19 07 02.152	22 58 56.07	.566 146	8	19 01 36.163	23 06 57.88	.051 395
25	19 06 52.281	22 59 11.84	.571 144	9	19 01 33.068	23 07 01.70	.066 579
26	19 06 42.483	−22 59 27.45	18.576 424	10	19 01 30.176	−23 07 05.21	19.081 896
27	19 06 32.763	22 59 42.90	.581 986	11	19 01 27.488	23 07 08.40	.097 343
28	19 06 23.121	22 59 58.18	.587 828	12	19 01 25.006	23 07 11.27	.112 915
29	19 06 13.562	23 00 13.30	.593 949	13	19 01 22.733	23 07 13.81	.128 608
30	19 06 04.085	23 00 28.25	.600 347	14	19 01 20.670	23 07 16.01	.144 417
31	19 05 54.695	−23 00 43.00	18.607 020	15	19 01 18.821	−23 07 17.88	19.160 339
Aug. 1	19 05 45.396	23 00 57.53	.613 965	16	19 01 17.186	23 07 19.42	.176 368
2	19 05 36.194	23 01 11.82	.621 180	17	19 01 15.768	23 07 20.64	.192 500
3	19 05 27.096	23 01 25.87	.628 663	18	19 01 14.568	23 07 21.53	.208 732
4	19 05 18.108	23 01 39.68	.636 410	19	19 01 13.587	23 07 22.12	.225 058
5	19 05 09.233	−23 01 53.24	18.644 420	20	19 01 12.824	−23 07 22.39	19.241 475
6	19 05 00.477	23 02 06.56	.652 689	21	19 01 12.279	23 07 22.37	.257 977
7	19 04 51.843	23 02 19.66	.661 214	22	19 01 11.951	23 07 22.04	.274 560
8	19 04 43.331	23 02 32.53	.669 994	23	19 01 11.839	23 07 21.40	.291 219
9	19 04 34.946	23 02 45.17	.679 024	24	19 01 11.943	23 07 20.44	.307 949
10	19 04 26.688	−23 02 57.58	18.688 303	25	19 01 12.264	−23 07 19.15	19.324 745
11	19 04 18.560	23 03 09.76	.697 827	26	19 01 12.805	23 07 17.50	.341 601
12	19 04 10.565	23 03 21.71	.707 594	27	19 01 13.568	23 07 15.49	.358 514
13	19 04 02.704	23 03 33.40	.717 600	28	19 01 14.556	23 07 13.14	.375 476
14	19 03 54.982	23 03 44.84	.727 843	29	19 01 15.770	23 07 10.43	.392 483
15	19 03 47.401	−23 03 56.01	18.738 321	30	19 01 17.211	−23 07 07.40	19.409 529
16	19 03 39.965	−23 04 06.91	18.749 029	Oct. 1	19 01 18.877	−23 07 04.05	19.426 609

GEOCENTRIC COORDINATES FOR 0h DYNAMICAL TIME

Date	Apparent Right Ascension	Apparent Declination	True Geocentric Distance	Date	Apparent Right Ascension	Apparent Declination	True Geocentric Distance
	h m s	° ′ ″			h m s	° ′ ″	
Oct. 1	19 01 18.877	−23 07 04.05	19.426 609	Nov. 16	19 06 22.399	−22 58 46.86	20.166 657
2	19 01 20.767	23 07 00.39	.443 719	17	19 06 33.413	22 58 28.93	.180 039
3	19 01 22.879	23 06 56.42	.460 852	18	19 06 44.580	22 58 10.72	.193 239
4	19 01 25.211	23 06 52.14	.478 005	19	19 06 55.900	22 57 52.21	.206 255
5	19 01 27.762	23 06 47.55	.495 172	20	19 07 07.370	22 57 33.40	.219 081
6	19 01 30.532	−23 06 42.64	19.512 347	21	19 07 18.991	−22 57 14.28	20.231 715
7	19 01 33.518	23 06 37.42	.529 528	22	19 07 30.763	22 56 54.87	.244 153
8	19 01 36.721	23 06 31.87	.546 707	23	19 07 42.683	22 56 35.17	.256 389
9	19 01 40.142	23 06 25.99	.563 882	24	19 07 54.750	22 56 15.20	.268 421
10	19 01 43.779	23 06 19.77	.581 048	25	19 08 06.958	22 55 54.97	.280 245
11	19 01 47.633	−23 06 13.22	19.598 198	26	19 08 19.305	−22 55 34.49	20.291 858
12	19 01 51.706	23 06 06.33	.615 330	27	19 08 31.785	22 55 13.77	.303 256
13	19 01 55.997	23 05 59.10	.632 439	28	19 08 44.394	22 54 52.81	.314 435
14	19 02 00.506	23 05 51.54	.649 519	29	19 08 57.129	22 54 31.61	.325 394
15	19 02 05.233	23 05 43.66	.666 567	30	19 09 09.985	22 54 10.16	.336 129
16	19 02 10.176	−23 05 35.47	19.683 578	Dec. 1	19 09 22.959	−22 53 48.47	20.346 636
17	19 02 15.334	23 05 26.98	.700 547	2	19 09 36.050	22 53 26.53	.356 914
18	19 02 20.706	23 05 18.18	.717 470	3	19 09 49.255	22 53 04.35	.366 960
19	19 02 26.287	23 05 09.08	.734 342	4	19 10 02.572	22 52 41.91	.376 771
20	19 02 32.075	23 04 59.68	.751 159	5	19 10 15.998	22 52 19.22	.386 346
21	19 02 38.070	−23 04 49.96	19.767 914	6	19 10 29.533	−22 51 56.28	20.395 680
22	19 02 44.269	23 04 39.92	.784 604	7	19 10 43.173	22 51 33.10	.404 773
23	19 02 50.673	23 04 29.54	.801 224	8	19 10 56.918	22 51 09.70	.413 623
24	19 02 57.283	23 04 18.82	.817 767	9	19 11 10.764	22 50 46.07	.422 226
25	19 03 04.099	23 04 07.75	.834 230	10	19 11 24.708	22 50 22.23	.430 582
26	19 03 11.122	−23 03 56.35	19.850 607	11	19 11 38.745	−22 49 58.20	20.438 687
27	19 03 18.352	23 03 44.61	.866 894	12	19 11 52.871	22 49 33.98	.446 541
28	19 03 25.785	23 03 32.57	.883 084	13	19 12 07.080	22 49 09.57	.454 141
29	19 03 33.420	23 03 20.22	.899 173	14	19 12 21.370	22 48 44.97	.461 485
30	19 03 41.252	23 03 07.59	.915 158	15	19 12 35.736	22 48 20.17	.468 570
31	19 03 49.278	−23 02 54.66	19.931 032	16	19 12 50.177	−22 47 55.17	20.475 396
Nov. 1	19 03 57.494	23 02 41.43	.946 791	17	19 13 04.691	22 47 29.94	.481 958
2	19 04 05.899	23 02 27.92	.962 432	18	19 13 19.277	22 47 04.51	.488 256
3	19 04 14.490	23 02 14.10	.977 950	19	19 13 33.934	22 46 38.86	.494 287
4	19 04 23.265	23 01 59.98	19.993 340	20	19 13 48.661	22 46 13.03	.500 049
5	19 04 32.222	−23 01 45.55	20.008 599	21	19 14 03.453	−22 45 47.01	20.505 541
6	19 04 41.361	23 01 30.81	.023 722	22	19 14 18.308	22 45 20.82	.510 760
7	19 04 50.679	23 01 15.75	.038 705	23	19 14 33.220	22 44 54.48	.515 705
8	19 05 00.177	23 01 00.38	.053 546	24	19 14 48.185	22 44 28.00	.520 374
9	19 05 09.852	23 00 44.70	.068 239	25	19 15 03.198	22 44 01.38	.524 766
10	19 05 19.705	−23 00 28.71	20.082 781	26	19 15 18.254	−22 43 34.63	20.528 880
11	19 05 29.733	23 00 12.43	.097 169	27	19 15 33.350	22 43 07.75	.532 715
12	19 05 39.933	22 59 55.86	.111 398	28	19 15 48.481	22 42 40.74	.536 269
13	19 05 50.304	22 59 39.01	.125 465	29	19 16 03.645	22 42 13.59	.539 543
14	19 06 00.841	22 59 21.90	.139 366	30	19 16 18.838	22 41 46.30	.542 535
15	19 06 11.541	−22 59 04.51	20.153 098	31	19 16 34.059	−22 41 18.89	20.545 245
16	19 06 22.399	−22 58 46.86	20.166 657	32	19 16 49.305	−22 40 51.34	20.547 673

GEOCENTRIC COORDINATES FOR 0^h DYNAMICAL TIME

Date	Apparent Right Ascension	Apparent Declination	True Geocentric Distance	Date	Apparent Right Ascension	Apparent Declination	True Geocentric Distance
	h m s	° ′ ″			h m s	° ′ ″	
Jan. 0	19 09 48.803	−21 43 58.98	31.169 655	Feb. 15	19 16 54.660	−21 31 40.11	30.969 075
1	19 09 58.484	21 43 43.47	.171 712	16	19 17 02.699	21 31 25.01	.958 506
2	19 10 08.182	21 43 27.87	.173 477	17	19 17 10.649	21 31 10.05	.947 710
3	19 10 17.895	21 43 12.21	.174 949	18	19 17 18.508	21 30 55.21	.936 692
4	19 10 27.620	21 42 56.47	.176 129	19	19 17 26.274	21 30 40.48	.925 453
5	19 10 37.352	−21 42 40.66	31.177 014	20	19 17 33.948	−21 30 25.86	30.913 998
6	19 10 47.090	21 42 24.74	.177 606	21	19 17 41.531	21 30 11.34	.902 329
7	19 10 56.825	21 42 08.55	.177 904	22	19 17 49.022	21 29 56.92	.890 448
8	19 11 06.516	21 41 52.64	.177 907	23	19 17 56.423	21 29 42.62	.878 360
9	19 11 16.248	21 41 36.94	.177 617	24	19 18 03.730	21 29 28.46	.866 066
10	19 11 25.975	−21 41 20.94	31.177 033	25	19 18 10.942	−21 29 14.44	30.853 571
11	19 11 35.692	21 41 04.82	.176 156	26	19 18 18.056	21 29 00.58	.840 878
12	19 11 45.397	21 40 48.63	.174 986	27	19 18 25.070	21 28 46.89	.827 990
13	19 11 55.092	21 40 32.38	.173 524	28	19 18 31.981	21 28 33.37	.814 912
14	19 12 04.776	21 40 16.06	.171 771	29	19 18 38.785	21 28 20.02	.801 646
15	19 12 14.447	−21 39 59.71	31.169 728	Mar. 1	19 18 45.481	−21 28 06.86	30.788 197
16	19 12 24.103	21 39 43.31	.167 395	2	19 18 52.067	21 27 53.88	.774 568
17	19 12 33.744	21 39 26.91	.164 775	3	19 18 58.539	21 27 41.08	.760 764
18	19 12 43.364	21 39 10.50	.161 866	4	19 19 04.898	21 27 28.45	.746 789
19	19 12 52.960	21 38 54.10	.158 672	5	19 19 11.143	21 27 15.99	.732 647
20	19 13 02.526	−21 38 37.71	31.155 193	6	19 19 17.272	−21 27 03.71	30.718 343
21	19 13 12.058	21 38 21.34	.151 430	7	19 19 23.286	21 26 51.59	.703 880
22	19 13 21.555	21 38 04.96	.147 383	8	19 19 29.185	21 26 39.66	.689 265
23	19 13 31.014	21 37 48.56	.143 055	9	19 19 34.968	21 26 27.91	.674 500
24	19 13 40.438	21 37 32.14	.138 445	10	19 19 40.636	21 26 16.35	.659 591
25	19 13 49.825	−21 37 15.70	31.133 555	11	19 19 46.186	−21 26 05.01	30.644 543
26	19 13 59.175	21 36 59.25	.128 385	12	19 19 51.617	21 25 53.88	.629 360
27	19 14 08.487	21 36 42.81	.122 937	13	19 19 56.926	21 25 42.98	.614 047
28	19 14 17.758	21 36 26.38	.117 212	14	19 20 02.111	21 25 32.32	.598 609
29	19 14 26.986	21 36 09.98	.111 212	15	19 20 07.168	21 25 21.90	.583 049
30	19 14 36.167	−21 35 53.62	31.104 938	16	19 20 12.095	−21 25 11.71	30.567 374
31	19 14 45.299	21 35 37.31	.098 391	17	19 20 16.891	21 25 01.75	.551 586
Feb. 1	19 14 54.378	21 35 21.05	.091 575	18	19 20 21.557	21 24 52.00	.535 691
2	19 15 03.401	21 35 04.85	.084 489	19	19 20 26.092	21 24 42.46	.519 692
3	19 15 12.364	21 34 48.71	.077 138	20	19 20 30.499	21 24 33.11	.503 594
4	19 15 21.265	−21 34 32.64	31.069 522	21	19 20 34.779	−21 24 23.98	30.487 401
5	19 15 30.103	21 34 16.62	.061 644	22	19 20 38.931	21 24 15.07	.471 118
6	19 15 38.874	21 34 00.66	.053 508	23	19 20 42.955	21 24 06.40	.454 748
7	19 15 47.578	21 33 44.75	.045 114	24	19 20 46.849	21 23 57.98	.438 297
8	19 15 56.213	21 33 28.90	.036 467	25	19 20 50.610	21 23 49.82	.421 768
9	19 16 04.779	−21 33 13.11	31.027 568	26	19 20 54.238	−21 23 41.92	30.405 167
10	19 16 13.276	21 32 57.39	.018 422	27	19 20 57.728	21 23 34.29	.388 498
11	19 16 21.701	21 32 41.74	31.009 030	28	19 21 01.081	21 23 26.93	.371 765
12	19 16 30.055	21 32 26.17	30.999 396	29	19 21 04.295	21 23 19.83	.354 975
13	19 16 38.335	21 32 10.70	.989 524	30	19 21 07.368	21 23 12.99	.338 131
14	19 16 46.538	−21 31 55.34	30.979 416	31	19 21 10.301	−21 23 06.41	30.321 240
15	19 16 54.660	−21 31 40.11	30.969 075	Apr. 1	19 21 13.093	−21 23 00.09	30.304 305

GEOCENTRIC COORDINATES FOR 0h DYNAMICAL TIME

Date	Apparent Right Ascension	Apparent Declination	True Geocentric Distance
	h m s	° ′ ″	
Apr. 1	19 21 13.093	−21 23 00.09	30.304 305
2	19 21 15.744	21 22 54.02	.287 332
3	19 21 18.256	21 22 48.19	.270 327
4	19 21 20.629	21 22 42.61	.253 294
5	19 21 22.863	21 22 37.29	.236 239
6	19 21 24.961	−21 22 32.23	30.219 167
7	19 21 26.920	21 22 27.44	.202 084
8	19 21 28.742	21 22 22.93	.184 994
9	19 21 30.423	21 22 18.71	.167 903
10	19 21 31.963	21 22 14.79	.150 817
11	19 21 33.360	−21 22 11.17	30.133 740
12	19 21 34.611	21 22 07.84	.116 677
13	19 21 35.717	21 22 04.79	.099 633
14	19 21 36.679	21 22 02.00	.082 612
15	19 21 37.498	21 21 59.47	.065 620
16	19 21 38.176	−21 21 57.19	30.048 661
17	19 21 38.716	21 21 55.17	.031 739
18	19 21 39.119	21 21 53.40	30.014 860
19	19 21 39.385	21 21 51.91	29.998 026
20	19 21 39.515	21 21 50.69	.981 243
21	19 21 39.508	−21 21 49.76	29.964 516
22	19 21 39.361	21 21 49.13	.947 849
23	19 21 39.074	21 21 48.79	.931 247
24	19 21 38.647	21 21 48.75	.914 715
25	19 21 38.079	21 21 48.99	.898 257
26	19 21 37.370	−21 21 49.52	29.881 878
27	19 21 36.520	21 21 50.32	.865 584
28	19 21 35.531	21 21 51.40	.849 379
29	19 21 34.404	21 21 52.74	.833 268
30	19 21 33.140	21 21 54.33	.817 255
May 1	19 21 31.741	−21 21 56.19	29.801 347
2	19 21 30.210	21 21 58.29	.785 548
3	19 21 28.548	21 22 00.65	.769 862
4	19 21 26.757	21 22 03.28	.754 296
5	19 21 24.837	21 22 06.19	.738 853
6	19 21 22.788	−21 22 09.38	29.723 539
7	19 21 20.609	21 22 12.85	.708 357
8	19 21 18.299	21 22 16.61	.693 314
9	19 21 15.857	21 22 20.64	.678 413
10	19 21 13.285	21 22 24.94	.663 658
11	19 21 10.584	−21 22 29.49	29.649 053
12	19 21 07.756	21 22 34.27	.634 603
13	19 21 04.805	21 22 39.28	.620 311
14	19 21 01.735	21 22 44.51	.606 181
15	19 20 58.547	21 22 49.97	.592 216
16	19 20 55.244	−21 22 55.65	29.578 421
17	19 20 51.827	−21 23 01.58	29.564 799
May 17	19 20 51.827	−21 23 01.58	29.564 799
18	19 20 48.296	21 23 07.75	.551 354
19	19 20 44.652	21 23 14.17	.538 089
20	19 20 40.893	21 23 20.84	.525 008
21	19 20 37.021	21 23 27.75	.512 116
22	19 20 33.036	−21 23 34.90	29.499 414
23	19 20 28.939	21 23 42.29	.486 909
24	19 20 24.731	21 23 49.90	.474 602
25	19 20 20.415	21 23 57.72	.462 499
26	19 20 15.992	21 24 05.74	.450 602
27	19 20 11.465	−21 24 13.96	29.438 915
28	19 20 06.837	21 24 22.38	.427 442
29	19 20 02.112	21 24 30.97	.416 188
30	19 19 57.291	21 24 39.75	.405 154
31	19 19 52.378	21 24 48.72	.394 346
June 1	19 19 47.375	−21 24 57.89	29.383 766
2	19 19 42.282	21 25 07.25	.373 419
3	19 19 37.101	21 25 16.83	.363 306
4	19 19 31.830	21 25 26.61	.353 433
5	19 19 26.470	21 25 36.58	.343 800
6	19 19 21.022	−21 25 46.74	29.334 412
7	19 19 15.489	21 25 57.05	.325 270
8	19 19 09.874	21 26 07.52	.316 377
9	19 19 04.181	21 26 18.12	.307 736
10	19 18 58.415	21 26 28.84	.299 348
11	19 18 52.579	−21 26 39.70	29.291 216
12	19 18 46.676	21 26 50.68	.283 341
13	19 18 40.709	21 27 01.80	.275 726
14	19 18 34.678	21 27 13.07	.268 372
15	19 18 28.585	21 27 24.47	.261 282
16	19 18 22.431	−21 27 36.02	29.254 456
17	19 18 16.217	21 27 47.71	.247 899
18	19 18 09.944	21 27 59.53	.241 610
19	19 18 03.613	21 28 11.48	.235 592
20	19 17 57.226	21 28 23.54	.229 847
21	19 17 50.787	−21 28 35.70	29.224 377
22	19 17 44.298	21 28 47.95	.219 183
23	19 17 37.762	21 29 00.29	.214 268
24	19 17 31.182	21 29 12.70	.209 633
25	19 17 24.563	21 29 25.18	.205 279
26	19 17 17.908	−21 29 37.72	29.201 209
27	19 17 11.221	21 29 50.34	.197 423
28	19 17 04.504	21 30 03.02	.193 925
29	19 16 57.759	21 30 15.77	.190 714
30	19 16 50.989	21 30 28.61	.187 792
July 1	19 16 44.193	−21 30 41.53	29.185 160
2	19 16 37.371	−21 30 54.53	29.182 819

GEOCENTRIC COORDINATES FOR 0^h DYNAMICAL TIME

Date	Apparent Right Ascension	Apparent Declination	True Geocentric Distance	Date	Apparent Right Ascension	Apparent Declination	True Geocentric Distance
	h m s	° ′ ″			h m s	° ′ ″	
July 1	19 16 44.193	−21 30 41.53	29.185 160	Aug. 16	19 11 48.305	−21 40 15.04	29.373 012
2	19 16 37.371	21 30 54.53	.182 819	17	19 11 43.163	21 40 25.47	.383 351
3	19 16 30.524	21 31 07.59	.180 770	18	19 11 38.112	21 40 35.72	.393 921
4	19 16 23.656	21 31 20.69	.179 013	19	19 11 33.156	21 40 45.82	.404 720
5	19 16 16.770	21 31 33.81	.177 549	20	19 11 28.297	21 40 55.75	.415 745
6	19 16 09.871	−21 31 46.94	29.176 377	21	19 11 23.539	−21 41 05.52	29.426 992
7	19 16 02.963	21 32 00.07	.175 497	22	19 11 18.882	21 41 15.13	.438 459
8	19 15 56.051	21 32 13.20	.174 910	23	19 11 14.328	21 41 24.58	.450 142
9	19 15 49.138	21 32 26.33	.174 615	24	19 11 09.879	21 41 33.90	.462 040
10	19 15 42.227	21 32 39.47	.174 611	25	19 11 05.533	21 41 43.06	.474 147
11	19 15 35.320	−21 32 52.62	29.174 900	26	19 11 01.291	−21 41 52.08	29.486 461
12	19 15 28.419	21 33 05.79	.175 479	27	19 10 57.154	21 42 00.94	.498 979
13	19 15 21.524	21 33 18.96	.176 349	28	19 10 53.121	21 42 09.62	.511 695
14	19 15 14.638	21 33 32.15	.177 510	29	19 10 49.197	21 42 18.10	.524 607
15	19 15 07.760	21 33 45.34	.178 961	30	19 10 45.385	21 42 26.38	.537 710
16	19 15 00.894	−21 33 58.53	29.180 702	31	19 10 41.689	−21 42 34.43	29.551 000
17	19 14 54.041	21 34 11.71	.182 732	Sept. 1	19 10 38.113	21 42 42.28	.564 472
18	19 14 47.203	21 34 24.86	.185 051	2	19 10 34.659	21 42 49.93	.578 121
19	19 14 40.385	21 34 37.97	.187 658	3	19 10 31.328	21 42 57.38	.591 945
20	19 14 33.588	21 34 51.04	.190 554	4	19 10 28.119	21 43 04.65	.605 938
21	19 14 26.817	−21 35 04.06	29.193 737	5	19 10 25.034	−21 43 11.74	29.620 096
22	19 14 20.075	21 35 17.02	.197 206	6	19 10 22.072	21 43 18.64	.634 414
23	19 14 13.366	21 35 29.92	.200 962	7	19 10 19.233	21 43 25.36	.648 889
24	19 14 06.694	21 35 42.76	.205 003	8	19 10 16.517	21 43 31.89	.663 516
25	19 14 00.061	21 35 55.54	.209 328	9	19 10 13.925	21 43 38.23	.678 292
26	19 13 53.470	−21 36 08.26	29.213 937	10	19 10 11.457	−21 43 44.36	29.693 211
27	19 13 46.923	21 36 20.93	.218 828	11	19 10 09.116	21 43 50.28	.708 269
28	19 13 40.421	21 36 33.56	.224 001	12	19 10 06.903	21 43 55.98	.723 463
29	19 13 33.963	21 36 46.14	.229 454	13	19 10 04.819	21 44 01.46	.738 788
30	19 13 27.551	21 36 58.67	.235 185	14	19 10 02.867	21 44 06.70	.754 240
31	19 13 21.186	−21 37 11.12	29.241 193	15	19 10 01.048	−21 44 11.72	29.769 815
Aug. 1	19 13 14.870	21 37 23.47	.247 475	16	19 09 59.365	21 44 16.50	.785 509
2	19 13 08.609	21 37 35.72	.254 030	17	19 09 57.820	21 44 21.06	.801 316
3	19 13 02.407	21 37 47.84	.260 856	18	19 09 56.412	21 44 25.40	.817 234
4	19 12 56.269	21 37 59.83	.267 948	19	19 09 55.143	21 44 29.53	.833 257
5	19 12 50.198	−21 38 11.71	29.275 306	20	19 09 54.013	−21 44 33.45	29.849 381
6	19 12 44.197	21 38 23.48	.282 927	21	19 09 53.020	21 44 37.16	.865 602
7	19 12 38.267	21 38 35.13	.290 807	22	19 09 52.164	21 44 40.67	.881 914
8	19 12 32.410	21 38 46.69	.298 945	23	19 09 51.443	21 44 43.97	.898 314
9	19 12 26.626	21 38 58.14	.307 338	24	19 09 50.858	21 44 47.05	.914 796
10	19 12 20.917	−21 39 09.49	29.315 984	25	19 09 50.410	−21 44 49.88	29.931 356
11	19 12 15.284	21 39 20.73	.324 879	26	19 09 50.100	21 44 52.45	.947 987
12	19 12 09.727	21 39 31.86	.334 022	27	19 09 49.933	21 44 54.77	.964 685
13	19 12 04.248	21 39 42.86	.343 409	28	19 09 49.910	21 44 56.82	.981 444
14	19 11 58.850	21 39 53.74	.353 038	29	19 09 50.033	21 44 58.62	.998 259
15	19 11 53.535	−21 40 04.46	29.362 907	30	19 09 50.303	−21 45 00.18	30.015 124
16	19 11 48.305	−21 40 15.04	29.373 012	Oct. 1	19 09 50.719	−21 45 01.51	30.032 035

GEOCENTRIC COORDINATES FOR 0^h DYNAMICAL TIME

Date	Apparent Right Ascension	Apparent Declination	True Geocentric Distance	Date	Apparent Right Ascension	Apparent Declination	True Geocentric Distance
	h m s	° ′ ″			h m s	° ′ ″	
Oct. 1	19 09 50.719	−21 45 01.51	30.032 035	Nov. 16	19 12 39.026	−21 41 43.09	30.774 075
2	19 09 51.279	21 45 02.61	.048 986	17	19 12 45.657	21 41 33.26	.787 640
3	19 09 51.982	21 45 03.49	.065 972	18	19 12 52.393	21 41 23.21	.801 025
4	19 09 52.827	21 45 04.16	.082 987	19	19 12 59.233	21 41 12.92	.814 228
5	19 09 53.813	21 45 04.59	.100 028	20	19 13 06.176	21 41 02.39	.827 243
6	19 09 54.939	−21 45 04.80	30.117 088	21	19 13 13.224	−21 40 51.62	30.840 066
7	19 09 56.205	21 45 04.77	.134 164	22	19 13 20.377	21 40 40.61	.852 694
8	19 09 57.611	21 45 04.49	.151 250	23	19 13 27.634	21 40 29.37	.865 122
9	19 09 59.158	21 45 03.97	.168 341	24	19 13 34.993	21 40 17.92	.877 346
10	19 10 00.845	21 45 03.19	.185 433	25	19 13 42.453	21 40 06.26	.889 362
11	19 10 02.675	−21 45 02.15	30.202 521	26	19 13 50.008	−21 39 54.41	30.901 166
12	19 10 04.647	21 45 00.86	.219 600	27	19 13 57.657	21 39 42.38	.912 755
13	19 10 06.763	21 44 59.31	.236 665	28	19 14 05.395	21 39 30.16	.924 125
14	19 10 09.024	21 44 57.50	.253 712	29	19 14 13.220	21 39 17.76	.935 273
15	19 10 11.428	21 44 55.46	.270 737	30	19 14 21.130	21 39 05.16	.946 196
16	19 10 13.976	−21 44 53.17	30.287 734	Dec. 1	19 14 29.122	−21 38 52.38	30.956 891
17	19 10 16.666	21 44 50.66	.304 698	2	19 14 37.194	21 38 39.39	.967 354
18	19 10 19.496	21 44 47.92	.321 626	3	19 14 45.347	21 38 26.21	.977 583
19	19 10 22.465	21 44 44.95	.338 512	4	19 14 53.578	21 38 12.83	.987 575
20	19 10 25.570	21 44 41.76	.355 351	5	19 15 01.886	21 37 59.24	30.997 327
21	19 10 28.810	−21 44 38.33	30.372 138	6	19 15 10.271	−21 37 45.46	31.006 838
22	19 10 32.185	21 44 34.65	.388 868	7	19 15 18.732	21 37 31.49	.016 103
23	19 10 35.695	21 44 30.71	.405 536	8	19 15 27.268	21 37 17.33	.025 122
24	19 10 39.341	21 44 26.50	.422 136	9	19 15 35.876	21 37 02.99	.033 891
25	19 10 43.126	21 44 22.01	.438 664	10	19 15 44.554	21 36 48.50	.042 408
26	19 10 47.050	−21 44 17.26	30.455 114	11	19 15 53.300	−21 36 33.86	31.050 672
27	19 10 51.113	21 44 12.26	.471 481	12	19 16 02.108	21 36 19.07	.058 680
28	19 10 55.313	21 44 07.01	.487 759	13	19 16 10.976	21 36 04.14	.066 430
29	19 10 59.649	21 44 01.53	.503 943	14	19 16 19.899	21 35 49.06	.073 918
30	19 11 04.118	21 43 55.83	.520 029	15	19 16 28.876	21 35 33.83	.081 144
31	19 11 08.717	−21 43 49.91	30.536 012	16	19 16 37.905	−21 35 18.42	31.088 105
Nov. 1	19 11 13.444	21 43 43.76	.551 886	17	19 16 46.986	21 35 02.85	.094 798
2	19 11 18.298	21 43 37.38	.567 648	18	19 16 56.119	21 34 47.11	.101 221
3	19 11 23.275	21 43 30.77	.583 293	19	19 17 05.304	21 34 31.19	.107 371
4	19 11 28.377	21 43 23.92	.598 816	20	19 17 14.540	21 34 15.12	.113 247
5	19 11 33.601	−21 43 16.84	30.614 213	21	19 17 23.824	−21 33 58.90	31.118 847
6	19 11 38.948	21 43 09.50	.629 479	22	19 17 33.154	21 33 42.55	.124 167
7	19 11 44.417	21 43 01.91	.644 611	23	19 17 42.527	21 33 26.09	.129 208
8	19 11 50.008	21 42 54.08	.659 605	24	19 17 51.938	21 33 09.51	.133 966
9	19 11 55.721	21 42 46.00	.674 455	25	19 18 01.384	21 32 52.83	.138 441
10	19 12 01.556	−21 42 37.67	30.689 159	26	19 18 10.861	−21 32 36.04	31.142 631
11	19 12 07.511	21 42 29.11	.703 711	27	19 18 20.367	21 32 19.15	.146 534
12	19 12 13.586	21 42 20.33	.718 110	28	19 18 29.899	21 32 02.15	.150 151
13	19 12 19.778	21 42 11.33	.732 349	29	19 18 39.454	21 31 45.05	.153 480
14	19 12 26.084	21 42 02.12	.746 426	30	19 18 49.032	21 31 27.83	.156 520
15	19 12 32.501	−21 41 52.71	30.760 336	31	19 18 58.629	−21 31 10.50	31.159 270
16	19 12 39.026	−21 41 43.09	30.774 075	32	19 19 08.247	−21 30 53.07	31.161 731

GEOCENTRIC POSITIONS FOR 0^h DYNAMICAL TIME

Date	Astrometric Right Ascension J2000.0	Astrometric Declination J2000.0	True Geocentric Distance	Date	Astrometric Right Ascension J2000.0	Astrometric Declination J2000.0	True Geocentric Distance
	h m s	° ′ ″			h m s	° ′ ″	
Jan. – 1	15 33 46.514	– 4 12 13.08	30.342 654	July 7	15 27 40.169	– 3 22 03.60	29.124 913
4	15 34 21.410	4 12 48.72	.280 322	12	15 27 24.787	3 23 36.65	.193 611
9	15 34 53.964	4 13 06.42	.213 452	17	15 27 12.170	3 25 26.57	.265 799
14	15 35 23.972	4 13 06.36	.142 560	22	15 27 02.439	3 27 32.84	.340 988
19	15 35 51.262	4 12 48.91	30.068 196	27	15 26 55.704	3 29 54.86	.418 671
24	15 36 15.690	– 4 12 14.50	29.990 902	Aug. 1	15 26 52.067	– 3 32 31.91	29.498 298
29	15 36 37.114	4 11 23.57	.911 211	6	15 26 51.598	3 35 23.03	.579 266
Feb. 3	15 36 55.393	4 10 16.72	.829 707	11	15 26 54.318	3 38 27.15	.660 984
8	15 37 10.415	4 08 54.76	.747 030	16	15 27 00.228	3 41 43.24	.742 901
13	15 37 22.106	4 07 18.69	.663 837	21	15 27 09.321	3 45 10.25	.824 478
18	15 37 30.433	– 4 05 29.62	29.580 773	26	15 27 21.586	– 3 48 47.09	29.905 169
23	15 37 35.385	4 03 28.65	.498 437	31	15 27 36.991	3 52 32.51	29.984 390
28	15 37 36.954	4 01 16.93	.417 424	Sept. 5	15 27 55.464	3 56 25.13	30.061 550
Mar. 4	15 37 35.152	3 58 55.78	.338 362	10	15 28 16.899	4 00 23.60	.136 118
9	15 37 30.033	3 56 26.69	.261 887	15	15 28 41.184	4 04 26.61	.207 604
14	15 37 21.692	– 3 53 51.26	29.188 610	20	15 29 08.206	– 4 08 32.90	30.275 541
19	15 37 10.262	3 51 11.07	.119 076	25	15 29 37.844	4 12 41.12	.339 452
24	15 36 55.882	3 48 27.63	29.053 772	30	15 30 09.951	4 16 49.83	.398 848
29	15 36 38.699	3 45 42.52	28.993 181	Oct. 5	15 30 44.340	4 20 57.55	.453 287
Apr. 3	15 36 18.889	3 42 57.43	.937 786	10	15 31 20.815	4 25 02.94	.502 401
8	15 35 56.667	– 3 40 14.11	28.888 034	15	15 31 59.181	– 4 29 04.70	30.545 872
13	15 35 32.285	3 37 34.29	.844 303	20	15 32 39.243	4 33 01.58	.583 399
18	15 35 06.004	3 34 59.55	.806 880	25	15 33 20.799	4 36 52.27	.614 680
23	15 34 38.079	3 32 31.36	.776 004	30	15 34 03.614	4 40 35.42	.639 439
28	15 34 08.774	3 30 11.22	.751 905	Nov. 4	15 34 47.431	4 44 09.78	.657 487
May 3	15 33 38.374	– 3 28 00.68	28.734 783	9	15 35 31.999	– 4 47 34.23	30.668 709
8	15 33 07.193	3 26 01.19	.724 772	14	15 36 17.075	4 50 47.76	.673 033
13	15 32 35.554	3 24 14.03	.721 912	19	15 37 02.420	4 53 49.38	.670 404
18	15 32 03.763	3 22 40.25	.726 161	24	15 37 47.781	4 56 38.05	.660 780
23	15 31 32.106	3 21 20.85	.737 461	29	15 38 32.880	4 59 12.83	.644 188
28	15 31 00.875	– 3 20 16.78	28.755 735	Dec. 4	15 39 17.439	– 5 01 32.95	30.620 749
June 2	15 30 30.373	3 19 28.91	.780 857	9	15 40 01.198	5 03 37.77	.590 645
7	15 30 00.906	3 18 57.95	.812 627	14	15 40 43.915	5 05 26.78	.554 092
12	15 29 32.760	3 18 44.36	.850 752	19	15 41 25.348	5 06 59.46	.511 310
17	15 29 06.186	3 18 48.38	.894 909	24	15 42 05.242	5 08 15.35	.462 558
22	15 28 41.413	– 3 19 10.23	28.944 774	29	15 42 43.338	– 5 09 14.15	30.408 183
27	15 28 18.671	3 19 50.07	29.000 003	Dec. 34	15 43 19.400	– 5 09 55.75	30.348 606
July 2	15 27 58.186	– 3 20 47.93	29.060 205				

HELIOCENTRIC POSITIONS FOR 0^h DYNAMICAL TIME

MEAN EQUINOX AND ECLIPTIC OF DATE

Date	Longitude	Latitude	Radius Vector	Date	Longitude	Latitude	Radius Vector
	h m s	° ′ ″			h m s	° ′ ″	
Jan. – 21	230 28 59.7	+ 14 54 39.3	29.685 84	Aug. 6	232 09 50.1	+ 14 39 40.9	29.705 22
Jan. 19	230 45 49.3	14 52 12.5	.688 75	Sept. 15	232 26 36.6	14 37 07.2	.708 89
Feb. 28	231 02 38.5	14 49 44.5	.691 79	Oct. 25	232 43 22.6	14 34 32.3	.712 69
Apr. 8	231 19 27.2	14 47 15.3	.694 95	Dec. 4	233 00 08.0	14 31 56.3	.716 62
May 18	231 36 15.3	14 44 45.0	.698 25	Dec. 44	233 16 52.8	+ 14 29 19.1	29.720 67
June 27	231 53 03.0	+ 14 42 13.6	29.701 67				

NOTES AND FORMULAS

Semidiameter and parallax

The apparent angular semidiameter of a planet is given by:

apparent S.D. = S.D. at unit distance / true distance

where the true distance is given in the daily geocentric ephemeris and the adopted semidiameter at unit distance is given by:

	″			″		″
Mercury	3.36	Jupiter:	equatorial	98.44	Uranus	35.02
Venus	8.34		polar	92.06	Neptune	33.50
Mars	4.68	Saturn:	equatorial	82.73	Pluto	2.07
			polar	73.82		

The difference in transit times of the limb and center of a planet in seconds of time is given approximately by:

difference in transit time = (apparent S.D. in seconds of arc) / 15 cos δ

where the sidereal motion of the planet is ignored.

The equatorial horizontal parallax of a planet is given by 8″.794 148 divided by its true geocentric distance; formulas for the corrections for diurnal parallax are given on page B61.

Time of transit of a planet

The transit times that are tabulated on pages E44–E51 are expressed in dynamical time (TDT) and refer to the transits over the ephemeris meridian; for most purposes this may be regarded as giving the universal time (UT) of transit over the Greenwich meridian.

The UT of transit over a local meridian is given by:

time of ephemeris transit − (λ/24) * first difference

with an error that is usually less than 1 second, where λ is the *east* longitude in hours and the first difference is about 24 hours.

Times of rising and setting

Approximate times of the rising and setting of a planet at a place with latitude φ may be obtained from the time of transit by applying the value of the hour angle h of the point on the horizon at the same declination as the planet; h is given by:

$$\cos h = -\tan\varphi \tan\delta$$

This ignores the sidereal motion of the planet during the interval between transit and rising or setting. Similarly, the time at which a planet reaches a zenith distance z may be obtained by determining the corresponding hour angle h from:

$$\cos h = -\tan\varphi \tan\delta + \sec\varphi \sec\delta \cos z$$

and applying h to the time of transit.

Date	Mercury	Venus	Mars	Jupiter	Saturn	Uranus	Neptune	Pluto
	h m s	h m s	h m s	h m s	h m s	h m s	h m s	h m
Jan. 0	10 28 20	9 14 42	10 53 36	4 28 46	13 55 20	12 21 45	12 32 08	8 56
1	10 29 27	9 15 42	10 52 52	4 24 50	13 51 52	12 18 04	12 28 22	8 53
2	10 30 45	9 16 43	10 52 09	4 20 53	13 48 23	12 14 24	12 24 36	8 49
3	10 32 11	9 17 45	10 51 25	4 16 55	13 44 55	12 10 43	12 20 50	8 45
4	10 33 45	9 18 48	10 50 42	4 12 57	13 41 27	12 07 03	12 17 03	8 41
5	10 35 27	9 19 53	10 49 59	4 08 58	13 37 59	12 03 22	12 13 17	8 37
6	10 37 15	9 20 58	10 49 17	4 04 58	13 34 31	11 59 42	12 09 31	8 33
7	10 39 09	9 22 05	10 48 35	4 00 58	13 31 03	11 56 01	12 05 45	8 30
8	10 41 09	9 23 13	10 47 53	3 56 57	13 27 36	11 52 21	12 01 58	8 26
9	10 43 14	9 24 22	10 47 11	3 52 55	13 24 08	11 48 40	11 58 12	8 22
10	10 45 24	9 25 32	10 46 29	3 48 53	13 20 41	11 45 00	11 54 26	8 18
11	10 47 38	9 26 43	10 45 48	3 44 49	13 17 13	11 41 19	11 50 40	8 14
12	10 49 56	9 27 55	10 45 06	3 40 45	13 13 46	11 37 38	11 46 53	8 10
13	10 52 18	9 29 07	10 44 25	3 36 41	13 10 18	11 33 58	11 43 07	8 07
14	10 54 44	9 30 21	10 43 44	3 32 36	13 06 51	11 30 17	11 39 21	8 03
15	10 57 12	9 31 36	10 43 04	3 28 30	13 03 24	11 26 36	11 35 35	7 59
16	10 59 44	9 32 51	10 42 23	3 24 23	12 59 57	11 22 56	11 31 48	7 55
17	11 02 18	9 34 07	10 41 42	3 20 15	12 56 30	11 19 15	11 28 02	7 51
18	11 04 54	9 35 24	10 41 02	3 16 07	12 53 03	11 15 34	11 24 15	7 47
19	11 07 34	9 36 42	10 40 21	3 11 59	12 49 36	11 11 53	11 20 29	7 44
20	11 10 15	9 38 01	10 39 41	3 07 49	12 46 09	11 08 13	11 16 43	7 40
21	11 12 58	9 39 20	10 39 01	3 03 39	12 42 42	11 04 32	11 12 56	7 36
22	11 15 43	9 40 39	10 38 21	2 59 29	12 39 15	11 00 51	11 09 10	7 32
23	11 18 30	9 41 59	10 37 40	2 55 17	12 35 48	10 57 10	11 05 23	7 28
24	11 21 18	9 43 20	10 37 00	2 51 05	12 32 22	10 53 29	11 01 37	7 24
25	11 24 08	9 44 41	10 36 20	2 46 53	12 28 55	10 49 47	10 57 50	7 20
26	11 26 59	9 46 02	10 35 40	2 42 40	12 25 28	10 46 06	10 54 03	7 17
27	11 29 52	9 47 24	10 35 00	2 38 26	12 22 01	10 42 25	10 50 17	7 13
28	11 32 45	9 48 46	10 34 20	2 34 11	12 18 34	10 38 43	10 46 30	7 09
29	11 35 40	9 50 08	10 33 40	2 29 56	12 15 08	10 35 02	10 42 43	7 05
30	11 38 35	9 51 31	10 32 59	2 25 41	12 11 41	10 31 21	10 38 57	7 01
31	11 41 31	9 52 53	10 32 19	2 21 25	12 08 14	10 27 39	10 35 10	6 57
Feb. 1	11 44 28	9 54 16	10 31 38	2 17 08	12 04 47	10 23 57	10 31 23	6 53
2	11 47 26	9 55 38	10 30 58	2 12 51	12 01 21	10 20 15	10 27 36	6 50
3	11 50 25	9 57 00	10 30 17	2 08 33	11 57 54	10 16 34	10 23 49	6 46
4	11 53 24	9 58 23	10 29 36	2 04 15	11 54 27	10 12 52	10 20 02	6 42
5	11 56 23	9 59 45	10 28 56	1 59 56	11 51 00	10 09 10	10 16 15	6 38
6	11 59 23	10 01 06	10 28 14	1 55 37	11 47 33	10 05 27	10 12 27	6 34
7	12 02 23	10 02 28	10 27 33	1 51 18	11 44 06	10 01 45	10 08 40	6 30
8	12 05 24	10 03 49	10 26 52	1 46 57	11 40 39	9 58 03	10 04 53	6 26
9	12 08 25	10 05 09	10 26 10	1 42 37	11 37 12	9 54 20	10 01 05	6 22
10	12 11 26	10 06 30	10 25 28	1 38 16	11 33 45	9 50 38	9 57 18	6 18
11	12 14 27	10 07 49	10 24 46	1 33 55	11 30 17	9 46 55	9 53 30	6 15
12	12 17 28	10 09 08	10 24 04	1 29 33	11 26 50	9 43 12	9 49 43	6 11
13	12 20 30	10 10 27	10 23 21	1 25 11	11 23 23	9 39 29	9 45 55	6 07
14	12 23 31	10 11 45	10 22 39	1 20 49	11 19 55	9 35 46	9 42 07	6 03
15	12 26 32	10 13 02	10 21 56	1 16 26	11 16 28	9 32 03	9 38 19	5 59

Date	Mercury	Venus	Mars	Jupiter	Saturn	Uranus	Neptune	Pluto
	h m s	h m s	h m s	h m s	h m s	h m s	h m s	h m
Feb. 15	12 26 32	10 13 02	10 21 56	1 16 26	11 16 28	9 32 03	9 38 19	5 59
16	12 29 32	10 14 18	10 21 12	1 12 03	11 13 00	9 28 20	9 34 31	5 55
17	12 32 32	10 15 34	10 20 29	1 07 40	11 09 33	9 24 36	9 30 43	5 51
18	12 35 31	10 16 49	10 19 45	1 03 16	11 06 05	9 20 53	9 26 55	5 47
19	12 38 30	10 18 03	10 19 01	0 58 52	11 02 37	9 17 09	9 23 07	5 43
20	12 41 26	10 19 16	10 18 16	0 54 28	10 59 09	9 13 25	9 19 19	5 39
21	12 44 21	10 20 28	10 17 32	0 50 04	10 55 41	9 09 41	9 15 30	5 36
22	12 47 14	10 21 40	10 16 47	0 45 40	10 52 13	9 05 57	9 11 42	5 32
23	12 50 04	10 22 50	10 16 01	0 41 15	10 48 44	9 02 13	9 07 53	5 28
24	12 52 50	10 24 00	10 15 16	0 36 50	10 45 16	8 58 28	9 04 05	5 24
25	12 55 32	10 25 09	10 14 30	0 32 25	10 41 47	8 54 44	9 00 16	5 20
26	12 58 09	10 26 16	10 13 44	0 28 00	10 38 19	8 50 59	8 56 27	5 16
27	13 00 39	10 27 23	10 12 57	0 23 35	10 34 50	8 47 14	8 52 38	5 12
28	13 03 02	10 28 28	10 12 10	0 19 10	10 31 21	8 43 29	8 48 49	5 08
29	13 05 17	10 29 33	10 11 23	0 14 45	10 27 52	8 39 44	8 45 00	5 04
Mar. 1	13 07 21	10 30 37	10 10 35	0 10 20	10 24 23	8 35 59	8 41 10	5 00
2	13 09 13	10 31 39	10 09 47	0 05 55	10 20 53	8 32 13	8 37 21	4 56
3	13 10 52	10 32 41	10 08 59	0 01 30	10 17 24	8 28 27	8 33 31	4 52
4	13 12 16	10 33 41	10 08 10	23 52 39	10 13 54	8 24 41	8 29 42	4 48
5	13 13 22	10 34 40	10 07 21	23 48 14	10 10 24	8 20 55	8 25 52	4 44
6	13 14 10	10 35 39	10 06 32	23 43 50	10 06 54	8 17 09	8 22 02	4 40
7	13 14 36	10 36 36	10 05 42	23 39 25	10 03 24	8 13 23	8 18 12	4 37
8	13 14 40	10 37 32	10 04 52	23 35 00	9 59 54	8 09 36	8 14 22	4 33
9	13 14 20	10 38 27	10 04 01	23 30 36	9 56 23	8 05 50	8 10 32	4 29
10	13 13 34	10 39 21	10 03 10	23 26 12	9 52 52	8 02 03	8 06 42	4 25
11	13 12 22	10 40 14	10 02 19	23 21 48	9 49 21	7 58 15	8 02 51	4 21
12	13 10 40	10 41 06	10 01 27	23 17 24	9 45 50	7 54 28	7 59 01	4 17
13	13 08 30	10 41 58	10 00 35	23 13 01	9 42 19	7 50 41	7 55 10	4 13
14	13 05 51	10 42 48	9 59 43	23 08 37	9 38 48	7 46 53	7 51 19	4 09
15	13 02 42	10 43 37	9 58 50	23 04 14	9 35 16	7 43 05	7 47 28	4 05
16	12 59 03	10 44 25	9 57 57	22 59 52	9 31 44	7 39 17	7 43 37	4 01
17	12 54 55	10 45 13	9 57 03	22 55 30	9 28 12	7 35 29	7 39 46	3 57
18	12 50 20	10 45 59	9 56 09	22 51 08	9 24 39	7 31 40	7 35 55	3 53
19	12 45 19	10 46 45	9 55 15	22 46 46	9 21 07	7 27 52	7 32 03	3 49
20	12 39 53	10 47 30	9 54 20	22 42 25	9 17 34	7 24 03	7 28 12	3 45
21	12 34 05	10 48 14	9 53 25	22 38 04	9 14 01	7 20 14	7 24 20	3 41
22	12 27 58	10 48 57	9 52 29	22 33 44	9 10 28	7 16 25	7 20 28	3 37
23	12 21 36	10 49 40	9 51 33	22 29 24	9 06 54	7 12 35	7 16 36	3 33
24	12 15 01	10 50 21	9 50 37	22 25 04	9 03 21	7 08 46	7 12 44	3 29
25	12 08 17	10 51 03	9 49 41	22 20 45	8 59 47	7 04 56	7 08 52	3 25
26	12 01 28	10 51 43	9 48 44	22 16 26	8 56 13	7 01 06	7 05 00	3 21
27	11 54 39	10 52 23	9 47 46	22 12 08	8 52 38	6 57 16	7 01 07	3 17
28	11 47 52	10 53 03	9 46 49	22 07 51	8 49 04	6 53 25	6 57 15	3 13
29	11 41 10	10 53 41	9 45 51	22 03 33	8 45 29	6 49 34	6 53 22	3 09
30	11 34 38	10 54 20	9 44 53	21 59 17	8 41 53	6 45 44	6 49 29	3 05
31	11 28 17	10 54 58	9 43 54	21 55 01	8 38 18	6 41 52	6 45 36	3 01
Apr. 1	11 22 10	10 55 35	9 42 55	21 50 45	8 34 42	6 38 01	6 41 43	2 57

Second transit: Jupiter, Mar. $3^d 23^h 57^d 04^s$.

Date	Mercury	Venus	Mars	Jupiter	Saturn	Uranus	Neptune	Pluto
	h m s	h m s	h m s	h m s	h m s	h m s	h m s	h m
Apr. 1	11 22 10	10 55 35	9 42 55	21 50 45	8 34 42	6 38 01	6 41 43	2 57
2	11 16 19	10 56 12	9 41 56	21 46 30	8 31 06	6 34 10	6 37 50	2 53
3	11 10 45	10 56 49	9 40 56	21 42 16	8 27 30	6 30 18	6 33 56	2 49
4	11 05 30	10 57 25	9 39 56	21 38 02	8 23 53	6 26 26	6 30 02	2 45
5	11 00 34	10 58 01	9 38 56	21 33 49	8 20 17	6 22 34	6 26 09	2 41
6	10 55 57	10 58 37	9 37 55	21 29 36	8 16 40	6 18 41	6 22 15	2 37
7	10 51 40	10 59 13	9 36 54	21 25 24	8 13 02	6 14 49	6 18 21	2 33
8	10 47 42	10 59 48	9 35 53	21 21 13	8 09 24	6 10 56	6 14 27	2 29
9	10 44 03	11 00 23	9 34 51	21 17 02	8 05 46	6 07 03	6 10 32	2 25
10	10 40 43	11 00 58	9 33 49	21 12 52	8 02 08	6 03 10	6 06 38	2 21
11	10 37 40	11 01 34	9 32 47	21 08 43	7 58 30	5 59 16	6 02 44	2 17
12	10 34 56	11 02 09	9 31 44	21 04 34	7 54 51	5 55 23	5 58 49	2 13
13	10 32 27	11 02 44	9 30 42	21 00 26	7 51 11	5 51 29	5 54 54	2 09
14	10 30 15	11 03 19	9 29 38	20 56 19	7 47 32	5 47 35	5 50 59	2 05
15	10 28 18	11 03 54	9 28 35	20 52 12	7 43 52	5 43 40	5 47 04	2 01
16	10 26 36	11 04 30	9 27 31	20 48 06	7 40 12	5 39 46	5 43 09	1 57
17	10 25 07	11 05 05	9 26 28	20 44 00	7 36 32	5 35 51	5 39 13	1 53
18	10 23 51	11 05 41	9 25 23	20 39 55	7 32 51	5 31 56	5 35 18	1 49
19	10 22 48	11 06 17	9 24 19	20 35 51	7 29 10	5 28 01	5 31 22	1 45
20	10 21 57	11 06 53	9 23 15	20 31 48	7 25 28	5 24 05	5 27 26	1 41
21	10 21 17	11 07 30	9 22 10	20 27 45	7 21 46	5 20 10	5 23 30	1 37
22	10 20 47	11 08 07	9 21 05	20 23 43	7 18 04	5 16 14	5 19 34	1 33
23	10 20 28	11 08 45	9 19 59	20 19 42	7 14 22	5 12 18	5 15 38	1 29
24	10 20 19	11 09 23	9 18 54	20 15 41	7 10 39	5 08 22	5 11 42	1 25
25	10 20 19	11 10 02	9 17 48	20 11 41	7 06 56	5 04 25	5 07 45	1 21
26	10 20 28	11 10 41	9 16 43	20 07 42	7 03 13	5 00 28	5 03 48	1 17
27	10 20 46	11 11 20	9 15 37	20 03 43	6 59 29	4 56 31	4 59 52	1 13
28	10 21 12	11 12 01	9 14 30	19 59 45	6 55 45	4 52 34	4 55 55	1 09
29	10 21 47	11 12 42	9 13 24	19 55 48	6 52 00	4 48 37	4 51 58	1 05
30	10 22 30	11 13 23	9 12 17	19 51 52	6 48 15	4 44 39	4 48 00	1 01
May 1	10 23 21	11 14 05	9 11 11	19 47 56	6 44 30	4 40 42	4 44 03	0 57
2	10 24 19	11 14 49	9 10 04	19 44 01	6 40 45	4 36 44	4 40 06	0 53
3	10 25 25	11 15 32	9 08 57	19 40 06	6 36 59	4 32 45	4 36 08	0 49
4	10 26 39	11 16 17	9 07 50	19 36 13	6 33 13	4 28 47	4 32 10	0 45
5	10 28 01	11 17 03	9 06 42	19 32 20	6 29 26	4 24 48	4 28 13	0 40
6	10 29 30	11 17 49	9 05 35	19 28 27	6 25 39	4 20 50	4 24 15	0 36
7	10 31 08	11 18 36	9 04 27	19 24 36	6 21 51	4 16 51	4 20 16	0 32
8	10 32 53	11 19 24	9 03 19	19 20 45	6 18 04	4 12 51	4 16 18	0 28
9	10 34 46	11 20 13	9 02 11	19 16 54	6 14 16	4 08 52	4 12 20	0 24
10	10 36 47	11 21 03	9 01 03	19 13 05	6 10 27	4 04 52	4 08 21	0 20
11	10 38 57	11 21 54	8 59 55	19 09 16	6 06 38	4 00 53	4 04 23	0 16
12	10 41 16	11 22 46	8 58 47	19 05 27	6 02 49	3 56 53	4 00 24	0 12
13	10 43 43	11 23 39	8 57 39	19 01 40	5 58 59	3 52 52	3 56 25	0 08
14	10 46 19	11 24 33	8 56 30	18 57 53	5 55 09	3 48 52	3 52 26	0 04
15	10 49 05	11 25 28	8 55 22	18 54 06	5 51 19	3 44 51	3 48 27	0 00
16	10 52 00	11 26 25	8 54 13	18 50 21	5 47 28	3 40 51	3 44 28	23 52
17	10 55 05	11 27 22	8 53 05	18 46 36	5 43 37	3 36 50	3 40 29	23 48

Second transit: Pluto, May $15^d 23^h 56^d$.

Date	Mercury	Venus	Mars	Jupiter	Saturn	Uranus	Neptune	Pluto
	h m s	h m s	h m s	h m s	h m s	h m s	h m s	h m
May 17	10 55 05	11 27 22	8 53 05	18 46 36	5 43 37	3 36 50	3 40 29	23 48
18	10 58 21	11 28 20	8 51 56	18 42 51	5 39 46	3 32 49	3 36 29	23 44
19	11 01 46	11 29 20	8 50 47	18 39 07	5 35 54	3 28 47	3 32 30	23 40
20	11 05 23	11 30 20	8 49 39	18 35 24	5 32 01	3 24 46	3 28 30	23 36
21	11 09 10	11 31 22	8 48 30	18 31 42	5 28 09	3 20 44	3 24 30	23 32
22	11 13 07	11 32 25	8 47 21	18 28 00	5 24 16	3 16 42	3 20 30	23 28
23	11 17 16	11 33 29	8 46 12	18 24 19	5 20 22	3 12 40	3 16 30	23 24
24	11 21 36	11 34 34	8 45 03	18 20 38	5 16 28	3 08 38	3 12 30	23 20
25	11 26 06	11 35 40	8 43 54	18 16 58	5 12 34	3 04 36	3 08 30	23 16
26	11 30 46	11 36 48	8 42 46	18 13 18	5 08 40	3 00 33	3 04 30	23 12
27	11 35 36	11 37 56	8 41 37	18 09 39	5 04 45	2 56 31	3 00 29	23 08
28	11 40 35	11 39 06	8 40 28	18 06 01	5 00 49	2 52 28	2 56 29	23 04
29	11 45 43	11 40 17	8 39 19	18 02 23	4 56 54	2 48 25	2 52 28	23 00
30	11 50 58	11 41 28	8 38 10	17 58 46	4 52 57	2 44 22	2 48 27	22 56
31	11 56 19	11 42 41	8 37 01	17 55 10	4 49 01	2 40 19	2 44 26	22 52
June 1	12 01 45	11 43 55	8 35 52	17 51 34	4 45 04	2 36 15	2 40 26	22 47
2	12 07 15	11 45 10	8 34 43	17 47 58	4 41 07	2 32 12	2 36 25	22 43
3	12 12 47	11 46 26	8 33 35	17 44 23	4 37 09	2 28 08	2 32 23	22 39
4	12 18 19	11 47 42	8 32 26	17 40 49	4 33 11	2 24 04	2 28 22	22 35
5	12 23 51	11 49 00	8 31 17	17 37 15	4 29 12	2 20 00	2 24 21	22 31
6	12 29 20	11 50 18	8 30 08	17 33 42	4 25 14	2 15 56	2 20 20	22 27
7	12 34 45	11 51 38	8 28 59	17 30 09	4 21 14	2 11 52	2 16 18	22 23
8	12 40 04	11 52 58	8 27 50	17 26 37	4 17 15	2 07 47	2 12 17	22 19
9	12 45 18	11 54 18	8 26 42	17 23 05	4 13 15	2 03 43	2 08 15	22 15
10	12 50 23	11 55 40	8 25 33	17 19 34	4 09 15	1 59 38	2 04 14	22 11
11	12 55 20	11 57 02	8 24 24	17 16 03	4 05 14	1 55 34	2 00 12	22 07
12	13 00 08	11 58 25	8 23 16	17 12 33	4 01 13	1 51 29	1 56 10	22 03
13	13 04 45	11 59 48	8 22 07	17 09 03	3 57 11	1 47 24	1 52 08	21 59
14	13 09 11	12 01 12	8 20 59	17 05 34	3 53 10	1 43 19	1 48 06	21 55
15	13 13 26	12 02 36	8 19 50	17 02 05	3 49 07	1 39 14	1 44 04	21 51
16	13 17 30	12 04 01	8 18 42	16 58 37	3 45 05	1 35 09	1 40 02	21 47
17	13 21 21	12 05 26	8 17 34	16 55 09	3 41 02	1 31 03	1 36 00	21 43
18	13 25 00	12 06 52	8 16 25	16 51 41	3 36 59	1 26 58	1 31 58	21 39
19	13 28 26	12 08 17	8 15 17	16 48 14	3 32 55	1 22 52	1 27 56	21 35
20	13 31 40	12 09 43	8 14 09	16 44 48	3 28 51	1 18 47	1 23 54	21 31
21	13 34 40	12 11 09	8 13 01	16 41 22	3 24 47	1 14 41	1 19 51	21 27
22	13 37 28	12 12 36	8 11 53	16 37 56	3 20 43	1 10 36	1 15 49	21 23
23	13 40 03	12 14 02	8 10 45	16 34 31	3 16 38	1 06 30	1 11 46	21 19
24	13 42 25	12 15 28	8 09 37	16 31 06	3 12 33	1 02 24	1 07 44	21 15
25	13 44 34	12 16 54	8 08 30	16 27 42	3 08 27	0 58 18	1 03 42	21 11
26	13 46 30	12 18 20	8 07 22	16 24 18	3 04 21	0 54 12	0 59 39	21 07
27	13 48 13	12 19 46	8 06 14	16 20 54	3 00 15	0 50 06	0 55 36	21 03
28	13 49 43	12 21 11	8 05 07	16 17 31	2 56 08	0 46 00	0 51 34	20 59
29	13 50 59	12 22 36	8 03 59	16 14 08	2 52 01	0 41 54	0 47 31	20 55
30	13 52 03	12 24 01	8 02 52	16 10 46	2 47 54	0 37 48	0 43 29	20 51
July 1	13 52 52	12 25 25	8 01 44	16 07 24	2 43 47	0 33 42	0 39 26	20 47
2	13 53 29	12 26 49	8 00 37	16 04 02	2 39 39	0 29 35	0 35 23	20 43

Date	Mercury	Venus	Mars	Jupiter	Saturn	Uranus	Neptune	Pluto
	h m s	h m s	h m s	h m s	h m s	h m s	h m s	h m
July 1	13 52 52	12 25 25	8 01 44	16 07 24	2 43 47	0 33 42	0 39 26	20 47
2	13 53 29	12 26 49	8 00 37	16 04 02	2 39 39	0 29 35	0 35 23	20 43
3	13 53 51	12 28 12	7 59 29	16 00 41	2 35 31	0 25 29	0 31 20	20 39
4	13 54 00	12 29 35	7 58 22	15 57 20	2 31 23	0 21 23	0 27 18	20 35
5	13 53 55	12 30 57	7 57 15	15 53 59	2 27 14	0 17 17	0 23 15	20 31
6	13 53 35	12 32 18	7 56 07	15 50 39	2 23 05	0 13 10	0 19 12	20 27
7	13 53 01	12 33 38	7 55 00	15 47 19	2 18 56	0 09 04	0 15 09	20 23
8	13 52 12	12 34 58	7 53 53	15 43 59	2 14 47	0 04 58	0 11 07	20 19
9	13 51 08	12 36 17	7 52 45	15 40 40	2 10 37	0 00 52	0 07 04	20 15
10	13 49 49	12 37 35	7 51 38	15 37 21	2 06 27	23 52 39	0 03 01	20 11
11	13 48 14	12 38 52	7 50 31	15 34 02	2 02 17	23 48 33	23 54 56	20 07
12	13 46 22	12 40 08	7 49 24	15 30 44	1 58 07	23 44 27	23 50 53	20 03
13	13 44 15	12 41 23	7 48 17	15 27 26	1 53 56	23 40 20	23 46 50	19 59
14	13 41 50	12 42 37	7 47 09	15 24 08	1 49 45	23 36 14	23 42 47	19 55
15	13 39 08	12 43 50	7 46 02	15 20 51	1 45 34	23 32 08	23 38 44	19 51
16	13 36 08	12 45 02	7 44 55	15 17 34	1 41 23	23 28 02	23 34 42	19 47
17	13 32 50	12 46 13	7 43 48	15 14 17	1 37 12	23 23 56	23 30 39	19 43
18	13 29 14	12 47 22	7 42 41	15 11 01	1 33 00	23 19 50	23 26 36	19 39
19	13 25 20	12 48 31	7 41 33	15 07 44	1 28 49	23 15 44	23 22 34	19 35
20	13 21 07	12 49 38	7 40 26	15 04 28	1 24 37	23 11 38	23 18 31	19 31
21	13 16 36	12 50 45	7 39 19	15 01 13	1 20 24	23 07 32	23 14 28	19 27
22	13 11 46	12 51 50	7 38 12	14 57 57	1 16 12	23 03 26	23 10 26	19 23
23	13 06 38	12 52 54	7 37 04	14 54 42	1 12 00	22 59 20	23 06 23	19 20
24	13 01 13	12 53 57	7 35 57	14 51 27	1 07 47	22 55 14	23 02 21	19 16
25	12 55 32	12 54 58	7 34 49	14 48 12	1 03 34	22 51 09	22 58 18	19 12
26	12 49 35	12 55 59	7 33 42	14 44 58	0 59 22	22 47 03	22 54 16	19 08
27	12 43 23	12 56 58	7 32 34	14 41 43	0 55 09	22 42 58	22 50 14	19 04
28	12 37 00	12 57 56	7 31 26	14 38 29	0 50 56	22 38 52	22 46 11	19 00
29	12 30 25	12 58 53	7 30 18	14 35 16	0 46 42	22 34 47	22 42 09	18 56
30	12 23 42	12 59 49	7 29 10	14 32 02	0 42 29	22 30 42	22 38 07	18 52
31	12 16 54	13 00 44	7 28 02	14 28 49	0 38 16	22 26 36	22 34 04	18 48
Aug. 1	12 10 02	13 01 37	7 26 54	14 25 35	0 34 02	22 22 31	22 30 02	18 44
2	12 03 11	13 02 29	7 25 45	14 22 22	0 29 49	22 18 26	22 26 00	18 40
3	11 56 22	13 03 21	7 24 36	14 19 10	0 25 35	22 14 22	22 21 58	18 36
4	11 49 40	13 04 11	7 23 27	14 15 57	0 21 22	22 10 17	22 17 56	18 32
5	11 43 07	13 05 00	7 22 18	14 12 45	0 17 08	22 06 12	22 13 54	18 28
6	11 36 46	13 05 48	7 21 09	14 09 33	0 12 55	22 02 08	22 09 53	18 24
7	11 30 41	13 06 34	7 19 59	14 06 21	0 08 41	21 58 03	22 05 51	18 20
8	11 24 55	13 07 20	7 18 50	14 03 09	0 04 27	21 53 59	22 01 49	18 16
9	11 19 29	13 08 05	7 17 40	13 59 57	0 00 14	21 49 55	21 57 48	18 13
10	11 14 25	13 08 49	7 16 29	13 56 46	23 51 46	21 45 51	21 53 46	18 09
11	11 09 47	13 09 32	7 15 19	13 53 34	23 47 33	21 41 47	21 49 45	18 05
12	11 05 36	13 10 14	7 14 08	13 50 23	23 43 19	21 37 43	21 45 43	18 01
13	11 01 52	13 10 55	7 12 57	13 47 12	23 39 05	21 33 40	21 41 42	17 57
14	10 58 37	13 11 35	7 11 46	13 44 01	23 34 52	21 29 36	21 37 41	17 53
15	10 55 51	13 12 14	7 10 34	13 40 50	23 30 39	21 25 33	21 33 40	17 49
16	10 53 35	13 12 53	7 09 22	13 37 40	23 26 25	21 21 30	21 29 39	17 45

Second transits: Uranus, July $9^{d}23^{h}56^{d}45^{s}$, Neptune, July $10^{d}23^{h}58^{d}58^{s}$,
Saturn, Aug. $9^{d}23^{h}56^{d}00^{s}$.

Date	Mercury	Venus	Mars	Jupiter	Saturn	Uranus	Neptune	Pluto
	h m s	h m s	h m s	h m s	h m s	h m s	h m s	h m
Aug. 16	10 53 35	13 12 53	7 09 22	13 37 40	23 26 25	21 21 30	21 29 39	17 45
17	10 51 49	13 13 31	7 08 10	13 34 29	23 22 12	21 17 27	21 25 38	17 41
18	10 50 33	13 14 08	7 06 57	13 31 19	23 17 59	21 13 24	21 21 37	17 37
19	10 49 45	13 14 44	7 05 44	13 28 09	23 13 46	21 09 21	21 17 36	17 33
20	10 49 26	13 15 20	7 04 31	13 24 59	23 09 33	21 05 19	21 13 35	17 30
21	10 49 34	13 15 55	7 03 17	13 21 49	23 05 20	21 01 16	21 09 35	17 26
22	10 50 09	13 16 30	7 02 03	13 18 39	23 01 07	20 57 14	21 05 34	17 22
23	10 51 08	13 17 04	7 00 48	13 15 29	22 56 54	20 53 12	21 01 34	17 18
24	10 52 31	13 17 38	6 59 33	13 12 20	22 52 42	20 49 10	20 57 34	17 14
25	10 54 15	13 18 11	6 58 17	13 09 10	22 48 29	20 45 08	20 53 34	17 10
26	10 56 19	13 18 44	6 57 01	13 06 01	22 44 17	20 41 07	20 49 34	17 06
27	10 58 41	13 19 16	6 55 45	13 02 51	22 40 05	20 37 06	20 45 34	17 02
28	11 01 17	13 19 48	6 54 28	12 59 42	22 35 53	20 33 04	20 41 34	16 58
29	11 04 07	13 20 20	6 53 10	12 56 33	22 31 42	20 29 03	20 37 34	16 55
30	11 07 07	13 20 52	6 51 52	12 53 24	22 27 30	20 25 03	20 33 35	16 51
31	11 10 17	13 21 23	6 50 33	12 50 15	22 23 19	20 21 02	20 29 35	16 47
Sept. 1	11 13 33	13 21 54	6 49 14	12 47 06	22 19 08	20 17 02	20 25 36	16 43
2	11 16 54	13 22 25	6 47 54	12 43 57	22 14 57	20 13 01	20 21 37	16 39
3	11 20 18	13 22 56	6 46 33	12 40 48	22 10 47	20 09 01	20 17 37	16 35
4	11 23 44	13 23 27	6 45 12	12 37 39	22 06 36	20 05 02	20 13 38	16 31
5	11 27 09	13 23 58	6 43 50	12 34 31	22 02 26	20 01 02	20 09 40	16 27
6	11 30 34	13 24 29	6 42 28	12 31 22	21 58 16	19 57 03	20 05 41	16 24
7	11 33 57	13 25 00	6 41 05	12 28 14	21 54 07	19 53 03	20 01 42	16 20
8	11 37 17	13 25 32	6 39 41	12 25 05	21 49 57	19 49 04	19 57 44	16 16
9	11 40 33	13 26 03	6 38 16	12 21 56	21 45 48	19 45 05	19 53 45	16 12
10	11 43 45	13 26 34	6 36 51	12 18 48	21 41 39	19 41 07	19 49 47	16 08
11	11 46 54	13 27 06	6 35 25	12 15 39	21 37 31	19 37 08	19 45 49	16 04
12	11 49 57	13 27 38	6 33 58	12 12 31	21 33 23	19 33 10	19 41 51	16 00
13	11 52 56	13 28 11	6 32 30	12 09 23	21 29 15	19 29 12	19 37 53	15 57
14	11 55 50	13 28 44	6 31 02	12 06 14	21 25 07	19 25 14	19 33 55	15 53
15	11 58 39	13 29 17	6 29 33	12 03 06	21 21 00	19 21 17	19 29 58	15 49
16	12 01 24	13 29 51	6 28 03	11 59 57	21 16 53	19 17 20	19 26 00	15 45
17	12 04 04	13 30 25	6 26 32	11 56 49	21 12 46	19 13 22	19 22 03	15 41
18	12 06 39	13 31 00	6 25 00	11 53 41	21 08 39	19 09 25	19 18 06	15 37
19	12 09 10	13 31 35	6 23 28	11 50 32	21 04 33	19 05 29	19 14 09	15 34
20	12 11 37	13 32 11	6 21 54	11 47 24	21 00 27	19 01 32	19 10 12	15 30
21	12 14 00	13 32 48	6 20 20	11 44 16	20 56 22	18 57 36	19 06 15	15 26
22	12 16 19	13 33 25	6 18 44	11 41 07	20 52 17	18 53 40	19 02 18	15 22
23	12 18 34	13 34 03	6 17 08	11 37 59	20 48 12	18 49 44	18 58 22	15 18
24	12 20 46	13 34 42	6 15 30	11 34 50	20 44 08	18 45 48	18 54 25	15 14
25	12 22 55	13 35 22	6 13 52	11 31 42	20 40 04	18 41 53	18 50 29	15 11
26	12 25 01	13 36 03	6 12 12	11 28 33	20 36 00	18 37 58	18 46 33	15 07
27	12 27 04	13 36 44	6 10 31	11 25 25	20 31 57	18 34 03	18 42 37	15 03
28	12 29 04	13 37 27	6 08 49	11 22 16	20 27 54	18 30 08	18 38 41	14 59
29	12 31 02	13 38 10	6 07 07	11 19 08	20 23 51	18 26 14	18 34 46	14 55
30	12 32 57	13 38 54	6 05 22	11 15 59	20 19 49	18 22 19	18 30 50	14 51
Oct. 1	12 34 51	13 39 39	6 03 37	11 12 50	20 15 47	18 18 25	18 26 55	14 48

Date	Mercury	Venus	Mars	Jupiter	Saturn	Uranus	Neptune	Pluto
	h m s	h m s	h m s	h m s	h m s	h m s	h m s	h m
Oct. 1	12 34 51	13 39 39	6 03 37	11 12 50	20 15 47	18 18 25	18 26 55	14 48
2	12 36 42	13 40 26	6 01 51	11 09 41	20 11 46	18 14 31	18 22 59	14 44
3	12 38 31	13 41 13	6 00 03	11 06 33	20 07 45	18 10 38	18 19 04	14 40
4	12 40 19	13 42 01	5 58 14	11 03 24	20 03 44	18 06 44	18 15 09	14 36
5	12 42 05	13 42 51	5 56 24	11 00 15	19 59 44	18 02 51	18 11 14	14 32
6	12 43 49	13 43 41	5 54 33	10 57 06	19 55 44	17 58 58	18 07 20	14 29
7	12 45 32	13 44 33	5 52 40	10 53 56	19 51 45	17 55 05	18 03 25	14 25
8	12 47 14	13 45 26	5 50 46	10 50 47	19 47 46	17 51 13	17 59 31	14 21
9	12 48 54	13 46 20	5 48 50	10 47 38	19 43 47	17 47 20	17 55 37	14 17
10	12 50 33	13 47 15	5 46 54	10 44 28	19 39 49	17 43 28	17 51 42	14 13
11	12 52 11	13 48 11	5 44 56	10 41 19	19 35 51	17 39 36	17 47 49	14 10
12	12 53 48	13 49 08	5 42 56	10 38 09	19 31 53	17 35 45	17 43 55	14 06
13	12 55 23	13 50 07	5 40 55	10 34 59	19 27 56	17 31 53	17 40 01	14 02
14	12 56 57	13 51 06	5 38 53	10 31 50	19 24 00	17 28 02	17 36 07	13 58
15	12 58 30	13 52 07	5 36 49	10 28 40	19 20 03	17 24 11	17 32 14	13 54
16	13 00 01	13 53 09	5 34 43	10 25 30	19 16 08	17 20 20	17 28 21	13 50
17	13 01 30	13 54 13	5 32 36	10 22 19	19 12 12	17 16 29	17 24 28	13 47
18	13 02 58	13 55 17	5 30 28	10 19 09	19 08 17	17 12 39	17 20 35	13 43
19	13 04 24	13 56 23	5 28 17	10 15 58	19 04 22	17 08 49	17 16 42	13 39
20	13 05 47	13 57 30	5 26 05	10 12 48	19 00 28	17 04 59	17 12 49	13 35
21	13 07 09	13 58 38	5 23 52	10 09 37	18 56 34	17 01 09	17 08 56	13 32
22	13 08 27	13 59 47	5 21 36	10 06 26	18 52 41	16 57 19	17 05 04	13 28
23	13 09 42	14 00 57	5 19 19	10 03 15	18 48 48	16 53 30	17 01 12	13 24
24	13 10 53	14 02 08	5 17 00	10 00 04	18 44 55	16 49 41	16 57 19	13 20
25	13 12 00	14 03 20	5 14 39	9 56 52	18 41 03	16 45 52	16 53 27	13 16
26	13 13 02	14 04 34	5 12 16	9 53 41	18 37 11	16 42 03	16 49 36	13 13
27	13 13 58	14 05 48	5 09 52	9 50 29	18 33 20	16 38 14	16 45 44	13 09
28	13 14 48	14 07 03	5 07 25	9 47 17	18 29 29	16 34 26	16 41 52	13 05
29	13 15 31	14 08 19	5 04 56	9 44 05	18 25 39	16 30 38	16 38 01	13 01
30	13 16 05	14 09 36	5 02 26	9 40 53	18 21 48	16 26 50	16 34 09	12 57
31	13 16 29	14 10 54	4 59 53	9 37 40	18 17 59	16 23 02	16 30 18	12 54
Nov. 1	13 16 42	14 12 13	4 57 18	9 34 27	18 14 09	16 19 14	16 26 27	12 50
2	13 16 42	14 13 32	4 54 41	9 31 14	18 10 20	16 15 27	16 22 36	12 46
3	13 16 28	14 14 52	4 52 02	9 28 01	18 06 32	16 11 40	16 18 45	12 42
4	13 15 57	14 16 12	4 49 20	9 24 48	18 02 44	16 07 53	16 14 54	12 38
5	13 15 08	14 17 33	4 46 37	9 21 34	17 58 56	16 04 06	16 11 04	12 35
6	13 13 58	14 18 55	4 43 51	9 18 20	17 55 09	16 00 19	16 07 13	12 31
7	13 12 24	14 20 17	4 41 02	9 15 06	17 51 22	15 56 33	16 03 23	12 27
8	13 10 24	14 21 39	4 38 12	9 11 52	17 47 35	15 52 46	15 59 32	12 23
9	13 07 55	14 23 01	4 35 18	9 08 37	17 43 49	15 49 00	15 55 42	12 20
10	13 04 54	14 24 24	4 32 23	9 05 23	17 40 03	15 45 14	15 51 52	12 16
11	13 01 18	14 25 47	4 29 25	9 02 07	17 36 17	15 41 28	15 48 02	12 12
12	12 57 04	14 27 10	4 26 24	8 58 52	17 32 32	15 37 43	15 44 13	12 08
13	12 52 11	14 28 33	4 23 21	8 55 37	17 28 48	15 33 57	15 40 23	12 04
14	12 46 36	14 29 56	4 20 15	8 52 21	17 25 03	15 30 12	15 36 33	12 01
15	12 40 20	14 31 19	4 17 06	8 49 04	17 21 19	15 26 27	15 32 44	11 57
16	12 33 22	14 32 41	4 13 55	8 45 48	17 17 36	15 22 42	15 28 55	11 53

Date	Mercury	Venus	Mars	Jupiter	Saturn	Uranus	Neptune	Pluto
	h m s	h m s	h m s	h m s	h m s	h m s	h m s	h m
Nov. 16	12 33 22	14 32 41	4 13 55	8 45 48	17 17 36	15 22 42	15 28 55	11 53
17	12 25 45	14 34 03	4 10 41	8 42 31	17 13 53	15 18 57	15 25 05	11 49
18	12 17 33	14 35 25	4 07 23	8 39 14	17 10 10	15 15 12	15 21 16	11 45
19	12 08 53	14 36 47	4 04 03	8 35 57	17 06 27	15 11 28	15 17 27	11 42
20	11 59 52	14 38 08	4 00 40	8 32 39	17 02 45	15 07 43	15 13 38	11 38
21	11 50 40	14 39 28	3 57 14	8 29 21	16 59 03	15 03 59	15 09 49	11 34
22	11 41 26	14 40 48	3 53 45	8 26 03	16 55 22	15 00 15	15 06 01	11 30
23	11 32 22	14 42 08	3 50 12	8 22 44	16 51 41	14 56 31	15 02 12	11 27
24	11 23 38	14 43 26	3 46 36	8 19 25	16 48 00	14 52 47	14 58 23	11 23
25	11 15 21	14 44 44	3 42 58	8 16 06	16 44 20	14 49 03	14 54 35	11 19
26	11 07 40	14 46 00	3 39 15	8 12 46	16 40 39	14 45 20	14 50 47	11 15
27	11 00 38	14 47 16	3 35 30	8 09 26	16 37 00	14 41 36	14 46 58	11 11
28	10 54 20	14 48 31	3 31 41	8 06 06	16 33 20	14 37 53	14 43 10	11 08
29	10 48 45	14 49 44	3 27 49	8 02 45	16 29 41	14 34 10	14 39 22	11 04
30	10 43 55	14 50 57	3 23 53	7 59 24	16 26 02	14 30 27	14 35 34	11 00
Dec. 1	10 39 47	14 52 08	3 19 54	7 56 02	16 22 24	14 26 44	14 31 46	10 56
2	10 36 20	14 53 18	3 15 51	7 52 40	16 18 46	14 23 01	14 27 58	10 53
3	10 33 30	14 54 27	3 11 45	7 49 18	16 15 08	14 19 18	14 24 11	10 49
4	10 31 14	14 55 34	3 07 35	7 45 55	16 11 30	14 15 36	14 20 23	10 45
5	10 29 30	14 56 39	3 03 21	7 42 32	16 07 53	14 11 53	14 16 35	10 41
6	10 28 13	14 57 44	2 59 05	7 39 08	16 04 16	14 08 11	14 12 48	10 37
7	10 27 22	14 58 46	2 54 44	7 35 44	16 00 39	14 04 29	14 09 00	10 34
8	10 26 54	14 59 48	2 50 20	7 32 20	15 57 03	14 00 46	14 05 13	10 30
9	10 26 46	15 00 47	2 45 52	7 28 55	15 53 27	13 57 04	14 01 26	10 26
10	10 26 56	15 01 45	2 41 21	7 25 29	15 49 51	13 53 22	13 57 38	10 22
11	10 27 22	15 02 41	2 36 46	7 22 04	15 46 16	13 49 40	13 53 51	10 18
12	10 28 03	15 03 36	2 32 07	7 18 37	15 42 40	13 45 59	13 50 04	10 15
13	10 28 56	15 04 28	2 27 25	7 15 11	15 39 05	13 42 17	13 46 17	10 11
14	10 30 00	15 05 19	2 22 39	7 11 44	15 35 30	13 38 35	13 42 30	10 07
15	10 31 15	15 06 09	2 17 50	7 08 16	15 31 56	13 34 54	13 38 43	10 03
16	10 32 39	15 06 56	2 12 57	7 04 48	15 28 22	13 31 12	13 34 56	10 00
17	10 34 11	15 07 42	2 08 01	7 01 19	15 24 48	13 27 31	13 31 09	9 56
18	10 35 51	15 08 25	2 03 01	6 57 50	15 21 14	13 23 49	13 27 23	9 52
19	10 37 38	15 09 07	1 57 58	6 54 21	15 17 40	13 20 08	13 23 36	9 48
20	10 39 31	15 09 47	1 52 52	6 50 50	15 14 07	13 16 27	13 19 49	9 44
21	10 41 30	15 10 26	1 47 43	6 47 20	15 10 34	13 12 46	13 16 03	9 41
22	10 43 34	15 11 02	1 42 30	6 43 49	15 07 01	13 09 05	13 12 16	9 37
23	10 45 43	15 11 37	1 37 15	6 40 17	15 03 29	13 05 24	13 08 29	9 33
24	10 47 56	15 12 09	1 31 56	6 36 45	14 59 56	13 01 43	13 04 43	9 29
25	10 50 13	15 12 40	1 26 35	6 33 12	14 56 24	12 58 02	13 00 56	9 25
26	10 52 35	15 13 09	1 21 12	6 29 38	14 52 52	12 54 21	12 57 10	9 22
27	10 55 00	15 13 35	1 15 46	6 26 05	14 49 20	12 50 40	12 53 23	9 18
28	10 57 28	15 14 00	1 10 17	6 22 30	14 45 49	12 46 59	12 49 37	9 14
29	11 00 00	15 14 23	1 04 47	6 18 55	14 42 17	12 43 18	12 45 51	9 10
30	11 02 35	15 14 44	0 59 15	6 15 19	14 38 46	12 39 37	12 42 04	9 06
31	11 05 12	15 15 03	0 53 41	6 11 43	14 35 15	12 35 57	12 38 18	9 02
32	11 07 52	15 15 20	0 48 06	6 08 06	14 31 44	12 32 16	12 34 32	8 59

EPHEMERIS FOR PHYSICAL OBSERVATIONS
FOR 0^h DYNAMICAL TIME

Date		Light-time	Magnitude	Surface Brightness	Diameter	Phase	Phase Angle	Defect of Illumination
		m			″		°	″
Jan.	−1	8.81	− 0.3	+3.0	6.35	0.686	68.2	1.99
	1	9.13	0.3	3.0	6.13	0.726	63.2	1.68
	3	9.43	0.3	3.0	5.93	0.760	58.7	1.43
	5	9.71	0.3	2.9	5.76	0.789	54.6	1.21
	7	9.97	0.3	2.9	5.61	0.815	50.9	1.04
	9	10.20	− 0.3	+2.9	5.48	0.837	47.5	0.89
	11	10.42	0.3	2.9	5.37	0.857	44.4	0.77
	13	10.63	0.3	2.9	5.26	0.875	41.5	0.66
	15	10.81	0.4	2.8	5.17	0.890	38.7	0.57
	17	10.98	0.4	2.8	5.10	0.904	36.1	0.49
	19	11.12	− 0.4	+2.7	5.03	0.917	33.6	0.42
	21	11.26	0.4	2.7	4.97	0.928	31.1	0.36
	23	11.37	0.5	2.6	4.92	0.938	28.7	0.30
	25	11.47	0.5	2.6	4.88	0.948	26.4	0.25
	27	11.55	0.6	2.5	4.84	0.957	24.0	0.21
	29	11.62	− 0.7	+2.4	4.81	0.965	21.6	0.17
	31	11.67	0.8	2.4	4.79	0.972	19.2	0.13
Feb.	2	11.70	0.8	2.3	4.78	0.979	16.7	0.10
	4	11.72	1.0	2.2	4.77	0.985	14.2	0.07
	6	11.71	1.1	2.0	4.78	0.990	11.6	0.05
	8	11.69	− 1.2	+1.9	4.79	0.994	8.9	0.03
	10	11.64	1.4	1.8	4.80	0.997	6.5	0.02
	12	11.58	1.5	1.7	4.83	0.998	5.0	0.01
	14	11.48	1.5	1.7	4.87	0.997	5.9	0.01
	16	11.37	1.5	1.7	4.92	0.994	8.9	0.03
	18	11.22	− 1.5	+1.8	4.99	0.987	13.0	0.06
	20	11.04	1.4	1.8	5.07	0.976	17.8	0.12
	22	10.82	1.4	1.9	5.17	0.959	23.4	0.21
	24	10.57	1.3	2.0	5.29	0.935	29.6	0.34
	26	10.28	1.2	2.1	5.44	0.902	36.4	0.53
	28	9.95	− 1.2	+2.1	5.62	0.859	44.0	0.79
Mar.	1	9.58	1.1	2.2	5.84	0.806	52.3	1.13
	3	9.17	1.0	2.3	6.10	0.741	61.2	1.58
	5	8.74	0.8	2.5	6.40	0.666	70.6	2.14
	7	8.29	0.7	2.6	6.75	0.583	80.4	2.81
	9	7.82	− 0.4	+2.8	7.15	0.496	90.5	3.60
	11	7.36	− 0.1	3.1	7.60	0.408	100.7	4.50
	13	6.92	+ 0.3	3.4	8.09	0.322	110.9	5.48
	15	6.50	0.8	3.7	8.61	0.242	121.1	6.52
	17	6.12	1.4	4.0	9.14	0.171	131.2	7.58
	19	5.79	+ 2.1	+4.4	9.67	0.111	141.1	8.60
	21	5.50	2.9	4.7	10.16	0.063	150.9	9.52
	23	5.28	3.8	4.8	10.60	0.029	160.2	10.29
	25	5.11	4.7	4.6	10.95	0.010	168.5	10.84
	27	5.00	5.1	4.3	11.19	0.005	172.2	11.13
	29	4.95	+ 4.6	+4.8	11.31	0.012	167.4	11.17
	31	4.95	+ 3.8	+5.1	11.31	0.030	159.9	10.96

EPHEMERIS FOR PHYSICAL OBSERVATIONS
FOR 0h DYNAMICAL TIME

Date		Sub-Earth Point		Sub-Solar Point			North Pole	
		Long.	Lat.	Long.	Dist.	P.A.	Dist.	P.A.
		°	°	°	″	°	″	°
Jan.	−1	312.07	− 4.93	20.15	2.95	100.27	− 3.16	12.27
	1	322.00	4.90	25.07	2.73	98.68	3.05	11.19
	3	331.79	4.88	30.36	2.53	97.01	2.96	10.01
	5	341.47	4.86	35.96	2.35	95.27	2.87	8.74
	7	351.05	4.85	41.82	2.18	93.45	2.80	7.41
	9	0.56	− 4.84	47.92	2.02	91.56	− 2.73	6.02
	11	10.01	4.83	54.20	1.88	89.61	2.67	4.58
	13	19.40	4.83	60.64	1.74	87.59	2.62	3.09
	15	28.75	4.83	67.20	1.62	85.50	2.58	1.57
	17	38.06	4.83	73.85	1.50	83.34	2.54	0.02
	19	47.32	− 4.83	80.57	1.39	81.11	− 2.51	358.45
	21	56.55	4.83	87.32	1.28	78.79	2.48	356.86
	23	65.74	4.84	94.09	1.18	76.37	2.45	355.27
	25	74.89	4.85	100.84	1.08	73.82	2.43	353.68
	27	84.00	4.85	107.54	0.98	71.09	2.41	352.09
	29	93.08	− 4.86	114.18	0.89	68.13	− 2.40	350.52
	31	102.10	4.87	120.72	0.79	64.82	2.39	348.97
Feb.	2	111.09	4.88	127.13	0.69	60.96	2.38	347.44
	4	120.02	4.89	133.37	0.59	56.20	2.38	345.95
	6	128.91	4.90	139.42	0.48	49.78	2.38	344.49
	8	137.75	− 4.92	145.24	0.37	40.00	− 2.38	343.08
	10	146.54	4.93	150.78	0.27	22.52	2.39	341.72
	12	155.27	4.96	155.99	0.21	348.73	2.41	340.42
	14	163.95	4.98	160.83	0.25	307.05	2.43	339.19
	16	172.58	5.01	165.25	0.38	282.16	2.45	338.02
	18	181.16	− 5.05	169.19	0.56	269.44	− 2.48	336.94
	20	189.71	5.10	172.59	0.78	262.00	2.52	335.94
	22	198.23	5.16	175.41	1.02	257.06	2.57	335.03
	24	206.75	5.24	177.62	1.31	253.49	2.64	334.22
	26	215.29	5.33	179.18	1.62	250.74	2.71	333.51
	28	223.90	− 5.45	180.13	1.95	248.54	− 2.80	332.91
Mar.	1	232.61	5.58	180.53	2.31	246.73	2.91	332.41
	3	241.50	5.74	180.48	2.67	245.18	3.03	332.02
	5	250.63	5.93	180.16	3.02	243.82	3.18	331.73
	7	260.09	6.15	179.75	3.33	242.56	3.36	331.53
	9	269.93	− 6.39	179.47	−3.58	241.33	− 3.55	331.39
	11	280.24	6.65	179.51	−3.73	240.04	3.77	331.30
	13	291.07	6.92	180.03	−3.78	238.60	4.01	331.25
	15	302.46	7.19	181.11	−3.69	236.86	4.27	331.21
	17	314.42	7.44	182.82	−3.44	234.65	4.53	331.19
	19	326.94	− 7.65	185.17	−3.03	231.60	− 4.79	331.19
	21	339.99	7.81	188.13	−2.47	226.96	5.04	331.19
	23	353.47	7.89	191.65	−1.79	218.59	5.25	331.22
	25	7.28	7.87	195.71	−1.09	198.59	5.42	331.28
	27	21.28	7.75	200.23	−0.76	143.59	5.54	331.38
	29	35.32	− 7.53	205.16	−1.23	97.59	− 5.60	331.50
	31	49.27	− 7.21	210.45	−1.94	81.81	− 5.61	331.63

EPHEMERIS FOR PHYSICAL OBSERVATIONS FOR 0^h DYNAMICAL TIME

Date		Light-time	Magnitude	Surface Brightness	Diameter	Phase	Phase Angle	Defect of Illumination
		m			″		°	″
Apr.	2	4.99	+ 3.1	+5.0	11.21	0.057	152.3	10.56
	4	5.08	2.6	4.9	11.02	0.091	145.0	10.02
	6	5.20	2.1	4.8	10.76	0.128	138.1	9.39
	8	5.35	1.7	4.6	10.46	0.167	131.8	8.72
	10	5.52	1.4	4.5	10.13	0.206	126.0	8.04
	12	5.71	+ 1.2	+4.4	9.79	0.246	120.6	7.39
	14	5.92	1.0	4.3	9.45	0.284	115.6	6.76
	16	6.14	0.8	4.2	9.11	0.321	110.9	6.18
	18	6.37	0.7	4.1	8.78	0.357	106.6	5.65
	20	6.61	0.6	4.0	8.47	0.391	102.5	5.15
	22	6.85	+ 0.5	+3.9	8.17	0.424	98.7	4.70
	24	7.10	0.4	3.8	7.88	0.456	95.0	4.28
	26	7.35	0.3	3.7	7.61	0.487	91.4	3.90
	28	7.61	0.3	3.6	7.35	0.518	87.9	3.54
	30	7.87	0.2	3.5	7.11	0.548	84.5	3.21
May	2	8.13	+ 0.1	+3.4	6.88	0.578	81.0	2.91
	4	8.39	+ 0.0	3.3	6.67	0.608	77.5	2.61
	6	8.65	− 0.1	3.2	6.47	0.638	73.9	2.34
	8	8.91	0.2	3.1	6.28	0.669	70.2	2.08
	10	9.17	0.3	3.0	6.10	0.701	66.3	1.82
	12	9.42	− 0.4	+2.9	5.94	0.734	62.1	1.58
	14	9.67	0.5	2.8	5.78	0.767	57.7	1.35
	16	9.91	0.6	2.6	5.64	0.802	52.9	1.12
	18	10.14	0.8	2.5	5.52	0.836	47.7	0.90
	20	10.35	1.0	2.3	5.40	0.871	42.1	0.70
	22	10.54	− 1.1	+2.1	5.31	0.905	36.0	0.51
	24	10.71	1.3	1.9	5.23	0.936	29.4	0.34
	26	10.84	1.6	1.7	5.16	0.963	22.3	0.19
	28	10.93	1.8	1.4	5.12	0.984	14.7	0.08
	30	10.98	2.1	1.2	5.09	0.996	6.9	0.02
June	1	10.99	− 2.3	+1.0	5.09	0.999	2.6	0.00
	3	10.95	2.0	1.2	5.11	0.992	10.1	0.04
	5	10.86	1.8	1.5	5.15	0.975	18.2	0.13
	7	10.72	1.5	1.7	5.22	0.949	26.0	0.26
	9	10.55	1.3	2.0	5.30	0.917	33.6	0.44
	11	10.35	− 1.1	+2.2	5.41	0.879	40.7	0.65
	13	10.12	0.9	2.3	5.53	0.839	47.3	0.89
	15	9.86	0.8	2.5	5.67	0.798	53.4	1.14
	17	9.60	0.6	2.7	5.83	0.757	59.1	1.42
	19	9.32	0.4	2.8	6.00	0.717	64.3	1.70
	21	9.04	− 0.3	+3.0	6.19	0.677	69.2	2.00
	23	8.76	0.2	3.1	6.39	0.640	73.8	2.30
	25	8.47	− 0.1	3.2	6.60	0.603	78.1	2.62
	27	8.19	+ 0.1	3.3	6.83	0.567	82.3	2.96
	29	7.90	0.2	3.5	7.08	0.533	86.2	3.31
July	1	7.63	+ 0.3	+3.6	7.34	0.499	90.1	3.68
	3	7.35	+ 0.4	+3.7	7.61	0.465	94.0	4.07

EPHEMERIS FOR PHYSICAL OBSERVATIONS
FOR 0^h DYNAMICAL TIME

Date		Sub-Earth Point		Sub-Solar Point			North Pole	
		Long.	Lat.	Long.	Dist.	P.A.	Dist.	P.A.
		°	°	°	″	°	″	°
Apr.	2	63.00	− 6.83	216.05	−2.61	74.89	− 5.56	331.76
	4	76.45	6.39	221.92	−3.16	71.04	5.47	331.87
	6	89.55	5.92	228.02	−3.59	68.58	5.35	331.95
	8	102.28	5.44	234.31	−3.90	66.86	5.21	332.00
	10	114.64	4.95	240.75	−4.10	65.59	5.05	332.01
	12	126.65	− 4.48	247.31	−4.22	64.61	− 4.88	331.98
	14	138.33	4.01	253.96	−4.26	63.83	4.71	331.92
	16	149.71	3.57	260.68	−4.26	63.19	4.55	331.84
	18	160.80	3.16	267.44	−4.21	62.67	4.39	331.74
	20	171.64	2.76	274.20	−4.13	62.25	4.23	331.65
	22	182.25	− 2.39	280.95	−4.04	61.92	− 4.08	331.56
	24	192.65	2.03	287.66	−3.92	61.66	3.94	331.49
	26	202.85	1.70	294.29	−3.80	61.48	3.80	331.44
	28	212.88	1.39	300.82	3.67	61.38	3.68	331.43
	30	222.74	1.10	307.23	3.54	61.37	3.55	331.47
May	2	232.44	− 0.82	313.47	3.40	61.44	− 3.44	331.56
	4	241.99	0.56	319.52	3.26	61.60	3.33	331.72
	6	251.40	0.31	325.33	3.11	61.86	3.23	331.94
	8	260.67	− 0.08	330.86	2.95	62.23	− 3.14	332.25
	10	269.80	+ 0.13	336.07	2.79	62.72	+ 3.05	332.65
	12	278.79	+ 0.34	340.91	2.62	63.33	+ 2.97	333.14
	14	287.64	0.54	345.32	2.44	64.10	2.89	333.74
	16	296.36	0.72	349.25	2.25	65.03	2.82	334.47
	18	304.93	0.90	352.64	2.04	66.16	2.76	335.32
	20	313.38	1.07	355.45	1.81	67.52	2.70	336.31
	22	321.69	+ 1.23	357.64	1.56	69.18	+ 2.65	337.45
	24	329.87	1.39	359.20	1.28	71.27	2.61	338.75
	26	337.94	1.55	0.14	0.98	74.06	2.58	340.22
	28	345.92	1.71	0.53	0.65	78.44	2.56	341.84
	30	353.82	1.88	0.48	0.31	89.39	2.54	343.61
June	1	1.69	+ 2.04	0.15	0.11	202.47	+ 2.54	345.52
	3	9.56	2.22	359.74	0.45	244.89	2.55	347.54
	5	17.46	2.40	359.47	0.80	252.28	2.57	349.64
	7	25.44	2.60	359.51	1.15	256.46	2.61	351.78
	9	33.51	2.81	0.03	1.47	259.71	2.65	353.94
	11	41.71	+ 3.03	1.13	1.76	262.55	+ 2.70	356.07
	13	50.04	3.26	2.85	2.03	265.14	2.76	358.14
	15	58.52	3.51	5.20	2.28	267.55	2.83	0.15
	17	67.16	3.77	8.17	2.50	269.81	2.91	2.06
	19	75.96	4.05	11.71	2.70	271.94	2.99	3.87
	21	84.92	+ 4.34	15.76	2.89	273.94	+ 3.08	5.58
	23	94.03	4.65	20.29	3.07	275.82	3.18	7.16
	25	103.31	4.98	25.23	3.23	277.60	3.29	8.63
	27	112.75	5.32	30.53	3.39	279.28	3.40	9.98
	29	122.36	5.68	36.13	3.53	280.86	3.52	11.22
July	1	132.15	+ 6.05	42.01	−3.67	282.37	+ 3.65	12.34
	3	142.12	+ 6.44	48.11	−3.79	283.82	+ 3.78	13.35

EPHEMERIS FOR PHYSICAL OBSERVATIONS
FOR 0^h DYNAMICAL TIME

Date		Light-time	Magnitude	Surface Brightness	Diameter	Phase	Phase Angle	Defect of Illumination
		m			″		°	″
July	1	7.63	+ 0.3	+3.6	7.34	0.499	90.1	3.68
	3	7.35	0.4	3.7	7.61	0.465	94.0	4.07
	5	7.09	0.5	3.8	7.89	0.432	97.8	4.48
	7	6.83	0.6	3.9	8.19	0.399	101.7	4.93
	9	6.58	0.7	4.0	8.51	0.365	105.7	5.40
	11	6.33	+ 0.8	+4.1	8.83	0.331	109.8	5.91
	13	6.10	1.0	4.2	9.17	0.296	114.0	6.45
	15	5.88	1.2	4.3	9.51	0.261	118.5	7.02
	17	5.68	1.4	4.5	9.85	0.226	123.2	7.62
	19	5.49	1.6	4.6	10.19	0.190	128.3	8.25
	21	5.32	+ 1.9	+4.7	10.51	0.155	133.6	8.88
	23	5.18	2.3	4.9	10.80	0.121	139.3	9.49
	25	5.06	2.7	5.0	11.05	0.089	145.2	10.07
	27	4.97	3.2	5.1	11.25	0.061	151.5	10.56
	29	4.92	3.7	5.2	11.36	0.037	157.7	10.94
	31	4.91	+ 4.3	+5.1	11.39	0.020	163.6	11.16
Aug.	2	4.94	4.7	4.9	11.32	0.012	167.6	11.19
	4	5.02	4.7	4.9	11.14	0.012	167.3	11.00
	6	5.15	4.1	5.0	10.87	0.023	162.4	10.61
	8	5.32	3.4	4.9	10.50	0.045	155.4	10.03
	10	5.55	+ 2.7	+4.7	10.08	0.078	147.6	9.29
	12	5.83	2.1	4.4	9.60	0.121	139.3	8.44
	14	6.15	1.5	4.1	9.10	0.173	130.8	7.53
	16	6.50	1.0	3.8	8.60	0.235	122.0	6.58
	18	6.90	0.5	3.5	8.11	0.304	113.1	5.64
	20	7.32	+ 0.1	+3.2	7.64	0.380	103.9	4.74
	22	7.76	− 0.2	3.0	7.20	0.460	94.6	3.89
	24	8.22	0.5	2.8	6.81	0.542	85.1	3.12
	26	8.67	0.7	2.6	6.45	0.624	75.6	2.42
	28	9.10	0.9	2.4	6.14	0.702	66.1	1.83
	30	9.52	− 1.0	+2.3	5.88	0.774	56.8	1.33
Sept.	1	9.90	1.2	2.1	5.65	0.835	47.9	0.93
	3	10.25	1.3	2.0	5.46	0.886	39.4	0.62
	5	10.56	1.4	1.9	5.30	0.926	31.6	0.39
	7	10.82	1.4	1.8	5.17	0.956	24.3	0.23
	9	11.04	− 1.5	+1.7	5.07	0.976	17.8	0.12
	11	11.23	1.6	1.6	4.98	0.989	12.0	0.05
	13	11.37	1.7	1.5	4.92	0.996	7.1	0.02
	15	11.48	1.7	1.5	4.87	0.999	4.3	0.01
	17	11.56	1.6	1.6	4.84	0.998	5.6	0.01
	19	11.62	− 1.4	+1.8	4.81	0.994	8.9	0.03
	21	11.65	1.2	1.9	4.80	0.988	12.3	0.06
	23	11.65	1.0	2.1	4.80	0.981	15.6	0.09
	25	11.64	0.9	2.2	4.81	0.973	18.8	0.13
	27	11.60	0.8	2.3	4.82	0.965	21.7	0.17
	29	11.55	− 0.7	+2.4	4.84	0.955	24.5	0.22
Oct.	1	11.48	− 0.6	+2.5	4.87	0.945	27.2	0.27

EPHEMERIS FOR PHYSICAL OBSERVATIONS
FOR 0^h DYNAMICAL TIME

Date		Sub-Earth Point		Sub-Solar Point			North Pole	
		Long.	Lat.	Long.	Dist.	P.A.	Dist.	P.A.
		°	°	°	″	°	″	°
July	1	132.15	+ 6.05	42.01	−3.67	282.37	+ 3.65	12.34
	3	142.12	6.44	48.11	−3.79	283.82	3.78	13.35
	5	152.27	6.85	54.40	−3.91	285.21	3.92	14.25
	7	162.64	7.27	60.84	−4.01	286.56	4.06	15.03
	9	173.22	7.72	67.41	−4.10	287.89	4.21	15.70
	11	184.03	+ 8.17	74.07	−4.16	289.23	+ 4.37	16.27
	13	195.10	8.64	80.79	−4.19	290.61	4.53	16.71
	15	206.44	9.11	87.54	−4.18	292.05	4.69	17.05
	17	218.07	9.58	94.31	−4.12	293.63	4.86	17.27
	19	230.02	10.05	101.06	−4.00	295.41	5.02	17.36
	21	242.29	+10.51	107.76	−3.81	297.51	+ 5.17	17.34
	23	254.90	10.93	114.39	−3.52	300.14	5.30	17.18
	25	267.83	11.30	120.93	−3.15	303.63	5.42	16.90
	27	281.07	11.60	127.33	−2.69	308.67	5.51	16.50
	29	294.57	11.81	133.57	−2.15	316.73	5.56	16.00
	31	308.27	+11.92	139.62	−1.61	331.23	+ 5.57	15.42
Aug.	2	322.07	11.90	145.42	−1.21	358.93	5.54	14.79
	4	335.87	11.76	150.95	−1.23	37.08	5.45	14.15
	6	349.56	11.50	156.15	−1.64	63.68	5.32	13.56
	8	3.02	11.13	160.98	−2.18	77.60	5.15	13.07
	10	16.16	+10.68	165.38	−2.70	85.46	+ 4.95	12.72
	12	28.90	10.16	169.30	−3.13	90.52	4.73	12.54
	14	41.19	9.61	172.69	−3.45	94.19	4.49	12.57
	16	52.99	9.05	175.49	−3.64	97.11	4.25	12.82
	18	64.31	8.50	177.67	−3.73	99.65	4.01	13.28
	20	75.14	+ 7.96	179.22	−3.71	101.98	+ 3.78	13.95
	22	85.50	7.46	180.15	−3.59	104.22	3.57	14.81
	24	95.43	7.00	180.53	3.39	106.44	3.38	15.82
	26	104.97	6.57	180.47	3.13	108.67	3.21	16.97
	28	114.17	6.20	180.14	2.81	110.96	3.05	18.19
	30	123.10	+ 5.86	179.73	2.46	113.30	+ 2.92	19.45
Sept.	1	131.82	5.56	179.46	2.10	115.76	2.81	20.71
	3	140.38	5.30	179.52	1.73	118.40	2.72	21.92
	5	148.85	5.07	180.05	1.39	121.36	2.64	23.07
	7	157.27	4.87	181.15	1.07	124.95	2.58	24.12
	9	165.68	+ 4.68	182.88	0.77	129.81	+ 2.52	25.06
	11	174.11	4.52	185.24	0.52	137.66	2.48	25.89
	13	182.58	4.37	188.21	0.30	154.18	2.45	26.61
	15	191.10	4.23	191.76	0.18	198.24	2.43	27.21
	17	199.67	4.10	195.83	0.24	251.07	2.41	27.69
	19	208.32	+ 3.97	200.36	0.37	271.80	+ 2.40	28.07
	21	217.03	3.86	205.30	0.51	280.47	2.40	28.35
	23	225.80	3.74	210.60	0.65	285.09	2.39	28.52
	25	234.64	3.63	216.21	0.77	287.91	2.40	28.61
	27	243.54	3.51	222.09	0.89	289.78	2.41	28.60
	29	252.51	+ 3.40	228.20	1.01	291.07	+ 2.42	28.51
Oct.	1	261.53	+ 3.29	234.49	1.11	291.97	+ 2.43	28.34

EPHEMERIS FOR PHYSICAL OBSERVATIONS
FOR 0h DYNAMICAL TIME

Date		Light-time	Magnitude	Surface Brightness	Diameter	Phase	Phase Angle	Defect of Illumination
		m			"		°	"
Oct.	1	11.48	− 0.6	+2.5	4.87	0.945	27.2	0.27
	3	11.40	0.5	2.6	4.91	0.934	29.8	0.33
	5	11.29	0.4	2.7	4.95	0.922	32.4	0.38
	7	11.18	0.4	2.8	5.00	0.910	34.9	0.45
	9	11.04	0.3	2.8	5.07	0.897	37.4	0.52
	11	10.90	− 0.3	+2.9	5.13	0.883	39.9	0.60
	13	10.73	0.2	2.9	5.21	0.869	42.5	0.68
	15	10.55	0.2	3.0	5.30	0.853	45.1	0.78
	17	10.36	0.2	3.0	5.40	0.835	47.9	0.89
	19	10.15	0.2	3.0	5.51	0.816	50.8	1.01
	21	9.93	− 0.2	+3.1	5.63	0.795	53.8	1.15
	23	9.69	0.2	3.1	5.77	0.772	57.1	1.32
	25	9.43	0.2	3.1	5.93	0.746	60.6	1.51
	27	9.16	0.2	3.1	6.11	0.716	64.4	1.74
	29	8.87	0.2	3.2	6.31	0.683	68.6	2.00
	31	8.56	− 0.2	+3.2	6.53	0.644	73.2	2.32
Nov.	2	8.24	0.1	3.2	6.79	0.601	78.3	2.71
	4	7.91	0.1	3.3	7.07	0.552	84.1	3.17
	6	7.57	− 0.0	3.3	7.39	0.495	90.5	3.73
	8	7.22	+ 0.1	3.4	7.75	0.432	97.8	4.40
	10	6.88	+ 0.3	+3.5	8.13	0.361	106.1	5.19
	12	6.55	0.6	3.6	8.54	0.285	115.5	6.11
	14	6.24	1.1	3.9	8.96	0.205	126.2	7.13
	16	5.98	1.8	4.1	9.36	0.128	138.1	8.16
	18	5.78	2.8	4.4	9.68	0.061	151.4	9.09
	20	5.66	+ 4.1	+4.3	9.88	0.016	165.6	9.73
	22	5.64	5.6	1.7	9.92	0.000	177.8	9.92
	24	5.72	3.9	4.3	9.77	0.020	163.7	9.58
	26	5.91	2.5	4.2	9.47	0.072	148.8	8.78
	28	6.18	1.4	3.9	9.05	0.149	134.6	7.70
	30	6.51	+ 0.7	+3.6	8.59	0.239	121.5	6.54
Dec.	2	6.89	+ 0.2	3.3	8.12	0.332	109.7	5.42
	4	7.29	− 0.1	3.2	7.67	0.421	99.1	4.44
	6	7.70	0.3	3.0	7.27	0.503	89.7	3.61
	8	8.10	0.4	3.0	6.91	0.575	81.3	2.93
	10	8.49	− 0.4	+2.9	6.59	0.638	74.0	2.39
	12	8.86	0.5	2.9	6.31	0.691	67.5	1.95
	14	9.21	0.5	2.8	6.08	0.737	61.7	1.60
	16	9.53	0.5	2.8	5.87	0.776	56.6	1.32
	18	9.83	0.5	2.8	5.69	0.808	51.9	1.09
	20	10.11	− 0.5	+2.8	5.53	0.836	47.7	0.91
	22	10.37	0.5	2.8	5.40	0.860	43.9	0.76
	24	10.60	0.5	2.7	5.28	0.880	40.5	0.63
	26	10.81	0.5	2.7	5.17	0.898	37.3	0.53
	28	11.00	0.5	2.7	5.08	0.913	34.3	0.44
	30	11.17	− 0.5	+2.7	5.01	0.926	31.5	0.37
	32	11.32	− 0.5	+2.6	4.94	0.938	28.8	0.31

EPHEMERIS FOR PHYSICAL OBSERVATIONS
FOR 0^h DYNAMICAL TIME

Date		Sub-Earth Point		Sub-Solar Point			North Pole	
		Long.	Lat.	Long.	Dist.	P.A.	Dist.	P.A.
		°	°	°	″	°	″	°
Oct.	1	261.53	+ 3.29	234.49	1.11	291.97	+ 2.43	28.34
	3	270.61	3.17	240.93	1.22	292.58	2.45	28.10
	5	279.75	3.06	247.50	1.33	292.97	2.47	27.78
	7	288.94	2.94	254.16	1.43	293.18	2.50	27.39
	9	298.18	2.82	260.88	1.54	293.25	2.53	26.93
	11	307.47	+ 2.70	267.63	1.65	293.19	+ 2.56	26.41
	13	316.82	2.58	274.40	1.76	293.02	2.60	25.83
	15	326.23	2.45	281.15	1.88	292.75	2.65	25.19
	17	335.69	2.32	287.85	2.00	292.40	2.70	24.50
	19	345.21	2.19	294.48	2.13	291.98	2.75	23.76
	21	354.81	+ 2.05	301.02	2.27	291.49	+ 2.82	22.98
	23	4.48	1.90	307.42	2.42	290.94	2.89	22.17
	25	14.23	1.75	313.66	2.58	290.34	2.96	21.33
	27	24.10	1.58	319.70	2.75	289.72	3.05	20.48
	29	34.08	1.41	325.50	2.94	289.07	3.15	19.63
	31	44.23	+ 1.23	331.02	3.13	288.42	+ 3.27	18.80
Nov.	2	54.55	1.04	336.22	3.32	287.79	3.39	18.01
	4	65.12	0.83	341.05	3.52	287.19	3.54	17.29
	6	75.97	0.60	345.44	−3.70	286.66	3.70	16.67
	8	87.18	0.35	349.36	−3.84	286.22	3.87	16.19
	10	98.84	+ 0.07	352.74	−3.91	285.89	+ 4.07	15.89
	12	111.04	− 0.24	355.53	−3.86	285.69	− 4.27	15.82
	14	123.86	0.57	357.70	−3.62	285.57	4.48	16.01
	16	137.36	0.93	359.24	−3.12	285.42	4.68	16.48
	18	151.54	1.31	0.16	−2.32	284.78	4.84	17.22
	20	166.27	− 1.69	0.53	−1.23	281.46	− 4.94	18.14
	22	181.30	2.05	0.47	−0.19	177.23	4.96	19.14
	24	196.28	2.36	0.13	−1.37	118.23	4.88	20.08
	26	210.86	2.62	359.73	−2.45	115.18	4.73	20.84
	28	224.80	2.82	359.46	−3.22	114.15	4.52	21.37
	30	237.97	− 2.97	359.52	−3.66	113.46	− 4.29	21.64
Dec.	2	250.38	3.08	0.06	−3.82	112.76	4.05	21.67
	4	262.10	3.16	1.18	−3.79	111.98	3.83	21.48
	6	273.25	3.23	2.91	3.63	111.08	3.63	21.11
	8	283.95	3.30	5.29	3.41	110.07	3.45	20.58
	10	294.30	− 3.35	8.27	3.17	108.93	− 3.29	19.91
	12	304.37	3.41	11.82	2.92	107.68	3.15	19.11
	14	314.24	3.46	15.90	2.68	106.31	3.03	18.20
	16	323.95	3.52	20.43	2.45	104.84	2.93	17.19
	18	333.55	3.58	25.38	2.24	103.27	2.84	16.09
	20	343.05	− 3.63	30.69	2.05	101.58	− 2.76	14.91
	22	352.49	3.69	36.31	1.87	99.80	2.69	13.66
	24	1.87	3.75	42.19	1.71	97.91	2.63	12.34
	26	11.21	3.81	48.29	1.57	95.91	2.58	10.96
	28	20.51	3.87	54.59	1.43	93.80	2.54	9.52
	30	29.79	− 3.93	61.03	1.31	91.56	− 2.50	8.04
	32	39.03	− 3.99	67.60	1.19	89.20	− 2.46	6.51

EPHEMERIS FOR PHYSICAL OBSERVATIONS FOR 0^h DYNAMICAL TIME

Date		Light-time	Magnitude	Surface Brightness	Diameter	Phase	Phase Angle	Defect of Illumination
		m			″		°	″
Jan.	−1	9.19	− 4.1	+1.2	15.10	0.743	61.0	3.89
	3	9.41	4.1	1.2	14.75	0.755	59.3	3.61
	7	9.62	4.1	1.2	14.42	0.767	57.7	3.36
	11	9.83	4.1	1.2	14.11	0.779	56.1	3.12
	15	10.04	4.0	1.1	13.82	0.790	54.5	2.90
	19	10.24	− 4.0	+1.1	13.55	0.801	53.0	2.70
	23	10.44	4.0	1.1	13.29	0.811	51.5	2.50
	27	10.64	4.0	1.1	13.04	0.822	50.0	2.33
	31	10.83	4.0	1.1	12.81	0.831	48.5	2.16
Feb.	4	11.02	4.0	1.1	12.59	0.841	47.0	2.00
	8	11.20	− 4.0	+1.1	12.39	0.850	45.6	1.86
	12	11.38	4.0	1.0	12.19	0.859	44.1	1.72
	16	11.56	4.0	1.0	12.00	0.868	42.7	1.59
	20	11.73	3.9	1.0	11.83	0.876	41.3	1.47
	24	11.90	3.9	1.0	11.66	0.884	39.9	1.36
	28	12.06	− 3.9	+1.0	11.50	0.891	38.5	1.25
Mar.	3	12.22	3.9	1.0	11.35	0.899	37.1	1.15
	7	12.38	3.9	1.0	11.21	0.906	35.7	1.05
	11	12.53	3.9	0.9	11.08	0.913	34.3	0.96
	15	12.67	3.9	0.9	10.95	0.920	32.9	0.88
	19	12.81	− 3.9	+0.9	10.83	0.926	31.5	0.80
	23	12.95	3.9	0.9	10.72	0.932	30.2	0.73
	27	13.08	3.9	0.9	10.61	0.938	28.8	0.66
	31	13.20	3.9	0.9	10.51	0.944	27.4	0.59
Apr.	4	13.33	3.9	0.9	10.41	0.949	26.0	0.53
	8	13.44	− 3.9	+0.9	10.33	0.955	24.6	0.47
	12	13.55	3.9	0.9	10.24	0.960	23.2	0.41
	16	13.65	3.9	0.8	10.16	0.964	21.8	0.36
	20	13.75	3.9	0.8	10.09	0.969	20.4	0.32
	24	13.85	3.9	0.8	10.02	0.973	18.9	0.27
	28	13.93	− 3.9	+0.8	9.96	0.977	17.5	0.23
May	2	14.01	3.9	0.8	9.90	0.980	16.1	0.19
	6	14.09	3.9	0.8	9.85	0.984	14.6	0.16
	10	14.15	3.9	0.8	9.80	0.987	13.1	0.13
	14	14.21	3.9	0.8	9.76	0.990	11.7	0.10
	18	14.27	− 3.9	+0.8	9.73	0.992	10.2	0.08
	22	14.31	3.9	0.8	9.69	0.994	8.7	0.06
	26	14.35	3.9	0.8	9.67	0.996	7.2	0.04
	30	14.39	3.9	0.8	9.65	0.998	5.6	0.02
June	3	14.41	3.9	0.7	9.63	0.999	4.1	0.01
	7	14.43	− 3.9	+0.7	9.62	0.999	2.6	0.00
	11	14.43	3.9	0.7	9.61	1.000	1.0	0.00
	15	14.44	3.9	0.7	9.61	1.000	0.6	0.00
	19	14.43	3.9	0.7	9.62	1.000	2.1	0.00
	23	14.41	3.9	0.7	9.63	0.999	3.7	0.01
	27	14.39	− 3.9	+0.7	9.64	0.998	5.3	0.02
July	1	14.36	− 3.9	+0.7	9.66	0.996	6.8	0.03

EPHEMERIS FOR PHYSICAL OBSERVATIONS
FOR 0^h DYNAMICAL TIME

Date	L_s	Sub-Earth Point		Sub-Solar Point				North Pole	
		Long.	Lat.	Long.	Lat.	Dist.	P.A.	Dist.	P.A.
	°	°	°	°	°	″	°	″	°
Jan. −1	300.29	243.24	− 1.03	182.25	− 2.30	6.60	105.38	− 7.55	13.32
3	306.76	253.98	0.97	194.65	2.13	6.34	103.60	7.37	11.70
7	313.22	264.75	0.91	207.03	1.94	6.10	101.69	7.21	9.97
11	319.67	275.53	0.83	219.41	1.72	5.86	99.66	7.06	8.14
15	326.11	286.32	0.74	231.77	1.48	5.63	97.53	6.91	6.24
19	332.54	297.13	− 0.65	244.13	− 1.23	5.41	95.32	− 6.77	4.27
23	338.96	307.94	0.55	256.47	0.96	5.20	93.03	6.64	2.25
27	345.38	318.77	0.45	268.80	0.67	4.99	90.70	6.52	0.20
31	351.78	329.61	0.35	281.13	0.38	4.80	88.35	6.41	358.15
Feb. 4	358.17	340.45	0.25	293.44	− 0.08	4.61	85.99	6.30	356.11
8	4.56	351.30	− 0.15	305.74	+ 0.21	4.42	83.65	− 6.19	354.10
12	10.93	2.15	− 0.05	318.04	0.50	4.24	81.36	− 6.09	352.14
16	17.30	13.01	+ 0.04	330.32	0.79	4.07	79.13	+ 6.00	350.25
20	23.66	23.87	0.12	342.60	1.07	3.90	76.98	5.91	348.46
24	30.01	34.73	0.20	354.87	1.33	3.74	74.93	5.83	346.76
28	36.35	45.59	+ 0.27	7.14	+ 1.58	3.58	73.00	+ 5.75	345.19
Mar. 3	42.69	56.46	0.34	19.40	1.80	3.42	71.19	5.68	343.74
7	49.03	67.33	0.39	31.66	2.01	3.27	69.52	5.61	342.42
11	55.36	78.19	0.43	43.92	2.19	3.12	67.99	5.54	341.25
15	61.68	89.06	0.46	56.17	2.34	2.98	66.62	5.48	340.22
19	68.01	99.92	+ 0.48	68.43	+ 2.47	2.83	65.41	+ 5.42	339.34
23	74.33	110.79	0.49	80.68	2.56	2.69	64.35	5.36	338.61
27	80.65	121.65	0.49	92.94	2.63	2.55	63.47	5.31	338.04
31	86.98	132.51	0.47	105.19	2.66	2.42	62.75	5.25	337.61
Apr. 4	93.30	143.37	0.45	117.45	2.66	2.28	62.20	5.21	337.34
8	99.63	154.23	+ 0.42	129.71	+ 2.62	2.15	61.82	+ 5.16	337.22
12	105.96	165.09	0.37	141.98	2.56	2.02	61.61	5.12	337.25
16	112.30	175.94	0.32	154.24	2.46	1.89	61.58	5.08	337.43
20	118.64	186.79	0.25	166.52	2.34	1.76	61.72	5.05	337.75
24	124.99	197.64	0.18	178.80	2.18	1.63	62.04	5.01	338.23
28	131.35	208.49	+ 0.11	191.08	+ 2.00	1.50	62.54	+ 4.98	338.86
May 2	137.72	219.34	+ 0.02	203.37	1.79	1.37	63.22	+ 4.95	339.63
6	144.09	230.19	− 0.07	215.67	1.56	1.24	64.08	− 4.93	340.54
10	150.47	241.03	0.16	227.97	1.31	1.11	65.13	4.90	341.60
14	156.86	251.87	0.26	240.29	1.05	0.99	66.37	4.88	342.80
18	163.27	262.72	− 0.36	252.61	+ 0.77	0.86	67.80	− 4.86	344.14
22	169.68	273.56	0.45	264.94	0.48	0.73	69.45	4.85	345.61
26	176.10	284.40	0.55	277.28	+ 0.18	0.60	71.34	4.83	347.20
30	182.53	295.24	0.65	289.63	− 0.12	0.47	73.54	4.82	348.90
June 3	188.97	306.08	0.74	301.99	0.41	0.34	76.22	4.82	350.71
7	195.42	316.93	− 0.83	314.36	− 0.71	0.22	80.00	− 4.81	352.61
11	201.87	327.77	0.91	326.73	0.99	0.09	89.26	4.81	354.58
15	208.34	338.61	0.98	339.12	1.26	0.05	237.76	4.81	356.61
19	214.81	349.45	1.05	351.51	1.52	0.18	255.80	4.81	358.68
23	221.29	0.29	1.11	3.92	1.76	0.31	260.53	4.81	0.76
27	227.77	11.14	− 1.15	16.33	− 1.97	0.44	263.80	− 4.82	2.84
July 1	234.26	21.98	− 1.19	28.74	− 2.16	0.57	266.61	− 4.83	4.88

EPHEMERIS FOR PHYSICAL OBSERVATIONS FOR 0^h DYNAMICAL TIME

Date		Light-time	Magnitude	Surface Brightness	Diameter	Phase	Phase Angle	Defect of Illumination
		m			″		°	″
July	1	14.36	− 3.9	+0.7	9.66	0.996	6.8	0.03
	5	14.32	3.9	0.8	9.69	0.995	8.4	0.05
	9	14.28	3.9	0.8	9.72	0.992	10.0	0.07
	13	14.22	3.9	0.8	9.76	0.990	11.6	0.10
	17	14.16	3.9	0.8	9.80	0.987	13.1	0.13
	21	14.09	− 3.9	+0.8	9.85	0.984	14.7	0.16
	25	14.02	3.9	0.8	9.90	0.980	16.3	0.20
	29	13.93	3.9	0.8	9.96	0.976	17.9	0.24
Aug.	2	13.84	3.9	0.8	10.03	0.971	19.4	0.29
	6	13.74	3.9	0.8	10.10	0.967	21.0	0.34
	10	13.64	− 3.9	+0.8	10.17	0.962	22.6	0.39
	14	13.53	3.9	0.8	10.26	0.956	24.1	0.45
	18	13.41	3.9	0.9	10.35	0.951	25.6	0.51
	22	13.29	3.9	0.9	10.44	0.945	27.2	0.58
	26	13.16	3.9	0.9	10.55	0.939	28.7	0.65
	30	13.02	− 3.9	+0.9	10.66	0.932	30.2	0.72
Sept.	3	12.88	3.9	0.9	10.77	0.925	31.7	0.81
	7	12.74	3.9	0.9	10.90	0.918	33.2	0.89
	11	12.58	3.9	0.9	11.03	0.911	34.7	0.98
	15	12.43	3.9	1.0	11.17	0.903	36.2	1.08
	19	12.27	− 3.9	+1.0	11.31	0.896	37.7	1.18
	23	12.10	3.9	1.0	11.47	0.888	39.2	1.29
	27	11.93	3.9	1.0	11.63	0.880	40.6	1.40
Oct.	1	11.76	3.9	1.0	11.80	0.871	42.1	1.52
	5	11.58	3.9	1.0	11.98	0.862	43.5	1.65
	9	11.40	− 3.9	+1.1	12.17	0.854	45.0	1.78
	13	11.21	4.0	1.1	12.38	0.845	46.4	1.92
	17	11.02	4.0	1.1	12.59	0.835	47.9	2.07
	21	10.83	4.0	1.1	12.81	0.826	49.4	2.23
	25	10.63	4.0	1.1	13.05	0.816	50.8	2.40
	29	10.43	− 4.0	+1.1	13.30	0.806	52.3	2.58
Nov.	2	10.23	4.0	1.2	13.56	0.796	53.8	2.77
	6	10.03	4.0	1.2	13.84	0.785	55.2	2.97
	10	9.82	4.0	1.2	14.14	0.774	56.7	3.19
	14	9.60	4.0	1.2	14.45	0.763	58.2	3.42
	18	9.39	− 4.1	+1.2	14.78	0.752	59.8	3.67
	22	9.17	4.1	1.2	15.13	0.740	61.3	3.93
	26	8.95	4.1	1.3	15.50	0.728	62.9	4.22
	30	8.73	4.1	1.3	15.90	0.715	64.5	4.52
Dec.	4	8.50	4.1	1.3	16.32	0.703	66.1	4.85
	8	8.27	− 4.2	+1.3	16.77	0.689	67.8	5.21
	12	8.04	4.2	1.3	17.25	0.676	69.4	5.60
	16	7.81	4.2	1.3	17.77	0.661	71.2	6.02
	20	7.58	4.2	1.4	18.32	0.647	73.0	6.47
	24	7.34	4.2	1.4	18.91	0.631	74.8	6.97
	28	7.10	− 4.3	+1.4	19.55	0.615	76.7	7.52
	32	6.86	− 4.3	+1.4	20.23	0.599	78.6	8.11

EPHEMERIS FOR PHYSICAL OBSERVATIONS
FOR 0^h DYNAMICAL TIME

Date	L_s	Sub-Earth Point		Sub-Solar Point				North Pole	
		Long.	Lat.	Long.	Lat.	Dist.	P.A.	Dist.	P.A.
	°	°	°	°	°	″	°	″	°
July 1	234.26	21.98	− 1.19	28.74	− 2.16	0.57	266.61	− 4.83	4.88
5	240.75	32.83	1.21	41.16	2.32	0.71	269.18	4.84	6.88
9	247.25	43.67	1.22	53.59	2.45	0.84	271.59	4.86	8.82
13	253.74	54.52	1.22	66.01	2.56	0.98	273.87	4.88	10.66
17	260.24	65.36	1.20	78.44	2.62	1.11	276.02	4.90	12.41
21	266.73	76.21	− 1.17	90.87	− 2.66	1.25	278.05	− 4.92	14.04
25	273.23	87.05	1.13	103.29	2.66	1.39	279.94	4.95	15.55
29	279.72	97.90	1.07	115.72	2.62	1.53	281.69	4.98	16.92
Aug. 2	286.20	108.74	1.00	128.13	2.56	1.67	283.30	5.01	18.16
6	292.68	119.59	0.91	140.54	2.46	1.81	284.77	5.05	19.25
10	299.16	130.43	− 0.81	152.95	− 2.32	1.95	286.09	− 5.09	20.20
14	305.62	141.28	0.70	165.35	2.16	2.10	287.26	5.13	21.00
18	312.08	152.12	0.57	177.73	1.98	2.24	288.27	5.17	21.65
22	318.54	162.96	0.43	190.11	1.76	2.39	289.13	5.22	22.16
26	324.98	173.79	0.28	202.48	1.53	2.53	289.84	5.27	22.51
30	331.41	184.63	− 0.12	214.84	− 1.27	2.68	290.40	− 5.33	22.72
Sept. 3	337.84	195.47	+ 0.05	227.18	1.00	2.83	290.80	+ 5.39	22.79
7	344.25	206.30	0.22	239.52	0.72	2.99	291.05	5.45	22.71
11	350.66	217.13	0.41	251.84	0.43	3.14	291.14	5.51	22.48
15	357.05	227.96	0.59	264.16	− 0.14	3.30	291.07	5.58	22.11
19	3.44	238.78	+ 0.78	276.46	+ 0.16	3.46	290.84	+ 5.66	21.59
23	9.82	249.61	0.97	288.76	0.45	3.62	290.46	5.73	20.93
27	16.19	260.43	1.16	301.05	0.74	3.79	289.92	5.81	20.13
Oct. 1	22.55	271.24	1.35	313.33	1.02	3.95	289.21	5.90	19.18
5	28.90	282.06	1.53	325.60	1.29	4.13	288.35	5.99	18.09
9	35.25	292.87	+ 1.70	337.87	+ 1.54	4.30	287.33	+ 6.08	16.86
13	41.59	303.67	1.87	350.13	1.77	4.48	286.15	6.18	15.50
17	47.92	314.47	2.03	2.40	1.98	4.67	284.83	6.29	14.00
21	54.25	325.26	2.18	14.65	2.16	4.86	283.36	6.40	12.38
25	60.58	336.05	2.31	26.91	2.32	5.06	281.76	6.52	10.65
29	66.90	346.83	+ 2.43	39.16	+ 2.45	5.26	280.04	+ 6.64	8.82
Nov. 2	73.23	357.61	2.52	51.42	2.55	5.47	278.22	6.77	6.91
6	79.55	8.37	2.61	63.67	2.62	5.69	276.30	6.91	4.92
10	85.87	19.13	2.66	75.93	2.66	5.91	274.32	7.06	2.89
14	92.20	29.87	2.70	88.19	2.66	6.14	272.29	7.22	0.84
18	98.53	40.60	+ 2.72	100.45	+ 2.63	6.39	270.24	+ 7.38	358.78
22	104.86	51.32	2.70	112.71	2.57	6.64	268.19	7.56	356.73
26	111.20	62.02	2.67	124.98	2.48	6.90	266.15	7.74	354.73
30	117.54	72.71	2.60	137.25	2.36	7.17	264.16	7.94	352.79
Dec. 4	123.89	83.37	2.51	149.53	2.21	7.46	262.23	8.15	350.93
8	130.24	94.00	+ 2.39	161.82	+ 2.03	7.76	260.38	+ 8.38	349.16
12	136.61	104.61	2.24	174.11	1.83	8.08	258.62	8.62	347.51
16	142.98	115.18	2.05	186.41	1.60	8.41	256.96	8.88	345.97
20	149.36	125.72	1.84	198.71	1.36	8.76	255.41	9.15	344.56
24	155.75	136.22	1.60	211.02	1.09	9.12	253.98	9.45	343.28
28	162.15	146.68	+ 1.32	223.34	+ 0.82	9.51	252.66	+ 9.77	342.13
32	168.56	157.08	+ 1.02	235.67	+ 0.53	9.92	251.45	+10.11	341.12

EPHEMERIS FOR PHYSICAL OBSERVATIONS FOR 0^h DYNAMICAL TIME

Date		Light-time	Magnitude	Surface Brightness	Diameter Eq.	Diameter Pol.	Phase	Phase Angle	Defect of Illumination
		m			″	″		°	″
Jan.	−1	20.12	+ 1.4	+4.1	3.87	3.85	0.992	10.1	0.03
	3	20.01	1.4	4.1	3.89	3.87	0.991	10.9	0.03
	7	19.88	1.4	4.1	3.91	3.89	0.990	11.6	0.04
	11	19.76	1.4	4.1	3.94	3.92	0.988	12.4	0.05
	15	19.63	1.4	4.1	3.96	3.94	0.987	13.1	0.05
	19	19.50	+ 1.4	+4.1	3.99	3.97	0.985	13.9	0.06
	23	19.37	1.4	4.1	4.02	4.00	0.984	14.6	0.07
	27	19.23	1.4	4.1	4.05	4.03	0.982	15.4	0.07
	31	19.10	1.4	4.1	4.08	4.05	0.980	16.1	0.08
Feb.	4	18.96	1.3	4.1	4.11	4.08	0.979	16.8	0.09
	8	18.82	+ 1.3	+4.1	4.14	4.12	0.977	17.6	0.10
	12	18.68	1.3	4.1	4.17	4.15	0.975	18.3	0.11
	16	18.53	1.3	4.1	4.20	4.18	0.973	19.0	0.11
	20	18.39	1.3	4.1	4.23	4.21	0.971	19.7	0.12
	24	18.24	1.3	4.1	4.27	4.25	0.969	20.4	0.13
	28	18.10	+ 1.3	+4.1	4.30	4.28	0.967	21.1	0.14
Mar.	3	17.95	1.3	4.1	4.33	4.31	0.964	21.8	0.15
	7	17.81	1.2	4.1	4.37	4.35	0.962	22.4	0.17
	11	17.66	1.2	4.1	4.41	4.39	0.960	23.1	0.18
	15	17.52	1.2	4.1	4.44	4.42	0.958	23.8	0.19
	19	17.37	+ 1.2	+4.1	4.48	4.46	0.955	24.4	0.20
	23	17.23	1.2	4.2	4.52	4.50	0.953	25.0	0.21
	27	17.08	1.2	4.2	4.56	4.54	0.951	25.7	0.22
	31	16.94	1.2	4.2	4.59	4.57	0.948	26.3	0.24
Apr.	4	16.80	1.2	4.2	4.63	4.61	0.946	26.9	0.25
	8	16.65	+ 1.1	+4.2	4.67	4.65	0.943	27.5	0.26
	12	16.51	1.1	4.2	4.71	4.69	0.941	28.1	0.28
	16	16.37	1.1	4.2	4.75	4.73	0.939	28.7	0.29
	20	16.23	1.1	4.2	4.80	4.78	0.936	29.3	0.31
	24	16.09	1.1	4.2	4.84	4.82	0.934	29.8	0.32
	28	15.95	+ 1.1	+4.2	4.88	4.86	0.931	30.4	0.34
May	2	15.81	1.1	4.2	4.92	4.90	0.929	30.9	0.35
	6	15.67	1.1	4.2	4.97	4.95	0.926	31.5	0.37
	10	15.53	1.0	4.2	5.01	4.99	0.924	32.0	0.38
	14	15.39	1.0	4.2	5.06	5.03	0.922	32.5	0.40
	18	15.25	+ 1.0	+4.2	5.10	5.08	0.919	33.0	0.41
	22	15.11	1.0	4.2	5.15	5.13	0.917	33.5	0.43
	26	14.98	1.0	4.2	5.20	5.17	0.915	34.0	0.44
	30	14.84	1.0	4.2	5.25	5.22	0.912	34.5	0.46
June	3	14.70	1.0	4.2	5.30	5.27	0.910	34.9	0.48
	7	14.56	+ 1.0	+4.2	5.35	5.32	0.908	35.4	0.49
	11	14.42	1.0	4.2	5.40	5.37	0.905	35.8	0.51
	15	14.27	0.9	4.2	5.45	5.43	0.903	36.3	0.53
	19	14.13	0.9	4.3	5.51	5.48	0.901	36.7	0.55
	23	13.99	0.9	4.3	5.56	5.54	0.899	37.1	0.56
	27	13.84	+ 0.9	+4.3	5.62	5.60	0.896	37.5	0.58
July	1	13.69	+ 0.9	+4.3	5.68	5.66	0.894	37.9	0.60

EPHEMERIS FOR PHYSICAL OBSERVATIONS
FOR 0^h DYNAMICAL TIME

Date	L_s	Sub-Earth Point		Sub-Solar Point				North Pole	
		Long.	Lat.	Long.	Lat.	Dist.	P.A.	Dist.	P.A.
	°	°	°	°	°	″	°	″	°
Jan. −1	167.28	307.52	+ 0.62	316.43	+ 5.43	0.34	91.73	+ 1.92	30.04
3	169.45	268.47	− 0.68	278.05	4.52	0.37	90.44	− 1.93	28.75
7	171.64	229.41	1.98	239.65	3.58	0.39	89.14	1.95	27.38
11	173.85	190.33	3.28	201.25	2.64	0.42	87.83	1.96	25.94
15	176.07	151.24	4.58	162.84	1.69	0.45	86.52	1.97	24.43
19	178.31	112.13	− 5.87	124.41	+ 0.73	0.48	85.22	− 1.97	22.86
23	180.56	73.00	7.15	85.98	− 0.24	0.51	83.92	1.98	21.24
27	182.82	33.84	8.42	47.52	1.21	0.54	82.64	1.99	19.55
31	185.11	354.66	9.67	9.06	2.19	0.57	81.38	2.00	17.82
Feb. 4	187.40	315.44	10.90	330.58	3.18	0.59	80.14	2.01	16.05
8	189.71	276.20	−12.11	292.08	− 4.16	0.62	78.93	− 2.01	14.23
12	192.04	236.92	13.29	253.56	5.15	0.65	77.75	2.02	12.38
16	194.38	197.61	14.44	215.03	6.13	0.68	76.61	2.02	10.49
20	196.74	158.26	15.55	176.47	7.11	0.71	75.51	2.03	8.58
24	199.11	118.88	16.62	137.89	8.09	0.74	74.45	2.03	6.65
28	201.49	79.45	−17.65	99.29	− 9.06	0.77	73.44	− 2.04	4.69
Mar. 3	203.89	39.98	18.64	60.67	10.03	0.80	72.48	2.05	2.72
7	206.30	0.48	19.57	22.02	10.98	0.83	71.57	2.05	0.74
11	208.72	320.93	20.45	343.34	11.92	0.86	70.72	2.06	358.75
15	211.15	281.35	21.27	304.64	12.85	0.89	69.93	2.06	356.75
19	213.60	241.73	−22.03	265.91	−13.76	0.93	69.19	− 2.07	354.76
23	216.05	202.07	22.72	227.15	14.65	0.96	68.52	2.08	352.78
27	218.52	162.38	23.35	188.36	15.53	0.99	67.90	2.08	350.80
31	221.00	122.65	23.90	149.54	16.38	1.02	67.35	2.09	348.84
Apr. 4	223.49	82.91	24.38	110.69	17.20	1.05	66.87	2.10	346.91
8	225.98	43.13	−24.78	71.81	−18.00	1.08	66.44	− 2.11	345.00
12	228.48	3.35	25.11	32.91	18.77	1.11	66.08	2.13	343.13
16	230.99	323.54	25.36	353.97	19.50	1.14	65.78	2.14	341.29
20	233.51	283.73	25.52	315.00	20.20	1.17	65.55	2.16	339.51
24	236.03	243.92	25.61	276.01	20.87	1.20	65.38	2.17	337.77
28	238.56	204.11	−25.61	236.99	−21.50	1.23	65.27	− 2.19	336.10
May 2	241.09	164.30	25.54	197.94	22.08	1.26	65.23	2.21	334.49
6	243.63	124.51	25.38	158.87	22.63	1.30	65.24	2.24	332.96
10	246.16	84.74	25.15	119.77	23.13	1.33	65.32	2.26	331.50
14	248.70	45.00	24.83	80.66	23.58	1.36	65.46	2.29	330.13
18	251.24	5.28	−24.45	41.52	−23.99	1.39	65.66	− 2.31	328.85
22	253.78	325.60	23.99	2.37	24.35	1.42	65.92	2.34	327.66
26	256.32	285.95	23.47	323.21	24.65	1.45	66.24	2.37	326.57
30	258.86	246.34	22.88	284.04	24.91	1.48	66.61	2.41	325.58
June 3	261.39	206.77	22.22	244.86	25.12	1.52	67.04	2.44	324.69
7	263.92	167.25	−21.51	205.67	−25.27	1.55	67.53	− 2.48	323.91
11	266.45	127.77	20.75	166.49	25.37	1.58	68.06	2.51	323.24
15	268.97	88.35	19.93	127.30	25.42	1.61	68.65	2.55	322.67
19	271.49	48.97	19.07	88.13	25.41	1.64	69.29	2.59	322.22
23	274.00	9.64	18.16	48.96	25.36	1.68	69.98	2.63	321.87
27	276.50	330.36	−17.22	9.80	−25.25	1.71	70.72	− 2.67	321.63
July 1	278.99	291.12	−16.24	330.66	−25.09	1.74	71.50	− 2.72	321.50

EPHEMERIS FOR PHYSICAL OBSERVATIONS
FOR 0^h DYNAMICAL TIME

Date		Light-time	Magnitude	Surface Brightness	Diameter		Phase	Phase Angle	Defect of Illumination
					Eq.	Pol.			
		m			″	″		°	″
July	1	13.69	+ 0.9	+4.3	5.68	5.66	0.894	37.9	0.60
	5	13.55	0.9	4.3	5.75	5.72	0.892	38.3	0.62
	9	13.39	0.9	4.3	5.81	5.78	0.890	38.7	0.64
	13	13.24	0.9	4.3	5.88	5.85	0.888	39.0	0.66
	17	13.08	0.8	4.3	5.95	5.92	0.886	39.4	0.67
	21	12.93	+ 0.8	+4.3	6.02	5.99	0.885	39.7	0.69
	25	12.76	0.8	4.3	6.10	6.07	0.883	40.0	0.71
	29	12.60	0.8	4.3	6.18	6.15	0.881	40.3	0.73
Aug.	2	12.43	0.8	4.3	6.26	6.23	0.879	40.6	0.75
	6	12.26	0.7	4.3	6.35	6.32	0.878	40.9	0.77
	10	12.09	+ 0.7	+4.4	6.44	6.41	0.876	41.2	0.80
	14	11.91	0.7	4.4	6.54	6.50	0.875	41.4	0.82
	18	11.73	0.7	4.4	6.64	6.60	0.874	41.6	0.84
	22	11.54	0.7	4.4	6.74	6.71	0.873	41.8	0.86
	26	11.35	0.6	4.4	6.85	6.82	0.872	42.0	0.88
	30	11.16	+ 0.6	+4.4	6.97	6.94	0.871	42.2	0.90
Sept.	3	10.96	0.6	4.4	7.10	7.06	0.870	42.3	0.92
	7	10.76	0.5	4.4	7.23	7.19	0.869	42.4	0.94
	11	10.56	0.5	4.4	7.37	7.33	0.869	42.4	0.96
	15	10.35	0.5	4.4	7.52	7.48	0.869	42.5	0.98
	19	10.14	+ 0.4	+4.4	7.67	7.64	0.869	42.5	1.00
	23	9.93	0.4	4.4	7.84	7.80	0.869	42.4	1.02
	27	9.71	0.4	4.5	8.02	7.98	0.870	42.3	1.04
Oct.	1	9.48	0.3	4.5	8.21	8.16	0.871	42.1	1.06
	5	9.26	0.3	4.5	8.40	8.36	0.872	41.9	1.08
	9	9.03	+ 0.2	+4.5	8.62	8.57	0.873	41.7	1.09
	13	8.80	0.2	4.5	8.84	8.80	0.875	41.3	1.10
	17	8.57	0.1	4.5	9.08	9.03	0.878	40.9	1.11
	21	8.34	+ 0.0	4.5	9.33	9.28	0.881	40.4	1.11
	25	8.11	− 0.0	4.5	9.60	9.55	0.884	39.8	1.11
	29	7.87	− 0.1	+4.5	9.88	9.83	0.888	39.1	1.11
Nov.	2	7.64	0.2	4.5	10.18	10.13	0.892	38.3	1.10
	6	7.41	0.2	4.5	10.50	10.45	0.897	37.4	1.08
	10	7.18	0.3	4.5	10.83	10.78	0.903	36.4	1.05
	14	6.96	0.4	4.5	11.18	11.13	0.909	35.2	1.02
	18	6.74	− 0.5	+4.5	11.55	11.49	0.915	33.8	0.98
	22	6.53	0.6	4.5	11.92	11.86	0.923	32.3	0.92
	26	6.32	0.7	4.4	12.31	12.25	0.931	30.5	0.85
	30	6.13	0.7	4.4	12.69	12.63	0.939	28.6	0.78
Dec.	4	5.95	0.8	4.4	13.08	13.02	0.947	26.5	0.69
	8	5.78	− 0.9	+4.4	13.46	13.39	0.956	24.2	0.59
	12	5.63	1.0	4.4	13.82	13.75	0.965	21.6	0.49
	16	5.50	1.1	4.3	14.15	14.07	0.973	18.9	0.38
	20	5.39	1.2	4.3	14.43	14.36	0.981	16.0	0.28
	24	5.31	1.3	4.3	14.67	14.59	0.987	12.9	0.18
	28	5.25	− 1.3	+4.2	14.84	14.76	0.993	9.7	0.11
	32	5.21	− 1.4	+4.2	14.93	14.85	0.997	6.4	0.05

EPHEMERIS FOR PHYSICAL OBSERVATIONS
FOR 0^h DYNAMICAL TIME

Date	L_s	Sub-Earth Point		Sub-Solar Point				North Pole	
		Long.	Lat.	Long.	Lat.	Dist.	P.A.	Dist.	P.A.
	°	°	°	°	°	″	°	″	°
July 1	278.99	291.12	−16.24	330.66	−25.09	1.74	71.50	− 2.72	321.50
5	281.48	251.94	15.23	291.54	24.88	1.78	72.32	2.76	321.47
9	283.95	212.80	14.19	252.44	24.62	1.81	73.18	2.80	321.54
13	286.42	173.71	13.13	213.36	24.32	1.85	74.08	2.85	321.71
17	288.87	134.66	12.06	174.31	23.97	1.89	75.01	2.89	321.98
21	291.31	95.66	−10.97	135.28	−23.58	1.92	75.97	− 2.94	322.34
25	293.75	56.70	9.86	96.29	23.14	1.96	76.97	2.99	322.78
29	296.17	17.78	8.75	57.32	22.67	2.00	77.98	3.04	323.31
Aug. 2	298.57	338.91	7.64	18.39	22.16	2.04	79.02	3.09	323.91
6	300.97	300.07	6.52	339.49	21.61	2.08	80.08	3.14	324.59
10	303.35	261.27	− 5.41	300.63	−21.03	2.12	81.14	− 3.19	325.34
14	305.71	222.51	4.31	261.80	20.41	2.16	82.22	3.24	326.15
18	308.07	183.79	3.21	223.00	19.77	2.20	83.30	3.30	327.02
22	310.40	145.10	2.12	184.24	19.10	2.25	84.38	3.35	327.94
26	312.73	106.45	− 1.05	145.51	18.40	2.29	85.47	− 3.41	328.91
30	315.04	67.84	+ 0.00	106.82	−17.68	2.34	86.54	+ 3.47	329.91
Sept. 3	317.33	29.26	1.03	68.16	16.93	2.39	87.60	3.53	330.95
7	319.61	350.72	2.04	29.54	16.17	2.44	88.64	3.59	332.02
11	321.87	312.23	3.02	350.95	15.39	2.49	89.67	3.66	333.11
15	324.12	273.77	3.97	312.39	14.59	2.54	90.67	3.73	334.22
19	326.35	235.35	+ 4.89	273.86	−13.78	2.59	91.65	+ 3.80	335.34
23	328.57	196.98	5.77	235.37	12.95	2.64	92.59	3.88	336.47
27	330.77	158.66	6.62	196.90	12.12	2.70	93.50	3.96	337.59
Oct. 1	332.96	120.40	7.42	158.46	11.27	2.75	94.36	4.05	338.70
5	335.13	82.19	8.18	120.04	10.42	2.81	95.19	4.14	339.79
9	337.29	44.04	+ 8.89	81.66	− 9.55	2.86	95.96	+ 4.23	340.87
13	339.43	5.96	9.56	43.29	8.69	2.92	96.69	4.34	341.91
17	341.56	327.95	10.18	4.95	7.82	2.97	97.36	4.45	342.92
21	343.67	290.03	10.74	326.63	6.94	3.03	97.97	4.56	343.89
25	345.77	252.19	11.24	288.33	6.07	3.07	98.52	4.69	344.81
29	347.86	214.46	+11.68	250.06	− 5.19	3.12	99.00	+ 4.82	345.66
Nov. 2	349.93	176.84	12.06	211.79	4.31	3.16	99.40	4.96	346.45
6	351.98	139.34	12.37	173.55	3.44	3.19	99.73	5.10	347.16
10	354.02	101.98	12.62	135.32	2.57	3.21	99.96	5.26	347.79
14	356.05	64.76	12.79	97.11	1.70	3.22	100.10	5.43	348.33
18	358.07	27.70	+12.88	58.90	− 0.83	3.21	100.13	+ 5.60	348.76
22	0.07	350.81	12.90	20.71	+ 0.03	3.18	100.05	5.78	349.07
26	2.06	314.10	12.83	342.53	0.89	3.13	99.82	5.97	349.27
30	4.04	277.60	12.67	304.35	1.74	3.04	99.43	6.16	349.33
Dec. 4	6.00	241.29	12.43	266.18	2.58	2.92	98.86	6.36	349.26
8	7.96	205.20	+12.10	228.02	+ 3.41	2.76	98.06	+ 6.55	349.05
12	9.90	169.31	11.69	189.86	4.24	2.55	96.98	6.73	348.70
16	11.83	133.62	11.19	151.70	5.06	2.29	95.52	6.90	348.21
20	13.75	98.13	10.62	113.54	5.87	1.98	93.52	7.06	347.59
24	15.66	62.81	9.98	75.38	6.66	1.63	90.65	7.19	346.85
28	17.56	27.63	+ 9.29	37.22	+ 7.45	1.25	86.11	+ 7.28	346.01
32	19.44	352.55	+ 8.56	359.06	+ 8.23	0.84	77.53	+ 7.34	345.10

EPHEMERIS FOR PHYSICAL OBSERVATIONS FOR 0^h DYNAMICAL TIME

Date		Light-time	Magnitude	Surface Brightness	Diameter Eq.	Diameter Pol.	Phase Angle	Defect of Illumination
		m			″	″	°	″
Jan.	−1	40.96	− 2.2	5.4	39.97	37.38	9.7	0.28
	3	40.49	2.3	5.4	40.45	37.83	9.3	0.27
	7	40.02	2.3	5.4	40.91	38.26	9.0	0.25
	11	39.58	2.3	5.4	41.37	38.69	8.5	0.23
	15	39.17	2.3	5.4	41.81	39.10	8.1	0.21
	19	38.77	− 2.4	5.4	42.23	39.50	7.5	0.18
	23	38.41	2.4	5.4	42.63	39.87	7.0	0.16
	27	38.07	2.4	5.4	43.01	40.23	6.3	0.13
	31	37.77	2.4	5.4	43.36	40.55	5.7	0.11
Feb.	4	37.50	2.4	5.4	43.67	40.84	5.0	0.08
	8	37.27	− 2.5	5.4	43.94	41.09	4.2	0.06
	12	37.07	2.5	5.4	44.17	41.31	3.5	0.04
	16	36.92	2.5	5.4	44.36	41.48	2.7	0.02
	20	36.80	2.5	5.4	44.50	41.61	1.9	0.01
	24	36.73	2.5	5.4	44.59	41.70	1.1	0.00
	28	36.69	− 2.5	5.4	44.63	41.74	0.3	0.00
Mar.	3	36.70	2.5	5.4	44.61	41.72	0.7	0.00
	7	36.75	2.5	5.4	44.55	41.67	1.5	0.01
	11	36.85	2.5	5.4	44.44	41.56	2.3	0.02
	15	36.98	2.5	5.4	44.28	41.41	3.1	0.03
	19	37.15	− 2.5	5.4	44.08	41.22	3.9	0.05
	23	37.36	2.4	5.4	43.83	40.99	4.6	0.07
	27	37.61	2.4	5.4	43.54	40.72	5.3	0.09
	31	37.89	2.4	5.4	43.22	40.42	6.0	0.12
Apr.	4	38.20	2.4	5.4	42.86	40.09	6.7	0.14
	8	38.55	− 2.4	5.4	42.48	39.73	7.3	0.17
	12	38.92	2.3	5.4	42.07	39.35	7.8	0.20
	16	39.32	2.3	5.4	41.65	38.95	8.3	0.22
	20	39.74	2.3	5.4	41.21	38.54	8.8	0.24
	24	40.18	2.3	5.4	40.75	38.11	9.2	0.26
	28	40.64	− 2.2	5.4	40.29	37.68	9.6	0.28
May	2	41.12	2.2	5.4	39.82	37.24	9.9	0.30
	6	41.61	2.2	5.4	39.35	36.80	10.1	0.31
	10	42.12	2.2	5.4	38.88	36.36	10.4	0.32
	14	42.63	2.1	5.4	38.41	35.93	10.5	0.32
	18	43.14	− 2.1	5.4	37.95	35.49	10.7	0.33
	22	43.67	2.1	5.4	37.50	35.07	10.7	0.33
	26	44.19	2.0	5.5	37.05	34.65	10.8	0.33
	30	44.72	2.0	5.5	36.62	34.25	10.8	0.32
June	3	45.24	2.0	5.5	36.19	33.85	10.7	0.32
	7	45.76	− 2.0	5.5	35.78	33.46	10.6	0.31
	11	46.27	1.9	5.5	35.39	33.09	10.5	0.30
	15	46.78	1.9	5.5	35.00	32.73	10.3	0.28
	19	47.28	1.9	5.4	34.63	32.39	10.1	0.27
	23	47.77	1.9	5.4	34.28	32.06	9.9	0.26
	27	48.24	− 1.9	5.4	33.94	31.74	9.7	0.24
July	1	48.71	− 1.8	5.4	33.62	31.44	9.4	0.22

EPHEMERIS FOR PHYSICAL OBSERVATIONS
FOR 0^h DYNAMICAL TIME

Date	L_s	Sub-Earth Point		Sub-Solar Point				North Pole	
		Long.	Lat.	Long.	Lat.	Dist.	P.A.	Dist.	P.A.
	°	°	°	°	°	″	°	″	°
Jan. −1	197.83	311.20	− 1.63	320.85	− 1.09	3.35	112.24	−18.69	24.94
3	198.14	193.64	1.65	202.97	1.11	3.28	112.13	18.91	24.94
7	198.45	76.13	1.67	85.08	1.13	3.19	112.00	19.12	24.94
11	198.76	318.66	1.69	327.19	1.15	3.07	111.85	19.34	24.93
15	199.07	201.23	1.70	209.27	1.17	2.93	111.68	19.54	24.91
19	199.38	83.83	− 1.71	91.35	− 1.18	2.77	111.47	−19.74	24.89
23	199.69	326.46	1.72	333.40	1.20	2.58	111.23	19.93	24.87
27	200.00	209.12	1.72	215.44	1.22	2.37	110.94	20.10	24.84
31	200.31	91.80	1.72	97.46	1.24	2.14	110.57	20.27	24.81
Feb. 4	200.61	334.50	1.72	339.46	1.26	1.89	110.11	20.41	24.77
8	200.92	217.22	− 1.72	221.44	− 1.27	1.62	109.50	−20.54	24.72
12	201.23	99.94	1.71	103.39	1.29	1.34	108.64	20.65	24.67
16	201.54	342.66	1.70	345.32	1.31	1.04	107.30	20.73	24.62
20	201.85	225.37	1.69	227.23	1.33	0.73	104.87	20.80	24.56
24	202.16	108.07	1.68	109.11	1.35	0.42	98.85	20.84	24.50
28	202.46	350.76	− 1.66	350.96	− 1.36	0.13	62.29	−20.86	24.44
Mar. 3	202.77	233.42	1.64	232.79	1.38	0.26	314.25	20.85	24.37
7	203.08	116.06	1.62	114.60	1.40	0.57	301.91	20.83	24.31
11	203.39	358.65	1.60	356.38	1.42	0.88	298.26	20.77	24.24
15	203.69	241.21	1.58	238.14	1.43	1.19	296.48	20.70	24.17
19	204.00	123.72	− 1.56	119.87	− 1.45	1.48	295.40	−20.60	24.11
23	204.31	6.18	1.53	1.58	1.47	1.76	294.67	20.49	24.04
27	204.62	248.59	1.51	243.27	1.49	2.02	294.12	20.35	23.98
31	204.92	130.94	1.48	124.93	1.50	2.26	293.69	20.20	23.92
Apr. 4	205.23	13.23	1.46	6.58	1.52	2.48	293.34	20.04	23.87
8	205.54	255.47	− 1.44	248.21	− 1.54	2.68	293.05	−19.86	23.83
12	205.84	137.64	1.42	129.82	1.56	2.86	292.81	19.67	23.79
16	206.15	19.74	1.40	11.42	1.57	3.01	292.60	19.47	23.75
20	206.46	261.79	1.38	253.00	1.59	3.15	292.42	19.26	23.73
24	206.77	143.77	1.36	134.57	1.61	3.26	292.27	19.05	23.71
28	207.07	25.70	− 1.35	16.13	− 1.62	3.35	292.13	−18.83	23.70
May 2	207.38	267.56	1.33	257.68	1.64	3.42	292.02	18.62	23.70
6	207.68	149.37	1.32	139.22	1.66	3.47	291.92	18.40	23.70
10	207.99	31.12	1.31	20.75	1.68	3.50	291.84	18.18	23.72
14	208.30	272.82	1.30	262.28	1.69	3.51	291.77	17.96	23.74
18	208.60	154.47	− 1.30	143.81	− 1.71	3.51	291.72	−17.74	23.76
22	208.91	36.07	1.29	25.33	1.73	3.49	291.67	17.53	23.80
26	209.22	277.62	1.29	266.85	1.74	3.46	291.63	17.32	23.84
30	209.52	159.13	1.29	148.37	1.76	3.42	291.61	17.12	23.89
June 3	209.83	40.61	1.30	29.89	1.78	3.37	291.59	16.92	23.94
7	210.13	282.04	− 1.30	271.42	− 1.79	3.30	291.57	−16.73	24.00
11	210.44	163.44	1.31	152.95	1.81	3.23	291.56	16.54	24.06
15	210.74	44.81	1.32	34.48	1.82	3.14	291.56	16.36	24.13
19	211.05	286.15	1.33	276.02	1.84	3.05	291.56	16.19	24.19
23	211.36	167.47	1.34	157.56	1.86	2.95	291.56	16.03	24.27
27	211.66	48.76	− 1.36	39.11	− 1.87	2.85	291.56	−15.87	24.34
July 1	211.97	290.03	− 1.38	280.67	− 1.89	2.74	291.55	−15.72	24.41

EPHEMERIS FOR PHYSICAL OBSERVATIONS
FOR 0^h DYNAMICAL TIME

Date		Light-time	Magnitude	Surface Brightness	Diameter		Phase Angle	Defect of Illumination
					Eq.	Pol.		
		m			″	″	°	″
July	1	48.71	− 1.8	5.4	33.62	31.44	9.4	0.22
	5	49.15	1.8	5.4	33.31	31.15	9.1	0.21
	9	49.59	1.8	5.4	33.02	30.88	8.7	0.19
	13	50.00	1.8	5.4	32.75	30.63	8.3	0.17
	17	50.40	1.8	5.4	32.49	30.39	8.0	0.16
	21	50.77	− 1.8	5.4	32.25	30.16	7.6	0.14
	25	51.13	1.7	5.4	32.03	29.95	7.1	0.12
	29	51.47	1.7	5.4	31.82	29.76	6.7	0.11
Aug.	2	51.78	1.7	5.4	31.62	29.58	6.2	0.09
	6	52.07	1.7	5.4	31.45	29.41	5.7	0.08
	10	52.33	− 1.7	5.4	31.29	29.26	5.2	0.07
	14	52.58	1.7	5.4	31.14	29.13	4.7	0.05
	18	52.79	1.7	5.4	31.02	29.01	4.2	0.04
	22	52.98	1.7	5.4	30.90	28.90	3.7	0.03
	26	53.15	1.7	5.4	30.81	28.81	3.2	0.02
	30	53.29	− 1.7	5.4	30.73	28.74	2.6	0.02
Sept.	3	53.40	1.7	5.4	30.66	28.68	2.1	0.01
	7	53.48	1.7	5.4	30.62	28.63	1.5	0.01
	11	53.54	1.7	5.4	30.58	28.60	1.0	0.00
	15	53.57	1.7	5.4	30.57	28.59	0.4	0.00
	19	53.57	− 1.7	5.4	30.57	28.59	0.3	0.00
	23	53.54	1.7	5.4	30.58	28.60	0.8	0.00
	27	53.49	1.7	5.4	30.61	28.63	1.3	0.00
Oct.	1	53.40	1.7	5.4	30.66	28.68	1.9	0.01
	5	53.29	1.7	5.4	30.73	28.74	2.4	0.01
	9	53.15	− 1.7	5.4	30.81	28.81	3.0	0.02
	13	52.98	1.7	5.4	30.91	28.90	3.5	0.03
	17	52.79	1.7	5.4	31.02	29.01	4.0	0.04
	21	52.57	1.7	5.4	31.15	29.13	4.5	0.05
	25	52.32	1.7	5.4	31.30	29.27	5.1	0.06
	29	52.05	− 1.7	5.4	31.46	29.43	5.6	0.07
Nov.	2	51.75	1.7	5.4	31.64	29.60	6.0	0.09
	6	51.42	1.7	5.4	31.84	29.78	6.5	0.10
	10	51.08	1.7	5.4	32.06	29.98	7.0	0.12
	14	50.71	1.8	5.4	32.29	30.20	7.4	0.13
	18	50.32	− 1.8	5.4	32.54	30.44	7.8	0.15
	22	49.91	1.8	5.5	32.81	30.69	8.2	0.17
	26	49.48	1.8	5.5	33.10	30.95	8.5	0.18
	30	49.03	1.8	5.5	33.40	31.24	8.9	0.20
Dec.	4	48.56	1.8	5.5	33.72	31.54	9.2	0.22
	8	48.08	− 1.9	5.5	34.06	31.85	9.5	0.23
	12	47.59	1.9	5.5	34.41	32.18	9.7	0.25
	16	47.08	1.9	5.5	34.78	32.53	9.9	0.26
	20	46.57	1.9	5.5	35.16	32.89	10.1	0.27
	24	46.04	2.0	5.5	35.56	33.26	10.2	0.28
	28	45.51	− 2.0	5.5	35.98	33.65	10.3	0.29
	32	44.98	− 2.0	5.5	36.41	34.05	10.4	0.30

EPHEMERIS FOR PHYSICAL OBSERVATIONS
FOR 0^h DYNAMICAL TIME

Date	L_s	Sub-Earth Point		Sub-Solar Point				North Pole	
		Long.	Lat.	Long.	Lat.	Dist.	P.A.	Dist.	P.A.
	°	°	°	°	°	″	°	″	°
July 1	211.97	290.03	− 1.38	280.67	− 1.89	2.74	291.55	−15.72	24.41
5	212.27	171.28	1.39	162.24	1.91	2.62	291.55	15.57	24.49
9	212.58	52.52	1.41	43.82	1.92	2.50	291.54	15.44	24.57
13	212.88	293.74	1.44	285.40	1.94	2.38	291.53	15.31	24.64
17	213.19	174.96	1.46	167.00	1.95	2.25	291.51	15.19	24.72
21	213.49	56.16	− 1.48	48.62	− 1.97	2.12	291.47	−15.08	24.79
25	213.80	297.35	1.51	290.24	1.99	1.99	291.43	14.97	24.86
29	214.10	178.54	1.54	171.88	2.00	1.85	291.37	14.87	24.93
Aug. 2	214.41	59.73	1.57	53.53	2.02	1.71	291.28	14.78	25.00
6	214.71	300.91	1.60	295.19	2.03	1.57	291.18	14.70	25.07
10	215.02	182.10	− 1.63	176.87	− 2.05	1.43	291.03	−14.63	25.13
14	215.32	63.29	1.66	58.56	2.06	1.29	290.84	14.56	25.19
18	215.62	304.48	1.69	300.27	2.08	1.14	290.58	14.50	25.24
22	215.93	185.68	1.73	181.99	2.09	1.00	290.24	14.45	25.29
26	216.23	66.88	1.76	63.73	2.11	0.85	289.75	14.40	25.34
30	216.54	308.09	− 1.80	305.49	− 2.13	0.70	289.03	−14.36	25.38
Sept. 3	216.84	189.32	1.83	187.26	2.14	0.55	287.91	14.33	25.41
7	217.15	70.55	1.87	69.05	2.16	0.41	285.95	14.31	25.44
11	217.45	311.80	1.91	310.86	2.17	0.26	281.70	14.29	25.47
15	217.75	193.06	1.95	192.68	2.19	0.12	266.61	14.29	25.49
19	218.06	74.34	− 1.98	74.52	− 2.20	0.07	162.11	−14.29	25.50
23	218.36	315.64	2.02	316.38	2.22	0.20	128.37	14.29	25.52
27	218.67	196.95	2.06	198.25	2.23	0.35	122.01	14.31	25.52
Oct. 1	218.97	78.28	2.10	80.14	2.25	0.50	119.42	14.33	25.52
5	219.27	319.64	2.14	322.05	2.26	0.65	118.02	14.36	25.51
9	219.58	201.02	− 2.18	203.97	− 2.27	0.79	117.13	−14.40	25.50
13	219.88	82.42	2.22	85.92	2.29	0.94	116.52	14.44	25.49
17	220.19	323.84	2.26	327.87	2.30	1.09	116.06	14.50	25.47
21	220.49	205.30	2.30	209.85	2.32	1.24	115.70	14.56	25.44
25	220.79	86.78	2.34	91.84	2.33	1.38	115.41	14.62	25.41
29	221.10	328.28	− 2.38	333.84	− 2.35	1.52	115.16	−14.70	25.38
Nov. 2	221.40	209.82	2.42	215.87	2.36	1.67	114.94	14.79	25.34
6	221.70	91.39	2.46	97.90	2.37	1.80	114.75	14.88	25.30
10	222.01	332.99	2.50	339.95	2.39	1.94	114.58	14.98	25.26
14	222.31	214.62	2.54	222.02	2.40	2.08	114.42	15.09	25.21
18	222.61	96.29	− 2.58	104.09	− 2.42	2.21	114.26	−15.20	25.16
22	222.92	337.99	2.62	346.18	2.43	2.34	114.12	15.33	25.11
26	223.22	219.73	2.66	228.28	2.44	2.46	113.98	15.46	25.06
30	223.52	101.51	2.70	110.40	2.46	2.58	113.85	15.60	25.01
Dec. 4	223.83	343.32	2.73	352.52	2.47	2.69	113.72	15.75	24.96
8	224.13	225.18	− 2.77	234.65	− 2.49	2.80	113.60	−15.91	24.90
12	224.43	107.07	2.80	116.79	2.50	2.90	113.47	16.07	24.85
16	224.73	349.01	2.84	358.94	2.51	3.00	113.35	16.25	24.80
20	225.04	230.99	2.87	241.09	2.53	3.08	113.23	16.43	24.75
24	225.34	113.01	2.91	123.25	2.54	3.16	113.12	16.61	24.71
28	225.64	355.07	− 2.94	5.41	− 2.55	3.23	113.00	−16.81	24.66
32	225.95	237.18	− 2.97	247.57	− 2.57	3.28	112.88	−17.01	24.62

SATURN, 1992

EPHEMERIS FOR PHYSICAL OBSERVATIONS
FOR 0^h DYNAMICAL TIME

Date		Light-time	Magnitude	Surface Brightness	Diameter Eq.	Diameter Pol.	Phase Angle	Defect of Illumination
		m			″	″	°	″
Jan.	−1	89.78	+ 0.7	6.9	15.33	13.86	2.7	0.01
	3	90.02	0.7	6.9	15.29	13.82	2.3	0.01
	7	90.23	0.6	6.9	15.25	13.79	2.0	0.00
	11	90.41	0.6	6.9	15.22	13.76	1.7	0.00
	15	90.55	0.6	6.9	15.20	13.73	1.3	0.00
	19	90.66	+ 0.6	6.9	15.18	13.71	1.0	0.00
	23	90.74	0.6	6.9	15.17	13.70	0.6	0.00
	27	90.78	0.6	6.8	15.16	13.69	0.3	0.00
	31	90.78	0.6	6.8	15.16	13.68	0.1	0.00
Feb.	4	90.76	0.6	6.8	15.16	13.69	0.5	0.00
	8	90.69	+ 0.6	6.9	15.17	13.69	0.8	0.00
	12	90.60	0.7	6.9	15.19	13.70	1.2	0.00
	16	90.46	0.7	6.9	15.21	13.72	1.5	0.00
	20	90.30	0.7	6.9	15.24	13.74	1.8	0.00
	24	90.10	0.7	6.9	15.27	13.77	2.2	0.01
	28	89.88	+ 0.7	6.9	15.31	13.80	2.5	0.01
Mar.	3	89.62	0.7	6.9	15.36	13.84	2.8	0.01
	7	89.33	0.8	7.0	15.40	13.88	3.1	0.01
	11	89.01	0.8	7.0	15.46	13.93	3.4	0.01
	15	88.66	0.8	7.0	15.52	13.98	3.7	0.02
	19	88.29	+ 0.8	7.0	15.59	14.04	4.0	0.02
	23	87.89	0.8	7.0	15.66	14.10	4.2	0.02
	27	87.47	0.8	7.0	15.73	14.17	4.5	0.02
	31	87.03	0.8	7.0	15.81	14.24	4.7	0.03
Apr.	4	86.56	0.8	7.0	15.90	14.31	4.9	0.03
	8	86.08	+ 0.8	7.0	15.99	14.39	5.1	0.03
	12	85.58	0.8	7.0	16.08	14.47	5.3	0.03
	16	85.07	0.8	7.0	16.18	14.56	5.4	0.04
	20	84.55	0.8	7.1	16.28	14.64	5.6	0.04
	24	84.01	0.8	7.1	16.38	14.74	5.7	0.04
	28	83.47	+ 0.8	7.1	16.49	14.83	5.7	0.04
May	2	82.92	0.8	7.1	16.60	14.93	5.8	0.04
	6	82.36	0.7	7.1	16.71	15.03	5.8	0.04
	10	81.81	0.7	7.1	16.82	15.13	5.8	0.04
	14	81.25	0.7	7.1	16.94	15.23	5.8	0.04
	18	80.70	+ 0.7	7.1	17.05	15.34	5.8	0.04
	22	80.16	0.7	7.1	17.17	15.44	5.7	0.04
	26	79.62	0.7	7.1	17.28	15.54	5.6	0.04
	30	79.10	0.6	7.1	17.40	15.65	5.5	0.04
June	3	78.58	0.6	7.0	17.51	15.75	5.4	0.04
	7	78.08	+ 0.6	7.0	17.62	15.85	5.2	0.04
	11	77.60	0.6	7.0	17.73	15.95	5.0	0.03
	15	77.14	0.6	7.0	17.84	16.05	4.8	0.03
	19	76.71	0.5	7.0	17.94	16.14	4.5	0.03
	23	76.29	0.5	7.0	18.04	16.23	4.2	0.02
	27	75.90	+ 0.5	7.0	18.13	16.31	4.0	0.02
July	1	75.55	+ 0.5	7.0	18.21	16.39	3.6	0.02

EPHEMERIS FOR PHYSICAL OBSERVATIONS FOR 0^h DYNAMICAL TIME

Date	L_s	Sub-Earth Point		Sub-Solar Point				North Pole	
		Long.	Lat.	Long.	Lat.	Dist.	P.A.	Dist.	P.A.
	°	°	°	°	°	″	°	″	°
Jan. −1	134.82	141.69	+24.02	139.06	+22.93	0.35	256.87	+ 6.44	6.90
3	134.95	144.29	23.84	141.98	22.88	0.31	256.92	6.43	6.90
7	135.07	146.89	23.64	144.91	22.83	0.26	257.04	6.43	6.89
11	135.19	149.51	23.45	147.87	22.78	0.22	257.26	6.42	6.89
15	135.32	152.15	23.25	150.84	22.73	0.17	257.67	6.42	6.88
19	135.44	154.80	+23.04	153.83	+22.68	0.13	258.49	+ 6.41	6.87
23	135.56	157.46	22.84	156.84	22.63	0.08	260.41	6.42	6.87
27	135.69	160.15	22.63	159.87	22.58	0.03	267.86	6.42	6.86
31	135.81	162.85	22.42	162.92	22.53	0.02	42.77	6.43	6.85
Feb. 4	135.93	165.58	22.21	165.99	22.48	0.06	66.75	6.43	6.84
8	136.06	168.32	+22.00	169.08	+22.43	0.11	69.94	+ 6.44	6.83
12	136.18	171.09	21.79	172.19	22.38	0.15	71.11	6.46	6.82
16	136.30	173.88	21.58	175.32	22.33	0.20	71.67	6.47	6.81
20	136.42	176.70	21.37	178.46	22.28	0.24	71.98	6.49	6.80
24	136.55	179.54	21.16	181.63	22.23	0.29	72.15	6.51	6.79
28	136.67	182.41	+20.95	184.81	+22.18	0.33	72.24	+ 6.53	6.78
Mar. 3	136.79	185.31	20.75	188.01	22.13	0.37	72.28	6.56	6.77
7	136.92	188.23	20.55	191.23	22.08	0.42	72.30	6.59	6.76
11	137.04	191.18	20.36	194.46	22.03	0.46	72.30	6.62	6.75
15	137.16	194.16	20.17	197.71	21.98	0.50	72.28	6.65	6.74
19	137.29	197.16	+19.99	200.97	+21.93	0.54	72.25	+ 6.68	6.73
23	137.41	200.19	19.81	204.25	21.88	0.57	72.22	6.71	6.71
27	137.54	203.26	19.64	207.54	21.82	0.61	72.19	6.75	6.70
31	137.66	206.35	19.48	210.85	21.77	0.64	72.16	6.79	6.69
Apr. 4	137.78	209.47	19.33	214.16	21.72	0.68	72.12	6.83	6.68
8	137.91	212.61	+19.19	217.49	+21.67	0.71	72.09	+ 6.87	6.67
12	138.03	215.79	19.05	220.82	21.62	0.73	72.07	6.92	6.67
16	138.15	218.99	18.93	224.17	21.57	0.76	72.04	6.96	6.66
20	138.28	222.22	18.81	227.52	21.51	0.78	72.03	7.01	6.65
24	138.40	225.48	18.71	230.87	21.46	0.80	72.01	7.06	6.64
28	138.52	228.76	+18.62	234.23	+21.41	0.82	72.01	+ 7.10	6.63
May 2	138.65	232.07	18.54	237.60	21.36	0.83	72.01	7.15	6.63
6	138.77	235.41	18.47	240.96	21.31	0.84	72.02	7.20	6.62
10	138.89	238.77	18.41	244.33	21.25	0.85	72.04	7.25	6.62
14	139.02	242.15	18.37	247.70	21.20	0.85	72.06	7.31	6.62
18	139.14	245.55	+18.34	251.06	+21.15	0.85	72.09	+ 7.36	6.61
22	139.27	248.98	18.32	254.42	21.10	0.85	72.14	7.41	6.61
26	139.39	252.42	18.32	257.78	21.04	0.84	72.19	7.46	6.61
30	139.51	255.89	18.33	261.13	20.99	0.83	72.26	7.51	6.61
June 3	139.64	259.37	18.36	264.47	20.94	0.81	72.34	7.55	6.61
7	139.76	262.87	+18.39	267.81	+20.88	0.79	72.43	+ 7.60	6.61
11	139.88	266.38	18.44	271.13	20.83	0.76	72.53	7.65	6.62
15	140.01	269.90	18.50	274.45	20.78	0.73	72.65	7.69	6.62
19	140.13	273.43	18.58	277.75	20.72	0.70	72.79	7.73	6.62
23	140.26	276.98	18.66	281.04	20.67	0.66	72.95	7.77	6.63
27	140.38	280.52	+18.76	284.31	+20.62	0.62	73.14	+ 7.81	6.64
July 1	140.50	284.08	+18.86	287.57	+20.56	0.57	73.36	+ 7.84	6.64

SATURN, 1992

EPHEMERIS FOR PHYSICAL OBSERVATIONS
FOR 0^h DYNAMICAL TIME

Date		Light-time	Magnitude	Surface Brightness	Diameter Eq.	Diameter Pol.	Phase Angle	Defect of Illumination
		m			″	″	°	″
July	1	75.55	+ 0.5	7.0	18.21	16.39	3.6	0.02
	5	75.22	0.4	7.0	18.29	16.46	3.3	0.02
	9	74.92	0.4	6.9	18.37	16.53	3.0	0.01
	13	74.66	0.4	6.9	18.43	16.59	2.6	0.01
	17	74.43	0.3	6.9	18.49	16.64	2.2	0.01
	21	74.24	+ 0.3	6.9	18.53	16.69	1.8	0.00
	25	74.09	0.3	6.9	18.57	16.72	1.4	0.00
	29	73.98	0.3	6.9	18.60	16.75	1.0	0.00
Aug.	2	73.90	0.2	6.8	18.62	16.77	0.6	0.00
	6	73.87	0.2	6.8	18.63	16.78	0.2	0.00
	10	73.87	+ 0.2	6.8	18.63	16.78	0.3	0.00
	14	73.91	0.2	6.8	18.62	16.78	0.7	0.00
	18	74.00	0.2	6.9	18.60	16.76	1.1	0.00
	22	74.12	0.3	6.9	18.56	16.73	1.5	0.00
	26	74.28	0.3	6.9	18.52	16.70	1.9	0.01
	30	74.48	+ 0.3	6.9	18.47	16.66	2.3	0.01
Sept.	3	74.72	0.3	6.9	18.42	16.61	2.7	0.01
	7	74.99	0.3	6.9	18.35	16.55	3.1	0.01
	11	75.30	0.4	7.0	18.28	16.48	3.4	0.02
	15	75.63	0.4	7.0	18.19	16.41	3.7	0.02
	19	76.00	+ 0.4	7.0	18.11	16.33	4.0	0.02
	23	76.40	0.4	7.0	18.01	16.25	4.3	0.03
	27	76.82	0.4	7.0	17.91	16.16	4.6	0.03
Oct.	1	77.26	0.5	7.0	17.81	16.07	4.8	0.03
	5	77.73	0.5	7.0	17.70	15.97	5.0	0.03
	9	78.22	+ 0.5	7.0	17.59	15.87	5.2	0.04
	13	78.72	0.5	7.0	17.48	15.77	5.4	0.04
	17	79.24	0.6	7.0	17.37	15.67	5.5	0.04
	21	79.77	0.6	7.1	17.25	15.56	5.6	0.04
	25	80.31	0.6	7.1	17.13	15.46	5.7	0.04
	29	80.86	+ 0.6	7.1	17.02	15.35	5.7	0.04
Nov.	2	81.41	0.6	7.1	16.90	15.25	5.8	0.04
	6	81.96	0.6	7.1	16.79	15.15	5.8	0.04
	10	82.51	0.7	7.1	16.68	15.04	5.7	0.04
	14	83.06	0.7	7.1	16.57	14.94	5.7	0.04
	18	83.60	+ 0.7	7.0	16.46	14.84	5.6	0.04
	22	84.13	0.7	7.0	16.36	14.75	5.5	0.04
	26	84.66	0.7	7.0	16.25	14.66	5.4	0.04
	30	85.17	0.7	7.0	16.16	14.57	5.2	0.03
Dec.	4	85.66	0.7	7.0	16.06	14.48	5.0	0.03
	8	86.14	+ 0.7	7.0	15.98	14.40	4.9	0.03
	12	86.59	0.7	7.0	15.89	14.32	4.6	0.03
	16	87.03	0.7	7.0	15.81	14.25	4.4	0.02
	20	87.45	0.7	7.0	15.74	14.18	4.2	0.02
	24	87.84	0.8	7.0	15.67	14.11	3.9	0.02
	28	88.20	+ 0.8	7.0	15.60	14.05	3.6	0.02
	32	88.54	+ 0.8	7.0	15.54	13.99	3.3	0.01

EPHEMERIS FOR PHYSICAL OBSERVATIONS
FOR 0^h DYNAMICAL TIME

Date	L_s	Sub-Earth Point		Sub-Solar Point				North Pole	
		Long.	Lat.	Long.	Lat.	Dist.	P.A.	Dist.	P.A.
	°	°	°	°	°	″	°	″	°
July 1	140.50	284.08	+18.86	287.57	+20.56	0.57	73.36	+ 7.84	6.64
5	140.63	287.63	18.98	290.81	20.51	0.52	73.62	7.87	6.65
9	140.75	291.19	19.10	294.03	20.45	0.47	73.93	7.90	6.66
13	140.88	294.74	19.23	297.23	20.40	0.41	74.33	7.92	6.67
17	141.00	298.28	19.37	300.41	20.35	0.35	74.83	7.94	6.67
21	141.12	301.82	+19.51	303.58	+20.29	0.29	75.54	+ 7.96	6.68
25	141.25	305.34	19.66	306.72	20.24	0.23	76.60	7.97	6.69
29	141.37	308.85	19.81	309.84	20.18	0.16	78.49	7.98	6.70
Aug. 2	141.50	312.35	19.96	312.94	20.13	0.09	82.95	7.98	6.71
6	141.62	315.83	20.12	316.01	20.07	0.03	107.85	7.98	6.72
10	141.74	319.28	+20.27	319.06	+20.02	0.05	232.08	+ 7.97	6.73
14	141.87	322.71	20.42	322.09	19.96	0.11	244.45	7.96	6.74
18	141.99	326.12	20.56	325.10	19.91	0.18	247.74	7.95	6.74
22	142.12	329.50	20.71	328.09	19.85	0.25	249.30	7.93	6.75
26	142.24	332.85	20.84	331.05	19.80	0.31	250.24	7.91	6.76
30	142.36	336.17	+20.97	334.00	+19.74	0.37	250.88	+ 7.88	6.77
Sept. 3	142.49	339.46	21.10	336.92	19.69	0.43	251.36	7.85	6.77
7	142.61	342.71	21.21	339.82	19.63	0.49	251.73	7.82	6.78
11	142.74	345.92	21.32	342.71	19.58	0.54	252.04	7.79	6.79
15	142.86	349.10	21.41	345.57	19.52	0.59	252.30	7.75	6.79
19	142.99	352.25	+21.50	348.42	+19.46	0.63	252.52	+ 7.71	6.80
23	143.11	355.35	21.57	351.25	19.41	0.67	252.71	7.67	6.80
27	143.23	358.42	21.64	354.07	19.35	0.71	252.88	7.62	6.80
Oct. 1	143.36	1.46	21.69	356.88	19.30	0.74	253.02	7.58	6.81
5	143.48	4.45	21.73	359.67	19.24	0.77	253.15	7.53	6.81
9	143.61	7.41	+21.76	2.45	+19.18	0.79	253.26	+ 7.48	6.81
13	143.73	10.34	21.77	5.22	19.13	0.81	253.36	7.43	6.81
17	143.86	13.23	21.77	7.99	19.07	0.83	253.45	7.38	6.81
21	143.98	16.09	21.76	10.74	19.01	0.84	253.52	7.34	6.81
25	144.11	18.91	21.74	13.50	18.96	0.84	253.58	7.29	6.81
29	144.23	21.71	+21.70	16.24	+18.90	0.84	253.63	+ 7.24	6.81
Nov. 2	144.35	24.48	21.65	18.99	18.84	0.84	253.67	7.19	6.80
6	144.48	27.22	21.59	21.74	18.79	0.84	253.70	7.14	6.80
10	144.60	29.94	21.51	24.48	18.73	0.83	253.73	7.10	6.80
14	144.73	32.64	21.43	27.23	18.67	0.81	253.74	7.06	6.79
18	144.85	35.31	+21.33	29.98	+18.61	0.80	253.76	+ 7.01	6.79
22	144.98	37.97	21.22	32.74	18.56	0.78	253.76	6.97	6.78
26	145.10	40.61	21.10	35.50	18.50	0.75	253.76	6.93	6.77
30	145.23	43.23	20.96	38.27	18.44	0.73	253.76	6.90	6.77
Dec. 4	145.35	45.85	20.82	41.05	18.38	0.70	253.76	6.86	6.76
8	145.48	48.46	+20.67	43.84	+18.33	0.67	253.76	+ 6.83	6.75
12	145.60	51.05	20.51	46.64	18.27	0.64	253.76	6.80	6.74
16	145.73	53.65	20.34	49.45	18.21	0.60	253.76	6.77	6.73
20	145.85	56.24	20.16	52.27	18.15	0.57	253.77	6.74	6.72
24	145.98	58.82	19.97	55.11	18.09	0.53	253.80	6.71	6.71
28	146.10	61.41	+19.77	57.96	+18.04	0.49	253.84	+ 6.69	6.70
32	146.23	64.01	+19.57	60.83	+17.98	0.45	253.90	+ 6.67	6.68

EPHEMERIS FOR PHYSICAL OBSERVATIONS
FOR 0^h DYNAMICAL TIME

Date		Light-time	Magnitude	Equatorial Diameter	Phase Angle	L_s	Sub-Earth Lat.	North Pole	
								Dist.	P.A.
		m		"	°	°	°	"	°
Jan.	−1	170.38	+ 5.8	3.42	0.3	296.31	−64.29	− 0.76	282.27
	9	170.42	5.8	3.42	0.2	296.43	63.74	0.77	281.67
	19	170.23	5.8	3.42	0.7	296.54	63.20	0.79	281.09
	29	169.80	5.8	3.43	1.1	296.65	62.67	0.80	280.54
Feb.	8	169.16	5.8	3.44	1.6	296.77	62.17	0.82	280.03
	18	168.32	+ 5.8	3.46	1.9	296.88	−61.70	− 0.83	279.57
	28	167.30	5.8	3.48	2.3	296.99	61.28	0.85	279.16
Mar.	9	166.13	5.8	3.51	2.6	297.11	60.92	0.87	278.82
	19	164.86	5.8	3.53	2.8	297.22	60.62	0.88	278.55
	29	163.50	5.7	3.56	2.9	297.33	60.39	0.89	278.35
Apr.	8	162.11	+ 5.7	3.59	2.9	297.45	−60.24	− 0.90	278.22
	18	160.72	5.7	3.62	2.9	297.56	60.17	0.91	278.16
	28	159.37	5.7	3.66	2.8	297.67	60.18	0.92	278.18
May	8	158.11	5.7	3.68	2.6	297.79	60.27	0.93	278.26
	18	156.96	5.6	3.71	2.3	297.90	60.43	0.93	278.42
	28	155.97	+ 5.6	3.73	1.9	298.01	−60.65	− 0.93	278.63
June	7	155.17	5.6	3.75	1.5	298.13	60.93	0.93	278.91
	17	154.57	5.6	3.77	1.1	298.24	61.25	0.92	279.22
	27	154.20	5.6	3.78	0.6	298.35	61.61	0.91	279.57
July	7	154.07	5.6	3.78	0.1	298.46	61.98	0.90	279.95
	17	154.19	+ 5.6	3.78	0.5	298.58	−62.35	− 0.89	280.33
	27	154.54	5.6	3.77	1.0	298.69	62.71	0.88	280.70
Aug.	6	155.13	5.6	3.76	1.4	298.80	63.04	0.87	281.04
	16	155.93	5.6	3.74	1.9	298.92	63.32	0.85	281.34
	26	156.92	5.6	3.71	2.2	299.03	63.56	0.84	281.59
Sept.	5	158.07	+ 5.7	3.69	2.5	299.14	−63.72	− 0.83	281.77
	15	159.35	5.7	3.66	2.7	299.25	63.82	0.82	281.88
	25	160.72	5.7	3.62	2.9	299.37	63.85	0.81	281.90
Oct.	5	162.14	5.7	3.59	2.9	299.48	63.79	0.81	281.83
	15	163.56	5.7	3.56	2.9	299.59	63.66	0.80	281.68
	25	164.96	+ 5.8	3.53	2.8	299.71	−63.45	− 0.80	281.45
Nov.	4	166.28	5.8	3.50	2.6	299.82	63.17	0.80	281.14
	14	167.49	5.8	3.48	2.3	299.93	62.82	0.81	280.77
	24	168.57	5.8	3.46	2.0	300.04	62.41	0.81	280.35
Dec.	4	169.47	5.8	3.44	1.6	300.16	61.96	0.82	279.88
	14	170.17	+ 5.8	3.42	1.2	300.27	−61.46	− 0.83	279.39
	24	170.66	5.8	3.41	0.7	300.38	60.93	0.84	278.88
	34	170.92	5.8	3.41	0.3	300.49	60.38	0.85	278.36
	44	170.95	+ 5.8	3.41	0.2	300.61	−59.83	− 0.87	277.86

EPHEMERIS FOR PHYSICAL OBSERVATIONS
FOR 0^h DYNAMICAL TIME

Date		Light-time	Magnitude	Equatorial Diameter	Phase Angle	L_s	Sub-Earth Lat.	North Pole Dist.	North Pole P.A.
		m		″	°	°	°	″	°
Jan.	−1	259.21	+ 8.0	2.15	0.3	248.38	−28.45	− 0.93	6.69
	9	259.30	8.0	2.15	0.1	248.44	28.54	0.93	6.35
	19	259.14	8.0	2.15	0.4	248.50	28.62	0.93	6.02
	29	258.74	8.0	2.15	0.7	248.56	28.71	0.93	5.69
Feb.	8	258.12	8.0	2.16	1.0	248.62	28.79	0.93	5.37
	18	257.29	+ 8.0	2.17	1.2	248.68	−28.86	− 0.93	5.08
	28	256.28	8.0	2.17	1.4	248.74	28.93	0.94	4.83
Mar.	9	255.11	8.0	2.18	1.6	248.80	28.98	0.94	4.60
	19	253.82	8.0	2.20	1.8	248.85	29.03	0.95	4.42
	29	252.45	7.9	2.21	1.9	248.91	29.07	0.95	4.29
Apr.	8	251.04	+ 7.9	2.22	1.9	248.97	−29.09	− 0.96	4.21
	18	249.63	7.9	2.23	1.9	249.03	29.10	0.96	4.17
	28	248.25	7.9	2.24	1.8	249.09	29.10	0.97	4.18
May	8	246.95	7.9	2.26	1.7	249.15	29.09	0.97	4.25
	18	245.77	7.9	2.27	1.5	249.21	29.06	0.98	4.35
	28	244.74	+ 7.9	2.28	1.3	249.27	−29.03	− 0.98	4.50
June	7	243.89	7.9	2.29	1.0	249.33	28.98	0.99	4.68
	17	243.25	7.9	2.29	0.7	249.39	28.93	0.99	4.89
	27	242.83	7.9	2.30	0.4	249.45	28.88	0.99	5.12
July	7	242.65	7.9	2.30	0.1	249.51	28.81	0.99	5.36
	17	242.71	+ 7.9	2.30	0.2	249.57	−28.75	− 0.99	5.61
	27	243.01	7.9	2.29	0.6	249.63	28.69	0.99	5.84
Aug.	6	243.54	7.9	2.29	0.9	249.68	28.63	0.99	6.06
	16	244.29	7.9	2.28	1.1	249.74	28.57	0.99	6.26
	26	245.23	7.9	2.27	1.4	249.80	28.52	0.98	6.42
Sept.	5	246.34	+ 7.9	2.26	1.6	249.86	−28.48	− 0.98	6.55
	15	247.59	7.9	2.25	1.7	249.92	28.45	0.97	6.63
	25	248.93	7.9	2.24	1.8	249.98	28.44	0.97	6.66
Oct.	5	250.33	7.9	2.23	1.9	250.04	28.44	0.96	6.65
	15	251.75	7.9	2.21	1.9	250.10	28.45	0.96	6.59
	25	253.15	+ 7.9	2.20	1.8	250.16	−28.47	− 0.95	6.48
Nov.	4	254.48	8.0	2.19	1.7	250.22	28.51	0.95	6.32
	14	255.71	8.0	2.18	1.5	250.28	28.56	0.94	6.12
	24	256.80	8.0	2.17	1.3	250.34	28.61	0.94	5.87
Dec.	4	257.72	8.0	2.16	1.1	250.40	28.68	0.93	5.60
	14	258.43	+ 8.0	2.16	0.8	250.46	−28.76	− 0.93	5.29
	24	258.93	8.0	2.15	0.5	250.52	28.83	0.93	4.97
	34	259.20	8.0	2.15	0.2	250.57	28.91	0.93	4.63
	44	259.22	+ 8.0	2.15	0.1	250.63	−29.00	− 0.93	4.29

EPHEMERIS FOR PHYSICAL OBSERVATIONS
FOR 0h DYNAMICAL TIME

Date		Light-time	Magnitude	Phase Angle	L_s	Sub-Earth Point Long.	Sub-Earth Point Lat.	North Pole P.A.
		m		°	°	°	°	°
Jan.	−1	252.35	+13.8	1.4	191.61	146.10	−11.44	84.90
	9	251.28	13.8	1.6	191.67	349.82	11.72	84.87
	19	250.07	13.8	1.7	191.74	193.52	11.96	84.85
	29	248.76	13.7	1.8	191.81	37.21	12.15	84.84
Feb.	8	247.40	13.7	1.9	191.88	240.87	12.29	84.83
	18	246.02	+13.7	1.9	191.95	84.52	−12.38	84.84
	28	244.66	13.7	1.8	192.01	288.14	12.41	84.85
Mar.	9	243.36	13.7	1.7	192.08	131.75	12.39	84.87
	19	242.18	13.7	1.6	192.15	335.34	12.32	84.90
	29	241.13	13.7	1.4	192.22	178.91	12.19	84.93
Apr.	8	240.25	+13.7	1.2	192.29	22.48	−12.03	84.96
	18	239.58	13.7	0.9	192.36	226.04	11.83	85.00
	28	239.12	13.7	0.7	192.42	69.59	11.60	85.03
May	8	238.90	13.7	0.5	192.49	273.15	11.35	85.07
	18	238.91	13.7	0.5	192.56	116.71	11.09	85.10
	28	239.15	+13.7	0.7	192.63	320.27	−10.83	85.12
June	7	239.63	13.7	1.0	192.70	163.85	10.58	85.15
	17	240.31	13.7	1.2	192.76	7.44	10.36	85.16
	27	241.19	13.7	1.4	192.83	211.05	10.16	85.17
July	7	242.22	13.7	1.6	192.90	54.67	10.00	85.18
	17	243.40	+13.7	1.8	192.97	258.31	− 9.88	85.18
	27	244.67	13.7	1.9	193.04	101.97	9.80	85.17
Aug.	6	246.00	13.7	1.9	193.10	305.65	9.78	85.16
	16	247.36	13.7	2.0	193.17	149.35	9.80	85.13
	26	248.71	13.7	1.9	193.24	353.06	9.88	85.11
Sept.	5	250.01	+13.8	1.8	193.31	196.79	−10.01	85.07
	15	251.23	13.8	1.7	193.38	40.54	10.19	85.03
	25	252.33	13.8	1.5	193.44	244.29	10.41	84.99
Oct.	5	253.27	13.8	1.3	193.51	88.06	10.67	84.94
	15	254.04	13.8	1.0	193.58	291.83	10.97	84.88
	25	254.61	+13.8	0.8	193.65	135.62	−11.30	84.82
Nov.	4	254.97	13.8	0.6	193.72	339.40	11.64	84.76
	14	255.10	13.8	0.5	193.78	183.18	12.01	84.70
	24	255.00	13.8	0.5	193.85	26.96	12.37	84.64
Dec.	4	254.66	13.8	0.7	193.92	230.73	12.73	84.58
	14	254.11	+13.8	1.0	193.99	74.50	−13.09	84.53
	24	253.35	13.8	1.2	194.06	278.25	13.42	84.48
	34	252.40	13.8	1.4	194.12	121.98	13.72	84.44
	44	251.30	+13.8	1.6	194.19	325.70	−13.99	84.40

FOR 0^h DYNAMICAL TIME

Date		Mars	Jupiter			Saturn	
			System I	System II	System III	System I	System III
		°	°	°	°	°	°
Jan.	0	297.76	315.09	277.84	101.80	23.47	232.34
	1	287.99	113.07	68.18	252.41	147.63	322.99
	2	278.23	271.04	218.53	43.03	271.78	53.63
	3	268.47	69.03	8.88	193.64	35.93	144.29
	4	258.70	227.01	159.23	344.26	160.09	234.94
	5	248.94	24.99	309.59	134.88	284.25	325.59
	6	239.17	182.98	99.95	285.51	48.40	56.24
	7	229.41	340.97	250.31	76.13	172.56	146.89
	8	219.64	138.97	40.67	226.76	296.72	237.55
	9	209.87	296.96	191.03	17.40	60.88	328.20
	10	200.10	94.96	341.40	168.03	185.04	58.86
	11	190.33	252.96	131.77	318.66	309.20	149.51
	12	180.56	50.97	282.14	109.30	73.36	240.17
	13	170.79	208.97	72.52	259.94	197.53	330.83
	14	161.01	6.98	222.90	50.59	321.69	61.49
	15	151.24	164.99	13.28	201.23	85.86	152.15
	16	141.46	323.00	163.66	351.88	210.02	242.81
	17	131.69	121.01	314.04	142.53	334.19	333.47
	18	121.91	279.03	104.42	293.18	98.36	64.13
	19	112.13	77.05	254.81	83.83	222.53	154.80
	20	102.35	235.07	45.20	234.49	346.70	245.46
	21	92.57	33.09	195.59	25.14	110.87	336.13
	22	82.78	191.11	345.98	175.80	235.04	66.79
	23	73.00	349.14	136.38	326.46	359.22	157.46
	24	63.21	147.16	286.78	117.13	123.39	248.13
	25	53.42	305.19	77.17	267.79	247.57	338.80
	26	43.63	103.22	227.57	58.46	11.75	69.47
	27	33.84	261.25	17.97	209.12	135.93	160.15
	28	24.05	59.28	168.38	359.79	260.11	250.82
	29	14.25	217.32	318.78	150.46	24.29	341.50
	30	4.45	15.35	109.18	301.13	148.47	72.17
	31	354.66	173.39	259.59	91.80	272.65	162.85
Feb.	1	344.86	331.43	50.00	242.48	36.84	253.53
	2	335.05	129.47	200.41	33.15	161.03	344.21
	3	325.25	287.51	350.82	183.83	285.21	74.89
	4	315.44	85.55	141.23	334.50	49.40	165.58
	5	305.64	243.59	291.64	125.18	173.59	256.26
	6	295.83	41.63	82.05	275.86	297.79	346.95
	7	286.01	199.67	232.46	66.54	61.98	77.63
	8	276.20	357.72	22.87	217.22	186.17	168.32
	9	266.38	155.76	173.29	7.90	310.37	259.01
	10	256.56	313.81	323.70	158.58	74.57	349.70
	11	246.74	111.85	114.12	309.26	198.77	80.40
	12	236.92	269.89	264.53	99.94	322.97	171.09
	13	227.10	67.94	54.94	250.62	87.17	261.79
	14	217.27	225.98	205.36	41.30	211.38	352.48
	15	207.44	24.03	355.77	191.98	335.58	83.18

FOR 0^h DYNAMICAL TIME

Date		Mars	Jupiter System I	Jupiter System II	Jupiter System III	Saturn System I	Saturn System III
		°	°	°	°	°	°
Feb.	15	207.44	24.03	355.77	191.98	335.58	83.18
	16	197.61	182.07	146.19	342.66	99.79	173.88
	17	187.78	340.11	296.60	133.33	224.00	264.59
	18	177.94	138.16	87.01	284.01	348.21	355.29
	19	168.10	296.20	237.43	74.69	112.42	85.99
	20	158.26	94.24	27.84	225.37	236.64	176.70
	21	148.42	252.28	178.25	16.05	0.85	267.41
	22	138.57	50.32	328.66	166.72	125.07	358.12
	23	128.73	208.36	119.07	317.40	249.29	88.83
	24	118.88	6.40	269.48	108.07	13.51	179.54
	25	109.02	164.44	59.88	258.75	137.73	270.26
	26	99.17	322.48	210.29	49.42	261.95	0.97
	27	89.31	120.51	0.70	200.09	26.18	91.69
	28	79.45	278.54	151.10	350.76	150.41	182.41
	29	69.59	76.58	301.50	141.43	274.63	273.13
Mar.	1	59.72	234.61	91.90	292.10	38.86	3.86
	2	49.85	32.64	242.30	82.76	163.10	94.58
	3	39.98	190.66	32.70	233.42	287.33	185.31
	4	30.11	348.69	183.09	24.08	51.57	276.03
	5	20.23	146.71	333.49	174.74	175.80	6.76
	6	10.36	304.73	123.88	325.40	300.04	97.50
	7	0.48	102.75	274.27	116.06	64.28	188.23
	8	350.59	260.77	64.65	266.71	188.53	278.96
	9	340.71	58.78	215.04	57.36	312.77	9.70
	10	330.82	216.80	5.42	208.01	77.02	100.44
	11	320.93	14.81	155.80	358.65	201.27	191.18
	12	311.04	172.81	306.18	149.30	325.52	281.92
	13	301.14	330.82	96.55	299.94	89.77	12.66
	14	291.25	128.82	246.92	90.58	214.02	103.41
	15	281.35	286.82	37.29	241.21	338.28	194.16
	16	271.44	84.81	187.66	31.84	102.53	284.90
	17	261.54	242.81	338.02	182.47	226.79	15.66
	18	251.63	40.80	128.38	333.10	351.05	106.41
	19	241.73	198.78	278.74	123.72	115.31	197.16
	20	231.81	356.77	69.09	274.34	239.58	287.92
	21	221.90	154.75	219.45	64.96	3.84	18.67
	22	211.99	312.73	9.79	215.57	128.11	109.43
	23	202.07	110.70	160.14	6.18	252.38	200.19
	24	192.15	268.67	310.48	156.79	16.65	290.96
	25	182.23	66.64	100.82	307.39	140.93	21.72
	26	172.30	224.60	251.15	97.99	265.20	112.49
	27	162.38	22.56	41.48	248.59	29.48	203.26
	28	152.45	180.52	191.81	39.18	153.76	294.03
	29	142.52	338.47	342.13	189.77	278.04	24.80
	30	132.59	136.42	132.45	340.36	42.32	115.57
	31	122.65	294.37	282.77	130.94	166.60	206.35
Apr.	1	112.72	92.31	73.08	281.52	290.89	297.12

FOR 0^h DYNAMICAL TIME

Date		Mars	Jupiter			Saturn	
			System I	System II	System III	System I	System III
		°	°	°	°	°	°
Apr.	1	112.72	92.31	73.08	281.52	290.89	297.12
	2	102.78	250.25	223.39	72.10	55.18	27.90
	3	92.85	48.18	13.70	222.67	179.47	118.68
	4	82.91	206.11	164.00	13.23	303.76	209.47
	5	72.97	4.04	314.29	163.80	68.05	300.25
	6	63.02	161.97	104.59	314.36	192.34	31.04
	7	53.08	319.88	254.88	104.91	316.64	121.82
	8	43.13	117.80	45.16	255.47	80.94	212.61
	9	33.19	275.71	195.45	46.01	205.24	303.40
	10	23.24	73.62	345.73	196.56	329.54	34.20
	11	13.29	231.52	136.00	347.10	93.85	124.99
	12	3.35	29.42	286.27	137.64	218.15	215.79
	13	353.40	187.32	76.54	288.17	342.46	306.59
	14	343.45	345.21	226.80	78.70	106.77	37.39
	15	333.50	143.10	17.06	229.22	231.08	128.19
	16	323.54	300.99	167.31	19.74	355.39	218.99
	17	313.59	98.87	317.56	170.26	119.70	309.80
	18	303.64	256.74	107.81	320.77	244.02	40.60
	19	293.69	54.62	258.06	111.28	8.34	131.41
	20	283.73	212.48	48.30	261.79	132.66	222.22
	21	273.78	10.35	198.53	52.29	256.98	313.03
	22	263.83	168.21	348.76	202.79	21.30	43.85
	23	253.87	326.07	138.99	353.28	145.63	134.66
	24	243.92	123.92	289.22	143.77	269.95	225.48
	25	233.97	281.77	79.44	294.26	34.28	316.30
	26	224.01	79.62	229.65	84.74	158.61	47.12
	27	214.06	237.46	19.87	235.22	282.94	137.94
	28	204.11	35.30	170.08	25.70	47.27	228.76
	29	194.16	193.14	320.28	176.17	171.61	319.59
	30	184.20	350.97	110.48	326.64	295.94	50.41
May	1	174.25	148.79	260.68	117.10	60.28	141.24
	2	164.30	306.62	50.88	267.56	184.62	232.07
	3	154.35	104.44	201.07	58.02	308.96	322.90
	4	144.41	262.26	351.25	208.47	73.30	53.74
	5	134.46	60.07	141.44	358.92	197.64	144.57
	6	124.51	217.88	291.62	149.37	321.99	235.41
	7	114.57	15.69	81.80	299.81	86.34	326.24
	8	104.63	173.49	231.97	90.25	210.69	57.08
	9	94.68	331.29	22.14	240.69	335.03	147.92
	10	84.74	129.08	172.31	31.12	99.39	238.77
	11	74.80	286.88	322.47	181.55	223.74	329.61
	12	64.87	84.67	112.63	331.97	348.09	60.45
	13	54.93	242.45	262.79	122.40	112.45	151.30
	14	45.00	40.24	52.94	272.82	236.81	242.15
	15	35.07	198.02	203.09	63.23	1.16	333.00
	16	25.14	355.79	353.24	213.65	125.52	63.85
	17	15.21	153.57	143.38	4.06	249.88	154.70

FOR 0^h DYNAMICAL TIME

Date		Mars	Jupiter			Saturn	
			System I	System II	System III	System I	System III
		°	°	°	°	°	°
May	17	15.21	153.57	143.38	4.06	249.88	154.70
	18	5.28	311.34	293.53	154.47	14.25	245.55
	19	355.36	109.10	83.66	304.87	138.61	336.41
	20	345.43	266.87	233.80	95.27	262.98	67.26
	21	335.51	64.63	23.93	245.67	27.34	158.12
	22	325.60	222.39	174.06	36.07	151.71	248.98
	23	315.68	20.15	324.19	186.46	276.08	339.84
	24	305.77	177.90	114.31	336.85	40.45	70.70
	25	295.85	335.65	264.43	127.24	164.82	161.56
	26	285.95	133.40	54.55	277.62	289.19	252.42
	27	276.04	291.14	204.67	68.00	53.57	343.29
	28	266.14	88.89	354.78	218.38	177.94	74.15
	29	256.23	246.63	144.89	8.76	302.32	165.02
	30	246.34	44.36	295.00	159.13	66.69	255.89
	31	236.44	202.10	85.11	309.51	191.07	346.76
June	1	226.55	359.83	235.21	99.87	315.45	77.63
	2	216.66	157.56	25.31	250.24	79.83	168.50
	3	206.77	315.29	175.41	40.61	204.21	259.37
	4	196.88	113.02	325.51	190.97	328.59	350.24
	5	187.00	270.74	115.60	341.33	92.98	81.12
	6	177.12	68.46	265.69	131.69	217.36	171.99
	7	167.25	226.18	55.78	282.04	341.74	262.87
	8	157.37	23.90	205.87	72.39	106.13	353.74
	9	147.50	181.61	355.95	222.75	230.52	84.62
	10	137.64	339.32	146.04	13.09	354.90	175.50
	11	127.77	137.03	296.12	163.44	119.29	266.38
	12	117.91	294.74	86.20	313.79	243.68	357.26
	13	108.05	92.45	236.28	104.13	8.07	88.14
	14	98.20	250.15	26.35	254.47	132.46	179.02
	15	88.35	47.86	176.42	44.81	256.85	269.90
	16	78.50	205.56	326.50	195.15	21.24	0.78
	17	68.65	3.26	116.57	345.49	145.63	91.67
	18	58.81	160.96	266.64	135.82	270.02	182.55
	19	48.97	318.65	56.70	286.15	34.42	273.43
	20	39.13	116.35	206.77	76.49	158.81	4.32
	21	29.30	274.04	356.83	226.81	283.21	95.20
	22	19.47	71.73	146.89	17.14	47.60	186.09
	23	9.64	229.42	296.95	167.47	171.99	276.98
	24	359.81	27.11	87.01	317.79	296.39	7.86
	25	349.99	184.80	237.07	108.12	60.78	98.75
	26	340.17	342.48	27.13	258.44	185.18	189.64
	27	330.36	140.17	177.18	48.76	309.58	280.52
	28	320.54	297.85	327.24	199.08	73.97	11.41
	29	310.73	95.53	117.29	349.40	198.37	102.30
	30	300.93	253.21	267.34	139.72	322.76	193.19
July	1	291.12	50.89	57.39	290.03	87.16	284.08
	2	281.32	208.57	207.44	80.35	211.56	14.97

FOR 0^h DYNAMICAL TIME

Date		Mars	Jupiter			Saturn	
			System I	System II	System III	System I	System III
		°	°	°	°	°	°
July	1	291.12	50.89	57.39	290.03	87.16	284.08
	2	281.32	208.57	207.44	80.35	211.56	14.97
	3	271.52	6.25	357.49	230.66	335.95	105.85
	4	261.73	163.92	147.53	20.97	100.35	196.74
	5	251.94	321.60	297.58	171.28	224.75	287.63
	6	242.15	119.27	87.62	321.60	349.14	18.52
	7	232.36	276.94	237.67	111.90	113.54	109.41
	8	222.58	74.62	27.71	262.21	237.94	200.30
	9	212.80	232.29	177.75	52.52	2.33	291.19
	10	203.02	29.96	327.79	202.83	126.73	22.07
	11	193.25	187.63	117.83	353.13	251.12	112.96
	12	183.48	345.30	267.87	143.44	15.52	203.85
	13	173.71	142.96	57.91	293.74	139.91	294.74
	14	163.94	300.63	207.95	84.05	264.31	25.62
	15	154.18	98.30	357.98	234.35	28.70	116.51
	16	144.42	255.96	148.02	24.65	153.09	207.39
	17	134.66	53.63	298.06	174.96	277.49	298.28
	18	124.91	211.30	88.09	325.26	41.88	29.16
	19	115.16	8.96	238.13	115.56	166.27	120.05
	20	105.41	166.62	28.16	265.86	290.66	210.93
	21	95.66	324.29	178.20	56.16	55.05	301.82
	22	85.92	121.95	328.23	206.46	179.44	32.70
	23	76.18	279.61	118.26	356.76	303.83	123.58
	24	66.44	77.27	268.29	147.06	68.22	214.46
	25	56.70	234.94	58.33	297.35	192.61	305.34
	26	46.97	32.60	208.36	87.65	316.99	36.22
	27	37.24	190.26	358.39	237.95	81.38	127.10
	28	27.51	347.92	148.42	28.25	205.76	217.98
	29	17.78	145.58	298.45	178.54	330.14	308.85
	30	8.06	303.24	88.48	328.84	94.53	39.73
	31	358.34	100.90	238.51	119.14	218.91	130.60
Aug.	1	348.62	258.56	28.54	269.43	343.29	221.48
	2	338.91	56.22	178.57	59.73	107.67	312.35
	3	329.19	213.88	328.60	210.03	232.04	43.22
	4	319.48	11.54	118.64	0.32	356.42	134.09
	5	309.78	169.20	268.67	150.62	120.79	224.96
	6	300.07	326.86	58.70	300.91	245.17	315.83
	7	290.37	124.52	208.73	91.21	9.54	46.69
	8	280.67	282.18	358.76	241.51	133.91	137.56
	9	270.97	79.84	148.79	31.80	258.28	228.42
	10	261.27	237.50	298.82	182.10	22.65	319.28
	11	251.58	35.16	88.85	332.40	147.01	50.14
	12	241.89	192.82	238.88	122.69	271.38	141.00
	13	232.20	350.48	28.91	272.99	35.74	231.86
	14	222.51	148.14	178.94	63.29	160.10	322.71
	15	212.83	305.80	328.97	213.59	284.46	53.57
	16	203.15	103.47	119.00	3.88	48.82	144.42

FOR 0h DYNAMICAL TIME

Date		Mars	Jupiter System I	Jupiter System II	Jupiter System III	Saturn System I	Saturn System III
		°	°	°	°	°	°
Aug.	16	203.15	103.47	119.00	3.88	48.82	144.42
	17	193.47	261.13	269.04	154.18	173.18	235.27
	18	183.79	58.79	59.07	304.48	297.53	326.12
	19	174.11	216.45	209.10	94.78	61.89	56.97
	20	164.44	14.11	359.13	245.08	186.24	147.81
	21	154.77	171.78	149.17	35.38	310.59	238.66
	22	145.10	329.44	299.20	185.68	74.94	329.50
	23	135.43	127.11	89.24	335.98	199.28	60.34
	24	125.77	284.77	239.27	126.28	323.62	151.18
	25	116.11	82.44	29.31	276.58	87.97	242.02
	26	106.45	240.10	179.34	66.88	212.31	332.85
	27	96.79	37.77	329.38	217.18	336.64	63.68
	28	87.14	195.43	119.42	7.49	100.98	154.51
	29	77.49	353.10	269.45	157.79	225.31	245.34
	30	67.84	150.77	59.49	308.09	349.64	336.17
	31	58.19	308.44	209.53	98.40	113.97	66.99
Sept.	1	48.54	106.11	359.57	248.70	238.30	157.82
	2	38.90	263.78	149.61	39.01	2.63	248.64
	3	29.26	61.45	299.65	189.32	126.95	339.46
	4	19.62	219.12	89.69	339.62	251.27	70.27
	5	9.99	16.79	239.73	129.93	15.59	161.09
	6	0.35	174.46	29.78	280.24	139.90	251.90
	7	350.72	332.14	179.82	70.55	264.22	342.71
	8	341.10	129.81	329.87	220.86	28.53	73.51
	9	331.47	287.49	119.91	11.17	152.84	164.32
	10	321.85	85.16	269.96	161.49	277.15	255.12
	11	312.23	242.84	60.01	311.80	41.45	345.92
	12	302.61	40.52	210.05	102.11	165.75	76.72
	13	292.99	198.20	0.10	252.43	290.05	167.52
	14	283.38	355.88	150.15	42.75	54.35	258.31
	15	273.77	153.56	300.20	193.06	178.65	349.10
	16	264.16	311.24	90.26	343.38	302.94	79.89
	17	254.56	108.93	240.31	133.70	67.23	170.68
	18	244.95	266.61	30.36	284.02	191.52	261.46
	19	235.35	64.29	180.42	74.34	315.81	352.25
	20	225.76	221.98	330.48	224.66	80.09	83.03
	21	216.16	19.67	120.53	14.99	204.37	173.81
	22	206.57	177.36	270.59	165.31	328.65	264.58
	23	196.98	335.05	60.65	315.64	92.93	355.35
	24	187.40	132.74	210.71	105.96	217.20	86.12
	25	177.82	290.43	0.77	256.29	341.48	176.89
	26	168.24	88.12	150.84	46.62	105.75	267.66
	27	158.66	245.82	300.90	196.95	230.01	358.42
	28	149.09	43.51	90.97	347.28	354.28	89.19
	29	139.52	201.21	241.03	137.61	118.54	179.94
	30	129.96	358.91	31.10	287.95	242.80	270.70
Oct.	1	120.40	156.61	181.17	78.28	7.06	1.46

FOR 0^h DYNAMICAL TIME

Date		Mars	Jupiter			Saturn	
			System I	System II	System III	System I	System III
		°	°	°	°	°	°
Oct.	1	120.40	156.61	181.17	78.28	7.06	1.46
	2	110.84	314.31	331.24	228.62	131.32	92.21
	3	101.28	112.01	121.31	18.96	255.57	182.96
	4	91.73	269.71	271.39	169.30	19.82	273.71
	5	82.19	67.42	61.46	319.64	144.07	4.45
	6	72.64	225.13	211.54	109.98	268.32	95.20
	7	63.11	22.83	1.62	260.33	32.56	185.94
	8	53.57	180.54	151.69	50.67	156.80	276.68
	9	44.04	338.25	301.78	201.02	281.04	7.41
	10	34.51	135.97	91.86	351.37	45.28	98.15
	11	24.99	293.68	241.94	141.72	169.52	188.88
	12	15.47	91.39	32.03	292.07	293.75	279.61
	13	5.96	249.11	182.11	82.42	57.98	10.34
	14	356.45	46.83	332.20	232.77	182.21	101.06
	15	346.95	204.55	122.29	23.13	306.44	191.79
	16	337.45	2.27	272.38	173.49	70.66	282.51
	17	327.95	159.99	62.47	323.84	194.88	13.23
	18	318.46	317.72	212.57	114.21	319.11	103.95
	19	308.98	115.45	2.67	264.57	83.32	194.66
	20	299.50	273.17	152.76	54.93	207.54	285.38
	21	290.03	70.90	302.86	205.30	331.76	16.09
	22	280.56	228.64	92.96	355.66	95.97	106.80
	23	271.10	26.37	243.07	146.03	220.18	197.50
	24	261.64	184.10	33.17	296.40	344.39	288.21
	25	252.19	341.84	183.28	86.78	108.59	18.91
	26	242.75	139.58	333.39	237.15	232.80	109.62
	27	233.32	297.32	123.50	27.53	357.00	200.32
	28	223.89	95.06	273.61	177.90	121.20	291.01
	29	214.46	252.81	63.72	328.28	245.40	21.71
	30	205.05	50.55	213.84	118.66	9.60	112.41
	31	195.64	208.30	3.95	269.05	133.80	203.10
Nov.	1	186.24	6.05	154.07	59.43	257.99	293.79
	2	176.84	163.80	304.19	209.82	22.18	24.48
	3	167.46	321.55	94.32	0.21	146.37	115.17
	4	158.08	119.31	244.44	150.60	270.56	205.85
	5	148.71	277.07	34.57	300.99	34.75	296.54
	6	139.34	74.83	184.70	91.39	158.94	27.22
	7	129.99	232.59	334.83	241.79	283.12	117.90
	8	120.64	30.35	124.96	32.19	47.30	208.58
	9	111.31	188.12	275.10	182.59	171.49	299.26
	10	101.98	345.88	65.24	332.99	295.67	29.94
	11	92.66	143.65	215.37	123.39	59.85	120.62
	12	83.35	301.43	5.52	273.80	184.02	211.29
	13	74.05	99.20	155.66	64.21	308.20	301.96
	14	64.76	256.98	305.80	214.62	72.37	32.64
	15	55.48	54.75	95.95	5.04	196.55	123.31
	16	46.21	212.53	246.10	155.45	320.72	213.98

PLANETARY CENTRAL MERIDIANS, 1992

FOR 0h DYNAMICAL TIME

Date		Mars	Jupiter System I	Jupiter System II	Jupiter System III	Saturn System I	Saturn System III
		°	°	°	°	°	°
Nov.	16	46.21	212.53	246.10	155.45	320.72	213.98
	17	36.95	10.32	36.25	305.87	84.89	304.64
	18	27.70	168.10	186.41	96.29	209.06	35.31
	19	18.46	325.89	336.56	246.71	333.23	125.98
	20	9.23	123.68	126.72	37.14	97.40	216.64
	21	0.01	281.47	276.88	187.56	221.56	307.31
	22	350.81	79.26	67.05	337.99	345.73	37.97
	23	341.61	237.06	217.21	128.42	109.89	128.63
	24	332.43	34.85	7.38	278.86	234.06	219.29
	25	323.26	192.66	157.55	69.29	358.22	309.95
	26	314.10	350.46	307.72	219.73	122.38	40.61
	27	304.96	148.26	97.89	10.17	246.54	131.27
	28	295.82	306.07	248.07	160.61	10.70	221.92
	29	286.70	103.88	38.25	311.06	134.86	312.58
	30	277.60	261.69	188.43	101.51	259.02	43.23
Dec.	1	268.50	59.51	338.62	251.96	23.18	133.89
	2	259.42	217.33	128.80	42.41	147.34	224.54
	3	250.35	15.14	278.99	192.87	271.49	315.20
	4	241.29	172.97	69.18	343.32	35.65	45.85
	5	232.25	330.79	219.38	133.78	159.80	136.50
	6	223.22	128.62	9.57	284.25	283.96	227.15
	7	214.20	286.45	159.77	74.71	48.11	317.80
	8	205.20	84.28	309.97	225.18	172.27	48.46
	9	196.20	242.11	100.18	15.65	296.42	139.11
	10	187.23	39.95	250.39	166.12	60.57	229.75
	11	178.26	197.79	40.59	316.60	184.73	320.40
	12	169.31	355.63	190.81	107.07	308.88	51.05
	13	160.37	153.48	341.02	257.55	73.03	141.70
	14	151.44	311.33	131.24	48.04	197.18	232.35
	15	142.52	109.18	281.46	198.52	321.34	323.00
	16	133.62	267.03	71.68	349.01	85.49	53.65
	17	124.73	64.88	221.90	139.50	209.64	144.29
	18	115.85	222.74	12.13	289.99	333.79	234.94
	19	106.98	20.60	162.36	80.49	97.94	325.59
	20	98.13	178.46	312.59	230.99	222.09	56.24
	21	89.28	336.33	102.83	21.49	346.24	146.88
	22	80.45	134.20	253.06	171.99	110.39	237.53
	23	71.62	292.07	43.31	322.50	234.54	328.18
	24	62.81	89.94	193.55	113.01	358.70	58.82
	25	54.00	247.82	343.80	263.52	122.85	149.47
	26	45.20	45.70	134.04	54.04	247.00	240.12
	27	36.41	203.58	284.30	204.55	11.15	330.77
	28	27.63	1.47	74.55	355.07	135.30	61.41
	29	18.85	159.36	224.81	145.60	259.45	152.06
	30	10.08	317.25	15.07	296.12	23.60	242.71
	31	1.31	115.14	165.33	86.65	147.76	333.36
	32	352.55	273.04	315.60	237.18	271.91	64.01

ROTATION ELEMENTS FOR MEAN EQUINOX AND EQUATOR OF DATE ON 1992 JANUARY 0 AT 0^h TDT

		North Pole Right Ascension	North Pole Declination	Argument of Prime Meridian at epoch	Argument of Prime Meridian var./day	Longitude of Central Meridian	Inclination of Equator to Orbit
		α_1	δ_1	W_0	$\dot{W}$	λ_e	
		°	°	°	°	°	°
Mercury		280.99	61.44	23.80	6.1385025	317.13	179.99
Venus		272.69	67.17	171.06	− 1.4813291	245.67	177.34
Mars		317.63	52.86	344.16	350.8919830	298.02	25.19
Jupiter	I	268.04	64.49	326.45	877.900	315.19	3.13
	II	268.04	64.49	288.95	870.270	277.91	3.13
	III	268.04	64.49	112.95	870.536	101.90	3.13
Saturn	III	40.23	83.51	282.93	810.7939024	232.00	26.72
Uranus		257.32	−15.09	145.34	−501.1600928	313.95	97.86
Neptune		298.43	42.73	113.85	483.7625981	308.18	30.19
Pluto		311.53	4.15	152.81	− 56.364	202.44	117.58

These data were derived from the "Report of the IAU/IAG/COSPAR Working Group on Cartographic Coordinates and Rotational Elements of the Planets and Satellites: 1988" (Davies *et al.*, *Celest. Mech.*, **46**, 187–204, 1989).

DEFINITIONS AND FORMULAS

α_1, δ_1 right ascension and declination of the north pole of the planet; variations during one year are negligible.

W_0 the angle measured from the planet's equator in the positive sense with respect to the planet's north pole from the ascending node of the planet's equator on the Earth's mean equator of date to the prime meridian of the planet.

$\dot{W}$ the daily rate of change of W_0. The sidereal periods of rotation are given on page E88.

α, δ, Δ apparent right ascension, declination and true distance of the planet at the time of observation (pages E14–E42).

W_1 argument of the prime meridian at the time of observation antedated by the light-time from the planet to the Earth.

$$W_1 = W_0 + \dot{W}\,(d - 0.005\,7755\,\Delta)$$

where d is the interval in days from Jan. 0 at 0^h TDT.

β_e planetocentric declination of the Earth, positive in the planet's northern hemisphere:

$$\sin\beta_e = -\sin\delta_1 \sin\delta - \cos\delta_1 \cos\delta \cos(\alpha_1 - \alpha), \text{ where } -90° < \beta_e < 90°.$$

p_n position angle of the central meridian, also called the position angle of the axis, measured eastwards from the north point:

$$\cos\beta_e \sin p_n = \cos\delta_1 \sin(\alpha_1 - \alpha)$$

$$\cos\beta_e \cos p_n = \sin\delta_1 \cos\delta - \cos\delta_1 \sin\delta \cos(\alpha_1 - \alpha), \text{ where } \cos\beta_e > 0.$$

λ_e planetographic longitude of the central meridian measured in the direction opposite to the direction of rotation:

$$\lambda_e = W_1 - K, \text{ if } \dot{W} \text{ is positive}$$

$$\lambda_e = K - W_1, \text{ if } \dot{W} \text{ is negative}$$

where K is given by

$$\cos\beta_e \sin K = -\cos\delta_1 \sin\delta + \sin\delta_1 \cos\delta \cos(\alpha_1 - \alpha)$$

$$\cos\beta_e \cos K = \cos\delta \sin(\alpha_1 - \alpha), \text{ where } \cos\beta_e > 0.$$

λ, φ planetographic longitude (measured in the direction opposite to the rotation) and latitude (measured positive to the planet's north) of a feature on the planet's surface.

s apparent semidiameter of the planet (see page E43).

$\Delta\alpha, \Delta\delta$ displacements in right ascension and declination of the feature (λ, φ) from the center of the planet:

$$\Delta\alpha \cos\delta = X \cos p_n + Y \sin p_n$$

$$\Delta\delta = -X \sin p_n + Y \cos p_n$$

where $X = s \cos\varphi \sin(\lambda - \lambda_e)$, if $\dot{W} > 0$; $X = -s \cos\varphi \sin(\lambda - \lambda_e)$, if $W < 0$;

$Y = s\,(\sin\varphi \cos\beta_e - \cos\varphi \sin\beta_e \cos(\lambda - \lambda_e))$.

PHYSICAL AND PHOTOMETRIC DATA

Planet	Mass[1]	Mean Equatorial Radius	Maximum Angular Diameter[2]	Minimum Geocentric Distance[3]	Flattening[4] (geometric)	Coefficients of the Potential		
						$10^3 J_2$	$10^6 J_3$	$10^8 J_4$
	$\times 10^{24}$ kg	km	″	au				
Mercury	0.330 22	2 439	11.0	0.613	0	—	—	—
Venus	4.869 0	6 052	60.2	0.277	0	0.027	—	—
Earth	5.974 2	6 378.140	—	—	0.003 352 81	1.082 63	− 2.54	− 1.61
(Moon)	0.073 483	1 738	1 864.8	0.002 57	0	0.202 7	—	—
Mars	0.641 91	3 397.4	17.9	0.524	0.005 186 5	1.964	36	—
Jupiter	1 898.8	71 492.4	46.8	4.203	0.064 808 8	14.75	—	− 580
Saturn	568.50	60 268.4	19.4	8.539	0.107 620 9	16.45	—	− 1 000
Uranus	86.625	25 559.4	3.9	18.182	0.030	12	—	—
Neptune	102.78	25 269.10	2.3	29.06	0.025 9	4	—	—
Pluto	0.015	1 500	0.1	38.44	0	—	—	—

Planet	Sidereal Period of Rotation[5]	Inclination of Equator to Orbit	Mean Density	Geometric Albedo[6]	Visual Magnitude[7]		Color Indices	
					$V(1,0)$	V_0	$B-V$	$U-B$
	d	°	g/cm³					
Mercury	58.646 2	0.0	5.43	0.106	− 0.42	—	0.93	0.41
Venus	− 243.01	177.3	5.24	0.65	− 4.40	—	0.82	0.50
Earth	0.997 269 68	23.45	5.515	0.367	− 3.86	—	—	—
(Moon)	27.321 66	6.68	3.34	0.12	+ 0.21	− 12.74	0.92	0.46
Mars	1.025 956 75	25.19	3.94	0.150	− 1.52	− 2.01	1.36	0.58
Jupiter	0.413 54 (System III)	3.12	1.33	0.52	− 9.40	− 2.70	0.83	0.48
Saturn	0.437 5 (System III)	26.73	0.70	0.47	− 8.88	+ 0.67	1.04	0.58
Uranus	− 0.65	97.86	1.30	0.51	− 7.19	+ 5.52	0.56	0.28
Neptune	0.768	29.56	1.76	0.41	− 6.87	+ 7.84	0.41	0.21
Pluto	− 6.386 7	118.?	1.1	0.3:	− 1.0	+ 15.12	0.80	0.31

[1] Values for the masses include the atmospheres but exclude satellites.

[2] The tabulated Maximum Angular Diameter is the diameter when the planet is at the tabulated Minimum Geocentric Distance.

[3] For Mercury and Venus the tabulated Minimum Geocentric Distance is the mean distance of the planet at inferior conjunction; for the outer planets it is the mean distance at opposition.

[4] The Flattening is the ratio of the difference of the equatorial and polar radii to the equatorial radius.

[5] Sidereal Period of Rotation is the rotation at the equator with respect to a fixed frame of reference. A negative sign indicates that the rotation is retrograde with respect to the pole that lies north of the invariable plane of the solar system. The period is measured in days of 86 400 SI seconds. Rotation elements are tabulated on page E87.

[6] The Geometric Albedo is the ratio of the illumination of the planet at zero phase angle to the illumination produced by a plane, absolutely white Lambert surface of the same radius and position as the planet.

[7] $V(1,0)$ is the visual magnitude of the planet reduced to a distance of 1 au from both the Sun and Earth and with phase angle zero. V_0 is the mean opposition magnitude. For Saturn the photometric quantities refer to the disk only.

Data for the Mean Equatorial Radius, Flattening, Sidereal Period of Rotation, and Inclination of Equator to Orbit are based on the "Report of the IAU/IAG/Cospar Working Group on Cartographic Coordinates and Rotational Elements of the Planets and Satellites: 1988" (M. E. Davies *et al.*, *Celest. Mech.*, **46**, 187–204, 1989). The data on this page are the best values available at the time of publication. They are not necessarily those used in preparing the ephemerides. The constants used in preparing the ephemerides are given on pages K6 and K7.

CONTENTS OF SECTION F

The satellite ephemerides were calculated using $\Delta T = 58$ seconds.

SATELLITES: ORBITAL DATA

Planet		Satellite	Orbital Period[1] (R = Retrogade)	Max. Elong. at Mean Opposition	Semimajor Axis	Orbital Eccentricity	Inclination of Orbit to Planet's Equator	Motion of Node on Fixed Plane[4]
			d	° ′ ″	×10³ km		°	°/yr
Earth		Moon	27.321 661		384.400	0.054 900 489	18.28–28.58	19.34[6]
Mars	I	Phobos	0.318 910 23	25	9.378	0.015	1.0	158.8
	II	Deimos	1.262 4407	1 02	23.459	0.000 5	0.9–2.7	6.614
Jupiter	I	Io	1.769 137 786	2 18	422	0.004	0.04	48.6
	II	Europa	3.551 181 041	3 40	671	0.009	0.47	12.0
	III	Ganymede	7.154 552 96	5 51	1 070	0.002	0.21	2.63
	IV	Callisto	16.689 018 4	10 18	1 883	0.007	0.51	0.643
	V	Amalthea	0.498 179 05	59	181	0.003	0.40	914.6
	VI	Himalia	250.566 2	1 02 46	11 480	0.157 98	27.63	
	VII	Elara	259.652 8	1 04 10	11 737	0.207 19	24.77	
	VIII	Pasiphae	735. R	2 08 26	23 500	0.378	145	
	IX	Sinope	758. R	2 09 31	23 700	0.275	153	
	X	Lysithea	259.22	1 04 04	11 720	0.107	29.02	
	XI	Carme	692. R	2 03 31	22 600	0.206 78	164	
	XII	Ananke	631. R	1 55 52	21 200	0.168 70	147	
	XIII	Leda	238.72	1 00 39	11 094	0.147 62	26.07	
	XIV	Thebe	0.674 5	1 13	222	0.015	0.8	
	XV	Adrastea	0.298 26	42	129			
	XVI	Metis	0.294 780	42	128			
Saturn	I	Mimas	0.942 421 813	30	185.52	0.020 2	1.53	365.0
	II	Enceladus	1.370 217 855	38	238.02	0.004 52	0.00	156.2[5]
	III	Tethys	1.887 802 160	48	294.66	0.000 00	1.86	72.25
	IV	Dione	2.736 914 742	1 01	377.40	0.002 230	0.02	30.85[5]
	V	Rhea	4.517 500 436	1 25	527.04	0.001 00	0.35	10.16
	VI	Titan	15.945 420 68	3 17	1 221.83	0.029 192	0.33	0.5213[5]
	VII	Hyperion	21.276 608 8	3 59	1 481.1	0.104	0.43	
	VIII	Iapetus	79.330 182 5	9 35	3 561.3	0.028 28	14.72	
	IX	Phoebe	550.48 R	34 51	12 952	0.163 26	177.[2]	
	X	Janus	0.694 5	24	151.472	0.007	0.14	
	XI	Epimetheus	0.694 2	24	151.422	0.009	0.34	
	XII	Helene	2.736 9	1 01	377.40	0.005	0.0	
	XIII	Telesto	1.887 8	48	294.66			
	XIV	Calypso	1.887 8	48	294.66			
	XV	Atlas	0.601 9	22	137.670	0.000	0.3	
	XVI	Prometheus	0.613 0	23	139.353	0.003	0.0	
	XVII	Pandora	0.628 5	23	141.700	0.004	0.0	
Uranus	I	Ariel	2.520 379 35	14	191.02	0.003 4	0.3	6.8
	II	Umbriel	4.144 177 2	20	266.30	0.005 0	0.36	3.6
	III	Titania	8.705 871 7	33	435.91	0.002 2	0.14	2.0
	IV	Oberon	13.463 238 9	44	583.52	0.000 8	0.10	1.4
	V	Miranda	1.413 479 25	10	129.39	0.002 7	4.2	19.8
	VI	Cordelia	0.335 033	4	49.77	<0.001	0.1	550.
	VII	Ophelia	0.376 409	4	53.79	0.010	0.1	419.
	VIII	Bianca	0.434 577	4	59.17	<0.001	0.2	229.
	IX	Cressida	0.463 570	5	61.78	<0.001	0.0	257.
	X	Desdemona	0.473 651	5	62.68	<0.001	0.2	245.
	XI	Juliet	0.493 066	5	64.35	<0.001	0.1	223.
	XII	Portia	0.513 196	5	66.09	<0.001	0.1	203.
	XIII	Rosalind	0.558 459	5	69.94	<0.001	0.3	129.
	XIV	Belinda	0.623 525	6	75.26	<0.001	0.0	167.
	XV	Puck	0.761 832	7	86.01	<0.001	0.31	81.
Neptune	I	Triton	5.876 854 1 R	17	354.76	0.000 016	157.345	0.5232
	II	Nereid	360.13619	4 21	5 513.4	0.751 2	27.6[3]	0.039
		1989N1	1.122316	6	117.65	<0.001	0.55	0.5232
		1989N2	0.554654	3	73.55	0.001 4	0.20	143.
		1989N3	0.334655	2	52.53	<0.001	0.07	466.
		1989N4	0.428745	3	61.95	<0.001	0.05	261.
		1989N5	0.311485	2	50.07	<0.001	0.21	551.
		1989N6	0.294396	2	48.23	<0.001	4.74	626.
Pluto	I	Charon	6.387 25	<1	19.6	<0.001	99.[3]	

[1] Sidereal periods, except that tropical periods are given for satellites of Saturn.
[2] Relative to ecliptic plane.
[3] Referred to equator of 1950.0.
[4] Rate of decrease (or increase) in the longitude of the ascending node.
[5] Rate of increase in the longitude of the apse.
[6] On the ecliptic plane.

Satellite		Mass (1/Planet)	Radius	Sidereal Period of Rotation[7]	Geometric Albedo (V)[9]	$V(1,0)$	V_0	$B-V$	$U-B$
			km	d					
	Moon	0.01230002	1738	S	0.12	+ 0.21	− 12.74	0.92	0.46
I	Phobos	1.5×10^{-8}	13.5 × 10.8 × 9.4	S	0.06	+11.8	11.3	0.6	
II	Deimos	3×10^{-9}	7.5 × 6.1 × 5.5	S	0.07	+12.89	12.40	0.65	0.18
I	Io	4.68×10^{-5}	1815	S	0.61	− 1.68	5.02	1.17	1.30
II	Europa	2.52×10^{-5}	1569	S	0.64	− 1.41	5.29	0.87	0.52
III	Ganymede	7.80×10^{-5}	2631	S	0.42	− 2.09	4.61	0.83	0.50
IV	Callisto	5.66×10^{-5}	2400	S	0.20	− 1.05	5.65	0.86	0.55
V	Amalthea	38×10^{-10}	135 × 83 × 75	S	0.05	+ 7.4	14.1	1.50	
VI	Himalia	50×10^{-10}	93	0.4	0.03	+ 8.14	14.84	0.67	0.30
VII	Elara	4×10^{-10}	38	0.5	0.03	+10.07	16.77	0.69	0.28
VIII	Pasiphae	1×10^{-10}	25			+10.33	17.03	0.63	0.34
IX	Sinope	0.4×10^{-10}	18			+11.6	18.3	0.7	
X	Lysithea	0.4×10^{-10}	18			+11.7	18.4	0.7	
XI	Carme	0.5×10^{-10}	20			+11.3	18.0	0.7	
XII	Ananke	0.2×10^{-10}	15			+12.2	18.9	0.7	
XIII	Leda	0.03×10^{-10}	8			+13.5	20.2	0.7	
XIV	Thebe	4×10^{-10}	55 × 45	S	0.05	+ 9.0	15.7	1.3	
XV	Adrastea	0.1×10^{-10}	12.510 × 7.5		0.05	+12.4	19.1		
XVI	Metis	0.5×10^{-10}	20		0.05	+10.8	17.5		
I	Mimas	8.0×10^{-8}	196	S	0.5	+ 3.3	12.9		
II	Enceladus	1.3×10^{-7}	250	S	1.0	+ 2.1	11.7	0.70	0.28
III	Tethys	1.3×10^{-6}	530	S	0.9	+ 0.6	10.2	0.73	0.30
IV	Dione	1.85×10^{-6}	560	S	0.7	+ 0.8	10.4	0.71	0.31
V	Rhea	4.4×10^{-6}	765	S	0.7	+ 0.1	9.7	0.78	0.38
VI	Titan	2.38×10^{-4}	2575	S	0.21	− 1.28	8.28	1.28	0.75
VII	Hyperion	3×10^{-8}	205 × 130 × 110		0.3	+ 4.63	14.19	0.78	0.33
VIII	Iapetus	3.3×10^{-6}	730	S	0.2[8]	+ 1.5	11.1	0.72	0.30
IX	Phoebe	7×10^{-10}	110	0.4	0.06	+ 6.89	16.45	0.70	0.34
X	Janus		110 × 100 × 80	S	0.8	+ 4.4 :	14.:		
XI	Epimetheus		70 × 60 × 50	S	0.8	+ 5.4 :	15.:		
XII	Helene		18 × 16 × 15		0.7	+ 8.4 :	18.:		
XIII	Telesto		17 × 14 × 13		0.5	+ 8.9 :	18.5 :		
XIV	Calypso		17 × 11 × 11		0.6	+ 9.1 :	18.7 :		
XV	Atlas		20 × 10		0.9	+ 8.4 :	18.:		
XVI	Prometheus		70 × 50 × 40		0.6	+ 6.4 :	16.:		
XVII	Pandora		55 × 45 × 35		0.9	+ 6.4 :	16.:		
I	Ariel	1.8×10^{-5}	579	S	0.34	+ 1.45	14.16	0.65	
II	Umbriel	1.2×10^{-5}	586	S	0.18	+ 2.10	14.81	0.68	
III	Titania	6.8×10^{-5}	790	S	0.27	+ 1.02	13.73	0.70	0.28
IV	Oberon	6.9×10^{-5}	762	S	0.24	+ 1.23	13.94	0.68	0.20
V	Miranda	0.2×10^{-5}	240	S	0.27	+ 3.6	16.3		
VI	Cordelia		13		0.07 :	+11.4	24.1		
VII	Ophelia		15		0.07 :	+11.1	23.8		
VIII	Bianca		21		0.07 :	+10.3	23.0		
IX	Cressida		31		0.07 :	+ 9.5	22.2		
X	Desdemona		27		0.07 :	+ 9.8	22.5		
XI	Juliet		42		0.07 :	+ 8.8	21.5		
XII	Portia		54		0.07 :	+ 8.3	21.0		
XIII	Rosalind		27		0.07 :	+ 9.8	22.5		
XIV	Belinda		33		0.07 :	+ 9.4	22.1		
XV	Puck		77		0.07 :	+ 7.5	20.2		
I	Triton	2.09×10^{-4}	1350	S	0.7	− 1.24	13.47	0.72	0.29
II	Nereid	2×10^{-7}	170		0.4	+ 4.0	18.7	0.65	
	1989N1		218 × 208 × 201		0.06	+ 5.6	20.3		
	1989N2		104 × 89		0.06	+ 7.3	22.0		
	1989N3		75		0.05	+ 7.9	22.6		
	1989N4		80 :		0.06 :	+ 7.6 :	22.3		
	1989N5		40 :		0.06 :	+ 9.1 :	23.8		
	1989N6		27 :		0.06 :	+10.0 :	24.7		
I	Charon	0.22	593	S	0.5	+ 0.9	16.8		

[7] S = Synchronous, rotation period same as orbital period.
[8] Bright side, 0.5; faint side, 0.05.
[9] V(Sun) = −26.8

APPARENT ORBITS OF THE SATELLITES ON JULY 1

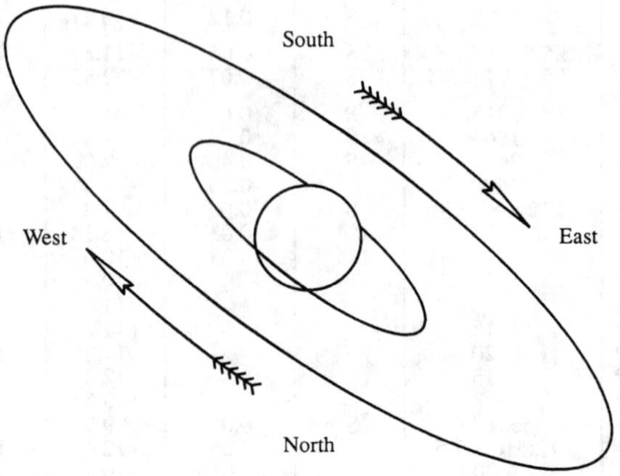

Name	Sidereal Period h m s
I Phobos	7 39 13.85
II Deimos	30 17 54.87

DEIMOS

UNIVERSAL TIME OF GREATEST EASTERN ELONGATION

Jan.	Feb.	Mar.	Apr.	May	June	July	Aug.	Sept.	Oct.	Nov.	Dec.
d h	d h	d h	d h	d h	d h	d h	d h	d h	d h	d h	d h
−1 21.1	1 18.7	1 21.3	1 06.5	1 15.8	1 01.1	1 10.2	2 01.4	1 10.1	1 18.5	1 02.7	1 10.2
1 03.5	3 01.1	3 03.7	2 12.9	2 22.2	2 07.4	2 16.6	3 07.8	2 16.5	3 00.9	2 09.0	2 16.5
2 09.9	4 07.5	4 10.1	3 19.3	4 04.6	3 13.9	3 23.0	4 14.2	3 22.8	4 07.2	3 15.3	3 22.8
3 16.2	5 13.8	5 16.5	5 01.6	5 11.0	4 20.2	5 05.3	5 20.5	5 05.1	5 13.6	4 21.6	5 05.1
4 22.6	6 20.2	6 22.8	6 08.0	6 17.4	6 02.6	6 11.7	7 02.9	6 11.5	6 19.9	6 04.0	6 11.4
6 05.0	8 02.6	8 05.2	7 14.5	7 23.7	7 08.9	7 18.1	8 09.2	7 17.9	8 02.2	7 10.3	7 17.7
7 11.3	9 09.0	9 11.6	8 20.8	9 06.1	8 15.4	9 00.4	9 15.6	9 00.2	9 08.6	8 16.6	8 23.9
8 17.7	10 15.3	10 18.0	10 03.2	10 12.5	9 21.7	10 06.8	10 22.0	10 06.6	10 14.9	9 22.9	10 06.2
10 00.0	11 21.7	12 00.4	11 09.6	11 18.9	11 04.1	11 13.2	12 04.3	11 12.9	11 21.3	11 05.3	11 12.5
11 06.4	13 04.1	13 06.8	12 16.0	13 01.3	12 10.5	12 19.6	13 10.7	12 19.3	13 03.6	12 11.6	12 18.8
12 12.8	14 10.5	14 13.2	13 22.4	14 07.6	13 16.9	14 01.9	14 17.1	14 01.6	14 09.9	13 17.9	14 01.1
13 19.2	15 16.8	15 19.5	15 04.7	15 14.1	14 23.3	15 08.3	15 23.4	15 08.0	15 16.3	15 00.2	15 07.4
15 01.5	16 23.2	17 01.9	16 11.2	16 20.5	16 05.6	16 14.7	17 05.8	16 14.3	16 22.6	16 06.5	16 13.6
16 07.9	18 05.6	18 08.3	17 17.5	18 02.8	17 12.0	17 21.0	18 12.2	17 20.7	18 05.0	17 12.8	17 19.9
17 14.3	19 12.0	19 14.7	18 23.9	19 09.2	18 18.4	19 03.4	19 18.5	19 03.0	19 11.3	18 19.2	19 02.2
18 20.6	20 18.3	20 21.0	20 06.3	20 15.6	20 00.8	20 09.8	21 00.9	20 09.4	20 17.7	20 01.5	20 08.5
20 03.0	22 00.7	22 03.4	21 12.7	21 22.0	21 07.1	21 16.2	22 07.2	21 15.7	22 00.0	21 07.8	21 14.7
21 09.4	23 07.1	23 09.8	22 19.1	23 04.4	22 13.5	22 22.5	23 13.6	22 22.1	23 06.3	22 14.1	22 21.0
22 15.8	24 13.5	24 16.2	24 01.5	24 10.7	23 19.9	24 04.9	24 19.9	24 04.4	24 12.6	23 20.4	24 03.3
23 22.1	25 19.8	25 22.6	25 07.8	25 17.2	25 02.3	25 11.3	26 02.3	25 10.8	25 19.0	25 02.7	25 09.5
25 04.5	27 02.2	27 04.9	26 14.3	26 23.5	26 08.6	26 17.6	27 08.6	26 17.1	27 01.3	26 09.0	26 15.8
26 10.9	28 08.6	28 11.4	27 20.6	28 05.9	27 15.1	28 00.0	28 15.0	27 23.5	28 07.7	27 15.3	27 22.1
27 17.2	29 15.0	29 17.7	29 03.0	29 12.3	28 21.4	29 06.3	29 21.4	29 05.8	29 14.0	28 21.6	29 04.3
28 23.6		31 00.1	30 09.4	30 18.7	30 03.8	30 12.7	31 03.7	30 12.2	30 20.3	30 03.9	30 10.6
30 06.0						31 19.1					31 16.8
31 12.4											32 23.1

PHOBOS

UNIVERSAL TIME OF EVERY THIRD GREATEST EASTERN ELONGATION

Jan.	Feb.	Mar.	Apr.	May	June	July	Aug.	Sept.	Oct.	Nov.	Dec.
d h	d h	d h	d h	d h	d h	d h	d h	d h	d h	d h	d h
0 03.9	1 17.0	1 10.3	1 01.6	1 16.8	1 08.2	1 00.4	1 14.6	1 05.7	1 20.8	1 11.7	1 03.6
1 02.8	2 16.0	2 09.2	2 00.5	2 15.8	2 07.1	1 23.4	2 13.5	2 04.7	2 19.7	2 10.7	2 02.6
2 01.8	3 14.9	3 08.2	2 23.5	3 14.8	3 06.1	2 22.4	3 12.5	3 03.6	3 18.7	3 09.7	3 01.5
3 00.8	4 13.9	4 07.2	3 22.5	4 13.8	4 05.1	3 21.3	4 11.5	4 02.6	4 17.7	4 08.6	4 00.5
3 23.7	5 12.9	5 06.2	4 21.4	5 12.8	5 04.1	4 20.3	5 10.5	5 01.6	5 16.6	5 07.6	4 23.5
4 22.7	6 11.9	6 05.2	5 20.4	6 11.7	6 03.0	5 19.3	6 09.4	6 00.5	6 15.6	6 06.6	5 22.4
5 21.7	7 10.9	7 04.1	6 19.4	7 10.7	7 02.0	6 18.3	7 08.4	6 23.5	7 14.6	7 05.5	6 21.4
6 20.7	8 09.8	8 03.1	7 18.4	8 09.7	8 01.0	7 17.2	8 07.4	7 22.5	8 13.5	8 04.5	7 20.3
7 19.6	9 08.8	9 02.1	8 17.4	9 08.7	9 00.0	8 16.2	9 06.4	8 21.5	9 12.5	9 03.5	8 19.3
8 18.6	10 07.8	10 01.1	9 16.3	10 07.7	9 22.9	9 15.2	10 05.3	9 20.4	10 11.5	10 02.4	9 18.3
9 17.6	11 06.8	11 00.0	10 15.3	11 06.6	10 21.9	10 14.2	11 04.3	10 19.4	11 10.5	11 01.4	10 17.2
10 16.6	12 05.7	11 23.0	11 14.3	12 05.6	11 20.9	11 13.1	12 03.3	11 18.4	12 09.4	12 00.4	11 16.2
11 15.5	13 04.7	12 22.0	12 13.3	13 04.6	12 19.9	12 12.1	13 02.2	12 17.3	13 08.4	12 23.3	12 15.1
12 14.5	14 03.7	13 21.0	13 12.3	14 03.6	13 18.8	13 11.1	14 01.2	13 16.3	14 07.4	13 22.3	13 14.1
13 13.5	15 02.7	14 19.9	14 11.2	15 02.6	14 17.8	14 10.1	15 00.2	14 15.3	15 06.3	14 21.3	14 13.1
14 12.5	16 01.6	15 18.9	15 10.2	16 01.5	15 16.8	15 09.0	15 23.2	15 14.3	16 05.3	15 20.2	15 12.0
15 11.4	17 00.6	16 17.9	16 09.2	17 00.5	16 15.8	16 08.0	16 22.1	16 13.2	17 04.3	16 19.2	16 11.0
16 10.4	17 23.6	17 16.9	17 08.2	17 23.5	17 14.8	17 07.0	17 21.1	17 12.2	18 03.2	17 18.2	17 09.9
17 09.4	18 22.6	18 15.9	18 07.2	18 22.5	18 13.7	18 06.0	18 20.1	18 11.2	19 02.2	18 17.1	18 08.9
18 08.4	19 21.5	19 14.8	19 06.1	19 21.4	19 12.7	19 04.9	19 19.0	19 10.1	20 01.2	19 16.1	19 07.8
19 07.4	20 20.5	20 13.8	20 05.1	20 20.4	20 11.7	20 03.9	20 18.0	20 09.1	21 00.1	20 15.0	20 06.8
20 06.3	21 19.5	21 12.8	21 04.1	21 19.4	21 10.7	21 02.9	21 17.0	21 08.1	21 23.1	21 14.0	21 05.7
21 05.3	22 18.5	22 11.8	22 03.1	22 18.4	22 09.6	22 01.9	22 16.0	22 07.1	22 22.1	22 13.0	22 04.7
22 04.3	23 17.4	23 10.7	23 02.1	23 17.3	23 08.6	23 00.8	23 14.9	23 06.0	23 21.0	23 11.9	23 03.7
23 03.3	24 16.4	24 09.7	24 01.0	24 16.3	24 07.6	23 23.8	24 13.9	24 05.0	24 20.0	24 10.9	24 02.6
24 02.2	25 15.4	25 08.7	25 00.0	25 15.3	25 06.6	24 22.8	25 12.9	25 04.0	25 19.0	25 09.9	25 01.6
25 01.2	26 14.4	26 07.7	25 23.0	26 14.3	26 05.5	25 21.7	26 11.9	26 02.9	26 17.9	26 08.8	26 00.5
26 00.2	27 13.3	27 06.7	26 22.0	27 13.3	27 04.5	26 20.7	27 10.8	27 01.9	27 16.9	27 07.8	26 23.5
26 23.1	28 12.3	28 05.6	27 20.9	28 12.2	28 03.5	27 19.7	28 09.8	28 00.9	28 15.9	28 06.7	27 22.4
27 22.1	29 11.3	29 04.6	28 19.9	29 11.2	29 02.5	28 18.7	29 08.8	28 23.8	29 14.8	29 05.7	28 21.4
28 21.1		30 03.6	29 18.9	30 10.2	30 01.4	29 17.6	30 07.7	29 22.8	30 13.8	30 04.7	29 20.3
29 20.1		31 02.6	30 17.9	31 09.2		30 16.6	31 06.7	30 21.8	31 12.8		30 19.3
30 19.0						31 15.6					31 18.3
31 18.0											32 17.2

PHOBOS

APPARENT DISTANCE AND POSITION ANGLE

Day (0^h UT)	Jan. a/Δ	Jan. p_2	Feb. a/Δ	Feb. p_2	Mar. a/Δ	Mar. p_2	Apr. a/Δ	Apr. p_2	May a/Δ	May p_2	June a/Δ	June p_2
	″	°	″	°	″	°	″	°	″	°	″	°
1	5.36	+68.5	5.64	+57.1	5.97	+43.9	6.37	+28.6	6.79	+14.7	7.29	+ 4.2
2	5.37	68.2	5.66	56.7	5.98	43.4	6.38	28.1	6.81	14.2	7.30	3.9
3	5.38	67.9	5.67	56.3	5.99	42.9	6.39	27.6	6.82	13.8	7.32	3.7
4	5.39	67.6	5.68	55.8	6.01	42.4	6.41	27.1	6.84	13.4	7.34	3.5
5	5.39	67.3	5.69	55.4	6.02	41.9	6.42	26.6	6.85	13.0	7.36	3.2
6	5.40	+67.0	5.70	+55.0	6.03	+41.4	6.43	+26.1	6.87	+12.6	7.37	+ 3.0
7	5.41	66.6	5.71	54.5	6.04	40.9	6.45	25.6	6.88	12.2	7.39	2.8
8	5.42	66.3	5.72	54.1	6.05	40.5	6.46	25.2	6.90	11.8	7.41	2.6
9	5.43	66.0	5.73	53.7	6.07	40.0	6.48	24.7	6.91	11.4	7.43	2.4
10	5.44	65.6	5.74	53.2	6.08	39.5	6.49	24.2	6.93	11.1	7.45	2.2
11	5.45	+65.3	5.75	+52.8	6.09	+39.0	6.50	+23.7	6.94	+10.7	7.46	+ 2.0
12	5.45	64.9	5.76	52.3	6.10	38.5	6.52	23.2	6.96	10.3	7.48	1.9
13	5.46	64.6	5.77	51.9	6.12	38.0	6.53	22.8	6.98	10.0	7.50	1.7
14	5.47	64.2	5.78	51.4	6.13	37.5	6.55	22.3	6.99	9.6	7.52	1.6
15	5.48	63.9	5.79	50.9	6.14	37.0	6.56	21.8	7.01	9.2	7.54	1.4
16	5.49	+63.5	5.81	+50.5	6.16	+36.5	6.57	+21.3	7.02	+ 8.9	7.56	+ 1.3
17	5.50	63.1	5.82	50.0	6.17	36.0	6.59	20.9	7.04	8.6	7.58	1.1
18	5.51	62.7	5.83	49.6	6.18	35.5	6.60	20.4	7.05	8.2	7.59	1.0
19	5.52	62.4	5.84	49.1	6.19	35.0	6.62	20.0	7.07	7.9	7.61	0.9
20	5.53	62.0	5.85	48.6	6.21	34.5	6.63	19.5	7.09	7.6	7.63	0.8
21	5.54	+61.6	5.86	+48.2	6.22	+34.0	6.64	+19.0	7.10	+ 7.2	7.65	+ 0.6
22	5.55	61.2	5.87	47.7	6.23	33.5	6.66	18.6	7.12	6.9	7.67	0.5
23	5.56	60.8	5.89	47.2	6.25	33.0	6.67	18.1	7.14	6.6	7.69	0.4
24	5.57	60.4	5.90	46.7	6.26	32.5	6.69	17.7	7.15	6.3	7.71	0.4
25	5.57	60.0	5.91	46.3	6.27	32.0	6.70	17.2	7.17	6.0	7.73	0.3
26	5.58	+59.6	5.92	+45.8	6.29	+31.5	6.72	+16.8	7.18	+ 5.7	7.75	+ 0.2
27	5.59	59.2	5.93	45.3	6.30	31.0	6.73	16.4	7.20	5.5	7.77	0.1
28	5.60	58.8	5.94	44.8	6.31	30.5	6.75	15.9	7.22	5.2	7.79	+ 0.1
29	5.61	58.4	5.96	+44.3	6.33	30.1	6.76	15.5	7.23	4.9	7.81	0.0
30	5.62	58.0			6.34	29.6	6.78	+15.1	7.25	4.7	7.84	0.0
31	5.63	+57.6			6.35	+29.1			7.27	+ 4.4		

Time from Eastern Elongation	F	p_1	Time from Eastern Elongation	F	p_1	Time from Eastern Elongation	F	p_1	Time from Eastern Elongation	F	p_1
h m		°	h m		°	h m		°	h m		°
0 00	1.000	51.0	2 00	0.299	307.3	4 00	0.991	228.6	6 00	0.354	104.3
0 10	0.991	48.7	2 10	0.352	285.1	4 10	0.965	226.2	6 10	0.438	89.6
0 20	0.966	46.3	2 20	0.435	270.0	4 20	0.922	223.7	6 20	0.534	79.8
0 30	0.924	43.8	2 30	0.530	260.1	4 30	0.865	220.8	6 30	0.630	73.0
0 40	0.867	40.9	2 40	0.626	253.2	4 40	0.794	217.5	6 40	0.721	68.0
0 50	0.796	37.7	2 50	0.717	248.2	4 50	0.711	213.5	6 50	0.802	64.1
1 00	0.714	33.7	3 00	0.799	244.2	5 00	0.619	208.3	7 00	0.872	60.8
1 10	0.623	28.5	3 10	0.870	240.9	5 10	0.523	201.3	7 10	0.928	58.0
1 20	0.526	21.6	3 20	0.926	238.1	5 20	0.428	191.0	7 20	0.969	55.5
1 30	0.431	11.5	3 30	0.967	235.6	5 30	0.346	175.5	7 30	0.993	53.1
1 40	0.349	356.2	3 40	0.992	233.2	5 40	0.297	152.8	7 40	1.000	50.8
1 50	0.298	333.8	3 50	1.000	230.9	5 50	0.300	126.3			

PHOBOS

APPARENT DISTANCE AND POSITION ANGLE

Day (0^h UT)	July		Aug.		Sept.		Oct.		Nov.		Dec.	
	a/Δ	p_2	a/Δ	p_2	a/Δ	p_2	a/Δ	p_2	a/Δ	p_2	a/Δ	p_2
	″	°	″	°	″	°	″	°	″	°	″	°
1	7.86	− 0.1	8.63	+ 1.9	9.73	+ 8.6	11.34	+17.1	13.97	+25.0	17.69	+28.3
2	7.88	0.1	8.66	2.1	9.77	8.8	11.41	17.4	14.08	25.2	17.82	28.3
3	7.90	0.1	8.68	2.2	9.81	9.1	11.48	17.6	14.19	25.4	17.95	28.3
4	7.92	0.1	8.71	2.4	9.86	9.4	11.55	17.9	14.30	25.6	18.09	28.3
5	7.94	0.1	8.75	2.5	9.90	9.7	11.62	18.2	14.41	25.7	18.22	28.2
6	7.97	− 0.1	8.78	+ 2.7	9.95	+ 9.9	11.69	+18.5	14.52	+25.9	18.35	+28.2
7	7.99	0.1	8.81	2.9	10.00	10.2	11.76	18.8	14.63	26.1	18.48	28.2
8	8.01	0.1	8.84	3.1	10.04	10.5	11.84	19.0	14.75	26.3	18.61	28.1
9	8.03	0.1	8.87	3.3	10.09	10.8	11.91	19.3	14.86	26.4	18.73	28.0
10	8.06	0.1	8.90	3.5	10.14	11.0	11.99	19.6	14.98	26.6	18.86	27.9
11	8.08	− 0.1	8.94	+ 3.7	10.19	+11.3	12.06	+19.9	15.10	+26.7	18.98	+27.9
12	8.10	0.0	8.97	3.9	10.24	11.6	12.14	20.1	15.22	26.9	19.10	27.8
13	8.13	0.0	9.00	4.1	10.29	11.9	12.22	20.4	15.34	27.0	19.22	27.7
14	8.15	+ 0.1	9.04	4.3	10.34	12.2	12.30	20.7	15.46	27.2	19.34	27.5
15	8.17	0.1	9.07	4.5	10.39	12.5	12.38	20.9	15.59	27.3	19.45	27.4
16	8.20	+ 0.2	9.10	+ 4.7	10.45	+12.7	12.47	+21.2	15.71	+27.4	19.56	+27.3
17	8.22	0.2	9.14	4.9	10.50	13.0	12.55	21.5	15.84	27.5	19.66	27.2
18	8.25	0.3	9.17	5.1	10.56	13.3	12.64	21.7	15.96	27.6	19.76	27.0
19	8.27	0.4	9.21	5.4	10.61	13.6	12.72	22.0	16.09	27.7	19.86	26.9
20	8.30	0.5	9.25	5.6	10.67	13.9	12.81	22.2	16.22	27.8	19.96	26.7
21	8.32	+ 0.6	9.28	+ 5.8	10.72	+14.2	12.90	+22.5	16.35	+27.9	20.04	+26.5
22	8.35	0.6	9.32	6.1	10.78	14.5	12.99	22.7	16.48	28.0	20.13	26.3
23	8.38	0.7	9.36	6.3	10.84	14.8	13.08	23.0	16.61	28.1	20.21	26.2
24	8.40	0.9	9.40	6.5	10.90	15.1	13.18	23.2	16.75	28.1	20.28	26.0
25	8.43	1.0	9.44	6.8	10.96	15.3	13.27	23.4	16.88	28.2	20.35	25.8
26	8.46	+ 1.1	9.48	+ 7.0	11.02	+15.6	13.37	+23.7	17.01	+28.2	20.41	+25.6
27	8.48	1.2	9.52	7.3	11.08	15.9	13.47	23.9	17.15	28.3	20.46	25.4
28	8.51	1.3	9.56	7.5	11.15	16.2	13.56	24.1	17.28	28.3	20.51	25.1
29	8.54	1.5	9.60	7.8	11.21	16.5	13.67	24.3	17.42	28.3	20.55	24.9
30	8.57	1.6	9.64	8.1	11.28	+16.8	13.77	24.5	17.55	+28.3	20.59	24.7
31	8.60	+ 1.8	9.68	+ 8.3			13.87	+24.8			20.62	+24.5

Apparent distance of satellite: $s = Fa/\Delta$

Position angle of satellite: $p = p_1 + p_2$

The differences of right ascension and declination, in the sense "satellite minus primary," are approximately

$$\Delta\alpha = s \sin p \sec(\delta + \Delta\delta)$$

$$\Delta\delta = s \cos p$$

DEIMOS

APPARENT DISTANCE AND POSITION ANGLE

Day (0ʰ UT)	Jan.		Feb.		Mar.		Apr.		May		June	
	a/Δ	p_2	a/Δ	p_2	a/Δ	p_2	a/Δ	p_2	a/Δ	p_2	a/Δ	p_2
	″	°	″	°	″	°	″	°	″	°	″	°
1	13.42	+65.6	14.12	+54.3	14.94	+41.5	15.93	+27.0	16.99	+14.1	18.23	+ 4.4
2	13.44	65.3	14.15	53.9	14.97	41.0	15.96	26.5	17.03	13.7	18.27	4.1
3	13.46	65.0	14.18	53.5	15.00	40.5	16.00	26.1	17.07	13.3	18.32	3.9
4	13.48	64.7	14.20	53.1	15.03	40.1	16.03	25.6	17.11	12.9	18.36	3.7
5	13.50	64.3	14.23	52.6	15.06	39.6	16.06	25.2	17.14	12.6	18.41	3.5
6	13.52	+64.0	14.25	+52.2	15.09	+39.1	16.10	+24.7	17.18	+12.2	18.45	+ 3.3
7	13.54	63.7	14.28	51.8	15.12	38.7	16.13	24.2	17.22	11.8	18.49	3.1
8	13.56	63.4	14.31	51.4	15.15	38.2	16.17	23.8	17.26	11.5	18.54	2.9
9	13.58	63.0	14.33	50.9	15.18	37.7	16.20	23.3	17.30	11.1	18.58	2.7
10	13.60	62.7	14.36	50.5	15.21	37.3	16.24	22.9	17.34	10.8	18.63	2.6
11	13.63	+62.3	14.39	+50.1	15.24	+36.8	16.27	+22.5	17.37	+10.4	18.67	+ 2.4
12	13.65	62.0	14.42	49.6	15.27	36.3	16.31	22.0	17.41	10.1	18.72	2.2
13	13.67	61.6	14.44	49.2	15.31	35.9	16.34	21.6	17.45	9.7	18.77	2.1
14	13.69	61.3	14.47	48.7	15.34	35.4	16.38	21.1	17.49	9.4	18.81	1.9
15	13.71	60.9	14.50	48.3	15.37	34.9	16.41	20.7	17.53	9.1	18.86	1.8
16	13.74	+60.6	14.53	+47.8	15.40	+34.4	16.45	+20.3	17.57	+ 8.8	18.91	+ 1.7
17	13.76	60.2	14.56	47.4	15.43	34.0	16.48	19.8	17.61	8.5	18.95	1.5
18	13.78	59.8	14.58	47.0	15.47	33.5	16.52	19.4	17.65	8.1	19.00	1.4
19	13.81	59.5	14.61	46.5	15.50	33.0	16.55	19.0	17.69	7.8	19.05	1.3
20	13.83	59.1	14.64	46.0	15.53	32.6	16.59	18.5	17.73	7.5	19.10	1.2
21	13.85	+58.7	14.67	+45.6	15.56	+32.1	16.63	+18.1	17.77	+ 7.2	19.15	+ 1.1
22	13.88	58.3	14.70	45.1	15.60	31.6	16.66	17.7	17.81	7.0	19.20	1.0
23	13.90	57.9	14.73	44.7	15.63	31.2	16.70	17.3	17.85	6.7	19.25	0.9
24	13.92	57.6	14.76	44.2	15.66	30.7	16.74	16.9	17.89	6.4	19.30	0.8
25	13.95	57.2	14.79	43.8	15.69	30.2	16.77	16.5	17.94	6.1	19.35	0.7
26	13.97	+56.8	14.82	+43.3	15.73	+29.8	16.81	+16.1	17.98	+ 5.9	19.40	+ 0.7
27	14.00	56.4	14.85	42.8	15.76	29.3	16.85	15.7	18.02	5.6	19.45	0.6
28	14.02	56.0	14.88	42.4	15.79	28.8	16.88	15.3	18.06	5.3	19.50	0.5
29	14.05	55.6	14.91	+41.9	15.83	28.4	16.92	14.9	18.10	5.1	19.55	0.5
30	14.07	55.1			15.86	27.9	16.96	+14.5	18.15	4.8	19.61	+ 0.4
31	14.10	+54.7			15.89	+27.4			18.19	+ 4.6		

Time from Eastern Elongation	F	p_1	Time from Eastern Elongation	F	p_1	Time from Eastern Elongation	F	p_1	Time from Eastern Elongation	F	p_1
h m		°	h m		°	h m		°	h m		°
0 00	1.000	53.0	8 00	0.306	306.3	16 00	0.986	230.0	24 00	0.386	100.4
0 40	0.991	50.7	8 40	0.364	284.9	16 40	0.955	227.5	24 40	0.476	87.6
1 20	0.965	48.2	9 20	0.450	270.6	17 20	0.908	224.9	25 20	0.574	79.1
2 00	0.923	45.6	10 00	0.547	261.1	18 00	0.846	221.8	26 00	0.669	73.0
2 40	0.865	42.7	10 40	0.643	254.5	18 40	0.771	218.3	26 40	0.757	68.4
3 20	0.793	39.3	11 20	0.733	249.6	19 20	0.684	213.9	27 20	0.834	64.7
4 00	0.709	35.2	12 00	0.814	245.7	20 00	0.590	208.1	28 00	0.899	61.6
4 40	0.617	29.9	12 40	0.882	242.5	20 40	0.493	200.1	28 40	0.948	58.9
5 20	0.520	22.6	13 20	0.936	239.6	21 20	0.401	188.2	29 20	0.982	56.4
6 00	0.425	12.0	14 00	0.974	237.1	22 00	0.327	170.2	30 00	0.998	54.0
6 40	0.344	355.9	14 40	0.995	234.7	22 40	0.294	145.3	30 40	0.997	51.7
7 20	0.298	332.7	15 20	0.999	232.4	23 20	0.318	119.6			

DEIMOS

APPARENT DISTANCE AND POSITION ANGLE

Day (0h UT)	July a/Δ	July p_2	Aug. a/Δ	Aug. p_2	Sept. a/Δ	Sept. p_2	Oct. a/Δ	Oct. p_2	Nov. a/Δ	Nov. p_2	Dec. a/Δ	Dec. p_2
	″	°	″	°	″	°	″	°	″	°	″	°
1	19.66	+ 0.4	21.58	+ 2.1	24.34	+ 8.2	28.38	+15.9	34.97	+23.0	44.25	+25.9
2	19.71	0.4	21.66	2.2	24.44	8.4	28.55	16.1	35.23	23.2	44.59	25.9
3	19.77	0.3	21.73	2.4	24.56	8.6	28.72	16.4	35.50	23.3	44.92	25.8
4	19.82	0.3	21.81	2.5	24.67	8.9	28.90	16.7	35.77	23.5	45.25	25.8
5	19.88	0.3	21.88	2.7	24.78	9.1	29.07	16.9	36.05	23.7	45.58	25.8
6	19.93	+ 0.3	21.96	+ 2.8	24.90	+ 9.4	29.25	+17.2	36.33	+23.8	45.91	+25.7
7	19.99	0.3	22.04	3.0	25.01	9.6	29.43	17.4	36.61	24.0	46.24	25.7
8	20.04	0.3	22.12	3.2	25.13	9.9	29.62	17.7	36.90	24.1	46.56	25.6
9	20.10	0.3	22.19	3.3	25.25	10.1	29.80	17.9	37.19	24.3	46.87	25.5
10	20.16	0.3	22.28	3.5	25.37	10.4	29.99	18.2	37.48	24.4	47.19	25.4
11	20.22	+ 0.3	22.36	+ 3.7	25.50	+10.7	30.19	+18.4	37.78	+24.6	47.49	+25.4
12	20.28	0.4	22.44	3.9	25.62	10.9	30.38	18.7	38.08	24.7	47.80	25.3
13	20.33	0.4	22.52	4.0	25.75	11.2	30.58	18.9	38.38	24.8	48.09	25.2
14	20.39	0.4	22.61	4.2	25.88	11.4	30.78	19.1	38.69	24.9	48.38	25.0
15	20.45	0.5	22.69	4.4	26.01	11.7	30.99	19.4	39.00	25.0	48.66	24.9
16	20.52	+ 0.5	22.78	+ 4.6	26.14	+12.0	31.19	+19.6	39.31	+25.1	48.93	+24.8
17	20.58	0.6	22.87	4.8	26.27	12.2	31.40	19.8	39.62	25.2	49.20	24.7
18	20.64	0.7	22.96	5.0	26.41	12.5	31.62	20.1	39.94	25.3	49.45	24.5
19	20.70	0.7	23.05	5.2	26.55	12.7	31.83	20.3	40.26	25.4	49.70	24.4
20	20.76	0.8	23.14	5.4	26.69	13.0	32.06	20.5	40.59	25.5	49.93	24.2
21	20.83	+ 0.9	23.23	+ 5.6	26.83	+13.3	32.28	+20.8	40.91	+25.6	50.15	+24.0
22	20.89	1.0	23.32	5.9	26.98	13.5	32.51	21.0	41.24	25.6	50.36	23.9
23	20.96	1.1	23.42	6.1	27.12	13.8	32.74	21.2	41.57	25.7	50.56	23.7
24	21.03	1.1	23.52	6.3	27.27	14.1	32.97	21.4	41.90	25.7	50.74	23.5
25	21.09	1.2	23.61	6.5	27.43	14.3	33.21	21.6	42.24	25.8	50.91	23.3
26	21.16	+ 1.4	23.71	+ 6.8	27.58	+14.6	33.45	+21.8	42.57	+25.8	51.06	+23.1
27	21.23	1.5	23.81	7.0	27.74	14.8	33.69	22.0	42.91	25.8	51.20	22.9
28	21.30	1.6	23.91	7.2	27.89	15.1	33.94	22.2	43.24	25.9	51.33	22.7
29	21.37	1.7	24.02	7.4	28.06	15.4	34.19	22.4	43.58	25.9	51.43	22.5
30	21.44	1.8	24.12	7.7	28.22	+15.6	34.45	22.6	43.92	+25.9	51.52	22.3
31	21.51	+ 2.0	24.23	+ 7.9			34.71	+22.8			51.59	+22.1

Apparent distance of satellite: $s = Fa/\Delta$

Position angle of satellite: $p = p_1 + p_2$

The differences of right ascension and declination, in the sense "satellite minus primary," are approximately

$$\Delta\alpha = s \sin p \sec(\delta + \Delta\delta)$$

$$\Delta\delta = s \cos p$$

APPARENT ORBITS OF SATELLITES I-V AT FEBRUARY 29

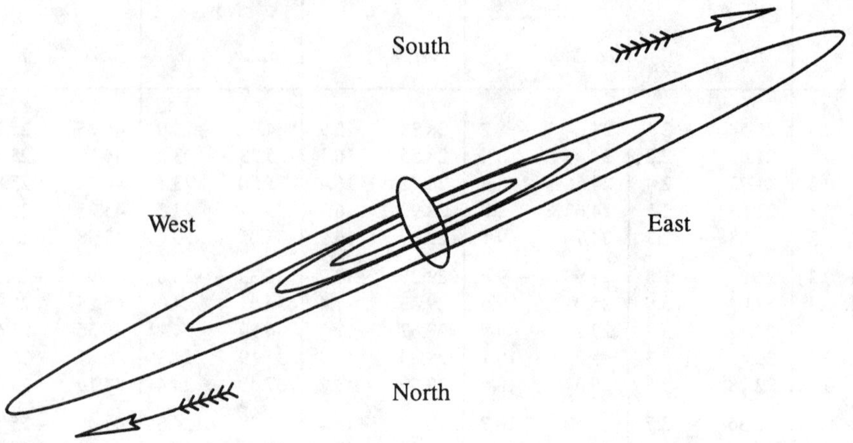

Orbits elongated in ratio of 3 to 1 in direction of minor axes.

	Name	Mean Synodic Period			Name	Sidereal Period
		d h m s	d			d
V	Amalthea	0 11 57 27.619 =	0.498 236 33	XIII	Leda	238.72
I	Io	1 18 28 35.946 =	1.769 860 49	X	Lysithea	259.22
II	Europa	3 13 17 53.736 =	3.554 094 17	XII	Ananke	631
III	Ganymede	7 03 59 35.856 =	7.166 387 22	XI	Carme	692
IV	Callisto	16 18 05 06.916 =	16.753 552 27	VIII	Pasiphae	735
VI	Himalia		266.00	IX	Sinope	758
VII	Elara		276.67			

SATELLITE V

UNIVERSAL TIME OF EVERY TWENTIETH GREATEST EASTERN ELONGATION

	d h		d h		d h		d h		d h
Jan.	0 23.7	Mar.	20 16.4	June	8 09.6	Aug.	27 03.1	Nov.	14 20.6
	10 22.8		30 15.5		18 08.7	Sept.	6 02.3		24 19.8
	20 21.9	Apr.	9 14.6		28 07.9		16 01.5	Dec.	4 19.0
	30 21.0		19 13.8	July	8 07.1		26 00.7		14 18.1
Feb.	9 20.1		29 12.9		18 06.3	Oct.	5 23.9		24 17.3
	19 19.2	May	9 12.0		28 05.5		15 23.1		34 16.4
	29 18.2		19 11.2	Aug.	7 04.7		25 22.3		
Mar.	10 17.3		29 10.4		17 03.9	Nov.	4 21.5		

MULTIPLES OF THE MEAN SYNODIC PERIOD

	d	h		d	h		d	h		d	h
1...........	0	12.0	6...........	2	23.7	11...........	5	11.5	16...........	7	23.3
2...........	0	23.9	7...........	3	11.7	12...........	5	23.5	17...........	8	11.3
3...........	1	11.9	8...........	3	23.7	13...........	6	11.4	18...........	8	23.2
4...........	1	23.8	9...........	4	11.6	14...........	6	23.4	19...........	9	11.2
5...........	2	11.8	10...........	4	23.6	15...........	7	11.4	20...........	9	23.2

DIFFERENTIAL COORDINATES FOR 0^h U.T.

Date	Satellite VI		Satellite VII		Date	Satellite VI		Satellite VII	
	Δα	Δδ	Δα	Δδ		Δα	Δδ	Δα	Δδ
	m s	′	m s	′		m s	′	m s	′
Jan. − 1	+ 3 18	− 0.2	− 2 58	+ 11.2	July 1	− 0 25	+ 25.2	− 3 15	− 8.6
3	3 28	2.9	2 44	12.2	5	− 0 11	24.5	3 20	7.8
7	3 37	5.6	2 28	13.0	9	+ 0 02	23.8	3 24	6.9
11	3 44	8.4	2 10	13.8	13	0 15	22.8	3 27	6.0
15	3 50	11.1	1 50	14.4	17	0 28	21.7	3 29	5.0
19	+ 3 54	− 13.8	− 1 29	+ 14.8	21	+ 0 40	+ 20.5	− 3 30	− 4.0
23	3 56	16.5	1 07	15.1	25	0 52	19.2	3 31	3.0
27	3 56	19.0	0 43	15.3	29	1 03	17.8	3 31	1.9
31	3 53	21.4	− 0 18	15.2	Aug. 2	1 13	16.3	3 30	− 0.8
Feb. 4	3 49	23.6	+ 0 08	15.0	6	1 23	14.7	3 28	+ 0.3
8	+ 3 42	− 25.6	+ 0 34	+ 14.5	10	+ 1 32	+ 13.0	− 3 26	+ 1.4
12	3 32	27.3	1 00	13.9	14	1 41	11.3	3 22	2.5
16	3 19	28.7	1 25	13.1	18	1 49	9.4	3 19	3.6
20	3 04	29.8	1 49	12.1	22	1 56	7.6	3 14	4.7
24	2 46	30.5	2 11	10.9	26	2 03	5.7	3 09	5.8
28	+ 2 26	− 30.8	+ 2 31	+ 9.6	30	+ 2 09	+ 3.7	− 3 03	+ 6.9
Mar. 3	2 04	30.7	2 48	8.2	Sept. 3	2 15	+ 1.7	2 56	8.0
7	1 39	30.2	3 01	6.6	7	2 20	− 0.3	2 49	9.0
11	1 13	29.2	3 10	5.0	11	2 24	2.3	2 40	10.1
15	0 46	27.8	3 15	3.3	15	2 28	4.4	2 31	11.0
19	+ 0 18	− 26.0	+ 3 16	+ 1.7	19	+ 2 30	− 6.4	− 2 22	+ 11.9
23	− 0 10	23.8	3 13	+ 0.0	23	2 33	8.5	2 11	12.8
27	0 37	21.3	3 05	− 1.6	27	2 34	10.5	2 00	13.5
31	1 03	18.5	2 54	3.1	Oct. 1	2 35	12.4	1 48	14.1
Apr. 4	1 28	15.4	2 40	4.6	5	2 35	14.4	1 35	14.7
8	− 1 50	− 12.1	+ 2 24	− 5.9	9	+ 2 34	− 16.2	− 1 21	+ 15.1
12	2 09	8.7	2 05	7.2	13	2 32	18.0	1 07	15.4
16	2 26	5.3	1 44	8.3	17	2 30	19.7	0 52	15.5
20	2 40	− 1.8	1 22	9.3	21	2 26	21.3	0 36	15.5
24	2 50	+ 1.6	1 00	10.2	25	2 22	22.7	0 20	15.2
28	− 2 57	+ 4.9	+ 0 38	− 10.9	29	+ 2 16	− 24.0	− 0 03	+ 14.7
May 2	3 00	8.0	+ 0 16	11.6	Nov. 2	2 09	25.1	+ 0 15	14.0
6	3 01	10.9	− 0 06	12.1	6	2 02	25.9	0 32	13.1
10	2 59	13.6	0 27	12.4	10	1 53	26.5	0 49	12.0
14	2 54	16.0	0 47	12.7	14	1 43	26.9	1 06	10.5
18	− 2 47	+ 18.2	− 1 06	− 12.9	18	+ 1 32	− 26.9	+ 1 22	+ 8.9
22	2 38	20.1	1 24	12.9	22	1 19	26.6	1 37	7.0
26	2 28	21.7	1 40	12.9	26	1 06	26.0	1 51	5.0
30	2 16	23.1	1 56	12.7	30	0 51	24.9	2 03	2.8
June 3	2 04	24.1	2 10	12.5	Dec. 4	0 36	23.5	2 14	+ 0.4
7	− 1 50	+ 25.0	− 2 23	− 12.1	8	+ 0 19	− 21.7	+ 2 23	− 2.0
11	1 37	25.5	2 34	11.7	12	+ 0 02	19.4	2 30	4.5
15	1 22	25.9	2 45	11.2	16	− 0 15	16.8	2 34	7.0
19	1 08	26.0	2 54	10.7	20	0 33	13.8	2 37	9.5
23	0 53	25.9	3 02	10.0	24	0 50	10.4	2 37	11.9
27	− 0 39	+ 25.6	− 3 09	− 9.3	28	− 1 07	− 6.8	+ 2 36	− 14.3
July 1	− 0 25	+ 25.2	− 3 15	− 8.6	32	− 1 23	− 3.0	+ 2 32	− 16.5

Differential coordinates are given in the sense "satellite minus planet."

DIFFERENTIAL COORDINATES FOR 0^h U.T.

Date		Satellite VIII		Satellite IX		Satellite X	
		$\Delta\alpha$	$\Delta\delta$	$\Delta\alpha$	$\Delta\delta$	$\Delta\alpha$	$\Delta\delta$
		m s	′	m s	′	m s	′
Jan.	− 1	+ 6 34	+ 0.9	− 5 56	+ 51.1	+ 2 20	− 43.2
	9	6 25	5.3	6 26	51.1	1 59	41.7
	19	6 11	9.9	6 53	50.5	1 29	37.5
	29	5 51	14.3	7 18	49.3	0 51	30.6
Feb.	8	5 25	18.5	7 38	47.3	+ 0 05	21.1
	18	+ 4 53	+ 22.3	− 7 55	+ 44.8	− 0 44	− 9.8
	28	4 15	25.4	8 06	41.7	1 30	+ 2.3
Mar.	9	3 31	27.7	8 14	38.1	2 09	14.1
	19	2 43	29.1	8 16	34.3	2 34	24.2
	29	1 51	29.6	8 14	30.3	2 44	31.7
Apr.	8	+ 0 57	+ 29.2	− 8 08	+ 26.3	− 2 38	+ 36.0
	18	+ 0 03	28.0	7 58	22.5	2 18	37.2
	28	− 0 49	26.2	7 45	18.8	1 48	35.5
May	8	1 39	23.9	7 28	15.3	1 12	31.5
	18	2 24	21.3	7 09	11.9	− 0 35	25.7
	28	− 3 04	+ 18.5	− 6 47	+ 8.7	+ 0 01	+ 18.9
June	7	3 38	15.6	6 22	5.5	0 34	11.5
	17	4 07	12.6	5 54	+ 2.3	1 01	+ 3.9
	27	4 29	9.5	5 24	− 0.9	1 24	− 3.5
July	7	4 46	6.4	4 51	4.2	1 42	10.4
	17	− 4 57	+ 3.2	− 4 15	− 7.5	+ 1 54	− 16.7
	27	5 03	− 0.1	3 38	10.9	2 02	22.3
Aug.	6	5 04	3.5	2 58	14.3	2 06	26.8
	16	5 01	7.1	2 17	17.7	2 07	30.4
	26	4 53	10.9	1 34	21.1	2 04	32.9
Sept.	5	− 4 42	− 14.8	− 0 51	− 24.3	+ 1 58	− 34.2
	15	4 28	18.9	− 0 07	27.2	1 50	34.3
	25	4 10	23.2	+ 0 36	29.8	1 39	33.0
Oct.	5	3 50	27.7	1 18	32.0	1 25	30.2
	15	3 27	32.3	1 57	33.6	1 09	26.0
	25	− 3 03	− 37.1	+ 2 35	− 34.7	+ 0 50	− 20.3
Nov.	4	2 36	42.0	3 09	35.0	0 28	13.3
	14	2 07	47.0	3 39	34.7	+ 0 04	− 5.2
	24	1 36	52.0	4 05	33.6	− 0 21	+ 3.7
Dec.	4	1 05	57.1	4 27	31.8	0 47	12.7
	14	− 0 32	− 62.1	+ 4 45	− 29.2	− 1 12	+ 21.2
	24	+ 0 02	67.1	4 58	26.1	1 34	28.6
	34	+ 0 36	− 72.0	+ 5 07	− 22.3	− 1 51	+ 34.2

Differential coordinates are given in the sense "satellite minus planet."

DIFFERENTIAL COORDINATES FOR 0^h U.T.

Date		Satellite XI Δα	Satellite XI Δδ	Satellite XII Δα	Satellite XII Δδ	Satellite XIII Δα	Satellite XIII Δδ
		m s	′	m s	′	m s	′
Jan.	− 1	+ 6 43	− 63.6	+ 4 17	+ 7.3	+ 2 49	+ 13.4
	9	6 34	65.0	4 21	7.1	2 15	18.1
	19	6 23	66.1	4 16	6.6	1 32	22.3
	29	6 10	66.8	4 03	5.6	+ 0 41	25.8
Feb.	8	5 55	66.8	3 42	4.3	− 0 15	28.1
	18	+ 5 37	− 66.3	+ 3 17	+ 2.5	− 1 13	+ 28.8
	28	5 15	65.0	2 47	+ 0.3	2 06	27.4
Mar.	9	4 50	63.0	2 14	− 2.0	2 48	23.9
	19	4 22	60.3	1 40	4.3	3 14	18.3
	29	3 51	56.8	1 04	6.6	3 20	11.1
Apr.	8	+ 3 16	− 52.7	+ 0 28	− 8.6	− 3 03	+ 3.0
	18	2 40	48.2	− 0 08	10.4	2 26	− 5.0
	28	2 01	43.2	0 44	11.7	1 34	11.9
May	8	1 22	37.8	1 19	12.6	− 0 34	16.9
	18	0 42	32.3	1 54	13.2	+ 0 26	19.5
	28	+ 0 02	− 26.5	− 2 27	− 13.4	+ 1 19	− 19.9
June	7	− 0 36	20.6	2 59	13.3	2 02	18.5
	17	1 13	14.6	3 29	13.0	2 33	15.9
	27	1 47	8.7	3 57	12.4	2 54	12.4
July	7	2 19	− 2.8	4 24	11.8	3 05	8.5
	17	− 2 46	+ 2.8	− 4 48	− 11.0	+ 3 08	− 4.6
	27	3 09	8.2	5 11	10.3	3 04	− 0.8
Aug.	6	3 26	13.1	5 31	9.5	2 54	+ 2.8
	16	3 38	17.6	5 50	8.9	2 40	6.0
	26	3 44	21.5	6 07	8.3	2 21	8.8
Sept.	5	− 3 44	+ 24.7	− 6 21	− 8.0	+ 1 59	+ 11.1
	15	3 37	27.1	6 34	7.8	1 33	12.8
	25	3 24	28.9	6 45	7.8	1 05	14.0
Oct.	5	3 05	29.8	6 54	8.1	0 34	14.6
	15	2 41	30.1	7 01	8.6	+ 0 02	14.5
	25	− 2 12	+ 29.7	− 7 06	− 9.3	− 0 33	+ 13.7
Nov.	4	1 39	28.7	7 09	10.3	1 06	12.1
	14	1 02	27.2	7 10	11.6	1 38	9.9
	24	− 0 23	25.3	7 10	13.0	2 06	7.0
Dec.	4	+ 0 20	22.9	7 07	14.6	2 24	+ 3.5
	14	+ 1 03	+ 20.4	− 7 03	− 16.3	− 2 31	− 0.2
	24	1 48	17.6	6 57	18.1	2 23	3.8
	34	+ 2 33	+ 14.6	− 6 48	− 19.8	− 1 58	− 7.1

Differential coordinates are given in the sense "satellite minus planet."

UNIVERSAL TIME OF SUPERIOR GEOCENTRIC CONJUNCTION

SATELLITE I

	d h m		d h m		d h m		d h m
Jan.	1 15 08	Mar.	21 04 53	June	8 19 44	Oct.	16 02 13
	3 09 35		22 23 20		10 14 13		17 20 43
	5 04 02		24 17 46		12 08 42		19 15 13
	6 22 30		26 12 13		14 03 12		21 09 43
	8 16 57		28 06 39		15 21 41		23 04 13
	10 11 24		30 01 06		17 16 10		24 22 43
	12 05 51		31 19 32		19 10 40		26 17 13
	14 00 18	Apr.	2 13 59		21 05 09		28 11 42
	15 18 44		4 08 26		22 23 39		30 06 12
	17 13 11		6 02 53		24 18 08	Nov.	1 00 42
	19 07 38		7 21 20		26 12 38		2 19 12
	21 02 04		9 15 46		28 07 08		4 13 41
	22 20 31		11 10 14		30 01 37		6 08 11
	24 14 58		13 04 41	July	1 20 07		8 02 41
	26 09 24		14 23 08		3 14 37		9 21 11
	28 03 50		16 17 35		5 09 07		11 15 40
	29 22 17		18 12 02		7 03 36		13 10 10
	31 16 43		20 06 30		8 22 06		15 04 39
Feb.	2 11 09		22 00 57		10 16 36		16 23 09
	4 05 36		23 19 25		12 11 06		18 17 38
	6 00 02		25 13 52		14 05 36		20 12 08
	7 18 28		27 08 20		16 00 06		22 06 37
	9 12 54		29 02 48		17 18 36		24 01 07
	11 07 20		30 21 15		19 13 06		25 19 36
	13 01 46	May	2 15 43		21 07 36		27 14 05
	14 20 12		4 10 11		23 02 06		29 08 35
	16 14 38		6 04 39		24 20 36	Dec.	1 03 04
	18 09 04		7 23 07		26 15 06		2 21 33
	20 03 30		9 17 36		28 09 36		4 16 02
	21 21 56		11 12 04		30 04 06		6 10 31
	23 16 22		13 06 32		31 22 36		8 05 00
	25 10 48		15 01 01	Aug.	2 17 06		9 23 29
	27 05 14		16 19 29		4 11 36		11 17 58
	28 23 40		18 13 57		6 06 06		13 12 27
Mar.	1 18 06		20 08 26		8 00 36		15 06 56
	3 12 32		22 02 55		9 19 07		17 01 25
	5 06 58		23 21 23		11 13 37		18 19 54
	7 01 24		25 15 52		13 08 07		20 14 22
	8 19 50		27 10 21		15 02 37		22 08 51
	10 14 16		29 04 50		16 21 07		24 03 20
	12 08 42		30 23 19		18 15 38		25 21 48
	14 03 08	June	1 17 48		20 10 08		27 16 17
	15 21 35		3 12 17		22 04 38		29 10 45
	17 16 01		5 06 46		.. .. ..		31 05 13
	19 10 27		7 01 15	Oct.	14 07 43		

UNIVERSAL TIME OF SUPERIOR GEOCENTRIC CONJUNCTION

SATELLITE II

	d h m		d h m		d h m		d h m
Jan.	0 10 37	Mar.	22 00 49	June	11 17 02	Oct.	17 19 14
	3 23 50		25 13 57		15 06 21		21 08 39
	7 13 02		29 03 06		18 19 41		24 22 04
	11 02 13	Apr.	1 16 14		22 09 02		28 11 29
	14 15 24		5 05 24		25 22 22	Nov.	1 00 52
	18 04 34		8 18 34		29 11 43		4 14 17
	21 17 44		12 07 44	July	3 01 05		8 03 40
	25 06 53		15 20 55		6 14 27		11 17 04
	28 20 02		19 10 06		10 03 49		15 06 27
Feb.	1 09 10		22 23 18		13 17 11		18 19 51
	4 22 18		26 12 31		17 06 34		22 09 13
	8 11 26		30 01 44		20 19 57		25 22 36
	12 00 33	May	3 14 57		24 09 20		29 11 58
	15 13 40		7 04 11		27 22 43	Dec.	3 01 20
	19 02 47		10 17 26		31 12 07		6 14 41
	22 15 54		14 06 41	Aug.	4 01 31		10 04 03
	26 05 00		17 19 57		7 14 54		13 17 23
	29 18 07		21 09 14		11 04 19		17 06 44
Mar.	4 07 14		24 22 30		14 17 43		20 20 03
	7 20 20		28 11 48		18 07 08		24 09 22
	11 09 27	June	1 01 06		21 20 32		27 22 41
	14 22 34		4 14 24		.. .. ..		31 11 59
	18 11 42		8 03 42	Oct.	14 05 50		

SATELLITE III

	d h m		d h m		d h m		d h m
Jan.	3 08 56	Mar.	29 01 25	June	22 23 15	Oct.	30 06 13
	10 12 33	Apr.	5 04 50		30 03 28	Nov.	6 10 34
	17 16 06		12 08 21	July	7 07 43		13 14 54
	24 19 34		19 11 55		14 12 01		20 19 12
	31 22 58		26 15 34		21 16 21		27 23 27
Feb.	8 02 18	May	3 19 17		28 20 42	Dec.	5 03 40
	15 05 37		10 23 05	Aug.	5 01 06		12 07 50
	22 08 54		18 02 57		12 05 30		19 11 58
	29 12 10		25 06 53		19 09 54		26 16 01
Mar.	7 15 27	June	1 10 53		.. .. ..		
	14 18 44		8 14 57	Oct.	15 21 25		
	21 22 03		15 19 05		23 01 49		

SATELLITE IV

	d h m		d h m		d h m		d h m
Jan.	6 21 48	Mar.	29 22 57	June	21 12 17	Oct.	17 10 09
	23 13 22	Apr.	15 14 12	July	8 07 42	Nov.	3 06 29
Feb.	9 04 07	May	2 06 22		25 03 35		20 02 29
	25 18 19		18 23 30	Aug.	10 23 50	Dec.	6 22 00
Mar.	13 08 26	June	4 17 31		.. .. ..		23 16 52

DYNAMICAL TIME OF GEOCENTRIC PHENOMENA

JANUARY

d	h m	
0	6 58	II.Ec.D.
	11 59	II.Oc.R.
	15 46	I.Sh.I.
	16 52	I.Tr.I.
	18 02	I.Sh.E.
	19 07	I.Tr.E.
1	12 52	I.Ec.D.
	16 16	I.Oc.R.
2	1 08	II.Sh.I.
	3 23	II.Tr.I.
	3 56	II.Sh.E.
	6 07	II.Tr.E.
	10 14	I.Sh.I.
	11 20	I.Tr.I.
	12 30	I.Sh.E.
	13 34	I.Tr.E.
3	2 42	III.Ec.D.
	6 16	III.Ec.R.
	7 13	III.Oc.D.
	7 21	I.Ec.D.
	10 39	III.Oc.R.
	10 44	I.Oc.R.
	20 15	II.Ec.D.
4	1 11	II.Oc.R.
	4 42	I.Sh.I.
	5 47	I.Tr.I.
	6 58	I.Sh.E.
	8 02	I.Tr.E.
5	1 49	I.Ec.D.
	5 11	I.Oc.R.
	14 25	II.Sh.I.
	16 37	II.Tr.I.
	17 13	II.Sh.E.
	19 20	II.Tr.E.
	23 10	I.Sh.I.
6	0 14	I.Tr.I.
	1 26	I.Sh.E.
	2 29	I.Tr.E.
	9 30	IV.Ec.D.
	13 48	IV.Ec.R.
	16 38	III.Sh.I.
	20 04	IV.Oc.D.
	20 10	III.Sh.E.
	20 17	I.Ec.D.
	20 57	III.Tr.I.
	23 32	IV.Oc.R.
	23 38	I.Oc.R.
7	0 21	III.Tr.E.
	9 32	II.Ec.D.
	14 23	II.Oc.R.
	17 39	I.Sh.I.
	18 41	I.Tr.I.
	19 55	I.Sh.E.

d	h m	
7	20 56	I.Tr.E.
8	14 46	I.Ec.D.
	18 05	I.Oc.R.
9	3 43	II.Sh.I.
	5 49	II.Tr.I.
	6 31	II.Sh.E.
	8 32	II.Tr.E.
	12 07	I.Sh.I.
	13 08	I.Tr.I.
	14 23	I.Sh.E.
	15 23	I.Tr.E.
10	6 39	III.Ec.D.
	9 14	I.Ec.D.
	10 14	III.Ec.R.
	10 50	III.Oc.D.
	12 32	I.Oc.R.
	14 16	III.Oc.R.
	22 49	II.Ec.D.
11	3 35	II.Oc.R.
	6 35	I.Sh.I.
	7 35	I.Tr.I.
	8 51	I.Sh.E.
	9 50	I.Tr.E.
12	3 42	I.Ec.D.
	6 59	I.Oc.R.
	17 00	II.Sh.I.
	19 01	II.Tr.I.
	19 48	II.Sh.E.
	21 45	II.Tr.E.
13	1 04	I.Sh.I.
	2 02	I.Tr.I.
	3 20	I.Sh.E.
	4 17	I.Tr.E.
	20 35	III.Sh.I.
	22 11	I.Ec.D.
14	0 07	III.Sh.E.
	0 32	III.Tr.I.
	1 26	I.Oc.R.
	3 55	III.Tr.E.
	12 06	II.Ec.D.
	16 45	II.Oc.R.
	18 55	IV.Sh.I.
	19 32	I.Sh.I.
	20 29	I.Tr.I.
	21 48	I.Sh.E.
	22 43	I.Tr.E.
	23 05	IV.Sh.E.
15	4 08	IV.Tr.I.
	7 29	IV.Tr.E.
	16 39	I.Ec.D.
	19 53	I.Oc.R.

d	h m	
16	6 18	II.Sh.I.
	8 12	II.Tr.I.
	9 06	II.Sh.E.
	10 56	II.Tr.E.
	14 00	I.Sh.I.
	14 55	I.Tr.I.
	16 16	I.Sh.E.
	17 10	I.Tr.E.
17	10 38	III.Ec.D.
	11 08	I.Ec.D.
	14 12	III.Ec.R.
	14 19	I.Oc.R.
	14 24	III.Oc.D.
	17 48	III.Oc.R.
18	1 23	II.Ec.D.
	5 56	II.Oc.R.
	8 28	I.Sh.I.
	9 22	I.Tr.I.
	10 44	I.Sh.E.
	11 37	I.Tr.E.
19	5 36	I.Ec.D.
	8 46	I.Oc.R.
	19 35	II.Sh.J.
	21 23	II.Tr.I.
	22 24	II.Sh.E.
20	0 07	II.Tr.E.
	2 57	I.Sh.I.
	3 48	I.Tr.I.
	5 13	I.Sh.E.
	6 03	I.Tr.E.
21	0 04	I.Ec.D.
	0 32	III.Sh.I.
	3 13	I.Oc.R.
	4 02	III.Tr.I.
	4 04	III.Sh.E.
	7 25	III.Tr.E.
	14 40	II.Ec.D.
	19 05	II.Oc.R.
	21 25	I.Sh.I.
	22 15	I.Tr.I.
	23 41	I.Sh.E.
22	0 30	I.Tr.E.
	18 33	I.Ec.D.
	21 39	I.Oc.R.
23	3 30	IV.Ec.D.
	7 43	IV.Ec.R.
	8 53	II.Sh.I.
	10 33	II.Tr.I.
	11 41	II.Sh.E.
	11 42	IV.Oc.D.
	13 17	II.Tr.E.
	15 04	IV.Oc.R.
	15 53	I.Sh.I.

d	h m	
23	16 41	I.Tr.I.
	18 09	I.Sh.E.
	18 56	I.Tr.E.
24	13 01	I.Ec.D.
	14 36	III.Ec.D.
	16 06	I.Oc.R.
	21 16	III.Oc.R.
25	3 57	II.Ec.D.
	8 15	II.Oc.R.
	10 22	I.Sh.I.
	11 08	I.Tr.I.
	12 38	I.Sh.E.
	13 23	I.Tr.E.
26	7 30	I.Ec.D.
	10 32	I.Oc.R.
	22 11	II.Sh.I.
	23 43	II.Tr.I.
27	0 59	II.Sh.E.
	2 27	II.Tr.E.
	4 50	I.Sh.I.
	5 34	I.Tr.I.
	7 06	I.Sh.E.
	7 49	I.Tr.E.
28	1 58	I.Ec.D.
	4 30	III.Sh.I.
	4 59	I.Oc.R.
	7 29	III.Tr.I.
	8 02	III.Sh.E.
	10 52	III.Tr.E.
	17 14	II.Ec.D.
	21 23	II.Oc.R.
	23 18	I.Sh.I.
29	0 01	I.Tr.I.
	1 34	I.Sh.E.
	2 16	I.Tr.E.
	20 27	I.Ec.D.
	23 25	I.Oc.R.
30	11 28	II.Sh.I.
	12 53	II.Tr.I.
	14 17	II.Sh.E.
	15 36	II.Tr.E.
	17 46	I.Sh.I.
	18 27	I.Tr.I.
	20 03	I.Sh.E.
	20 42	I.Tr.E.
31	12 53	IV.Sh.I.
	14 55	I.Ec.D.
	16 58	IV.Sh.E.
	17 51	I.Oc.R.
	18 34	III.Ec.D.
	19 20	IV.Tr.I.
	22 38	IV.Tr.E.

I. Jan. 15	II. Jan. 14	III. Jan. 17	IV. Jan. 23
$x_1 = -1.8,\ y_1 = -0.1$	$x_1 = -2.3,\ y_1 = -0.3$	$x_1 = -3.0,\ y_1 = -0.3$	$x_1 = -4.0,\ y_1 = -0.7$
		$x_2 = -1.0,\ y_2 = -0.3$	$x_2 = -2.3,\ y_2 = -0.7$

NOTE.—I. denotes ingress; E., egress; D., disappearance; R., reappearance; Ec., eclipse; Oc., occultation; Tr., transit of the satellite; Sh., transit of the shadow.

CONFIGURATIONS OF SATELLITES I–IV FOR JANUARY

UNIVERSAL TIME

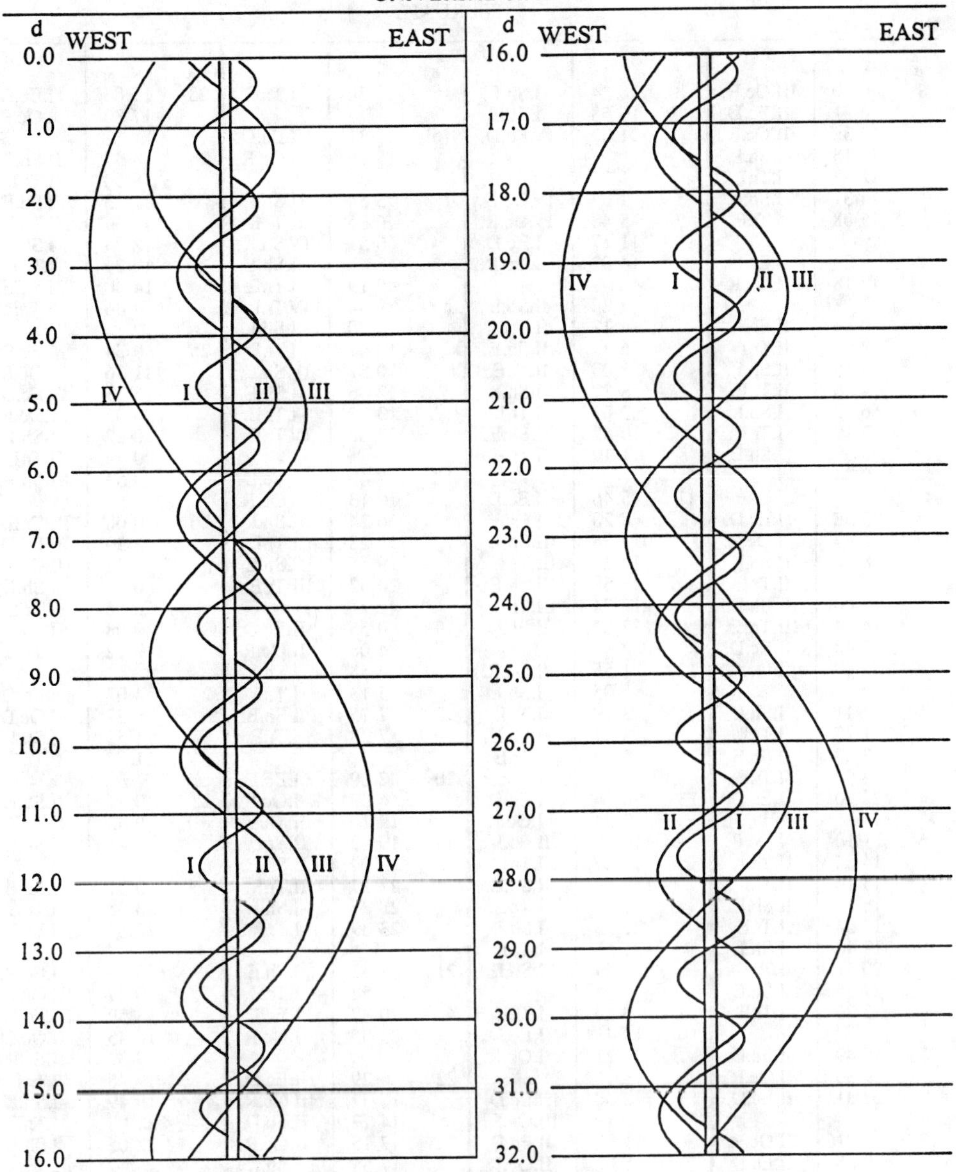

PHASES OF THE ECLIPSES

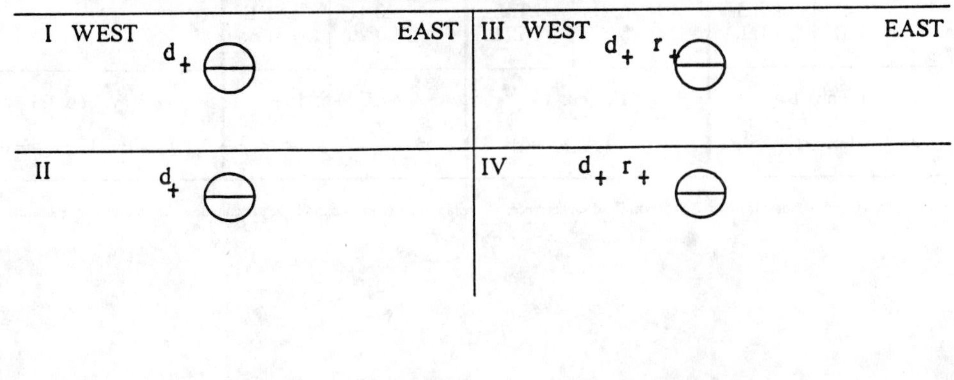

DYNAMICAL TIME OF GEOCENTRIC PHENOMENA

FEBRUARY

d	h m	
1	0 40	III.Oc.R.
	6 31	II.Ec.D.
	10 32	II.Oc.R.
	12 15	I.Sh.I.
	12 53	I.Tr.I.
	14 31	I.Sh.E.
	15 08	I.Tr.E.
2	9 23	I.Ec.D.
	12 18	I.Oc.R.
3	0 46	II.Sh.I.
	2 02	II.Tr.I.
	3 35	II.Sh.E.
	4 46	II.Tr.E.
	6 43	I.Sh.I.
	7 19	I.Tr.I.
	8 59	I.Sh.E.
	9 34	I.Tr.E.
4	3 52	I.Ec.D.
	6 44	I.Oc.R.
	8 28	III.Sh.I.
	10 51	III.Tr.I.
	12 00	III.Sh.E.
	14 14	III.Tr.E.
	19 48	II.Ec.D.
	23 40	II.Oc.R.
5	1 11	I.Sh.I.
	1 45	I.Tr.I.
	3 28	I.Sh.E.
	4 00	I.Tr.E.
	22 20	I.Ec.D.
6	1 10	I.Oc.R.
	14 04	II.Sh.I.
	15 10	II.Tr.I.
	16 52	II.Sh.E.
	17 54	II.Tr.E.
	19 40	I.Sh.I.
	20 11	I.Tr.I.
	21 56	I.Sh.E.
	22 26	I.Tr.E.
7	16 49	I.Ec.D.
	19 36	I.Oc.R.
	22 31	III.Ec.D.
8	4 01	III.Oc.R.
	9 05	II.Ec.D.
	12 47	II.Oc.R.
	14 08	I.Sh.I.
	14 37	I.Tr.I.
8	16 24	I.Sh.E.
	16 53	I.Tr.E.
	21 32	IV.Ec.D.
9	1 39	IV.Ec.R.
	2 27	IV.Oc.D.
	5 48	IV.Oc.R.
	11 17	I.Ec.D.
	14 02	I.Oc.R.
10	3 22	II.Sh.I.
	4 19	II.Tr.I.
	6 11	II.Sh.E.
	7 03	II.Tr.E.
	8 37	I.Sh.I.
	9 03	I.Tr.I.
	10 53	I.Sh.E.
	11 19	I.Tr.E.
11	5 46	I.Ec.D.
	8 28	I.Oc.R.
	12 26	III.Sh.I.
	14 11	III.Tr.I.
	15 58	III.Sh.E.
	17 34	III.Tr.E.
	22 22	II.Ec.D.
12	1 55	II.Oc.R.
	3 05	I.Sh.I.
	3 29	I.Tr.I.
	5 21	I.Sh.E.
	5 45	I.Tr.E.
13	0 14	I.Ec.D.
	2 54	I.Oc.R.
	16 40	II.Sh.I.
	17 27	II.Tr.I.
	19 28	II.Sh.E.
	20 11	II.Tr.E.
	21 33	I.Sh.I.
	21 55	I.Tr.I.
	23 49	I.Sh.E.
14	0 10	I.Tr.E.
	18 43	I.Ec.D.
	21 21	I.Oc.R.
15	2 30	III.Ec.D.
	7 19	III.Oc.R.
	11 38	II.Ec.D.
	15 02	II.Oc.R.
	16 02	I.Sh.I.
	16 21	I.Tr.I.
	18 18	I.Sh.E.
15	18 36	I.Tr.E.
16	13 11	I.Ec.D.
	15 46	I.Oc.R.
17	5 58	II.Sh.I.
	6 35	II.Tr.I.
	6 51	IV.Sh.I.
	8 47	II.Sh.E.
	9 19	II.Tr.E.
	9 48	IV.Tr.I.
	10 30	I.Sh.I.
	10 47	I.Tr.I.
	10 51	IV.Sh.E.
	12 46	I.Sh.E.
	13 02	I.Tr.E.
	13 08	IV.Tr.E.
18	7 40	I.Ec.D.
	10 13	I.Oc.R.
	16 24	III.Sh.I.
	17 28	III.Tr.I.
	19 55	III.Sh.E.
	20 52	III.Tr.E.
19	0 55	II.Ec.D.
	4 08	II.Oc.R.
	4 58	I.Sh.I.
	5 13	I.Tr.I.
	7 15	I.Sh.E.
	7 28	I.Tr.E.
20	2 09	I.Ec.D.
	4 39	I.Oc.R.
	19 16	II.Sh.I.
	19 42	II.Tr.I.
	22 04	II.Sh.E.
	22 27	II.Tr.E.
	23 27	I.Sh.I.
	23 39	I.Tr.I.
21	1 43	I.Sh.E.
	1 54	I.Tr.E.
	20 37	I.Ec.D.
	23 05	I.Oc.R.
22	6 29	III.Ec.D.
	10 37	III.Oc.R.
	14 12	II.Ec.D.
	17 15	II.Oc.R.
	17 55	I.Sh.I.
	18 05	I.Tr.I.
	20 11	I.Sh.E.
	20 20	I.Tr.E.
23	15 06	I.Ec.D.
	17 30	I.Oc.R.
24	8 35	II.Sh.I.
	8 50	II.Tr.I.
	11 23	II.Sh.E.
	11 35	II.Tr.E.
	12 24	I.Sh.I.
	12 31	I.Tr.I.
	14 40	I.Sh.E.
	14 46	I.Tr.E.
25	9 34	I.Ec.D.
	11 56	I.Oc.R.
	15 33	IV.Ec.D.
	20 03	IV.Oc.R.
	20 22	III.Sh.I.
	20 44	III.Tr.I.
	23 53	III.Sh.E.
26	0 08	III.Tr.E.
	3 29	II.Ec.D.
	6 22	II.Oc.R.
	6 52	I.Sh.I.
	6 56	I.Tr.I.
	9 08	I.Sh.E.
	9 12	I.Tr.E.
27	4 03	I.Ec.D.
	6 22	I.Oc.R.
	21 53	II.Sh.I.
	21 57	II.Tr.I.
28	0 41	II.Sh.E.
	0 42	II.Tr.E.
	1 21	I.Sh.I.
	1 22	I.Tr.I.
	3 37	I.Sh.E.
	3 38	I.Tr.E.
	22 31	I.Ec.D.
29	0 49	I.Ec.R.
	10 27	III.Oc.D.
	13 59	III.Ec.R.
	16 45	II.Oc.D.
	19 32	II.Ec.R.
	19 48	I.Tr.I.
	19 49	I.Sh.I.
	22 03	I.Tr.E.
	22 05	I.Sh.E.

I. Feb. 14	II. Feb. 15	III. Feb. 15	IV. Feb. 25
$x_1 = -1.3,\ y_1 = -0.1$	$x_1 = -1.4,\ y_1 = -0.3$	$x_1 = -1.7,\ y_1 = -0.3$	$x_1 = -1.1,\ y_1 = -0.7$

NOTE.—I. denotes ingress; E., egress; D., disappearance; R., reappearance; Ec., eclipse; Oc., occultation; Tr., transit of the satellite; Sh., transit of the shadow.

CONFIGURATIONS OF SATELLITES I–IV FOR FEBRUARY

UNIVERSAL TIME

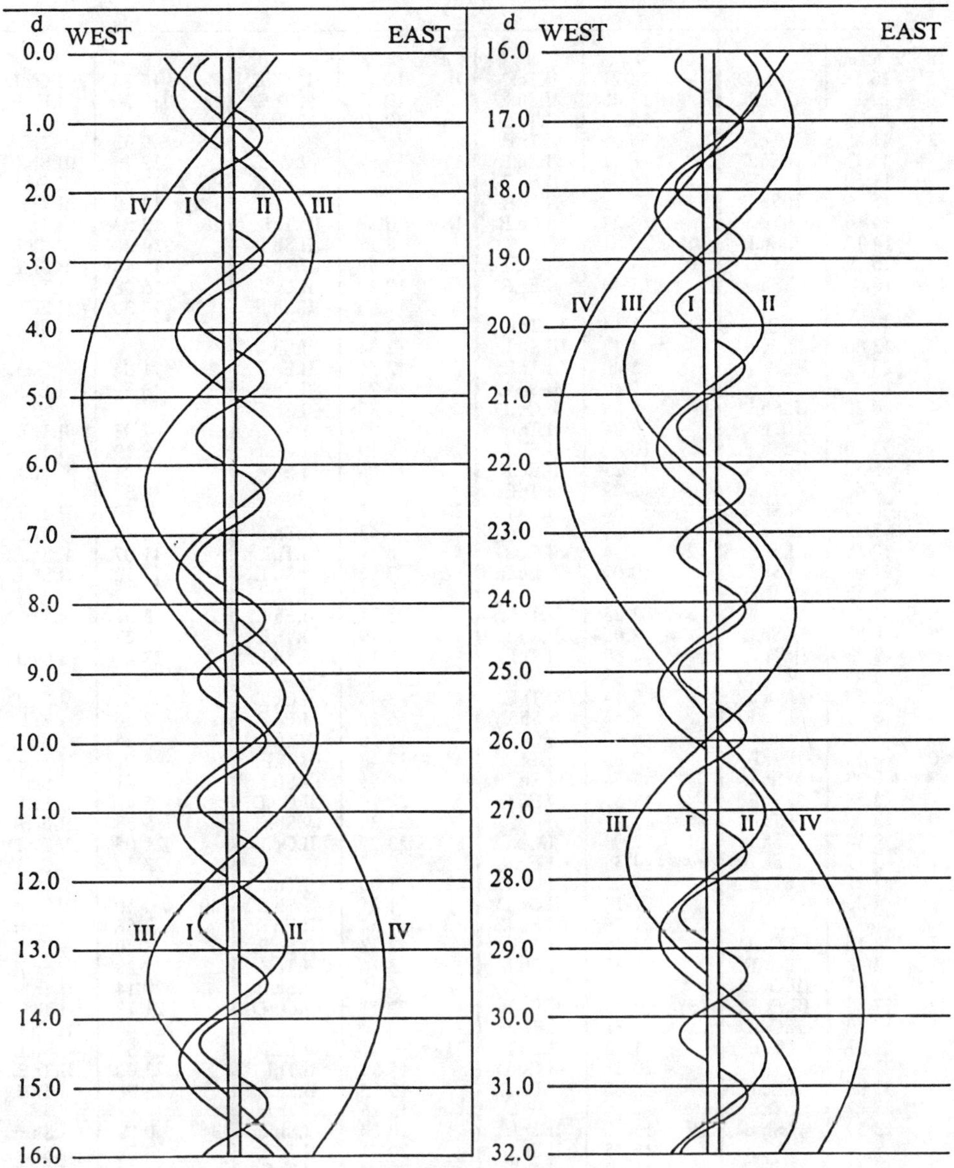

PHASES OF THE ECLIPSES

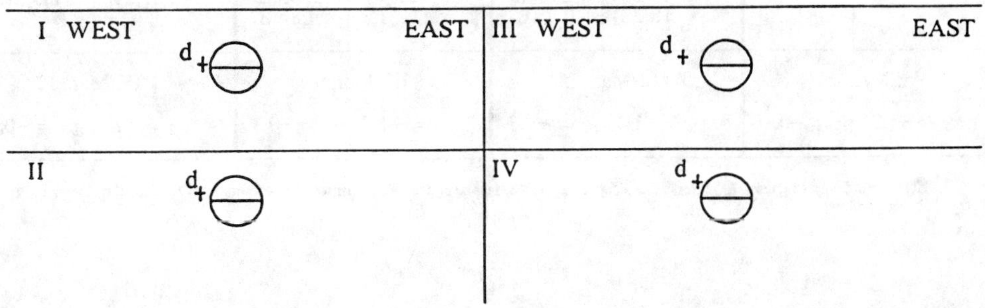

DYNAMICAL TIME OF GEOCENTRIC PHENOMENA

MARCH

d	h m	
1	16 58	I.Oc.D.
	19 17	I.Ec.R.
2	11 05	II.Tr.I.
	11 11	II.Sh.I.
	13 50	II.Tr.E.
	13 59	II.Sh.E.
	14 14	I.Tr.I.
	14 17	I.Sh.I.
	16 29	I.Tr.E.
	16 34	I.Sh.E.
3	11 24	I.Oc.D.
	13 46	I.Ec.R.
	23 59	III.Tr.I.
4	0 20	III.Sh.I.
	3 23	III.Tr.E.
	3 50	III.Sh.E.
	5 52	II.Oc.D.
	8 40	I.Tr.I.
	8 46	I.Sh.I.
	8 49	II.Ec.R.
	10 55	I.Tr.E.
	11 02	I.Sh.E.
	23 55	IV.Tr.I.
5	0 51	IV.Sh.I.
	3 23	IV.Tr.E.
	4 45	IV.Sh.E.
	5 50	I.Oc.D.
	8 15	I.Ec.R.
6	0 12	II.Tr.I.
	0 29	II.Sh.I.
	2 58	II.Tr.E.
	3 06	I.Tr.I.
	3 14	I.Sh.I.
	3 17	II.Sh.E.
	5 21	I.Tr.E.
	5 30	I.Sh.E.
7	0 16	I.Oc.D.
	2 43	I.Ec.R.
	13 44	III.Oc.D.
	17 58	III.Ec.R.
	18 59	II.Oc.D.
	21 32	I.Tr.I.
	21 43	I.Sh.I.
	22 06	II.Ec.R.
	23 47	I.Tr.E.
	23 59	I.Sh.E.
8	18 42	I.Oc.D.
	21 12	I.Ec.R.
9	13 21	II.Tr.I.
	13 48	II.Sh.I.
	15 58	I.Tr.I.
	16 06	II.Tr.E.
	16 11	I.Sh.I.
	16 36	II.Sh.E.
	18 13	I.Tr.E.
	18 27	I.Sh.E.
10	13 08	I.Oc.D.
	15 40	I.Ec.R.
11	3 14	III.Tr.I.
	4 18	III.Sh.I.
	6 40	III.Tr.E.
	7 48	III.Sh.E.
	8 05	II.Oc.D.
	10 24	I.Tr.I.
	10 40	I.Sh.I.
	11 23	II.Ec.R.
	12 39	I.Tr.E.
	12 56	I.Sh.E.
12	7 34	I.Oc.D.
	10 09	I.Ec.R.
13	2 28	II.Tr.I.
	3 06	II.Sh.I.
	4 50	I.Tr.I.
	5 08	I.Sh.I.
	5 14	II.Tr.E.
	5 54	II.Sh.E.
	6 40	IV.Oc.D.
	7 05	I.Tr.E.
	7 24	I.Sh.E.
	13 31	IV.Ec.R.
14	2 00	I.Oc.D.
	4 38	I.Ec.R.
	17 01	III.Oc.D.
	21 12	II.Oc.D.
	21 56	III.Ec.R.
	23 16	I.Tr.I.
	23 37	I.Sh.I.
15	0 39	II.Ec.R.
	1 31	I.Tr.E.
	1 53	I.Sh.E.
	20 26	I.Oc.D.
	23 06	I.Ec.R.
16	15 37	II.Tr.I.
	16 25	II.Sh.I.
	17 42	I.Tr.I.
	18 05	I.Sh.I.
	18 23	II.Tr.E.
16	19 13	II.Sh.E.
	19 57	I.Tr.E.
	20 21	I.Sh.E.
17	14 52	I.Oc.D.
	17 35	I.Ec.R.
18	6 33	III.Tr.I.
	8 17	III.Sh.I.
	9 59	III.Tr.E.
	10 20	II.Oc.D.
	11 46	III.Sh.E.
	12 08	I.Tr.I.
	12 34	I.Sh.I.
	13 56	II.Ec.R.
	14 23	I.Tr.E.
	14 50	I.Sh.E.
19	9 19	I.Oc.D.
	12 04	I.Ec.R.
20	4 45	II.Tr.I.
	5 43	II.Sh.I.
	6 34	I.Tr.I.
	7 02	I.Sh.I.
	7 31	II.Tr.E.
	8 31	II.Sh.E.
	8 49	I.Tr.E.
	9 18	I.Sh.E.
21	3 45	I.Oc.D.
	6 32	I.Ec.R.
	14 10	IV.Tr.I.
	17 47	IV.Tr.E.
	18 51	IV.Sh.I.
	20 19	III.Oc.D.
	22 39	IV.Sh.E.
	23 27	II.Oc.D.
22	1 00	I.Tr.I.
	1 31	I.Sh.I.
	1 54	III.Ec.R.
	3 13	II.Ec.R.
	3 16	I.Tr.E.
	3 47	I.Sh.E.
	22 11	I.Oc.D.
23	1 01	I.Ec.R.
	17 54	II.Tr.I.
	19 02	II.Sh.I.
	19 26	I.Tr.I.
	20 00	I.Sh.I.
	20 41	II.Tr.E.
	21 42	I.Tr.E.
	21 49	II.Sh.E.
	22 15	I.Sh.E.
24	16 38	I.Oc.D.
	19 30	I.Ec.R.
25	9 53	III.Tr.I.
	12 16	III.Sh.I.
	12 35	II.Oc.D.
	13 20	III.Tr.E.
	13 53	I.Tr.I.
	14 28	I.Sh.I.
	15 44	III.Sh.E.
	16 08	I.Tr.E.
	16 30	II.Ec.R.
	16 44	I.Sh.E.
26	11 04	I.Oc.D.
	13 59	I.Ec.R.
27	7 03	II.Tr.I.
	8 19	I.Tr.I.
	8 21	II.Sh.I.
	8 57	I.Sh.I.
	9 50	II.Tr.E.
	10 35	I.Tr.E.
	11 07	II.Sh.E.
	11 12	I.Sh.E.
28	5 31	I.Oc.D.
	8 27	I.Ec.R.
	23 41	III.Oc.D.
29	1 43	II.Oc.D.
	2 46	I.Tr.I.
	3 25	I.Sh.I.
	5 01	I.Tr.E.
	5 41	I.Sh.E.
	5 47	II.Ec.R.
	5 53	III.Ec.R.
	21 05	IV.Oc.D.
	23 57	I.Oc.D.
30	0 50	IV.Oc.R.
	2 56	I.Ec.R.
	3 40	IV.Ec.D.
	7 28	IV.Ec.R.
	20 14	II.Tr.I.
	21 12	I.Tr.I.
	21 40	II.Sh.I.
	21 54	I.Sh.I.
	23 00	II.Tr.E.
	23 28	I.Tr.E.
31	0 09	I.Sh.E.
	0 26	II.Sh.E.
	18 24	I.Oc.D.
	21 25	I.Ec.R.

I. Mar. 15	II. Mar. 15	III. Mar. 14	IV. Mar. 13
$x_2 = +1.3,\ y_2 = -0.1$	$x_2 = +1.4,\ y_2 = -0.3$	$x_2 = +1.7,\ y_2 = -0.3$	$x_2 = +2.1,\ y_2 = -0.6$

NOTE.—I. denotes ingress; E., egress; D., disappearance; R., reappearance; Ec., eclipse; Oc., occultation; Tr., transit of the satellite; Sh., transit of the shadow.

CONFIGURATIONS OF SATELLITES I–IV FOR MARCH

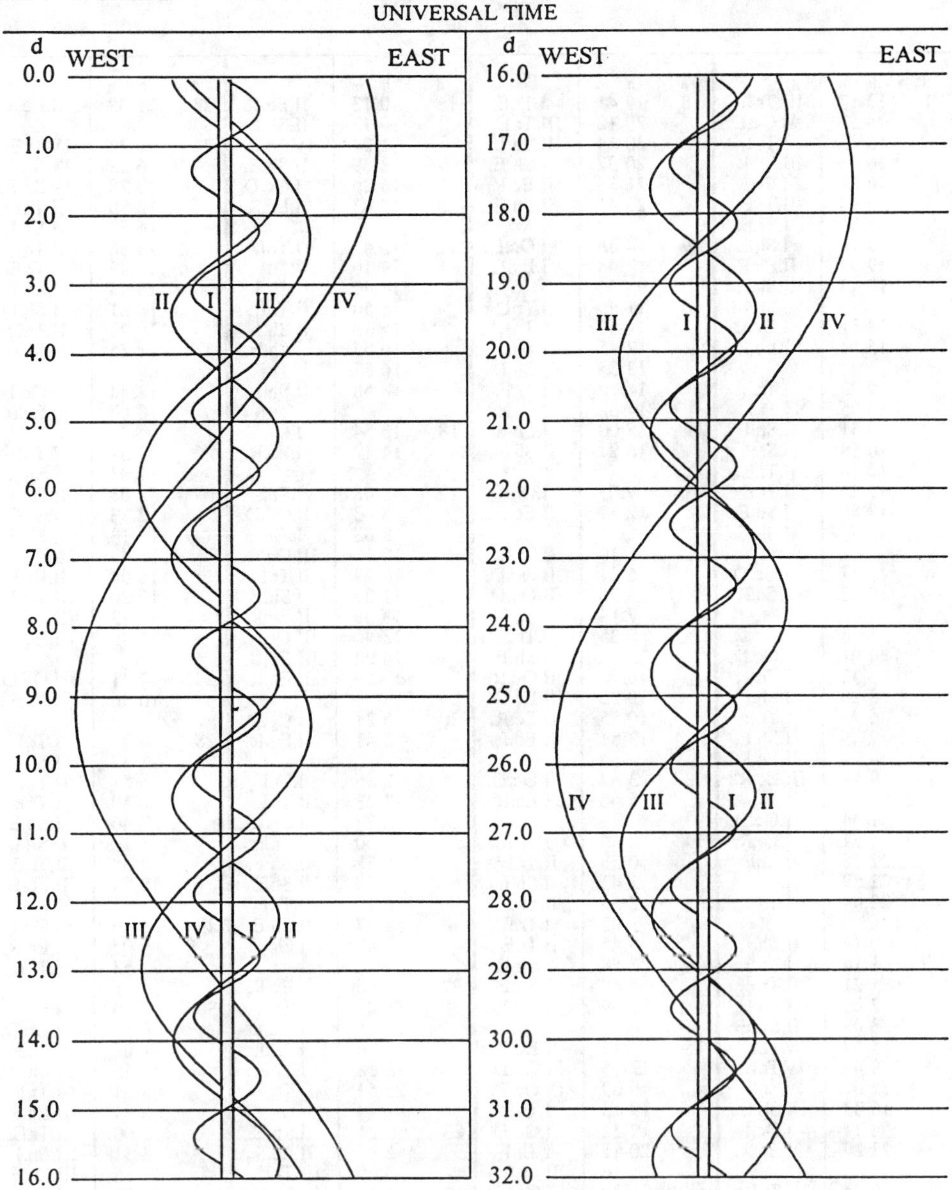

PHASES OF THE ECLIPSES

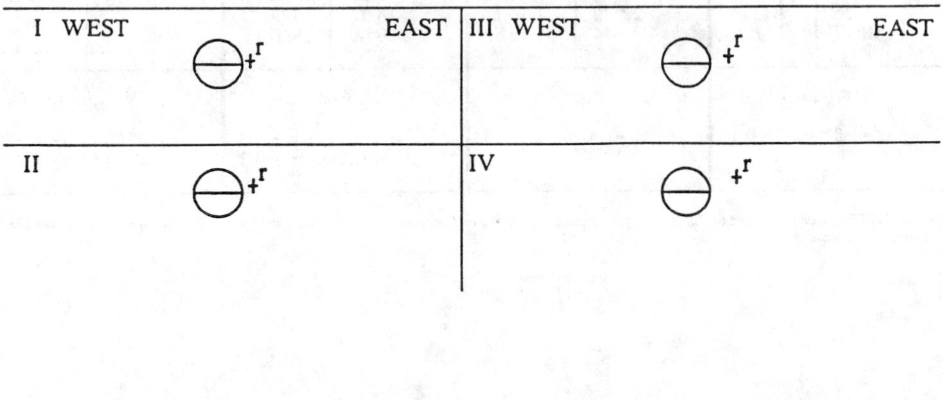

DYNAMICAL TIME OF GEOCENTRIC PHENOMENA

APRIL

d	h m	
1	13 17	III.Tr.I.
	14 52	II.Oc.D.
	15 39	I.Tr.I.
	16 16	III.Sh.I.
	16 22	I.Sh.I.
	16 44	III.Tr.E.
	17 54	I.Tr.E.
	18 38	I.Sh.E.
	19 04	II.Ec.R.
	19 43	III.Sh.E.
2	12 51	I.Oc.D.
	15 53	I.Ec.R.
3	9 23	II.Tr.I.
	10 05	I.Tr.I.
	10 51	I.Sh.I.
	10 58	II.Sh.I.
	12 10	II.Tr.E.
	12 21	I.Tr.E.
	13 06	I.Sh.E.
	13 44	II.Sh.E.
4	7 17	I.Oc.D.
	10 22	I.Ec.R.
5	3 06	III.Oc.D.
	4 01	II.Oc.D.
	4 32	I.Tr.I.
	5 20	I.Sh.I.
	6 47	I.Tr.E.
	7 35	I.Sh.E.
	8 21	II.Ec.R.
	9 51	III.Ec.R.
6	1 44	I.Oc.D.
	4 51	I.Ec.R.
	22 35	II.Tr.I.
	22 59	I.Tr.I.
	23 48	I.Sh.I.
7	0 17	II.Sh.I.
	1 14	I.Tr.E.
	1 21	II.Tr.E.
	2 03	I.Sh.E.
	3 03	II.Sh.E.
	4 59	IV.Tr.I.
	8 45	IV.Tr.E.
	12 52	IV.Sh.I.
	16 33	IV.Sh.E.
	20 11	I.Oc.D.
	23 20	I.Ec.R.
8	16 44	III.Tr.I.
	17 11	II.Oc.D.
	17 26	I.Tr.I.
	18 17	I.Sh.I.

d	h m	
8	19 41	I.Tr.E.
	20 12	III.Tr.E.
	20 14	III.Sh.I.
	20 32	I.Sh.E.
	21 38	II.Ec.R.
	23 41	III.Sh.E.
9	14 38	I.Oc.D.
	17 48	I.Ec.R.
10	11 45	II.Tr.I.
	11 52	I.Tr.I.
	12 45	I.Sh.I.
	13 35	II.Sh.I.
	14 08	I.Tr.E.
	14 32	II.Tr.E.
	15 00	I.Sh.E.
	16 21	II.Sh.E.
11	9 05	I.Oc.D.
	12 17	I.Ec.R.
12	6 19	I.Tr.I.
	6 21	II.Oc.D.
	6 36	III.Oc.D.
	7 14	I.Sh.I.
	8 35	I.Tr.E.
	9 29	I.Sh.E.
	10 06	III.Oc.R.
	10 22	III.Ec.D.
	10 55	II.Ec.R.
	13 51	III.Ec.R.
13	3 32	I.Oc.D.
	6 46	I.Ec.R.
14	0 46	I.Tr.I.
	0 58	II.Tr.I.
	1 43	I.Sh.I.
	2 55	II.Sh.I.
	3 02	I.Tr.E.
	3 45	II.Tr.E.
	3 58	I.Sh.E.
	5 40	II.Sh.E.
	21 59	I.Oc.D.
15	1 15	I.Ec.R.
	12 15	IV.Oc.D.
	16 09	IV.Oc.R.
	19 13	I.Tr.I.
	19 32	II.Oc.D.
	20 11	I.Sh.I.
	20 16	III.Tr.I.
	21 29	I.Tr.E.
	21 44	IV.Ec.D.
	22 26	I.Sh.E.
	23 44	III.Tr.E.

d	h m	
16	0 13	II.Ec.R.
	0 13	III.Sh.I.
	1 24	IV.Ec.R.
	3 39	III.Sh.E.
	16 26	I.Oc.D.
	19 43	I.Ec.R.
17	13 41	I.Tr.I.
	14 10	II.Tr.I.
	14 40	I.Sh.I.
	15 56	I.Tr.E.
	16 13	II.Sh.I.
	16 55	I.Sh.E.
	16 57	II.Tr.E.
	18 58	II.Sh.E.
18	10 54	I.Oc.D.
	14 12	I.Ec.R.
19	8 08	I.Tr.I.
	8 43	II.Oc.D.
	9 08	I.Sh.I.
	10 10	III.Oc.D.
	10 23	I.Tr.E.
	11 23	I.Sh.E.
	13 30	II.Ec.R.
	13 40	III.Oc.R.
	14 22	III.Ec.D.
	17 50	III.Ec.R.
20	5 21	I.Oc.D.
	8 41	I.Ec.R.
21	2 35	I.Tr.I.
	3 23	II.Tr.I.
	3 37	I.Sh.I.
	4 50	I.Tr.E.
	5 32	II.Sh.I.
	5 52	I.Sh.E.
	6 10	II.Tr.E.
	8 17	II.Sh.E.
	23 49	I.Oc.D.
22	3 10	I.Ec.R.
	21 03	I.Tr.I.
	21 55	II.Oc.D.
	22 06	I.Sh.I.
	23 18	I.Tr.E.
	23 51	III.Tr.I.
23	0 21	I.Sh.E.
	2 47	II.Ec.R.
	3 20	III.Tr.E.
	4 12	III.Sh.I.
	7 37	III.Sh.E.
	18 16	I.Oc.D.
	20 40	IV.Tr.I.

d	h m	
23	21 39	I.Ec.R.
24	0 32	IV.Tr.E.
	6 55	IV.Sh.I.
	10 28	IV.Sh.E.
	15 30	I.Tr.I.
	16 34	I.Sh.I.
	16 36	II.Tr.I.
	17 45	I.Tr.E.
	18 49	I.Sh.E.
	18 51	II.Sh.I.
	19 23	II.Tr.E.
	21 35	II.Sh.E.
25	12 44	I.Oc.D.
	16 07	I.Ec.R.
26	9 57	I.Tr.I.
	11 03	I.Sh.I.
	11 08	II.Oc.D.
	12 13	I.Tr.E.
	13 18	I.Sh.E.
	13 49	III.Oc.D.
	16 04	II.Ec.R.
	17 20	III.Oc.R.
	18 22	III.Ec.D.
	21 49	III.Ec.R.
27	7 11	I.Oc.D.
	10 36	I.Ec.R.
28	4 25	I.Tr.I.
	5 32	I.Sh.I.
	5 51	II.Tr.I.
	6 40	I.Tr.E.
	7 46	I.Sh.E.
	8 10	II.Sh.I.
	8 38	II.Tr.E.
	10 54	II.Sh.E.
29	1 39	I.Oc.D.
	5 05	I.Ec.R.
	22 53	I.Tr.I.
30	0 00	I.Sh.I.
	0 21	II.Oc.D.
	1 08	I.Tr.E.
	2 15	I.Sh.E.
	3 31	III.Tr.I.
	5 21	II.Ec.R.
	7 00	III.Tr.E.
	8 11	III.Sh.I.
	11 35	III.Sh.E.
	20 07	I.Oc.D.
	23 34	I.Ec.R.

I. Apr. 15	II. Apr. 16	III. Apr. 12	IV. Apr. 15-16
$x_2 = +1.8,\ y_2 = -0.1$	$x_2 = +2.3,\ y_2 = -0.3$	$x_1 = +1.1,\ y_1 = -0.3$ $x_2 = +3.0,\ y_2 = -0.3$	$x_1 = +3.1,\ y_1 = -0.6$ $x_2 = +4.6,\ y_2 = -0.5$

NOTE.—I. denotes ingress; E., egress; D., disappearance; R., reappearance; Ec., eclipse; Oc., occultation; Tr., transit of the satellite; Sh., transit of the shadow.

CONFIGURATIONS OF SATELLITES I–IV FOR APRIL

UNIVERSAL TIME

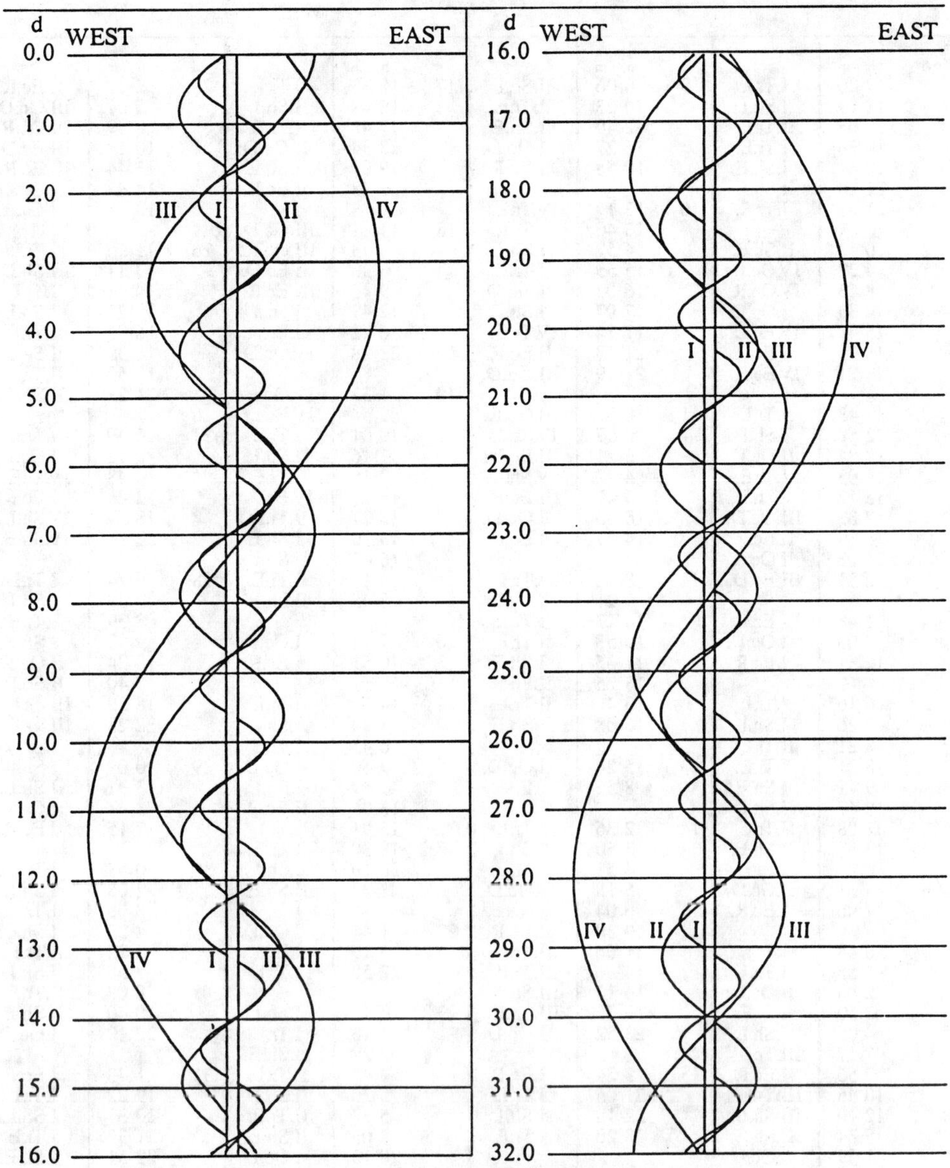

PHASES OF THE ECLIPSES

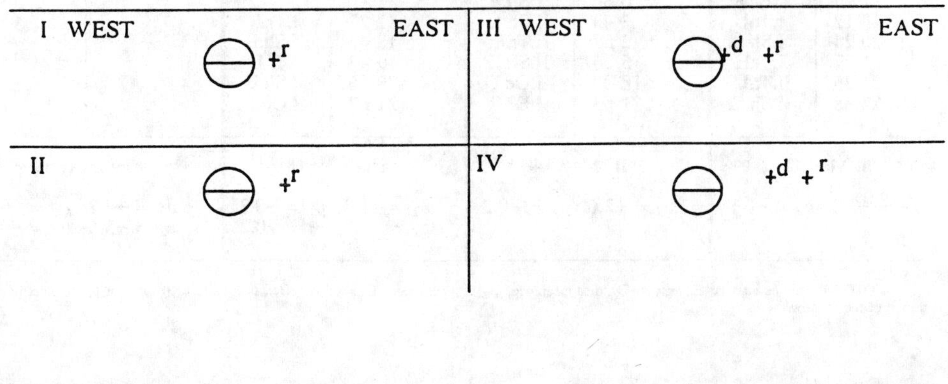

DYNAMICAL TIME OF GEOCENTRIC PHENOMENA

MAY

d	h m		d	h m		d	h m		d	h m	
1	17 20	I.Tr.I.	9	0 06	II.Sh.I.	17	15 33	I.Tr.I.	25	2 23	II.Ec.R.
	18 29	I.Sh.I.		0 23	II.Tr.E.		16 47	I.Sh.I.		5 07	III.Oc.D.
	19 05	II.Tr.I.		2 49	II.Sh.E.		17 48	I.Tr.E.		8 39	III.Oc.R.
	19 36	I.Tr.E.		16 27	I.Oc.D.		18 34	II.Oc.D.		10 19	III.Ec.D.
	20 44	I.Sh.E.		19 58	I.Ec.R.		19 01	I.Sh.E.		13 44	III.Ec.R.
	21 28	II.Sh.I.					23 48	II.Ec.R.		14 43	I.Oc.D.
	21 52	II.Tr.E.	10	13 17	IV.Tr.I.					18 17	I.Ec.R.
				13 40	I.Tr.I.	18	1 11	III.Oc.D.			
2	0 12	II.Sh.E.		14 52	I.Sh.I.		4 43	III.Oc.R.	26	11 56	I.Tr.I.
	4 22	IV.Oc.D.		15 55	I.Tr.E.		6 20	III.Ec.D.		13 11	I.Sh.I.
	8 23	IV.Oc.R.		16 03	II.Oc.D.		9 45	III.Ec.R.		14 11	I.Tr.E.
	14 35	I.Oc.D.		17 07	I.Sh.E.		12 49	I.Oc.D.		15 25	I.Sh.E.
	15 49	IV.Ec.D.		17 14	IV.Tr.E.		16 22	I.Ec.R.		16 03	II.Tr.I.
	18 03	I.Ec.R.		21 13	II.Ec.R.		21 28	IV.Oc.D.		18 39	II.Sh.I.
	19 21	IV.Ec.R.		21 19	III.Oc.D.					18 50	II.Tr.E.
						19	1 32	IV.Oc.R.		21 22	II.Sh.E.
3	11 48	I.Tr.I.	11	0 51	III.Oc.R.		9 54	IV.Ec.D.			
	12 58	I.Sh.I.		0 57	IV.Sh.I.		10 01	I.Tr.I.	27	6 49	IV.Tr.I.
	13 34	II.Oc.D.		2 21	III.Ec.D.		11 16	I.Sh.I.		9 12	I.Oc.D.
	14 03	I.Tr.E.		4 22	IV.Sh.E.		12 16	I.Tr.E.		10 48	IV.Tr.E.
	15 12	I.Sh.E.		5 47	III.Ec.R.		13 19	IV.Ec.R.		12 46	I.Ec.R.
	17 32	III.Oc.D.		10 55	I.Oc.D.		13 27	II.Tr.I.		18 59	IV.Sh.I.
	18 39	II.Ec.R.		14 27	I.Ec.R.		13 30	I.Sh.E.		22 16	IV.Sh.E.
	21 03	III.Oc.R.					16 02	II.Sh.I.			
	22 21	III.Ec.D.	12	8 08	I.Tr.I.		16 14	II.Tr.E.	28	6 24	I.Tr.I.
				9 21	I.Sh.I.		18 45	II.Sh.E.		7 39	I.Sh.I.
4	1 48	III.Ec.R.		10 23	I.Tr.E.					8 39	I.Tr.E.
	9 03	I.Oc.D.		10 53	II.Tr.I.	20	7 17	I.Oc.D.		9 53	I.Sh.E.
	12 31	I.Ec.R.		11 35	I.Sh.E.		10 51	I.Ec.R.		10 24	II.Oc.D.
				13 25	II.Sh.I.					15 40	II.Ec.R.
5	6 16	I.Tr.I.		13 40	II.Tr.E.	21	4 30	I.Tr.I.		18 59	III.Tr.I.
	7 26	I.Sh.I.		16 08	II.Sh.E.		5 44	I.Sh.I.		22 28	III.Tr.E.
	8 21	II.Tr.I.					6 45	I.Tr.E.			
	8 31	I.Tr.E.	13	5 24	I.Oc.D.		7 50	II.Oc.D.	29	0 09	III.Sh.I.
	9 41	I.Sh.E.		8 55	I.Ec.R.		7 59	I.Sh.E.		3 31	III.Sh.E.
	10 47	II.Sh.I.					13 05	II.Ec.R.		3 41	I.Oc.D.
	11 08	II.Tr.E.	14	2 36	I.Tr.I.		15 01	III.Tr.I.		7 15	I.Ec.R.
	13 31	II.Sh.E.		3 50	I.Sh.I.		18 30	III.Tr.E.			
				4 51	I.Tr.E.		20 10	III.Sh.I.	30	0 53	I.Tr.I.
6	3 31	I.Oc.D.		5 18	II.Oc.D.		23 32	III.Sh.E.		2 08	I.Sh.I.
	7 00	I.Ec.R.		6 04	I.Sh.E.					3 08	I.Tr.E.
				10 30	II.Ec.R.	22	1 46	I.Oc.D.		4 22	I.Sh.E.
7	0 44	I.Tr.I.		11 06	III.Tr.I.		5 19	I.Ec.R.		5 22	II.Tr.I.
	1 55	I.Sh.I.		14 36	III.Tr.E.		22 58	I.Tr.I.		7 58	II.Sh.I.
	2 48	II.Oc.D.		16 10	III.Sh.I.					8 09	II.Tr.E.
	2 59	I.Tr.E.		19 33	III.Sh.E.	23	0 13	I.Sh.I.		10 40	II.Sh.E.
	4 09	I.Sh.E.		23 52	I.Oc.D.		1 13	I.Tr.E.		22 10	I.Oc.D.
	7 17	III.Tr.I.					2 27	I.Sh.E.			
	7 56	II.Ec.R.	15	3 24	I.Ec.R.		2 45	II.Tr.I.	31	1 43	I.Ec.R.
	10 46	III.Tr.E.		21 05	I.Tr.I.		5 20	II.Sh.I.		19 22	I.Tr.I.
	12 10	III.Sh.I.		22 18	I.Sh.I.		5 32	II.Tr.E.		20 37	I.Sh.I.
	15 34	III.Sh.E.		23 20	I.Tr.E.		8 03	II.Sh.E.		21 37	I.Tr.E.
	21 59	I.Oc.D.					20 15	I.Oc.D.		22 51	I.Sh.E.
			16	0 09	II.Tr.I.		23 48	I.Ec.R.		23 42	II.Oc.D.
8	1 29	I.Ec.R.		0 33	I.Sh.E.						
	19 12	I.Tr.I.		2 43	II.Sh.I.	24	17 27	I.Tr.I.			
	20 24	I.Sh.I.		2 57	II.Tr.E.		18 42	I.Sh.I.			
	21 27	I.Tr.E.		5 26	II.Sh.E.		19 42	I.Tr.E.			
	21 36	II.Tr.I.		18 20	I.Oc.D.		20 56	I.Sh.E.			
	22 38	I.Sh.E.		21 53	I.Ec.R.		21 07	II.Oc.D.			

I. May 15	II. May 14	III. May 18	IV. May 19
$x_2 = +2.0,\ y_2 = -0.1$	$x_2 = +2.6,\ y_2 = -0.2$	$x_1 = +1.8,\ y_1 = -0.2$ $x_2 = +3.7,\ y_2 = -0.2$	$x_1 = +4.2,\ y_1 = -0.5$ $x_2 = +5.6,\ y_2 = -0.5$

NOTE.—I. denotes ingress; E., egress; D., disappearance; R., reappearance; Ec., eclipse; Oc., occultation; Tr., transit of the satellite; Sh., transit of the shadow.

CONFIGURATIONS OF SATELLITES I–IV FOR MAY

UNIVERSAL TIME

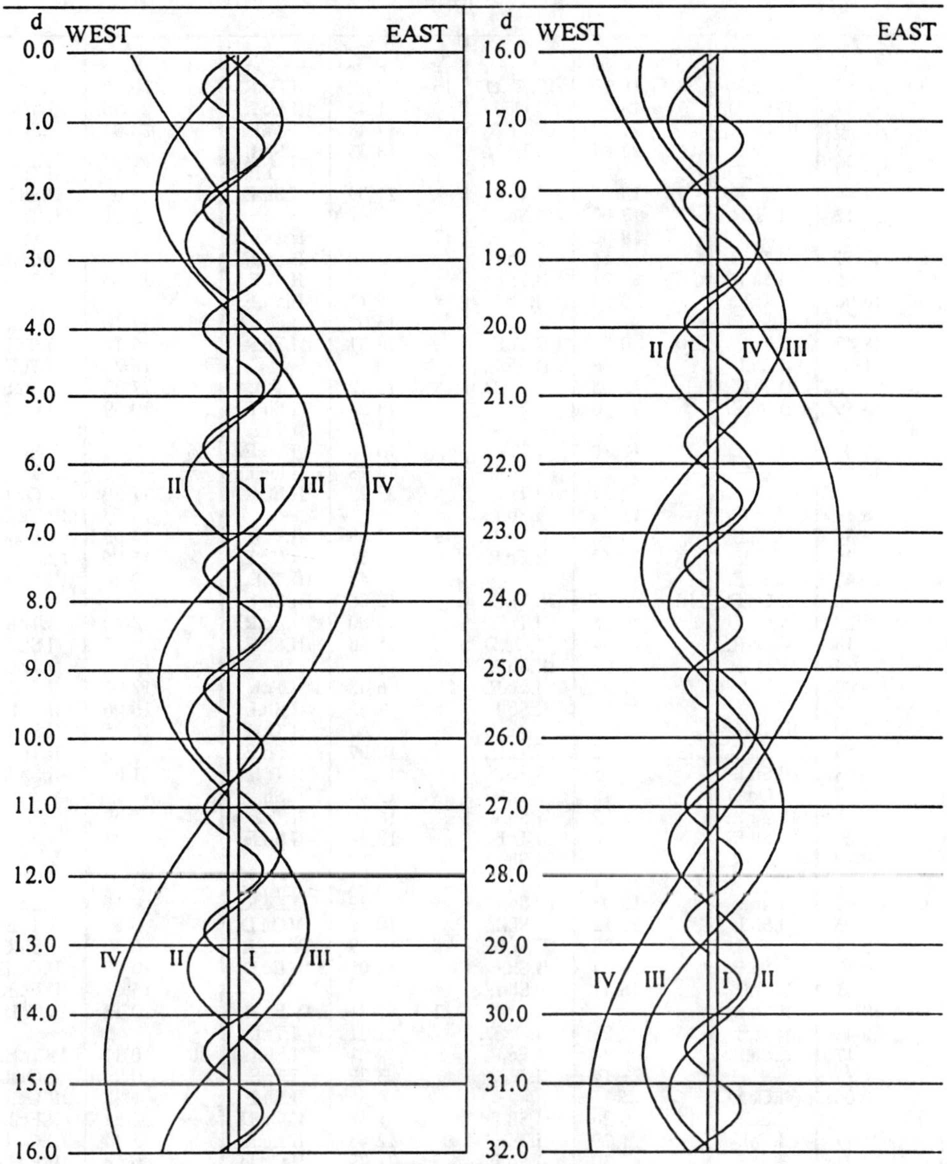

PHASES OF THE ECLIPSES

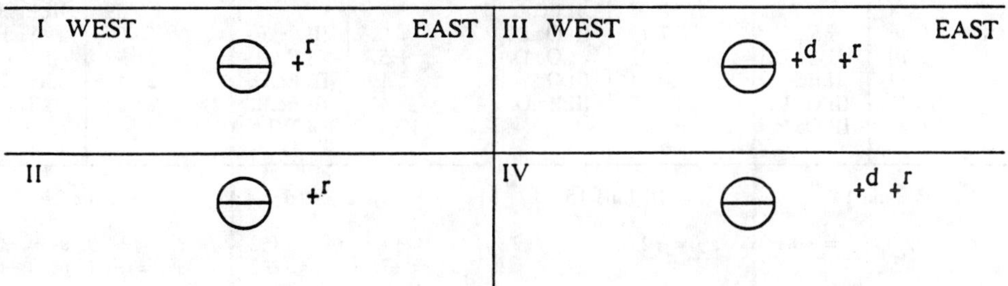

DYNAMICAL TIME OF GEOCENTRIC PHENOMENA

JUNE

d	h m	
1	4 57	II.Ec.R.
	9 07	III.Oc.D.
	12 40	III.Oc.R.
	14 19	III.Ec.D.
	16 39	I.Oc.D.
	17 44	III.Ec.R.
	20 12	I.Ec.R.
2	13 51	I.Tr.I.
	15 05	I.Sh.I.
	16 06	I.Tr.E.
	17 20	I.Sh.E.
	18 42	II.Tr.I.
	21 17	II.Sh.I.
	21 28	II.Tr.E.
	23 59	II.Sh.E.
3	11 08	I.Oc.D.
	14 41	I.Ec.R.
4	8 19	I.Tr.I.
	9 34	I.Sh.I.
	10 35	I.Tr.E.
	11 48	I.Sh.E.
	13 00	II.Oc.D.
	15 29	IV.Oc.D.
	18 15	II.Ec.R.
	19 34	IV.Oc.R.
	23 00	III.Tr.I.
5	2 30	III.Tr.E.
	3 59	IV.Ec.D.
	4 08	III.Sh.I.
	5 37	I.Oc.D.
	7 15	IV.Ec.R.
	7 29	III.Sh.E.
	9 10	I.Ec.R.
6	2 48	I.Tr.I.
	4 03	I.Sh.I.
	5 04	I.Tr.E.
	6 17	I.Sh.E.
	8 01	II.Tr.I.
	10 35	II.Sh.I.
	10 48	II.Tr.E.
	13 17	II.Sh.E.
7	0 06	I.Oc.D.
	3 39	I.Ec.R.
	21 17	I.Tr.I.
	22 31	I.Sh.I.
	23 33	I.Tr.E.
8	0 46	I.Sh.E.
	2 19	II.Oc.D.
	7 32	II.Ec.R.
	13 11	III.Oc.D.
	16 44	III.Oc.R.

d	h m	
8	18 19	III.Ec.D.
	18 35	I.Oc.D.
	21 43	III.Ec.R.
	22 07	I.Ec.R.
9	15 47	I.Tr.I.
	17 00	I.Sh.I.
	18 02	I.Tr.E.
	19 14	I.Sh.E.
	21 21	II.Tr.I.
	23 54	II.Sh.I.
10	0 08	II.Tr.E.
	2 36	II.Sh.E.
	13 05	I.Oc.D.
	16 36	I.Ec.R.
11	10 16	I.Tr.I.
	11 29	I.Sh.I.
	12 31	I.Tr.E.
	13 43	I.Sh.E.
	15 38	II.Oc.D.
	20 50	II.Ec.R.
12	3 05	III.Tr.I.
	6 35	III.Tr.E.
	7 34	I.Oc.D.
	8 07	III.Sh.I.
	11 05	I.Ec.R.
	11 28	III.Sh.E.
13	1 12	IV.Tr.I.
	4 45	I.Tr.I.
	5 10	IV.Tr.E.
	5 57	I.Sh.I.
	7 00	I.Tr.E.
	8 12	I.Sh.E.
	10 41	II.Tr.I.
	13 03	IV.Sh.I.
	13 12	II.Sh.I.
	13 28	II.Tr.E.
	15 54	II.Sh.E.
	16 11	IV.Sh.E.
14	2 03	I.Oc.D.
	5 34	I.Ec.R.
	23 14	I.Tr.I.
15	0 26	I.Sh.I.
	1 29	I.Tr.E.
	2 40	I.Sh.E.
	4 58	II.Oc.D.
	10 08	II.Ec.R.
	17 19	III.Oc.D.
	20 32	I.Oc.D.
	20 51	III.Oc.R.
	22 19	III.Ec.D.

d	h m	
16	0 02	I.Ec.R.
	1 42	III.Ec.R.
	17 43	I.Tr.I.
	18 55	I.Sh.I.
	19 59	I.Tr.E.
	21 09	I.Sh.E.
17	0 02	II.Tr.I.
	2 30	II.Sh.I.
	2 49	II.Tr.E.
	5 12	II.Sh.E.
	15 02	I.Oc.D.
	18 31	I.Ec.R.
18	12 12	I.Tr.I.
	13 24	I.Sh.I.
	14 28	I.Tr.E.
	15 38	I.Sh.E.
	18 18	II.Oc.D.
	23 25	II.Ec.R.
19	7 13	III.Tr.I.
	9 31	I.Oc.D.
	10 43	III.Tr.E.
	12 06	III.Sh.I.
	13 00	I.Ec.R.
	15 26	III.Sh.E.
20	6 42	I.Tr.I.
	7 52	I.Sh.I.
	8 57	I.Tr.E.
	10 07	I.Sh.E.
	13 23	II.Tr.I.
	15 49	II.Sh.I.
	16 10	II.Tr.E.
	18 30	II.Sh.E.
21	4 01	I.Oc.D.
	7 29	I.Ec.R.
	10 16	IV.Oc.D.
	14 19	IV.Oc.R.
	22 05	IV.Ec.D.
22	1 11	IV.Ec.R.
	1 11	I.Tr.I.
	2 21	I.Sh.I.
	3 27	I.Tr.E.
	4 35	I.Sh.E.
	7 38	II.Oc.D.
	12 43	II.Ec.R.
	21 29	III.Oc.D.
	22 30	I.Oc.D.
23	1 01	III.Oc.R.
	1 57	I.Ec.R.
	2 19	III.Ec.D.
	5 41	III.Ec.R.
	19 41	I.Tr.I.

d	h m	
23	20 50	I.Sh.I.
	21 56	I.Tr.E.
	23 04	I.Sh.E.
24	2 45	II.Tr.I.
	5 07	II.Sh.I.
	5 31	II.Tr.E.
	7 49	II.Sh.E.
	17 00	I.Oc.D.
	20 26	I.Ec.R.
25	14 10	I.Tr.I.
	15 18	I.Sh.I.
	16 26	I.Tr.E.
	17 33	I.Sh.E.
	20 59	II.Oc.D.
26	2 00	II.Ec.R.
	11 25	III.Tr.I.
	11 30	I.Oc.D.
	14 55	III.Tr.E.
	14 55	I.Ec.R.
	16 06	III.Sh.I.
	19 26	III.Sh.E.
27	8 40	I.Tr.I.
	9 47	I.Sh.I.
	10 55	I.Tr.E.
	12 02	I.Sh.E.
	16 06	II.Tr.I.
	18 25	II.Sh.I.
	18 53	II.Tr.E.
	21 06	II.Sh.E.
28	5 59	I.Oc.D.
	9 24	I.Ec.R.
29	3 09	I.Tr.I.
	4 16	I.Sh.I.
	5 25	I.Tr.E.
	6 30	I.Sh.E.
	10 20	II.Oc.D.
	15 18	II.Ec.R.
	20 15	IV.Tr.I.
30	0 11	IV.Tr.E.
	0 29	I.Oc.D.
	1 42	III.Oc.D.
	3 52	I.Ec.R.
	5 14	III.Oc.R.
	6 18	III.Ec.D.
	7 06	IV.Sh.I.
	9 39	III.Ec.R.
	10 05	IV.Sh.E.
	21 39	I.Tr.I.
	22 44	I.Sh.I.
	23 54	I.Tr.E.

I. June 14	II. June 15	III. June 15-16	IV. June 21-22
$x_2 = +2.0,\ y_2 = -0.1$	$x_2 = +2.6,\ y_2 = -0.2$	$x_1 = +1.8,\ y_1 = -0.2$ $x_2 = +3.6,\ y_2 = -0.2$	$x_1 = +4.0,\ y_1 = -0.5$ $x_2 = +5.2,\ y_2 = -0.5$

NOTE.—I. denotes ingress; E., egress; D., disappearance; R., reappearance; Ec., eclipse; Oc., occultation; Tr., transit of the satellite; Sh., transit of the shadow.

CONFIGURATIONS OF SATELLITES I–IV FOR JUNE

UNIVERSAL TIME

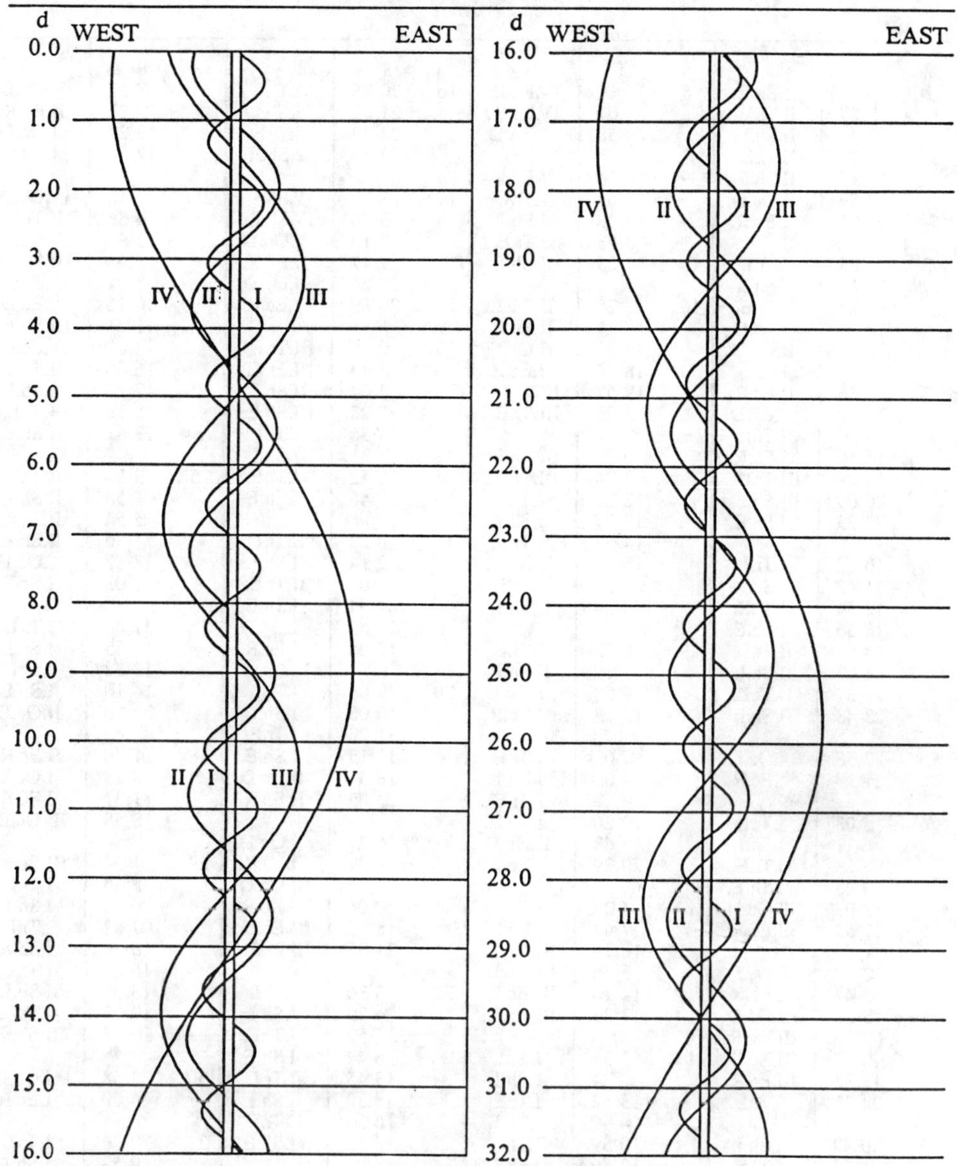

PHASES OF THE ECLIPSES

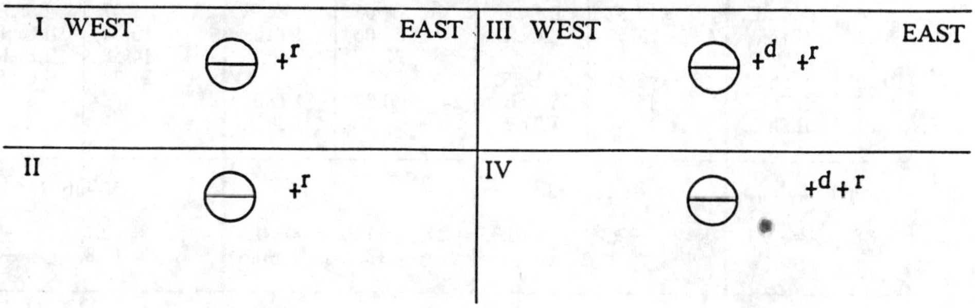

DYNAMICAL TIME OF GEOCENTRIC PHENOMENA

JULY

d	h m	
1	0 59	I.Sh.E.
	5 29	II.Tr.I.
	7 44	II.Sh.I.
	8 15	II.Tr.E.
	10 25	II.Sh.E.
	18 59	I.Oc.D.
	22 21	I.Ec.R.
2	16 08	I.Tr.I.
	17 13	I.Sh.I.
	18 24	I.Tr.E.
	19 28	I.Sh.E.
	23 42	II.Oc.D.
3	4 36	II.Ec.R.
	13 28	I.Oc.D.
	15 39	III.Tr.I.
	16 50	I.Ec.R.
	19 09	III.Tr.E.
	20 05	III.Sh.I.
	23 25	III.Sh.E.
4	10 38	I.Tr.I.
	11 42	I.Sh.I.
	12 54	I.Tr.E.
	13 56	I.Sh.E.
	18 51	II.Tr.I.
	21 02	II.Sh.I.
	21 37	II.Tr.E.
	23 43	II.Sh.E.
5	7 58	I.Oc.D.
	11 18	I.Ec.R.
6	5 08	I.Tr.I.
	6 11	I.Sh.I.
	7 23	I.Tr.E.
	8 25	I.Sh.E.
	13 03	II.Oc.D.
	17 53	II.Ec.R.
7	2 28	I.Oc.D.
	5 47	I.Ec.R.
	5 58	III.Oc.D.
	9 29	III.Oc.R.
	10 17	III.Ec.D.
	13 38	III.Ec.R.
	23 37	I.Tr.I.
8	0 39	I.Sh.I.
	1 53	I.Tr.E.
	2 54	I.Sh.E.
	5 44	IV.Oc.D.
	8 13	II.Tr.I.
	9 41	IV.Oc.R.
	10 20	II.Sh.I.
	10 59	II.Tr.E.
	13 01	II.Sh.E.

d	h m	
8	16 11	IV.Ec.D.
	19 07	IV.Ec.R.
	20 58	I.Oc.D.
9	0 16	I.Ec.R.
	18 07	I.Tr.I.
	19 08	I.Sh.I.
	20 23	I.Tr.E.
	21 23	I.Sh.E.
10	2 25	II.Oc.D.
	7 11	II.Ec.R.
	15 28	I.Oc.D.
	18 44	I.Ec.R.
	19 57	III.Tr.I.
	23 26	III.Tr.E.
11	0 05	III.Sh.I.
	3 24	III.Sh.E.
	12 37	I.Tr.I.
	13 37	I.Sh.I.
	14 53	I.Tr.E.
	15 51	I.Sh.E.
	21 36	II.Tr.I.
	23 38	II.Sh.I.
12	0 22	II.Tr.E.
	2 19	II.Sh.E.
	9 58	I.Oc.D.
	13 13	I.Ec.R.
13	7 07	I.Tr.I.
	8 05	I.Sh.I.
	9 22	I.Tr.E.
	10 20	I.Sh.E.
	15 48	II.Oc.D.
	20 29	II.Ec.R.
14	4 27	I.Oc.D.
	7 42	I.Ec.R.
	10 16	III.Oc.D.
	13 46	III.Oc.R.
	14 16	III.Ec.D.
	17 36	III.Ec.R.
15	1 37	I.Tr.I.
	2 34	I.Sh.I.
	3 52	I.Tr.E.
	4 49	I.Sh.E.
	10 59	II.Tr.I.
	12 56	II.Sh.I.
	13 45	II.Tr.E.
	15 37	II.Sh.E.
	22 57	I.Oc.D.
16	2 11	I.Ec.R.
	15 53	IV.Tr.I.
	19 42	IV.Tr.E.

d	h m	
16	20 06	I.Tr.I.
	21 03	I.Sh.I.
	22 22	I.Tr.E.
	23 17	I.Sh.E.
17	1 08	IV.Sh.I.
	3 58	IV.Sh.E.
	5 10	II.Oc.D.
	9 47	II.Ec.R.
	17 27	I.Oc.D.
	20 39	I.Ec.R.
18	0 15	III.Tr.I.
	3 44	III.Tr.E.
	4 04	III.Sh.I.
	7 22	III.Sh.E.
	14 36	I.Tr.I.
	15 31	I.Sh.I.
	16 52	I.Tr.E.
	17 46	I.Sh.E.
19	0 22	II.Tr.I.
	2 14	II.Sh.I.
	3 07	II.Tr.E.
	4 54	II.Sh.E.
	11 57	I.Oc.D.
	15 08	I.Ec.R.
20	9 06	I.Tr.I.
	10 00	I.Sh.I.
	11 22	I.Tr.E.
	12 15	I.Sh.E.
	18 34	II.Oc.D.
	23 05	II.Ec.R.
21	6 27	I.Oc.D.
	9 37	I.Ec.R.
	14 36	III.Oc.D.
	18 06	III.Oc.R.
	18 16	III.Ec.D.
	21 35	III.Ec.R.
22	3 36	I.Tr.I.
	4 29	I.Sh.I.
	5 52	I.Tr.E.
	6 44	I.Sh.E.
	13 45	II.Tr.I.
	15 32	II.Sh.I.
	16 30	II.Tr.E.
	18 12	II.Sh.E.
23	0 57	I.Oc.D.
	4 05	I.Ec.R.
	22 06	I.Tr.I.
	22 57	I.Sh.I.
24	0 22	I.Tr.E.
	1 12	I.Sh.E.

d	h m	
24	7 57	II.Oc.D.
	12 22	II.Ec.R.
	19 27	I.Oc.D.
	22 34	I.Ec.R.
25	1 42	IV.Oc.D.
	4 36	III.Tr.I.
	5 30	IV.Oc.R.
	8 03	III.Sh.I.
	8 04	III.Tr.E.
	10 16	IV.Ec.D.
	11 21	III.Sh.E.
	13 01	IV.Ec.R.
	16 36	I.Tr.I.
	17 26	I.Sh.I.
	18 52	I.Tr.E.
	19 41	I.Sh.E.
26	3 09	II.Tr.I.
	4 50	II.Sh.I.
	5 54	II.Tr.E.
	7 30	II.Sh.E.
	13 57	I.Oc.D.
	17 03	I.Ec.R.
27	11 06	I.Tr.I.
	11 55	I.Sh.I.
	13 22	I.Tr.E.
	14 10	I.Sh.E.
	21 20	II.Oc.D.
28	1 40	II.Ec.R.
	8 28	I.Oc.D.
	11 31	I.Ec.R.
	18 58	III.Oc.D.
29	1 34	III.Ec.R.
	5 36	I.Tr.I.
	6 23	I.Sh.I.
	7 52	I.Tr.E.
	8 38	I.Sh.E.
	16 32	II.Tr.I.
	18 08	II.Sh.I.
	19 17	II.Tr.E.
	20 48	II.Sh.E.
30	2 58	I.Oc.D.
	6 00	I.Ec.R.
31	0 06	I.Tr.I.
	0 52	I.Sh.I.
	2 22	I.Tr.E.
	3 07	I.Sh.E.
	10 44	II.Oc.D.
	14 58	II.Ec.R.
	21 28	I.Oc.D.

I. July 14	II. July 13	III. July 14	IV. July 25
$x_2 = +1.8,\ y_2 = -0.1$	$x_2 = +2.3,\ y_2 = -0.3$	$x_1 = +1.2,\ y_1 = -0.3$ $x_2 = +3.0,\ y_2 = -0.3$	$x_1 = +2.7,\ y_1 = -0.6$ $x_2 = +3.8,\ y_2 = -0.6$

NOTE.—I. denotes ingress; E., egress; D., disappearance; R., reappearance; Ec., eclipse; Oc., occultation; Tr., transit of the satellite; Sh., transit of the shadow.

CONFIGURATIONS OF SATELLITES I–IV FOR JULY

UNIVERSAL TIME

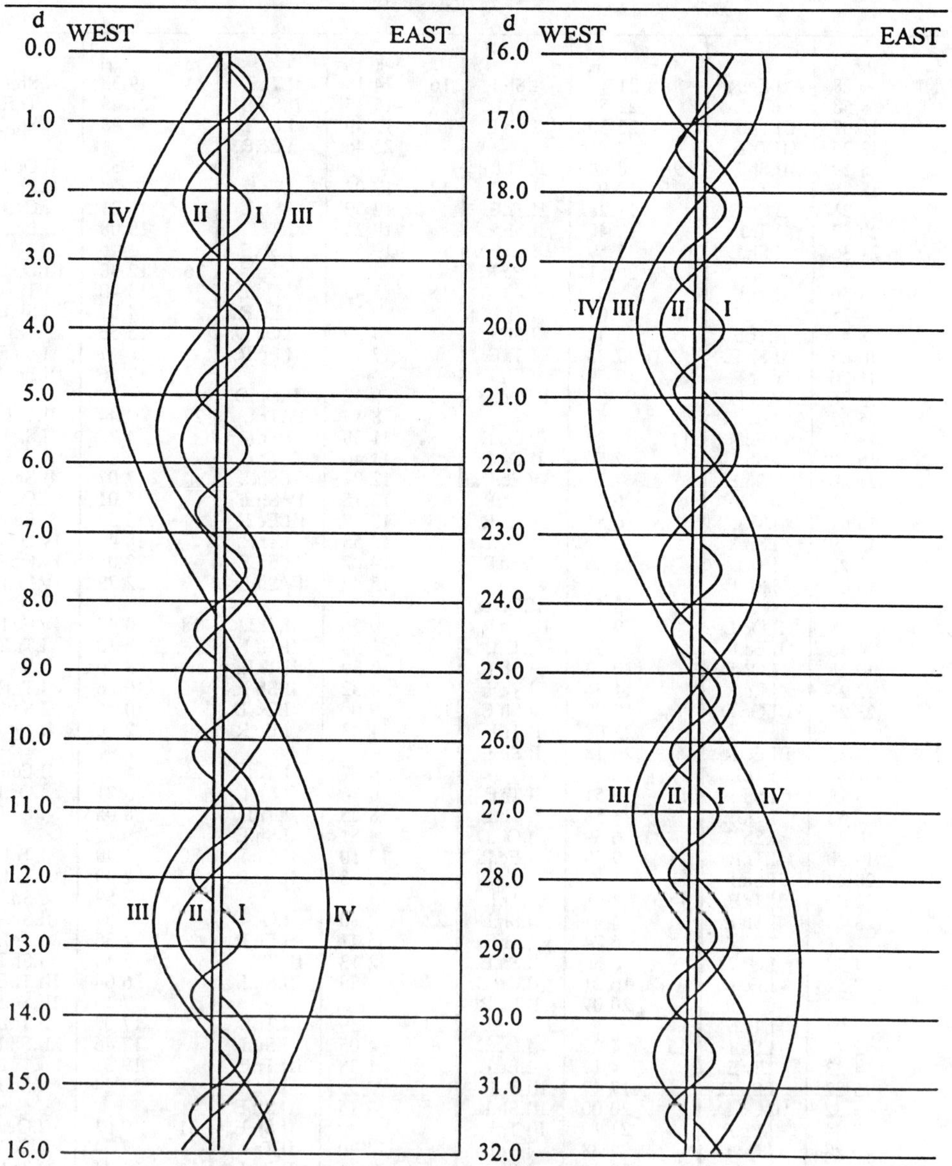

PHASES OF THE ECLIPSES

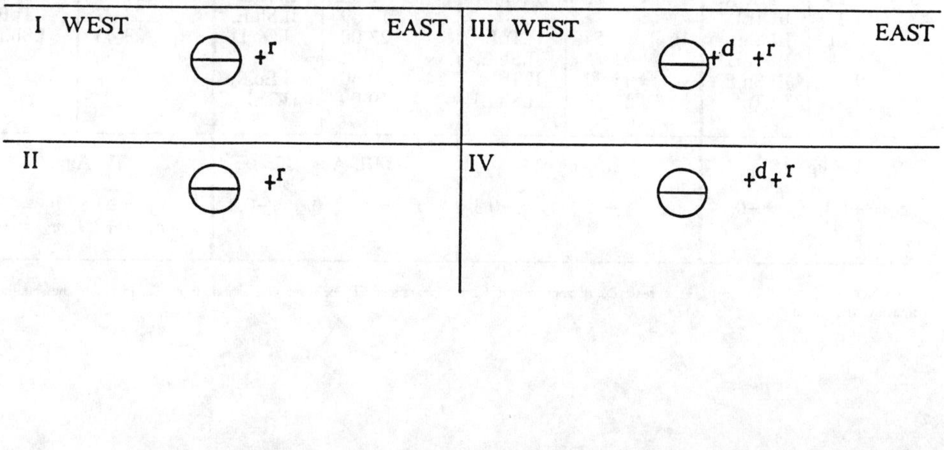

DYNAMICAL TIME OF GEOCENTRIC PHENOMENA

AUGUST

d	h m	
1	0 28	I.Ec.R.
	8 58	III.Tr.I.
	12 02	III.Sh.I.
	12 25	III.Tr.E.
	15 19	III.Sh.E.
	18 36	I.Tr.I.
	19 21	I.Sh.I.
	20 52	I.Tr.E.
	21 36	I.Sh.E.
2	5 56	II.Tr.I.
	7 25	II.Sh.I.
	8 40	II.Tr.E.
	10 05	II.Sh.E.
	11 56	IV.Tr.I.
	15 36	IV.Tr.E.
	15 58	I.Oc.D.
	18 57	I.Ec.R.
	19 12	IV.Sh.I.
	21 50	IV.Sh.E.
3	13 07	I.Tr.I.
	13 49	I.Sh.I.
	15 23	I.Tr.E.
	16 04	I.Sh.E.
4	0 08	II.Oc.D.
	4 16	II.Ec.R.
	10 28	I.Oc.D.
	13 26	I.Ec.R.
	23 22	III.Oc.D.
5	5 33	III.Ec.R.
	7 37	I.Tr.I.
	8 18	I.Sh.I.
	9 53	I.Tr.E.
	10 33	I.Sh.E.
	19 20	II.Tr.I.
	20 43	II.Sh.I.
	22 04	II.Tr.E.
	23 23	II.Sh.E.
6	4 58	I.Oc.D.
	7 54	I.Ec.R.
7	2 07	I.Tr.I.
	2 47	I.Sh.I.
	4 23	I.Tr.E.
	5 02	I.Sh.E.
	13 32	II.Oc.D.
	17 34	II.Ec.R.
	23 28	I.Oc.D.
8	2 23	I.Ec.R.
	13 21	III.Tr.I.
	16 01	III.Sh.I.
	16 48	III.Tr.E.
	19 17	III.Sh.E.
	20 37	I.Tr.I.
8	21 15	I.Sh.I.
	22 53	I.Tr.E.
	23 30	I.Sh.E.
9	8 44	II.Tr.I.
	10 01	II.Sh.I.
	11 28	II.Tr.E.
	12 40	II.Sh.E.
	17 59	I.Oc.D.
	20 51	I.Ec.R.
10	15 07	I.Tr.I.
	15 44	I.Sh.I.
	17 23	I.Tr.E.
	17 59	I.Sh.E.
	22 03	IV.Oc.D.
11	1 39	IV.Oc.R.
	2 56	II.Oc.D.
	4 22	IV.Ec.D.
	6 52	II.Ec.R.
	6 54	IV.Ec.R.
	12 29	I.Oc.D.
	15 20	I.Ec.R.
12	3 46	III.Oc.D.
	9 31	III.Ec.R.
	9 37	I.Tr.I.
	10 13	I.Sh.I.
	11 54	I.Tr.E.
	12 28	I.Sh.E.
	22 08	II.Tr.I.
	23 18	II.Sh.I.
13	0 51	II.Tr.E.
	1 58	II.Sh.E.
	6 59	I.Oc.D.
	9 49	I.Ec.R.
14	4 08	I.Tr.I.
	4 41	I.Sh.I.
	6 24	I.Tr.E.
	6 56	I.Sh.E.
	16 21	II.Oc.D.
	20 09	II.Ec.R.
15	1 29	I.Oc.D.
	4 17	I.Ec.R.
	17 47	III.Tr.I.
	20 00	III.Sh.I.
	21 12	III.Tr.E.
	22 38	I.Tr.I.
	23 10	I.Sh.I.
	23 16	III.Sh.E.
16	0 54	I.Tr.E.
	1 25	I.Sh.E.
	11 32	II.Tr.I.
	12 36	II.Sh.I.
16	14 15	II.Tr.E.
	15 15	II.Sh.E.
	19 59	I.Oc.D.
	22 46	I.Ec.R.
17	17 08	I.Tr.I.
	17 39	I.Sh.I.
	19 24	I.Tr.E.
	19 54	I.Sh.E.
18	5 45	II.Oc.D.
	9 28	II.Ec.R.
	14 30	I.Oc.D.
	17 14	I.Ec.R.
19	8 12	III.Oc.D.
	8 19	IV.Tr.I.
	11 38	I.Tr.I.
	11 46	IV.Tr.E.
	12 07	I.Sh.I.
	13 15	IV.Sh.I.
	13 28	III.Ec.R.
	13 55	I.Tr.E.
	14 22	I.Sh.E.
	15 42	IV.Sh.E.
20	0 56	II.Tr.I.
	1 53	II.Sh.I.
	3 39	II.Tr.E.
	4 32	II.Sh.E.
	9 00	I.Oc.D.
	11 43	I.Ec.R.
21	6 09	I.Tr.I.
	6 36	I.Sh.I.
	8 25	I.Tr.E.
	8 51	I.Sh.E.
	19 10	II.Oc.D.
	22 45	II.Ec.R.
22	3 30	I.Oc.D.
	6 11	I.Ec.R.
	22 13	III.Tr.I.
	23 59	III.Sh.I.
23	0 39	I.Tr.I.
	1 05	I.Sh.I.
	1 38	III.Tr.E.
	2 55	I.Tr.E.
	3 15	III.Sh.E.
	3 20	I.Sh.E.
	14 20	II.Tr.I.
	15 11	II.Sh.I.
	17 03	II.Tr.E.
	17 49	II.Sh.E.
	22 00	I.Oc.D.
24	0 40	I.Ec.R.
	19 09	I.Tr.I.
24	19 33	I.Sh.I.
	21 25	I.Tr.E.
	21 48	I.Sh.E.
25	8 35	II.Oc.D.
	12 04	II.Ec.R.
	16 31	I.Oc.D.
	19 08	I.Ec.R.
26	12 38	III.Oc.D.
	13 40	I.Tr.I.
	14 02	I.Sh.I.
	15 56	I.Tr.E.
	16 17	I.Sh.E.
	17 26	III.Ec.R.
27	3 44	II.Tr.I.
	4 28	II.Sh.I.
	6 27	II.Tr.E.
	7 07	II.Sh.E.
	11 01	I.Oc.D.
	13 37	I.Ec.R.
	18 41	IV.Oc.D.
	22 01	IV.Oc.R.
	22 29	IV.Ec.D.
28	0 47	IV.Ec.R.
	8 10	I.Tr.I.
	8 30	I.Sh.I.
	10 26	I.Tr.E.
	10 46	I.Sh.E.
	22 00	II.Oc.D.
29	1 21	II.Ec.R.
	5 31	I.Oc.D.
	8 05	I.Ec.R.
30	2 40	I.Tr.I.
	2 40	III.Tr.I.
	2 59	I.Sh.I.
	3 59	III.Sh.I.
	4 56	I.Tr.E.
	5 14	I.Sh.E.
	6 04	III.Tr.E.
	7 14	III.Sh.E.
	17 09	II.Tr.I.
	17 46	II.Sh.I.
	19 51	II.Tr.E.
	20 24	II.Sh.E.
31	0 01	I.Oc.D.
	2 34	I.Ec.R.
	21 11	I.Tr.I.
	21 28	I.Sh.I.
	23 27	I.Tr.E.
	23 43	I.Sh.E.

I. Aug. 15	II. Aug. 14	III. Aug. 12	IV. Aug. 11
$x_2 = +1.4,\ y_2 = -0.1$	$x_2 = +1.7,\ y_2 = -0.3$	$x_2 = +2.2,\ y_2 = -0.3$	$x_1 = +1.8,\ y_1 = -0.6$ $x_2 = +2.9,\ y_2 = -0.6$

NOTE.—I. denotes ingress; E., egress; D., disappearance; R., reappearance; Ec., eclipse; Oc., occultation; Tr., transit of the satellite; Sh., transit of the shadow.

CONFIGURATIONS OF SATELLITES I–IV FOR AUGUST

UNIVERSAL TIME

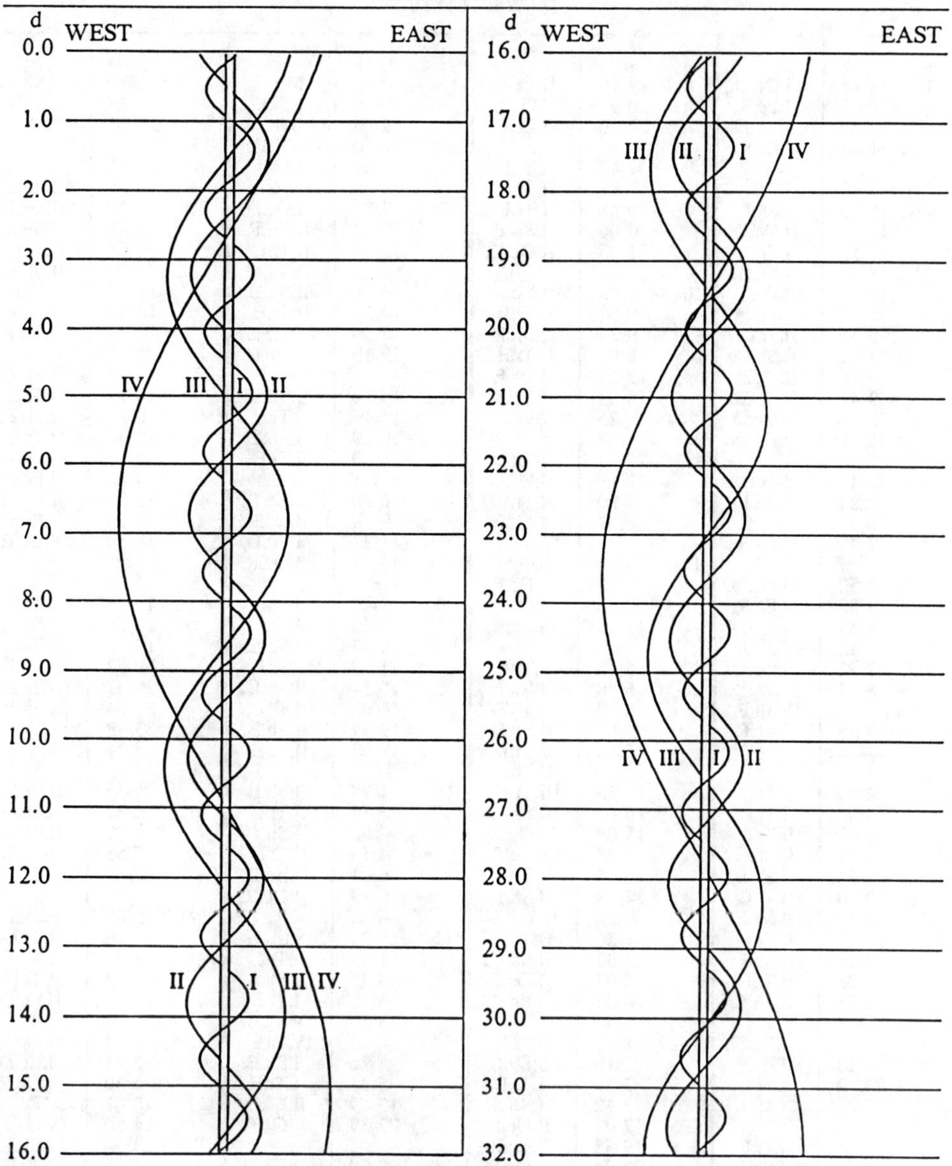

PHASES OF THE ECLIPSES

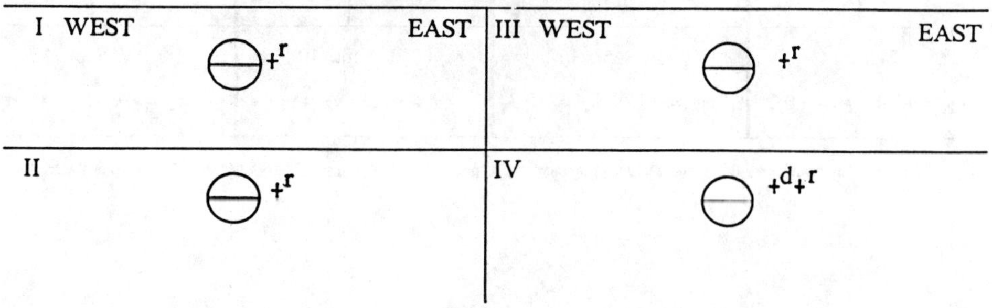

DYNAMICAL TIME OF GEOCENTRIC PHENOMENA

SEPTEMBER

d	h m	
1	11 25	II.Oc.D.
	14 40	II.Ec.R.
	18 32	I.Oc.D.
	21 02	I.Ec.R.
2	15 41	I.Tr.I.
	15 56	I.Sh.I.
	17 05	III.Oc.D.
	17 57	I.Tr.E.
	18 11	I.Sh.E.
	21 24	III.Ec.R.
3	6 33	II.Tr.I.
	7 03	II.Sh.I.
	9 14	II.Tr.E.
	9 41	II.Sh.E.
	13 02	I.Oc.D.
	15 31	I.Ec.R.
4	10 11	I.Tr.I.
	10 25	I.Sh.I.
	12 27	I.Tr.E.
	12 40	I.Sh.E.
5	0 50	II.Oc.D.
	3 57	II.Ec.R.
	4 55	IV.Tr.I.
	7 18	IV.Sh.I.
	7 32	I.Oc.D.
	8 04	IV.Tr.E.
	9 32	IV.Sh.E.
	9 59	I.Ec.R.
6	4 42	I.Tr.I.
	4 53	I.Sh.I.
	6 58	I.Tr.E.
	7 08	III.Tr.I.
	7 09	I.Sh.E.
	7 57	III.Sh.I.
	10 30	III.Tr.E.
	11 11	III.Sh.E.
	19 57	II.Tr.I.
	20 20	II.Sh.I.
	22 38	II.Tr.E.
	22 58	II.Sh.E.
7	2 02	I.Oc.D.
	4 28	I.Ec.R.
	23 12	I.Tr.I.
	23 22	I.Sh.I.
8	1 28	I.Tr.E.
	1 37	I.Sh.E.
	14 16	II.Oc.D.

d	h m	
8	17 16	II.Ec.R.
	20 33	I.Oc.D.
	22 56	I.Ec.R.
9	17 42	I.Tr.I.
	17 51	I.Sh.I.
	19 58	I.Tr.E.
	20 06	I.Sh.E.
	21 33	III.Oc.D.
10	1 22	III.Ec.R.
	9 21	II.Tr.I.
	9 37	II.Sh.I.
	12 02	II.Tr.E.
	12 15	II.Sh.E.
	15 03	I.Oc.D.
	17 25	I.Ec.R.
11	12 13	I.Tr.I.
	12 19	I.Sh.I.
	14 29	I.Tr.E.
	14 34	I.Sh.E.
12	3 41	II.Oc.D.
	6 33	II.Ec.R.
	9 33	I.Oc.D.
	11 53	I.Ec.R.
13	6 43	I.Tr.I.
	6 48	I.Sh.I.
	8 59	I.Tr.E.
	9 03	I.Sh.E.
	11 35	III.Tr.I.
	11 56	III.Sh.I.
	14 56	III.Tr.E.
	15 09	III.Sh.E.
	15 29	IV.Oc.D.
	18 38	IV.Ec.R.
	22 45	II.Tr.I.
	22 54	II.Sh.I.
14	1 26	II.Tr.E.
	1 32	II.Sh.E.
	4 03	I.Oc.D.
	6 22	I.Ec.R.
15	1 13	I.Tr.I.
	1 16	I.Sh.I.
	3 29	I.Tr.E.
	3 32	I.Sh.E.
	17 07	II.Oc.D.
	19 52	II.Ec.R.
	22 34	I.Oc.D.

d	h m	
16	0 50	I.Ec.R.
	19 44	I.Tr.I.
	19 45	I.Sh.I.
	22 00	I.Tr.E.
	22 00	I.Sh.E.
17	2 01	III.Oc.D.
	5 21	III.Oc.R.
	12 10	II.Tr.I.
	12 11	II.Sh.I.
	14 49	II.Sh.E.
	14 50	II.Tr.E.
	17 04	I.Oc.D.
	19 19	I.Oc.R.
18	14 14	I.Sh.I.
	14 14	I.Tr.I.
	16 29	I.Sh.E.
	16 30	I.Tr.E.
19	6 30	II.Ec.D.
	9 14	II.Oc.R.
	11 33	I.Ec.D.
	13 49	I.Oc.R.
20	8 42	I.Sh.I.
	8 44	I.Tr.I.
	10 57	I.Sh.E.
	11 00	I.Tr.E.
	15 54	III.Sh.I.
	16 03	III.Tr.I.
	19 07	III.Sh.E.
	19 22	III.Tr.E.
21	1 28	II.Sh.I.
	1 34	II.Tr.I.
	4 05	II.Sh.E.
	4 13	II.Tr.E.
	6 01	I.Ec.D.
	8 19	I.Oc.R.
22	1 21	IV.Sh.I.
	1 39	IV.Tr.I.
	3 11	I.Sh.I.
	3 15	I.Tr.I.
	3 20	IV.Sh.E.
	4 23	IV.Tr.E.
	5 26	I.Sh.E.
	5 31	I.Tr.E.
	19 49	II.Ec.D.
	22 39	II.Oc.R.
23	0 29	I.Ec.D.
	2 49	I.Oc.R.

d	h m	
23	21 39	I.Sh.I.
	21 45	I.Tr.I.
	23 54	I.Sh.E.
24	0 01	I.Tr.E.
	6 05	III.Ec.D.
	9 47	III.Oc.R.
	14 45	II.Sh.I.
	14 58	II.Tr.I.
	17 22	II.Sh.E.
	17 37	II.Tr.E.
	18 58	I.Ec.D.
	21 20	I.Oc.R.
25	16 08	I.Sh.I.
	16 15	I.Tr.I.
	18 23	I.Sh.E.
	18 31	I.Tr.E.
26	9 06	II.Ec.D.
	12 04	II.Oc.R.
	13 26	I.Ec.D.
	15 50	I.Oc.R.
27	10 36	I.Sh.I.
	10 46	I.Tr.I.
	12 51	I.Sh.E.
	13 01	I.Tr.E.
	19 52	III.Sh.I.
	20 31	III.Tr.I.
	23 04	III.Sh.E.
	23 48	III.Tr.E.
28	4 02	II.Sh.I.
	4 22	II.Tr.I.
	6 39	II.Sh.E.
	7 00	II.Tr.E.
	7 55	I.Ec.D.
	10 20	I.Oc.R.
29	5 05	I.Sh.I.
	5 16	I.Tr.I.
	7 20	I.Sh.E.
	7 32	I.Tr.E.
	22 25	II.Ec.D.
30	1 30	II.Oc.R.
	2 23	I.Ec.D.
	4 50	I.Oc.R.
	10 41	IV.Ec.D.
	14 50	IV.Oc.R.
	23 34	I.Sh.I.
	23 47	I.Tr.I.

I. Sept. 14	II. Sept. 15	III. Sept. 10	IV. Sept. 13
x_2 = +1.0, y_2 = −0.2	x_2 = +1.0, y_2 = −0.3	x_2 = +1.2, y_2 = −0.4	x_2 = +0.7, y_2 = −0.8

NOTE.—I. denotes ingress; E., egress; D., disappearance; R., reappearance; Ec., eclipse; Oc., occultation; Tr., transit of the satellite; Sh., transit of the shadow.

CONFIGURATIONS OF SATELLITES I–IV FOR SEPTEMBER

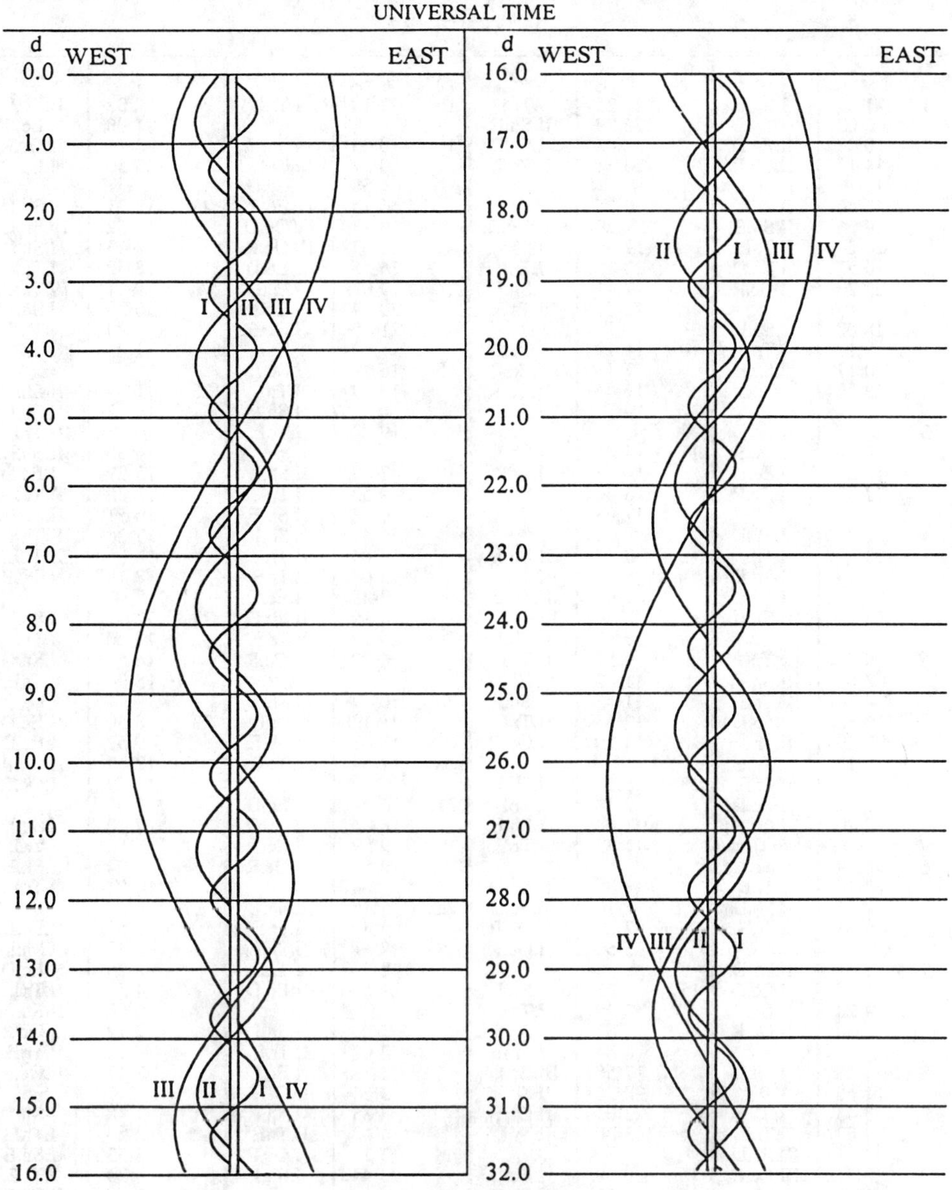

PHASES OF THE ECLIPSES

I WEST EAST

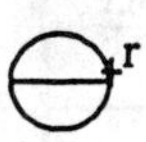

III WEST EAST

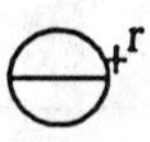

II

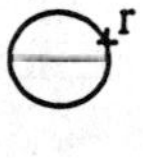

IV

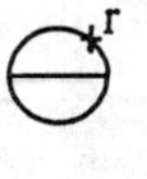

DYNAMICAL TIME OF GEOCENTRIC PHENOMENA

OCTOBER

d	h m	
1	1 49	I.Sh.E.
	2 02	I.Tr.E.
	10 03	III.Ec.D.
	14 13	III.Oc.R.
	17 19	II.Sh.I.
	17 46	II.Tr.I.
	19 56	II.Sh.E.
	20 24	II.Tr.E.
	20 52	I.Ec.D.
	23 20	I.Oc.R.
2	18 02	I.Sh.I.
	18 17	I.Tr.I.
	20 17	I.Sh.E.
	20 32	I.Tr.E.
3	11 43	II.Ec.D.
	14 54	II.Oc.R.
	15 20	I.Ec.D.
	17 50	I.Oc.R.
4	12 31	I.Sh.I.
	12 47	I.Tr.I.
	14 46	I.Sh.E.
	15 03	I.Tr.E.
	23 51	III.Sh.I.
5	0 59	III.Tr.I.
	3 03	III.Sh.E.
	4 14	III.Tr.E.
	6 36	II.Sh.I.
	7 09	II.Tr.I.
	9 12	II.Sh.E.
	9 47	II.Tr.E.
	9 48	I.Ec.D.
	12 20	I.Oc.R.
6	6 59	I.Sh.I.
	7 17	I.Tr.I.
	9 14	I.Sh.E.
	9 33	I.Tr.E.
7	1 01	II.Ec.D.
	4 17	I.Ec.D.
	4 20	II.Oc.R.
	6 50	I.Oc.R.
8	1 28	I.Sh.I.
	1 48	I.Tr.I.
	3 43	I.Sh.E.
	4 03	I.Tr.E.
	14 01	III.Ec.D.
	18 37	III.Oc.R.
	19 25	IV.Sh.I.
	19 53	II.Sh.I.
	20 33	II.Tr.I.
	21 07	IV.Sh.E.

d	h m	
8	22 25	IV.Tr.I.
	22 29	II.Sh.E.
	22 45	I.Ec.D.
	23 11	II.Tr.E.
9	0 37	IV.Tr.E.
	1 20	I.Oc.R.
	19 56	I.Sh.I.
	20 18	I.Tr.I.
	22 11	I.Sh.E.
	22 33	I.Tr.E.
10	14 19	II.Ec.D.
	17 13	I.Ec.D.
	17 44	II.Oc.R.
	19 50	I.Oc.R.
11	14 25	I.Sh.I.
	14 48	I.Tr.I.
	16 40	I.Sh.E.
	17 03	I.Tr.E.
12	3 50	III.Sh.I.
	5 26	III.Tr.I.
	7 00	III.Sh.E.
	8 39	III.Tr.E.
	9 10	II.Sh.I.
	9 57	II.Tr.I.
	11 42	I.Ec.D.
	11 45	II.Sh.E.
	12 34	II.Tr.E.
	14 20	I.Oc.R.
13	8 53	I.Sh.I.
	9 18	I.Tr.I.
	11 08	I.Sh.E.
	11 33	I.Tr.E.
14	3 38	II.Ec.D.
	6 10	I.Ec.D.
	7 09	II.Oc.R.
	8 50	I.Oc.R.
15	3 22	I.Sh.I.
	3 49	I.Tr.I.
	5 37	I.Sh.E.
	6 04	I.Tr.E.
	17 58	III.Ec.D.
	22 26	II.Sh.I.
	23 01	III.Oc.R.
	23 20	II.Tr.I.
16	0 39	I.Ec.D.
	1 02	II.Sh.E.
	1 57	II.Tr.E.
	3 20	I.Oc.R.
	21 50	I.Sh.I.

d	h m	
16	22 19	I.Tr.I.
17	0 05	I.Sh.E.
	0 34	I.Tr.E.
	4 50	IV.Ec.D.
	6 15	IV.Ec.R.
	9 18	IV.Oc.D.
	11 02	IV.Oc.R.
	16 55	II.Ec.D.
	19 07	I.Ec.D.
	20 34	II.Oc.R.
	21 50	I.Oc.R.
18	16 19	I.Sh.I.
	16 49	I.Tr.I.
	18 34	I.Sh.E.
	19 04	I.Tr.E.
19	7 49	III.Sh.I.
	9 53	III.Tr.I.
	10 58	III.Sh.E.
	11 43	II.Sh.I.
	12 44	II.Tr.I.
	13 04	III.Tr.E.
	13 35	I.Ec.D.
	14 19	II.Sh.E.
	15 20	II.Tr.E.
	16 20	I.Oc.R.
20	10 47	I.Sh.I.
	11 19	I.Tr.I.
	13 02	I.Sh.E.
	13 34	I.Tr.E.
21	6 14	II.Ec.D.
	8 04	I.Ec.D.
	9 59	II.Oc.R.
	10 50	I.Oc.R.
22	5 16	I.Sh.I.
	5 49	I.Tr.I.
	7 31	I.Sh.E.
	8 04	I.Tr.E.
	21 56	III.Ec.D.
23	1 00	II.Sh.I.
	2 07	II.Tr.I.
	2 32	I.Ec.D.
	3 24	III.Oc.R.
	3 35	II.Sh.E.
	4 42	II.Tr.E.
	5 20	I.Oc.R.
	23 44	I.Sh.I.
24	0 19	I.Tr.I.
	1 59	I.Sh.E.
	2 34	I.Tr.E.

d	h m	
24	19 32	II.Ec.D.
	21 00	I.Ec.D.
	23 23	II.Oc.R.
	23 50	I.Oc.R.
25	13 31	IV.Sh.I.
	14 52	IV.Sh.E.
	18 13	I.Sh.I.
	18 50	I.Tr.I.
	19 11	IV.Tr.I.
	20 28	I.Sh.E.
	20 33	IV.Tr.E.
	21 04	I.Tr.E.
26	11 47	III.Sh.I.
	14 17	II.Sh.I.
	14 18	III.Tr.I.
	14 56	III.Sh.E.
	15 29	I.Ec.D.
	15 30	II.Tr.I.
	16 51	II.Sh.E.
	17 26	III.Tr.E.
	18 05	II.Tr.E.
	18 19	I.Oc.R.
27	12 41	I.Sh.I.
	13 20	I.Tr.I.
	14 56	I.Sh.E.
	15 34	I.Tr.E.
28	8 50	II.Ec.D.
	9 57	I.Ec.D.
	12 47	II.Oc.R.
	12 49	I.Oc.R.
29	7 10	I.Sh.I.
	7 50	I.Tr.I.
	9 24	I.Sh.E.
	10 04	I.Tr.E.
30	1 55	III.Ec.D.
	3 33	II.Sh.I.
	4 25	I.Ec.D.
	4 53	II.Tr.I.
	6 08	II.Sh.E.
	7 19	I.Oc.R.
	7 27	II.Tr.E.
	7 47	III.Oc.R.
31	1 38	I.Sh.I.
	2 20	I.Tr.I.
	3 53	I.Sh.E.
	4 34	I.Tr.E.
	22 08	II.Ec.D.
	22 53	I.Ec.D.

I. Oct. 14	II. Oct. 14	III. Oct. 15	IV. Oct. 17
$x_1 = -1.3,\ y_1 = -0.2$	$x_1 = -1.5,\ y_1 = -0.4$	$x_1 = -1.9,\ y_1 = -0.4$	$x_1 = -2.2,\ y_1 = -0.9$ $x_2 = -1.6,\ y_2 = -0.9$

NOTE.—I. denotes ingress; E., egress; D., disappearance; R., reappearance; Ec., eclipse; Oc., occultation; Tr., transit of the satellite; Sh., transit of the shadow.

CONFIGURATIONS OF SATELLITES I–IV FOR OCTOBER

UNIVERSAL TIME

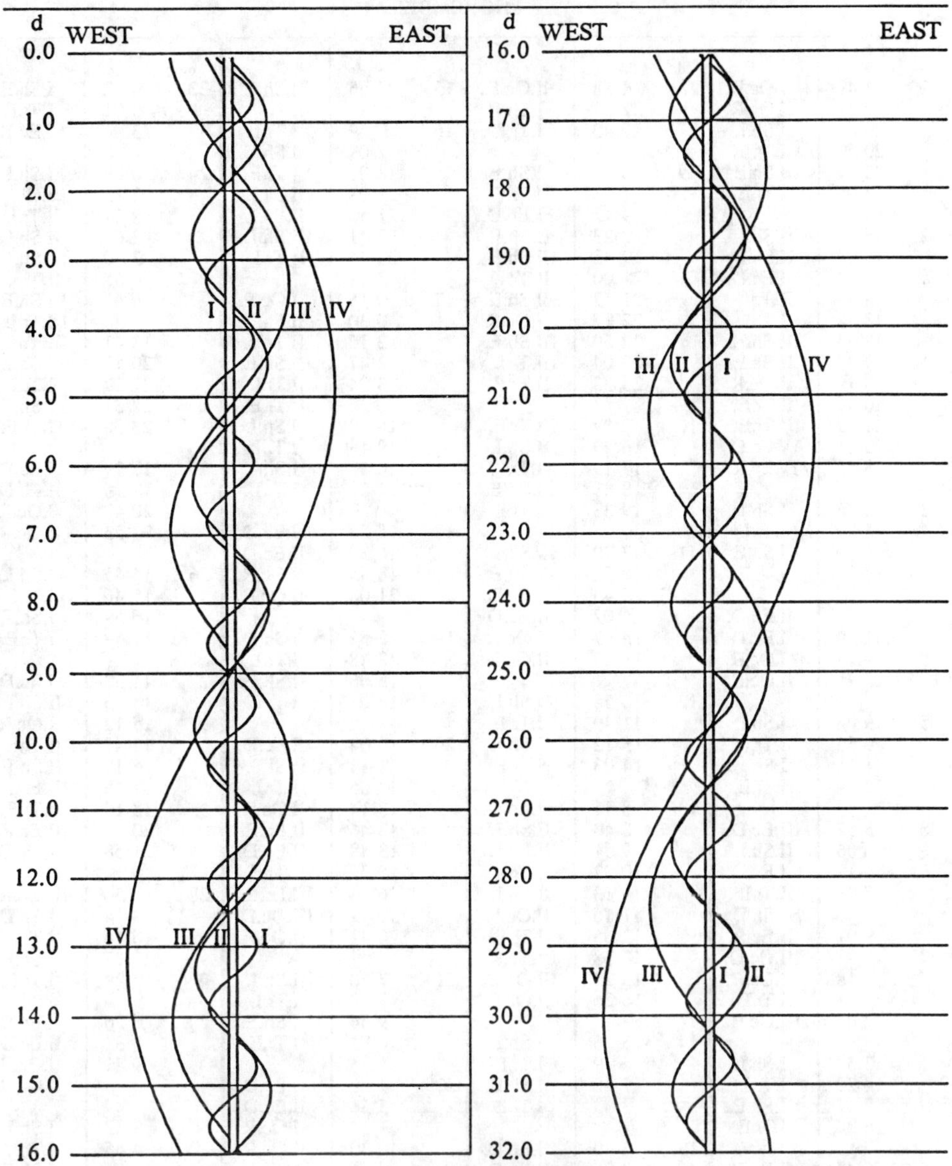

PHASES OF THE ECLIPSES

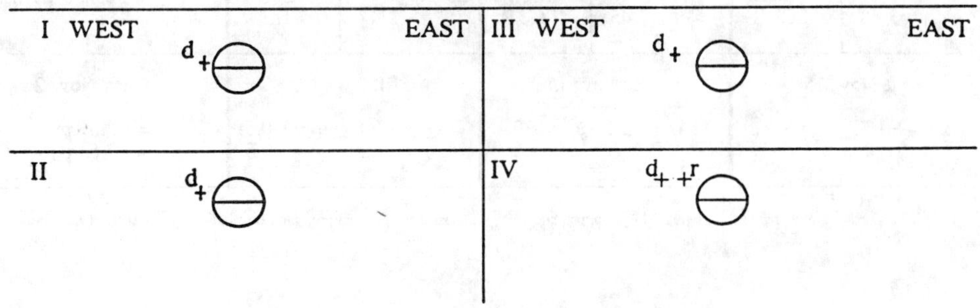

DYNAMICAL TIME OF GEOCENTRIC PHENOMENA

NOVEMBER

d	h m	
1	1 49	I.Oc.R.
	2 11	II.Oc.R.
	20 07	I.Sh.I.
	20 50	I.Tr.I.
	22 21	I.Sh.E.
	23 04	I.Tr.E.
2	15 45	III.Sh.I.
	16 50	II.Sh.I.
	17 22	I.Ec.D.
	18 15	II.Tr.I.
	18 42	III.Tr.I.
	18 53	III.Sh.E.
	19 24	II.Sh.E.
	20 19	I.Oc.R.
	20 49	II.Tr.E.
	21 47	III.Tr.E.
	23 01	IV.Ec.D.
	23 58	IV.Ec.R.
3	14 35	I.Sh.I.
	15 20	I.Tr.I.
	16 50	I.Sh.E.
	17 34	I.Tr.E.
4	11 26	II.Ec.D.
	11 50	I.Ec.D.
	14 48	I.Oc.R.
	15 35	II.Oc.R.
5	9 04	I.Sh.I.
	9 50	I.Tr.I.
	11 18	I.Sh.E.
	12 04	I.Tr.E.
6	5 52	III.Ec.D.
	6 06	II.Sh.I.
	6 18	I.Ec.D.
	7 38	II.Tr.I.
	8 41	II.Sh.E.
	9 01	III.Ec.R.
	9 01	III.Oc.D.
	9 18	I.Oc.R.
	10 11	II.Tr.E.
	12 07	III.Oc.R.
7	3 32	I.Sh.I.
	4 20	I.Tr.I.
	5 47	I.Sh.E.
	6 34	I.Tr.E.
8	0 44	II.Ec.D.
	0 47	I.Ec.D.
	3 48	I.Oc.R.
8	4 58	II.Oc.R.
	22 01	I.Sh.I.
	22 49	I.Tr.I.
9	0 15	I.Sh.E.
	1 03	I.Tr.E.
	19 15	I.Ec.D.
	19 23	II.Sh.I.
	19 43	III.Sh.I.
	21 00	II.Tr.I.
	21 57	II.Sh.E.
	22 17	I.Oc.R.
	22 50	III.Sh.E.
	23 04	III.Tr.I.
	23 33	II.Tr.E.
10	2 07	III.Tr.E.
	16 29	I.Sh.I.
	17 19	I.Tr.I.
	18 43	I.Sh.E.
	19 33	I.Tr.E.
11	7 39	IV.Sh.I.
	8 32	IV.Sh.E.
	13 43	I.Ec.D.
	14 02	II.Ec.D.
	16 47	I.Oc.R.
	18 22	II.Oc.R.
12	10 58	I.Sh.I.
	11 49	I.Tr.I.
	13 12	I.Sh.E.
	14 03	I.Tr.E.
13	8 11	I.Ec.D.
	8 40	II.Sh.I.
	9 51	III.Ec.D.
	10 22	II.Tr.I.
	11 13	II.Sh.E.
	11 16	I.Oc.R.
	12 55	II.Tr.E.
	12 58	III.Ec.R.
	13 23	III.Oc.D.
	16 26	III.Oc.R.
14	5 26	I.Sh.I.
	6 19	I.Tr.I.
	7 40	I.Sh.E.
	8 33	I.Tr.E.
15	2 40	I.Ec.D.
	3 20	II.Ec.D.
	5 46	I.Oc.R.
	7 45	II.Oc.R.
15	23 55	I.Sh.I.
16	0 49	I.Tr.I.
	2 09	I.Sh.E.
	3 02	I.Tr.E.
	21 08	I.Ec.D.
	21 56	II.Sh.I.
	23 41	III.Sh.I.
	23 44	II.Tr.I.
17	0 15	I.Oc.R.
	0 30	II.Sh.E.
	2 16	II.Tr.E.
	2 47	III.Sh.E.
	3 24	III.Tr.I.
	6 25	III.Tr.E.
	18 23	I.Sh.I.
	19 18	I.Tr.I.
	20 37	I.Sh.E.
	21 32	I.Tr.E.
18	15 36	I.Ec.D.
	16 39	II.Ec.D.
	18 45	I.Oc.R.
	21 08	II.Oc.R.
19	12 52	I.Sh.I.
	13 48	I.Tr.I.
	15 06	I.Sh.E.
	16 01	I.Tr.E.
20	10 04	I.Ec.D.
	11 13	II.Sh.I.
	13 06	II.Tr.I.
	13 14	I.Oc.R.
	13 46	II.Sh.E.
	13 48	III.Ec.D.
	15 37	II.Tr.E.
	16 54	III.Ec.R.
	17 42	III.Oc.D.
	20 42	III.Oc.R.
21	7 20	I.Sh.I.
	8 18	I.Tr.I.
	9 34	I.Sh.E.
	10 31	I.Tr.E.
22	4 33	I.Ec.D.
	5 56	II.Ec.D.
	7 44	I.Oc.R.
	10 30	II.Oc.R.
23	1 49	I.Sh.I.
	2 47	I.Tr.I.
23	4 02	I.Sh.E.
	5 01	I.Tr.E.
	23 01	I.Ec.D.
24	0 29	II.Sh.I.
	2 13	I.Oc.R.
	2 27	II.Tr.I.
	3 03	II.Sh.E.
	3 40	III.Sh.I.
	4 58	II.Tr.E.
	6 45	III.Sh.E.
	7 44	III.Tr.I.
	10 41	III.Tr.E.
	20 17	I.Sh.I.
	21 17	I.Tr.I.
	22 31	I.Sh.E.
	23 30	I.Tr.E.
25	17 29	I.Ec.D.
	19 15	II.Ec.D.
	20 42	I.Oc.R.
	23 53	II.Oc.R.
26	14 45	I.Sh.I.
	15 46	I.Tr.I.
	16 59	I.Sh.E.
	17 59	I.Tr.E.
27	11 57	I.Ec.D.
	13 46	II.Sh.I.
	15 12	I.Oc.R.
	15 48	II.Tr.I.
	16 19	II.Sh.E.
	17 45	III.Ec.D.
	18 18	II.Tr.E.
	20 51	III.Ec.R.
	21 58	III.Oc.D.
28	0 56	III.Oc.R.
	9 14	I.Sh.I.
	10 16	I.Tr.I.
	11 27	I.Sh.E.
	12 29	I.Tr.E.
29	6 26	I.Ec.D.
	8 32	II.Ec.D.
	9 41	I.Oc.R.
	13 15	II.Oc.R.
30	3 42	I.Sh.I.
	4 45	I.Tr.I.
	5 56	I.Sh.E.
	6 58	I.Tr.E.

I. Nov. 15	II. Nov. 15	III. Nov. 13	IV. Nov. 2
$x_1 = -1.7,\ y_1 = -0.2$	$x_1 = -2.1,\ y_1 = -0.4$	$x_1 = -2.8,\ y_1 = -0.5$	$x_1 = -3.0,\ y_1 = -0.9$
		$x_2 = -1.1,\ y_2 = -0.5$	$x_2 = -2.6,\ y_2 = -0.9$

NOTE.—I. denotes ingress; E., egress; D., disappearance; R., reappearance; Ec., eclipse; Oc., occultation; Tr., transit of the satellite; Sh., transit of the shadow.

CONFIGURATIONS OF SATELLITES I–IV FOR NOVEMBER

UNIVERSAL TIME

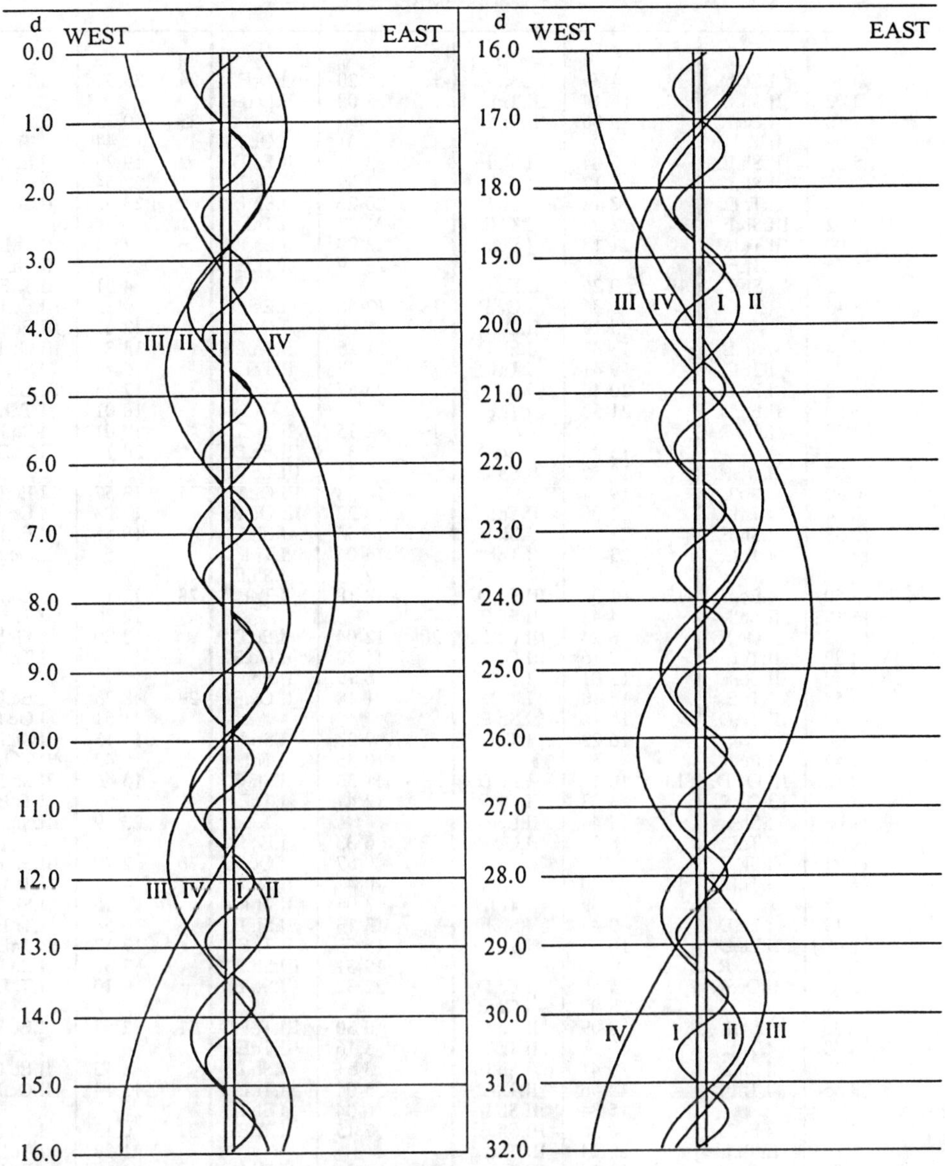

PHASES OF THE ECLIPSES

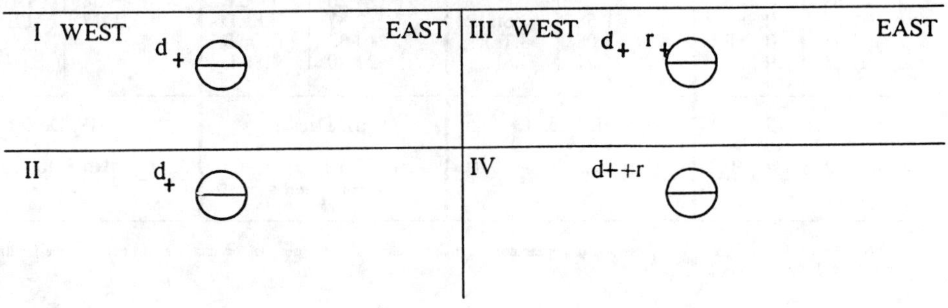

DYNAMICAL TIME OF GEOCENTRIC PHENOMENA

DECEMBER

d	h m	
1	0 54	I.Ec.D.
	3 02	II.Sh.I.
	4 10	I.Oc.R.
	5 09	II.Tr.I.
	5 35	II.Sh.E.
	7 38	III.Sh.I.
	7 38	II.Tr.E.
	10 42	III.Sh.E.
	12 00	III.Tr.I.
	14 55	III.Tr.E.
	22 11	I.Sh.I.
	23 15	I.Tr.I.
2	0 24	I.Sh.E.
	1 27	I.Tr.E.
	19 22	I.Ec.D.
	21 50	II.Ec.D.
	22 39	I.Oc.R.
3	2 37	II.Oc.R.
	16 39	I.Sh.I.
	17 44	I.Tr.I.
	18 53	I.Sh.E.
	19 57	I.Tr.E.
4	13 50	I.Ec.D.
	16 19	II.Sh.I.
	17 08	I.Oc.R.
	18 29	II.Tr.I.
	18 52	II.Sh.E.
	20 58	II.Tr.E.
	21 42	III.Ec.D.
5	0 47	III.Ec.R.
	2 12	III.Oc.D.
	5 07	III.Oc.R.
	11 08	I.Sh.I.
	12 13	I.Tr.I.
	13 21	I.Sh.E.
	14 26	I.Tr.E.
6	8 19	I.Ec.D.
	11 08	II.Ec.D.
	11 38	I.Oc.R.
	15 57	II.Oc.R.
7	5 36	I.Sh.I.
	6 43	I.Tr.I.
	7 49	I.Sh.E.
	8 55	I.Tr.E.
8	2 47	I.Ec.D.
	5 35	II.Sh.I.
	6 07	I.Oc.R.
	7 49	II.Tr.I.
	8 08	II.Sh.E.
	10 18	II.Tr.E.
	11 36	III.Sh.I.
8	14 39	III.Sh.E.
	16 14	III.Tr.I.
	19 06	III.Tr.E.
9	0 04	I.Sh.I.
	1 12	I.Tr.I.
	2 18	I.Sh.E.
	3 24	I.Tr.E.
	21 15	I.Ec.D.
10	0 26	II.Ec.D.
	0 36	I.Oc.R.
	5 19	II.Oc.R.
	18 33	I.Sh.I.
	19 41	I.Tr.I.
	20 46	I.Sh.E.
	21 53	I.Tr.E.
11	15 43	I.Ec.D.
	18 52	II.Sh.I.
	19 04	I.Oc.R.
	21 09	II.Tr.I.
	21 24	II.Sh.E.
	23 37	II.Tr.E.
12	1 39	III.Ec.D.
	4 43	III.Ec.R.
	6 24	III.Oc.D.
	9 16	III.Oc.R.
	13 01	I.Sh.I.
	14 10	I.Tr.I.
	15 14	I.Sh.E.
	16 22	I.Tr.E.
13	10 11	I.Ec.D.
	13 33	I.Oc.R.
	13 44	II.Ec.D.
	18 38	II.Oc.R.
14	7 30	I.Sh.I.
	8 39	I.Tr.I.
	9 43	I.Sh.E.
	10 51	I.Tr.E.
15	4 40	I.Ec.D.
	8 02	I.Oc.R.
	8 09	II.Sh.I.
	10 28	II.Tr.I.
	10 41	II.Sh.E.
	12 56	II.Tr.E.
	15 34	III.Sh.I.
	18 36	III.Sh.E.
	20 24	III.Tr.I.
	23 13	III.Tr.E.
16	1 58	I.Sh.I.
	3 08	I.Tr.I.
	4 11	I.Sh.E.
16	5 20	I.Tr.E.
	23 08	I.Ec.D.
17	2 31	I.Oc.R.
	3 02	II.Ec.D.
	7 59	II.Oc.R.
	20 26	I.Sh.I.
	21 37	I.Tr.I.
	22 39	I.Sh.E.
	23 49	I.Tr.E.
18	17 36	I.Ec.D.
	21 00	I.Oc.R.
	21 25	II.Sh.I.
	23 47	II.Tr.I.
	23 57	II.Sh.E.
19	2 15	II.Tr.E.
	5 37	III.Ec.D.
	8 41	III.Ec.R.
	10 33	III.Oc.D.
	13 22	III.Oc.R.
	14 55	I.Sh.I.
	16 06	I.Tr.I.
	17 08	I.Sh.E.
	18 18	I.Tr.E.
20	12 04	I.Ec.D.
	15 28	I.Oc.R.
	16 19	II.Ec.D.
	21 18	II.Oc.R.
21	9 23	I.Sh.I.
	10 35	I.Tr.I.
	11 36	I.Sh.E.
	12 47	I.Tr.E.
22	6 33	I.Ec.D.
	9 57	I.Oc.R.
	10 42	II.Sh.I.
	13 06	II.Tr.I.
	13 13	II.Sh.E.
	15 33	II.Tr.E.
	19 31	III.Sh.I.
	22 32	III.Sh.E.
23	0 30	III.Tr.I.
	3 16	III.Tr.E.
	3 51	I.Sh.I.
	5 04	I.Tr.I.
	6 04	I.Sh.E.
	7 15	I.Tr.E.
24	1 01	I.Ec.D.
	4 26	I.Oc.R.
	5 37	II.Ec.D.
	10 37	II.Oc.R.
	22 20	I.Sh.I.
24	23 32	I.Tr.I.
25	0 33	I.Sh.E.
	1 44	I.Tr.E.
	19 29	I.Ec.D.
	22 54	I.Oc.R.
	23 58	II.Sh.I.
26	2 24	II.Tr.I.
	2 30	II.Sh.E.
	4 51	II.Tr.E.
	9 35	III.Ec.D.
	12 37	III.Ec.R.
	14 38	III.Oc.D.
	16 48	I.Sh.I.
	17 25	III.Oc.R.
	18 01	I.Tr.I.
	19 01	I.Sh.E.
	20 13	I.Tr.E.
27	13 57	I.Ec.D.
	17 23	I.Oc.R.
	18 55	II.Ec.D.
	23 55	II.Oc.R.
28	11 17	I.Sh.I.
	12 30	I.Tr.I.
	13 29	I.Sh.E.
	14 41	I.Tr.E.
29	8 25	I.Ec.D.
	11 51	I.Oc.R.
	13 15	II.Sh.I.
	15 42	II.Tr.I.
	15 46	II.Sh.E.
	18 08	II.Tr.E.
	23 29	III.Sh.I.
30	2 29	III.Sh.E.
	4 33	III.Tr.I.
	5 45	I.Sh.I.
	6 58	I.Tr.I.
	7 17	III.Tr.E.
	7 58	I.Sh.E.
	9 10	I.Tr.E.
31	2 54	I.Ec.D.
	6 19	I.Oc.R.
	8 13	II.Ec.D.
	13 14	II.Oc.R.
32	0 13	I.Sh.I.
	1 27	I.Tr.I.
	2 26	I.Sh.E.
	3 38	I.Tr.E.
	21 22	I.Ec.D.

I. Dec. 15	II. Dec. 13	III. Dec. 12	IV. Dec.
$x_1 = -1.9,\ y_1 = -0.2$	$x_1 = -2.5,\ y_1 = -0.5$	$x_1 = -3.3,\ y_1 = -0.6$ $x_2 = -1.7,\ y_2 = -0.6$	No Eclipse

NOTE.—I. denotes ingress; E., egress; D., disappearance; R., reappearance; Ec., eclipse; Oc., occultation; Tr., transit of the satellite; Sh., transit of the shadow.

CONFIGURATIONS OF SATELLITES I–IV FOR DECEMBER

UNIVERSAL TIME

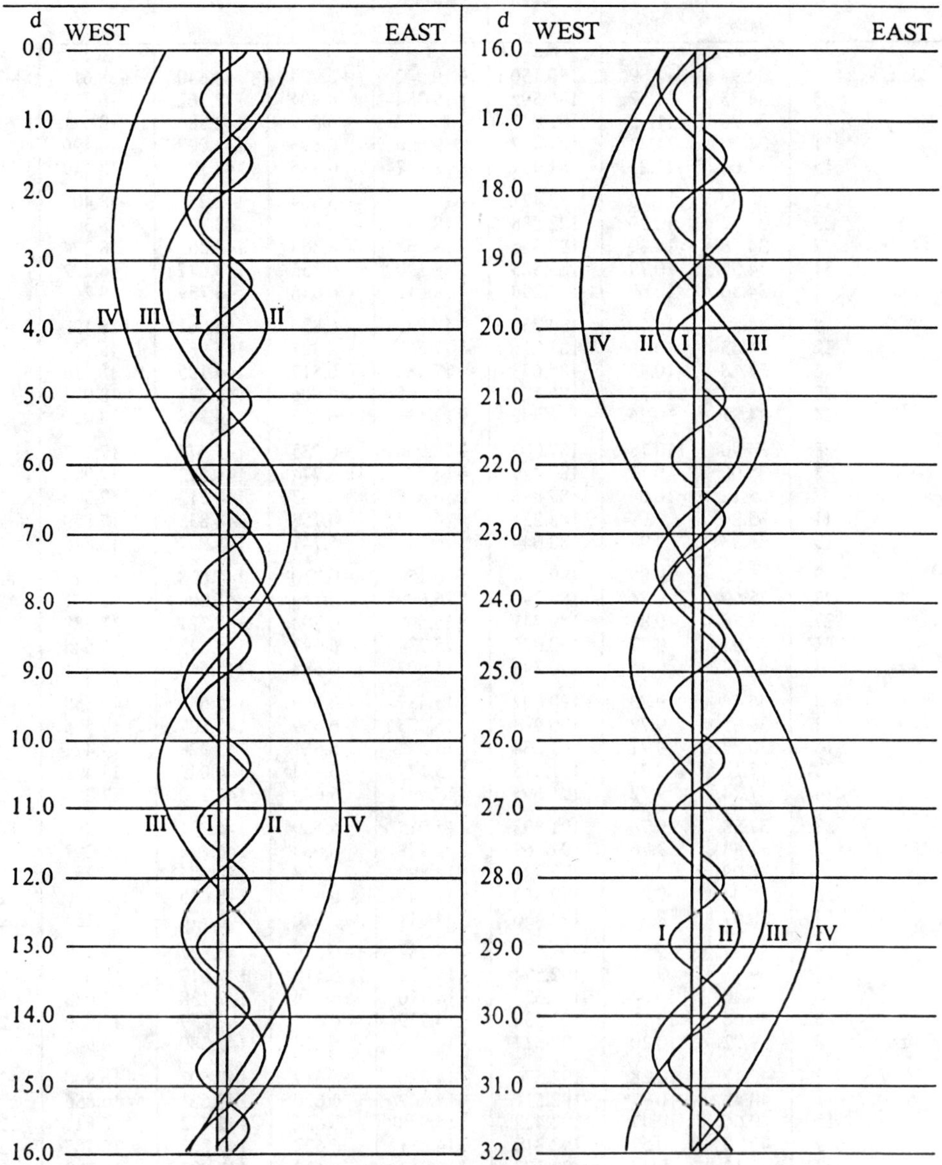

PHASES OF THE ECLIPSES

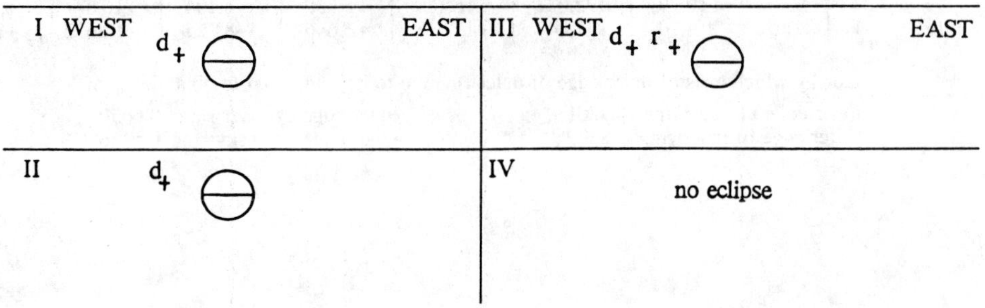

FOR 0^h UNIVERSAL TIME

Date		Axes of outer edge of outer ring: Major	Axes of outer edge of outer ring: Minor	*U*	*B*	*P*	*U′*	*B′*	*P′*
		″	″	°	°	°	°	°	°
Jan.	−1	34.94	11.69	180.150	+19.542	+6.909	142.640	+18.614	+21.966
	3	34.85	11.57	180.592	19.384	6.904	142.762	18.573	22.003
	7	34.76	11.45	181.041	19.222	6.899	142.884	18.531	22.041
	11	34.70	11.33	181.497	19.056	6.893	143.007	18.490	22.078
	15	34.64	11.21	181.957	18.887	6.886	143.129	18.449	22.115
	19	34.60	11.10	182.421	+18.716	+6.879	143.251	+18.407	+22.153
	23	34.57	10.99	182.888	18.542	6.871	143.373	18.365	22.190
	27	34.56	10.89	183.356	18.367	6.863	143.495	18.324	22.227
	31	34.55	10.79	183.823	18.190	6.855	143.617	18.282	22.264
Feb.	4	34.56	10.69	184.290	18.012	6.846	143.739	18.240	22.300
	8	34.59	10.59	184.755	+17.835	+6.837	143.861	+18.198	+22.337
	12	34.63	10.50	185.216	17.657	6.827	143.983	18.156	22.374
	16	34.68	10.42	185.673	17.481	6.817	144.105	18.114	22.410
	20	34.74	10.33	186.123	17.305	6.806	144.227	18.071	22.446
	24	34.81	10.26	186.567	17.132	6.796	144.348	18.029	22.483
	28	34.90	10.18	187.003	+16.960	+6.785	144.470	+17.987	+22.519
Mar.	3	35.00	10.11	187.429	16.792	6.774	144.592	17.944	22.555
	7	35.12	10.05	187.846	16.626	6.763	144.713	17.902	22.591
	11	35.24	9.99	188.251	16.465	6.752	144.835	17.859	22.626
	15	35.38	9.94	188.644	16.308	6.741	144.957	17.816	22.662
	19	35.53	9.89	189.024	+16.156	+6.730	145.078	+17.773	+22.698
	23	35.69	9.84	189.390	16.009	6.719	145.200	17.731	22.733
	27	35.86	9.81	189.741	15.869	6.708	145.321	17.688	22.768
	31	36.05	9.77	190.075	15.734	6.698	145.443	17.645	22.804
Apr.	4	36.24	9.75	190.393	15.607	6.688	145.564	17.601	22.839
	8	36.44	9.73	190.692	+15.487	+6.679	145.685	+17.558	+22.874
	12	36.65	9.72	190.973	15.375	6.669	145.807	17.515	22.909
	16	36.87	9.71	191.234	15.272	6.661	145.928	17.471	22.944
	20	37.10	9.71	191.475	15.177	6.653	146.049	17.428	22.978
	24	37.34	9.72	191.695	15.091	6.645	146.170	17.384	23.013
	28	37.58	9.74	191.893	+15.015	+6.639	146.291	+17.341	+23.047
May	2	37.83	9.76	192.069	14.948	6.633	146.412	17.297	23.082
	6	38.09	9.79	192.221	14.892	6.628	146.534	17.253	23.116
	10	38.34	9.82	192.350	14.846	6.623	146.655	17.210	23.150
	14	38.61	9.87	192.456	14.811	6.620	146.776	17.166	23.184
	18	38.87	9.92	192.537	+14.787	+6.617	146.897	+17.122	+23.218
	22	39.13	9.98	192.594	14.773	6.615	147.017	17.077	23.252
	26	39.40	10.04	192.627	14.770	6.614	147.138	17.033	23.285
	30	39.66	10.12	192.634	14.779	6.615	147.259	16.989	23.319
June	3	39.92	10.20	192.617	14.798	6.616	147.380	16.945	23.352
	7	40.17	10.28	192.576	+14.828	+6.617	147.501	+16.900	+23.386
	11	40.42	10.37	192.511	14.869	6.620	147.621	16.856	23.419
	15	40.66	10.47	192.422	14.920	6.624	147.742	16.811	23.452
	19	40.90	10.57	192.310	14.981	6.628	147.863	16.767	23.485
	23	41.12	10.68	192.177	15.051	6.633	147.983	16.722	23.518
	27	41.33	10.79	192.022	+15.130	+6.639	148.104	+16.677	+23.551
July	1	41.52	10.90	191.847	+15.218	+6.646	148.225	+16.632	+23.584

Factor by which axes of outer edge of outer ring are to be multiplied to obtain axes of:

Inner edge of outer ring 0.8801
Outer edge of inner ring 0.8599
Inner edge of inner ring 0.6650
Inner edge of dusky ring 0.5486

FOR 0^h UNIVERSAL TIME

Date		Axes of outer edge of outer ring Major	Minor	*U*	*B*	*P*	*U′*	*B′*	*P′*
		″	″	°	°	°	°	°	°
July	1	41.52	10.90	191.847	+15.218	+6.646	148.225	+16.632	+23.584
	5	41.70	11.01	191.652	15.313	6.653	148.345	16.587	23.616
	9	41.87	11.13	191.441	15.416	6.661	148.466	16.542	23.649
	13	42.02	11.25	191.214	15.525	6.669	148.586	16.497	23.681
	17	42.14	11.36	190.973	15.639	6.677	148.706	16.452	23.713
	21	42.25	11.47	190.719	+15.757	+6.686	148.827	+16.407	+23.745
	25	42.34	11.58	190.455	15.879	6.695	148.947	16.361	23.777
	29	42.40	11.69	190.183	16.004	6.704	149.067	16.316	23.809
Aug.	2	42.45	11.79	189.905	16.130	6.714	149.188	16.270	23.841
	6	42.47	11.89	189.623	16.257	6.723	149.308	16.225	23.873
	10	42.47	11.98	189.340	+16.383	+6.732	149.428	+16.179	+23.904
	14	42.44	12.06	189.057	16.508	6.740	149.548	16.134	23.936
	18	42.39	12.13	188.778	16.630	6.749	149.668	16.088	23.967
	22	42.32	12.20	188.505	16.749	6.757	149.788	16.042	23.998
	26	42.23	12.25	188.239	16.863	6.765	149.908	15.996	24.029
	30	42.12	12.29	187.983	+16.973	+6.772	150.028	+15.950	+24.060
Sept.	3	41.98	12.33	187.739	17.076	6.779	150.148	15.904	24.091
	7	41.83	12.35	187.510	17.172	6.785	150.268	15.858	24.122
	11	41.66	12.36	187.297	17.261	6.791	150.388	15.812	24.152
	15	41.48	12.36	187.102	17.342	6.796	150.508	15.765	24.183
	19	41.28	12.35	186.927	+17.414	+6.800	150.628	+15.719	+24.213
	23	41.06	12.33	186.772	17.476	6.804	150.747	15.672	24.244
	27	40.84	12.30	186.640	17.530	6.808	150.867	15.626	24.274
Oct.	1	40.60	12.26	186.530	17.573	6.810	150.987	15.579	24.304
	5	40.36	12.21	186.446	17.606	6.812	151.107	15.533	24.334
	9	40.10	12.15	186.386	+17.629	+6.814	151.226	+15.486	+24.364
	13	39.85	12.08	186.352	17.641	6.814	151.346	15.439	24.394
	17	39.59	12.00	186.344	17.642	6.815	151.465	15.392	24.423
	21	39.32	11.91	186.361	17.633	6.814	151.585	15.345	24.453
	25	39.06	11.82	186.405	17.612	6.813	151.704	15.298	24.482
	29	38.79	11.72	186.475	+17.581	+6.811	151.824	+15.251	+24.511
Nov.	2	38.53	11.61	186.572	17.540	6.808	151.943	15.204	24.540
	6	38.27	11.50	186.693	17.488	6.805	152.063	15.157	24.570
	10	38.02	11.39	186.840	17.426	6.801	152.182	15.110	24.598
	14	37.77	11.26	187.011	17.353	6.797	152.301	15.062	24.627
	18	37.52	11.14	187.206	+17.271	+6.791	152.421	+15.015	+24.656
	22	37.28	11.01	187.424	17.180	6.785	152.540	14.967	24.685
	26	37.05	10.88	187.665	17.078	6.779	152.659	14.920	24.713
	30	36.83	10.75	187.926	16.968	6.771	152.779	14.872	24.741
Dec.	4	36.62	10.61	188.208	16.849	6.763	152.898	14.824	24.770
	8	36.42	10.48	188.509	+16.722	+6.755	153.017	+14.776	+24.798
	12	36.23	10.34	188.828	16.587	6.745	153.136	14.729	24.826
	16	36.04	10.20	189.164	16.445	6.735	153.255	14.681	24.854
	20	35.87	10.07	189.516	16.295	6.724	153.374	14.633	24.881
	24	35.71	9.93	189.883	16.138	6.712	153.493	14.584	24.909
	28	35.57	9.79	190.264	+15.975	+6.700	153.612	+14.536	+24.937
	32	35.43	9.65	190.657	+15.806	+6.687	153.731	+14.488	+24.964

Factor by which axes of outer edge of outer ring are to be multiplied to obtain axes of:

Inner edge of outer ring 0.8801

Outer edge of inner ring 0.8599

Inner edge of inner ring 0.6650

Inner edge of dusky ring 0.5486

APPARENT ORBITS OF SATELLITES I–VII, AT DATE OF OPPOSITION, AUGUST 7

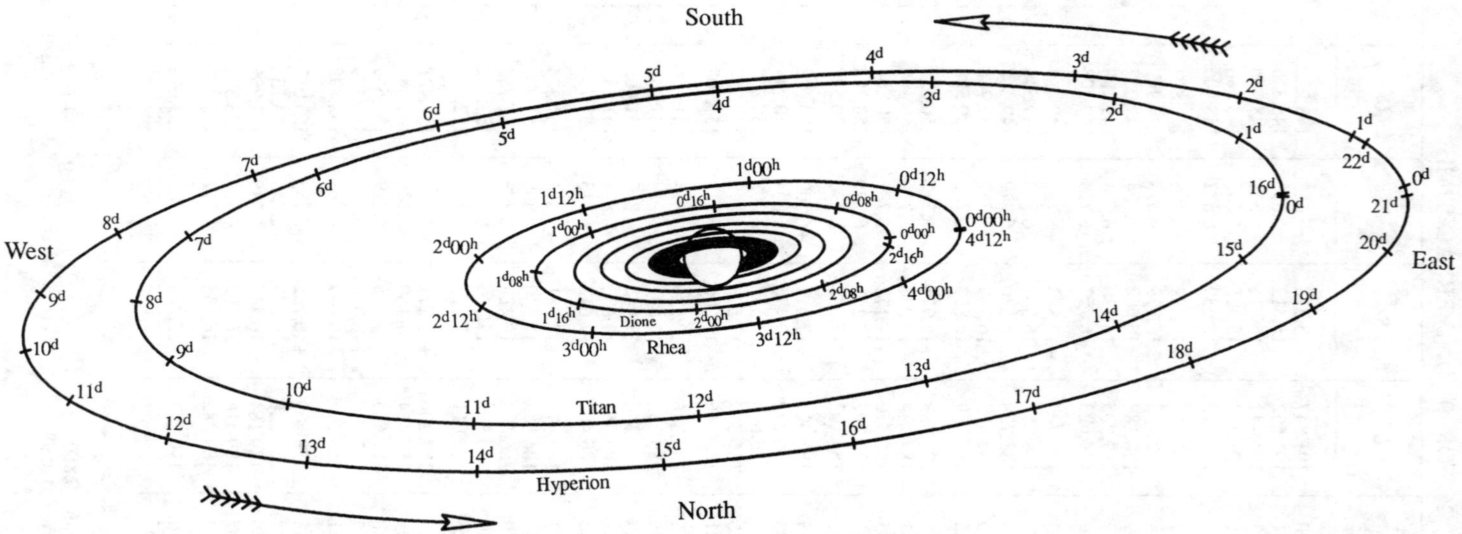

	Name	Mean Synodic Period d	h		Name	Mean Synodic Period d	h
I	Mimas	0	22.6	VI	Titan	15	23.3
II	Enceladus	1	08.9	VII	Hyperion................	21	07.6
III	Tethys	1	21.3	VIII	Iapetus	79	22.1
IV	Dione	2	17.7	IX	Phoebe	523	15.6
V	Rhea	4	12.5				

UNIVERSAL TIME OF GREATEST EASTERN ELONGATION

Jan.	Feb.	Mar.	Apr.	May	June	July	Aug.	Sept.	Oct.	Nov.	Dec.
					MIMAS						
d h	d h	d h	d h	d h	d h	d h	d h	d h	d h	d h	d h
0 08.4	1 09.7	1 15.1	1 17.6	1 21.4	1 01.2	1 04.9	1 07.1	1 09.4	1 13.1	1 15.6	1 19.6
1 07.1	2 08.3	2 13.7	2 16.2	2 20.0	1 23.8	2 03.5	2 05.7	2 08.0	2 11.8	2 14.2	2 18.2
2 05.7	3 07.0	3 12.3	3 14.8	3 18.7	2 22.4	3 02.1	3 04.3	3 06.6	3 10.4	3 12.9	3 16.8
3 04.3	4 05.6	4 10.9	4 13.5	4 17.3	3 21.0	4 00.7	4 03.0	4 05.2	4 09.0	4 11.5	4 15.4
4 02.9	5 04.2	5 09.5	5 12.1	5 15.9	4 19.6	4 23.3	5 01.6	5 03.9	5 07.6	5 10.1	5 14.1
5 01.6	6 02.8	6 08.2	6 10.7	6 14.5	5 18.3	5 21.9	6 00.2	6 02.5	6 06.2	6 08.7	6 12.7
6 00.2	7 01.5	7 06.8	7 09.3	7 13.1	6 16.9	6 20.5	6 22.8	7 01.1	7 04.9	7 07.4	7 11.3
6 22.8	8 00.1	8 05.4	8 07.9	8 11.8	7 15.5	7 19.2	7 21.4	7 23.7	8 03.5	8 06.0	8 09.9
7 21.4	8 22.7	9 04.0	9 06.6	9 10.4	8 14.1	8 17.8	8 20.0	8 22.3	9 02.1	9 04.6	9 08.6
8 20.1	9 21.3	10 02.7	10 05.2	10 09.0	9 12.7	9 16.4	9 18.6	9 20.9	10 00.7	10 03.2	10 07.2
9 18.7	10 20.0	11 01.3	11 03.8	11 07.6	10 11.3	10 15.0	10 17.3	10 19.6	10 23.3	11 01.8	11 05.8
10 17.3	11 18.6	11 23.9	12 02.4	12 06.2	11 10.0	11 13.6	11 15.9	11 18.2	11 22.0	12 00.5	12 04.4
11 15.9	12 17.2	12 22.5	13 01.0	13 04.8	12 08.6	12 12.2	12 14.5	12 16.8	12 20.6	12 23.1	13 03.1
12 14.6	13 15.8	13 21.2	13 23.7	14 03.5	13 07.2	13 10.8	13 13.1	13 15.4	13 19.2	13 21.7	14 01.7
13 13.2	14 14.4	14 19.8	14 22.3	15 02.1	14 05.8	14 09.5	14 11.7	14 14.0	14 17.8	14 20.3	15 00.3
14 11.8	15 13.1	15 18.4	15 20.9	16 00.7	15 04.4	15 08.1	15 10.3	15 12.6	15 16.4	15 19.0	15 22.9
15 10.4	16 11.7	16 17.0	16 19.5	16 23.3	16 03.0	16 06.7	16 08.9	16 11.3	16 15.1	16 17.6	16 21.6
16 09.1	17 10.3	17 15.6	17 18.1	17 21.9	17 01.6	17 05.3	17 07.6	17 09.9	17 13.7	17 16.2	17 20.2
17 07.7	18 08.9	18 14.3	18 16.8	18 20.6	18 00.3	18 03.9	18 06.2	18 08.5	18 12.3	18 14.8	18 18.8
18 06.3	19 07.6	19 12.9	19 15.4	19 19.2	18 22.9	19 02.5	19 04.8	19 07.1	19 10.9	19 13.4	19 17.4
19 04.9	20 06.2	20 11.5	20 14.0	20 17.8	19 21.5	20 01.1	20 03.4	20 05.7	20 09.5	20 12.1	20 16.1
20 03.6	21 04.8	21 10.1	21 12.6	21 16.4	20 20.1	20 23.8	21 02.0	21 04.3	21 08.2	21 10.7	21 14.7
21 02.2	22 03.4	22 08.8	22 11.2	22 15.0	21 18.7	21 22.4	22 00.6	22 03.0	22 06.8	22 09.3	22 13.3
22 00.8	23 02.1	23 07.4	23 09.9	23 13.6	22 17.3	22 21.0	22 23.2	23 01.6	23 05.4	23 07.9	23 11.9
22 23.5	24 00.7	24 06.0	24 08.5	24 12.3	23 15.9	23 19.6	23 21.9	24 00.2	24 04.0	24 06.6	24 10.6
23 22.1	24 23.3	25 04.6	25 07.1	25 10.9	24 14.6	24 18.2	24 20.5	24 22.8	25 02.6	25 05.2	25 09.2
24 20.7	25 21.9	26 03.2	26 05.7	26 09.5	25 13.2	25 16.8	25 19.1	25 21.4	26 01.3	26 03.8	26 07.8
25 19.3	26 20.6	27 01.9	27 04.3	27 08.1	26 11.8	26 15.4	26 17.7	26 20.1	26 23.9	27 02.4	27 06.4
26 17.9	27 19.2	28 00.5	28 03.0	28 06.7	27 10.4	27 14.1	27 16.3	27 18.7	27 22.5	28 01.1	28 05.1
27 16.6	28 17.8	28 23.1	29 01.6	29 05.3	28 09.0	28 12.7	28 14.9	28 17.3	28 21.1	28 23.7	29 03.7
28 15.2	29 16.4	29 21.7	30 00.2	30 03.9	29 07.6	29 11.3	29 13.6	29 15.9	29 19.8	29 22.3	30 02.3
29 13.8		30 20.4	30 22.8	31 02.6	30 06.2	30 09.9	30 12.2	30 14.5	30 18.4	30 20.9	31 00.9
30 12.4		31 19.0				31 08.5	31 10.8		31 17.0		31 23.6
31 11.1											32 22.2
					ENCELADUS						
d h	d h	d h	d h	d h	d h	d h	d h	d h	d h	d h	d h
−1 20.4	1 18.0	1 12.9	2 01.5	2 05.0	1 08.5	1 11.8	1 24.0	1 03.3	1 06.7	1 19.1	1 22.8
1 05.3	3 02.9	2 21.8	3 10.3	3 13.9	2 17.4	2 20.7	3 08.9	2 12.2	2 15.6	3 04.0	3 07.7
2 14.2	4 11.8	4 06.7	4 19.2	4 22.8	4 02.2	4 05.6	4 17.7	3 21.0	4 00.5	4 12.9	4 16.6
3 23.1	5 20.7	5 15.6	6 04.1	6 07.7	5 11.1	5 14.5	6 02.6	5 05.9	5 09.3	5 21.8	6 01.5
5 08.0	7 05.6	7 00.5	7 13.0	7 16.6	6 20.0	6 23.3	7 11.5	6 14.8	6 18.2	7 06.7	7 10.4
6 16.9	8 14.5	8 09.4	8 21.9	9 01.5	8 04.9	8 08.2	8 20.4	7 23.7	8 03.1	8 15.6	8 19.3
8 01.8	9 23.4	9 18.3	10 06.8	10 10.3	9 13.8	9 17.1	10 05.2	9 08.6	9 12.0	10 00.5	10 04.2
9 10.7	11 08.3	11 03.2	11 15.7	11 19.2	10 22.6	11 02.0	11 14.1	10 17.4	10 20.9	11 09.4	11 13.1
10 19.6	12 17.2	12 12.0	13 00.6	13 04.1	12 07.5	12 10.8	12 23.0	12 02.3	12 05.8	12 18.3	12 22.0
12 04.5	14 02.1	13 20.9	14 09.5	14 13.0	13 16.4	13 19.7	14 07.9	13 11.2	13 14.7	14 03.2	14 06.9
13 13.4	15 11.0	15 05.8	15 18.4	15 21.9	15 01.3	15 04.6	15 16.7	14 20.1	14 23.5	15 12.0	15 15.8
14 22.3	16 19.9	16 14.7	17 03.3	17 06.8	16 10.2	16 13.5	17 01.6	16 05.0	16 08.4	16 20.9	17 00.7
16 07.2	18 04.8	17 23.6	18 12.1	18 15.7	17 19.0	17 22.3	18 10.5	17 13.8	17 17.3	18 05.8	18 09.6
17 16.1	19 13.7	19 08.5	19 21.0	20 00.5	19 03.9	19 07.2	19 19.4	18 22.7	19 02.2	19 14.7	19 18.5
19 01.0	20 22.6	20 17.4	21 05.9	21 09.4	20 12.8	20 16.1	21 04.3	20 07.6	20 11.1	20 23.6	21 03.4
20 09.9	22 07.5	22 02.3	22 14.8	22 18.3	21 21.7	22 01.0	22 13.1	21 16.5	21 20.0	22 08.5	22 12.3
21 18.8	23 16.4	23 11.2	23 23.7	24 03.2	23 06.6	23 09.8	23 22.0	23 01.4	23 04.9	23 17.4	23 21.2
23 03.7	25 01.3	24 20.1	25 08.6	25 12.1	24 15.4	24 18.7	25 06.9	24 10.3	24 13.8	25 02.3	25 06.1
24 12.6	26 10.2	26 05.0	26 17.5	26 21.0	26 00.3	26 03.6	26 15.8	25 19.1	25 22.7	26 11.2	26 15.0
25 21.5	27 19.1	27 13.9	28 02.4	28 05.8	27 09.2	27 12.5	28 00.6	27 04.0	27 07.5	27 20.1	27 23.9
27 06.4	29 04.0	28 22.8	29 11.3	29 14.7	28 18.1	28 21.3	29 09.5	28 12.9	28 16.4	29 05.0	29 08.8
28 15.3		30 07.7	30 20.1	30 23.6	30 02.9	30 06.2	30 18.4	29 21.8	30 01.3	30 13.9	30 17.7
30 00.2		31 16.6				31 15.1			31 10.2		32 02.6
31 09.1											33 11.5

UNIVERSAL TIME OF GREATEST EASTERN ELONGATION

TETHYS

Jan.	Feb.	Mar.	Apr.	May	June	July	Aug.	Sept.	Oct.	Nov.	Dec.
d h	d h	d h	d h	d h	d h	d h	d h	d h	d h	d h	d h
−1 19.1	2 19.1	2 03.1	1 08.3	1 13.4	2 15.6	2 20.4	2 01.0	1 05.7	1 10.5	2 12.8	2 18.0
1 16.4	4 16.4	4 00.4	3 05.6	3 10.7	4 12.9	4 17.7	3 22.3	3 03.0	3 07.8	4 10.1	4 15.3
3 13.7	6 13.8	5 21.8	5 03.0	5 08.0	6 10.2	6 15.0	5 19.6	5 00.3	5 05.1	6 07.4	6 12.6
5 11.1	8 11.1	7 19.1	7 00.3	7 05.3	8 07.5	8 12.2	7 16.9	6 21.6	7 02.4	8 04.8	8 10.0
7 08.4	10 08.4	9 16.4	8 21.6	9 02.6	10 04.8	10 09.5	9 14.2	8 18.9	8 23.7	10 02.1	10 07.3
9 05.7	12 05.8	11 13.7	10 18.9	10 24.0	12 02.1	12 06.8	11 11.5	10 16.2	10 21.0	11 23.4	12 04.6
11 03.1	14 03.1	13 11.1	12 16.2	12 21.3	13 23.4	14 04.1	13 08.8	12 13.5	12 18.3	13 20.7	14 02.0
13 00.4	16 00.4	15 08.4	14 13.6	14 18.6	15 20.7	16 01.4	15 06.1	14 10.8	14 15.6	15 18.0	15 23.3
14 21.8	17 21.8	17 05.7	16 10.9	16 15.9	17 18.0	17 22.7	17 03.3	16 08.1	16 13.0	17 15.4	17 20.6
16 19.1	19 19.1	19 03.0	18 08.2	18 13.2	19 15.3	19 20.0	19 00.6	18 05.4	18 10.3	19 12.7	19 18.0
18 16.4	21 16.4	21 00.4	20 05.5	20 10.5	21 12.6	21 17.3	20 21.9	20 02.7	20 07.6	21 10.0	21 15.3
20 13.7	23 13.8	22 21.7	22 02.8	22 07.8	23 09.9	23 14.6	22 19.2	21 24.0	22 04.9	23 07.3	23 12.6
22 11.1	25 11.1	24 19.0	24 00.1	24 05.1	25 07.2	25 11.9	24 16.5	23 21.3	24 02.2	25 04.7	25 09.9
24 08.4	27 08.4	26 16.3	25 21.5	26 02.4	27 04.5	27 09.2	26 13.8	25 18.6	25 23.5	27 02.0	27 07.3
26 05.8	29 05.8	28 13.7	27 18.8	27 23.7	29 01.8	29 06.4	28 11.1	27 15.9	27 20.8	28 23.3	29 04.6
28 03.1		30 11.0	29 16.1	29 21.0	30 23.1	31 03.7	30 08.4	29 13.2	29 18.2	30 20.7	31 01.9
30 00.4				31 18.3					31 15.5		32 23.3
31 21.8											34 20.6

DIONE

Jan.	Feb.	Mar.	Apr.	May	June	July	Aug.	Sept.	Oct.	Nov.	Dec.
d h	d h	d h	d h	d h	d h	d h	d h	d h	d h	d h	d h
−2 12.1	3 02.8	1 12.2	3 08.9	3 11.7	2 14.3	2 16.6	1 18.8	3 14.7	3 17.0	2 19.7	2 22.6
1 05.9	5 20.6	4 05.9	6 02.7	6 05.4	5 08.0	5 10.3	4 12.5	6 08.3	6 10.7	5 13.4	5 16.3
3 23.6	8 14.3	6 23.7	8 20.4	8 23.1	8 01.7	8 04.0	7 06.1	9 02.0	9 04.4	8 07.1	8 10.0
6 17.4	11 08.0	9 17.4	11 14.1	11 16.8	10 19.3	10 21.6	9 23.8	11 19.7	11 22.1	11 00.8	11 03.8
9 11.1	14 01.8	12 11.2	14 07.8	14 10.5	13 13.0	13 15.3	12 17.4	14 13.3	14 15.8	13 18.5	13 21.5
12 04.8	16 19.5	15 04.9	17 01.5	17 04.2	16 06.7	16 08.9	15 11.1	17 07.0	17 09.5	16 12.3	16 15.2
14 22.6	19 13.3	17 22.6	19 19.2	19 21.9	19 00.3	19 02.6	18 04.7	20 00.7	20 03.2	19 06.0	19 09.0
17 16.3	22 07.0	20 16.3	22 12.9	22 15.6	21 18.0	21 20.2	20 22.4	22 18.3	22 20.9	21 23.7	22 02.7
20 10.1	25 00.7	23 10.1	25 06.6	25 09.3	24 11.7	24 13.9	23 16.0	25 12.0	25 14.6	24 17.4	24 20.5
23 03.8	27 18.5	26 03.8	28 00.3	28 03.0	27 05.3	27 07.5	26 09.7	28 05.7	28 08.3	27 11.1	27 14.2
25 21.6		28 21.5	30 18.0	30 20.6	29 23.0	30 01.2	29 03.3	30 23.4	31 02.0	30 04.9	30 07.9
28 15.3		31 15.2					31 21.0				33 01.7
31 09.1											35 19.4

RHEA

Jan.	Feb.	Mar.	Apr.	May	June	July	Aug.	Sept.	Oct.	Nov.	Dec.
d h	d h	d h	d h	d h	d h	d h	d h	d h	d h	d h	d h
−2 22.9	4 03.5	2 06.9	2 22.7	4 14.1	5 05.1	2 07.3	2 21.6	3 11.9	5 02.6	1 05.2	2 20.7
3 11.5	8 16.1	6 19.5	7 11.2	9 02.6	9 17.5	6 19.6	7 09.9	8 00.3	9 15.0	5 17.7	7 09.2
8 00.0	13 04.7	11 08.0	11 23.7	13 15.0	14 05.8	11 08.0	11 22.3	12 12.6	14 03.4	10 06.2	11 21.8
12 12.6	17 17.2	15 20.6	16 12.2	18 03.4	18 18.2	15 20.3	16 10.6	17 01.0	18 15.8	14 18.6	16 10.3
17 01.2	22 05.8	20 09.1	21 00.7	22 15.9	23 06.6	20 08.6	20 22.9	21 13.4	23 04.3	19 07.2	20 22.9
21 13.8	26 18.4	24 21.6	25 13.2	27 04.3	27 18.9	24 21.0	25 11.3	26 01.8	27 16.7	23 19.7	25 11.4
26 02.4		29 10.2	30 01.6	31 16.7		29 09.3	29 23.6	30 14.2		28 08.2	29 24.0
30 14.9											34 12.6

UNIVERSAL TIME OF CONJUNCTIONS AND ELONGATIONS

TITAN

Eastern Elongation		Inferior Conjunction		Western Elongation		Superior Conjunction	
	d h		d h		d h		d h
						Jan.	− 3 06.5
Jan.	1 02.2	Jan.	5 04.5	Jan.	9 09.1		13 07.1
	17 02.8		21 05.2		25 09.8		29 07.6
Feb.	2 03.5	Feb.	6 06.0	Feb.	10 10.5	Feb.	14 08.2
	18 04.1		22 06.8		26 11.2	Mar.	1 08.7
Mar.	5 04.6	Mar.	9 07.4	Mar.	13 11.6		17 09.0
	21 04.9		25 07.7		29 11.8	Apr.	2 09.0
Apr.	6 05.0	Apr.	10 07.8	Apr.	14 11.8		18 08.8
	22 04.7		26 07.5		30 11.4	May	4 08.2
May	8 04.1	May	12 06.9	May	16 10.6		20 07.3
	24 03.1		28 05.8	June	1 09.3	June	5 06.0
June	9 01.7	June	13 04.2		17 07.7		21 04.3
	24 23.9		29 02.3	July	3 05.7	July	7 02.3
July	10 21.8	July	15 00.0		19 03.5		23 00.0
	26 19.4		30 21.6	Aug.	4 01.0	Aug.	7 21.6
Aug.	11 16.9	Aug.	15 19.0		19 22.5		23 19.2
	27 14.5		31 16.5	Sept.	4 20.1	Sept.	8 16.9
Sept.	12 12.3	Sept.	16 14.3		20 18.0		24 14.9
	28 10.3	Oct.	2 12.4	Oct.	6 16.2	Oct.	10 13.3
Oct.	14 08.8		18 10.9		22 14.9		26 12.1
	30 07.7	Nov.	3 10.0	Nov.	7 14.0	Nov.	11 11.3
Nov.	15 07.0		19 09.4		23 13.6		27 10.9
Dec.	1 06.7	Dec.	5 09.3	Dec.	9 13.5	Dec.	13 10.8
	17 06.7		21 09.5		25 13.7		29 11.0
	33 07.1						

HYPERION

Eastern Elongation		Inferior Conjunction		Western Elongation		Superior Conjunction	
	d h		d h		d h		d h
				Jan.	− 4 21.7	Jan.	2 11.8
Jan.	8 14.1	Jan.	13 15.6		18 07.9		23 23.0
	30 00.6	Feb.	4 00.9	Feb.	8 17.7	Feb.	14 09.7
Feb.	20 10.2		25 09.5	Mar.	1 02.7	Mar.	6 19.5
Mar.	12 19.0	Mar.	17 17.3		22 11.0		28 04.4
Apr.	3 02.8	Apr.	8 00.1	Apr.	12 18.1	Apr.	18 12.0
	24 09.2		29 05.9	May	4 00.1	May	9 18.2
May	15 14.6	May	20 10.6		25 04.9		30 23.1
June	5 18.7	June	10 14.2	June	15 08.7	June	21 02.8
	26 21.9	July	1 17.0	July	6 11.4	July	12 05.4
July	18 00.4		22 19.2		27 13.6	Aug.	2 07.6
Aug.	8 02.5	Aug.	12 21.3	Aug.	17 15.6		23 09.7
	29 04.8	Sept.	2 23.7	Sept.	7 18.0	Sept.	13 12.6
Sept.	19 07.9		24 02.9		28 21.5	Oct.	4 16.6
Oct.	10 12.3	Oct.	15 07.3	Oct.	20 02.2		25 22.3
	31 18.2	Nov.	5 13.1	Nov.	10 08.6	Nov.	16 05.7
Nov.	22 01.8		26 20.2	Dec.	1 16.5	Dec.	7 14.8
Dec.	13 10.5	Dec.	18 04.7		23 01.6		29 01.3
	34 20.6						

IAPETUS

Eastern Elongation		Inferior Conjunction		Western Elongation		Superior Conjunction	
	d h		d h		d h		d h
				Jan.	−12 04.1	Jan.	8 18.7
Jan.	28 02.8	Feb.	17 05.0	Mar.	9 13.6	Mar.	29 22.9
Apr.	18 02.1	May	8 00.2	May	28 22.5	June	17 18.7
July	6 10.9	July	25 21.4	Aug.	15 09.8	Sept.	4 02.6
Sept.	22 18.6	Oct.	12 07.4	Nov.	2 04.7	Nov.	22 09.6
Dec.	11 12.9	Dec.	31 14.8	Dec.	52 23.8		

APPARENT DISTANCE AND POSTION ANGLE

Time from Eastern Elongation	MIMAS F	MIMAS p_1	Time from Eastern Elongation	ENCELADUS F	ENCELADUS p_1	TETHYS F	TETHYS p_1	Time from Eastern Elongation	DIONE F	DIONE p_1
h		°	d h		°		°	d h		°
0.0	1.000	98.0	0 00	1.000	97.0	1.000	97.0	0 00	1.000	97.0
0.5	0.991	100.1	0 01	0.983	100.1	0.991	99.4	0 02	0.983	100.1
1.0	0.964	102.4	0 02	0.934	103.4	0.965	101.8	0 04	0.934	103.4
1.5	0.921	104.8	0 03	0.854	107.3	0.923	104.5	0 06	0.854	107.3
2.0	0.861	107.4	0 04	0.747	112.0	0.864	107.4	0 08	0.747	112.1
2.5	0.787	110.6	0 05	0.621	118.6	0.792	110.9	0 10	0.620	118.7
3.0	0.701	114.4	0 06	0.485	128.8	0.708	115.1	0 12	0.484	129.0
3.5	0.605	119.5	0 07	0.358	146.7	0.616	120.5	0 14	0.357	147.0
4.0	0.504	126.4	0 08	0.283	178.4	0.519	127.8	0 16	0.283	178.8
4.5	0.405	136.9	0 09	0.314	215.1	0.424	138.6	0 18	0.316	215.4
5.0	0.320	153.6	0 10	0.425	238.6	0.345	154.9	0 20	0.427	238.8
5.5	0.271	178.9	0 11	0.561	251.5	0.300	178.3	0 22	0.563	251.6
6.0	0.284	207.8	0 12	0.694	259.3	0.310	204.5	1 00	0.696	259.4
6.5	0.349	229.8	0 13	0.810	264.8	0.369	225.5	1 02	0.812	264.8
7.0	0.442	243.6	0 14	0.902	269.0	0.456	239.6	1 04	0.903	269.0
7.5	0.543	252.5	0 15	0.965	272.5	0.552	249.0	1 06	0.966	272.5
8.0	0.643	258.6	0 16	0.997	275.6	0.648	255.6	1 08	0.997	275.7
8.5	0.735	263.1	0 17	0.995	278.7	0.738	260.5	1 10	0.994	278.8
9.0	0.817	266.7	0 18	0.960	281.9	0.819	264.4	1 12	0.958	282.0
9.5	0.886	269.6	0 19	0.893	285.5	0.886	267.6	1 14	0.891	285.6
10.0	0.939	272.2	0 20	0.797	289.8	0.939	270.5	1 16	0.795	289.9
10.5	0.977	274.5	0 21	0.679	295.4	0.976	273.0	1 18	0.676	295.6
11.0	0.997	276.7	0 22	0.545	303.7	0.996	275.5	1 20	0.542	304.0
11.5	0.999	278.8	0 23	0.411	317.4	0.999	277.8	1 22	0.407	318.0
12.0	0.983	281.0	1 00	0.305	342.5	0.984	280.2	2 00	0.304	343.6
12.5	0.950	283.3	1 01	0.287	19.9	0.952	282.7	2 02	0.289	21.1
13.0	0.900	285.7	1 02	0.371	50.0	0.904	285.5	2 04	0.375	50.6
13.5	0.835	288.6	1 03	0.500	66.6	0.841	288.5	2 06	0.505	67.0
14.0	0.756	292.0	1 04	0.636	76.3	0.764	292.2	2 08	0.641	76.5
14.5	0.665	296.2	1 05	0.761	82.6	0.677	296.8	2 10	0.765	82.8
15.0	0.567	301.8	1 06	0.864	87.2	0.582	302.7	2 12	0.868	87.4
15.5	0.466	309.9	1 07	0.941	91.0	0.485	311.1	2 14	0.943	91.1
16.0	0.370	322.3	1 08	0.987	94.3	0.394	323.5	2 16	0.988	94.4
16.5	0.295	342.2	1 09	1.000	97.4	0.324	342.3	2 18	1.000	97.5
17.0	0.268	10.2	1 10	0.979	100.5	0.297	7.5	2 20	0.978	100.6
17.5	0.304	37.2	1 11			0.326	32.6			
18.0	0.383	55.8	1 12			0.397	51.1			
18.5	0.481	67.5	1 13			0.489	63.3			
19.0	0.582	75.1	1 14			0.586	71.5			
19.5	0.679	80.5	1 15			0.681	77.4			
20.0	0.768	84.6	1 16			0.767	81.9			
20.5	0.845	87.9	1 17			0.844	85.6			
21.0	0.908	90.6	1 18			0.906	88.7			
21.5	0.956	93.1	1 19			0.954	91.4			
22.0	0.986	95.3	1 20			0.985	93.9			
22.5	1.000	97.5	1 21			0.999	96.3			
23.0	0.995	99.6	1 22			0.996	98.6			

Apparent distance of satellite is Fa/Δ
Position angle of satellite is $p_1 + p_2$

APPARENT DISTANCE AND POSTION ANGLE

Time from Eastern Elongation	RHEA		Time from Eastern Elongation	TITAN		HYPERION		Time from Eastern Elongation	IAPETUS	
	F	p_1		F	p_1	F	p_1		F	p_1
d h		°	d h		°		°	d		°
0 00	1.000	97.0	0 00	0.991	96.0	0.996	96.0	0	0.977	90.0
0 03	0.986	99.9	0 10	0.983	98.6	0.980	97.9	2	0.967	90.4
0 06	0.945	102.9	0 20	0.951	101.4	0.949	99.9	4	0.931	90.8
0 09	0.878	106.3	1 06	0.895	104.4	0.904	102.1	6	0.870	91.3
0 12	0.789	110.3	1 16	0.818	108.0	0.845	104.6	8	0.787	91.8
0 15	0.681	115.6	2 02	0.724	112.3	0.774	107.4	10	0.684	92.5
0 18	0.560	123.0	2 12	0.617	118.1	0.691	110.9	12	0.564	93.4
0 21	0.437	134.5	2 22	0.503	126.5	0.599	115.5	14	0.429	95.0
1 00	0.332	154.3	3 08	0.393	139.7	0.501	121.7	16	0.285	98.0
1 03	0.284	185.8	3 18	0.309	161.5	0.403	131.0	18	0.136	107.5
1 06	0.326	217.8	4 04	0.285	192.3	0.314	146.0	20	0.049	211.4
1 09	0.429	238.4	4 14	0.337	220.2	0.255	170.1	22	0.185	257.2
1 12	0.551	250.4	5 00	0.435	238.1	0.254	200.4	24	0.333	263.2
1 15	0.673	258.0	5 10	0.549	249.0	0.310	225.0	26	0.475	265.5
1 18	0.782	263.4	5 20	0.662	256.1	0.398	240.4	28	0.606	266.8
1 21	0.873	267.5	6 06	0.765	261.2	0.496	249.9	30	0.724	267.6
2 00	0.941	270.9	6 16	0.855	265.2	0.594	256.3	32	0.825	268.3
2 03	0.984	273.9	7 02	0.926	268.4	0.686	260.9	34	0.907	268.7
2 06	1.000	276.8	7 12	0.976	271.3	0.769	264.5	36	0.968	269.2
2 09	0.988	279.7	7 22	1.003	273.9	0.841	267.4	38	1.007	269.5
2 12	0.949	282.7	8 08	1.006	276.5	0.900	269.8	40	1.022	269.9
2 15	0.884	286.0	8 18	0.984	279.1	0.945	272.0	42	1.014	270.3
2 18	0.796	290.0	9 04	0.939	281.9	0.976	274.1	44	0.982	270.6
2 21	0.689	295.1	9 14	0.870	285.0	0.992	276.0	46	0.926	271.0
3 00	0.569	302.3	10 00	0.781	288.9	0.994	277.9	48	0.848	271.5
3 03	0.445	313.5	10 10	0.674	293.8	0.982	279.8	50	0.750	272.1
3 06	0.338	332.6	10 20	0.555	300.7	0.957	281.8	52	0.633	272.9
3 09	0.284	3.3	11 06	0.433	311.5	0.919	284.0	54	0.500	274.0
3 12	0.320	35.9	11 16	0.325	330.2	0.869	286.3	56	0.355	276.1
3 15	0.420	57.3	12 02	0.269	1.2	0.810	289.0	58	0.203	281.2
3 18	0.543	69.7	12 12	0.302	34.8	0.741	292.1	60	0.056	315.9
3 21	0.664	77.6	12 22	0.401	56.7	0.665	295.9	62	0.128	72.0
4 00	0.775	83.0	13 08	0.521	69.1	0.585	300.7	64	0.282	82.2
4 03	0.867	87.2	13 18	0.641	76.9	0.502	307.1	66	0.431	85.2
4 06	0.937	90.7	14 04	0.751	82.3	0.422	316.0	68	0.569	86.7
4 09	0.982	93.7	14 14	0.844	86.3	0.351	328.8	70	0.692	87.7
4 12	1.000	96.6	15 00	0.916	89.7	0.301	346.8	72	0.796	88.4
4 15	0.990	99.4	15 10	0.965	92.6	0.286	9.1	74	0.878	88.9
			15 20	0.989	95.3	0.312	30.6	76	0.936	89.3
			16 06	0.988	97.9	0.369	47.1	78	0.969	89.7
			16 16			0.443	58.6	80	0.977	90.1
			17 02			0.525	66.7	82	0.957	90.5
			17 12			0.607	72.6			
			17 22			0.687	77.1			
			18 08			0.761	80.7			
			18 18			0.828	83.7			
			19 04			0.885	86.3			
			19 14			0.932	88.5			
			20 00			0.967	90.6			
			20 10			0.990	92.6			
			20 20			0.999	94.5			
			21 06			0.994	96.3			
			21 16			0.975	98.3			

Apparent distance of satellite is Fa/Δ
Position angle of satellite is $p_1 + p_2$

APPARENT DISTANCE AND POSTION ANGLE

Day (0^h UT)		MIMAS a/Δ	MIMAS p_2	ENCELADUS a/Δ	ENCELADUS p_2	TETHYS a/Δ	TETHYS p_2	DIONE a/Δ	DIONE p_2
		″	°	″	°	″	°	″	°
Jan.	−1	23.7	−2.6	30.4	−0.1	37.6	−0.6	48.2	−0.1
	3	23.6	2.6	30.3	0.1	37.5	0.6	48.1	0.1
	7	23.6	2.7	30.3	0.1	37.5	0.6	48.0	0.1
	11	23.5	2.7	30.2	0.1	37.4	0.6	47.9	0.1
	15	23.5	2.7	30.2	0.1	37.3	0.6	47.8	0.1
	19	23.5	−2.7	30.1	−0.1	37.3	−0.5	47.7	−0.1
	23	23.5	2.7	30.1	0.1	37.2	0.5	47.7	0.1
	27	23.4	2.7	30.1	0.1	37.2	0.5	47.7	0.1
	31	23.4	2.7	30.1	0.1	37.2	0.5	47.7	0.2
Feb.	4	23.4	2.7	30.1	0.1	37.2	0.5	47.7	0.2
	8	23.5	−2.6	30.1	−0.1	37.3	−0.5	47.7	−0.2
	12	23.5	2.6	30.1	0.2	37.3	0.4	47.8	0.2
	16	23.5	2.5	30.2	0.2	37.4	0.4	47.9	0.2
	20	23.6	2.5	30.2	0.2	37.4	0.4	47.9	0.2
	24	23.6	2.4	30.3	0.2	37.5	0.4	48.0	0.2
	28	23.7	−2.3	30.4	−0.2	37.6	−0.4	48.2	−0.2
Mar.	3	23.7	2.2	30.5	0.2	37.7	0.4	48.3	0.2
	7	23.8	2.2	30.6	0.2	37.8	0.4	48.5	0.2
	11	23.9	2.1	30.7	0.2	38.0	0.4	48.6	0.3
	15	24.0	2.0	30.8	0.2	38.1	0.3	48.8	0.3
	19	24.1	−1.9	30.9	−0.2	38.3	−0.3	49.0	−0.3
	23	24.2	1.8	31.1	0.3	38.5	0.3	49.3	0.3
	27	24.3	1.7	31.2	0.3	38.6	0.3	49.5	0.3
	31	24.5	1.6	31.4	0.3	38.8	0.3	49.7	0.3
Apr.	4	24.6	1.5	31.5	0.3	39.0	0.3	50.0	0.3
	8	24.7	−1.3	31.7	−0.3	39.3	−0.3	50.3	−0.3
	12	24.9	1.2	31.9	0.3	39.5	0.3	50.6	0.3
	16	25.0	1.1	32.1	0.3	39.7	0.2	50.9	0.3
	20	25.2	1.0	32.3	0.3	40.0	0.2	51.2	0.4
	24	25.3	0.9	32.5	0.3	40.2	0.2	51.5	0.4
	28	25.5	−0.8	32.7	−0.3	40.5	−0.2	51.9	−0.4
May	2	25.7	0.7	32.9	0.3	40.8	0.2	52.2	0.4
	6	25.8	0.6	33.1	0.3	41.0	0.2	52.6	0.4
	10	26.0	0.5	33.4	0.4	41.3	0.2	52.9	0.4
	14	26.2	0.4	33.6	0.4	41.6	0.2	53.3	0.4
	18	26.4	−0.4	33.8	−0.4	41.9	−0.1	53.6	−0.4
	22	26.5	0.3	34.1	0.4	42.2	0.1	54.0	0.4
	26	26.7	0.2	34.3	0.4	42.4	0.1	54.4	0.4
	30	26.9	0.1	34.5	0.4	42.7	0.1	54.7	0.4
June	3	27.1	−0.1	34.7	0.4	43.0	0.1	55.1	0.4
	7	27.3	0.0	35.0	−0.4	43.3	−0.1	55.4	−0.4
	11	27.4	0.0	35.2	0.4	43.6	0.0	55.8	0.4
	15	27.6	+0.1	35.4	0.4	43.8	0.0	56.1	0.4
	19	27.7	0.1	35.6	0.3	44.1	0.0	56.4	0.4
	23	27.9	0.2	35.8	0.3	44.3	0.0	56.7	0.4
	27	28.0	+0.2	36.0	−0.3	44.5	0.0	57.0	−0.4
July	1	28.2	+0.2	36.1	−0.3	44.7	0.0	57.3	−0.4

APPARENT DISTANCE AND POSTION ANGLE

Day (0h UT)		MIMAS		ENCELADUS		TETHYS		DIONE	
		a/Δ	p_2	a/Δ	p_2	a/Δ	p_2	a/Δ	p_2
		″	°	″	°	″	°	″	°
July	1	28.2	+0.2	36.1	−0.3	44.7	0.0	57.3	−0.4
	5	28.3	0.2	36.3	0.3	44.9	+0.1	57.6	0.4
	9	28.4	0.2	36.4	0.3	45.1	0.1	57.8	0.3
	13	28.5	0.2	36.6	0.3	45.3	0.1	58.0	0.3
	17	28.6	0.2	36.7	0.3	45.4	0.1	58.2	0.3
	21	28.7	+0.2	36.8	−0.3	45.5	+0.1	58.3	−0.3
	25	28.7	0.2	36.9	0.3	45.6	0.2	58.4	0.3
	29	28.8	0.2	36.9	0.3	45.7	0.2	58.5	0.3
Aug.	2	28.8	0.2	36.9	0.3	45.7	0.2	58.6	0.3
	6	28.8	0.1	37.0	0.3	45.8	0.2	58.6	0.3
	10	28.8	+0.1	37.0	−0.2	45.8	+0.2	58.6	−0.3
	14	28.8	0.0	36.9	0.2	45.7	0.2	58.6	0.3
	18	28.8	0.0	36.9	0.2	45.7	0.3	58.5	0.3
	22	28.7	−0.1	36.8	0.2	45.6	0.3	58.4	0.2
	26	28.6	0.1	36.8	0.2	45.5	0.3	58.3	0.2
	30	28.6	−0.2	36.7	−0.2	45.4	+0.3	58.1	−0.2
Sept.	3	28.5	0.3	36.5	0.2	45.2	0.3	57.9	0.2
	7	28.4	0.4	36.4	0.2	45.1	0.3	57.7	0.2
	11	28.3	0.5	36.3	0.2	44.9	0.4	57.5	0.2
	15	28.1	0.5	36.1	0.2	44.7	0.4	57.2	0.2
	19	28.0	−0.6	35.9	−0.2	44.5	+0.4	57.0	−0.2
	23	27.9	0.7	35.7	0.2	44.2	0.4	56.7	0.2
	27	27.7	0.8	35.5	0.2	44.0	0.4	56.4	0.2
Oct.	1	27.5	0.9	35.3	0.2	43.7	0.4	56.0	0.2
	5	27.4	1.0	35.1	0.2	43.5	0.5	55.7	0.2
	9	27.2	−1.2	34.9	−0.2	43.2	+0.5	55.3	−0.2
	13	27.0	1.3	34.7	0.2	42.9	0.5	55.0	0.2
	17	26.9	1.4	34.5	0.2	42.7	0.5	54.6	0.2
	21	26.7	1.5	34.2	0.2	42.4	0.5	54.3	0.2
	25	26.5	1.6	34.0	0.2	42.1	0.5	53.9	0.2
	29	26.3	−1.7	33.8	−0.2	41.8	+0.5	53.5	−0.2
Nov.	2	26.1	1.8	33.5	0.2	41.5	0.5	53.2	0.2
	6	26.0	1.9	33.3	0.2	41.2	0.6	52.8	0.2
	10	25.8	2.0	33.1	0.2	41.0	0.6	52.5	0.2
	14	25.6	2.1	32.9	0.2	40.7	0.6	52.1	0.2
	18	25.5	−2.2	32.7	−0.2	40.4	+0.6	51.8	−0.2
	22	25.3	2.3	32.5	0.2	40.2	0.6	51.5	0.2
	26	25.1	2.4	32.3	0.2	39.9	0.6	51.1	0.2
	30	25.0	2.5	32.1	0.2	39.7	0.6	50.8	0.2
Dec.	4	24.8	2.6	31.9	0.2	39.5	0.6	50.5	0.2
	8	24.7	−2.6	31.7	−0.2	39.2	+0.6	50.3	−0.2
	12	24.6	2.7	31.5	0.3	39.0	0.6	50.0	0.3
	16	24.5	2.7	31.4	0.3	38.8	0.6	49.7	0.3
	20	24.3	2.8	31.2	0.3	38.7	0.6	49.5	0.3
	24	24.2	2.8	31.1	0.3	38.5	0.6	49.3	0.3
	28	24.1	−2.9	31.0	−0.3	38.3	+0.6	49.1	−0.3
	32	24.0	−2.9	30.8	−0.3	38.2	+0.6	48.9	−0.3

APPARENT DISTANCE AND POSTION ANGLE

Day (0^h UT)	RHEA a/Δ	RHEA p_2	TITAN a/Δ	TITAN p_2	HYPERION a/Δ	HYPERION p_2	IAPETUS a/Δ	IAPETUS p_2
	″	°	″	°	″	°	″	°
Jan. −1	67.3	+0.1	156	+0.7	190	+0.7	455	+2.8
3	67.2	0.1	156	0.7	189	0.7	453	2.7
7	67.0	0.1	155	0.7	189	0.6	452	2.5
11	66.9	0.1	155	0.7	188	0.6	452	2.4
15	66.8	0.1	155	0.7	188	0.6	451	2.3
19	66.7	+0.1	155	+0.7	188	+0.6	450	+2.2
23	66.6	0.1	154	0.7	187	0.6	450	2.0
27	66.6	0.1	154	0.7	187	0.6	450	1.9
31	66.6	0.1	154	0.6	187	0.6	450	1.8
Feb. 4	66.6	+0.1	154	0.6	187	0.5	450	1.7
8	66.7	0.0	154	+0.6	187	+0.5	450	+1.5
12	66.7	0.0	155	0.6	188	0.5	451	1.4
16	66.8	0.0	155	0.6	188	0.5	451	1.3
20	66.9	0.0	155	0.6	188	0.5	452	1.2
24	67.1	0.0	155	0.6	188	0.5	453	1.1
28	67.3	0.0	156	+0.6	189	+0.4	454	+0.9
Mar. 3	67.5	0.0	156	0.5	189	0.4	456	0.8
7	67.7	0.0	157	0.5	190	0.4	457	0.7
11	67.9	0.0	157	0.5	191	0.4	459	0.6
15	68.2	0.0	158	0.5	191	0.4	460	0.5
19	68.5	0.0	159	+0.5	192	+0.4	462	+0.4
23	68.8	−0.1	159	0.5	193	0.3	464	0.3
27	69.1	0.1	160	0.5	194	0.3	467	0.2
31	69.5	0.1	161	0.5	195	0.3	469	+0.1
Apr. 4	69.8	0.1	162	0.4	196	0.3	472	0.0
8	70.2	−0.1	163	+0.4	197	+0.3	474	−0.1
12	70.6	0.1	164	0.4	198	0.3	477	0.1
16	71.1	0.1	165	0.4	199	0.2	480	0.2
20	71.5	0.1	166	0.4	201	0.2	483	0.3
24	72.0	0.1	167	0.4	202	0.2	486	0.3
28	72.4	−0.1	168	+0.4	203	+0.2	489	−0.4
May 2	72.9	0.1	169	0.4	204	0.2	492	0.4
6	73.4	0.1	170	0.4	206	0.2	496	0.5
10	73.9	0.1	171	0.4	207	0.2	499	0.5
14	74.4	0.1	172	0.4	209	0.2	502	0.5
18	74.9	−0.1	174	+0.4	210	+0.2	506	−0.6
22	75.4	0.1	175	0.4	211	0.2	509	0.6
26	75.9	0.1	176	0.4	213	0.2	513	0.6
30	76.4	0.1	177	0.4	214	0.2	516	0.6
June 3	76.9	0.1	178	0.4	216	0.2	520	0.6
7	77.4	−0.1	179	+0.4	217	+0.2	523	−0.6
11	77.9	0.1	181	0.4	218	0.2	526	0.5
15	78.4	0.1	182	0.4	220	0.2	529	0.5
19	78.8	0.1	183	0.4	221	0.2	532	0.5
23	79.2	0.1	184	0.4	222	0.2	535	0.5
27	79.6	−0.1	185	+0.4	223	+0.2	538	−0.4
July 1	80.0	−0.1	185	+0.4	224	+0.2	540	−0.4

APPARENT DISTANCE AND POSTION ANGLE

Day (0h UT)	RHEA		TITAN		HYPERION		IAPETUS	
	a/Δ	p_2	a/Δ	p_2	a/Δ	p_2	a/Δ	p_2
	″	°	″	°	″	°	″	°
July 1	80.0	−0.1	185	+0.4	224	+0.2	540	−0.4
5	80.4	0.1	186	0.4	225	0.2	543	0.3
9	80.7	0.1	187	0.4	226	0.2	545	0.3
13	81.0	0.1	188	0.4	227	0.2	547	0.2
17	81.2	0.1	188	0.4	228	0.3	548	0.1
21	81.4	−0.1	189	+0.4	228	+0.3	550	−0.1
25	81.6	−0.1	189	0.5	229	0.3	551	0.0
29	81.7	0.0	189	0.5	229	0.3	552	+0.1
Aug. 2	81.8	0.0	190	0.5	229	0.3	552	0.2
6	81.8	0.0	190	0.5	229	0.3	553	0.2
10	81.8	0.0	190	+0.5	229	+0.3	553	+0.3
14	81.8	0.0	190	0.5	229	0.4	552	0.4
18	81.7	0.0	189	0.5	229	0.4	552	0.5
22	81.6	0.0	189	0.5	229	0.4	551	0.5
26	81.4	0.0	189	0.5	228	0.4	550	0.6
30	81.2	0.0	188	+0.5	228	+0.4	548	+0.7
Sept. 3	80.9	0.0	187	0.5	227	0.4	546	0.7
7	80.6	0.0	187	0.6	226	0.4	544	0.8
11	80.3	0.0	186	0.6	225	0.4	542	0.9
15	79.9	0.0	185	0.6	224	0.4	540	0.9
19	79.5	0.0	184	+0.6	223	+0.4	537	+1.0
23	79.1	0.0	183	0.6	222	0.4	534	1.0
27	78.7	+0.1	182	0.6	221	0.5	531	1.0
Oct. 1	78.2	0.1	181	0.6	220	0.5	528	1.1
5	77.8	0.1	180	0.6	218	0.5	525	1.1
9	77.3	+0.1	179	+0.6	217	+0.5	522	+1.1
13	76.8	0.1	178	0.6	216	0.5	519	1.1
17	76.3	0.1	177	0.6	214	0.5	515	1.1
21	75.8	0.1	176	0.6	213	0.5	512	1.1
25	75.3	0.1	174	0.6	211	0.5	508	1.1
29	74.8	+0.1	173	+0.6	210	+0.5	505	+1.1
Nov. 2	74.3	0.1	172	0.6	209	0.5	501	1.1
6	73.8	0.1	171	0.6	207	0.4	498	1.0
10	73.3	+0.1	170	0.6	206	0.4	495	1.0
14	72.8	0.0	169	0.6	205	0.4	491	0.9
18	72.3	0.0	168	+0.6	203	+0.4	488	+0.9
22	71.9	0.0	167	0.6	202	0.4	485	0.8
26	71.4	0.0	165	0.5	201	0.4	482	0.8
30	71.0	0.0	164	0.5	200	0.4	479	0.7
Dec. 4	70.6	0.0	164	0.5	198	0.4	477	0.6
8	70.2	0.0	163	+0.5	197	+0.4	474	+0.5
12	69.8	0.0	162	0.5	196	0.4	471	0.4
16	69.5	0.0	161	0.5	195	0.3	469	0.4
20	69.1	0.0	160	0.5	195	0.3	467	0.3
24	68.8	0.0	159	0.5	194	0.3	465	0.2
28	68.5	0.0	159	+0.5	193	+0.3	463	+0.1
32	68.3	0.0	158	+0.4	192	+0.3	461	−0.1

ORBITAL POSITIONS FOR 0^h UNIVERSAL TIME

Date		MIMAS *L*	MIMAS *M*	MIMAS *θ*	ENCELADUS *L*	ENCELADUS *M*	TETHYS *L*	TETHYS *θ*	DIONE *L*	DIONE *M*
		°	°	°	°	°	°	°	°	°
Jan.	−1	137.617	38.6	201.5	63.177	162.3	130.649	117.9	343.390	19.9
	3	225.588	122.5	197.5	34.107	131.9	173.441	117.1	149.530	185.7
	7	313.560	206.5	193.5	5.036	101.5	216.232	116.3	315.669	351.5
	11	41.531	290.5	189.5	335.965	71.1	259.024	115.5	121.808	157.3
	15	129.502	14.4	185.5	306.895	40.6	301.815	114.7	287.947	323.1
	19	217.473	98.4	181.5	277.824	10.2	344.607	113.9	94.087	128.9
	23	305.444	182.4	177.5	248.754	339.8	27.399	113.1	260.226	294.7
	27	33.415	266.4	173.5	219.683	309.4	70.190	112.3	66.365	100.5
	31	121.386	350.3	169.5	190.613	279.0	112.982	111.5	232.504	266.3
Feb.	4	209.357	74.3	165.5	161.543	248.5	155.773	110.7	38.644	72.1
	8	297.328	158.3	161.5	132.472	218.1	198.565	109.9	204.783	237.9
	12	25.298	242.2	157.5	103.402	187.7	241.357	109.1	10.922	43.7
	16	113.269	326.2	153.5	74.332	157.3	284.148	108.4	177.061	209.5
	20	201.240	50.2	149.5	45.261	126.9	326.940	107.6	343.201	15.3
	24	289.211	134.1	145.5	16.191	96.4	9.731	106.8	149.340	181.1
	28	17.181	218.1	141.5	347.121	66.0	52.523	106.0	315.479	346.9
Mar.	3	105.152	302.1	137.5	318.051	35.6	95.315	105.2	121.618	152.7
	7	193.123	26.0	133.5	288.980	5.2	138.106	104.4	287.758	318.5
	11	281.093	110.0	129.5	259.910	334.7	180.898	103.6	93.897	124.3
	15	9.064	194.0	125.5	230.840	304.3	223.689	102.8	260.036	290.1
	19	97.034	277.9	121.5	201.770	273.9	266.481	102.0	66.175	95.9
	23	185.005	1.9	117.5	172.700	243.5	309.273	101.2	232.314	261.7
	27	272.975	85.9	113.5	143.630	213.1	352.064	100.4	38.454	67.5
	31	0.946	169.8	109.5	114.560	182.6	34.856	99.6	204.593	233.3
Apr.	4	88.916	253.8	105.5	85.489	152.2	77.647	98.9	10.732	39.1
	8	176.886	337.8	101.5	56.419	121.8	120.439	98.1	176.871	204.9
	12	264.857	61.7	97.5	27.349	91.4	163.231	97.3	343.011	10.7
	16	352.827	145.7	93.5	358.279	61.0	206.022	96.5	149.150	176.5
	20	80.797	229.7	89.5	329.209	30.5	248.814	95.7	315.289	342.4
	24	168.767	313.6	85.5	300.139	0.1	291.606	94.9	121.428	148.2
	28	256.738	37.6	81.5	271.069	329.7	334.397	94.1	287.567	314.0
May	2	344.708	121.6	77.5	241.999	299.3	17.189	93.3	93.707	119.8
	6	72.678	205.5	73.5	212.929	268.8	59.980	92.5	259.846	285.6
	10	160.648	289.5	69.5	183.859	238.4	102.772	91.7	65.985	91.4
	14	248.618	13.5	65.5	154.789	208.0	145.564	90.9	232.124	257.2
	18	336.588	97.4	61.5	125.719	177.6	188.355	90.1	38.263	63.0
	22	64.558	181.4	57.5	96.649	147.2	231.147	89.4	204.403	228.8
	26	152.528	265.4	53.5	67.579	116.7	273.939	88.6	10.542	34.6
	30	240.498	349.3	49.5	38.509	86.3	316.730	87.8	176.681	200.4
June	3	328.467	73.3	45.5	9.439	55.9	359.522	87.0	342.820	6.2
	7	56.437	157.3	41.5	340.368	25.5	42.314	86.2	148.960	172.0
	11	144.407	241.2	37.5	311.298	355.1	85.105	85.4	315.099	337.8
	15	232.377	325.2	33.5	282.228	324.6	127.897	84.6	121.238	143.6
	19	320.347	49.2	29.5	253.158	294.2	170.688	83.8	287.377	309.4
	23	48.316	133.1	25.5	224.088	263.8	213.480	83.0	93.516	115.2
	27	136.286	217.1	21.5	195.018	233.4	256.272	82.2	259.656	281.0
July	1	224.255	301.1	17.5	165.948	202.9	299.063	81.4	65.795	86.8
4^d motion		1527.970	1524.0	−4.0	1050.930	1049.6	762.792	−0.8	526.139	525.8

ORBITAL POSITIONS FOR 0^h UNIVERSAL TIME

Date		MIMAS			ENCELADUS		TETHYS		DIONE	
		L	*M*	*θ*	*L*	*M*	*L*	*θ*	*L*	*M*
		°	°	°	°	°	°	°	°	°
July	1	224.255	301.1	17.5	165.948	202.9	299.063	81.4	65.795	86.8
	5	312.225	25.0	13.5	136.877	172.5	341.855	80.6	231.934	252.6
	9	40.194	109.0	9.5	107.807	142.1	24.647	79.8	38.073	58.4
	13	128.164	193.0	5.5	78.737	111.7	67.438	79.1	204.213	224.2
	17	216.133	276.9	1.5	49.667	81.3	110.230	78.3	10.352	30.0
	21	304.103	0.9	357.5	20.596	50.8	153.022	77.5	176.491	195.8
	25	32.072	84.9	353.5	351.526	20.4	195.813	76.7	342.630	1.6
	29	120.042	168.8	349.5	322.456	350.0	238.605	75.9	148.770	167.4
Aug.	2	208.011	252.8	345.5	293.386	319.6	281.397	75.1	314.909	333.2
	6	295.980	336.8	341.5	264.315	289.2	324.188	74.3	121.048	139.0
	10	23.949	60.7	337.5	235.245	258.7	6.980	73.5	287.187	304.8
	14	111.919	144.7	333.5	206.174	228.3	49.772	72.7	93.326	110.6
	18	199.888	228.7	329.5	177.104	197.9	92.563	71.9	259.466	276.4
	22	287.857	312.6	325.5	148.033	167.5	135.355	71.1	65.605	82.2
	26	15.826	36.6	321.5	118.963	137.0	178.147	70.3	231.744	248.0
	30	103.795	120.5	317.5	89.892	106.6	220.938	69.6	37.883	53.8
Sept.	3	191.764	204.5	313.5	60.822	76.2	263.730	68.8	204.023	219.7
	7	279.733	288.5	309.5	31.751	45.8	306.522	68.0	10.162	25.5
	11	7.702	12.4	305.5	2.680	15.4	349.314	67.2	176.301	191.3
	15	95.671	96.4	301.5	333.610	344.9	32.105	66.4	342.440	357.1
	19	183.640	180.4	297.5	304.539	314.5	74.897	65.6	148.580	162.9
	23	271.609	264.3	293.5	275.468	284.1	117.689	64.8	314.719	328.7
	27	359.577	348.3	289.5	246.397	253.7	160.480	64.0	120.858	134.5
Oct.	1	87.546	72.3	285.5	217.326	223.2	203.272	63.2	286.998	300.3
	5	175.515	156.2	281.5	188.256	192.8	246.064	62.4	93.137	106.1
	9	263.484	240.2	277.5	159.185	162.4	288.855	61.6	259.276	271.9
	13	351.452	324.2	273.5	130.114	132.0	331.647	60.8	65.415	77.7
	17	79.421	48.1	269.5	101.043	101.6	14.439	60.1	231.555	243.5
	21	167.389	132.1	265.5	71.971	71.1	57.230	59.3	37.694	49.3
	25	255.358	216.1	261.5	42.900	40.7	100.022	58.5	203.833	215.1
	29	343.327	300.0	257.5	13.829	10.3	142.814	57.7	9.973	20.9
Nov.	2	71.295	24.0	253.5	344.758	339.9	185.606	56.9	176.112	186.7
	6	159.263	108.0	249.5	315.687	309.4	228.397	56.1	342.251	352.5
	10	247.232	191.9	245.5	286.615	279.0	271.189	55.3	148.390	158.3
	14	335.200	275.9	241.5	257.544	248.6	313.981	54.5	314.530	324.1
	18	63.169	359.9	237.5	228.472	218.2	356.772	53.7	120.669	129.9
	22	151.137	83.8	233.5	199.401	187.8	39.564	52.9	286.808	295.7
	26	239.105	167.8	229.5	170.329	157.3	82.356	52.1	92.948	101.5
	30	327.073	251.7	225.5	141.258	126.9	125.148	51.3	259.087	267.3
Dec.	4	55.042	335.7	221.5	112.186	96.5	167.939	50.6	65.226	73.1
	8	143.010	59.7	217.5	83.114	66.1	210.731	49.8	231.366	238.9
	12	230.978	143.6	213.5	54.042	35.6	253.523	49.0	37.505	44.7
	16	318.946	227.6	209.5	24.971	5.2	296.314	48.2	203.644	210.5
	20	46.914	311.6	205.5	355.899	334.8	339.106	47.4	9.784	16.3
	24	134.882	35.5	201.5	326.827	304.4	21.898	46.6	175.923	182.1
	28	222.850	119.5	197.5	297.755	273.9	64.690	45.8	342.062	347.9
	32	310.818	203.5	193.5	268.683	243.5	107.481	45.0	148.202	153.7
4^d motion		1527.969	1524.0	−4.0	1050.929	1049.6	762.792	−0.8	526.139	525.8

ORBITAL POSITIONS FOR 0h UNIVERSAL TIME

Date		RHEA				TITAN			
		L	M	θ	γ	L	M	θ	γ
		°	°	°	°	°	°	°	°
Jan.	−1	278.966	81.2	56.7	0.311	221.445	21.36	240.03	0.364
	3	237.726	40.0	56.5	0.311	311.753	111.66	240.03	0.364
	7	196.485	358.7	56.4	0.311	42.060	201.96	240.04	0.364
	11	155.245	317.4	56.3	0.311	132.368	292.26	240.04	0.364
	15	114.005	276.1	56.2	0.311	222.675	22.57	240.05	0.364
	19	72.765	234.9	56.0	0.311	312.983	112.87	240.05	0.364
	23	31.525	193.6	55.9	0.311	43.290	203.17	240.05	0.364
	27	350.285	152.3	55.8	0.311	133.598	293.47	240.06	0.364
	31	309.045	111.0	55.7	0.311	223.906	23.77	240.06	0.364
Feb.	4	267.804	69.7	55.5	0.311	314.213	114.08	240.07	0.364
	8	226.564	28.5	55.4	0.311	44.521	204.38	240.07	0.364
	12	185.324	347.2	55.3	0.311	134.828	294.68	240.08	0.364
	16	144.084	305.9	55.2	0.311	225.136	24.98	240.08	0.364
	20	102.844	264.6	55.0	0.311	315.443	115.28	240.08	0.364
	24	61.604	223.4	54.9	0.311	45.751	205.58	240.09	0.363
	28	20.364	182.1	54.8	0.311	136.059	295.89	240.09	0.363
Mar.	3	339.123	140.8	54.7	0.311	226.366	26.19	240.10	0.363
	7	297.883	99.5	54.5	0.311	316.674	116.49	240.10	0.363
	11	256.643	58.3	54.4	0.311	46.981	206.79	240.10	0.363
	15	215.403	17.0	54.3	0.311	137.289	297.09	240.11	0.363
	19	174.163	335.7	54.2	0.311	227.596	27.39	240.11	0.363
	23	132.923	294.4	54.0	0.311	317.904	117.70	240.11	0.363
	27	91.683	253.1	53.9	0.311	48.212	208.00	240.12	0.363
	31	50.442	211.9	53.8	0.311	138.519	298.30	240.12	0.363
Apr.	4	9.202	170.6	53.7	0.311	228.827	28.60	240.12	0.363
	8	327.962	129.3	53.5	0.311	319.134	118.90	240.13	0.363
	12	286.722	88.0	53.4	0.311	49.442	209.20	240.13	0.363
	16	245.482	46.8	53.3	0.311	139.749	299.51	240.13	0.363
	20	204.242	5.5	53.2	0.311	230.057	29.81	240.14	0.363
	24	163.001	324.2	53.0	0.311	320.365	120.11	240.14	0.363
	28	121.761	282.9	52.9	0.311	50.672	210.41	240.14	0.363
May	2	80.521	241.7	52.8	0.311	140.980	300.71	240.15	0.363
	6	39.281	200.4	52.7	0.311	231.287	31.01	240.15	0.363
	10	358.041	159.1	52.5	0.311	321.595	121.32	240.15	0.363
	14	316.801	117.8	52.4	0.311	51.902	211.62	240.16	0.362
	18	275.561	76.5	52.3	0.311	142.210	301.92	240.16	0.362
	22	234.320	35.3	52.2	0.311	232.517	32.22	240.16	0.362
	26	193.080	354.0	52.0	0.311	322.825	122.52	240.17	0.362
	30	151.840	312.7	51.9	0.311	53.133	212.82	240.17	0.362
June	3	110.600	271.4	51.8	0.311	143.440	303.12	240.17	0.362
	7	69.360	230.2	51.7	0.311	233.748	33.43	240.17	0.362
	11	28.120	188.9	51.6	0.311	324.055	123.73	240.18	0.362
	15	346.880	147.6	51.4	0.311	54.363	214.03	240.18	0.362
	19	305.639	106.3	51.3	0.311	144.670	304.33	240.18	0.362
	23	264.399	65.1	51.2	0.311	234.978	34.63	240.18	0.362
	27	223.159	23.8	51.1	0.311	325.286	124.93	240.19	0.362
July	1	181.919	342.5	50.9	0.311	55.593	215.24	240.19	0.362
4d motion		318.760	318.7	. . .	. . . .	90.308	90.30	. . . . .	. . . .

ORBITAL POSITIONS FOR 0^h UNIVERSAL TIME

Date		RHEA L	RHEA M	RHEA θ	RHEA γ	TITAN L	TITAN M	TITAN θ	TITAN γ
		°	°	°	°	°	°	°	°
July	1	181.919	342.5	50.9	0.311	55.593	215.24	240.19	0.362
	5	140.679	301.2	50.8	0.311	145.901	305.54	240.19	0.362
	9	99.439	259.9	50.7	0.311	236.208	35.84	240.19	0.362
	13	58.199	218.7	50.6	0.311	326.516	126.14	240.20	0.362
	17	16.958	177.4	50.4	0.311	56.823	216.44	240.20	0.362
	21	335.718	136.1	50.3	0.311	147.131	306.74	240.20	0.362
	25	294.478	94.8	50.2	0.311	237.438	37.05	240.20	0.362
	29	253.238	53.6	50.1	0.311	327.746	127.35	240.21	0.362
Aug.	2	211.998	12.3	49.9	0.311	58.054	217.65	240.21	0.361
	6	170.758	331.0	49.8	0.311	148.361	307.95	240.21	0.361
	10	129.517	289.7	49.7	0.311	238.669	38.25	240.21	0.361
	14	88.277	248.4	49.6	0.311	328.976	128.55	240.21	0.361
	18	47.037	207.2	49.4	0.311	59.284	218.86	240.21	0.361
	22	5.797	165.9	49.3	0.311	149.591	309.16	240.22	0.361
	26	324.557	124.6	49.2	0.311	239.899	39.46	240.22	0.361
	30	283.317	83.3	49.1	0.311	330.206	129.76	240.22	0.361
Sept.	3	242.077	42.1	48.9	0.311	60.514	220.06	240.22	0.361
	7	200.836	0.8	48.8	0.311	150.822	310.36	240.22	0.361
	11	159.596	319.5	48.7	0.311	241.129	40.67	240.23	0.361
	15	118.356	278.2	48.6	0.311	331.437	130.97	240.23	0.361
	19	77.116	237.0	48.4	0.311	61.744	221.27	240.23	0.361
	23	35.876	195.7	48.3	0.311	152.052	311.57	240.23	0.361
	27	354.636	154.4	48.2	0.311	242.359	41.87	240.23	0.361
Oct.	1	313.396	113.1	48.1	0.311	332.667	132.17	240.23	0.361
	5	272.155	71.8	47.9	0.311	62.974	222.47	240.23	0.361
	9	230.915	30.6	47.8	0.311	153.282	312.78	240.23	0.361
	13	189.675	349.3	47.7	0.311	243.590	43.08	240.24	0.361
	17	148.435	308.0	47.6	0.311	333.897	133.38	240.24	0.361
	21	107.195	266.7	47.4	0.311	64.205	223.68	240.24	0.360
	25	65.955	225.5	47.3	0.311	154.512	313.98	240.24	0.360
	29	24.715	184.2	47.2	0.311	244.820	44.28	240.24	0.360
Nov.	2	343.474	142.9	47.1	0.311	335.127	134.59	240.24	0.360
	6	302.234	101.6	46.9	0.311	65.435	224.89	240.24	0.360
	10	260.994	60.3	46.8	0.311	155.742	315.19	240.24	0.360
	14	219.754	19.1	46.7	0.311	246.050	45.49	240.24	0.360
	18	178.514	337.8	46.6	0.311	336.357	135.79	240.24	0.360
	22	137.274	296.5	46.4	0.311	66.665	226.09	240.25	0.360
	26	96.033	255.2	46.3	0.311	156.973	316.40	240.25	0.360
	30	54.793	214.0	46.2	0.311	247.280	46.70	240.25	0.360
Dec.	4	13.553	172.7	46.1	0.311	337.588	137.00	240.25	0.360
	8	332.313	131.4	45.9	0.311	67.895	227.30	240.25	0.360
	12	291.073	90.1	45.8	0.311	158.203	317.60	240.25	0.360
	16	249.833	48.8	45.7	0.311	248.510	47.90	240.25	0.360
	20	208.593	7.6	45.6	0.311	338.818	138.20	240.25	0.360
	24	167.352	326.3	45.4	0.311	69.125	228.51	240.25	0.360
	28	126.112	285.0	45.3	0.311	159.433	318.81	240.25	0.360
	32	84.872	243.7	45.2	0.311	249.741	49.11	240.25	0.360
4^d motion		318.760	318.7	. . .	. . . .	90.308	90.30	. . . . .	. . . .

ORBITAL POSITIONS FOR 0^h UNIVERSAL TIME

Date		HYPERION						IAPETUS		
		L	M	θ	γ	e	a	L	M	γ
		°	°	°	°		″	°	°	°
Jan.	−1	118.428	92.25	256.41	0.936	0.07913	2045.9	139.056	257.40	15.322
	3	186.038	160.12	256.39	0.936	0.07897	2045.6	157.207	275.55	15.322
	7	253.662	228.00	256.38	0.936	0.07881	2045.3	175.359	293.70	15.322
	11	321.300	295.90	256.37	0.936	0.07864	2045.1	193.510	311.85	15.322
	15	28.953	3.81	256.35	0.936	0.07848	2044.8	211.662	330.00	15.322
	19	96.619	71.74	256.34	0.937	0.07832	2044.5	229.813	348.15	15.323
	23	164.299	139.68	256.32	0.937	0.07815	2044.2	247.965	6.30	15.323
	27	231.993	207.64	256.31	0.937	0.07799	2043.9	266.116	24.45	15.323
	31	299.701	275.61	256.30	0.937	0.07783	2043.7	284.268	42.60	15.323
Feb.	4	7.424	343.60	256.28	0.937	0.07767	2043.4	302.419	60.75	15.323
	8	75.160	51.60	256.27	0.937	0.07751	2043.1	320.571	78.91	15.323
	12	142.910	119.61	256.25	0.937	0.07735	2042.8	338.722	97.06	15.323
	16	210.674	187.64	256.24	0.937	0.07720	2042.6	356.874	115.21	15.324
	20	278.451	255.69	256.22	0.937	0.07704	2042.3	15.025	133.36	15.324
	24	346.242	323.75	256.21	0.938	0.07689	2042.0	33.177	151.51	15.324
	28	54.047	31.82	256.20	0.938	0.07674	2041.8	51.328	169.66	15.324
Mar.	3	121.864	99.90	256.18	0.938	0.07660	2041.5	69.480	187.81	15.324
	7	189.694	168.00	256.17	0.938	0.07645	2041.3	87.631	205.96	15.324
	11	257.537	236.12	256.15	0.938	0.07631	2041.0	105.783	224.11	15.325
	15	325.392	304.24	256.14	0.938	0.07618	2040.8	123.934	242.26	15.325
	19	33.259	12.38	256.13	0.938	0.07604	2040.5	142.086	260.41	15.325
	23	101.139	80.53	256.11	0.938	0.07591	2040.3	160.237	278.56	15.325
	27	169.029	148.70	256.10	0.938	0.07579	2040.1	178.389	296.71	15.325
	31	236.931	216.87	256.08	0.938	0.07567	2039.9	196.540	314.86	15.325
Apr.	4	304.844	285.06	256.07	0.939	0.07555	2039.7	214.692	333.01	15.325
	8	12.767	353.26	256.06	0.939	0.07544	2039.5	232.843	351.16	15.326
	12	80.700	61.46	256.04	0.939	0.07533	2039.3	250.995	9.31	15.326
	16	148.643	129.68	256.03	0.939	0.07523	2039.1	269.146	27.46	15.326
	20	216.596	197.91	256.01	0.939	0.07513	2038.9	287.297	45.61	15.326
	24	284.557	266.15	256.00	0.939	0.07504	2038.7	305.449	63.76	15.326
	28	352.526	334.39	255.99	0.939	0.07495	2038.6	323.600	81.92	15.326
May	2	60.504	42.65	255.97	0.939	0.07487	2038.4	341.752	100.07	15.327
	6	128.489	110.91	255.96	0.939	0.07479	2038.3	359.903	118.22	15.327
	10	196.481	179.18	255.94	0.940	0.07472	2038.1	18.055	136.37	15.327
	14	264.480	247.46	255.93	0.940	0.07466	2038.0	36.206	154.52	15.327
	18	332.485	315.74	255.92	0.940	0.07460	2037.9	54.358	172.67	15.327
	22	40.495	24.03	255.90	0.940	0.07454	2037.8	72.509	190.82	15.327
	26	108.510	92.32	255.89	0.940	0.07450	2037.7	90.661	208.97	15.327
	30	176.530	160.62	255.87	0.940	0.07446	2037.6	108.812	227.12	15.328
June	3	244.554	228.93	255.86	0.940	0.07442	2037.5	126.964	245.27	15.328
	7	312.581	297.23	255.85	0.940	0.07440	2037.5	145.115	263.42	15.328
	11	20.612	5.54	255.83	0.940	0.07438	2037.4	163.267	281.57	15.328
	15	88.644	73.86	255.82	0.940	0.07436	2037.4	181.418	299.72	15.328
	19	156.679	142.17	255.80	0.941	0.07435	2037.3	199.570	317.87	15.328
	23	224.714	210.49	255.79	0.941	0.07435	2037.3	217.721	336.02	15.328
	27	292.751	278.80	255.78	0.941	0.07436	2037.3	235.873	354.17	15.329
July	1	0.787	347.12	255.76	0.941	0.07437	2037.3	254.024	12.32	15.329

ORBITAL POSITIONS FOR 0^h UNIVERSAL TIME

Date		Hyperion L	Hyperion M	Hyperion θ	Hyperion γ	Hyperion e	Hyperion a	Iapetus L	Iapetus M	Iapetus γ
		°	°	°	°		″	°	°	°
July	1	0.787	347.12	255.76	0.941	0.07437	2037.3	254.024	12.32	15.329
	5	68.824	55.43	255.75	0.941	0.07439	2037.3	272.176	30.47	15.329
	9	136.859	123.75	255.73	0.941	0.07441	2037.3	290.327	48.62	15.329
	13	204.893	192.06	255.72	0.941	0.07445	2037.4	308.479	66.77	15.329
	17	272.924	260.37	255.71	0.941	0.07448	2037.4	326.630	84.93	15.329
	21	340.953	328.68	255.69	0.941	0.07453	2037.5	344.782	103.08	15.330
	25	48.979	36.99	255.68	0.942	0.07458	2037.5	2.933	121.23	15.330
	29	117.002	105.29	255.66	0.942	0.07464	2037.6	21.085	139.38	15.330
Aug.	2	185.020	173.59	255.65	0.942	0.07470	2037.7	39.236	157.53	15.330
	6	253.033	241.88	255.64	0.942	0.07477	2037.8	57.388	175.68	15.330
	10	321.041	310.16	255.62	0.942	0.07485	2037.9	75.539	193.83	15.330
	14	29.044	18.44	255.61	0.942	0.07493	2038.0	93.691	211.98	15.330
	18	97.040	86.72	255.59	0.942	0.07502	2038.1	111.842	230.13	15.331
	22	165.029	154.98	255.58	0.942	0.07511	2038.3	129.994	248.28	15.331
	26	233.011	223.24	255.57	0.942	0.07521	2038.4	148.145	266.43	15.331
	30	300.985	291.49	255.55	0.942	0.07532	2038.6	166.297	284.58	15.331
Sept.	3	8.952	359.73	255.54	0.943	0.07543	2038.8	184.448	302.73	15.331
	7	76.909	67.97	255.52	0.943	0.07554	2038.9	202.600	320.88	15.331
	11	144.858	136.19	255.51	0.943	0.07567	2039.1	220.751	339.03	15.332
	15	212.797	204.40	255.49	0.943	0.07579	2039.3	238.903	357.18	15.332
	19	280.726	272.61	255.48	0.943	0.07592	2039.5	257.054	15.33	15.332
	23	348.645	340.80	255.47	0.943	0.07606	2039.7	275.206	33.48	15.332
	27	56.553	48.98	255.45	0.943	0.07620	2039.9	293.357	51.63	15.332
Oct.	1	124.451	117.15	255.44	0.943	0.07634	2040.1	311.509	69.78	15.332
	5	192.337	185.31	255.42	0.943	0.07649	2040.4	329.660	87.94	15.332
	9	260.211	253.45	255.41	0.944	0.07665	2040.6	347.812	106.09	15.333
	13	328.074	321.58	255.40	0.944	0.07680	2040.8	5.963	124.24	15.333
	17	35.924	29.70	255.38	0.944	0.07696	2041.1	24.115	142.39	15.333
	21	103.762	97.81	255.37	0.944	0.07713	2041.3	42.266	160.54	15.333
	25	171.587	165.90	255.35	0.944	0.07729	2041.6	60.418	178.69	15.333
	29	239.399	233.98	255.34	0.944	0.07746	2041.8	78.569	196.84	15.333
Nov.	2	307.197	302.05	255.33	0.944	0.07763	2042.1	96.721	214.99	15.334
	6	14.983	10.10	255.31	0.944	0.07781	2042.4	114.872	233.14	15.334
	10	82.755	78.13	255.30	0.944	0.07798	2042.6	133.024	251.29	15.334
	14	150.513	146.16	255.28	0.944	0.07816	2042.9	151.175	269.44	15.334
	18	218.258	214.17	255.27	0.945	0.07834	2043.2	169.327	287.59	15.334
	22	285.988	282.16	255.26	0.945	0.07853	2043.4	187.478	305.74	15.334
	26	353.705	350.14	255.24	0.945	0.07871	2043.7	205.630	323.89	15.334
	30	61.407	58.10	255.23	0.945	0.07889	2044.0	223.781	342.04	15.335
Dec.	4	129.096	126.05	255.21	0.945	0.07908	2044.3	241.933	0.19	15.335
	8	196.770	193.98	255.20	0.945	0.07926	2044.6	260.084	18.34	15.335
	12	264.431	261.90	255.19	0.945	0.07945	2044.8	278.236	36.49	15.335
	16	332.077	329.81	255.17	0.945	0.07963	2045.1	296.387	54.64	15.335
	20	39.710	37.70	255.16	0.945	0.07982	2045.4	314.539	72.79	15.335
	24	107.328	105.57	255.14	0.945	0.08000	2045.7	332.690	90.95	15.335
	28	174.933	173.43	255.13	0.946	0.08019	2045.9	350.842	109.10	15.336
	32	242.523	241.28	255.12	0.946	0.08037	2046.2	8.993	127.25	15.336

DIFFERENTIAL COORDINATES OF HYPERION FOR 0^h U.T.

Date	Δα	Δδ	Date	Δα	Δδ	Date	Δα	Δδ
	s	′		s	′		s	′
Jan. −1	−10	+0.9	Apr. 30	−4	−0.6	Sept. 1	+9	−1.0
1	−5	+1.1	May 2	−11	−0.2	3	−1	−1.0
3	+3	+1.0	4	−14	+0.4	5	−10	−0.5
5	+9	+0.6	6	−12	+0.8	7	−15	+0.2
7	+13	+0.1	8	−6	+1.0	9	−15	+0.8
9	+13	−0.5	10	+2	+0.9	11	−9	+1.2
11	+9	−0.9	12	+9	+0.5	13	−1	+1.2
13	+2	−1.0	14	+14	0.0	15	+7	+0.9
15	−6	−0.6	16	+14	−0.5	17	+13	+0.3
17	−12	0.0	18	+10	−0.8	19	+16	−0.3
19	−12	+0.6	20	+1	−0.8	21	+13	−0.9
21	−9	+1.0	22	−7	−0.5	23	+5	−1.1
23	−3	+1.1	24	−13	+0.1	25	−5	−0.8
25	+4	+0.9	26	−14	+0.6	27	−13	−0.2
27	+10	+0.4	28	−10	+0.9	29	−15	+0.5
29	+13	−0.1	30	−3	+1.0	Oct. 1	−13	+1.0
31	+12	−0.6	June 1	+5	+0.8	3	−6	+1.2
Feb. 2	+7	−0.9	3	+12	+0.3	5	+2	+1.1
4	−1	−0.8	5	+15	−0.2	7	+10	+0.7
6	−8	−0.4	7	+13	−0.7	9	+14	+0.1
8	−12	+0.2	9	+7	−0.9	11	+15	−0.6
10	−12	+0.7	11	−3	−0.7	13	+10	−1.0
12	−7	+1.0	13	−11	−0.3	15	+1	−1.0
14	−1	+1.0	15	−15	+0.3	17	−8	−0.6
16	+6	+0.7	17	−14	+0.8	19	−14	+0.1
18	+11	+0.3	19	−8	+1.0	21	−14	+0.7
20	+13	−0.3	21	0	+1.0	23	−10	+1.1
22	+11	−0.7	23	+8	+0.6	25	−3	+1.2
24	+5	−0.9	25	+14	+0.1	27	+5	+0.9
26	−3	−0.7	27	+15	−0.4	29	+11	+0.4
28	−10	−0.2	29	+11	−0.8	31	+14	−0.2
Mar. 1	−13	+0.3	July 1	+3	−0.9	Nov. 2	+13	−0.7
3	−11	+0.8	3	−7	−0.6	4	+6	−1.0
5	−6	+1.0	5	−14	0.0	6	−3	−0.8
7	+1	+0.9	7	−15	+0.6	8	−11	−0.3
9	+8	+0.6	9	−12	+1.0	10	−14	+0.3
11	+12	+0.1	11	−5	+1.1	12	−13	+0.8
13	+13	−0.4	13	+4	+0.9	14	−8	+1.1
15	+10	−0.8	15	+11	+0.5	16	0	+1.0
17	+2	−0.8	17	+16	−0.1	18	+7	+0.7
19	−6	−0.6	19	+14	−0.7	20	+12	+0.2
21	−12	0.0	21	+8	−1.0	22	+14	−0.4
23	−13	+0.5	23	−2	−0.9	24	+11	−0.8
25	−10	+0.9	25	−11	−0.4	26	+3	−0.9
27	−4	+1.0	27	−15	+0.3	28	−6	−0.6
29	+4	+0.8	29	−15	+0.8	30	−12	−0.1
31	+10	+0.4	31	−9	+1.1	Dec. 2	−14	+0.5
Apr. 2	+13	−0.1	Aug. 2	−1	+1.1	4	−11	+0.9
4	+13	−0.6	4	+8	+0.8	6	−5	+1.0
6	+8	−0.8	6	+14	+0.2	8	+2	+0.9
8	−1	−0.8	8	+16	−0.4	10	+9	+0.5
10	−9	−0.4	10	+12	−0.9	12	+13	0.0
12	−13	+0.2	12	+4	−1.0	14	+13	−0.5
14	−13	+0.7	14	−6	−0.7	16	+8	−0.8
16	−8	+0.9	16	−14	−0.1	18	0	−0.8
18	−1	+0.9	18	−16	+0.5	20	−8	−0.5
20	+6	+0.7	20	−12	+1.0	22	−13	+0.1
22	+12	+0.2	22	−5	+1.2	24	−13	+0.6
24	+14	−0.3	24	+3	+1.0	26	−9	+0.9
26	+12	−0.7	26	+11	+0.6	28	−3	+1.0
28	+5	−0.8	28	+15	−0.1	30	+4	+0.8
30	−4	−0.6	30	+15	−0.7	32	+10	+0.4

Differential coordinates are given in the sense "satellite minus planet."

DIFFERENTIAL COORDINATES OF IAPETUS FOR 0^h U.T.

Date	Δα	Δδ	Date	Δα	Δδ	Date	Δα	Δδ
	s	′		s	′		s	′
Jan. −1	− 22	+ 0.7	Apr. 30	+ 20	− 0.1	Sept. 1	− 9	+ 0.5
1	18	0.7	May 2	15	0.1	3	− 3	0.5
3	14	0.7	4	10	0.2	5	+ 3	0.5
5	9	0.7	6	+ 5	0.2	7	9	0.4
7	− 4	0.6	8	0	0.2	9	15	0.4
9	+ 1	0.6	10	− 5	0.2	11	20	0.3
11	+ 6	+ 0.5	12	− 11	− 0.2	13	+ 25	+ 0.3
13	11	0.4	14	16	0.2	15	29	0.2
15	15	0.3	16	20	0.2	17	32	+ 0.1
17	19	0.2	18	25	0.2	19	35	0.0
19	23	0.1	20	28	0.2	21	36	− 0.1
21	26	+ 0.1	22	32	0.2	23	37	0.2
23	+ 28	0.0	24	− 34	− 0.1	25	+ 36	− 0.2
25	30	− 0.1	26	36	0.1	27	34	0.3
27	31	0.2	28	36	0.1	29	32	0.4
29	31	0.3	30	36	− 0.1	Oct. 1	29	0.4
31	30	0.3	June 1	35	0.0	3	24	0.5
Feb. 2	29	0.4	3	34	0.0	5	20	0.5
4	+ 26	− 0.4	5	− 31	0.0	7	+ 15	− 0.5
6	24	0.5	7	28	+ 0.1	9	9	0.5
8	20	0.5	9	24	0.1	11	+ 4	0.5
10	16	0.5	11	19	0.1	13	− 2	0.5
12	12	0.5	13	14	0.2	15	8	0.5
14	7	0.5	15	8	0.2	17	13	0.4
16	+ 3	− 0.4	17	− 2	+ 0.2	19	− 18	− 0.4
18	− 2	0.4	19	+ 4	0.2	21	23	0.3
20	7	0.4	21	10	0.2	23	27	0.2
22	11	0.3	23	15	0.2	25	30	0.2
24	16	0.3	25	21	0.2	27	33	− 0.1
26	20	0.2	27	25	0.2	29	35	0.0
28	− 24	− 0.2	29	+ 29	+ 0.2	31	− 36	+ 0.1
Mar. 1	27	0.1	July 1	33	0.2	Nov. 2	36	0.2
3	29	− 0.1	3	35	0.1	4	35	0.2
5	31	0.0	5	37	+ 0.1	6	34	0.3
7	32	0.0	7	37	0.0	8	32	0.3
9	33	+ 0.1	9	36	0.0	10	29	0.4
11	− 32	+ 0.1	11	+ 35	− 0.1	12	− 25	+ 0.4
13	32	0.2	13	32	0.1	14	21	0.4
15	30	0.2	15	29	0.2	16	17	0.4
17	27	0.2	17	25	0.2	18	12	0.4
19	24	0.2	19	20	0.3	20	6	0.4
21	21	0.2	21	14	0.3	22	− 1	0.4
23	− 17	+ 0.3	23	+ 9	− 0.3	24	+ 4	+ 0.4
25	12	0.3	25	+ 3	0.3	26	10	0.4
27	7	0.3	27	− 4	0.3	28	15	0.3
29	− 2	0.3	29	10	0.3	30	19	0.3
31	+ 3	0.2	31	15	0.3	Dec. 2	23	0.2
Apr. 2	8	0.2	Aug. 2	21	0.3	4	26	0.1
4	+ 13	+ 0.2	4	− 26	− 0.3	6	+ 29	+ 0.1
6	18	0.2	6	30	0.2	8	31	0.0
8	22	0.2	8	34	0.2	10	32	0.0
10	25	0.1	10	36	− 0.1	12	32	− 0.1
12	28	0.1	12	38	0.0	14	32	0.1
14	31	0.1	14	39	0.0	16	30	0.2
16	+ 32	+ 0.1	16	− 39	+ 0.1	18	+ 28	− 0.2
18	33	0.0	18	38	0.2	20	25	0.2
20	32	0.0	20	37	0.2	22	22	0.2
22	31	0.0	22	34	0.3	24	18	0.3
24	30	0.0	24	30	0.3	26	13	0.3
26	27	− 0.1	26	26	0.4	28	9	0.3
28	+ 24	− 0.1	28	− 21	+ 0.4	30	+ 4	− 0.3
30	+ 20	− 0.1	30	− 15	+ 0.5	32	− 1	− 0.3

Differential coordinates are given in the sense "satellite minus planet."

DIFFERENTIAL COORDINATES OF PHOEBE FOR 0^h U.T.

Date		Δα	Δδ	Date		Δα	Δδ	Date		Δα	Δδ
		m s	′			m s	′			m s	′
Jan.	−1	+ 1 15	+ 1.5	Apr.	30	− 1 36	− 6.3	Sept.	1	− 0 42	+ 0.3
	1	1 12	1.4	May	2	1 38	6.4		3	0 40	0.4
	3	1 09	1.3		4	1 39	6.4		5	0 37	0.6
	5	1 07	1.2		6	1 40	6.4		7	0 35	0.7
	7	1 04	1.0		8	1 42	6.4		9	0 32	0.9
	9	1 01	0.9		10	1 43	6.4		11	0 30	1.0
	11	+ 0 58	+ 0.8		12	− 1 44	− 6.4		13	− 0 27	+ 1.1
	13	0 55	0.7		14	1 45	6.4		15	0 24	1.3
	15	0 52	0.5		16	1 45	6.4		17	0 22	1.4
	17	0 49	0.4		18	1 46	6.4		19	0 19	1.6
	19	0 46	0.3		20	1 47	6.3		21	0 17	1.7
	21	0 43	+ 0.1		22	1 47	6.3		23	0 14	1.8
	23	+ 0 39	0.0		24	− 1 48	− 6.3		25	− 0 11	+ 2.0
	25	0 36	− 0.2		26	1 48	6.2		27	0 09	2.1
	27	0 33	0.3		28	1 48	6.2		29	0 06	2.2
	29	0 30	0.4		30	1 49	6.1	Oct.	1	0 03	2.3
	31	0 27	0.6	June	1	1 49	6.1		3	− 0 01	2.5
Feb.	2	0 23	0.7		3	1 49	6.0		5	+ 0 02	2.6
	4	+ 0 20	− 0.9		5	− 1 49	− 5.9		7	+ 0 05	+ 2.7
	6	0 17	1.1		7	1 49	5.8		9	0 07	2.8
	8	0 14	1.2		9	1 48	5.7		11	0 10	3.0
	10	0 10	1.4		11	1 48	5.6		13	0 13	3.1
	12	0 07	1.5		13	1 48	5.5		15	0 15	3.2
	14	0 04	1.7		15	1 47	5.4		17	0 18	3.3
	16	+ 0 01	− 1.8		17	− 1 47	− 5.3		19	+ 0 21	+ 3.4
	18	− 0 03	2.0		19	1 46	5.2		21	0 23	3.6
	20	0 06	2.2		21	1 46	5.1		23	0 26	3.7
	22	0 09	2.3		23	1 45	5.0		25	0 28	3.8
	24	0 12	2.5		25	1 44	4.9		27	0 31	3.9
	26	0 16	2.7		27	1 43	4.7		29	0 33	4.0
	28	− 0 19	− 2.8		29	− 1 42	− 4.6		31	+ 0 36	+ 4.1
Mar.	1	0 22	3.0	July	1	1 41	4.5	Nov.	2	0 38	4.2
	3	0 25	3.1		3	1 40	4.3		4	0 41	4.4
	5	0 28	3.3		5	1 39	4.2		6	0 43	4.5
	7	0 32	3.4		7	1 38	4.1		8	0 46	4.6
	9	0 35	3.6		9	1 36	3.9		10	0 48	4.7
	11	− 0 38	− 3.8		11	− 1 35	− 3.8		12	+ 0 51	+ 4.8
	13	0 41	3.9		13	1 33	3.6		14	0 53	4.9
	15	0 44	4.1		15	1 32	3.5		16	0 55	5.0
	17	0 47	4.2		17	1 30	3.3		18	0 58	5.1
	19	0 50	4.3		19	1 29	3.1		20	1 00	5.2
	21	0 52	4.5		21	1 27	3.0		22	1 02	5.3
	23	− 0 55	− 4.6		23	− 1 25	− 2.8		24	+ 1 05	+ 5.4
	25	0 58	4.8		25	1 24	2.7		26	1 07	5.6
	27	1 01	4.9		27	1 22	2.5		28	1 09	5.7
	29	1 03	5.0		29	1 20	2.4		30	1 11	5.8
	31	1 06	5.1		31	1 18	2.2	Dec.	2	1 13	5.9
Apr.	2	1 08	5.2	Aug.	2	1 16	2.0		4	1 15	6.0
	4	− 1 11	− 5.4		4	− 1 14	− 1.9		6	+ 1 17	+ 6.1
	6	1 13	5.5		6	1 12	1.7		8	1 19	6.2
	8	1 16	5.6		8	1 10	1.6		10	1 21	6.3
	10	1 18	5.7		10	1 08	1.4		12	1 23	6.4
	12	1 20	5.8		12	1 06	1.2		14	1 25	6.5
	14	1 22	5.8		14	1 04	1.1		16	1 27	6.6
	16	− 1 24	− 5.9		16	− 1 01	− 0.9		18	+ 1 29	+ 6.7
	18	1 26	6.0		18	0 59	0.8		20	1 31	6.8
	20	1 28	6.1		20	0 57	0.6		22	1 32	6.9
	22	1 30	6.1		22	0 54	0.5		24	1 34	7.0
	24	1 32	6.2		24	0 52	0.3		26	1 36	7.1
	26	1 33	6.2		26	0 50	− 0.2		28	1 37	7.1
	28	− 1 35	− 6.3		28	− 0 47	0.0		30	+ 1 39	+ 7.2
	30	− 1 36	− 6.3		30	− 0 45	+ 0.1		32	+ 1 41	+ 7.3

Differential coordinates are given in the sense “satellite minus planet.”

TRUE ORBITAL LONGITUDE AND RADIUS VECTOR

The formulae and constants for obtaining the true orbital longitude u and the radius vector r of Satellites I–VIII (where γ is the inclination) follow. Quantities that change during the year are tabulated separately.

Mimas

$r/a = 1.0002 - 0.0201 \cos M - 0.0002 \cos 2M$ $\quad a = 255''.9 \quad \gamma = 1°31'.0$

$u = L + 2°.303 \sin M + 0°.029 \sin 2M$

Enceladus

$r/a = 1 - 0.0044 \cos M$ $\quad a = 328''.3 \quad \gamma = 1'.4$

$u = L + 0°.509 \sin M$ $\quad u - \theta = 36° + 263.15\ (\mathrm{JD} - 243\ 6000.5)$

Tethys

$r/a = 1 \quad u = L$ $\quad a = 406''.4 \quad \gamma = 1°05'.56$

Dione

$r/a = 1 - 0.0022 \cos M$ $\quad a = 520''.5 \quad \gamma = 1'.4$

$u = L + 0°.253 \sin M$ $\quad u - \theta = 214° + 131.62\ (\mathrm{JD} - 243\ 6000.5)$

Rhea, Titan, Hyperion

$r/a = 1 + e^2/2 - e \cos M - e^2/2 \cos 2M - \ldots$

$u = L + 2e \sin M + \ldots$

Rhea

$a = 726''.9$

e	=	0.00127	January 0 – March 3
	=	0.00128	March 4 – December 32

Titan

$a = 1684''.4$

e	=	0.02888	January 0 – September 3
	=	0.02889	September 4 – December 32

Iapetus

$r/a = 1.0004 - 0.0283 \cos M - 0.0004 \cos 2M$ $\quad a = 4908''.6$

$u = L + 3°.240 \sin M + 0°.057 \sin 2M + 0°.001 \sin 3M$

θ	=	254.37	January 0 – February 20
	=	254.36	February 21 – May 18
	=	254.35	May 19 – August 14
	=	254.34	August 15 – November 10
	=	254.33	November 11 – December 32

SATURNICENTRIC RECTANGULAR COORDINATES

Apparent rectangular coordinates, with the x-axis in the plane of the rings, positive toward the east, and the y-axis positive toward the north pole of Saturn, are given by

$$x = \xi / (1 + \zeta) \qquad y = \eta / (1 + \zeta)$$

$$\xi = (a/\Delta)\,(r/a)\,[\cos b \sin (l - U)]$$

$$\eta = (a/\Delta)\,(r/a)\,[\cos b \sin B \cos (l - U) + \sin b \cos B]$$

$$\zeta = (a/\Delta)\,(r/a)\,[\cos b \cos B \cos (l - U) - \sin b \sin B]$$

$$\sin b = \sin (u - \theta) \sin \gamma$$

$$\cos b \sin (l - \theta) = \sin (u - \theta) \cos \gamma$$

$$\cos b \cos (l - \theta) = \cos (u - \theta)$$

For Satellites I–V, apparent rectangular coordinates may be obtained with sufficient acccuracy from

$$x = (a/\Delta)\,(r/a)\,[1/(1 + \zeta)] \sin (u - U) = s \sin (p - P)$$

$$y = (a/\Delta)\,(r/a)\,[1/(1 + \zeta)]\,[\sin B \cos (u - U) + \cos B \sin \gamma \sin (u - \theta)]$$

$$= s \cos (p - P)$$

and the tables given below. In critical cases ascend.

Mimas

$u - U$	$\frac{1}{1+\zeta}$	$u - U$
°		°
0.0		360.0
	0.9999	
67.3		292.7
	1.0000	
112.6		247.4
	1.0001	
247.3		112.7

Tethys

$u - U$	$\frac{1}{1+\zeta}$	$u - U$
°		°
0.0		360.0
	9.9998	
43.4		316.6
	0.9999	
75.9		284.1
	1.0000	
104.0		256.0
	1.0001	
136.5		223.5
	1.0002	
223.4		136.6

Rhea

$u - U$	$\frac{1}{1+\zeta}$	$u - U$
°		°
0.0		360.0
	0.9996	
18.6		341.4
	0.9997	
47.4		312.6
	0.9998	
66.0		294.0
	0.9999	
82.2		277.8
	1.0000	
97.7		262.3
	1.0001	
113.9		246.1
	1.0002	
132.5		227.5
	1.0003	
161.3		198.7
	1.0004	
198.6		161.4

Enceladus

$u - U$	$\frac{1}{1+\zeta}$	$u - U$
°		°
0.0		360.0
	0.9998	
25.9		334.1
	0.9999	
72.5		287.5
	1.0000	
107.4		252.6
	1.0001	
154.0		206.0
	1.0002	
205.9		154.1

Dione

$u - U$	$\frac{1}{1+\zeta}$	$u - U$
°		°
0.0		360.0
	0.9997	
19.0		341.0
	0.9998	
55.4		304.6
	0.9999	
79.1		280.9
	1.0000	
100.8		259.2
	1.0001	
124.5		235.5
	1.0002	
160.9		199.1
	1.0003	
199.0		161.0

APPARENT ORBITS OF SATELLITES I–IV AT DATE OF OPPOSITION, JULY 7

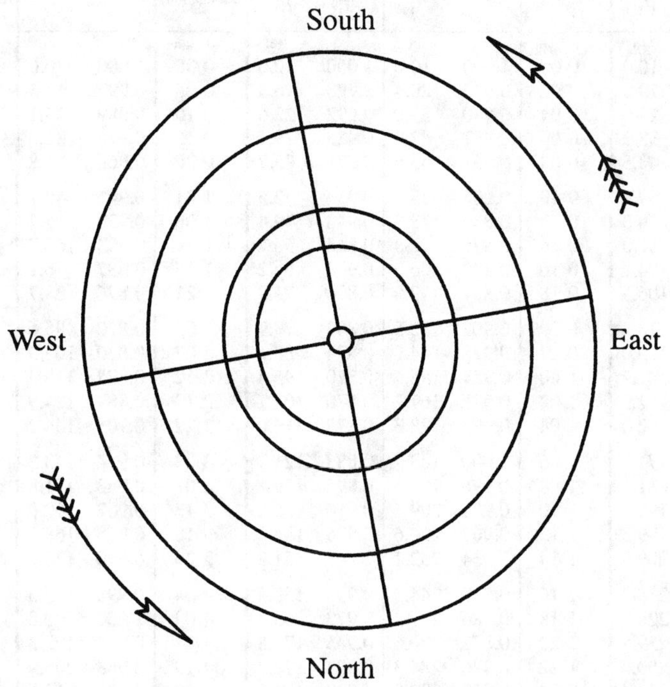

	Name	Sidereal Period	
		d	h
V	Miranda	1.4	
I	Ariel	2	12.489
II	Umbriel	4	03.460
III	Titania	8	16.941
IV	Oberon	13	11.118

RINGS OF URANUS

Ring	Semimajor Axis	Eccentricity	Azimuth of Periapse	Precession Rate
	km		°	°/d
6	41870	0.0014	236	2.77
5	42270	0.0018	182	2.66
4	42600	0.0012	120	2.60
α	44750	0.0007	331	2.18
β	45700	0.0005	231	2.03
η	47210	— —	—	—
γ	47660	— —	—	—
δ	48330	0.0005	140	—
ε	51180	0.0079	216	1.36

Epoch: 1977 March 10, 20^h UT (J.D. 244 3213.33)

APPARENT DISTANCE AND POSTION ANGLE

Time from Northern Elongation	Miranda F	Miranda p_1	Time from Northern Elongation	Ariel F	Ariel p_1	Umbriel F	Umbriel p_1	Time from Northern Elongation	Titania F	Titania p_1	Time from Northern Elongation	Oberon F	Oberon p_1
d h		°	d h		°		°	d h		°	d h		°
0 00	1.000	10.0	0 00	1.000	10.0	1.000	10.0	0 00	1.000	10.0	0 00	1.000	10.0
0 01	0.996	19.2	0 02	0.995	20.4	0.998	16.3	0 05	0.997	17.5	0 08	0.997	17.8
0 02	0.984	28.6	0 04	0.980	31.0	0.992	22.6	0 10	0.989	25.1	0 16	0.988	25.6
0 03	0.965	38.3	0 06	0.957	42.0	0.983	29.1	0 15	0.976	32.8	1 00	0.975	33.6
0 04	0.943	48.5	0 08	0.931	53.6	0.971	35.7	0 20	0.960	40.8	1 08	0.958	41.9
0 05	0.919	59.1	0 10	0.905	65.9	0.956	42.5	1 01	0.941	49.1	1 16	0.938	50.5
0 06	0.896	70.3	0 12	0.883	78.8	0.940	49.4	1 06	0.922	57.7	2 00	0.918	59.5
0 07	0.879	82.0	0 14	0.871	92.3	0.924	56.7	1 11	0.903	66.7	2 08	0.899	68.9
0 08	0.870	94.1	0 16	0.870	106.0	0.908	64.2	1 16	0.887	76.1	2 16	0.884	78.7
0 09	0.870	106.3	0 18	0.881	119.5	0.894	71.9	1 21	0.876	85.7	3 00	0.873	88.7
0 10	0.880	118.4	0 20	0.902	132.5	0.882	79.9	2 02	0.870	95.6	3 08	0.869	98.9
0 11	0.897	130.0	0 22	0.928	144.9	0.874	88.1	2 07	0.870	105.5	3 16	0.872	109.2
0 12	0.919	141.2	1 00	0.954	156.6	0.870	96.4	2 12	0.877	115.3	4 00	0.881	119.3
0 13	0.943	151.9	1 02	0.977	167.7	0.870	104.7	2 17	0.889	124.9	4 08	0.895	129.1
0 14	0.966	162.0	1 04	0.993	178.3	0.875	113.0	2 22	0.905	134.2	4 16	0.914	138.6
0 15	0.984	171.7	1 06	1.000	188.7	0.883	121.1	3 03	0.924	143.2	5 00	0.934	147.6
0 16	0.996	181.1	1 08	0.996	199.1	0.895	129.1	3 08	0.943	151.8	5 08	0.954	156.3
0 17	1.000	190.3	1 10	0.982	209.6	0.910	136.8	3 13	0.962	160.0	5 16	0.972	164.7
0 18	0.996	199.5	1 12	0.961	220.6	0.926	144.3	3 18	0.978	168.0	6 00	0.986	172.8
0 19	0.983	209.0	1 14	0.934	232.1	0.943	151.5	3 23	0.990	175.7	6 08	0.996	180.6
0 20	0.965	218.7	1 16	0.908	244.3	0.958	158.5	4 04	0.998	183.3	6 16	1.000	188.4
0 21	0.942	228.8	1 18	0.886	257.2	0.973	165.2	4 09	1.000	190.8	7 00	0.998	196.2
0 22	0.918	239.5	1 20	0.872	270.6	0.985	171.8	4 14	0.997	198.3	7 08	0.991	204.0
0 23	0.896	250.7	1 22	0.870	284.3	0.993	178.2	4 19	0.988	205.9	7 16	0.978	212.0
1 00	0.879	262.4	2 00	0.879	297.8	0.999	184.5	5 00	0.975	213.6	8 00	0.961	220.2
1 01	0.870	274.5	2 02	0.899	311.0	1.000	190.8	5 05	0.958	221.6	8 08	0.942	228.7
1 02	0.871	286.7	2 04	0.924	323.4	0.998	197.1	5 10	0.939	230.0	8 16	0.922	237.7
1 03	0.880	298.7	2 06	0.951	335.2	0.991	203.5	5 15	0.920	238.6	9 00	0.903	247.0
1 04	0.898	310.4	2 08	0.975	346.3	0.982	210.0	5 20	0.901	247.7	9 08	0.886	256.7
1 05	0.920	321.6	2 10	0.992	357.0	0.969	216.6	6 01	0.886	257.0	9 16	0.875	266.7
1 06	0.944	332.2	2 12	1.000	7.4	0.954	223.4	6 06	0.875	266.7	10 00	0.870	276.8
1 07	0.967	342.3	2 14	0.997	17.8	0.938	230.4	6 11	0.870	276.6	10 08	0.871	287.1
1 08	0.985	352.0	2 16			0.922	237.7	6 16	0.871	286.5	10 16	0.879	297.2
1 09	0.996	1.4	2 18			0.906	245.2	6 21	0.878	296.3	11 00	0.892	307.1
1 10	1.000	10.6	2 20			0.892	253.0	7 02	0.890	305.9	11 08	0.910	316.7
1 11	0.995	19.8	2 22			0.881	261.0	7 07	0.907	315.1	11 16	0.930	325.8
			3 00			0.873	269.2	7 12	0.926	324.1	12 00	0.950	334.6
			3 02			0.869	277.5	7 17	0.945	332.6	12 08	0.968	343.0
			3 04			0.870	285.8	7 22	0.964	340.8	12 16	0.983	351.1
			3 06			0.876	294.1	8 03	0.979	348.8	13 00	0.994	359.0
			3 08			0.885	302.2	8 08	0.991	356.5	13 08	1.000	6.8
			3 10			0.897	310.1	8 13	0.998	4.0	13 16	0.999	14.6
			3 12			0.912	317.8	8 18	1.000	11.5			
			3 14			0.928	325.2						
			3 16			0.945	332.4						
			3 18			0.960	339.4						
			3 20			0.974	346.1						
			3 22			0.986	352.6						
			4 00			0.994	359.0						
			4 02			0.999	5.4						
			4 04			1.000	11.7						

Apparent distance of satellite is Fa/Δ
Position angle of satellite is $p_1 + p_2$

APPARENT DISTANCE AND POSITION ANGLE

Day (0h UT)		a/Δ Miranda	a/Δ Ariel	a/Δ Umbriel	a/Δ Titania	a/Δ Oberon	p_2	Day (0h UT)		a/Δ Miranda	a/Δ Ariel	a/Δ Umbriel	a/Δ Titania	a/Δ Oberon	p_2
		″	″	″	″	″	°			″	″	″	″	″	°
Jan.	−1	8.7	12.9	17.9	29.4	39.3	+ 2.4	July	7	9.7	14.2	19.8	32.5	43.4	0.0
	9	8.7	12.8	17.9	29.4	39.3	1.7		17	9.7	14.2	19.8	32.4	43.4	+ 0.3
	19	8.7	12.9	17.9	29.4	39.3	1.1		27	9.6	14.2	19.7	32.4	43.3	0.6
	29	8.8	12.9	18.0	29.5	39.4	+ 0.5	Aug.	6	9.6	14.1	19.7	32.3	43.1	0.9
Feb.	8	8.8	12.9	18.0	29.6	39.6	0.0		16	9.6	14.0	19.6	32.1	42.9	1.2
	18	8.8	13.0	18.1	29.7	39.8	− 0.5		26	9.5	14.0	19.4	31.9	42.6	+ 1.4
	28	8.9	13.1	18.2	29.9	40.0	0.9	Sept.	5	9.4	13.9	19.3	31.7	42.3	1.6
Mar.	9	9.0	13.2	18.4	30.1	40.3	1.3		15	9.3	13.7	19.1	31.4	42.0	1.7
	19	9.0	13.3	18.5	30.3	40.6	1.5		25	9.3	13.6	19.0	31.1	41.6	1.7
	29	9.1	13.4	18.7	30.6	40.9	1.7	Oct.	5	9.2	13.5	18.8	30.9	41.3	1.7
Apr.	8	9.2	13.5	18.8	30.9	41.3	− 1.8		15	9.1	13.4	18.6	30.6	40.9	+ 1.6
	18	9.3	13.6	19.0	31.1	41.6	1.8		25	9.0	13.3	18.5	30.3	40.6	1.4
	28	9.3	13.7	19.1	31.4	42.0	1.7	Nov.	4	9.0	13.2	18.3	30.1	40.2	1.1
May	8	9.4	13.8	19.3	31.6	42.3	1.6		14	8.9	13.1	18.2	29.9	40.0	0.8
	18	9.5	14.0	19.4	31.9	42.6	1.5		24	8.8	13.0	18.1	29.7	39.7	+ 0.4
	28	9.5	14.0	19.6	32.1	42.9	− 1.2	Dec.	4	8.8	12.9	18.0	29.5	39.5	− 0.1
June	7	9.6	14.1	19.7	32.2	43.1	1.0		14	8.8	12.9	17.9	29.4	39.3	0.6
	17	9.6	14.2	19.7	32.4	43.3	0.7		24	8.7	12.8	17.9	29.3	39.2	1.1
	27	9.7	14.2	19.8	32.5	43.4	− 0.4		34	8.7	12.8	17.8	29.3	39.2	− 1.6

UNIVERSAL TIME OF GREATEST EASTERN ELONGATION

MIRANDA

Jan.	Feb.	Mar.	Apr.	May	June	July	Aug.	Sept.	Oct.	Nov.	Dec.
d h	d h	d h	d h	d h	d h	d h	d h	d h	d h	d h	d h
−1 16.7	1 04.1	1 20.6	1 23.6	1 15.2	1 16.9	1 10.3	1 13.2	1 14.5	1 06.9	1 10.2	1 02.1
1 03.0	2 14.2	3 06.2	3 09.3	3 01.5	3 02.9	2 20.0	2 23.1	3 00.7	2 16.6	2 19.9	2 12.5
2 13.0	4 00.5	4 16.0	4 19.0	4 11.8	4 13.0	4 05.5	4 08.9	4 11.2	4 02.4	4 05.4	3 22.6
3 22.9	5 10.9	6 01.9	6 04.5	5 21.8	5 23.2	5 15.3	5 18.6	5 21.4	5 12.4	5 15.1	5 08.5
5 08.5	6 20.9	7 11.9	7 14.2	7 07.6	7 09.7	7 01.1	7 04.1	7 07.3	6 22.5	7 00.8	6 18.3
6 18.2	8 06.8	8 22.1	9 00.0	8 17.4	8 19.8	8 11.0	8 13.8	8 17.2	8 09.1	8 10.6	8 03.8
8 03.7	9 16.5	10 08.5	10 09.8	10 02.9	10 05.8	9 21.0	9 23.5	10 02.7	9 19.3	9 20.5	9 13.6
9 13.4	11 02.3	11 18.7	11 19.9	11 12.6	11 15.7	11 07.5	11 09.2	11 12.4	11 05.6	11 06.8	10 23.3
10 23.1	12 11.8	13 04.8	13 06.2	12 22.3	13 01.5	12 17.7	12 19.1	12 22.1	12 15.5	12 17.0	12 08.9
12 08.9	13 21.4	14 14.7	14 16.4	14 08.0	14 11.1	14 04.0	14 05.3	14 07.6	14 01.4	14 03.5	13 18.7
13 18.9	15 07.1	16 00.5	16 02.6	15 17.9	15 20.8	15 14.1	15 15.5	15 17.4	15 11.0	15 13.6	15 04.7
15 05.2	16 16.9	17 10.1	16 02.7	17 04.0	17 06.3	17 00.0	17 01.8	17 03.3	16 20.7	16 23.6	16 14.7
16 15.5	18 02.9	18 19.8	17 12.8	18 14.1	18 16.1	18 09.6	18 12.2	18 13.4	18 06.1	18 09.4	18 00.9
18 01.7	19 13.1	20 05.3	18 22.7	20 00.3	20 02.0	19 19.3	19 22.3	19 23.6	19 15.9	19 19.2	19 11.4
19 12.0	20 23.3	21 15.1	20 08.4	21 10.7	21 12.0	21 04.8	21 08.1	21 10.2	21 01.6	21 04.6	20 21.5
20 22.0	22 09.7	23 00.9	21 18.2	22 20.8	22 22.1	22 14.5	22 17.9	22 20.4	22 11.5	22 14.3	22 07.6
22 07.7	23 19.8	24 10.9	23 03.7	24 06.7	24 08.6	24 00.3	24 03.4	24 06.5	23 21.5	24 00.0	23 17.4
23 17.4	25 05.8	25 20.9	24 13.4	25 16.6	25 18.8	25 10.1	25 13.1	25 16.4	25 07.9	25 09.7	25 03.0
25 02.9	26 15.7	27 07.4	25 23.2	27 02.3	27 05.0	26 20.1	26 22.8	27 02.2	26 18.2	26 19.6	26 12.8
26 12.6	28 01.4	28 17.6	27 08.9	28 11.8	28 14.9	28 06.4	28 08.4	28 11.7	28 04.6	28 05.7	27 22.5
27 22.3	29 10.9	30 03.8	28 18.8	29 21.5	30 00.7	29 16.6	29 18.2	29 21.4	29 14.6	29 15.9	29 08.0
29 07.9		31 13.8	30 05.1	31 07.1		31 03.1	31 04.3		31 00.5		30 17.8
30 17.9											32 03.7

UNIVERSAL TIME OF GREATEST NORTHERN ELONGATION

ARIEL

Jan.	Feb.	Mar.	Apr.	May	June	July	Aug.	Sept.	Oct.	Nov.	Dec.
d h	d h	d h	d h	d h	d h	d h	d h	d h	d h	d h	d h
−1 08.1	1 02.3	2 07.9	1 13.6	1 19.4	1 01.3	1 07.2	3 01.8	2 07.8	2 13.7	1 19.5	2 01.4
1 20.6	3 14.7	4 20.4	4 02.1	4 07.9	3 13.8	3 19.7	5 14.3	4 20.2	5 02.2	4 08.0	4 13.9
4 09.1	6 03.1	7 08.9	6 14.6	6 20.4	6 02.3	6 08.2	8 02.8	7 08.7	7 14.7	6 20.5	7 02.3
6 21.5	8 15.6	9 21.3	9 03.1	9 08.9	8 14.8	8 20.7	10 15.3	9 21.2	10 03.2	9 09.0	9 14.8
9 10.0	11 04.1	12 09.8	11 15.6	11 21.3	11 03.3	11 09.2	13 03.8	12 09.7	12 15.7	11 21.5	12 03.3
11 22.4	13 16.6	14 22.3	14 04.1	14 09.8	13 15.8	13 21.7	15 16.3	14 22.2	15 04.2	14 10.0	14 15.8
14 10.9	16 05.1	17 10.7	16 16.6	16 22.3	16 04.3	16 10.2	18 04.8	17 10.7	17 16.6	16 22.5	17 04.3
16 23.4	18 17.6	19 23.2	19 05.1	19 10.8	18 16.8	18 22.7	20 17.3	19 23.2	20 05.1	19 10.9	19 16.7
19 11.8	21 06.0	22 11.7	21 17.5	21 23.3	21 05.3	21 11.2	23 05.8	22 11.7	22 17.6	21 23.5	22 05.2
22 00.3	23 18.5	25 00.2	24 06.0	24 11.8	23 17.8	23 23.8	25 18.2	25 00.2	25 06.1	24 11.9	24 17.6
24 12.8	26 07.0	27 12.7	26 18.5	27 00.3	26 06.2	26 12.2	28 06.8	27 12.7	27 18.6	27 00.4	27 06.1
27 01.3	28 19.5	30 01.2	29 07.0	29 12.8	28 18.7	29 00.8	30 19.2	30 01.2	30 07.0	29 12.9	29 18.6
29 13.7						31 13.3					32 07.1

UMBRIEL

Jan.	Feb.	Mar.	Apr.	May	June	July	Aug.	Sept.	Oct.	Nov.	Dec.
d h	d h	d h	d h	d h	d h	d h	d h	d h	d h	d h	d h
0 08.7	2 11.9	2 11.9	4 15.6	3 15.6	1 15.7	4 19.6	2 20.2	5 00.0	4 00.2	2 00.5	1 00.5
4 12.1	6 15.3	6 15.4	8 19.0	7 19.0	5 19.1	8 23.2	6 23.7	9 03.6	8 03.7	6 03.7	5 04.0
8 15.5	10 18.7	10 18.8	12 22.4	11 22.4	9 22.6	13 02.6	11 03.2	13 07.0	12 07.1	10 07.2	9 07.4
12 18.9	14 22.2	14 22.3	17 01.8	16 01.9	14 02.1	17 06.1	15 06.7	17 10.5	16 10.5	14 10.7	13 10.9
16 22.3	19 01.7	19 01.8	21 05.4	20 05.4	18 05.6	21 09.6	19 10.2	21 13.9	20 14.0	18 14.2	17 14.3
21 01.7	23 04.9	23 05.2	25 08.8	24 08.8	22 09.0	25 13.2	23 13.7	25 17.3	24 17.5	22 17.7	21 17.8
25 05.1	27 08.4	27 08.6	29 12.2	28 12.2	26 12.6	29 16.7	27 17.1	29 20.8	28 21.0	26 21.1	25 21.1
29 08.5		31 12.1			30 16.1		31 20.6				30 00.6
											34 04.1

TITANIA

Jan.	Feb.	Mar.	Apr.	May	June	July	Aug.	Sept.	Oct.	Nov.	Dec.
d h	d h	d h	d h	d h	d h	d h	d h	d h	d h	d h	d h
−6 08.0	6 19.9	3 22.3	7 17.6	3 20.4	7 16.3	3 19.4	7 15.6	2 18.8	7 14.8	2 17.3	7 12.4
3 00.8	15 12.7	12 15.1	16 10.5	12 13.3	16 09.3	12 12.4	16 08.7	11 11.8	16 07.6	11 10.1	16 05.1
11 17.6	24 05.5	21 07.9	25 03.4	21 06.3	25 02.3	21 05.5	25 01.7	20 04.8	25 00.5	20 02.9	24 21.9
20 10.3		30 00.7		29 23.2		29 22.5		28 21.9		28 19.6	33 14.7
29 03.1											

OBERON

Jan.	Feb.	Mar.	Apr.	May	June	July	Aug.	Sept.	Oct.	Nov.	Dec.
d h	d h	d h	d h	d h	d h	d h	d h	d h	d h	d h	d h
−7 02.4	2 10.0	13 18.5	9 16.2	6 14.4	2 13.1	12 23.1	8 22.0	4 20.3	1 18.5	11 03.7	8 01.3
6 12.9	15 20.7	27 05.5	23 03.3	20 01.8	16 00.3	26 10.5	22 09.3	18 07.5	15 05.6	24 14.5	21 12.2
19 23.3	29 07.6				29 11.7				28 16.7		34 23.1

APPARENT ORBIT OF TRITON AT DATE OF OPPOSITION, JULY 9

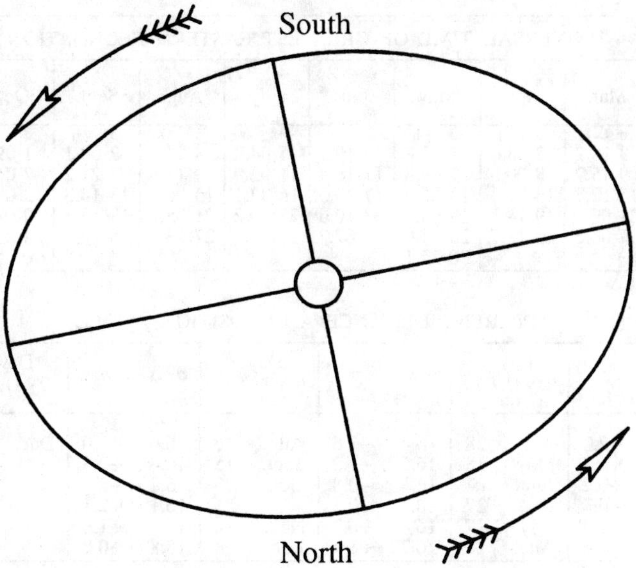

Name		Sidereal Period
I	Triton	$5^d\ 21^h.044$
II	Nereid	$360^d.2$

DIFFERENTIAL COORDINATES OF NEREID FOR 0^h U.T.

Date		Δα cosδ	Δδ	Date		Δα cosδ	Δδ	Date		Δα cosδ	Δδ
		′ ″	″			′ ″	″			′ ″	″
Jan.	−1	+3 41.0	+65.4	May	8	−0 13.0	+14.1	Sept.	15	+4 36.7	+64.4
	9	3 29.7	64.1		18	0 37.9	+06.9		25	4 41.5	67.1
	19	3 17.5	62.5		28	1 01.7	−00.6	Oct.	5	4 43.0	69.1
	29	3 04.4	60.5	June	7	1 22.3	08.0		15	4 42.0	70.5
Feb.	8	2 50.2	58.1		17	1 34.2	14.4		25	4 38.9	71.4
	18	+2 34.9	+55.2		27	−1 19.8	−16.5	Nov.	4	+4 34.1	+71.9
	28	2 18.4	51.9	July	7	+0 04.6	−03.3		14	4 27.8	72.1
Mar.	9	2 00.6	48.1		17	1 39.6	+16.8		24	4 20.4	71.9
	19	1 41.4	43.7		27	2 41.9	31.5	Dec.	4	4 11.9	71.5
	29	1 20.9	38.8	Aug.	6	3 24.0	42.2		14	4 02.5	70.7
Apr.	8	+0 59.1	+33.4		16	+3 53.5	+50.2		24	+3 52.2	+69.6
	18	0 36.0	27.5		26	4 14.0	56.3		34	3 41.1	68.2
	28	+0 11.8	+21.0	Sept.	5	+4 27.9	+60.9		44	+3 29.0	+66.4

TRITON

UNIVERSAL TIME OF GREATEST EASTERN ELONGATION

Jan.	Feb.	Mar.	Apr.	May	June	July	Aug.	Sept.	Oct.	Nov.	Dec.
d h	d h	d h	d h	d h	d h	d h	d h	d h	d h	d h	d h
−1 04.5	3 10.1	3 18.8	2 03.6	1 12.8	5 19.2	5 04.9	3 14.5	2 00.1	1 09.5	5 15.6	5 00.5
5 01.4	9 07.0	9 15.7	8 00.6	7 09.8	11 16.4	11 02.0	9 11.6	7 21.2	7 06.5	11 12.6	10 21.4
10 22.4	15 03.9	15 12.7	13 21.7	13 06.9	17 13.5	16 23.1	15 08.8	13 18.3	13 03.6	17 09.6	16 18.3
16 19.3	21 00.9	21 09.7	19 18.7	19 04.0	23 10.6	22 20.2	21 05.9	19 15.3	19 00.6	23 06.5	22 15.3
22 16.2	26 21.8	27 06.6	25 15.7	25 01.0	29 07.7	28 17.4	27 03.0	25 12.4	24 21.6	29 03.5	28 12.2
28 13.1				30 22.1					30 18.6		34 09.1

APPARENT DISTANCE AND POSITION ANGLE

Day (0^h U.T.)	a/Δ	p_2	Day (0^h U.T.)	a/Δ	p_2	Day (0^h U.T.)	a/Δ	p_2	Day (0^h U.T.)	a/Δ	p_2
	″	°		″	°		″	°		″	°
Jan. −1	15.7	+2.2	Apr. 28	16.4	−1.6	Aug. 26	16.6	+2.0	Dec. 24	15.7	−0.2
19	15.7	+1.2	May 18	16.5	−1.3	Sept. 15	16.4	+2.3	44	15.7	−1.2
Feb. 8	15.8	+0.2	June 7	16.7	−0.8	Oct. 5	16.2	+2.4			
28	15.9	−0.7	27	16.7	−0.1	25	16.1	+2.1			
Mar. 19	16.0	−1.3	July 17	16.8	+0.7	Nov. 14	15.9	+1.6			
Apr. 8	16.2	−1.6	Aug. 6	16.7	+1.4	Dec. 4	15.8	+0.8			

Time from Eastern Elongation	F	p_1	Time from Eastern Elongation	F	p_1	Time from Eastern Elongation	F	p_1	Time from Eastern Elongation	F	p_1
d h		°	d h		°	d h		°	d h		°
0 00	1.000	101.0	1 12	0.737	193.6	3 00	0.999	283.8	4 12	0.739	18.7
0 03	0.996	106.7	1 15	0.745	203.9	3 03	0.991	289.5	4 15	0.753	28.8
0 06	0.984	112.4	1 18	0.763	213.8	3 06	0.975	295.3	4 18	0.775	38.5
0 09	0.964	118.3	1 21	0.789	223.2	3 09	0.953	301.4	4 21	0.804	47.6
0 12	0.939	124.6	2 00	0.821	231.9	3 12	0.924	307.8	5 00	0.838	55.9
0 15	0.908	131.2	2 03	0.856	239.9	3 15	0.891	314.6	5 03	0.873	63.6
0 18	0.874	138.3	2 06	0.891	247.3	3 18	0.856	322.0	5 06	0.907	70.7
0 21	0.839	146.0	2 09	0.924	254.1	3 21	0.822	330.0	5 09	0.938	77.4
1 00	0.805	154.3	2 12	0.952	260.6	4 00	0.790	338.7	5 12	0.964	83.6
1 03	0.775	163.4	2 15	0.975	266.6	4 03	0.763	348.1	5 15	0.984	89.5
1 06	0.753	173.0	2 18	0.991	272.4	4 06	0.745	358.0	5 18	0.996	95.3
1 09	0.739	183.2	2 21	0.999	278.1	4 09	0.737	8.3	5 21	1.000	100.9

Apparent distance of satellite is Fa/Δ
Position angle of satellite is $p_1 + p_2$

APPARENT ORBIT OF CHARON AT DATE OF OPPOSITION, MAY 12

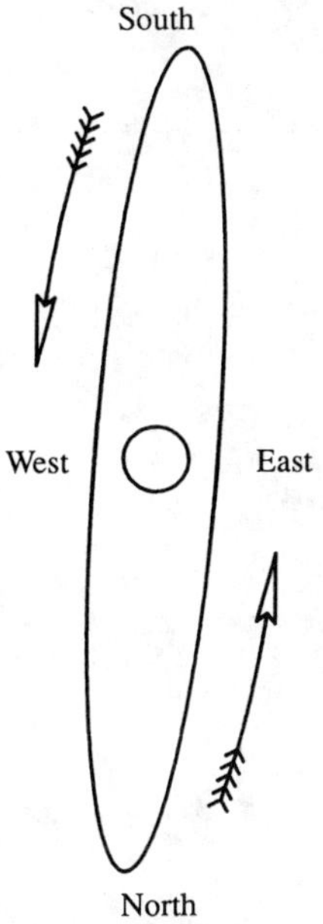

Sidereal Period: $6^d 09^h.29$

UNIVERSAL TIME OF NORTHERN ELONGATION

	d h		d h		d h
Jan.	0 20.4	May	1 05.1	Sept.	5 23.0
	7 05.7		7 14.4		12 08.3
	13 15.0		13 23.7		18 17.5
	20 00.3		20 09.1		25 02.8
	26 09.5		26 18.4	Oct.	1 12.1
Feb.	1 18.8	June	2 03.7		7 21.3
	8 04.1		8 13.0		14 06.6
	14 13.4		14 22.3		20 15.8
	20 22.7		21 07.6		27 01.1
	27 08.0		27 16.9	Nov.	2 10.3
Mar.	4 17.3	July	4 02.2		8 19.6
	11 02.6		10 11.5		15 04.9
	17 11.9		16 20.8		21 14.1
	23 21.2		23 06.1		27 23.4
	30 06.5		29 15.4	Dec.	4 08.6
Apr.	5 15.8	Aug.	5 00.6		10 17.9
	12 01.2		11 09.9		17 03.2
	18 10.5		17 19.2		23 12.4
	24 19.8		24 04.5		29 21.7
			30 13.8		36 07.0

APPARENT DISTANCE AND POSITION ANGLE

Day (0^h UT)	a/Δ	p_2	Day (0^h UT)	a/Δ	p_2	Day (0^h UT)	a/Δ	p_2	Day (0^h UT)	a/Δ	p_2
	″	°		″	°		″	°		″	°
Jan. − 1	0.9	0.0	Apr. 28	0.9	+ 0.1	Aug. 26	0.9	+ 0.2	Dec. 24	0.9	− 0.3
19	0.9	0.0	May 18	0.9	0.2	Sept. 15	0.9	0.1	44	0.9	− 0.4
Feb. 8	0.9	0.0	June 7	0.9	0.2	Oct. 5	0.9	+ 0.1			
28	0.9	0.0	27	0.9	0.2	25	0.9	0.0			
Mar. 19	0.9	0.0	July 17	0.9	0.2	Nov. 14	0.9	− 0.1			
Apr. 8	0.9	+ 0.1	Aug. 6	0.9	+ 0.2	Dec. 4	0.9	− 0.2			

Time from Northern Elongation	F	p_1	Time from Northern Elongation	F	p_1	Time from Northern Elongation	F	p_1	Time from Northern Elongation	F	p_1
d h		°	d h		°	d h		°	d h		°
0 00	1.000	355.0	1 16	0.163	109.9	3 08	0.991	176.2	5 00	0.251	319.6
0 04	0.987	356.4	1 20	0.272	142.9	3 12	0.956	177.6	5 04	0.387	334.1
0 08	0.948	357.9	2 00	0.410	155.5	3 16	0.896	179.3	5 08	0.525	340.9
0 12	0.884	359.5	2 04	0.546	161.7	3 20	0.813	181.2	5 12	0.653	345.0
0 16	0.798	1.5	2 08	0.672	165.5	4 00	0.709	183.6	5 16	0.765	347.7
0 20	0.691	4.0	2 12	0.782	168.1	4 04	0.588	186.9	5 20	0.859	349.9
1 00	0.568	7.6	2 16	0.872	170.2	4 08	0.455	192.1	6 00	0.930	351.6
1 04	0.432	13.2	2 20	0.939	171.9	4 12	0.316	201.8	6 04	0.977	353.1
1 08	0.294	24.2	3 00	0.982	173.4	4 16	0.191	225.5	6 08	0.999	354.5
1 12	0.176	52.2	3 04	1.000	174.8	4 20	0.154	280.9	6 12	0.994	355.9

Apparent distance of satellite is $F\,a/\Delta$.
Position angle of satellite is $p_1 + p_2$.

CONTENTS OF SECTION G

Notes

The osculating elements for periodic comets returning to perihelion in 1992 have been supplied by B. G. Marsden, Smithsonian Astrophysical Observatory, and are intended for use in the generation of ephemerides by numerical integration.

The geocentric ephemerides of the four principal minor planets (1 Ceres; 2 Pallas; 3 Juno; 4 Vesta) give, at an interval of 2 days, the astrometric right ascension and declination, referred to the mean equator and equinox of J2000·0, the geometric distance and the time of ephemeris transit. Linear interpolation is sufficient for the distance and ephemeris transit, but for the astrometric right ascension and declination second differences are significant. The tabulations are similar to those for Pluto, and the use of the data is similar to that for major planets.

Opposition dates (in right ascension) and visual magnitudes in 1992, and osculating elements for epoch 1992 June 27·0 TDT (JD 244 8800·5) and ecliptic and equinox J2000·0, for 143 of the larger minor planets are given on pages G10–G12; in these tabulations *H* is the absolute visual magnitude at zero phase angle and *G* is the slope parameter which depends on the albedo. The data were supplied by the Institute of Theoretical Astronomy, Leningrad.

PERIODIC COMETS, 1992

OSCULATING ELEMENTS FOR ECLIPTIC AND EQUINOX OF B1950·0

Name	Perihelion Time T	Perihelion Distance q	Eccentricity e	Period P	Arg. of Perihelion ω	Long of Asc. Node Ω	Inclination i	Osculation Date Epoch
		au		years	°	°	°	
Chernykh	Jan. 27·821 73	2·356 3219	0·593 7580	13·97	263·196 95	129·745 80	5·082 40	Jan. 19
Giacobini-Zinner	Apr. 13·227 90	1·034 0027	0·706 4891	6·61	172·521 03	194·682 21	31·827 96	Apr. 8
Tsuchinshan 2	May 20·093 55	1·782 1573	0·504 4087	6·82	203·160 68	287·593 74	6·715 06	May 18
Kowal 1	June 12·881 38	4·673 3559	0·234 4131	15·08	181·287 65	28·059 44	4·353 51	June 27
Grigg-Skjellerup	July 22·137 96	0·994 6894	0·664 3254	5·10	359·276 08	212·631 49	21·104 16	Aug. 6
Smirnova-Chernykh	Aug. 5·934 95	3·572 1440	0·147 3143	8·57	88·938 37	76·838 22	6·628 23	Aug. 6
Shoemaker 2	Aug. 6·875 42	1·327 0989	0·664 6408	7·87	317·626 73	54·748 89	21·548 86	Aug. 6
du Toit-Hartley	Aug. 27·614 80	1·197 1614	0·601 6141	5·21	251·565 16	308·562 30	2·938 06	Sept. 15
Wolf	Aug. 28·128 99	2·427 7044	0·405 5992	8·25	162·292 28	203·440 50	27·483 39	Sept. 15
Daniel	Sept. 1·670 95	1·649 4513	0·551 9641	7·06	10·991 68	68·370 00	20·133 98	Sept. 15
Schuster	Sept. 6·388 07	1·539 2363	0·589 5205	7·26	355·718 27	49·918 92	20·130 32	Sept. 15
Giclas	Sept. 13·091 90	1·846 7947	0·493 2117	6·96	276·440 52	111·872 56	7·286 46	Sept. 15
Singer Brewster	Oct. 28·152 65	2·026 7137	0·413 8999	6·43	46·673 80	191·904 25	9·199 05	Oct. 25
Gale	Dec. 18·231 57	1·213 8685	0·758 1421	11·24	215·405 98	59·286 91	10·731 78	Dec. 4

GEOCENTRIC POSITIONS FOR 0^h DYNAMICAL TIME

Date		Astrometric J2000·0 R.A.	Astrometric J2000·0 Dec.	True Distance	Ephemeris Transit
		h m s	° ′ ″		h m
Jan.	1	17 49 44·1	−23 48 08	3·784	11 08·6
	3	17 53 17·5	23 52 15	3·780	11 04·3
	5	17 56 50·6	23 56 06	3·774	10 59·9
	7	18 00 23·4	23 59 39	3·768	10 55·6
	9	18 03 55·7	24 02 55	3·762	10 51·3
	11	18 07 27·6	−24 05 54	3·755	10 46·9
	13	18 10 59·0	24 08 37	3·747	10 42·6
	15	18 14 29·8	24 11 03	3·739	10 38·2
	17	18 17 59·9	24 13 13	3·730	10 33·8
	19	18 21 29·4	24 15 08	3·721	10 29·4
	21	18 24 58·3	−24 16 47	3·711	10 25·0
	23	18 28 26·3	24 18 12	3·700	10 20·6
	25	18 31 53·6	24 19 21	3·689	10 16·2
	27	18 35 20·1	24 20 16	3·678	10 11·8
	29	18 38 45·7	24 20 57	3·666	10 07·3
	31	18 42 10·3	−24 21 24	3·653	10 02·8
Feb.	2	18 45 34·0	24 21 38	3·640	9 58·3
	4	18 48 56·5	24 21 39	3·626	9 53·8
	6	18 52 18·0	24 21 28	3·612	9 49·3
	8	18 55 38·2	24 21 05	3·598	9 44·8
	10	18 58 57·2	−24 20 31	3·583	9 40·2
	12	19 02 14·9	24 19 46	3·567	9 35·6
	14	19 05 31·2	24 18 51	3·551	9 31·0
	16	19 08 46·2	24 17 45	3·534	9 26·4
	18	19 11 59·7	24 16 31	3·517	9 21·7
	20	19 15 11·7	−24 15 08	3·500	9 17·0
	22	19 18 22·3	24 13 36	3·482	9 12·3
	24	19 21 31·2	24 11 57	3·463	9 07·6
	26	19 24 38·6	24 10 11	3·445	9 02·8
	28	19 27 44·2	24 08 19	3·425	8 58·1
Mar.	1	19 30 48·1	−24 06 21	3·406	8 53·2
	3	19 33 50·2	24 04 18	3·386	8 48·4
	5	19 36 50·3	24 02 10	3·365	8 43·5
	7	19 39 48·6	23 59 59	3·344	8 38·6
	9	19 42 44·8	23 57 46	3·323	8 33·7
	11	19 45 38·9	−23 55 30	3·301	8 28·7
	13	19 48 30·9	23 53 13	3·279	8 23·7
	15	19 51 20·8	23 50 55	3·257	8 18·6
	17	19 54 08·4	23 48 37	3·234	8 13·5
	19	19 56 53·7	23 46 20	3·211	8 08·4
	21	19 59 36·7	−23 44 04	3·188	8 03·2
	23	20 02 17·3	23 41 50	3·165	7 58·0
	25	20 04 55·5	23 39 40	3·141	7 52·8
	27	20 07 31·1	23 37 33	3·117	7 47·5
	29	20 10 04·1	23 35 32	3·092	7 42·2
	31	20 12 34·5	−23 33 36	3·068	7 36·8
Apr.	2	20 15 02·0	−23 31 47	3·043	7 31·4

Date		Astrometric J2000·0 R.A.	Astrometric J2000·0 Dec.	True Distance	Ephemeris Transit
		h m s	° ′ ″		h m
Apr.	2	20 15 02·0	−23 31 47	3·043	7 31·4
	4	20 17 26·7	23 30 05	3·018	7 25·9
	6	20 19 48·4	23 28 32	2·993	7 20·4
	8	20 22 07·2	23 27 08	2·967	7 14·8
	10	20 24 22·8	23 25 54	2·942	7 09·2
	12	20 26 35·3	−23 24 51	2·916	7 03·5
	14	20 28 44·6	23 24 01	2·890	6 57·8
	16	20 30 50·5	23 23 22	2·864	6 52·0
	18	20 32 53·1	23 22 58	2·838	6 46·2
	20	20 34 52·2	23 22 47	2·812	6 40·3
	22	20 36 47·8	−23 22 52	2·786	6 34·3
	24	20 38 39·7	23 23 14	2·759	6 28·3
	26	20 40 27·8	23 23 53	2·733	6 22·2
	28	20 42 12·1	23 24 50	2·707	6 16·1
	30	20 43 52·4	23 26 06	2·681	6 09·9
May	2	20 45 28·6	−23 27 43	2·654	6 03·6
	4	20 47 00·5	23 29 40	2·628	5 57·3
	6	20 48 28·2	23 32 00	2·602	5 50·8
	8	20 49 51·4	23 34 42	2·576	5 44·4
	10	20 51 10·1	23 37 47	2·550	5 37·8
	12	20 52 24·3	−23 41 16	2·525	5 31·1
	14	20 53 33·7	23 45 09	2·499	5 24·4
	16	20 54 38·3	23 49 28	2·474	5 17·6
	18	20 55 38·0	23 54 12	2·449	5 10·7
	20	20 56 32·7	23 59 22	2·424	5 03·8
	22	20 57 22·2	−24 04 59	2·400	4 56·7
	24	20 58 06·5	24 11 04	2·376	4 49·6
	26	20 58 45·3	24 17 36	2·352	4 42·4
	28	20 59 18·7	24 24 36	2·328	4 35·0
	30	20 59 46·5	24 32 04	2·305	4 27·6
June	1	21 00 08·5	−24 40 00	2·283	4 20·1
	3	21 00 24·8	24 48 24	2·261	4 12·5
	5	21 00 35·2	24 57 15	2·239	4 04·8
	7	21 00 39·7	25 06 33	2·218	3 57·0
	9	21 00 38·2	25 16 18	2·197	3 49·1
	11	21 00 30·7	−25 26 27	2·177	3 41·1
	13	21 00 17·2	25 37 02	2·158	3 33·0
	15	20 59 57·7	25 48 00	2·139	3 24·9
	17	20 59 32·2	25 59 20	2·121	3 16·6
	19	20 59 00·5	26 11 02	2·104	3 08·2
	21	20 58 22·9	−26 23 03	2·087	2 59·7
	23	20 57 39·3	26 35 22	2·072	2 51·1
	25	20 56 49·7	26 47 57	2·057	2 42·4
	27	20 55 54·3	27 00 46	2·042	2 33·6
	29	20 54 53·2	27 13 47	2·029	2 24·7
July	1	20 53 46·5	−27 26 57	2·017	2 15·7
	3	20 52 34·5	−27 40 13	2·005	2 06·7

GEOCENTRIC POSITIONS FOR 0^h DYNAMICAL TIME

Date		Astrometric J2000·0 R.A.	Astrometric J2000·0 Dec.	True Distance	Ephemeris Transit
		h m s	° ′ ″		h m
July	1	20 53 46·5	−27 26 57	2·017	2 15·7
	3	20 52 34·5	27 40 13	2·005	2 06·7
	5	20 51 17·4	27 53 33	1·995	1 57·5
	7	20 49 55·4	28 06 53	1·985	1 48·3
	9	20 48 28·9	28 20 11	1·977	1 39·0
	11	20 46 58·1	−28 33 24	1·969	1 29·6
	13	20 45 23·4	28 46 29	1·963	1 20·2
	15	20 43 45·2	28 59 23	1·958	1 10·7
	17	20 42 03·8	29 12 03	1·953	1 01·2
	19	20 40 19·5	29 24 27	1·950	0 51·6
	21	20 38 32·9	−29 36 31	1·948	0 42·0
	23	20 36 44·4	29 48 14	1·947	0 32·3
	25	20 34 54·4	29 59 32	1·947	0 22·6
	27	20 33 03·3	30 10 24	1·948	0 12·9
	29	20 31 11·8	30 20 46	1·950	0 03·2
	31	20 29 20·4	−30 30 37	1·954	23 48·6
Aug.	2	20 27 29·5	30 39 55	1·958	23 38·9
	4	20 25 39·8	30 48 38	1·964	23 29·3
	6	20 23 51·6	30 56 45	1·970	23 19·6
	8	20 22 05·5	31 04 16	1·978	23 10·0
	10	20 20 22·0	−31 11 10	1·987	23 00·5
	12	20 18 41·4	31 17 26	1·996	22 51·0
	14	20 17 04·3	31 23 04	2·007	22 41·5
	16	20 15 30·8	31 28 04	2·019	22 32·1
	18	20 14 01·6	31 32 27	2·032	22 22·8
	20	20 12 36·8	−31 36 12	2·045	22 13·6
	22	20 11 16·7	31 39 20	2·060	22 04·4
	24	20 10 01·8	31 41 53	2·075	21 55·4
	26	20 08 52·2	31 43 49	2·092	21 46·4
	28	20 07 48·3	31 45 11	2·109	21 37·5
	30	20 06 50·1	−31 45 59	2·127	21 28·7
Sept.	1	20 05 58·1	31 46 13	2·146	21 20·1
	3	20 05 12·1	31 45 56	2·165	21 11·5
	5	20 04 32·5	31 45 09	2·185	21 03·0
	7	20 03 59·1	31 43 52	2·206	20 54·6
	9	20 03 32·2	−31 42 08	2·227	20 46·4
	11	20 03 11·6	31 39 56	2·250	20 38·2
	13	20 02 57·4	31 37 19	2·272	20 30·2
	15	20 02 49·5	31 34 17	2·295	20 22·2
	17	20 02 48·0	31 30 52	2·319	20 14·4
	19	20 02 52·7	−31 27 03	2·343	20 06·6
	21	20 03 03·7	31 22 53	2·368	19 59·0
	23	20 03 20·8	31 18 22	2·393	19 51·4
	25	20 03 44·0	31 13 30	2·418	19 44·0
	27	20 04 13·3	31 08 19	2·444	19 36·7
	29	20 04 48·4	−31 02 49	2·470	19 29·4
Oct.	1	20 05 29·4	−30 57 00	2·497	19 22·3

Date		Astrometric J2000·0 R.A.	Astrometric J2000·0 Dec.	True Distance	Ephemeris Transit
		h m s	° ′ ″		h m
Oct.	1	20 05 29·4	−30 57 00	2·497	19 22·3
	3	20 06 16·0	30 50 55	2·524	19 15·2
	5	20 07 08·1	30 44 32	2·550	19 08·3
	7	20 08 05·7	30 37 54	2·578	19 01·4
	9	20 09 08·4	30 30 59	2·605	18 54·6
	11	20 10 16·2	−30 23 50	2·633	18 47·9
	13	20 11 29·0	30 16 25	2·660	18 41·3
	15	20 12 46·5	30 08 46	2·688	18 34·7
	17	20 14 08·6	30 00 53	2·716	18 28·2
	19	20 15 35·3	29 52 45	2·744	18 21·8
	21	20 17 06·3	−29 44 24	2·772	18 15·5
	23	20 18 41·6	29 35 48	2·800	18 09·3
	25	20 20 21·0	29 26 59	2·828	18 03·1
	27	20 22 04·4	29 17 57	2·856	17 56·9
	29	20 23 51·7	29 08 41	2·884	17 50·9
	31	20 25 42·7	−28 59 12	2·912	17 44·9
Nov.	2	20 27 37·2	28 49 30	2·940	17 38·9
	4	20 29 35·1	28 39 35	2·967	17 33·1
	6	20 31 36·3	28 29 28	2·995	17 27·2
	8	20 33 40·7	28 19 08	3·023	17 21·4
	10	20 35 48·0	−28 08 36	3·050	17 15·7
	12	20 37 58·3	27 57 51	3·077	17 10·0
	14	20 40 11·3	27 46 54	3·104	17 04·4
	16	20 42 27·0	27 35 44	3·131	16 58·8
	18	20 44 45·2	27 24 22	3·157	16 53·2
	20	20 47 06·0	−27 12 47	3·184	16 47·7
	22	20 49 29·1	27 01 00	3·210	16 42·2
	24	20 51 54·4	26 49 01	3·235	16 36·8
	26	20 54 22·0	26 36 49	3·261	16 31·4
	28	20 56 51·6	26 24 25	3·286	16 26·0
	30	20 59 23·2	−26 11 49	3·311	16 20·7
Dec.	2	21 01 56·7	25 59 02	3·335	16 15·4
	4	21 04 31·9	25 46 03	3·360	16 10·1
	6	21 07 08·8	25 32 52	3·384	16 04·9
	8	21 09 47·2	25 19 30	3·407	15 59·6
	10	21 12 27·2	−25 05 57	3·430	15 54·4
	12	21 15 08·6	24 52 12	3·453	15 49·3
	14	21 17 51·3	24 38 16	3·475	15 44·1
	16	21 20 35·4	24 24 08	3·497	15 39·0
	18	21 23 20·6	24 09 50	3·519	15 33·8
	20	21 26 07·0	−23 55 21	3·540	15 28·8
	22	21 28 54·6	23 40 41	3·560	15 23·7
	24	21 31 43·2	23 25 50	3·581	15 18·6
	26	21 34 32·8	23 10 49	3·600	15 13·6
	28	21 37 23·2	22 55 38	3·620	15 08·5
	30	21 40 14·5	−22 40 17	3·638	15 03·5
	32	21 43 06·6	−22 24 47	3·657	14 58·5

Second Transit: July $29^d\ 23^h\ 58^m.3$

PALLAS, 1992

GEOCENTRIC POSITIONS FOR 0^h DYNAMICAL TIME

Date		Astrometric J2000·0 R.A.	Astrometric J2000·0 Dec.	True Distance	Ephemeris Transit
		h m s	° ′ ″		h m
Jan.	**1**	16 35 21·4	+ 3 10 45	3·587	9 54·3
	3	16 38 26·1	3 15 28	3·576	9 49·5
	5	16 41 29·6	3 20 39	3·566	9 44·7
	7	16 44 31·8	3 26 20	3·555	9 39·9
	9	16 47 32·7	3 32 30	3·543	9 35·0
	11	16 50 32·1	+ 3 39 10	3·532	9 30·1
	13	16 53 30·2	3 46 19	3·520	9 25·2
	15	16 56 26·7	3 53 57	3·507	9 20·2
	17	16 59 21·6	4 02 05	3·494	9 15·3
	19	17 02 14·9	4 10 41	3·481	9 10·3
	21	17 05 06·6	+ 4 19 47	3·467	9 05·3
	23	17 07 56·5	4 29 23	3·454	9 00·2
	25	17 10 44·6	4 39 28	3·439	8 55·1
	27	17 13 30·8	4 50 02	3·425	8 50·0
	29	17 16 15·2	5 01 06	3·410	8 44·9
	31	17 18 57·5	+ 5 12 39	3·395	8 39·7
Feb.	**2**	17 21 37·7	5 24 41	3·380	8 34·5
	4	17 24 15·7	5 37 12	3·364	8 29·3
	6	17 26 51·4	5 50 12	3·348	8 24·0
	8	17 29 24·8	6 03 41	3·332	8 18·6
	10	17 31 55·8	+ 6 17 38	3·315	8 13·3
	12	17 34 24·4	6 32 03	3·298	8 07·9
	14	17 36 50·3	6 46 55	3·282	8 02·4
	16	17 39 13·6	7 02 14	3·264	7 56·9
	18	17 41 34·3	7 18 00	3·247	7 51·4
	20	17 43 52·2	+ 7 34 13	3·230	7 45·8
	22	17 46 07·2	7 50 52	3·212	7 40·2
	24	17 48 19·3	8 07 57	3·194	7 34·5
	26	17 50 28·4	8 25 28	3·176	7 28·8
	28	17 52 34·4	8 43 24	3·158	7 23·0
Mar.	**1**	17 54 37·2	+ 9 01 44	3·140	7 17·2
	3	17 56 36·7	9 20 28	3·122	7 11·3
	5	17 58 32·8	9 39 35	3·103	7 05·3
	7	18 00 25·3	9 59 04	3·085	6 59·3
	9	18 02 14·3	10 18 55	3·066	6 53·3
	11	18 03 59·7	+10 39 07	3·048	6 47·1
	13	18 05 41·2	10 59 39	3·029	6 41·0
	15	18 07 19·0	11 20 29	3·011	6 34·7
	17	18 08 52·8	11 41 38	2·992	6 28·4
	19	18 10 22·6	12 03 04	2·973	6 22·0
	21	18 11 48·4	+12 24 46	2·955	6 15·6
	23	18 13 09·9	12 46 43	2·937	6 09·0
	25	18 14 27·2	13 08 54	2·918	6 02·5
	27	18 15 40·1	13 31 18	2·900	5 55·8
	29	18 16 48·6	13 53 54	2·882	5 49·1
	31	18 17 52·4	+14 16 39	2·864	5 42·2
Apr.	**2**	18 18 51·6	+14 39 34	2·846	5 35·4

Date		Astrometric J2000·0 R.A.	Astrometric J2000·0 Dec.	True Distance	Ephemeris Transit
		h m s	° ′ ″		h m
Apr.	**2**	18 18 51·6	+14 39 34	2·846	5 35·4
	4	18 19 45·9	15 02 35	2·828	5 28·4
	6	18 20 35·5	15 25 41	2·810	5 21·3
	8	18 21 20·0	15 48 51	2·793	5 14·2
	10	18 21 59·6	16 12 02	2·776	5 07·0
	12	18 22 34·2	+16 35 13	2·759	4 59·7
	14	18 23 03·6	16 58 22	2·742	4 52·3
	16	18 23 27·8	17 21 26	2·726	4 44·8
	18	18 23 46·9	17 44 24	2·710	4 37·3
	20	18 24 00·6	18 07 13	2·694	4 29·6
	22	18 24 09·0	+18 29 52	2·679	4 21·9
	24	18 24 12·1	18 52 19	2·664	4 14·1
	26	18 24 09·7	19 14 29	2·649	4 06·2
	28	18 24 01·9	19 36 22	2·635	3 58·2
	30	18 23 48·6	19 57 54	2·621	3 50·1
May	**2**	18 23 29·9	+20 19 02	2·608	3 41·9
	4	18 23 05·7	20 39 44	2·595	3 33·7
	6	18 22 36·2	20 59 57	2·583	3 25·3
	8	18 22 01·3	21 19 37	2·571	3 16·9
	10	18 21 21·3	21 38 42	2·559	3 08·3
	12	18 20 36·0	+21 57 09	2·549	2 59·7
	14	18 19 45·8	22 14 55	2·538	2 51·0
	16	18 18 50·7	22 31 58	2·529	2 42·2
	18	18 17 50·8	22 48 15	2·520	2 33·4
	20	18 16 46·3	23 03 44	2·511	2 24·4
	22	18 15 37·4	+23 18 21	2·503	2 15·4
	24	18 14 24·1	23 32 03	2·496	2 06·3
	26	18 13 06·8	23 44 49	2·490	1 57·2
	28	18 11 45·7	23 56 36	2·484	1 48·0
	30	18 10 21·0	24 07 21	2·478	1 38·7
June	**1**	18 08 53·0	+24 17 01	2·474	1 29·4
	3	18 07 22·1	24 25 36	2·470	1 20·0
	5	18 05 48·4	24 33 02	2·467	1 10·6
	7	18 04 12·5	24 39 19	2·465	1 01·2
	9	18 02 34·6	24 44 24	2·463	0 51·7
	11	18 00 55·1	+24 48 19	2·462	0 42·2
	13	17 59 14·4	24 51 01	2·462	0 32·6
	15	17 57 32·7	24 52 30	2·463	0 23·1
	17	17 55 50·5	24 52 46	2·464	0 13·5
	19	17 54 08·1	24 51 48	2·467	0 04·0
	21	17 52 25·9	+24 49 38	2·470	23 49·6
	23	17 50 44·2	24 46 14	2·473	23 40·1
	25	17 49 03·3	24 41 38	2·478	23 30·6
	27	17 47 23·8	24 35 50	2·483	23 21·1
	29	17 45 45·8	24 28 52	2·489	23 11·6
July	**1**	17 44 09·7	+24 20 45	2·496	23 02·2
	3	17 42 36·0	+24 11 30	2·503	22 52·8

Second Transit: June 19^d 23^h $59^m.2$

GEOCENTRIC POSITIONS FOR 0^h DYNAMICAL TIME

Date		Astrometric J2000·0 R.A. h m s	Astrometric J2000·0 Dec. ° ′ ″	True Dist-ance	Ephem-eris Transit h m
July	**1**	17 44 09·7	+24 20 45	2·496	23 02·2
	3	17 42 36·0	24 11 30	2·503	22 52·8
	5	17 41 04·9	24 01 09	2·512	22 43·4
	7	17 39 36·7	23 49 45	2·521	22 34·1
	9	17 38 11·8	23 37 20	2·530	22 24·9
	11	17 36 50·2	+23 23 57	2·541	22 15·7
	13	17 35 32·4	23 09 37	2·552	22 06·5
	15	17 34 18·4	22 54 25	2·564	21 57·5
	17	17 33 08·4	22 38 21	2·576	21 48·5
	19	17 32 02·7	22 21 30	2·589	21 39·6
	21	17 31 01·4	+22 03 53	2·603	21 30·7
	23	17 30 04·5	21 45 33	2·618	21 22·0
	25	17 29 12·3	21 26 34	2·633	21 13·3
	27	17 28 24·9	21 06 57	2·649	21 04·6
	29	17 27 42·3	20 46 46	2·665	20 56·1
	31	17 27 04·6	+20 26 04	2·682	20 47·7
Aug.	**2**	17 26 32·0	20 04 54	2·699	20 39·3
	4	17 26 04·4	19 43 17	2·717	20 31·0
	6	17 25 41·8	19 21 19	2·736	20 22·8
	8	17 25 24·3	18 59 00	2·755	20 14·7
	10	17 25 11·8	+18 36 25	2·774	20 06·6
	12	17 25 04·4	18 13 35	2·794	19 58·7
	14	17 25 01·9	17 50 32	2·815	19 50·8
	16	17 25 04·3	17 27 20	2·835	19 43·0
	18	17 25 11·6	17 03 59	2·857	19 35·3
	20	17 25 23·7	+16 40 33	2·878	19 27·7
	22	17 25 40·6	16 17 02	2·900	19 20·1
	24	17 26 02·2	15 53 30	2·923	19 12·6
	26	17 26 28·5	15 29 58	2·945	19 05·2
	28	17 26 59·5	15 06 27	2·968	18 57·9
	30	17 27 34·9	+14 43 00	2·991	18 50·7
Sept.	**1**	17 28 14·9	14 19 39	3·015	18 43·5
	3	17 28 59·2	13 56 24	3·039	18 36·4
	5	17 29 47·8	13 33 19	3·063	18 29·4
	7	17 30 40·5	13 10 23	3·087	18 22·4
	9	17 31 37·4	+12 47 39	3·111	18 15·5
	11	17 32 38·2	12 25 07	3·136	18 08·7
	13	17 33 42·9	12 02 48	3·160	18 01·9
	15	17 34 51·4	11 40 45	3·185	17 55·2
	17	17 36 03·6	11 18 56	3·210	17 48·6
	19	17 37 19·4	+10 57 24	3·235	17 42·0
	21	17 38 38·8	10 36 09	3·260	17 35·4
	23	17 40 01·6	10 15 12	3·285	17 29·0
	25	17 41 27·9	9 54 34	3·311	17 22·6
	27	17 42 57·4	9 34 16	3·336	17 16·2
	29	17 44 30·1	+ 9 14 19	3·361	17 09·9
Oct.	**1**	17 46 06·0	+ 8 54 42	3·386	17 03·6

Date		Astrometric J2000·0 R.A. h m s	Astrometric J2000·0 Dec. ° ′ ″	True Dist-ance	Ephem-eris Transit h m
Oct.	**1**	17 46 06·0	+ 8 54 42	3·386	17 03·6
	3	17 47 44·9	8 35 28	3·411	16 57·4
	5	17 49 26·7	8 16 36	3·436	16 51·3
	7	17 51 11·3	7 58 07	3·461	16 45·2
	9	17 52 58·7	7 40 01	3·486	16 39·1
	11	17 54 48·8	+ 7 22 19	3·511	16 33·1
	13	17 56 41·4	7 05 01	3·536	16 27·1
	15	17 58 36·5	6 48 07	3·560	16 21·1
	17	18 00 34·1	6 31 37	3·585	16 15·2
	19	18 02 34·0	6 15 33	3·609	16 09·4
	21	18 04 36·2	+ 5 59 53	3·633	16 03·5
	23	18 06 40·6	5 44 40	3·657	15 57·8
	25	18 08 47·2	5 29 52	3·680	15 52·0
	27	18 10 55·9	5 15 30	3·704	15 46·3
	29	18 13 06·6	5 01 34	3·727	15 40·6
	31	18 15 19·2	+ 4 48 05	3·749	15 34·9
Nov.	**2**	18 17 33·7	4 35 02	3·772	15 29·3
	4	18 19 49·9	4 22 26	3·794	15 23·7
	6	18 22 07·9	4 10 17	3·816	15 18·2
	8	18 24 27·4	3 58 35	3·838	15 12·6
	10	18 26 48·6	+ 3 47 19	3·859	15 07·1
	12	18 29 11·2	3 36 30	3·880	15 01·6
	14	18 31 35·2	3 26 07	3·901	14 56·1
	16	18 34 00·7	3 16 12	3·921	14 50·7
	18	18 36 27·5	3 06 43	3·941	14 45·3
	20	18 38 55·6	+ 2 57 41	3·960	14 39·9
	22	18 41 24·9	2 49 06	3·979	14 34·5
	24	18 43 55·3	2 40 57	3·998	14 29·1
	26	18 46 26·9	2 33 16	4·016	14 23·8
	28	18 48 59·5	2 26 01	4·034	14 18·5
	30	18 51 33·1	+ 2 19 12	4·051	14 13·2
Dec.	**2**	18 54 07·5	2 12 51	4·068	14 07·9
	4	18 56 42·8	2 06 55	4·084	14 02·6
	6	18 59 18·9	2 01 26	4·100	13 57·3
	8	19 01 55·8	1 56 22	4·115	13 52·0
	10	19 04 33·3	+ 1 51 45	4·130	13 46·8
	12	19 07 11·4	1 47 33	4·144	13 41·6
	14	19 09 50·1	1 43 46	4·158	13 36·3
	16	19 12 29·4	1 40 25	4·171	13 31·1
	18	19 15 09·1	1 37 28	4·184	13 25·9
	20	19 17 49·3	+ 1 34 57	4·196	13 20·7
	22	19 20 30·0	1 32 50	4·208	13 15·5
	24	19 23 10·9	1 31 08	4·219	13 10·3
	26	19 25 52·2	1 29 51	4·229	13 05·1
	28	19 28 33·7	1 28 57	4·239	12 59·9
	30	19 31 15·3	+ 1 28 28	4·248	12 54·8
	32	19 33 57·1	+ 1 28 22	4·257	12 49·6

GEOCENTRIC POSITIONS FOR 0^h DYNAMICAL TIME

Date		Astrometric J2000·0 R.A.	Dec.	True Distance	Ephemeris Transit
		h m s	° ′ ″		h m
Jan.	1	21 10 46·3	−12 31 56	3·187	14 29·4
	3	21 14 16·4	12 22 25	3·197	14 25·0
	5	21 17 47·6	12 12 32	3·207	14 20·6
	7	21 21 20·0	12 02 16	3·216	14 16·3
	9	21 24 53·5	11 51 38	3·225	14 12·0
	11	21 28 27·9	−11 40 38	3·233	14 07·7
	13	21 32 03·3	11 29 17	3·241	14 03·4
	15	21 35 39·7	11 17 34	3·248	13 59·1
	17	21 39 16·9	11 05 31	3·255	13 54·9
	19	21 42 55·0	10 53 07	3·261	13 50·7
	21	21 46 33·9	−10 40 24	3·267	13 46·4
	23	21 50 13·5	10 27 21	3·273	13 42·2
	25	21 53 54·0	10 13 58	3·278	13 38·0
	27	21 57 35·2	10 00 17	3·282	13 33·8
	29	22 01 17·1	9 46 16	3·286	13 29·7
	31	22 04 59·8	− 9 31 58	3·290	13 25·5
Feb.	2	22 08 43·1	9 17 22	3·293	13 21·3
	4	22 12 27·1	9 02 28	3·296	13 17·2
	6	22 16 11·7	8 47 18	3·298	13 13·1
	8	22 19 56·9	8 31 52	3·300	13 08·9
	10	22 23 42·7	− 8 16 10	3·301	13 04·8
	12	22 27 29·1	8 00 13	3·302	13 00·7
	14	22 31 15·9	7 44 01	3·303	12 56·6
	16	22 35 03·3	7 27 35	3·303	12 52·5
	18	22 38 51·2	7 10 55	3·302	12 48·5
	20	22 42 39·6	− 6 54 02	3·302	12 44·4
	22	22 46 28·5	6 36 56	3·300	12 40·3
	24	22 50 17·9	6 19 37	3·299	12 36·3
	26	22 54 07·8	6 02 07	3·296	12 32·2
	28	22 57 58·2	5 44 25	3·294	12 28·2
Mar.	1	23 01 49·1	− 5 26 33	3·291	12 24·2
	3	23 05 40·6	5 08 30	3·288	12 20·2
	5	23 09 32·5	4 50 18	3·284	12 16·1
	7	23 13 24·8	4 31 57	3·280	12 12·1
	9	23 17 17·6	4 13 27	3·275	12 08·1
	11	23 21 10·9	− 3 54 50	3·270	12 04·1
	13	23 25 04·6	3 36 06	3·265	12 00·2
	15	23 28 58·8	3 17 15	3·259	11 56·2
	17	23 32 53·4	2 58 18	3·253	11 52·2
	19	23 36 48·4	2 39 15	3·246	11 48·3
	21	23 40 44·0	− 2 20 08	3·239	11 44·3
	23	23 44 40·1	2 00 56	3·232	11 40·4
	25	23 48 36·6	1 41 40	3·224	11 36·4
	27	23 52 33·7	1 22 21	3·216	11 32·5
	29	23 56 31·3	1 02 59	3·208	11 28·6
	31	0 00 29·4	− 0 43 36	3·199	11 24·7
Apr.	2	0 04 28·0	− 0 24 11	3·190	11 20·8

Date		Astrometric J2000·0 R.A.	Dec.	True Distance	Ephemeris Transit
		h m s	° ′ ″		h m
Apr.	2	0 04 28·0	− 0 24 11	3·190	11 20·8
	4	0 08 27·2	− 0 04 45	3·181	11 16·9
	6	0 12 26·8	+ 0 14 40	3·171	11 13·0
	8	0 16 27·0	0 34 05	3·161	11 09·1
	10	0 20 27·7	0 53 28	3·151	11 05·3
	12	0 24 28·9	+ 1 12 49	3·140	11 01·4
	14	0 28 30·7	1 32 08	3·129	10 57·6
	16	0 32 33·0	1 51 23	3·118	10 53·7
	18	0 36 35·8	2 10 35	3·106	10 49·9
	20	0 40 39·3	2 29 41	3·095	10 46·1
	22	0 44 43·3	+ 2 48 43	3·083	10 42·3
	24	0 48 48·0	3 07 39	3·070	10 38·5
	26	0 52 53·3	3 26 29	3·058	10 34·7
	28	0 56 59·1	3 45 12	3·045	10 30·9
	30	1 01 05·6	4 03 46	3·031	10 27·1
May	2	1 05 12·7	+ 4 22 12	3·018	10 23·4
	4	1 09 20·5	4 40 29	3·004	10 19·6
	6	1 13 28·8	4 58 35	2·990	10 15·9
	8	1 17 37·7	5 16 31	2·976	10 12·1
	10	1 21 47·2	5 34 15	2·962	10 08·4
	12	1 25 57·3	+ 5 51 47	2·947	10 04·7
	14	1 30 07·9	6 09 06	2·932	10 01·0
	16	1 34 19·2	6 26 12	2·917	9 57·3
	18	1 38 31·1	6 43 03	2·902	9 53·6
	20	1 42 43·7	6 59 40	2·887	9 50·0
	22	1 46 56·8	+ 7 16 02	2·871	9 46·3
	24	1 51 10·6	7 32 07	2·855	9 42·7
	26	1 55 25·0	7 47 56	2·839	9 39·0
	28	1 59 40·0	8 03 27	2·823	9 35·4
	30	2 03 55·5	8 18 39	2·807	9 31·8
June	1	2 08 11·6	+ 8 33 32	2·790	9 28·2
	3	2 12 28·2	8 48 06	2·773	9 24·6
	5	2 16 45·3	9 02 19	2·757	9 21·0
	7	2 21 02·9	9 16 10	2·739	9 17·4
	9	2 25 20·9	9 29 40	2·722	9 13·8
	11	2 29 39·4	+ 9 42 47	2·705	9 10·2
	13	2 33 58·3	9 55 31	2·688	9 06·7
	15	2 38 17·6	10 07 51	2·670	9 03·1
	17	2 42 37·3	10 19 48	2·652	8 59·6
	19	2 46 57·3	10 31 19	2·634	8 56·0
	21	2 51 17·7	+10 42 24	2·617	8 52·5
	23	2 55 38·4	10 53 03	2·598	8 48·9
	25	2 59 59·4	11 03 15	2·580	8 45·4
	27	3 04 20·6	11 13 00	2·562	8 41·9
	29	3 08 41·9	11 22 16	2·544	8 38·3
July	1	3 13 03·3	+11 31 03	2·525	8 34·8
	3	3 17 24·7	+11 39 21	2·507	8 31·3

JUNO, 1992

GEOCENTRIC POSITIONS FOR 0^h DYNAMICAL TIME

Date	Astrometric J2000·0 R.A.	Astrometric J2000·0 Dec.	True Distance	Ephemeris Transit	Date	Astrometric J2000·0 R.A.	Astrometric J2000·0 Dec.	True Distance	Ephemeris Transit
	h m s	° ′ ″		h m		h m s	° ′ ″		h m
July 1	3 13 03·3	+11 31 03	2·525	8 34·8	**Oct. 1**	6 12 29·6	+ 8 19 14	1·632	5 31·6
3	3 17 24·7	11 39 21	2·507	8 31·3	**3**	6 15 14·2	8 03 05	1·614	5 26·5
5	3 21 46·1	11 47 09	2·488	8 27·8	**5**	6 17 53·5	7 46 36	1·595	5 21·3
7	3 26 07·4	11 54 27	2·469	8 24·3	**7**	6 20 27·4	7 29 50	1·576	5 15·9
9	3 30 28·6	12 01 13	2·450	8 20·7	**9**	6 22 55·7	7 12 47	1·558	5 10·5
11	3 34 49·6	+12 07 29	2·431	8 17·2	**11**	6 25 18·3	+ 6 55 30	1·540	5 05·0
13	3 39 10·4	12 13 12	2·412	8 13·7	**13**	6 27 34·9	6 37 59	1·522	4 59·4
15	3 43 30·9	12 18 24	2·393	8 10·1	**15**	6 29 45·4	6 20 17	1·504	4 53·7
17	3 47 51·1	12 23 02	2·374	8 06·6	**17**	6 31 49·5	6 02 24	1·486	4 47·9
19	3 52 10·9	12 27 08	2·355	8 03·0	**19**	6 33 47·2	5 44 24	1·468	4 42·0
21	3 56 30·1	+12 30 41	2·336	7 59·5	**21**	6 35 38·0	+ 5 26 18	1·451	4 35·9
23	4 00 48·9	12 33 39	2·316	7 55·9	**23**	6 37 21·9	5 08 09	1·434	4 29·8
25	4 05 07·0	12 36 04	2·297	7 52·3	**25**	6 38 58·7	4 49 58	1·417	4 23·5
27	4 09 24·3	12 37 53	2·278	7 48·7	**27**	6 40 28·1	4 31 49	1·401	4 17·1
29	4 13 40·9	12 39 08	2·258	7 45·1	**29**	6 41 50·0	4 13 44	1·384	4 10·6
31	4 17 56·5	+12 39 48	2·239	7 41·5	**31**	6 43 04·3	+ 3 55 46	1·368	4 04·0
Aug. 2	4 22 11·0	12 39 53	2·219	7 37·8	**Nov. 2**	6 44 10·8	3 37 57	1·353	3 57·2
4	4 26 24·4	12 39 22	2·199	7 34·2	**4**	6 45 09·4	3 20 20	1·337	3 50·3
6	4 30 36·6	12 38 16	2·180	7 30·5	**6**	6 46 00·0	3 02 58	1·322	3 43·3
8	4 34 47·5	12 36 35	2·160	7 26·8	**8**	6 46 42·5	2 45 54	1·308	3 36·1
10	4 38 57·0	+12 34 18	2·140	7 23·1	**10**	6 47 16·8	+ 2 29 11	1·294	3 28·8
12	4 43 05·0	12 31 26	2·121	7 19·3	**12**	6 47 42·8	2 12 53	1·280	3 21·3
14	4 47 11·4	12 27 58	2·101	7 15·5	**14**	6 48 00·5	1 57 03	1·266	3 13·8
16	4 51 16·2	12 23 55	2·081	7 11·7	**16**	6 48 09·6	1 41 43	1·254	3 06·0
18	4 55 19·2	12 19 16	2·062	7 07·9	**18**	6 48 10·3	1 26 59	1·241	2 58·2
20	4 59 20·3	+12 14 01	2·042	7 04·0	**20**	6 48 02·5	+ 1 12 53	1·229	2 50·2
22	5 03 19·4	12 08 12	2·022	7 00·1	**22**	6 47 46·3	0 59 30	1·218	2 42·1
24	5 07 16·4	12 01 47	2·002	6 56·2	**24**	6 47 21·7	0 46 54	1·207	2 33·8
26	5 11 11·0	11 54 48	1·983	6 52·2	**26**	6 46 48·9	0 35 08	1·197	2 25·4
28	5 15 03·2	11 47 13	1·963	6 48·2	**28**	6 46 08·2	0 24 16	1·188	2 16·8
30	5 18 52·9	+11 39 05	1·943	6 44·2	**30**	6 45 19·8	+ 0 14 22	1·179	2 08·1
Sept. 1	5 22 39·8	11 30 23	1·923	6 40·1	**Dec. 2**	6 44 23·9	+ 0 05 30	1·171	1 59·4
3	5 26 23·9	11 21 08	1·904	6 35·9	**4**	6 43 21·0	− 0 02 19	1·163	1 50·4
5	5 30 05·1	11 11 20	1·884	6 31·7	**6**	6 42 11·4	0 09 00	1·157	1 41·4
7	5 33 43·2	11 01 00	1·864	6 27·5	**8**	6 40 55·5	0 14 31	1·151	1 32·3
9	5 37 18·1	+10 50 08	1·845	6 23·2	**10**	6 39 33·9	− 0 18 50	1·146	1 23·1
11	5 40 49·7	10 38 45	1·825	6 18·8	**12**	6 38 07·0	0 21 54	1·141	1 13·8
13	5 44 17·9	10 26 52	1·805	6 14·4	**14**	6 36 35·4	0 23 41	1·138	1 04·4
15	5 47 42·4	10 14 29	1·786	6 09·9	**16**	6 34 59·5	0 24 10	1·135	0 54·9
17	5 51 03·2	10 01 37	1·767	6 05·4	**18**	6 33 20·1	0 23 20	1·133	0 45·4
19	5 54 20·2	+ 9 48 17	1·747	6 00·8	**20**	6 31 37·8	− 0 21 09	1·132	0 35·9
21	5 57 33·0	9 34 29	1·728	5 56·1	**22**	6 29 53·3	0 17 37	1·132	0 26·3
23	6 00 41·6	9 20 14	1·709	5 51·4	**24**	6 28 07·4	0 12 45	1·133	0 16·7
25	6 03 45·7	9 05 35	1·689	5 46·5	**26**	6 26 20·8	− 0 06 34	1·135	0 07·0
27	6 06 45·2	8 50 30	1·670	5 41·6	**28**	6 24 34·3	+ 0 00 55	1·138	23 52·6
29	6 09 39·8	+ 8 35 03	1·651	5 36·7	**30**	6 22 48·7	+ 0 09 40	1·141	23 43·0
Oct. 1	6 12 29·6	+ 8 19 14	1·632	5 31·6	**32**	6 21 04·7	+ 0 19 38	1·146	23 33·4

Second Transit: December 27^d 23^h $57^m.4$

VESTA, 1992

GEOCENTRIC POSITIONS FOR 0h DYNAMICAL TIME

Date		Astrometric J2000·0 R.A.	Astrometric J2000·0 Dec.	True Distance	Ephemeris Transit
		h m s	° ′ ″		h m
Jan.	1	11 47 23·1	+ 8 23 11	1·926	5 06·7
	3	11 48 48·4	8 23 13	1·899	5 00·2
	5	11 50 08·6	8 23 59	1·873	4 53·7
	7	11 51 23·5	8 25 28	1·847	4 47·0
	9	11 52 32·9	8 27 42	1·822	4 40·3
	11	11 53 36·7	+ 8 30 41	1·796	4 33·5
	13	11 54 34·7	8 34 27	1·772	4 26·6
	15	11 55 26·8	8 38 59	1·747	4 19·6
	17	11 56 12·8	8 44 18	1·723	4 12·5
	19	11 56 52·5	8 50 24	1·699	4 05·3
	21	11 57 25·9	+ 8 57 18	1·676	3 57·9
	23	11 57 52·8	9 05 00	1·654	3 50·5
	25	11 58 12·9	9 13 30	1·631	3 43·0
	27	11 58 26·3	9 22 48	1·610	3 35·3
	29	11 58 32·6	9 32 54	1·589	3 27·6
	31	11 58 32·0	+ 9 43 45	1·569	3 19·7
Feb.	2	11 58 24·1	9 55 23	1·549	3 11·7
	4	11 58 09·1	10 07 44	1·530	3 03·6
	6	11 57 46·9	10 20 47	1·512	2 55·3
	8	11 57 17·5	10 34 31	1·494	2 47·0
	10	11 56 41·0	+10 48 51	1·478	2 38·5
	12	11 55 57·5	11 03 47	1·462	2 29·9
	14	11 55 07·1	11 19 13	1·447	2 21·2
	16	11 54 10·1	11 35 08	1·433	2 12·4
	18	11 53 06·5	11 51 27	1·419	2 03·5
	20	11 51 56·6	+12 08 07	1·407	1 54·4
	22	11 50 40·7	12 25 03	1·396	1 45·3
	24	11 49 19·0	12 42 12	1·385	1 36·1
	26	11 47 51·9	12 59 28	1·376	1 26·8
	28	11 46 19·8	13 16 46	1·368	1 17·4
Mar.	1	11 44 43·2	+13 34 01	1·360	1 07·9
	3	11 43 02·5	13 51 07	1·354	0 58·4
	5	11 41 18·4	14 07 59	1·349	0 48·8
	7	11 39 31·4	14 24 32	1·345	0 39·2
	9	11 37 42·2	14 40 39	1·342	0 29·5
	11	11 35 51·3	+14 56 16	1·340	0 19·8
	13	11 33 59·6	15 11 19	1·339	0 10·1
	15	11 32 07·5	15 25 41	1·339	0 00·4
	17	11 30 15·8	15 39 20	1·340	23 45·8
	19	11 28 25·1	15 52 11	1·343	23 36·1
	21	11 26 35·9	+16 04 13	1·346	23 26·5
	23	11 24 49·0	16 15 20	1·350	23 16·8
	25	11 23 04·8	16 25 31	1·355	23 07·3
	27	11 21 23·9	16 34 44	1·361	22 57·8
	29	11 19 47·0	16 42 56	1·369	22 48·3
	31	11 18 14·4	+16 50 07	1·377	22 39·0
Apr.	2	11 16 46·8	+16 56 14	1·386	22 29·7
Apr.	2	11 16 46·8	+16 56 14	1·386	22 29·7
	4	11 15 24·5	17 01 17	1·395	22 20·5
	6	11 14 07·9	17 05 16	1·406	22 11·4
	8	11 12 57·5	17 08 11	1·417	22 02·4
	10	11 11 53·5	17 10 02	1·429	21 53·6
	12	11 10 56·1	+17 10 51	1·442	21 44·8
	14	11 10 05·5	17 10 39	1·456	21 36·2
	16	11 09 21·9	17 09 26	1·470	21 27·6
	18	11 08 45·3	17 07 15	1·485	21 19·2
	20	11 08 15·8	17 04 07	1·500	21 10·9
	22	11 07 53·4	+17 00 04	1·516	21 02·7
	24	11 07 38·2	16 55 06	1·532	20 54·7
	26	11 07 30·1	16 49 16	1·549	20 46·7
	28	11 07 29·1	16 42 35	1·567	20 38·9
	30	11 07 35·1	16 35 05	1·584	20 31·2
May	2	11 07 48·2	+16 26 47	1·603	20 23·6
	4	11 08 08·2	16 17 43	1·621	20 16·1
	6	11 08 34·9	16 07 54	1·640	20 08·7
	8	11 09 08·4	15 57 23	1·659	20 01·4
	10	11 09 48·4	15 46 11	1·679	19 54·3
	12	11 10 34·7	+15 34 19	1·699	19 47·2
	14	11 11 27·1	15 21 50	1·719	19 40·3
	16	11 12 25·6	15 08 46	1·739	19 33·4
	18	11 13 29·8	14 55 07	1·759	19 26·7
	20	11 14 39·6	14 40 54	1·780	19 20·0
	22	11 15 54·9	+14 26 10	1·801	19 13·4
	24	11 17 15·5	14 10 55	1·822	19 06·9
	26	11 18 41·2	13 55 10	1·843	19 00·5
	28	11 20 11·9	13 38 56	1·864	18 54·2
	30	11 21 47·4	13 22 14	1·885	18 47·9
June	1	11 23 27·7	+13 05 06	1·906	18 41·8
	3	11 25 12·4	12 47 32	1·927	18 35·7
	5	11 27 01·6	12 29 33	1·949	18 29·6
	7	11 28 54·9	12 11 11	1·970	18 23·7
	9	11 30 52·3	11 52 26	1·991	18 17·8
	11	11 32 53·6	+11 33 20	2·013	18 12·0
	13	11 34 58·7	11 13 52	2·034	18 06·2
	15	11 37 07·3	10 54 06	2·055	18 00·5
	17	11 39 19·5	10 34 00	2·077	17 54·8
	19	11 41 34·9	10 13 35	2·098	17 49·3
	21	11 43 53·6	+ 9 52 53	2·119	17 43·7
	23	11 46 15·5	9 31 53	2·140	17 38·2
	25	11 48 40·4	9 10 37	2·161	17 32·8
	27	11 51 08·3	8 49 06	2·182	17 27·4
	29	11 53 39·0	8 27 18	2·203	17 22·0
July	1	11 56 12·6	+ 8 05 16	2·224	17 16·7
	3	11 58 48·8	+ 7 43 00	2·244	17 11·5

Second Transit: March 15d 23h 55m·5

GEOCENTRIC POSITIONS FOR 0^h DYNAMICAL TIME

Date		Astrometric J2000·0 R.A.	Astrometric J2000·0 Dec.	True Distance	Ephemeris Transit
		h m s	° ′ ″		h m
July	1	11 56 12·6	+ 8 05 16	2·224	17 16·7
	3	11 58 48·8	7 43 00	2·244	17 11·5
	5	12 01 27·7	7 20 31	2·265	17 06·3
	7	12 04 09·1	6 57 49	2·285	17 01·1
	9	12 06 52·9	6 34 56	2·305	16 56·0
	11	12 09 39·0	+ 6 11 51	2·325	16 50·9
	13	12 12 27·4	5 48 36	2·345	16 45·8
	15	12 15 18·0	5 25 10	2·365	16 40·8
	17	12 18 10·8	5 01 36	2·385	16 35·8
	19	12 21 05·6	4 37 52	2·404	16 30·9
	21	12 24 02·5	+ 4 14 00	2·423	16 25·9
	23	12 27 01·4	3 49 59	2·442	16 21·1
	25	12 30 02·2	3 25 52	2·461	16 16·2
	27	12 33 05·0	3 01 37	2·480	16 11·4
	29	12 36 09·7	2 37 16	2·499	16 06·6
	31	12 39 16·3	+ 2 12 48	2·517	16 01·8
Aug.	2	12 42 24·7	1 48 16	2·535	15 57·1
	4	12 45 34·9	1 23 39	2·553	15 52·4
	6	12 48 46·8	0 58 59	2·571	15 47·7
	8	12 52 00·5	0 34 15	2·588	15 43·1
	10	12 55 15·7	+ 0 09 28	2·606	15 38·5
	12	12 58 32·6	− 0 15 21	2·623	15 33·9
	14	13 01 51·1	0 40 11	2·640	15 29·3
	16	13 05 11·2	1 05 03	2·656	15 24·8
	18	13 08 32·9	1 29 55	2·672	15 20·3
	20	13 11 56·2	− 1 54 48	2·689	15 15·8
	22	13 15 21·0	2 19 40	2·705	15 11·4
	24	13 18 47·4	2 44 32	2·720	15 06·9
	26	13 22 15·3	3 09 22	2·736	15 02·5
	28	13 25 44·9	3 34 10	2·751	14 58·1
	30	13 29 15·9	− 3 58 56	2·766	14 53·8
Sept.	1	13 32 48·5	4 23 38	2·780	14 49·5
	3	13 36 22·5	4 48 16	2·795	14 45·2
	5	13 39 58·0	5 12 50	2·809	14 40·9
	7	13 43 35·0	5 37 18	2·823	14 36·6
	9	13 47 13·4	− 6 01 41	2·836	14 32·4
	11	13 50 53·2	6 25 57	2·850	14 28·2
	13	13 54 34·5	6 50 07	2·863	14 24·0
	15	13 58 17·2	7 14 09	2·876	14 19·8
	17	14 02 01·4	7 38 04	2·888	14 15·7
	19	14 05 47·0	− 8 01 50	2·900	14 11·6
	21	14 09 34·0	8 25 27	2·912	14 07·5
	23	14 13 22·5	8 48 55	2·924	14 03·4
	25	14 17 12·5	9 12 12	2·935	13 59·4
	27	14 21 03·9	9 35 19	2·946	13 55·4
	29	14 24 56·8	− 9 58 14	2·957	13 51·4
Oct.	1	14 28 51·0	−10 20 57	2·968	13 47·4

Date		Astrometric J2000·0 R.A.	Astrometric J2000·0 Dec.	True Distance	Ephemeris Transit
		h m s	° ′ ″		h m
Oct.	1	14 28 51·0	−10 20 57	2·968	13 47·4
	3	14 32 46·7	10 43 27	2·978	13 43·5
	5	14 36 43·7	11 05 44	2·988	13 39·5
	7	14 40 42·2	11 27 47	2·997	13 35·6
	9	14 44 41·9	11 49 35	3·007	13 31·8
	11	14 48 43·1	−12 11 08	3·016	13 27·9
	13	14 52 45·6	12 32 25	3·024	13 24·1
	15	14 56 49·5	12 53 27	3·033	13 20·3
	17	15 00 54·7	13 14 11	3·041	13 16·5
	19	15 05 01·3	13 34 39	3·048	13 12·7
	21	15 09 09·2	−13 54 48	3·056	13 09·0
	23	15 13 18·5	14 14 39	3·063	13 05·2
	25	15 17 29·2	14 34 10	3·070	13 01·5
	27	15 21 41·2	14 53 22	3·076	12 57·9
	29	15 25 54·4	15 12 14	3·082	12 54·2
	31	15 30 08·9	−15 30 45	3·088	12 50·6
Nov.	2	15 34 24·6	15 48 54	3·093	12 47·0
	4	15 38 41·6	16 06 41	3·098	12 43·4
	6	15 42 59·7	16 24 06	3·103	12 39·8
	8	15 47 19·0	16 41 08	3·108	12 36·2
	10	15 51 39·4	−16 57 46	3·112	12 32·7
	12	15 56 01·0	17 14 01	3·115	12 29·2
	14	16 00 23·6	17 29 51	3·119	12 25·7
	16	16 04 47·4	17 45 16	3·122	12 22·2
	18	16 09 12·2	18 00 15	3·125	12 18·7
	20	16 13 38·1	−18 14 49	3·127	12 15·3
	22	16 18 05·1	18 28 57	3·129	12 11·9
	24	16 22 32·9	18 42 38	3·131	12 08·5
	26	16 27 01·8	18 55 52	3·132	12 05·1
	28	16 31 31·4	19 08 38	3·133	12 01·7
	30	16 36 02·0	−19 20 56	3·134	11 58·3
Dec.	2	16 40 33·3	19 32 46	3·134	11 54·9
	4	16 45 05·3	19 44 08	3·134	11 51·6
	6	16 49 38·0	19 55 01	3·134	11 48·3
	8	16 54 11·4	20 05 25	3·133	11 44·9
	10	16 58 45·4	−20 15 19	3·132	11 41·6
	12	17 03 19·9	20 24 44	3·130	11 38·3
	14	17 07 55·1	20 33 39	3·129	11 35·0
	16	17 12 30·7	20 42 04	3·126	11 31·8
	18	17 17 06·8	20 49 59	3·124	11 28·5
	20	17 21 43·2	−20 57 23	3·121	11 25·2
	22	17 26 20·1	21 04 17	3·118	11 21·9
	24	17 30 57·2	21 10 41	3·114	11 18·7
	26	17 35 34·5	21 16 34	3·110	11 15·4
	28	17 40 11·9	21 21 56	3·106	11 12·2
	30	17 44 49·4	−21 26 48	3·102	11 08·9
	32	17 49 27·0	−21 31 09	3·097	11 05·6

OPPOSITION DATES, MAGNITUDES AND OSCULATING ELEMENTS FOR EPOCH 1992 JUNE 27·0 TDT, ECLIPTIC AND EQUINOX J2000·0

Name	No.	Magnitude Parameters *H*	*G*	Opposition Date	Mag.	Dia-meter km	Inclin-ation *i* °	Long of Asc. Node *Ω* °	Argument of Peri-helion *ω* °	Mean Distance *a*	Daily Motion *n* °	Eccen-tricity *e*	Mean Anomaly *M* °
Ceres	1	3·32	0·11	July 28	7·5	1003	10·600	80·677	71·116	2·7674	0·21409	0·0766	141·462
Pallas	2	4·13	0·15	June 19	9·4	608	34·812	173·307	309·700	2·7703	0·21375	0·2346	127·630
Juno	3	5·31	0·30	Dec. 27	7·7	247	12·992	170·530	246·775	2·6694	0·22598	0·2571	341·381
Vesta	4	3·16	0·34	Mar. 13	5·9	538	7·135	103·988	149·919	2·3615	0·27159	0·0896	315·201
Astraea	5	6·91	0·25	June 12	10·6	117	5·358	141·788	356·313	2·5742	0·23864	0·1917	106·575
Flora	8	6·48	0·33	Jan. 26	8·7	151	5·886	111·109	285·094	2·2021	0·30161	0·1562	117·823
Metis	9	6·32	0·29	May 27	9·9	151	5·584	69·072	5·103	2·3860	0·26742	0·1226	178·407
Parthenope	11	6·62	0·27	May 1	9·7	150	4·621	125·688	194·853	2·4521	0·25668	0·0995	285·180
Victoria	12	7·23	0·24	Apr. 12	10·1	126	8·370	235·674	69·400	2·3344	0·27633	0·2189	305·225
Irene	14	6·27	0·09	Feb. 18	9·0	158	9·110	86·839	94·811	2·5869	0·23689	0·1666	5·430
Eunomia	15	5·22	0·20	Apr. 16	9·9	272	11·758	293·569	97·482	2·6427	0·22942	0·1868	197·076
Psyche	16	5·98	0·22	Mar. 1	10·4	250	3·093	150·491	228·608	2·9269	0·19683	0·1345	154·590
Thetis	17	7·77	0·13	Apr. 20	10·4	109	5·585	125·671	135·898	2·4693	0·25400	0·1354	335·715
Fortuna	19	7·09	0·10	Apr. 13	10·9	215	1·571	211·640	182·084	2·4414	0·25837	0·1593	185·919
Massalia	20	6·52	0·26	Jan. 21	8·5	131	0·702	206·955	255·194	2·4093	0·26356	0·1444	55·489
Lutetia	21	7·33	0·16	July 5	9·7	115	3·067	80·969	250·241	2·4354	0·25933	0·1625	322·735
Thalia	23	7·07	0·37	Sept. 20	11·2	111	10·151	67·331	59·274	2·6303	0·23105	0·2301	228·876
Themis	24	7·07	0·10	Mar. 29	10·6	234	0·763	36·048	109·959	3·1278	0·17817	0·1356	48·892
Phocaea	25	7·78	0·09	Dec. 30	12·4	72	21·574	214·347	90·443	2·3999	0·26511	0·2561	94·686
Euterpe	27	7·07	0·25	June 10	10·6	108	1·584	94·820	356·072	2·3468	0·27415	0·1718	169·491
Amphitrite	29	5·84	0·21	Jan. 3	8·9	195	6·111	356·624	63·477	2.5552	0·24130	0·0716	78·652
Urania	30	7·74	0·41	June 27	10·6	91	2·096	308·111	86·039	2·3653	0·27094	0·1278	255·485
Euphrosyne	31	6·53	0·15	June 1	12·2	370	26·341	31·308	62·919	3·1448	0·17674	0·2284	157·854
Circe	34	8·37	0·03	June 7	12·1	111	5·483	184·769	327·946	2·6862	0·22388	0·1071	96·300
Atalante	36	8·35	0·15	Dec. 3	10·9	118	18·482	358·923	46·284	2·7466	0·21652	0·3042	341·050
Fides	37	7·28	0·25	Aug. 5	11·1	95	3·077	7·814	61·555	2·6414	0·22959	0·1776	253·775
Laetitia	39	6·16	0·25	Nov. 14	9·6	163	10·369	157·411	207·987	2·7685	0·21396	0·1143	6·176
Harmonia	40	7·14	0·31	Jan. 23	9·8	100	4·257	94·381	269·009	2·2680	0·28857	0·0463	159·047
Daphne	41	7.34	0.15	Dec. 4	12·4	204	15·777	178·380	46·213	2·7611	0·21482	0·2752	189·820
Isis	42	7·50	0·25	Jan. 24	11·6	97	8·538	84·724	235·736	2·4417	0·25832	0·2244	193·205
Ariadne	43	8·01	0·25	Nov. 3	10·9	85	3·466	265·151	15·889	2·2035	0·30132	0·1679	64·816
Nysa	44	7·05	0·44	Jan. 16	9·1	82	3·703	131·674	342·278	2·4236	0·26122	0·1503	43·800
Eugenia	45	7·27	0·15	July 15	10·8	226	6·605	148·028	87·046	2·7211	0·21957	0·0804	47·268
Hestia	46	8·38	0·11	Feb. 28	12·3	133	2·333	181·259	176·084	2·5270	0·24536	0·1713	184·560
Aglaja	47	7·86	0·13	Mar. 5	12·4	158	4·994	3·465	314·586	2·8842	0·20122	0·1301	236·282
Europa	52	6·29	0·15	Oct. 25	10·6	289	7·439	129·272	337·512	3·1023	0·18037	0·0997	272·429
Alexandra	54	7·70	0·15	Aug. 30	10·4	180	11·827	313·585	345·309	2·7095	0·22099	0·1997	13·251
Melete	56	8·30	0·15	Dec. 23	12·9	146	8·079	193·840	102·924	2·5969	0·23552	0·2352	100·246
Mnemosyne	57	6·95	0·07	June 7	12·1	109	15·211	199·451	215·799	3·1470	0·17655	0·1194	209·993
Echo	60	8·68	0·33	May 9	11·9	51	3·597	192·103	269·540	2·3931	0·26623	0·1832	121·450
Ausonia	63	7·52	0·25	Sept. 29	10·4	91	5·789	338·222	295·104	2·3959	0·26577	0·1252	55·025
Maja	66	9·39	0·15	Sept. 10	12·3	85	3·053	8·045	42·790	2·6456	0·22904	0·1742	296·176
Asia	67	8·36	0·25	Feb. 10	12·3	58	6·014	202·982	105·417	2·4222	0·26145	0·1854	236·413
Leto	68	6·84	0·11	Sept. 7	9·6	126	7·963	44·528	304·103	2·7812	0·21250	0·1875	340·311
Panopaea	70	7·99	0·15	Apr. 25	11·3	151	11·581	48·029	255·617	2·6144	0·23316	0·1832	306·109
Niobe	71	7·26	0.37	Jan. 25	11·0	115	23·292	316·476	265·900	2·7558	0·21545	0·1750	313·197
Feronia	72	9·00	0·23	Nov. 26	11·7	96	5·411	208·247	102·569	2·2662	0·28890	0·1211	55·791
Frigga	77	8·57	0·26	July 4	12·6	67	2·434	1·572	60·686	2·6676	0·22622	0·1339	229·238
Eurynome	79	7·83	0·18	June 3	11·7	76	4·626	207·024	200·199	2·4436	0·25802	0·1938	222·881
Sappho	80	8·10	0·30	Mar. 9	11·8	83	8·664	218·958	139·171	2·2965	0·28321	0·2000	201·099

OPPOSITION DATES, MAGNITUDES AND OSCULATING ELEMENTS FOR EPOCH 1992 JUNE 27·0 TDT, ECLIPTIC AND EQUINOX J2000·0

Name	No.	Magnitude Parameters H	G	Opposition Date		Mag.	Diameter	Inclination i	Long of Asc. Node Ω	Argument of Perihelion ω	Mean Distance a	Daily Motion n	Eccentricity e	Mean Anomaly M
							km	°	°	°		°		°
Alkmene	82	8·51	0·34	Sept.	30	12·7	65	2·837	25·909	110·300	2·7638	0·21451	0·2205	232·523
Beatrix	83	8·89	0·30	Dec.	4	12·1	123	4·975	27·878	166·659	2·4317	0·25992	0·0825	205·080
Klio	84	9·26	0·15	July	25	10·9	90	9·341	327·890	14·698	2·3617	0·27156	0·2376	326·792
Julia	89	6·57	0·14	June	8	10·2	155	16·136	311·752	44·898	2·5501	0·24202	0·1826	287·063
Klotho	97	7·70	0·25	July	25	11·9	95	11·758	160·181	267·809	2·6676	0·22622	0·2589	256·930
Hera	103	7·59	0·11	Aug.	3	10·8	96	5·420	136·381	189·510	2·7017	0·22195	0·0818	339·495
Artemis	105	8·89	0·29	Aug.	11	11·8	126	21·485	188·545	56·023	2·3731	0·26961	0·1766	46·124
Dione	106	7·42	0·17	Sept.	10	11·1	139	4·623	62·604	330·500	3·1632	0·17519	0·1800	313·379
Camilla	107	7·09	0·15	Nov.	19	11·9	211	9·926	174·149	296·648	3·4825	0·15166	0·0841	290·580
Felicitas	109	8·87	0·11	June	20	13·7	75	7·887	3·508	56·397	2·6937	0·22294	0·2993	231·785
Amalthea	113	8·63	0·26	Apr.	11	11·0	47	5·034	123·655	79·056	2·3756	0·26918	0·0876	18·169
Kassandra	114	8·24	0·10	Feb.	4	11·1	136	4·941	164·514	351·848	2·6758	0·22518	0·1407	16·833
Thyra	115	7·51	0·14	July	19	10·5	93	11·588	309·320	95·873	2·3798	0·26846	0·1927	266·978
Brunhild	123	8·93	0·25	Oct.	16	12·1	47	6·416	308·232	123·735	2·6956	0·22270	0·1203	298·752
Alkeste	124	8·13	0·31	Oct.	27	11·8	67	2·949	188·451	62·513	2·6299	0·23110	0·0786	108·945
Antigone	129	7·05	0·37	Nov.	4	11·9	115	12·219	136·529	109·024	2·8675	0·20298	0·2137	113·416
Elektra	130	7·06	0·15	Nov.	24	11·0	173	22·875	145·962	234·605	3·1148	0·17929	0·2174	356·667
Vala	131	9·99	0·25	Nov.	27	13·3	35	4·955	65·848	158·466	2·4320	0·25988	0·0672	164·715
Hertha	135	8·21	0·19	May	16	11·1	78	2·303	344·132	339·708	2·4288	0·26038	0·2046	306·559
Meliboea	137	8·04	0·10	Dec.	16	13·2	150	13·436	202·571	107·873	3·1110	0·17962	0·2232	82·488
Juewa	139	7·79	0·15	Jan.	7	11·3	163	10·943	2·226	165·848	2·7798	0·21266	0·1780	350·405
Siwa	140	8·20	0·15	Dec.	12	12·6	103	3·189	107·409	195·878	2·7308	0·21841	0·2173	81·217
Lumen	141	8·56	0·15	Jan.	3	11·9	133	11·912	319·128	56·927	2·6668	0·22631	0·2140	101·086
Vibilia	144	7·87	0·08	July	2	11·1	130	4·815	76·731	293·294	2·6542	0·22793	0·2344	296·031
Adeona	145	8·05	0·01	Feb.	22	11·2	195	12·619	77·701	44·969	2·6725	0·22559	0·1470	47·790
Lucina	146	8·15	0·13	Aug.	4	11·9	141	13·086	84·359	145·522	2·7192	0·21980	0·0636	64·915
Aemilia	159	8·07	0·15	Mar.	2	11·9	140	6·124	134·439	338·689	3·1037	0·18025	0·1102	60·160
Baucis	172	8·80	0·25	Aug.	3	11·0	67	10·033	332·316	359·452	2·3798	0·26846	0·1144	332·519
Elsa	182	9·30	0·30	June	25	13·0	39	2·002	107·335	309·397	2·4161	0·26244	0·1869	232·721
Celuta	186	9·08	0·29	Dec.	7	12·3	49	13·186	14·954	315·184	2·3616	0·27158	0·1513	46·162
Phthia	189	9·51	0·25	May	14	12·7	41	5·177	203·845	167·123	2·4499	0·25703	0·0370	236·639
Nausikaa	192	7·13	0·03	June	19	10·3	94	6·827	343·640	29·789	2·4019	0·26477	0·2482	285·560
Philomela	196	6·64	0·47	Dec.	9	11·0	161	7·260	72·731	216·582	3·1127	0·17948	0·0272	116·857
Kallisto	204	9·00	0·25	Aug.	28	12·2	50	8·265	205·426	56·124	2·6717	0·22570	0·1739	43·359
Lacrimosa	208	9·05	0·25	Oct.	13	13·1	42	1·754	4·859	114·706	2·8936	0·20024	0·0104	240·458
Isolda	211	7·84	0·03	Jan.	25	11·5	166	3·873	264·157	175·747	3·0478	0·18524	0·1546	62·423
Thusnelda	219	9·43	0·25	Oct.	22	11·0	39	10·846	201·047	142·229	2·3537	0·27294	0·2250	357·144
Athamantis	230	7·47	0·35	July	15	10·4	121	9·439	240·059	139·254	2·3822	0·26806	0·0618	277·402
Hypatia	238	8·10	0·15	May	25	12·7	154	12·403	184·280	209·528	2·9050	0·19906	0·0912	221·439
Eukrate	247	8·00	0·07	Oct	24	10·9	142	25·021	0·537	54·365	2·7410	0·21719	0·2439	325·486
Libussa	264	8·40	0·25	Apr.	28	12·7	63	10·431	49·903	340·338	2·7969	0·21072	0·1368	202·499
Unitas	306	9·05	0·25	Jan.	23	12·7	43	7·264	142·124	166·879	2·3588	0·27206	0·1507	215·113
Polyxo	308	8·18	0·28	June	10	11·7	138	4·355	182·064	115·558	2·7491	0·21624	0·0377	328·046
Chaldaea	313	8·86	0·05	Mar.	6	10·9	160	11·622	176·997	315·060	2·3764	0·26905	0·1798	54·706
Tamara	326	9·35	0·15	Nov.	10	13·0	80	23·724	32·402	238·596	2·3181	0·27926	0·1894	87·754
Dembowska	349	5·98	0·32	Dec.	2	9·8	144	8·264	32·807	343·489	2·9244	0·19708	0·0897	15·938
Eleonora	354	6·32	0·32	May	3	10·2	153	18·430	140·721	4·951	2·7959	0·21083	0·1169	71·512
Liguria	356	8·17	0·15	Sept.	26	11·0	150	8·238	355·519	77·128	2·7578	0·21521	0·2390	295·646
Carlova	360	8·41	0·15	Mar.	26	13·0	130	11·704	132·788	289·740	3·0020	0·18949	0·1793	118·832
Isara	364	9·85	0·25	Feb.	3	12·0	28	6·003	105·742	312·668	2·2212	0·29772	0·1493	102·177

OPPOSITION DATES, MAGNITUDES AND OSCULATING ELEMENTS FOR EPOCH 1992 JUNE 27·0 TDT, ECLIPTIC AND EQUINOX J2000·0

Name	No.	Magnitude Parameters H	G	Opposition Date	Mag.	Diameter km	Inclination i °	Long of Asc. Node Ω °	Argument of Perihelion ω °	Mean Distance a	Daily Motion n °	Eccentricity e	Mean Anomaly M °
Corduba	365	9·27	0·30	Apr. 28	13·8	99	12·802	185·563	215·796	2·8013	0·21022	0·1570	185·954
Amicitia	367	10·95	0·25	Apr. 20	13·1	20	2·944	83·590	55·011	2·2192	0·29813	0·0956	81·200
Myrrha	381	8·50	0·15	Mar. 10	13·1	126	12·527	125·559	133·685	3·2170	0·17082	0·1037	298·752
Siegena	386	7·42	0·23	Jan. 20	11·6	191	20·269	167·169	219·439	2·8994	0·19964	0·1684	106·784
Industria	389	7·88	0·25	July 9	11·2	81	8·144	282·706	264·045	2·6087	0·23391	0·0635	91·051
Lampetia	393	8·40	0·15	Dec. 31	13·7	129	14·883	213·409	89·367	2·7753	0·21317	0·3324	100·786
Vienna	397	9·36	0·22	Sept. 19	11·6	50	12·845	228·495	139·277	2·6351	0·23041	0·2465	336·129
Arsinoe	404	9·05	0·19	Oct. 12	13·6	102	14·120	92·750	121·422	2·5936	0·23597	0·2000	123·884
Chloris	410	8·26	0·08	June 19	10·3	134	10·929	97·339	172·528	2·7282	0·21872	0·2368	0·890
Palatia	415	9·38	0·32	Sept. 7	12·3	93	8·133	127·546	296·031	2·7904	0·21145	0·3041	297·043
Hungaria	434	11·47	0·38	Feb. 16	13·9	11	22·506	175·476	123·720	1·9445	0·36350	0·0736	261·861
Bathilde	441	8·40	0·25	Nov. 29	11·8	66	8·137	254·050	200·629	2·8070	0·20957	0·0789	304·225
Gyptis	444	7·85	0·23	June 9	11·6	165	10·265	196·001	154·647	2·7691	0·21389	0·1761	291·488
Patientia	451	6·65	0·20	Feb. 25	11·0	276	15·219	89·592	344·134	3·0650	0·18368	0·0724	92·438
Bruchsalia	455	8·96	0·15	Dec. 9	11·9	105	12·035	76·806	272·236	2·6567	0·22761	0·2927	18·256
Papagena	471	6·61	0·29	Jan. 6	9·9	143	14·933	84·506	313·602	2·8922	0·20038	0·2288	79·479
Hedwig	476	8·71	0·15	Oct. 16	12·5	113	10·945	286·640	0·590	2·6495	0·22853	0·0756	66·072
Iva	497	10·01	0·11	Apr. 28	14·9	31	4·840	6·840	2·985	2·8500	0·20485	0·3008	243·040
Tokio	498	8·95	0·15	July 20	11·2	71	9·527	97·744	240·370	2·6495	0·22854	0·2246	329·022
Davida	511	6·17	0·02	Feb. 5	10·0	323	15·927	107·972	339·377	3·1762	0·17412	0·1793	58·482
Amherstia	516	8·25	0·25	Dec. 22	12·8	63	12·959	329·432	257·417	2·6752	0·22525	0·2777	208·368
Herculina	532	5·78	0·25	May 31	9·1	150	16·347	108·058	74·713	2·7722	0·21353	0·1756	53·267
Carmen	558	9·07	0·25	Oct. 17	13·0	64	8·355	144·143	310·630	2·9089	0·19866	0·0396	268·746
Suleika	563	8·61	0·25	Nov. 20	10·8	51	10·230	85·632	335·518	2·7134	0·22051	0·2355	325·232
Semiramis	584	8·74	0·34	Aug. 1	10·6	55	10·712	282·564	84·472	2·3732	0·26958	0·2352	315·094
Scheila	596	8·89	0·15	June 7	11·8	134	14·632	71·034	175·882	2·9320	0·19631	0·1591	11·190
Patroclus	617	8·17	0·15	Jan. 8	15·5	147	22·037	44·419	307·260	5·2335	0·08232	0·1384	114·174
Hektor	624	7·47	0·15	May 24	14·6	179	18·234	342·787	181·126	5·1871	0·08343	0·0247	83·387
Zelinda	654	8·43	0·05	Apr. 28	11·4	128	18·128	278·680	213·659	2·2970	0·28311	0·2306	80·667
Rachele	674	7·43	0·25	Aug. 4	12·4	102	13·538	58·821	40·240	2·9226	0·19726	0·1954	215·008
Pax	679	9·01	0·15	Jan. 28	12·5	73	24·374	112·829	265·887	2·5882	0·23670	0·3110	108·700
Ekard	694	9·01	0·15	Nov. 27	11·7	101	15·848	230·818	110·477	2·6707	0·22583	0·3229	14·225
Winchester	747	7·68	0·15	May 9	13·6	205	18·178	130·261	275·898	2·9923	0·19042	0·3443	184·989
Montefiore	782	11·53	0·25	Aug. 6	14·1	15	5·261	80·635	81·088	2·1798	0·30625	0·0389	136·759
Zwetana	785	9·58	0·25	Jan. 2	12·9	49	12·706	72·557	128·885	2·5721	0·23893	0·2082	326·365
Hispania	804	7·87	0·22	Apr. 11	12·3	141	15·386	347·956	343·763	2·8380	0·20615	0·1397	264·775
Benkoela	863	9·13	0·40	Mar. 16	13·7	33	25·412	117·211	95·287	3·1997	0·17220	0·0436	338·329
Alinda	887	14·20	0·25	Aug. 1	19·0	4	9·250	110·892	349·739	2·4928	0·25042	0·5584	246·008
Laodamia	1011	12·85	0·25	July 20	17·3	7	5·476	132·788	353·181	2·3955	0·26583	0·3471	159·037
Belgica	1052	12·02	0·25	Dec. 21	13·9	12	4·694	99·811	297·014	2·2361	0·29475	0·1434	348·357
Anchises	1173	8·91	0·15	Jan. 9	16·3	92	6·906	283·954	39·364	5·3291	0·08012	0·1356	149·026
Icarus	1566	16·65	0·25	Aug. 5	18·5	20	22·883	88·165	31·217	1·0780	0·88056	0·8268	209·505
Alikoski	1567	9·57	0·15	Feb. 5	14·1	72	17·262	51·839	117·765	3·2137	0·17108	0·0866	351·899

CONTENTS OF SECTION H

Except for the tables of radio sources and pulsars, positions tabulated in Section H are referred to the mean equator and equinox of J1992.5 = 1992 July 2.125 = JD 244 8805.625. Positions of radio sources are referred to J2000.0 = JD 245 1545.0.

Name	B.S.	Right Ascension	Declination	Notes	V	U–B	B–V	Spectral Type
		h m s	° ′ ″					
θ Oct	9084	0 01 13.0	−77 06 26	fv	4.78	+1.41	+1.27	K2 III
30 Psc	9089	0 01 34.6	− 6 03 21	fv	4.41	+1.83	+1.63	M3 III
2 Cet	9098	0 03 21.4	−17 22 40	fv	4.55	−0.12	−0.05	B9 IV
33 Psc	3	0 04 57.1	− 5 44 58	fvd6	4.61	+0.89	+1.04	K0 III–IV
21 α And	15	0 07 59.9	+29 02 57	fvd6	2.06	−0.46	−0.11	B9p Hg Mn
11 β Cas	21	0 08 46.5	+59 06 30	fsvd6	2.27	+0.11	+0.34	F2 III–IV
ε Phe	25	0 09 01.9	−45 47 20	f	3.88	+0.84	+1.03	K0 III
22 And	27	0 09 55.7	+46 01 50	f	5.03	+0.25	+0.40	F2 II
θ Scl	35	0 11 21.2	−35 10 30	f	5.25		+0.44	F3/5 V
88 γ Peg	39	0 12 50.9	+15 08 31	fsvd6	2.83	−0.87	−0.23	B2 IV
89 χ Peg	45	0 14 12.8	+20 09 54	fsv	4.80	+1.93	+1.57	M2^{+} III
7 Cet	48	0 14 15.6	−18 58 27	v	4.44	+1.99	+1.66	M1 III
25 σ And	68	0 17 56.1	+36 44 37	fv6	4.52	+0.07	+0.05	A2 Va
8 ι Cet	74	0 19 02.8	− 8 51 56	fvd	3.56	+1.25	+1.22	K1 IIIb
ζ Tuc	77	0 19 41.0	−64 55 08	f	4.23	+0.02	+0.58	F9 V
41 Psc	80	0 20 12.7	+ 8 08 55	fv	5.37	+1.55	+1.34	gK3
27 ρ And	82	0 20 43.5	+37 55 38	f	5.18	+0.05	+0.42	F6 IV
R And	90	0 23 38.1	+38 32 09	svd	7.39	+1.25	+1.97	S5/4.5e
β Hyi	98	0 25 21.9	−77 17 47	fv	2.80	+0.11	+0.62	G1 IV
κ Phe	100	0 25 50.1	−43 43 17		3.94	+0.11	+0.17	A6 Vn
α Phe	99	0 25 54.9	−42 20 48	fd67	2.39	+0.88	+1.09	K0 IIIb
	118	0 30 00.2	−23 49 45	f6	5.19		+0.12	A5 Vn
λ^{1} Phe	125	0 31 03.4	−48 50 42	fd6	4.77	+0.04	+0.02	A1 Va
β^{1} Tuc	126	0 31 12.3	−62 59 58	d6	4.37	−0.17	−0.07	B9 V
15 κ Cas	130	0 32 34.1	+62 53 26	fsv6	4.16	−0.80	+0.14	B1 Ia
29 π And	154	0 36 28.7	+33 40 41	fvd6	4.36	−0.55	−0.14	B5 V
17 ζ Cas	153	0 36 33.0	+53 51 21	fv6	3.66	−0.87	−0.20	B2 IV
	157	0 36 57.0	+35 21 30	s	5.48	+0.48	+0.88	G2 Ib–II
30 ε And	163	0 38 09.5	+29 16 16	f	4.37	+0.47	+0.87	G8 IIIp
31 δ And	165	0 38 55.5	+30 49 12	fsd6	3.27	+1.48	+1.28	K3 III
18 α Cas	168	0 40 04.7	+56 29 47	fvd	2.23	+1.13	+1.17	K0^{-} IIIa
μ Phe	180	0 40 58.4	−46 07 34	f	4.59	+0.72	+0.97	G8 III
η Phe	191	0 43 01.1	−57 30 15	fd	4.36	−0.02	0.00	A0.5 IV
16 β Cet	188	0 43 12.8	−18 01 40	fv	2.04	+0.87	+1.02	G9.5 III CN−1
22 ο Cas	193	0 44 18.3	+48 14 36	fvd6	4.54	−0.51	−0.07	B5 III
34 ζ And	215	0 46 56.4	+24 13 35	fvd6	4.06	+0.90	+1.12	K0 III
63 δ Psc	224	0 48 17.6	+ 7 32 40	fd	4.43	+1.86	+1.50	K4.5 IIIb
λ Hyi	236	0 48 19.9	−74 57 51	f	5.07	+1.68	+1.37	K5 III
64 Psc	225	0 48 35.0	+16 54 01	fd6	5.07		+0.51	F8 V
24 η Cas	219	0 48 38.6	+57 46 34	sd6	3.44	+0.03	+0.57	G0 V
35 ν And	226	0 49 23.9	+41 02 17	f6	4.53	−0.58	−0.15	B5 V
19 ϕ^{2} Cet	235	0 49 45.1	−10 41 05	fv	5.19	−0.02	+0.50	F8 V
	233	0 50 16.0	+64 12 25	fcv6	5.39		+0.49	G0 III–IV + B9.5 V
20 Cet	248	0 52 37.5	− 1 11 06	f	4.77	+1.93	+1.57	M0^{-} IIIa
λ^{2} Tuc	270	0 54 43.6	−69 34 03	f	5.45	+1.00	+1.09	K2 III
27 γ Cas	264	0 56 15.1	+60 40 34	fvd6	2.47	−1.08	−0.15	B0.5 IVe
37 μ And	269	0 56 20.1	+38 27 32	fd	3.87	+0.15	+0.13	A5 V
38 η And	271	0 56 48.4	+23 22 39	d6	4.42	+0.69	+0.94	G8^{-} IIIb
α Scl	280	0 58 14.7	−29 23 52	fsv6	4.31	−0.56	−0.16	B7 III (C II)
71 ε Psc	294	1 02 33.2	+ 7 50 59	fd	4.28	+0.70	+0.96	G9 III Fe−2

Name	B.S.	Right Ascension	Declination	Notes	*V*	*U–B*	*B–V*	Spectral Type
		h m s	° ′ ″					
β Phe	322	1 05 45.0	−46 45 31	vd7	3.31	+0.57	+0.89	G8 III
ι Tuc	332	1 07 00.9	−61 48 55	f	5.37		+0.88	G5 III
υ Phe	331	1 07 27.4	−41 31 37	fd	5.21	+0.09	+0.16	A3 IV/V
	285	1 07 36.4	+86 13 02	f	4.25	+1.33	+1.21	K2 III
30 μ Cas	321	1 07 46.2	+54 53 02	fvd6	5.17	+0.09	+0.69	G5 Vb
ζ Phe	338	1 08 04.3	−55 17 09	vd6	3.92	−0.41	−0.08	B7 V
31 η Cet	334	1 08 12.8	−10 13 19	fd	3.45	+1.19	+1.16	K2–III CN 0.5
42 φ And	335	1 09 03.9	+47 12 07	d7	4.25	−0.34	−0.07	B7 III
43 β And	337	1 09 18.6	+35 34 51	fvd	2.06	+1.96	+1.58	$M0^+$ IIIa
33 θ Cas	343	1 10 38.6	+55 06 36	vd6	4.33	+0.12	+0.17	A7 V
84 χ Psc	351	1 11 03.0	+20 59 42	f	4.66	+0.82	+1.03	G8.5 III
83 τ Psc	352	1 11 14.8	+30 03 00	f6	4.51	+1.01	+1.09	K0.5 IIIb
86 ζ Psc	361	1 13 20.3	+ 7 32 09	fd67	4.86	+0.09	+0.32	F0 Vn
κ Tuc	377	1 15 31.0	−68 54 57	vd7	4.86	+0.03	+0.47	F6 IV
89 Psc	378	1 17 24.7	+ 3 34 30	f6	5.16	+0.08	+0.07	A3 V
90 υ Psc	383	1 19 03.2	+27 13 29	f6	4.76	+0.10	+0.03	A2 IV
34 φ Cas	382	1 19 36.3	+58 11 33	smd6	4.98		+0.68	F0 Ia
46 ξ And	390	1 21 53.8	+45 29 23	f6	4.88	+0.99	+1.08	$K0^-$ IIIb
45 θ Cet	402	1 23 38.9	− 8 13 19	fd	3.60	+0.93	+1.06	K0 IIIb
37 δ Cas	403	1 25 19.2	+60 11 47	fsvd6	2.68	+0.12	+0.13	A5 III–IV
36 ψ Cas	399	1 25 23.8	+68 05 28	fd	4.74	+0.94	+1.05	K0 III
94 Psc	414	1 26 17.3	+19 12 06	f	5.50	+1.05	+1.11	gK1
48 ω And	417	1 27 12.3	+45 22 06	fd	4.83	0.00	+0.42	F5 V
γ Phe	429	1 28 02.4	−43 21 23	fv6	3.41	+1.85	+1.57	$M0^-$ IIIa
48 Cet	433	1 29 14.6	−21 40 05	fd7	5.12	+0.04	+0.02	A1 Va
δ Phe	440	1 30 56.4	−49 06 41	f	3.95	+0.70	+0.99	G9 III
99 η Psc	437	1 31 04.9	+15 18 26	fvd	3.62	+0.75	+0.97	G7 IIIa
50 υ And	458	1 36 21.3	+41 22 05	fd6	4.09	+0.06	+0.54	F8 V
α Eri	472	1 37 26.1	−57 16 29	fv	0.46	−0.66	−0.16	B3 Vp
51 And	464	1 37 31.8	+48 35 26	f	3.57	+1.45	+1.28	$K3^-$ III
40 Cas	456	1 37 54.2	+73 00 08	fvd	5.28		+0.96	G8 IIIa
106 ν Psc	489	1 41 02.4	+ 5 26 59	f	4.44	+1.57	+1.36	K3 IIIb Ba 0.1
	490	1 41 37.4	+35 12 29	f	5.40	−0.20	−0.09	B9 IV–V
π Scl	497	1 41 48.3	−32 21 52	fm	5.26		+1.04	K1 II/III
	500	1 42 20.7	− 3 43 40	f	4.99	+1.58	+1.38	K3 II–III
φ Per	496	1 43 11.2	+50 39 04	fv6	4.07	−0.93	−0.04	B2 Vep
52 τ Cet	509	1 43 43.2	−15 58 36	fd	3.50	+0.21	+0.72	G8 V
110 ο Psc	510	1 44 59.8	+ 9 07 13	fsv	4.26	+0.71	+0.96	G8 III
ε Scl	514	1 45 17.7	−25 05 24	fd7	5.31	+0.02	+0.39	F0 V
	513	1 45 36.6	− 5 46 15	s	5.34	+1.88	+1.52	K4 III
53 χ Cet	531	1 49 13.0	−10 43 24	fd	4.67	+0.03	+0.33	F2 V
55 ζ Cet	539	1 51 05.4	−10 22 19	fvd6	3.73	+1.07	+1.14	K0 III
2 α Tri	544	1 52 39.2	+29 32 33	fd6	3.41	+0.06	+0.49	F6 IV
111 ξ Psc	549	1 53 10.0	+ 3 09 03	fv6	4.62	+0.72	+0.94	K0 III
ψ Phe	555	1 53 20.7	−46 20 21	fv6	4.41	+1.70	+1.59	M4 III
45 ε Cas	542	1 53 51.0	+63 38 00	fv	3.38	−0.60	−0.15	B3 Vp
φ Phe	558	1 54 03.4	−42 32 01	f6	5.11	−0.15	−0.06	Ap Hg
6 β Ari	553	1 54 13.5	+20 46 18	fv6	2.64	+0.10	+0.13	A5 V
η^2 Hyi	570	1 54 44.7	−67 41 03	f	4.69	+0.64	+0.95	G8.5 III
χ Eri	566	1 55 40.0	−51 38 46	fvd7	3.70	+0.46	+0.85	G8 IIIb CN IV

Name	B.S.	Right Ascension	Declination	Notes	V	U−B	B−V	Spectral Type
		h m s	° ′ ″					
α Hyi	591	1 58 32.0	−61 36 22	f	2.86	+0.14	+0.28	F0 V
59 υ Cet	585	1 59 39.1	−21 06 50	f	4.00	+1.91	+1.57	M0 IIIb
113 α Psc	596	2 01 39.5	+ 2 43 40	vmd68	3.79	−0.05	+0.03	A0p Si Sr
4 Per	590	2 01 47.9	+54 27 06	f6	5.04	−0.32	−0.08	B8 III
50 Cas	580	2 02 46.9	+72 23 07	f6	3.98	+0.03	−0.01	A1 Va
57 γ[1] And	603	2 03 26.2	+42 17 38	fd6	2.26	+1.58	+1.37	K3⁻ IIb
ν For	612	2 04 09.3	−29 19 57	fv	4.69	−0.51	−0.17	B9.5p Si
13 α Ari	617	2 06 45.0	+23 25 38	fv6	2.00	+1.12	+1.15	K2⁻ IIIab Ca−1
4 β Tri	622	2 09 05.7	+34 57 08	f6	3.00	+0.10	+0.14	A5 III
μ For	652	2 12 34.7	−30 45 32	f	5.28	−0.06	−0.02	A2 Vn
65 ξ[1] Cet	649	2 12 36.1	+ 8 48 42	fvd6	4.37	+0.60	+0.89	G7 II−III Fe−1
	645	2 13 06.2	+51 01 53	fvd6	5.31	+0.62	+0.93	K0 III
	641	2 13 09.4	+58 31 35	s	6.44	+0.23	+0.60	A3 Iab
φ Eri	674	2 16 14.5	−51 32 48	fd	3.56	−0.39	−0.12	B8 V
67 Cet	666	2 16 36.6	− 6 27 23	f	5.51	+0.76	+0.96	G8.5 III
9 γ Tri	664	2 16 52.0	+33 48 46	f	4.01	+0.02	+0.02	A0 IV−Vn
62 And	670	2 18 47.6	+47 20 44	f	5.30		−0.01	A1 V
68 o Cet	681	2 18 58.0	− 3 00 41	vd	2−10	+1.09	+1.42	M5.5e + pec
δ Hyi	705	2 21 36.8	−68 41 36	f	4.09	+0.05	+0.03	A1 Va
κ For	695	2 22 12.0	−23 51 01	f	5.20		+0.60	G0 Va
κ Hyi	715	2 22 49.3	−73 40 47	f	5.01	+1.04	+1.09	K1 III
1 α UMi	424	2 23 45.7	+89 13 51	fvd6	2.02	+0.38	+0.60	F5−8 Ib
λ Hor	714	2 24 41.3	−60 20 43	f	5.35		+0.39	F2 III
72 ρ Cet	708	2 25 35.3	−12 19 27	f	4.89	−0.07	−0.03	A0 III−IVn
κ Eri	721	2 26 42.6	−47 44 14	f6	4.25	−0.50	−0.14	B5 IV
12 Tri	717	2 27 43.5	+29 38 10	fm	5.28			F0 III
73 ξ[2] Cet	718	2 27 45.6	+ 8 25 36	f6	4.28	−0.12	−0.06	A0 III⁻
ι Cas	707	2 28 26.4	+67 22 09	vd	4.52	+0.06	+0.12	A5p Sr
14 Tri	736	2 31 38.6	+36 06 52	f	5.15	+1.78	+1.47	K5 III
76 σ Cet	740	2 31 43.9	−15 16 38	f	4.75	−0.02	+0.45	F5 IV−Vs
μ Hyi	776	2 31 49.2	−79 08 32	f	5.28	+0.73	+0.98	G8 III
78 ν Cet	754	2 35 28.8	+ 5 33 39	fd67	4.86	+0.53	+0.87	G8 III
	753	2 35 40.2	+ 6 51 05	fsd6	5.82	+0.81	+0.98	K3⁻ V
	743	2 37 18.4	+72 47 10	f	5.16	+0.58	+0.88	G8 III
32 ν Ari	773	2 38 23.4	+21 55 45	f6	5.30	+0.18	+0.16	A7 V
82 δ Cet	779	2 39 05.9	+ 0 17 47	fv6	4.07	−0.87	−0.22	B2 IV
ε Hyi	806	2 39 28.3	−68 17 56	f	4.11	−0.14	−0.06	B9 V
ι Eri	794	2 40 22.3	−39 53 14	f	4.11	+0.74	+1.02	K0 III
ζ Hor	802	2 40 25.6	−54 34 55	f6	5.21	−0.01	+0.40	F4 IV
86 γ Cet	804	2 42 54.7	+ 3 12 16	d7	3.47	+0.07	+0.09	A2 Va
35 Ari	801	2 43 00.6	+27 40 32	f6	4.66	−0.62	−0.13	B3 V
14 Per	800	2 43 35.7	+44 15 56	f	5.43	+0.65	+0.90	G0 Ib Ca 1
13 θ Per	799	2 43 41.1	+49 11 50	fvd	4.12	0.00	+0.49	F7 V
89 π Cet	811	2 43 45.9	−13 53 25	f6	4.25	−0.45	−0.14	B7 V
87 μ Cet	813	2 44 32.2	+10 04 58	fvd6	4.27	+0.08	+0.31	F0 IV
1 τ[1] Eri	818	2 44 45.1	−18 36 14	6	4.47	0.00	+0.48	F5 V
β For	841	2 48 46.6	−32 26 14	fd	4.46	+0.69	+0.99	G8.5 IIIb Fe−0.5
41 Ari	838	2 49 32.5	+27 13 48	fvd6	3.63	−0.37	−0.10	B8 Vn
16 Per	840	2 50 06.5	+38 17 17	vd	4.23	+0.08	+0.34	F2 III
15 η Per	834	2 50 08.7	+55 51 53	fd6	3.76	+1.89	+1.68	K3⁻ Ib−IIa

Name	B.S.	Right Ascension	Declination	Notes	V	U–B	B–V	Spectral Type
		h m s	° ′ ″					
2 τ^2 Eri	850	2 50 41.9	−21 02 05	fd	4.75	+0.63	+0.91	K0 III
43 σ Ari	847	2 51 04.7	+15 03 05	f	5.49	−0.43	−0.09	B7 V
R Hor	868	2 53 37.9	−49 55 14	vm	4.00			M6.5e:
18 τ Per	854	2 53 43.3	+52 43 56	fcvd6	3.95	+0.46	+0.74	G5 III + A4 V
3 η Eri	874	2 56 03.6	− 8 55 40	fv	3.89	+1.00	+1.11	K1 IIIb
	875	2 56 14.8	− 3 44 32	f6	5.17	+0.05	+0.08	A3 Vn
θ^1 Eri	897	2 57 58.6	−40 20 04	fvmd68	3.42	+0.12	+0.12	A5 III
θ^2 Eri	898	2 57 59.2	−40 20 04	vmd8	4.42	+0.12	+0.12	A1 Va
24 Per	882	2 58 35.7	+35 09 12	f	4.93	+1.29	+1.23	K2 III
91 λ Cet	896	2 59 18.7	+ 8 52 40	f	4.70	−0.45	−0.12	B6 III
92 α Cet	911	3 01 53.2	+ 4 03 38	fv	2.53	+1.94	+1.64	M1.5 IIIa
11 τ^3 Eri	919	3 02 03.7	−23 39 13	f	4.09	+0.08	+0.16	A5 V
θ Hyi	939	3 02 14.2	−71 55 54	fd7	5.53	−0.51	−0.14	B9 IVp
μ Hor	934	3 03 26.2	−59 46 00	f	5.11		+0.34	F0 IV
23 γ Per	915	3 04 15.0	+53 28 39	fcd6	2.93	+0.45	+0.70	G5 III + A2 V
25 ρ Per	921	3 04 41.6	+38 48 42	fv	3.39	+1.79	+1.65	M4 II
	881	3 05 06.0	+79 23 23	fd6	5.49		+1.57	M2 IIIab
26 β Per	936	3 07 40.7	+40 55 38	fvd6	2.12	−0.37	−0.05	B8 V
ι Per	937	3 08 31.3	+49 35 06	fd	4.05	+0.10	+0.61	G0 V
27 κ Per	941	3 08 59.3	+44 49 46	vd6	3.80	+0.83	+0.98	K0 III
57 δ Ari	951	3 11 12.0	+19 41 55	fv	4.35	+0.87	+1.03	K0 III
α For	963	3 11 45.1	−29 00 59	vd7	3.87	+0.02	+0.52	F8 V
	977	3 12 21.7	−57 20 58	fsv	5.74	+2.83	+2.28	C6:,2.5 Ba2 Y4
94 Cet	962	3 12 23.4	− 1 13 26	fd7	5.06	+0.12	+0.57	F8 V
58 ζ Ari	972	3 14 28.2	+21 01 01	f	4.89	−0.01	−0.01	A0.5 Va^+
13 ζ Eri	984	3 15 28.1	− 8 50 50	fv6	4.80	+0.09	+0.23	A5m:
29 Per	987	3 18 05.6	+50 11 42	sm6	5.15		−0.05	B3 V
96 κ Cet	996	3 18 58.1	+ 3 20 35	fsvd	4.83	+0.19	+0.68	G5 V
16 τ^4 Eri	1003	3 19 11.0	−21 47 05	vd	3.69	+1.81	+1.62	M3 III
	961	3 19 21.3	+77 42 29	fvd	5.45	+0.11	+0.19	A5 III:
	1008	3 19 37.7	−43 05 53	f	4.27	+0.22	+0.71	G8 V
	999	3 19 53.1	+29 01 18		4.47	+1.79	+1.55	K4 III
61 τ Ari	1005	3 20 47.6	+21 07 13	fd	5.28	−0.52	−0.07	B5 IV
33 α Per	1017	3 23 47.1	+49 50 06	fsvmd	1.80		+0.48	F5 Ib
	1009	3 24 01.1	+64 33 35	fv	5.23		+2.08	M0 II
1 ο Tau	1030	3 24 24.5	+ 9 00 10	fv6	3.60	+0.61	+0.89	G6 IIIa Fe−1
	1029	3 25 25.3	+49 05 42	sm	6.07		−0.08	B7 V
2 ξ Tau	1038	3 26 45.7	+ 9 42 25	f6	3.74	−0.33	−0.09	B9 Vn
	1035	3 28 27.4	+59 54 53	fvd	4.21	−0.24	+0.41	B9 Ia
κ Ret	1083	3 29 14.7	−62 57 50	fd	4.72	−0.04	+0.40	F5 IV−V
	1040	3 29 18.6	+58 51 12	s6	4.54	−0.11	+0.56	A0 Ia
35 σ Per	1052	3 30 02.6	+47 58 11	fvm	4.35		+1.37	K3 III
17 Eri	1070	3 30 14.7	− 5 06 02	f	4.73	−0.27	−0.09	B9 Vs
5 Tau	1066	3 30 27.5	+12 54 41	fd6	4.11	+1.02	+1.12	K0 II−III
18 ε Eri	1084	3 32 34.6	− 9 29 00	fsd	3.73	+0.59	+0.88	K2 V
19 τ^5 Eri	1088	3 33 27.4	−21 39 28	fm6	4.26		−0.10	B8 V
20 Eri	1100	3 35 56.9	−17 29 30	fv	5.23		−0.13	B9p Si
37 ψ Per	1087	3 35 57.2	+48 10 06	vm	4.23		−0.06	B5 Ve
10 Tau	1101	3 36 29.4	+ 0 22 42	f	4.28	+0.07	+0.58	F9 IV−V
	1106	3 36 49.5	−40 17 56	f	4.58	+0.77	+1.04	K1 III

Name	B.S.	Right Ascension	Declination	Notes	V	U–B	B–V	Spectral Type
		h m s	° ′ ″					
	1105	3 41 29.9	+63 11 35	fv6	5.10		+1.63	S3.5/2
δ For	1134	3 41 57.0	−31 57 43	f6	5.00	−0.60	−0.16	B5 IV
39 δ Per	1122	3 42 23.3	+47 45 51	fvd6	3.01	−0.51	−0.13	B5 III
23 δ Eri	1136	3 42 53.3	− 9 47 18	fv	3.54	+0.69	+0.92	$K0^+$ IV
38 o Per	1131	3 43 50.8	+32 15 54	vd6	3.83	−0.75	+0.05	B1 III
β Ret	1175	3 44 06.2	−64 49 50	fd6	3.85	+1.10	+1.13	K2 III
24 Eri	1146	3 44 07.6	− 1 11 11	f6	5.25	−0.39	−0.10	B7 V
17 Tau	1142	3 44 25.7	+24 05 25	fmd6	3.70		−0.11	B6 III
41 ν Per	1135	3 44 40.9	+42 33 19	fvd	3.77	+0.31	+0.42	F5 II
19 Tau	1145	3 44 45.6	+24 26 39	vmd6	4.30		−0.11	B6 IV
29 Tau	1153	3 45 16.5	+ 6 01 37	fd6	5.35	−0.61	−0.12	B3 V
20 Tau	1149	3 45 22.8	+24 20 41	svmd6	3.88		−0.07	B7 III (Fe II)
26 π Eri	1162	3 45 47.2	−12 07 30	v	4.42	+2.01	+1.63	$M2^-$ IIIab
23 Tau	1156	3 45 52.8	+23 55 31	vmd	4.18		−0.06	B6 IV
27 τ^6 Eri	1173	3 46 31.5	−23 16 18	f	4.23	0.00	+0.42	F3 III
25 η Tau	1165	3 47 02.3	+24 04 56	fmd	2.87		−0.09	B7 III
γ Hyi	1208	3 47 20.9	−74 15 43	f	3.24	+1.99	+1.62	M2 III
27 Tau	1178	3 48 42.9	+24 01 51	fvmd6	3.63		−0.08	B8 III
	1155	3 48 49.7	+65 30 13	v	4.47	+2.13	+1.88	$M2^+$ IIab
	1195	3 49 10.4	−36 13 22	f	4.17	+0.69	+0.95	G8 III
γ Cam	1148	3 49 33.4	+71 18 36	fd	4.63	+0.07	+0.03	A1 IIIn
44 ζ Per	1203	3 53 39.6	+31 51 42	fsvd67	2.85	−0.77	+0.12	B1 Ib
45 ε Per	1220	3 57 20.9	+39 59 20	fsvd67	2.89	−0.99	−0.18	B0.5 III
34 γ Eri	1231	3 57 40.7	−13 31 46	fvd	2.95	+1.96	+1.59	M0.5 IIIb Ca−1
46 ξ Per	1228	3 58 28.6	+35 46 12	fv6	4.04	−0.92	+0.01	O7.5
δ Ret	1247	3 58 37.5	−61 25 17	f	4.56	+1.96	+1.62	M2 IIIab
35 λ Tau	1239	4 00 15.8	+12 28 10	fv6	3.47	−0.62	−0.12	B3 IV
35 Eri	1244	4 01 09.2	− 1 34 13	f	5.28	−0.55	−0.15	B5 V
38 ν Tau	1251	4 02 45.4	+ 5 58 08	f	3.91	+0.07	+0.03	A1 Va
37 Tau	1256	4 04 15.0	+22 03 43	fd	4.36	+0.95	+1.07	K0 III
47 λ Per	1261	4 06 01.4	+50 19 53	f	4.29	−0.04	−0.01	A0 IIIn
	1279	4 07 16.5	+15 08 35	svmd6	6.01		+0.40	F3 V
48 Per	1273	4 08 06.9	+47 41 35	fv	4.04	−0.55	−0.03	B3 Ve
43 Tau	1283	4 08 43.7	+19 35 23	f	5.50		+1.07	K1 III
	1270	4 08 49.3	+59 53 19	s	6.28	+0.92	+1.14	G8 IIa
44 Tau	1287	4 10 22.4	+26 27 43	fv	5.41	+0.06	+0.34	F2 IV−V
38 o^1 Eri	1298	4 11 29.9	− 6 51 24	fv	4.04	+0.13	+0.33	F3 V
α Hor	1326	4 13 45.2	−42 18 46	f	3.86	+1.00	+1.10	K2 III
α Ret	1336	4 14 19.6	−62 29 33	fd6	3.35	+0.63	+0.91	G8 II−III
51 μ Per	1303	4 14 20.7	+48 23 27	fvd67	4.14	+0.64	+0.95	G0 Ib
40 o^2 Eri	1325	4 14 55.6	− 7 39 51	vd	4.43	+0.45	+0.82	K0.5 V
49 μ Tau	1320	4 15 07.6	+ 8 52 26	fm6	4.29		−0.05	B3 IV
48 Tau	1319	4 15 20.7	+15 22 56	svmd	6.32		+0.40	F3 V
γ Dor	1338	4 15 49.8	−51 30 19	fv	4.25	+0.03	+0.30	F0 V
ε Ret	1355	4 16 21.1	−59 19 12	d	4.44	+1.07	+1.08	K2 IV
41 Eri	1347	4 17 36.6	−33 48 59	vd67	3.56	−0.37	−0.12	Ap Mn
54 γ Tau	1346	4 19 21.9	+15 36 36	fvm6	3.63		+0.99	$K0^-$ IIIab CN 1
57 Tau	1351	4 19 32.3	+14 01 03	svmd6	5.59		+0.28	F0 IV
54 Per	1343	4 19 55.3	+34 32 57	fd	4.93	+0.69	+0.94	G8 III CN 0.5
	1327	4 19 57.7	+65 07 23	s	5.27	+0.47	+0.81	G5 IIb

Name	B.S.	Right Ascension	Declination	Notes	V	U−B	B−V	Spectral Type
		h m s	° ′ ″					
	1367	4 20 19.4	−20 39 26	fm	5.38		−0.02	A1 V
η Ret	1395	4 21 48.4	−63 24 15	fm	5.23		+0.95	G8 III
61 δ Tau	1373	4 22 30.1	+17 31 31	fvmd6	3.76		+0.98	G9.5 III CN 0.5
63 Tau	1376	4 22 59.2	+16 45 37	csmd6	5.63		+0.30	A2 m F3 III
42 ξ Eri	1383	4 23 18.4	− 3 45 45	fv6	5.17	+0.08	+0.08	A2 V
43 Eri	1393	4 23 45.3	−34 02 02	fm	3.95		+1.49	K5 III
65 κ Tau	1387	4 24 55.3	+22 16 38	vmd6	4.22		+0.14	A7 V
68 Tau	1389	4 25 03.3	+17 54 41	vmd6	4.30		+0.05	A2 IV−Vs
69 υ Tau	1392	4 25 51.5	+22 47 49	vmd6	4.29		+0.26	A8 Vn
71 Tau	1394	4 25 55.1	+15 36 06	vmd6	4.49		+0.25	A8 Vn
77 θ^1 Tau	1411	4 28 08.7	+15 56 45	d6	3.85	+0.76	+0.96	G9 III CN−0.5
74 ε Tau	1409	4 28 10.7	+19 09 51	fd	3.54	+0.88	+1.02	G9.5 III CN 0.5
78 θ^2 Tau	1412	4 28 14.0	+15 51 17	svd6	3.42	+0.15	+0.18	A7 III
δ Cae	1443	4 30 36.3	−44 58 11	fv	5.07		−0.19	B2 IV−V
1 Cam	1417	4 31 26.0	+53 53 43	fd6	5.77	−0.73	+0.18	B0 IIIn
50 υ^1 Eri	1453	4 33 13.0	−29 46 53	v	4.51	+0.72	+0.98	K0$^+$ III CN−1
86 ρ Tau	1444	4 33 25.3	+14 49 45	fvm6	4.65		+0.24	A8 Vn
α Dor	1465	4 33 50.0	−55 03 37	fvd7	3.27	−0.35	−0.10	A0 IIIp Si
88 Tau	1458	4 35 14.5	+10 08 45	vd6	4.25	+0.11	+0.18	A5m:
52 υ^2 Eri	1464	4 35 15.5	−30 34 39	f	3.82	+0.72	+0.98	G8.5 IIIa
87 α Tau	1457	4 35 29.4	+16 29 41	fsvd6	0.85	+1.90	+1.54	K5$^+$ III
48 ν Eri	1463	4 35 56.6	− 3 22 03	fvd6	3.93	−0.89	−0.21	B2 III
58 Per	1454	4 36 10.1	+41 15 00	cd6	4.25	+0.82	+1.22	K0 II−III + B9 V
R Dor	1492	4 36 40.3	−62 05 31	svd	5.40	+0.86	+1.58	M8e III:
90 Tau	1473	4 37 44.3	+12 29 47	md6	4.27		+0.13	A6 Vn
53 Eri	1481	4 37 50.2	−14 19 06	fd67	3.87	+1.01	+1.09	K2 IIIb
54 Eri	1496	4 40 06.8	−19 41 08	vd	4.32	+1.81	+1.61	M3/4 III
α Cae	1502	4 40 19.2	−41 52 40	fvd	4.45	+0.01	+0.34	F1 V
β Cae	1503	4 41 47.5	−37 09 31	f	5.05	+0.04	+0.37	F1 V
94 τ Tau	1497	4 41 47.6	+22 56 35	fd67	4.28	−0.57	−0.13	B3 V
57 μ Eri	1520	4 45 07.6	− 3 16 05	f6	4.02	−0.60	−0.15	B4 IV
4 Cam	1511	4 47 22.6	+56 44 40	fmd	5.26		+0.25	A3m
	1533	4 49 24.3	+37 28 32	f	4.88	+1.70	+1.44	K3.5 III Ba 0.2:
1 π^3 Ori	1543	4 49 26.0	+ 6 56 55	fvd6	3.19	−0.01	+0.45	F6 V
2 π^2 Ori	1544	4 50 12.2	+ 8 53 16	6	4.36	0.00	+0.01	A0.5 IVn
3 π^4 Ori	1552	4 50 48.4	+ 5 35 34	fsv6	3.69	−0.81	−0.17	B2 III
97 Tau	1547	4 50 56.1	+18 49 39	fvd	5.13	+0.12	+0.21	A9 IIIn
4 o^1 Ori	1556	4 52 06.5	+14 14 19	fcv	4.74		+1.84	S3.5/1$^-$
61 ω Eri	1560	4 52 31.5	− 5 27 54	6	4.39	+0.16	+0.25	A9 IV
9 α Cam	1542	4 53 18.0	+66 19 51	f	4.29	−0.88	+0.03	O9.5 Ia
8 π^5 Ori	1567	4 53 51.6	+ 2 25 44	fv6	3.72	−0.83	−0.18	B2 III
η Men	1629	4 55 23.8	−74 56 55	f	5.47	+1.83	+1.52	K4 III
9 o^2 Ori	1580	4 55 56.9	+13 30 11	d	4.07	+1.11	+1.15	K2− III Fe−1
3 ι Aur	1577	4 56 30.3	+33 09 17	fv	2.69	+1.78	+1.53	K3 II
7 Cam	1568	4 56 41.0	+53 44 27	d67	4.47	−0.01	−0.02	A0m A1 III
10 π^6 Ori	1601	4 58 09.5	+ 1 42 11	v	4.47	+1.55	+1.40	K2$^-$ II
7 ε Aur	1605	5 01 25.8	+43 48 46	fvd6	2.99	+0.33	+0.54	F0 Iape
8 ζ Aur	1612	5 01 57.2	+41 03 56	fcv6	3.75	+0.38	+1.22	K5 II + B5 V
102 ι Tau	1620	5 02 38.8	+21 34 47	fmd	4.64		+0.15	A7 V
10 β Cam	1603	5 02 44.9	+60 25 55	fd	4.03	+0.63	+0.92	G1 Ib−IIa

Name			B.S.	Right Ascension	Declination	Notes	V	U−B	B−V	Spectral Type
				h m s	° ′ ″					
11		Ori	1638	5 04 08.4	+15 23 39	fv	4.68	−0.09	−0.06	A0p Si
	η^2	Pic	1663	5 04 46.3	−49 35 16	fv	5.03	+1.88	+1.49	K5 III
2	ε	Lep	1654	5 05 08.6	−22 22 51	fv	3.19	+1.78	+1.46	K4 III
	ζ	Dor	1674	5 05 22.9	−57 28 58	f	4.72	−0.04	+0.52	F7 V
10	η	Aur	1641	5 05 59.3	+41 13 30	fv	3.17	−0.67	−0.18	B3 V
67	β	Eri	1666	5 07 28.8	− 5 05 45	fvd	2.79	+0.10	+0.13	A3 IVn
69	λ	Eri	1679	5 08 47.2	− 8 45 48	fv	4.27	−0.90	−0.19	B2 IVn
16		Ori	1672	5 08 54.9	+ 9 49 13	fvmd6	5.43		+0.24	A2m
3	ι	Lep	1696	5 11 56.9	−11 52 40	d	4.45	−0.40	−0.10	B9 V:
5	μ	Lep	1702	5 12 35.7	−16 12 51	fsv	3.31	−0.39	−0.11	B9 IIIp Mn
4	κ	Lep	1705	5 12 53.1	−12 57 00	d7	4.36	−0.37	−0.10	B9 V:
17	ρ	Ori	1698	5 12 53.9	+ 2 51 09	vd67	4.46	+1.16	+1.19	K1 III CN 0.5
11	μ	Aur	1689	5 12 54.9	+38 28 34	f	4.86	+0.09	+0.18	A4m:
	θ	Dor	1744	5 13 45.7	−67 11 38	f	4.83	+1.39	+1.28	K2.5 IIIa
19	β	Ori	1713	5 14 10.6	− 8 12 36	fsvd6	0.12	−0.66	−0.03	B8 Ia
13	α	Aur	1708	5 16 08.1	+45 59 28	fcvd67	0.08	+0.44	+0.80	G6 III + G2 III
	ο	Col	1743	5 17 12.8	−34 54 08	f	4.83	+0.80	+1.00	K0/1 III/IV
20	τ	Ori	1735	5 17 14.5	− 6 51 08	fsd6	3.60	−0.47	−0.11	B5 III
15	λ	Aur	1729	5 18 36.8	+40 05 35	fd	4.71	+0.12	+0.63	G1.5 IV−V Fe−1
	ζ	Pic	1767	5 19 11.1	−50 36 50	f	5.45	+0.01	+0.51	F7 III−IV
6	λ	Lep	1756	5 19 13.8	−13 11 03	f	4.29	−1.03	−0.26	B0.5 IV
			1686	5 21 18.7	+79 13 26	fd	5.05		+0.47	F7 Vs
22		Ori	1765	5 21 22.8	− 0 23 22	f6	4.73	−0.79	−0.17	B2 IV−V
29		Ori	1784	5 23 35.1	− 7 48 52		4.14	+0.69	+0.96	G8 III Fe−0.5
28	η	Ori	1788	5 24 06.0	− 2 24 13	vd6	3.36	−0.92	−0.17	B0.5 Vnn
24	γ	Ori	1790	5 24 43.7	+ 6 20 36	fvd6	1.64	−0.87	−0.22	B2 III
112	β	Tau	1791	5 25 49.0	+28 36 06	fsd	1.65	−0.49	−0.13	B7 III
115		Tau	1808	5 26 43.8	+17 57 22	fd	5.42	−0.53	−0.10	B5 V
9	β	Lep	1829	5 27 55.4	−20 45 54	fvd	2.84	+0.46	+0.82	G5 II
			1856	5 29 57.1	−47 04 58	fd7	5.46	+0.21	+0.62	G3 IV
32		Ori	1839	5 30 23.0	+ 5 56 34	d7	4.20	−0.55	−0.14	B5 V
	ε	Col	1862	5 30 56.8	−35 28 33		3.87	+1.08	+1.14	K1 II/III
34	δ	Ori	1851	5 31 37.4	− 0 17 23	sd	6.85	−0.71	−0.16	B2 Vh
34	δ	Ori	1852	5 31 37.4	− 0 18 15	fvd6	2.23	−1.05	−0.22	O9.5 II
119		Tau	1845	5 31 46.4	+18 35 21	v	4.38	+2.21	+2.07	M2 Iab−Ib
	γ	Men	1953	5 32 10.6	−76 20 48	fd	5.19	+1.19	+1.13	K2 III
25	χ	Aur	1843	5 32 14.4	+32 11 13	fv6	4.76	−0.46	+0.34	B5 Iab
11	α	Lep	1865	5 32 23.9	−17 49 38	fsvd	2.58	+0.23	+0.21	F0 Ib
	β	Dor	1922	5 33 33.6	−62 29 41	fvm	3.40		+0.80	F9 Ib
37	ϕ^1	Ori	1876	5 34 24.5	+ 9 29 06	f6	4.41	−0.97	−0.16	B0.5 IV−V
39	λ	Ori	1879	5 34 43.5	+ 9 55 46	vd8	3.66	−1.03	−0.18	O8
			1890	5 34 59.5	− 4 29 53	svm6	6.55		−0.14	B2 Vh
			1891	5 35 00.2	− 4 25 47	smd	6.25		−0.16	B2.5 V
44	ι	Ori	1899	5 35 03.9	− 5 54 52	fsmd6	2.76		−0.23	O9 III
46	ε	Ori	1903	5 35 50.0	− 1 12 23	fsvd6	1.70	−1.04	−0.19	B0 Ia
40	ϕ^2	Ori	1907	5 36 29.6	+ 9 17 13	s	4.09	+0.64	+0.95	K0 IIIb Fe−2
123	ζ	Tau	1910	5 37 11.8	+21 08 18	fsvd6	3.00	−0.67	−0.19	B1 IV:((e)) (shell)
48	σ	Ori	1931	5 38 22.1	− 2 36 14	d6	3.81	−1.01	−0.24	O9.5 V
	α	Col	1956	5 39 22.6	−34 04 40	fvd	2.64	−0.46	−0.12	B7 IV
50	ζ	Ori	1948	5 40 22.8	− 1 56 45	vmd68	2.05	−1.07	−0.21	O9.5 Ib

Name			B.S.	Right Ascension	Declination	Notes	*V*	*U−B*	*B−V*	Spectral Type
				h m s	° ′ ″					
50	ζ	Ori	1949	5 40 22.8	− 1 56 48	vmd68	4.21	−1.07	−0.21	B0 III
13	γ	Lep	1983	5 44 09.0	−22 27 02	fd	3.60	0.00	+0.47	F7 V
	δ	Dor	2015	5 44 45.5	−65 44 18	f	4.35	+0.12	+0.21	A7 IV
27	ο	Aur	1971	5 45 19.1	+49 49 25	f	5.47	+0.07	+0.03	A0p Cr
14	ζ	Lep	1998	5 46 36.9	−14 49 28	f6	3.55	+0.07	+0.10	A2 Van
130		Tau	1990	5 46 59.9	+17 43 36	f	5.49	+0.27	+0.30	F0 III
	β	Pic	2020	5 47 06.4	−51 04 08		3.85	+0.10	+0.17	A5 V
53	κ	Ori	2004	5 47 24.0	− 9 40 19	fv	2.06	−1.03	−0.17	B0.5 Ia
	γ	Pic	2042	5 49 41.5	−56 10 06	f	4.51	+0.98	+1.10	K1 III
	β	Col	2040	5 50 41.7	−35 46 15	f	3.12	+1.21	+1.16	K1.5 III
			2049	5 50 43.0	−52 06 38	f	5.17	+0.72	+0.99	G8 III
32	ν	Aur	2012	5 50 58.2	+39 08 49	fd	3.97	+1.09	+1.13	K0 III CN 0.5
15	δ	Lep	2035	5 50 59.9	−20 52 46	f	3.81	+0.68	+0.99	K0 III Fe−1.5 CH 0.5
136		Tau	2034	5 52 51.3	+27 36 40	fvd6	4.58	+0.03	−0.02	A0 IV
54	χ^1	Ori	2047	5 53 56.3	+20 16 31	d6	4.41	+0.07	+0.59	G0$^-$ V Ca 0.5
30	ξ	Aur	2029	5 54 13.0	+55 42 21	f	4.99	+0.12	+0.05	A1 Va
58	α	Ori	2061	5 54 45.9	+ 7 24 22	fvd6	0.50	+2.06	+1.85	M1−M2 Ia−Iab
16	η	Lep	2085	5 56 03.8	−14 10 07	f	3.71	+0.01	+0.33	F0 IV
	γ	Col	2106	5 57 16.2	−35 17 02	fd	4.36	−0.66	−0.18	B2.5 IV
60		Ori	2103	5 58 26.4	+ 0 33 10	fd6	5.22	+0.01	+0.01	A1 Vs
33	δ	Aur	2077	5 58 54.6	+54 17 05	fd	3.72	+0.87	+1.00	K0$^-$ III
	η	Col	2120	5 58 55.0	−42 48 55	f	3.96	+1.08	+1.14	G8/K1 II
34	β	Aur	2088	5 58 58.7	+44 56 50	fvd6	1.90	+0.05	+0.03	A1 IV
37	θ	Aur	2095	5 59 12.6	+37 12 45	vd67	2.62	−0.18	−0.08	B9.5p Si
35	π	Aur	2091	5 59 22.7	+45 56 12	v	4.26	+1.83	+1.72	M3 II
61	μ	Ori	2124	6 01 58.2	+ 9 38 52	vmd6	4.12		+0.15	A2m
62	χ^2	Ori	2135	6 03 28.5	+20 08 20	svd	4.63	−0.68	+0.28	B2 Ia
1		Gem	2134	6 03 39.8	+23 15 51	fd67	4.16	+0.52	+0.82	G5 III−IV
17		Lep	2148	6 04 39.0	−16 29 01	sv6	4.93	+0.12	+0.24	A(shell)
67	ν	Ori	2159	6 07 08.6	+14 46 11	f6	4.42	−0.66	−0.17	B3 IV
			2180	6 08 38.9	−22 25 33	f	5.50		−0.01	A0 IV
	ν	Dor	2221	6 08 47.1	−68 50 31	f	5.06	−0.21	−0.08	B8 V
	δ	Pic	2212	6 10 09.1	−54 58 01	fv6	4.81	−1.03	−0.23	B0.5 IV
	α	Men	2261	6 10 27.9	−74 45 03	f	5.09	+0.33	+0.72	G5 V
70	ξ	Ori	2199	6 11 30.8	+14 12 39	d6	4.48	−0.65	−0.18	B3 IV
36		Cam	2165	6 12 05.8	+65 43 15	f6	5.32	+1.44	+1.34	K2 II−III
7	η	Gem	2216	6 14 25.5	+22 30 34	vd6	3.28	+1.66	+1.60	M3$^-$ IIIab
5	γ	Mon	2227	6 14 29.4	− 6 16 19	d	3.98	+1.41	+1.32	K1 III Ba 0.5
44	κ	Aur	2219	6 14 54.0	+29 30 05	fv	4.35	+0.80	+1.02	G9 IIIb
74		Ori	2241	6 16 01.3	+12 16 29	fd	5.04	−0.02	+0.42	F5 IV−V
	κ	Col	2256	6 16 17.1	−35 08 16	fv	4.37	+0.83	+1.00	K0.5 IIIa
			2209	6 18 01.3	+69 19 24	f6	4.80	0.00	+0.03	A0 IV+nn
2		Lyn	2238	6 18 57.7	+59 00 52	fv	4.48	+0.03	+0.01	A1 Va
7		Mon	2273	6 19 21.1	− 7 49 10	f6	5.27	−0.75	−0.19	B2.5 V
1	ζ	CMa	2282	6 20 01.5	−30 03 35	fd6	3.02	−0.72	−0.19	B2.5 V
	δ	Col	2296	6 21 50.3	−33 25 57	6	3.85	+0.52	+0.88	G7 II
2	β	CMa	2294	6 22 22.1	−17 57 07	fsvd6	1.98	−0.98	−0.23	B1 II−III
13	μ	Gem	2286	6 22 30.4	+22 31 05	fsvd	2.88	+1.85	+1.64	M3 IIIab
8		Mon	2298	6 23 22.2	+ 4 35 50	fmd68	4.44	+0.12	+0.20	A5 IV
	α	Car	2326	6 23 47.1	−52 41 29	f	−0.72	+0.10	+0.15	A9 II

Name			B.S.	Right Ascension	Declination	Notes	V	U–B	B–V	Spectral Type
				h m s	° ′ ″					
			2305	6 23 49.3	−11 31 33	f	5.22		+1.24	K3 III
46	ψ^1	Aur	2289	6 24 19.3	+49 17 33	fv6	4.91	+2.29	+1.97	K5–M0 Iab–Ib
10		Mon	2344	6 27 35.3	− 4 45 26	fmd	5.05		−0.18	B2 V
	λ	CMa	2361	6 27 53.5	−32 34 30		4.48	−0.61	−0.17	B4 V
18	ν	Gem	2343	6 28 31.1	+20 13 02	fd6	4.15	−0.48	−0.13	B6 III
4	ζ^1	CMa	2387	6 31 32.6	−23 24 45	vmd6	4.34		−0.25	B1 III
			2392	6 32 25.8	−11 09 38	sv	6.24	+0.78	+1.11	G9.5 III: Ba 3
13		Mon	2385	6 32 29.9	+ 7 20 20	f	4.50	−0.18	0.00	A0 Ib–II
			2395	6 33 15.1	− 1 12 51	f	5.10	−0.56	−0.14	B5 Vn
5	ξ^2	CMa	2414	6 34 44.5	−22 57 31	f	4.54	−0.03	−0.05	A0 III
			2435	6 34 48.6	−52 58 09		4.39	−0.15	−0.02	A0 II
7	ν^2	CMa	2429	6 36 21.4	−19 14 58	v	3.95	+1.01	+1.06	K1.5 III–IV Fe 1
24	γ	Gem	2421	6 37 16.7	+16 24 22	fd6	1.93	+0.04	0.00	A1 IVs
	ν	Pup	2451	6 37 31.9	−43 11 21	fv6	3.17	−0.41	−0.11	B8 III
8	ν^3	CMa	2443	6 37 33.6	−18 13 50		4.43	+1.04	+1.15	K0.5 II–III
15		Mon	2456	6 40 33.9	+ 9 54 11	svmd6	4.65		−0.25	O7 V
27	ε	Gem	2473	6 43 28.3	+25 08 20	fsvd6	2.98	+1.46	+1.40	G8 Ib
30		Gem	2478	6 43 33.9	+13 14 09	d	4.49	+1.16	+1.16	K0 III CN 1 Ca 1
9	α	CMa	2491	6 44 49.1	−16 42 20	fd6	−1.46	−0.06	0.00	A0m A1 Va
31	ξ	Gem	2484	6 44 52.1	+12 54 15	fv	3.36	+0.06	+0.43	F5 IV
			2401	6 44 57.9	+79 34 28	f6	5.45	−0.02	+0.50	F8 V
			2513	6 45 15.7	−52 11 35	sm	6.32			G6 Iab
56	ψ^5	Aur	2483	6 46 11.9	+43 35 08	fd	5.25	+0.05	+0.56	G0 V
57	ψ^6	Aur	2487	6 47 05.3	+48 47 53	f	5.22	+1.04	+1.12	K0 III
			2518	6 47 06.0	−37 55 16	fd	5.26	−0.25	−0.08	B8/9 V
18		Mon	2506	6 47 28.2	+ 2 25 15	f6	4.47	+1.04	+1.11	$K0^+$ IIIa
	α	Pic	2550	6 48 06.8	−61 56 00	f	3.27	+0.13	+0.21	A7 Vn
13	κ	CMa	2538	6 49 33.6	−32 29 58	fv	3.96	−0.92	−0.23	B1.5 IVne
			2554	6 49 41.6	−53 36 47	6	4.40	+0.61	+0.92	G3: III
	τ	Pup	2553	6 49 45.0	−50 36 20	f6	2.93	+1.21	+1.20	K1 III
			2534	6 50 20.7	− 8 01 55	sv	6.29	+0.02	0.00	A2: V:p Sr–Cr–Eu
	ι	Vol	2602	6 51 32.2	−70 57 15	f	5.40	−0.38	−0.11	B7 IV
34	θ	Gem	2540	6 52 17.7	+33 58 15	fd6	3.60	+0.14	+0.10	A3 III
43		Cam	2511	6 52 53.9	+68 53 53	f	5.12	−0.43	−0.13	B7 III
16	o^1	CMa	2580	6 53 49.2	−24 10 27	svm	3.86		+1.73	K2 Iab
14	θ	CMa	2574	6 53 50.5	−12 01 44	f	4.07	+1.70	+1.43	K4 III
			2591	6 54 12.5	−42 21 20	sv	6.32	+2.79	+2.24	C5,2.5
20	ι	CMa	2596	6 55 48.1	−17 02 38	vm	4.38		−0.07	B3 II
15		Lyn	2560	6 56 37.7	+58 25 59	d7	4.35	+0.52	+0.85	G5 III–IV
21	ε	CMa	2618	6 58 19.9	−28 57 42	fd	1.50	−0.93	−0.21	B2 II
			2527	6 58 58.9	+76 59 17	f6	4.55	+1.66	+1.36	K4 III
22	σ	CMa	2646	7 01 25.2	−27 55 26	fvmd	3.46		+1.73	K7 Ib
42	ω	Gem	2630	7 01 57.4	+24 13 36	fsv	5.18	+0.68	+0.94	G5 Ib–II
24	o^2	CMa	2653	7 02 42.7	−23 49 19	fsm6	3.03		−0.09	B3 Iab
23	γ	CMa	2657	7 03 25.1	−15 37 19	f	4.11	−0.49	−0.12	B8 II
43	ζ	Gem	2650	7 03 39.9	+20 34 54	fvd6	3.79		+0.79	F9 Ib
			2666	7 03 48.5	−42 19 33	fv6	5.20	+0.15	+0.20	A3m A3–F0
			2683	7 04 09.9	−56 44 18	fm	5.17		−0.04	Ap Si
25	δ	CMa	2693	7 08 05.2	−26 22 51	fsv6	1.86	+0.57	+0.65	F8 Ia
	γ^1	Vol	2735	7 08 46.2	−70 29 06	vmd68	5.67	+0.60	+0.91	F0/3

Name	B.S.	Right Ascension	Declination	Notes	*V*	*U–B*	*B–V*	Spectral Type
		h m s	° ′ ″					
γ^{2} Vol	2736	7 08 48.8	−70 29 13	fmd8	3.78	+0.60	+0.91	G9 III
20 Mon	2701	7 09 51.3	− 4 13 30	fd	4.92	+0.78	+1.03	K0 III
46 τ Gem	2697	7 10 39.8	+30 15 29	vd7	4.41	+1.41	+1.26	K2 III
63 Aur	2696	7 11 08.4	+39 20 00	f6	4.90	+1.74	+1.45	K4– III–IIIa
22 δ Mon	2714	7 11 28.9	− 0 28 48	fd	4.15	+0.02	−0.01	A1 III^{+}
48 Gem	2706	7 11 59.1	+24 08 29	s	5.85	+0.09	+0.36	F5 III–IV
	2740	7 12 20.8	−46 44 48	f	4.49	−0.01	+0.32	F0 IV
51 Gem	2717	7 12 56.4	+16 10 20	fvd	5.00	+1.82	+1.66	M4 IIIab
	2748	7 13 18.6	−44 37 38	vd	5.10		+1.56	M5 IIIe
27 CMa	2745	7 13 56.8	−26 20 21	vd6	4.66	−0.71	−0.19	B3 IIIe
28 ω CMa	2749	7 14 30.4	−26 45 34	v	3.85	−0.73	−0.17	B2 IV/Ve
δ Vol	2803	7 16 50.1	−67 56 37	f	3.98	+0.45	+0.79	F9 Ib
π Pup	2773	7 16 52.7	−37 05 02	fvd	2.70	+1.24	+1.62	K3 Ib
54 λ Gem	2763	7 17 39.7	+16 33 16	fvd67	3.58	+0.10	+0.11	A3 V
30 τ CMa	2782	7 18 23.8	−24 56 25	vmd6	4.39		−0.15	O9 Ib
55 δ Gem	2777	7 19 40.5	+21 59 48	fd67	3.53	+0.04	+0.34	F0 IV
66 Aur	2805	7 23 37.4	+40 41 14	f6	5.19	+1.24	+1.23	K1 IIIa CN 1
31 η CMa	2827	7 23 47.9	−29 17 17	fsmd	2.44		−0.07	B5 Ia
60 ι Gem	2821	7 25 15.7	+27 48 48	f	3.79	+0.85	+1.03	G9 IIIb
3 β CMi	2845	7 26 44.7	+ 8 18 17	fvd6	2.90	−0.28	−0.09	B8 V
4 γ CMi	2854	7 27 45.3	+ 8 56 29	d6	4.32	+1.54	+1.43	K3^{-} III Fe−1
62 ρ Gem	2852	7 28 37.8	+31 47 59	fd6	4.18	−0.03	+0.32	F1 V
σ Pup	2878	7 28 59.6	−43 17 10	fd6	3.25	+1.78	+1.51	K5 III
6 CMi	2864	7 29 22.8	+12 01 21	fv	4.54	+1.37	+1.28	K1 III
	2906	7 33 43.9	−22 16 47	f	4.45	+0.06	+0.51	F6 IV
66 α Gem	2891	7 34 07.2	+31 54 19	fvmd68	1.99	+0.01	+0.04	A1m A2 Va
66 α Gem	2890	7 34 07.4	+31 54 20	fmd68	2.85	+0.01	+0.04	A2m A5 V:
69 υ Gem	2905	7 35 27.7	+26 54 46	fvd	4.06	+1.94	+1.54	M0 III–IIIb
	2934	7 35 28.6	−52 31 01	fv6	4.94	+1.63	+1.40	K3 III
25 Mon	2927	7 36 54.3	− 4 05 38	fvd	5.13	+0.12	+0.44	F6 III
	2937	7 37 05.4	−34 57 05	fd7	4.53	−0.31	−0.09	B8 V
	2609	7 37 12.6	+87 02 16	fv	5.07	+1.97	+1.63	M2^{-} IIIab
	2948	7 38 30.9	−26 47 04	vmd8	4.50	−0.57	−0.17	B6 V
10 α CMi	2943	7 38 54.5	+ 5 14 40	fsvd67	0.38	+0.02	+0.42	F5 IV–V
R Pup	2974	7 40 35.3	−31 38 35	sv	6.65	+0.85	+1.20	G2 0–Ia
26 α Mon	2970	7 40 53.3	− 9 32 00	f	3.93	+0.88	+1.02	G9 III Fe−1
ζ Vol	3024	7 41 55.0	−72 35 17	fd7	3.95	+0.83	+1.04	G9 III
24 Lyn	2946	7 42 22.5	+58 43 43	fd	4.99	+0.08	+0.08	A2 IVn
75 σ Gem	2973	7 42 50.7	+28 54 08	vd6	4.28	+0.97	+1.12	K1 III
3 Pup	2996	7 43 30.4	−28 56 12	6	3.96	−0.09	+0.18	A2 Ib
77 κ Gem	2985	7 43 59.7	+24 24 59	fd7	3.57	+0.69	+0.93	G8 III
78 β Gem	2990	7 44 51.5	+28 02 41	fvd	1.14	+0.85	+1.00	K0 IIIb
	3017	7 44 59.2	−37 57 01	vm	3.59		+1.72	K5 IIa
4 Pup	3015	7 45 36.1	−14 32 43	f	5.04	+0.09	+0.33	A6n
81 Gem	3003	7 45 41.4	+18 31 43	fd6	4.88	+1.75	+1.45	K4 III
11 CMi	3008	7 45 51.5	+10 47 13	fv6	5.30	−0.02	+0.01	A0.5 IV–nn
	2999	7 46 09.3	+37 32 10	fvm	5.18		+1.58	M2^{+} IIIb
80 π Gem	3013	7 47 01.4	+33 26 04	fvd7	5.14	+1.95	+1.60	M1^{+} IIIa
	3037	7 47 17.9	−46 35 23	fm6	5.23		−0.14	B1.5 IV
o Pup	3034	7 47 46.4	−25 55 06	vd	4.50	−1.02	−0.05	B1 IV:nne

Name			B.S.	Right Ascension	Declination	Notes	V	U–B	B–V	Spectral Type
				h m s	° ′ ″					
7	ξ	Pup	3045	7 48 58.7	−24 50 26	fd6	3.34	+1.16	+1.24	G6 Iab–Ib
			3055	7 49 00.6	−46 21 15	d	4.11	−1.01	−0.18	B0 III
13	ζ	CMi	3059	7 51 18.7	+ 1 47 11	f	5.14	−0.49	−0.12	B8 II
			3080	7 51 57.6	−40 33 23	fc6	3.73	+0.78	+1.04	K1/2 II + A0
			3084	7 52 22.7	−38 50 36	vd	4.49	−0.69	−0.19	B2.5 V
83	φ	Gem	3067	7 53 02.3	+26 47 08	f6	4.97	+0.10	+0.09	A3 V
			3090	7 53 05.0	−48 05 00		4.24	−1.00	−0.14	B0.5 Ib
11		Pup	3102	7 56 32.2	−22 51 35		4.20	+0.42	+0.72	F8 II
	χ	Car	3117	7 56 35.3	−52 57 43	fv	3.47	−0.67	−0.18	B3 IVp
			3113	7 57 22.2	−30 18 51	fv	4.79	+0.18	+0.15	A7 III
	V	Pup	3129	7 58 01.4	−49 13 28	cvd6	4.41	−0.96	−0.17	B1 Vp + B2:
			3075	7 59 18.5	+73 56 20	f	5.41	+1.64	+1.42	K3 III
27		Mon	3122	7 59 21.7	− 3 39 32	f	4.93	+1.21	+1.21	K2 III
			3153	7 59 30.0	−60 33 59	sm	5.16		+1.72	M1.5 IIa
			3131	7 59 31.9	−18 22 42	fm	4.62		+0.08	A2 IVn
			3145	8 01 52.5	+ 2 21 20	d	4.39	+1.28	+1.25	K2 III
	χ	Gem	3149	8 03 03.5	+27 48 57	fd6	4.94	+1.09	+1.12	K2 III
	ζ	Pup	3165	8 03 19.2	−39 58 55	fs	2.25	−1.11	−0.26	O5 If
15	ρ	Pup	3185	8 07 13.5	−24 16 57	fvd6	2.81	+0.19	+0.43	F5 IIp
27		Lyn	3173	8 07 53.7	+51 31 44	fd	4.84	0.00	+0.05	A1 Va
	ε	Vol	3223	8 07 54.6	−68 35 43	d67	4.35	−0.46	−0.11	B6 IV
29	ζ	Mon	3188	8 08 13.0	− 2 57 42	d	4.38	+0.69	+0.97	G2 Ib
16		Pup	3192	8 08 41.5	−19 13 22	6	4.40	−0.60	−0.15	B5 IV
	γ^1	Vel	3206	8 09 15.4	−47 19 24	vd6	4.27	−0.92	−0.23	B1 IV
	γ^2	Vel	3207	8 09 18.1	−47 18 52	fcvmd68	1.82	−0.99	−0.22	WC8
			3225	8 11 05.4	−39 35 46	v6	4.45	+1.86	+1.62	K3 II
			3182	8 12 04.4	+68 29 49	f	5.32	+0.81	+1.04	G7 II
20		Pup	3229	8 12 59.3	−15 45 55	f	4.99		+1.07	G5 II
			3243	8 13 46.9	−40 19 29	d6	4.44	+1.09	+1.17	K1 II–III
17	β	Cnc	3249	8 16 06.5	+ 9 12 32	fvd	3.52	+1.77	+1.48	K4 III Ba 0.5
			3270	8 18 16.5	−36 38 09	f	4.45	+0.11	+0.22	A7 III
	α	Cha	3318	8 18 43.9	−76 53 46		4.07	−0.02	+0.39	F4 IV
18	χ	Cnc	3262	8 19 36.6	+27 14 33	f	5.14	−0.06	+0.47	F6 V
	θ	Cha	3340	8 20 52.6	−77 27 38	fd	4.35	+1.20	+1.16	K2 III CN 0.5
			3282	8 21 05.3	−33 01 49	f	4.83	+1.60	+1.45	$K3^-$ II
31		Lyn	3275	8 22 19.4	+43 12 46	fv	4.25	+1.90	+1.55	K4.5 III
	ε	Car	3307	8 22 21.6	−59 29 07	fcv	1.86	+0.19	+1.28	K3: III + B2: V
			3315	8 24 44.3	−24 01 18	fmd6	5.28		+1.48	K4.5 III CN 1
			3314	8 25 17.2	− 3 52 54	f	3.90	−0.02	−0.02	A0 Va
	β	Vol	3347	8 25 39.4	−66 06 43	fv	3.77	+1.14	+1.13	K2 III
1	o	UMa	3323	8 29 38.8	+60 44 38	fsvd	3.36	+0.52	+0.84	G5 IIIa
33	η	Cnc	3366	8 32 16.5	+20 28 01	f	5.33	+1.39	+1.25	K3 III
4	δ	Hya	3410	8 37 15.6	+ 5 43 49	fd6	4.16	+0.01	0.00	A1 IVnn
			3426	8 37 22.8	−42 57 46	f	4.14	+0.16	+0.11	A7 II
5	σ	Hya	3418	8 38 21.9	+ 3 22 05	f	4.44	+1.28	+1.21	$K1^+$ III
6		Hya	3431	8 39 40.1	−12 26 55	f	4.98		+1.42	K4 III
	β	Pyx	3438	8 39 48.6	−35 16 53	d6	3.97	+0.65	+0.94	G4 III
	o	Vel	3447	8 40 04.7	−52 53 43	fvm6	3.62		−0.18	B3 IV
			3445	8 40 22.6	−46 37 19	fvd	3.84	+0.30	+0.71	F3 Ia
			3457	8 40 27.2	−59 44 03	vd6	4.33	−0.80	−0.11	B1.5 III

Name			B.S.	Right Ascension	Declination	Notes	V	U–B	B–V	Spectral Type
				h m s	° ′ ″					
34		Lyn	3422	8 40 30.1	+45 51 39	f	5.37	+0.75	+0.99	G8 IV
	η	Cha	3502	8 41 35.6	−78 56 11	f	5.47	−0.35	−0.10	B8 V
7	η	Hya	3454	8 42 50.0	+ 3 25 33	v6	4.30	−0.74	−0.20	B4 V
43	γ	Cnc	3449	8 42 51.2	+21 29 45	fmd6	4.67		+0.01	A1 Va
	α	Pyx	3468	8 43 17.4	−33 09 33	fv	3.68	−0.88	−0.18	B1.5 III
			3477	8 44 07.9	−42 37 19	md	4.05		+0.87	G8 IIIab
47	δ	Cnc	3461	8 44 15.6	+18 10 56	fd	3.94	+0.99	+1.08	K0 IIIb
	δ	Vel	3485	8 44 29.8	−54 40 50	d7	1.96	+0.07	+0.04	A1 Va
			3487	8 45 46.4	−46 00 50		3.91	−0.05	0.00	A1 II
12		Hya	3484	8 46 01.3	−13 31 12	d6	4.32	+0.62	+0.90	G8 III Fe−1
48	ι	Cnc	3475	8 46 14.7	+28 47 16	fvd	4.02	+0.78	+1.01	G8 II–III
11	ε	Hya	3482	8 46 22.8	+ 6 26 48	cvd67	3.38	+0.36	+0.68	K III: + F0 V
			3498	8 46 31.1	−56 44 31	v	4.49	−0.73	−0.17	B3 Vne
13	ρ	Hya	3492	8 48 02.2	+ 5 51 57	d6	4.36	−0.04	−0.04	A0 Vn
14		Hya	3500	8 48 59.1	− 3 24 54	fv	5.31	−0.35	−0.09	B9p Hg Mn
	γ	Pyx	3518	8 50 12.8	−27 40 55	f	4.01	+1.40	+1.27	K3⁻ III
			3571	8 54 52.7	−60 36 57	fd	3.84	−0.45	−0.10	B7 II/III
16	ζ	Hya	3547	8 54 59.9	+ 5 58 28	f	3.11	+0.80	+1.00	G9 IIIa
			3582	8 56 47.4	−59 12 01	fvmd8	5.08	−0.77	−0.19	B2 IV–V
	ζ	Oct	3678	8 57 53.0	−85 38 03	f	5.42	+0.07	+0.31	F0 III
65	α	Cnc	3572	8 58 04.6	+11 53 14	fvd6	4.25	+0.15	+0.14	A3m
9	ι	UMa	3569	8 58 41.8	+48 04 18	fvd6	3.14	+0.07	+0.19	A7 IV
64	σ^3	Cnc	3575	8 59 05.1	+32 26 53	fd	5.20		+0.93	G8 III
			3591	8 59 48.6	−41 13 28	fcv6	4.45	+0.38	+0.65	G8/K1 III + A
			3579	9 00 09.3	+41 48 46	fd67	3.97	+0.05	+0.44	F5 V
8	ρ	UMa	3576	9 01 52.6	+67 39 34	fv	4.76	+1.88	+1.53	M3 IIIb Ca 1
	α	Vol	3615	9 02 19.8	−66 21 58	f6	4.00	+0.13	+0.14	A3m A7
12	κ	UMa	3594	9 03 07.0	+47 11 12	fvd7	3.60	+0.01	0.00	A0 IIIn
			3614	9 03 53.8	−47 04 04	f	3.75	+1.22	+1.20	K2 III
			3643	9 05 08.3	−72 34 21		4.48	+0.22	+0.61	F8 II
			3612	9 06 03.3	+38 28 57	f	4.56	+0.82	+1.04	G7 Ib–II
76	κ	Cnc	3623	9 07 20.5	+10 41 55	fvd6	5.24	−0.43	−0.11	B8 IIIp Hg Mn
	λ	Vel	3634	9 07 43.2	−43 24 08	fvd	2.21	+1.81	+1.66	K4.5 Ib
15		UMa	3619	9 08 20.7	+51 38 07		4.48	+0.12	+0.27	A1m
77	ξ	Cnc	3627	9 08 55.7	+22 04 34	fd6	5.14	+0.80	+0.97	G9 IIIa Fe−0.5 CH−1
13	σ^2	UMa	3616	9 09 44.1	+67 09 56	vd7	4.80	+0.02	+0.49	F7 IV–V
	a	Car	3659	9 10 46.2	−58 56 10	v6	3.44	−0.70	−0.19	B2 IV–V
			3663	9 11 06.5	−62 17 10		3.97	−0.67	−0.18	B3 III
	β	Car	3685	9 13 07.2	−69 41 11	f	1.68	+0.03	0.00	A1 III
36		Lyn	3652	9 13 18.9	+43 14 57	f	5.32	−0.45	−0.14	B8 IIIp Mn
22	θ	Hya	3665	9 13 58.5	+ 2 20 46	fvd6	3.88	−0.12	−0.06	B9.5 IV (C II)
			3696	9 15 59.5	−57 30 35	v	4.34	+1.98	+1.63	K7 III
	ι	Car	3699	9 16 53.4	−59 14 37	fv	2.25	+0.16	+0.18	A9 Ib
38		Lyn	3690	9 18 22.8	+36 50 05	d67	3.82	+0.06	+0.06	A2 IV⁻
40	α	Lyn	3705	9 20 36.0	+34 25 29	fv	3.13	+1.94	+1.55	K7 IIIab
	θ	Pyx	3718	9 21 09.7	−25 56 00	f	4.72	+2.02	+1.63	M0.5 III
	κ	Vel	3734	9 21 52.9	−54 58 43	f6	2.50	−0.75	−0.18	B2 IV–V
1	κ	Leo	3731	9 24 13.1	+26 12 54	fvd7	4.46	+1.31	+1.23	K2 III
30	α	Hya	3748	9 27 13.1	− 8 37 33	fvd	1.98	+1.72	+1.44	K3 IIIa
	ε	Ant	3765	9 28 56.1	−35 55 06	f6	4.51	+1.68	+1.44	K3 III

Name	B.S.	Right Ascension	Declination	Notes	*V*	*U–B*	*B–V*	Spectral Type
		h m s	° ′ ″					
ψ Vel	3786	9 30 24.2	–40 26 01	vd7	3.60	–0.03	+0.36	F2 IV
23 UMa	3757	9 30 56.6	+63 05 42	fvd	3.67	+0.10	+0.33	F0 III–IV
N Vel	3803	9 30 59.6	–57 00 04	fv	3.13	+1.89	+1.55	K5 III
4 λ Leo	3773	9 31 17.6	+23 00 05	v	4.31	+1.89	+1.54	K4.5 IIIb
5 ξ Leo	3782	9 31 32.5	+11 20 00	fvd	4.97	+0.86	+1.05	G9.5 III
	3821	9 31 33.1	–73 02 52	f	5.47	+1.75	+1.56	K4 III
R Car	3816	9 32 03.4	–62 45 20	vmd	4.00			M5e
25 θ UMa	3775	9 32 21.5	+51 42 43	fdv6	3.17	+0.02	+0.46	F6 IV
	3808	9 32 51.7	–21 04 56	f	5.01		+1.02	K0 III
10 LMi	3800	9 33 45.9	+36 25 52	f	4.55	+0.62	+0.92	G7.5 III Fe–0.5
24 UMa	3771	9 33 49.7	+69 51 50	fv	4.56	+0.34	+0.77	G5 III–IV
	3825	9 34 13.6	–59 11 45		4.08	–0.56	+0.01	B5 II
26 UMa	3799	9 34 18.8	+52 05 07		4.50	+0.04	+0.01	A1 Va
	3751	9 36 03.7	+81 21 37	f	4.29	+1.72	+1.48	K3 IIIa
	3836	9 36 33.6	–49 19 17	d	4.35	+0.13	+0.17	A5 V
	3834	9 38 03.8	+ 4 41 00	f	4.68	+1.46	+1.32	K3 III
35 ι Hya	3845	9 39 28.4	– 1 06 31	fv	3.91	+1.46	+1.32	K2.5 III
38 κ Hya	3849	9 39 56.8	–14 17 53	fv	5.06	–0.57	–0.15	B5 V
14 ο Leo	3852	9 40 45.1	+ 9 55 36	fcd6	3.52	+0.21	+0.49	F6 II + A1 V
16 ψ Leo	3866	9 43 19.4	+14 03 23	fvd	5.35		+1.63	M2$^+$ IIIab
θ Ant	3871	9 43 52.0	–27 44 06	fcd7	4.79	+0.35	+0.51	F7 II–III + A8 V
l Car	3884	9 45 02.4	–62 28 24	fvm	3.40		+1.20	G8 Ia (CNII)
17 ε Leo	3873	9 45 25.6	+23 48 32	fv	2.98	+0.47	+0.80	G1 II
υ Car	3890	9 46 54.9	–65 02 13	md8	3.15	+0.13	+0.27	A8 Ib
R Leo	3882	9 47 09.3	+11 27 50	vm	4.40			M7e
	3881	9 48 06.5	+46 03 22	f	5.09	+0.10	+0.62	G0.5 Va
29 υ UMa	3888	9 50 27.7	+59 04 28	fvd	3.80	+0.10	+0.29	F0 IV
39 υ^1 Hya	3903	9 51 07.0	–14 48 40		4.12	+0.65	+0.92	G8.5 IIIa
24 μ Leo	3905	9 52 20.3	+26 02 33	fs	3.88	+1.39	+1.22	K2 III CN 1 Ca 1
	3923	9 54 31.0	–18 58 25	f6	4.94	+1.93	+1.57	K5 III
φ Vel	3940	9 56 35.9	–54 31 55	fd	3.54	–0.62	–0.08	B5 Ib
19 LMi	3928	9 57 13.6	+41 05 30	f6	5.14	0.00	+0.46	F5 V
η Ant	3947	9 58 32.9	–35 51 18	fd	5.23	+0.08	+0.31	F1 III–IV
29 π Leo	3950	9 59 49.1	+ 8 04 49	fv	4.70	+1.93	+1.60	M2$^-$ IIIab
20 LMi	3951	10 00 34.8	+31 57 39	fd	5.36	+0.27	+0.66	G3 Va H 1
40 υ^2 Hya	3970	10 04 45.5	–13 01 41	fv6	4.60	–0.27	–0.09	B9 III–IV
30 η Leo	3975	10 06 55.5	+16 47 58	fsvd	3.52	–0.21	–0.03	A0 Ib
21 LMi	3974	10 06 59.3	+35 16 54	v	4.48	+0.08	+0.18	A7 V
31 Leo	3980	10 07 30.4	+10 02 05	d	4.37	+1.75	+1.45	K4 III Ca 0.5
15 α Sex	3981	10 07 33.3	– 0 20 05		4.49	–0.07	–0.04	A0 III
32 α Leo	3982	10 07 58.4	+12 00 15	fvd6	1.35	–0.36	–0.11	B7 Vn
41 λ Hya	3994	10 10 13.3	–12 19 01	fd6	3.61	+0.92	+1.01	K0 III CN 0.5
ω Car	4037	10 13 33.5	–70 00 02	f	3.32	–0.33	–0.08	B8 III
	4023	10 14 25.2	–42 05 05	f6	3.85	+0.06	+0.05	A2 Va
36 ζ Leo	4031	10 16 16.4	+23 27 18	fsvd6	3.44	+0.20	+0.31	F0 III
33 λ UMa	4033	10 16 38.8	+42 57 07	fsm	3.45		+0.03	A1 IV
	4050	10 16 49.9	–61 17 41	fvd	3.40	+1.72	+1.54	K2.5 II
22 ε Sex	4042	10 17 15.4	– 8 01 53	f	5.24	+0.13	+0.31	F4 V:
	4049	10 17 46.9	–28 57 16	fm	5.34		+0.24	A0 Ib–IIp
41 γ^1 Leo	4057	10 19 33.6	+19 52 47	vmd68	2.61	+1.00	+1.15	K1$^-$ IIIb Fe–0.5

Name	B.S.	Right Ascension	Declination	Notes	V	U–B	B–V	Spectral Type
		h m s	° ′ ″					
41 γ^2 Leo	4058	10 19 33.9	+19 52 43	md8	3.80	+1.00	+1.15	G7 III Fe–1
	4074	10 20 38.0	–56 00 19	d7	4.50	–0.58	–0.12	B3 III
34 μ UMa	4069	10 21 53.1	+41 32 15	fv6	3.05	+1.89	+1.59	M0 III
	4080	10 22 00.2	–41 36 44	fv	4.83	+1.08	+1.12	K1 III
	4086	10 23 09.5	–37 58 18	f	5.33		+0.25	A8 V
	4072	10 23 35.8	+65 36 16	fv6	4.97	–0.13	–0.06	A0p Hg
	4102	10 24 14.8	–73 59 36	fv6	4.00	–0.01	+0.35	F3 V
42 μ Hya	4094	10 25 43.6	–16 47 52	fm	3.81		+1.48	K4$^+$ III
α Ant	4104	10 26 48.5	–31 01 46	fv6	4.25	+1.63	+1.45	K4.5 III
31 β LMi	4100	10 27 27.1	+36 44 45	fd67	4.21	+0.64	+0.90	G9 IIIab
	4114	10 27 36.1	–58 42 04	fv	3.82	+0.24	+0.31	F0 Ib
29 δ Sex	4116	10 29 05.9	– 2 42 02	f	5.21	–0.12	–0.06	B9.5 V
36 UMa	4112	10 30 09.0	+56 01 09	f	4.84	–0.01	+0.52	F8 V
	4084	10 30 12.6	+82 35 50	fv	5.26	–0.05	+0.37	F4 V
	4140	10 31 45.4	–61 38 48	fv	3.32	–0.72	–0.09	B4 Vne
47 ρ Leo	4133	10 32 25.0	+ 9 20 43	fvd6	3.85	–0.96	–0.14	B1 Iab
	4143	10 32 37.8	–46 57 53	fd7	5.02	+0.59	+1.04	K1/2 III
44 Hya	4145	10 33 39.5	–23 42 23	fd	5.08		+1.60	K5 III
	4126	10 34 28.2	+75 45 07	f	4.84		+0.96	G8.5 III
37 UMa	4141	10 34 41.0	+57 07 17	f	5.16	–0.02	+0.34	F2 V
	4159	10 35 17.9	–57 31 07	v6	4.45	+1.79	+1.62	K5 II
γ Cha	4174	10 35 23.0	–78 34 08	fv	4.11	+1.95	+1.58	M0 III
	4167	10 36 59.1	–48 11 12	d67	3.84	+0.07	+0.30	A3m F0–F2
37 LMi	4166	10 38 18.0	+32 00 55	f	4.71	+0.54	+0.81	G2.5 IIa
	4180	10 39 00.4	–55 33 51	fd	4.28	+0.75	+1.04	G2 II
	4181	10 42 32.3	+69 06 56	f	5.00		+1.38	K3 III
θ Car	4199	10 42 41.2	–64 21 18	fm6	2.76		–0.23	B0.5 Vp
41 LMi	4192	10 43 00.5	+23 13 40	f	5.08	+0.05	+0.04	A2 IV
	4191	10 43 06.6	+46 14 36	fd6	5.18	+0.01	+0.33	F5 III
η Car	4210	10 44 46.1	–59 38 41	vmd	6.22		+0.62	pec.
42 LMi	4203	10 45 27.0	+30 43 19	fd6	5.24	–0.14	–0.06	A1 Vn
δ^2 Cha	4234	10 45 43.0	–80 30 02	f	4.45	–0.70	–0.19	B2.5 IV
51 Leo	4208	10 46 00.3	+18 55 52	f	5.49		+1.12	gK3
μ Vel	4216	10 46 26.8	–49 22 49	d67	2.69	+0.57	+0.90	G5 IIIa
53 Leo	4227	10 48 51.8	+10 35 06	f6	5.25	+0.05	+0.01	A2 V
ν Hya	4232	10 49 15.3	–16 09 16	f	3.11	+1.30	+1.25	K2 III
46 LMi	4247	10 52 53.6	+34 15 20	fv	3.83	+0.91	+1.04	K0$^+$ III–IV
	4257	10 53 11.3	–58 48 48	vd6	3.78	+0.65	+0.95	K0 IIIb
54 Leo	4259	10 55 12.5	+24 47 24	md8	4.51	+0.01	+0.02	A1 IIIn + A1 IVn
ι Ant	4273	10 56 22.0	–37 05 51	f	4.60	+0.84	+1.03	K0 III
47 UMa	4277	10 59 02.9	+40 28 14	f	5.05	+0.13	+0.61	G0 V
7 α Crt	4287	10 59 24.5	–18 15 32	f	4.08	+1.00	+1.09	K0$^+$ III
	4293	10 59 48.6	–42 11 08	f	4.39	+0.12	+0.11	A3 IV
58 Leo	4291	11 00 10.4	+ 3 39 28	fd	4.84	+1.12	+1.16	K1 IIIp Ba
48 β UMa	4295	11 01 23.5	+56 25 22	fv6	2.37	+0.01	–0.02	A0m A1 IV–V
60 Leo	4300	11 01 55.8	+20 13 12		4.42	+0.05	+0.05	A0.5m A3 V
50 α UMa	4301	11 03 16.2	+61 47 29	fvd6	1.79	+0.92	+1.07	K0$^-$ IIIa
63 χ Leo	4310	11 04 37.8	+ 7 22 36	fvd7	4.63	+0.08	+0.33	Fp
χ^1 Hya	4314	11 04 58.2	–27 15 11	fd7	4.94	+0.04	+0.36	F3 IV
	4337	11 08 16.0	–58 56 04	fcv6	3.91	+0.94	+1.23	G4 0–Ia

Name	B.S.	Right Ascension	Declination	Notes	V	U–B	B–V	Spectral Type
		h m s	° ′ ″					
52 ψ UMa	4335	11 09 14.6	+44 32 21	f	3.01	+1.11	+1.14	K1 III
11 β Crt	4343	11 11 17.3	–22 47 05	f6	4.48	+0.06	+0.03	A2 IV
	4350	11 12 12.5	–49 03 37	fd6	5.36		+0.18	A3 IV/V
68 δ Leo	4357	11 13 42.6	+20 33 54	fvd	2.56	+0.12	+0.12	A4 IV
70 θ Leo	4359	11 13 50.8	+15 28 14	fv	3.34	+0.06	–0.01	A2 IV
74 φ Leo	4368	11 16 16.8	– 3 36 38	fd	4.47	+0.14	+0.21	A7 IVn
	4369	11 16 35.3	– 7 05 39	svd67	6.14	+0.15	+0.20	A7: III:p Sr–Cr
53 ξ UMa	4375	11 17 47.0	+31 34 17	cvmd68	4.41	+0.04	+0.59	F8.5 V
54 ν UMa	4377	11 18 04.5	+33 08 07	fd6	3.48	+1.55	+1.40	K3⁻ III
55 UMa	4380	11 18 43.5	+38 13 36	f6	4.78	+0.03	+0.12	A1 Va
12 δ Crt	4382	11 18 57.9	–14 44 16	f6	3.56	+0.97	+1.12	G8 III–IV
π Cen	4390	11 20 39.8	–54 27 00	fd7	3.89	–0.59	–0.15	B5 Vn
77 σ Leo	4386	11 20 45.0	+ 6 04 14	f6	4.05	–0.12	–0.06	A0 III⁺
78 ι Leo	4399	11 23 32.1	+10 34 14	vd67	3.94	+0.07	+0.41	F2 IV
15 γ Crt	4405	11 24 30.4	–17 38 34	fd	4.08	+0.11	+0.21	A7 IV–V
84 τ Leo	4418	11 27 33.1	+ 2 53 51	fd	4.95	+0.79	+1.00	G7.5 IIIa
1 λ Dra	4434	11 30 57.9	+69 22 21	fv	3.84	+1.97	+1.62	M0 III Ca–1
ξ Hya	4450	11 32 37.9	–31 48 58	fd	3.54	+0.71	+0.94	G7 III
λ Cen	4467	11 35 25.9	–62 58 42	fd	3.13	–0.17	–0.04	B9.5 IIn
	4466	11 35 33.7	–47 36 00	f	5.25	+0.12	+0.25	A7m
21 θ Crt	4468	11 36 18.1	– 9 45 39	f6	4.70	–0.18	–0.08	B9.5 Vn
91 υ Leo	4471	11 36 33.9	– 0 46 56	fd	4.30	+0.75	+1.00	G8⁺ IIIb
ο Hya	4494	11 39 50.4	–34 42 11	f	4.70	–0.22	–0.07	B9 V
61 UMa	4496	11 40 39.4	+34 14 38	fsvd	5.33	+0.25	+0.72	G8 V
3 Dra	4504	11 42 03.6	+66 47 11	f	5.30		+1.28	K3 III
	4511	11 43 09.6	–62 26 52	sv	5.05	+0.35	+0.80	G0 0–Ia Fe 1
27 ζ Crt	4514	11 44 22.9	–18 18 33	f	4.73	+0.74	+0.97	G8 IIIa
λ Mus	4520	11 45 14.9	–66 41 14	fd	3.64	+0.15	+0.16	A7 V
3 ν Vir	4517	11 45 28.4	+ 6 34 17	fv	4.03	+1.79	+1.51	M1 III
63 χ UMa	4518	11 45 39.4	+47 49 16	fv	3.71	+1.16	+1.18	K0.5 IIIb
	4522	11 46 08.9	–61 08 12	fvd	4.11	+0.58	+0.90	G3 II
93 Leo	4527	11 47 36.0	+20 15 38	fcvd6	4.53	+0.28	+0.55	G4 III–IV + A7 V
	4532	11 48 22.3	–26 42 29	fv	5.11	+1.67	+1.60	M4⁺ III
94 β Leo	4534	11 48 40.7	+14 36 50	fvd	2.14	+0.07	+0.09	A3 Va
	4537	11 49 19.0	–63 44 48	v	4.32	–0.59	–0.15	B3 V
5 β Vir	4540	11 50 18.3	+ 1 48 25	fd	3.61	+0.11	+0.55	F9 V
	4546	11 50 46.1	–45 07 54	f	4.46	+1.46	+1.30	K3 III
β Hya	4552	11 52 31.7	–33 51 58	vd7	4.28	–0.33	–0.10	Ap Si
64 γ UMa	4554	11 53 26.3	+53 44 11	fv6	2.44	+0.02	0.00	A0 Van
95 Leo	4564	11 55 17.4	+15 41 19	fd6	5.53	+0.12	+0.11	A3 V
30 η Crt	4567	11 55 38.0	–17 06 33	fv	5.18	0.00	–0.02	A0 Va
8 π Vir	4589	12 00 29.3	+ 6 39 22	fd6	4.66	+0.11	+0.13	A5 V
θ^1 Cru	4599	12 02 38.5	–63 16 16	d6	4.33	+0.04	+0.27	A3m A8–A8
	4600	12 03 16.2	–42 23 32	f	5.15	–0.03	+0.41	F6 V
9 ο Vir	4608	12 04 49.6	+ 8 46 28	fs	4.12	+0.63	+0.98	G8 IIIa CN–1 Ba 1 CH 1
η Cru	4616	12 06 29.2	–64 34 19	d6	4.15	+0.03	+0.34	F2 III
	4618	12 07 41.8	–50 37 10	d	4.47	–0.67	–0.15	B2 IIIne
δ Cen	4621	12 07 58.0	–50 40 50	fvd	2.60	–0.90	–0.12	B2 IVne
1 α Crv	4623	12 08 01.5	–24 41 13		4.02	–0.02	+0.32	F2 V
2 ε Crv	4630	12 09 44.3	–22 34 41	fv	3.00	+1.47	+1.33	K2.5 IIIa Ba 0.2:

Name			B.S.	Right Ascension	Declination	Notes	V	U−B	B−V	Spectral Type
				h m s	° ′ ″					
	ρ	Cen	4638	12 11 15.5	−52 19 37		3.96	−0.62	−0.15	B3 V
			4646	12 11 51.2	+77 39 28	f6	5.14		+0.33	A5m
	δ	Cru	4656	12 14 44.6	−58 42 26	fv	2.80	−0.91	−0.23	B2 IV
69	δ	UMa	4660	12 15 03.4	+57 04 27	fvd	3.31	+0.07	+0.08	A2 Van
4	γ	Crv	4662	12 15 25.2	−17 30 01	fv6	2.59	−0.34	−0.11	B8 IIIp Hg Mn
	ε	Mus	4671	12 17 09.6	−67 55 09	v6	4.11	+1.55	+1.58	M5 III
	β	Cha	4674	12 17 53.6	−79 16 14	fv	4.26	−0.51	−0.12	B5 Vn
	ζ	Cru	4679	12 18 01.6	−63 57 41	d	4.04	−0.69	−0.17	B2.5 V
3		CVn	4690	12 19 26.7	+49 01 33	f	5.29	+1.97	+1.66	M1^{+} IIIab
15	η	Vir	4689	12 19 31.3	− 0 37 31	fvd6	3.89	+0.06	+0.02	A1 IV^{+}
16		Vir	4695	12 19 58.1	+ 3 21 15	fvd	4.96	+1.15	+1.16	K0.5 IIIb Fe−0.5
	ε	Cru	4700	12 20 57.1	−60 21 35	v	3.59	+1.63	+1.42	K3/4 III
12		Com	4707	12 22 07.7	+25 53 16	fcvd6	4.79	+0.26	+0.49	G5 III + A5
6		CVn	4728	12 25 28.8	+39 03 37	f	5.02	+0.73	+0.96	G9 III
	α^1	Cru	4730	12 26 10.6	−63 03 27	fcmd68	1.58	−0.96	−0.26	B0.5 IV
	α^2	Cru	4731	12 26 11.3	−63 03 29	cmd8	2.09	−0.96	−0.26	B1 V
15	γ	Com	4737	12 26 33.9	+28 18 36	m	4.35		+1.13	K1 III Fe 0.5
	σ	Cen	4743	12 27 37.9	−50 11 21	f	3.91	−0.78	−0.19	B2 V
			4748	12 27 58.4	−38 59 59	fm	5.44		−0.08	B8/9 V
7	δ	Crv	4757	12 29 28.5	−16 28 25	fvd7	2.95	−0.09	−0.05	B9.5 IV− n
74		UMa	4760	12 29 36.4	+58 26 49	fm	5.32		+0.20	δ Del
	γ	Cru	4763	12 30 44.8	−57 04 17	fvd	1.63	+1.78	+1.59	M3.5 III
8	η	Crv	4775	12 31 41.0	−16 09 17	v6	4.31	+0.01	+0.38	F0 IV
	γ	Mus	4773	12 32 00.6	−72 05 30	f	3.87	−0.62	−0.15	B5 V
5	κ	Dra	4787	12 33 09.9	+69 49 46	fv6	3.87	−0.57	−0.13	B6 IIIpe
			4783	12 33 16.8	+33 17 20	f	5.42	+0.83	+1.00	K0 III CN−1
8	β	CVn	4785	12 33 23.2	+41 23 53	fsv6	4.26	+0.05	+0.59	G0 V
9	β	Crv	4786	12 33 59.5	−23 21 19	fv	2.65	+0.60	+0.89	G5 IIb
23		Com	4789	12 34 28.7	+22 40 14	fd6	4.81	−0.01	0.00	A0m A1 IV
24		Com	4792	12 34 45.2	+18 25 06	fvd	5.02	+1.11	+1.15	K2 III
	α	Mus	4798	12 36 43.8	−69 05 40	fvd	2.69	−0.83	−0.20	B2 IV−V
	τ	Cen	4802	12 37 17.4	−48 30 00		3.86	+0.03	+0.05	A1 IVnn
26	χ	Vir	4813	12 38 51.5	− 7 57 16	fd	4.66	+1.39	+1.23	K2 III CN 1.5
	γ	Cen	4819	12 41 06.0	−48 55 07	d67	2.17	−0.01	−0.01	A1 IV
29	γ	Vir	4826	12 41 16.7	− 1 24 30	cvmd8	3.68	−0.03	+0.36	F0m F2 V
29	γ	Vir	4825	12 41 16.9	− 1 24 30	cvmd68	3.65	−0.03	+0.36	F1 V
30	ρ	Vir	4828	12 41 30.3	+10 16 37	fv6	4.88	+0.03	+0.09	A0 Va
			4839	12 43 36.5	−28 16 58	fm	5.48		+1.34	K3 III
	Y	CVn	4846	12 44 46.8	+45 28 52	fv	4.99	+6.33	+2.54	C5,5
32	d^2	Vir	4847	12 45 14.3	+ 7 42 51	fv6	5.22	+0.15	+0.33	A8m
	β	Mus	4844	12 45 49.0	−68 04 02	d7	3.05	−0.74	−0.18	B2 V
	β	Cru	4853	12 47 16.7	−59 38 52	fvd6	1.25	−1.00	−0.23	B0.5 III
			4874	12 50 16.7	−33 57 31	fd	4.91	−0.11	−0.04	A0 IV
31		Com	4883	12 51 20.0	+27 34 53	fs	4.94	+0.20	+0.67	G0 IIIp
			4888	12 52 41.2	−48 54 09	6	4.33	+1.58	+1.37	K3/4 III
			4889	12 53 01.1	−40 08 18	f	4.27	+0.12	+0.21	A7 III
77	ε	UMa	4905	12 53 42.0	+56 00 02	fv6	1.77	+0.02	−0.02	A0p IV: Cr
40	ψ	Vir	4902	12 53 57.7	− 9 29 54	fvd	4.79	+1.53	+1.60	M3^{-} III Ca−1
	ι	Oct	4870	12 54 07.6	−85 04 58	fd	5.46	+0.79	+1.02	K0 III
	μ^1	Cru	4898	12 54 09.0	−57 08 14	d	4.03	−0.76	−0.17	B2 IV−V

Name	B.S.	Right Ascension	Declination	Notes	V	U–B	B–V	Spectral Type
		h m s	° ′ ″					
8 Dra	4916	12 55 10.7	+65 28 45	f	5.24	+0.02	+0.28	F0 V
43 δ Vir	4910	12 55 13.5	+ 3 26 17	fvd	3.38	+1.78	+1.58	$M3^+$ III
12 α^2 CVn	4915	12 55 40.7	+38 21 32	fvd	2.90	–0.32	–0.12	A0p III: Si Eu
78 UMa	4931	13 00 24.6	+56 24 24	svd7	4.93	+0.01	+0.36	F2 V
δ Mus	4923	13 01 44.7	–71 30 31	f6	3.62	+1.26	+1.18	K2 III
47 ε Vir	4932	13 01 48.2	+10 59 58	fsvd	2.83	+0.73	+0.94	G8 IIIab
14 CVn	4943	13 05 23.5	+35 50 20	fv	5.25	–0.20	–0.08	B9 V
ξ^2 Cen	4942	13 06 28.2	–49 51 58	fd6	4.27	–0.79	–0.19	B1.5 V
51 θ Vir	4963	13 09 33.7	– 5 29 57	fd6	4.38	–0.01	–0.01	A1 IV
43 β Com	4983	13 11 31.4	+27 54 58	fd6	4.26	+0.07	+0.57	F9.5 V
η Mus	4993	13 14 44.0	–67 51 18	fvd6	4.80	–0.35	–0.08	B7 V
	5006	13 16 28.1	–31 28 00	fm	5.10		+0.96	K0 III
20 CVn	5017	13 17 12.4	+40 36 43	fsv	4.73	+0.21	+0.30	F3 III
60 σ Vir	5015	13 17 13.6	+ 5 30 33	fv	4.80	+1.95	+1.67	M1 III
61 Vir	5019	13 18 00.7	–18 16 11	fd	4.74	+0.26	+0.71	G6 V
46 γ Hya	5020	13 18 30.8	–23 07 56	fvd	3.00	+0.66	+0.92	G8 IIIa
ι Cen	5028	13 20 10.4	–36 40 23	f	2.75	+0.03	+0.04	A2 Va
	5035	13 22 08.5	–60 56 57	fd	4.53	–0.60	–0.13	B3 V
79 ζ UMa	5054	13 23 37.5	+54 57 52	fvmd68	2.27	+0.03	+0.02	A1 Va
79 ζ UMa	5055	13 23 38.4	+54 57 39	vmd6	3.95	+0.09	+0.13	A1m A7 IV–V
67 α Vir	5056	13 24 47.8	–11 07 20	fvmd6	0.97	–0.93	–0.24	B1 IV
80 UMa	5062	13 24 55.5	+55 01 37	v6	4.01	+0.08	+0.16	A5 V
68 Vir	5064	13 26 19.3	–12 40 08	fv	5.25	+1.75	+1.52	M0 III
70 Vir	5072	13 28 03.8	+13 49 07	fd	4.98	+0.26	+0.71	G4 V
	5085	13 28 10.6	+59 59 04	fmd	5.40	–0.02	–0.01	A1 Vn
	5089	13 30 36.4	–39 22 08	vd67	3.88	+1.03	+1.17	G8 III
78 Vir	5105	13 33 45.1	+ 3 41 50	fv6	4.94	0.00	+0.03	A1p
79 ζ Vir	5107	13 34 18.6	– 0 33 28	f	3.37	+0.10	+0.11	A2 IV^-
	5110	13 34 27.7	+37 13 14	fv6	4.98	+0.06	+0.40	F2 IV
ε Cen	5132	13 39 24.5	–53 25 43	fvd	2.30	–0.92	–0.22	B1 III
	5134	13 39 31.7	–49 54 44	svm	6.00	+1.15	+1.50	M6 III
82 Vir	5150	13 41 13.1	– 8 39 55	fv	5.01	+1.95	+1.63	M1.5 III
1 Cen	5168	13 45 15.6	–33 00 22	fv6	4.23	0.00	+0.38	F3 IV
	5171	13 46 39.0	–62 33 10	svd	6.51	+1.19	+1.98	K0 0–Ia
4 τ Boo	5185	13 46 54.4	+17 29 38	fvd7	4.50	+0.04	+0.48	F7 V
85 η UMa	5191	13 47 14.7	+49 21 02	fv6	1.86	–0.67	–0.19	B3 V
2 Cen	5192	13 49 00.5	–34 24 49	v	4.19	+1.45	+1.50	M4.5 III
ν Cen	5190	13 49 03.2	–41 39 02	v6	3.41	–0.84	–0.22	B2 IV
5 υ Boo	5200	13 49 06.9	+15 50 06	v	4.06	+1.89	+1.52	K5.5 III
μ Cen	5193	13 49 09.7	–42 26 12	fsvd6	3.04	–0.72	–0.17	B2 V:e
89 Vir	5196	13 49 27.8	–18 05 49	f	4.97	+0.92	+1.06	K0.5 III–IIIb
10 Dra	5226	13 51 12.8	+64 45 37	fvd	4.65	+1.89	+1.58	M3.5 III
8 η Boo	5235	13 54 19.6	+18 26 06	fsd6	2.68	+0.20	+0.58	G0 IV
ζ Cen	5231	13 55 04.1	–47 15 06	f6	2.55	–0.92	–0.22	B2.5 IV
	5241	13 57 05.9	–63 39 01	f	4.71	+1.04	+1.11	K1.5 III
φ Cen	5248	13 57 48.8	–42 03 52	v	3.83	–0.83	–0.21	B2 IV
47 Hya	5250	13 58 05.8	–24 56 09	f6	5.15	–0.39	–0.10	B8n
υ^1 Cen	5249	13 58 12.8	–44 46 02		3.87	–0.80	–0.20	B2 IV–V
υ^2 Cen	5260	14 01 15.2	–45 34 03	6	4.34	+0.27	+0.60	F6 II
93 τ Vir	5264	14 01 15.9	+ 1 34 50	fd6	4.26	+0.12	+0.10	A3 V

Name	B.S.	Right Ascension	Declination	Notes	V	U–B	B–V	Spectral Type
		h m s	° ′ ″					
	5270	14 02 09.7	+ 9 43 21	sv	6.20	+0.38	+0.90	K1 CN–5 Fe–4
β Cen	5267	14 03 17.3	–60 20 14	fvmd6	0.61	–0.98	–0.23	B1 III
11 α Dra	5291	14 04 11.1	+64 24 42	fsv6	3.65	–0.08	–0.05	A0 III
θ Aps	5261	14 04 34.9	–76 45 40	fsvm	5.50	+1.05	+1.55	M6.5 III:
χ Cen	5285	14 05 35.2	–41 08 38	v	4.36	–0.77	–0.19	B2 V
49 π Hya	5287	14 05 56.6	–26 38 47	f	3.27	+1.04	+1.12	K2$^-$ III Fe–0.5
5 θ Cen	5288	14 06 14.3	–36 20 00	fd	2.06	+0.87	+1.01	K0$^-$ IIIb
	5299	14 07 37.8	+43 53 24	fv	5.27	+1.66	+1.59	M4.5: III
4 UMi	5321	14 08 52.1	+77 34 58	f6	4.82		+1.36	K3 III
12 d Boo	5304	14 10 03.4	+25 07 37	f6	4.83	+0.07	+0.54	F8 IV
98 κ Vir	5315	14 12 29.7	–10 14 21	f	4.19	+1.47	+1.33	K2.5 IIIb Fe–0.5
16 α Boo	5340	14 15 19.1	+19 13 16	fv	–0.04	+1.27	+1.23	K1.5 III Fe–0.5
99 ι Vir	5338	14 15 37.3	– 5 57 54	fv	4.08	+0.04	+0.52	F7 III–IV
21 ι Boo	5350	14 15 54.0	+51 24 06	fvd6	4.75	+0.06	+0.20	A9 V
19 λ Boo	5351	14 16 05.9	+46 07 21	fv	4.18	+0.05	+0.08	A0 Va
	5361	14 17 40.8	+35 32 38	f6	4.81	+0.92	+1.06	K0 III
100 λ Vir	5359	14 18 42.2	–13 20 13	fvd6	4.52	+0.12	+0.13	A2m A7 V
18 Boo	5365	14 18 54.5	+13 02 19	fd	5.41	–0.03	+0.38	F3 V
ι Lup	5354	14 18 55.2	–46 01 25	v	3.55	–0.72	–0.18	B2.5 IV
	5358	14 19 47.8	–56 21 08	f	4.33	–0.43	+0.12	B6 Ib
ψ Cen	5367	14 20 05.9	–37 51 04	fd	4.05	–0.11	–0.03	A0 III
	5378	14 22 34.4	–39 28 41	v	4.42	–0.75	–0.18	B7 IIIp
	5392	14 23 48.9	+ 5 51 14	f6	5.10	+0.10	+0.12	A5 V
	5390	14 24 22.9	–24 46 21	fm	5.32		+0.96	K0 III
23 θ Boo	5404	14 24 56.5	+51 53 07	fvd	4.05	+0.01	+0.50	F7 V
δ Oct	5339	14 25 38.8	–83 38 03	v	4.32	+1.45	+1.31	K2 III
τ^1 Lup	5395	14 25 39.2	–45 11 16	fvd	4.56	–0.79	–0.15	B2 IV
τ^2 Lup	5396	14 25 41.7	–45 20 45	cd67	4.35	+0.19	+0.43	F4 IV + A7:
22 Boo	5405	14 26 06.4	+19 15 37	f	5.39	+0.23	+0.23	F0m
5 UMi	5430	14 27 32.1	+75 43 46	fvd	4.25	+1.70	+1.44	K4$^-$ III
52 Hya	5407	14 27 44.0	–29 27 30	fvd	4.97	–0.41	–0.07	B8 IV
105 φ Vir	5409	14 27 48.9	– 2 11 41	fsvd67	4.81	+0.21	+0.70	G2 IV
25 ρ Boo	5429	14 31 30.4	+30 24 15	fvd	3.58	+1.44	+1.30	K3$^-$ III
27 γ Boo	5435	14 31 46.6	+38 20 27	fvd	3.03	+0.12	+0.19	A7 IV$^+$
σ Lup	5425	14 32 06.4	–50 25 27		4.42	–0.84	–0.19	B2 III
28 σ Boo	5447	14 34 21.2	+29 46 39	fvd	4.46	–0.08	+0.36	F2 V
η Cen	5440	14 35 01.7	–42 07 31	fvd67	2.31	–0.83	–0.19	B1.5 Vn
ρ Lup	5453	14 37 22.7	–49 23 36	v	4.05	–0.56	–0.15	B5 V
33 Boo	5468	14 38 33.5	+44 26 12	fm6	5.39	–0.04	0.00	A1 V
α^2 Cen	5460	14 39 04.2	–60 48 26	fmd	1.33	+0.63	+0.88	K1 V
α^1 Cen	5459	14 39 05.7	–60 48 11	fmd6	–0.01	+0.33	+0.71	G2 V
30 ζ Boo	5478	14 40 47.4	+13 45 37	cvmd68	4.43	+0.05	+0.05	A2 Va
30 ζ Boo	5477	14 40 47.5	+13 45 37	cvmd8	4.83	+0.05	+0.05	A2 Va
α Lup	5469	14 41 25.7	–47 21 23	fvd6	2.30	–0.89	–0.20	B1.5 III
	5471	14 41 29.5	–37 45 42		4.00	–0.70	–0.17	B3 V
α Cir	5463	14 41 53.6	–64 56 35	fvd6	3.19	+0.12	+0.24	F1 Vp Sr Eu
107 μ Vir	5487	14 42 39.9	– 5 37 33	f6	3.88	–0.02	+0.38	F3 IV
34 Boo	5490	14 43 05.6	+26 33 34	fv	4.81	+1.94	+1.66	M3$^-$ III
	5485	14 43 11.8	–35 08 30	f	4.05	+1.53	+1.35	K3 IIIb
36 ε Boo	5506	14 44 39.6	+27 06 20	md	2.40		+0.96	K0$^-$ II–III

Name			B.S.	Right Ascension	Declination	Notes	*V*	*U−B*	*B−V*	Spectral Type
				h m s	° ′ ″					
109		Vir	5511	14 45 52.2	+ 1 55 27	fv	3.72	−0.03	−0.01	A0 IVnn
			5495	14 46 29.6	−52 21 08	fmd	5.21		+0.98	G8 III
	α	Aps	5470	14 46 54.1	−79 00 49	f	3.83	+1.68	+1.43	K3 III CN 0.5
56		Hya	5516	14 47 18.5	−26 03 23	f	5.24	+0.65	+0.94	G8/K0
58		Hya	5526	14 49 50.8	−27 55 46		4.41	+1.49	+1.40	K3 III
8	α^1	Lib	5530	14 50 16.3	−15 57 59	fd	5.15	−0.03	+0.41	F3 V
9	α^2	Lib	5531	14 50 27.8	−16 00 39	fvd6	2.75	+0.09	+0.15	A3 III−IV
7	β	UMi	5563	14 50 43.3	+74 11 10	fvd	2.08	+1.78	+1.47	$K4^-$ III
	ο	Lup	5528	14 51 08.8	−43 32 41	d67	4.32	−0.61	−0.15	B5 IV
			5552	14 51 15.0	+59 19 28	f	5.46	+1.60	+1.36	K4 III
			5558	14 55 17.0	−33 49 33	fmd6	5.34			A0 V
15	ξ^2	Lib	5564	14 56 21.6	−11 22 47	fm	5.46		+1.49	gK4
16		Lib	5570	14 56 47.4	− 4 18 59		4.49	+0.05	+0.32	F0 IV
			5589	14 57 27.7	+65 57 44	fv6	4.60	+1.59	+1.59	M4.5 III
	β	Lup	5571	14 58 02.3	−43 06 15	f6	2.68	−0.87	−0.22	B2 IV
	κ	Cen	5576	14 58 40.3	−42 04 28	fvd6	3.13	−0.79	−0.20	B2 IV
19	δ	Lib	5586	15 00 34.3	− 8 29 22	fvd6	4.92	−0.10	0.00	B9.5 V
42	β	Boo	5602	15 01 39.8	+40 25 12	fv	3.50	+0.72	+0.97	G8 IIIa Fe−0.5
110		Vir	5601	15 02 31.3	+ 2 07 13		4.40	+0.88	+1.04	K0+ IIIb Fe−0.5
20	σ	Lib	5603	15 03 37.8	−25 15 10	fv	3.29	+1.94	+1.70	M2.5 III
43	ψ	Boo	5616	15 04 07.5	+26 58 36	f	4.54	+1.33	+1.24	K2 III
			5635	15 06 03.8	+54 35 06	f	5.25	+0.64	+0.96	G8 III Fe−1
45		Boo	5634	15 06 58.3	+24 53 53	fvd	4.93	−0.02	+0.43	F5 V
	λ	Lup	5626	15 08 20.1	−45 15 05	d67	4.05	−0.68	−0.18	B3 V
	κ	Lup	5646	15 11 24.6	−48 42 35	fd	3.87	−0.13	−0.05	B9.5 IVnn
	ζ	Lup	5649	15 11 44.6	−52 04 16	fd	3.41	+0.66	+0.92	G8 III
24	ι	Lib	5652	15 11 47.6	−19 45 49	fvd6	4.54	−0.35	−0.08	A0p Si
1		Lup	5660	15 14 09.7	−31 29 30	f	4.91	+0.28	+0.37	F1 II
			5691	15 14 33.0	+67 22 31	f	5.13	+0.08	+0.53	F8 V
3		Ser	5675	15 14 49.0	+ 4 58 01	f	5.33	+0.91	+1.09	gK0
49	δ	Boo	5681	15 15 12.0	+33 20 33	fvd6	3.49	+0.68	+0.95	G8 III Fe−1
27	β	Lib	5685	15 16 36.2	− 9 21 20	fv6	2.61	−0.36	−0.11	B8 IVn
	β	Cir	5670	15 16 55.3	−58 46 25	f	4.07	+0.09	+0.09	A3 V
2		Lup	5686	15 17 22.4	−30 07 17	v	4.34	+1.07	+1.10	K0 IIIa Fe 1
	μ	Lup	5683	15 18 00.5	−47 50 53	d7	4.27	−0.37	−0.08	B8 V
	γ	TrA	5671	15 18 12.0	−68 39 09	fv	2.89	−0.02	0.00	A1 III
13	γ	UMi	5735	15 20 44.2	+71 51 39	fv	3.05	+0.12	+0.05	A3 III
	δ	Lup	5695	15 20 52.7	−40 37 15	fv	3.22	−0.89	−0.22	B1.5 IV
	ϕ^1	Lup	5705	15 21 19.7	−36 14 05	fd	3.56	+1.88	+1.54	K4 III
	ε	Lup	5708	15 22 10.2	−44 39 46	d67	3.37	−0.75	−0.18	B2 IV/V
	ϕ^2	Lup	5712	15 22 40.5	−36 49 55	f	4.54	−0.63	−0.15	B4 V
	γ	Cir	5704	15 22 46.6	−59 17 40	c7d	4.51	−0.35	+0.19	B5 IV
51	μ^1	Boo	5733	15 24 12.4	+37 24 12	fvd6	4.31	+0.07	+0.31	F2 III−IV
12	ι	Dra	5744	15 24 45.7	+58 59 32	fvd	3.29	+1.22	+1.16	K2 III
9	τ^1	Ser	5739	15 25 26.5	+15 27 15	fv	5.17	+1.95	+1.66	M1 IIIa
3	β	CrB	5747	15 27 31.2	+29 07 53	fvd6	3.68	+0.11	+0.28	F0p Cr Eu
52	ν^1	Boo	5763	15 30 39.6	+40 51 30	f	5.02	+1.90	+1.59	K4.5 IIIb Ba 0.5
	κ^1	Aps	5730	15 30 41.0	−73 21 51	fvd	5.49	−0.77	−0.12	B1 pne
4	θ	CrB	5778	15 32 37.6	+31 23 03	fvd	4.14	−0.54	−0.13	B6 Vnn
37		Lib	5777	15 33 46.1	−10 02 21	f	4.62	+0.86	+1.01	K1 III−IV

Name	B.S.	Right Ascension	Declination	Notes	*V*	*U–B*	*B–V*	Spectral Type
		h m s	° ′ ″					
5 α CrB	5793	15 34 22.2	+26 44 23	fv6	2.23	−0.02	−0.02	A0 IV
13 δ Ser	5789	15 34 26.6	+10 33 50	vmd	4.23			F0 IV
γ Lup	5776	15 34 38.3	−41 08 31	d7	2.78	−0.82	−0.20	B2 IV
38 γ Lib	5787	15 35 06.4	−14 45 54	fd	3.91	+0.74	+1.01	G8.5 III
	5784	15 35 41.1	−44 22 20	f	5.43		+1.50	K4/5 III
ε TrA	5771	15 36 01.6	−66 17 33	fd	4.11	+1.16	+1.17	K1/2 III
39 υ Lib	5794	15 36 34.1	−28 06 38	fd	3.58	+1.58	+1.38	K3.5 III
ω Lup	5797	15 37 32.8	−42 32 35	d6	4.33	+1.72	+1.42	K4.5 III
54 φ Boo	5823	15 37 33.4	+40 22 39	f	5.24	+0.53	+0.88	G7 III–IV Fe−2
40 τ Lib	5812	15 38 11.7	−29 45 13	6	3.66	−0.70	−0.17	B2.5 V
	5798	15 38 15.9	−52 20 55	fd	5.44		0.00	B9 V
43 κ Lib	5838	15 41 30.8	−19 39 18	fvd6	4.74	+1.95	+1.57	M0$^-$ IIIb
8 γ CrB	5849	15 42 25.7	+26 19 08	vd67	3.84	−0.04	0.00	A0 IV
24 α Ser	5854	15 43 53.9	+ 6 26 56	fd	2.65	+1.24	+1.17	K2 IIIb CN 1
16 ζ UMi	5903	15 44 18.9	+77 49 04	fv	4.32	+0.05	+0.04	A2 III–IVn
28 β Ser	5867	15 45 50.5	+15 26 42	fd	3.67	+0.08	+0.06	A2 IV
27 λ Ser	5868	15 46 04.7	+ 7 22 35	v6	4.43	+0.11	+0.60	G0$^-$ V
	5886	15 46 33.0	+62 37 21	f	5.19		+0.04	A2 IV
35 κ Ser	5879	15 48 24.1	+18 09 52	fv	4.09	+1.95	+1.62	M0.5 IIIab
32 μ Ser	5881	15 49 13.7	− 3 24 28	f6	3.54	−0.11	−0.04	A0 III
10 δ CrB	5889	15 49 16.8	+26 05 27	s	4.63	+0.37	+0.80	G5 III–IV Fe−1
37 ε Ser	5892	15 50 26.5	+ 4 30 00	f	3.71	+0.11	+0.15	A2m
5 χ Lup	5883	15 50 28.9	−33 36 17	f6	3.95	−0.13	−0.04	B9.5p III Hg
11 κ CrB	5901	15 50 57.0	+35 40 49	fsd	4.82	+0.87	+1.00	K1 IVa
1 ξ Her	5914	15 52 25.0	+42 28 20	f	4.62	0.00	+0.56	F9 V
45 λ Lib	5902	15 52 53.9	−20 08 42	fd6	5.03	−0.56	−0.01	B2.5 V
46 θ Lib	5908	15 53 23.9	−16 42 28		4.15	+0.81	+1.02	G8.5 IIIb
β TrA	5897	15 54 28.6	−63 24 29	fd	2.85	+0.05	+0.29	F0 IV
41 γ Ser	5933	15 56 06.4	+15 41 09	fvd	3.85	−0.03	+0.48	F6 V
5 ρ Sco	5928	15 56 25.3	−29 11 33	d	3.88	−0.82	−0.20	B2 IV–V
13 ε CrB	5947	15 57 16.6	+26 53 57	fsd	4.15	+1.28	+1.23	K2 IIIab
	5960	15 57 36.7	+54 46 15	fv6	4.95	+0.05	+0.26	F0 IV
48 Lib	5941	15 57 46.1	−14 15 29	fv6	4.88	−0.20	−0.10	B5 IIIpe
6 π Sco	5944	15 58 23.8	−26 05 35	fvd6	2.89	−0.91	−0.19	B1 V
	5943	15 58 59.5	−41 43 24	f	4.99		+1.00	K0 II/III
T CrB	5958	15 59 11.3	+25 56 28	vmd6	2–11			gM3 + pec
η Lup	5948	15 59 37.4	−38 22 33	d8	3.43	−0.83	−0.22	B2.5 IV
7 δ Sco	5953	15 59 53.3	−22 36 03	fd6	2.32	−0.91	−0.12	B0.5 IV
49 Lib	5954	15 59 54.3	−16 30 42	fd6	5.47	+0.03	+0.52	F8 V
13 θ Dra	5986	16 01 44.9	+58 35 07	f6	4.01	+0.10	+0.52	F8 IV–V
ξ Sco	5977	16 03 57.3	−11 21 11	cd7	4.16	+0.03	+0.45	F6 IV
8 β^1 Sco	5984	16 05 00.0	−19 47 07	fvd6	2.62	−0.87	−0.07	B0.5 V
8 β^2 Sco	5985	16 05 00.3	−19 46 55	svd	4.92	−0.70	−0.02	B2 V
δ Nor	5980	16 05 57.5	−45 09 12	f	4.72	+0.15	+0.23	A1m A7–F3
θ Lup	5987	16 06 05.9	−36 46 57	f	4.23	−0.70	−0.17	B2.5 Vn
9 ω^1 Sco	5993	16 06 22.1	−20 38 57	s	3.96	−0.81	−0.04	B1 V
10 ω^2 Sco	5997	16 06 57.9	−20 50 56	v	4.32	+0.50	+0.84	G4 II–III
7 κ Her	6008	16 07 44.2	+17 04 00	fvd	5.00	+0.61	+0.95	G5 III
11 φ Her	6023	16 08 32.0	+44 57 16	fv6	4.26	−0.28	−0.07	B9p Hg Mn
16 τ CrB	6018	16 08 41.8	+36 30 35	fvd6	4.76	+0.86	+1.01	K1$^-$ III–IV

Name			B.S.	Right Ascension	Declination	Notes	*V*	*U–B*	*B–V*	Spectral Type
				h m s	° ′ ″					
19		UMi	6079	16 11 01.9	+75 53 48	f	5.48	−0.45	−0.15	B8 V
14	ν	Sco	6027	16 11 33.5	−19 26 30	d6	4.01	−0.65	+0.04	B2 IVp
	κ	Nor	6024	16 12 53.1	−54 36 42	fd	4.94	+0.78	+1.04	G8 III
1	δ	Oph	6056	16 13 57.1	− 3 40 32	fvd	2.74	+1.96	+1.58	M0.5 III
	δ	TrA	6030	16 14 45.0	−63 40 02	fd	3.85	+0.86	+1.11	G2 Ib–IIa
21	η	UMi	6116	16 17 43.1	+75 46 22	fd	4.95	+0.08	+0.37	F5 V
2	ε	Oph	6075	16 17 55.4	− 4 40 29	fd	3.24	+0.75	+0.96	G9.5 IIIb Fe−0.5
			6077	16 19 04.2	−30 53 21	fd6	5.49	−0.01	+0.47	F6 III
	δ^1	Aps	6020	16 19 12.4	−78 40 41	fvd	4.68	+1.69	+1.69	M5 IIb
	γ^2	Nor	6072	16 19 16.6	−50 08 16	fd	4.02	+1.16	+1.08	K0 III
22	τ	Her	6092	16 19 30.9	+46 19 52	fvd	3.89	−0.56	−0.15	B5 IV
20	σ	Sco	6084	16 20 43.9	−25 34 31	fvd6	2.89	−0.70	+0.13	B1 III
20	γ	Her	6095	16 21 35.4	+19 10 13	fvd6	3.75	+0.18	+0.27	A9 III
50	σ	Ser	6093	16 21 41.5	+ 1 02 47	f	4.82	+0.04	+0.34	F3 V
4	ψ	Oph	6104	16 23 39.8	−20 01 13		4.50	+0.82	+1.01	K0$^-$ II–III
14	η	Dra	6132	16 23 53.3	+61 31 52	vd67	2.74	+0.70	+0.91	G8$^-$ IIIab
24	ω	Her	6117	16 25 04.2	+14 03 01	fvd	4.57	−0.04	0.00	B9p Cr
7	χ	Oph	6118	16 26 35.3	−18 26 23	v6	4.42	−0.75	+0.28	B1.5 Ve
	ε	Nor	6115	16 26 38.0	−47 32 19	d67	4.47	−0.54	−0.07	B4 V
	ζ	TrA	6098	16 27 39.3	−70 04 06	f6	4.91	+0.04	+0.55	F9 V
15		Dra	6161	16 27 59.7	+68 47 04	f	5.00	−0.12	−0.06	B9.5 III
21	α	Sco	6134	16 28 56.8	−26 24 57	fvd6	0.96	+1.34	+1.83	M1.5: Iab–Ib
27	β	Her	6148	16 29 53.8	+21 30 20	fvd6	2.77	+0.69	+0.94	G7 IIIa Fe−0.5
10	λ	Oph	6149	16 30 32.1	+ 2 00 00	vd67	3.82	+0.01	+0.01	A1 IV
8	φ	Oph	6147	16 30 42.5	−16 35 48	d	4.28	+0.72	+0.92	G8$^+$ IIIa
			6143	16 30 53.5	−34 41 19	f6	4.23	−0.80	−0.16	B2 III–IV
9	ω	Oph	6153	16 31 41.4	−21 27 03	v	4.45	+0.13	+0.13	Ap Sr Cr
	γ	Aps	6102	16 32 16.9	−78 52 53	f6	3.89	+0.62	+0.91	G8/K0 III
35	σ	Her	6168	16 33 51.7	+42 27 08	fvd	4.20	−0.10	−0.01	A0 IIIn
23	τ	Sco	6165	16 35 24.9	−28 12 03	fs	2.82	−1.03	−0.25	B0 V
			6166	16 35 52.8	−35 14 27	v	4.16	+1.94	+1.57	K5 III
13	ζ	Oph	6175	16 36 44.7	−10 33 08	fv	2.56	−0.86	+0.02	O9.5 Vn
42		Her	6200	16 38 32.6	+48 56 34	fvd	4.90		+1.55	M3$^-$ IIIab
40	ζ	Her	6212	16 41 00.2	+31 36 58	vd67	2.81	+0.21	+0.65	G0 IV
			6196	16 41 08.3	−17 43 41	f	4.96	+0.87	+1.11	G7.5 II–III CN 1 Ba 0.5
	β	Aps	6163	16 41 59.7	−77 30 10	d	4.24	+0.95	+1.06	K0 III
44	η	Her	6220	16 42 38.3	+38 56 11	fvd	3.53	+0.60	+0.92	G7.5 IIIb Fe−1
			6237	16 45 09.2	+56 47 42	f6	4.85	−0.06	+0.38	F5 V
22	ε	UMi	6322	16 46 43.0	+82 03 02	fvd6	4.23	+0.55	+0.89	G5 III
	α	TrA	6217	16 47 51.9	−69 00 53	f	1.92	+1.56	+1.44	K2 IIb–IIIa
	η	Ara	6229	16 49 08.1	−59 01 43	fd	3.76	+1.94	+1.57	K5 III
20		Oph	6243	16 49 25.1	−10 46 13	f6	4.65	+0.07	+0.47	F7 III
26	ε	Sco	6241	16 49 40.6	−34 16 49	fv	2.29	+1.27	+1.15	K2 IIIb
	μ^1	Sco	6247	16 51 21.7	−38 02 06	fv6	3.08	−0.87	−0.20	B1.5 IV
51		Her	6270	16 51 26.6	+24 40 07	f	5.04	+1.29	+1.25	K0.5 IIIa Ca 0.5
	μ^2	Sco	6252	16 51 49.6	−38 00 19	d	3.57	−0.85	−0.21	B2 IV
53		Her	6279	16 52 41.0	+31 42 50	fd	5.32	−0.02	+0.29	F2 V
25	ι	Oph	6281	16 53 39.2	+10 10 38	fv6	4.38	−0.32	−0.08	B8 V
	ζ^2	Sco	6271	16 54 03.3	−42 20 57	v	3.62	+1.65	+1.37	K4 III
27	κ	Oph	6299	16 57 18.8	+ 9 23 11	fsv	3.20	+1.18	+1.15	K2 III

Name	B.S.	Right Ascension	Declination	Notes	*V*	*U–B*	*B–V*	Spectral Type
		h m s	° ′ ″					
ζ Ara	6285	16 57 59.9	−55 58 44	f	3.13	+1.97	+1.60	K4 III
ε^{1} Ara	6295	16 58 59.1	−53 08 59	f	4.06	+1.71	+1.45	K4 IIIab
58 ε Her	6324	17 00 00.2	+30 56 14	f6	3.92	−0.10	−0.01	A0 IV+
30 Oph	6318	17 00 39.8	− 4 12 43	fvd	4.82	+1.83	+1.48	K4 III
59 Her	6332	17 01 19.7	+33 34 44	f	5.25	+0.02	+0.02	A3 IV
60 Her	6355	17 05 01.8	+12 45 03	fd	4.91	+0.05	+0.12	A4 IV
22 ζ Dra	6396	17 08 45.8	+65 43 26	f	3.17	−0.43	−0.12	B6 IIIp
35 η Oph	6378	17 09 56.9	−15 42 58	d67	2.43	+0.09	+0.06	A2 Va^{+}
η Sco	6380	17 11 36.9	−43 13 48	f	3.33	+0.09	+0.41	F2 V
64 α^{1} Her	6406	17 14 18.4	+14 23 55	svd	3.08	+1.01	+1.44	M5 Ib–II
65 δ Her	6410	17 14 43.4	+24 50 52	fvd6	3.14	+0.08	+0.08	A1 Vann
67 π Her	6418	17 14 47.2	+36 49 02	fv	3.16	+1.66	+1.44	K3 II
	6452	17 19 59.0	+18 03 52	fv	5.00		+1.62	$M1^{+}$ IIIab
72 Her	6458	17 20 22.7	+32 28 37	fvd	5.39	+0.07	+0.62	G0 V
53 ν Ser	6446	17 20 24.3	−12 50 23	d7	4.33	+0.05	+0.03	A1.5 IV
40 ξ Oph	6445	17 20 33.2	−21 06 19	d7	4.39	−0.05	+0.39	F1 Vs
ι Aps	6411	17 21 15.5	−70 06 58	fvd7	5.41	−0.23	−0.04	B8/9 Vn
42 θ Oph	6453	17 21 32.9	−24 59 33	fvd6	3.27	−0.86	−0.22	B2 IV
β Ara	6461	17 24 40.5	−55 31 25	f	2.85	+1.56	+1.46	K3 Ib–IIa
γ Ara	6462	17 24 45.7	−56 22 16	d	3.34	−0.96	−0.13	B1 Ib
44 Oph	6486	17 25 54.7	−24 10 08	fv	4.17	+0.12	+0.28	A9m:
49 σ Oph	6498	17 26 08.5	+ 4 08 47	fsv	4.34	+1.62	+1.50	K2 II
	6493	17 26 14.0	− 5 04 49	fv6	4.54	−0.03	+0.39	F3 IIIs
45 Oph	6492	17 26 52.5	−29 51 39	f	4.29	+0.09	+0.40	δ Del
34 υ Sco	6508	17 30 15.2	−37 17 26	f6	2.69	−0.82	−0.22	B2 IV
23 β Dra	6536	17 30 15.8	+52 18 24	fsd	2.79	+0.64	+0.98	G2 Ib–IIa
δ Ara	6500	17 30 25.2	−60 40 42	fd	3.62	−0.31	−0.10	B8 Vn
76 λ Her	6526	17 30 26.1	+26 06 57	fv	4.41	+1.68	+1.44	K3.5 III
α Ara	6510	17 31 15.7	−49 52 15	fvd6	2.95	−0.69	−0.17	B2 Vne
27 Dra	6566	17 31 59.6	+68 08 23	fd6	5.05	+0.92	+1.08	G9 III–IIIb
24 ν^{1} Dra	6554	17 32 01.7	+55 11 21	fvd6	4.88	+0.04	+0.26	Am
25 ν^{2} Dra	6555	17 32 07.1	+55 10 40	fvd6	4.87	+0.06	+0.28	Am
35 λ Sco	6527	17 33 06.0	−37 05 56	fvd6	1.63	−0.89	−0.22	B1.5 IV
55 α Oph	6556	17 34 35.2	+12 33 54	fvd6	2.08	+0.10	+0.15	A5 Vnn
23 δ UMi	6789	17 34 36.7	+86 35 28	f	4.36	+0.03	+0.02	A1 Van
	6546	17 36 01.8	−38 37 50	v	4.29	+0.90	+1.09	G8/K0 III/IV
θ Sco	6553	17 36 46.8	−42 59 37	fv	1.87	+0.22	+0.40	F0 II
28 ω Dra	6596	17 36 59.7	+68 45 41	fd6	4.80	−0.01	+0.43	F4 V
55 ξ Ser	6561	17 37 09.4	−15 23 40	fd6	3.54	+0.14	+0.26	F0 IIIb
85 ι Her	6588	17 39 15.2	+46 00 36	fsvd6	3.80	−0.69	−0.18	B3 IV
56 ο Ser	6581	17 40 59.6	−12 52 18	v6	4.26	+0.10	+0.08	A2 Va
κ Sco	6580	17 41 58.1	−39 01 36	fv6	2.41	−0.89	−0.22	B1.5 III
31 ψ Dra	6636	17 42 04.3	+72 09 10	fd	4.58	+0.01	+0.42	F5 V
58 Oph	6595	17 42 58.8	−21 40 48	fd	4.87	−0.03	+0.47	F7 V:
84 Her	6608	17 43 03.1	+24 19 50	s	5.71	+0.27	+0.65	G2 IIIb
60 β Oph	6603	17 43 06.1	+ 4 34 12	f	2.77	+1.24	+1.16	K2 III CN 0.5
μ Ara	6585	17 43 32.9	−51 49 51	f	5.15		+0.70	G5 V
η Pav	6582	17 44 59.7	−64 43 16	f	3.62	+1.17	+1.19	K2 II
86 μ Her	6623	17 46 09.9	+27 43 29	fsd	3.42	+0.39	+0.75	G5 IV
ι^{1} Sco	6615	17 47 03.6	−40 07 29	fsd6	3.03	+0.27	+0.51	F2 Ia

Name	B.S.	Right Ascension	Declination	Notes	V	U−B	B−V	Spectral Type
		h m s	° ′ ″					
3 X Sgr	6616	17 47 05.3	−27 49 43	fvm	4.20		+0.70	F3 II
62 γ Oph	6629	17 47 31.0	+ 2 42 35	f6	3.75	+0.04	+0.04	A0 Van
	6630	17 49 20.9	−37 02 29	fd	3.21	+1.19	+1.17	K0/1 III
35 Dra	6701	17 49 47.1	+76 57 51	f	5.04	+0.08	+0.49	F7 IV
32 ξ Dra	6688	17 53 23.9	+56 52 25	fd	3.75	+1.21	+1.18	K2 III
89 Her	6685	17 55 07.0	+26 03 03	fsv6	5.46	+0.27	+0.34	F2 Ib
91 θ Her	6695	17 55 59.7	+37 15 04	fv	3.86	+1.46	+1.35	K1 IIa CN 2
33 γ Dra	6705	17 56 25.9	+51 29 23	fsd	2.23	+1.87	+1.52	K5 III
92 ξ Her	6703	17 57 28.4	+29 14 54	fv	3.70	+0.70	+0.94	G8.5 III
94 ν Her	6707	17 58 12.9	+30 11 23	v	4.41	+0.15	+0.39	F2 II
64 ν Oph	6698	17 58 36.8	− 9 46 23	f	3.34	+0.88	+0.99	K0$^-$ IIIa CN−1
93 Her	6713	17 59 43.4	+16 45 03	f	4.67	+1.22	+1.26	K0.5 IIb
67 Oph	6714	18 00 16.2	+ 2 55 53	fsd	3.97	−0.62	+0.02	B5 Ib
68 Oph	6723	18 01 22.4	+ 1 18 18	vd67	4.45	0.00	+0.02	A0.5 Van
W Sgr	6742	18 04 32.5	−29 34 51	vmd6	4.30		+0.80	G0 Ib/II
70 Oph	6752	18 05 04.6	+ 2 30 03	d67	4.03	+0.54	+0.86	K0 V
10 γ Sgr	6746	18 05 19.6	−30 25 29	fv6	2.99	+0.77	+1.00	K0$^+$ III
θ Ara	6743	18 06 02.8	−50 05 34	f	3.66	−0.85	−0.08	B2 Ib
72 Oph	6771	18 06 59.6	+ 9 33 45	fd6	3.73	+0.10	+0.12	A4 IVs
103 o Her	6779	18 07 15.0	+28 45 40	fv6	3.83	−0.07	−0.03	A0 II−III
	6791	18 07 15.2	+43 27 38	s6	5.00	+0.71	+0.91	G8 III CN−1 CH−3
π Pav	6745	18 07 51.6	−63 40 10	v6	4.35	+0.18	+0.22	A7p Sr
102 Her	6787	18 08 26.3	+20 48 46	d	4.36	−0.81	−0.16	B2 IV
ε Tel	6783	18 10 40.4	−45 57 23	fd	4.53	+0.78	+1.01	K0 III
13 μ Sgr	6812	18 13 18.9	−21 03 41	fvd6	3.86	−0.49	+0.23	B9 Ia
36 Dra	6850	18 13 51.2	+64 23 41	f	5.03	−0.04	+0.38	F5 V
	6819	18 16 29.6	−56 01 35	f6	5.33	−0.69	−0.05	B3 IIIep
η Sgr	6832	18 17 07.2	−36 45 52	fvd7	3.11	+1.71	+1.56	M3.5 IIIab
1 κ Lyr	6872	18 19 35.9	+36 03 39	fv	4.33	+1.19	+1.17	K2$^-$ III CN 0.5
74 Oph	6866	18 20 29.6	+ 3 22 24	fd	4.86	+0.62	+0.91	G8 III
19 δ Sgr	6859	18 20 30.9	−29 49 55	fd	2.70	+1.55	+1.38	K2.5 IIIa CN 0.5
43 φ Dra	6920	18 20 51.9	+71 20 03	vd67	4.22	−0.33	−0.10	A0p Si
58 η Ser	6869	18 20 55.3	− 2 54 04	fvd	3.26	+0.66	+0.94	K0 III−IV
44 χ Dra	6927	18 21 11.5	+72 43 47	fvd6	3.57	−0.06	+0.49	F7 V
ξ Pav	6855	18 22 32.2	−61 29 53	fvd67	4.36	+1.55	+1.48	K4 III
109 Her	6895	18 23 22.7	+21 45 57	fsvd	3.84	+1.17	+1.18	K2 III
20 ε Sgr	6879	18 23 40.5	−34 23 19	fd	1.85	−0.13	−0.03	A0 II$^-$
39 Dra	6923	18 23 48.0	+58 47 46	d6	4.98	+0.04	+0.08	A2 Va
α Tel	6897	18 26 25.1	−45 58 23	f	3.51	−0.64	−0.17	B3 IV
22 λ Sgr	6913	18 27 30.5	−25 25 35	f	2.81	+0.89	+1.04	K1 IIIb
ζ Tel	6905	18 28 15.3	−49 04 32		4.13	+0.82	+1.02	G8/K0 III
γ Sct	6930	18 28 46.2	−14 34 16	f	4.70	+0.06	+0.06	A2 III$^-$
60 Ser	6935	18 29 17.6	− 1 59 26	f6	5.39	+0.76	+0.96	K0 III
θ Cra	6951	18 32 58.1	−42 19 07	f	4.64	+0.76	+1.01	G8 III
α Sct	6973	18 34 48.0	− 8 14 59	fv	3.85	+1.54	+1.33	K3 III
	6985	18 36 06.4	+ 9 06 58	f	5.39	−0.02	+0.37	F5 IIIs
3 α Lyr	7001	18 36 41.1	+38 46 35	fsvd	0.03	−0.01	0.00	A0 Va
δ Sct	7020	18 41 51.8	− 9 03 37	fvd6	4.72	+0.14	+0.35	F3 IIIp
ζ Pav	6982	18 42 09.7	−71 26 08	fd	4.01	+1.02	+1.14	K0 III
ε Sct	7032	18 43 06.8	− 8 16 59	fd	4.90	+0.87	+1.12	G8 IIb

Name	B.S.	Right Ascension	Declination	Notes	*V*	*U–B*	*B–V*	Spectral Type
		h m s	° ′ ″					
6 ζ^1 Lyr	7056	18 44 30.9	+37 35 49	vd6	4.36	+0.16	+0.19	A4m
27 ϕ Sgr	7039	18 45 11.3	−26 59 56	fd6	3.17	−0.36	−0.11	B8.5 III
110 Her	7061	18 45 20.4	+20 32 20	fvd	4.19	+0.01	+0.46	F6 V
	7064	18 45 46.3	+26 39 14	f	4.83	+1.23	+1.20	K2 III
50 Dra	7124	18 46 36.9	+75 25 31	fm6	5.35		+0.05	A1 Vn
111 Her	7069	18 46 41.4	+18 10 22	fd6	4.36	+0.07	+0.13	A5 III
β Sct	7063	18 46 46.6	− 4 45 23	f6	4.22	+0.81	+1.10	G4 IIa
R Sct	7066	18 47 04.9	− 5 42 49	sv	5.20	+1.64	+1.47	K0 Ib:p Ca−1
η^1 Cra	7062	18 48 18.1	−43 41 20	f	5.49		+0.13	A2 Vn
10 β Lyr	7106	18 49 48.2	+33 21 13	fcvd6	3.45	−0.56	0.00	Bpe
χ Oct	6721	18 50 30.5	−87 36 54	f	5.28	+1.60	+1.28	K3 III
47 o Dra	7125	18 51 05.5	+59 22 45	fd6	4.66	+1.04	+1.19	G9 III Fe−0.5
λ Pav	7074	18 51 31.4	−62 11 49	fvd	4.22	−0.89	−0.14	B2 II–III
12 δ^2 Lyr	7139	18 54 14.5	+36 53 20	vd	4.30	+1.65	+1.68	M4 II
52 υ Dra	7180	18 54 29.5	+71 17 14	f6	4.82	+1.10	+1.15	K0$^+$ III–IIIa Ba 0.2
34 σ Sgr	7121	18 54 48.0	−26 18 24	fd	2.02	−0.75	−0.22	B2.5 V
13 R Lyr	7157	18 55 06.4	+43 56 10	fsv6	4.04	+1.41	+1.59	M5 IIIv
63 θ^1 Ser	7141	18 55 50.8	+ 4 11 36	fvd	4.61	+0.10	+0.16	A5 V
κ Pav	7107	18 56 10.9	−67 14 38	vm	3.90		+0.60	F5 I–II
37 ξ^2 Sgr	7150	18 57 17.0	−21 07 01	f	3.51	+1.13	+1.18	K1 III
λ Tel	7134	18 57 51.8	−52 56 57	fm6	5.03			A0 III+
14 γ Lyr	7178	18 58 39.8	+32 40 44	fvd	3.24	−0.09	−0.05	B9 II
13 ϵ Aql	7176	18 59 16.9	+15 03 28	fd6	4.02	+1.04	+1.08	K1$^-$ III CN 0.5
12 Aql	7193	19 01 16.8	− 5 45 00	v	4.02	+1.04	+1.09	K1 III
38 ζ Sgr	7194	19 02 08.1	−29 53 30	d67	2.60	+0.06	+0.08	A2 IV–V
39 o Sgr	7217	19 04 14.1	−21 45 11	vd	3.77	+0.85	+1.01	G9 IIIb
17 ζ Aql	7235	19 05 03.9	+13 51 07	fvd6	2.99	−0.01	+0.01	A0 Vann
16 λ Aql	7236	19 05 51.1	− 4 53 39	f	3.44	−0.27	−0.09	B9 V
γ CrA	7226	19 05 54.7	−37 04 29	md68	5.01	+0.02	+0.52	F7 IV–V
40 τ Sgr	7234	19 06 28.4	−27 40 55	f6	3.32	+1.15	+1.19	K1.5 IIIb
18 ι Lyr	7262	19 07 02.1	+36 05 17	f	5.28	−0.51	−0.11	B6 IV
α CrA	7254	19 08 57.8	−37 55 00	f	4.11	+0.08	+0.04	A2 IV
41 π Sgr	7264	19 09 19.1	−21 02 10	fvd7	2.89	+0.22	+0.35	F2 II–III
β CrA	7259	19 09 30.8	−39 21 12		4.11	+1.07	+1.20	K0 II/III CN Ib
20 Aql	7279	19 12 16.3	− 7 57 09	fv	5.34	−0.44	+0.13	B3 V
57 δ Dra	7310	19 12 33.3	+67 38 54	fd	3.07	+0.78	+1.00	G9 III
20 η Lyr	7298	19 13 30.2	+39 07 58	vd6	4.39	−0.65	−0.15	B2.5 IV
60 τ Dra	7352	19 15 41.9	+73 20 30	fv6	4.45	+1.45	+1.25	K2$^+$ IIIb CN 1
21 θ Lyr	7314	19 16 06.5	+38 07 12	fvd	4.36	+1.23	+1.26	K0 II
1 κ Cyg	7328	19 16 55.8	+53 21 16	fv6	3.77	+0.74	+0.96	G9 III
43 Sgr	7304	19 17 11.8	−18 58 00	fmd	4.96		+1.02	G8 II–III
25 ω Aql	7315	19 17 27.9	+11 34 53	f	5.28	+0.22	+0.20	F0 IV
44 ρ^1 Sgr	7340	19 21 14.3	−17 51 42	vd	3.93	+0.13	+0.22	F0 IV–V
46 υ Sgr	7342	19 21 17.9	−15 58 10	fvd6	4.61	−0.53	+0.10	Apep
β^1 Sgr	7337	19 22 06.0	−44 28 25	fd	4.01	−0.39	−0.10	B8 V
β^2 Sgr	7343	19 22 40.7	−44 48 52		4.29	+0.07	+0.34	F2 III
α Sgr	7348	19 23 22.1	−40 37 50	f6	3.97	−0.33	−0.10	B8 V
31 Aql	7373	19 24 36.8	+11 55 41	fvd	5.16	+0.42	+0.77	G7 IV H 1
30 δ Aql	7377	19 25 07.2	+ 3 05 58	fvd6	3.36	+0.04	+0.32	F0 IV
6 α Vul	7405	19 28 23.6	+24 38 58	fvd	4.44	+1.81	+1.50	M1 IIIb

Name			B.S.	Right Ascension	Declination	Notes	V	U–B	B–V	Spectral Type
				h m s	° ′ ″					
10	ι	Cyg	7420	19 29 31.0	+51 42 49	f	3.79	+0.11	+0.14	A5 Vn
36		Aql	7414	19 30 16.3	− 2 48 18	fv	5.03	+2.05	+1.75	M1 IIIab
6	β	Cyg	7417	19 30 25.1	+27 56 37	fcvmd8	3.24	+0.62	+1.13	K3 II + B9.5 V
8		Cyg	7426	19 31 29.6	+34 26 12	f	4.74	−0.65	−0.14	B3 IV
61	σ	Dra	7462	19 32 22.6	+69 38 54	svd	4.68	+0.38	+0.79	K0 V
38	μ	Aql	7429	19 33 43.4	+ 7 21 45	fvd	4.45	+1.26	+1.17	K3⁻ IIIb CN 0.5
	ι	Tel	7424	19 34 39.7	−48 06 57	f	4.90		+1.09	K0 III
13	θ	Cyg	7469	19 36 14.4	+50 12 13	fd	4.48	−0.03	+0.38	F4 V
52		Sgr	7440	19 36 15.1	−24 54 02	fvd	4.60	−0.15	−0.07	B9
41	ι	Aql	7447	19 36 20.0	− 1 18 13	vd	4.36	−0.44	−0.08	B5 III
39	κ	Aql	7446	19 36 29.3	− 7 02 40	fv	4.95	−0.87	0.00	B0.5 IIIn
5	α	Sge	7479	19 39 45.7	+17 59 46	d	4.37	+0.43	+0.78	G1 II
54		Sgr	7476	19 40 17.6	−16 18 39	fvd	5.30	+1.06	+1.13	K2 III
			7495	19 40 36.4	+45 30 24	svd	5.06	+0.15	+0.40	F5 II–III
6	β	Sge	7488	19 40 42.7	+17 27 30	f	4.37	+0.89	+1.05	G8 IIIa CN 0.5
16		Cyg	7503	19 41 37.0	+50 30 27	sd	5.96	+0.19	+0.64	G1.5 Vb
16		Cyg	7504	19 41 40.0	+50 30 00	sd	6.20	+0.20	+0.66	G3 V
55		Sgr	7489	19 42 05.4	−16 08 31	fd6	5.06	+0.09	+0.33	F0 IVn:
10		Vul	7506	19 43 24.2	+25 45 13	f	5.49	+0.67	+0.93	G8 III
15		Cyg	7517	19 44 00.4	+37 20 09	f	4.89	+0.69	+0.95	G7⁺ III
18	δ	Cyg	7528	19 44 44.5	+45 06 44	vd67	2.87	−0.10	−0.03	B9.5 III
50	γ	Aql	7525	19 45 54.2	+10 35 41	fd	2.72	+1.68	+1.52	K3 II
56		Sgr	7515	19 45 55.5	−19 46 46	fm	4.86		+0.93	K1 III
7	δ	Sge	7536	19 47 03.2	+18 30 55	fcvd6	3.82	+0.96	+1.41	M2 II + A0 V
	ν	Tel	7510	19 47 24.6	−56 22 52	f	5.35		+0.20	A9 Vn
63	ε	Dra	7582	19 48 12.1	+70 14 56	vd67	3.83	+0.52	+0.89	G7 IIIb CN–2
	χ	Cyg	7564	19 50 16.6	+32 53 42	vd	4.23	+0.96	+1.82	S6+/1e
53	α	Aql	7557	19 50 25.1	+ 8 50 53	fd	0.77	+0.08	+0.22	A7 Vn
			7552	19 51 20.2	−39 53 38	sv6	5.33	−0.22	−0.06	A0: IV:p Cr Eu
			7589	19 51 45.5	+47 00 28	sv	5.62	−0.97	−0.07	O9.5 Iab
9		Sge	7574	19 52 01.7	+18 39 09	sv6	6.23	−0.92	+0.01	O8 If
55	η	Aql	7570	19 52 05.5	+ 0 59 10	fv6	3.90	+0.51	+0.89	F6–G1 Ib
			7575	19 52 55.2	− 3 08 03	fsv	5.65	+0.10	+0.20	A3: IV:p Sr Cr Eu
	ι	Sgr	7581	19 54 44.7	−41 53 19	f	4.13	+0.90	+1.08	G8 III
60	β	Aql	7602	19 54 56.7	+ 6 23 16	fvd	3.71	+0.48	+0.86	G8 IV
21	η	Cyg	7615	19 56 01.5	+35 03 48	fvd	3.89	+0.89	+1.02	K0 III
61		Sgr	7614	19 57 31.5	−15 30 43	fm	5.02		+0.05	A3 IV
12	γ	Sge	7635	19 58 25.4	+19 28 17	fsv	3.47	+1.93	+1.57	M0⁻ III
	θ^1	Sgr	7623	19 59 15.0	−35 17 49	f6	4.37	−0.67	−0.15	B2.5 IV
	ε	Pav	7590	19 59 44.1	−72 55 52	fv	3.96	−0.05	−0.03	A0 Va
15		Vul	7653	20 00 47.5	+27 43 57	fv6	4.64	+0.16	+0.18	A4 III
62		Sgr	7650	20 02 11.9	−27 43 52	fv	4.58	+1.80	+1.65	M4.5 III
	ξ	Tel	7673	20 06 48.8	−52 54 10	fv6	4.94	+1.84	+1.62	M1 IIab
	δ	Pav	7665	20 07 59.8	−66 12 07	fv	3.56	+0.45	+0.76	G6/8 IV
28		Cyg	7708	20 09 08.9	+36 49 02	fv6	4.93	−0.77	−0.13	B2.5 V
1	κ	Cep	7750	20 09 09.1	+77 41 21	fd7	4.39	−0.11	−0.05	B9 III
65	θ	Aql	7710	20 10 55.1	− 0 50 39	fd6	3.23	−0.14	−0.07	B9.5 III⁺
33		Cyg	7740	20 13 13.4	+56 32 41	fv6	4.30	+0.08	+0.11	A3 IV–Vn
31	o^1	Cyg	7735	20 13 23.7	+46 43 06	fcvd6	3.79	+0.42	+1.28	K2 II + B4 V
67	ρ	Aql	7724	20 13 55.8	+15 10 28	f6	4.95	+0.01	+0.08	A1 Va

Name	B.S.	Right Ascension	Declination	Notes	V	$U-B$	$B-V$	Spectral Type
		h m s	° ′ ″					
32 o^2 Cyg	7751	20 15 14.4	+47 41 28	cvd6	3.98	+1.03	+1.52	K3 II + B9: V
24 Vul	7753	20 16 27.8	+24 38 52	fm	5.32		+0.95	G8 III
5 α^1 Cap	7747	20 17 14.0	−12 31 54	fd6	4.24	+0.78	+1.07	G3 Ib
34 P Cyg	7763	20 17 30.6	+38 00 34	sv	4.81	−0.58	+0.42	B1pe
6 α^2 Cap	7754	20 17 38.3	−12 34 07	fmd6	3.56		+0.94	G8 IIIb
9 β Cap	7776	20 20 35.4	−14 48 20	fcd67	3.08	+0.28	+0.79	G5 II + A0 V:
37 γ Cyg	7796	20 21 57.5	+40 13 57	fsvd	2.20	+0.53	+0.68	F8 Ib
	7794	20 22 48.4	+ 5 19 07	f	5.31	+0.77	+0.97	G8 III–IV
39 Cyg	7806	20 23 33.6	+32 09 57	s	4.43	+1.50	+1.33	K2.5 III Fe−0.5
α Pav	7790	20 25 03.5	−56 45 35	fvd6	1.94	−0.71	−0.20	B2.5 V
41 Cyg	7834	20 29 05.3	+30 20 36	fv	4.01	+0.27	+0.40	F5 II
69 Aql	7831	20 29 15.5	− 2 54 39	f	4.91	+1.22	+1.15	K2 III
2 θ Cep	7850	20 29 27.4	+62 58 08	f6	4.22	+0.16	+0.20	A7 III
73 Dra	7879	20 31 36.7	+74 55 44	fv6	5.20	+0.11	+0.07	A0p Sr Cr Eu
2 ε Del	7852	20 32 51.3	+11 16 39	fv	4.03	−0.47	−0.13	B6 III
α Ind	7869	20 37 02.5	−47 19 05	fd	3.11	+0.79	+1.00	K0 III CN III/IV
6 β Del	7882	20 37 11.9	+14 34 08	d6	3.63	+0.08	+0.44	F5 IV
71 Aql	7884	20 37 57.0	− 1 07 54	vd6	4.32	+0.69	+0.95	G7.5 IIIa
29 Vul	7891	20 38 11.2	+21 10 29	f	4.82	−0.08	−0.02	A0 Va (shell)
7 κ Del	7896	20 38 45.9	+10 03 34	fd	5.05	+0.24	+0.71	G2 IV
9 α Del	7906	20 39 17.4	+15 53 07	fvd6	3.77	−0.21	−0.06	B9 IV
15 υ Cap	7900	20 39 37.4	−18 09 56	f	5.10	+1.99	+1.66	M1 III
49 Cyg	7921	20 40 44.3	+32 16 49	sd6	5.51		+0.88	G8 IIb
50 α Cyg	7924	20 41 10.6	+45 15 12	fsvd	1.25	−0.24	+0.09	A2 Ia
11 δ Del	7928	20 43 06.5	+15 02 50	fv6	4.43	+0.10	+0.32	F0 IVp
η Ind	7920	20 43 29.5	−51 56 54	f	4.51	+0.09	+0.27	A9 III–IV
β Pav	7913	20 44 17.4	−66 13 50	f	3.42	+0.12	+0.16	A7 III
3 η Cep	7957	20 45 08.3	+61 48 34	fd	3.43	+0.62	+0.92	K0 IV
	7955	20 45 10.0	+57 33 10	fd6	4.51	+0.10	+0.54	F8 IV–V
52 Cyg	7942	20 45 21.2	+30 41 32	d	4.22	+0.89	+1.05	K0 IIIa
16 ψ Cap	7936	20 45 39.2	−25 17 54	f	4.14	+0.02	+0.43	F4 V
53 ε Cyg	7949	20 45 54.5	+33 56 31	fd6	2.46	+0.87	+1.03	K0 III
12 γ^2 Del	7948	20 46 18.6	+16 05 49	fd	4.27	+0.97	+1.04	K1 IV
54 λ Cyg	7963	20 47 07.0	+36 27 47	d67	4.53	−0.49	−0.11	B6 IV
2 ε Aqr	7950	20 47 16.2	− 9 31 25	f	3.77	+0.02	0.00	A1 III^-
3 Aqr	7951	20 47 20.5	− 5 03 20	fv	4.42	+1.92	+1.65	M3 III
ι Mic	7943	20 47 58.8	−44 00 59	fd7	5.11	+0.06	+0.35	F1 IV
55 Cyg	7977	20 48 41.0	+46 05 10	svd	4.84	−0.45	+0.41	B3 Ia
18 ω Cap	7980	20 51 22.5	−26 56 51	fv	4.11	+1.93	+1.64	M0 III Ba 0.5
6 μ Aqr	7990	20 52 15.0	− 9 00 42	f6	4.73	+0.11	+0.32	A3m
β Ind	7986	20 54 13.7	−58 28 59	fd	3.65	+1.23	+1.25	K1 II
32 Vul	8008	20 54 14.4	+28 01 44	fv	5.01	+1.79	+1.48	K4 III
	8023	20 56 18.9	+44 53 45	s6	5.96	−0.85	+0.05	O6
58 ν Cyg	8028	20 56 53.6	+41 08 17	f6	3.94	0.00	+0.02	A0.5 IIIn
33 Vul	8032	20 57 56.2	+22 17 48	f	5.31		+1.40	K3.5 III
20 Cap	8033	20 59 10.6	−19 03 52	svm	6.23			A0 IIIp Si
59 Cyg	8047	20 59 34.2	+47 29 29	fvd6	4.74	−0.94	−0.05	B1.5 Venn
γ Mic	8039	21 00 50.0	−32 17 15	fd	4.67	+0.54	+0.89	G8 III
σ Oct	7228	21 01 57.1	−88 59 12	fv	5.47	+0.13	+0.27	F0 III
ζ Mic	8048	21 02 29.3	−38 39 40	fm	5.35			F3 V

Name	B.S.	Right Ascension	Declination	Notes	V	$U-B$	$B-V$	Spectral Type
		h m s	° ′ ″					
α Oct	8021	21 03 49.7	−77 03 11	fc6	5.15	+0.13	+0.49	G2 III + A7 III
62 ξ Cyg	8079	21 04 39.5	+43 53 52	fsv6	3.72	+1.83	+1.65	K4.5 Ib–II
23 θ Cap	8075	21 05 31.6	−17 15 47	f6	4.07	+0.01	−0.01	A1 V
61 a Cyg	8085	21 06 33.7	+38 42 44	fsvd	5.21	+1.11	+1.18	K5 V
61 b Cyg	8086	21 06 35.1	+38 42 19	svd	6.03	+1.23	+1.37	K7 V
24 Cap	8080	21 06 41.4	−25 02 10	fd	4.50	+1.93	+1.61	M1$^-$ III
13 ν Aqr	8093	21 09 11.2	−11 24 09	f	4.51	+0.70	+0.94	G8 III
5 γ Equ	8097	21 09 58.6	+10 06 04	fvd	4.69	+0.10	+0.26	F0p Sr Eu
64 ζ Cyg	8115	21 12 37.0	+30 11 45	fsd6	3.20	+0.76	+0.99	G8$^+$ III–IIIa Ba 0.5
ο Pav	8092	21 12 38.9	−70 09 27	f6	5.02	+1.56	+1.58	M1/2 III
	8110	21 12 50.7	−27 39 01	f	5.42		+1.42	K5 III
7 δ Equ	8123	21 14 06.9	+ 9 58 35	d67	4.49	−0.01	+0.50	G1 V
65 τ Cyg	8130	21 14 29.5	+38 00 48	vd67	3.72	+0.02	+0.39	F3 IV
8 α Equ	8131	21 15 27.0	+ 5 13 00	fcd6	3.92	+0.29	+0.53	G2 II–III + A4 V
67 σ Cyg	8143	21 17 07.3	+39 21 47	fv6	4.23	−0.39	+0.12	B9 Ia
ε Mic	8135	21 17 29.1	−32 12 15	f	4.71	+0.02	+0.06	A1m A2 Va$^+$
66 υ Cyg	8146	21 17 36.6	+34 51 55	fvd6	4.43	−0.82	−0.11	B2 Ve
5 α Cep	8162	21 18 24.1	+62 33 13	fvd	2.44	+0.11	+0.22	A7 Van
θ Ind	8140	21 19 20.1	−53 28 53	d7	4.39	+0.12	+0.19	A5 V
θ^1 Mic	8151	21 20 17.0	−40 50 30	fv	4.82	−0.07	+0.02	Ap Cr Eu
1 Peg	8173	21 21 44.4	+19 46 20	fd6	4.08	+1.06	+1.11	K1 III
32 ι Cap	8167	21 21 49.8	−16 52 01	f	4.28	+0.58	+0.90	G7 III Fe−1.5
18 Aqr	8187	21 23 47.0	−12 54 38	fvd	5.49		+0.29	F1 V
69 Cyg	8209	21 25 28.6	+36 38 06	sd	5.94	−0.94	−0.08	B0 Ib
γ Pav	8181	21 25 49.8	−65 24 02	fv	4.22	−0.12	+0.49	F6 Vp
34 ζ Cap	8204	21 26 14.4	−22 26 39	fd6	3.74	+0.59	+1.00	G4p Ba
36 Cap	8213	21 28 17.8	−21 50 25		4.51	+0.60	+0.91	G7 IIIb CN−0.5 Fe−1
8 β Cep	8238	21 28 33.9	+70 31 40	fvd6	3.23	−0.95	−0.22	B1 III
71 Cyg	8228	21 29 10.3	+46 30 26	f	5.24	+0.80	+0.97	K0$^-$ III
2 Peg	8225	21 29 36.5	+23 36 21	fd	4.57	+1.93	+1.62	M1$^+$ III
22 β Aqr	8232	21 31 09.9	− 5 36 16	fsd	2.91	+0.56	+0.83	G0 Ib
73 ρ Cyg	8252	21 33 41.9	+45 33 31	fv	4.02	+0.56	+0.89	G8 III Fe−0.5
74 Cyg	8266	21 36 38.9	+40 22 47	f	5.01		+0.18	A5 V
23 ξ Aqr	8264	21 37 21.2	− 7 53 17	fd6	4.69	+0.13	+0.17	A7 V
5 Peg	8267	21 37 24.4	+19 17 05	f	5.45	+0.14	+0.30	F1 IV
9 Cep	8279	21 37 43.2	+62 02 53	sv	4.73	−0.53	+0.30	B2 Ib
40 γ Cap	8278	21 39 40.6	−16 41 47	f6	3.68	+0.20	+0.32	Am
75 Cyg	8284	21 39 53.4	+43 14 23	svd	5.11	+1.90	+1.60	M1 IIIab
ν Oct	8254	21 40 39.8	−77 25 26	fd6	3.76	+0.89	+1.00	K0 III
11 Cep	8317	21 41 48.8	+71 16 37	f	4.56	+1.10	+1.10	K1.5 III
μ Cep	8316	21 43 16.7	+58 44 44	svd	4.08	+2.42	+2.35	M2$^-$ Ia
8 ε Peg	8308	21 43 49.1	+ 9 50 25	fsvd	2.39	+1.70	+1.53	K2 Ib–II
9 Peg	8313	21 44 09.4	+17 18 55	sv	4.34	+1.00	+1.17	G5 Ib
10 κ Peg	8315	21 44 18.3	+25 36 37	d67	4.13	+0.03	+0.43	F5 IV
9 ι PsA	8305	21 44 30.1	−33 03 37	fd6	4.34	−0.11	−0.05	A0 IV
10 ν Cep	8334	21 45 13.9	+61 05 10	fv	4.29	+0.13	+0.52	A2 Ia
81 π^2 Cyg	8335	21 46 30.9	+49 16 29	f6	4.23	−0.71	−0.12	B2.5 III
49 δ Cap	8322	21 46 37.7	−16 09 42	fvd6	2.87	+0.09	+0.29	A5m F2 IV:
14 Peg	8343	21 49 30.8	+30 08 21	f6	5.04	+0.03	−0.03	A1 Vs
ο Ind	8333	21 50 09.7	−69 39 53	f	5.53	+1.63	+1.37	K2/3 III

Name	B.S.	Right Ascension	Declination	Notes	V	$U-B$	$B-V$	Spectral Type
		h m s	° ′ ″					
16 Peg	8356	21 52 43.3	+25 53 23	f6	5.08	−0.67	−0.17	B3 V
51 μ Cap	8351	21 52 53.3	−13 35 14	f	5.08	−0.01	+0.37	F2 V
γ Gru	8353	21 53 28.6	−37 24 01	f	3.01	−0.37	−0.12	B8 III
13 Cep	8371	21 54 38.0	+56 34 32	sv	5.80	−0.02	+0.73	B8 Ib
δ Ind	8368	21 57 24.7	−55 01 43	fd7	4.40	+0.10	+0.28	F0 IV
ε Ind	8387	22 02 47.5	−56 49 02	f	4.69	+0.99	+1.06	K4/5 V
17 ξ Cep	8417	22 03 34.4	+64 35 28	d6	4.29	+0.09	+0.34	A3m
20 Cep	8426	22 04 46.8	+62 44 56	fv	5.27	+1.78	+1.41	K4 III
19 Cep	8428	22 04 55.0	+62 14 36	sd	5.11	−0.84	+0.08	O9 Ib
34 α Aqr	8414	22 05 23.9	− 0 21 23	fsd	2.96	+0.74	+0.98	G2 Ib
λ Gru	8411	22 05 39.9	−39 34 47	f	4.46	+1.66	+1.37	K3 III
33 ι Aqr	8418	22 06 02.0	−13 54 23	f6	4.27	−0.29	−0.07	B9 IV−V
24 ι Peg	8430	22 06 39.7	+25 18 30	fvd6	3.76	−0.04	+0.44	F5 V
α Gru	8425	22 07 45.8	−46 59 51	fvd	1.74	−0.47	−0.13	B7 IV
14 μ PsA	8431	22 07 56.9	−33 01 31	f	4.50	+0.05	+0.05	A1 IVnn
29 π Peg	8454	22 09 39.2	+33 08 28	f	4.29	+0.18	+0.46	F3 III
24 Cep	8468	22 09 39.9	+72 18 15	f	4.79	+0.61	+0.92	G7 II−III
26 θ Peg	8450	22 09 49.3	+ 6 09 39	fv6	3.53	+0.10	+0.08	A2m A1 IV−V
21 ζ Cep	8465	22 10 35.6	+58 09 51	fv6	3.35	+1.71	+1.57	K1.5 Ib
22 λ Cep	8469	22 11 15.4	+59 22 39	sv	5.04	−0.74	+0.25	O6 If
	8485	22 13 33.4	+39 40 39	fvd6	4.49	+1.45	+1.39	K2.5 III
	8546	22 13 52.8	+86 04 14	f6	5.27	−0.11	−0.03	B9.5 Vn
16 λ PsA	8478	22 13 53.3	−27 48 15	f	5.43	−0.55	−0.16	B8 III
23 ε Cep	8494	22 14 45.4	+57 00 22	vd6	4.19	+0.04	+0.28	F0 IV
1 Lac	8498	22 15 38.5	+37 42 41		4.13	+1.63	+1.46	$K3^-$ II−III
43 θ Aqr	8499	22 16 26.3	− 7 49 15	f	4.16	+0.81	+0.98	G9 III
α Tuc	8502	22 17 59.6	−60 17 50	fd6	2.86	+1.54	+1.39	K3 III
ε Oct	8481	22 19 13.0	−80 28 39	fv	5.10	+1.09	+1.47	M6 III
31 Peg	8520	22 21 08.9	+12 10 02	fv	5.01	−0.81	−0.13	B2 IV−V
47 Aqr	8516	22 21 10.9	−21 38 10	fm	5.13		+1.07	K0 III
48 γ Aqr	8518	22 21 16.2	− 1 25 31	fvd6	3.84	−0.12	−0.05	B9.5 III−IV
3 β Lac	8538	22 23 15.9	+52 11 29	f	4.43	+0.77	+1.02	G9 IIIb Ca 1
52 π Aqr	8539	22 24 53.7	+ 1 20 21	fv	4.66	−0.98	−0.03	B1 Ve
δ Tuc	8540	22 26 48.5	−65 00 18	vd7	4.48	−0.07	−0.03	B9.5 IVn
ν Gru	8552	22 28 13.0	−39 10 12	fd	5.47		+0.95	G8 III
55 ζ^1 Aqr	8558	22 28 26.6	− 0 03 32	cmd8	4.59	−0.01	+0.40	F3 IV
55 ζ^2 Aqr	8559	22 28 26.9	− 0 03 31	cmd8	4.42	−0.01	+0.40	F3 IV
δ^1 Gru	8556	22 28 49.4	−43 32 03	fvd	3.97	+0.80	+1.03	G6/8 III
27 δ Cep	8571	22 28 53.5	+58 22 36	fvd6	3.75		+0.60	F5 Ibv
5 Lac	8572	22 29 13.1	+47 40 06	cv6	4.36	+1.11	+1.68	M0 II + B8 V
δ^2 Gru	8560	22 29 18.7	−43 47 16	vd	4.11	+1.71	+1.57	M4.5 IIIa
38 Peg	8574	22 29 41.2	+32 32 03	f	5.47	−0.25	−0.10	B9.5 V
6 Lac	8579	22 30 09.8	+43 05 06	6	4.51	−0.74	−0.09	B2 IV
57 σ Aqr	8573	22 30 15.0	−10 42 59	f6	4.82	−0.11	−0.06	A0 IV
7 α Lac	8585	22 30 58.9	+50 14 38	fd	3.77	0.00	+0.01	A1 Va
17 β PsA	8576	22 31 04.9	−32 23 05	fd7	4.29	+0.02	+0.01	A1 Va
59 υ Aqr	8592	22 34 17.1	−20 44 48	f	5.20	0.00	+0.44	F5 V
62 η Aqr	8597	22 34 58.3	− 0 09 23	f	4.02	−0.26	−0.09	B9 IV−V:n
31 Cep	8615	22 35 35.0	+73 36 15	f	5.08	+0.16	+0.39	F3 III−IV
63 κ Aqr	8610	22 37 22.1	− 4 16 01	fd	5.03	+1.16	+1.14	K2 III

Name	B.S.	Right Ascension	Declination	Notes	V	$U-B$	$B-V$	Spectral Type
		h m s	° ′ ″					
30 Cep	8627	22 38 23.0	+63 32 43	f6	5.19		+0.06	A3 IV
10 Lac	8622	22 38 55.4	+39 00 40	fd	4.88	−1.04	−0.20	O9 V
	8626	22 39 13.9	+37 33 13	sd	6.03		+0.86	G3 Ib–II CN−1 CH 2 Fe−1
11 Lac	8632	22 40 11.1	+44 14 13		4.46	+1.36	+1.33	K2.5 III
18 ε PsA	8628	22 40 14.5	−27 04 58	fv	4.17	−0.37	−0.11	B8 Ve
42 ζ Peg	8634	22 41 05.3	+10 47 31	fd	3.40	−0.25	−0.09	B8.5 III–IV
β Gru	8636	22 42 13.4	−46 55 26	fv	2.11	+1.60	+1.62	M4.5 III
44 η Peg	8650	22 42 39.0	+30 10 55	fcvd6	2.94	+0.55	+0.86	G8 II + F0 V
13 Lac	8656	22 43 45.4	+41 46 47	fd	5.08		+0.96	K0 III
β Oct	8630	22 45 19.4	−81 25 16	f6	4.15	+0.11	+0.20	A9 IV/V
47 λ Peg	8667	22 46 10.2	+23 31 34	f	3.95	+0.91	+1.07	G8 IIIa CN 0.5
46 ξ Peg	8665	22 46 19.1	+12 08 03	d	4.19	−0.03	+0.50	F7 V
68 Aqr	8670	22 47 09.0	−19 39 09	fm	5.26		+0.94	G8 III
ε Gru	8675	22 48 06.3	−51 21 23	fv	3.49	+0.10	+0.08	A2 Va
71 τ Aqr	8679	22 49 11.7	−13 37 56	fvd	4.01	+1.95	+1.57	M0 III
32 ι Cep	8694	22 49 24.7	+66 09 39	fs	3.52	+0.90	+1.05	K0⁻ III
48 μ Peg	8684	22 49 38.4	+24 33 43	fs	3.48	+0.68	+0.93	G8⁺ III
	8685	22 50 36.7	−39 11 48	fm	5.42		+1.43	K3 III
22 γ PsA	8695	22 52 06.6	−32 54 56	vd7	4.46	−0.14	−0.04	A0m A1 III–IV
73 λ Aqr	8698	22 52 13.4	− 7 37 11	fv	3.74	+1.74	+1.64	M2.5 IIIa Fe−1
76 δ Aqr	8709	22 54 15.2	−15 51 39	fv	3.27	+0.08	+0.05	A3 V
	8748	22 54 29.7	+84 18 22	f	4.71	+1.69	+1.43	K4 III
23 δ PsA	8720	22 55 32.1	−32 34 48	d	4.21	+0.69	+0.97	G8 III
	8726	22 56 06.2	+49 41 36	sv	4.95	+1.96	+1.78	K5 Ib
24 α PsA	8728	22 57 14.3	−29 39 44	fv	1.16	+0.08	+0.09	A3 V
	8732	22 58 10.1	−35 33 48	s	6.13		+0.58	F8 III–IV
	8752	22 59 46.0	+56 54 19	sv	5.00	+1.16	+1.42	G4v 0
ζ Gru	8747	23 00 26.4	−52 47 40	f6	4.12	+0.70	+0.98	G8/K0 III
1 o And	8762	23 01 34.5	+42 17 08	fvd6	3.62	−0.53	−0.09	B6pe
π PsA	8767	23 03 05.0	−34 47 24	fv6	5.11	+0.02	+0.29	F0 III
53 β Peg	8775	23 03 24.6	+28 02 31	fvd	2.42	+1.96	+1.67	M2.5: II–III
4 β Psc	8773	23 03 29.7	+ 3 46 46	fv	4.53	−0.49	−0.12	B6 Ve
54 α Peg	8781	23 04 23.2	+15 09 53	fv6	2.49	−0.05	−0.04	A0 III–IV
86 Aqr	8789	23 06 16.8	−23 47 02	d	4.47	+0.58	+0.90	G6 IIIb
θ Gru	8787	23 06 27.6	−43 33 40	d7	4.28	+0.16	+0.42	δ Del
55 Peg	8795	23 06 37.6	+ 9 22 08	fv	4.52	+1.90	+1.57	M1 IIIab
33 π Cep	8819	23 07 39.5	+75 20 49	d67	4.41	+0.46	+0.80	G2 III
88 Aqr	8812	23 09 02.9	−21 12 47	f	3.66	+1.24	+1.22	K1.5 III
ι Gru	8820	23 09 56.2	−45 17 15	f6	3.90	+0.86	+1.02	K1 III
59 Peg	8826	23 11 21.5	+ 8 40 45	f	5.16	+0.08	+0.13	A5 Vn
90 φ Aqr	8834	23 13 56.1	− 6 05 22	f	4.22	+1.90	+1.56	M1.5 III
91 ψ^1 Aqr	8841	23 15 29.9	− 9 07 43	fd	4.21	+0.99	+1.11	K0 III
6 γ Psc	8852	23 16 46.6	+ 3 14 28	fs	3.69	+0.58	+0.92	G9 III: Fe−2
γ Tuc	8848	23 16 59.7	−58 16 37	f	3.99	−0.02	+0.40	F1 III
93 ψ^2 Aqr	8858	23 17 30.9	− 9 13 25	d	4.39	−0.56	−0.15	B5 Vn
γ Scl	8863	23 18 25.2	−32 34 22	f	4.41	+1.06	+1.13	K1 III
95 ψ^3 Aqr	8865	23 18 34.3	− 9 39 06	fvd	4.98		−0.02	A0 Va
62 τ Peg	8880	23 20 15.9	+23 41 57	fv	4.60	+0.10	+0.17	A5 V
98 Aqr	8892	23 22 34.6	−20 08 30	f	3.97	+0.95	+1.10	K0 III
4 Cas	8904	23 24 30.1	+62 14 30	fvmd	4.97		+1.68	M2⁻ IIIab

Name			B.S.	Right Ascension	Declination	Notes	*V*	*U–B*	*B–V*	Spectral Type
				h m s	° ′ ″					
68	υ	Peg	8905	23 25 00.3	+23 21 46	fsm	4.41		+0.61	F8 III
99		Aqr	8906	23 25 39.2	–20 40 59	vm	4.39		+1.48	K5 III
8	κ	Psc	8911	23 26 32.9	+ 1 12 52	fvd	4.94	–0.02	+0.03	A0p Cr Sr
	τ	Oct	8862	23 27 08.1	–87 31 25	f	5.49	+1.43	+1.27	K2 III
10	θ	Psc	8916	23 27 35.3	+ 6 20 16	f	4.28	+1.01	+1.07	K1 III
70		Peg	8923	23 28 46.5	+12 43 09	f	4.55	+0.73	+0.94	G7$^+$ III
			8924	23 29 08.8	– 4 34 25	sv	6.25	+1.16	+1.09	K3$^-$ IIIb
	β	Scl	8937	23 32 34.2	–37 51 36	fv	4.37	–0.36	–0.09	B9.5 IVp Hg Mn
			8952	23 34 39.4	+71 36 02	s	5.84	+1.73	+1.80	G9 Ib
	ι	Phe	8949	23 34 40.5	–42 39 24	fvd	4.71	+0.07	+0.08	Ap Sr
16	λ	And	8961	23 37 11.7	+46 25 03	fvd6	3.82	+0.69	+1.01	G8 III–IV
			8959	23 37 26.9	–45 32 02	f6	4.74	+0.09	+0.08	A1/2 V
17	ι	And	8965	23 37 46.0	+43 13 36	fv6	4.29	–0.29	–0.10	B8 V
35	γ	Cep	8974	23 39 02.1	+77 35 26	fsv	3.21	+0.94	+1.03	K1 III–IV CN 1
17	ι	Psc	8969	23 39 33.9	+ 5 35 08	fvd	4.13	0.00	+0.51	F7 V
19	κ	And	8976	23 40 02.2	+44 17 33	fd	4.14	–0.26	–0.08	B9 IVn
	μ	Scl	8975	23 40 14.6	–32 06 53	fv	5.31	+0.66	+0.97	K0 III
18	λ	Psc	8984	23 41 39.8	+ 1 44 19	f6	4.50	+0.08	+0.20	A7 V
105	ω^2	Aqr	8988	23 42 20.0	–14 35 11	fd6	4.49	–0.12	–0.04	B9.5 IV
106		Aqr	8998	23 43 48.8	–18 19 07	f	5.24	–0.27	–0.08	B9 Vn
20	ψ	And	9003	23 45 39.6	+46 22 43	fd	4.95	+0.82	+1.11	G5 Ib
			9013	23 47 33.0	+67 45 54	f6	5.04		–0.01	A1 Vn
20		Psc	9012	23 47 33.4	– 2 48 12	fd	5.49	+0.70	+0.94	gG8
	δ	Scl	9016	23 48 32.2	–28 10 18	fd	4.57	–0.03	+0.01	A0 Va+n
81	φ	Peg	9036	23 52 06.4	+19 04 43	fv	5.08	+1.86	+1.60	M3$^-$ IIIb
82		Peg	9039	23 52 14.1	+10 54 20	fvm	5.29			A4 Vn
7	ρ	Cas	9045	23 54 00.4	+57 27 28	fv	4.54	+1.12	+1.22	G2v 0
84	ψ	Peg	9064	23 57 22.5	+25 05 59	fv	4.66	+1.68	+1.59	M3 III
27		Psc	9067	23 58 17.4	– 3 35 51	fvd6	4.86	+0.70	+0.93	G9 III
	π	Phe	9069	23 58 32.6	–52 47 16	fv	5.13	+1.03	+1.13	K0 III
28	ω	Psc	9072	23 58 55.6	+ 6 49 18	fv6	4.01	+0.06	+0.42	F3 V
	ε	Tuc	9076	23 59 31.8	65 37 08	fv	4.50	–0.28	–0.08	B9 IV

Notes

f	FK4 position and proper motion
s	MK standard
c	composite or combined spectrum
v	variable star
m	magnitude and color from Yale Bright Star Catalogue 3rd ed.
d	double star data given in Yale Bright Star Catalogue 3rd ed.
1	companion is optical
2	visual binary
3	common proper motion components
4	fixed–seperation companion
5	two spectra are indicated on radial velocity plates
7	magnitude and colors refer to combined light of two or more stars
8	colors but not magnitudes refer to combined light of two or more stars

BS=HR No.	Name	Right Ascension	Declination	Standards Code	V	$U-B$	$B-V$	$V-R$	$V-I$	Spectral Class
		h m s	° ′ ″							
21	11 β Cas	0 08 46.5	+59 06 30		2.27	+0.12	+0.34	+0.31	+0.51	F2 III–IV
39	88 γ Peg	0 12 50.9	+15 08 31	1	2.84	−0.86	−0.23	−0.10	−0.29	B2 IV
45	89 ξ Peg	0 14 12.8	+20 09 54	1	4.80	+1.93	+1.57	+1.34	+2.47	M2 III
63	24 θ And	0 16 41.9	+38 38 24		4.61	+0.05	+0.06	+0.08	+0.09	A2 V
113		0 29 54.6	+59 56 10		5.94	−0.36	+0.01			B9 IIIn
130	15 κ Cas	0 32 34.1	+62 53 26		4.16	−0.80	+0.14	+0.14	+0.20	B1 Ia
321	30 μ Cas	1 07 46.2	+54 53 02		5.18	+0.09	+0.69	+0.63	+1.04	G5 Vp
437	99 η Psc	1 31 04.9	+15 18 26		3.62	+0.74	+0.97	+0.72	+1.22	G8 III
493	107 Psc	1 42 05.3	+20 13 56	1	5.24	+0.49	+0.84	+0.69	+1.12	K1 V
553	6 β Ari	1 54 13.5	+20 46 18		2.65	+0.10	+0.13	+0.14	+0.22	A5 V
617	13 α Ari	2 06 45.0	+23 25 38	2	2.00	+1.13	+1.15	+0.84	+1.46	K2$^-$ IIIab Ca−1
718	73 ξ^2 Cet	2 27 45.6	+ 8 25 36	1	4.29	−0.11	−0.06	+0.02	−0.03	B9 III
753		2 35 40.2	+ 6 51 05	1	5.82	+0.79	+0.97	+0.83	+1.36	K3 V
875		2 56 14.8	− 3 44 32	2	5.17	+0.05	+0.08	+0.11	+0.16	A1 V
996	96 κ Cet	3 18 58.1	+ 3 20 35		4.84	+0.19	+0.68	+0.57	+0.93	G5 V
1034		3 27 30.9	+49 02 14		4.98	−0.55	−0.10	+0.01	−0.09	B3 V
1046		3 29 25.7	+55 25 35		5.10	+0.05	+0.04	+0.09	+0.08	A1 V
1084	18 ε Eri	3 32 34.6	− 9 29 00	1	3.73	+0.58	+0.88	+0.72	+1.19	K2 V
1131	38 o Per	3 43 50.8	+32 15 54		3.83	−0.75	+0.05	+0.12	+0.12	B1 III
1144	18 Tau	3 44 42.8	+24 48 58	1	5.65	−0.36	−0.07	+0.03	−0.04	B8 V
1165	25 η Tau	3 47 02.3	+24 04 56	1	2.87	−0.35	−0.09	+0.03	−0.01	B7 III
1172		3 47 54.1	+23 23 55		5.45	−0.32	−0.07	+0.05	−0.01	B8 V
1228	46 ξ Per	3 58 28.6	+35 46 12		4.04	−0.93	+0.02	+0.16	+0.15	O7.5
1346	54 γ Tau	4 19 21.9	+15 36 36		3.65	+0.81	+0.99	+0.73	+1.20	K0$^-$ IIIab
1373	61 δ Tau	4 22 30.1	+17 31 31		3.76	+0.82	+0.99	+0.73	+1.20	K1 III
1409	74 ε Tau	4 28 10.7	+19 09 51	1	3.54	+0.87	+1.01	+0.73	+1.23	G9.5 III CN 0.5
1411	77 θ^1 Tau	4 28 08.7	+15 56 45	1	3.83	+0.72	+0.95	+0.71	+1.18	G9 III CN−0.5
1412	78 θ^2 Tau	4 28 14.0	+15 51 17	1	3.39	+0.12	+0.18	+0.18	+0.27	A7 III
1543	1 π^3 Ori	4 49 26.0	+ 6 56 55	1	3.19	−0.01	+0.46	+0.42	+0.68	F6 V
1552	3 π^4 Ori	4 50 48.4	+ 5 35 34		3.68	−0.81	−0.16	−0.05	−0.21	B2 III
1641	10 η Aur	5 05 59.3	+41 13 30	1	3.18	−0.67	−0.18	−0.05	−0.22	B3 V
1666	67 β Eri	5 07 28.8	− 5 05 45		2.79	+0.10	+0.13	+0.14	+0.22	A3 III
1781		5 23 19.2	− 0 09 59	1	5.70	−0.88	−0.21	−0.08	−0.27	B2 V
1791	112 β Tau	5 25 49.0	+28 36 06		1.65	−0.49	−0.13	−0.01	−0.11	B7 III
1855	36 υ Ori	5 31 34.0	− 7 18 24	1	4.62	−1.07	−0.26	−0.12	−0.38	B0 V
1861		5 32 18.5	− 1 35 49	1	5.35	−0.93	−0.19	−0.05	−0.24	B1 V
1938		5 40 06.9	+31 21 17	1	6.04	−0.21	+0.05	+0.11	+0.16	B7 V
2010	134 Tau	5 49 07.7	+12 38 57		4.91	−0.16	−0.07	+0.02	−0.06	B9 IV
2047	54 ξ^1 Ori	5 53 56.3	+20 16 31		4.41	+0.08	+0.59	+0.51	+0.82	G0$^-$ V Ca 0.5
2382	12 Mon	6 31 55.3	+ 4 51 42		5.83	+0.78	+1.00	+0.72	+1.25	K0 III
2421	24 γ Gem	6 37 16.7	+16 24 22		1.92	+0.05	0.00	+0.06	+0.05	A0 IV
2693	25 δ CMa	7 08 05.2	−26 22 51		1.84	+0.54	+0.67	+0.51	+0.84	F8 Ia
2763	54 λ Gem	7 17 39.7	+16 33 16		3.58	+0.09	+0.12	+0.12	+0.17	A3 V
2782	30 τ CMa	7 18 23.8	−24 56 25		4.40	−0.99	−0.15	−0.04	−0.22	O9 Ib
2787		7 18 02.4	−36 43 12		4.67	−0.79	−0.10	+0.10	+0.05	B3 Ve

BS=HR No.	Name	Right Ascension	Declination	Standards Code	*V*	*U−B*	*B−V*	*V−R*	*V−I*	Spectral Class
		h m s	° ′ ″							
2852	62 ρ Gem	7 28 37.8	+31 47 59	1	4.18	−0.02	+0.32	+0.32	+0.51	F1 V
2990	78 β Gem	7 44 51.5	+28 02 41		1.14	+0.86	+1.00	+0.75	+1.25	K0 IIIb
3249	17 β Cnc	8 16 06.5	+ 9 12 32	2	3.53	+1.77	+1.48	+1.12	+1.90	K4 III Ba 0.5
3314		8 25 17.2	− 3 52 54		3.90	−0.03	−0.02	+0.03	−0.02	A0 Va
3427	39 Cnc	8 39 40.6	+20 02 04	1	6.39	+0.83	+0.98	+0.72	+1.19	K0 III
3454	7 η Hya	8 42 50.0	+ 3 25 33	2	4.30	−0.74	−0.20	−0.07	−0.26	B4 V
3569	9 ι UMa	8 58 41.8	+48 04 18		3.14	+0.07	+0.19	+0.22	+0.29	A7 IV
3579		9 00 09.3	+41 48 46		3.97	+0.06	+0.43	+0.40	+0.62	F5 V
3815	11 LMi	9 35 12.7	+35 50 40	1	5.41	+0.44	+0.77	+0.62	+0.99	G8 IV−V
3974	21 LMi	10 06 59.3	+35 16 54	1	4.49	+0.07	+0.18	+0.18	+0.25	A7 V
3982	32 α Leo	10 07 58.4	+12 00 15	1	1.35	−0.36	−0.11	−0.02	−0.12	B7 V
4031	36 ζ Leo	10 16 16.4	+23 27 18		3.44	+0.19	+0.31	+0.31	+0.50	F0 III
4033	33 λ UMa	10 16 38.8	+42 57 07		3.45	+0.06	+0.03	+0.08	+0.07	A2 IV
4054	40 Leo	10 19 19.7	+19 30 33		4.80	+0.01	+0.45	+0.45	+0.68	F6 IV
4112	36 UMa	10 30 09.0	+56 01 09		4.84	−0.01	+0.52	+0.48	+0.76	F8 V
4133	47 ρ Leo	10 32 25.0	+ 9 20 43		3.85	−0.95	−0.14	−0.05	−0.21	B1 Iab
4456	90 Leo	11 34 19.1	+16 50 19	1	5.95	−0.65	−0.16	−0.06	−0.24	B3 V
4534	94 β Leo	11 48 40.7	+14 36 50		2.14	+0.08	+0.08	+0.06	+0.08	A3 Va
4550		11 52 32.9	+37 46 21	1	6.45	+0.17	+0.75	+0.66	+1.11	G8 V P
4554	64 γ UMa	11 53 26.3	+53 44 11		2.44	+0.03	0.00	0.00	−0.03	A0 Van
4623	1 α Crv	12 08 01.5	−24 41 13		4.02	−0.02	+0.32	+0.30	+0.48	F2 V
4660	69 δ UMa	12 15 03.4	+57 04 27		3.31	+0.07	+0.08	+0.06	+0.06	A3 V
4662	4 γ Crv	12 15 25.2	−17 30 01	1	2.58	−0.35	−0.11	−0.04	−0.13	B8 IIIp
4707	12 Com	12 22 07.7	+25 53 16	1	4.81	+0.27	+0.49	+0.47	+0.80	G5 III + A5
4751		12 28 22.1	+25 56 27	1	6.65	+0.08	+0.22	+0.15	+0.23	A0p
4752	17 Com	12 28 32.3	+25 57 15	1	5.29	−0.10	−0.06	+0.02	−0.06	A0p (Si)
4785	8 β CVn	12 33 23.2	+41 23 53		4.27	+0.05	+0.59	+0.54	+0.85	G0 V
4983	43 β Com	13 11 31.4	+27 54 58	1	4.26	+0.08	+0.58	+0.49	+0.79	F9.5 V
5019	61 Vir	13 18 00.7	−18 16 11	1	4.74	+0.26	+0.71	+0.58	+0.94	G6 V
5062	80 UMa	13 24 55.5	+55 01 37		4.02	+0.08	+0.16	+0.17	+0.24	A5 V
5185	4 τ Boo	13 46 54.4	+17 29 38		4.50	+0.05	+0.48	+0.41	+0.65	F7 V
5235	8 η Boo	13 54 19.6	+18 26 06		2.68	+0.20	+0.58	+0.44	+0.73	G0 IV
5264	93 τ Vir	14 01 15.9	+ 1 34 50		4.26	+0.13	+0.10	+0.15	+0.21	A3 III
5340	16 α Boo	14 15 19.1	+19 13 16		−0.05	+1.28	+1.23	+0.97	+1.62	K2 IIIp
5359	100 λ Vir	14 18 42.2	−13 20 13		4.52	+0.09	+0.13	+0.10	+0.14	A2m
5447	28 σ Boo	14 34 21.2	+29 46 39		4.47	−0.08	+0.37	+0.34	+0.53	F2 V
5511	109 Vir	14 45 52.2	+ 1 55 27		3.73	−0.03	−0.01	+0.07	+0.05	A0 IVnn
5570	16 Lib	14 56 47.4	− 4 18 59		4.49	+0.04	+0.32	+0.32	+0.49	F0 IV
5634	45 Boo	15 06 58.3	+24 53 53		4.93	−0.02	+0.43	+0.40	+0.61	F5 V
5685	27 β Lib	15 16 36.2	− 9 21 20	2	2.61	−0.37	−0.11	−0.04	−0.14	B8 IVn
5854	24 α Ser	15 43 53.9	+ 6 26 56	2	2.64	+1.25	+1.17	+0.81	+1.37	K2 IIIb CN 1
5868	27 λ Ser	15 46 04.7	+ 7 22 35		4.43	+0.10	+0.60	+0.51	+0.83	G0⁻ V
5933	41 γ Ser	15 56 06.4	+15 41 09		3.86	−0.03	+0.48	+0.49	+0.73	F6 V
5947	13 ε CrB	15 57 16.6	+26 53 57	2	4.15	+1.28	+1.23	+0.89	+1.51	K2 IIIab
6092	22 τ Her	16 19 30.9	+46 19 52	2	3.90	−0.57	−0.15	−0.09	−0.26	B5 IV

BS=HR No.	Name	Right Ascension	Declination	Standards Code	*V*	*U−B*	*B−V*	*V−R*	*V−I*	Spectral Class
		h m s	° ′ ″							
6175	13 ζ Oph	16 36 44.7	−10 33 08		2.56	−0.85	+0.02	+0.10	+0.06	O9.5 Vn
6603	60 β Oph	17 43 06.1	+ 4 34 12	1	2.77	+1.24	+1.17	+0.82	+1.39	K2 III CN 0.5
6629	62 γ Oph	17 47 31.0	+ 2 42 35	1	3.75	+0.04	+0.04	+0.04	+0.04	A0 Van
6705	33 γ Dra	17 56 25.9	+51 29 23		2.22	+1.88	+1.52	+1.14	+1.99	K5 III
7178	14 γ Lyr	18 58 39.8	+32 40 44		3.24	−0.08	−0.05	−0.03	−0.04	B9 II
7235	17 ζ Aql	19 05 03.9	+13 51 07		2.99	−0.01	+0.01	+0.01	+0.01	A0 Vann
7377	30 δ Aql	19 25 07.2	+ 3 05 58		3.36	+0.04	+0.32	+0.25	+0.41	F0 IV
7446	39 κ Aql	19 36 29.3	− 7 02 41	1	4.96	−0.87	0.00	+0.06	+0.02	B0.5 IIIn
7602	60 β Aql	19 54 56.7	+ 6 23 16	1	3.72	+0.49	+0.86	+0.66	+1.15	G8 IV
7906	9 α Del	20 39 17.4	+15 53 07	1	3.77	−0.21	−0.06	0.00	−0.04	B9 IV
7950	2 ε Aqr	20 47 16.2	− 9 31 25		3.77	+0.02	0.00	+0.07	+0.07	A1 V
8085†	61 Cyg A	21 06 34.4	+38 42 32		5.22	+1.11	+1.17	+1.03	+1.68	K5 V
8086†	61 Cyg B	21 06 34.4	+38 42 32		6.03	+1.23	+1.37	+1.17	+2.00	K7 V
8469	22 λ Cep	22 11 15.4	+59 22 39		5.05	−0.74	+0.24	+0.28	+0.43	O6 If
8622	10 Lac	22 38 55.4	+39 00 40	2	4.88	−1.05	−0.20	−0.09	−0.30	O9 V
8781	54 α Peg	23 04 23.2	+15 09 53		2.48	−0.06	−0.04	+0.01	−0.02	A0 III–IV
8832		23 12 55.2	+57 07 37	2	5.57	+0.89	+1.00	+0.83	+1.36	K3 V

†Center of gravity position; see Bright Stars list for orbital position.

BS=HR No.	Name			Right Ascension	Declination	Spectral Type	V	$b-y$	m_1	c_1	β	Type
				h m s	° ′ ″							
9088	85		Peg	0 01 46.7	+27 02 33	G2 V	5.75	+0.430	0.187	0.214	2.558	AF
9091		ζ	Scl	0 01 56.9	−29 45 45	B5 V	5.04	−0.063	0.106	0.450	2.712	
9107				0 04 30.0	+34 37 05	G2 V	6.10	+0.412	0.169	0.312		
15	21	α	And	0 07 59.9	+29 02 57	B9p Hg Mn	2.06*	−0.046	0.120	0.520	2.743	
21	11	β	Cas	0 08 46.5	+59 06 30	F2 III−IV	2.27*	+0.216	0.177	0.785		
27	22		And	0 09 55.7	+46 01 50	F2 II	5.04	+0.273	0.123	1.082	2.666	AF
63	24	θ	And	0 16 41.9	+38 38 24	A2 V	4.62	+0.026	0.180	1.049	2.880	AF
100		κ	Phe	0 25 50.1	−43 43 17	A6 Vn	3.95	+0.098	0.194	0.918	2.846	
114	28		And	0 29 43.5	+29 42 37	Am	5.23*	+0.169	0.165	0.869		
184	20	π	Cas	0 43 03.2	+46 59 01	A5 V	4.96	+0.086	0.226	0.901		
193	22	o	Cas	0 44 18.3	+48 14 36	B5 III	4.62*	+0.007	0.076	0.479	2.667	
233				0 50 16.0	+64 12 25	G0 III−IV + B9.5 V	5.39	+0.355	0.127	0.696		
269	37	μ	And	0 56 20.1	+38 27 32	A5 V	3.87	+0.068	0.194	1.056	2.865	AF
343	33	θ	Cas	1 10 38.6	+55 06 36	A7 V	4.34*	+0.087	0.213	0.997		
373	39		Cet	1 16 13.4	− 2 32 23	gG5	5.41*	+0.554	0.285	0.335		
413	93	ρ	Psc	1 25 51.0	+19 08 00	F2 V:	5.35	+0.259	0.146	0.481		
458	50	υ	And	1 36 21.3	+41 22 05	F8 V	4.10	+0.344	0.179	0.409	2.629	AF
493	107		Psc	1 42 05.3	+20 13 56	K1 V	5.24	+0.493	0.364	0.298		
531	53	χ	Cet	1 49 13.0	−10 43 24	F2 V	4.66	+0.209	0.188	0.649	2.737	
617	13	α	Ari	2 06 45.0	+23 25 38	K2⁻ IIIab Ca−1	2.00	+0.696	0.526	0.395		
623	14		Ari	2 08 59.6	+25 54 17	F2 III	4.98	+0.210	0.185	0.874	2.723	AF
635	64		Cet	2 10 57.3	+ 8 32 06	G0 IV	5.64	+0.361	0.180	0.469	2.627	
660	8	δ	Tri	2 16 35.7	+34 11 25	G0 V	4.86	+0.390	0.187	0.259		
672				2 17 38.0	+ 1 43 21	G0.5 IVb	5.60	+0.370	0.188	0.405	2.619	
675	10		Tri	2 18 30.8	+28 36 30	A2 V	5.03	+0.011	0.161	1.145		
685	9		Per	2 21 49.8	+55 48 42	A2 IA	5.17*	+0.321	−0.038	0.753		
717	12		Tri	2 27 43.5	+29 38 10	F0 III	5.29	+0.178	0.211	0.780		
773	32	ν	Ari	2 38 23.4	+21 55 45	A7 V	5.30	+0.092	0.182	1.095	2.829	
784				2 39 50.4	− 9 29 06	F6 V	5.79	+0.330	0.168	0.362	2.627	
801	35		Ari	2 43 00.6	+27 40 32	B3 V	4.65	−0.052	0.097	0.333	2.684	B
811	89	π	Cet	2 43 45.9	−13 53 25	B7 V	4.25	−0.052	0.105	0.599	2.718	
812	38		Ari	2 44 33.0	+12 24 52	A7 IV	5.18*	+0.136	0.186	0.842	2.798	AF
813	87	μ	Cet	2 44 32.2	+10 04 58	F0 IV	4.27*	+0.189	0.188	0.756	2.751	
870				2 55 49.7	+ 8 21 06	F7 IV	5.97	+0.306	0.175	0.505	2.662	
913				3 01 47.0	− 6 31 26	G0 IV−V	6.20	+0.373	0.205	0.394	2.621	
937		ι	Per	3 08 31.3	+49 35 06	G0 V	4.05	+0.376	0.201	0.376		
962	94		Cet	3 12 23.4	− 1 13 26	F8 V	5.06	+0.363	0.186	0.425		
1006		ζ^1	Ret	3 17 36.3	−62 36 14	G3−5 V	5.51	+0.403	0.204	0.284		
1010		ζ^2	Ret	3 18 03.0	−62 32 05	G2 V	5.23	+0.381	0.183	0.297		
1017	33	α	Per	3 23 47.1	+49 50 06	F5 Ib	1.79	+0.302	0.195	1.074	2.677	AF
1024				3 22 55.8	− 7 49 12	G2 V	6.20	+0.449	0.198	0.295		
1030	1	o	Tau	3 24 24.5	+ 9 00 10	G6 IIIa Fe−1	3.61	+0.547	0.333	0.426		
1089				3 34 25.2	+ 6 23 35	G0	6.49	+0.408	0.183	0.452	2.613	
1140	16		Tau	3 44 21.4	+24 15 59	B7 IV	5.46	+0.005	0.097	0.650	2.750	
1144	18		Tau	3 44 42.8	+24 48 58	B8 V	5.67	−0.021	0.107	0.638	2.750	B

* *V* magnitude may be or is variable.

BS=HR No.	Name			Right Ascension	Declination	Spectral Type	V	$b-y$	m_1	c_1	β	Type
				h m s	° ′ ″							
1178	27		Tau	3 48 42.9	+24 01 51	B8 III	3.62	−0.019	0.092	0.708	2.696	B
1201				3 52 44.3	+17 18 19	F4 V	5.97	+0.221	0.166	0.610	2.712	
1269	42	ψ	Tau	4 06 32.6	+28 58 53	F1 V	5.23	+0.226	0.159	0.588		
1292	45		Tau	4 10 56.3	+ 5 30 14	F1 IV−V	5.71	+0.231	0.164	0.597	2.710	
1303	51	μ	Per	4 14 20.7	+48 23 27	G0 Ib	4.15*	+0.614	0.268	0.551		
1321				4 15 01.8	+ 6 10 54	G5 IV	6.94	+0.425	0.240	0.297	2.580	
1322				4 15 05.2	+ 6 10 07	G0 IV	6.32	+0.369	0.185	0.331	2.606	
1327				4 19 57.7	+65 07 23	G5 IIb	5.26	+0.513	0.286	0.402		
1329	50	ω	Tau	4 16 49.2	+20 33 38	A3m	4.94	+0.146	0.235	0.745		
1331	51		Tau	4 17 56.5	+21 33 41	A8 V	5.64	+0.171	0.191	0.784		
1341	56		Tau	4 19 10.0	+21 45 21	A0p	5.38	−0.094	0.197	0.536	2.768	
1346	54	γ	Tau	4 19 21.9	+15 36 36	K0$^-$ IIIab CN 1	3.64*	+0.596	0.422	0.385		
1373	61	δ	Tau	4 22 30.1	+17 31 31	G9.5 III CN 0.5	3.76*	+0.597	0.424	0.405		
1376	63		Tau	4 22 59.2	+16 45 37	A2 m F3 III	5.63	+0.179	0.244	0.731	2.785	
1387	65	κ	Tau	4 24 55.3	+22 16 38	A7 V	4.22*	+0.070	0.200	1.054	2.864	
1388	67		Tau	4 24 58.1	+22 10 59	A5 N	5.28*	+0.149	0.193	0.840		
1394	71		Tau	4 25 55.1	+15 36 06	A8 Vn	4.49	+0.153	0.183	0.933		
1409	74	ε	Tau	4 28 10.7	+19 09 51	G9.5 III CN 0.5	3.53	+0.616	0.449	0.417		
1411	77	θ^1	Tau	4 28 08.7	+15 56 45	G9 III CN−0.5	3.85	+0.584	0.394	0.393		
1412	78	θ^2	Tau	4 28 14.0	+15 51 17	A7 III	3.41*	+0.101	0.199	1.014	2.831	AF
1414	79		Tau	4 28 25.0	+13 01 53	A5m	5.02	+0.116	0.225	0.907	2.836	
1430	83		Tau	4 30 12.0	+13 42 31	F0 V N	5.40	+0.154	0.200	0.813		
1444	86	ρ	Tau	4 33 25.3	+14 49 45	A8 Vn	4.65	+0.146	0.199	0.829	2.797	
1457	87	α	Tau	4 35 29.4	+16 29 41	K5$^+$ III	0.86*	+0.955	0.814	0.373		
1543	1	π^3	Ori	4 49 26.0	+ 6 56 55	F6 V	3.18*	+0.299	0.162	0.416	2.652	AF
1552	3	π^4	Ori	4 50 48.4	+ 5 35 34	B2 III	3.68	−0.056	0.073	0.135	2.606	B
1577	3	ι	Aur	4 56 30.3	+33 09 17	K3 II	2.69*	+0.937	0.775	0.307		
1620	102	ι	Tau	5 02 38.8	+21 34 47	A7 V	4.63	+0.078	0.203	1.034	2.847	
1641	10	η	Aur	5 05 59.3	+41 13 30	B3 V	3.16*	−0.085	0.104	0.318	2.685	B
1656	104		Tau	5 07 00.4	+18 38 07	G4 V	4.91	+0.410	0.201	0.328		
1662	13		Ori	5 07 13.6	+ 9 27 48	G1 IV	6.17	+0.398	0.185	0.350	2.590	
1672	16		Ori	5 08 54.9	+ 9 49 13	A2m	5.42	+0.136	0.251	0.835	2.828	
1729	15	λ	Aur	5 18 36.8	+40 05 35	G1.5 IV−V Fe−1	4.71	+0.389	0.206	0.363	2.598	
1861				5 32 18.5	− 1 35 49	B1 V	5.34*	−0.074	0.073	0.002	2.615	B
1865	11	α	Lep	5 32 23.9	−17 49 38	F0 Ib	2.57	+0.142	0.150	1.496		
1905	122		Tau	5 36 37.6	+17 02 10	F0 V	5.53	+0.132	0.203	0.856		
2034	136		Tau	5 52 51.3	+27 36 40	A0 IV	4.56	+0.001	0.133	1.152		
2047	54	χ^1	Ori	5 53 56.3	+20 16 31	G0$^-$ V Ca 0.5	4.41	+0.378	0.194	0.307	2.599	AF
2056		λ	Col	5 52 50.5	−33 48 10	B5 V	4.89*	−0.070	0.115	0.413	2.718	
2106		γ	Col	5 57 16.2	−35 17 02	B2.5 IV	4.36	−0.073	0.093	0.362	2.644	
2143	40		Aur	6 06 04.1	+38 29 02	A4m	5.35*	+0.139	0.222	0.923		
2233				6 15 11.3	− 0 30 32	F6 V	5.62	+0.325	0.154	0.446	2.633	
2236				6 15 30.7	+ 1 10 19	F5 IV:	6.36	+0.299	0.148	0.476	2.645	
2264	45		Aur	6 21 09.7	+53 27 22	F5 III	5.33	+0.285	0.170	0.627		
2313				6 24 53.4	− 0 56 28	F8 V	5.88	+0.361	0.170	0.395	2.613	

* *V* magnitude may be or is variable.

BS=HR No.	Name	Right Ascension	Declination	Spectral Type	*V*	*b−y*	m_1	c_1	β	Type
		h m s	° ′ ″							
2473	27 ε Gem	6 43 28.3	+25 08 21	G8 Ib	3.00	+0.868	0.656	0.282		
2483	56 ψ^5 Aur	6 46 11.9	+43 35 08	G0 V	5.25	+0.359	0.184	0.376		
2484	31 ξ Gem	6 44 52.1	+12 54 15	F5 IV	3.36*	+0.288	0.167	0.552		
2585	16 Lyn	6 57 04.3	+45 06 16	A2 V	4.91	+0.014	0.159	1.109		
2622		6 59 55.9	− 5 21 23	G0 III–IV	6.29	+0.359	0.192	0.402		
2657	23 γ CMa	7 03 25.1	−15 37 19	B8 II	4.11	−0.046	0.099	0.556	2.689	
2707	21 Mon	7 11 00.6	− 0 17 21	A8n	5.44*	+0.185	0.184	0.875		
2763	54 λ Gem	7 17 39.7	+16 33 16	A3 V	3.58*	+0.048	0.198	1.055		
2777	55 δ Gem	7 19 40.5	+21 59 48	F0 IV	3.53	+0.221	0.156	0.696	2.712	
2779		7 19 23.5	+ 7 09 27	F8 V	5.92	+0.339	0.169	0.469	2.628	
2798		7 20 55.3	− 8 51 49	F5	6.55	+0.343	0.174	0.390		
2807		7 21 55.9	− 2 57 52	F5	6.24	+0.432	0.216	0.588		
2845	3 β CMi	7 26 44.7	+ 8 18 17	B8 V	2.89*	−0.038	0.113	0.799	2.731	B
2852	62 ρ Gem	7 28 37.8	+31 47 59	F1 V	4.18	+0.214	0.155	0.613	2.713	AF
2857	64 Gem	7 28 52.4	+28 08 03	A6 V	5.05	+0.062	0.202	1.013		
2866		7 29 03.8	− 7 32 08	F8 V	5.86	+0.311	0.155	0.392		
2880	7 δ^1 CMi	7 31 42.5	+ 1 55 51	F0 III	5.25	+0.128	0.173	1.198		
2883		7 31 44.2	− 8 51 52	F5 V	5.93	+0.355	0.124	0.335	2.595	
2886	68 Gem	7 33 10.8	+15 50 36	A1 V	5.28	+0.037	0.143	1.178		
2918		7 36 10.8	+ 5 52 43	G0 V	5.90	+0.375	0.188	0.387	2.610	
2927	25 Mon	7 36 54.3	− 4 05 38	F6 III	5.14	+0.283	0.180	0.643		
2930	71 ο Gem	7 38 40.6	+34 36 08	F3 III	4.89	+0.270	0.173	0.654		
2948/9		7 38 30.9	−26 47 04	B6 V	3.83	−0.076	0.121	0.400		
2961		7 39 11.5	−38 17 27	B2.5 V	4.84	−0.084	0.103	0.303		
2985	77 κ Gem	7 43 59.7	+24 24 59	G8 III	3.57	+0.573	0.379	0.398		
3003	81 Gem	7 45 41.4	+18 31 43	K4 III	4.85	+0.895	0.735	0.451		
3084		7 52 22.7	−38 50 36	B2.5 V	4.50*	−0.083	0.104	0.244		
3131		7 59 31.9	−18 22 42	A2 IVn	4.61	+0.048	0.161	1.122	2.837	
3173	27 Lyn	8 07 53.7	+51 31 44	A1 Va	4.81	+0.017	0.151	1.105		
3249	17 β Cnc	8 16 06.5	+ 9 12 32	K4 III Ba 0.5	3.52	+0.914	0.758	0.371		
3262	18 χ Cnc	8 19 36.6	+27 14 33	F6 V	5.14	+0.314	0.146	0.384		
3271		8 19 50.2	− 0 53 07	F9 V	6.17	+0.385	0.193	0.414	2.612	
3297	1 Hya	8 24 12.6	− 3 43 36	F3 V	5.60	+0.311	0.138	0.400	2.631	
3314		8 25 17.2	− 3 52 54	A0 Va	3.90	−0.006	0.156	1.024	2.898	B
3410	4 δ Hya	8 37 15.6	+ 5 43 49	A1 IVnn	4.15	+0.009	0.152	1.091	2.855	B
3454	7 η Hya	8 42 50.0	+ 3 25 33	B4 V	4.30*	−0.087	0.093	0.241	2.653	B
3459		8 43 18.3	− 7 12 23	G2 IB	4.63	+0.517	0.294	0.472		
3538		8 53 55.8	− 5 24 21	G3 V	6.01	+0.410	0.239	0.325	2.597	
3555	59 σ^2 Cnc	8 56 28.9	+32 56 23	A7 IV	5.45	+0.084	0.205	0.972		
3619	15 UMa	9 08 20.7	+51 38 07	A1m	4.46	+0.165	0.248	0.762		
3624	14 τ UMa	9 10 18.3	+63 32 40	Am	4.65	+0.214	0.253	0.711		
3657		9 13 11.6	+21 18 52	A2 V	6.48	+0.017	0.164	1.094		
3662	18 UMa	9 15 39.2	+54 03 12	A5 V	4.84*	+0.113	0.196	0.892		
3665	22 θ Hya	9 13 58.5	+ 2 20 46	B9.5 IV (C II)	3.88	−0.028	0.145	0.944		
3757	23 UMa	9 30 56.6	+63 05 42	F0 III–IV	3.67*	+0.211	0.180	0.752		

* *V* magnitude may be or is variable.

BS=HR No.	Name	Right Ascension	Declination	Spectral Type	*V*	*b−y*	m_1	c_1	β	Type
		h m s	° ′ ″							
3759	31 τ^1 Hya	9 28 46.1	− 2 44 09	F6 V	4.60	+0.295	0.164	0.453		
3775	25 θ UMa	9 32 21.5	+51 42 43	F6 IV	3.18	+0.314	0.153	0.463		
3800	10 LMi	9 33 45.9	+36 25 52	G7.5 III Fe−0.5	4.55	+0.561	0.349	0.375		
3815	11 LMi	9 35 12.7	+35 50 40	G8 IV−V	5.41	+0.473	0.304	0.372		
3849	38 κ Hya	9 39 56.8	−14 17 53	B5 V	5.07	−0.070	0.110	0.407	2.704	B
3852	14 ο Leo	9 40 45.1	+ 9 55 36	F6 II + A1 V	3.52	+0.306	0.234	0.615		
3856		9 39 08.5	−61 17 38	B9 V	4.51*	−0.034	0.140	0.821		
3881		9 48 06.5	+46 03 22	G0.5 Va	5.10	+0.390	0.203	0.382		
3893	4 Sex	9 50 06.7	+ 4 22 44	F7 Vn	6.24	+0.306	0.161	0.419	2.646	
3901		9 50 59.2	− 6 08 48	F8 V	6.43	+0.363	0.185	0.412		
3906	7 Sex	9 51 49.0	+ 2 29 21	A0 Vs	6.03	−0.015	0.136	1.040		
3928	19 LMi	9 57 13.6	+41 05 30	F5 V	5.14	+0.300	0.165	0.457		
3951	20 LMi	10 00 34.8	+31 57 39	G3 Va Hδ 1	5.35	+0.416	0.234	0.388	2.599	
3974	21 LMi	10 06 59.3	+35 16 54	A7 V	4.49*	+0.106	0.201	0.876	2.837	AF
3975	30 η Leo	10 06 55.5	+16 47 58	A0 Ib	3.53	+0.030	0.068	0.966		
4031	36 ζ Leo	10 16 16.4	+23 27 18	F0 III	3.44	+0.196	0.169	0.986	2.722	AF
4054	40 Leo	10 19 19.7	+19 30 33	F6 IV	4.79*	+0.299	0.166	0.462		
4057/8	41 γ^1 Leo	10 19 33.6	+19 52 47	$K1^-$ IIIb Fe−0.5	1.98*	+0.689	0.457	0.373		
4090	30 LMi	10 25 29.2	+33 50 04	F0 V	4.73	+0.150	0.196	0.959		
4101	45 Leo	10 27 15.2	+ 9 48 04	A0p	6.04	−0.036	0.180	0.956		
4119	30 β Sex	10 29 54.5	− 0 35 54	B6 V	5.08	−0.061	0.113	0.479	2.730	B
4133	47 ρ Leo	10 32 25.0	+ 9 20 43	B1 Iab	3.86*	−0.027	0.040	−0.040	2.552	B
4166	37 LMi	10 38 18.0	+32 00 55	G2.5 IIa	4.72	+0.512	0.297	0.477	2.595	AF
4277	47 UMa	10 59 02.9	+40 28 14	G0 V	5.05	+0.392	0.203	0.337		
4288	49 UMa	11 00 25.3	+39 15 09	F0 M	5.07	+0.142	0.198	1.012		
4293		10 59 48.6	−42 11 08	A3 IV	4.38	+0.059	0.179	1.116		
4300	60 Leo	11 01 55.8	+20 13 12	A0.5m A3 V	4.42	+0.022	0.194	1.019		
4343	11 β Crt	11 11 17.3	−22 47 05	A2 IV	4.47	+0.011	0.164	1.190	2.877	
4378		11 17 57.5	+12 01 33	A2 V	6.66	+0.024	0.190	1.052		
4386	77 σ Leo	11 20 45.0	+ 6 04 14	A0 III^+	4.05	−0.020	0.127	1.014		
4392	56 UMa	11 22 25.0	+43 31 26	G8 II	4.99	+0.610	0.416	0.396		
4405	15 γ Crt	11 24 30.4	−17 38 34	A7 IV−V	4.07	+0.118	0.195	0.895	2.823	AF
4456	90 Leo	11 34 19.1	+16 50 19	B3 V	5.95	−0.066	0.095	0.323	2.687	B
4501	62 UMa	11 41 10.9	+31 47 15	F4 V	5.74	+0.312	0.118	0.401		
4515	2 ξ Vir	11 44 53.9	+ 8 18 00	A4 V	4.85	+0.090	0.196	0.928	2.855	
4527	93 Leo	11 47 36.0	+20 15 38	G4 III−IV + A7 V	4.53*	+0.352	0.186	0.725		
4534	94 β Leo	11 48 40.7	+14 36 50	A3 Va	2.14*	+0.044	0.210	0.975	2.900	AF
4540	5 β Vir	11 50 18.3	+ 1 48 25	F9 V	3.60	+0.354	0.186	0.415	2.629	AF
4550		11 52 32.9	+37 46 21	G8 V P	6.43	+0.483	0.225	0.153		
4554	64 γ UMa	11 53 26.3	+53 44 11	A0 Van	2.44	+0.006	0.153	1.113	2.884	B
4618		12 07 41.8	−50 37 10	B2 IIIne	4.47	−0.076	0.108	0.254	2.682	
4689	15 η Vir	12 19 31.3	− 0 37 31	A1 IV^+	3.90*	+0.017	0.163	1.130		
4695	16 Vir	12 19 58.1	+ 3 21 15	K0.5 IIIb Fe−0.5	4.97	+0.717	0.485	0.516		
4705		12 21 48.2	+24 48 55	A0 V	6.20	−0.002	0.169	1.034		
4707	12 Com	12 22 07.7	+25 53 16	G5 III + A5	4.81	+0.322	0.175	0.779	2.701	

* *V* magnitude may be or is variable.

BS=HR No.	Name	Right Ascension	Declination	Spectral Type	*V*	*b−y*	m_1	c_1	β	Type
		h m s	° ′ ″							
4753	18 Com	12 29 04.5	+24 09 01	F5 III	5.48	+0.289	0.170	0.609		
4775	8 η Crv	12 31 41.0	−16 09 17	F0 IV	4.30*	+0.245	0.167	0.543	2.700	
4789	23 Com	12 34 28.7	+22 40 14	A0m A1 IV	4.81	+0.008	0.144	1.090		
4802	τ Cen	12 37 17.4	−48 30 00	A1 IVnn	3.86	+0.026	0.159	1.086	2.870	
4861	28 Com	12 47 51.8	+13 35 38	A1 V	6.56	+0.012	0.167	1.052		
4865	29 Com	12 48 31.7	+14 09 48	A1 V	5.70	+0.020	0.156	1.130		
4869	30 Com	12 48 55.6	+27 35 35	A2 V	5.78	+0.025	0.169	1.074		
4883	31 Com	12 51 20.0	+27 34 53	G0 IIIp	4.93	+0.437	0.186	0.416	2.592	AF
4889		12 53 01.1	−40 08 18	A7 III	4.26	+0.125	0.185	0.971	2.816	
4914	12 α^1 CVn	12 55 39.4	+38 21 19	F0 V	5.60	+0.230	0.152	0.578		
4931	78 UMa	13 00 24.6	+56 24 24	F2 V	4.92*	+0.244	0.170	0.575	2.707	AF
4983	43 β Com	13 11 31.4	+27 54 58	F9.5 V	4.26	+0.370	0.191	0.337	2.608	AF
5011	59 Vir	13 16 24.2	+ 9 27 48	F8 V	5.19	+0.372	0.191	0.385	2.614	
5017	20 CVn	13 17 12.4	+40 36 43	F3 III	4.72*	+0.174	0.238	0.915		
5062	80 UMa	13 24 55.5	+55 01 37	A5 V	4.02*	+0.097	0.192	0.928	2.847	AF
5072	70 Vir	13 28 03.7	+13 49 03	G4 V	4.97	+0.446	0.232	0.350		
5163		13 43 30.8	− 5 27 41	A1 V	6.53	+0.028	0.172	0.980		
5168	1 Cen	13 45 15.6	−33 00 22	F3 IV	4.23*	+0.247	0.164	0.548	2.700	
5235	8 η Boo	13 54 19.6	+18 26 06	G0 IV	2.68	+0.376	0.203	0.476	2.627	AF
5270		14 02 09.7	+ 9 43 21	K1 CN−5 Fe−4	6.21	+0.638	0.087	0.541	2.533	AF
5280		14 02 43.0	+51 00 28	A2 V	6.15	+0.020	0.181	1.016		
5285	χ Cen	14 05 35.2	−41 08 38	B2 V	4.36*	−0.094	0.102	0.161	2.661	
5304	12 D Boo	14 10 03.4	+25 07 37	F8 IV	4.82	+0.347	0.172	0.443		
5414		14 28 11.6	+28 19 21	A1 V	7.62	+0.014	0.168	1.018		
5415		14 28 13.5	+28 19 27	A1 V	7.12	+0.008	0.146	1.020		
5447	28 σ Boo	14 34 21.2	+29 46 39	F2 V	4.47*	+0.253	0.135	0.484	2.675	AF
5511	109 Vir	14 45 52.2	+ 1 55 27	A0 IVnn	3.74	+0.006	0.137	1.078	2.846	B
5522		14 48 31.0	− 0 49 00	B9 Vp:v	6.16	−0.007	0.132	0.996		
5530	8 α^1 Lib	14 50 16.3	−15 57 59	F3 V	5.16	+0.265	0.156	0.494	2.681	AF
5531	9 α^2 Lib	14 50 27.8	−16 00 39	A3 III–IV	2.75	+0.074	0.192	0.996	2.860	AF
5626	λ Lup	15 08 20.1	−45 15 05	B3 V	4.06	−0.077	0.105	0.265	2.687	
5633		15 06 59.7	+18 28 13	A3 V	6.02	+0.032	0.190	1.017		
5634	45 Boo	15 06 58.3	+24 53 53	F5 V	4.93	+0.287	0.161	0.448		
5660	1 Lup	15 14 09.7	−31 29 30	F1 II	4.92	+0.246	0.132	1.367	2.741	
5681	49 δ Boo	15 15 12.0	+33 20 33	G8 III Fe−1	3.49	+0.587	0.346	0.410		
5685	27 β Lib	15 16 36.2	− 9 21 20	B8 IVn	2.61	−0.040	0.100	0.750	2.706	B
5717	7 Ser	15 22 01.9	+12 35 38	A0 V	6.28	+0.008	0.136	1.044		
5752		15 28 30.0	+47 13 37	Am	6.15	+0.046	0.194	1.142		
5754		15 27 32.7	+62 18 05	A5 IV	6.40	+0.062	0.210	0.982		
5793	5 α CrB	15 34 22.2	+26 44 23	A0 IV	2.24*	+0.000	0.144	1.060		
5825		15 40 40.2	−44 38 12	F5 IV–V	4.64	+0.270	0.152	0.458	2.678	
5854	24 α Ser	15 43 53.9	+ 6 26 56	K2 IIIb CN 1	2.64	+0.715	0.572	0.445		
5868	27 λ Ser	15 46 04.7	+ 7 22 35	G0$^-$ V	4.43	+0.383	0.193	0.366	2.605	
5885	1 Sco	15 50 31.6	−25 43 44	B1.5 V N	4.65	+0.006	0.070	0.122	2.639	
5933	41 γ Ser	15 56 06.4	+15 41 09	F6 V	3.86	+0.319	0.151	0.401	2.632	AF

* *V* magnitude may be or is variable.

BS=HR No.	Name			Right Ascension	Declination	Spectral Type	V	$b-y$	m_1	c_1	β	Type
				h m s	° ′ ″							
5936	12	λ	CrB	15 55 31.3	+37 58 06	F2	5.44	+0.230	0.161	0.654		
5947	13	ε	CrB	15 57 16.6	+26 53 57	K2 IIIab	4.15	+0.751	0.570	0.414		
5968	15	ρ	CrB	16 00 45.5	+33 19 34	G2 V	5.40	+0.396	0.176	0.331		
5993	9	ω^1	Sco	16 06 22.1	−20 38 57	B1 V	3.94	+0.037	0.042	0.009	2.617	B
5997	10	ω^2	Sco	16 06 57.9	−20 50 56	G4 II–III	4.32	+0.522	0.285	0.448	2.577	AF
6027	14	ν	Sco	16 11 33.5	−19 26 30	B2 IVp	3.99	+0.080	0.051	0.137	2.663	
6092	22	τ	Her	16 19 30.9	+46 19 52	B5 IV	3.88*	−0.056	0.089	0.440	2.702	B
6141	22		Sco	16 29 45.1	−25 05 57	B2 V	4.79	−0.047	0.092	0.191	2.665	B
6175	13	ζ	Oph	16 36 44.7	−10 33 08	O9.5 Vn	2.56	+0.088	0.014	−0.069	2.583	
6243	20		Oph	16 49 25.1	−10 46 13	F7 III	4.64	+0.311	0.164	0.532	2.647	
6332	59		Her	17 01 19.7	+33 34 44	A3 IV	5.28	+0.001	0.172	1.102	2.885	
6355	60		Her	17 05 01.8	+12 45 03	A4 IV	4.90	+0.064	0.207	0.992	2.877	AF
6378	35	η	Oph	17 09 56.9	−15 42 58	A2 Va+	2.42	+0.029	0.186	1.076	2.894	
6458	72		Her	17 20 22.7	+32 28 37	G0 V	5.39*	+0.405	0.178	0.312	2.588	
6536	23	β	Dra	17 30 15.8	+52 18 24	G2 Ib–IIa	2.78	+0.610	0.323	0.423	2.599	
6581	56	ο	Ser	17 40 59.6	−12 52 18	A2 Va	4.25*	+0.049	0.168	1.108	2.874	
6588	85	ι	Her	17 39 15.2	+46 00 36	B3 IV	3.80	−0.064	0.078	0.294	2.661	B
6595	58		Oph	17 42 58.8	−21 40 48	F7 V:	4.87	+0.304	0.150	0.408	2.645	
6603	60	β	Oph	17 43 06.1	+ 4 34 12	K2 III CN 0.5	2.76	+0.719	0.553	0.451		
6629	62	γ	Oph	17 47 31.0	+ 2 42 35	A0 Van	3.75	+0.024	0.165	1.055	2.905	B
6714	67		Oph	18 00 16.2	+ 2 55 53	B5 Ib	3.97	+0.081	0.020	0.302	2.585	B
6723	68		Oph	18 01 22.4	+ 1 18 18	A0.5 Van	4.44*	+0.029	0.137	1.087	2.842	
6743		θ	Ara	18 06 02.8	−50 05 34	B2 Ib	3.67	+0.007	0.037	0.006	2.582	
6775	99		Her	18 06 44.4	+30 33 38	F7 V	5.06	+0.356	0.136	0.321		
6930		γ	Sct	18 28 46.2	−14 34 16	A2 III−	4.69	+0.045	0.147	1.208	2.846	
7069	111		Her	18 46 41.4	+18 10 22	A5 III	4.36	+0.061	0.216	0.942	2.895	AF
7119				18 54 17.3	−15 36 46	B5 II	5.09	+0.175	0.026	0.468	2.626	
7152		ε	CrA	18 58 13.1	−37 07 03	F0 V	4.85*	+0.253	0.161	0.617		
7178	14	γ	Lyr	18 58 39.8	+32 40 44	B9 II	3.24	+0.001	0.093	1.219	2.751	B
7235	17	ζ	Aql	19 05 03.9	+13 51 07	A0 Vann	2.99	+0.012	0.147	1.080	2.873	B
7253				19 06 19.9	+28 36 59	F0 III	5.53	+0.176	0.189	0.747	2.756	
7254		α	CrA	19 08 57.8	−37 55 00	A2 IV	4.11	+0.024	0.181	1.057	2.890	
7328	1	κ	Cyg	19 16 55.8	+53 21 16	G9 III	3.76	+0.579	0.390	0.430		
7340	44	ρ^1	Sgr	19 21 14.3	−17 51 42	F0 IV–V	3.93*	+0.130	0.194	0.950	2.809	
7377	30	δ	Aql	19 25 07.2	+ 3 05 58	F0 IV	3.37*	+0.203	0.170	0.711	2.733	AF
7446	39	κ	Aql	19 36 29.3	− 7 02 41	B0.5 IIIn	4.95	+0.085	−0.024	−0.031	2.563	B
7447	41	ι	Aql	19 36 20.0	− 1 18 13	B5 III	4.36	−0.017	0.087	0.574	2.704	B
7462	61	σ	Dra	19 32 22.6	+69 38 54	K0 V	4.67	+0.472	0.324	0.266		
7469	13	θ	Cyg	19 36 14.4	+50 12 13	F4 V	4.49	+0.262	0.157	0.502	2.689	
7479	5	α	Sge	19 39 45.7	+17 59 46	G1 II	4.39	+0.489	0.259	0.471		
7503	16		Cyg	19 41 37.0	+50 30 27	G1.5 Vb	5.98	+0.410	0.212	0.368		
7504				19 41 40.0	+50 30 00	G3 V	6.23	+0.417	0.223	0.349		
7525	50	γ	Aql	19 45 54.2	+10 35 41	K3 II	2.71	+0.936	0.762	0.292		
7534	17		Cyg	19 46 08.5	+33 42 36	F5 V	5.01	+0.312	0.155	0.436		
7557	53	α	Aql	19 50 25.1	+ 8 50 53	A7 Vn	0.76	+0.137	0.178	0.880		

* V magnitude may be or is variable.

BS=HR No.	Name	Right Ascension	Declination	Spectral Type	*V*	*b−y*	m_1	c_1	β	Type
		h m s	° ′ ″							
7560	54 ο Aql	19 50 40.1	+10 23 47	F8 V	5.13	+0.356	0.182	0.415		
7602	60 β Aql	19 54 56.7	+ 6 23 16	G8 IV	3.72*	+0.522	0.303	0.345		
7610	61 φ Aql	19 55 52.9	+11 24 13	A1 V	5.29	−0.006	0.178	1.021		
7773	8 ν Cap	20 20 14.9	−12 47 00	B9 V	4.76	−0.020	0.135	1.011	2.853	
7796	37 γ Cyg	20 21 57.5	+40 13 57	F8 Ib	2.23	+0.396	0.296	0.885	2.641	AF
7858	3 η Del	20 33 35.8	+13 00 04	A2 V	5.40	+0.023	0.207	0.983	2.918	
7906	9 α Del	20 39 17.4	+15 53 07	B9 IV	3.77	−0.019	0.125	0.893	2.799	B
7936	16 ψ Cap	20 45 39.2	−25 17 54	F4 V	4.14	+0.278	0.161	0.465	2.673	
7949	53 ε CYG	20 45 54.5	+33 56 31	K0−III	2.46	+0.627	0.415	0.425		
7977	55 Cyg	20 48 41.0	+46 05 10	B3 Ia	4.86*	+0.356	−0.067	0.153	2.530	B
7984	56 Cyg	20 49 48.9	+44 01 51	A4m	5.04	+0.108	0.209	0.897	2.844	
8060	22 η Cap	21 03 58.7	−19 53 06	A5 V	4.86	+0.090	0.191	0.946	2.861	
8085†	61 Cyg A	21 06 34.4	+38 42 32	K5 V	5.21	+0.656	0.677	0.136		
8086†	61 Cyg B	21 06 34.4	+38 42 32	K7 V	6.04	+0.792	0.673	0.063		
8143	67 σ Cyg	21 17 07.3	+39 21 47	B9 Ia	4.23	+0.138	0.027	0.571	2.583	B
8162	5 α Cep	21 18 24.1	+62 33 13	A7 Van	2.45*	+0.125	0.190	0.936	2.808	
8181	γ Pav	21 25 49.8	−65 24 02	F6 Vp	4.23	+0.333	0.118	0.315	2.613	
8267	5 Peg	21 37 24.4	+19 17 05	F1 IV	5.47*	+0.199	0.172	0.890	2.734	
8279	9 Cep	21 37 43.2	+62 02 53	B2 Ib	4.73*	+0.275	−0.051	0.135	2.558	B
8313	9 Peg	21 44 09.4	+17 18 55	G5 Ib	4.34	+0.706	0.479	0.346		
8344	13 Peg	21 49 47.3	+17 15 02	F2 III−IV	5.29*	+0.263	0.156	0.545	2.688	
8353	γ Gru	21 53 28.6	−37 24 01	B8 III	3.01	−0.045	0.106	0.726		
8425	α Gru	22 07 45.8	−46 59 51	B7 IV	1.74	−0.058	0.107	0.568	2.729	
8431	14 μ PsA	22 07 56.9	−33 01 31	A1 IVnn	4.50	+0.032	0.167	1.070	2.872	
8454	29 π Peg	22 09 39.2	+33 08 28	F3 III	4.29	+0.304	0.177	0.778		
8494	23 ε Cep	22 14 45.4	+57 00 22	F0 IV	4.19*	+0.169	0.192	0.787	2.758	AF
8551	35 Peg	22 27 28.8	+ 4 39 28	K0 III−IV	4.79	+0.640	0.420	0.418		
8585	7 α Lac	22 30 58.9	+50 14 38	A1 Va	3.77	+0.001	0.173	1.030	2.906	B
8613	9 Lac	22 37 03.8	+51 30 23	A7 IV	4.65	+0.149	0.172	0.935	2.784	
8622	10 Lac	22 38 55.4	+39 00 40	O9 V	4.89	−0.066	0.037	−0.117	2.587	B
8630	β Oct	22 45 19.4	−81 25 16	A9 IV/V	4.14	+0.124	0.191	0.915	2.817	
8634	42 ζ Peg	22 41 05.3	+10 47 31	B8.5 III−IV	3.40	−0.035	0.114	0.867	2.768	B
8665	46 ξ Peg	22 46 19.1	+12 08 03	F7 V	4.19	+0.330	0.147	0.407		
8675	ε Gru	22 48 06.3	−51 21 23	A2 Va	3.49	+0.051	0.161	1.143	2.856	
8709	76 δ Aqr	22 54 15.2	−15 51 39	A3 V	3.28	+0.036	0.167	1.157	2.890	
8728	24 α PsA	22 57 14.3	−29 39 44	A3 V	1.16	+0.039	0.208	0.985	2.906	
8729	51 Peg	22 57 05.8	+20 43 42	G5 V	5.45	+0.415	0.233	0.372		
8781	54 α Peg	23 04 23.2	+15 09 53	A0 III−IV	2.48	−0.012	0.130	1.128	2.840	B
8826	59 Peg	23 11 21.5	+ 8 40 45	A5 Vn	5.16	+0.076	0.164	1.091	2.820	
8830	7 And	23 12 12.3	+49 21 55	F0 V	4.53	+0.188	0.169	0.713		
8848	γ Tuc	23 16 59.7	−58 16 37	F1 III	3.99	+0.271	0.143	0.564	2.665	
8880	62 τ Peg	23 20 15.9	+23 41 57	A5 V	4.60*	+0.105	0.166	1.009		
8899		23 23 25.3	+32 29 24	F4 Vw	6.69	+0.321	0.121	0.404		
8954	16 PsC	23 36 00.3	+ 2 03 38	F6 Vbvw	5.69	+0.306	0.122	0.386		
8965	17 ι And	23 37 46.0	+43 13 36	B8 V	4.29	−0.031	0.100	0.784	2.728	B
8969	17 ι Psc	23 39 33.9	+ 5 35 08	F7 V	4.13	+0.331	0.161	0.398	2.621	AF
8976	19 κ And	23 40 02.2	+44 17 33	B9 IVn	4.14	−0.035	0.131	0.831	2.833	B
9072	28 ω Psc	23 58 55.6	+ 6 49 18	F3 V	4.03	+0.271	0.154	0.631	2.667	
9076	ε Tuc	23 59 31.8	−65 37 08	B9 IV	4.50	−0.023	0.098	0.881	2.722	

* *V* magnitude may be or is variable.

†Center of gravity position; see Bright Stars list for orbital position

HD No.	BS=HR No.	Name	Right Ascension	Declination	Vis. Mag.	Spectral Type	Radial Velocity
			h m s	° ′ ″			km/sec
693‡	33	6 Cet	0 10 53.0	−15 30 33	4.89	F6 V	+ 14.7±0.2
3712	168	18 α Cas	0 40 04.7	+56 29 47	2.23	K0$^-$ IIIa	− 3.9 0.1
3765			0 40 24.4	+40 08 50	7.36	dK5	− 63.0 0.2
4128‡	188	16 β Cet	0 43 12.8	−18 01 40	2.04	G9.5 CN−1	+ 13.1 0.1
4388			0 46 02.8	+30 54 39	7.51	K3 III	− 28.3 0.6
8779‡	416		1 26 04.2	− 0 26 15	6.41	gK0	− 5.0±0.6
9138	434	98 μ Psc	1 29 47.5	+ 6 06 19	4.84	K4 III	+ 35.4 0.5
12029			1 58 16.0	+29 20 36	7.80	K2 III	+ 38.6 0.5
12929	617	13 α Ari	2 06 45.0	+23 25 38	2.00	K2$^-$ IIIab Ca−1	− 14.3 0.2
18884‡	911	92 α Cet	3 01 53.2	+ 4 03 38	2.53	M1.5 IIIa	− 25.8 0.1
22484‡	1101	10 Tau	3 36 29.4	+ 0 22 42	4.28	F9 IV−V	+ 27.9±0.1
23169			3 43 25.9	+25 42 08	8.75	G2 V	+ 13.3 0.2
26162‡	1283	43 Tau	4 08 43.7	+19 35 23	5.50	K1 III	+ 23.9 0.6
29139‡	1457	87 α Tau	4 35 29.4	+16 29 41	0.85	K5$^+$ III	+ 54.1 0.1
29587			4 41 04.4	+42 06 19	7.29	dG2	+112.4 0.2
32963			5 07 27.9	+26 19 08	7.72	G2 V	− 63.1±0.4
36079‡	1829	9 β Lep	5 27 55.4	−20 45 54	2.84	G5 II	− 13.5 0.1
42397			6 11 07.0	+25 00 43	8.03	G0 IV	+ 37.4 0.4
51250*	2593	18 μ CMa	6 55 46.0	−14 02 00	5.00	K2 III + B9 V:	+ 19.6 0.5
62509	2990	78 β Gem	7 44 51.5	+28 02 41	1.14	K0 IIIb	+ 3.3 0.1
65583*			8 00 04.4	+29 14 09	7.00	dG7	+ 12.5±0.4
65934			8 01 43.7	+26 39 34	7.94	G8 III	+ 35.0 0.3
66141‡	3145		8 01 52.5	+ 2 21 20	4.39	K2 III	+ 70.9 0.3
75935			8 53 23.3	+26 56 31	8.63	G8 V	− 18.9 0.3
80170	3694		9 16 39.5	−39 22 11	5.33	K5 III−IV	0.0 0.2
81797‡	3748	30 α Hya	9 27 13.1	− 8 37 33	1.98	K3 IIIa	− 4.4±0.2
84441	3873	17 ε Leo	9 45 25.6	+23 48 32	2.98	G1 IIab	+ 4.8 0.1
86801			10 01 08.5	+28 36 10	8.88	G0 V	− 14.5 0.4
89449‡	4054	40 Leo	10 19 19.7	+19 30 33	4.79	F6 IV	+ 6.5 0.5
90861			10 29 28.6	+28 37 11	7.20	K2 III	+ 36.3 0.4
92588‡	4182	33 Sex	10 41 01.3	− 1 42 07	6.26	sgK1	+ 42.8±0.1
102494			11 47 33.1	+27 22 55	7.44	G8 IV	− 22.9 0.3
102870	4540	5 β Vir	11 50 18.3	+ 1 48 25	3.61	F9 V	+ 5.0 0.2
103095	4550		11 52 32.9	+37 46 21	6.45	G8 Vp	− 99.1 0.3
107328‡	4695	16 Vir	12 19 58.1	+ 3 21 15	4.96	K0.5 IIIb Fe−0.5	+ 35.7 0.3
108903	4763	γ Cru	12 30 44.8	−57 04 17	1.63	F0 IV	+ 21.3±0.1
109379	4786	9 β Crv	12 33 59.5	−23 21 19	2.65	G5 IIb	− 7.0 0.0
112299			12 55 06.5	+25 46 43	8.66	F8 V	+ 3.4 0.5
114762‡			13 11 58.0	+17 33 24	7.31	dF7	+ 49.9 0.5
122693			14 02 31.3	+24 35 52	8.21	F8 V	− 6.3 0.2
123782	5300	13 Boo	14 08 00.5	+49 29 37	5.25	M2 IIIab	− 13.4±0.3
124897‡	5340	16 α Boo	14 15 19.1	+19 13 16	−0.04	K1.5 III Fe−0.5	− 5.3 0.1
126053	5384		14 22 52.3	+ 1 16 35	6.27	G1 V	− 18.5 0.4
132737			14 59 33.1	+27 11 23	8.02	K0 III	− 24.1 0.3
136202‡	5694	5 Ser	15 18 55.8	+ 1 47 36	5.06	F8 IV−V	+ 53.5 0.2

*Radial velocity is now known to vary.

‡Candidate for inclusion in revised list of primary standard stars currently in preparation by IAU Commission 30; recommended for intensive observation.

HD No.	BS=HR No.	Name			Right Ascension	Declination	Vis. Mag.	Spectral Type	Radial Velocity
					h m s	° ′ ″			km/sec
140913					15 44 49.0	+28 29 36	8.21	G0 V	− 20.8±0.4
144579					16 04 41.3	+39 10 36	6.66	dG8	− 60.0 0.3
145001	6008	7	κ	Her	16 07 44.2	+17 04 00	5.00	G5 III	− 9.5 0.2
146051‡	6056	1	δ	Oph	16 13 57.1	− 3 40 32	2.74	M0.5 III	− 19.8 0.0
149803					16 35 36.6	+29 45 36	8.40	F7 V	− 7.6 0.4
150798	6217		α	TrA	16 47 51.9	−69 00 53	1.92	K2 IIb−IIIa	− 3.7±0.2
154417	6349				17 04 54.0	+ 0 42 47	6.01	G0 V	− 17.4 0.3
157457	6468		κ	Ara	17 25 24.8	−50 37 38	5.23	G8 III	+ 17.4 0.2
161096‡	6603	60	β	Oph	17 43 06.1	+ 4 34 12	2.77	K2 III CN 0.5	− 12.0 0.1
168454	6859	19	δ	Sgr	18 20 30.9	−29 49 55	2.70	K2.5 IIIa CN 0.5	− 20.0 0.0
171232					18 32 17.7	+25 29 00	7.73	G8 III	− 35.9±0.5
171391‡	6970				18 34 37.4	−10 59 01	5.14	G8 III	+ 6.9 0.2
182572‡	7373	31		Aql	19 24 36.8	+11 55 41	5.16	G7 IV Hδ1	−100.5 0.4
†					19 34 42.4	+29 04 12	9.05	F7 V	− 36.6 0.5
186791	7525	50	γ	Aql	19 45 54.2	+10 35 41	2.72	K3 II	− 2.1 0.2
187691‡	7560	54	ο	Aql	19 50 40.1	+10 23 47	5.11	F8 V	+ 0.1±0.3
194071					20 22 18.9	+28 13 20	8.13	G8 III	− 9.8 0.1
203638‡	8183	33		Cap	21 23 44.2	−20 53 03	5.77	K0 III	+ 21.9 0.1
204867‡	8232	22	β	Aqr	21 31 09.9	− 5 36 16	2.91	G0 Ib	+ 6.7 0.1
206778*	8308	8	ε	Peg	21 43 49.1	+ 9 50 25	2.39	K2 Ib−II	+ 5.2 0.2
212943‡	8551	35		Peg	22 27 28.8	+ 4 39 28	4.79	K0 III−IV	+ 54.3±0.3
213014‡					22 27 49.7	+17 13 30	7.70	dG8	− 39.7 0.0
213947					22 34 15.4	+26 33 33	7.53	K4 III	+ 16.7 0.3
222368	8969	17	ι	Psc	23 39 33.9	+ 5 35 08	4.13	F7 V	+ 5.3 0.2
223094					23 46 02.7	+28 39 42	7.45	K5 III	+ 19.6 0.3
223311	9014				23 48 09.4	− 6 25 20	6.07	gK4	− 20.4±0.1
223647	9032		γ	1Oct	23 51 41.5	−82 03 38	5.11	G5 III	+ 13.8 0.4

*Radial velocity is now known to vary.

†BD 28°3402

‡Candidate for inclusion in revised list of primary standard stars currently in preparation by IAU Commission 30; recommended for intensive observation.

NGC or IC	Right Ascension	Declination	Revised Morphological Type	T	L	Log D_{25}	Log R_{25}	B_T^w	$B-V$	$U-B$	V	V_0
	h m	° ′									km/sec	km/sec
W–L–M	0 01.58	−15 30.4	IB(s)m	+10	8	2.01	0.39	11.27	0.40	−0.24	− 120	− 38
N7814	0 02.87	+16 06.2	SA(s)ab: sp	+ 2		1.80	0.38	11.49	0.90		+1047	+1249
N0045	0 13.69	−23 13.4	SA(s)dm	+ 8	8	1.91	0.15	11.11	0.69	−0.04	+ 468	+ 508
N0055	0 14.53	−39 13.8	SB(s)m: sp	+ 9		2.51	0.70	7.84			+ 131	+ 98
N0134	0 30.0	−33 18	SAB(s)bc	+ 4		1.91	0.50	11.05	0.88	+0.29	+1545	+1531
N0147	0 32.78	+48 27.9	E5p	− 5		2.11	0.20	10.24	0.94		− 263	− 11
N0185	0 38.54	+48 17.7	E3p	− 5		2.06	0.07	10.03	0.90		− 245	+ 4
N0205	0 39.96	+41 38.9	E5p	− 5		2.24	0.25	8.83	0.84		− 239	+ 1
N0221	0 42.29	+40 49.5	cE2	− 6		1.88	0.12	9.09	0.94	+0.47	− 217	+ 21
N0224	0 42.33	+41 13.7	SA(s)b	+ 3	2	3.25	0.45	4.37	0.91	+0.50	− 299	− 61
N0247	0 46.77	−20 48.1	SAB(s)d	+ 7	7	2.30	0.43	9.53	0.65		+ 150	+ 180
N0253	0 47.22	−25 19.8	SAB(s)c	+ 5	4	2.40	0.53	8.13	0.97		+ 249	+ 259
SMC	0 52.5	−72 52	SB(s)mp	+ 9	7	3.45	0.25	2.79	0.50		+ 150	− 30
N0300	0 54.53	−37 43.6	SA(s)d	+ 7	6	2.30	0.13	8.71			+ 145	+ 97
I1613	1 04.41	+ 2 04.7	IAB(s)m	+10	9	2.08	0.03	9.93	0.60		− 238	− 125
N0488	1 21.39	+ 5 13.1	SA(r)b	+ 3	1	1.72	0.11	11.15	0.86		+2180	+2292
N0578	1 30.12	−22 42.4	SAB(rs)c	+ 5	3	1.68	0.18	11.48	0.57	−0.14	+1696	+1689
N0598	1 33.45	+30 37.0	SA(s)cd	+ 6	4	2.79	0.20	6.27	0.55	−0.10	− 183	+ 2
N0613	1 33.95	−29 27.3	SB(rs)bc	+ 4	2	1.76	0.10	10.76	0.76	+0.15	+1500	+1462
N0628	1 36.30	+15 44.6	SA(s)c	+ 5	1	2.01	0.03	9.77	0.58		+ 655	+ 793
N0672	1 47.49	+27 23.8	SB(s)cd	+ 6	5	1.82	0.39	11.41	0.59	−0.11	+ 412	+ 578
N0772	1 58.93	+18 58.4	SA(s)b	+ 3	1	1.85	0.20	11.10	0.77		+2431	+2562
N0891	2 22.08	+42 18.8	SA(s)b? sp	+ 3		2.13	0.68	10.95	0.92	+0.27	+ 524	+ 706
N0908	2 22.73	−21 16.1	SA(s)c	+ 5	1	1.74	0.30	10.87	0.67		+1511	+1470
N0925	2 26.83	+33 32.8	SAB(s)d	+ 7	4	1.99	0.21	10.59	0.59		+ 560	+ 716
N0936	2 27.25	− 1 11.3	LB(rs)0$^+$	− 1		1.72	0.08	11.11	0.96	+0.55	+1317	+1350
FORNX	2 39.60	−34 33.5	E0p	− 5		2.30	0.16	9.04			+ 53	− 51
N1023	2 39.94	+39 01.8	LB(rs)0$^-$	− 3		1.94	0.42	10.38	1.00	+0.55	+ 614	+ 776
N1055	2 41.37	+ 0 24.6	SBb: sp	+ 3	4	1.88	0.40	11.42	0.85	+0.20	+1050	+1077
N1068	2 42.30	− 0 02.7	(R)SA(rs)b	+ 3		1.84	0.07	9.55	0.70	+0.08	+1109	+1134
N1073	2 43.28	+ 1 20.7	SB(rs)c	+ 5	3	1.69	0.03	11.50	0.53	−0.05	+1216	+1245
N1097	2 46.00	−30 18.4	SB(s)b	+ 3	2	1.97	0.15	10.21	1.00		+1320	+1227
N1232	3 09.42	−20 36.5	SAB(rs)c	+ 5	1	1.89	0.05	10.48	0.63		+1720	+1644
N1291	3 17.03	−41 09.2	(R)SB(s)0/a	0		2.02	0.06	9.45	0.93	+0.44	+ 824	+ 674
N1313	3 18.17	−66 31.5	SB(s)d	+ 7		1.93	0.11	9.77			+ 448	+ 241
N1300	3 19.34	−19 26.3	SB(rs)bc	+ 4	1	1.81	0.18	11.13	0.68	+0.13	+1502	+1422
N1316	3 22.40	−37 14.0	LAB(s)0p	− 2		1.85	0.11	9.75	0.90	+0.49	+1774	+1632
N1332	3 25.95	−21 21.6	L(s)0$^-$: sp	− 3		1.66	0.42	11.25	0.90		+1564	+1471
N1365	3 33.32	−36 09.8	SB(s)b	+ 3	2	1.99	0.25	10.20	0.62	+0.05	+1649	+1502
N1380	3 36.16	−35 00.0	LA0	− 2		1.69	0.41	11.21			+1809	+1664
N1398	3 38.55	−26 21.6	(R′)SB(r)ab	+ 2	1	1.82	0.10	10.62	0.95		+1419	+1299
N1433	3 41.78	−47 14.7	SB(r)a	+ 1		1.83	0.05	10.67	0.69	+0.23	+ 984	+ 802
N1448	3 44.28	−44 40.0	SAcd: sp	+ 6	5	1.91	0.65	11.38			+1182	+1005
I0342	3 46.08	+68 04.3	SAB(rs)cd	+ 6	2	2.25	0.01	9.16			+ 32	+ 228
I0356	4 06.99	+69 47.5	SA(s)abp	+ 2		1.72	0.11	11.43			+ 822	+1015

NGC or IC	Right Ascension	Declination	Revised Morphological Type	T	L	Log D_{25}	Log R_{25}	B_T^w	$B-V$	$U-B$	V	V_0
	h m	° ′									km/sec	km/sec
N1566	4 19.83	−54 57.4	SAB(s)bc	+ 4		1.88	0.09	10.23	0.85	0.00	+1394	+1178
N1617	4 31.49	−54 37.0	SB(s)a	+ 1		1.67	0.29	11.30	0.95	+0.42	+1000	+ 778
N1672	4 45.59	−59 15.6	SB(s)b	+ 3	2	1.68	0.09	11.02			+1309	+1076
N1808	5 07.45	−37 31.4	(R)SAB(s)a	+ 1		1.86	0.25	10.72	0.81	+0.30	+ 981	+ 769
LMC	5 23.6	−69 46	SB(s)m	+ 9	6	3.81	0.07	0.63	0.55		+ 260	+ 13
N2146	6 17.52	+78 21.6	SB(s)ab	+ 2		1.78	0.20	11.24	0.74		+ 838	+1028
N2217	6 21.36	−27 13.8	(R)LB(rs)0$^+$	− 1		1.68	0.04	11.48	1.03	+0.54	+1476	+1243
N2336	7 25.79	+80 11.6	SAB(r)bc	+ 4	1	1.84	0.24	11.16	0.66		+2199	+2389
N2366	7 28.13	+69 13.9	IB(s)m	+10	8	1.88	0.33	11.46	0.55		+ 107	+ 252
N2403	7 36.15	+65 37.0	SAB(s)cd	+ 6	5	2.25	0.21	8.90	0.50		+ 131	+ 259
N2442	7 36.42	−69 30.8	SB(s)b	+ 3	2	1.78	0.04	11.12			+ 657	+ 384
HLMII	8 18.15	+70 44.4	Im	+10	8	1.88	0.09	11.11	0.52		+ 158	+ 305
N2613	8 33.04	−22 56.7	SA(s)b	+ 3	3	1.86	0.53	11.35	0.93	+0.36	+1712	+1444
N2683	8 52.22	+33 26.9	SA(rs)b	+ 3	4	1.97	0.57	10.61	0.89	+0.29	+ 284	+ 242
N2655	8 54.67	+78 15.1	SAB(s)0/a	0		1.71	0.07	10.95	0.86		+1445	+1623
N2775	9 09.94	+ 7 04.1	SA(r)ab	+ 2		1.65	0.10	11.20	0.87	+0.38	+1135	+ 965
N2768	9 11.05	+60 04.1	E6	− 5		1.80	0.35	10.96	0.93		+1408	+1502
N2784	9 11.99	−24 08.4	LA(s)0:	− 2		1.71	0.35	11.35	1.15	+0.72	+ 708	+ 435
N2841	9 21.51	+51 00.4	SA(r)b:	+ 3	1	1.91	0.33	10.17	0.85	+0.41	+ 652	+ 700
N2903	9 31.75	+21 31.9	SAB(rs)bc	+ 4	2	2.10	0.28	9.56	0.64	+0.05	+ 569	+ 467
N2997	9 45.32	−31 09.4	SAB(rs)c	+ 5	1	1.91	0.10	10.36			+1089	+ 805
N2976	9 46.65	+67 57.1	SAcp	+ 5		1.69	0.29	10.86	0.70		+ 42	+ 175
N3031	9 54.99	+69 06.2	SA(s)ab	+ 2	2	2.41	0.26	7.86	0.93		− 44	+ 95
N3034	9 55.21	+69 42.8	I0 sp	0		2.05	0.39	9.28	0.87		+ 246	+ 388
N3079	10 01.45	+55 43.1	SB(s)c sp	+ 5	3	1.88	0.65	11.27	0.64		+1137	+1212
N3077	10 02.75	+68 46.2	I0p	0		1.66	0.10	10.67	0.80	+0.15	+ 10	+ 148
N3115	10 04.86	− 7 40.9	L0$^-$ sp	− 3		1.92	0.42	10.00	0.95	+0.57	+ 698	+ 476
LEO I	10 08.05	+12 20.7	E3	− 5		2.03	0.11	10.81	0.97			
N3166	10 13.36	+ 3 27.8	SAB(rs)0/a	0		1.72	0.29	11.50	0.91		+1381	+1203
N3169	10 13.84	+ 3 30.5	SA(s)ap	+ 1		1.68	0.18	11.30	0.80		+1229	+1051
N3184	10 17.82	+41 27.2	SAB(rs)cd	+ 6	3	1.84	0.01	10.40	0.65		+ 589	+ 593
N3198	10 19.46	+45 35.2	SB(rs)c	+ 5	3	1.92	0.35	10.92	0.54		+ 665	+ 691
I2574	10 27.81	+68 27.1	SAB(s)m	+ 9	8	2.09	0.32	10.84	0.47		+ 46	+ 185
N3338	10 41.74	+13 47.2	SA(s)c	+ 5	3	1.74	0.17	11.34	0.55		+1316	+1191
N3344	10 43.11	+24 57.7	(R)SAB(r)bc	+ 4	3	1.84	0.03	10.51	0.55		+ 585	+ 513
N3351	10 43.56	+11 44.7	SB(r)b	+ 3	3	1.87	0.16	10.54	0.79	+0.20	+ 807	+ 673
N3359	10 46.14	+63 15.8	SB(rs)c	+ 5	3	1.83	0.20	11.02	0.55		+1007	+1124
N3368	10 46.37	+11 51.7	SAB(rs)ab	+ 2		1.85	0.14	10.11	0.86	+0.27	+ 905	+ 773
N3379	10 47.44	+12 37.3	E1	− 5		1.65	0.05	10.20	0.94	+0.52	+ 885	+ 756
N3384	10 47.89	+12 40.2	LB(s)0$^-$:	− 3		1.77	0.35	10.91	0.91	+0.46	+ 770	+ 642
N3486	11 00.02	+29 00.9	SAB(r)c	+ 5	3	1.84	0.11	10.93	0.52		+ 720	+ 674
N3521	11 05.44	+ 0 00.4	SAB(rs)bc	+ 4	3	1.98	0.28	9.99	0.84		+ 815	+ 640
N3556	11 11.09	+55 42.8	SB(s)cd sp	+ 6		1.92	0.52	10.71	0.61	−0.01	+ 685	+ 772
N3623	11 18.53	+13 08.0	SAB(rs)a	+ 1	3	2.00	0.48	10.24	0.90	+0.41	+ 780	+ 666
N3627	11 19.86	+13 01.8	SAB(s)b	+ 3	3	1.94	0.30	9.74	0.70	+0.22	+ 697	+ 583

NGC or IC	Right Ascension	Declination	Revised Morphological Type	T	L	Log D_{25}	Log R_{25}	B_T^w	$B-V$	$U-B$	V	V_0
	h m	° ′									km/sec	km/sec
N3628	11 19.88	+13 38.1	Sbp sp	+ 3		2.17	0.61	10.31	0.80		+ 839	+ 728
N3631	11 20.62	+53 12.7	SA(s)c	+ 5	1	1.66	0.05	11.04	0.60		+1167	+1245
N3675	11 25.73	+43 37.7	SA(s)b	+ 3	3	1.77	0.26	11.09			+ 701	+ 735
N3718	11 32.17	+53 06.6	SB(s)ap	+ 1		1.94	0.29	11.21	0.73		+1014	+1095
N3726	11 32.93	+47 04.3	SAB(r)c	+ 5	2	1.78	0.13	10.95	0.51		+ 765	+ 818
N3938	11 52.43	+44 09.8	SA(s)c	+ 5	1	1.73	0.04	10.90	0.52		+ 792	+ 838
N3945	11 52.84	+60 43.1	LB(rs)0$^+$	− 1		1.74	0.18	11.44	0.92		+1220	+1340
N3953	11 53.44	+52 22.3	SB(r)bc	+ 4	1	1.82	0.26	10.81	0.70		+ 959	+1043
N3992	11 57.21	+53 25.1	SB(rs)bc	+ 4	1	1.88	0.19	10.64	0.79		+1059	+1149
N4036	12 01.07	+61 56.3	L0$^-$	− 3		1.65	0.34	11.48	0.90	+0.55	+1382	+1510
N4051	12 02.78	+44 34.5	SAB(rs)bc	+ 4	3	1.70	0.10	10.93	0.67	0.00	+ 674	+ 726
N4088	12 05.21	+50 35.0	SAB(rs)bc	+ 4	2	1.76	0.36	11.14	0.60		+ 742	+ 822
N4096	12 05.65	+47 31.2	SAB(rs)c	+ 5	3	1.81	0.50	11.13	0.44		+ 494	+ 561
N4125	12 07.70	+65 13.1	E6p	− 5		1.71	0.20	10.73	0.88		+1339	+1482
N4151	12 10.15	+39 26.7	(R′)SAB(rs)ab:	+ 2		1.77	0.13	11.12	0.75	0.00	+ 970	+1002
N4192	12 13.42	+14 56.6	SAB(s)ab	+ 2	2	1.98	0.48	10.91	0.79		− 129	− 206
N4214	12 15.27	+36 22.3	IAB(s)m	+10	6	1.90	0.10	10.22	0.46	−0.30	+ 289	+ 309
N4216	12 15.51	+13 11.2	SAB(s)b:	+ 3	3	1.92	0.58	10.97	0.99	+0.55	+ 15	− 69
N4236	12 16.33	+69 30.8	SB(s)dm	+ 8	7	2.27	0.43	10.09	0.40		− 1	+ 160
N4244	12 17.12	+37 51.0	SA(s)cd: sp	+ 6	7	2.21	0.81	10.60	0.44		+ 242	+ 270
N4254	12 18.45	+14 27.5	SA(s)c	+ 5	1	1.73	0.05	10.43	0.58	−0.02	+2400	+2324
N4258	12 18.59	+47 20.9	SAB(s)bc	+ 4		2.26	0.36	9.01	0.68		+ 465	+ 537
N4274	12 19.46	+29 39.2	(R)SB(r)ab	+ 2	4	1.84	0.39	11.30	0.93	+0.41	+ 722	+ 715
N4293	12 20.84	+18 25.6	(R)SB(s)0/a	0		1.78	0.31	11.22			+ 882	+ 825
N4303	12 21.53	+ 4 31.0	SAB(rs)bc	+ 4	1	1.78	0.04	10.17	0.54		+1599	+1483
N4314	12 22.18	+29 56.0	SB(rs)a	+ 1		1.68	0.05	11.32	0.84	+0.30	+ 883	+ 879
N4321	12 22.53	+15 51.9	SAB(s)bc	+ 4	1	1.84	0.05	10.11	0.73		+1610	+1543
N4365	12 24.10	+ 7 21.6	E3	− 5		1.79	0.13	10.51			+1177	+1074
N4374	12 24.68	+12 55.7	E1	− 5		1.70	0.06	10.26	0.97	+0.58	+ 933	+ 8[illegible]
N4382	12 25.03	+18 13.9	LA(s)0$^+$p	− 1		1.85	0.13	10.09	0.88		+ 773	+ 7[illegible]
N4395	12 25.44	+33 35.4	SA(s)m:	+ 9	8	2.11	0.07	10.39	0.54		+ 294	+ 307
N4406	12 25.81	+12 59.3	E3	− 5		1.87	0.13	10.07	0.93	+0.52	− 341	− 419
N4429	12 27.06	+11 09.0	La(r)0$^+$	− 1		1.74	0.33	11.15	0.94	+0.54	+1114	+1029
N4438	12 27.38	+13 03.0	SA(s)0/ap:	0		1.97	0.38	10.91	0.83		+ 259	+ 182
N4442	12 27.69	+ 9 50.6	LB(s)0	− 2		1.66	0.37	11.40	0.92	+0.55	+ 580	+ 490
N4449	12 27.85	+44 08.2	IBm	+10	5	1.71	0.14	9.94	0.41	−0.30	+ 200	+ 262
N4450	12 28.11	+17 07.6	SA(s)ab	+ 2		1.68	0.14	10.93	0.82		+2048	+1990
N4472	12 29.40	+ 8 02.6	E2	− 5		1.95	0.08	9.30	0.94		+ 914	+ 817
N4473	12 29.43	+13 28.3	E5	− 5		1.65	0.24	11.07	0.88	+0.53	+2279	+2205
N4490	12 30.23	+41 40.8	SB(s)dp	+ 7	5	1.77	0.28	10.24	0.44	−0.18	+ 577	+ 629
N4486	12 30.44	+12 26.0	E$^+$0−1p	− 4		1.86	0.03	9.58	0.94	+0.57	+1257	+1180
N4494	12 31.03	+25 49.0	E1−2	− 5		1.68	0.10	10.73	0.89	+0.48	+1307	+1289
N4501	12 31.61	+14 27.6	SA(rs)b	+ 3	1	1.84	0.25	10.27	0.75	+0.25	+2057	+1989
N4517	12 32.38	+ 0 09.2	SA(s)cd: sp	+ 6		2.01	0.73	11.20	0.73		+1128	+1001
N4526	12 33.67	+ 7 44.4	LAB(s)0:	− 2		1.86	0.49	10.61	0.94	+0.54	+ 450	+ 355

NGC or IC	Right Ascension	Declination	Revised Morphological Type	T	L	Log D_{25}	Log R_{25}	B_T^w	$B-V$	$U-B$	V	V_0
	h m	° ′									km/sec	km/sec
N4527	12 33.76	+ 2 41.6	SAB(s)bc	+ 4	3	1.80	0.44	11.33	0.88		+1730	+1614
N4535	12 33.96	+ 8 14.6	SAB(s)c	+ 5	1	1.83	0.13	10.52	0.70		+1946	+1853
N4536	12 34.07	+ 2 13.7	SAB(rs)cd	+ 4	3	1.87	0.33	11.01	0.60		+1927	+1810
N4548	12 35.06	+14 32.4	SB(rs)b	+ 3		1.73	0.09	10.98	0.79	+0.30	+ 468	+ 403
N4559	12 35.59	+28 00.1	SAB(rs)bc	+ 6	4	2.02	0.33	10.34	0.45		+ 807	+ 802
N4565	12 35.97	+26 01.6	SA(s)b? sp	+ 3	1	2.21	0.77	10.33	0.83		+1136	+1122
N4569	12 36.45	+13 12.4	SAB(rs)ab	+ 2		1.98	0.31	10.25	0.75	+0.30	− 312	− 382
N4579	12 37.35	+11 51.6	SAB(rs)b	+ 3		1.73	0.09	10.56	0.83	+0.32	+1805	+1730
N4594	12 39.59	−11 35.0	SA(s)a sp	+ 1		1.95	0.34	9.29	0.97		+1128	+ 963
N4605	12 39.66	+61 39.1	SB(s)cp	+ 5		1.74	0.38	10.95			+ 148	+ 286
N4621	12 41.66	+11 41.2	E5	− 5		1.71	0.18	10.80	0.96		+ 414	+ 341
N4631	12 41.75	+32 34.8	SB(s)d sp	+ 7	5	2.18	0.66	9.81	0.54		+ 620	+ 638
N4636	12 42.46	+ 2 43.7	E0−1	− 5		1.79	0.09	10.48	0.94	+0.50	+ 979	+ 869
N4649	12 43.29	+11 35.5	E2	− 5		1.86	0.07	9.82	1.00		+1200	+1128
N4654	12 43.58	+13 10.0	SAB(rs)cd	+ 6	3	1.67	0.20	11.14	0.64	−0.07	+1036	+ 970
N4656	12 43.60	+32 12.6	SB(s)mp	+ 9	7	2.14	0.62	10.86	0.43		+ 645	+ 662
N4697	12 48.21	− 5 45.6	E6	− 5		1.78	0.20	10.20	0.93	+0.39	+1308	+1170
N4725	12 50.08	+25 32.6	SAB(r)abp	+ 2	1	2.04	0.14	9.99	0.74		+1138	+1131
N4736	12 50.53	+41 09.7	(R)SA(r)ab	+ 2	3	2.04	0.08	8.90	0.75	+0.16	+ 269	+ 329
N4754	12 51.92	+11 21.3	LB(r)0$^-$:	− 3		1.67	0.26	11.46	0.95	+0.50	+1461	+1393
N4753	12 51.98	− 1 09.5	I0	0		1.73	0.27	10.82	0.95	+0.48	+1255	+1137
N4762	12 52.56	+11 16.3	LB(r)0? sp	− 2		1.94	0.73	11.19	0.90	+0.40	+ 945	+ 878
N4826	12 56.37	+21 43.3	(R)SA(rs)ab	+ 2		1.97	0.24	9.37	0.84	+0.21	+ 397	+ 377
N4856	12 58.94	−15 00.0	SB(s)0/a	0		1.66	0.46	11.48	0.97		+1251	+1088
N4945	13 05.01	−49 25.7	SB(s)cd: sp	+ 6		2.30	0.66	9.63			+ 594	+ 356
N5005	13 10.58	+37 05.9	SAB(rs)bc	+ 4	3	1.73	0.30	10.68	0.82	+0.30	+1015	+1069
N5033	13 13.10	+36 38.3	SA(s)c	+ 5	2	2.02	0.27	10.63	0.54		+ 907	+ 961
N5055	13 15.49	+42 04.4	SA(rs)bc	+ 4	3	2.09	0.21	9.36	0.73		+ 509	+ 587
N5102	13 21.54	−36 35.4	LA0$^-$	− 3		1.97	0.43	10.30	0.70	+0.27	+ 454	+ 247
N5128	13 25.02	−42 58.8	L0p	− 2		2.26	0.10	7.83	0.98		+ 541	+ 323
N5194	13 29.55	+47 14.2	SA(s)bcp	+ 4	1	2.04	0.15	9.00	0.60		+ 460	+ 565
N5195	13 29.66	+47 18.7	I0p	0		1.73	0.10	10.52	0.90	+0.40	+ 552	+ 658
N5236	13 36.57	−29 49.8	SAB(s)c	+ 5	2	2.05	0.04	8.21			+ 518	+ 337
N5248	13 37.17	+ 8 55.5	SAB(rs)bc	+ 4	1	1.81	0.12	10.67	0.63	0.00	+1146	+1102
N5247	13 37.65	−17 50.8	SA(s)bc	+ 4	2	1.73	0.06	11.13	0.59	−0.10	+1655	+1511
N5322	13 49.01	+60 13.8	E3−4	− 5		1.74	0.15	10.86	0.88	+0.44	+1902	+2061
N5364	13 55.83	+ 5 03.1	SA(rs)bcp	+ 4	1	1.85	0.15	11.05	0.65		+1393	+1349
N5457	14 02.96	+54 23.4	SAB(rs)cd	+ 6	1	2.43	0.01	8.16	0.46		+ 241	+ 388
N5474	14 04.77	+53 41.9	SA(s)cdp	+ 6	7	1.65	0.03	11.31	0.50		+ 270	+ 416
N5585	14 19.55	+56 45.8	SAB(s)d	+ 7	7	1.74	0.17	11.37	0.50		+ 300	+ 462
N5566	14 19.96	+ 3 58.0	SB(r)ab	+ 2	4	1.81	0.43	11.43	0.86	+0.42	+1518	+1489
N5643	14 32.21	−44 08.4	SAB(rs)c	+ 5	5	1.66	0.05	10.84			+1142	+ 962
N5866	15 06.29	+55 47.5	LA0$^+$ sp	− 1		1.72	0.35	10.96	0.85	+0.40	+ 692	+ 874
N5907	15 15.72	+56 21.1	SA(s)c: sp	+ 5	3	2.09	0.84	11.15	0.77		+ 592	+ 780
N5921	15 21.57	+ 5 05.8	SB(r)bc	+ 4	2	1.69	0.07	11.47	0.63	+0.02	+1475	+1503

NGC or IC	Right Ascen-sion	Declina-tion	Revised Morphological Type	T	L	Log D_{25}	Log R_{25}	B_T^w	$B-V$	$U-B$	V	V_0
	h m	° ′									km/sec	km/sec
N6300	17 16.26	−62 48.7	SB(r)b	+ 3		1.73	0.18	11.13			+1140	+ 988
N6384	17 32.04	+ 7 04.0	SAB(r)bc	+ 4	1	1.78	0.15	11.32	0.73		+1660	+1801
N6503	17 49.52	+70 08.9	SA(s)cd	+ 6	5	1.79	0.42	10.94	0.67	+0.05	+ 62	+ 315
N6744	19 09.06	−63 52.2	SAB(r)bc	+ 4	3	2.19	0.18	9.26			+ 644	+ 519
N6822	19 44.53	−14 49.5	IB(s)m	+10	8	2.01	0.03	9.31			− 56	+ 65
N6946	20 34.69	+60 07.8	SAB(rs)cd	+ 6	1	2.04	0.05	9.68	0.80		+ 46	+ 338
N7331	22 36.74	+34 22.7	SA(s)bc	+ 4	2	2.03	0.43	10.39	0.84	+0.25	+ 826	+1105
N7410	22 54.58	−39 42.1	SB(s)a	+ 1		1.74	0.43	11.40	0.92	+0.49	+1638	+1634
I5267	22 56.80	−43 26.3	LA(rs)0:	− 2		1.70	0.09	11.35	0.93	+0.30	+1715	+1691
N7424	22 56.88	−41 06.8	SAB(rs)cd	+ 6	3	1.88	0.05	10.99			+ 862	+ 850
N7582	23 17.98	−42 24.8	(R′)SB(s)ab	+ 2		1.66	0.32	11.47	0.77	+0.16	+1452	+1427
N7640	23 21.76	+40 48.2	SB(s)c	+ 5	3	2.03	0.63	11.44	0.54		+ 369	+ 642
I5332	23 34.06	−36 08.5	SA(s)d	+ 7		1.82	0.11	11.25	0.66	−0.10	+ 702	+ 701
N7793	23 57.45	−32 37.9	SA(s)d	+ 7	6	1.96	0.14	9.64	0.59	−0.10	+ 209	+ 214

W−L−M	= A2359−15	= Wolf−Lundmark−Melotte neb. = DDO 221
SMC	= A0051−73	= Small Magellanic Cloud
FORNX	= A0237−34	= Fornax System
LMC	= A0524−69	= Large Magellanic Cloud
HLMII	= A0813+70	= Holmberg II = DDO 50
LEO I	= A1005+12	= Regulus System = DDO 74
NGC 224	= M31	= Andromeda Nebula
NGC 221	= M32	
NGC 598	= M33	= Triangulum Nebula
NGC 5194	= M51	= Whirlpool Nebula
NGC 5457	= M101	= Pinwheel Nebula
NGC 4594	= M104	= Sombrero Nebula
NGC 5128		= Centaurus A

IAU Desig.	Name	R.A.	Dec.	Ang. Diam.	Dist.	Trumpler Class	Tot. Mag.	Spec-trum	Mag.*	Log age	Log Fe / H	A_v
		h m	° ′	′	pc							
C0001–302	Blanco 1	0 03.9	–29 59	89	240	IV 3 m			8	7.85	+0.03	0.27
C0022+610	NGC 103	0 24.9	+61 18	5	2900	II 1 m	5.8	B3	11	7.46		1.62
C0027+599	NGC 129	0 29.5	+60 11	21	1600	III 2 m	9.8	B3	11	7.67		1.77
C0029+628	King 14	0 31.4	+63 07	7	2800	III 1 p		B2	10	7.61		1.65
C0030+630	NGC 146	0 32.6	+63 15	6	3200	II 2 p	9.6	B3		7.61		2.07
C0036+608	NGC 189	0 39.2	+61 02	3	750	III 1 p	11.1	A0		7.00		1.62
C0039+850	NGC 188	0 43.6	+85 18	13	1500	I 2 r	9.3	F2	10	9.91	–0.06	0.12
C0040+615	NGC 225	0 43.0	+61 45	12	610	III 1 pn	8.9	A2		8.11		0.84
C0112+585	NGC 436	1 15.2	+58 46	5	2100	I 2 m	9.3	B5	10	8.15		0.45
C0115+580	NGC 457	1 18.6	+58 17	13	3100	II 3 r	5.1	B2	6	7.00		1.41
C0126+630	NGC 559	1 29.0	+63 16	4	1100	I 1 m	7.4		9	8.66	–0.76	2.70
C0129+604	NGC 581	1 32.7	+60 40	6	2500	II 2 m	6.9	B2	9	7.46		1.14
C0132+610	Trump 1	1 35.2	+61 15	4	2300	II 2 p	8.9	B2	10	7.61		1.29
C0139+637	NGC 637	1 42.4	+63 58	3	2400	I 2 m	7.3	B0	8	8.34		1.17
C0140+616	NGC 654	1 43.5	+61 51	4	2000	II 2 r	8.2		10			2.58
C0140+604	NGC 659	1 43.7	+60 40	4	2500	I 2 m	7.2		10	7.89		1.74
C0142+610	NGC 663	1 45.5	+61 13	16	2200	II 3 r	6.4	B1	9	7.00		2.34
C0144+717	Coll 463	1 47.8	+71 55	36	660	III 2 M	5.8			7.89		0.57
C0149+615	IC 166	1 52.0	+61 48	4	3100	II 1 r			17	9.38		2.31
C0154+374	NGC 752	1 57.3	+37 38	50	360	II 2 r	6.6	F0	8	9.38	–0.21	0.06
C0155+552	NGC 744	1 57.9	+55 26	11	1400	III 1 p	7.8	B7	10	8.30		1.14
C0211+590	Stock 2	2 14.4	+59 14	60	300	I 2 m		B8		8.23		1.26
C0215+569	NGC 869	2 18.5	+57 07	29	2200	I 3 r	4.3	B1	7	7.26		1.62
C0218+568	NGC 884	2 21.9	+57 05	29	2200	I 3 r	4.4	B1	7			1.62
C0225+604	Mark 6	2 29.1	+60 37	4	510	III 1 P			8			1.56
C0228+612	IC 1805	2 32.1	+61 25	21	2200	II 3 mn	4.8	O6	9	7.00		2.46
C0233+557	Trump 2	2 36.8	+55 57	20	560	II 2 p	9.0	B9		8.30		0.90
C0238+613	NGC 1027	2 42.1	+61 31	20	1200	II 3 mn	7.4	B3	9	7.89		0.96
C0238+425	NGC 1039	2 41.5	+42 45	35	450	II 3 r	5.8	B8	9	8.03	–0.26	0.12
C0247+602	IC 1848	2 50.6	+60 25	12	2300	I 3 pn	7.0	O7		7.00		1.89
C0311+470	NGC 1245	3 14.1	+47 13	10	2200	II 2 r	7.7		12	9.03	+0.14	0.81
C0318+484	Mel 20	3 21.5	+48 35	185	160	III 3 m	2.3	B1	3	7.61	+0.10	0.27
C0328+371	NGC 1342	3 31.1	+37 19	14	540	III 2 m	7.2	A1	8	8.80	–0.13	0.81
C0344+239	Pleiades	3 46.5	+24 06	110	130	I 3 m	1.5	B5	3	8.11	+0.11	0.15
C0403+622	NGC 1502	4 07.1	+62 19	7	810	I 3 m	4.1	B0	7	7.00		2.13
C0411+511	NGC 1528	4 14.8	+51 13	23	750	II 2 m	6.4	B8	10	8.34	–0.10	0.87
C0417+448	Berk 11	4 20.0	+44 54	5	2200	II 2 m			15	7.00		
C0417+501	NGC 1545	4 20.3	+50 14	18	760	IV 2 p	4.6		9	8.30		1.02
C0424+157	Hyades 1	4 26.4	+15 51	330	40		0.8	A2	4	8.85	+0.12	
C0443+189	NGC 1647	4 45.6	+19 04	45	520	II 2 r	6.2	B7	9	8.11		1.11
C0445+108	NGC 1662	4 48.1	+10 55	20	380	II 3 m	8.0	A0	9	8.11	–0.20	0.96
C0447+436	NGC 1664	4 50.5	+43 41	18	1100		7.2	A0	10	8.38		0.72
C0504+369	NGC 1778	5 07.6	+37 02	6	1500	III 2 p	8.5			8.11		0.93
C0509+166	NGC 1817	5 11.7	+16 41	15	1800	IV 2 r	7.8		9	8.90	–0.26	0.99
C0518–685	NGC 1901	5 17.8	–68 27	40	420	III 3 m				8.92		0.15

* Magnitude of brightest cluster member

IAU Desig.	Name	R.A.	Dec.	Ang. Diam.	Dist.	Trumpler Class	Tot. Mag.	Spec-trum	Mag.*	Log age	Log Fe / H	A_v
		h m	° ′	′	pc							
C0519+333	NGC 1893	5 22.2	+33 23	11	4000	II 3 m	7.8			7.00		1.62
C0520+295	Berk 19	5 23.6	+29 35	6	4800	II 1 m			15	9.49	−0.50	
C0524+352	NGC 1907	5 27.5	+35 19	6	1300	I 1 mn	10.2		11	8.11	−0.10	1.35
C0525+358	NGC 1912	5 28.2	+35 50	21	1200	II 2 r	6.8	B5	8	8.19	−0.11	0.69
C0532+341	NGC 1960	5 35.6	+34 08	12	1200	I 3 r	6.5	B3	9	7.61		0.63
C0532−054	Trapez.	5 35.0	− 5 23	47	460					7.61		
C0546+336	King 8	5 48.9	+33 38	7	3400	II 2 m			15	8.26	−0.50	2.49
C0548+217	Berk 21	5 51.3	+21 47	6	3600	I 2			6	7.00		
C0549+325	NGC 2099	5 51.9	+32 33	23	1300	I 2 r	6.2	B9	11	8.30	+0.01	0.96
C0600+104	NGC 2141	6 02.7	+10 26	10	4200	I 2 r	10.8		15	9.60	−0.46	
C0601+240	IC 2157	6 04.5	+24 00	6	1900	II 1 p	9.1		12	7.85		1.65
C0604+241	NGC 2158	6 07.0	+24 06	4	3900		12.1		15	9.27	−0.60	1.23
C0605+139	NGC 2169	6 08.0	+13 58	6	920	III 3 m	7.0	B1		7.46		0.36
C0605+243	NGC 2168	6 08.4	+24 21	28	850	III 3 r	5.6	B4	8	8.07		0.66
C0606+203	NGC 2175	6 09.3	+20 20	18	2700	III 3 m	6.8		8	7.00		1.14
C0609+054	NGC 2186	6 11.8	+ 5 27	4	1800	II 2 m	9.2		12	8.30		0.90
C0611+128	NGC 2194	6 13.4	+12 48	10	2700	II 2 r	10.0		13	8.57		
C0613−186	NGC 2204	6 15.4	−18 39	12	4300	II 2 r	9.3		13	9.27	−0.58	
C0618−072	NGC 2215	6 20.7	− 7 17	11	980	II 2 m	8.6		11	8.80		0.93
C0624−047	NGC 2232	6 26.2	− 4 45	29	360	III 2 p	4.2			7.61	−0.20	
C0627−312	NGC 2243	6 29.5	−31 17	4	4100	I 2 r	10.5			9.49	−0.55	
C0629+049	NGC 2244	6 32.0	+ 4 52	23	1500	II 3 m	5.2	O5	7	7.00		1.35
C0632+084	NGC 2251	6 34.3	+ 8 22	10	1600	III 2 m	8.8			8.66		0.69
C0634+094	Trump 5	6 36.3	+ 9 27	7	980	III 1 m	10.9		17	9.86		2.01
C0638+099	NGC 2264	6 40.6	+ 9 54	20	790	III 3 mn	4.1	O8	5	7.00	−0.15	0.18
C0644−206	NGC 2287	6 46.7	−20 44	38	640	I 3 r	5.0	B5	8	8.30	+0.09	
C0645+411	NGC 2281	6 48.8	+41 04	14	460	I 3 m	7.2	A0	8	8.72	−0.04	0.24
C0649+005	NGC 2301	6 51.4	+ 0 29	12	760	I 3 r	6.3		8	8.11	+0.04	0.12
C0649−070	NGC 2302	6 51.6	− 7 03	2	1100	III 2 m			12	7.89		0.72
C0649+030	Biur 10	6 51.8	+ 2 57	4	6200	I 1 p			15	7.46		
C0652−245	Coll 121	6 53.9	−24 37	50	630	IV 3 p	5.8			7.61		0.03
C0700−082	NGC 2323	7 02.8	− 8 20	16	1000	II 3 r	7.2	B8	9	7.89		0.90
C0701+011	NGC 2324	7 03.8	+ 1 04	7	3200	II 2 r	7.9		12	7.00	−0.13	0.15
C0704−100	NGC 2335	7 06.2	−10 04	12	1000	III 2 mn	9.3		10	8.03	+0.20	1.14
C0705−105	NGC 2343	7 07.9	−10 38	6	870	II 2 pn	7.5		8	7.89	−0.20	0.57
C0706−130	NGC 2345	7 08.0	−13 09	12	1800	II 3 r	8.1		9	7.89		1.74
C0712−102	NGC 2353	7 14.2	−10 17	20	1000	III 3 p	5.2	B0	9	7.89		0.27
C0712−256	NGC 2354	7 13.9	−25 43	20	1800	III 2 r	8.9			8.92		0.39
C0712−310	Coll 132	7 14.1	−31 09	95	410	III 3 p	3.8			7.89		
C0715−155	NGC 2360	7 17.4	−15 37	12	1100	I 3 r	9.1	B8		9.27	−0.13	0.18
C0716−248	NGC 2362	7 18.5	−24 56	8	1600	I 3 r	3.8	O8	8	7.00		0.33
C0717−130	Haff 6	7 19.8	−13 07	4	1100	IV 2 m			16	9.03		
C0721−131	NGC 2374	7 23.7	−13 15	19	1200	IV 2 p	7.3			8.34		
C0722−321	Coll 140	7 23.6	−32 11	42	380	III 3 m	4.2			7.73	+0.05	
C0722−209	NGC 2384	7 24.7	−21 01	2	3300	IV 3 p	8.2			7.00		0.90

* Magnitude of brightest cluster member

IAU Desig.	Name	R.A.	Dec.	Ang. Diam.	Dist.	Trumpler Class	Tot. Mag.	Spectrum	Mag.*	Log age	Log Fe / H	A_v
		h m	° ′	′	pc							
C0724–476	Mel 66	7 26.1	–47 43	10	2900	II 1 r	10.7			9.69	–0.36	0.48
C0731–153	NGC 2414	7 32.9	–15 26	4	4200	I 3 m	8.2			7.08		1.59
C0734–205	NGC 2421	7 36.0	–20 36	10	1900	I 2 r	9.0		11	7.61		1.35
C0734–143	NGC 2422	7 36.3	–14 29	29	470	I 3 m	4.3	B3	5	7.89		0.21
C0734–137	NGC 2423	7 36.8	–13 51	19	750	II 2 m	7.0			8.60	–0.04	0.36
C0735–119	Mel 71	7 37.2	–12 03	9	2700	II 2 r	9.0			8.57	–0.29	
C0735+216	NGC 2420	7 38.0	+21 35	10	1900	I 1 r	10.0		11	9.47	–0.61	
C0738–334	Boch 15	7 39.8	–33 32	3	3800	IV 2 pn				7.00		
C0738–315	NGC 2439	7 40.5	–31 38	10	3600	II 3 r	7.1	B1	9	7.61		0.72
C0739–147	NGC 2437	7 41.5	–14 48	27	1600	II 2 r	6.6	B9	10	8.50		0.15
C0742–237	NGC 2447	7 44.3	–23 51	22	1100	I 3 r	6.5	B9	9	8.50	0.00	0.15
C0743–378	NGC 2451	7 45.1	–37 57	45	320	II 2 m	3.7		6	7.61	–0.45	0.12
C0745–271	NGC 2453	7 47.4	–27 13	4	2100	I 3 m	9.0			7.00		1.38
C0746–261	Rup 36	7 48.2	–26 16	4	2100	IV 1 m			12	8.30		
C0750–384	NGC 2477	7 52.0	–38 32	27	1200	I 2 r	5.7		12	9.10	+0.04	0.87
C0752–241	NGC 2482	7 54.6	–24 17	12	750	IV 1 m	8.8			8.66	+0.12	0.09
C0754–299	NGC 2489	7 55.9	–30 03	8	1200	I 2 m	9.3		11	8.50		1.02
C0757–284	Rup 44	7 58.7	–28 34	4	4200	IV 2 m			12	7.00		2.01
C0757–607	NGC 2516	7 58.2	–60 51	29	410	I 3 r	3.3	B3	7	7.85	–0.23	0.27
C0757–106	NGC 2506	7 59.8	–10 46	6	3200	I 2 r	8.9		11	9.36	–0.55	0.27
C0803–280	NGC 2527	8 04.9	–28 08	22	570	II 2 m	8.3			8.77	0.00	0.27
C0805–297	NGC 2533	8 06.7	–29 52	3	1300	II 2 r	10.0			7.89		0.75
C0808–126	NGC 2539	8 10.4	–12 49	21	1200	III 2 m	8.0	A0	9	8.66	–0.20	0.30
C0809–491	NGC 2547	8 10.5	–49 15	20	440	I 3 m	5.0		7	7.61	–0.13	0.06
C0810–374	NGC 2546	8 12.2	–37 37	40	990	III 2 m	5.2	B0	7	8.11	+0.30	0.30
C0811–056	NGC 2548	8 13.4	– 5 47	54	630	I 3 r	5.5	A0	8	8.50	+0.10	0.15
C0816–304	NGC 2567	8 18.3	–30 37	10	1600	II 2 m	8.4		11	8.46	+0.20	0.66
C0816–295	NGC 2571	8 18.6	–29 43	13	1200	II 3 m	7.4			7.08	+0.08	0.90
C0837+201	Praesepe	8 39.6	+20 01	95	160	II 3 m	3.9	A0	6	8.92	+0.07	
C0837–460	Pisms 6	8 39.0	–46 11	1	1700	II 3 p			9	7.08		1.14
C0838–459	Wtrloo 6	8 40.1	–46 07	2	1800	II 3 p				7.61		
C0838–528	IC 2391	8 40.0	–53 02	50	150	II 3 m	2.6	B5	4	7.73	–0.04	0.09
C0839–480	IC 2395	8 40.9	–48 10	7	950	II 3 m	4.6	B5		7.00		0.36
C0839–461	Pisms 8	8 41.2	–46 15	2	1500	II 2 p			10	7.61		2.13
C0840–469	NGC 2660	8 42.0	–47 07	4	2900	I 1 r	10.8		13	9.22	+0.06	1.05
C0843–527	NGC 2669	8 44.6	–52 56	12	1000	III 3 m	6.0			7.61		0.54
C0846–423	Trump 10	8 47.5	–42 27	14	420	II 3 m	5.0	B3		7.61	–0.08	0.12
C0847+120	NGC 2682	8 50.0	+11 50	29	790	II 3 r	7.4	B8	9	9.70	–0.10	0.15
C0922–515	Rup 76	9 23.9	–51 42	5	1300	IV 2 p			13	8.30		
C0925–549	Rup 77	9 26.8	–55 05	2	4900	II 1 m			14	7.54		
C0927–534	Rup 78	9 29.0	–53 38	0	3300	II 2 m			15	7.67		
C0939–536	Rup 79	9 40.7	–53 48	11	2400	III 2 p			11	7.54		2.37
C1001–598	NGC 3114	10 02.5	–60 04	35	910		4.5	B9	9	7.89	+0.01	
C1022–575	Wester 2	10 23.7	–57 43	1	5000	IV 1 pn	11.3			7.00		4.86
C1024–576	NGC 3247	10 25.7	–57 54	6	1400	III 2 p	10.3			8.19		0.90

* Magnitude of brightest cluster member

IAU Desig.	Name	R.A.	Dec.	Ang. Diam.	Dist.	Trumpler Class	Tot. Mag.	Spec- trum	Mag.*	Log age	Log Fe/H	A_v
		h m	° ′	′	pc							
C1025–573	IC 2581	10 27.1	−57 36	7	2300	II 2 pn	5.3	B0		7.08		1.23
C1033–579	NGC 3293	10 35.5	−58 11	5	2500		6.2		8	7.00		0.87
C1035–583	NGC 3324	10 37.0	−58 35	5	3100					7.00		1.29
C1036–538	NGC 3330	10 38.3	−54 06	6	1400	III 2 m	8.4			7.89		0.51
C1040–588	Boch 10	10 41.9	−59 06	20	3200	II 3 mn				7.00		
C1041–597	Coll 228	10 42.7	−59 58	14	2300		4.9			7.00		1.44
C1041–641	IC 2602	10 42.9	−64 21	50	150	I 3 r	1.6	B0	3	7.46	−0.20	0.09
C1041–593	Trump 14	10 43.6	−59 31	4	2900		6.8			7.00		1.50
C1042–591	Trump 15	10 44.5	−59 19	3	2000	III 2 pn	9.0			7.00		1.53
C1043–594	Trump 16	10 44.9	−59 40	10	2600		6.7	O5		7.00		1.41
C1045–598	Boch 11	10 47.0	−60 03	21	3600	IV 3 pn				7.00		
C1055–614	Boch 12	10 57.1	−61 42	10	2200	III 3 p				7.61		0.72
C1057–600	NGC 3496	10 59.5	−60 18	9	1000	II 1 r	9.2			8.50		1.47
C1059–595	Pisms 17	11 00.8	−59 47	0	4100				9	7.00		1.47
C1104–584	NGC 3532	11 06.1	−58 38	55	480	II 3 r	3.4	B5	8	8.46		
C1108–599	NGC 3572	11 10.1	−60 12	6	2800	II 3 mn	4.5		7	7.00		1.44
C1108–601	Hogg 10	11 10.4	−60 20	3	2600					7.26		1.32
C1109–600	Coll 240	11 10.9	−60 15	25	2600	III 2 mn				7.00		
C1109–604	Trump 18	11 11.1	−60 38	12	1300	II 3 m	8.2			7.85		0.99
C1110–605	NGC 3590	11 12.6	−60 45	4	2100	I 2 p	6.8			7.54		1.14
C1110–586	Stock 13	11 12.8	−58 53	3	2700	I 3 pn			10	7.00		0.69
C1112–609	NGC 3603	11 14.7	−61 13	2	5200	II 3 mn	9.2			7.00		3.96
C1115–624	IC 2714	11 17.5	−62 40	12	1100	II 2 r	8.2		10	8.30		1.29
C1117–632	Mel 105	11 19.1	−63 28	4	2000	I 2 r	9.4			8.50		1.08
C1123–429	NGC 3680	11 25.3	−43 12	12	790	I 2 m	8.6		10	9.60	−0.07	0.09
C1133–613	NGC 3766	11 35.8	−61 34	12	1800	I 3 r	4.6	B0	8	7.08		0.51
C1134–627	IC 2944	11 36.3	−62 59	14	1900	III 3 mn	2.8	O6		7.00		0.99
C1141–622	Stock 14	11 43.6	−62 27	4	2700	III 3 p			10	7.00		0.75
C1148–554	NGC 3960	11 50.5	−55 39	6	1700	I 2 m	8.8			9.03	−0.30	
C1154–623	Rup 97	11 56.9	−62 36	3	3100	IV 1 p			12	8.72		0.60
C1204–609	NGC 4103	12 06.3	−61 12	6	1500	I 2 m	7.4		10	7.61		0.81
C1221–616	NGC 4349	12 24.1	−61 51	15	860	II 2 m	8.0	B8	11	8.50		0.69
C1222+263	Coma Ber	12 24.7	+26 09	275	80	III 3 r	2.9	A0	5	8.66	−0.03	
C1225–598	NGC 4439	12 28.0	−60 03	4	1400		8.7			7.79		0.93
C1226–604	Harvd 5	12 28.6	−60 43	5	1000		8.8			7.79		
C1232+365	Upgren 1	12 34.7	+36 21	18	140	IV 2 p		F3				0.25
C1239–627	NGC 4609	12 41.9	−62 56	4	1300	II 2 m	4.5		10	7.61		0.87
C1250–600	κ Cru	12 53.1	−60 18	10	1500		5.2	B3	7	7.38		0.90
C1315–623	Stock 16	13 18.6	−62 31	3	1900	III 3 pn			10	7.00		1.53
C1317–646	Rup 107	13 20.1	−64 54	5	2000	III 2 p			12	8.11		
C1324–587	NGC 5138	13 26.8	−58 58	7	1400	II 2 m	9.8			7.89		0.78
C1327–606	NGC 5168	13 30.7	−60 54	4	1300	I 2 m	11.5			7.89		1.02
C1328–625	Trump 21	13 31.7	−62 45	4	1100	I 2 p	9.6			7.89		0.60
C1343–626	NGC 5281	13 46.1	−62 52	4	1300	I 3 m	8.2		10	7.61		0.75
C1350–616	NGC 5316	13 53.4	−61 50	13	1100	II 2 r	8.8	B8	11	8.30	+0.01	0.51

* Magnitude of brightest cluster member

IAU Desig.	Name	R.A.	Dec.	Ang. Diam.	Dist.	Trumpler Class	Tot. Mag.	Spec-trum	Mag.*	Log age	Log Fe/H	A_v
		h m	° ′	′	pc							
C1404–480	NGC 5460	14 07.1	–48 17	25	770	I 3 m	6.1		9	8.11		0.39
C1420–611	Lynga 2	14 23.5	–61 22	12	1100	II 3 m				7.61		0.54
C1424–594	NGC 5606	14 27.2	–59 36	3	1900	I 3 p	10.0			7.08		1.32
C1426–605	NGC 5617	14 29.2	–60 41	10	1600	I 3 r	8.5	B3	10	7.98	–0.51	1.47
C1427–609	Trump 22	14 30.6	–61 08	6	1600	III 2 m	9.9	B4	12	7.61		1.59
C1431–563	NGC 5662	14 34.7	–56 31	12	620	II 3 r	7.7		10	7.89		
C1440+697	Ursa Maj	14 40.8	+69 36		30				2	8.30		
C1445–543	NGC 5749	14 48.4	–54 30	7	840	II 2 m	8.8			7.89		1.17
C1501–541	NGC 5822	15 04.6	–54 19	39	730	II 2 r	6.5	B9	10	9.08	–0.07	0.39
C1502–554	NGC 5823	15 05.2	–55 34	10	1100	II 2 r	8.6		13	8.80	–0.13	0.78
C1511–588	Pisms 20	15 14.9	–59 02	4	2500					7.00		3.39
C1559–603	NGC 6025	16 03.0	–60 29	12	770	II 3 r	6.0		7	7.85	+0.23	
C1601–517	Lynga 6	16 04.2	–51 54	4	1600					7.46		2.94
C1603–539	NGC 6031	16 07.0	–54 03	2	1600	I 3 p	12.2			8.07		1.23
C1609–540	NGC 6067	16 12.6	–54 12	12	1700	I 3 r	6.5		10	8.11	–0.05	1.05
C1614–577	NGC 6087	16 18.2	–57 53	12	900	II 2 m	6.0	B5	8	7.85		0.36
C1622–405	NGC 6124	16 25.1	–40 39	29	470	I 3 r	6.3	B8	9	8.30		2.07
C1623–261	Antares	16 25.6	–26 13	505		III 3 p	1.0					
C1624–490	NGC 6134	16 27.2	–49 08	6	650		8.8		11	8.50	+0.25	1.29
C1632–455	NGC 6178	16 35.2	–45 37	4	900	III 3 p	7.2			7.08		
C1637–486	NGC 6193	16 40.7	–48 45	14	1300		5.4	O7		7.00		1.50
C1642–469	NGC 6204	16 45.9	–47 01	4	890	I 3 m	8.4	O6		7.73		1.47
C1645–537	NGC 6208	16 48.9	–53 48	15	990	III 2 r	9.5			9.16		0.51
C1650–417	NGC 6231	16 53.5	–41 47	14	1900		3.4	O9	6	7.00		1.35
C1652–394	NGC 6242	16 55.1	–39 29	9	1100		8.2	B5		7.73		1.11
C1653–405	Trump 24	16 56.5	–40 39	60	1900		8.6			7.00		1.08
C1654–447	NGC 6249	16 57.1	–44 46	6	1000	II 2 m	9.3			7.61		
C1654–457	NGC 6250	16 57.4	–45 56	7	1000	II 3 r	8.0			7.38		
C1657–446	NGC 6259	17 00.2	–44 40	10	1900	II 2 r	8.6		11	8.30	+0.29	1.86
C1714–355	Boch 13	17 16.8	–35 33	14	1600	III 3 m				7.08		
C1714–429	NGC 6322	17 17.9	–42 57	10	1200	I 3 m	6.5	B0		7.00		1.53
C1720–499	IC 4651	17 24.1	–49 56	12	730	II 2 r	8.0		10	9.40	–0.16	0.30
C1731–325	NGC 6383	17 34.3	–32 34	4	1300	II 3 mn	5.4			7.54		0.75
C1732–334	Trump 27	17 35.7	–33 29	6	1400	III 3 m	9.1			7.89		4.05
C1733–324	Trump 28	17 36.3	–32 29	7	1400	III 2 mn	9.4			8.30		2.13
C1734–362	Rup 127	17 37.2	–36 16	8	1500	II 2 p			11	7.08		2.97
C1736–321	NGC 6405	17 39.6	–32 12	14	490	II 3 r	4.6	B5	7	7.89	+0.10	0.48
C1741–323	NGC 6416	17 43.9	–32 21	18	770	III 2 m	8.7			8.50		0.93
C1743+057	IC 4665	17 45.9	+ 5 43	40	340	III 2 m	5.3	B4	6	7.89		0.48
C1747–302	NGC 6451	17 50.2	–30 13	7	560	I 2 m	8.2		12	9.62		0.21
C1750–348	NGC 6475	17 53.4	–34 48	80	240	I 3 r	3.3	B5	7	8.11		0.09
C1753–190	NGC 6494	17 56.4	–19 01	27	640	II 2 r	5.9	B9	10	8.30		0.78
C1758–237	Boch 14	18 01.5	–23 42	2	1100	III 1 pn				7.00		
C1800–279	NGC 6520	18 02.9	–27 54	6	1600	I 2 m	7.6		9	8.66		0.90
C1801–225	NGC 6531	18 04.2	–22 30	13	1200	I 3 r	7.2	B0	8	7.61		0.87

* Magnitude of brightest cluster member

IAU Desig.	Name	R.A.	Dec.	Ang. Diam.	Dist.	Trumpler Class	Tot. Mag.	Spectrum	Mag.*	Log age	Log Fe/H	A_v
		h m	° ′	′	pc							
C1801–243	NGC 6530	18 04.3	–24 20	14	1500		5.1	O5	6	7.00		0.90
C1804–233	NGC 6546	18 06.8	–23 20	13	1200	II 1 r	8.2			7.08		
C1815–122	NGC 6604	18 17.7	–12 14	2	2100	I 3 mn	7.5			7.00		2.88
C1816–138	NGC 6611	18 18.4	–13 47	6	2800		6.5	O7	11	7.00		2.13
C1817–171	NGC 6613	18 19.5	–17 08	9	1200	II 3 pn	7.5			7.61		1.26
C1825+065	NGC 6633	18 27.4	+ 6 34	27	310	III 2 m	5.6	B6	8	8.80	–0.11	0.51
C1828–192	IC 4725	18 31.2	–19 15	32	710	I 3 m	6.2	B4	8	7.61	–0.06	1.44
C1830–104	NGC 6649	18 33.0	–10 24	5	1600	I 3 m	10.0		13	8.03		3.48
C1834–082	NGC 6664	18 36.3	– 8 14	16	1400	III 2 m	8.5	B3	9	8.11		1.83
C1836+054	IC 4756	18 38.6	+ 5 26	52	390	II 3 r	5.4		8	8.92	+0.04	0.63
C1840–041	Trump 35	18 42.5	– 4 08	9	1800	I 2 m	10.0			7.89		3.42
C1842–094	NGC 6694	18 44.8	– 9 24	14	1500	II 3 m	9.0	B8	11	7.94		2.13
C1848–052	NGC 6704	18 50.5	– 5 13	5	1900	I 2 m	9.3		12	7.54		3.06
C1848–063	NGC 6705	18 50.7	– 6 17	13	1700		6.1	B8	11	8.30	+0.10	1.08
C1850–204	Coll 394	18 53.0	–20 24	22	640		6.3			7.89		
C1851–199	NGC 6716	18 54.1	–19 54	6	610	IV 1 p	7.5			8.03	–0.28	0.36
C1851+368	Steph 1	18 53.3	+36 54	20	320	IV 3 p				7.61	–0.10	
C1905+041	NGC 6755	19 07.4	+ 4 13	14	1700	II 2 r	8.6		11	7.08		3.42
C1906+046	NGC 6756	19 08.3	+ 4 40	4	1500	I 1 m	10.6		13	7.79		4.26
C1919+377	NGC 6791	19 20.5	+37 50	15	4700	I 2 r			15	0.00	+0.04	0.63
C1936+464	NGC 6811	19 38.0	+46 33	12	1100	III 1 r	9.0	A3	11	8.92		0.42
C1939+400	NGC 6819	19 41.1	+40 10	5	2100		9.5	A0	11	9.36	–0.11	1.35
C1941+231	NGC 6823	19 42.8	+23 17	12	2600	I 3 mn		O7		7.00		2.46
C1948+229	NGC 6830	19 50.7	+23 03	12	1700	II 2 p	8.9		10	7.73		1.68
C1950+292	NGC 6834	19 51.9	+29 24	4	2200	II 2 m	9.7		11	7.61		1.77
C1950+182	Harvd 20	19 52.8	+18 19	6	2700	IV 2 p	9.6			8.11		
C2002+438	NGC 6866	20 03.5	+43 58	6	1300	II 2 r	9.1	A2	10	8.75		0.39
C2002+290	Roslnd 4	20 04.6	+29 11	5	2700	II 3 mn				7.08		
C2004+356	NGC 6871	20 05.6	+35 45	20	1700	II 2 pn	5.8	O9		7.08		1.14
C2007+353	Biur 2	20 08.9	+35 28	12	1500	III 2 p			16	7.00		
C2009+263	NGC 6885	20 11.7	+26 28	7	600	III 2 m	5.7	B8	6	9.16	–0.16	0.21
C2014+374	IC 4996	20 16.2	+37 37	5	1600	II 3 pn	7.1	B0	8	7.00		2.07
C2018+385	Berk 86	20 20.2	+38 40	7	1100	IV 2 mn			13	7.61		
C2019+372	Berk 87	20 21.4	+37 20	12	840	III 2 m			13	7.00		
C2021+406	NGC 6910	20 22.8	+40 45	7	1500	I 3 mn	7.3	B0		7.00		2.79
C2022+383	NGC 6913	20 23.7	+38 30	6	1300	II 3 mn	7.5	B0	9	7.00		2.79
C2030+604	NGC 6939	20 31.3	+60 37	7	1200	II 1 r	10.1			9.20	–0.11	1.44
C2032+281	NGC 6940	20 34.3	+28 17	31	810	III 2 r	7.2	A2	11	9.27	+0.04	0.72
C2109+454	NGC 7039	21 10.9	+45 37	25	810	IV 2 m	6.8			7.54		0.36
C2121+461	NGC 7062	21 22.9	+46 21	6	1700	II 2 m	8.3			8.69		1.29
C2122+478	NGC 7067	21 23.9	+47 59	3	3700	II 1 p	8.3			7.61		2.49
C2122+362	NGC 7063	21 24.1	+36 28	7	630	III 1 p	8.9			8.11		0.21
C2127+468	NGC 7082	21 29.2	+47 03	25	1300					7.89	+0.03	0.81
C2130+482	NGC 7092	21 31.9	+48 24	31	290	III 2 m	5.3	A0	7	8.30		0.15
C2144+655	NGC 7142	21 45.7	+65 46	4	3000	I 2 r	10.0		11	9.49	–0.11	0.54

* Magnitude of brightest cluster member

IAU Desig.	Name	R.A.	Dec.	Ang. Diam.	Dist.	Trumpler Class	Tot. Mag.	Spec-trum	Mag.*	Log age	Log Fe/H	A_v
		h m	° ′	′	pc							
C2151+470	IC 5146	21 53.1	+47 14	9	960	III 2 pn	8.3	B1		8.30		1.98
C2152+623	NGC 7160	21 53.5	+62 34	7	810	I 3 p	6.4	B2		7.61		1.56
C2203+462	NGC 7209	22 04.9	+46 27	25	900	III 1 m	7.8	A0	9	8.50		0.60
C2208+551	NGC 7226	22 10.2	+55 23	1	2100	I 2 m	13.3			8.34		1.77
C2210+570	NGC 7235	22 12.3	+57 15	4	3200	II 3 m	9.2			7.00		2.82
C2213+496	NGC 7243	22 15.0	+49 51	21	760	II 2 m	6.7	B6	8	8.03		0.54
C2213+540	NGC 7245	22 15.0	+54 18	4	1800	II 2 m				8.19		1.77
C2218+578	NGC 7261	22 20.1	+58 03	5	2100	II 3 m	9.8			7.46	–0.46	2.88
C2227+551	Berk 96	22 29.1	+55 22	2	4900	I 2 p			13	7.00		
C2245+578	NGC 7380	22 46.7	+58 03	12	3000	III 2 mn	8.8	O9	10	7.00		1.86
C2306+602	King 19	23 08.0	+60 29	6	1200	III 2 p			12	7.61		2.37
C2309+603	NGC 7510	23 11.2	+60 32	4	3100	II 3 m	9.3	B2	10			3.18
C2313+602	Mark 50	23 14.9	+60 26	4	3400	III 1 pn				7.00		2.49
C2322+613	NGC 7654	23 23.9	+61 33	12	1600	II 2 r	8.2	B7	11	8.23		1.80
C2350+616	King 12	23 52.6	+61 55	2	2500	II 1 p			10	7.00		
C2354+611	NGC 7788	23 56.3	+61 21	9	2300	I 2 p	9.4	B1		7.89		0.81
C2354+564	NGC 7789	23 56.6	+56 41	15	1800	II 2 r	7.5	B9	10	9.25	–0.10	0.81
C2355+609	NGC 7790	23 58.1	+61 10	17	3000	II 2 m	7.2		10	7.46		1.83

* Magnitude of brightest cluster member

C0001–302 = ζ Scl Cluster
C0129+604 = M103
C0215+569 = h Per
C0218+568 = χ Per
C0238+425 = M34
C0344+239 = M45
C0525+358 = M38
C0532+341 = M36
C0549+325 = M37
C0605+243 = M35
C0629+049 = Rosette Cluster
C0638+099 = S Mon Cluster
C0644–206 = M41
C0700–082 = M50
C0716–248 = τ CMa Cluster
C0734–143 = M47
C0739–147 = M46
C0742–237 = M93
C0811–056 = M48
C0837+201 = M44 = NGC 2632
C0838–528 = o Vel Cluster
C0847+120 = M67
C1041–641 = θ Car Cluster
C1043–594 = η Car Cluster
C1239–627 = Coal–Sack Cluster
C1250–600 = NGC 4755 = the Jewel Box Cluster
C1736–321 = M6
C1750–348 = M7
C1753–190 = M23
C1801–225 = M21
C1816–138 = M16
C1817–171 = M18
C1828–192 = M25
C1842–094 = M26
C1848–063 = M11
C2022+383 = M29
C2130+482 = M39
C2322+613 = M52

Berk = Berkeley; Biur = Biurakan; Boch = Bochum; Coll = Collinder; Haff= Haffner; Harvd = Harvard; Mark = Markarian; Mel = Melotte; Pisms = Pismis; Roslnd = Roslund; Rup = Ruprecht; Trump = Trumpler; Steph = Stephenson; Wtrloo = Waterloo; Wester = Westerlund

IAU Desig.	Name	Right Ascension	Declination	V	$B-V$	$U-B$	$V-I$	Type	m/H	$E_{(B-V)}$	$(m-M)_v$	D_o	R	Remarks
		h m	° ′										kpc	
C0021−723	NGC 104	00 23.8	−72 07	4.04	0.86	0.34	1.42	G4	−0.75	0.04	13.46	4.6	7.8	47 Tuc (X−ray)
C0050−268	NGC 288	00 52.4	−26 38	8.56	0.66	0.11		((F6))	−1.39	0.00	14.70	8.7	12.2	
C0100−711	NGC 362	01 03.0	−70 53	6.42	0.76	0.13	1.31	F9	−1.39	0.04	14.90	8.9	9.8	D 62
C0310−554	NGC 1261	03 12.0	−55 15	8.64	0.70	0.12		F7	−1.17	0.02	15.70	13.4	15.8	
C0325+794	Pal 1	03 32.3	+79 33						−1.01	0.12	18.7	45.8	51.4	
C0344−718	NGC 1466	03 44.6	−71 42						−2.15	0.07	Wb	39.4	38.4	SL 1
C0354−498	AM 1	03 54.8	−49 38						−1.68	0.00	Wb	39.4	38.4	E 1, ESO 201−SC 10
C0422−213	Erid 1	04 24.4	−21 12						−1.22	0.00	Wb	84.7	90.2	
C0435−590	Reticulum	04 36.1	−58 52						−2.01	0.02	Wb	50.4	51.3	ESO 118−G 31
C0443+313	Pal 2	04 45.6	+31 22		1.5:				−1.68	1.45:	Wb	13.6	22.2	
C0444−840	NGC 1841	04 46.3	−84 01						−1.56	0.07	Wb	40.9	38.3	
C0512−400	NGC 1851	05 13.9	−40 03	6.70	0.77	0.14	1.35	F7	−1.25	0.07	15.40	10.8	15.9	D 508, MX 0513−40
C0522−245	NGC 1904	05 23.9	−24 32	7.84	0.60	0.04	1.19	F5	−1.47	0.00	15.65	13.5	19.8	M 79
C0647−359	NGC 2298	06 48.7	−36 00	9.44	0.74	0.22	1.44	F5	−2.06	0.11	15.80	12.2	17.5	
C0734+390	NGC 2419	07 37.6	+38 54	10.80	0.68	0.11	1.20	(F5.5)	−1.98	0.03	19.94	92.9	100.7	
C0737−337	AM 2	07 39.1	−33 50							0.53	Wb	57.7	61.5	ESO 368−SC 07
C0911−646	NGC 2808	09 11.9	−64 50	6.13	0.91	0.27	1.69	F7	−1.47	0.21	15.52	9.2	11.2	
C0921−770	E 3	09 21.0	−77 15						−0.96	0.30	Wb	8.3	9.7	ESO 037−SC 01
C0923−545	UKS 2	09 25.1	−54 41						−0.37	0.74	Wb	9.0	11.9	
C1003+003	Pal 3	10 05.1	+00 06	14.50	0.6:				−1.68	0.03	20.0	95.5	99.0	Sex C
C1015−461	NGC 3201	10 17.3	−46 22	7.10	0.97	0.38	1.64	F6	−1.60	0.28	14.15	4.4	9.1	D 445
C1117−649		11 19.4	−65 11							0.79	Wb	59.5	56.6	ESO 093−SC?08
C1126+292	Pal 4	11 28.9	+29 01	14.50	0.6:				−1.30	0.00	19.85	93.3	96.1	
C1207+188	NGC 4147	12 09.7	+18 35	10.28	0.62	0.06	1.06	F2/3	−1.68	0.02	16.28	17.5	19.9	
C1223−724	NGC 4372	12 25.3	−72 37	8.00	0.87	: 0.28	:	F5	−1.77	0.45	14.90	4.8	7.3	
C1235−509	Rup 106	12 38.2	−51 07							0.24	Wb	26.7	5.4	
C1236−264	NGC 4590	12 39.1	−26 42	8.25	0.66	0.03	1.18	F2/3	−1.85	0.03	15.01	9.6	10.0	M 68
C1256−706	NGC 4833	12 59.1	−70 50	7.36	0.96	0.31	1.66	F3	−1.98	0.38	14.90	5.4	7.1	
C1310+184	NGC 5024	13 12.5	+18 12	7.71	0.65	0.06	1.11	F6	−1.89	0.05	16.34	17.2	17.9	M 53
C1313+179	NGC 5053	13 16.1	+17 44	9.98	0.63		1.16	((F3))	−2.02	0.03	16.00	15.1	16.0	
C1323−472	NGC 5139	13 26.3	−47 26	3.65	0.79	0.19	1.36	F5	−1.60	0.11	13.92	5.2	6.7	Omega Cen
C1339+286	NGC 5272	13 41.8	+28 25	6.41	0.69	0.10	1.15	F6	−1.30	0.00	15.00	10.0	12.2	M 3
C1343−511	NGC 5286	13 46.0	−51 20	7.48	0.90	0.29	1.51	F5	−1.60	0.27	15.61	8.8	7.2	D 388
C1353−269	AM 4	13 55.9	−27 08						−2.23	0.06	Wb	30.3	25.6	
C1403+287	NGC 5466	14 05.1	+28 34	9.35	0.75		1.05	((F5))	−1.85	0.05	15.96	14.4	15.1	
C1427−057	NGC 5634	14 29.2	−05 57	9.58	0.68	0.13	1.25	F3/4	−1.77	0.07	16.90	21.6	17.6	
C1436−263	NGC 5694	14 39.2	−26 30	10.17	0.72	0.07	1.27	F4	−1.89	0.08	17.60	29.3	23.5	
C1452−820	IC 4499	14 59.1	−82 11	10.70	.8:				−1.77	0.24	17.12	18.4	15.3	
C1500−328	NGC 5824	15 03.5	−33 02	8.96	0.76	0.15	1.38	F4	−1.98	0.14	17.32	23.5	17.2	
C1513+000	Pal 5	15 15.7	−00 05	11.6:					−1.43	0.02	17.2	26.7	21.7	
C1514−208	NGC 5897	15 17.0	−20 59	8.59	0.75	0.05	: 1.28	F7	−1.47	0.06	15.60	12.0	6.9	
C1516+022	NGC 5904	15 18.2	+02 07	6.03	0.71	0.12	1.19	F7	−1.60	0.07	14.51	7.2	6.4	M 5
C1524−505	NGC 5927	15 27.5	−50 39	7.95	1.31	0.83	2.09	G2	−0.67	0.55	16.10	7.2	4.7	
C1531−504	NGC 5946	15 34.9	−50 38	9.11	1.19	0.43		F7/8	−1.34	0.56	16.7	9.3	5.1	
C1535−499	BH 176	15 38.6	−50 01							0.73	Wb	85.7	78.3	
C1542−376	NGC 5986	15 45.6	−37 46	7.53	0.89	0.30	1.61	F5	−1.72	0.27	15.90	10.0	4.5	D 552
C1608+150	Pal 14	16 10.7	+14 59						−1.34	0.03	Wb	75.3	69.9	Arp 1, AvdB
C1614−228	NGC 6093	16 16.6	−22 57	7.31	0.84	0.20	1.44	F6	−2.15	0.21	15.28	8.3	3.0	M 80
C1620−264	NGC 6121	16 23.1	−26 30	5.96	1.04	0.44	1.84	F8	−1.09	0.31	12.90	2.4	6.3	M 4
C1620−720	NGC 6101	16 25.0	−72 11	8.9:	1.0:			F5:	−1.68	0.08	15.70	12.2	8.6	
C1624−259	NGC 6144	16 26.8	−26 00	9.07	0.94		1.42	F5/6	−1.81	0.36	15.6	7.6	2.6	
C1624−387	NGC 6139	16 27.2	−38 50	8.99	1.38	0.68	2.45	F6/7:	−1.60	0.68	17.0	8.9	2.9	
C1625−352	Ter 3	16 28.2	−35 20							0.32	Wb	27.2	19.0	
C1629−129	NGC 6171	16 32.1	−13 02	8.17	1.13	0.52	1.88	G0:	−0.88	0.37	15.03	5.8	3.9	M 107
C1636−283		16 39.0	−28 23						−1.01	0.31	Wb	10.3	2.8	ESO 452−SC 11

IAU Desig.	Name	Right Ascension	Declination	V	$B-V$	$U-B$	$V-I$	Type	m/H	$E_{(B-V)}$	$(m-M)_v$	D_o	R	Remarks
		h m	° ′										kpc	
C1639+365	NGC 6205	16 41.4	+36 28	5.86	0.69	0.06	1.12	(F5.4)	−1.60	0.02	14.35	7.2	8.7	M 13
C1645+476	NGC 6229	16 46.8	+47 32	9.39	0.74	0.09	1.25	(F7.3)	−1.39	0.01	17.2	27.1	26.6	
C1644−018	NGC 6218	16 46.9	−01 56	6.88	0.86	0.20	1.46	F8	−1.89	0.19	14.30	5.4	4.7	M 12
C1650−220	NGC 6235	16 53.0	−22 10	10.23	0.88			F9:	−1.60	0.38	16.6	11.7	4.0	
C1654−040	NGC 6254	16 56.8	−04 05	6.63	0.92	0.24	1.60	F3	−1.51	0.26	14.05	4.4	5.1	M 10
C1656−370	NGC 6256	16 59.0	−37 07						−1.56	0.88:	Wb	9.1	2.0	
C1657−004	Pal 15	16 59.7	−00 32						−1.26	0.12	Wb	69.7	62.3	
C1658−300	NGC 6266	17 00.7	−30 06	6.53	1.14	0.52	2.10	F9	−1.26	0.46	15.38	5.9	2.8	M 62
C1659−262	NGC 6273	17 02.2	−26 15	6.83	1.00	0.35	1.73	F7	−2.40	0.38	16.35	10.5	2.5	M 19
C1701−246	NGC 6284	17 04.0	−24 45	9.03	0.97	0.36	1.72	F9	−1.34	0.27	16.0	10.5	2.6	
C1702−226	NGC 6287	17 04.7	−22 42	9.44	1.26		2.33	F5	−1.72	0.36	15.8	8.4	1.6	
C1707−265	NGC 6293	17 09.7	−26 34	8.39	0.97	0.27	1.68	F3	−1.85	0.34	15.5	7.5	1.5	
C1708−271	TJ 5	17 10.8	−27 11											
C1711−294	NGC 6304	17 14.1	−29 27	8.38	1.32	0.82	2.26	G3	−0.54	0.58	15.50	5.2	3.4	
C1713−280	NGC 6316	17 16.2	−28 08	9.00	1.30	0.57	2.42	G2	−0.62	0.48	16.7	10.6	2.3	
C1715+432	NGC 6341	17 16.9	+43 09	6.50	0.63	0.02	1.10	(F2.8)	−1.89	0.01	14.50	7.8	9.7	M 92
C1714−237	NGC 6325	17 17.5	−23 45	10.73	1.69		2.82	G0	−2.02	0.80	16.70	6.5	2.3	
C1715−277	TJ 16	17 18.0	−27 46											TBJ 2
C1715−262	TJ 15	17 18.1	−26 17											
C1715−278	TJ 17	17 18.2	−27 50						<−0.07					TBJ 1
C1716−184	NGC 6333	17 18.8	−18 31	7.75	0.96	0.30	1.75	F5/6	−1.77	0.36	15.7	8.0	1.8	M 9
C1718−195	NGC 6342	17 20.7	−19 35	10.10	1.36		1.95	G3/4	−0.75	0.49	17.5	15.0	6.9	
C1720−177	NGC 6356	17 23.1	−17 48	8.28	1.14	0.58	1.85	G3	−1.17	0.21	17.07	18.9	10.7	
C1720−263	NGC 6355	17 23.5	−26 21	9.76	1.58		2.36	G0	−1.34	0.76	16.6	6.6	2.0	
C1721−484	NGC 6352	17 24.9	−48 25	8.40	1.03	0.61		G4	−0.07	0.25	14.47	5.4	3.9	
C1724−307	Ter 2	17 27.1	−30 47						−0.54	1.31	Wb	10.0	1.4	HP 3 (X−ray)
C1725−050	NGC 6366	17 27.3	−05 04	10.09	1.60	0.99			−0.71	0.65	15.4	4.5	4.8	
C1727−315	Ter 4	17 30.2	−31 35						−0.29	1.55	Wb	16.1	7.4	HP 4
C1727−299	HP 1	17 30.6	−29 59		2.0:			((G5):)	−1.68	1.41	Wb	9.5	0.9	
C1726−670	NGC 6362	17 31.1	−67 03					G3	−0.71	0.12	14.65	7.1	5.3	D 225
C1728−338	Grindlay	1 17 31.5	−33 50							3.2 :	WB	11.8	3.2	4U/MXB 1728−34
C1730−333	Liller 1	17 32.9	−33 23						−0.29	2.91	Wb	7.9	1.2	(X−ray)
C1731−390	NGC 6380	17 33.9	−39 04						−1.30	1.38	WB	4.0	5.0	Pismis 25, Ton 1
C1732−304	Ter 1	17 35.3	−30 28						+0.10	1.52	Wb	10.6	1.9	HP 2 (X−ray)
C1733−390	Pismis 26	17 35.6	−38 33							0.91	Wb	8.7	1.5	Ton 2
C1732−447	NGC 6388	17 35.7	−44 44	6.64	1.16	0.62		G2	−0.62	0.32	16.83	14.3	6.5	
C1735−032	NGC 6402	17 37.2	−03 14	7.49	1.24	0.60	2.10	F4	−2.19	0.58	16.90	9.9	4.3	M 14
C1735−238	NGC 6401	17 38.2	−23 54	9.44	1.32		2.52	F9	−1.01	0.79	16.6	6.3	2.3	
C1736−536	NGC 6397	17 40.1	−53 40	5.90	0.76	0.15		F4	−2.02	0.13	12.30	2.4	6.4	D 366
C1740−247		17 43.1	−24 43											ESO 520−SC?20
C1740−262	Pal 6	17 43.2	−26 13	13.6:	3.4:				+0.22	1.80	18.0	2.6	5.9	
C1742+031	NGC 6426	17 44.5	+03 10	11.48	0.99	0.32	1.66	G1::	−1.94	0.40	17.3	15.7	9.6	
C1741−328	TJ 23	17 44.7	−32 46											
C1745−247	Ter 5	17 47.7	−24 47	13.50	4.00		5.90		−0.71	2.14	Wb	7.1	1.8	XB 1745−25
C1746−203	NGC 6440	17 48.4	−20 21	9.39	1.97	0.97	3.23	G4:	−0.54	1.01	16.4	4.1	4.5	MX 1746−20
C1746−370	NGC 6441	17 49.7	−37 03	7.24	1.25	0.79	2.16	G2	−0.07	0.45	16.50	10.1	2.1	3U 1746−37
C1747−312	Ter 6	17 50.2	−31 17							1.46	Wb	12.8	4.0	HP 5
C1748−346	NGC 6453	17 50.4	−34 36	9.77	1.17		2.53	F8	−1.51	0.61	Wb	10.7	2.1	
C1751−241	UKS 1	17 54.0	−24 09						−1.22	3.07	Wb	10.4	1.8	
C1755−442	NGC 6496	17 58.5	−44 16	8.80	1.1:			G4	−0.71	0.07	14.3	6.5	2.8	
C1758−268	Ter 9	18 01.3	−26 51						−0.45	1.71	Wb	7.0	1.9	
C1759−089	NGC 6517	18 01.4	−08 58	10.29	1.81	0.99	3.04	F8	−1.47	1.14	18.1	7.4	3.0	
C1800−260	Ter 10	18 02.8	−26 04							1.71	Wb	14.6	5.9	
C1800−300	NGC 6522	18 03.1	−30 02	8.75	1.20	0.64	1.95	F7/8	−1.56	0.50	15.64	6.3	2.3	
C1801−003	NGC 6535	18 03.5	−00 18	10.62	0.96	0.34		G0	−1.56	0.36	16.1	9.6	4.7	

IAU Desig.	Name	Right Ascension	Declination	V	$B-V$	$U-B$	$V-I$	Type	m/H	$E_{(B-V)}$	$(m-M)_v$	D_o	R	Remarks
		h m	° ′										kpc	
C1801−300	NGC 6528	18 04.3	−30 03	9.67	1.43	0.95	2.28	G3	−0.96	0.63	16.4	7.3	1.3	
C1802−075	NGC 6539	18 04.4	−07 35	9.62	1.91	1.16	2.39	G4::	−1.05	1.22	15.7	2.2	6.5	
C1804−250	NGC 6544	18 06.9	−25 00	8.30	1.46	0.68	2.50	F9:	−2.15	0.63	15.2	4.2	4.3	
C1804−437	NGC 6541	18 07.5	−43 42	6.91	0.76	0.14		F6	−2.02	0.13	14.60	6.8	2.6	D 473
C1806−259	NGC 6553	18 08.8	−25 55	8.13	1.63	1.06	2.89	G4	−0.41	0.79	16.35	5.6	3.0	
C1807−317	NGC 6558	18 09.8	−31 46					F7	−1.51	0.40	16.1	9.0	1.1	
C1808−072	IC 1276	18 10.3	−07 13						−0.84	0.92	18.5	12.4	5.6	Pal 7
C1809−227	Ter 11	18 12.0	−22 45							1.57	Wb	23.7	15.1	
C1810−318	NGC 6569	18 13.2	−31 50	8.76	1.29	0.54	2.12	G1:	−1.01	0.63	16.5	7.7	1.3	
C1812−121	Kodaira 1	18 14.7	−12 03											
C1814−522	NGC 6584	18 18.0	−52 13	8.87	0.79	0.17		F6	−1.56	0.11	16.4	16.1	9.1	D 376
C1820−303	NGC 6624	18 23.2	−30 22	8.31	1.10	0.57	1.84	G4/5	−0.84	0.25	15.20	7.5	1.5	4U/MXB 1820−30
C1821−249	NGC 6626	18 24.1	−24 52	6.99	1.09	0.45	1.82	F8	−1.81	0.33	14.9	5.8	3.0	M 28
C1827−255	NGC 6638	18 30.5	−25 30	9.03	1.12	0.54	1.92	G0	−0.92	0.36	16.8	13.3	5.2	
C1828−323	NGC 6637	18 30.9	−32 21	7.79	1.02	0.48	1.68	G2/3	−0.92	0.17	15.7	10.4	2.6	M 69
C1828−235	NGC 6642	18 31.5	−23 29	8.80	1.12	0.47	1.86	F8	−1.30	0.36	14.8	5.3	3.5	
C1832−330	NGC 6652	18 35.3	−33 00	8.93	0.89	0.36	1.44	G3	−0.92	0.11	17.0	21.3	13.0	
C1833−239	NGC 6656	18 35.9	−23 55	5.07	1.00	0.28	1.83	F5	−1.81	0.35	13.55	3.0	5.6	M 22
C1838−198	Pal 8	18 41.0	−19 50						−0.50	0.30	18.4	30.3	22.3	
C1840−323	NGC 6681	18 42.7	−32 18	8.18	0.72	0.14	1.31	F5	−0.92	0.07	15.40	10.8	3.2	M 70
C1850−087	NGC 6712	18 52.7	−08 43	8.13	1.14	0.53	2.00	F5	−1.26	0.35	15.51	7.4	3.7	A 1850−08 (X−ray)
C1851−305	NGC 6715	18 54.6	−30 29	7.61	0.84	0.24	1.44	F7/8	−1.85	0.14	17.11	21.4	13.3	M 54
C1852−227	NGC 6717	18 54.6	−22 43					F6	−2.19	0.18	16.5	15.2	7.5	Pal 9
C1856−367	NGC 6723	18 59.0	−36 39	7.26	0.74	0.26	1.36	F9	−1.09	0.01	14.80	9.0	2.7	D 523
C1902+017	NGC 6749	19 04.8	+01 52	11.07	1.63			((F8))	−0.37	0.96:	Wb	12.8	7.7	
C1906−600	NGC 6752	19 10.2	−60 00	5.76	0.66	0.08		F4/5	−1.64	0.01	13.20	4.3	5.5	D 295
C1908+009	NGC 6760	19 10.8	+01 01	9.08	1.68	0.8:	2.80	G5	−0.84	0.91	15.9	3.8	5.6	
C1914+300	NGC 6779	19 16.3	+30 10	8.21	0.87	0.21	1.48	(F4.6)	−2.32	0.22	15.60	9.4	9.4	M 56
C1914−347	Ter 7	19 17.2	−34 40							0.06	Wb	36.4	28.4	ESO 397−SC14
C1916+184	Pal 10	19 17.8	+18 33							1.20	18.6	8.5	7.5	
C1925−304	Arp 2	19 28.2	−30 22						−1.85	0.11	Wb	28.3	20.4	ESO 460−SC06
C1936−310	NGC 6809	19 39.5	−30 59	6.33	0.69	0.12	1.27	F4	−1.56	0.07	14.00	5.7	4.1	M 55
C1938−341	Ter 8	19 41.3	−34 01							0.12	Wb	48.2	40.4	
C1942−081	Pal 11	19 44.9	−08 02						−0.92	0.15	17.6	26.4	20.0	
C1951+186	NGC 6838	19 53.4	+18 45	8.28	1.12	0.53	1.81	G1	−0.45	0.28	13.90	3.9	7.2	M 71
C2003−220	NGC 6864	20 05.6	−21 57	8.52	0.86	0.28	1.52	F9	−1.68	0.17	16.85	18.1	11.8	M 75
C2031+072	NGC 6934	20 33.8	+07 23	9.03	0.77	0.20	1.24	F7/8	−1.30	0.12	16.22	14.6	11.9	
C2050−127	NGC 6981	20 53.0	−12 34	9.35	0.74	0.11	1.28	F7	−1.56	0.03	16.29	17.3	13.0	M 72
C2059+160	NGC 7006	21 01.1	+16 09	10.67	0.74	0.15	1.22	F6	−1.72	0.13	18.12	34.5	31.9	
C2127+119	NGC 7078	21 29.6	+12 08	6.48	0.68	0.06	1.18	F3/4	−2.06	0.07	15.26	10.1	1.5	M 15, 3U 2131+11
C2130−010	NGC 7089	21 33.1	−00 51	6.50	0.68	0.08	1.17	F4	−1.81	0.06	15.45	11.2	10.3	M 2
C2137−234	NGC 7099	21 39.9	−23 13	7.56	0.60	0.04	1.10	F3	−2.19	0.01	14.60	8.2	7.4	M 30
C2139−472		21 42.3	−47 03											ESO 287−?53
C2143−214	Pal 12	21 46.2	−21 17						−1.13	0.02	19.0	61.2	56.7	
C2304+124	Pal 13	23 06.4	+12 44	14.50	0.7:			((F6))	−0.96	0.05	17.10	24.4	25.5	
C2305−159	NGC 7492	23 08.1	−15 39	11.48	0.40	0.22		((F5))	−1.81	0.00	16.70	21.9	21.3	
C2346−732	AM 3	23 48.5	−72 59											

IAU Desig.	Name	Right Ascension J2000.0	Declination J2000.0	ID	$M(V)$	Z	S 5GHz	Code
		h m s	° ′ ″				Jy	
0003−066		0 06 13.895	− 6 23 35.34	G	19.7		1.5	
0008−264		0 11 01.247	−26 12 33.38	Q	19.0			
0016+731		0 19 45.789	+73 27 30.07	Q	18		1.7	
0019−000	4C+00.02	0 22 25.428	+ 0 14 56.14	G	21.1		1.1	
0022−423		0 24 42.993	−42 02 03.58	?	22		1.5	
0026+346	OB 343	0 29 14.242	+34 56 32.22	G	20.2		1.2	
0056−001	4C−00.06	0 59 05.514	+ 0 06 51.66	Q	17.7	0.717	1.4	
0104−408		1 06 45.108	−40 34 19.96	Q	18.1			
0106+013	4C 01.02	1 08 38.771	+ 1 35 00.32	Q	18.5	2.107		
0111+021		1 13 43.144	+ 2 22 17.30	G	16.3	0.047		
0112−017	UM 310	1 15 17.095	− 1 27 04.59	Q	17	1.365	0.9	
0113−118		1 16 12.522	−11 36 15.44	G	18.5			
0116+319	4C 31.04	1 19 34.998	+32 10 50.04	G	15.7	0.059	1.5	
0119+041	OC 033	1 21 56.860	+ 4 22 24.7	Q	19.5	0.637	1.1	
0133+476	OC 457	1 36 58.595	+47 51 29.10	L	18	0.860	2.0	A
0135−247	OC−259	1 37 38.346	−24 30 53.83	Q	16.9	0.831	0.7	
0138−097		1 41 25.832	− 9 28 43.69	L	18		1.2	
0146+056	OC 079	1 49 22.374	+ 5 55 53.55	Q?	20	2.345	0.9	
0149+218		1 52 18.055	+22 07 07.69	Q	18		1.4	
0153+744		1 57 34.976	+74 42 43.26	Q	16		1.1	
0202+149	4C 15.05	2 04 50.414	+15 14 11.05	Q	21.9			
0202+319		2 05 04.925	+32 12 30.01	Q	18	1.466	1.2	
0202−172		2 04 57.676	−17 01 19.78	Q	18.5	1.740	1.2	
0208−512		2 10 46.199	−51 01 01.94	Q	17.5	1.003		
0212+735		2 17 30.820	+73 49 32.63	L	19		2.2	A
0224+671	4C 67.05	2 28 50.052	+67 21 03.03	Q	19.5			
0234+285	CTD20	2 37 52.406	+28 48 08.99	Q	18.5	1.207		A
0235+164	OD 160	2 38 38.930	+16 36 59.28	L	19		1.4	A,a
0237−233	PHL 8462	2 40 08.176	−23 09 15.75	Q	17	2.224	0.9	A
0256+075	OD 094.7	2 59 27.083	+ 7 47 39.6	Q?	18		0.8	
0300+470	4C 47.08	3 03 35.242	+47 16 16.28	L	18			A
0316+413	3C84	3 19 48.160	+41 30 42.11	G	15.1			A
0319+121	OE 131	3 21 53.104	+12 21 14.00	Q	19		1.1	
0332−403		3 34 13.654	−40 08 25.39	Q	18.5	1.445	1.5	A
0333+321	NRAO 140	3 36 30.108	+32 18 29.35	Q	17.0	1.253	2.4	A
0336−019	CTA26	3 39 30.938	− 1 46 35.80	Q	17.9	0.852	2.6	A
0355+508	NRAO 150	3 59 29.748	+50 57 50.17	EF				
0400+258	OF 200	4 03 05.580	+26 00 01.51	Q	18	2.109	1.2	
0402−362		4 03 53.750	−36 05 01.91	Q	16.0	1.417		
0406+121		4 09 22.009	+12 17 39.84	Q	22			A
0414−189		4 16 36.546	−18 51 08.3	Q	19	1.536	1.3	
0420−014	OF−035	4 23 15.801	− 1 20 33.06	Q	18	0.915	3.1	A
0420+417	VR41.04.01	4 23 56.010	+41 50 02.72	Q	19			
0428+205	OF 247	4 31 03.755	+20 37 34.25	G	20	0.219	2.3	
0434−188		4 37 01.483	−18 44 48.62	Q	20.0			
0438−436		4 40 17.180	−43 33 08.60	Q	19.8	2.852	3.9	
0440−003	NRAO 190	4 42 38.661	0 17 43.42	Q	18.5	0.844	1.5	A
0451−282		4 53 14.646	−28 07 37.32	Q	19.0			
0454+844		5 08 42.363	+84 32 04.56	L	16.5		1.6	A
0457+024	OF 097	4 59 52.049	+ 2 29 31.08	Q	18.0	2.370	1.2	

IAU Desig.	Name	Right Ascension J2000.0	Declination J2000.0	ID	*M(V)*	*Z*	*S* 5GHz	Code
		h m s	° ′ ″				Jy	
0458−020	4C−02.19	5 01 12.805	− 1 59 13.8	Q	18.4	2.286	1.9	
0500+019		5 03 21.194	+ 2 03 04.55	?	20			
0528−250		5 30 07.964	−25 03 29.80	Q	17	2.765	0.8	
0528+134	OG 147	5 30 56.417	+13 31 55.15	Q	20.3			A
0529+075	OG 050	5 32 38.997	+ 7 32 43.30	Q	19.0			b
0537−441		5 38 50.361	−44 05 08.94	Q	15.5	0.894		A
0539−057		5 41 38.082	− 5 41 49.50	EF				
0552+398	DA 193	5 55 30.806	+39 48 49.17	Q	18	2.365	4.7	A
0605−085		6 07 59.700	− 8 34 49.99	Q	18.0			A,c
0607−157		6 09 40.950	−15 42 40.67	Q	17	0.324	2.4	A
0609+607		6 14 23.859	+60 46 21.81	Q	20		1.1	
0615+820		6 26 02.917	+82 02 25.64	Q	17.5		1.0	
0637−752		6 35 46.517	−75 16 16.86	Q	15.8	0.651		
0642+449	OH 471	6 46 32.017	+44 51 16.61	Q	19	3.40	0.7	
0710+439	OI 417	7 13 38.177	+43 49 17.01	G	19.8		1.6	
0711+356	OI 318	7 14 24.819	+35 34 39.77	Q	19.0	1.62	1.2	
0716+714		7 21 53.448	+71 20 36.44	L	13.2		1.1	
0723−008	OI−039	7 25 50.640	− 0 54 56.54	G	18.0	0.128		A
0727−115		7 30 19.113	−11 41 12.61	EF	(?)		3.0	A
0733−174		7 35 45.814	−17 35 48.39	EF	(?)		1.9	
0735+178	OI 158	7 38 07.394	+17 42 19.00	L	16.5	(0.424)	2.1	A
0736+017	OI 061	7 39 18.032	+ 1 37 04.64	Q	18	0.191	2.2	
0738+313	OI 363	7 41 10.704	+31 12 00.22	Q	17.5	0.630	1.6	A
0742+103	DW0742	7 45 33.060	+10 11 12.69	EF			3.6	A
0743−673		7 43 31.518	−67 26 25.96	Q	17.0	1.51		
0748+126	OI 280	7 50 52.046	+12 31 04.83	Q	18	0.889	1.5	A
0804+499	OJ 508	8 08 39.665	+49 50 36.55	G	18.4	0.351	1.1	
0814+425	OJ 425	8 18 16.000	+42 22 45.41	Q	18.5		1.6	A
0823+033		8 25 50.338	+ 3 09 24.51	Q	18		1.0	A,c
0826−373		8 28 04.785	−37 31 06.19	Q	16		1.8	
0827+243	OJ 248	8 30 52.087	+24 10 59.81	Q	17.5	2.046		
0828+493	OJ 448	8 32 23.214	+49 13 21.04	Q	17.5	1.046	1.5	
0831+557	4C 55.16	8 34 54.903	+55 34 21.09	G	18.5	0.242	5.5	d
0833+585		8 37 22.409	+58 25 01.86	Q	18	2.101	1.2	
0836+710	4C 71.01	8 41 24.368	+70 53 42.18	Q	16.5		2.5	A
0839+187		8 42 05.095	+18 35 40.98	Q	16.5	0.259	1.0	
0851+202	OJ 287	8 54 48.875	+20 06 30.63	L	14.5	(0.306)	2.8	A
0859−140	OJ−199	9 02 16.832	−14 15 30.90	Q	17.8	1.327	2.1	A
0859+470	OJ 499	9 03 03.991	+46 51 04.13	Q	18.7	1.462		
0906+015	4C 01.24	9 09 10.100	+ 1 21 35.4	Q	17.5	1.012	1.4	
0917+624	OK 630	9 21 36.236	+62 15 52.14	Q	19.5		1.2	
0919−260	OK−232	9 21 29.357	−26 18 43.36	Q	19	2.300	2.1	
0923+392	4C 39.25	9 27 03.014	+39 02 20.85	Q	17.8	0.699		A
0941−080		9 43 36.945	− 8 19 30.87	G	19		1.0	
0952+179	VR17.09.04	9 54 56.823	+17 43 31.22	Q	18.0	1.472		
0954+658		9 58 47.247	+65 33 54.81	Q	18		0.6	
1004+141	OL 108.1	10 07 41.498	+13 56 29.59	Q	18.0	2.707		
1015−314		10 18 09.278	−31 44 14.08	G	21		1.3	
1030+415	VR10.41.03	10 33 03.711	+41 16 06.16	Q	18.2	1.120	0.6	
1031+567	OL 553	10 35 07.047	+56 28 46.76	Q	20.3		1.2	

IAU Desig.	Name	Right Ascension J2000.0	Declination J2000.0	ID	M(V)	Z	S 5GHz	Code
		h m s	° ′ ″				Jy	
1032−199		10 35 02.156	−20 11 34.35	Q	19	2.198	0.9	
1034−293	OL−259	10 37 16.080	−29 34 02.82	L	18		1.9	A
1038+064	OL 064.5	10 41 17.162	+ 6 10 16.94	Q	16.5	1.270		
1039+811		10 44 23.086	+80 54 39.45	Q	16.5			c
1040+123	4C 12.37	10 42 44.606	+12 03 31.25	Q	17.3	1.029		
1055+018	4C 01.28	10 58 29.605	+ 1 33 58.81	Q	18.0	0.888		
1104−445		11 07 08.695	−44 49 07.61	Q	18.0	1.598		A
1111+149	OM 118	11 13 58.695	+14 42 26.94	Q	18.0	0.869		
1116+128	4C 12.39	11 18 57.302	+12 34 41.71	Q	19.3	2.118		
1117+146	4C 14.41	11 20 27.803	+14 20 54.95	Q	20		1.1	
1123+264.	PB2704	11 25 53.712	+26 10 19.97	Q	18.5	2.341		A
1127−145	OM−146	11 30 07.052	−14 49 27.39	Q	16.9	1.187	4.7	A
1128+385		11 30 53.283	+38 15 18.54	Q	19.0			
1130+009		11 33 20.056	+ 0 40 52.83	Q	18.5			
1143−245	OM−272	11 46 08.108	−24 47 32.95	Q	18.5	1.95		c
1144−379		11 47 01.370	−38 12 11.03	Q	16.2			
1145−071		11 47 51.559	− 7 24 41.18	Q	18.5		1.0	
1148−001	4C−00.47	11 50 43.870	− 0 23 54.21	Q	17.6	1.982	1.9	A
1150+812		11 53 12.514	+80 58 29.09	Q	18.6	1.25	1.2	
1155+251		11 58 25.790	+24 50 17.93	G	18		0.9	
1213+350	4C 35.28	12 15 55.600	+34 48 15.04	Q	20.0		0.9	
1216+487	ON 428	12 19 06.419	+48 29 56.09	Q	18.5	1.073	1.0	
1219+285	W COM	12 21 31.681	+28 13 58.44	L	15			
1222+037	4C 03.23	12 24 52.422	+ 3 30 50.28	Q	19.0	0.957		
1226+023	3C 273	12 29 06.6997*	+ 2 03 08.59	Q	12.86	0.1584	5.8	A
1228+126	3C 274	12 30 49.423	+12 23 28.04	G	9.6	0.004		e
1237−101	ON−162	12 39 43.065	−10 23 28.77	Q	18.2	0.753	1.0	
1243−072	ON−073	12 46 04.235	− 7 30 46.62	Q	18.5	(0.267)	1.4	
1244−255		12 46 46.802	−25 47 49.30	Q	18.0	0.633		
1245−197		12 48 23.900	−19 59 18.66	Q	20.5		2.3	
1252+119	ON 187	12 54 38.253	+11 41 05.83	Q	16.6	0.870	1.0	
1253−055	3C 279	12 56 11.167	− 5 47 21.53	Q	16.8	0.536		
1255−316		12 57 59.071	−31 55 16.90	Q	19.5		1.0	
1302−102	OP−106	13 05 33.016	−10 33 19.6	Q	15.2	0.286	1.0	
1308+326	OP 313	13 10 28.663	+32 20 43.78	L	19	0.996	2.5	A
1311+678	4C 67.22	13 13 27.984	+67 35 50.36	EF			0.9	
1313−333	OP−322	13 16 07.986	−33 38 59.18	Q	20.0	2.21		A
1323+321	4C 32.44	13 26 16.512	+31 54 09.40	G	19.0		2.3	A
1328+254	3C 287	13 30 37.691	+25 09 10.85	Q	18.0	1.055	3.2	c
1328+307	3C 286	13 31 08.284	+30 30 32.94	Q	17.0	0.846	7.4	
1334−127		13 37 39.783	−12 57 24.70	Q	18.5		1.9	A
1342+663		13 44 08.679	+66 06 11.64	Q	19.0			
1345+125	4C 12.50	13 47 33.359	+12 17 24.21	G	17.0	0.122	2.7	
1349−439		13 52 56.535	−44 12 40.40	L	18.5	0.053		
1354−152		13 57 11.240	−15 27 28.73	Q	18.5		1.5	
1354+195	4C 19.44	13 57 04.436	+19 19 07.37	Q	16.5	0.720	1.8	A
1404+286	OQ 208	14 07 00.394	+28 27 14.67	G	14.0	0.077	3.0	A
1418+546	OQ 530	14 19 46.598	+54 23 14.78	L	14.5		0.7	A
1430−178	OQ-151	14 32 57.690	−18 01 35.24	Q	19.0	2.331		
1435+638		14 36 45.800	+63 36 37.86	Q	15.0	2.060	0.9	

*Reference for origin of right ascension.

IAU Desig.	Name	Right Ascension J2000.0	Declination J2000.0	ID	*M(V)*	*Z*	*S* 5GHz	Code
		h m s	° ′ ″				Jy	
1442+101	OQ 172	14 45 16.461	+ 9 58 36.05	Q	18.4	3.53	1.1	
1502+106	OR 103	15 04 24.980	+10 29 39.19	Q	18.9	1.833	2.1	A
1504−166		15 07 04.791	−16 52 30.16	Q	18.5	0.876		
1510−089		15 12 50.533	− 9 05 59.84	Q	16.5	0.361		
1511+238	4C 23.41	15 13 40.186	+23 38 35.18	?	20		0.8	
1519−273		15 22 37.676	−27 30 10.78	Q	19		2.0	A
1546+027	OR 078	15 49 29.435	+ 2 37 01.15	Q	18.0	0.412	1.3	
1547+507		15 49 17.468	+50 38 05.76	Q	18.5		0.7	
1555+001	DW 1555	15 57 51.434	− 0 01 50.42	Q	19.0	1.770	1.2	A, f
1607+268	CTD93	16 09 13.315	+26 41 28.98	Q	19		1.7	
1610−771		16 17 49.260	−77 17 18.47	Q	19.0	1.710		
1611+343	DA 406	16 13 41.064	+34 12 47.91	Q	17.5	1.404	2.2	A
1633+382	4C 38.41	16 35 15.493	+38 08 04.50	Q	18.0	1.81	1.9	A
1637+574	OS 562	16 38 13.462	+57 20 23.94	Q	17.0	(0.745)	1.6	
1638+398	NRAO 512	16 40 29.633	+39 46 46.03	L	18.5			A
1641+399	3C 345	16 42 58.810	+39 48 36.99	Q	16.3	0.595		A
1652+398	4C 39.49	16 53 52.227	+39 45 36.45	L	14.0	0.033	1.2	
1656+053	OS 094	16 58 33.447	+ 5 15 16.44	Q	16.5	0.879		
1717+178	OT 129	17 19 13.049	+17 45 06.44	Q	18.5			g
1730−130	NRAO 530	17 33 02.706	−13 04 49.55	Q	18.5	0.902		
1732+389		17 34 20.577	+38 57 51.41	G	19.5		1.3	
1738+476	OT 465	17 39 57.127	+47 37 58.37	Q	17.5			
1739+522	4C 51.37	17 40 36.980	+52 11 43.43	Q	18.5	1.375	1.9	
1741−038		17 43 58.857	− 3 50 04.62	Q	18.5		2.2	A
1748−253		17 51 51.265	−25 23 59.80	Q	18.4		0.5	
1749+701		17 48 32.839	+70 05 50.77	L	16.5	(0.76)	1.2	A
1749+096	OT 081	17 51 32.816	+ 9 39 00.68	L	16.5		1.6	
1751+288		17 53 42.474	+28 48 04.91	Q	20.0		0.8	
1803+784		18 00 45.669	+78 28 04.00	L	16.4		2.5	
1807+698	3C 371	18 06 50.680	+69 49 28.11	G	14.2	0.51		
1821+107		18 24 02.855	+10 44 23.77	Q	16.0	1.036	1.1	A
1908−202		19 11 09.654	−20 06 55.03	?	22			
1921−293	OV−236	19 24 51.056	−29 14 30.11	Q	17	0.352	6.8	A
1928+738	4C 73.18	19 27 48.490	+73 58 01.55	Q	15.5	0.36	3.0	
1933−400		19 37 16.208	−39 58 00.88	Q	19.0		0.7	A
1934−638		19 39 25.006	−63 42 45.68	G	18.4	0.183		e
1936−155		19 39 26.655	−15 25 43.06	Q	20.5			
1947+079	OV 080	19 50 05.536	+ 8 07 13.93	Q?	21		1.2	
1958−179	OV−198	20 00 57.090	−17 48 57.67	Q	18.5	0.65	1.2	A
2007+776		20 05 31.001	+77 52 43.22	L	16.5		1.0	
2008−068		20 11 14.214	− 6 44 03.65	EF				
2021+614	OW 637	20 22 06.681	+61 36 58.82	Q	19.0		2.3	A
2029+121		20 31 54.994	+12 19 41.34	Q	18.5			
2030+547	OW 551	20 31 47.959	+54 55 03.15	?	18.7			
2037+511	3C 418	20 38 37.030	+51 19 12.59	Q	21.0	1.686		
2106−413		21 09 33.184	−41 10 20.48	Q	18.4		2.2	
2113+293		21 15 29.414	+29 33 38.37	Q	19.5		0.8	A
2128+048	3CR 433	21 30 32.874	+ 5 02 17.45	EF			2.1	
2128−123		21 31 35.260	−12 07 04.81	Q	16.0	0.501	2.4	
2131−021	4C−02.81	21 34 10.313	− 1 53 17.28	L	19.0	0.556	1.9	

IAU Desig.	Name	Right Ascension J2000.0	Declination J2000.0	ID	*M(V)*	*Z*	*S* 5GHz	Code
		h m s	° ′ ″				Jy	
2134+004	OX 057	21 36 38.586	+ 0 41 54.21	Q	18	1.936	10.3	A
2136+141	OX 161	21 39 01.303	+14 23 35.97	Q	18.5	2.427	1.2	
2144+092	OX 074	21 47 10.159	+ 9 29 46.65	Q	18.6	(1.609)	0.7	
2145+067	4C 06.69	21 48 05.459	+ 6 57 38.61	Q	17.5	0.990	2.5	A
2150+173		21 52 24.816	+17 34 37.8	G	21.0		0.7	
2155−152	OX−192	21 58 06.282	−15 01 09.32	L	18.0			
2200+420	BL LAC	22 02 43.291	+42 16 39.98	L	14	0.070	2.4	A
2201+315	4C 31.63	22 03 14.968	+31 45 38.29	Q	14.5	0.298		
2203−188	MSH 22−101	22 06 10.413	−18 35 38.77	Q	19.5	0.614	4.1	
2210−257		22 13 02.499	−25 29 30.17	Q	19.5		0.9	
2216−038	4C−03.79	22 18 52.036	− 3 35 36.91	Q	17	0.901	3.2	
2227−088		22 29 40.082	− 8 32 54.43	Q	18		1.2	
2227−399		22 30 40.276	−39 42 52.02	Q	18	0.323	0.6	
2230+114	CTA 102	22 32 36.409	+11 43 50.90	Q	17.3	1.037	3.6	A
2234+282	CTD 135	22 36 22.471	+28 28 57.42	Q	19	0.795	1.3	A
2243−123		22 46 18.232	−12 06 51.28	Q	17	0.63	2.4	A
2245−328		22 48 38.686	−32 35 52.17	Q	18.6	2.268	1.8	A
2251+158	3C 454.3	22 53 57.748	+16 08 53.56	L	16.1	0.859		A
2253+417	OY 489	22 55 36.708	+42 02 52.54	Q	18.8	1.476		
2254+074	OY 091	22 57 17.304	+ 7 43 12.27	L	16.4		0.5	
2255−282		22 58 05.862	−27 58 23.04	Q	17	0.926	1.6	
2318+049	OZ 031	23 20 44.854	+ 5 13 49.94	Q	19	0.623	0.8	
2319+272	4C 27.50	23 21 59.859	+27 32 46.42	Q	20	1.26	0.8	
2320−035		23 23 31.954	− 3 17 05.02	Q	18.0	1.411		
2326−477		23 29 17.707	−47 30 19.19	Q	17.0	1.299		
2328+107	4C 10.73	23 30 40.849	+11 00 18.68	Q	18.1	1.489	1.1	
2329−162		23 31 38.655	−15 56 56.99	Q	20		0.9	
2331−240	OZ−252	23 33 55.275	−23 43 40.74	G	16.5	0.048	1.0	
2337+264		23 40 29.029	+26 41 56.79	Q	20.0		0.8	
2344+092	4C 09.74	23 46 36.839	+ 9 30 45.49	Q	17.5	0.677	1.9	
2345−167	OZ−176	23 48 02.609	−16 31 12.02	Q	18.5	0.6	2.7	A
2351+456	4C 45.51	23 54 21.677	+45 53 04.16	G	19.9		1.2	
2352+495	DA 611	23 55 09.460	+49 50 08.33	G	19	0.237		A

Identification: Q=Quasar, G=Galaxy, L=BL Lac object, EF=Empty field, ?=uncertain.

Code: A—source observed by VLA and JPL
a—nebulous extension
b—extended HII
c—optical double
d—optical multiple
e—optically diffuse
f—nebulous (POSS—E)
g—diffuse (POSS—O)

RADIO FLUX CALIBRATORS, J2000.0

Source	Right Ascension	Declination	S_{400}	S_{750}	S_{1400}	S_{1665}	S_{2700}	S_{5000}
	h m s	° ′ ″	Jy	Jy	Jy	Jy	Jy	Jy
3C 48[e]	1 37 41.299	+33 09 35.41	39.4	25.6	15.9	13.9	9.20	5.24
3C 123	4 37 04.4	+29 40 15	119.2	77.7	48.7	42.4	28.5	16.5
3C 147[e, g]	5 42 36.127	+49 51 07.23	48.2	33.9	22.4	19.8	13.6	7.98
3C 161	6 27 10.0	− 5 53 07	41.2	28.9	19.0	16.8	11.4	6.62
3C 218	9 18 06.0	−12 05 45	134.6	76.0	43.1	36.8	23.7	13.5
3C 227	9 47 46.4	+ 7 25 12	20.3	12.1	7.21	6.25	4.19	2.52
3C 249.1	11 04 11.5	+76 59 01	6.1	4.0	2.48	2.14	1.40	0.77
3C 274[f]	12 30 49.6	+12 23 21	625	365	214	184	122	71.9
3C 286[e]	13 31 08.284	+30 30 32.94	25.1	19.7	14.8	13.6	10.5	7.30
3C 295	14 11 20.7	+52 12 09	54.1	36.3	22.3	19.2	12.2	6.36
3C 348	16 51 08.3	+ 4 59 26	168.1	86.8	45.0	37.5	22.6	11.8
3C 353	17 20 29.5	− 0 58 52	131.1	88.2	57.3	50.5	35.0	21.2
DR 21	20 39 01.2	+42 19 45	—	—	—	—	—	—
NGC 7027[d]	21 07 01.6	+42 14 10	—	—	1.35	1.65	3.5	5.7

Source	S_{8000}	S_{10700}	S_{15000}	S_{22235}	Spec.	Ident.	Polarization (at 5 GHz)	Angular Size (at 1.4 GHz)
	Jy	Jy	Jy	Jy			%	″
3C 48[e]	3.31	2.46	1.72	1.11	C−	QSS	5	< 1
3C 123	10.6	7.94	5.63	3.71	C−	GAL	2	20
3C 147[e, g]	5.10	3.80	2.65	1.71	C−	QSS	< 1	< 1
3C 161	4.18	3.09	2.14	—	C−	GAL	5	< 3
3C 218	8.81	6.77	—	—	S	GAL	1	core 25, halo 220
3C 227	1.71	1.34	1.02	0.73	S	GAL	7	180
3C 249.1	0.47	0.34	0.23	—	S	QSS	—	15
3C 274[f]	48.1	37.5	28.1	—	S	GAL	1	halo 400[a]
3C 286[e]	5.38	4.40	3.44	2.55	C−	QSS	11	< 5
3C 295	3.65	2.53	1.61	0.92	C−	GAL	0.1	4
3C 348	7.19	5.30	—	—	S	GAL	8	115[b]
3C 353	14.2	10.9	—	—	C−	GAL	5	150
DR 21	21.6	20.8	20.0	19.0	Th	HII	—	20[c]
NGC 7027[d]	—	6.43	6.16	5.86	Th	PN	< 1	10

a) Halo has steep spectral index, so for $\lambda \leq 6$ cm, more than 90% of the flux is in the core. Spectrum curves positively above 20 GHz.
b) Angular distance between the two components.
c) Angular size at 2 cm, but consists of 5 smaller components.
d) Data up to 5 GHz are the direct measurements, not calculated from fit.
e) Suitable for calibration of interferometers and synthesis telescopes.
f) Virgo A.
g) Indications of time variability above 5 GHz.

Designation Discovery	4U	Right Ascension	Declination	Flux[1]	Mag.[2]	Identified Counterpart	Type
		h m s	° ′ ″				
4U0005+20	0005+20	0 05 56.4	+20 09 41	5	14.0*	Mkn 335	G
Cep XR−1	0022+63	0 24 51	+64 06.0	16		Tycho's SNR	R
		0 28 50.4	+13 13 35		14.8	PG0026+129	Q
3U0026−09	0037−10	0 41 27.6	− 9 20.1	5	15.7	Abell 85	C
2U0022+42	0037+39	0 42 20	+41 13.7	4	4.8	M31=NGC 224	G
		0 48 22.8	+31 54 58		15.5*	Mkn 348	G
		0 53 11.3	+12 39 10		14.3*	I Zw 1	G
		0 59 28.3	+31 47 14		15.0*	Mkn 352	G
		1 02 49	+ 2 19.1		16.0	UMT301	Q
SMC X−1	0115−73	1 16 53.2	−73 28 57	66	13.2	Sanduleak 160	S
		1 21 36.8	− 1 04 46		15.1*	II Zw 1	G
2A0120−591	0106−59	1 23 28.6	−58 50 42	6	13.2	Fairall 9	G
		1 35 59.8	+20 55 10		18.1V	3CR47	Q
		2 14 10.1	− 0 48 13		14.5*	Mkn 590	G
		2 18	+62 37			HB3	R
4U0223+31	0223+31	2 27 48.6	+31 16 45	6	13.9*	NGC 931	G
4U0241+61	0241+61	2 44 21.9	+62 26 13	6	16.4		Q
GX146−15	0253+41	2 54.0	+41 33	9	14.5	NGC 1129	G+C
2U0528+13	0254+13	2 58.5	+13 33		15.6	Abell 401	C
2A0311−227		3 13 53.0	−22 37 22		14.8*V	EF Eri	S
Per XR−1	0316+41	3 19.3	+41 27	86	12.7	NGC 1275	G+C
H0324+28		3 26 07.7	+28 41 26		6.5V	UX Ari	S
4U0336+01	0336+01	3 36 24.2	+ 0 33 49	180	5.7	HR 1099	S
2A0335+096	0344+11	3 37.4	+10 04	3	12.2	Zw0335.1+0956	C
H0349+17		3 49 58	+17 14.1		9.2V	V471 Tau	S
2U0352+30	0352+30	3 54 54.8	+31 01 27	55	6.1V	X Per	S
2U0410+10	0410+10	4 13 00.4	+10 27 05	6	17.4	Abell 478	C
H0405−08		4 14 55.6	− 7 39 51		4.4	40 Eri	S
H0415+38	0407+37?	4 17 51.1	+38 00 31	5	18.0	3C111	G
	0432+05	4 32 47.2	+ 5 20 19	5	14.2	3C120	G
2A0431−136		4 33.3	−13 16		15.3	Abell 496	C
		4 36 00.8	−10 23 28		14.5*	Mkn 618	G
H0457+46		5 00	+46 34			HB 9	R
MX0513−40	0513−40	5 13 52.0	−40 03 14	33	8.1	NGC 1851	A
H0523−00		5 15 48.4	− 0 09 29		14.6*	Akn 120	G
LMC X−2	0520−72	5 20 36.5	−71 58 01	29	17		S
		5 25 06	−69 38 52			N132D	R
		5 25 25	−65 59 32			(N49)	R
		5 25 59	−66 05 41			N49	R
2A0526−328		5 29 08.9	−32 49 23		13.5V	TV Col	S
		5 32 06	−71 00 53			N206	R
LMC X−4	0532−66	5 32 49	−66 22 32	7	14.0		S
Tau XR−1	0531+21	5 34 04	+22 00 35	1730	8.4	Crab Nebula	R+P
		5 34 22	−70 33 34			DEM 238	R
		5 35 43	−66 02 24			N63A	R

Designation Discovery	4U	Right Ascension	Declination	Flux[1]	Mag.[2]	Identified Counterpart	Type
		h m s	° ′ ″				
A0538−66		5 35 44.9	−66 50 40		12.8V		S
		5 36 19.7	−70 39 10			DEM 249	R
		5 37 51	−69 10 14			N157B	R
A0535+26	0538+26?	5 38 26.6	+26 18 43	4	9.1	HDE 245770	S
LMC X−3	0538−64	5 38 53.9	−64 05 15	46	16.9		S
		5 40 15.8	−69 20 02			N158A	R
		5 45 43.1	−32 18 33		5.2	μ Col	S
		5 47 14	−69 42 32			N135	R
3U0545−32	0543−31	5 50 23.7	−32 16 28	7	16.1	PKS0548−322	G
2S0549−074		5 51 49.6	− 7 27 32		14.0	NGC 2110	G
MX0600+46	0558+46	5 54 20.2	+46 26 22	5	14.5	MCG 8−11−11	G
		6 15 17	+28 34.5		11.3V	KR Aur	S
2U0613+09	0614+09	6 16 42.5	+ 9 08 10	220	18.5V	V1055 Ori	S
2U0601+21	0617+23?	6 16.9	+22 33	6		IC 443	R
A0620−00		6 22 21	− 0 20 29		16.4V	V616 Mon	T
4U0720+55	0720+55	7 20 55	+55 47.2	5	13.6	Abell 576	C
		7 36 19.1	+58 47 22		14.5*	Mkn 9	G
		7 41 58.7	+65 11 46		15.0*	Mkn 78	G
		7 54 39	+22 01.5		8.8V	U Gem	S
		7 54 55.0	+39 12 25		15.5*	Mkn 382	G
	0821−42	8 22 46	−43 00.2	14		Puppis A	R
Vela XR−2	0833−45	8 35 05	−45 09 01	17	20.0	PSR0833−45	R+P
		8 50 27.3	+15 23 57		17.7V	LB8755	Q
Vela XR−1	0900−40	9 01 50	−40 31 30	450	6.7	HD 77581	S
3U0901−09	0900−09	9 08.5	− 9 37	9	15.2	Abell 754	C
		10 31 09.2	+28 49 20		16.6	Ton 524A	G
		10 44 46.1	−59 38 41		6.2V	η Car	S+H
A1044−59	1053−58	10 47	−59 36	5		G287.8−0.5	R
2A1052+606		10 55 15.5	+60 30 35		8.8	SAO 015338	S
		11 03.3	+28 55		15.2	IC 3510	G
A1103+38		11 04 02.4	+38 14 58		13.5*	Mkn 421	G
Cen XR−3	1118−60	11 20 56	−60 34 59	360	13.3V	V779 Cen	S
H1122−59		11 24.1	−59 24			MSH 11−54	R
		11 25 11.6	+54 25 28		16.0*	Mkn 40	G
A1136−37	1136−37	11 38 39.4	−37 41 49	5	12.8*	NGC 3783	G
2U1134−61	1137−65	11 39 08.4	−65 21 22	17	5.2	HD 101379	S
2U1144+19	1143+19	11 44 20	+19 47.5	5	13.5	Abell 1367	G+C
Cen X−5	1145−61	11 47 38	−62 09 54	130	9.2	HD 102567	S
		12 04 19.2	+27 56 42		15.5V	GQ Com	Q
2U1207+39	1206+39	12 10 09	+39 26.7	8	11.2*	NGC 4151	G
H1209−52		12 11.8	−52 56			PKS1209−52	R
		12 18 03.8	+29 51 18		14.0*	Mkn 766	G
		12 21 24.6	+75 21 07		14.5	Mkn 205	Q
		12 24 40	+12 55.7		9.3	M84=NGC 4374	G
		12 25.5	+12 42		12.7*	NGC 4388	G

Designation Discovery	4U	Right Ascension	Declination	Flux[1]	Mag.[2]	Identified Counterpart	Type
		h m s	° ′ ″				
		12 25 49	+12 59.3		9.7	M86=NGC 4406	G
GX301−2	1223−62	12 26 12	−62 43 43	73	10.0V	BP Cru	S
		12 26.8	+ 9 28		12.5	NGC 4424	G
		12 27 23	+13 03.0		11.3*	NGC 4438	G
		12 28 02.5	+31 31 07		15.9	B2 1225+317	Q
		12 28.6	+14 01		12.0	NGC 4459	G
	1226+02	12 28 43.6	+ 2 05 37	5	13.0	3C273	Q
1E1227.0+1403		12 29 11.0	+13 48 56		17.4		Q
		12 29 26	+13 28.3		11.6*	NGC 4473	G
		12 29.7	+13 41		11.5	NGC 4477	G
Vir XR−1	1228+12	12 30 26	+12 26.0	40	9.2	M87=NGC 4486	G+C
		12 33 53	+11 07		15.2	IC 3510	G
		12 35 12.2	−39 52 05		12.9	NGC 4507	G
4U1240−05	1240−05	12 39.2	− 5 18	4	12.0	NGC 4593	G
2U1247−41	1246−41	12 48.4	−41 16	9	12.4*	NGC 4696	G+C
2U1253−28	1249−28	12 52 00	−29 13 08	8	11.5V	EX Hya	S
Coma XR−1	1257+28	12 59.6	+27 57.7	27	10.7	Coma Cluster	C
GX304−1	1258−61	13 00 48.2	−61 33 37	100	14.7		S
MX1313+29		13 16.0	+29 09		12.5*	HZ 43	S
	1322−42	13 25 02	−42 58.8	15	7.2*	Cen A=NGC 5128	G
4U1326+11	1326+11	13 29.0	+11 47	4	14.2	NGC 5171	G+C
2U1348+24	1348+25	13 48 31.4	+26 37.7	8	16.0	Abell 1795	C
2A1347−300		13 48 53.6	−30 16 23		12.8*	IC 4329A	G+C
		13 53 02.4	+63 47 57		14.8	PG1351+640	Q
		14 11 24.3	+52 14 16		20.5	3C295	C
TWX−1	1410−03	14 12.9	− 3 10	3	13.6*	NGC 5506	G
2A1415+255	1414+25	14 17 39.4	+25 10 14	6	13.1*	NGC 5548	G
		14 29 12	−62 38.9		12.4V	PROXIMA CEN	S
		14 34 36.7	+48 41 43		16.5*	Mkn 474	G
Cen XR−4		14 57 54.6	−31 38 20		19.0		T
	1458−41	15 01 52	−41 42.1	4	19.9	SN 1006	R
GX9+50		15 10.6	+ 5 47		16.0	Abell 2029	C
Cir XR−1	1516−56	15 20 06.0	−57 08 24	1300	22.5*V	BR Cir	S
2A1519+082		15 21 29.4	+ 7 43 59		15.5	NGC 5920	G+C
		15 26 24.4	+10 00 40		18.0	4C10.43	Q
A1524−61		15 27 38.8	−61 51 26		19.0	Nova TrA 1974	T
		15 35 43.2	+57 55 40		15.0*	Mkn 290	G
2U1537−52	1538−52	15 41 49.5	−52 21 45	33	14.5V	QV Nor	S
		15 54 48.1	+19 12 58		15.0*	Mkn 291	G
		15 55 39	−37 54 48		12.0	The 12	S
3U1555+27	1556+27	15 58.0	+27 15		16.0	Abell 2142	C
3U1551+15	1601+15	16 01 53.4	+16 02 28	6	13.8	Abell 2147	G+C
		16 04 35.9	+23 56 18		15.0	NGC 6051	G+C
MX1608−52	1608−52	16 12 08.3	−52 24 16	73	21 V	QX Nor	S
H1615−51		16 17 01.4	−51 01 20			RCW 103	R

Designation Discovery	4U	Right Ascension	Declination	Flux[1]	Mag.[2]	Identified Counterpart	Type
		h m s	° ′ ″				
Sco XR−1	1617−15	16 19 29.3	−15 37 19	31000	12.4V	V818 Sco	S
3U1639+40	1627+39	16 28.4	+39 32	7	13.9	Abell 2199	C
2U1626−67	1626−67	16 31 30.8	−67 26 44	33	18.5V	KZ TrA	S
2A1630+057		16 32.4	+ 5 36		17.1	Abell 2204	C
2U1637−53	1636−53	16 40 18.5	−53 24 11	460	17.5V	V801 Ara	S
Ara XR−1	1642−45	16 45 17	−45 37	820		G339.6−0.1	H
4U1651+39	1651+39	16 53 37.2	+39 46 20	4	13.5*	Mkn 501	G
Her X−1	1656+35	16 57 34	+35 21 11	180	13.0V	HZ Her	S
		17 00 50.4	+29 25 10		17.0*	Mkn 504	G
GX339−4	1658−48	17 02 15.2	−48 46 44	630	15.4V	V821 Ara	S
2U1700−37	1700−37	17 03 26	−37 50 02	180	6.7	HD 153919	S
2U1706+78	1707+78	17 04 19.8	+78 39.1	7	15.3	Abell 2256	C
H1705−25		17 07 47.0	−25 04 55		21*	Nova Oph 1977	T
		17 22 22.8	+24 36 44		16.4V	V396 Her	Q
		17 22 23.0	+30 53 11		15.5*	Mkn 506	G
4U1722−30	1722−30	17 27 04.0	−30 47 47	13	17	Terzan 2	A
		17 30.1	−21 28.7		19	Kepler's SNR	R
GX9+9	1728−16	17 31 18.0	−16 57 26	470	16.6		S
GX1+4	1728−24	17 31 34.5	−24 44 35	110	18.7V	V2116 Oph	S
MXB1730−335		17 32 55	−33 23 03		17.5	Liller 1	A
GX346−7	1735−44	17 38 25.1	−44 26 45	380	17.5V	V926 Sco	S
MX1746−20		17 48 26.0	−20 21 27		12.0	NGC 6440	A
		17 48 40.6	+68 42 08		16.0*	Mkn 507	G
L10	1746−37	17 49 42.1	−37 02 57	73	8.4*	NGC 6441	A
2U1808+50	1813+50	18 16 02.2	+49 51 53	10	12.3V	AM Her	S
Sgr XR−4	1820−30	18 23 11.5	−30 21 53	580	8.6*	NGC 6624	A
H1832+32		18 34 46.7	+32 41 24		14.7	3C382	G
Ser XR−1	1837+04	18 39 35.2	+ 5 01 52	510	15.1V	MM Ser	S
2U1828+81	1847+78	18 42 41	+79 45 46	5	15.0	3C390.3	G
A1850−08	1850−08	18 52 39.7	− 8 42 56	16	8.9	NGC 6712	A
4U1849−31	1849−31	18 54 34	−31 10.5	7	14.7V	V1223 Sgr	S
2U1907+02	1901+03?	18 55 45	+ 1 18.4	160		Westerhout 44	R
	1907+09	19 09 17.1	+ 9 49 03	36	16.4		S
Aql XR−1	1908+00	19 10 53.1	+ 0 34 20	360	16.0V	V1333 Aql	S
A1909+04	1908+05	19 11 27.4	+ 4 58 13	7	14.2V	SS433	S
2A1914−58	9 1924−59	19 20 35.8	−58 41 05	4	14.1	ESO 141−G55	G
2U1926+43	1919+44	19 20 59	+43 55.5	7	15.4	Abell 2319	C
		19 33	+31 16			G65.2+5.7	R
H1938+16		19 41 50	+17 04.7		14.7V	UU Sge	S
Cyg XR−1	1956+35	19 58 04.6	+35 10 51	2100	8.9	HDE 226868	S
3U1956+11	1957+11	19 59 02.8	+11 41 16	32	18.7		S
2U1957+40	1957+40	19 59 12	+40 43	7	16.2	Cyg A	C
1E2014.1+3702		20 15 46.8	+37 10 43			G74.9+1.2	R
Cyg X−3	2030+40	20 32 10	+40 55 56	700		V1521 Cyg	S
2A2040−115		20 43 45.3	−10 45 02		13.0*	Mkn 509	G

Designation Discovery	4U	Right Ascension	Declination	Flux[1]	Mag.[2]	Identified Counterpart	Type
		h m s	° ′ ″				
Vul XR−1?	2046+31?	20 51 31	+31 03	3		Cygnus Loop	R
2U2134+11	2129+12	21 29 36.7	+12 08 02	8	6.0	M15=NGC 7078	A
2U2130+47	2129+47	21 31 09.7	+47 15 24	36	16.2V	V1727 Cyg	S
		21 42 25	+43 33.0		8.2V	SS Cyg	S
Cyg XR−2	2142+38	21 44 23	+38 17 13	1000	15.5V	V1341 Cyg	S
2A2151−316		21 58 25.6	−30 15 41		17.0	PKS2155−304	G
H2208−47		22 08.9	−47 12		11.8	NGC 7213	G
		22 16 49.9	+14 12 15		15.5*	Mkn 304	G
2S2251−178		22 53 41.9	−17 37 19		17.0	MR2251−178	Q
GF2259+586		23 00 49.0	+58 50 20		15.0	G109.1−1.0	R+S
2A2259+085	2300+08	23 02 53.0	+ 8 50 02	5	13.0*	NGC 7469	G
		23 03 40.7	+22 34 58		15.0*	Mkn 315	G
2A2302−088	2305−07	23 04 20.1	− 8 43 34	4	13.9	MCG−2−58−22	G
2A2315−428		23 17 58	−42 24.8		11.8	NGC 7582	G
Cas XR−1	2321+58	23 23 05	+58 46	97	19.6	Cas A	R
3U2346+26	2345+27	23 50 38	+27 06.8	4	13.8	Abell 2666	C
4U2351+06	2351+06	23 55 38.9	+ 7 28 52	7	15.5*	Mkn 541	G

[1] (2−6) kev flux, units are 10^{-11} ergs $cm^{-2}s^{-1}$.
[2] *V* magnitude unless followed by *, then *B* magnitude. V designates variable magnitude.

Type Designation:
A – Globular Cluster
C – Cluster of Galaxies
G – Galaxy
H – HII Region
P – Pulsar
Q – Quasi−Stellar Object
R – Supernova Remnant
S – Stellar
T – Transient (Nova-like optically).

ECLIPSING VARIABLES

Name		H.D.	Right Ascension	Declination	Type	Magnitude Max	Magnitude Min	Mag. Type	Epoch (2400000+)	Period	Spectrum
			h m s	° ′						d	
U	Cep	5679	1 01 37	+81 50.1	EA	6.75	9.2	V	44541.603	2.493	B7Ve + G8III–IV
ζ	Phe	6882	1 08 04	−55 17.2	EA	3.92	4.4	V	41643.689	1.669	B6V + B9V
RZ	Cas	17138	2 48 14	+69 36.2	EA	6.18	7.7	V	43200.306	1.195	A3V
β	Per	19356	3 07 41	+40 55.6	EA	2.12	3.4	V	40953.465	2.867	B8V + G5IV + Am
λ	Tau	25204	4 00 16	+12 28.2	EA	3.3	3.8	p	35089.204	3.952	B3V + A4IV
HU	Tau	29365	4 37 50	+20 40.2	EA	5.92	6.7	V	42412.456	2.056	A8V
ε	Aur	31964	5 01 25	+43 48.7	EA	2.92	3.8	V	35629	9892	A8Ia–F2Iaep
AR	Aur	34364	5 17 49	+33 45.6	EA	6.15	6.8	V	38402.183	4.134	ApHgMn + B9V
TZ	Men	39780	5 31 41	−84 47.5	EA	6.2	6.9	p	38196.370	8.569	B9.5IV–V
WW	Aur	46052	6 31 58	+32 27.6	EA	5.79	6.5	V	41399.305	2.525	A3m: + A3m:
R	CMa	57167	7 19 07	−16 22.8	EA	5.70	6.3	V	44289.361	1.135	F1V
V	Pup	65818	7 58 02	−49 13.5	EB	4.7	5.2	p	28648.304	1.454	B1Vp + B3IV:
TY	Pyx	77137	8 59 25	−27 47.2	E	6.87	7.4	V	43187.230	3.198	G5 + G5
CV	Vel	77464	9 00 24	−51 31.5	EA	6.5	7.3	p	42048.668	6.889	B2V + B2V
S	Ant	82610	9 31 59	−28 35.7	EW	6.4	6.9	V	35139.929	0.648	A9Vn
W	UMa	83950	9 43 14	+55 59.2	EW	7.9	8.6	V	41004.397	0.333	dF8p + F8p
δ	Lib	132742	15 00 34	− 8 29.3	EA	4.92	5.9	V	42937.423	2.327	B9.5V
i	Boo	133640	15 03 34	+47 40.9	EW	6.5	7.1	v	39370.422	0.267	G2V + G2V
GG	Lup	135876	15 18 27	−40 45.8	EB	5.4	6.0	p	34532.325	2.164	B5 + A0
R	Ara	149730	16 39 07	−56 58.8	EA	6.0	6.9	p	25818.028	4.425	B9IV–V
V1010	Oph	151676	16 49 02	−15 39.4	EB	6.1	7.0	v	38937.771	0.661	A5V
V861	Sco	152667	16 56 04	−40 48.7	EB	6.07	6.6	V		7.848	B0.5Iae
U	Oph	156247	17 16 09	+ 1 13.1	EA	5.88	6.5	V	36727.424	1.677	B5Vnn + B5V
u	Her	156633	17 17 03	+33 06.5	EB	4.6	5.3	p	44069.386	2.051	B1.5Vp + B5III
V539	Ara	161783	17 49 52	−53 36.7	EA	5.66	6.1	V	39314.342	3.169	B2V + B3V
RS	Sgr	167647	18 17 06	−34 06.6	EA	6.0	6.9	p	20586.387	2.415	B3V + A
β	Lyr	174639	18 49 48	+33 21.2	EB	3.34	4.3	V	45342.39	12.935	B7Ve + A8p
RS	Vul	180939	19 17 21	+22 25.7	EA	6.9	7.6	p	32808.257	4.477	B5V + A2
U	Sge	181182	19 18 29	+19 35.8	EA	6.58	9.1	V	40774.463	3.380	B8III + K:
V505	Sgr	187949	19 52 41	−14 37.3	EA	6.48	7.5	V	40087.336	1.182	A0V + F8IV
EE	Peg	206155	21 39 40	+ 9 09.0	EA	6.9	7.6	v	40286.432	2.628	A3Vm + F4:
VV	Cep	208816	21 56 27	+63 35.4	EA	4.80	5.3	V	43360	7430	M2Ia–Iabep + B8:Ve
AR	Lac	210334	22 08 23	+45 42.3	E	6.11	6.7	V	39376.495	1.983	G2Iv + K0III

See page H74 for Type codes.

Designation Discovery	4U	Right Ascension	Declination	Flux[1]	Mag.[2]	Identified Counterpart	Type
		h m s	° ′ ″				
Vul XR−1?	2046+31?	20 51 31	+31 03	3		Cygnus Loop	R
2U2134+11	2129+12	21 29 36.7	+12 08 02	8	6.0	M15=NGC 7078	A
2U2130+47	2129+47	21 31 09.7	+47 15 24	36	16.2V	V1727 Cyg	S
		21 42 25	+43 33.0		8.2V	SS Cyg	S
Cyg XR−2	2142+38	21 44 23	+38 17 13	1000	15.5V	V1341 Cyg	S
2A2151−316		21 58 25.6	−30 15 41		17.0	PKS2155−304	G
H2208−47		22 08.9	−47 12		11.8	NGC 7213	G
		22 16 49.9	+14 12 15		15.5*	Mkn 304	G
2S2251−178		22 53 41.9	−17 37 19		17.0	MR2251−178	Q
GF2259+586		23 00 49.0	+58 50 20		15.0	G109.1−1.0	R+S
2A2259+085	2300+08	23 02 53.0	+ 8 50 02	5	13.0*	NGC 7469	G
		23 03 40.7	+22 34 58		15.0*	Mkn 315	G
2A2302−088	2305−07	23 04 20.1	− 8 43 34	4	13.9	MCG−2−58−22	G
2A2315−428		23 17 58	−42 24.8		11.8	NGC 7582	G
Cas XR−1	2321+58	23 23 05	+58 46	97	19.6	Cas A	R
3U2346+26	2345+27	23 50 38	+27 06.8	4	13.8	Abell 2666	C
4U2351+06	2351+06	23 55 38.9	+ 7 28 52	7	15.5*	Mkn 541	G

[1] (2−6) kev flux, units are 10^{-11} ergs $cm^{-2}s^{-1}$.
[2] *V* magnitude unless followed by *, then *B* magnitude. V designates variable magnitude.

Type Designation:
- A − Globular Cluster
- C − Cluster of Galaxies
- G − Galaxy
- H − HII Region
- P − Pulsar
- Q − Quasi−Stellar Object
- R − Supernova Remnant
- S − Stellar
- T − Transient (Nova-like optically).

ECLIPSING VARIABLES

Name		H.D.	Right Ascension	Declination	Type	Magnitude Max	Magnitude Min	Mag. Type	Epoch (2400000+)	Period	Spectrum
			h m s	° ′						d	
U	Cep	5679	1 01 37	+81 50.1	EA	6.75	9.2	V	44541.603	2.493	B7Ve + G8III–IV
ζ	Phe	6882	1 08 04	−55 17.2	EA	3.92	4.4	V	41643.689	1.669	B6V + B9V
RZ	Cas	17138	2 48 14	+69 36.2	EA	6.18	7.7	V	43200.306	1.195	A3V
β	Per	19356	3 07 41	+40 55.6	EA	2.12	3.4	V	40953.465	2.867	B8V + G5IV + Am
λ	Tau	25204	4 00 16	+12 28.2	EA	3.3	3.8	p	35089.204	3.952	B3V + A4IV
HU	Tau	29365	4 37 50	+20 40.2	EA	5.92	6.7	V	42412.456	2.056	A8V
ε	Aur	31964	5 01 25	+43 48.7	EA	2.92	3.8	V	35629	9892	A8Ia–F2Iaep
AR	Aur	34364	5 17 49	+33 45.6	EA	6.15	6.8	V	38402.183	4.134	ApHgMn + B9V
TZ	Men	39780	5 31 41	−84 47.5	EA	6.2	6.9	p	38196.370	8.569	B9.5IV–V
WW	Aur	46052	6 31 58	+32 27.6	EA	5.79	6.5	V	41399.305	2.525	A3m: + A3m:
R	CMa	57167	7 19 07	−16 22.8	EA	5.70	6.3	V	44289.361	1.135	F1V
V	Pup	65818	7 58 02	−49 13.5	EB	4.7	5.2	p	28648.304	1.454	B1Vp + B3IV:
TY	Pyx	77137	8 59 25	−27 47.2	E	6.87	7.4	V	43187.230	3.198	G5 + G5
CV	Vel	77464	9 00 24	−51 31.5	EA	6.5	7.3	p	42048.668	6.889	B2V + B2V
S	Ant	82610	9 31 59	−28 35.7	EW	6.4	6.9	V	35139.929	0.648	A9Vn
W	UMa	83950	9 43 14	+55 59.2	EW	7.9	8.6	V	41004.397	0.333	dF8p + F8p
δ	Lib	132742	15 00 34	− 8 29.3	EA	4.92	5.9	V	42937.423	2.327	B9.5V
i	Boo	133640	15 03 34	+47 40.9	EW	6.5	7.1	v	39370.422	0.267	G2V + G2V
GG	Lup	135876	15 18 27	−40 45.8	EB	5.4	6.0	p	34532.325	2.164	B5 + A0
R	Ara	149730	16 39 07	−56 58.8	EA	6.0	6.9	p	25818.028	4.425	B9IV–V
V1010	Oph	151676	16 49 02	−15 39.4	EB	6.1	7.0	v	38937.771	0.661	A5V
V861	Sco	152667	16 56 04	−40 48.7	EB	6.07	6.6	V		7.848	B0.5Iae
U	Oph	156247	17 16 09	+ 1 13.1	EA	5.88	6.5	V	36727.424	1.677	B5Vnn + B5V
u	Her	156633	17 17 03	+33 06.5	EB	4.6	5.3	p	44069.386	2.051	B1.5Vp + B5III
V539	Ara	161783	17 49 52	−53 36.7	EA	5.66	6.1	V	39314.342	3.169	B2V + B3V
RS	Sgr	167647	18 17 06	−34 06.6	EA	6.0	6.9	p	20586.387	2.415	B3V + A
β	Lyr	174639	18 49 48	+33 21.2	EB	3.34	4.3	V	45342.39	12.935	B7Ve + A8p
RS	Vul	180939	19 17 21	+22 25.7	EA	6.9	7.6	p	32808.257	4.477	B5V + A2
U	Sge	181182	19 18 29	+19 35.8	EA	6.58	9.1	V	40774.463	3.380	B8III + K:
V505	Sgr	187949	19 52 41	−14 37.3	EA	6.48	7.5	V	40087.336	1.182	A0V + F8IV
EE	Peg	206155	21 39 40	+ 9 09.0	EA	6.9	7.6	v	40286.432	2.628	A3Vm + F4:
VV	Cep	208816	21 56 27	+63 35.4	EA	4.80	5.3	V	43360	7430	M2Ia–Iabep + B8:Ve
AR	Lac	210334	22 08 23	+45 42.3	E	6.11	6.7	V	39376.495	1.983	G2Iv + K0III

See page H74 for Type codes.

PULSATING VARIABLES

Name		H.D.	Right Ascension	Declination	Type	Magnitude Max	Magnitude Min	Mag. Type	Epoch (2400000+)	Period	Spectrum
			h m s	° ′						d	
S	Scl	1115	0 15 00	−32 05.2	M	5.5	13.6	v	42343	365.32	M3e–M8e
T	Cet	1760	0 21 23	−20 06.0	SRc	5.0	6.9	v	40562	158.9	M5–M6SIIe
R	And	1967	0 23 41	+38 32.2	M	5.8	14.9	v	43135	409.33	S3,5e–S8,8e(M7e)
TV	Psc	2411	0 27 39	+17 51.1	SR	4.65	5.4	V		70	M3IIIv
KK	Per	13136	2 09 45	+56 31.4	Lc	6.6	7.7	V			M1–M3.5 Iab–Ib
o	Cet	14386	2 18 58	− 3 00.5	M	2.0	10.1	v	44839	331.96	M5e–M9e
U	Cet	15971	2 33 22	−13 10.9	M	6.8	13.4	v	42137	234.76	M2e–M6e
R	Tri	16210	2 36 35	+34 13.9	M	5.4	12.6	v	42014	266.48	M4IIIe
R	Hor	18242	2 53 36	−49 55.3	M	4.7	14.3	v	41490	403.97	M7IIIe
ρ	Per	19058	3 04 41	+38 48.8	SRb	3.30	4.0	V		50:	M4II
R	Dor	29712	4 36 41	−62 05.4	SRb	4.8	6.6	v		338:	M8IIIq:e
R	Cae	29844	4 40 15	−38 15.1	M	6.7	13.7	v	40645	390.95	M6e
R	Pic	30551	4 45 57	−49 15.5	SRa	6.7	10.0	v	38091	164.2	M1IIe–M4IIe
R	Lep	31996	4 59 16	−14 49.0	M	5.5	11.7	v	40800	432.13	C6IIe
RX	Lep	33664	5 11 02	−11 51.6	Lb	5.0	7.0	v			M6III
β	Dor	37350	5 33 33	−62 29.7	δ Cep	3.46	4.0	V	35206.44	9.842	F4Ia–G4Iab
α	Ori	39801	5 54 46	+ 7 24.3	SRc	0.40	1.3	V		2110	M1–M2 Ia–Iab
U	Ori	39816	5 55 23	+20 10.5	M	4.8	12.6	v	42280	372.40	M6.5IIIe
η	Gem	42995	6 14 25	+22 30.5	SRb	3.2	3.9	v	37725	232.9	M3III
T	Mon	44990	6 24 49	+ 7 05.4	δ Cep	5.59	6.6	V	36137.090	27.020	F7Iab–K1Iab
RT	Aur	45412	6 28 05	+30 29.9	δ Cep	5.00	5.8	V	42361.155	3.728	F4Ib–G1Ib
IS	Gem	49380	6 49 12	+32 37.0	SRd	6.6	7.3	p		47:	K3II
ζ	Gem	52973	7 03 40	+20 34.9	δ Cep	3.66	4.1	V	36791.922	10.150	F7Ib–G3Ib
L_2	Pup	56096	7 13 19	−44 37.8	SRb	2.6	6.2	v	40813	140.42	M5IIIe
AK	Hya	73844	8 39 33	−17 16.5	SRb	6.33	6.9	V		112:	M4III
R	Car	82901	9 32 03	−62 45.4	M	3.9	10.5	v	42000	308.71	M4e–M8e
R	Leo	84748	9 47 09	+11 27.8	M	4.4	11.3	v	41688	312.43	M8IIIe
S	Car	88366	10 09 08	−61 30.8	M	4.5	9.9	v	42112	149.49	K5e–M6e
VY	UMa	92839	10 44 33	+67 27.0	Lb	5.89	6.5	V			C5II
VW	UMa	94902	10 58 31	+70 01.8	SR	6.85	7.7	V		125	M2
U	Car	95109	10 57 30	−59 41.5	δ Cep	5.72	7.0	V	37320.055	38.768	F6–G7Iab
S	Mus	106111	12 12 23	−70 06.6	δ Cep	5.90	6.4	V	35837.992	9.660	F6Ib
RY	UMa	107397	12 20 06	+61 21.1	SRb	6.68	8.5	V	40810	311	M2–M3IIIe
SS	Vir	108105	12 24 51	+ 0 48.6	M	6.0	9.6	v	40653	354.66	Ne(C5,3e)
BO	Mus	109372	12 34 27	−67 42.9	Lb	6.0	6.7	v			Mb
R	Vir	109914	12 38 07	+ 7 01.8	M	6.0	12.1	v	42512	145.64	M4.5IIIe
R	Mus	110311	12 41 37	−69 22.0	δ Cep	5.93	6.7	V	40896.13	7.476	F7Ib
SW	Vir	114961	13 13 41	− 2 46.0	SRb	6.85	7.8	V	40709	150:	M7III
FH	Vir	115322	13 16 02	+ 6 32.7	SRb	6.92	7.4	V	40740	70:	M6III
V	CVn	115898	13 19 08	+45 34.0	SRa	6.52	8.5	V	43929	191.89	M4e–M6eIIIa:
R	Hya	117287	13 29 19	−23 14.6	M	4.5	9.5	v	41676	389.61	M7IIIe
T	Cen	119090	13 41 20	−33 33.6	SRa	5.5	9.0	v	43242	90.44	K0:e–M4II:e
V412	Cen	121518	13 56 58	−57 40.5	Lb	7.1	9.6	B			M3IAB–IB – M7
θ	Aps	122250	14 04 36	−76 45.6	SRb	6.4	8.6	p		119	M7III

See page H74 for Type codes.

PULSATING VARIABLES

Name		H.D.	Right Ascension	Declination	Type	Magnitude Max	Magnitude Min	Mag. Type	Epoch (2400000+)	Period	Spectrum
			h m s	° ′						d	
R	Cen	124601	14 16 02	−59 52.8	M	5.3	11.8	v	41942	546.2	M4e–M8IIe
τ^4	Ser	139216	15 36 07	+15 07.5	Lb	7.5	8.9	p			M5IIb–IIIa
R	Ser	141850	15 50 21	+15 09.4	M	5.16	14.4	v	42315	356.41	M7IIIe
AT	Dra	147232	16 17 08	+59 46.4	Lb	6.8	7.5	p			M4IIIa
α	Sco	148478	16 28 56	−26 24.9	SRc	0.88	1.8	V	08600	1733	M1.5Iab–Ib + B4Ve
g	Her	148783	16 28 24	+41 53.9	SRb	5.7	7.2	p		70:	M6III
RS	Sco	152476	16 55 04	−45 05.5	M	6.2	13.0	v	42134	320.06	M5e–M8e
α^1	Her	156014	17 14 18	+14 23.9	SRc	3	4	v			M5Ib–II
VW	Dra	156947	17 16 24	+60 40.7	SRd	6.0	6.5	v		170:	K1.5IIIb
BM	Sco	160371	17 40 29	−32 12.6	SRd	6.8	8.7	p		850:	K2.5Ib
X	Sgr	161592	17 47 05	−27 49.7	δ Cep	4.24	4.8	V	36968.852	7.012	F7II
OP	Her	163990	17 56 35	+45 21.1	Lb	7.7	8.3	p			M5IIb–IIIa
W	Sgr	164975	18 04 32	−29 34.9	δ Cep	4.30	5.0	V	37678.578	7.594	F4–G1Ib
VX	Sgr	165674	18 07 37	−22 13.4	SRc	6.5	12.5	v	36493	732	M4Iae–M9.5
Y	Sgr	168608	18 20 56	−18 51.8	δ Cep	5.40	6.1	V	36230.180	5.773	F8I
T	Lyr	———	18 32 04	+36 59.6	Lb	7.8	9.6	v			R6(C5,3)
X	Oph	172171	18 37 59	+ 8 49.6	M	5.9	9.2	v	41478	334.39	M6IIIe + K1III
XY	Lyr	172380	18 37 51	+39 39.7	Lc	7.3	7.8	p			M4–M5 Ib–II
κ	Pav	174694	18 56 10	−67 14.6	CWa	3.94	4.7	V	40858.53	9.088	F5I–II
R	Lyr	175865	18 55 07	+43 56.1	SRb	3.88	5.0	V	35920	46.0	M5III
FF	Aql	176155	18 57 55	+17 21.1	δ Cep	5.18	5.6	V	41576.428	4.470	F5Ia–F8Ia
MT	Tel	176387	19 01 40	−46 39.6	RRc	8.68	9.2	V	38479.332	0.316	A0
R	Aql	177940	19 06 00	+ 8 13.2	M	5.5	12.0	v	43458	284.2	M5e–M9e
RR	Lyr	182989	19 25 14	+42 46.3	RRab	7.06	8.1	V	42995.405	0.566	A8–F7
UX	Dra	183556	19 21 52	+76 32.7	SRa	5.94	7.1	V		168:	C7,3
χ	Cyg	187796	19 50 16	+32 53.7	M	3.3	14.2	v	42143	406.93	S6,2e–S10,4e
η	Aql	187929	19 52 06	+ 0 59.1	δ Cep	3.48	4.3	V	36084.656	7.176	F6Ib–G4Ib
V449	Cyg	188344	19 53 04	+33 55.7	Lb	7.4	9.0	p			M1–M4
RR	Sgr	188378	19 55 29	−29 12.6	M	5.6	14.0	v	41133	334.58	M5e–M7e
S	Sge	188727	19 55 41	+16 36.9	δ Cep	5.28	6.0	V	36082.168	8.382	F6Ib–G5Ibv
EU	Del	196610	20 37 34	+18 14.5	SRb	5.8	6.9	v	35794	59.5	M6IIIFe
X	Cyg	197572	20 43 06	+35 33.6	δ Cep	5.87	6.8	V	35915.918	16.386	F7Ib–G8Ibv
T	Vul	198726	20 51 09	+28 13.3	δ Cep	5.44	6.0	V	35934.758	4.435	F5Ib–G0Ib
T	Cep	202012	21 09 27	+68 27.7	M	5.2	11.3	v	44177	388.14	M5.5e–M8.8e
W	Cyg	205730	21 35 45	+45 20.4	SRb	6.8	8.9	p	38659.73	126.26	M5IIIae
V460	Cyg	206570	21 41 42	+35 28.5	Lb	5.6	7.0	v			N1(C6,5)
μ	Cep	206936	21 43 17	+58 44.8	SRc	3.43	5.1	V		730	M2Iae
π^1	Gru	212087	22 22 16	−45 59.1	SRb	5.41	6.7	V		150:	S5
δ	Cep	213306	22 28 53	+58 22.6	δ Cep	3.48	4.3	V	36075.445	5.366	F5Ib–G1Ib
ER	Aqr	218074	23 05 01	−22 31.7	Lb	7.14	7.8	V			M3
R	Aqr	222800	23 43 26	−15 19.5	M	5.8	12.4	v	42398	386.96	M5e–M8.5e + P
TX	Psc	223075	23 46 01	+ 3 26.7	Lb	6.9	7.7	p			N0(C6,2)

See page H74 for Type codes.

ERUPTIVE VARIABLES

Name	H.D.	Right Ascension	Declination	Type	Magnitude Max	Magnitude Min	Mag. Type	Epoch (2400000+)	Period	Spectrum
		h m s	° ′						d	
WW Cet	———	0 11 02	−11 31.3	Z Cam	9.3	16.8	p		31.2:	P(UG)
RX And	———	1 04 10	+41 15.6	Z Cam	10.3	14.0	v		14.3:	
KT Per	———	1 36 36	+54 48.4	Z Cam	10.7	15	p		12:	
WX Hyi	———	2 09 38	−63 21.1	UG	9.6	14.7	v			
VW Hyi	———	4 09 11	−71 18.8	UG	8.4	14.4	v		27.8:	
SS Aur	———	6 12 48	+47 44.6	UG	10.5	15.0	v		55.8	
IR Gem	———	6 47 11	+28 05.3	UG	10.7	14.5			75:	
U Gem	64511	7 54 39	+22 01.4	UG	8.2	14.9	v		103:	M4.5 + WD
Z Cam	———	8 24 24	+73 08.2	Z Cam	10.2	14.5	v		22:	
SW UMa	———	8 36 09	+53 30.2	UG	10.8	16	v		459:	
BZ UMa	———	8 53 20	+59 29.2	UG	10.5	16	p		110:	
CU Vel	———	8 58 12	−41 46.3	UG	10.7	15.5	p		150:	
SY Cnc	———	9 00 37	+17 56.0	Z Cam	10.6	13.7	p		27.3	
T Pyx	———	9 04 22	−32 20.8	Nr	6.3	14.0	v	39501	7000:	P
CH UMa	———	10 06 50	+68 33.0	UG	10.7	15.9	p		204:	
T Leo	———	11 38 04	+ 3 24.8	UG	10	15.4	p			
BC UMa	———	11 51 57	+56 19.1	UG	10.9	17.5	p			
BV Cen	———	13 30 52	−54 56.2	UG	10.7	13.6	v		149.4:	
Z Aps	———	14 06 16	−71 20.1	Z Cam	10.7	12.7	v		19:	
T CrB	143454	15 59 11	+25 56.4	Nr	2.0	10.8	v	31860	9000:	M3III + P(Q)
U Sco	———	16 22 05	−17 51.6	Nr	8.8	19	p	44048	3400:	
AH Her	———	16 43 51	+25 15.9	Z Cam	10.2	14.7	v		19.8:	
RS Oph	162214	17 49 49	− 6 42.4	Nr	5.3	12.3	p	39791		0cp + M2ep
MV Lyr	———	19 07 03	+44 00.4	NL	10.5	14.0	p			
WZ Sge	———	20 07 15	+17 40.9	Nr(E)	7.0	15.5	p	32001	1900:	P(Q)
P Cyg	193237	20 17 31	+38 00.6	S Dor	3.0	6.0	v			B2pe
V Sge	———	20 19 55	+21 04.8	NL	9.5	13.9	v			
AE Aqr	———	20 39 45	− 0 53.9	UG?	10.4	12.0	B			
VY Aqr	———	21 11 46	− 8 51.6	UG	8.0	16.6	p	45667		
SS Cyg	206697	21 42 25	+43 33.1	UG	8.2	12.4	v		50.1:	A1−dGep
RU Peg	———	22 13 41	+12 39.9	UG	9.0	13.1	v		67.8	sdBe + G8IVn

See Page H74 for Type codes.

OTHER VARIABLES

Name		H.D.	Right Ascension	Declination	Type	Magnitude Max	Magnitude Min	Mag. Type	Epoch (2400000+)	Period	Spectrum
			h m s	° ′						d	
EG	And	4174	0 44 12	+40 38.3	Z And	7.08	7.8	V			M2IIIep
SU	Tau	247925	5 48 39	+19 03.9	RCB	9.1	16.0	v			G0ep(C1,0)
U	Mon	59693	7 30 26	− 9 45.7	RVb	6.1	8.1	p	37395	92.26	F8e−K0Ib:p
AR	Pup	———	8 02 44	−36 34.5	RVb	8.7	10.9	p		75	cF0−cF8
AI	Vel	69213	8 13 50	−44 33.2	δ Sct	6.4	7.1	v		0.111	A2p−F2p
VZ	Cnc	73857	8 40 28	+ 9 51.2	δ Sct	7.18	7.9	V	41304.364	0.178	A7III−F2III
WY	Vel	81137	9 21 44	−52 31.9	Z And	8.8	10.2	p			M3Ib:ep + B
IW	Car	82085	9 26 42	−63 35.9	RVb	7.9	9.6	p	29401	67.5	F7−F8
RU	Cen	105578	12 09 01	−45 23.0	RV	8.7	10.7	p	28015.51	64.727	A7Ib−G2pe
UW	Cen	———	12 42 51	−54 29.2	RCB	9.1	14.5	v			K
TX	CVn	———	12 44 20	+36 48.3	Z And	9.2	11.8	p			B1−B9Veq + K0III−M4
S	Aps	———	15 08 37	−72 01.8	RCB	9.6	15.2	v			R3
R	CrB	141527	15 48 16	+28 10.8	RCB	5.71	14.8	V			C0,0(F8pep)
AG	Dra	———	16 01 39	+66 49.4	Z And	8.8	11.8	p			Gep
RY	Ara	———	17 20 30	−51 06.8	RV	9.2	12.1	p	30220	143.5	G5−K0
V703	Sco	160589	17 41 48	−32 31.3	δ Sct	7.82	8.5	B	37186.365	0.115	F0−F5
RS	Tel	———	18 18 18	−46 33.0	RCB	9.3	13.0	p			R8
AC	Her	170756	18 29 57	+21 51.7	RVa	7.43	9.7	B	35052	75.461	F2Ibp−K4e
V	CrA	173539	18 47 02	−38 09.9	RCB	8.3	16.5	v			C(r0)
R	Sct	173819	18 47 05	− 5 42.8	RVa	4.45	8.2	V	32078.3	140.05	G0Iae−K0Ibpv
FN	Sgr	———	18 53 27	−19 00.3	Z And	9.0	13.9	p			P
RY	Sgr	180093	19 16 03	−33 32.1	RCB	6.0	15.0	v			G0Ipe(C1,0)
BF	Cyg	———	19 23 36	+29 39.6	Z And	9.3	13.4	p			Bep + M5III
CH	Cyg	182917	19 24 21	+50 13.6	Z And	6.4	8.7	V		97	M7IIIab + B
CI	Cyg	———	19 49 56	+35 40.0	Z And	9.0	11.6	v			
RR	Tel	———	20 03 40	−55 44.6	Z And	6.5	16.5	p			F5ep
RS	Gru	206379	21 42 35	−48 13.4	δ Sct	7.93	8.4	V	41599.999	0.147	A6−F0
AG	Peg	207757	21 50 41	+12 35.4	Z And	6.0	9.4	v		830.14	WN6 + M1−M3II−III
Z	And	221650	23 33 18	+48 46.7	Z And	8.0	12.4	p			M2III + B1eq
SX	Phe	223065	23 46 08	−41 36.7	δ Sct	6.78	7.5	V	38636.617	0.054	A2V

TYPE OF VARIATION

E	eclipsing	EA	eclipsing, Algol type
EB	eclipsing, β Lyr type	EW	eclipsing, W UMa type
δ Cep	cepheid, classical type	CWa	cepheid, W Vir type
Lb	slow irregular type	Lc	irregular supergiants of late spectral type
RV	RV Tauri type	RVa	RV Tauri type with constant mean brightness
M	Mira, long period variable	RVb	RV Tauri type with varying mean brightness
Nr	recurrent novae	δ Sct	δ Sct type, pulsating stars of spectral class A and F
Nr(E)	recurrent novae and eclipsing variable	NL	nova like stars
UG	U Gem, SS Cyg type, outbursts	Z Cam	Z Cam type, UG type variations with standstills
SR	semi-regular	Z And	Z And type (symbiotic stars)

SRa	semi-regular, late spectral class, strong periodicities
SRb	semi-regular, late spectral class, weak periodicities
SRc	semi-regular, late spectral class, disk component stars
SRd	semi-regular, spectrum F, G, or K
RRab	RR Lyr with sharp asymmetric light curves
RRc	RR Lyr with symmetric sinusoidal light curves
RCB	R CrB type, high luminosity stars with non-periodic drops in brightness
S Dor	high luminosity stars of spectral classes Bpeq-Fpeq, irregular variations

TYPE OF MAGNITUDE

p	photographic magnitudes	V	photoelectric visual magnitudes
v	visual magnitudes	B	photoelectric blue magnitudes

Name	Right Ascension	Declination	Flux 6 cm	Flux 11 cm	V	$B-V$	z	M(abs)	Code
	h m s	° ′ ″	Jy	Jy					
UM 18	0 04 57.2	+ 5 21 40	0.257	0.17	16.00		1.890	−30.3	R
S5 0014+81	0 16 40.5	+81 32 39	0.551	0.61	16.50		3.410	−30.8	R
PG 0026+12	0 28 50.3	+13 13 36	0.002		14.78	0.26	0.142	−24.8	O
UM 281	0 50 39.4	− 1 05 10			16.00		1.870	−30.2	
PG 0052+251	0 54 27.9	+25 23 13			15.42		0.155		
DHM0054−284	0 56 03.5	−27 46 12			19.55		3.610	−28.1	
UM 294	0 58 01.6	+ 0 38 49			16.00		1.920	−30.3	
PHL 957	1 02 47.6	+13 13 53			16.57	0.40	2.690	−30.5	O
PKS 0106+01	1 08 15.6	+ 1 32 36	3.730	2.04	18.39	0.15	2.107	−28.2	O
UM 100	1 22 32.9	+ 2 55 12			18.00		3.272	−29.3	
Q 0122−380	1 23 57.1	−37 46 45			16.50		2.190	−30.2	
Q 0130−403	1 32 42.2	−40 08 46			17.02	0.66	3.015	−30.5	
UM 673	1 44 54.9	− 9 47 27			17.00		2.720	−30.2	
UM 141	1 48 55.5	+ 1 55 10			17.00		2.909	−30.5	
UM 148	1 56 12.6	+ 4 43 25			17.00		2.990	−30.6	
UM 154	2 01 36.3	+ 3 48 33		0.01	16.00		2.440	−30.7	
NAB 0205+02	2 07 26.5	+ 2 40 48	0.002		15.41	0.26	0.155		O
UM 402	2 09 27.9	+ 0 31 08			16.00		2.840	−31.4	
Q 0207−398	2 09 09.6	−39 40 58			17.15	0.20	2.805	−30.2	
UM 678	2 51 20.5	−22 02 10			18.40		3.200	−28.8	
UM 679	2 51 27.7	−18 15 54			18.60		3.210	−28.6	
Q 0254−334	2 56 29.5	−33 17 10			16.00		1.849	−30.2	
Q 0347−383	3 49 26.7	−38 11 48			17.30		3.230	−29.9	
Q 0401−350B	4 02 54.4	−34 58 10			19.50		3.250	−27.7	
PKS 0405−12	4 07 27.3	−12 12 47	1.990	2.36	14.57	0.18	0.574	−28.4	O
Q 0420−388	4 21 59.1	−38 45 54	0.107	0.14	16.90		3.120	−30.4	O
PKS 0438−43	4 40 03.0	−43 33 58	7.580	6.17	18.80		2.852	−28.6	O
3C 138.0	5 20 43.8	+16 37 57	4.160	5.99	18.84	0.53	0.759	−24.9	O
PKS0537−441	5 38 36.9	−44 05 23	3.960	3.84	16.48	0.52	0.894	−27.5	O
3C 147.0	5 42 01.2	+49 50 56	8.180	12.98	17.80	0.65	0.545	−25.1	O
B2 0552+39A	5 54 59.4	+39 48 46	4.814	3.53	18.00		2.365	−28.6	O
PKS 0637−75	6 36 01.0	−75 15 54	6.190	4.51	15.75	0.33	0.651	−27.6	O
OH 471	6 45 59.2	+44 51 47	0.778	1.27	18.49	1.08	3.400	−28.8	O
MARK 380	7 18 54.0	+74 28 47			17.00		2.737	−30.2	O
1E0754+3928	7 57 29.7	+39 21 40			14.36	0.38	0.096	−24.5	
PG 0804+761	8 10 01.5	+76 04 02			15.15		0.100	−23.7	
3C 196.0	8 13 03.7	+48 14 25	4.360	7.66	17.79	0.57	0.871	−26.2	O
PG 0844+349	8 47 14.2	+34 46 44			14.00		0.064	−23.9	
0846+51W1	8 49 25.8	+51 10 10	0.258	0.25	15.72	0.56	1.860	−30.5	O
PG 0906+48	9 09 39.5	+48 15 32			16.06	0.40	0.118		O
B2 0923+39	9 26 34.9	+39 04 19	7.570	4.54	17.86	0.06	0.698	−25.7	O
PG 0953+415	9 56 25.0	+41 17 50			14.50		0.239	−26.3	
PKS 1004+13	10 07 02.0	+12 51 09	0.420	0.64	15.15	0.13	0.240		O
TON 34	10 19 29.5	+27 48 11			15.69	0.37	1.924	−30.6	
EX 1059+730	11 02 06.4	+72 49 10			14.70	1.7	0.089	−24.0	
Q 1101−264	11 03 03.4	−26 42 50			16.02	0.06	2.145	−30.6	O
PG 1114+445	11 16 41.8	+44 16 05			16.05		0.144		
PG 1115+080	11 17 53.6	+ 7 48 28			15.80		1.722	−30.2	
PG 1116+215	11 18 44.8	+21 21 46			15.17		0.177		

Code: O=Optical position, R=Radio position with an accuracy better than one arc second.

Name	Right Ascension	Declination	Flux 6 cm	Flux 11 cm	*V*	*B−V*	*z*	*M*(abs)	Code
	h m s	° ′ ″	Jy	Jy					
PKS 1127−14	11 29 44.3	−14 46 58	7.310	6.43	16.90	0.27	1.187	−28.0	O
PG 1138+040	11 40 53.4	+ 3 49 29			16.05		1.876	−30.2	
Q 1159+124	12 01 10.6	+12 09 48					3.510		
PG 1211+143	12 13 54.6	+14 05 42			14.63		0.085	−23.9	
B2 1225+31	12 28 02.4	+31 31 07	0.330	0.33	15.87	0.28	2.200	−30.7	O
PG 1241+176	12 43 48.4	+17 23 32			15.38		1.273		
PG 1247+268	12 49 43.7	+26 33 07			15.80		2.038	−30.7	
3C 279	12 55 47.8	− 5 44 56	5.340	11.96	17.75	0.26	0.536	−25.1	O
PKS1302−102	13 05 09.4	−10 30 55	1.280	1.23	15.23	−0.05	0.286		O
PG 1307+085	13 09 24.3	+ 8 22 13			15.28		0.155		
3C 286.0	13 30 47.5	+30 32 51	7.480	10.26	17.25	0.26	0.846	−26.6	O
Q 1346+001	13 48 54.6	+ 0 03 38					3.268		O
PG 1351+64	13 53 02.4	+63 47 57	0.032		14.84	0.34	0.088	−23.8	O
PKS1402+044	14 04 38.4	+ 4 17 44	0.710	0.58	18.50		3.202	−28.7	O
PG 1404+226	14 06 01.3	+22 25 51			15.82		0.098	−23.1	
PG 1411+442	14 13 30.6	+44 02 19			14.99		0.089	−23.7	
PG 1415+451	14 16 43.1	+44 58 11			15.74		0.114		
PG 1416−129	14 18 39.4	−13 08 41			15.40		0.129		
S4 1435+63	14 36 35.3	+63 38 34	1.240	1.41	15.00		2.060	−31.5	O
PG 1435−067	14 37 52.5	− 6 56 22			15.54		0.129	−23.8	
OQ 172	14 44 54.6	+10 00 29	1.150	1.77	17.78	0.80	3.530	−29.7	O
3C 309.1	14 59 05.8	+71 42 06	3.760	5.30	16.78	0.46	0.905	−27.3	O
PG 1519+226	15 20 54.7	+22 29 14			16.09		0.137		
PG 1552+085	15 54 22.7	+ 8 23 39			16.02		0.119		
1601+182	16 02 58.6	+18 10 17					3.280		
PKS 1610−77	16 16 46.3	−77 16 12	5.550	3.80	19.00		1.710	−26.9	O
TON 256	16 13 54.5	+26 05 23		0.02	15.41	0.65	0.131		O
PKS1614+051	16 16 15.2	+ 5 00 37	0.850	0.67	19.50		3.208	−27.7	R
1623.7+268B	16 25 29.8	+26 48 10			16.00		2.518	−30.8	
PG 1626+554	16 27 46.4	+55 23 29			16.17		0.133		
B2 1633+38	16 34 59.7	+38 08 57	4.080	2.57	18.00		1.814	−28.1	O
PG 1634+706	16 34 32.3	+70 32 27			14.90		1.334	−30.4	
MC 1635+119	16 37 25.3	+11 50 42	0.080	0.11	16.57	0.49	0.146		O
3C 345.0	16 42 43.5	+39 49 27	5.650	6.01	15.96	0.29	0.594	−27.1	O
PG 1700+518	17 01 14.2	+51 49 59			15.43		0.292		
3C 351.0	17 04 35.7	+60 45 04	1.210	2.03	15.28	0.13	0.371		O
PG 1718+481	17 19 26.1	+48 04 39			15.33		1.084		
NRAO 530	17 32 37.2	−13 04 32	4.220	4.90	18.50		0.902	−25.6	O
PKS2000−330	20 02 55.3	−32 53 04	1.030	0.62	19.00		3.780	−28.8	O
20 05+40	20 07 29.1	+40 28 28	4.450	4.60	19.00		1.736	−27.0	O
3C 418.0	20 38 23.4	+51 17 37	3.790	4.71	21.00		1.687	−24.9	O
PKS 2126−15	21 28 47.3	−15 40 40	1.240	1.17	17.30		3.275	−30.0	O
PKS 2145+06	21 47 43.1	+ 6 55 33	4.410	3.46	16.47	0.38	0.990	−27.8	O
PKS 2203−18	22 05 45.7	−18 37 52	4.240	5.25	18.50		0.618	−24.7	
PKS2204−573	22 07 23.2	−57 09 47	0.360	0.43	16.60		2.725	−30.6	
3C 446	22 25 23.7	− 4 59 18	4.070	4.28	18.39	0.44	1.404	−27.1	O
	22 30 06.7	−39 15 28			18.80		3.450	−28.5	
CTA 102	22 32 14.0	+11 41 32	3.650	4.93	17.33	0.42	1.037	−27.2	O
Q 2313−423	23 15 42.6	−42 07 24			19.50		3.360	−27.7	

Code: O=Optical position, R=Radio position with an accuracy better than one arc second.

PSR	Right Ascension	Declination	Period	$\dot{P}$	Epoch	DM	S_{400}
	h m s	° ′ ″	s	10^{-15} ss^{-1}	24	$cm^{-3}pc$	Jy
0031–07	0 34 08.9	− 7 22 01	0.94295078486	0.40	40690	10.8	25
0136+57	1 39 19.9	+58 14 32.0	0.27244563408	10.68	43890	72.2	50
0138+59	1 41 40.0	+60 09 29.8	1.22294826723	0.39	41794	34.8	55
0329+54	3 32 59.2	+54 34 42.9	0.71451866398	2.04	40621	26.7	1400
0355+54	3 58 53.6	+54 13 14.5	0.15638005591	4.38	41593	57.0	60
0450+55	4 54 07.6	+55 43 39.5	0.34072820672	2.36	43890	15.6	40
0450–18	4 52 34.4	−17 59 25.5	0.5489353684	5.74	41536	39.9	55
0525+21	5 28 52.1	+22 00 22.8	3.74549702902	40.05	41994	50.9	93
0531+21	5 34 31.9	+22 00 52.0	0.03313404075	422.43	41351	56.7	800
0538–75	5 36 30.8	−75 43 59.3	1.2458554349	0.57	43555	18.3	75
0611+22	6 14 16.2	+22 25 10.4	0.33492505401	59.63	42881	96.7	25
0628–28	6 30 49.5	−28 34 43.8	1.2444170726	7.10	44124	34.3	90
0655+64*	7 00 37.6	+64 18 11.0	0.19567094486	0.00	43987	8.7	40
0736–40	7 38 32.4	−40 42 39.4	0.37491871098	1.61	42554	160.8	190
0740–28	7 42 48.9	−28 22 52	0.16675244661	16.83	42554	73.7	195
0808–47	8 09 43.9	−47 53 55.4	0.54719837196	3.08	43556	228.3	46
0809+74	8 14 59.4	+74 29 06.7	1.29224132384	0.16	40689	5.7	50
0818–13	8 20 26.3	−13 50 54.9	1.23812810723	2.10	41006	40.9	90
0818–41	8 20 15.5	−41 14 36.6	0.5454455279	0.02	43557	111	65
0820+02*	8 23 09.7	+ 1 59 12.9	0.86487275	<0.5	43419	22.2	22
0823+26	8 26 51.2	+26 37 25.2	0.53065995906	1.72	42717	19.4	70
0833–45	8 35 20.6	−45 10 35.8	0.08924726825	124.68	43892	69.0	5000
0834+06	8 37 05.5	+ 6 10 13.7	1.27376417152	6.79	41708	12.8	65
0835–41	8 37 21.1	−41 35 14.3	0.75162112843	3.54	43557	147.6	197
0919+06	9 22 13.8	+ 6 38 21.6	0.43061431165	13.72	43890	27.2	40
0940–55	9 42 15.6	−55 52 55.4	0.66436112446	22.73	43555	180.2	55
0950+08	9 53 09.3	+ 7 55 35.7	0.25306506819	0.22	41501	2.9	900
0959–54	10 01 37.9	−55 07 08.8	1.4365681929	51.66	43558	130.6	80
1054–62	10 56 25.5	−62 58 47.6	0.42244618669	3.57	43556	323.4	45
1055–52	10 57 58.7	−52 26 56.5	0.19710760818	5.83	43556	30.1	80
1112+50	11 15 38.3	+50 30 13.9	1.65643808033	2.49	41536	9.1	20
1133+16	11 36 03.3	+15 51 00.8	1.18791153608	3.73	41665	4.8	340
1154–62	11 57 15.2	−62 24 50.6	0.40052094651	3.93	43556	325.2	145
1221–63	12 24 22.2	−64 07 53.7	0.21647481429	4.95	43556	96.9	48
1237+25	12 39 40.5	+24 53 49.1	1.38244861210	0.95	40611	9.2	160
1240–64	12 43 17.2	−64 23 23.4	0.38847933086	4.50	42710	297.4	110
1323–58	13 26 58.3	−58 59 30.1	0.4779896901	3.21	43556	283	120
1323–62	13 27 17.2	−62 22 43	0.5299062943	18.89	43556	318.4	135
1356–60	13 59 58.2	−60 38 08.2	0.12750077685	6.33	43556	295.0	105
1426–66	14 30 40.9	−66 23 04.5	0.78543998083	2.77	43556	65.3	130
1449–64	14 53 32.7	−64 13 15.0	0.17948389392	2.74	43177	71.0	230
1451–68	14 56 00.2	−68 43 38.9	0.26337677865	0.09	42554	8.6	350
1508+55	15 09 25.9	+55 31 35.2	0.73967789896	5.03	40625	19.5	125
1509–58	15 13 56	−59 08 08	0.15021718	520.	45042	235	2
1530–53	15 34 08.3	−53 34 19.1	1.3688805090	1.42	43559	24.8	70
1540–06	15 43 30.1	− 6 20 44.0	0.70906364986	0.88	43890	18.6	50
1541+09	15 43 38.8	+ 9 29 16.9	0.74844817748	0.43	42304	34.9	100

*Member of a binary system

PSR	Right Ascension	Declination	Period	$\dot{P}$	Epoch	DM	S_{400}
	h m s	° ′ ″	s	10^{-15} ss^{-1}	24	$cm^{-3}pc$	Jy
1556–44	15 59 41.5	−44 38 45.9	0.25705572352	1.01	42554	58.8	110
1558–50	16 02 18.9	−51 00 04.4	0.8642020784	69.57	43559	169.5	45
1600–49	16 04 23.0	−49 09 57.3	0.32741728797	1.01	43557	140.8	44
1604–00	16 07 12.1	+ 0 32 40.1	0.42181611020	0.30	42307	10.7	45
1641–45	16 44 49.3	−45 59 09.2	0.45505464292	20.13	43634	475	375
1642–03	16 45 02.0	− 3 17 58.5	0.38768879135	1.78	40622	35.6	300
1706–16	17 09 26.4	−16 40 57	0.65305047326	6.38	40622	24.8	60
1727–47	17 31 41.9	−47 44 33.1	0.82972364148	163.67	43494	121.9	190
1737+13	17 40 07.4	+ 3 11 57.6	0.80304971623	1.45	43893	48.4	70
1738–08	17 41 22.5	− 8 40 33	2.0430815106	2.27	43891	74	40
1747–46	17 51 42.2	−46 57 24.0	0.74235203333	1.29	43557	21.7	70
1749 28	17 52 58.6	−28 06 37.8	0.56255316830	8.15	40128	50.8	1300
1804–08	18 07 38.0	− 8 47 42.8	0.16372736083	0.02	43891	112.8	55
1818–04	18 20 52.6	− 4 27 40	0.59807263930	6.33	40622	84.3	170
1821–19	18 24 00.4	−19 45 51	0.18933213477	5.23	43557	226	52
1831–04	18 34 25.3	− 4 26 35	0.29010815629	0.19	44054	79	75
1844–04	18 47 23	− 4 02 12	0.5977390	51.9	42004	141.9	100
1845–01	18 48 24	− 1 24 05	0.65942849	5.2	42005	163	60
1857–26	19 00 48	−26 00 38	0.61220908	0.16	42005	38.1	120
1859+03	19 01 31.8	+ 3 31 06.2	0.65544511516	7.48	42100	402.9	125
1900+01	19 03 29.9	+ 1 35 38.2	0.72930163274	4.03	42346	243.4	60
1907+10	19 09 48.7	+11 02 03.3	0.28363867083	2.63	42540	144	55
1911–04	19 13 54.1	− 4 40 47.6	0.82593368968	4.06	40624	89.4	120
1913+16*	19 15 28.0	+16 06 27.4	0.05902999526	0.00	42321	167	6
1914+13	19 16 58.7	+13 12 50.7	0.28184027865	3.61	42847	230	45
1919+21	19 21 44.8	+21 53 01.8	1.33730119226	1.34	40689	12.4	240
1920+21	19 22 53.5	+21 10 42.2	1.07791915514	8.18	42547	220	45
1929+10	19 32 13.8	+10 59 31.4	0.22651715301	1.15	41704	3.1	130
1929+20	19 32 08.1	+20 20 45.3	0.26821490434	4.17	43029	210.0	50
1933+16	19 35 47.8	+16 16 40.6	0.35873624827	6.00	42265	158.5	260
1937+21	19 39 38.5	+21 34 59.1	0.00155780649	0.00	45303	71.2	200
1944+17	19 46 53.0	+18 05 41.5	0.44061846173	0.02	41501	16.3	60
1946+35	19 48 25.0	+35 40 11.1	0.71730676525	7.05	42221	129.1	120
1952+29	19 54 22.6	+29 23 17.9	0.42667678559	0.00	42434	7.9	20
1953+29*	19 55 27.9	+29 08 41.9	0.00613317	< 0.04		104.5	15
2016+28	20 18 03.8	+28 39 54.2	0.55795340728	0.14	40689	14.1	150
2020+28	20 22 37.0	+28 54 23.5	0.34340079150	1.89	41348	24.6	250
2021+51	20 22 49.9	+51 54 49	0.52919532782	3.05	40625	22.5	60
2045–16	20 48 35.3	−16 16 45	1.96156687985	10.96	40695	11.5	130
2111+46	21 13 24.3	+46 44 08.4	1.01468444504	0.71	41006	141.5	190
2217+47	22 19 48.1	+47 54 53.9	0.53846739454	2.76	40624	43.5	63
2255+58	22 57 57.7	+59 09 14.6	0.36824365392	5.75	42629	148	60
2303+30	23 05 58.2	+31 00 01.6	1.57588474427	2.89	42341	49.9	25
2310+42	23 13 08.5	+42 53 12.5	0.34943363975	0.11	43891	17.3	40
2319+60	23 21 55.3	+60 24 29.5	2.2564837049	7.03	41535	96	70

*Member of a binary system

CONTENTS OF SECTION J

Pages J6–J16 contain as complete a list as possible of observatories that are currently engaged in professional programs of astronomical observations. Entries are in alphabetical order according to geographical location. If only the formal name of an observatory is known, its location can be found in the Index List (pp. J2–J5). In the General List observatories with radio or infrared telescopes are designated with an 'R' or 'I', respectively, in the Description column. The column labelled Ref. specifies the year in which an observatory appeared in the Instrumentation Lists of the 1981–1984 editions. East longitudes and north latitudes are considered to be positive.

INDEX LIST

INDEX LIST

INDEX LIST

Name	Place
Michigan University, Central	Mount Pleasant
Michigan, Univ. of	Dexter
Millimeter Radio Ast. Institute	Pico Veleta
Millimeter Radio Ast. Institute	Plateau de Bure
Millimeter Wave	Mount Locke
Mills	Dundee
Millstone Hill Atm. Sci. Fac.	Westford
Millstone Hill Radar	Westford
MIRA Oliver Observing Station	Chews Ridge
Mitchell, Maria	Nantucket
MMT (Multiple-Mirror Tel.)	Mount Hopkins
Molongo	Hoskinstown
Monterey Inst. for Radio Ast. (MIRA)	Chews Ridge
Moore	Brownsboro
Moraes, Abrahão de	Valinhos
Morehead	Chapel Hill
Morrison	Fayette
Morro Santana	Porto Alegre
Mountain	Alma-Ata
Mount Cuba	Greenville
Mullard	Cambridge
Multiple-Mirror Telescope (MMT)	Mount Hopkins
Nagoya University (Branch)	Toyokawa
Nagoya University (Branch)	Kamiku Isshiki
Nassau	Montville
National Central University	Chung-li
National Obs. of Athens (Branch)	Pentele
National Obs. of Brazil	Rio de Janeiro
National Obs. of Chile	Cerro Calan
National Obs. of Colombia	Bogotá
National Obs. of Cosmic Physics	San Miguel
National Obs. of Japan	Mitaka
National Obs. of Mexico (Branch)	San Pedro Martir
National Obs. of Mexico	Tonantzintla
National Obs. of Spain	Madrid
National Obs. of Spain (Branch)	Yebes
National Obs. of Uruguay	Montevideo
National Radio	Green Bank
National Radio (Branch)	Kitt Peak
National Radio	Socorro
National Solar	Sunspot
Naval	Buenos Aires
Naval	San Fernando
Naval Research Lab. (Branch)	Sugar Grove
Naval Research Laboratory (NRL)	Washington
Naval, U.S. (Branch)	Black Birch Mountain
Naval, U.S. (Branch)	Flagstaff
Naval, U.S. (Branch)	Miami
Naval, U.S.	Washington
New Mexico State Univ. (Branch)	Blue Mesa
New Mexico State Univ. (Branch)	Tortugas
Nice	Mont Gros
Nicholas Copernicus	Brno
Nizamiah	Hyderabad
Norikura Solar	Mount Norikura
Northern Arizona University	Flagstaff
North Gornergrat	Zermatt
NRL (Branch)	Sugar Grove
NRL (Naval Research Lab.)	Washington
Nuffield Radio Ast. Labs.	Jodrell Bank
Oak Ridge	Harvard
O'Brien	Marine-on-St. Croix
Ohio State	Delaware
Okayama Astrophysical	Mount Chikurin
Oklahoma, Univ. of	Norman
Ole Rømer	Århus
Oliver Observing Station, MIRA	Chews Ridge
Ooty (Ootacamund)	Udhagamandalam
Owens Valley	Big Pine
Pagasa	Quezon City
Paris (Branch)	Nançay
Perkins	Delaware
Perth	Bickley
Pico dos Dias	Itajuba
Piedade	Belo Horizonte
Prof. Manuel de Barros	Vila Nova de Gaia
Pulkovo (Branch)	Kislovodsk
Québec	Dolbeau
Radio Astronomy Center	Udhagamandalam
Radio Astronomy Institute	Kharkov
Radio Astronomy Institute	Stanford
Radio Astronomy Institute	Villa Elisa
Ramon Maria Aller	Santiago de Compostela
Remeis	Bamberg
Research Inst. of Magnetosphere	Kasugai
Ritter	Toledo
Riverview College	Lane Cove
Rome	Monte Mario
Rømer, Ole	Århus
Roque de los Muchachos	La Palma Island
Rosemary Hill	Bronson
Rothney Astrophysical	Priddis
Royal Greenwich (Branch)	Herstmonceux
Royal Obs. Edinburgh (Branch)	Siding Spring Mountain
Royal Obs. of Belgium (Branch)	Humain
Royal Obs. of Belgium	Uccle
Royal Obs. of Scotland	Edinburgh
Royal Radar Establishment	Malvern
Rutherfurd	New York
Sacramento Peak	Sunspot
Sagamore Hill	Hamilton
San Vittore	Bologna
Schwarzschild, Karl	Tautenburg
Shaanxi	Lintong
Shanghai (Branch)	Sheshan
Shanghai	Xujiahui
Shattuck	Hanover
Simon Stevin	Hoeven
Slovak Technical University	Bratislava
Smithsonian Astrophysical	Cambridge

INDEX LIST

Place	Description*		East Longitude	Latitude	Height (Sea Level)	Ref.
			° ′	° ′	m	
Algonquin Prov. Park, Ontario	Algonquin Radio Obs.	R	− 78 04.4	+45 57.3	260	81
Alma-Ata, Kazakh S.S.R.	Mountain Obs.		+ 76 57.4	+43 11.3	1450	84
Alston, England	Wilfred Hall Obs.		− 2 35.6	+53 48.1	63	82
Anacapri, Italy	Capri Obs.		+ 14 11.8	+40 33.5	137	82
Anderson Mesa, Arizona	Lowell Obs. Sta.		−111 32.2	+35 05.8	2200	
Ankara, Turkey	Ankara Univ. Obs.	R	+ 32 46.8	+39 50.6	1266	82
Arcetri, Italy	Arcetri Astrophysical Obs.		+ 11 15.3	+43 45.2	184	82
Arecibo, Puerto Rico	Arecibo Obs.	R	− 66 45.2	+18 20.6	496	81
Århus, Denmark	Ole Rømer Obs.		+ 10 11.8	+56 07.7	50	82
Armagh, Northern Ireland	Armagh Obs.		− 6 38.9	+54 21.2	64	82
Arosa, Switzerland	Arosa Astrophysical Obs.		+ 9 40.1	+46 47.0	2050	82
Ashkhabad, Turkmen S.S.R.	Ashkhabad Astrophysical Lab.		+ 58 21.2	+37 57.4	234	84
Asiago, Italy	Asiago Astrophysical Obs.		+ 11 31.7	+45 51.7	1045	81
Athens, Greece	National Obs. of Athens		+ 23 43.2	+37 58.4	110	81
Atibaia, Brazil	Itapetinga Radio Obs.	R	− 46 33.5	−23 11.1	806	83
Atlanta, Georgia	Fernbank Obs.		− 84 19.1	+33 46.7	320	81
Auckland, New Zealand	Auckland Obs.		+174 46.7	−36 54.4	80	82
Bailey, Colorado	Chamberlin Obs. Dick Mtn. Sta.		−105 26.2	+39 25.6	2675	83
Bamberg, German Fed. Rep.	Remeis Obs.		+ 10 53.4	+49 53.1	288	82
Barcelona, Spain	Fabra Obs.		+ 2 07.6	+41 25.0	420	82
Barcelona, Spain	Faculty of Physics Obs.		+ 2 07.1	+41 23.2	97	82
Beijing, China	Beijing Normal Univ. Obs.	R	+116 21.6	+39 57.4	70	
Belo Horizonte, Brazil	Piedade Obs.		− 43 30.7	−19 49.3	1746	82
Beloit, Wisconsin	Thompson Obs.		− 89 01.9	+42 30.3	255	82
Bergedorf, German Fed. Rep.	Hamburg Obs.		+ 10 14.5	+53 28.9	45	83
Berlin, German Dem. Rep.	Archenhold Obs.		+ 13 28.7	+52 29.2	41	81
Berlin, German Fed. Rep.	Wilhelm-Foerster Obs.		+ 13 21.2	+52 27.5	78	82
Besançon, France	Besançon Obs.		+ 5 59.2	+47 15.0	312	81
Bickley, Western Australia	Perth Obs.		+116 08.1	−32 00.5	391	83
Big Bear Lake, California	Big Bear Solar Obs.		−116 54.9	+34 15.2	2067	82
Big Pine, California	Owens Valley Radio Obs.	R	−118 16.9	+37 13.9	1236	81
Binningen, Switzerland	Univ. of Basle Ast. Inst.		+ 7 35.0	+47 32.5	318	82
Björnstorp, Sweden	Lund Obs. Jävan Sta.		+ 13 26.0	+55 37.4	145	82
Black Birch Mtn., New Zealand	Carter Obs. Sta.		+173 48.2	−41 44.9	1396	
Black Birch Mtn., New Zealand	U.S. Naval Obs. Sta.		+173 48.2	−41 44.7	1366	
Blue Mesa, New Mexico	N.M. State Univ. Obs. Sta.		−107 09.9	+32 29.5	2025	82
Bochum, German Fed. Rep.	Bochum Public Obs.		+ 7 13.4	+51 27.9	132	82
Bogota, Colombia	National Ast. Obs.		− 74 04.9	+ 4 35.9	2640	82
Bologna, Italy	San Vittore Obs.		+ 11 20.5	+44 28.1	280	83
Boone, Iowa	Erwin W. Fick Obs.		− 93 56.5	+42 00.3	332	82
Bornova, Turkey	Ege Univ. Obs.		+ 27 16.5	+38 23.9	795	82
Borowiec, Poland	Astronomical Latitude Obs.	L	+ 17 04.5	+52 16.6	80	84
Bosque Alegre, Argentina	Córdoba Obs. Astrophys. Sta.		− 64 32.8	−31 35.9	1250	81
Boulder, Colorado	Sommers-Bausch Obs.		−105 15.8	+40 00.2	1653	82
Bouzaréa, Algeria	Algier Obs.		+ 3 02.1	+36 48.1	345	82

* 'R' denotes an observatory with radio instruments; 'I' denotes an observatory with infrared instruments; 'L' denotes an observatory with laser instruments.

Place	Description*		East Longitude	Latitude	Height (Sea Level)	Ref.
			° ′	° ′	m	
Brannenburg, German Fed. Rep.	Wendelstein Solar Obs.		+ 12 00.8	+47 42.5	1838	82
Bratislava, Czechoslovakia	Slovak Technical Univ. Obs.		+ 17 07.2	+48 09.3	171	82
Bristol Springs, New York	C.E. Kenneth Mees Obs.		− 77 24.5	+42 42.0	701	82
Brno, Czechoslovakia	Nicholas Copernicus Obs.		+ 16 35.3	+49 12.3	310	81
Bro, Sweden	Kvistaberg Obs.		+ 17 36.4	+59 30.1	—	81
Bronson, Florida	Rosemary Hill Obs.		− 82 35.2	+29 24.0	44	82
Brooklyn, Indiana	Goethe Link Obs.		− 86 23.7	+39 33.0	300	81
Brorfelde, Denmark	Copenhagen Univ. Obs.		+ 11 40.0	+55 37.5	90	82
Brownsboro, Kentucky	Moore Obs.		− 85 31.8	+38 20.1	216	82
Brussels, Belgium	Ast. and Astrophys. Inst.		+ 4 23.0	+50 48.8	147	81
Bucharest, Romania	Bucharest Ast. Obs.		+ 26 05.8	+44 24.8	81	82
Budapest, Hungary	Konkoly Obs.		+ 18 57.9	+47 30.0	474	82
Budapest, Hungary	Urania Obs.		+ 19 03.9	+47 29.1	166	82
Buenos Aires, Argentina	Naval Obs.		− 58 21.3	−34 37.3	6	82
Calar Alto, Spain	German Spanish Ast. Center		− 2 32.2	+37 13.8	2168	81
Calern, France	Cote d'Azur Obs.	I,L	+ 6 55.6	+43 44.9	1270	84
Cambridge, England	Cambridge Univ. Observatories		+ 0 05.7	+52 12.8	30	
Cambridge, England	Mullard Radio Ast. Obs.	R	+ 0 02.6	+52 10.2	17	81
Cambridge, Massachusetts	Harvard College Obs.	R	− 71 07.8	+42 22.8	24	82
Cambridge, Massachusetts	Smithsonian Astrophysical Obs.		− 71 07.8	+42 22.8	24	82
Cape Town, South Africa	South African Ast. Obs.		+ 18 28.7	−33 56.1	18	81
Capilla de Caleu, Chile	Cerro El Roble Ast. Obs.		− 71 01.2	−32 58.9	2220	81
Capilla Peak, New Mexico	Capilla Peak Obs.		−106 24.3	+34 41.8	2842	82
Capoterra, Italy	Cagliari Ast. Obs.	L	+ 8 58.6	+39 08.2	205	82
Caracas, Venezuela	Cagigal Obs.		− 66 55.7	+10 30.4	1026	82
Carloforte, Italy	International Latitude Obs.		+ 8 18.7	+39 08.2	22	82
Cassel, California	Hat Creek Radio Ast. Obs.	R	−121 28.4	+40 49.1	1043	81
Castleknock, Ireland	Dunsink Obs.		− 6 20.2	+53 23.3	85	82
Catania, Italy	Catania Astrophysical Obs.		+ 15 05.2	+37 30.2	47	82
Cebreros, Spain	Deep Space Sta.	R	− 4 22.0	+40 27.3	789	84
Cerro Calán, Chile	Cerro Calán National Ast. Obs.		− 70 32.8	−33 23.8	860	82
Cerro Las Campanas, Chile	Las Campanas Obs.		− 70 42.0	−29 00.5	2282	83
Cerro La Silla, Chile	European Southern Obs.	R	− 70 43.8	−29 15.4	2347	81
Cerro Tololo, Chile	Cerro Tololo Inter-Amer. Obs.	R	− 70 48.9	−30 09.9	2215	82
Chapel Hill, North Carolina	Morehead Obs.		− 79 03.0	+35 54.8	161	
Charlottesville, Virginia	Leander McCormick Obs.		− 78 31.4	+38 02.0	264	82
Chavannes-des-Bois, Switz.	Univ. of Lausanne Obs.		+ 6 08.2	+46 18.4	465	83
Chews Ridge, California	MIRA Oliver Observing Sta.		−121 34.2	+36 18.3	1525	
Chilbolton, England	Chilbolton Obs.	R	− 1 26.2	+51 08.7	92	81
Chions, Italy	Chaonis Obs.		+ 12 42.7	+45 50.6	15	
Chung-li, Taiwan	National Central Univ. Obs.		+121 11.2	+24 58.2	152	82
Cincinnati, Ohio	Cincinnati Obs.	R	− 84 25.4	+39 08.3	247	82
Cluj-Napoca, Romania	Cluj-Napoca Ast. Obs.		+ 23 35.9	+46 42.8	750	82
Cocoa, Florida	Brevard Community College Obs.		− 80 45.7	+28 23.1	17	82
Coimbra, Portugal	Coimbra Ast. Obs.		− 8 25.8	+40 12.4	99	82

* 'R' denotes an observatory with radio instruments; 'I' denotes an observatory with infrared instruments; 'L' denotes an observatory with laser instruments.

Place	Description*	East Longitude	Latitude	Height (Sea Level)	Ref.
		° ′	° ′	m	
College Park, Maryland	Univ. of Maryland Obs. R	− 76 57.4	+39 00.1	53	82
Columbia, South Carolina	Melton Memorial Obs.	− 81 01.6	+33 59.8	98	82
Columbia, South Carolina	Univ. of S.C. Radio Obs. R	− 81 01.9	+33 59.8	127	
Copenhagen, Denmark	Copenhagen Univ. Obs.	+ 12 34.6	+55 41.2	—	82
Córdoba, Argentina	Córdoba Ast. Obs.	− 64 11.8	−31 25.3	434	82
Cracow, Poland	Jagellonian Obs. Ft. Skala Sta. R	+ 19 49.6	+50 03.3	314	81
Cracow, Poland	Jagellonian Univ. Ast. Obs.	+ 19 57.6	+50 03.9	225	82
Culgoora, New South Wales	Australian Tel. Natl. Facility R	+149 33.7	−30 18.9	217	81
Daegang, South Korea	Sobaeksan Ast. Obs.	+128 27.4	+36 56.0	1390	82
Daejeon, South Korea	Daeduk Radio Ast. Obs. R	+127 22.3	+36 23.9	120	
Danbury, Connecticut	Western Conn. State Univ. Obs.	− 73 26.7	+41 24.0	128	82
Debrecen, Hungary	Heliophysical Obs.	+ 21 37.4	+47 33.6	132	84
Decatur, Georgia	Bradley Obs.	− 84 17.6	+33 45.9	316	82
Dehra Dun, India	Haig Obs.	+ 78 02.9	+30 18.9	681	82
Delaware, Ohio	Ohio State Radio Obs. R	− 83 02.9	+40 15.1	282	81
Delaware, Ohio	Perkins Obs.	− 83 03.3	+40 15.1	280	81
Denver, Colorado	Chamberlin Obs.	−104 57.2	+39 40.6	1644	83
Devon, Alberta	Devon Ast. Obs.	−113 45.5	+53 23.4	708	82
Dexter, MIchigan	Univ. of Mich. Radio Ast. Obs. R	− 83 56.2	+42 23.9	345	81
Dolbeau, Québec	Québec Ast. Obs.	− 72 15.3	+48 50.2	110	82
Dresden, German Dem. Rep.	Lohrmann Obs.	+ 13 52.3	+51 03.0	324	83
Dundee, Scotland	Mills Obs.	− 3 00.7	+56 27.9	152	82
Dushanbe, Tadzhik S.S.R.	Inst. of Astrophysics	+ 68 46.9	+38 33.7	820	83
Dwingeloo, Netherlands	Dwingeloo Radio Obs. R	+ 6 23.8	+52 48.8	25	81
East Lansing, Michigan	Michigan State Univ. Obs.	− 84 29.0	+42 42.4	274	82
Edinburgh, Scotland	City Obs.	− 3 10.8	+55 57.4	107	82
Edinburgh, Scotland	Royal Obs.	− 3 11.0	+55 55.5	146	81
Effelsberg, German Fed. Rep.	Max Planck Inst. for Radio Ast. R	+ 6 53.1	+50 31.6	369	81
El Leoncito, Argentina	El Leoncito Ast. Complex	− 69 18.0	−31 48.0	2552	
El Leoncito, Argentina	Yale-San Juan Southern Sta.	− 69 19.8	−31 48.1	2348	
Eschweiler, German Fed. Rep.	Stockert Radio Obs. R	+ 6 43.4	+50 34.2	435	81
Evanston, Illinois	Dearborn Obs.	− 87 40.5	+42 03.4	195	82
Evanston, Illinois	Lindheimer Ast. Research Center	− 87 40.3	+42 03.6	205	82
Fan Mountain, Virginia	Leander McCormick Obs. Sta.	− 78 41.6	+37 52.7	566	82
Fayette, Missouri	Morrison Obs.	− 92 41.8	+39 09.1	228	82
Flagstaff, Arizona	Lowell Obs.	−111 39.9	+35 12.2	2219	81
Flagstaff, Arizona	Northern Arizona Univ. Obs.	−111 39.2	+35 11.1	2110	82
Flagstaff, Arizona	U.S. Naval Obs. Sta.	−111 44.4	+35 11.0	2316	81
Floirac, France	Bordeaux Univ. Obs. R	− 0 31.7	+44 50.1	73	82
Fort Davis, Texas	George R. Agassiz Sta. R	−103 56.8	+30 38.1	1603	82
Fort Irwin, California	Goldstone Complex R	−116 50.9	+35 23.4	1036	81
Gauribidanur, India	Gauribidanur Radio Obs. R	+ 77 26.1	+13 36.2	686	
Glasgow, Scotland	Glasgow Univ. Obs.	− 4 18.3	+55 54.1	53	81
Göttingen, German Fed. Rep.	Göttingen Univ. Obs.	+ 9 56.6	+51 31.8	159	
Graz, Austria	Univ. of Graz Obs.	+ 15 26.9	+47 04.6	375	82

* ‘R’ denotes an observatory with radio instruments; ‘I’ denotes an observatory with infrared instruments; ‘L’ denotes an observatory with laser instruments.

Place	Description*		East Longitude	Latitude	Height (Sea Level)	Ref.
			° ′	° ′	m	
Green Bank, West Virginia	National Radio Ast. Obs.	R	− 79 50.5	+38 25.8	836	83
Greenbelt, Maryland	GSFC Optical Test Site	L	− 76 49.6	+39 01.3	53	81
Greenville, Delaware	Mount Cuba Ast. Obs.		− 75 38.0	+39 47.1	92	82
Gyula, Hungary	Heliophysical Obs. Sta.		+ 21 16.2	+46 39.2	135	84
Haleakala, Hawaii	C.E.K. Mees Solar Obs.		−156 15.4	+20 42.4	3054	
Haleakala, Hawaii	LURE Obs.	L	−156 15.5	+20 42.6	3048	82
Hamilton, Massachusetts	Sagamore Hill Radio Obs.	R	− 70 49.3	+42 37.9	53	81
Hannover, German Fed. Rep.	Inst. of Geodesy Ast. Obs.		+ 9 42.8	+52 23.3	71	82
Hanover, New Hampshire	Shattuck Obs.		− 72 17.0	+43 42.3	183	82
Hartebeeshoek, South Africa	Hartebeeshoek Radio Ast. Obs.	R	+ 27 41.1	−25 53.4	1391	
Hartebeespoort, South Africa	Leiden Obs. Southern Sta.		+ 27 52.6	−25 46.4	1220	81
Harvard, Massachusetts	Oak Ridge Obs.	R	− 71 33.5	+42 30.3	185	81
Haverford, Pennsylvania	Strawbridge Obs.	R	− 75 18.2	+40 00.7	116	82
Helsinki, Finland	Univ. of Helsinki Obs.		+ 24 57.3	+60 09.7	33	82
Helwân, Egypt	Helwân Obs.		+ 31 22.8	+29 51.5	116	82
Herstmonceux, England	Royal Greenwich Obs. Sta.	L	+ 0 20.9	+50 52.2	50	83
Hobart, Tasmania	Univ. of Tasmania Obs.	R	+147 32.0	−42 50.0	300	81
Hoeven, Netherlands	Simon Stevin Obs.	R	+ 4 33.8	+51 34.0	9	82
Hoher List, German Fed. Rep.	Hoher List Obs.		+ 6 51.0	+50 09.8	533	81
Holmdel, New Jersey	Crawford Hill Obs.	R	− 74 11.2	+40 23.5	114	81
Hoskinstown, New South Wales	Molongo Radio Obs.	R	+149 25.4	−35 22.3	732	84
Humain, Belgium	Royal Obs. Radio Ast. Sta.	R	+ 5 15.3	+50 11.5	293	84
Hvar, Yugoslavia	Hvar Obs.		+ 16 26.9	+43 10.7	238	
Hyderabad, India	Nizamiah Obs.		+ 78 27.2	+17 25.9	554	82
Incline Village, Nevada	Maclean Obs.		−119 55.7	+39 17.7	2546	82
Irkutsk, R.S.F.S.R.	Irkutsk Ast. Obs.		+104 20.7	+52 16.7	468	83
Istanbul, Turkey	Istanbul Univ. Obs.		+ 28 57.9	+41 00.7	65	82
Istanbul, Turkey	Kandilli Obs.		+ 29 03.7	+41 03.8	120	82
Itajubá, Brazil	Pico dos Dias Obs.		− 45 35.0	−22 32.1	1870	82
Ithaca, New York	Hartung-Boothroyd Obs.		− 76 23.1	+42 27.5	534	82
Izaña, Tenerife Is., Canaries	Teide Obs.	R,I	− 16 29.8	+28 17.5	2395	
Japal, India	Japal-Rangapur Obs.	R	+ 78 43.7	+17 05.9	695	83
Jelm Mountain, Wyoming	Wyoming Infrared Obs.	I	−105 58.6	+41 05.9	2943	
Jena, German Dem. Rep.	Friedrich-Schiller Univ. Obs.		+ 11 29.2	+50 55.8	356	82
Jodrell Bank, England	Nuffield Radio Ast. Labs.	R	− 2 18.4	+53 14.2	78	81
Jungfraujoch, Switzerland	High Alpine Research Obs.		+ 7 59.1	+46 32.9	3576	81
Kaliningrad, R.S.F.S.R.	Kaliningrad Univ. Obs.		+ 20 29.7	+54 42.8	24	83
Kamiku Isshiki, Japan	Nagoya Univ. Fujigane Sta.	R	+138 36.7	+35 25.6	1015	81
Kamitakara, Japan	Hida Obs.		+137 18.5	+36 14.9	1276	84
Kanzelhöhe, Austria	Kanzelhöhe Solar Obs.		+ 13 54.4	+46 40.7	1526	82
Kashima, Japan	Kashima Space Commun. Center	R	+140 39.8	+35 57.3	32	81
Kasugai, Japan	Research Inst. of Magnetosphere	R	+137 00.6	+35 16.6	800	81
Kavalur, India	Vainu Bappu Obs.		+ 78 49.6	+12 34.6	725	
Kazan, R.S.F.S.R.	Engelhardt Ast. Obs.		+ 48 48.9	+55 50.3	98	81
Kazan, R.S.F.S.R.	Kazan University Obs.		+ 49 07.3	+55 47.4	79	83

* 'R' denotes an observatory with radio instruments; 'I' denotes an observatory with infrared instruments; 'L' denotes an observatory with laser instruments.

Place	Description*		East Longitude	Latitude	Height (Sea Level)	Ref.
			° ′	° ′	m	
Kemps Creek, New South Wales	Fleurs Radio Obs.	R	+150 46.5	−33 51.8	45	81
Kharkov, Ukrainian S.S.R.	Inst. of Radio Ast.	R	+ 36 56.0	+49 38.0	150	
Kharkov, Ukrainian S.S.R.	Kharkov Univ. Ast. Obs.		+ 36 13.9	+50 00.2	138	82
Kiev, Ukrainian S.S.R.	Kiev Univ. Obs.		+ 30 29.9	+50 27.2	184	83
Kiev, Ukrainian S.S.R.	Main Ast. Obs.		+ 30 30.4	+50 21.9	188	83
Kirkkonummi, Finland	Metsahovi Obs.		+ 24 23.8	+60 13.2	60	81
Kirkkonummi, Finland	Metsahovi Obs. Radio Rsch. Sta.	R	+ 24 23.6	+60 13.1	61	81
Kisarazu, Japan	Kisarazu Technical College Obs.	R	+139 57.6	+35 22.8	35	81
Kislovodsk, R.S.F.S.R.	Pulkovo Obs. Sta.		+ 42 31.8	+43 44.0	2130	84
Kiso, Japan	Kiso Obs.		+137 37.7	+35 47.6	1130	81
Kitab, Uzbek S.S.R.	Kitab Ast. Sta.		+ 66 52.9	+39 08.0	658	83
Kitt Peak, Arizona	Kitt Peak National Obs.		−111 36.0	+31 57.8	2120	81
Kitt Peak, Arizona	McGraw-Hill Obs.		−111 37.0	+31 57.0	1925	81
Kitt Peak, Arizona	National Radio Ast. Obs.	R	−111 36.9	+31 57.2	1938	81
Kitt Peak, Arizona	Steward Obs. Sta.		−111 36.0	+31 57.8	2071	81
Kitt Peak, Arizona	Warner and Swasey Obs. Sta.		−111 35.9	+31 57.6	2084	83
Kodaikanal, India	Kodaikanal Solar Obs.		+ 77 28.1	+10 13.8	2343	
Königstuhl, German Fed. Rep.	State Obs.		+ 8 43.3	+49 23.9	570	82
Kottamia, Egypt	Kottamia Obs.		+ 31 49.5	+29 55.9	476	81
Kunming, China	Yunnan Obs.	R	+102 47.3	+25 01.5	1940	82
Kurashiki, Japan	Kurashiki Hydrographic Obs.		+133 46.3	+34 35.4	6	
Kutztown, Pennsylvania	Kutztown Univ. Obs		− 75 47.1	+40 30.9	158	82
Kyoto, Japan	Kwasan Obs.		+135 47.6	+34 59.7	221	84
Kyoto, Japan	Kyoto Univ. Ast. Dept. Obs.		+135 47.2	+35 01.7	86	81
Kyoto, Japan	Kyoto Univ. Physics Dept. Obs.		+135 47.2	+35 01.7	80	
Lafayette, California	Leuschner Obs.		−122 09.4	+37 55.1	304	82
La Mariquita Mtn., Mexico	Cananea Astrophysical Obs.		−110 23.0	+31 03.2	2480	
Lane Cove, New South Wales	Riverview College Obs.		+151 09.5	−33 49.8	25	82
La Palma Island, Canary Islands	Roque de los Muchachos Obs.	R	− 17 52.9	+28 45.6	2326	
La Plata, Argentina	La Plata Ast. Obs.		− 57 55.9	−34 54.5	17	81
Las Cruces, New Mexico	Corralitos Obs.		−107 02.6	+32 22.8	1453	82
Lawrence, Kansas	Clyde W. Tombaugh Obs.		− 95 15.0	+38 57.6	323	82
Legon, Ghana	Univ. of Ghana Obs.	R	− 0 11.4	+ 5 38.9	96	81
Leiden, Netherlands	Leiden Obs.		+ 4 29.1	+52 09.3	12	81
Lembang, (Java), Indonesia	Bosscha Obs.		+107 37.0	− 6 49.5	1300	84
Leningrad, R.S.F.S.R.	Leningrad Univ. Obs.		+ 30 17.7	+59 56.5	3	83
Liège, Belgium	Cointe Obs.		+ 5 33.9	+50 37.1	127	81
Lintong, China	Shaanxi Ast. Obs.	R	+109 33.1	+34 56.7	468	
Lisbon, Portugal	Lisbon Ast. Obs.		− 9 11.2	+38 42.7	111	82
Loiano, Italy	Bologna Univ. Obs.		+ 11 20.2	+44 15.5	785	81
Lomnický Štít, Czechoslovakia	Lomnický Štít Coronal Obs.		+ 20 13.2	+49 11.8	2632	82
Los Angeles, California	Griffith Obs.		−118 17.9	+34 07.1	357	82
Lund, Sweden	Lund Obs.		+ 13 11.2	+55 41.9	34	82
Lustbühel, Austria	Lustbühel Obs.		+ 15 29.7	+47 03.9	480	82
Lvov, Ukrainian S.S.R.	Lvov Polytechnic Inst. Obs.		+ 24 00.9	+49 50.2	340	83

* 'R' denotes an observatory with radio instruments; 'I' denotes an observatory with infrared instruments; 'L' denotes an observatory with laser instruments.

Place	Description*		East Longitude	Latitude	Height (Sea Level)	Ref.
			° ′	° ′	m	
Lvov, Ukrainian S.S.R.	Lvov Univ. Obs.		+ 24 01.8	+49 50.0	330	83
Madison, Wisconsin	Washburn Obs.		− 89 24.5	+43 04.6	292	82
Madrid, Spain	National Ast. Obs.		− 3 41.1	+40 24.6	670	82
Maipu, Chile	Maipu Radio Ast. Obs.	R	− 70 51.5	−33 30.1	446	81
Malvern, England	Royal Radar Establishment	R	− 2 20.0	+52 08.1	93	81
Malvern, Pennsylvania	Flower and Cook Obs.		− 75 29.6	+40 00.0	155	81
Manastash Ridge, Washington	Manastash Ridge Obs.		−120 43.4	+46 57.1	1198	82
Manchester, England	Godlee Obs.		− 2 14.0	+53 28.6	77	82
Manora Peak, India	Uttar Pradesh State Obs.		+ 79 27.4	+29 21.7	1927	82
Marine-on-St. Croix, Minnesota	O'Brien Obs.		− 92 46.6	+45 10.9	308	82
Mauna Kea, Hawaii	Caltech Submillimeter Obs.	R	−155 28.3	+19 49.6	4060	
Mauna Kea, Hawaii	Joint Astronomy Centre	R,I	−155 28.4	+19 49.5	4194	
Mauna Kea, Hawaii	Mauna Kea Obs.	I	−155 28.3	+19 49.6	4215	81
Mazelspoort, South Africa	Boyden Obs.		+ 26 24.3	−29 02.3	1387	81
Mead, Nebraska	Behlen Obs.		− 96 26.8	+41 10.3	362	82
Medicina, Italy	Medicina Radio Ast. Sta.	R	+ 11 38.7	+44 31.2	41	
Merate, Italy	Brera-Milan Ast. Obs.		+ 9 25.7	+45 42.0	340	81
Mérida, Venezuela	Llano del Hato Obs.		− 70 52.0	+ 8 47.4	3610	81
Meudon, France	Meudon Obs.		+ 2 13.9	+48 48.3	162	84
Miami, Florida	U.S. Naval Obs. Time Sta.	R	− 80 23.1	+25 36.8	7	81
Middletown, Connecticut	Van Vleck Obs.		− 72 39.6	+41 33.3	65	82
Milan, Italy	Brera-Milan Ast. Obs.		+ 9 11.5	+45 28.0	146	82
Mill Hill, England	Univ. of London Obs.		− 0 14.4	+51 36.8	81	82
Mitaka, Japan	National Ast. Obs.	R	+139 32.5	+35 40.3	58	81
Miyun, China	Beijing Obs. Sta.	R	+116 45.9	+40 33.4	160	84
Mizusawa, Japan	Mizusawa Astrogeodynamics Obs.		+141 07.9	+39 08.1	61	82
Monte Mario, Italy	Rome Obs.		+ 12 27.1	+41 55.3	152	82
Montevideo, Uruguay	National Obs.		− 56 12.8	−34 54.6	24	84
Mont Gros, France	Nice Obs.		+ 7 18.1	+43 43.4	372	81
Mont Mégantic, Québec	Mont Mégantic Ast. Obs.		− 71 09.2	+45 27.3	1114	81
Montville, Ohio	Nassau Ast. Obs.		− 81 04.5	+41 35.5	390	83
Moscow, R.S.F.S.R.	Sternberg State Ast. Inst.		+ 37 32.7	+55 42.0	195	83
Mount Abu, India	Gurushikhar Infrared Obs.	I	+ 72 46.8	+24 39.1	1700	
Mount Aragatz, Armenian S.S.R.	Byurakan Astrophysical Obs.	R	+ 44 17.5	+40 20.1	1500	84
Mount Bigelow, Arizona	Steward Obs. Catalina Sta.		−110 43.9	+32 25.0	2510	81
Mount Chikurin, Japan	Okayama Astrophysical Obs.		+133 35.8	+34 34.4	372	81
Mount Dodaira, Japan	Dodaira Obs.	L	+139 11.8	+36 00.2	879	81
Mount Ekar, Italy	Mount Ekar Obs.		+ 11 34.3	+45 50.6	1350	81
Mount Evans, Colorado	Mount Evans Obs.		−105 38.4	+39 35.2	4313	82
Mount Hamilton, California	Lick Obs.		−121 38.2	+37 20.6	1290	84
Mount Hopkins, Arizona	Fred L. Whipple (MMT) Obs.		−110 53.1	+31 41.3	2608	84
Mount John, New Zealand	Mount John Univ. Obs.		+170 27.9	−43 59.2	1027	83
Mount Kanobili, Georgian S.S.R.	Abastumani Astrophysical Obs.	R	+ 42 49.3	+41 45.3	1583	83
Mount Killíni, Greece	Kryonerion Ast. Obs.		+ 22 37.3	+37 58.4	905	82
Mount Laguna, California	Mount Laguna Obs.		−116 25.6	+32 50.4	1859	82

* 'R' denotes an observatory with radio instruments; 'I' denotes an observatory with infrared instruments; 'L' denotes an observatory with laser instruments.

Place	Description*		East Longitude	Latitude	Height (Sea Level)	Ref.
			° ′	° ′	m	
Mount Lemmon, Arizona	Mount Lemmon Infrared Obs.	I	−110 47.5	+32 26.5	2776	81
Mount Lemmon, Arizona	Steward Obs. Catalina Sta.		−110 47.3	+32 26.6	2790	81
Mount Locke, Texas	McDonald Obs.	L	−104 01.3	+30 40.3	2075	84
Mount Locke, Texas	Millimeter Wave Obs.	R	−104 01.7	+30 40.3	2031	81
Mount Norikura, Japan	Norikura Solar Obs.	I	+137 33.3	+36 06.8	2876	82
Mount Pleasant, Michigan	Central Michigan Univ. Obs.		− 84 46.5	+43 35.3	258	82
Mount Stromlo, A.C.T., Australia	Mount Stromlo Obs.		+149 00.5	−35 19.2	767	83
Mount Wilson, California	Mount Wilson Obs.	R	−118 03.6	+34 13.0	1742	82
Mount Zin, Israel	Florence and George Wise Obs.		+ 34 45.8	+30 35.8	874	81
Munich, German Fed. Rep.	Munich Univ. Obs.		+ 11 36.5	+48 08.7	529	82
Nagoya, Japan	Nagoya Univ. Radio Ast. Lab.	R	+136 58.4	+35 08.9	75	81
Nakaminato, Japan	Hiraiso Solar Terr. Rsch. Center	R	+140 37.5	+36 22.0	27	82
Nançay, France	Paris Obs. Radio Ast. Sta.	R	+ 2 11.8	+47 22.8	150	81
Nantucket, Massachusetts	Maria Mitchell Obs.		− 70 06.3	+41 16.8	20	82
Naples, Italy	Capodimonte Ast. Obs.		+ 14 15.3	+40 51.8	150	81
Nashville, Tennessee	Arthur J. Dyer Obs.		− 86 48.3	+36 03.1	345	82
Neuchatel, Switzerland	Cantonal Obs.		+ 6 57.5	+46 59.9	488	82
New Salem, Massachusetts	Five College Radio Ast. Obs.	R	− 72 20.7	+42 23.5	314	
New York, New York	Rutherfurd Obs.		− 73 57.5	+40 48.6	25	82
Nijmegen, Netherlands	Catholic Univ. Ast. Inst.		+ 5 52.1	+51 49.5	62	82
Nikolaev, Ukrainian S.S.R.	Nikolaev Ast. Obs.		+ 31 58.5	+46 58.3	54	83
Nobeyama, Japan	Nobeyama Cosmic Radio Obs.	R	+138 29.0	+35 56.0	1350	
Nobeyama, Japan	Nobeyama Solar Radio Obs.	R	+138 28.8	+35 56.3	1350	81
Norman, Oklahoma	Univ. of Oklahoma Obs.		− 97 26.6	+35 12.1	363	82
North Liberty, Iowa	North Liberty Radio Obs.	R	− 91 34.5	+41 46.3	241	81
Oakland, California	Chabot Obs.		−122 10.6	+37 47.2	100	82
Odessa, Ukrainian S.S.R.	Odessa Obs.		+ 30 45.5	+46 28.6	60	83
Old Town, Florida	Univ. of Florida Radio Obs.	R	− 83 02.1	+29 31.7	8	81
Ondřejov, Czechoslovakia	Ondřejov Obs.	R	+ 14 47.0	+49 54.6	533	
Onsala, Sweden	Onsala Space Obs.	R	+ 11 55.1	+57 23.6	24	81
Ostrowik, Poland	Warsaw Univ. Obs.		+ 21 25.2	+52 05.4	138	82
Ottawa, Ontario	Ottawa River Solar Obs.		− 75 53.6	+45 23.2	58	82
Padua, Italy	Padua Ast. Obs.		+ 11 52.3	+45 24.0	38	82
Palermo, Sicily	Palermo Univ. Ast. Obs.		+ 13 21.5	+38 06.7	72	82
Palo Alto, California	Stanford Center for Radar Ast.	R	−122 10.7	+37 27.5	172	
Palomar Mountain, California	Palomar Obs.		−116 51.8	+33 21.4	1706	81
Paris, France	Paris Obs.		+ 2 20.2	+48 50.2	67	81
Parkes, New South Wales	Austral. Natl. Radio Ast. Obs.	R	+148 15.7	−33 00.0	392	81
Partizanskoye, Ukrainian S.S.R.	Crimean Astrophysical Obs.		+ 34 01.0	+44 43.7	550	
Pentele, Greece	National Obs. Sta.	R	+ 23 51.8	+38 02.9	509	82
Penticton, British Columbia	Dominion Radio Astrophys. Obs.	R	−119 37.2	+49 19.2	545	81
Philadelphia, Pennsylvania	The Franklin Inst. Obs.		− 75 10.4	+39 57.5	30	82
Pic du Midi, France	Pic du Midi Obs.		+ 0 08.7	+42 56.2	2861	82
Pico Veleta, Spain	Millimeter Radio Ast. Inst.	R	− 3 24.0	+37 04.1	2870	
Piikkiö, Finland	Turku Univ. Obs.		+ 22 26.8	+60 25.0	40	83

* 'R' denotes an observatory with radio instruments; 'I' denotes an observatory with infrared instruments; 'L' denotes an observatory with laser instruments.

Place	Description*		East Longitude	Latitude	Height (Sea Level)	Ref.
			° ′	° ′	m	
Pine Bluff, Wisconsin	Pine Bluff Obs.		− 89 41.1	+43 04.7	366	82
Pino Torinese, Italy	Turin Ast. Obs.		+ 7 46.5	+45 02.3	622	83
Piszkéstetö, Hungary	Konkoly Obs. Mountain Sta.		+ 19 53.7	+47 55.1	958	81
Pittsburgh, Pennsylvania	Allegheny Obs.		− 80 01.3	+40 29.0	380	82
Piwnice, Poland	Piwnice Ast. Obs.	R	+ 18 33.3	+53 05.8	91	82
Plateau de Bure, France	Grenoble Obs.	R	+ 5 54.5	+44 38.0	2552	
Plateau de Bure, France	Millimeter Radio Ast. Inst.	R	+ 5 54.4	+44 38.0	2552	
Poltava, Ukrainian S.S.R.	Gravimetric Obs.		+ 34 32.8	+49 36.3	151	84
Porto Alegre, Brazil	Morro Santana Obs.		− 51 07.6	−30 03.2	300	
Potsdam, German Dem. Rep.	Central Inst. for Earth Physics		+ 13 04.0	+52 22.9	91	82
Potsdam, German Dem. Rep.	Einstein Tower Solar Obs.	R	+ 13 03.9	+52 22.8	100	83
Potsdam, German Dem. Rep.	Potsdam Astrophysical Obs.		+ 13 04.0	+52 22.9	107	
Poznań, Poland	Poznań Univ. Ast. Obs.	L	+ 16 52.7	+52 23.8	85	82
Prague, Czechoslovakia	Charles Univ. Ast. Inst.		+ 14 23.7	+50 04.6	267	82
Preston, England	Jeremiah Horrocks Obs.		− 2 42.2	+53 46.5	34	82
Priddis, Alberta	Rothney Astrophysical Obs.	I	−114 17.3	+50 52.1	1272	82
Princeton, New Jersey	FitzRandolph Obs.		− 74 38.8	+40 20.7	43	81
Prostějov, Czechoslovakia	Prostějov Obs.		+ 17 09.8	+49 29.2	225	83
Providence, Rhode Island	Ladd Obs.		− 71 24.0	+41 50.3	69	82
Pulkovo, R.S.F.S.R.	Main Ast. Obs.	R	+ 30 19.6	+59 46.4	75	83
Purple Mountain, China	Purple Mountain Obs.	R	+118 49.3	+32 04.0	367	83
Quezon City, Philippines	Manila Obs.	R	+121 04.6	+14 38.2	58	82
Quezon City, Philippines	Pagasa Ast. Obs.		+121 04.3	+14 39.2	70	82
Quito, Ecuador	Quito Ast. Obs.		− 78 29.9	− 0 13.0	2818	82
Rattlesnake Mtn., Pennsylvania	Black Moshannon Obs.		− 78 00.3	+40 55.3	738	
Richmond Hill, Ontario	David Dunlap Obs.		− 79 25.3	+43 51.8	244	81
Riga, Latvian S.S.R.	Latvian State Univ. Ast. Obs.	L	+ 24 07.0	+56 57.1	39	84
Riga, Latvian S.S.R.	Riga Radio-Astrophysical Obs.	R	+ 24 24.0	+56 47.0	75	
Rio de Janeiro, Brazil	National Obs.		− 43 13.4	−22 53.7	33	81
Rio de Janeiro, Brazil	Valongo Obs.		− 43 11.2	−22 53.9	52	82
Riverside, Iowa	Univ. Of Iowa Obs.		− 91 33.6	+41 30.9	221	82
Riverside, Maryland	Maryland Point Obs.	R	− 77 13.9	+38 22.4	20	81
Robledo, Spain	Deep Space Sta.	R	− 4 14.9	+40 25.8	774	
Roden, Netherlands	Kapteyn Obs.		+ 6 26.6	+53 07.7	12	82
Rome (Castel Gandolfo), Italy	Vatican Obs.		+ 12 39.1	+41 44.8	450	82
Roquetas, Spain	Ebro Obs.	R	+ 0 29.6	+40 49.2	50	82
Rostov-on-Don, R.S.F.S.R.	Rostov Univ. Ast. Obs.		+ 39 40.0	+47 15.0	100	82
St. Andrews, Scotland	St. Andrews Univ. Obs.		− 2 48.9	+56 20.2	30	82
St. Corona at Schopfl, Austria	L. Figl Astrophysical Obs.		+ 15 55.4	+48 05.0	890	82
St. Genis Laval, France	Lyon Univ. Obs.		+ 4 47.1	+45 41.7	299	82
St. Michel, France	Obs. of Haute-Provence		+ 5 42.8	+43 55.9	665	81
Saltsjöbaden, Sweden	Stockholm Obs.		+ 18 18.5	+59 16.3	60	82
San Fernando, California	San Fernando Obs.	R	−118 29.5	+34 18.5	371	81
San Fernando, Spain	Naval Obs.	L	− 6 12.2	+36 28.0	27	81
San Juan, Argentina	Félix Aguilar Obs.		− 68 37.2	−31 30.6	700	82

* ‘R’ denotes an observatory with radio instruments; ‘I’ denotes an observatory with infrared instruments; ‘L’ denotes an observatory with laser instruments.

Place	Description*		East Longitude	Latitude	Height (Sea Level)	Ref.
			° ′	° ′	m	
San Miguel, Argentina	National Obs. of Cosmic Physics		− 58 43.9	−34 33.4	37	82
San Pedro Martir, Mexico	National Ast. Obs.		−115 27.8	+31 02.6	2830	81
Santiago, Chile	Manuel Foster Astrophys. Obs.		− 70 37.8	−33 25.1	840	82
Santiago de Compostela, Spain	Ramon Maria Aller Obs.		− 8 33.6	+42 52.5	240	82
Sauverny, Switzerland	Geneva Obs.		+ 6 08.2	+46 18.4	465	81
Saxapahaw, North Carolina	Three College Obs.		− 79 24.4	+35 56.7	183	
Schauinsland Mtn., German F.R.	Schauinsland Obs.		+ 7 54.4	+47 54.9	1240	82
Sendai, Japan	Sendai Ast. Obs.		+140 51.9	+38 15.4	45	82
Sendai, Japan	Tohoku Univ. Obs.		+140 50.6	+38 15.4	153	82
Serra la Nave, Sicily	Catania Obs. Stellar Sta.		+ 14 58.4	+37 41.5	1735	81
Shahe, China	Beijing Obs. Sta.	R,L	+116 19.7	+40 06.1	40	84
Sheshan, China	Shanghai Obs. Sheshan Sta.	L	+121 11.2	+31 05.8	100	
Siding Spring Mtn., New So. Wales	Royal Obs. Edinburgh Sta.		+149 04.2	−31 16.5	1145	
Siding Spring Mtn., New So. Wales	Siding Spring Obs.	R	+149 03.7	−31 16.4	1149	83
Simeis, Ukrainian S.S.R.	Crimean Astrophysical Obs.	R	+ 34 01.0	+44 32.1	676	84
Simosato, Japan	Simosato Hydrographic Obs.	R,L	+135 56.4	+33 34.5	63	
Sirahama, Japan	Sirahama Hydrographic Obs.		+138 59.3	+34 42.8	172	
Skalnaté Pleso, Czechoslovakia	Skalnaté Pleso Obs.		+ 20 14.7	+49 11.3	1783	82
Skibotn, Norway	Skibotn Obs.		+ 20 21.9	+69 20.9	157	82
Socorro, New Mexico	Joint Obs. for Cometary Rsch.		−107 11.3	+33 59.1	3235	82
Socorro, New Mexico	National Radio Ast. Obs.	R	−107 37.1	+34 04.7	2124	81
Sondre Stromfjiord, Greenland	Incoherent Scatter Facility	R	− 50 57.0	+66 59.2	180	
Sonneberg, German Dem. Rep.	Sonneberg Obs.		+ 11 11.5	+50 22.7	640	81
South Park, Colorado	Tiara Obs.		−105 31.0	+38 58.2	2679	82
Stanford, California	Radio Ast. Inst.	R	−122 11.3	+37 23.9	80	81
Stanford, California	SRI Radio Ast. Obs.	R	−122 10.6	+37 24.3	168	
Stephanion, Greece	Stephanion Obs.		+ 22 49.7	+37 45.3	800	
Strasbourg, France	Strasbourg Obs.		+ 7 46.2	+48 35.0	142	81
Sugar Grove, West Virginia	Naval Research Lab. Radio Sta.	R	− 79 16.4	+38 31.2	705	81
Sunspot, New Mexico	Sacr. Peak Natl. Solar Obs.		−105 49.2	+32 47.2	2811	82
Sutherland, South Africa	South African Ast. Obs. Sta.		+ 20 48.7	−32 22.7	1771	81
Swarthmore, Pennsylvania	Sproul Obs.		− 75 21.4	+39 54.3	63	82
Syracuse, New York	Syracuse Univ. Obs.		− 76 08.3	+43 02.2	160	82
Table Mountain, California	Table Mountain Obs.	R	−117 40.9	+34 22.9	2286	82
Taipei, Taiwan	Taipei Obs.		+121 31.6	+25 04.7	31	
Tartu, Estonian S.S.R.	Wilhelm Struve Astrophys. Obs.		+ 26 28.0	+58 16.0	—	83
Tashkent, Uzbek S.S.R.	Tashkent Obs.		+ 69 17.6	+41 19.5	477	83
Tautenburg, German Dem. Rep.	Karl Schwarzschild Obs.		+ 11 42.8	+50 58.9	331	83
Teramo, Italy	Collurania Ast. Obs.		+ 13 44.0	+42 39.5	388	82
Thessaloníki, Greece	Univ. of Thessaloníki Obs.		+ 22 57.5	+40 37.0	28	82
Tianjing, China	Beijing Obs. Latitude Sta.		+117 03.5	+39 08.0	5	84
Tidbinbilla, A.C.T., Australia	Deep Space Sta.	R	+148 58.8	−35 24.1	656	82
Tokyo, Japan	Tokyo Hydrographic Obs.		+139 46.2	+35 39.7	41	
Toledo, Ohio	Ritter Obs.		− 83 36.8	+41 39.7	201	81
Tomsk, R.S.F.S.R.	Tomsk Univ. Obs.		+ 84 56.8	+56 28.1	130	84

* 'R' denotes an observatory with radio instruments; 'I' denotes an observatory with infrared instruments; 'L' denotes an observatory with laser instruments.

Place	Description*		East Longitude	Latitude	Height (Sea Level)	Ref.
			° ′	° ′	m	
Tonantzintla, Mexico	National Ast. Obs.	R	− 98 18.8	+19 02.0	2150	82
Topeka, Kansas	Zenas Crane Obs.		− 95 41.8	+39 02.2	306	82
Tortugas Mountain, New Mexico	N.M. State Univ. Obs. Sta.		−106 41.8	+32 17.6	1505	82
Toulouse, France	Toulouse Univ. Obs.		+ 1 27.8	+43 36.7	195	82
Toyokawa, Japan	Nagoya Univ. Sugadaira Sta.	R	+138 19.3	+36 31.2	1280	81
Toyokawa, Japan	Nagoya Univ. Toyokawa Sta.	R	+137 22.2	+34 50.1	25	81
Toyokawa, Japan	Toyokawa Obs.	R	+137 22.3	+34 50.2	18	
Tremsdorf, German Dem. Rep.	Tremsdorf Radio Ast. Obs.	R	+ 13 08.2	+52 17.1	35	83
Trieste, Italy	Trieste Ast. Obs.	R	+ 13 52.5	+45 38.5	400	81
Tübingen, German Fed. Rep.	Tübingen Univ. Ast. Obs.	R	+ 9 03.5	+48 32.3	470	82
Tucson, Arizona	Steward Obs.		−110 56.9	+32 14.0	757	81
Tumamoc Hill, Arizona	Steward Obs. Catalina Sta.		−111 00.3	+32 12.8	950	81
Uccle, Belgium	Royal Obs. of Belgium	R	+ 4 21.5	+50 47.9	105	81
Uchinoura, Japan	Kagoshima Space Center	R	+131 04.0	+31 13.7	228	82
Udhagamandalam (Ooty), India	Radio Ast. Center	R	+ 76 40.0	+11 22.9	2150	81
U.S. Air Force Academy, Col.	U.S. Air Force Academy Obs.		−104 52.5	+39 00.4	2187	82
University, Alabama	Univ. of Alabama Obs.		− 87 32.5	+33 12.6	87	82
Utrecht, Netherlands	Utrecht Univ. Ast. Inst.		+ 5 07.8	+52 05.2	14	82
Valašské Meziříčí, Czech.	Valašské Meziříčí Obs.		+ 17 58.5	+49 27.8	338	82
Valinhos, Brazil	Abrahao de Moraes Obs.	R	− 46 58.0	−23 00.1	850	
Victoria, British Columbia	Climenhaga Obs.		−123 18.5	+48 27.8	74	82
Victoria, British Columbia	Dominion Astrophysical Obs.		−123 25.0	+48 31.2	238	84
Vienna, Austria	Kuffner Obs.		+ 16 17.8	+48 12.8	302	82
Vienna, Austria	Urania Obs.		+ 16 23.1	+48 12.7	193	82
Vienna, Austria	Vienna Univ. Obs.		+ 16 20.2	+48 13.9	241	82
Vila Nova de Gaia, Portugal	Prof. Manuel de Barros Obs.	R	− 8 35.3	+41 06.5	232	82
Villa Elisa, Argentina	Argentine Radio Ast. Inst.	R	− 58 08.2	−34 52.1	11	81
Villanova, Pennsylvania	Villanova Univ. Obs.	R	− 75 20.5	+40 02.4	—	82
Vilnius, Lithuanian S.S.R.	Vilnius Ast. Obs.		+ 25 17.2	+54 41.0	122	83
Washington, D.C.	NRL Radio Ast. Obs.	R	− 77 01.6	+38 49.3	30	81
Washington, D.C.	U.S. Naval Obs.		− 77 04.0	+38 55.3	92	81
Wasosz, Poland	Wroclaw Univ. Sta.		+ 16 39.6	+51 28.5	140	
Wellesley, Massachusetts	Whitin Obs.		− 71 18.2	+42 17.7	32	82
Wellington, New Zealand	Carter Obs.		+174 46.0	−41 17.2	129	83
Westerbork, Netherlands	Westerbork Radio Ast. Obs.	R	+ 6 36.3	+52 55.0	16	81
Westford, Massachusetts	George R. Wallace Jr. Aph. Obs.		− 71 29.1	+42 36.6	107	82
Westford, Massachusetts	Haystack Obs.	R	− 71 29.3	+42 37.4	146	81
Westford, Massachusetts	Millstone Hill Atm. Sci. Fac.	R	− 71 29.7	+42 36.6	146	
Westford, Massachusetts	Millstone Hill Radar Obs.	R	− 71 29.5	+42 37.0	156	81
Westford, Massachusetts	Westford Antenna Facility	R	− 71 29.7	+42 36.8	115	
Williams Bay, Wisconsin	Yerkes Obs.		− 88 33.4	+42 34.2	334	81
Williamstown, Massachusetts	Hopkins Obs.	R	− 73 12.1	+42 42.7	215	82
Wroclaw, Poland	Wroclaw Univ. Ast. Obs.		+ 17 05.3	+51 06.7	115	82
Wuhan, China	Wucheng Time Obs.	L	+114 20.7	+30 32.5	28	
Xinglong, China	Beijing Obs. Sta.	I	+117 34.5	+40 23.7	870	84

* 'R' denotes an observatory with radio instruments; 'I' denotes an observatory with infrared instruments; 'L' denotes an observatory with laser instruments.

Place	Description*		East Longitude	Latitude	Height (Sea Level)	Ref.
			° ′	° ′	m	
Xujiahui, China	Shanghai Obs. Xujiahui Sta.	R	+121 25.6	+31 11.4	5	
Yebes, Spain	National Obs. Ast. Center	R	− 3 06.0	+40 31.5	914	82
Zagreb, Yugoslavia	Geodetical Faculty Obs.		+ 16 01.3	+45 49.5	146	82
Zelenchukskaya, R.S.F.S.R.	Special Astrophysical Obs.	R	+ 41 26.5	+43 39.2	2100	81
Zermatt, Switzerland	Gornergrat North and South Obs.	R,I	+ 7 47.1	+45 59.1	3135	81
Zimmerwald, Switzerland	Zimmerwald Obs.		+ 7 27.9	+46 52.6	929	81
Zürich, Switzerland	Swiss Federal Obs.		+ 8 33.1	+47 22.6	469	81

* 'R' denotes an observatory with radio instruments; 'I' denotes an observatory with infrared instruments; 'L' denotes an observatory with laser instruments.

CONTENTS OF SECTION K

JULIAN DAY NUMBER, 1950–2000

OF DAY COMMENCING AT GREENWICH NOON ON:

Year	Jan. 0	Feb. 0	Mar. 0	Apr. 0	May 0	June 0	July 0	Aug. 0	Sept. 0	Oct. 0	Nov. 0	Dec. 0
1950	243 3282	3313	3341	3372	3402	3433	3463	3494	3525	3555	3586	3616
1951	3647	3678	3706	3737	3767	3798	3828	3859	3890	3920	3951	3981
1952	4012	4043	4072	4103	4133	4164	4194	4225	4256	4286	4317	4347
1953	4378	4409	4437	4468	4498	4529	4559	4590	4621	4651	4682	4712
1954	4743	4774	4802	4833	4863	4894	4924	4955	4986	5016	5047	5077
1955	243 5108	5139	5167	5198	5228	5259	5289	5320	5351	5381	5412	5442
1956	5473	5504	5533	5564	5594	5625	5655	5686	5717	5747	5778	5808
1957	5839	5870	5898	5929	5959	5990	6020	6051	6082	6112	6143	6173
1958	6204	6235	6263	6294	6324	6355	6385	6416	6447	6477	6508	6538
1959	6569	6600	6628	6659	6689	6720	6750	6781	6812	6842	6873	6903
1960	243 6934	6965	6994	7025	7055	7086	7116	7147	7178	7208	7239	7269
1961	7300	7331	7359	7390	7420	7451	7481	7512	7543	7573	7604	7634
1962	7665	7696	7724	7755	7785	7816	7846	7877	7908	7938	7969	7999
1963	8030	8061	8089	8120	8150	8181	8211	8242	8273	8303	8334	8364
1964	8395	8426	8455	8486	8516	8547	8577	8608	8639	8669	8700	8730
1965	243 8761	8792	8820	8851	8881	8912	8942	8973	9004	9034	9065	9095
1966	9126	9157	9185	9216	9246	9277	9307	9338	9369	9399	9430	9460
1967	9491	9522	9550	9581	9611	9642	9672	9703	9734	9764	9795	9825
1968	9856	9887	9916	9947	9977	*0008	*0038	*0069	*0100	*0130	*0161	*0191
1969	244 0222	0253	0281	0312	0342	0373	0403	0434	0465	0495	0526	0556
1970	244 0587	0618	0646	0677	0707	0738	0768	0799	0830	0860	0891	0921
1971	0952	0983	1011	1042	1072	1103	1133	1164	1195	1225	1256	1286
1972	1317	1348	1377	1408	1438	1469	1499	1530	1561	1591	1622	1652
1973	1683	1714	1742	1773	1803	1834	1864	1895	1926	1956	1987	2017
1974	2048	2079	2107	2138	2168	2199	2229	2260	2291	2321	2352	2382
1975	244 2413	2444	2472	2503	2533	2564	2594	2625	2656	2686	2717	2747
1976	2778	2809	2838	2869	2899	2930	2960	2991	3022	3052	3083	3113
1977	3144	3175	3203	3234	3264	3295	3325	3356	3387	3417	3448	3478
1978	3509	3540	3568	3599	3629	3660	3690	3721	3752	3782	3813	3843
1979	3874	3905	3933	3964	3994	4025	4055	4086	4117	4147	4178	4208
1980	244 4239	4270	4299	4330	4360	4391	4421	4452	4483	4513	4544	4574
1981	4605	4636	4664	4695	4725	4756	4786	4817	4848	4878	4909	4939
1982	4970	5001	5029	5060	5090	5121	5151	5182	5213	5243	5274	5304
1983	5335	5366	5394	5425	5455	5486	5516	5547	5578	5608	5639	5669
1984	5700	5731	5760	5791	5821	5852	5882	5913	5944	5974	6005	6035
1985	244 6066	6097	6125	6156	6186	6217	6247	6278	6309	6339	6370	6400
1986	6431	6462	6490	6521	6551	6582	6612	6643	6674	6704	6735	6765
1987	6796	6827	6855	6886	6916	6947	6977	7008	7039	7069	7100	7130
1988	7161	7192	7221	7252	7282	7313	7343	7374	7405	7435	7466	7496
1989	7527	7558	7586	7617	7647	7678	7708	7739	7770	7800	7831	7861
1990	244 7892	7923	7951	7982	8012	8043	8073	8104	8135	8165	8196	8226
1991	8257	8288	8316	8347	8377	8408	8438	8469	8500	8530	8561	8591
1992	8622	8653	8682	8713	8743	8774	8804	8835	8866	8896	8927	8957
1993	8988	9019	9047	9078	9108	9139	9169	9200	9231	9261	9292	9322
1994	9353	9384	9412	9443	9473	9504	9534	9565	9596	9626	9657	9687
1995	244 9718	9749	9777	9808	9838	9869	9899	9930	9961	9991	*0022	*0052
1996	245 0083	0114	0143	0174	0204	0235	0265	0296	0327	0357	0388	0418
1997	0449	0480	0508	0539	0569	0600	0630	0661	0692	0722	0753	0783
1998	0814	0845	0873	0904	0934	0965	0995	1026	1057	1087	1118	1148
1999	1179	1210	1238	1269	1299	1330	1360	1391	1422	1452	1483	1513
2000	245 1544	1575	1604	1635	1665	1696	1726	1757	1788	1818	1849	1879

OF DAY COMMENCING AT GREENWICH NOON ON:

Year	Jan. 0	Feb. 0	Mar. 0	Apr. 0	May 0	June 0	July 0	Aug. 0	Sept. 0	Oct. 0	Nov. 0	Dec. 0
2000	245 1544	1575	1604	1635	1665	1696	1726	1757	1788	1818	1849	1879
2001	1910	1941	1969	2000	2030	2061	2091	2122	2153	2183	2214	2244
2002	2275	2306	2334	2365	2395	2426	2456	2487	2518	2548	2579	2609
2003	2640	2671	2699	2730	2760	2791	2821	2852	2883	2913	2944	2974
2004	3005	3036	3065	3096	3126	3157	3187	3218	3249	3279	3310	3340
2005	245 3371	3402	3430	3461	3491	3522	3552	3583	3614	3644	3675	3705
2006	3736	3767	3795	3826	3856	3887	3917	3948	3979	4009	4040	4070
2007	4101	4132	4160	4191	4221	4252	4282	4313	4344	4374	4405	4435
2008	4466	4497	4526	4557	4587	4618	4648	4679	4710	4740	4771	4801
2009	4832	4863	4891	4922	4952	4983	5013	5044	5075	5105	5136	5166
2010	245 5197	5228	5256	5287	5317	5348	5378	5409	5440	5470	5501	5531
2011	5562	5593	5621	5652	5682	5713	5743	5774	5805	5835	5866	5896
2012	5927	5958	5987	6018	6048	6079	6109	6140	6171	6201	6232	6262
2013	6293	6324	6352	6383	6413	6444	6474	6505	6536	6566	6597	6627
2014	6658	6689	6717	6748	6778	6809	6839	6870	6901	6931	6962	6992
2015	245 7023	7054	7082	7113	7143	7174	7204	7235	7266	7296	7327	7357
2016	7388	7419	7448	7479	7509	7540	7570	7601	7632	7662	7693	7723
2017	7754	7785	7813	7844	7874	7905	7935	7966	7997	8027	8058	8088
2018	8119	8150	8178	8209	8239	8270	8300	8331	8362	8392	8423	8453
2019	8484	8515	8543	8574	8604	8635	8665	8696	8727	8757	8788	8818
2020	245 8849	8880	8909	8940	8970	9001	9031	9062	9093	9123	9154	9184
2021	9215	9246	9274	9305	9335	9366	9396	9427	9458	9488	9519	9549
2022	9580	9611	9639	9670	9700	9731	9761	9792	9823	9853	9884	9914
2023	9945	9976	*0004	*0035	*0065	*0096	*0126	*0157	*0188	*0218	*0249	*0279
2024	246 0310	0341	0370	0401	0431	0462	0492	0523	0554	0584	0615	0645
2025	246 0676	0707	0735	0766	0796	0827	0857	0888	0919	0949	0980	1010
2026	1041	1072	1100	1131	1161	1192	1222	1253	1284	1314	1345	1375
2027	1406	1437	1465	1496	1526	1557	1587	1618	1649	1679	1710	1740
2028	1771	1802	1831	1862	1892	1923	1953	1984	2015	2045	2076	2106
2029	2137	2168	2196	2227	2257	2288	2318	2349	2380	2410	2441	2471
2030	246 2502	2533	2561	2592	2622	2653	2683	2714	2745	2775	2806	2836
2031	2867	2898	2926	2957	2987	3018	3048	3079	3110	3140	3171	3201
2032	3232	3263	3292	3323	3353	3384	3414	3445	3476	3506	3537	3567
2033	3598	3629	3657	3688	3718	3749	3779	3810	3841	3871	3902	3932
2034	3963	3994	4022	4053	4083	4114	4144	4175	4206	4236	4267	4297
2035	246 4328	4359	4387	4418	4448	4479	4509	4540	4571	4601	4632	4662
2036	4693	4724	4753	4784	4814	4845	4875	4906	4937	4967	4998	5028
2037	5059	5090	5118	5149	5179	5210	5240	5271	5302	5332	5363	5393
2038	5424	5455	5483	5514	5544	5575	5605	5636	5667	5697	5728	5758
2039	5789	5820	5848	5879	5909	5940	5970	6001	6032	6062	6093	6123
2040	246 6154	6185	6214	6245	6275	6306	6336	6367	6398	6428	6459	6489
2041	6520	6551	6579	6610	6640	6671	6701	6732	6763	6793	6824	6854
2042	6885	6916	6944	6975	7005	7036	7066	7097	7128	7158	7189	7219
2043	7250	7281	7309	7340	7370	7401	7431	7462	7493	7523	7554	7584
2044	7615	7646	7675	7706	7736	7767	7797	7828	7859	7889	7920	7950
2045	246 7981	8012	8040	8071	8101	8132	8162	8193	8224	8254	8285	8315
2046	8346	8377	8405	8436	8466	8497	8527	8558	8589	8619	8650	8680
2047	8711	8742	8770	8801	8831	8862	8892	8923	8954	8984	9015	9045
2048	9076	9107	9136	9167	9197	9228	9258	9289	9320	9350	9381	9411
2049	9442	9473	9501	9532	9562	9593	9623	9654	9685	9715	9746	9776
2050	246 9807	9838	9866	9897	9927	9958	9988	*0019	*0050	*0080	*0111	*0141

K4 JULIAN DATES OF GREGORIAN CALENDAR DATES

The Julian date (JD) corresponding to any instant is the interval in mean solar days elapsed since 4713 BC January 1 at Greenwich mean noon (12^h UT). To determine the JD at 0^h UT for a given Gregorian calendar date, sum the values from Table A for century, Table B for year and Table C for month; then add the day of the month. Julian dates for the current year are given on page B4.

A. Julian date at January 0^d 0^h UT of centurial year

Year	1600†	1700	1800	1900	2000†	2100
Julian date	230 5447·5	234 1971·5	237 8495·5	241 5019·5	245 1544·5	248 8068·5

† Centurial years that are exactly divisible by 400 are leap years in the Gregorian calendar. To determine the JD for any date in such a year, subtract 1 from the JD in Table A and use the leap year portion of Table C. (For 1600 and 2000 the JDs tabulated in Table A are actually for January 1^d 0^h.)

B. Addition to give Julian date for January 0^d 0^h UT of year

Year	Add	Year	Add	Year	Add	Year	Add
0	0	25	9131	50	18262	75	27393
1	365	26	9496	51	18627	76*	27758
2	730	27	9861	52*	18992	77	28124
3	1095	28*	10226	53	19358	78	28489
4*	1460	29	10592	54	19723	79	28854
5	1826	30	10957	55	20088	80*	29219
6	2191	31	11322	56*	20453	81	29585
7	2556	32*	11687	57	20819	82	29950
8*	2921	33	12053	58	21184	83	30315
9	3287	34	12418	59	21549	84*	30680
10	3652	35	12783	60*	21914	85	31046
11	4017	36*	13148	61	22280	86	31411
12*	4382	37	13514	62	22645	87	31776
13	4748	38	13879	63	23010	88*	32141
14	5113	39	14244	64*	23375	89	32507
15	5478	40*	14609	65	23741	90	32872
16*	5843	41	14975	66	24106	91	33237
17	6209	42	15340	67	24471	92*	33602
18	6574	43	15705	68*	24836	93	33968
19	6939	44*	16070	69	25202	94	34333
20*	7304	45	16436	70	25567	95	34698
21	7670	46	16801	71	25932	96*	35063
22	8035	47	17166	72*	26297	97	35429
23	8400	48*	17531	73	26663	98	35794
24*	8765	49	17897	74	27028	99	36159

* Leap years

Examples

a. 1981 November 14

Table A	
1900 Jan. 0	241 5019·5
+ Table B	+ 2 9585
1981 Jan. 0	244 4604·5
+ Table C (n.y.)	+ 304
1981 Nov. 0	244 4908·5
+ Day of Month	+ 14
1981 Nov. 14	244 4922·5

b. 2000 September 24

Table A	
2000 Jan. 1	245 1544·5
− 1 (for 2000)	− 1
2000 Jan. 0	245 1543·5
+ Table B	+ 0
2000 Jan. 0	245 1543·5
+ Table C (l.y.)	+ 244
2000 Sept. 0	245 1787·5
+ Day of Month	+ 24
2000 Sept. 24	245 1811·5

c. 2001 June 21

Table A	
2000 Jan. 1	245 1544·5
+ Table B	+ 365
2001 Jan. 0	245 1909·5
+ Table C (n.y.)	+ 151
2001 June 0	245 2060·5
+ Day of Month	+ 21
2001 June 21	245 2081·5

C. Addition to give Julian date for beginning of month (0^d 0^h UT)

	Jan.	Feb.	Mar.	Apr.	May	June	July	Aug.	Sept.	Oct.	Nov.	Dec.
Normal year	0	31	59	90	120	151	181	212	243	273	304	334
Leap year	0	31	60	91	121	152	182	213	244	274	305	335

WARNING: prior to 1925 Greenwich mean noon (i.e. 12^h UT) was usually denoted by 0^h GMT in astronomical publications.

IAU (1964) System of Astronomical Constants

This system of constants was replaced for the 1984 edition of the *Astronomical Almanac* by the IAU (1976) System of Astronomical Constants given on pages K6 to K7.

Defining constants

Number of ephemeris seconds in one tropical year (1900)	$s = 31\ 556\ 925{\cdot}974\ 7$
Gaussian gravitational constant	$k = 0{\cdot}017\ 202\ 098\ 950\ 000$
	$= 3\ 548''{\cdot}187\ 606\ 965\ 1$

Primary constants

Astronomical unit	$149\ 600 \times 10^6$ m
Velocity of light	$299\ 792{\cdot}5 \times 10^3$ m / sec
Equatorial radius of the Earth	6 378 160 m
Dynamical form-factor for Earth	0·001 082 7
Geocentric gravitational constant	$398\ 603 \times 10^9\ m^3\ s^{-2}$
Mass ratio: Earth / Moon	81·30
General precession in longitude per tropical century (1900)	$5025''{\cdot}64$
Constant of nutation (1900)	$9''{\cdot}210$

Derived constants

Solar parallax	$8''{\cdot}794$
Light-time for unit distance	$499^s{\cdot}012$
Constant of aberration	$20''{\cdot}496$
Flattening factor for Earth	1 / 298·25
	= 0·003 352 89
Heliocentric gravitational constant	$132\ 718 \times 10^{15}\ m^3\ s^{-2}$
Mass ratio: Sun / Earth	332 958
Mass ratio: Sun / (Earth + Moon)	328 912
Mean distance of the Moon	$384\ 400 \times 10^3$ m
Constant of sine parallax for Moon	$3\ 422''{\cdot}451$

Constants related to the Figure of the Earth

Equatorial radius (primary)	$a = 6\ 378\ 160$ m
Polar radius	$a\,(1 - f) = 6\ 356\ 774{\cdot}7$ m
Square of eccentricity	$e^2 = 0{\cdot}006\ 694\ 54$

Reduction from geodetic latitude ϕ to geocentric latitude ϕ'

$$\phi' - \phi = -11'\ 32''{\cdot}743\ 0 \sin 2\phi + 1''{\cdot}163\ 3 \sin 4\phi - 0''{\cdot}002\ 6 \sin 6\phi$$

Radius vector

$$\rho = a\,(0{\cdot}998\ 327\ 073 + 0{\cdot}001\ 676\ 438 \cos 2\phi - 0{\cdot}000\ 003\ 519 \cos 4\phi + 0{\cdot}000\ 000\ 008 \cos 6\phi)$$

One degree of latitude (m)

$$111\ 133{\cdot}35 - 559{\cdot}84 \cos 2\phi + 1{\cdot}17 \cos 4\phi \quad (\phi = \text{mid-latitude of arc})$$

One degree of longitude (m)

$$111\ 413{\cdot}28 \cos \phi - 93{\cdot}51 \cos 3\phi + 0{\cdot}12 \cos 5\phi$$

The complete system of astronomical constants is given in *Supplement to the A.E. 1968* (pages 4s–7s).

Old Constants

The IAU (1964) system was introduced into the planetary ephemerides in 1968 except that those for the Sun and inner planets continued to be based on the following values of the constants that were in use immediately prior to the introduction of the IAU (1964) System.

Solar parallax	$8''{\cdot}80$
Light-time for unit distance	$498^s{\cdot}38$
Constant of aberration	$20''{\cdot}47$
Mass ratio	
Sun / (Earth + Moon)	329 390
Earth / Moon (planetary theory)	81·45

IAU (1976) System of Astronomical Constants

Units:

The units meter (m), kilogram (kg), and second (s) are the units of length, mass, and time in the International System of Units (SI).

The astronomical unit of time is a time interval of one day (D) of 86400 seconds. An interval of 36525 days is one Julian century.

The astronomical unit of mass is the mass of the Sun (S).

The astronomical unit of length is that length (A) for which the Gaussian gravitational constant (k) takes the value 0·017 202 098 95 when the units of measurement are the astronomical units of length, mass, and time. The dimensions of k^2 are those of the constant of gravitation (G), i.e., $L^3 M^{-1} T^{-2}$. The term "unit distance" is also used for the length A.

In the preparation of the ephemerides and the fitting of the ephemerides to all the observational data available, it was necessary to modify some of the constants and planetary masses. The modified values of the constants are indicated in brackets following the (1976) System values.

Defining constants:

1. Gaussian gravitational constant	$k = 0{\cdot}017\ 202\ 098\ 95$
2. Speed of light	$c = 299\ 792\ 458$ m s^{-1}

Primary constants:

3. Light-time for unit distance	$\tau_A = 499{\cdot}004\ 782$ s [499·004 7837 . . .]
4. Equatorial radius for Earth	$a_e = 6378\ 140$ m
[IUGG value	$a_e = 6378\ 137$ m]
5. Dynamical form-factor for Earth	$J_2 = 0{\cdot}001\ 082\ 63$
6. Geocentric gravitational constant	$GE = 3{\cdot}986\ 005 \times 10^{14}$ m^3 s^{-2} [$3{\cdot}986\ 004\ 48 \ldots \times 10^{14}$]
7. Constant of gravitation	$G = 6{\cdot}672 \times 10^{-11}$ m^3 kg^{-1} s^{-2}
8. Ratio of mass of Moon to that of Earth	$\mu = 0{\cdot}012\ 300\ 02$ [0·012 300 034]
9. General precession in longitude, per Julian century, at standard epoch 2000	$\rho = 5029''{\cdot}0966$
10. Obliquity of the ecliptic, at standard epoch 2000	$\varepsilon = 23° 26' 21''{\cdot}448$ [23° 26′ 21″·4119]

Derived constants:

11. Constant of nutation, at standard epoch 2000	$N = 9''{\cdot}2025$
12. Unit distance	$c\tau_A = A = 1{\cdot}495\ 978\ 70 \times 10^{11}$ m [$1{\cdot}495\ 978\ 706\ 6 \times 10^{11}$]
13. Solar parallax	$\arcsin(a_e/A) = \pi_\odot = 8''{\cdot}794\ 148$
14. Constant of aberration, for standard epoch 2000	$\kappa = 20''{\cdot}49\ 552$
15. Flattening factor for the Earth	$f = 0{\cdot}003\ 352\ 81$ $= 1/298{\cdot}257$
16. Heliocentric gravitational constant	$A^3k^2/D^2 = GS = 1{\cdot}327\ 124\ 38 \times 10^{20}$ m^3 s^{-2} [$1{\cdot}327\ 124\ 40 \ldots \times 10^{20}$]
17. Ratio of mass of Sun to that of the Earth	$(GS)/(GE) = S/E = 332\ 946{\cdot}0$ [332 946·038 . . .]
18. Ratio of mass of Sun to that of Earth + Moon	$(S/E)/(1+\mu) = 328\ 900{\cdot}5$ [328 900·55]
19. Mass of the Sun	$(GS)/G = S = 1{\cdot}9891 \times 10^{30}$ kg

IAU (1976) System of Astronomical Constants (continued)

20. System of planetary masses

Ratios of mass of Sun to masses of the planets

Mercury	6 023 600	Jupiter	1 047·355 [1 047·350]
Venus	408 523·5	Saturn	3 498·5 [3 498·0]
Earth + Moon	328 900·5	Uranus	22 869 [22 960]
Mars	3 098 710	Neptune	19 314
		Pluto	3 000 000 [130 000 000]

Other Quantities for Use in the Preparation of Ephemerides

It is recommended that the values given in the following list should normally be used in the preparation of new ephemerides.

21. Masses of minor planets

Minor planet	Mass in solar mass
(1) Ceres	$5{\cdot}9 \times 10^{-10}$
(2) Pallas	1.1×10^{-10} [$1{\cdot}081 \times 10^{-10}$]
(4) Vesta	$1{\cdot}2 \times 10^{-10}$ [$1{\cdot}379 \times 10^{-10}$]

22. Masses of satellites

Planet	Satellite	Satellite / Planet
Jupiter	Io	$4{\cdot}70 \times 10^{-5}$
	Europa	$2{\cdot}56 \times 10^{-5}$
	Ganymede	$7{\cdot}84 \times 10^{-5}$
	Callisto	$5{\cdot}6 \times 10^{-5}$
Saturn	Titan	$2{\cdot}41 \times 10^{-4}$
Neptune	Triton	2×10^{-3}

23. Equatorial radii in km

Mercury	2 439	Jupiter	71 398	Pluto	2 500
Venus	6 052	Saturn	60 000		
Earth	6 378·140	Uranus	25 400	Moon	1 738
Mars	3 397·2	Neptune	24 300	Sun	696 000

24. Gravity fields of planets

Planet	J_2	J_3	J_4
Earth	+0·001 082 63	$-0{\cdot}254 \times 10^{-5}$	$-0{\cdot}161 \times 10^{-5}$
Mars	+0·001 964	$+0{\cdot}36 \times 10^{-4}$	
Jupiter	+0·014 75		$-0{\cdot}58 \times 10^{-3}$
Saturn	+0·016 45		$-0{\cdot}10 \times 10^{-2}$
Uranus	+0·012		
Neptune	+0·004		

(Mars: $C_{22} = -0{\cdot}000\,055$, $S_{22} = +0{\cdot}000\,031$, $S_{31} = +0{\cdot}000\,026$)

25. Gravity field of the Moon

$\gamma = (B - A)/C = 0{\cdot}000\,2278$ $\qquad C/MR^2 = 0{\cdot}392$

$\beta = (C - A)/B = 0{\cdot}000\,6313$ $\qquad I = 5552''{\cdot}7 = 1° 32' 32''{\cdot}7$

$C_{20} = -0{\cdot}000\,2027$	$C_{30} = -0{\cdot}000\,006$	$C_{32} = +0{\cdot}000\,0048$
$C_{22} = +0{\cdot}000\,0223$	$C_{31} = +0{\cdot}000\,029$	$S_{32} = +0{\cdot}000\,0017$
	$S_{31} = +0{\cdot}000\,004$	$C_{33} = +0{\cdot}000\,0018$
		$S_{33} = -0{\cdot}000\,001$

$\Delta T = \mathrm{ET} - \mathrm{UT}$

Year	ΔT	Year	ΔT	Year	ΔT	Year	ΔT	Year	ΔT
	s		s		s		s		s
1620·0	+124	1660·0	+37	1700·0	+ 9	1740·0	+12	1780·0	+17
1621	119	1661	36	1701	9	1741	12	1781	17
1622	115	1662	35	1702	9	1742	12	1782	17
1623	110	1663	34	1703	9	1743	12	1783	17
1624	106	1664	33	1704	9	1744	13	1784	17
1625·0	+102	1665·0	+32	1705·0	+ 9	1745·0	+13	1785·0	+17
1626	98	1666	31	1706	9	1746	13	1786	17
1627	95	1667	30	1707	9	1747	13	1787	17
1628	91	1668	28	1708	10	1748	13	1788	17
1629	88	1669	27	1709	10	1749	13	1789	17
1630·0	+ 85	1670·0	+26	1710·0	+10	1750·0	+13	1790·0	+17
1631	82	1671	25	1711	10	1751	14	1791	17
1632	79	1672	24	1712	10	1752	14	1792	16
1633	77	1673	23	1713	10	1753	14	1793	16
1634	74	1674	22	1714	10	1754	14	1794	16
1635·0	+ 72	1675·0	+21	1715·0	+10	1755·0	+14	1795·0	+16
1636	70	1676	20	1716	10	1756	14	1796	15
1637	67	1677	19	1717	11	1757	14	1797	15
1638	65	1678	18	1718	11	1758	15	1798	14
1639	63	1679	17	1719	11	1759	15	1799	14
1640·0	+ 62	1680·0	+16	1720·0	+11	1760·0	+15	1800·0	+13·7
1641	60	1681	15	1721	11	1761	15	1801	13·4
1642	58	1682	14	1722	11	1762	15	1802	13·1
1643	57	1683	14	1723	11	1763	15	1803	12·9
1644	55	1684	13	1724	11	1764	15	1804	12·7
1645·0	+ 54	1685·0	+12	1725·0	+11	1765·0	+16	1805·0	+12·6
1646	53	1686	12	1726	11	1766	16	1806	12·5
1647	51	1687	11	1727	11	1767	16	1807	12·5
1648	50	1688	11	1728	11	1768	16	1808	12·5
1649	49	1689	10	1729	11	1769	16	1809	12·5
1650·0	+ 48	1690·0	+10	1730·0	+11	1770·0	+16	1810·0	+12·5
1651	47	1691	10	1731	11	1771	16	1811	12·5
1652	46	1692	9	1732	11	1772	16	1812	12·5
1653	45	1693	9	1733	11	1773	16	1813	12·5
1654	44	1694	9	1734	12	1774	16	1814	12·5
1655·0	+ 43	1695·0	+ 9	1735·0	+12	1775·0	+17	1815·0	+12·5
1656	42	1696	9	1736	12	1776	17	1816	12·5
1657	41	1697	9	1737	12	1777	17	1817	12·4
1658	40	1698	9	1738	12	1778	17	1818	12·3
1659·0	+ 38	1699·0	+ 9	1739·0	+12	1779·0	+17	1819·0	+12·2

This table is based on an adopted value of $-26''/\mathrm{cy}^2$ for the tidal term ($\dot{n}$) in the mean motion of the Moon from the results of analyses of observations of lunar occultations of stars, eclipses of the Sun, and transits of Mercury. (See F. R. Stephenson and L. V. Morrison, 1984, *Phil. Trans. R. Soc. London,* Ser. A, **313**, 47–70).

To calculate the values of ΔT for a different value of the tidal term ($\dot{n}'$), add

$$-0{\cdot}000\,091\,(\dot{n}' + 26)\,(\mathrm{year} - 1955)^2 \text{ seconds}$$

to the tabulated values of ΔT.

1820–1983, ΔT = ET − UT. FROM 1984, ΔT = TDT − UT.

Year	ΔT	Year	ΔT	Year	ΔT	Year	ΔT	Year	ΔT
	s		s		s		s		s
1820·0	+12·0	1860·0	+7·88	1900·0	− 2·72	1940·0	+24·33	1980·0	+50·54
1821	11·7	1861	7·82	1901	1·54	1941	24·83	1981	51·38
1822	11·4	1862	7·54	1902	− 0·02	1942	25·30	1982	52·17
1823	11·1	1863	6·97	1903	+ 1·24	1943	25·70	1983	52·96
1824	10·6	1864	6·40	1904	2·64	1944	26·24	1984	53·79
1825·0	+10·2	1865·0	+6·02	1905·0	+ 3·86	1945·0	+26·77	1985·0	+54·34
1826	9·6	1866	5·41	1906	5·37	1946	27·28	1986	54·87
1827	9·1	1867	4·10	1907	6·14	1947	27·78	1987	55·32
1828	8·6	1868	2·92	1908	7·75	1948	28·25	1988	55·82
1829	8·0	1869	1·82	1909	9·13	1949	28·71	1989	56·30
1830·0	+ 7·5	1870·0	+1·61	1910·0	+10·46	1950·0	+29·15	1990	+56·86
1831	7·0	1871	+0·10	1911	11·53	1951	29·57		
1832	6·6	1872	−1·02	1912	13·36	1952	29·97		
1833	6·3	1873	1·28	1913	14·65	1953	30·36		
1834	6·0	1874	2·69	1914	16·01	1954	30·72	Extrapolated	
1835·0	+ 5·8	1875·0	−3·24	1915·0	+17·20	1955·0	+31·07	1991·0	+57·4
1836	5·7	1876	3·64	1916	18·24	1956	31·35	1992	58·0
1837	5·6	1877	4·54	1917	19·06	1957	31·68	1993	58·6
1838	5·6	1878	4·71	1918	20·25	1958	32·18		
1839	5·6	1879	5·11	1919	20·95	1959	32·68		
1840·0	+ 5·7	1880·0	−5·40	1920·0	+21·16	1960·0	+33·15		
1841	5·8	1881	5·42	1921	22·25	1961	33·59		
1842	5·9	1882	5·20	1922	22·41	1962	34·00		
1843	6·1	1883	5·46	1923	23·03	1963	34·47		
1844	6·2	1884	5·46	1924	23·49	1964	35·03		
1845·0	+ 6·3	1885·0	−5·79	1925·0	+23·62	1965·0	+35·73		
1846	6·5	1886	5·63	1926	23·86	1966	36·54		
1847	6·6	1887	5·64	1927	24·49	1967	37·43		
1848	6·8	1888	5·80	1928	24·34	1968	38·29		
1849	6·9	1889	5·66	1929	24·08	1969	39·20		
1850·0	+ 7·1	1890·0	−5·87	1930·0	+24·02	1970·0	+40·18		
1851	7·2	1891	6·01	1931	24·00	1971	41·17		
1852	7·3	1892	6·19	1932	23·87	1972	42·23		
1853	7·4	1893	6·64	1933	23·95	1973	43·37		
1854	7·5	1894	6·44	1934	23·86	1974	44·49		
1855·0	+ 7·6	1895·0	−6·47	1935·0	+23·93	1975·0	+45·48		
1856	7·7	1896	6·09	1936	23·73	1976	46·46		
1857	7·7	1897	5·76	1937	23·92	1977	47·52		
1858	7·8	1898	4·66	1938	23·96	1978	48·53		
1859·0	+ 7·8	1899·0	−3·74	1939·0	+24·02	1979·0	+49·59		

Difference TAI–UTC

Date	ΔAT
	s
1972 Jan. 1	+10·00
1972 July 1	+11·00
1973 Jan. 1	+12·00
1974 Jan. 1	+13·00
1975 Jan. 1	+14·00
1976 Jan. 1	+15·00
1977 Jan. 1	+16·00
1978 Jan. 1	+17·00
1979 Jan. 1	+18·00
1980 Jan. 1	+19·00
1981 July 1	+20·00
1982 July 1	+21·00
1983 July 1	+22·00
1985 July 1	+23·00
1988 Jan. 1	+24·00
1990 Jan. 1	+25·00

In critical cases descend

$$\left.\begin{matrix}\Delta ET\\ \Delta TT\end{matrix}\right\} = \Delta AT + 32^{s}{\cdot}184$$

See page B4 for a summary of the notation for time-scales.

1979 BIH SYSTEM

Date	*x*	*y*	Date	*x*	*y*	Date	*x*	*y*
1970	″	″	1978	″	″	1986	″	″
Jan. 1	−0·140	+0·144	Jan. 1	+0·007	+0·015	Jan. 1	+0·187	+0·072
Apr. 1	−0·097	+0·397	Apr. 1	−0·231	+0·240	Apr. 1	−0·041	+0·139
July 1	+0·139	+0·405	July 1	−0·042	+0·483	July 1	−0·075	+0·324
Oct. 1	+0·174	+0·125	Oct. 1	+0·236	+0·353	Oct. 1	+0·062	+0·395
1971			1979			1987		
Jan. 1	−0·081	+0·026	Jan. 1	+0·140	+0·076	Jan. 1	+0·146	+0·315
Apr. 1	−0·199	+0·313	Apr. 1	−0·107	+0·133	Apr. 1	+0·096	+0·212
July 1	+0·050	+0·523	July 1	−0·117	+0·351	July 1	−0·003	+0·208
Oct. 1	+0·249	+0·263	Oct. 1	+0·092	+0·408	Oct. 1	−0·053	+0·295
1972			1980			1988		
Jan. 1	+0·045	+0·050	Jan. 1	+0·129	+0·251	Jan. 1	−0·023	+0·414
Apr. 1	−0·180	+0·174	Apr. 1	+0·014	+0·189	Apr. 1	+0·134	+0·407
July 1	−0·031	+0·409	July 1	−0·044	+0·280	July 1	+0·171	+0·253
Oct. 1	+0·142	+0·344	Oct. 1	−0·006	+0·338	Oct. 1	+0·012	+0·132
1973			1981			1989		
Jan. 1	+0·129	+0·139	Jan. 1	+0·056	+0·361	Jan. 1	−0·159	+0·316
Apr. 1	−0·035	+0·129	Apr. 1	+0·088	+0·285	Apr. 1	+0·028	+0·482
July 1	−0·075	+0·286	July 1	+0·075	+0·209	July 1	+0·238	+0·368
Oct. 1	+0·035	+0·347	Oct. 1	−0·045	+0·210	Oct. 1	+0·167	+0·106
1974			1982					
Jan. 1	+0·115	+0·252	Jan. 1	−0·091	+0·378			
Apr. 1	+0·037	+0·185	Apr. 1	+0·093	+0·431			
July 1	+0·014	+0·216	July 1	+0·231	+0·239			
Oct. 1	+0·002	+0·225	Oct. 1	+0·036	+0·060			
1975			1983					
Jan. 1	−0·055	+0·281	Jan. 1	−0·211	+0·249			
Apr. 1	+0·027	+0·344	Apr. 1	−0·069	+0·538			
July 1	+0·151	+0·249	July 1	+0·269	+0·436			
Oct. 1	+0·063	+0·115	Oct. 1	+0·235	+0·069			
1976			1984					
Jan. 1	−0·145	+0·204	Jan. 1	−0·125	+0·089			
Apr. 1	−0·091	+0·399	Apr. 1	−0·211	+0·410			
July 1	+0·159	+0·390	July 1	+0·119	+0·543			
Oct. 1	+0·227	+0·158	Oct. 1	+0·313	+0·246			
1977			1985					
Jan. 1	−0·065	+0·076	Jan. 1	+0·051	+0·025			
Apr. 1	−0·226	+0·362	Apr. 1	−0·196	+0·220			
July 1	+0·085	+0·500	July 1	−0·044	+0·482			
Oct. 1	+0·281	+0·230	Oct. 1	+0·214	+0·404			

The angles *x*, *y*, are defined on page B60.
From 1988 the values of *x* and *y* have been taken from the IERS Bulletin B, published by the Bureau Central de L'IERS, Observatoire de Paris, 61 Avenue de l'Observatoire, 75014 Paris, France.

Introduction

In the reduction of astrometric observations of high precision it is necessary to distinguish between several different systems of terrestrial coordinates that are used to specify the positions of points on or near the surface of the Earth. The formulae on page B60 for the reduction for polar motion give the relationships between the representations of a geocentric vector referred to the celestial reference frame of the true equator and equinox of date and to the current conventional terrestrial reference frame. The pole and prime meridian of this terrestrial reference frame are the same as those of the current international geodetic reference system (IUGG, 1980), which is based on a spheroid whose size, shape and gravity field are specified by a set of adopted parameters, and whose centre coincides with the centre of mass of the Earth. (The term "spheroid" is here used in the sense of an ellipsoid whose equatorial section is a circle and for which each meridional section is an ellipse.) A spheroid provides a suitable approximation to mean sea level (i.e. to the geoid) and so is used as a reference surface for the specification of geodetic coordinates. Many different spheroids are in use for regional mapping, but in general the centres of these spheroids do not coincide with the centre of mass of the Earth. A few have been used for global purposes, and the associated gravity fields for which the spheroid is an equipotential surface have also been specified. The geoid may, however, differ from a global reference spheroid by as much as 100 m in some regions. The parameters for a selection of reference spheroids are given in the table on page K13.

Reduction from geodetic to geocentric coordinates

The position of a point relative to a terrestrial reference frame may be expressed in three ways:

(i) geocentric equatorial rectangular coordinates, x, y, z;
(ii) geocentric longitude, latitude and radius, λ, ϕ', ρ;
(iii) geodetic longitude, latitude and height, λ, ϕ, h.

The geodetic and geocentric longitudes of a point are the same, while the relationship between the geodetic and geocentric latitudes of a point is illustrated in the figure on page K13, which represents a meridional section through the reference spheroid. The geocentric radius ρ is usually expressed in units of the equatorial radius of the reference spheroid. The following relationships hold between the geocentric and geodetic coordinates:

$$x = a\rho \cos\phi' \cos\lambda = (aC + h)\cos\phi \cos\lambda$$
$$y = a\rho \cos\phi' \sin\lambda = (aC + h)\cos\phi \sin\lambda$$
$$z = a\rho \sin\phi' = (aS + h)\sin\phi$$

where a is the equatorial radius of the spheroid and C and S are auxiliary functions that depend on the geodetic latitude and on the flattening f of the reference spheroid. The polar radius b and the eccentricity of e of the ellipse are given by:

$$b = a(1 - f) \qquad e^2 = 2f - f^2 \qquad \text{or} \qquad 1 - e^2 = (1 - f)^2$$

It follows from the geometrical properties of the ellipse that:

$$C = \{\cos^2\phi + (1 - f)^2 \sin^2\phi\}^{-1/2} \qquad S = (1 - f)^2 C$$

Geocentric coordinates may be calculated directly from geodetic coordinates but an iterative procedure must be used for the inverse calculation. Series expressions and tables are available for certain values of f for the calculation of C and S and also of ρ and $\phi - \phi'$ for points on the spheroid ($h = 0$). The quantity $\phi - \phi'$ is sometimes known as the "reduction of the latitude" or the "angle of the vertical", and it is of the order of 10′ in mid-latitudes. To a first approximation when h is small the geocentric radius is increased by h/a and the angle of the vertical is unchanged. The height h refers to a height above the reference spheroid and differs from the height above mean sea level (i.e. above the geoid) by the "undulation of the geoid" at the point.

Reduction from geodetic to geocentric coordinates (continued)

An iterative procedure for calculating λ, ϕ, h from x, y, z is as follows:

Calculate: $\lambda = \tan^{-1}(y/x)$ $\quad r = (x^2 + y^2)^{1/2}$ $\quad e^2 = 2f - f^2$

Calculate the first approximation to ϕ from: $\phi = \tan^{-1}(z/r)$

Then perform the following iteration until ϕ is unchanged to the required precision:

$$\phi_1 = \phi \qquad C = (1 - e^2 \sin^2 \phi_1)^{-1/2} \qquad \phi = \tan^{-1}((z + aCe^2 \sin \phi_1)/r)$$

Then: $h = r/\cos\phi - aC$

Other geodetic reference systems

In practice, most geodetic positions are referred either (a) to a regional geodetic datum that is represented by a spheroid that approximates to the geoid in the region considered or (b) to a global reference system that is defined for a particular system of measurement (e.g. a satellite navigation system). Data for the reduction of such geodetic coordinates to the conventional reference frame are not readily available, but it is hoped that the following notes, formulae and data will be useful and that more detailed and more up-to-date information will be included in future volumes.

(a) Each regional geodetic datum is specified by the size and shape of an adopted spheroid and by the coordinates of an "origin point". The principal axis of the spheroid is generally close to the mean axis of rotation of the Earth, but the centre of the spheroid may not coincide with the centre of mass of the Earth. The offset is usually represented by the geocentric rectangular coordinates (x_0, y_0, z_0) of the centre of the regional spheroid. The reduction from regional geodetic coordinates (λ, ϕ, h) to geocentric rectangular coordinates referred to the conventional reference frame (and hence to geodetic coordinates relative to a reference spheroid) may then be made by using the expressions:

$$x = x_0 + (aC + h)\cos\phi\cos\lambda$$
$$y = y_0 + (aC + h)\cos\phi\sin\lambda$$
$$z = z_0 + (aS + h)\sin\phi$$

Current estimates of x_0, y_0, z_0 are given, to an accuracy of about 10 m, for some regional datums in the lower table on page K13.

(b) The global reference systems for the various space techniques of measurement differ slightly, although all give good approximations (in some cases to within 1 m) to the conventional reference frame. The transformations from one system to another involve translation, rotation and scaling (i.e. 7 parameters in all) and are still the subject of specialist study.

Astronomical coordinates

Many astrometric observations that are used in the determination of the terrestrial coordinates of the point of observation use the local vertical, which defines the zenith, as a principal reference axis; the coordinates so obtained are called "astronomical coordinates". The local vertical is in the direction of the vector sum of the acceleration due to the gravitational field of the Earth and of the apparent acceleration due to the rotation of the Earth on its axis. The vertical is normal to the equipotential (or level) surface at the point, but it is inclined to the normal to the geodetic reference spheroid; the angle of inclination is known as the "deflection of the vertical".

The astronomical coordinates of an observatory may differ significantly (e.g. by as much as 1′) from its geodetic coordinates, which are required for the determination of the geocentric coordinates of the observatory for use in computing, for example, parallax corrections for solar system observations. The size and direction of the deflection may be estimated by studying the gravity field in the region concerned. The deflection may affect both the latitude and the longitude, and hence local time. Astronomical coordinates also vary with time because they are affected by polar motion (see page B60).

GEODETIC REFERENCE SPHEROIDS

Name and Date	Equatorial Radius, a m	Reciprocal of Flattening, $1/f$	Gravitational Constant, GM 10^{14} m^3 s^{-2}	Dynamical Form Factor, J_2	Ang. Velocity of Earth, ω 10^{-5} rad s^{-1}
MERIT 1983	637 8137	298·257	—	—	—
GRS 80 (IUGG, 1980)	8137	298·257 222	3·986 005	0·001 082 63	7·292 115
IAU 1976	8140	298·257	3·986 005	0·001 082 63	—
South American 1969	8160	298·25	—	—	—
GRS 67 (IUGG, 1967)	8160	298·247 167	3·986 03	0·001 082 7	7·292 115 146 7
Australian National 1965	8160	298·25	—	—	—
IAU 1964	8160	298·25	3·986 03	0·001 082 7	7·292 1
Krassovski 1942	8245	298·3	—	—	—
International 1924 (Hayford)	8388	297	—	—	—
Clarke 1880 mod.	8249·145	293·466 3	—	—	—
Clarke 1866	8206·4	294·978 698	—	—	—
Bessel 1841	7397·155	299·152 813	—	—	—
Everest 1830	7276·345	300·801 7	—	—	—
Airy 1830	7563·396	299·324 964	—	—	—

IUGG, 1980. See H. Moritz, Geodetic Reference System 1980, *Bull. Géodésique*, **58** (3), 388–398, 1984.

REGIONAL GEODETIC DATUMS

Geodetic Datum	Spheroid	Origin	Latitude ° ′ ″	Longitude ° ′ ″	Centre Offset x_0 m	y_0 m	z_0 m
New Arc 1950	Clarke 1880 mod.	Buffelsfontein	−33 59 32·000	25 30 44·622		—	
Australian Geodetic 1966	Australian National 1965	Johnston Memorial Cairn	−25 56 54·55	133 12 30·08	−122	−43	138
European 1950	International 1924	Helmert Tower	52 22 51·45	13 03 58·74	−84	−105	−126
Indian 1938	Everest 1830	Kalianpur	24 07 11·26	77 39 17·57		—	
North American 1927	Clarke 1866	Meades Ranch	39 13 26·686	261 27 29·494	−22	158	176
Ordnance Survey GB SN80	Airy 1830	Herstmonceux	50 51 55·271	00 20 45·882	372	−127	433
Pico de las Nieves (Canaries)	International 1924	Pico de las Nieves	27 57 41·273	344 25 49·476		—	
Potsdam	Bessel 1841	Helmert Tower	52 22 53·954	13 04 01·153		—	
Pulkovo 1942	Krassovski 1942	Pulkovo Obs.	59 46 18·55	30 19 42·09		—	
South American 1969	S.American 1969	Chua	−19 45 41·653	311 53 55·936	−75	5	−43
Tokyo	Bessel 1841	Tokyo Obs. (old)	35 39 17·51	139 44 40·50	−143	514	675

GEODETIC AND GEOCENTRIC COORDINATES

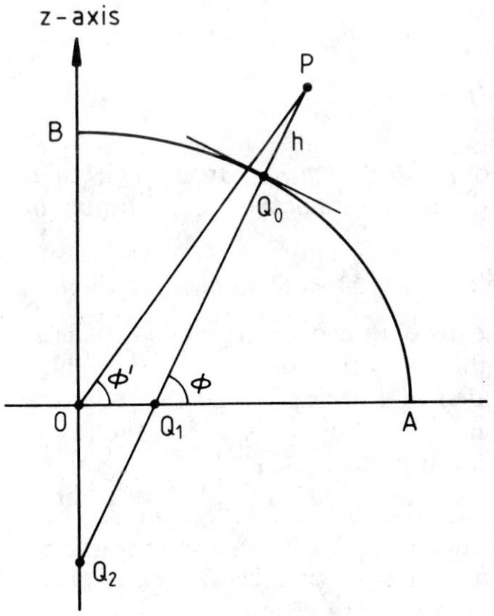

O is centre of Earth

OA = equatorial radius, a

OB = polar radius, $b = a(1-f)$

OP = geocentric radius, ap

PQ_0 is normal to the reference spheroid

$Q_0Q_1 = aS$

$Q_0Q_2 = aC$

ϕ = geodetic latitude

ϕ' = geocentric latitude

INTRODUCTION AND NOTATION

The interpolation methods described in this section, together with the accompanying tables, are usually sufficient to interpolate to full precision the ephemerides in this volume. Additional notes, formulae and tables are given in the booklets *Interpolation and Allied Tables* and *Subtabulation* (see p. ix) and in many textbooks on numerical analysis. It is recommended that interpolated values of the Moon's right ascension, declination and horizontal parallax are derived from the daily polynomial coefficients that are provided for this purpose on pages D23–D45.

f_p denotes the value of the function $f(t)$ at the time $t = t_0 + ph$, where h is the interval of tabulation, t_0 is a tabular argument, and $p = (t - t_0)/h$ is known as the interpolating factor. The notation for the differences of the tabular values is shown in the following table; it is derived from the use of the central-difference operator δ, which is defined by:

$$\delta f_p = f_{p+1/2} - f_{p-1/2}$$

The symbol for the function is usually omitted in the notation for the differences. Tables are given for use with Bessel's interpolation formula for p in the range 0 to +1. The differences may be expressed in terms of function values for convenience in the use of programmable calculators or computers.

Arg.	Function	Differences 1st	2nd	3rd	4th
t_{-2}	f_{-2}		δ^2_{-2}		
		$\delta_{-3/2}$		$\delta^3_{-3/2}$	
t_{-1}	f_{-1}		δ^2_{-1}		δ^4_{-1}
		$\delta_{-1/2}$		$\delta^3_{-1/2}$	
t_0	f_0		δ^2_0		δ^4_0
		$\delta_{1/2}$		$\delta^3_{1/2}$	
t_{+1}	f_{+1}		δ^2_1		δ^4_1
		$\delta_{3/2}$		$\delta^3_{3/2}$	
t_{+2}	f_{+2}		δ^2_2		

$$\delta_{1/2} = f_1 - f_0$$
$$\delta^2_0 = \delta_{1/2} - \delta_{-1/2} = f_1 - 2f_0 + f_{-1}$$
$$\delta^2_0 + \delta^2_1 = f_2 - f_1 - f_0 + f_{-1}$$
$$\delta^3_{1/2} = \delta^2_1 - \delta^2_0 = f_2 - 3f_1 + 3f_0 - f_{-1}$$
$$\delta^4_0 = \delta^3_{1/2} - \delta^3_{-1/2} = f_2 - 4f_1 + 6f_0 - 4f_{-1} + f_{-2}$$
$$\delta^4_0 + \delta^4_1 = f_3 - 3f_2 + 2f_1 + 2f_0 - 3f_{-1} + f_{-2}$$

$$p \equiv \text{the interpolating factor} = (t - t_0)/(t_1 - t_0) = (t - t_0)/h$$

BESSEL'S INTERPOLATION FORMULA

In this notation Bessel's interpolation formula is:

$$f_p = f_0 + p\,\delta_{1/2} + B_2\,(\delta^2_0 + \delta^2_1) + B_3\,\delta^3_{1/2} + B_4\,(\delta^4_0 + \delta^4_1) + \cdots$$

where
$$B_2 = p\,(p-1)/4 \qquad B_3 = p\,(p-1)\,(p-\tfrac{1}{2})/6$$
$$B_4 = (p+1)\,p\,(p-1)\,(p-2)/48$$

The maximum contribution to the truncation error of f_p, for $0 < p < 1$, from neglecting each order of difference is less than 0·5 in the unit of the end figure of the tabular function if

$$\delta^2 < 4 \qquad \delta^3 < 60 \qquad \delta^4 < 20 \qquad \delta^5 < 500.$$

The critical table of B_2 opposite provides a rapid means of interpolating when δ^2 is less than 500 and higher-order differences are negligible or when full precision is not required. The interpolating factor p should be rounded to 4 decimals, and the required value of B_2 is then the tabular value opposite the interval in which p lies, or it is the value above and to the right of p if p exactly equals a tabular argument. B_2 is always negative. The effects of the third and fourth differences can be estimated from the values of B_3 and B_4 given in the last column.

INVERSE INTERPOLATION

Inverse interpolation to derive the interpolating factor p, and hence the time, for which the function takes a specified value f_p is carried out by successive approximations. The first estimate p_1 is obtained from:

$$p_1 = (f_p - f_0) / \delta_{1/2}$$

This value of p is used to obtain an estimate of B_2, from the critical table or otherwise, and hence an improved estimate of p from:

$$p = p_1 - B_2 (\delta_0^2 + \delta_1^2) / \delta_{1/2}$$

This last step is repeated until there is no further change in B_2 or p; the effects of higher-order differences may be taken into account in this step.

CRITICAL TABLE FOR B_2

p	B_2	p	B_2	p	B_2	p	B_2	p	B_2
0·0000	−	0·1101	−	0·2719	−	0·7280	−	0·8898	−
	·000		·025		·050		·049		·024
0·0020		0·1152		0·2809		0·7366		0·8949	
	·001		·026		·051		·048		·023
0·0060		0·1205		0·2902		0·7449		0·9000	
	·002		·027		·052		·047		·022
0·0101		0·1258		0·3000		0·7529		0·9049	
	·003		·028		·053		·046		·021
0·0142		0·1312		0·3102		0·7607		0·9098	
	·004		·029		·054		·045		·020
0·0183		0·1366		0·3211		0·7683		0·9147	
	·005		·030		·055		·044		·019
0·0225		0·1422		0·3326		0·7756		0·9195	
	·006		·031		·056		·043		·018
0·0267		0·1478		0·3450		0·7828		0·9242	
	·007		·032		·057		·042		·017
0·0309		0·1535		0·3585		0·7898		0·9289	
	·008		·033		·058		·041		·016
0·0352		0·1594		0·3735		0·7966		0·9335	
	·009		·034		·059		·040		·015
0·0395		0·1653		0·3904		0·8033		0·9381	
	·010		·035		·060		·039		·014
0·0439		0·1713		0·4105		0·8098		0·9427	
	·011		·036		·061		·038		·013
0·0483		0·1775		0·4367		0·8162		0·9472	
	·012		·037		·062		·037		·012
0·0527		0·1837		0·5632		0·8224		0·9516	
	·013		·038		·061		·036		·011
0·0572		0·1901		0·5894		0·8286		0·9560	
	·014		·039		·060		·035		·010
0·0618		0·1966		0·6095		0·8346		0·9604	
	·015		·040		·059		·034		·009
0·0664		0·2033		0·6264		0·8405		0·9647	
	·016		·041		·058		·033		·008
0·0710		0·2101		0·6414		0·8464		0·9690	
	·017		·042		·057		·032		·007
0·0757		0·2171		0·6549		0·8521		0·9732	
	·018		·043		·056		·031		·006
0·0804		0·2243		0·6673		0·8577		0·9774	
	·019		·044		·055		·030		·005
0·0852		0·2316		0·6788		0·8633		0·9816	
	·020		·045		·054		·029		·004
0·0901		0·2392		0·6897		0·8687		0·9857	
	·021		·046		·053		·028		·003
0·0950		0·2470		0·7000		0·8741		0·9898	
	·022		·047		·052		·027		·002
0·1000		0·2550		0·7097		0·8794		0·9939	
	·023		·048		·051		·026		·001
0·1050		0·2633		0·7190		0·8847		0·9979	
	·024		·049		·050		·025		·000
0·1101		0·2719		0·7280		0·8898		1·0000	

p	B_3
0·0	0·000
0·1	+0·006
0·2	0·008
0·3	0·007
0·4	+0·004
0·5	0·000
0·6	−0·004
0·7	0·007
0·8	0·008
0·9	−0·006
1·0	0·000

p	B_4
0·0	0·000
0·1	+0·004
0·2	0·007
0·3	0·010
0·4	0·011
0·5	+0·012
0·6	0·011
0·7	0·010
0·8	0·007
0·9	+0·004
1·0	0·000

In critical cases ascend. B_2 is always negative.

POLYNOMIAL REPRESENTATIONS

It is sometimes convenient to construct a simple polynomial representation of the form

$$f_p = a_0 + a_1 p + a_2 p^2 + a_3 p^3 + a_4 p^4 + \cdots$$

which may be evaluated in the nested form

$$f_p = (((a_4 p + a_3) p + a_2) p + a_1) p + a_0$$

Expressions for the coefficients $a_0, a_1, \ldots$ may be obtained from Stirling's interpolation formula, neglecting fifth-order differences:

$$a_4 = \delta_0^4 / 24 \qquad a_2 = \delta_0^2 / 2 - a_4 \qquad a_0 = f_0$$

$$a_3 = (\delta_{1/2}^3 + \delta_{-1/2}^3) / 12 \qquad a_1 = (\delta_{1/2} + \delta_{-1/2}) / 2 - a_3$$

This is suitable for use in the range $-\frac{1}{2} \leqslant p \leqslant 1\frac{1}{2}$, and it may be adequate in the range $-2 \leqslant p \leqslant 2$, but it should not normally be used outside this range. Techniques are available in the literature for obtaining polynomial representations which give smaller errors over similar or larger intervals. The coefficients may be expressed in terms of function values rather than differences.

EXAMPLES

To find (a) the declination of the Sun at $16^{h}\ 23^{m}\ 14^{s}{\cdot}8$ TDT on 1984 January 19, (b) the right ascension of Mercury at $17^{h}\ 21^{m}\ 16^{s}{\cdot}8$ TDT on 1984 January 8, and (c) the time on 1984 January 8 when Mercury's right ascension is exactly $18^{h}\ 04^{m}$.

Difference tables for the Sun and Mercury are constructed as shown below, where the differences are in units of the end figures of the function. Second-order differences are sufficient for the Sun, but fourth-order differences are required for Mercury.

1984 Jan	Sun Dec.	δ	δ^2
18	$-20°44'\ 48''{\cdot}3$		
		+7212	
19	−20 32 47·1		+233
		+7445	
20	−20 20 22·6		+230
		+7675	
21	−20 07 35·1		

1984 Jan	Mercury R.A.	δ	δ^2	δ^3	δ^4
6	$18^{h}\ 10^{m}\ 10{\cdot}12$				
		−18709			
7	18 07 03·03		+4299		
		−14410		−16	
8	18 04 38·93		+4283		−104
		−10127		−120	
9	18 02 57·66		+4163		−76
		−5964		−196	
10	18 01 58·02		+3967		
		−1997			
11	18 01 38·05				

(a) *Use of Bessel's formula*

The tabular interval is one day, hence the interpolating factor is 0·68281. From the critical table, $B_2 = -0{\cdot}054$, and

$$f_p = -20°\ 32'\ 47''{\cdot}1 + 0{\cdot}68281\,(+744''{\cdot}5) - 0{\cdot}054\,(+23''{\cdot}3 + 23''{\cdot}0)$$
$$= -20°\ 24'\ 21''{\cdot}2$$

(b) *Use of polynomial formula*

Using the polynomial method, the coefficients are:

$a_4 = -1^{s}{\cdot}04/24 = -0^{s}{\cdot}043$

$a_3 = (-1^{s}{\cdot}20 - 0^{s}{\cdot}16)/12 = -0^{s}{\cdot}113$

$a_2 = +42^{s}{\cdot}83/2 + 0^{s}{\cdot}043 = +21^{s}{\cdot}458$

$a_1 = (-101^{s}{\cdot}27 - 144^{s}{\cdot}10)/2 + 0^{s}{\cdot}113 = -122^{s}{\cdot}572$

$a_0 = 18^{h} + 278^{s}{\cdot}93$

where an extra decimal place has been kept as a guarding figure.

Then $f_p = 18^{h} + 278^{s}{\cdot}93 - 122^{s}{\cdot}572p + 21^{s}{\cdot}458p^2 - 0^{s}{\cdot}113p^3 - 0^{s}{\cdot}043p^4$
The interpolating factor $p = 0{\cdot}72311$, hence $f_p = 18^{h}\ 03^{m}\ 21^{s}{\cdot}46$

(c) *Inverse interpolation*

Since $f_p = 18^{h}\ 04^{m}$ the first estimate for p is:

$$p_1 = (18^{h}\ 04^{m} - 18^{h}\ 04^{m}\ 38^{s}{\cdot}93)/(-101^{s}{\cdot}27) = 0{\cdot}38442$$

From the critical table, with $p = 0{\cdot}3844$, $B_2 = -0{\cdot}059$. Also

$$(\delta_0^2 + \delta_1^2)/\delta_{1/2} = (+42{\cdot}83 + 41{\cdot}63)/(-101{\cdot}27) = -0{\cdot}834$$

The second approximation to p is:

$$p = 0{\cdot}38442 + 0{\cdot}059\,(-0{\cdot}834) = 0{\cdot}33521 \text{ which gives } t = 8^{h}\ 02^{m}\ 42^{s};$$

as a check, using the polynomial found in (b) with $p = 0{\cdot}33521$ gives

$$f_p = 18^{h}\ 04^{m}\ 00^{s}{\cdot}25.$$

The next approximation is $B_2 = -0{\cdot}056$ and $p = 0{\cdot}38442 + 0{\cdot}056\,(-0{\cdot}834) = 0{\cdot}33772$ which gives $t = 8^{h}\ 06^{m}\ 19^{s}$; using the polynomial in (b) with $p = 0{\cdot}33772$ gives

$$f_p = 18^{h}\ 03^{m}\ 59^{s}{\cdot}98.$$

SUBTABULATION

Coefficients for use in the systematic interpolation of an ephemeris to a smaller interval are given in the following table for certain values of the ratio of the two intervals. The table is entered for each of the appropriate multiples of this ratio to give the corresponding decimal value of the interpolating factor p and the Bessel coefficients. The values of p are exact or recurring decimal numbers. The values of the coefficients may be rounded to suit the maximum number of figures in the differences.

BESSEL COEFFICIENTS FOR SUBTABULATION

Ratio of intervals												Bessel Coefficients		
$\frac{1}{2}$	$\frac{1}{3}$	$\frac{1}{4}$	$\frac{1}{5}$	$\frac{1}{6}$	$\frac{1}{8}$	$\frac{1}{10}$	$\frac{1}{12}$	$\frac{1}{20}$	$\frac{1}{24}$	$\frac{1}{40}$	p	B_2	B_3	B_4
										1	0·025	−0·006094	0·00193	0·0010
									1		0·0416	−0·009983	0·00305	0·0017
								1		2	0·050	−0·011875	0·00356	0·0020
										3	0·075	−0·017344	0·00491	0·0030
							1		2		0·0833	−0·019097	0·00530	0·0033
						1		2		4	0·100	−0·022500	0·00600	0·0039
					1				3	5	0·125	−0·027344	0·00684	0·0048
								3		6	0·150	−0·031875	0·00744	0·0057
				1			2		4		0·1666	−0·034722	0·00772	0·0062
										7	0·175	−0·036094	0·00782	0·0064
			1			2		4		8	0·200	−0·040000	0·00800	0·0072
									5		0·2083	−0·041233	0·00802	0·0074
										9	0·225	−0·043594	0·00799	0·0079
		1			2		3	5	6	10	0·250	−0·046875	0·00781	0·0085
										11	0·275	−0·049844	0·00748	0·0091
									7		0·2916	−0·051649	0·00717	0·0095
						3		6		12	0·300	−0·052500	0·00700	0·0097
										13	0·325	−0·054844	0·00640	0·0101
	1			2			4		8		0·3333	−0·055556	0·00617	0·0103
								7		14	0·350	−0·056875	0·00569	0·0106
					3				9	15	0·375	−0·058594	0·00488	0·0109
			2			4		8		16	0·400	−0·060000	0·00400	0·0112
							5		10		0·4166	−0·060764	0·00338	0·0114
										17	0·425	−0·061094	0·00305	0·0114
								9		18	0·450	−0·061875	0·00206	0·0116
									11		0·4583	−0·062066	0·00172	0·0116
										19	0·475	−0·062344	0·00104	0·0117
1		2		3	4	5	6	10	12	20	0·500	−0·062500	0·00000	0·0117
										21	0·525	−0·062344	−0·00104	0·0117
									13		0·5416	−0·062066	−0·00172	0·0116
								11		22	0·550	−0·061875	−0·00206	0·0116
										23	0·575	−0·061094	−0·00305	0·0114
							7		14		0·5833	−0·060764	−0·00338	0·0114
			3			6		12		24	0·600	−0·060000	−0·00400	0·0112
					5				15	25	0·625	−0·058594	−0·00488	0·0109
								13		26	0·650	−0·056875	−0·00569	0·0106
	2			4			8		16		0·6666	−0·055556	−0·00617	0·0103
										27	0·675	−0·054844	−0·00640	0·0101
						7		14		28	0·700	−0·052500	−0·00700	0·0097
									17		0·7083	−0·051649	−0·00717	0·0095
										29	0·725	−0·049844	−0·00748	0·0091
		3			6		9	15	18	30	0·750	−0·046875	−0·00781	0·0085
										31	0·775	−0·043594	−0·00799	0·0079
									19		0·7916	−0·041233	−0·00802	0·0074
			4			8		16		32	0·800	−0·040000	−0·00800	0·0072
										33	0·825	−0·036094	−0·00782	0·0064
				5			10		20		0·8333	−0·034722	−0·00772	0·0062
								17		34	0·850	−0·031875	−0·00744	0·0057
					7				21	35	0·875	−0·027344	−0·00684	0·0048
						9		18		36	0·900	−0·022500	−0·00600	0·0039
							11		22		0·9166	−0·019097	−0·00530	0·0033
										37	0·925	−0·017344	−0·00491	0·0030
								19		38	0·950	−0·011875	−0·00356	0·0020
									23		0·9583	−0·009983	−0·00305	0·0017
										39	0·975	−0·006094	−0·00193	0·0010

This explanation specifies the sources for the theories and data used in constructing the ephemerides in this volume, explains basic concepts required to use the ephemerides, and where appropriate states the precise meaning of tabulated quantities. Definitions of individual terms are given in the Glossary (Section M).

The IAU (1976) System of Astronomical Constants was adopted by the General Assembly of the IAU at Grenoble. These constants are given on page K6 of this volume. Additional resolutions concerning time scales and the astronomical reference system were adopted by the IAU in 1979 at Montreal and in 1982 at Patras. A complete list of these resolutions, with constants, formulae and explanatory notes, is given in the *Supplement to the Astronomical Almanac for 1984*, which is published in the 1984 *Astronomical Almanac.*

Fundamental ephemerides of the Sun, Moon and planets were calculated by a simultaneous numerical integration at the Jet Propulsion Laboratory. These ephemerides, designated DE200/LE200, cover the period 1800–2050. Optical, radar, laser, and spacecraft observations were analyzed to determine starting conditions for the numerical integration. In order to obtain the best fit of the ephemerides to the observational data, some modifications to the IAU (1976) System of Astronomical Constants were necessary. These modifications are listed on pages K6 and K7. A satisfactory ephemeris for Uranus for the 1980's could be computed only by excluding observations made before 1900. Additional information about the new ephemerides is included in the *Supplement to the Astronomical Almanac for 1984.* DE200/LE200 is available on magnetic tape.

Reference Frame

Beginning in 1984 the standard epoch of the fundamental astronomical coordinate system is 2000 January 1, 12^h TDB (JD 245 1545.0), which is denoted J2000.0. The numerical integration used as the basis for ephemerides in this volume is in a reference frame defined by the mean equator and dynamical equinox of J2000.0. Rigorous reduction methods presented in Section B were used to construct the published tabular ephemerides.

In practice, the dynamical equinox, defined by the ascending node of the ecliptic on the mean equator at epoch J2000.0, differs from the origin of right ascension (the catalog equinox) of the FK5 star catalog. Although the exact value of the difference is uncertain, it is thought to be less than $0''.04$ at the current time. Likewise, the radio reference frame may differ from optical reference frames.

Time Scales

Terrestrial dynamical time (TDT) is the tabular argument of the fundamental geocentric ephemerides. For ephemerides referred to the barycenter of the solar system, the argument is barycentric dynamical time (TDB). In the terminology of the general theory of relativity, TDT corresponds to a proper time, while TDB corresponds to a coordinate time. These scales are defined so that the difference between them is purely periodic, with an amplitude less than $0^s.002$. Like their predecessor, ephemeris time (ET), TDT and TDB are independent of the Earth's rotation.

In the astronomical system of units, the unit of time is the day of 86400 seconds of barycentric dynamical time (TDB). For long periods, however, the

Julian century of 36525 days is used. Use of the tropical year and Besselian epochs was discontinued in 1984.

International atomic time (TAI) is the most precisely determined time scale that is now available for astronomical use. This scale results from analyses by the Bureau International des Poids et Mesures in Paris of data from atomic time standards of many countries. Although TAI was not introduced until 1972 January 1, atomic time scales have been available since 1956. Therefore, TAI may be extrapolated backwards for the period 1956–1971. The fundamental unit of TAI is the unit of time in the international system of units, the SI second; it is defined as the duration of 9 192 631 770 periods of the radiation corresponding to the transition between two hyperfine levels of the ground state of the cesium 133 atom.

Universal time (UT), which serves as the basis of civil timekeeping, is formally defined by a mathematical formula which relates UT to Greenwich mean sidereal time. Thus UT is determined from observations of the diurnal motions of the stars. It implicitly contains nonuniformities due to variations in the rotation of the Earth. A UT scale determined directly from stellar observations is dependent on the place of observation; these scales are designated UT0. A time scale that is independent of the location of the observer is established by removing from UT0 the effect of the variation of the observer's meridian due to the observed motion of the geographic pole; this time scale is designated UT1. A tabulation of the quantity ΔT = TDT – UT1 is given on page K9.

Since 1972 January 1, the time scale distributed by most broadcast time services has been based on the redefined coordinated universal time (UTC), which differs from TAI by an integral number of seconds. UTC is maintained within $0^{s}.90$ of UT1 by the introduction of one second steps (leap seconds) when necessary, normally at the end of June or December. DUT1, an approximation to the difference UT1 minus UTC, is transmitted in code on broadcast time signals. Beginning in 1962, an increasing number of broadcast time services cooperated to provide a consistent time standard, until most broadcast signals were synchronized to the redefined UTC in 1972. For a while prior to 1972, broadcast time signals were kept within $0^{s}.1$ of UT2 (UT1 corrected by an adopted formula for the seasonal variation) by the introduction of step adjustments, normally of $0^{s}.1$, and occasionally by changes in the duration of the second. Since the table on page K9 is based on the signals broadcast by WWV, special corrections may be required to derive UT1 times from other signals broadcast prior to 1972.

Universal time and UT are commonly used to mean UT0, UT1 or UTC, according to context. In this volume, UT1 is always implied where the differences are significant.

Greenwich mean sidereal time (GMST) is defined as the Greenwich hour angle of the mean equinox of date. The defining relation between sidereal and universal time is:

$$\text{GMST of } 0^{h}\text{ UT1} = 6^{h}41^{m}50^{s}.54841 + 8640\,184^{s}.812\,866\,T + 0^{s}.093\,104\,T^{2} - 6^{s}.2\times10^{-6}\,T^{3}$$

where T is measured in Julian centuries of 36525 days of UT1 from 2000 January 1, 12^{h} UT1 (JD 245 1545.0 UT1). (S. Aoki *et al.*, *Astron. Astrophys.*, **105**, 359, 1982).

To provide continuity with pre-1984 practices, the difference between TDT and TAI was set to the current estimate of the difference between ET and TAI:

$$TDT = TAI + 32^s.184.$$

Thus procedures analogous to those used with ephemerides tabulated as functions of ET are generally applicable to ephemerides based on TDT. The tabulations for 0^h TDT may be converted to 0^h UT1 by interpolation to time ΔT.

Since TDT is independent of the Earth's rotation, calculations of hour angles and phenomena referred to a geographic meridian are provisionally referred to the ephemeris meridian, which is 1.002 738 ΔT east of the Greenwich meridian. Only when ΔT is specified can quantities be referred to the Greenwich meridian.

Section A: Summary of Principal Phenomena

The lunations given on page A1 are numbered in continuation of E. W. Brown's series, of which No. 1 commenced on 1923 January 16 (*Mon. Not. Roy. Astron. Soc.*, **93**, 603, 1933).

The list of occultations of planets and bright stars by the Moon on page A2 gives the approximate times and areas of visibility for the major planets, except Neptune and Pluto, and the bright stars Aldebaran, Antares, Pollux, Regulus and Spica. More detailed information about these events and of other occultations by the Moon may be obtained from the International Lunar Occultation Centre at the address given at the foot of page A2.

Times tabulated on page A3 for the stationary points of the planets are the instants at which the planet is stationary in apparent geocentric right ascension; but for elongations of the planets from the Sun, the tabular times are for the geometric configurations. From inferior conjunction to superior conjunction for Mercury or Venus, or from conjunction to opposition for a superior planet, the elongation from the Sun is west; from superior to inferior conjunction, or from opposition to conjunction, the elongation is east. Because planetary orbits do not lie exactly in the ecliptic plane, elongation passages from west to east or from east to west do not in general coincide with oppositions and conjunctions.

Dates of heliocentric phenomena are given on page A3. Since they are determined from the actual perturbed motion, these dates generally differ from dates obtained by using the elements of the mean orbit. The date on which the radius vector is a minimum may differ considerably from the date on which the heliocentric longitude of a planet is equal to the longitude of perihelion of the mean orbit. Similarly, when the heliocentric latitude of a planet is zero, the heliocentric longitude may not equal the longitude of the mean node.

Configurations of the Sun, Moon and Planets (pages A9–A11) is a chronological listing, with times to the nearest hour, of geocentric phenomena. Included are eclipses; lunar perigees, apogees and phases; phenomena in apparent geocentric longitude of the planets and of the minor planets Ceres, Pallas, Juno and Vesta; times when the planets and minor planets are stationary in right ascension and when the geocentric distance to Mars is a minimum; and geocentric conjunctions in apparent right ascension of the planets with the Moon, with each other, and with the bright stars Aldebaran, Pollux, Regulus, Spica and Antares, provided these conjunctions are considered to occur sufficiently far from the Sun to permit observation. Thus conjunctions in right ascension are excluded if they occur within 15° of the Sun for the Moon, Mars and Saturn; within 10° for Venus and Jupiter; and within approximately 10° for

Mercury, depending on Mercury's brightness. The occurrence of occultations of planets and bright stars is indicated by "Occn."; the areas of visibility are given in the list on page A2. Geocentric phenomena differ from the actually observed configurations by the effects of the geocentric parallax at the place of observation, which for configurations with the Moon may be quite large.

The explanation for the tables of sunrise and sunset, twilight, moonrise and moonset is given on page A12; examples are given on page A13.

Eclipses

The elements and circumstances are computed according to Bessel's method from apparent right ascensions and declinations of the Sun and Moon. From 1986 onwards, positions of the Sun and Moon are derived from DE200/LE200. Semidiameters of the Sun and Moon used in the calculation of eclipses do not include irradiation. The adopted semidiameter of the Sun at unit distance is 15′59″.63 (A. Auwers, *Astronomische Nachrichten,* No. 3068, 367, 1891), the same as in the ephemeris of the Sun. The apparent semidiameter of the Moon is equal to arcsin (k sin π), where π is the Moon's horizontal parallax and k is an adopted constant. In 1982, to be consistent with the System of Astronomical Constants (1976) and the ephemeris based on DE200/LE200, the IAU adopted k = 0.272 5076, corresponding to the mean radius of Watts' datum as determined by observations of occultations and to the adopted radius of the Earth. This value is introduced for 1986 onwards. Corrections to the ephemerides, if any, are noted in the beginning of the eclipse section.

In calculating lunar eclipses the radius of the geocentric shadow of the Earth is increased by one-fiftieth part to allow for the effect of the atmosphere. Refraction is neglected in calculating solar and lunar eclipses. Because the circumstances of eclipses are calculated for the surface of the ellipsoid, refraction is not included in Besselian elements. For local predictions, corrections for refraction are unnecessary; they are required only in precise comparisons of theory with observation in which many other refinements are also necessary.

The solar eclipse maps show the path of the eclipse, beginning and ending times of the eclipse, and the region of visibility, including restrictions due to rising and setting of the Sun. The short-dash and long-dash lines show, respectively, the progress of the leading and trailing edge of the penumbra; thus, at a given location, times of first and last contact may be interpolated.

Besselian elements characterize the geometric position of the shadow of the Moon relative to the Earth. The exterior tangents to the surfaces of the Sun and Moon form the umbral cone; the interior tangents form the penumbral cone. The common axis of these two cones is the axis of the shadow. To form a system of geocentric rectangular coordinates, the geocentric plane perpendicular to the axis of the shadow is taken as the xy-plane. This is called the fundamental plane. The x-axis is the intersection of the fundamental plane with the plane of the equator; it is positive toward the east. The y-axis is positive toward the north. The z-axis is parallel to the axis of the shadow and is positive toward the Moon. The tabular values of x and y are the coordinates, in units of the Earth's equatorial radius, of the intersection of the axis of the shadow with the fundamental plane. The direction of the axis of the shadow is specified by the declination d and hour angle μ of the point on the celestial sphere toward which the axis is directed.

The radius of the umbral cone is regarded as positive for an annular eclipse and negative for a total eclipse. The angles f_1 and f_2 are the angles at which the tangents that form the penumbral and umbral cones, respectively, intersect the axis of the shadow.

Section B: Time Scales and Coordinate Systems

Calendar

Over extended intervals, civil time is ordinarily reckoned according to conventional calendar years and adopted historical eras; in constructing and regulating civil calendars and fixing ecclesiastical calendars, a number of auxiliary cycles and periods are used.

To facilitate chronological reckoning, the system of Julian day (JD) numbers maintains a continuous count of astronomical days, beginning with JD 0 on 1 January 4713 B.C., Julian proleptic calendar. Julian day numbers for the current year are given on page B4 and in the Universal and Sidereal Times table, pages B8–B15. To determine JD numbers for other years on the Gregorian calendar, consult the Julian Day Number tables, pages K2–K4.

Note that the Julian day begins at noon, whereas the calendar day begins at the preceding midnight. Thus the Julian day system is consistent with astronomical practice before 1925, with the astronomical day being reckoned from noon. For critical applications, the Julian date should include a specification as to whether UT, TDT or TDB is used.

Universal and Sidereal Times

The tabulations of Greenwich mean sidereal time (GMST) at 0^h UT are calculated from the defining relation between sidereal time and universal time (see the introductory discussion of Time Scales in this Explanation). The tabulation of Greenwich apparent sidereal time (GAST) is calculated by adding the equation of the equinoxes (the total nutation in longitude, multiplied by the cosine of the true obliquity of the ecliptic) to GMST. Following the general practice of this volume, UT implies UT1 in critical applications. Useful formulae and examples are given on pages B6–B7.

Reduction of Astronomical Coordinates

Formulae and tables for a variety of methods of apparent place reduction are presented. Choice of a particular method should be made according to accuracy requirements.

Reduction to apparent place from mean place for standard epoch J2000.0 is most accurately accomplished by formulae given on pages B36–B41. These require the rectangular position and velocity components of the Earth with respect to the solar system barycenter (even pages B44–B58) and the precession and nutation matrix (odd pages B45–B59). The Earth's position and velocity components are derived from the simultaneous numerical integration DE200/LE200 described on page L1. In critical applications the tabular argument of the barycentric ephemeris is barycentric dynamical time (TDB).

Beginning in 1984 the standard epoch of the stellar data tabulations (Section H) is the middle of the Julian year, rather than the beginning of the Besselian year. Thus the Besselian and second-order day numbers are referred to the mean equator and equinox of the middle of the current Julian year. Formulae for precessing positions from the standard epoch J2000.0 to the current year are given on page B18. These formulae are based on expressions for annual rates of precession given by J. H. Lieske *et al.* (*Astron. Astrophys.*, **58**, 1–16, 1977), in conformance with the IAU (1976) value of the constant of precession.

Section C: The Sun

Apparent geocentric coordinates of the Sun are given on even pages C4–C18; geocentric rectangular coordinates referred to the mean equator and equinox of J2000.0 are given on pages C20–C23. These ephemerides are based on the simultaneous numerical integration DE200/LE200 described on page L1. The tabular argument of the solar ephemerides is terrestrial dynamical time (TDT). Although the apparent right ascension and declination are antedated for light-time, the true geocentric distance in astronomical units is the geometric distance at the tabular time.

The rotation elements listed on page C3, as well as the daily tabulations of rotational parameters (odd pages C5–C19), are due to R. C. Carrington (*Observations of the Spots on the Sun,* 1863). The synodic rotation numbers are in continuation of Carrington's Greenwich photoheliographic series, of which Number 1 commenced on 1853 November 9. The tabular values of the semidiameter are computed by taking the arc sine of the quantity formed by dividing the IAU solar radius (696 000 km) by the true distance in km.

Formulae for geocentric and heliographic coordinates are given on pages C1–C3.

Section D: The Moon

The geocentric ephemerides of the Moon are based on the numerical integration DE200/LE200 described on page L1. The tabular argument is terrestrial dynamical time (TDT).

For high precision calculations the polynomial ephemeris on pages D23–D45 should be used; procedures for evaluating the polynomials are given on page D22. A daily geocentric ephemeris to lower precision is given on the even numbered pages D6–D20. Although the tabular apparent right ascension and declination are antedated for light-time, the horizontal parallax is the geometric value for the tabular time. It is derived from $\arcsin(1/r)$, where r is the true distance in units of the Earth's equatorial radius. The semidiameter s is computed from $s = \arcsin(k \sin \pi)$, where $k = 0.272\,493$ is the ratio of the equatorial radius of the Moon to the equatorial radius of the Earth and π is the horizontal parallax. No correction is made for irradiation.

Beginning in 1985 the physical ephemeris (odd pages D7–D21) is based on the formulae and constants for physical librations given by D. Eckhardt (*The Moon and the Planets*, **25**, 3, 1981; *High Precision Earth Rotation and Earth-Moon Dynamics,* ed. O. Calame, pages 193–198, 1982), but with the IAU value of 1°32′32″.7 for the inclination of the mean lunar equator to the ecliptic. Although values of Eckhardt's constants differ slightly from those of the IAU, this is of no consequence to the precision of the tabulation. Optical librations are first calculated from rigorous formulae; then the total librations (optical and physical) are calculated from the rigorous formulae by replacing I with $I + \rho$, Ω with $\Omega + \sigma$ and ☾ with ☾ $+ \tau$. Included in the calculations are perturbations for all terms greater than 0°.0001 in solution 500 of the first Eckhardt reference and in Table I of the second reference. Since apparent coordinates of the Sun and Moon

are used in the calculations, aberration is fully included, except for the inappreciable difference between the light-time from the Sun to the Moon and from the Sun to the Earth.

The selenographic coordinates of the Earth and Sun specify the point on the lunar surface where the Earth and Sun are in the selenographic zenith. The selenographic longitude and latitude of the Earth are the total geocentric, optical and physical librations in longitude and latitude, respectively. When the longitude is positive, the mean central point of the disk is displaced eastward on the celestial sphere, exposing to view a region on the west limb. When the latitude is positive, the mean central point is displaced toward the south, exposing to view the north limb. If the principal moment of inertia axis toward the Earth is used as the origin for measuring librations, rather than the traditional origin in the mean direction of the Earth from the Moon, there is a constant offset of 214″.2 in τ, or equivalently a correction of –0°.059 to the Earth's selenographic longitude.

The tabulated selenographic colongitude of the Sun is the east selenographic longitude of the morning terminator. It is calculated by subtracting the selenographic longitude of the Sun from 90° or 450°. Colongitudes of 270°, 0°, 90° and 180° correspond to New Moon, First Quarter, Full Moon and Last Quarter, respectively.

The position angles of the axis of rotation and the midpoint of the bright limb are measured counterclockwise around the disk from the north point. The position angle of the terminator may be obtained by adding 90° to the position angle of the bright limb before Full Moon and by subtracting 90° after Full Moon.

For precise reductions of observations, the tabular data should be reduced to topocentric values. Formulae for this purpose by R. d'E. Atkinson (*Mon. Not. Roy. Astron. Soc.*, **111** , 448, 1951) are given on page D5.

Additional formulae and data pertaining to the Moon are given on pages D2–D5, D46.

Section E: Major Planets

The heliocentric and geocentric ephemerides of the planets are based on the numerical integration DE200/LE200 described on page L1. Terrestrial dynamical time (TDT) is the tabular argument of the geocentric ephemerides. The argument of the heliocentric ephemerides is barycentric dynamical time (TDB).

Although the apparent right ascension and declination are antedated for light-time, the true geocentric distance in astronomical units is the geometric distance for the tabular time. For Pluto the astrometric ephemeris is comparable with observations referred to catalog mean places of comparison stars (corrected for proper motion and annual parallax, if significant, to the epoch of observation), provided the catalog is referred to the J2000.0 reference frame and the observations are corrected for geocentric parallax.

Ephemerides for Physical Observations of the Planets

The physical ephemerides of the planets have been calculated from the fundamental solar system ephemerides used elsewhere in this volume. Except where otherwise noted, physical data are based on the "Report of the IAU Working Group on Cartographic Coordinates and Rotational Elements of the Planets and Satellites" (M. E. Davies *et al., Celest. Mech.*, **29**, 309–321, 1983;

hereafter referred to as the IAU Report on Cartographic Coordinates).

All tabulated quantities are corrected for light-time, so the given values apply to the disk that is visible at the tabular time. Except for planetographic longitudes, all tabulated quantities vary so slowly that they remain unchanged if the time argument is considered to be universal time rather than dynamical time. Conversion from dynamical to universal time affects the tabulated planetographic longitudes by several tenths of a degree for all planets except Mercury, Venus and Pluto.

The tabulated light-time is the travel time for light arriving at the Earth at the tabular time. Expressions for the visual magnitudes of the planets are due to D. L. Harris (*Planets and Satellites,* ed. G. P. Kuiper and B. L. Middlehurst, page 272, 1961), except that values for $V(1,0)$, the visual magnitude at unit distance, are those given on page E88 of this volume. The tabulated surface brightness is the average visual magnitude of an area of one square arc-second of the illuminated portion of the apparent disk. For a few days around inferior and superior conjunctions, the tabulated magnitude and surface brightness of Mercury and Venus are only approximate; surface brightness is not tabulated near inferior conjunction. For Saturn the magnitude includes the contribution due to the rings, but the surface brightness applies only to the disk of the planet.

The apparent disk of an oblate planet is always an ellipse, with an oblateness less than or equal to the oblateness of the planet itself, depending on the apparent tilt of the planet's axis. For planets with significant oblateness, the apparent equatorial and polar diameters are separately tabulated.

The phase is the ratio of the apparent illuminated area of the disk to the total area of the disk, as seen from the Earth. The phase angle is the planetocentric elongation of the Earth from the Sun. In the accompanying diagram of the apparent disk of a planet, the defect of illumination is designated by q. It is the length of the unilluminated section of the diameter passing through the sub-Earth point e (the center of the disk) and the sub-solar point s. The position angle of the defect of illumination can be computed by adding 180° to the tabulated position angle of the sub-solar point. Calculations of phase and defect of illumination are based on the geometric terminator, which is defined by the plane crossing through the planet's center of mass, orthogonal to the direction of the Sun.

The tabulated quantity L_s is the planetocentric orbital longitude of the Sun, measured eastward in the planet's orbital plane from the planet's vernal equinox. Instantaneous orbital and equatorial planes are used in computing L_s. Values of L_s of 0°, 90°, 180° and 270° correspond to the beginning of spring, summer, autumn and winter, respectively, for the planet's northern hemisphere.

The orientation of the pole of a planet is specified by the right ascension α_0 and declination δ_0 of the north pole, with respect to the Earth's mean equator and equinox of J2000.0. According to the IAU definition, the north pole is the pole that lies on the north side of the invariable plane of the solar system. Because of precession of a planet's axis, α_0 and δ_0 may vary slowly with time; values for the current year are given on page E87.

The angle W of the prime meridian is measured counterclockwise (when viewed from above the planet's north pole) along the planet's equator from the ascending node of the planet's equator on the Earth's mean equator of J2000.0. For a planet with direct rotation (counterclockwise as viewed from the planet's north pole), W increases with time. Values of W and its rate of change are given on page E87.

Except for Saturn and Pluto, expressions for the pole and prime meridian of each planet are based on the IAU Report on Cartographic Coordinates. The

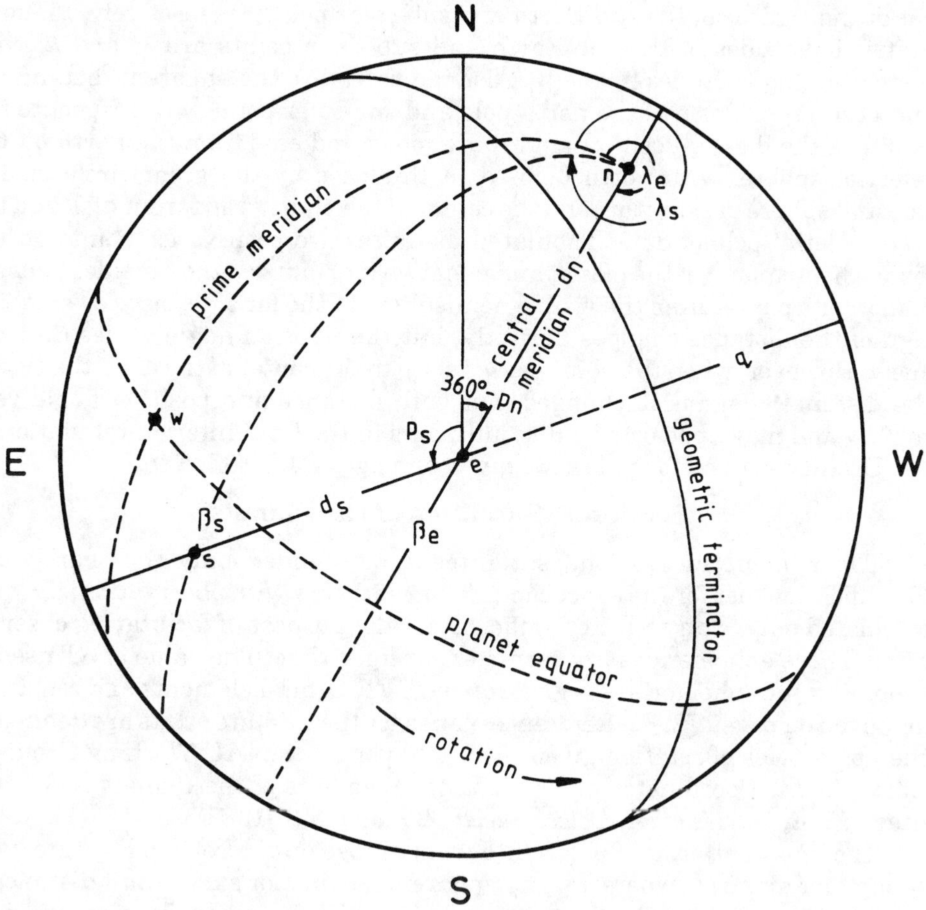

rotation of Saturn was provided by M. D. Desch and M. L. Kaiser. For Pluto the pole and rotation are due to R. S. Harrington and J. W. Christy (*Astron. Jour.*, **86**, 442, 1981), under the assumptions that the orbital motion of Pluto's satellite is synchronous with Pluto's rotation and that the satellite's orbital plane is coincident with Pluto's equatorial plane.

Tabulated longitudes and latitudes of the sub-Earth and sub-solar points are in planetographic coordinates. Planetographic longitude is reckoned from the prime meridian and increases from 0° to 360° in the direction opposite rotation. The planetographic latitude of a point is the angle between the planet's equator and the normal to the reference spheroid at the point. Latitudes north of the equator are positive. For Jupiter and Saturn multiple longitude systems are defined, each system corresponding to a different apparent rate of rotation. System I applies to the visible cloud layer in the equatorial region of each planet. System II applies to the visible cloud layer at higher latitudes of Jupiter; there is no System II for Saturn. System III applies to the origin of the radio emissions of both Jupiter and Saturn. Since rotational periods of Uranus and Neptune are not well known, no planetographic longitudes are given for these planets.

Planetographic coordinates are illustrated in the diagram. At the center of the apparent disk is the sub-Earth point *e*; other reference points are the sub-solar point *s* and the north pole *n*. For an oblate planet, the Earth and Sun are not at the zeniths of the sub-Earth and sub-solar points, respectively. Planetocentric longitudes of the sub-Earth and sub-solar points are λ_e and λ_s with corresponding latitudes β_e and β_s. Also indicated are the apparent distances *d* and position angles *p* of the north pole and sub-solar point with respect to the center of the disk *e*. Position angles are measured east from the north on the celestial sphere, with north defined in this case by the great circle on the celestial sphere passing through the center of the planet's apparent disk and the true celestial pole of date. Tabulated distances are positive for points on the visible hemisphere of the planet and negative for points on the far side. Thus, as point *n* or *s* passes from the visible hemisphere to the far side, or vice versa, the sign of the distance changes abruptly, but the position angle varies continuously. However, when the point passes close to the center of the disk, the sign of the distance remains unchanged, but both distance and position angle vary rapidly and may appear to be discontinuous in the fixed interval tabulations.

Useful data and formulae are given on pages E43, E87, E88.

Section F: Satellites of the Planets

The ephemerides of the satellites are intended only for search and identification, not for the exact comparison of theory with observation; they are calculated only to an accuracy sufficient for the purpose of facilitating observations. These ephemerides are corrected for light-time. The value of ΔT used in preparing the ephemerides is given on page F1. Orbital elements and constants are given on pages F2, F3. Reference planes for the satellite orbits are defined by the north poles of rotation given in the "Report of the IAU Working Group on Cartographic Coordinates and Rotational Elements of the Planets and Satellites" (M. E. Davies *et al.*, *Celest. Mech.*, **29**, 309–321, 1983).

The apparent orbit of a satellite is an ellipse on the celestial sphere, with semimajor axis a/Δ, where a is the apparent semimajor axis at unit distance in seconds of arc and Δ is the geocentric distance of the primary. In calculating the tables for finding the position angle and apparent angular distance with respect to the primary, the value of the eccentricity of the apparent orbit at opposition is used. The apparent geocentric distance s is measured from the central point of the geometric disk of the primary, and the position angle p is measured eastward from the north celestial pole. Neglected in the calculations are the effect of the eccentricity of the actual orbit upon its projection onto the apparent orbit and the variation of the eccentricity of the apparent orbit. Approximately, therefore, $s = F(a/\Delta)$, where F is the ratio of s to the apparent distance at greatest elongation. At greatest elongations $p = P \pm 90°$, where P is the position angle of the extremity of the minor axis of the apparent orbit that is directed toward the pole of the orbit from which motion appears counterclockwise. With P_0 denoting an arbitrary fixed integral number of degrees, usually the approximate value of P at opposition, the value of p at any time is expressed in the form $p_1 + p_2$, where p_1 is the sum of $P_0 + 90°$ plus the amount of motion in position angle since elongation, and p_2 denotes the correction $P - P_0$. In the tables of p_1 the tabular entry for argument 0^h00^m is the value of $P_0 + 90°$.

Approximate formulae for calculating differential coordinates of satellites are given with the relevant tables.

Satellites of Mars

The ephemerides of the satellites of Mars are computed from the orbital elements given by H. Struve (*Sitzungsberichte der Königlich Preuss. Akad. der Wiss.*, p. 1073, 1911).

Satellites of Jupiter

The ephemerides of Satellites I–IV are based on the theory of J. H. Lieske (*Astron. Astrophys.*, **65**, 83, 1978), with constants due to J.-E. Arlot (*Astron. Astrophys.*, **107**, 305, 1982).

Elongations of Satellite V are computed from circular orbital elements determined by A. J. J. Van Woerkom (*Astron. Pap. Amer. Ephem.*, Vol, XIII, Pt. I, pages 8, 14, 16, 1950). The differential coordinates of Satellites VI–XIII are computed from numerical integrations, using starting coordinates and velocities calculated at the U. S. Naval Observatory.

The actual geocentric phenomena of Satellites I–IV are not instantaneous. Since the tabulated times are for the middle of the phenomena, a satellite is usually observable after the tabulated time of eclipse disappearance (EcD) and before the time of eclipse reappearance (EcR). In the case of Satellite IV the difference is sometimes quite large. Light curves of eclipse phenomena are discussed by D. L. Harris (*Planets and Satellites,* ed. G. P. Kuiper and B. M. Middlehurst, pages 327–340, 1961).

To facilitate identification, approximate configurations of Satellites I–IV are shown in graphical form on pages facing the tabular ephemerides of the geocentric phenomena. Time is shown by the vertical scale, with horizontal lines denoting 0^h UT. For any time the curves specify the relative positions of the satellites in the equatorial plane of Jupiter. The width of the central band, which represents the disk of Jupiter, is scaled to the planet's equatorial diameter.

For eclipses the points d of immersion into the shadow and points r of emersion from the shadow are shown pictorially at the foot of the right-hand pages for the superior conjunctions nearest the middle of each month. At the foot of the left-hand pages, rectangular coordinates of these points are given in units of the equatorial radius of Jupiter. The x-axis lies in Jupiter's equatorial plane, positive toward the east; the y-axis is positive toward the north pole of Jupiter. The subscript 1 refers to the beginning of an eclipse, subscript 2 to the end of an eclipse.

Satellites and Rings of Saturn

Since the rings of Saturn lie in the equatorial plane of the planet, the inclination and ascending node of the ring plane are determined from the position of the north pole of Saturn, as defined by M. E. Davies *et al.* (*Celest. Mech.,* **29**, 309–321, 1983). The apparent dimensions of the outer ring and factors for computing relative dimensions of the rings are from L. W. Esposito *et al.* (*Saturn,* eds. T. Gehrels and M. S. Matthews, pages 468–478, 1984).

Since the appearance of the rings depends upon the Saturnicentric positions of the Earth and Sun, the following quantities are tabulated in the ephemeris:

U, the geocentric longitude of Saturn, measured in the plane of the rings eastward from its ascending node on the mean equator of the Earth; the Saturnicentric longitude of the Earth, measured in the same way, is $U + 180°$.

B, the Saturnicentric latitude of the Earth, referred to the plane of the rings, positive toward the north; when B is positive the visible surface of the rings is the northern surface.

P, the geocentric position angle of the northern semiminor axis of the apparent ellipse of the rings, measured eastward from north.

U', the heliocentric longitude of Saturn, measured in the plane of the rings eastward from its ascending node on the ecliptic; the Saturnicentric longitude of the Sun, measured in the same way, is $U' + 180°$.

B', the Saturnicentric latitude of the Sun, referred to the plane of the rings, positive toward the north; when B' is positive the northern surface of the rings is illuminated.

P', the heliocentric position angle of the northern semiminor axis of the rings on the heliocentric celestial sphere, measured eastward from the great circle that passes through Saturn and the poles of the ecliptic.

The ephemeris of the rings is corrected for light-time.

The ephemerides of Satellites I–VI and of Iapetus are computed from the orbital elements determined by G. Struve (*Veröff. der Universitätssternwarte zu Berlin-Babelsberg,* Vol. VI, Pt. 4, 1930, and Pt. 5, 1933). The ephemeris of Hyperion is computed from the elements given by J. Woltjer, Jr. (*Annalen van de Sterrewacht te Leiden,* Vol. XVI, Pt. 3, p. 64, 1928), and of Phoebe from the theory by F. E. Ross (*Annals of Harvard College Obs.*, Vol. LIII, No. VI, 1905).

For Satellites I–V times of eastern elongation are tabulated; for Satellites VI–VIII times of all elongations and conjunctions are tabulated. Tables for finding approximate distance s and position angle p are given for Satellites I–VIII. On the diagram of the orbits of Satellites I–VII, points of eastern elongation are marked "0". From the tabular times of these elongations the apparent position of a satellite at any other time can be marked on the diagram by setting off on the orbit the elapsed interval since last eastern elongation. For Hyperion and Iapetus ephemerides of differential coordinates are also included. An ephemeris of differential coordinates is given for Phoebe.

Solar perturbations are not included in calculating the tables of elongations and conjunctions, distances and position angles for Satellites I–VIII. For Satellites I–IV, the orbital eccentricity e is neglected. However, the 4-day tabulations of mean orbital longitude L and mean anomaly M for Satellites I–VIII are calculated from accurate values of the orbital elements; in the case of Titan, solar perturbations are included. Also tabulated are values of the elements that have large variations. The tabular values of the elements can thus be used to obtain perturbed orbital positions. Therefore, using the Saturnicentric position of the Earth, referred to the orbital plane of the satellite, one can calculate the perturbed apparent distance and position angle, and hence differential coordinates in right ascension and declination.

The ascending node of the ring-plane on the mean equator of the Earth serves as the origin for measuring mean orbital longitude L, true longitude u, and the longitude of the ascending node θ of the orbit on the plane of the rings. L and u are reckoned along the ring-plane to the node of the orbit, then along the orbit. Tabulated values of L and M are geometric values at the tabular times, not corrected for light-time.

Satellites and Rings of Uranus

Data for the Uranian Rings are from the analysis of J. L. Elliot *et al.* (*Astron. Jour.*, **86**, 444, 1981). Ephemerides of the satellites are calculated from orbital elements determined by J. Laskar and R. A. Jacobson (*Astron. Astrophys.*, **188**, 212–224, 1987).

Satellites of Neptune

The ephemerides of Triton and Nereid are calculated from elements by R. A. Jacobson (*Astron. Astrophys.*, **231**, 241–250, 1990).

Satellite of Pluto

The ephemeris of Charon is calculated from the elements of D. J. Tholen (*Astron. Jour.*, **90**, 2353, 1985).

Section G: Minor Planets

The ephemerides of Ceres, Pallas, Juno and Vesta give astrometric right ascensions and declinations, referred to the mean equator and equinox of J2000.0, geometric distances from the Earth, and times of ephemeris transit. Astrometric positions are obtained by adding planetary aberration to the geometric positions, referred to the origin of the FK5 system, and then subtracting stellar aberration. Thus these positions are comparable with observations that are referred to catalog mean places of reference stars on the FK5 system, provided the observations are corrected for geocentric parallax and the star positions are corrected for proper motion and annual parallax, if significant, to the epoch of observation. These ephemerides are based on the heliocentric ephemerides of R. L. Duncombe (*Astron. Pap. Amer. Ephem.*, Vol. XX, Pt. II, 1969).

Orbital elements and opposition dates for the larger minor planets are based on data from the Minor Planet Center and the Institute of Theoretical Astronomy. Data concerning physical characteristics are from D. Morrison (*Icarus*, **31**, 185, 1977).

Section H: Stellar Data

Except for the positions of radio sources and pulsars (pages H57–H62, H75, H76) all positions in this section are mean places for the middle of the current Julian year, referred to the origin of the FK5 system. The positions of radio sources and pulsars are mean places for J2000.0.

Bright Stars

Included in the list of bright stars are 1482 stars chosen according to the following criteria:

a. all stars of visual magnitude 4.5 or brighter, as listed in the third revised edition of the *Yale Bright Star Catalogue* (BSC);
b. all FK5 stars brighter than 5.5;
c. all MK atlas standards in the BSC (W. W. Morgan *et al.*, *Revised MK Spectral Atlas for Stars Earlier Than the Sun,* 1978; and P. C. Keenan and R. C. McNeil, *Atlas of Spectra of the Cooler Stars: Types G, K, M, S, and C,*

1976).

Flamsteed and Bayer designations are given with the constellation name and the BSC number. For FK5 stars, positions referred to J2000.0 are taken directly from the "basic" FK5 and precessed to the equator and equinox of the middle of the current year. For the remainder of the stars, B1950.0 positions are taken from the SAO Catalog and converted to J2000.0 by precepts given on page B42, then precessed to the equator and equinox of the middle of the current year. Orbital positions are given for these binary stars: BS 1948/49, 2890/91, 4825/26, 5477/ 78, 8085/86, 2491, 3579, 5459/60, 6134. Orbital elements were taken from the *Third Catalog of Orbits of Visual Binary Stars* (W. S. Finsen and C. E. Worley, *Republic Obs. Circ.* 129, 1970). Whenever possible V magnitudes and color indices *U–B* and *B–V* are due to B. Nicolet (*Astron. Astrophys. Supp.*, **34**, 1, 1978); otherwise these data are taken from the BSC. Spectral types were provided by W. P. Bidelman. Codes in the Notes column are explained at the end of the table (page H31).

Photometric Standards

A selection of 107 stars to serve as standards of the *UBVRI* photometric system was supplied by H. L. Johnson. Photometric data for these stars are due to Johnson *et al.* (*Comm. Lunar Planetary Lab*, Vol. 4, Pt. 3, Table 2, 1966). Primary standards of the *UBVRI* and *UBV* systems are specified by 1 and 2, respectively, in the Standards Code column. As given in the above reference, the filter bands have the following effective wavelengths: *U*, 3600 Å; *B*, 4400 Å; *V*, 5500 Å; *R*, 7000 Å; *I*, 9000 Å.

The selection and photometric data for standards on the Strömgren four-color and Hβ systems are those of C. L. Perry, E. H. Olsen and D. L. Crawford (*Pub. Astron. Soc. Pac.*, **99**, 1184, 1987). Only the 319 stars which have both four-color and Hβ data are included. The *u* band is centered at 3500 Å; *v* at 4100 Å; *b* at 4700 Å; and *y* at 5500 Å. Four indices are tabulated: $b–y$, $m_1 = (v–b) – (b–y)$, $c_1 = (u–v) – (v–b)$ and Hβ.

In both photometric tables, star names and numbers are taken from the *Yale Bright Star Catalogue*. Spectral types are taken from the Bright Stars list (pages H2–H31) or from the photometric references cited above. Positions are obtained by the procedures used for the Bright Stars list.

Radial Velocity Standards

The selection of radial velocity standard stars is based on a list of bright standards taken from the report of IAU Sub-Commission 30a On Standard Velocity Stars (*Trans. IAU*, **IX**, 442, 1957) and list of faint standards (*Trans. IAU*, **XVA**, 409, 1973). These two lists are quite dated relative to modern accuracies and are being revised. Nine stars have been dropped from these lists whose velocities have been found to be variable by more than 1 km/sec and flags have also been used to indicate those stars whose velocities have been found to differ by more than ±1 km/sec. The error bars have also been dropped since thay are considered unrealistic below ±0.5 km/sec. Those stars which are candidates for inclusion in the revised list of primary standards, currently being prepared by IAU Commission 30, are flagged. Those stars are recommended for intensive observation. Positions are obtained by the procedures used for the Bright Stars list. *V* magnitudes are due to B. Nicolet (*Astron. Astrophys. Supp.*, **34**, 1, 1978) when possible; otherwise they are estimated from the given photographic magnitudes and spectral types. The spectral types are taken from the Bright Stars list (pages H2–H31), the *Yale Bright Star Catalogue*, or the original IAU list, in that order of preference.

Bright Galaxies

The list of galaxies is comprised of the brightest ($B_T^w \leq 11.50$) and largest (log $D_{25} \leq 1.65$) galaxies, as compiled by G. de Vaucouleurs from the *Second Reference Catalog of Bright Galaxies* (G. de Vaucouleurs *et al.*, 1976), hereafter referred to as RC2. In addition to serving as an identification list, the precision of the diameters, photometric data and velocities makes these data suitable for calibration purposes.

Identification numbers from the *New General Catalog* (NGC) have the prefix N; those from the *Index Catalog* (IC) have the prefix I. Other abbreviations are interpreted at the end of the table (page H48). Morphological types are based on the revised Hubble system (see G. de Vaucouleurs, *Handbuch der Physik*, **53**, 275, 1959; *Astrophys. Jour. Supp.*, **8**, 31, 1963).

The column headed *T* gives a numerical index to the stages of the Hubble sequence:

T	−6	−5	−4	−3	−2	−1	0	+1	+2	+3	+4	+5	+6	+7	+8	+9	+10	+11
type	cE	E	E+	L−	L	L+	S0/a	Sa	Sab	Sb	Sbc	Sc	Scd	Sd	Sdm	Sm	Im	cI

where E=elliptical, L=lenticular, S=spiral, I=irregular, c=compact.

The column headed *L* gives the luminosity class in the David Dunlap Observatory system:

L	1	2	3	4	5	6	7	8	9
class	I	I–II	II	II–III	III	III–IV	IV	IV–V	V

Columns headed Log D_{25} and Log R_{25} give logarithms to base 10 of the apparent isophotol diameter (D_{25}) and the ratio of the major diameter to the minor diameter ($R_{25} = D_{25} / d_{25}$), measured at or reduced to the surface brightness level $\mu_s = 25.0$ mag/arcsec^2.

The total magnitude in the *B* system is given in the column headed B_T^w. This quantity is a revision of B_T in RC2 (see G. de Vaucouleurs and G. Bollinger, *Astrophys. Jour. Supp.*, **34**, 469, 1977). Total (asymptotic) color indices in the standard *UBV* system are given in the columns *B–V* and *U–B*. The tabulated values are derived by extrapolation from photoelectric color-aperture data or from precise photographic surface photometry with photoelectric zero point.

Observed radial velocities *V* in km/s are weighted means, corrected for systematic errors, of all optical and radio observations. A few values not listed in RC2 are from new or revised determinations. Radial velocities, corrected for solar motion relative to the Local Group of galaxies, are calculated from $V_0 = V + 300 \cos b \sin l$, as recommended by IAU Commission 28 (*Trans, IAU*, **XVIB**, 201, 1977).

Star Clusters

The list of 288 open clusters was supplied by G. Lyngå. It is a selection from the 5th (1987) edition of the Lund-Strasbourg catalogue (original edition described by G. Lyngå, *Astron. Data Cen. Bul.*, 2, 1981). The latest edition is available from the NASA World Data Center A, Greenbelt, MD or from Centre de Données Stellaires, Strasbourg. For each cluster, two identifications are given. First is the designation adopted by the IAU, while the second is the traditional name (G. Alter *et al.*, *Catalogue of Star Clusters and Associations,* 2nd ed., 1970).

Positions are referred to the mean equator and equinox of the middle of the Julian year. The tabulated angular diameter of a cluster pertains to the cluster's nucleus. Trumpler classification is defined by R. S. Trumpler (*Lick Obs. Bul.*, **XIV**, 154, 1930).

The total magnitude of the cluster usually refers to the integrated blue magnitude. The tabulated spectrum refers to the hottest member of the cluster. Under the heading Mag. is tabulated the magnitude of the brightest cluster member.

The logarithm to the base 10 of the cluster age is determined from the turnoff point on the main sequence. Log (Fe/H) is mostly determined from photometric narrow band or intermediate band studies.

Extinction in V is tabulated under A_V. This is determined using the assumption that $A_V = 3E_{(B-V)}$, where the color excess $E_{(B-V)}$ is derived from some of the brightest cluster members.

The table of 157 galactic globular clusters was supplied by R. E. White. It is based on a list circulated some years ago by IAU Commission 37. The entire table is sorted, first, according to the cluster's 1950.0 right ascension (first four digits) and second, according to its diminishing (north to south) declination. Both coordinates are folded into the IAU enumeration.

In keeping with IAU recommendations, at least two indentifications are provided, the IAU designation and the most common catalog name. Additional common aliases are given under Remarks. The positions are referred to the mean equator and equinox of the middle of the Julian year and are based on the 1950.0 coordinates of S. J. Shawl and R. E. White (*Astron. Jour.*, **91**, 312, 1986) or R. F. Webbink ("Coordinates of Reference Stars in the Fields of Globular Clusters", Univ. of Illinois preprint, 1982).

The integrated V magnitude and $(B–V)$, $(U–B)$ and $(V–I)$ color indicies are taken exclusively from Table F in the compilation of B. V. Kukarkin (*Globular Star Clusters*, 1974). Integrated spectral types are due to J. E. Hesser and S. J. Shawl (*Pub. Astron. Soc. Pac.*, **97**, 465, 1985) except for types enclosed in parentheses, which are from the Kukarkin compilation. The m/H values are from R. F. Webbink (*IAU Symposium No. 113, Dynamics of Star Clusters*, eds. J. Goodman and P. Hut, pages 541–577, Table IIa, 1985).

Values for $E_{(B-V)}$, $(m-M)_V$, distance from the Sun, D_0 and galactocentric distance, R (assuming $R_0 = 8.5$ kpc) are due to W. E. Harris (*Astron. Jour.*, **81**, 1095, Tables II and III, 1976) or to R. F. Webbink (denoted by Wb in column 11) from the previous reference. In addition to alternate names, the Remarks column also indicates which clusters have X-ray sources.

Radio Source Standards

The list of 233 radio source positions is that of Argue *et al.* (*Astron. Astrophys.* **130**, 191, 1984). This list was compiled by a working group under IAU Commission 24 as a first step in defining a catalog of extragalactic objects that have both radio and optical counterparts. Positions in the list were compiled from a number of previously published catalogs. The origin of right ascension is defined by the right ascension of 1226+023 (3C273B) at epoch J2000.0, $12^h 26^m 06^s.6997$, as computed by Kaplan *et al.* (*Astron. Jour.* **87**, 570, 1982), based on the B1950.0 position determined for the source by C. Hazard *et* *al.* (*Nature Phys. Sci.*, **233**, 89, 1971). An indication of the uncertainty of a position is given by the number of digits in the tabulated coordinates; the end figures may be subject to revision. The column headed S_{5GHz} gives the flux density in Janskys at 5 GHz. Fluxes of many of the sources vary, however, and the tabulated flux is meant to serve only as a rough guide.

Data for the list of flux standards are due to J. W. M. Baars *et al.* (*Astron. Astrophys.*, **61**, 99, 1977), as updated by the authors. Flux densities S. measured in Janskys, are given for ten frequencies ranging from 400 to 22235 MHz. Positions are referred to the mean equinox and equator of J2000.0. For flux calibration of interferometers, positions of three sources are given with increased precision. Positions of 3C 48 and 3C 147 are due to B. Elsmore and M. Ryle (*Mon. Not. Roy. Astron. Soc.*, **174**, 411, 1976); the position of 3C 286 is from the list of astrometric radio sources, pages H57–H61. Positions of the other sources are due to Baars *et al.*, as cited above.

Selected Identified X-Ray Sources

The X-ray sources were selected by J. F. Dolan from his unpublished survey file. Two common designations of X-ray sources are tabulated: the discovery designation, usually taken from the first published detection of the source, and the designation in the *Fourth Uhuru Catalog* (4U) of W. Forman *et al.* (*Astrophys. Jour. Supp.*, **38**, 357, 1978). When no discovery designation is listed, the source is consistently referred to by the common name of the identified counterpart. Although the listed counterparts are usually optical, the common designation of the radio or infrared counterpart is given in the absence of an optical counterpart. When no identified counterpart is listed, the counterpart has no common designation.

Tabulated positions are based on published positions of identified counterparts. The (2–6) kev flux, in units of 10^{-11} erg cm^{-2} s^{-1} (10^{-14} watts m^{-2}), is taken from the 4U catalog. For sources with variable X-ray intensities, the maximum observed flux from the 4U catalog is tabulated. The tabulated magnitude is the optical magnitude of the counterpart in the *V* filter, unless marked by an asterisk, in which case the *B* magnitude is given. Variable magnitude objects are denoted by V; for these objects the tabulated magnitude pertains to maximum brightness. Codes specifying the type of the identified counterpart are explained at the end of the table (page H67).

Variable Stars

The list containing 181 variable stars has been compiled by J. A. Mattei using as reference the Third Edition of the *General Catalogue of Variable Stars* and its three *Supplements,* the *Sky Catalog 2000.0, Volume 2,* and the data files of the American Association of Variable Star Observers. The brightest stars for each class with amplitude of 0.5 magnitude or more have been selected. The following magnitude criteria at maximum brightness have been used:

a. eclipsing variables brighter than magnitude 7.0;
b. pulsating variables:
    RR Lyrae stars brighter than magnitude 9.0;
    Cepheids brighter than 6.0;
    Mira variables brighter than 7.0;
    Semiregular variables brighter than 7.0;
    Irregular variables brighter than 8.0;

c. eruptive variables:

U Geminorum, Z Camelopardalis, recurrent novae, and nova-like variables brighter than magnitude 11.0;

d. other types:

RV Tauri variables brighter than magnitude 9.0;
R Coronae Borealis variables brighter than 10.0;
Symbiotic stars (Z Andromedae) brighter than 10.0;
d Scuti variables brighter than 9.0.

Selected Quasars

With the collaboration of T. M. Heckman, a set of 99 quasars was selected from the catalog of M.-P. Véron-Cetty and P. Véron (*A Catalogue of Quasars and Active Nuclei*, ESO Scientific Report No. 1, 1984). The following selection criteria were used:

$V \leq 15.45$ (24 quasars);
$M(abs) \leq -30.2$ (27 quasars);
z (redshift) ≤ 0.150 (19 quasars) or $z \geq 3.200$ (18 quasars);
6 cm flux density ≥ 3.6 Janskys (23 quasars).

No objects classified as Seyfert, BL Lac or HII were included.

Positions are given for the equator and equinox of the middle of the current year. Flux densities are given for 6 cm and 7 cm. The authors of the catalog caution that many of the V magnitudes are inaccurate and, in any case, variable. However, the $(B-V)$ color indices do not vary much and should be more accurate. Absolute magnitudes are computed assuming H_0=50 kms $^{-1}$Mpc^{-1}, q_0=0, and an optical spectral index of 0.7.

Selected Pulsars

A selection of 92 pulsars was provided by J. H. Taylor. The selection criterion is that S_{400}, the mean flux density at 400 MHz, be greater than 40 milli-Janskys. In addition about a dozen pulsars of special interest are included, such as the binary and millisecond pulsars.

Positions are referred to the equator and equinox of J2000.0. For each pulsar the period P in seconds and the time rate of change $\dot{P}$ in units of 10^{-15} ss^{-1} are given for the specified epoch. The dispersion measure DM is in pc cm^{-3}.

aberration: the apparent angular displacement of the observed position of a celestial object from its **geometric position**, caused by the finite velocity of light in combination with the motions of the observer and of the observed object. (See **aberration, planetary**.)

aberration, annual: the component of stellar aberration (see **aberration, stellar**) resulting from the motion of the Earth about the Sun.

aberration, diurnal: the component of stellar aberration (see **aberration, stellar**) resulting from the observer's diurnal motion about the center of the Earth.

aberration, E-terms of: terms of annual aberration (see **aberration, annual**) depending on the **eccentricity** and longitude of **perihelion** (see **longitude of pericenter**) of the Earth.

aberration, elliptic: see **aberration, E-terms of.**

aberration, planetary: the apparent angular displacement of the observed position of a celestial body produced by motion of the observer (see **aberration, stellar**) and the actual motion of the observed object (see **correction for light-time**).

aberration, secular: the component of stellar aberration (see **aberration, stellar**) resulting from the essentially uniform and rectilinear motion of the entire solar system in space. Secular aberration is usually disregarded.

aberration, stellar: the apparent angular displacement of the observed position of a celestial body resulting from the motion of the observer. Stellar aberration is divided into diurnal, annual and secular components (see **aberration, diurnal; aberration, annual; aberration, secular**).

altitude: the angular distance of a celestial body above or below the horizon, measured along the great circle passing through the body and the **zenith**. Altitude is 90° minus **zenith distance.**

anomaly: angular measurement of a body in its **orbit** from its **perihelion.**

aphelion: the point in a planetary **orbit** that is at the greatest distance from the Sun.

apogee: the point at which a body in **orbit** around the Earth reaches its farthest distance from the Earth.

apparent place: the position on a **celestial sphere**, centered at the Earth, determined by removing from the directly observed position of a celestial body the effects that depend on the **topocentric** location of the observer; i.e., **refraction**, diurnal aberration (see **aberration, diurnal**) and geocentric (diurnal) **parallax**. Thus the position at which the object would actually be seen from the center of the Earth, displaced by planetary aberration (except the diurnal part – see **aberration, planetary & aberration, diurnal**) and referred to the **true equinox and equator**.

apparent solar time: the measure of time based on the diurnal motion of the true Sun. The rate of diurnal motion undergoes seasonal variation because of the **obliquity** of the **ecliptic** and because of the **eccentricity** of the Earth's **orbit**. Additional small variations result from irregularities in the rotation of the Earth on its axis.

aspect: the apparent position of any of the planets, or the Moon relative to the Sun, as seen from Earth.

astrometric ephemeris: an **ephemeris** of a solar system body in which the tabulated positions are essentially comparable to catalog **mean places** of stars at a **standard epoch**. An astrometric position is obtained by adding to the geometric position, computed from gravitational theory, the correction for **light-time**. Prior to 1984, the E-terms of annual aberration (see **aberration, annual & aberration, E-terms of**) were also added to the geometric position.

astronomical coordinates: the longitude and latitude of a point on the Earth relative to the **geoid**. These coordinates are influenced by local gravity anomalies. (See **zenith; longitude, terrestrial; latitude, terrestrial.**)

astronomical unit (a.u.): the radius of a circular **orbit** in which a body of negligible mass, and free of **perturbations**, would revolve around the Sun in $2\pi/k$ days, where k is the **Gaussian gravitational constant**. This is slightly less than the **semimajor axis** of the Earth's **orbit**.

atomic second: see **second, Système International.**

augmentation: the amount by which the apparent **semidiameter** of a celestial body, as observed from the surface of the Earth, is greater than the semidiameter that would be observed from the center of the Earth.

azimuth: the angular distance measured clockwise along the **horizon** from a specified reference point (usually north) to the intersection with the great circle drawn from the **zenith** through a body on the **celestial sphere**.

barycenter: the center of mass of a system of bodies; e.g., the center of mass of the solar system or the Earth-Moon system.

barycentric dynamical time (TDB): the independent argument of ephemerides and equations of motion that are referred to the **barycenter** of the solar system. A family of time scales results from the transformation by various theories and metrics of relativistic theories of **terrestrial dynamical time (TDT)**. TDB differs from TDT only by periodic variations. In the terminology of the general theory of relativity, TDB may be considered to be a coordinate time. (See **dynamical time.**)

Brilliancy: for Mercury and Venus the quantity ks^2/r^2, where $k = 0.5(1 + \cos i)$, i is the phase angle, s is the apparent **semidiameter**, and r is the heliocentric distance.

calendar: a system of reckoning time in which days are enumerated according to their position in cyclic patterns.

catalog equinox: the intersection of the **hour circle** of zero **right ascension** of a star catalog with the **celestial equator**. (See **dynamical equinox & equator.**)

celestial ephemeris pole: the reference pole for **nutation** and **polar motion**; the axis of figure for the mean surface of a model Earth in which the free motion has zero amplitude. This pole has no nearly-diurnal **nutation** with respect to a space-fixed or Earth-fixed coordinate system.

celestial equator: the projection onto the **celestial sphere** of the Earth's **equator**. (See **mean equator and equinox & true equator and equinox.**)

celestial pole: either of the two points projected onto the **celestial sphere** by the extension of the Earth's axis of rotation to infinity.

celestial sphere: an imaginary sphere of arbitrary radius upon which celestial bodies may be considered to be located. As circumstances require, the celestial sphere may be centered at the observer, at the Earth's center, or at any other location.

conjunction: the phenomenon in which two bodies have the same apparent celestial longitude (see **longitude, celestial**) or **right ascension** as viewed from a third body. Conjunctions are usually tabulated as **geocentric** phenomena, however. For Mercury and Venus, geocentric inferior conjunction occurs when the planet is between the Earth and Sun, and superior conjunction occurs when the Sun is between the planet and Earth.

constellation: a grouping of stars, usually with pictorial or mythical associations, that serves to identify an area of the **celestial sphere**. Also one of the precisely defined areas of the celestial sphere, associated with a grouping of stars, that the International Astronomical Union has designated as a constellation.

coordinated universal time (UTC): the time scale available from broadcast time signals. UTC differs from TAI (see **international atomic time**) by an integral number of seconds; it is maintained within +0.90 second of UT1 (see **universal time**) by the introduction of one second steps (leap seconds). (See **leap second**.)

culmination: passage of a celestial object across the observer's **meridian**; also called "meridian passage". More precisely, culmination is the passage through the point of greatest altitude in the diurnal path. Upper culmination (also called "culmination above pole" for circumpolar stars and the Moon) or transit is the crossing closer to the observer's **zenith**. Lower culmination (also called "culmination below pole" for circumpolar stars and the Moon) is the crossing farther from the zenith.

day: an interval of 86 400 SI seconds (see **second, Système International**), unless otherwise indicated.

day numbers: quantities that facilitate hand calculations of the reduction of **mean place to apparent place**. Besselian day numbers depend solely on the Earth's position and motion; second-order day numbers, used in higher precision reductions, depend on the positions of both the Earth and the star.

declination: angular distance on the **celestial sphere** north or south of the **celestial equator**. It is measured along the **hour circle** passing through the celestial object. Declination is usually given in combination with **right ascension** or **hour angle**.

defect of illumination: the angular amount of the observed lunar or planetary disk that is not illuminated to an observer on the Earth.

deflection of light: the angle by which the apparent path of a photon is altered from a straight line by the gravitational field of the Sun. The path is deflected radially away from the Sun by up to 1″.75 at the Sun's limb. Correction for this effect, which is independent of wavelength, is included in the reduction from **mean place to apparent place**.

deflection of the vertical: the angle between the astronomical vertical and the geodetic vertical. (See **zenith**; **astronomical coordinates**; **geodetic coordinates**.)

Delta T (ΔT): the difference between **dynamical time** and **universal time**; specifically the difference between **terrestrial dynamical time (TDT)** and UT1: $\Delta T = TDT - UT1$.

direct motion: for orbital motion in the solar system, motion that is counterclockwise in the orbit as seen from the north pole of the **ecliptic**; for an object observed on the celestial sphere, motion that is from west to east, resulting from the relative motion of the object and the Earth.

diurnal motion: the apparent daily motion, caused by the Earth's rotation, of celestial bodies across the sky from east to west.

DUT1: the predicted value of the difference between UT1 and UTC, transmitted in code on broadcast time signals: DUT1 = UT1 − UTC. (See **universal time & coordinated universal time**.)

dynamical equinox: the ascending **node** of the Earth's mean **orbit** on the Earth's **equator**; i.e., the intersection of the **ecliptic** with the celestial equator at which the Sun's **declination** is changing from south to north. (See **catalog equinox & equinox**.)

dynamical time: the family of time scales introduced in 1984 to replace **ephemeris time** as the independent argument of dynamical theories and ephemerides. (See **barycentric dynamical time & terrestrial dynamical time**.)

eccentric anomaly: in undisturbed elliptic motion, the angle measured at the center of the ellipse from **pericenter** to the point on the circumscribing auxiliary circle from which a perpendicular to the major axis would intersect the orbiting body. (See **mean anomaly & true anomaly**.)

eccentricity: a parameter that specifies the shape of a conic section; one of the standard elements used to describe an elliptic **orbit** (see **elements, orbital**).

eclipse: the obscuration of a celestial body caused by its passage through the shadow cast by another body.

eclipse, annular: a solar **eclipse** (see **eclipse, solar**) in which the solar disk is never completely covered but is seen as an annulus or ring at maximum eclipse. An annular eclipse occurs when the apparent disk of the Moon is smaller than that of the Sun.

eclipse, lunar: an **eclipse** in which the Moon passes through the shadow cast by the Earth. The eclipse may be total (the Moon passing completely through the Earth's **umbra**), partial (the Moon passing partially through the Earth's umbra at maximum eclipse), or penumbral (the Moon passing only through the Earth's **penumbra**).

eclipse, solar: an **eclipse** in which the Earth passes through the shadow cast by the Moon. It may be total (observer in the Moon's **umbra**), partial (observer in the Moon's **penumbra**), or annular (see **eclipse, annular**).

ecliptic: the mean plane of the Earth's **orbit** around the Sun.

elements, Besselian: quantities tabulated for the calculation of accurate predictions of an **eclipse** or **occultation** for any point on or above the surface of the Earth.

elements, orbital: parameters that specify the position and motion of a body in **orbit**. (See **osculating elements & mean elements**.)

elongation, greatest: the instants when the **geocentric** angular distances of Mercury and Venus are at a maximum from the Sun.

elongation (planetary): the **geocentric** angle between a planet and the Sun, measured in the plane of the planet, Earth and Sun. Planetary elongations are measured from 0° to 180°, east or west of the Sun.

elongation (satellite): the **geocentric** angle between a satellite and its primary, measured in the plane of the satellite, planet and Earth. Satellite elongations are measured from 0° east or west of the planet.

epact: the age of the Moon; the number of days since new Moon, diminished by one day, on January 1 in the Gregorian ecclesiastical lunar cycle. (See **Gregorian calendar** and **lunar phases.**)

ephemeris: a tabulation of the positions of a celestial object in an orderly sequence for a number of dates.

ephemeris hour angle: an **hour angle** referred to the **ephemeris meridian.**

ephemeris longitude: longitude (see **longitude, terrestrial**) measured eastward from the **ephemeris meridian.**

ephemeris meridian: a fictitious **meridian** that rotates independently of the Earth at the uniform rate implicitly defined by **terrestrial dynamical time (TDT).** The ephemeris meridian is 1.002 738 ΔT east of the Greenwich meridian, where $\Delta T = TDT - UT1$

ephemeris time (ET): the time scale used prior to 1984 as the independent variable in gravitational theories of the solar system. In 1984, ET was replaced by **dynamical time.**

ephemeris transit: the passage of a celestial body or point across the **ephemeris meridian.**

epoch: an arbitrary fixed instant of time or date used as a chronological reference datum for calendars (see **calendar**), celestial reference systems, star catalogues, or orbital motions (see **orbit**).

equation of center: in elliptic motion the **true anomaly** minus the **mean anomaly.** It is the difference between the actual angular position in the elliptic **orbit** and the position the body would have if its angular motion were uniform.

equation of the equinoxes: the **right ascension** of the mean **equinox** (see **mean equator and equinox**) referred to the **true equator and equinox**; apparent **sidereal time** minus mean sidereal time. (See **apparent place** & **mean place.**)

equation of time: the **hour angle** of the true Sun minus the hour angle of the **fictitious mean sun**; alternatively, **apparent solar time** minus **mean solar time.**

equator: the great circle on the surface of a body formed by the intersection of the surface with the plane passing through the center of the body perpendicular to the axis of rotation. (See **celestial equator.**)

equinox: either of the two points on the **celestial sphere** at which the **ecliptic** intersects the **celestial equator**; also the time at which the Sun passes through either of these intersection points; ie., when the apparent longitude (see **apparent place** & **longitude, celestial**) of the Sun is 0° or 180°. (See **catalog equinox** & **dynamical equinox** for precise usage.)

era: a system of chronological notation reckoned from a given date.

fictitious mean sun: an imaginary body introduced to define **mean solar time**; essentially the name of a mathematical formula that defined mean solar time. This concept is no longer used in high precision work.

flattening: a parameter that specifies the degree by which a planet's figure differs from that of a sphere; the ratio $f = (a-b)/a$, where a is the equatorial radius and b is the polar radius.

frequency: the number of cycles or complete alternations per unit time of a carrier wave, band, or oscillation.

frequency standard: a generator whose output is used as a precise frequency reference; a primary frequency standard is one whose frequency corresponds to the adopted definition of the second (see **second, Système International**), with its specified accuracy achieved without calibration of the device.

Gaussian gravitational constant (k=0.0172 0209 895): the constant defining the astronomical system of units of length (**astronomical unit**), mass (solar mass) and time (day), by means of Kepler's third law. The dimensions of k^2 are those of Newton's constant of gravitation: $L^3M^{-1}T^{-2}$.

gegenshein: faint nebulous light about 20° across near the **ecliptic** and opposite the Sun, best seen in September and October. Also called counterglow.

geocentric: with reference to, or pertaining to, the center of the Earth.

geocentric coordinates: the latitude and longitude of a point on the Earth's surface relative to the center of the Earth; also celestial coordinates given with respect to the center of the Earth. (See **zenith; latitude, terrestrial; longitude, terrestrial.**)

geodetic coordinates: the latitude and longitude of a point on the Earth's surface determined from the geodetic vertical (normal to the specified spheroid). (See **zenith; latitude, terrestrial; longitude, terrestrial.**)

geoid: an equipotential surface that coincides with mean sea level in the open ocean. On land it is the level surface that would be assumed by water in an imaginary network of frictionless channels connected to the ocean.

geometric position: the **geocentric** position of an object on the **celestial sphere referred** to the **true equator and equinox**, but without the displacement due to planetary aberration. (See **apparent place; mean place; aberration, planetary.**)

Greenwich sidereal date (GSD): the number of **sidereal days** elapsed at Greenwich since the beginning of the Greenwich sidereal day that was in progress at **Julian date** 0.0.

Greenwich sidereal day number: the integral part of the **Greenwich sidereal date.**

Gregorian calendar: the calendar introduced by Pope Gregory XIII in 1582 to replace the **Julian calendar**; the calendar now used as the civil calendar in most countries. Every year that is exactly divisible by four is a leap year, except for centurial years, which must be exactly divisible by 400 to be leap years. Thus 2000 is a leap year, but 1900 and 2100 are not leap years.

height: elevation above ground or distance upwards from a given level (esp. sea level) to a fixed point. (See **altitude.**)

heliocentric: with reference to, or pertaining to, the center of the Sun.

horizon: a plane perpendicular to the line from an observer to the zenith. The great circle formed by the intersection of the **celestial sphere** with a plane perpendicular to the line from an observer to the zenith is called the astronomical horizon.

horizontal parallax: the difference between the **topocentric** and **geocentric** positions of an object, when the object is on the astronomical **horizon.**

hour angle: angular distance on the **celestial sphere** measured westward along the **celestial equator** from the **meridian** to the **hour circle** that passes through a celestial object.

hour circle: a great circle on the **celestial sphere** that passes through the **celestial poles** and is therefore perpendicular to the **celestial equator**.

inclination: the angle between two planes or their poles; usually the angle between an orbital plane and a reference plane; one of the standard orbital elements (see **elements, orbital**) that specifies the orientation of an **orbit**.

international atomic time (TAI): the continuous scale resulting from analyses by the Bureau International des Poids et Mesures of atomic time standards in many countries. The fundamental unit of TAI is the SI second (see **second, Système International**), and the epoch is 1958 January 1.

invariable plane: the plane through the center of mass of the solar system perpendicular to the angular momentum vector of the solar system.

irradiation: an optical effect of contrast that makes bright objects viewed against a dark background appear to be larger than they really are.

Julian calendar: the calendar introduced by Julius Caesar in 46 B.C. to replace the Roman calendar. In the Julian calendar a common year is defined to comprise 365 days, and every fourth year is a leap year comprising 366 days. The Julian calendar was superseded by the **Gregorian calendar**.

Julian date (JD): the interval of time in days and fraction of a day since 1 January 4713 B.C., Greenwich noon, **Julian proleptic calendar**. In precise work the time scale, e.g., **dynamical time** or **universal time**, should be specified.

Julian date, modified (MJD): the Julian date minus 240 0000.5.

Julian day number (JD): the integral part of the **Julian date**.

Julian proleptic calendar: the calendric system employing the rules of the **Julian calendar**, but extended and applied to dates preceding the introduction of the Julian calendar.

Julian year: a period of 365.25 days. This period served as the basis for the **Julian calendar**.

Laplacian plane: for planets see **invariable plane**; for a system of satellites, the fixed plane relative to which the vector sum of the disturbing forces has no orthogonal component.

latitude, celestial: angular distance on the **celestial sphere** measured north or south of the **ecliptic** along the great circle passing through the poles of the ecliptic and the celestial object.

latitude, terrestrial: angular distance on the Earth measured north or south of the **equator** along the **meridian** of a geographic location.

leap second: a second (see **second, Système International**) added between 60^s and 0^s at announced times to keep UTC within $0^s.90$ of UT1. Generally, leap seconds are added at the end of June or December.

librations: variations in the orientation of the Moon's surface with respect to an observer on the Earth. Physical librations are due to variations in the rate at which the Moon rotates on its axis. The much larger optical librations are due to variations in the rate of the Moon's orbital motion, the obliquity of the Moon's equator to its orbital plane, and the diurnal changes of geometric perspective of an observer on the Earth's surface.

light, deflection of: the bending of the beam of light due to gravity. It is observable when the light from a star or planet passes a massive object such as the Sun.

light-time: the interval of time required for light to travel from a celestial body to the Earth. During this interval the motion of the body in space causes an angular displacement of its **apparent place** from its geometric place (see **geometric position**). (See **aberration, planetary**.)

light-year: the distance that light traverses in a vacuum during one year.

limb: the apparent edge of the Sun, Moon, or a planet or any other celestial body with a detectable disc.

limb correction: correction which must be made to the distance between the center of mass of the Moon and its limb. These corrections are due to the irregular surface of the Moon and are a function of the **librations** in longitude (see **longitude, celestial**) and latitude (see **latitude, celestial**) and the position angle from the central **meridian**.

local sidereal time: the local **hour angle of a catalog equinox**.

longitude, celestial: angular distance on the **celestial sphere** measured eastward along the **ecliptic** from the **dynamical equinox** to the great circle passing through the poles of the ecliptic and the celestial object.

longitude, terrestrial: angular distance measured along the Earth's **equator** from the Greenwich **meridian** to the meridian of a geographic location.

luminosity class: distinctions among stars of the same spectral class. (See **Spectral types and classes**.)

lunar phases: cyclically recurring apparent forms of the Moon. New Moon, First Quarter, Full Moon and Last Quarter are defined as the times at which the excess of the apparent celestial longitude (see **longitude, celestial**) of the Moon over that of the Sun is 0°, 90°, 180° and 270°, respectively.

lunation: the period of time between two consecutive New Moons.

magnitude, stellar: a measure on a logarithmic scale of the brightness of a celestial object considered as a point source.

magnitude of a lunar eclipse: the fraction of the lunar diameter obscured by the shadow of the Earth at the greatest phase of a lunar eclipse (see **eclipse, lunar**), measured along the common diameter.

magnitude of a solar eclipse: the fraction of the solar diameter obscured by the Moon at the greatest phase of a solar eclipse (see **eclipse, solar**), measured along the common diameter.

mean anomaly: in undisturbed elliptic motion, the product of the **mean motion** of an orbiting body and the interval of time since the body passed **pericenter**. Thus the mean anomaly is the angle from pericenter of a hypothetical body moving with a constant angular speed that is equal to the mean motion. (See **true anomaly & eccentric anomaly**.)

mean distance: **the semimajor axis** of an elliptic **orbit**.

mean elements: elements of an adopted reference **orbit** (see **elements, orbital**) that approximates the actual, perturbed orbit. Mean elements may serve as the basis for calculating **perturbations**.

mean equator and equinox: the celestial reference system determined by ignoring small variations of short period in the motions of the **celestial equator**. Thus the mean equator and equinox are affected only by **precession**. Positions in star catalogs are normally referred to the mean catalog equator and equinox (see **catalog equinox**) of a **standard epoch**.

mean motion: in undisturbed elliptic motion, the constant angular speed required for a body to complete one revolution in an **orbit** of a specified **semimajor axis.**

mean place: the coordinates, referred to the **mean equator and equinox** of a **standard epoch**, of an object on the **celestial sphere** centered at the Sun. A mean place is determined by removing from the directly observed position the effects of **refraction**, geocentric and stellar parallax, and stellar aberration (see **aberration, stellar**), and by referring the coordinates to the mean equator and equinox of a standard epoch. In compiling star catalogs it has been the practice not to remove the secular part of stellar aberration (see **aberration, secular**). Prior to 1984, it was additionally the practice not to remove the elliptic part of annual aberration (see **aberration, annual** & **aberration, E-terms of**).

mean solar time: a measure of time based conceptually on the diurnal motion of the **fictitious mean sun**, under the assumption that the Earth's rate of rotation is constant.

meridian: a great circle passing through the **celestial poles** and through the **zenith** of any location on Earth. For planetary observations a meridian is half the great circle passing through the planet's poles and through any location on the planet.

month: the period of one complete synodic or sidereal revolution of the Moon around the Earth; also a calendrical unit that approximates the period of revolution.

moonrise, moonset: the times at which the apparent upper limb of the Moon is on the astronomical **horizon**; i.e., when the true **zenith distance**, referred to the center of the Earth, of the central point of the disk is $90° 34' + s - \pi$, where s is the Moon's **semidiameter**, π is the **horizontal parallax**, and $34'$ is the adopted value of horizontal **refraction**.

nadir: the point on the **celestial sphere** diametrically opposite to the **zenith**.

node: either of the points on the **celestial sphere** at which the plane of an **orbit** intersects a reference plane. The position of a node is one of the standard orbital elements (see **elements, orbital**) used to specify the orientation of an orbit.

nutation: the short-period oscillations in the motion of the pole of rotation of a freely rotating body that is undergoing torque from external gravitational forces. Nutation of the Earth's pole is discussed in terms of components in **obliquity** and longitude (see **longitude, celestial**.)

obliquity: in general the angle between the equatorial and orbital planes of a body or, equivalently, between the rotational and orbital poles. For the Earth the obliquity of the **ecliptic** is the angle between the planes of the **equator** and the ecliptic.

occultation: the obscuration of one celestial body by another of greater apparent diameter; especially the passage of the Moon in front of a star or planet, or the disappearance of a satellite behind the disk of its primary. If the primary source of illumination of a reflecting body is cut off by the occultation, the phenomenon is also called an **eclipse**. The occultation of the Sun by the Moon is a solar eclipse (see **eclipse, solar**.)

opposition: a configuration of the Sun, Earth and a planet in which the apparent **geocentric** longitude (see **longitude, celestial**) of the planet differs by 180° from the apparent geocentric longitude of the Sun.

orbit: the path in space followed by a celestial body.

osculating elements: a set of parameters (see **elements, orbital**) that specifies the instantaneous position and velocity of a celestial body in its perturbed orbit. Osculating elements describe the unperturbed (two-body) orbit that the body would follow if **perturbations** were to cease instantaneously.

parallax: the difference in apparent direction of an object as seen from two different locations; conversely the angle at the object that is subtended by the line joining two designated points. Geocentric (diurnal) parallax is the difference in direction between a **topocentric** observation and a hypothetical **geocentric** observation. Heliocentric or annual parallax is the difference between hypothetical geocentric and heliocentric observations; it is the angle subtended at the observed object by the **semimajor axis** of the Earth's **orbit**. (See also **horizontal parallax**.)

parsec: the distance at which one **astronomical unit** subtends an angle of one second of arc; equivalently the distance to an object having an annual **parallax** of one second of arc.

penumbra: the portion of a shadow in which light from an extended source is partially but not completely cut off by an intervening body; the area of partial shadow surrounding the **umbra**.

pericenter: the point in an **orbit** that is nearest to the center of force. (See **perigee & perihelion**.)

perigee: the point at which a body in **orbit** around the Earth most closely approaches the Earth. Perigee is sometimes used with reference to the apparent orbit of the Sun around the Earth.

perihelion: the point at which a body in **orbit** around the Sun most closely approaches the Sun.

period: the interval of time required to complete one revolution in an **orbit** or one cycle of a periodic phenomenon, such as a cycle of phases (see **phase**).

perturbations: deviations between the actual **orbit** of a celestial body and an assumed reference orbit; also the forces that cause deviations between the actual and reference orbits. Perturbations, according to the first meaning, are usually calculated as quantities to be added to the coordinates of the reference orbit to obtain the precise coordinates.

phase: the ratio of the illuminated area of the apparent disk of a celestial body to the area of the entire apparent disk taken as a circle. For the Moon, phase designations (see **lunar phases**) are defined by specific configurations of the Sun, Earth and Moon. For eclipses phase designations (total, partial, penumbral, etc.) provide general descriptions of the phenomena. (See **eclipse, solar**; **eclipse, annular**; **eclipse, lunar**.)

phase angle: the angle measured at the center of an illuminated body between the light source and the observer.

photometry: a measurement of the intensity of light usually specified for a specific frequency range.

Planetocentric coordinates: coordinates for general use, where the z-axis is the mean axis of rotation, the x-axis is the intersection of the planetary

equator (normal to the z-axis through the center of mass) and an arbitrary prime **meridian**, and the y-axis completes a right hand coordinate system. Longitude (see **longitude, celestial**) of a point is measured positive to the prime meridian as defined by rotational elements. Latitude (see **latitude, celestial**) of a point is the angle between the planetary equator and a line to the center of mass. The radius is measured from the center of mass to the surface point.

Planetographic coordinates: coordinates for cartographic purposes dependent on an equipotential surface as a reference surface. Longitude (see **longitude, celestial**) of a point is measured in the direction opposite to the rotation (positive to the west for direct rotation) from the cartographic position of the prime **meridian** defined by a clearly observable surface feature. Latitude (see **latitude, celestial**) of a point is the angle between the planetary **equator** (normal to the z-axis and through the center of mass) and normal to the reference surface at the point. The height of a point is specified as the distance above a point with the same longitude and latitude on the reference surface.

polar motion: the irregularly varying motion of the Earth's pole of rotation with respect to the Earth's crust. (See **celestial ephemeris pole.**)

precession: the uniformly progressing motion of the pole of rotation of a freely rotating body undergoing torque from external gravitational forces. In the case of the Earth, the component of precession caused by the Sun and Moon acting on the Earth's equatorial bulge is called lunisolar precession; the component caused by the action of the planets is called planetary precession. The sum of lunisolar and planetary precession is called general precession. (See **nutation.**)

proper motion: the projection onto the **celestial sphere** of the space motion of a star relative to the solar system; thus the transverse component of the space motion of a star with respect to the solar system. Proper motion is usually tabulated in star catalogs as changes in **right ascension** and **declination** per year or century.

quadrature: a configuration in which two celestial bodies have apparent longitudes (see **longitude, celestial**) that differ by 90° as viewed from a third body. Quadratures are usually tabulated with respect to the Sun as viewed from the center of the Earth.

radial velocity: the rate of change of the distance to an object.

refraction, astronomical: the change in direction of travel (bending) of a light ray as it passes obliquely through the atmosphere. As a result of refraction the observed **altitude** of a celestial object is greater than its geometric altitude. The amount of refraction depends on the altitude of the object and on atmospheric conditions.

retrograde motion: for orbital motion in the solar system, motion that is clockwise in the **orbit** as seen from the north pole of the **ecliptic**; for an object observed on the **celestial sphere**, motion that is from east to west, resulting from the relative motion of the object and the Earth. (See **direct motion.**)

right ascension: angular distance on the **celestial sphere** measured eastward along the **celestial equator** from the **equinox** to the **hour circle** passing through the celestial object. Right ascension is usually given in combination with **declination**.

Saros: a Babylonian lunar cycle of 6585.32 days, or 18 years 11.33 days, or 223 lunations, at the end of which the centers of the Sun and Moon return so nearly to the relative positions of the beginning that all the eclipses (see **eclipse**) of the period recur approximately as before, but in longitudes (see **longitude, terrestrial**) approximately 120° to the west. (See **lunar phases**.)

satellite: natural body revolving around a planet.

satellite, artificial: device launched into a closed orbit around the Earth, another planet, the Sun, ect.

second, Système International (SI): the duration of 9192 631 770 cycles of radiation corresponding to the transition between two hyperfine levels of the ground state of cesium 133.

selenocentric: with reference to, or pertaining to, the center of the Moon.

semidiameter: the angle at the observer subtended by the equatorial radius of the Sun, Moon or a planet.

semimajor axis: half the length of the major axis of an ellipse; a standard element used to describe an elliptical **orbit** (see **elements, orbital**.)

sidereal day: the interval of time between two consecutive **transits** of the **catalog equinox**. (See **sidereal time**.)

sidereal hour angle: angular distance on the **celestial sphere** measured westward along the **celestial equator** from **the catalog equinox** to the **hour circle** passing through the celestial object. It is equal to 360° minus **right ascension** in degrees.

sidereal time: the measure of time defined by the apparent diurnal motion of the **catalog equinox**; hence a measure of the rotation of the Earth with respect to the stars rather than the Sun.

solstice: either of the two points on the **ecliptic** at which the apparent longitude (see **longitude, celestial**) of the Sun is 90° or 270°; also the time at which the Sun is at either point.

Spectral types or classes: catagorization of stars according to their spectra, primarily due to differing temperatures of the stellar atmosphere. From hottest to coolest, the spectral types are O, B, A, F, G, K and M.

standard epoch: a date and time that specifies the reference system to which celestial coordinates are referred. Prior to 1984 coordinates of star catalogs were commonly referred to the **mean equator and equinox** of the beginning of a Besselian year (see **year, Besselian**). Beginning with 1984 the **Julian year** has been used, as denoted by the prefix J, e.g., J2000.0.

stationary point (of a planet): the position at which the rate of change of the apparent **right ascension** (see **apparent place**) of a planet is momentarily zero.

sunrise, sunset: the times at which the apparent upper limb of the Sun is on the astronomical horizon; i.e., when the true zenith distance, referred to the center of the Earth, of the central point of the disk is 90° 50′, based on adopted values of 34′ for horizontal **refraction** and 16′ for the Sun's **semidiameter**.

surface brightness (of a planet): the visual magnitude of an average square arc-second area of the illuminated portion of the apparent disk.

synodic period: for planets, the mean interval of time between successive **conjunctions** of a pair of planets, as observed from the Sun; for satellites, the mean interval between successive conjunctions of a satellite with the Sun, as observed from the satellite's primary.

synodic time: pertaining to successive conjunctions; successive returns of a planet to the same **aspect** as determined by Earth.

terrestrial dynamical time (TDT): the independent argument for apparent **geocentric** ephemerides. At 1977 January $1^{d}00^{h}00^{m}00^{s}$ TAI, the value of TDT was exactly 1977 January $1^{d}.000\ 3725$. The unit of TDT is 86 400 SI seconds at mean sea level. For practical purposes TDT = TAI+$32^{s}.184$. (See **barycentric dynamical time; dynamical time; international atomic time.**)

terminator: the boundary between the illuminated and dark areas of the apparent disk of the Moon, a planet or a planetary satellite.

topocentric: with reference to, or pertaining to, a point on the surface of the Earth, usually with reference to a coordinate system.

transit: the passage of a celestial object across a **meridian**; also the passage of one celestial body in front of another of greater apparent diameter (e.g., the passage of Mercury or. Venus across the Sun or Jupiter's satellites across its disk); however, the passage of the Moon in front of the larger apparent Sun is called an annular eclipse (see **eclipse, annular**). The passage of a body's shadow across another body is called a shadow transit; however, the passage of the Moon's shadow across the Earth is called a solar eclipse (see **eclipse, solar**).

true anomaly: the angle, measured at the focus nearest the **pericenter** of an elliptical orbit, between the pericenter and the radius vector from the focus to the orbiting body; one of the standard orbital elements (see **elements, orbital**). (See also **eccentric anomaly, mean anomaly.**)

true equator and equinox: the celestial coordinate system determined by the instantaneous positions of the **celestial equator** and **ecliptic**. The motion of this system is due to the progressive effect of **precession** and the short-term, periodic variations of **nutation**. (See **mean equator and equinox.**)

twilight: the interval of time preceding sunrise and following sunset (see **sunrise, sunset**) during which the sky is partially illuminated. Civil twilight comprises the interval when the **zenith distance**, referred to the center of the Earth, of the central point of the Sun's disk is between 90° 50′ and 96°, nautical twilight comprises the interval from 96° to 102°, astronomical twilight comprises the interval from 102° to 108°.

umbra: the portion of a shadow cone in which none of the light from an extended light source (ignoring **refraction**) can be observed.

universal time (UT): a measure of time that conforms, within a close approximation, to the mean diurnal motion of the Sun and serves as the basis of all civil timekeeping. UT is formally defined by a mathematical formula as a function of **sidereal time**. Thus UT is determined from observations of the diurnal motions of the stars. The time scale determined directly from such observations is designated UT0; it is slightly dependent on the place of observation. When UT0 is corrected for the shift in longitude (see **longitude, terrestrial**) of the observing station caused by **polar motion**, the time scale UT1 is obtained. Whenever the designation UT is used in this volume, UT1 is implied.

vernal equinox: the ascending **node** of the **ecliptic** on the **celestial equator**; also the time at which the apparent longitude (see **apparent place** & **longitude, celestial**) of the Sun is 0°. (See **equinox**.)

vertical: apparent direction of gravity at the point of observation (normal to the plane of a free level surface.)

week: an arbitrary period of days, usually seven days; approximately equal to the number of days counted between the four phases of the Moon. (See **lunar phases**.)

year: a period of time based on the revolution of the Earth around the Sun. The calendar year (see **Gregorian calendar**) is an approximation to the tropical year (see **year, tropical**). The anomalistic year is the mean interval between successive passages of the Earth through **perihelion**. The sidereal year is the mean period of revolution with respect to the background stars. (See **Julian year** & **year, Besselian**.)

year, Besselian: the period of one complete revolution in **right ascension** of the **fictitious mean sun**, as defined by Newcomb. The beginning of a Besselian year, traditionally used as as **standard epoch**, is denoted by the suffix ".0". Since 1984 standard epochs have been defined by the **Julian year** rather that the Besselian year. For distinction, the beginning of the Besselian year is now identified by the prefix B (e.g., B1950.0).

year, tropical: the period of one complete revolution of the mean longitude of the sun with respect to the **dynamical equinox**. The tropical year is longer than the Besselian year (see **year, Besselian**) by $0^s.148T$, where T is centuries from B1900.0.

zenith: in general, the point directly overhead on the **celestial sphere**. The astronomical zenith is the extension to infinity of a plumb line. The geocentric zenith is defined by the line from the center of the Earth through the observer. The geodetic zenith is the normal to the geodetic ellipsoid at the observer's location. (See **deflection of the vertical**.)

zenith distance: angular distance on the **celestial sphere** measured along the great circle from the **zenith** to the celestial object. Zenith distance is 90° minus **altitude**.

zodiacal light: a nebulous light seen in the east before twilight and in the west after twilight. It is triangular in shape along the **ecliptic** with the base on the horizon and its apex at varying altitudes. It is best seen in middle latitudes (see **latitude, terrestrial**) on spring evenings and autumn mornings.

Definitions of astronomical terms are provided in the Glossary, Section M. Entries in the Glossary are not cited in the Index.

Definitions of astronomical terms are provided in the Glossary, Section M. Entries in the Glossary are not cited in the Index.

Definitions of astronomical terms are provided in the Glossary, Section M. Entries in the Glossary are not cited in the Index.

Definitions of astronomical terms are provided in the Glossary, Section M. Entries in the Glossary are not cited in the Index.

Definitions of astronomical terms are provided in the Glossary, Section M. Entries in the Glossary are not cited in the Index.

Definitions of astronomical terms are provided in the Glossary, Section M. Entries in the Glossary are not cited in the Index.

Definitions of astronomical terms are provided in the Glossary, Section M. Entries in the Glossary are not cited in the Index.

☆U.S. GOVERNMENT PRINTING OFFICE: 1991 249-891